AF697792

Other Titles in This Series

43 **Luis A. Caffarelli and Xavier Cabré,** Fully nonlinear elliptic equations, 1995
42 **Victor Guillemin and Shlomo Sternberg,** Variations on a theme by Kepler, 1990
41 **Alfred Tarski and Steven Givant,** A formalization of set theory without variables, 1987
40 **R. H. Bing,** The geometric topology of 3-manifolds, 1983
39 **N. Jacobson,** Structure and representations of Jordan algebras, 1968
38 **O. Ore,** Theory of graphs, 1962
37 **N. Jacobson,** Structure of rings, 1956
36 **W. H. Gottschalk and G. A. Hedlund,** Topological dynamics, 1955
35 **A. C. Schaeffer and D. C. Spencer,** Coefficient regions for Schlicht functions, 1950
34 **J. L. Walsh,** The location of critical points of analytic and harmonic functions, 1950
33 **J. F. Ritt,** Differential algebra, 1950
32 **R. L. Wilder,** Topology of manifolds, 1949
31 **E. Hille and R. S. Phillips,** Functional analysis and semi-groups, 1957
30 **T. Radó,** Length and area, 1948
29 **A. Weil,** Foundations of algebraic geometry, 1946
28 **G. T. Whyburn,** Analytic topology, 1942
27 **S. Lefschetz,** Algebraic topology, 1942
26 **N. Levinson,** Gap and density theorems, 1940
25 **Garrett Birkhoff,** Lattice theory, 1940
24 **A. A. Albert,** Structure of algebras, 1939
23 **G. Szegö,** Orthogonal polynomials, 1939
22 **C. N. Moore,** Summable series and convergence factors, 1938
21 **J. M. Thomas,** Differential systems, 1937
20 **J. L. Walsh,** Interpolation and approximation by rational functions in the complex domain, 1935
19 **R. E. A. C. Paley and N. Wiener,** Fourier transforms in the complex domain, 1934
18 **M. Morse,** The calculus of variations in the large, 1934
17 **J. M. Wedderburn,** Lectures on matrices, 1934
16 **G. A. Bliss,** Algebraic functions, 1933
15 **M. H. Stone,** Linear transformations in Hilbert space and their applications to analysis, 1932
14 **J. F. Ritt,** Differential equations from the algebraic standpoint, 1932
13 **R. L. Moore,** Foundations of point set theory, 1932
12 **S. Lefschetz,** Topology, 1930
11 **D. Jackson,** The theory of approximation, 1930
10 **A. B. Coble,** Algebraic geometry and theta functions, 1929
9 **G. D. Birkhoff,** Dynamical systems, 1927
8 **L. P. Eisenhart,** Non-Riemannian geometry, 1922
7 **E. T. Bell,** Algebraic arithmetic, 1927
6 **G. C. Evans,** The logarithmic potential, discontinuous Dirichlet and Neumann problems, 1927
5.1 **G. C. Evans,** Functionals and their applications; selected topics, including integral equations, 1918
5.2 **O. Veblen,** Analysis situs, 1922
4 **L. E. Dickson,** On invariants and the theory of numbers
W. F. Osgood, Topics in the theory of functions of several complex variables, 1914

(*Continued in the back of this publication*)

American Mathematical Society
Colloquium Publications
Volume 31

Functional Analysis and Semi-Groups

Einar Hille
Ralph S. Phillips

American Mathematical Society
Providence, Rhode Island

2000 *Mathematics Subject Classification.* Primary 47-02;
Secondary 46-02, 47Dxx.

Library of Congress Card Number 58-102
International Standard Book Number 0-8218-1031-6
International Standard Serial Number 0065-9258

Revised and Expanded Edition 1957.

♾ The paper used in this book is acid-free and falls within the guidelines
established to ensure permanence and durability.
Visit the AMS home page at URL: `http://www.ams.org/`

13 12 11 10 9 8   05 04 03 02 01 00

TO THE MEMORY OF

J. D. TAMARKIN

WHO ENCOURAGED AND INSPIRED

US BOTH

FOREWORD TO THE REVISED EDITION

Seven years have now elapsed since the publication of the first edition of this treatise in October 1948. The friendly interest which the mathematical community has shown for this work has been most gratifying; in fact, the edition was out of print by 1954. We can also report the appearance of a Russian translation of the book in 1951 [*Funkcional'nyj Analiz i Polugruppy*, Izdatel'stvo Innostrannoj Literatury, Moskva, 1951, 636 pages].

Since 1948 both the analytical theory of semi-groups and its applications have made vigorous progress. K. Yosida found the basic generation theorem independently of Hille in 1948 and proceeded to apply it to the diffusion equation in a series of important papers. Inspired by Yosida's work, Hille made a new attack on Cauchy's problem with the aid of semi-group theory, starting in 1949. Soon thereafter, W. Feller became interested in the possibilities of the new approach and he and his students have contributed much to the theory; we mention in particular his penetrating investigation of the singular boundary value problem for the diffusion equation. Another of the early workers in the general theory of semi-groups of linear operators was R. S. Phillips who filled in many of the gaps which Hille had left behind and then went on to broaden the theory using representation theory for semi-group algebras, perturbation methods, extended classes of semi-groups, and adjoint semi-groups. In a different direction, in 1948 Hille laid the foundations of a Lie theory of semi-groups.

When early in 1952 it became obvious that a new printing of the treatise would be needed and it was clear that the new advances in the theory called for extensive revision, Hille asked Phillips to collaborate with him on a new edition. The resulting treatise is now offered to the public. The original has been completely rewritten, mostly by Phillips; the old framework is still there together with most of the old results, but much has been added, very much indeed. The changes are partly a matter of exposition, and partly a matter of methods and results. Thus in keeping with the spirit of the times the algebraic tools now play a major role and are introduced early in the book; they lead to a more satisfactory operational calculus and spectral theory in Chapters XV and XVI. On the other hand, the Laplace-Stieltjes transform methods, used by Hille for such purposes, have not been replaced but rather supplemented by the new tools.

Part I on Functional Analysis has been augmented and rearranged. The old Chapter IV (Functions on Vectors to Vectors) has been relegated to the end of the book except for certain parts essential for the body of the work which have been incorporated with Chapter III. The new chapter IV contains a treatment of the Gelfand representation theory for commutative Banach algebras together with substantial portions of the old Chapters V (Analysis in a Banach Algebra) and XXII (Notes on Banach Algebras). The new Chapter V contains a modified version of the operational calculus for Banach algebras which was presented in the old Chapter V and, in addition, it contains an operational calculus for closed unbounded linear operators. The discussion of Laplace integrals has been moved from Chapter X to Chapter VI. Part II is now called Basic Properties of Semi-Groups but contains most of the material to be found in the first half of the old Part II. However important additions have been made; we note Phillips' classification of semi-groups in Chapters X and XI and his solution of the generation problem in Chapter XII. The new part III, Advanced Analytic Theory of Semi-Groups, is based on the latter half of the old Part II. Nevertheless except for the last two chapters, the material here is largely new. It contains Phillips's main contributions to the theory (perturbation theory, adjoint theory, operational calculus, and spectral theory). Part IV, Special Semi-Groups and Applications, corresponds to the old Part III. Here the main change, sad to report, is the omission of the discussion of the applications to partial differential equations. These applications had grown so tremendously that an adequate treatment now requires a treatise of its own. All we could do was to insert a discussion of the abstract Cauchy problem in Chapter XXIII. Finally the old Appendix has become a fifth part entitled Extensions of the Theory. Here the main addition is Hille's development of the Lie theory of semi-groups.

As usual the revision has taken more work and more time than originally planned. Both authors have been liberally supported by grants from public funds, Hille through the Office of Scientific Research of the Air Research and Development Command, United States Air Force, Contract No. AF 18 (600)–469, Phillips through the Office of Ordnance Research, United States Army, Contract No. DA–04–495–Ord–406. This support is gratefully acknowledged. On the personal side it is a great pleasure for us to express our gratitude to many friends who have aided us in preparing the manuscript of the revised edition. Advice, aid, and valuable suggestions have been received from R. P. Agnew, A. V. Balakrishnan, J. Brooks, N. Dunford, H. A. Dye, N. Jacobson, S. Kakutani, T. Kato, D. G. Kendall, R. A. Moore, J. Schwartz, and K. Yosida. To all helpers, named and unnamed, we extend our warmest thanks.

New Haven, Conn. and Los Angeles, Calif. EINAR HILLE, R. S. PHILLIPS
September 1955

CONVENTIONS

Each Part of the book starts with a Summary, each Chapter with an Orientation. The chapters are divided into sections and the sections, except for orientations, are grouped into paragraphs. Cross references are normally to sections, rarely to paragraphs. Section 3.10 is the tenth section in Chapter III; it belongs to §2 which is referred to as §3.2 when necessary. The page headings show the numbers of current section and paragraph, the integral part of the former is the number of the chapter. Definitions, formulas, lemmas, and theorems are numbered separately within each section; thus Theorem 9.4.2 is the second theorem in section 9.4. References to the literature give the author's name followed by numerals in brackets referring to his book or paper by that number in the Bibliography at the close of the book. Such references are given in the text when needed; collected references for a whole chapter occur after the orientation to the chapter in question except for chapters with heterogeneous subject matter in which case they are given at the end of each paragraph.

TABLE OF CONTENTS

PART ONE

FUNCTIONAL ANALYSIS

PART TWO

BASIC PROPERTIES OF SEMI-GROUPS

PART ONE

FUNCTIONAL ANALYSIS

Summary. The first part of this treatise is devoted to an exposition of the basic ideas of modern functional analysis. There are six chapters entitled: *Abstract Spaces, Linear Operations, Vector-Valued Functions, Banach Algebras, Analysis in a Banach Algebra,* and *Laplace Integrals and Binomial Series.*

We start with a description of the algebraic and topological properties of the spaces to be considered and then proceed to a study of linear operations on one such space to another. We are mostly concerned with Banach spaces, but other topologies will occur besides the normed one. The abstract function theory starts in Chapter III where we shall study functions on scalars to vectors, on vectors to scalars, and on vectors to vectors. Additional material on the latter topic is to be found at the end of the book in Chapter XXVI. Banach algebras is one of our central topics. The discussion in Chapter IV leads up to the Gelfand representation theory for a commutative (B)-algebra with a unit element of which several important illustrations are given. In Chapter V a firm foundation is laid for the operational calculus, another of our main themes. Supplementary material on noncommutative Banach algebras without unit element will be found in Chapter XXIV. Finally Chapter VI is devoted to the analytic representation of abstract-valued functions which are holomorphic in a half-plane and satisfy a suitable boundedness condition. Again this is one of the principal tools in the later discussion.

CHAPTER I

ABSTRACT SPACES

1.1. Orientation. This chapter is intended to give a brief review of the basic notions in the theory of abstract spaces which will be needed in what follows. The material is grouped under four paragraph headings: *Topological Concepts*, *Additive Spaces*, *Linear Spaces*, and *Algebraic Spaces*. The abstract spaces occurring in this treatise normally have a definite algebraic structure in addition to being topological spaces of one type or another. This fact underlies the choice of the headings. As practically all the material is taken from current literature, proofs are usually omitted. The reader who needs further explanations is referred to the literature quoted at the end of the paragraphs.

1. Topological Concepts

1.2. Sets. Let $\mathfrak{S}$ be an abstract set with elements $x, y, \cdots; X, Y, X_\alpha$ stand for subsets of $\mathfrak{S}$. The symbols $X \cup Y$ and $\bigcup_\alpha X_\alpha$ denote *unions*, $X \cap Y$ and $\bigcap_\alpha X_\alpha$ *intersections* of the indicated subsets. X and Y are said to be *disjoint* if $X \cap Y = \emptyset$, the *empty set*. The complement of X with respect to $\mathfrak{S}$ is written as $\tilde{X}$ or $\mathfrak{S} \ominus X$; more generally the complement of X with respect to Y is $Y \cap \tilde{X} = Y \ominus X$. We denote the set consisting of all elements x possessing the property P by $[x; \text{P}]$. Often when the property P is clear from the context this symbol is abbreviated as simply $[x]$, or in the case of a sequence of elements as $\{x_n\}$.

A set $\mathfrak{S}$ is *partially ordered* if for some pairs of elements x, y an *ordering relation* $x < y$ exists (also denoted by $y > x$) which satisfies

O_1. *For all* x, $x < x$ *(reflexive)*,
O_2. *If* $x < y$ *and* $y < x$, *then* $x = y$ *(proper)*,
O_3. *If* $x < y$ *and* $y < z$, *then* $x < z$ *(transitive)*.

If $x < y$ we say that x is contained in y or that y contains x. The set $\mathfrak{S}$ is *simply ordered* if all pairs x, y are ordered. From a given statement about a partially ordered set we obtain a *dual* statement by reversing all of the order relations involved. It will be noticed that our postulates are self-dual so that any theorem about a partially ordered set will imply its dual.

In a partially ordered set $\mathfrak{S}$ the subset Y is said to have y_0 for an *upper bound* if $y < y_0$ for every $y \in Y$. The element x_0 is *maximal* for $\mathfrak{S}$ if $x > x_0$ implies $x_0 > x$. The notions of *lower bound* and *minimal element* are defined dually. For the existence of maximal elements we have the convenient *maximal principle* of Max Zorn, one form of which is

MAXIMAL PRINCIPLE. *If $\mathfrak{S}$ is a partially ordered set in which each simply ordered subset has an upper bound in $\mathfrak{S}$, then $\mathfrak{S}$ contains at least one maximal element.*

The maximal principle is equivalent to the axiom of choice; its use has been found essential to many parts of the theory of abstract spaces.

For a partially ordered set we define a *least upper bound* (l.u.b.) of a subset Y to be an upper bound which is contained in every other upper bound. A *greatest lower bound* (g.l.b.) is defined dually. It is clear from O_2 that a given subset can have at most one l.u.b. and g.l.b. A *lattice* is a partially ordered set $\mathfrak{S}$ any two of whose elements x, y have a l.u.b. $x \vee y$ and a g.l.b. $x \wedge y$. We see by induction that any finite subset of a lattice has a l.u.b. and a g.l.b. For a given subset X we denote by $\vee [x; x \in X]$ and $\wedge [x; x \in X]$ the l.u.b. and the g.l.b., respectively, of the set X if these bounds exist.

1.3. Hausdorff spaces. A set $\mathfrak{X}$ of *points* and a collection of *subsets* Σ defines a *Hausdorff space* if

H_1. *For any two distinct points x and y of $\mathfrak{X}$ there exist disjoint sets N_1 and N_2 of Σ such that $x \in N_1$ and $y \in N_2$;*

H_2. *For any two sets N_1 and N_2 of Σ which contain the same point x, there exists a set N_3 of Σ such that $x \in N_3 \subset N_1 \cap N_2$.*

A subset $N \in \Sigma$ is called a *neighborhood* for each x belonging to N, and the system Σ is called a *neighborhood basis* (or a *basis*) for $\mathfrak{X}$. For a detailed discussion of these topological concepts the reader is referred to Alexandroff-Hopf [1] or Kuratowski [1].

A set G is *open* if each point $x \in G$ is contained in a neighborhood contained in G. A set F is *closed* if $\tilde{F}$ is open. *Open sets have an open union and closed sets a closed intersection.* The intersection of a countable number of open sets is called a G_δ set, and the union of a countable number of closed sets an F_σ set.

The set of all open sets or, equivalently, the set of all closed sets, determines by definition the topology of $\mathfrak{X}$. Hence any system Σ' of open sets such that each open subset of $\mathfrak{X}$ can be obtained as the union of open sets belonging to Σ' can serve as a *neighborhood basis* for $\mathfrak{X}$, determining the same topology. Such a system Σ' will automatically satisfy the postulates H_1 and H_2 in a Hausdorff space. For this reason we call any open set containing the point x a neighborhood of x, denoted by $N(x)$. A neighborhood basis for a given point x is a system $\Sigma'(x)$ of neighborhoods of x such that each open set containing x also contains some $N(x) \in \Sigma'(x)$.

For an arbitrary subset X, the *closure* of X, denoted by $\bar{X}$, is defined as the set of all x such that every $N(x)$ has a non-empty intersection with X. Equivalently, $\bar{X}$ is the intersection of all closed sets containing X. It is easy to show that the closure operation satisfies the following conditions.

C_1. $\overline{X \cup Y} = \bar{X} \cup \bar{Y}$,

C_2. $\overline{\overline{X}} = \bar{X}$,
C_3. $\bar{X} = X$ *if* X *is empty or is a single point.*
From this it follows that $X \subset \bar{X}$. Further $X = \bar{X}$ if and only if X is closed.

REMARK. A set $\mathfrak{X}$ together with a closure operation defined on all subsets and satisfying the conditions C_1, C_2, and C_3 is called a *topological space*. Translated in terms of a basis this means that a topological space satisfies the condition H_2 but only a weakened form of the *separation axiom* H_1, namely,
H_1'. *For any two distinct points x and y of $\mathfrak{X}$ there exists a set $N \in \Sigma$ such that $x \in N$ and $y \notin N$.*

We shall not have occasion to make use of the added generality afforded by a topological space.

We say that a topological space satisfies the *first countability axiom* if each of its points possesses a countable basis. If the entire space has a countable basis, then we say that the space satisfies the *second countability axiom.*

The point x_0 is a *limit point* of the set X if $x_0 \in \overline{X \ominus x_0}$ or equivalently if every $N(x_0)$ contains points of X distinct from x_0. The set X' of limit points of X is called the *derived set* of X; $\bar{X} = X \cup X'$ and hence X is closed if $X' \subset X$. X is *dense in itself* if $X' \supset X$ and *perfect* if $X' = X$. A set X is *dense* in $\mathfrak{X}$ if $\bar{X} = \mathfrak{X}$, dense in Y if $Y \subset \overline{X \cap Y}$. X is a *separable set* if there is a countable set which is dense in X. In particular, $\mathfrak{X}$ is a *separable space* if there is a countable set which is dense in $\mathfrak{X}$. If $\mathfrak{X}$ satisfies the second countability axiom then it is necessarily separable.

The union of all open sets contained in X is the *interior* of X, denoted by $\text{Int}(X)$. If $x \in \text{Int}(X)$, then x is an *interior point* of X. The *boundary* of X is the intersection of the closures of X and of $\tilde{X}$. X is *nowhere dense* if $\text{Int}(\bar{X}) = \emptyset$. X is of the *first category* in $\mathfrak{X}$ if X is the union of a countable number of sets each of which is nowhere dense in $\mathfrak{X}$; otherwise X is of the second category. If X is of the second category, then $\text{Int}(\bar{X})$ is not empty.

If X_0 is a subset of $\mathfrak{X}$, then a *relative topology* can be introduced in X_0 by means of the neighborhood basis Σ_0 consisting of all sets $N_0 = N \cap X_0$ where $N \in \Sigma$. It is clear that in this topology the *relative closure* of a subset $Y \subset X_0$ is $\bar{Y} \cap X_0$.

A topological space is *connected* if it is not the union of two open non-void disjoint sets. In a connected space the empty set and the whole space are the only sets which are simultaneously open and closed (*clopen*). A subset X_0 of $\mathfrak{X}$ is connected if it is not the union of two relatively open non-void disjoint subsets of X_0. A *component* of a set is a connected subset which is not contained in any larger connected subset. Each point of a set X determines a unique component containing it and X is the union of its components.

A system of sets $[A_\alpha]$ is called a *covering* of the set X if each point of X is an interior point of at least one set A_α. A set X has the *Borel property* if every system of open sets $[G_\alpha]$ which covers X contains a finite sub-system also covering X. A system of sets $[B_\alpha]$ has the *finite intersection property* if every finite subsystem has a non-empty intersection.

A subset X is called *compact* if it has the Borel property. In particular, $\mathfrak{X}$ is a *compact space* if it has the Borel property. It is easily shown by way of complementation that a set X is compact if and only if each system of relatively closed subsets of X having the finite intersection property has a non-empty intersection. Compact subsets of a Hausdorff space are necessarily closed and each closed subset of a compact set is itself compact. A subset is called *conditionally compact* if its closure is compact. A space is said to be *locally compact* if each point of the space has a conditionally compact neighborhood. A locally compact space can be embedded in a compact space by appropriately adjoining one other point.

One can also introduce more restrictive separation axioms. A topological space is called *regular* if for any point x and closed set F, $x \cap F = \emptyset$, there exist disjoint open sets G_1 and G_2 such that $x \in G_1$ and $F \subset G_2$. A topological space is called *normal* if for any two disjoint closed sets F_1 and F_2 there exist disjoint open sets G_1 and G_2 such that $F_1 \subset G_1$ and $F_2 \subset G_2$. It is clear that a normal space is regular and that a regular space is Hausdorff. On the other hand a compact Hausdorff space is necessarily normal.

We next define the notion of a *directed set*; this is a partially ordered system of indices $\mathfrak{A}$ such that any two elements of $\mathfrak{A}$ have a common successor; this means that given $\alpha, \beta \in \mathfrak{A}$ there exists a $\gamma \in \mathfrak{A}$ such that $\alpha < \gamma$ and $\beta < \gamma$. A directed set of points $[x_\alpha]$ is a function assigning to each $\alpha \in \mathfrak{A}$ a point $x_\alpha \in \mathfrak{X}$. Following E. H. Moore and H. L. Smith we say that the directed set $[x_\alpha]$ contained in a topological space *converges to a limit point* $x \in \mathfrak{X}$ (in symbols, $\lim_\alpha x_\alpha = x$ or $x_\alpha \to x$) if for each neighborhood $N(x)$ of x there exists a $\beta(N) \in \mathfrak{A}$ such that $x_\alpha \in N(x)$ for all $\alpha > \beta(N)$. In a Hausdorff space a convergent directed set of points has a unique limit point.

Another very useful notion is that of a *Cartesian product*. Corresponding to each element α of an index set $\mathfrak{A}$, we suppose given a topological (Hausdorff) space $\mathfrak{X}_\alpha$. Then the Cartesian product $\prod_{\alpha \in \mathfrak{A}} \mathfrak{X}_\alpha$ is defined as the set of all functions p with domain $\mathfrak{A}$ such that $p(\alpha) \in \mathfrak{X}_\alpha$ for every $\alpha \in \mathfrak{A}$. When $\mathfrak{A}$ consists simply of the integers $(1, 2, \cdots, n)$, $\prod_\alpha \mathfrak{X}_\alpha$ is also denoted by $\mathfrak{X}_1 \times \mathfrak{X}_2 \times \cdots \times \mathfrak{X}_n$. We introduce a topology in the product set $\prod_\alpha \mathfrak{X}_\alpha$ as follows. Let Σ_α be a neighborhood basis for $\mathfrak{X}_\alpha$. Then a neighborhood basis Σ for $\prod_\alpha \mathfrak{X}_\alpha$ is given by the system of all product sets of the form $\prod_\alpha N_\alpha$ where $N_\alpha \in \Sigma_\alpha$ for a finite subset of indices and $N_\alpha = \mathfrak{X}_\alpha$ for the remaining indices. It is easy to show that Σ satisfies the postulates H_1' and H_2 (H_1 and H_2) and hence that $\prod_\alpha \mathfrak{X}_\alpha$ is a topological (Hausdorff) space. According to a theorem due to A. Tychonoff, *the Cartesian product of a family of compact topological spaces is compact.* For a proof see S. Lefschetz [1, p. 19].

1.4. Continuous transformations. Let $\mathfrak{X}$ and $\mathfrak{Y}$ be two Hausdorff spaces which may be identical or distinct. Let $y = T(x)$ be a single-valued function with domain $\mathfrak{D}$ in $\mathfrak{X}$ and range $\mathfrak{R}$ in $\mathfrak{Y}$, that is, to each $x \in \mathfrak{D}$ corresponds a unique $y \in \mathfrak{R}$ and each y in $\mathfrak{R}$ is the image of at least one x in $\mathfrak{D}$. Then $y = T(x)$ is said to define a *mapping* (*transformation*) *of* (*on*) $\mathfrak{D}$ *onto* $\mathfrak{R}$. In order to indicate merely that the range $\mathfrak{R}$ is contained in $\mathfrak{Y}$, the mapping or transformation is said to be *into* $\mathfrak{Y}$ or simply *to* $\mathfrak{Y}$.

If $X \subset \mathfrak{D}$, the symbol $T(X)$ denotes the set $[T(x); x \in X]$ and $T(X)$ is called the *image* of X under the mapping T. If $Y \subset \mathfrak{Y}$ then the set $[x; T(x) \in Y]$ is called the *inverse image* of Y and is denoted by $T^{-1}(Y)$. It is clear that

$$Y = T[T^{-1}(Y)] \quad \text{for all} \quad Y \subset \mathfrak{R},$$

$$X \subset T^{-1}[T(X)] \quad \text{for all} \quad X \subset \mathfrak{D},$$

and, in general, the inclusion is proper. The mapping is *one-to-one* if equality holds, that is, if $Y = T(X)$ implies $X = T^{-1}(Y)$ for every $X \subset \mathfrak{D}$. In this case there exists a unique single-valued function $x = T^{-1}(y)$ with domain $\mathfrak{R}$ and range $\mathfrak{D}$ such that

$$T[T^{-1}(y)] = y \quad \text{for all} \quad y \in \mathfrak{R}$$

and

$$T^{-1}[T(x)] = x \quad \text{for all} \quad x \in \mathfrak{D}.$$

We call $x = T^{-1}(y)$ the *inverse* of $y = T(x)$.

The point set $[(x, T(x)); x \in \mathfrak{D}]$ in the (Cartesian) product space $\mathfrak{X} \times \mathfrak{Y}$ is known as the *graph* of $T(x)$.

We come now to the concept of a continuous transformation. Here we shall take $\mathfrak{D} = \mathfrak{X}$; this is merely a notational convenience which avoids the use of the relative topology for $\mathfrak{D}$. Further we shall denote closure in both $\mathfrak{X}$ and $\mathfrak{Y}$ by the same symbol.

The transformation $y = T(x)$ is continuous at $x = x_0$ if for every set X the condition $x_0 \in \bar{X}$ implies that $T(x_0) \in \overline{T(X)}$. $T(x)$ is a continuous transformation if $T(x)$ is continuous at all points.

An equivalent formulation in the neighborhood topology for the continuity of $y = T(x)$ at $x = x_0$ is the following: For each $N[T(x_0)]$ there exists an $N(x_0)$ such that $T[N(x_0)] \subset N[T(x_0)]$. Necessary and sufficient conditions in order that $T(x)$ be a continuous transformation are:

(1) $T(\bar{X}) \subset \overline{T(X)}$ for every set X, or

(2) $T^{-1}(G)$ is open for every open set G, or

(3) $T^{-1}(F)$ is closed for every closed set F.

Functions of two or more variables are discussed in the same manner by introducing the product space on which they are defined. In the case of Hausdorff spaces $\mathfrak{X}$ and $\mathfrak{Y}$ satisfying the first countability axiom, $T(x)$ is continuous at $x = x_0$ if and only if $x_n \to x_0$ implies that $T(x_n) \to T(x_0)$.

If $y = T(x)$ is a one-to-one mapping of $\mathfrak{X}$ onto $\mathfrak{Y}$ and if both T and T^{-1} are continuous, then $T(x)$ is called a homeomorphic mapping or simply a homeomorphism.

A continuous transformation maps compact subsets onto compact subsets. Since compact subsets of a Hausdorff space are closed, it follows that a one-to-one continuous mapping of a compact topological space X onto a Hausdorff space Y has a continuous inverse and is therefore a homeomorphism.

A numerically-valued function on an abstract space is known as a *functional*. Concerning the existence of continuous functionals, Urysohn has established the following result: If F_0 and F_1 are disjoint closed subsets of a normal space $\mathfrak{X}$, then there exists a continuous real-valued functional f on $\mathfrak{X}$ such that $f(x) = 0$ on F_0, $f(x) = 1$ on F_1, and $0 \leqq f(x) \leqq 1$ elsewhere.

Let $\mathfrak{F}$ be a family of continuous complex-valued functionals defined on a topological space $\mathfrak{X}$ and let Σ' be the system of open sets $[f^{-1}(U);\ U$ an open subset of the complex plane, $f \in \mathfrak{F}]$. If the functionals in $\mathfrak{F}$ collectively separate (i.e. distinguish between) points of $\mathfrak{X}$, then Σ' is a neighborhood basis for a Hausdorff topology on the point set $\mathfrak{X}$. It is clear that Σ' defines a *weaker topology* than the original topology in the sense that the open sets arising from Σ' are also open in the original topology.

THEOREM 1.4.1. *If $\mathfrak{F}$ is a family of continuous complex-valued functionals on a compact Hausdorff space, separating the points of $\mathfrak{X}$, then the topology induced by $\mathfrak{F}$ is equivalent with the original topology.*

PROOF. Let $\mathfrak{X}'$ be the induced topology on $\mathfrak{X}$. Then the identity mapping on $\mathfrak{X}$ to $\mathfrak{X}'$ is clearly one-to-one and continuous; it is therefore a homeomorphism.

We conclude this section with a result typical of the category theorems which appear throughout the study of abstract spaces.

THEOREM 1.4.2. *Let $\mathfrak{F}$ be a family of lower semi-continuous real-valued functionals on $\mathfrak{X}$ having the property that* $\sup\ [f(x); f \in \mathfrak{F}] < \infty$ *for each x belonging to a given set X_0 of the second category. Then there exists an open set $G \subset \mathrm{Int}(\overline{X_0})$ and a constant M such that $f(x) \leqq M$ for all $x \in G$ and for all $f \in \mathfrak{F}$.*

PROOF. Set $Y_n = [x; f(x) \leqq n, x \in \bar{X}_0, f \in \mathfrak{F}]$. Then each Y_n is closed and $X_0 \subset \bigcup Y_n \subset \overline{X_0}$. Since X_0 is of the second category, it cannot be the union of a denumerable number of sets of the first category. Thus one of the Y_n's, say Y_{n_0}, is also of the second category. Consequently there exists an open set $G \subset \mathrm{Int}(Y_{n_0})$ and setting $M = n_0$ we see that G fulfills the requirements of the theorem.

1.5. Metric spaces. A set $\mathfrak{X}$ is called a *metric space* if for each pair of points x, y in $\mathfrak{X}$ there is defined a real-valued function $d(x, y)$, called the *distance* from x to y, subject to the two postulates of Lindenbaum:

D_1. $d(x, y) = 0$ *if and only if* $x = y$,

D_2. $d(x, y) \leqq d(z, x) + d(z, y)$ *for any three points* x, y, *and* z.

These two properties imply:

D_3. $d(x, y) \geqq 0$,

D_4. $d(x, y) = d(y, x)$,

and in view of D_4 the *triangle axiom* D_2 may be written as

D_2'. $d(x, y) \leqq d(x, z) + d(z, y)$.

In a metric space $\mathfrak{X}$ the set of points $[x; d(x_0, x) < \epsilon]$ is called the *sphere* of radius ϵ about x_0. The aggregate of all spheres Σ satisfies the postulates H_1 and H_2 and constitutes a basis for the topology of $\mathfrak{X}$. Thus topologized, a metric space is normal and satisfies the first countability axiom. A topological space in which an equivalent neighborhood basis can be defined in this way is said to be *metrizable*.

A metric space is separable if and only if it satisfies the second countability axiom. Conversely, Urysohn has shown that a normal topological space which satisfies the second countability axiom is metrizable.

In a metric space a sequence $\{x_n\}$ converges to a limit x_0 if and only if $d(x_0, x_n) \rightarrow 0$. A convergent sequence $\{x_n\}$ satisfies the *Cauchy condition*: $\lim_{m,n\rightarrow\infty} d(x_m, x_n) = 0$; any sequence satisfying this condition is called a *Cauchy sequence*. A metric space is said to be *complete* if every Cauchy sequence converges to a limit. Each metric space can be embedded in a complete metric space, uniquely defined as the set of all Cauchy sequences, $\{x_n\}, \{y_n\}, \cdots$, with $d(\{x_n\}, \{y_n\}) = \lim_{n\rightarrow\infty} d(x_n, y_n)$ and where the sequences whose distance apart is zero are identified.

A complete metric space is of the second category. In fact if a set X is of the first category in a complete metric space, then $\tilde{X}$ is dense in the space. On the other hand if X is a set of the second category contained in any metric space, then the closure of X contains a sphere. It follows that if $\mathfrak{X}$ is a complete metric space and if $\mathfrak{X}$ is represented as the union of a countable number of subsets, $\mathfrak{X} = \bigcup_n X_n$, then for at least one value of n the closure of X_n contains a sphere.

A set X is *bounded* if it is contained in a sphere. The *diameter* of a bounded set X is sup $[d(x, y); x, y \in X]$. A set X is *totally bounded* if for each $\delta > 0$ the set may be covered by a finite number of spheres of diameter $<\delta$. A totally bounded set is separable. In a complete metric space a totally bounded set is conditionally compact, that is, its closure is compact. Conversely, a compact subset of a metric space is totally bounded and a compact metric space is complete. A topological space $\mathfrak{X}$ is said to be *sequentially compact* if every sequence $\{x_n\}$ contains a subsequence which converges to a point in $\mathfrak{X}$. In a metric space a set is sequentially compact if and only if it is compact.

If $\mathfrak{X}$ and $\mathfrak{Y}$ are complete metric spaces, if $\{T_n(x)\}$, $n = 1, 2, \cdots$, are transformations with domain $\mathfrak{D}_n \subset \mathfrak{X}$ and range $\mathfrak{R}_n \subset \mathfrak{Y}$, and if $\lim_{n\rightarrow\infty} T_n(x)$ exists for each x in $\mathfrak{D} \subset \bigcap_n \mathfrak{D}_n$, then the limit defines a transformation $T(x)$ with domain $\mathfrak{D}$ and the sequence $\{T_n\}$ is said to converge to T in $\mathfrak{D}$. If $\mathfrak{X}$ and $\mathfrak{Y}$ are complete metric spaces, if $y = T(x)$ has $\mathfrak{X}$ as its domain and has its range in $\mathfrak{Y}$, and if $T(x)$ takes bounded sets in $\mathfrak{X}$ into totally bounded sets in $\mathfrak{Y}$, then $T(x)$ is called a *compact transformation*.

References. Alexandroff and Hopf [1], G. Birkhoff [5], Kuratowski [1], Lefschetz [1], Moore and Smith [1].

2. Additive Spaces

1.6. Algebraic systems. The topological spaces which are most useful in functional analysis are at the same time *algebraic systems*. By this we mean that one or more algebraic operations are defined in the space which is closed under the operations in question. There are essentially three *binary operations*, conventionally referred to as *addition, multiplication*, and *scalar multiplication*, which come into consideration in this connection.

The simplest and most primitive systems are obtained when there is only one operation defined, either addition or multiplication. The system is then a *semigroup* or possibly a *group* depending upon the stringency of the postulates. When two operations are defined there are two useful alternatives: addition and multiplication give rise to a *ring*, whereas addition and scalar multiplication give rise to a *linear system*. Finally if all three operations are defined, the system is an *algebra*. Formal definitions will be given when needed; here only an orientation is desired.

The systems discussed in the remainder of this chapter are determined by three sets of postulates: the first defining the algebraic operations which are permissible, the second the topology of the space, and the third the relations between operations and topology. The purpose of the third set is to ensure continuity of the operations in the particular topology under consideration.

From this enumeration the reader will perceive that there are five basic possibilities of constructing algebraico-topological systems. We shall, however, omit multiplicative spaces as well as ring spaces. The three remaining alternatives, additive, linear, and algebraic spaces, will be highly specialized since we shall have no use for the general theory of such spaces.

Definition 1.6.1. *Let $\mathfrak{X}$ and $\mathfrak{X}'$ be two algebraic systems of the same kind such as two groups, two linear systems over the same field, or two algebras over the same field. A mapping $x \to x'$ of $\mathfrak{X}$ onto $\mathfrak{X}'$ is called a homomorphism if the defined algebraic combinations are preserved. We denote this relationship by $\mathfrak{X} \sim \mathfrak{X}'$. If the mapping is one-to-one, it is called an isomorphism and we write $\mathfrak{X} \cong \mathfrak{X}'$. An isomorphism of $\mathfrak{X}$ onto itself is called an automorphism. If the two systems are also topological algebraic spaces, they are then called isomorphic only if an algebraic isomorphism exists which is at the same time a homeomorphism.*

1.7. Additive groups. Let $\mathfrak{X}$ be a set, containing at least two distinct elements, in which a binary operation called addition is defined.

Definition 1.7.1. *$\mathfrak{X}$ is an additive group if the following conditions hold:*

A_1. *Every ordered pair of elements x, y has a uniquely defined sum $x + y$;*

A_2. *Addition is associative, that is, $(x + y) + z = x + (y + z)$;*

A_3. *There is a zero element θ such that $x + \theta = \theta + x = x$ for all x;*

A_4. *To every x there is a negative, written $-x$, such that $x + (-x) = \theta$.*

The postulates imply that the zero element and the negative are unique and that $(-x) + x = \theta$ for all x, whence it follows that $-(-x) = x$. The properties of the negative give the *law of cancellation*:

$$x + z = y + z \quad \text{or} \quad z + x = z + y \quad \text{implies} \quad x = y.$$

If addition satisfies the further postulate

A_5. *Addition is commutative*: $x + y = y + x$,

the group is called a *commutative* or *abelian* group.

A subset X of $\mathfrak{X}$ is called a *subgroup* if X is an additive group under the same operation.

If X and Y are subsets of $\mathfrak{X}$, the symbols $-X$, $X + Y$, and $X - Y$ denote the sets $[-x; x \in X]$, $[x + y; x \in X, y \in Y]$, and $[x - y; x \in X, y \in Y]$ respectively. In particular $X + y = [x + y; x \in X]$ and $y + X = [y + x; x \in X]$.

A homomorphic mapping of a group X onto a group X' is an isomorphism if and only if $x \to \theta$ implies that $x = \theta$.

1.8. Additive group spaces. Let $\mathfrak{X}$ be an additive group and suppose that a topology has been introduced in $\mathfrak{X}$ making $\mathfrak{X}$ a Hausdorff space.

Definition 1.8.1. *An additive group $\mathfrak{X}$ is a topological additive group if $\mathfrak{X}$ is a Hausdorff space and to every pair of elements x, y and every neighborhood $N(x - y)$ of $x - y$ there are neighborhoods $N(x)$ of x and $N(y)$ of y such that $N(x) - N(y) \subset N(x - y)$.*

Stated in other words this postulate requires that $x - y$ be a continuous function of (x, y). This postulate is equivalent to the two conditions: $-x$ continuous in x and $x + y$ continuous in (x, y). This of course implies that all topological properties of subsets of X are left invariant by the group operations of translation and inversion. In particular we see that $\mathfrak{X}$ possesses a *uniform topology*, that is to say, the translates of the neighborhoods of $x = \theta$ form a basis for the space. It also follows that the mappings $x \to x + a$, $a + x$, and $-x$ are homeomorphisms of $\mathfrak{X}$ onto $\mathfrak{X}$ for each a. Actually a somewhat weaker postulate suffices for this purpose (see E. Hille [8]).

References. Banach [2], van Dantzig [1], Hille [8], Pontrjagin [1], A. Weil [1, 2].

3. Linear Spaces

1.9. Linear systems. In an additive group $\mathfrak{X}$ we introduce a second operation, that of *scalar multiplication*. The scalars could form any *domain of integrity* without essential modification of the subsequent postulates, but for the sake of simplicity we restrict ourselves to the most important case in which the scalars are either *real* or *complex numbers*. The scalar field will be denoted by Φ.

DEFINITION 1.9.1. *$\mathfrak{X}$ is a linear system (or module) over the scalar field Φ if its elements admit of the two operations of addition and scalar multiplication, subject to the following conditions:*

Addition satisfies postulates A_1 *to* A_4 *of Definition* 1.7.1 *together with* A_5.

Scalar multiplication satisfies:

S_1. *To every scalar $\alpha \in \Phi$ and every element $x \in \mathfrak{X}$, there is a uniquely defined scalar product $\alpha x = x\alpha$ in $\mathfrak{X}$;*

S_2. $(\alpha + \beta)x = \alpha x + \beta x$;

S_3. $\alpha(x + y) = \alpha x + \alpha y$;

S_4. $\alpha(\beta x) = (\alpha\beta)x$;

S_5. $1 \cdot x = x$.

From S_2, S_5, and A_3 it follows that $-x = (-1)x$ and that $0 \cdot x = \theta$ for all x.

If X is a subset of $\mathfrak{X}$ and A a subset of Φ, the then symbol AX stands for the set $[\alpha x; \alpha \in A, x \in X]$.

A linear system, with or without an imposed topology, is usually referred to as a *linear vector space* and the elements are called *vectors*. A system of n vectors $x_1, x_2, \cdots, x_n$ is said to be *linearly independent* if the equation

$$\alpha_1 x_1 + \alpha_2 x_2 + \cdots + \alpha_n x_n = \theta, \qquad \alpha_k \in \Phi,$$

implies $\alpha_1 = \alpha_2 = \cdots = \alpha_n = 0$. They are *linearly dependent* if such an equal tion holds in which at least one coefficient is different from zero. If $\mathfrak{X}$ contains n linearly independent vectors, but every system of $(n + 1)$ vectors is linearly dependent, then $\mathfrak{X}$ is said to be of *dimension n*. If the number of linearly independent vectors is not finite, then $\mathfrak{X}$ is said to be of *infinite dimension*. Any set of n linearly independent vectors $[y_k]$ in an n-dimensional linear system $\mathfrak{X}$ constitutes a *basis* for $\mathfrak{X}$ and each vector $x \in \mathfrak{X}$ has a unique representation of the form $x = \sum_1^n \alpha_k y_k$, $\alpha_k \in \Phi$.

A linear system $\mathfrak{L}$ contained in $\mathfrak{X}$ is called a *linear subspace* of $\mathfrak{X}$ or a *linear manifold* in $\mathfrak{X}$. To each set X in $\mathfrak{X}$ there is a *least linear subspace* $\mathfrak{L}(X)$ which contains X; $\mathfrak{L}(X)$ consists of all elements of the form $\sum_1^n \alpha_k x_k$ where $\alpha_k \in \Phi$, $x_k \in X$, and n is finite; $\mathfrak{L}(X)$ is often referred to as the *linear extension* of X. A set of the form $a + \mathfrak{L}$ is called a *flat space*. A linear subspace $\mathfrak{M}$ is said to be *maximal* if it is not equal to the entire space and if it is not properly contained in any other linear subspace. A set of the form $a + \mathfrak{M}$ is called a *maximal flat space*. Finally we note that a complex linear system $\mathfrak{X}$ can also be treated as a real linear system since multiplication by real scalars is well defined. In this context we shall speak of *real linear subspaces* of $\mathfrak{X}$.

A subset C of $\mathfrak{X}$ is said to be *convex* if $x_1, x_2 \in C$ implies that the *line segment* $[x_1, x_2] \equiv [\alpha x_1 + (1 - \alpha)x_2; 0 \leqq \alpha \leqq 1] \subset C$. To each set X in $\mathfrak{X}$ there is a least convex set $\mathfrak{C}(X)$ which contains X; $\mathfrak{C}(X)$ consists of all elements of the form $\sum_1^n \alpha_k x_k$ where $\alpha_k \geqq 0$, $\sum_1^n \alpha_k = 1$, $x_k \in X$, and n is finite; the set $\mathfrak{C}(X)$ is also called the *convex extension* of X.

A subset X of a complex linear system $\mathfrak{X}$ is called *circular* if $x \in X$ implies that $e^{i\varphi}x \in X$ for $0 < \varphi < 2\pi$.

Under a homomorphism linear manifolds, convex sets, and circular sets go into their respective varieties. Two finite dimensional linear systems over the same scalar field are isomorphic if and only if they are of the same dimension.

Suppose $\mathfrak{X}_1, \mathfrak{X}_2, \cdots, \mathfrak{X}_n$ are n linear systems over the same scalar field Φ. The *product* of these systems, denoted by $\mathfrak{X}_1 \times \mathfrak{X}_2 \times \cdots \times \mathfrak{X}_n$, is the set of all ordered n-tuples $(x_1, x_2, \cdots, x_n)$ where $x_k \in \mathfrak{X}_k$; the arithmetic operations are defined as

$$(x_1, x_2, \cdots, x_n) + (y_1, y_2, \cdots, y_n) = (x_1 + y_1, x_2 + y_2, \cdots, x_n + y_n),$$

$$\alpha(x_1, x_2, \cdots, x_n) = (\alpha x_1, \alpha x_2, \cdots, \alpha x_n).$$

The product system is also referred to as the *direct sum*, in which case it is denoted by $\mathfrak{X}_1 \oplus \mathfrak{X}_2 \oplus \cdots \oplus \mathfrak{X}_n$. The notion can be generalized in an obvious fashion to any indexed family of linear systems over Φ.

We conclude this section with a result on convex sets due to S. Kakutani [1] and J. W. Tukey [1] which will be basic for our treatment of separation theorems in Chapter II.

THEOREM 1.9.1. *Suppose that A and B are disjoint convex sets in a linear system $\mathfrak{X}$. Then there exist complementary convex sets C and D such that $C \supset A$ and $D \supset B$.*

PROOF. Let $\mathfrak{F}$ be the family of all pairs of convex sets (X, Y) such that $X \supset A$, $Y \supset B$, and $X \cap Y = \emptyset$. Write $(X, Y) < (X', Y')$ if $X \subset X'$ and $Y \subset Y'$. Then $\mathfrak{F}$ is partially ordered by this relation and each simply ordered subfamily has an upper bound in $\mathfrak{F}$. By the maximal principle $\mathfrak{F}$ contains a maximal element (C, D). It is required to show that $C \cup D = \mathfrak{X}$. Consider an arbitrary point $p \in \mathfrak{X}$. We cannot have simultaneously $c \in [p, d]$ and $d' \in [p, c']$ where $c, c' \in C$ and $d, d' \in D$; for this would imply that $[c, c']$ intersects $[d, d']$, contrary to $C \cap D = \emptyset$. Suppose then that $c \in [p, d]$ does not occur. In this case the convex extension D' of $D \cup p$ will be disjoint from C so that $(C, D') \in \mathfrak{F}$. From the maximality of (C, D) it follows that $D = D'$ and hence that $C \cup D = \mathfrak{X}$.

1.10. Linear spaces. We suppose that $\mathfrak{X}$ is a linear system and proceed to introduce a topology in $\mathfrak{X}$ in such a manner that the arithmetic operations become continuous.

DEFINITION 1.10.1. *A linear system $\mathfrak{X}$ is a topological linear space if $\mathfrak{X}$ is a topological additive group and to every $\alpha \in \Phi$, $x \in \mathfrak{X}$ and every neighborhood $N(\alpha x)$ of αx there are neighborhoods $N(\alpha)$ of α and $N(x)$ of x such that $N(\alpha)N(x) \subset N(\alpha x)$.*

The added effect of this postulate is to require that αx be a continuous function of (α, x). This suffices to make the mapping $T(x) = \alpha x$ of $\mathfrak{X}$ onto $\mathfrak{X}$ homeomorphic for each $\alpha \neq 0$.

The most familiar example of a topological linear space is the real (or complex) n-dimensional euclidean space E_n (or Z_n) consisting of all n-tuples $\alpha = (\alpha_1, \cdots, \alpha_n)$ of real (or complex) numbers α_k. Addition and scalar multiplication are defined by

$$\alpha + \beta = (\alpha_1 + \beta_1, \cdots, \alpha_n + \beta_n), \qquad \gamma\alpha = (\gamma\alpha_1, \cdots, \gamma\alpha_n);$$

and the topology is defined by the metric function

$$d(\alpha, \beta) = \{\sum_1^n | \alpha_k - \beta_k |^2\}^{\frac{1}{2}}.$$

Verification of the postulates is left to the reader. The euclidean space is typical of all finite dimensional topological linear spaces, as the following result due to A. Tychonoff [1] shows.

THEOREM 1.10.1. *Each n-dimensional topological linear space $\mathfrak{X}_n$ over the real (or complex) numbers is isomorphic with E_n (or Z_n).*

PROOF. Let $[y_k; k = 1, 2, \cdots, n]$ be a basis for $\mathfrak{X}_n$. Then the mapping $T(\{\alpha_k\}) = \sum_1^n \alpha_k y_k = x$ defines an algebraic isomorphism between the linear systems E_n (or Z_n) and $\mathfrak{X}_n$. Further the mapping is continuous; for given $N(x)$ there exist neighborhoods $N(\alpha_k)$ of $\alpha_k \in \Phi$ such that $\sum_1^n N(\alpha_k)y_k \subset N(x)$. The topologies being uniform, it is therefore sufficient to show that $T(S_\epsilon)$, $S_\epsilon \equiv [\{\alpha_k\}; d(\{\alpha_k\}, \theta) < \epsilon]$, contains an open subset of $\mathfrak{X}_n$. Now the boundary C_ϵ of S_ϵ is compact and consequently so is $T(C_\epsilon)$; hence there exists a neighborhood N_ϵ of θ disjoint from $T(C_\epsilon)$. The relation $0\theta = \theta$ implies the existence of neighborhoods $N(0)$ of 0 and $N(\theta)$ of θ such that $N(0)N(\theta) \subset N_\epsilon$. In other words $\alpha N(\theta) \subset N_\epsilon$ if only $|\alpha| < \delta_\epsilon$. If for any $\{\alpha_k\} \in T^{-1}[N(\theta)]$ we have $\gamma = d(\{\alpha_k\}, \theta) > \epsilon/\delta_\epsilon$, then $\epsilon\gamma^{-1}\{\alpha_k\} \in C_\epsilon$; this is contrary to $\epsilon\gamma^{-1}N(\theta) \cap T(C_\epsilon) = \emptyset$. Consequently the θ-neighborhood, $\frac{1}{2}\delta_\epsilon N(\theta)$, is contained in $T(S_\epsilon)$ and this proves that T is a homeomorphism.

A *closed linear subspace* is a linear subspace which is closed. For any subspace X, the *least closed linear subspace* containing X is simply the closure of the least linear subspace $\mathfrak{L} \equiv \mathfrak{L}(X)$ containing X. To prove this it suffices to show that $\bar{\mathfrak{L}}$ is linear. Suppose that $x \in \mathfrak{L}$ and $y \in \bar{\mathfrak{L}}$. Then $\alpha x + \beta y \in \alpha x + \beta\bar{\mathfrak{L}} \subset \alpha x + \overline{\beta\mathfrak{L}} \subset \overline{\alpha x + \beta\mathfrak{L}} = \bar{\mathfrak{L}}$. Thus $\alpha\mathfrak{L} + \beta y \subset \bar{\mathfrak{L}}$ and hence $\alpha\bar{\mathfrak{L}} + \beta y \subset \overline{\alpha\mathfrak{L} + \beta y} \subset \bar{\mathfrak{L}}$. Consequently for $x, y \in \bar{\mathfrak{L}}$ we have $\alpha x + \beta y \in \bar{\mathfrak{L}}$. The set $\overline{\mathfrak{L}(X)}$ is also called the *closed linear extension* of X. A set X is called *fundamental* in $\mathfrak{X}$ if $\overline{\mathfrak{L}(X)} = \mathfrak{X}$.

The closure of the convex extension $\mathfrak{C}(X)$ of a subset X can be shown in like manner to be the *least closed convex set* containing X; the proof is modified only to the extent that $\alpha, \beta \geqq 0$ and $\alpha + \beta = 1$. $\overline{\mathfrak{C}(X)}$ is also called the *closed convex extension* of X. A *convex body* is a convex set having interior points. If C is a convex body, then $\overline{\text{Int}(C)} \supset C$.

REMARK. The above proofs that $\overline{\mathfrak{L}(X)}$ is linear and $\overline{\mathfrak{C}(X)}$ convex required only that the mappings $x \to a + x$ and αx, a and α fixed, be continuous.

Since a topological linear space is in particular a Hausdorff space, $\lim_\pi x_\pi$ is uniquely defined for the family $\mathfrak{F}$ of all convergent directed point sets defined on a given directed set $[\pi]$. Further this limit operation is linear on $\mathfrak{F}$ in the sense that

$$\lim_\pi(\alpha x_\pi + \beta y_\pi) = \alpha \lim_\pi x_\pi + \beta \lim_\pi y_\pi$$

for all $[x_\pi]$, $[y_\pi] \in \mathfrak{F}$.

DEFINITION 1.10.2. *A subset D of a real (or complex) linear system $\mathfrak{X}$ is said to be finitely open if for each choice of elements $y_1, y_2, \cdots, y_n \in \mathfrak{X}$, the elements $\sum_1^n \alpha_k y_k$ which are in D correspond to an open subset of the space E_n (or Z_n) with elements $(\alpha_1, \alpha_2, \cdots, \alpha_n)$.*

If $\mathfrak{X}$ is a topological linear space, then, according to Theorem 1.10.1 a subset D is finitely open if and only if D intersects every finite dimensional subspace in a relatively open set. Further if $\mathfrak{X}$ is complex linear, the property of a set being finitely open is independent of whether we treat $\mathfrak{X}$ as a real or a complex space.

The following example, due to E. G. Begle and H. Pollard, shows that there are sets which are finitely open without being open. Let $\{x_n\}$ be a sequence of points in a topological linear space $\mathfrak{X}$ such that (i) for each n, the point x_{n+1} is not in the linear vector space spanned by $x_1, x_2, \cdots, x_n$, and (ii) $\lim x_n = x_0$ exists. Let D be the complement of the set $\{x_n\}$. Since D contains x_0, it cannot be open. On the other hand, any k-dimensional subspace of $\mathfrak{X}$ can contain at most k points x_n; its intersection with D is consequently relatively open and hence D is finitely open.

Given an arbitrary linear system $\mathfrak{X}$, the collection of all finitely open subsets is a suitable neighborhood basis and defines a Hausdorff topology in $\mathfrak{X}$ which we shall call the *finite topology*. If $\mathfrak{X}$ is finite-dimensional, this topology is equivalent to the euclidean topology. For arbitrary $\mathfrak{X}$, it is clear that postulate H_2 is satisfied. Postulate H_1 follows from the fact that the relative topology for each n-dimensional subspace $\mathfrak{L}$ is equivalent with the euclidean topology. This is proved as follows: By definition the intersection of $\mathfrak{L}$ with any finitely open subset of $\mathfrak{X}$ is finitely open in $\mathfrak{L}$. Conversely suppose that $N_\mathfrak{L}$ is a finitely open subset of $\mathfrak{L}$. Employing the maximal principle it can be shown that there exists a subspace $\mathfrak{M}$ such that $\mathfrak{L} \cap \mathfrak{M} = \theta$ and $\mathfrak{L} + \mathfrak{M} = \mathfrak{X}$. We now define $N = N_\mathfrak{L} + \mathfrak{M}$. Then clearly $N_\mathfrak{L} = N \cap \mathfrak{L}$. It remains to show that N is finitely open in $\mathfrak{X}$. Let $\mathfrak{F}$ be any finite dimensional subspace; without loss of generality we may suppose that $\mathfrak{L} \subset \mathfrak{F}$. Then $\mathfrak{F} = \mathfrak{L} + \mathfrak{M} \cap \mathfrak{F}$ and $\mathfrak{F} \cap N = N_\mathfrak{L} + \mathfrak{M} \cap \mathfrak{F}$. Hence $\mathfrak{F} \cap N$ is finitely open in $\mathfrak{F}$ and this proves the assertion. As a consequence of this assertion and the fact that any two points of $\mathfrak{X}$ may be joined by a line segment, we see that $\mathfrak{X}$ is connected. We also note that the mappings $x \to -x$, $a + x$, and αx for fixed a and $\alpha \neq 0$ are homeomorphisms of $\mathfrak{X}$ onto $\mathfrak{X}$ in the finite topology. Kakutani and Klee [1] have shown that if the dimension of $\mathfrak{X}$ is countable then the finite topology defines a linear topological space. However this is no longer true if the dimension of $\mathfrak{X}$ is uncountable for in this case vector addition is continuous in each variable separately but not jointly as required by Definition 1.8.1. We shall denote the *finite closure* of a set X by $\bar{X}^\circ$ and the *finite interior* of a set X by $\text{Int}^\circ(X)$. It is clear that $\overline{\mathfrak{L}(X)^\circ}$ is linear and the $\overline{\mathfrak{C}(X)^\circ}$ is convex for any subset X.

THEOREM 1.10.2. *If C and D are complementary convex subsets of a linear system $\mathfrak{X}$ and $\mathfrak{M} = \bar{C}^\circ \cap \bar{D}^\circ$, then $\mathfrak{M}$ is either a maximal real flat space or else $\mathfrak{M} = \mathfrak{X}$.*

PROOF. Since $\bar{C}^\circ$ and $\bar{D}^\circ$ are both finitely closed and convex the same is true of $\mathfrak{M}$. Now let x and y be distinct points of $\mathfrak{M}$ and suppose that z is such that $y \in [x, z]$. If $z \notin \mathfrak{M}$, then z belongs to, say, $\text{Int}^\circ(C)$. In this case it is easy to see that $y \in \text{Int}^\circ(C)$ which is impossible. Thus each real line determined by any two points of $\mathfrak{M}$ belongs to $\mathfrak{M}$ and hence $\mathfrak{M}$ is a flat space. We may suppose without loss of generality that $\theta \in \mathfrak{M}$. Thus if $p \in \text{Int}^\circ(C)$, then $-p \in \text{Int}^\circ(D)$. For arbitrary $x \in C$, $[x, -p]$ intersects $\mathfrak{M}$ whereas for arbitrary $y \in D$, $[y, p]$ intersects $\mathfrak{M}$. Therefore $\mathfrak{X}$ is contained in the real linear extension of $\mathfrak{M}$ and p; it follows that $\mathfrak{M}$ is either a maximal real flat space or all of $\mathfrak{X}$.

The Tychonoff [1] fixed point theorem is one of the most beautiful results in this field. We list it here even though we shall not have occasion to make use of it.

THEOREM 1.10.3. *Let $\mathfrak{X}$ be a topological linear space with a convex neighborhood basis and let C be a convex compact subset of $\mathfrak{X}$. If $F(x)$ is a continuous mapping of C into itself, then $F(x_0) = x_0$ for some $x_0 \in C$.*

1.11. Partially ordered linear systems. It is sometimes convenient to introduce a partial ordering into a real linear system. To be appropriate to a linear system the order relation must satisfy further conditions besides O_1, O_2, and O_3 of Section 1.2.

DEFINITION 1.11.1. *A partially ordered linear system $\mathfrak{X}$ is a real linear system, partially ordered so that the transformations $x \to x + a$ and $x \to \alpha x$ $(\alpha > 0)$ induce order automorphisms in the system.*

This postulate requires that if $x > y$ then $x + a > y + a$ and $\alpha x > \alpha y$ for all a and all $\alpha > 0$. In particular $x > y$ and $\alpha > 0$ implies $x - y > y - y = \theta$ and hence that $-\alpha y = \alpha(x - y) - \alpha x > \theta - \alpha x = -\alpha x$. From this we see that the mapping $x \to \beta x$ $(\beta < 0)$ is an order anti-automorphism.

DEFINITION 1.11.2. *An element x of a partially ordered linear system $\mathfrak{X}$ is said to be positive if $x > \theta$. The set of all positive elements forms the positive cone $\mathfrak{X}^+$ of $\mathfrak{X}$.*

The next theorem shows that the positive cone completely determines the ordering in $\mathfrak{X}$.

THEOREM 1.11.1. *Let $\mathfrak{X}$ be a partially ordered linear system. Then*

(i) $\theta \in \mathfrak{X}^+$,

(ii) *If $x \in \mathfrak{X}^+$ and $\alpha \geqq 0$, then $\alpha x \in \mathfrak{X}^+$,*

(iii) *If* $x \in \mathfrak{X}^+$ *and* $-x \in \mathfrak{X}^+$, *then* $x = \theta$,
(iv) *If* $x \in \mathfrak{X}^+$ *and* $y \in \mathfrak{X}^+$, *then* $x + y \in \mathfrak{X}^+$.
Conversely if $\mathfrak{X}^+$ *is a subset of* $\mathfrak{X}$ *satisfying the above conditions* (i) *to* (iv) *and if* $x > y$ *is defined as* $x - y \in \mathfrak{X}^+$, *then this defines* $\mathfrak{X}$ *as a partially ordered linear system.*

For a proof see G. Birkhoff [5, pp. 214–215]. When $\mathfrak{X}$ is a partially ordered linear topological space we shall always assume that $\mathfrak{X}^+$ is a closed subset of $\mathfrak{X}$.

DEFINITION 1.11.3. *A linear lattice is a partially ordered linear system which is also a lattice.*

By hypothesis $x \to x + a$ is an order automorphism. As a consequence we have $(x \vee y) + a = (x + a) \vee (y + a)$ and in particular $x \vee y = (x - y) \vee \theta + y$. Also since $x \to -x$ is an order anti-automorphism we have $x \wedge y = -(-x \vee -y)$. Hence the existence of $x \vee \theta$ for all x belonging to a partially ordered linear system $\mathfrak{X}$ is enough to ensure that $\mathfrak{X}$ is a lattice.

A linear lattice is said to be *complete* if every subset X with an upper bound possesses an l.u.b., denoted by $\vee [x; x \in X]$. If $\mathfrak{X}$ is complete and the subset Y is bounded from below then a g.l.b. for Y exists and is, in fact, $\wedge [x; x \in Y] = - \vee [-x; x \in Y]$.

1.12. Banach spaces. By far the most important class of linear spaces are the *Banach spaces*, (B)-spaces for short. Here the metric topology is based upon a *norm*.

DEFINITION 1.12.1. *A linear system* $\mathfrak{X}$ *is called a (B)-space if*
(1) *with every element* x *there is associated a real number* $\| x \|$, *called the norm of* x, *with the properties:*

$$N_1 .\ \| x \| \geqq 0 \quad \text{and} \quad \| x \| = 0 \quad \text{if and only if } x = \theta,$$

$$N_2 .\ \| \alpha x \| = | \alpha | \, \| x \| ,$$

$$N_3 .\ \| x + y \| \leqq \| x \| + \| y \| ;$$

(2) $d(x, y) = \| x - y \|$;
(3) $\mathfrak{X}$ *is complete in the resulting topology.*
$\mathfrak{X}$ *is a real or a complex (B)-space according as* Φ *is the real or the complex number field.*

From N_1, N_2, and N_3 we see that $d(x, y) = \| x - y \|$ is a suitable metric function. The continuity properties of the arithmetic operations listed in Definitions 1.8.1 and 1.10.1 follow directly from the relations

$$\| (x_1 - x_2) - (y_1 - y_2) \| \leqq \| x_1 - y_1 \| + \| x_2 - y_2 \|$$

and

$$\| \alpha x - \beta y \| \leqq | \alpha - \beta | \, \| x \| + | \beta | \, \| x - y \| .$$

Finally $| \| x \| - \| y \| | \leqq \| x - y \|$ shows that $\| x \|$ is a continuous function of x.

A linear system satisfying just the conditions (1) and (2) of Definition 1.12.1 is called a *normed linear space*. A normed linear space $\mathfrak{X}_0$ can always be embedded in the Banach space obtained from the class of all Cauchy sequences $[\{x_n\}]$, $x_n \in \mathfrak{X}_0$, by setting $\| \{x_n\} \| = \lim_n \| x_n \|$ and then identifying any two sequences whose distance apart is zero.

Numerous examples of (B)-spaces are to be found in Banach's treatise [2]. Here we note the following:

1. $l_p (1 \leqq p < \infty)$: The set of all sequences $\{\alpha_n\}$, $\alpha_n \in \Phi$, such that $\sum_1^\infty | \alpha_n |^p < \infty$; $\| \{\alpha_n\} \| = [\sum_1^\infty | \alpha_n |^p]^{1/p}$.

2. m: Bounded sequences $\{\alpha_n\}$, $\alpha_n \in \Phi$, with $\| \{\alpha_n\} \| = \sup | \alpha_n |$.

c: The linear subspace of m consisting of all convergent sequences.

c_0: The linear subspace of c consisting of all sequences converging to 0.

3. $C(\mathfrak{S})$: The set of all bounded continuous Φ-valued functions $f(\sigma)$ on a Hausdorff space $\mathfrak{S}$ with $\| f \| = \sup | f(\sigma) |$.

In Examples 4 and 5, $\mathfrak{S}$ is an abstract set, $\mathfrak{E}$ is a σ-ring of subsets of $\mathfrak{S}$ and $m(E)$ is a σ-finite measure defined on $\mathfrak{E}$ (cf. P. R. Halmos [1]).

4. $L_p(\mathfrak{S}; m) (1 \leqq p < \infty)$: The set of all measurable Φ-valued functions $f(\sigma)$ such that $\int | f(\sigma) |^p \, dm < \infty$. The corresponding (B)-space is made up of classes of such functions, each class X consisting of all functions which differ from a representative function only on sets of measure zero; $\| X \| = [\int | f(\sigma) |^p \, dm]^{1/p}$ where $f \in X$. $L_p(\mathfrak{S}; m)$ includes l_p as a special instance.

5. $L_\infty(\mathfrak{S}; m)$: The set of all essentially bounded, measurable, Φ-valued functions. The corresponding (B)-space again consists of classes X and $\| X \| = \text{ess. l.u.b.} \, | f(\sigma) |$ where $f \in X$.

6. Hilbert space $\mathfrak{H}$: This is a linear system over Φ with an inner product (x, y) defined on $\mathfrak{H} \times \mathfrak{H}$ to Φ such that $(x, y) = \overline{(y, x)}$, $(\alpha x + \beta y, z) = \alpha(x, z) + \beta(y, z)$, $(x, x) \geqq 0$, and $(x, x) = 0$ if and only if $x = \theta$. The norm is defined as $\| x \| = [(x, x)]^{\frac{1}{2}}$; employing the *Schwarz inequality*, $| (x, y)| \leqq \| x \| \, \| y \|$, it can be shown that conditions N_1, N_2, and N_3 are satisfied. $\mathfrak{H}$ is assumed complete in the resulting topology. The n-dimensional euclidean space Z_n is a Hilbert space as is $L_2(\mathfrak{S}; m)$. Additional material on $\mathfrak{H}$ can be found in M. H. Stone's treatise [3] and B. Sz.-Nagy [3].

If X is an arbitrary separable subset of a normed linear space $\mathfrak{X}$, then the least closed linear subspace $\mathfrak{L}$ containing X is also separable. For taking a sequence $\{x_n\}$ dense in X, the set of all finite sums $\sum \alpha_k x_k$, α_k having rational real and imaginary parts, is again denumerable and dense in $\overline{\mathfrak{L}(\{x_n\})} = \mathfrak{L}$.

Lemma 1.12.1. *If $\mathfrak{X}$ is a finite dimensional normed linear space, then each closed bounded subset is compact.*

Proof. By Theorem 1.10.1, $\mathfrak{X}$ is isomorphic with E_n (or Z_n). Since the latter is locally compact, the same is true of $\mathfrak{X}$. Thus for some $\epsilon > 0$, the sphere $S_\epsilon \equiv [x; \| x \| < \epsilon]$ is conditionally compact. The mapping $x \to \alpha x$ being continuous, it follows that each closed bounded subset of $\mathfrak{X}$ is compact.

Lemma 1.12.2. *Suppose that $y_1, y_2, \cdots, y_n$ are linearly independent vectors of a normed linear space and define $\mathfrak{L}_k = \mathfrak{L}[y_i ; i \leqq k]$. Then there is a vector $x_n \in \mathfrak{L}_n$, $\| x_n \| = 1$, such that $\| x_n - x \| \geqq 1$ for all $x \in \mathfrak{L}_{n-1}$.*

PROOF. Let $d = \inf [\| y_n - x \|; x \in \mathfrak{L}_{n-1}]$. By the previous lemma relatively closed and bounded subsets of $\mathfrak{L}_{n-1}$ are compact and hence there exists an $x_0 \in \mathfrak{L}_{n-1}$ such that $\| y_n - x_0 \| = d$. Further $d > 0$ since by hypothesis $x_0 \in \mathfrak{L}_{n-1}$ and y_n are linearly independent. Set $x_n = d^{-1}(y_n - x_0)$. Then for any $x \in \mathfrak{L}_{n-1}$ we have $\| x_n - x \| = d^{-1} \| y_n - (x_0 + dx) \| \geqq 1$.

THEOREM 1.12.2. *Every bounded closed subset of a normed linear space $\mathfrak{X}$ is compact if and only if $\mathfrak{X}$ is finite dimensional.*

PROOF. Lemma 1.12.1 establishes the necessity. Conversely suppose that $\mathfrak{X}$ is infinite dimensional and that the closed unit sphere is compact. Then there exists a denumerable set of vectors $\{y_n\}$ such that any finite subset is linearly independent. Setting $\mathfrak{L}_k = \mathfrak{L}[y_i ; i \leqq k]$ and applying Lemma 1.12.2 we obtain vectors $x_k \in L_k$, $\| x_k \| = 1$, such that $\| x_k - x \| \geqq 1$ for $x \in \mathfrak{L}_{k-1}$; as a consequence $\| x_k - x_j \| \geqq 1$ for $k \neq j$. Clearly $\{x_k\}$ can contain no convergent subsequence and hence the closed unit sphere in $\mathfrak{X}$ cannot be compact.

Before concluding this section we shall consider the notion of a quotient space for linear systems. If $\mathfrak{L}$ is a linear subspace of $\mathfrak{X}$, we can define a congruence relation in $\mathfrak{X}$ modulo $\mathfrak{L}$. We say that x and $y \in \mathfrak{X}$ are *congruent modulo* $\mathfrak{L}$, $x \equiv y \pmod{\mathfrak{L}}$, if $x - y \in \mathfrak{L}$. This relation is reflexive, symmetric, and transitive. The elements of $\mathfrak{X}$ congruent to a given element x (mod $\mathfrak{L}$) form a *residue class* X which is determined and represented by x; in this way $\mathfrak{X}$ is divided into a system of mutually exclusive residue classes $X, Y, \cdots$. Now if $x_1 - x_2 \in \mathfrak{L}$ and $y_1 - y_2 \in \mathfrak{L}$, then

$$(\alpha x_1 + \beta y_1) - (\alpha x_2 + \beta y_2) = \alpha(x_1 - x_2) + \beta(y_1 - y_2) \in \mathfrak{L}.$$

As a consequence we may define $\alpha X + \beta Y$ to be the residue class containing $\alpha x + \beta y$ where x and y are representative elements of X and Y respectively. The collection of residue classes then becomes a linear system which we denote by $\mathfrak{X} \div \mathfrak{L}$ and refer to as *a quotient space*; the zero element of $\mathfrak{X} \div \mathfrak{L}$ is $\Theta = \mathfrak{L}$. The mapping which takes x into the residue class containing x is a homomorphism of $\mathfrak{X}$ onto $\mathfrak{X} \div \mathfrak{L}$.

THEOREM 1.12.3. *If $\mathfrak{L}$ is a closed linear subspace of a normed linear space $\mathfrak{X}$, then $\mathfrak{X} \div \mathfrak{L}$ becomes a normed linear space if we define*

$$\| X \| = \inf [\, \| x \|; x \in X]. \tag{1.12.2}$$

If $\mathfrak{X}$ is complete then so is $\mathfrak{X} \div \mathfrak{L}$.

PROOF. We have only to verify the norm properties of $\mathfrak{X} \div \mathfrak{L}$. $\| X \| = 0$ if and only if there exists a sequence $\{x_n\} \subset X$ with $\| x_n \| \to 0$. Since X is closed this means that $\theta \in X$ and hence that $X = \mathfrak{L} = \Theta$. $\| \alpha X \| = \inf [\, \| \alpha x \|; x \in X] = | \alpha | \, \| X \|$. Likewise $\| X + Y \| = \inf [\| x + y \|; x \in X, y \in Y] \leqq \inf [\| x \| + \| y \|; x \in X, y \in Y] = \| X \| + \| Y \|$. Thus $\mathfrak{X} \div \mathfrak{L}$ is a normed space.

Suppose next that $\mathfrak{X}$ is complete and that $\{X_n\}$ is a Cauchy sequence in $\mathfrak{X} \div \mathfrak{L}$. Without restricting the generality we may assume that $\sum_{n=1}^{\infty} \| X_{n+1} - X_n \| < \infty$ since this condition will clearly be satisfied by a suitably chosen subsequence. We then successively choose a point x_n from each X_n so that $\| x_{n-1} - x_n \| \leqq 2 \| X_{n-1} - X_n \|$. Then the sequence $\{x_n\}$ forms a Cauchy sequence having a limit point x_0 in $\mathfrak{X}$. If X_0 is the residue class determined by x_0, then $\| X_n - X_0 \| \to 0$ and hence $\mathfrak{X} \div \mathfrak{L}$ is complete.

Because of condition N_3, the unit sphere in a (B)-space is necessarily convex. A (B)-space is said to be *uniformly convex* if to each $\epsilon > 0$ there corresponds a $\delta_\epsilon > 0$ such that if $\| x \| = \| y \| = 1$ and $\| x - y \| \geqq \epsilon$, then $\| \frac{1}{2}(x + y) \| \leqq 1 - \delta_\epsilon$. This notion is due to J. A. Clarkson [1]. The spaces $L_p(\mathfrak{S}; m)$, $1 < p < \infty$, and $\mathfrak{H}$ are uniformly convex.

PROBLEMS.

1. A bounded subset X of l_p $(1 \leqq p < \infty)$ is conditionally compact if and only if

$$\limsup_{n\to\infty} [\sum_{n}^{\infty} | \alpha_k |^p; \{\alpha_k\} \in X] = 0.$$

2. Let $[\mathfrak{X}_\alpha ; \alpha \in \mathfrak{A}]$ be a collection of (B)-spaces. The aggregate of elements $[x_\alpha ; \alpha \in \mathfrak{A}]$ belonging to the product space for which $\sum_\alpha \| x_\alpha \|^p < \infty$ $(1 \leqq p < \infty)$ forms a (B)-space with $\| [x_\alpha] \| = [\sum_\alpha \| x_\alpha \|^p]^{1/p}$.

3. A real (B)-space $\mathfrak{X}$ can be embedded in the complex (B)-space of all ordered pairs $[(x, y); x, y \in \mathfrak{X}]$ in which $(x_1, y_1) + (x_2, y_2) = (x_1 + x_2, y_1 + y_2)$, $(\alpha + i\beta)(x, y) = (\alpha x - \beta y, \alpha y + \beta x)$, and $\| (x, y) \| = \sup [\| \cos \theta\, x + \sin \theta\, y \|; 0 \leqq \theta < 2\pi]$.

4. If C is a closed convex subset of a uniformly convex (B)-space and if $x_0 \notin C$, then there exists a unique $y_0 \in C$ such that $\| x_0 - y_0 \| = \inf [\| x_0 - y \|; y \in C]$.

References. Banach [2], G. Birkhoff [5], Clarkson [1], Halmos [1], Kakutani [1], Stone [3], Sz.-Nagy [3], Tukey [1], Tychonoff [1].

4. ALGEBRAIC SPACES

1.13. Algebras. An algebraic system in which all three operations are defined is known as an *algebra*.

DEFINITION 1.13.1. *$\mathfrak{A}$ is an algebra over the scalar field Φ if its elements admit of the three operations of addition, multiplication, and scalar multiplication, subject to the following conditions.*

$\mathfrak{A}$ is a linear system in the sense of Definition 1.9.1.

The multiplication satisfies:

M_1. *Every ordered pair of elements x, y has a unique product xy;*

M_2. *Multiplication is associative: $(xy)z = x(yz)$.*

D. *Addition and multiplication are distributive*:

$$x(y + z) = xy + xz, \qquad (y + z)\,x = yx + zx.$$

S_6. *Multiplication and scalar multiplication commute*:

$$\alpha x\beta y = \alpha\beta xy.$$

Further conditions which will sometimes be imposed are:

M_3. *There exists a unit element e such that* $ex = xe = x$ *for each* x;

M_4. *Multiplication is commutative*: $xy = yx$.

We speak of an *algebra with a unit element* if M_3 holds, a *commutative* or *abelian algebra* if M_4 holds.

If X and Y are two given subsets of $\mathfrak{A}$, the symbol XY will denote the set $[xy;\, x \in X,\, y \in Y]$.

If $\mathfrak{A}_1 \subset \mathfrak{A}$ and $\mathfrak{A}_1$ is an algebra, then $\mathfrak{A}_1$ is called a *subalgebra* of $\mathfrak{A}$. To every subset X of $\mathfrak{A}$ there corresponds a *least subalgebra* $\mathfrak{A}(X)$ containing X; it consists of all finite multinomials obtained by adding and multiplying elements of X, the coefficients being in Φ.

Definition 1.13.2. *Let $\mathfrak{A}$ be an algebra. A linear set $\mathfrak{i}$ of $\mathfrak{A}$ is called a right ideal (left ideal) if $xy \in \mathfrak{i}$ ($yx \in \mathfrak{i}$) whenever $x \in \mathfrak{i}$ and $y \in \mathfrak{A}$. A linear set which is both a right ideal and a left ideal is called a two sided ideal. An ideal is said to be proper if $\mathfrak{i} \neq [\theta]$, $\mathfrak{A}$.*

The ideals $[\theta]$ and $\mathfrak{A}$, known as the *zero* and *unit ideals* respectively, occur in every algebra and are the *improper ideals*. In a commutative algebra all ideals are two sided.

The *product* of the algebras $\mathfrak{A}_1, \mathfrak{A}_2, \cdots, \mathfrak{A}_n$ over Φ, denoted by $\mathfrak{A}_1 \times \mathfrak{A}_2 \times \cdots \times \mathfrak{A}_n$, is defined as the product of the $\mathfrak{A}_k$'s considered as linear systems with the further operation

$$(x_1, x_2, \cdots, x_n)\,(y_1, y_2, \cdots, y_n) = (x_1y_1, x_2y_2, \cdots, x_ny_n).$$

Definition 1.13.3. *Let $\mathfrak{A}$ be an algebra with a unit element e. An element x is said to be regular if there is an element x^{-1}, called the inverse of x, such that $xx^{-1} = x^{-1}x = e$. A non-regular element is called singular. An element y (or z) is said to be a right (left) inverse of x if $xy = e$ (or $zx = e$).*

There is no mention of inverses in our postulates; however if $\mathfrak{A}$ has a unit element it is *a priori* evident that some elements have inverses. The distinction between elements which have inverses and those which do not is fundamental. An element may have any number of either right inverses or left inverses; nevertheless, if $xy = e = zx$, then $zxy = ze = ey$ and $y = z$. *Thus in an associative algebra the existence of a right inverse implies that the element has either no left inverse or that the right inverse is also the left inverse and the element is regular.* In non-associative systems this is not necessarily true. If $\mathfrak{A}$ has no divisors of zero (see Section 4.5), then right and left inverses are also unique when they exist.

DEFINITION 1.13.4. *For a given subset X of an algebra $\mathfrak{A}$, the set of all elements of $\mathfrak{A}$ which commute with every element of X is called the commutant of X and is denoted by X^c.*

We note that $X^c \subset Y^c$ if $X \supset Y$ and that $X^{cc} \supset X$. As a consequence $X^{ccc} \subset X^c$. On the other hand, since each element of X^{cc} by definition commutes with every element of X^c, we also have $X^{ccc} \supset X^c$. It follows that $X^{ccc} = X^c$ and hence that $X^{c^{n+2}} = X^{c^n}$ for all $n \geqq 1$.

THEOREM 1.13.1. *X^c is a subalgebra of $\mathfrak{A}$. If X is abelian, then so is X^{cc}. Further if $\mathfrak{A}$ possesses a unit element e then $e \in X^c$ and if $x \in X^c$ is regular in $\mathfrak{A}$ then $x^{-1} \in X^c$.*

PROOF. If $x, y \in X^c$ and $z \in X$, then $xz = zx$ and $yz = zy$. This implies that $(x + y)z = z(x + y)$, $(\alpha x)z = z(\alpha x)$, and $xyz = xzy = zxy$. Hence $x + y$, αx, and xy all belong to X^c and this suffices to show that X^c is a subalgebra.

If X is abelian, then $X^c \supset X$ and hence $X^{cc} \subset X^c$. Since each element of X^{cc} by definition commutes with every element of X^c, it follows that X^{cc} is itself abelian.

If $\mathfrak{A}$ possesses a unit element, then clearly $e \in X^c$. Further if $x \in X^c$ is regular and if $z \in X$, then $x^{-1}z = x^{-1}zxx^{-1} = x^{-1}xzx^{-1} = zx^{-1}$ and hence x^{-1} also belongs to X^c.

1.14. Algebraic spaces. As in the previous cases we again introduce a topology.

DEFINITION 1.14.1. *An algebra $\mathfrak{A}$ is a topological algebra (or topological algebraic space) if $\mathfrak{A}$ is a topological linear space and to every $x, y \in \mathfrak{A}$ and every neighborhood $N(xy)$ of xy there are neighborhoods $N(x)$ of x and $N(y)$ of y such that $xN(y) \subset N(xy)$ and $N(x)y \subset N(xy)$.*

The added effect of this postulate is to require that xy be continuous in x and y *separately*, that is, that the mappings $T_1(x) = ax$ and $T_2(x) = xa$ of $\mathfrak{A}$ *into* $\mathfrak{A}$ be continuous for each $a \in \mathfrak{A}$; in general T_1 and T_2 will not be homeomorphisms.

For any subset X, the *least closed subalgebra* containing X is the closure of the least subalgebra $\mathfrak{A}(X)$ containing X. If X is abelian then so is $\overline{\mathfrak{A}(X)}$.

THEOREM 1.14.1. *Let X be a subset of a topological algebra $\mathfrak{A}$. Then the commutant X^c of X is closed.*

PROOF. Suppose that X^c is not closed. Then for any $y \in \overline{X^c} \ominus X^c$ there exists a $z \in X$ such that $yz - zy \neq \theta$. We can then find a neighborhood $N(yz - zy)$ of $yz - zy$ which does not contain θ. Applying the postulates of Definitions 1.8.1 and 1.14.1 successively we see that there exists a neighborhood $N(y)$ such that $N(y)z - zN(y) \subset N(yz - zy)$. Since $y \in \overline{X^c}$ there is an $x \in N(y) \cap X^c$ and hence $\theta = xz - zx \in N(yz - zy)$ which is impossible.

1.15. Banach algebras. We shall be principally concerned with a class of topological algebras called *Banach algebras*, (B)-algebras for short.

Definition 1.15.1. *$\mathfrak{B}$ is a (B)-algebra if $\mathfrak{B}$ is an algebra as well as a (B)-space and if, in addition,*

$$\| xy \| \leqq \| x \| \| y \| .$$

It is a real or a complex Banach algebra according as Φ is the real or complex number field.

The inequality

$$\| x_1y_1 - x_2y_2 \| \leqq \| x_1 \| \| y_1 - y_2 \| + \| y_2 \| \| x_1 - x_2 \|$$

shows that xy is a continuous function of both variables together. It follows, in particular, that $\mathfrak{B}$ is a topological algebra.

Many examples of (B)-algebras appear throughout this book. The classical example of a commutative (B)-algebra is $C(\mathfrak{S})$ where $\mathfrak{S}$ is a Hausdorff space and the arithmetic operations are given by

$$(f_1 + f_2)(\sigma) = f_1(\sigma) + f_2(\sigma), \qquad (\alpha f)(\sigma) = \alpha f(\sigma), \qquad (f_1f_2)(\sigma) = f_1(\sigma)f_2(\sigma).$$

An algebra $\mathfrak{A}$ satisfying all of the postulates for a (B)-algebra except completeness is called a *normed algebra*; such an algebra can be embedded in the usual way in the complete (B)-algebra $\mathfrak{B}$ consisting of the Cauchy sequences in $\mathfrak{A}$ with identifications. If $\mathfrak{A}$ satisfies M_3 or M_4, then the same will be true of $\mathfrak{B}$.

If a (B)-algebra $\mathfrak{B}$ has a unit element e (distinct from θ), then necessarily $\| e \| \geqq 1$. However for most purposes one may assume that $\| e \| = 1$ since $\mathfrak{B}$ is always isomorphic with an algebra $\mathfrak{B}'$ having a unit element with norm one. This will be proved in Theorem 2.14.3.

Definition 1.15.2. *A (B)-algebra $\mathfrak{B}$ is called a (*)-algebra if it admits an involution operator $x \to x^*$ defined on all $\mathfrak{B}$ with the properties:*

$$(1) \quad (x^*)^* = x; \qquad (2) \quad (xy)^* = y^*x^*; \qquad (3) \quad (\alpha x + \beta y)^* = \bar{\alpha}x^* + \bar{\beta}y^*.$$

The element x^* in a (*)-algebra is called the *adjoint* of x; x is said to be *self-adjoint* if $x = x^*$; and x is said to be *normal* if $xx^* = x^*x$. It is clear that x^*x, xx^*, $x + x^*$, and $i(x - x^*)$ are all self-adjoint. Further $\theta^* = (\theta\theta^*)^* = \theta\theta^* = \theta$ so that θ is self-adjoint. If $\mathfrak{B}$ has a unit element e, then $e^* = ee^* = (ee^*)^* = (e^*)^* = e$ and hence e is also self-adjoint. Finally we note that $x = \frac{1}{2}(x + x^*) + i(1/2i)(x - x^*)$ is the unique representation of x as the sum of a self-adjoint element plus i times another.

Definition 1.15.3. *An (A*)-algebra is a (*)-algebra with the property that the spectrum (see Definition 4.7.1) of each of its self-adjoint elements is real.*

Definition 1.15.4. *A (B*)-algebra is a (*)-algebra in which, for all elements in the algebra, $\| xx^* \| = \| x \| \| x^* \|$.*

References. Gelfand [4], Loomis [1].

CHAPTER II

LINEAR TRANSFORMATIONS

2.1. Orientation. In this chapter we shall give a survey of the theory of linear transformations. The presentation is self-contained for those parts of the theory which have a bearing on our main problem. Although the material is far from complete in other respects, the reader will find some indications of other directions of research; and the References are recommended for supplementary reading. The chapter is grouped under four paragraph headings: *Additive Transformations*, *Linear Functionals*, *Linear Transformations*, and *Spaces of Endomorphisms*.

1. Additive Transformations

2.2. Additive transformations. In this chapter we are concerned with transformations possessing certain algebraic properties and defined on an algebraic system. For the sake of simplicity we use the same symbols for corresponding operations in the domain and range spaces and do not distinguish between the two zero elements.

DEFINITION 2.2.1. *Let $\mathfrak{X}$ and $\mathfrak{Y}$ be two topological additive groups and $y = T(x)$ a transformation on $\mathfrak{X}$ into $\mathfrak{Y}$. $T(x)$ is said to be additive if for all x_1 and x_2 we have*

$$T(x_1 + x_2) = T(x_1) + T(x_2).$$

The definition implies that

$$T(\theta) = \theta \quad \text{and} \quad T(-x) = -T(x).$$

THEOREM 2.2.1. *An additive transformation is continuous everywhere if it is continuous at a single point.*

PROOF. Suppose that $T(x)$ is continuous at $x = x_0$, take $x_1 \neq x_0$, and let X be any set with $x_1 \in \bar{X}$. Then

$$x_0 \in \bar{X} - x_1 + x_0 = \overline{X - x_1 + x_0}$$

and continuity at x_0 implies

$$T(x_0) \in \overline{T(X - x_1 + x_0)} = \overline{T(X) - T(x_1) + T(x_0)} = \overline{T(X)} - T(x_1) + T(x_0),$$

whence $T(x_1) \in \overline{T(X)}$. Hence $T(x)$ is continuous at the arbitrary point x_0.

2.3. Linear transformations. We turn now to linear spaces. We recall that the scalar field Φ is restricted to be either the real or the complex field.

DEFINITION 2.3.1. *Let $\mathfrak{X}$ and $\mathfrak{Y}$ be two topological linear spaces with the same scalar field Φ and let $y = T(x)$ be a transformation on $\mathfrak{X}$ into $\mathfrak{Y}$. If $T(\alpha x) = \alpha T(x)$ for all x and all real α, then $T(x)$ is said to be real-homogeneous; $T(x)$ is homogeneous if the relation holds for all $\alpha \in \Phi$.*

The two notions obviously coincide if Φ is the real field.

DEFINITION 2.3.2. *An additive and homogeneous transformation is said to be linear.*

The reader should observe that we are not following the usage of the Polish school according to which a linear transformation is also continuous.

THEOREM 2.3.1. *An additive continuous transformation on one topological linear space to another with the same scalar field is linear if either* (i) *Φ is the real field or* (ii) *Φ is the complex field and, in addition, $T(ix) = iT(x)$ for all x.*

PROOF. The additivity obviously implies $T(\rho x) = \rho T(x)$ for each rational ρ. Now the relative topology for any one-dimensional real subspace of a topological linear space is homeomorphic with E_1. Hence for α irrational and a sequence of rational numbers $\rho_n \to \alpha$, we have $\rho_n x \to \alpha x$ and $\rho_n T(x) \to \alpha T(x)$. Continuity then implies that $T(\alpha x) = \alpha T(x)$. Thus $T(x)$ is real-homogeneous and the remainder of the proof is trivial.

2.4. Bounded transformations. In normed linear spaces the notion of boundedness is fundamental in the theory of additive transformations.

DEFINITION 2.4.1. *A transformation $y = T(x)$ on one normed linear space to another is said to be bounded if it takes bounded sets into bounded sets. For an additive bounded transformation*

$$\| T \| \equiv \sup [\, \| T(x) \| \,;\, \| x \| \leqq 1]$$

is called the bound or norm of T.

THEOREM 2.4.1. *An additive transformation $y = T(x)$ on one normed linear space $\mathfrak{X}$ to another is continuous if and only if it is bounded. In this case*

$$\| T(x) \| \leqq \| T \| \, \| x \|, \qquad \textit{for all } x \in \mathfrak{X}. \tag{2.4.1}$$

PROOF. If $T(x)$ is additive and bounded, then for any rational number $\rho > \| x \|$ we have $\| T(x) \| = \rho \| T(x/\rho) \| \leqq \rho \| T \|$. Since this holds for all $\rho > \| x \|$, we obtain (2.4.1). Consequently

$$\| T(x) - T(x_0) \| = \| T(x - x_0) \| \leqq \| T \| \, \| x - x_0 \|$$

and continuity follows. On the other hand if $T(x)$ is continuous, then given

$\epsilon > 0$ there exists a $\delta_\epsilon > 0$ such that $\| T(x) \| \leqq \epsilon$ if only $\| x \| \leqq \delta_\epsilon$. Thus for any $z \in \mathfrak{X}$, $\| T(\delta_\epsilon z/\| z \|) \| \leqq \epsilon$ and, since T is real-homogeneous by Theorem 2.3.1, we have $\| T(z) \| \leqq (\epsilon/\delta_\epsilon) \| z \|$; this implies boundedness.

2.5. Subadditive functionals. We now introduce a new notion, that of *subadditivity.*

Definition 2.5.1. *A finite real-valued functional $F(x)$ defined on a normed linear space $\mathfrak{X}$ is said to be subadditive if for all $x_1, x_2 \in \mathfrak{X}$ we have*

$$F(x_1 + x_2) \leqq F(x_1) + F(x_2). \tag{2.5.1}$$

Subadditive functionals play an important role in the study of bounded additive transformations on one normed linear space $\mathfrak{X}$ to another. For if $T(x)$ is such a transformation, then $F(x) \equiv \| T(x) \|$ is clearly subadditive on $\mathfrak{X}$. As we shall see later, subadditive functions also appear in other connections in the study of semi-groups of transformations.

Theorem 2.5.1. *If a subadditive functional $F(x)$ defined on a normed linear space $\mathfrak{X}$ is non-negative outside of a sphere, then it is non-negative for all $x \in \mathfrak{X}$.*

Proof. Suppose $F(x) \geqq 0$ for all x, $\| x \| \geqq R$. Then for a given x, $\| x \| < R$, there is a positive integer n with $n \| x \| \geqq R$ unless, of course, $x = \theta$. Hence $0 \leqq F(nx) \leqq nF(x)$ and $F(x) \geqq 0$. For $x = \theta$ we have $F(\theta) \leqq 2F(\theta)$ so that $F(\theta) \geqq 0$.

Remark. If we do not restrict $F(x)$ to be finite-valued, then the conditions of the above theorem are satisfied by one other functional: $F(x) = +\infty$ for $x \neq 0$ and $F(\theta) = -\infty$.

Theorem 2.5.2. *A subadditive functional defined on a normed linear space, continuous at $x = \theta$ with $F(\theta) = 0$, is continuous for all x.*

Proof. The double inequality

$$F(x) - F(-h) \leqq F(x + h) \leqq F(x) + F(h)$$

together with continuity at $x = \theta$ shows that $\lim_{h \to \theta} F(x + h) = F(x)$.

Theorem 2.5.3. *If $F(x)$ is a continuous subadditive functional, then*

$$0 \leqq M_F \equiv \sup [F(x); \| x \| \leqq 1] < \infty$$

and

$$-M_F \| x \| \leqq F(x) \leqq M_F(\| x \| + 1), \qquad x \in \mathfrak{X}. \tag{2.5.2}$$

Proof. Since $F(x)$ is clearly bounded above in some ϵ-sphere about θ, $F(x) \leqq nF(x/n)$ implies boundedness above in every sphere so that $M_F < \infty$. From $F(\theta) \leqq 2F(\theta)$ we see that $0 \leqq F(\theta) \leqq M_F$. If $n - 1 \leqq \| x \| < n$, we have $F(x) \leqq nF(x/n) \leqq nM_F \leqq M_F(\| x \| + 1)$. In order to establish the first inequality in (2.5.2), let $\epsilon > 0$ be given; then from $0 \leqq F(\theta) \leqq F(x) + F(-x)$ we see that

$-F(x) \leqq F(-x) \leqq (M_F + \epsilon) \| x \|$ for $\| x \| \geqq M_F/\epsilon$. Now, $F(x) + (M_F + \epsilon) \| x \|$ being the sum of two subadditive functionals, is itself subadditive and, since it is non-negative for $\| x \| \geqq M_F/\epsilon$, it must be non-negative for all $x \in \mathfrak{X}$. Finally because this holds for each $\epsilon > 0$, we have $F(x) \geqq -M_F \| x \|$ and the theorem is proved.

REMARK. The subadditive functional $F(x) = \| x \|^\alpha, 0 < \alpha < 1$, shows that an inequality of the form $F(x) \leqq M \| x \|$ need not hold for all x even when $F(\theta) = 0$.

We come next to the *principle of uniform boundedness*; here we require that $\mathfrak{X}$ be complete.

THEOREM 2.5.4. *Let* $[F_\alpha(x); \alpha \in \mathfrak{A}]$ *be a family of lower semi-continuous subadditive functionals defined on a* (B)*-space* $\mathfrak{X}$. *If* $\sup_\alpha F_\alpha(x) < \infty$ *for each* $x \in \mathfrak{X}$, *then* $\sup_\alpha M_{F_\alpha} < \infty$.

PROOF. According to Theorem 1.4.2, a sphere $S \equiv [x; \| x - x_0 \| < r]$ and a constant M exist such that $F_\alpha(x) \leqq M$ for all $x \in S$ and all $\alpha \in \mathfrak{A}$. Set $M_0 = \sup_\alpha F_\alpha(-x_0)$. Then for $\| y \| < r$, $x_0 + y \in S$ so that

$$F_\alpha(y) \leqq F_\alpha(-x_0) + F_\alpha(x_0 + y) \leqq M_0 + M = M', \qquad \alpha \in \mathfrak{A}.$$

Consequently for arbitrary x, $\| x \| \leqq 1$, and $n_0 > 1/r$ we have $F_\alpha(x) \leqq n_0 F_\alpha(x/n_0) \leqq n_0 M'$ and this proves the theorem.

The following somewhat sharper result holds for bounded additive transformations; this is the so-called *uniform boundedness theorem*.

THEOREM 2.5.5. *Let* $[T_\alpha(x); \alpha \in \mathfrak{A}]$ *be a family of bounded additive transformations on one normed linear space* $\mathfrak{X}$ *to another. If* $\sup_\alpha \| T_\alpha(x) \| < \infty$ *for each* x *belonging to a given subset* $X_0 \subset \mathfrak{X}$ *of the second category, then* $\sup_\alpha \| T_\alpha \| < \infty$.

PROOF. The proof proceeds as in Theorem 2.5.4. In this case, however, the boundedness of $\sup_\alpha \| T_\alpha(-x_0) \|$ is not postulated but results from the fact that $\sup_\alpha \| T_\alpha(-x_0) \| = \sup_\alpha \| T_\alpha(x_0) \| \leqq M$.

References. Banach [2].

2. LINEAR FUNCTIONALS

2.6. Geometrical aspects. We come now to the study of linear functionals; these constitute by far the most important class of linear transformations. The notions of additive, homogeneous, linear, and continuous functionals are obtained from the preceding paragraph by specializing the range space $\mathfrak{Y}$ to be E_1 or Z_1. In this section we shall be concerned with some geometrical aspects of linear functionals.

If $\mathfrak{X}$ is a complex linear system we can consider both real- and complex-linear functionals on $\mathfrak{X}$. There is, in fact, a rather simple relationship between these two. On the one hand if $F(x)$ is complex-linear, then $f(x) = \Re[F(x)]$ is clearly real-linear and $\Im[F(x)] = \Re[-iF(x)] = -\Re[F(ix)] = -f(ix)$; thus $F(x) = f(x) - if(ix)$ is uniquely determined by its real part. On the other hand if $f(x)$ is a given real-linear functional, then $f(x)$ is easily shown to be the real part of the complex-linear functional $F(x) = f(x) - if(ix)$. We also note that $F(x)$ is continuous if and only if $f(x) = \Re[F(x)]$ is continuous.

Each real (or complex)-linear functional $f(x)$ defines a maximal real (or complex) subspace, namely, the *kernel* of $f(x)$, $\mathfrak{M}_0 \equiv [x; f(x) = 0]$; likewise $\mathfrak{M}_\alpha \equiv [x; f(x) = \alpha]$ is a maximal real (or complex) flat space. Conversely, given a maximal real (or complex) subspace $\mathfrak{M}$ and $p \notin \mathfrak{M}$, then each element of $\mathfrak{X}$ has a unique representation $x = y + \alpha p$ where $y \in \mathfrak{M}$ and α is real (or complex); the functional $f(x) = \alpha$ is a real (or complex)-linear functional with kernel $\mathfrak{M}$. Further $f(x)$ is uniquely determined by $\mathfrak{M}$ to within a constant multiple. For if $f'(x)$ has the same kernel as $f(x)$, then $f(x)p - x \in \mathfrak{M}$ so that $f'[f(x)p - x] = 0$; hence $f'(x) = [f'(p)]f(x)$.

If $\mathfrak{M}$ is a maximal real subspace and $f(x)$ is a real-linear functional with kernel $\mathfrak{M}$, then the set A is said to be *bounded* by $\mathfrak{M} + p$ if either $A \subset [x; f(x) \leqq f(p)]$ or $A \subset [x; f(x) \geqq f(p)]$. The sets A and B are said to be *separated* by $\mathfrak{M} + p$ if they are bounded by and lie on opposite sides of $\mathfrak{M} + p$. It is clear that these concepts do not depend on the particular real-linear functional $f(x)$ employed as long as $\mathfrak{M}$ is its kernel. We are now in a position to prove the *separation theorem.*

THEOREM 2.6.1. *Let A and B be disjoint convex subsets of a linear system and suppose that* Int° $(A) \neq \emptyset$. *Then A and B can be separated by a maximal real flat space disjoint from* Int° (A).

PROOF. By Theorem 1.9.1 there exist complementary convex sets $C \supset A$ and $D \supset B$. Set $\mathfrak{M} = \bar{C}^\circ \cap \bar{D}^\circ$. Since Int° $(A) \cap \mathfrak{M} = \emptyset$, it follows from Theorem 1.10.2 that $\mathfrak{M}$ is a maximal real flat space. We may suppose without loss of generality that $\theta \in \mathfrak{M}$. Let $f(x)$ be a real-linear functional with kernel $\mathfrak{M}$, $f(\mathfrak{M}) = 0$. If $f(p) > 0$, then $p \notin \mathfrak{M}$ and hence $p \in$ Int° (C) or Int° (D). If $f(p) > 0$ and $f(q) < 0$, then it is clear that $[p, q] \cap \mathfrak{M} \neq \emptyset$. If, in addition, p and $q \in C$, then it is easy to see that $[p, q] \subset$ Int° (C); since this is impossible it follows that $\mathfrak{M}$ bounds C (and likewise D). Finally suppose that $c \in C$, $d \in D$, $f(c) > 0$, and $f(d) > 0$. Then $[c, d] \cap \mathfrak{M} = \emptyset$. This is contrary to the fact that the relatively open non-empty sets Int° $(C) \cap [c, d]$ and Int° $(D) \cap [c, d]$ cannot fill out the connected segment $[c, d]$. Consequently $\mathfrak{M}$ separates C and D and *a fortiori* A and B.

THEOREM 2.6.2. *A real- or complex-linear functional defined on a topological linear space is continuous if and only if there exists a non-empty convex open set disjoint from its kernel.*

PROOF. If $F(x)$ is continuous then $X_- \equiv [x; f(x) \equiv \Re[F(x)] < 0]$ is clearly an open convex set, disjoint from the kernel of $F(x)$. In proving the converse we first suppose that the functional $f(x)$ is real-linear and that the convex open set C is contained in $[x; f(x) \leqq 0]$. Choose $q \in C$. Then if $p \in X_-$ there exists an $\alpha > 0$ such that $f(p - \alpha q) < 0$ and hence $f(p - \alpha q + \alpha C) < 0$. Now $p - \alpha q + \alpha C$ is open and $p \in p - \alpha q + \alpha C \subset X_-$ so that X_- is open. Similarly working with $-C$ we find that $X_+ \equiv [x; f(x) > 0]$ is also open. If we choose p_0 so that $f(p_0) = 1$ we see that $f^{-1}[(\alpha, \beta)] = (X_- + \beta p_0) \cap (X_+ + \alpha p_0)$ is open and consequently $f(x)$ is continuous. Finally if $F(x)$ is complex-linear, then $F(C)$ is a convex subset of Z_1 disjoint from 0. Hence $F(C)$ is bounded by a line of the form $E_1 F(p_1)$, and if $\arg [F(p_1)] = \varphi$ then C is bounded by the kernel of $f(x) \equiv \Re[e^{-i\varphi}F(x)]$. It follows as above that $f(x)$ is continuous and hence the same is true of $F(x)$.

Corollary. *A real- or complex-linear functional defined on a linear topological space with a convex neighborhood basis is continuous if and only if its kernel is a closed point set.*

The most useful version of the separation theorem is embodied in the following theorem due in its original form to S. Mazur [1] and M. Eidelheit [1], and in its present form to V. L. Klee, Jr. [1].

Theorem 2.6.3. *If $\mathfrak{X}$ is a topological linear space, A a convex body, and B a convex set such that* Int $(A) \cap B = \emptyset$, *then there exists a continuous real-linear functional $f(x)$ and a constant γ such that $f(\text{Int}(A)) < \gamma$ and $f(A) \leqq \gamma \leqq f(B)$.*

Proof. Since Int (A) is convex and finitely open, Theorem 2.6.1 asserts the existence of a maximal real flat space $\mathfrak{M}$ which separates Int (A) and B and which is disjoint from Int (A). The associated real-linear functional $f(x)$ is continuous by Theorem 2.6.2 and $f(\text{Int}(A)) < \gamma$ implies that $f(A) \leqq \gamma$.

As an immediate consequence of Theorems 2.6.2 and 2.6.3 we see that a necessary and sufficient condition for the existence of non-trivial real-linear (and hence complex-linear) continuous functionals is the existence of open convex proper subsets of $\mathfrak{X}$ (cf. J. P. LaSalle [1]). Topological spaces possessing convex open sets have been the object of much research (cf. A. Kolmogoroff [3] and J. V. Wehausen [1]). Examples of topological linear spaces possessing no non-trivial continuous linear functionals are given by the space of measurable functions (S. Banach [2, p. 234]) and the spaces L_p with $0 < p < 1$ (M. M. Day [1]).

Another very useful result on convex sets is the Krein-Milman [1] theorem on extreme points. By an extreme point of a convex set C is meant a point which does not belong to the interior of any line segment contained in C. We state the following extension of this theorem due to J. L. Kelley [1].

Theorem 2.6.4. *Let $\mathfrak{X}$ be a topological linear space with the property that for each $x \neq \theta$ there is a continuous linear functional for which $f(x) \neq 0$. Let C be a compact convex subset of $\mathfrak{X}$. Then C is the closed convex extension of the set of its extreme points.*

2.7. Extension of linear functionals. The extent to which values of a linear functional can be preassigned is fundamental in the theory of linear spaces. The basic result in this direction was found by H. Hahn [1]; this is Theorem 2.7.3

for real $\mathfrak{X}$. S. Banach [1] (cf. [2, pp. 27–29]) later rediscovered the result and gave it the following more general formulation.

THEOREM 2.7.1. *Given a real linear system $\mathfrak{X}$ and*
(1) *a subadditive positive-homogeneous functional $p(x)$ defined on $\mathfrak{X}$:*

$$p(x + y) \leqq p(x) + p(y), \qquad p(\alpha x) = \alpha p(x) \quad \text{for} \quad \alpha \geqq 0;$$

(2) *a real-linear functional $f_0(x)$ defined on a linear subspace $\mathfrak{L}$ of $\mathfrak{X}$ such that $f_0(x) \leqq p(x)$ for all $x \in \mathfrak{L}$.*
Then there exists a linear functional $f(x)$ defined on $\mathfrak{X}$ such that $f(x) \leqq p(x)$ for all $x \in \mathfrak{X}$ and $f(x) = f_0(x)$ in $\mathfrak{L}$.

PROOF. Although the Hahn-Banach original argument is quite elegant, it is not very illuminating. The following proof provides the theorem with a geometrical setting (cf. Jean Dieudonné [1] and V. L. Klee, Jr. [1]). Let $\mathfrak{X}_1 = \mathfrak{X} \times E_1$ and consider the following subsets in $\mathfrak{X}_1$: $A_1 = [(x, \alpha); \alpha > p(x)]$, $A_2 = [(x, \alpha); \alpha > f_0(x), x \in \mathfrak{L}]$, $A = \mathfrak{C}(A_1 \cup A_2)$, and $B = [(x, \alpha); \alpha < f_0(x), x \in \mathfrak{L}]$. It is easy to show that $p(x)$ is a convex function on $\mathfrak{X}$ and hence that $A_1 \subset A$ is a finitely open convex subset of $\mathfrak{X}_1$; we note that $(\theta, 1) \in A_1$. Likewise it is easy to verify that B is convex, that $A \cap B = \emptyset$, and that $(\theta, 0) \in M_0 \equiv [(x, \alpha); \alpha = f_0(x), x \in \mathfrak{L}] \subset \bar{A}^\circ \cap \bar{B}^\circ$. Now according to Theorem 2.6.1 there exists a maximal real linear subspace $\mathfrak{M}$ which separates A and B, and contains $\bar{A}^\circ \cap \bar{B}^\circ$. Geometrically $\mathfrak{M}$ is a plane of support for A at $(\theta, 0)$. Since $(\theta, 1) \notin \mathfrak{M}$, there exists a real-linear functional $g[(x, \alpha)]$ with kernel $\mathfrak{M}$ of the form $g[(x, \alpha)] = f(x) - \alpha$, where $f(x)$ is real-linear on $\mathfrak{X}$. It is clear that A_1 lies "above" the plane $\alpha = f(x)$ and hence that $f(x) \leqq p(x)$ for all $x \in \mathfrak{X}$. Finally since $M_0 \subset \mathfrak{M}$ we have $f(x) = f_0(x)$ in $\mathfrak{L}$.

For the complex analogue of this theorem we use an argument due to H. F. Bohnenblust and A. Sobczyk [1].

THEOREM 2.7.2. *Given a complex linear system $\mathfrak{X}$ and*
(1) *a subadditive positive-homogeneous and circular functional $c(x)$ defined on $\mathfrak{X}$:*

$$c(x + y) \leqq c(x) + c(y), \qquad c(\alpha x) = |\alpha| \, c(x) \quad \text{for all complex } \alpha;$$

(2) *a complex-linear functional $F_0(x)$ defined on a complex linear subspace $\mathfrak{L}$ of $\mathfrak{X}$ such that $|F_0(x)| \leqq c(x)$, $x \in \mathfrak{L}$.*
Then there exists a complex-linear functional $F(x)$ defined on $\mathfrak{X}$ such that $|F(x)| \leqq c(x)$ for all $x \in \mathfrak{X}$ and $F(x) = F_0(x)$ in $\mathfrak{L}$.

PROOF. Let $f_0(x) = \mathfrak{R}[F_0(x)]$. Then $f_0(x)$ is a real-linear functional defined in $\mathfrak{L}$, $f_0(x) \leqq |f_0(x)| \leqq c(x)$ in $\mathfrak{L}$, and $F_0(x) = f_0(x) - if_0(ix)$. The previous theorem may be employed to obtain a real-linear extension $f(x)$ of $f_0(x)$ such that $f(x) \leqq c(x)$ for all $x \in \mathfrak{X}$. It is clear that $F(x) \equiv f(x) - if(ix)$ is a complex-linear extension of $F_0(x)$. Further, if $\varphi = \arg [F(x)]$, we have

$$|F(x)| = e^{-i\varphi}F(x) = F(e^{-i\varphi}x) = f(e^{-i\varphi}x) \leqq c(x),$$

and this completes the proof.

For the validity of Theorem 2.7.2 it is essential that the initial functional $F_0(x)$ be defined on a complex linear subspace $\mathfrak{L}$ rather than a real linear subspace. Indeed, Bohnenblust and Sobczyk have shown that in every complex (B)-space $\mathfrak{X}$ of infinite dimension there always exists a real linear subspace on which there is a bounded complex linear functional which does not admit of a bounded extension to $\mathfrak{X}$.

The two preceding theorems have very important applications to normed linear spaces. In such spaces it is customary to denote the linear bounded functionals by the generic symbol $x^* = x^*(x)$. We recall that x^* is *linear* if

$$x^*(\alpha x + \beta y) = \alpha x^*(x) + \beta x^*(y) \tag{2.7.1}$$

for all $\alpha, \beta \in \Phi$ and $x, y \in \mathfrak{X}$; it is *bounded* if

$$| x^*(x) | \leqq M \| x \| \tag{2.7.2}$$

for all $x \in \mathfrak{X}$; and its *norm* is given by

$$\| x^* \| = \sup [| x^*(x) | ; \| x \| \leqq 1]. \tag{2.7.3}$$

THEOREM 2.7.3. *Given a normed linear space $\mathfrak{X}$, a linear subspace $\mathfrak{L}$, and a linear bounded functional $f(x)$ defined on $\mathfrak{L}$. Then there exists a linear bounded functional $x^*(x)$ defined on $\mathfrak{X}$ such that $x^*(x) = f(x)$ in $\mathfrak{L}$ and the norm of $x^*(x)$ on $\mathfrak{X}$ is the same as the norm of $f(x)$ on $\mathfrak{L}$.*

PROOF. If

$$M = \| f \|_{\mathfrak{L}} = \sup [| f(x) | ; \| x \| \leqq 1, x \in \mathfrak{L}],$$

we take $p(x) = c(x) = M\| x \|$ and apply Theorem 2.7.1 in the real case and Theorem 2.7.2 in the complex case. Let $x^*(x)$ be the resulting extension. In the real case $\pm x^*(x) = x^*(\pm x) \leqq M \| x \|$. Thus in both the real and the complex case $| x^*(x) | \leqq M \| x \|$. Consequently $\| f \|_{\mathfrak{L}} \leqq \sup [| x^*(x) | ; \| x \| \leqq 1] = \| x^* \| \leqq \sup [M \| x \| ; \| x \| \leqq 1] = M = \| f \|_{\mathfrak{L}}$, and hence $\| x^* \| = \| f \|_{\mathfrak{L}}$.

On the basis of this theorem we can now prove various existence theorems for linear bounded functionals. The next two theorems are due to H. Hahn [1].

THEOREM 2.7.4. *To each point x_0 of the normed linear space $\mathfrak{X}$ there exists a linear bounded functional on $\mathfrak{X}$ such that $x^*(x_0) = \| x_0 \|$ and $\| x^* \| = 1$.*

PROOF. We may assume that $x_0 \neq \theta$. Then the elements of the form αx_0, $\alpha \in \Phi$, form a linear subspace $\mathfrak{L}$ on which we define $f(\alpha x_0) = \alpha \| x_0 \|$ so that $\| f \|_{\mathfrak{L}} = 1$. The desired result then follows from the preceding theorem.

THEOREM 2.7.5. *Given a normed linear space $\mathfrak{X}$, a linear subspace $\mathfrak{L}_0$, and a vector x_1 at a positive distance d from $\mathfrak{L}_0$. Then there is a linear bounded functional x^* such that* (i) $x^*(\mathfrak{L}_0) = 0$, (ii) $x^*(x_1) = 1$, *and* (iii) $\| x^* \| = 1/d$.

PROOF. Set $\mathfrak{L} = \mathfrak{L}_0 + \Phi x_1$. Then $\mathfrak{L}$ is a linear subspace and each $x \in \mathfrak{L}$ has a unique representation of the form $x = x_0 + \alpha x_1$ where $x_0 \in \mathfrak{L}_0$ and $\alpha \in \Phi$. We define $f(x) = \alpha$ on $\mathfrak{L}$. It is obvious that f is a linear functional on $\mathfrak{L}$ such that $f(x) = 0$ in $\mathfrak{L}_0$ and $f(x_1) = 1$. Moreover

$$\| f \|_{\mathfrak{L}} = \sup \left[\frac{|\alpha|}{\| x_0 + \alpha x_1 \|} ;\ x_0 \in \mathfrak{L}_0 ,\ \alpha \neq 0 \right]$$

$$= \sup \left[\frac{1}{\| x_1 + (1/\alpha) x_0 \|} ;\ x_0 \in \mathfrak{L}_0 ,\ \alpha \neq 0 \right] = 1/d.$$

The extension theorem is now used to complete the proof.

We recall that a set X_0 is called *fundamental* if the closed linear extension of X_0 is the entire space $\mathfrak{X}$. As an immediate corollary of Theorem 2.7.5 we have

THEOREM 2.7.6. *A necessary and sufficient condition that a subset X_0 of a normed linear space $\mathfrak{X}$ be fundamental is that every linear bounded functional which vanishes on X_0 vanishes identically.*

The problem of solving infinite systems of linear equations in functionals was attacked in special but typical cases by F. Riesz [1, 2] and E. Helly [1], and in the general case by H. Hahn [1].

THEOREM 2.7.7. *Given a normed linear space $\mathfrak{X}$, a collection of elements $[x_\alpha ; \alpha \in \mathfrak{A}] \subset \mathfrak{X}$, and a collection of numbers $[c_\alpha ; \alpha \in \mathfrak{A}] \subset \Phi$. A necessary and sufficient condition for the existence of a linear bounded functional x^* such that* (i) $x^*(x_\alpha) = c_\alpha$ *for all* $\alpha \in \mathfrak{A}$, *and* (ii) $\| x^* \| \leqq M$ *is that the inequality*

$$| \textstyle\sum_\pi \beta_\alpha c_\alpha | \leqq M \| \textstyle\sum_\pi \beta_\alpha x_\alpha \| \tag{2.7.4}$$

hold for each finite subset π of $\mathfrak{A}$ and each choice of numbers $[\beta_\alpha] \subset \Phi$.

PROOF. The necessity is obvious. To prove the sufficiency we define $f(\sum_\pi \beta_\alpha x_\alpha) = \sum_\pi \beta_\alpha c_\alpha$ on $\mathfrak{L} \equiv \mathfrak{L}([x_\alpha])$. The inequality (2.7.4) implies that f is single-valued, f is clearly linear on $\mathfrak{L}$, and (2.7.4) asserts that $\| f \|_{\mathfrak{L}} \leqq M$. The result now follows from Theorem 2.7.3.

The dual problem of solving an infinite system of linear equations for an *element* of the space does not in general have a solution. The best result here is due to E. Helly [2]; the following proof of Helly's theorem was devised by Yukio Mimura.

THEOREM 2.7.8. *Given a normed linear space $\mathfrak{X}$, a finite set of linear bounded functionals $[x_i^* ; i = 1, \cdots, n]$, and the numbers $[c_i ; i = 1, \cdots, n] \subset \Phi$. A necessary and sufficient condition that there exist for each $\epsilon > 0$ a vector $x_\epsilon \in \mathfrak{X}$ such that* (i) $x_i^*(x_\epsilon) = c_i$ *for* $i = 1, \cdots, n$, *and* (ii) $\| x_\epsilon \| \leqq M + \epsilon$ *is that the inequality*

$$\left| \sum_{i=1}^{n} \beta_i c_i \right| \leqq M \left\| \sum_{i=1}^{n} \beta_i x_i^* \right\| \tag{2.7.5}$$

hold for each choice of numbers $[\beta_i] \subset \Phi$.

PROOF. The necessity being obvious, we shall prove only the sufficiency. We may suppose without loss of generality that the x_i^* are linearly independent. Otherwise we can work with a linearly independent subsystem which spans the same subspace as the original system; the inequality (2.7.5) implies that an x_ϵ satisfying the conditions (i) and (ii) for this subsystem will also satisfy these conditions for the original system. Consider the mapping $x \to T(x) = (x_1^*(x), \cdots, x_n^*(x))$ which maps $\mathfrak{X}$ onto E_n (or Z_n). Let $S_\epsilon \equiv [x; \| x \| \leqq M + \epsilon]$. Then $T(S_\epsilon)$ is a convex subset of E_n (or Z_n). The linear independence of the x_i^* assures us that θ is an interior point of $T(S_\epsilon)$. Suppose for a given $\epsilon > 0$ that there is no $x_\epsilon \in \mathfrak{X}$ satisfying (i) and (ii). Then the point $p = (c_1, \cdots, c_n) \notin T(S_\epsilon)$. As a consequence there exists a maximal real flat space in E_n (or Z_n) which contains p and bounds $T(S_\epsilon)$. In other words there exist complex numbers $[\alpha_i]$ such that $\Re[\sum_1^n \alpha_i x_i^*(x)] \leqq \Re[\sum_1^n \alpha_i c_i] > 0$ for all $x \in S_\epsilon$. Since S_ϵ is circular we obtain $| \sum_1^n \alpha_i x_i^*(x) | \leqq | \sum_1^n \alpha_i c_i | > 0$ for all $x \in S_\epsilon$. Now

$$\sup \left[\left| \sum_1^n \alpha_i x_i^*(x) \right| ; x \in S_\epsilon \right] = (M + \epsilon) \left\| \sum_1^n \alpha_i x_i^* \right\|$$

and therefore $(M + \epsilon) \| \sum_1^n \alpha_i x_i^* \| \leqq | \sum_1^n \alpha_i c_i | > 0$ which is contrary to (2.7.5). It follows that there is an $x_\epsilon \in \mathfrak{X}$ satisfying (i) and (ii).

The indeterminancy of the solution in the preceding theorem is further illustrated by

THEOREM 2.7.9. *If the dimension of the linear system $\mathfrak{X}$ exceeds n and if $f_1, \cdots, f_n$ are n given linear functionals on $\mathfrak{X}$, then the homogeneous system of equations*

$$f_j(x) = 0, \qquad j = 1, 2, \cdots, n,$$

always has a non-trivial solution.

PROOF. Choose $n + 1$ linearly independent vectors $x_1, \cdots, x_{n+1}$ in $\mathfrak{X}$. If $x = \sum_1^{n+1} \xi_i x_i$ and $\alpha_{jk} = f_j(x_k)$, then the system

$$f_j(x) \equiv \alpha_{j,1}\xi_1 + \cdots + \alpha_{j,n+1}\xi_{n+1} = 0, \qquad j = 1, 2, \cdots, n,$$

has a non-trivial solution $(\xi_1^0, \cdots, \xi_{n+1}^0)$ which defines a non-trivial solution $x^0 = \sum_1^{n+1} \xi_i^0 x_i$ of the original system.

2.8. The adjoint space. Let $\mathfrak{X}$ be a real or complex (B)-space and consider the set $\mathfrak{X}^*$ of linear bounded functionals on $\mathfrak{X}$. If $x_1^*, x_2^* \in \mathfrak{X}^*$ and $\alpha_1, \alpha_2 \in \Phi$, then it is clear that $\alpha_1 x_1^* + \alpha_2 x_2^*$ also belongs to $\mathfrak{X}^*$. It follows that addition and scalar multiplication are defined in $\mathfrak{X}^*$ and it is easily shown that $\mathfrak{X}^*$ is a linear system in the sense of Definition 1.9.1. The zero element of $\mathfrak{X}^*$ is the functional which vanishes for all x. We introduce a normed topology in $\mathfrak{X}^*$ by setting $\| x^* \| = \sup [| x^*(x) | ; \| x \| \leqq 1]$ as in formula (2.7.3). It is now easy to show that this defines a proper norm in the sense of Definition 1.12.1 and that $\mathfrak{X}^*$ is complete in the resulting topology. Thus $\mathfrak{X}^*$ is again a (B)-space; it is called the *conjugate* or *adjoint space* of $\mathfrak{X}$.

Numerous examples of (B)-spaces and their adjoints are to be found in Banach's monograph [2]. We note that $c^* = l_1$, $l_1^* = m$, $l_p^* = l_q$, and $L_p(\mathfrak{S}; m)^* = L_q(\mathfrak{S}; m)$; here $1/p + 1/q = 1$. The reader is referred to T. H. Hildebrandt [2] for a discussion of m^* and to S. Kakutani [6] for $C(\mathfrak{S})^*$ and $M(\mathfrak{S}; m)^*$.

THEOREM 2.8.1. *To each bounded linear functional x_0^* on a Hilbert space $\mathfrak{H}$ there is a unique $x_0 \in \mathfrak{H}$ such that $x_0^*(x) = (x, x_0)$ for all $x \in \mathfrak{H}$. The mapping $x_0^* \to x_0$ is an isometry of $\mathfrak{X}^*$ onto $\mathfrak{H}$ taking $\alpha x_0^* + \beta y_0^*$ into $\bar{\alpha} x_0 + \bar{\beta} y_0$ if $x_0^* \to x_0$ and $y_0^* \to y_0$.*

PROOF. The kernel $\mathfrak{M} = [x; x_0^*(x) = 0]$ is a closed linear subspace of $\mathfrak{H}$. Choose a $y_0 \notin \mathfrak{M}$. By Problem 4 of Section 1.12 there is a unique $z_0 \in \mathfrak{M}$ such that $\| y_0 - z_0 \| = \inf [\| y_0 - z \|; z \in \mathfrak{M}]$. Thus for $w_0 = y_0 - z_0$ and $z \in \mathfrak{M}$ we see that the first variation of $\| w_0 + \epsilon z \|$ vanishes and hence that $(z, w_0) = 0$ for all $z \in \mathfrak{M}$. Now for any $x \in \mathfrak{H}$, $z = x_0^*(x) w_0 - x_0^*(w_0) x \in$

$\mathfrak{M}$ so that $x_0^*(x)(w_0, w_0) - x_0^*(w_0)(x, w_0) = 0$. Thus $x_0^*(x) = (x, x_0)$ where

$$x_0 = [\overline{x_0^*(w_0)/(w_0, w_0)}]w_0,$$

and the rest of the theorem readily follows.

If x and y are distinct elements of $\mathfrak{X}$, then according to Theorem 2.7.4 there is at least one $x^* \in \mathfrak{X}^*$ such that $x^*(x) \neq x^*(y)$. In other words

THEOREM 2.8.2. *If $x^*(x) = x^*(y)$ for all $x^* \in \mathfrak{X}^*$, then $x = y$.*

The value of the functional $x^* \in \mathfrak{X}^*$ at the point $x \in \mathfrak{X}$ is a number $B(x, x^*) \in \Phi$. We may consider $B(x, x^*)$ as a *bilinear functional* since obviously

$$B(\alpha x + \beta y, x^*) = \alpha B(x, x^*) + \beta B(y, x^*), \tag{2.8.1}$$

$$B(x, \gamma x^* + \delta y^*) = \gamma B(x, x^*) + \delta B(x, y^*), \tag{2.8.2}$$

$$| B(x, x^*) | \leqq \| x \| \| x^* \| . \tag{2.8.3}$$

The adjoint space $\mathfrak{X}^*$, being a (B)-space, also has an adjoint space $\mathfrak{X}^{**}$. This is made up of the linear bounded functionals defined on $\mathfrak{X}^*$. Formulas (2.8.2) and and (2.8.3) show for fixed $x_0 \in \mathfrak{X}$ that $B(x_0, x^*)$ is a linear bounded functional on $\mathfrak{X}^*$, that is, there exists an $x_0^{**} \in \mathfrak{X}^{**}$ such that

$$x_0^{**}(x^*) = x^*(x_0) \tag{2.8.4}$$

for all $x^* \in \mathfrak{X}^*$. By formula (2.8.3) we have $\| x_0^{**} \| \leqq \| x_0 \|$. Actually the equality must hold; for by Theorem 2.7.4 there is a functional x_0^* of unit norm for which $x_0^{**}(x_0^*) = x_0^*(x_0) = \| x_0 \|$.

The correspondence $x_0 \to x_0^{**}$ establishes a *natural mapping* of $\mathfrak{X}$ onto a subset $\mathfrak{X}_0^{**}$ of $\mathfrak{X}^{**}$. This mapping is one-to-one by Theorem 2.8.2; if $x \to x^{**}$, $y \to y^{**}$, then $\alpha x + \beta y \to \alpha x^{**} + \beta y^{**}$, and $\| x \| = \| x^{**} \|$. Thus under this natural mapping, $\mathfrak{X}$ is *isometrically isomorphic* with $\mathfrak{X}_0^{**}$. It follows that the space $\mathfrak{X}$ may be embedded in this way in the space $\mathfrak{X}^{**}$ without change of algebraic or metric relations. This is the sense to be attached to

THEOREM 2.8.3. $\mathfrak{X} \subset \mathfrak{X}^{**}$.

The case in which $\mathfrak{X} = \mathfrak{X}^{**}$ *under the natural embedding* is of particular interest. Such spaces were originally called *regular* by H. Hahn in 1927; following E. R. Lorch (1939) we shall use the more suggestive term *reflexive*. We note that it is possible for a (B)-space to be isometrically isomorphic with its second adjoint without being reflexive (cf. R. C. James [1]). The (B)-spaces l_p, $L_p(\mathfrak{S}; m)$ $(1 < p < \infty)$, and $\mathfrak{H}$ are reflexive. More generally D. Milman [1] and B. J. Pettis [3] (cf. S. Kakutani [3]) have shown that every uniformly convex space is reflexive.

The adjoint space to a separable (B)-space need not be separable. For example, l_1 is separable whereas $m = l_1^*$ is not. In the other direction we have

THEOREM 2.8.4. *If $\mathfrak{X}^*$ is separable, then $\mathfrak{X}$ is separable.*

PROOF. Since $\mathfrak{X}^*$ is separable, there is a countable set $\{x_n^*\}$ dense on the surface of the unit sphere in $\mathfrak{X}^*$. As $\| x_n^* \| = 1$, it follows from (2.7.3) that there exists an $x_n \in \mathfrak{X}$ such that $\| x_n \| \leqq 1$ and $| x_n^*(x_n) | \geqq \frac{1}{2}$. If the set $\{x_n\}$ is not fundamental, then Theorem 2.7.5 asserts the existence of a functional x^* with $\| x^* \| = 1$ and $x^*(x_n) = 0$ for all n. We then have

$$\| x^* - x_n^* \| \geqq | (x^* - x_n^*)(x_n) | = | x_n^*(x_n) | \geqq \tfrac{1}{2},$$

so that $\{x_n^*\}$ cannot be dense on the unit sphere in $\mathfrak{X}^*$ contrary to our assumption. It follows that $\{x_n\}$ is fundamental and hence that $\mathfrak{X} = \overline{\mathfrak{L}(\{x_n\})}$ is separable.

DEFINITION 2.8.1. *A subset $\Lambda \subset \mathfrak{X}^*$ is called a determining set for $\mathfrak{X}$ if $\| x \| = \sup [\, | x^*(x) | \,;\, x^* \in \Lambda]$ for all $x \in \mathfrak{X}$.*

It is clear that the elements of a determining set for $\mathfrak{X}$ must be of norm $\leqq 1$.

THEOREM 2.8.5. *If $\mathfrak{X}$ is separable, then both $\mathfrak{X}$ and $\mathfrak{X}^*$ possess denumerable determining sets.*

PROOF. We first obtain a denumerable determining set for $\mathfrak{X}$. By assumption there exists a denumerable dense subset $\{y_n\}$ in $\mathfrak{X}$. Applying Theorem 2.7.4 we obtain a denumerable set Λ of functionals $\{x_n^*\}$ such that $| x_n^*(y_n) | = \| y_n \|$, $\| x_n^* \| = 1$. For arbitrary $x \in \mathfrak{X}$ and $\epsilon > 0$ there is a y_{n_ϵ} with $\epsilon \geqq \| x - y_{n_\epsilon} \| \geqq | \| x \| - \| y_{n_\epsilon} \| |$ so that $\| x \| \geqq | x_{n_\epsilon}^*(x) | \geqq | x_{n_\epsilon}^*(y_{n_\epsilon}) | - | x_{n_\epsilon}^*(x - y_{n_\epsilon}) | \geqq \| y_{n_\epsilon} \| - \epsilon \geqq \| x \| - 2\epsilon$. It follows that Λ is a denumerable determining set for $\mathfrak{X}$. In the case of $\mathfrak{X}^*$, let Λ_0 be a denumerable set $\{x_n\}$ dense in the unit sphere in $\mathfrak{X}$. Then clearly

$$\sup [\, | x^*(x_n) | \,;\, x_n \in \Lambda_0] = \sup [\, | x^*(x) | \,;\, \| x \| \leqq 1] = \| x^* \|.$$

Thus if $x_n \to x_n^{**}$ under the natural embedding of $\mathfrak{X}$ in $\mathfrak{X}^{**}$, then it is clear that $\Lambda_1 \equiv \{x_n^{**}\}$ is a determining set for $\mathfrak{X}^*$.

A concept closely related to that of a determining set is given by

DEFINITION 2.8.2. *A closed linear subspace $\Gamma \subset \mathfrak{X}^*$ is called a determining manifold for $\mathfrak{X}$ if $\| x \| = \sup [\, | x^*(x) |; \| x^* \| \leqq 1, x^* \in \Gamma]$ for all $x \in \mathfrak{X}$.*

Any closed linear subspace Γ of $\mathfrak{X}^*$ is, of course, a (B)-space and each $x \in \mathfrak{X}$ defines a linear bounded functional on Γ. If, in addition, Γ is a determining manifold then the norm of the functional on Γ corresponding to x is simply $\| x \|$. Thus the following result, which is due to N. Dunford [3], is an immediate corollary of Theorem 2.5.5.

THEOREM 2.8.6. *Let Γ be a determining manifold for $\mathfrak{X}$ and suppose $[x_\alpha\,; \alpha \in \mathfrak{A}] \subset \mathfrak{X}$ is such that $\sup_\alpha | x^*(x_\alpha) | < \infty$ for each $x^* \in \Gamma$. Then $\sup_\alpha \| x_\alpha \| < \infty$.*

PROBLEM. Let $\mathfrak{Y} = \prod_\alpha \mathfrak{X}_\alpha$ be the product of the (B)-spaces $[\mathfrak{X}_\alpha\,; \alpha \in \mathfrak{A}]$, defined as in Problem 2 of Section 1.12 with $\| [x_\alpha] \| = [\sum_\alpha \| x_\alpha \|^p]^{1/p}$, $1 < p < \infty$. Show that $\mathfrak{Y}^* = \prod_\alpha \mathfrak{X}_\alpha^*$ with $\| [x_\alpha^*] \| = [\sum_\alpha \| x_\alpha^* \|^q]^{1/q}$, where $1/p + 1/q = 1$. For the case $p = 1$, show that $\| [x_\alpha^*] \| = \sup_\alpha \| x_\alpha^* \|$.

2.9. The weak topology of $\mathfrak{X}$. The normed topology of a (B)-space is often called the *strong topology*. In this terminology a Cauchy sequence $\{x_n\}$ is said to be *strongly convergent* and its limit is called the *strong limit* of $\{x_n\}$. In addition to the strong topology, we shall make use of another topology, called the *weak topology*, in terms of which the system is a *topological linear space* in the sense of Definition 1.10.1.

The neighborhoods are defined as follows. Let $x_0 \in \mathfrak{X}$, $\epsilon > 0$, and $x_1^*, x_2^*, \cdots, x_n^*$ be any finite set of functionals in $\mathfrak{X}^*$. Then

$$N(x_0\,;x_1^*, \cdots, x_n^*\,;\epsilon) \equiv [x;\, |\, x_k^*(x - x_0)\,| < \epsilon,\, k = 1, \cdots, n] \tag{2.9.1}$$

constitutes a neighborhood. The set of all such neighborhoods is the system $[N_\alpha]$ which defines the topology. Actually the set of all neighborhoods of the form (2.9.1) suffices to form a neighborhood basis for the point x_0. For, if $x_0 \in N(y_0) = N(y_0\,;x_1^*, \cdots, x_n^*\,;\epsilon)$, then $N(x_0\,;x_1^*, \cdots, x_n^*\,;\delta) \subset N(y_0)$ provided that $\delta < \epsilon - \max_k |\, x_k^*(y_0 - x_0)\,|$. In order to verify the Hausdorff axiom H_1, we take

$$N(x_0) = N(x_0\,;x^*;\epsilon), \qquad N(y_0) = N(y_0\,;x^*;\epsilon)$$

where $x^* \in \mathfrak{X}^*$ is chosen to satisfy the condition $x^*(y_0 - x_0) = 1$. This gives $x^*(x - x_0) - x^*(x - y_0) = 1$ and hence if $\epsilon < \frac{1}{2}$, then we have $N(x_0) \cap N(y_0) = \emptyset$. Further if N_1 and N_2 are two neighborhoods of x_0:

$$N_1 = N(x_0\,;x_1^*, \cdots, x_m^*\,;\epsilon_1), \qquad N_2 = N(x_0\,;x_{m+1}^*, \cdots, x_{m+n}^*\,;\epsilon_2)$$

then

$$N_3 = N(x_0\,;x_1^*, \cdots, x_m^*, x_{m+1}^*, \cdots, x_{m+n}^*\,;\epsilon_3),$$

with $\epsilon_3 < \min(\epsilon_1, \epsilon_2)$, has the property $N_3 \subset N_1 \cap N_2$ so that H_2 is satisfied. It is a simple matter to verify that the axioms of Definitions 1.8.1 and 1.10.1 are also satisfied. It follows that $\mathfrak{X}$ is a topological linear space relative to this weak topology. Thus $x + y$ and αx are continuous functions of (x, y) and (α, x) respectively, just as in the strong topology.

The weak topology of Hilbert space was introduced by J. von Neumann in 1929; however the corresponding notion of weak convergence is much older. For the case of the Lebesgue spaces it goes back to F. Riesz (1910).

DEFINITION 2.9.1. *A directed set $[x_\pi]$ in a (B)-space $\mathfrak{X}$ is said to be weakly convergent if $\lim_\pi x^*(x_\pi)$ exists for each $x^* \in \mathfrak{X}^*$; it is said to be weakly convergent to an element x_0 if $\lim_\pi x^*(x_\pi) = x^*(x_0)$ for all x^*.*

DEFINITION 2.9.2. *A subset X of a (B)-space is said to be weakly (sequentially) complete if every weakly convergent directed set (sequence) is weakly convergent to an element of X.*

DEFINITION 2.9.3. *A subset X of a (B)-space is said to be (conditionally) sequentially weakly compact if every sequence in X contains a subsequence which converges weakly to an element (in $\mathfrak{X}$) in X.*

Strong convergence implies weak convergence, but not conversely. However, for weakly convergent sequences we do have the following result which is a special instance of Theorem 2.8.6.

THEOREM 2.9.1. *If a sequence* $\{x_n\}$ *converges weakly, then* $\sup_n \| x_n \| < \infty$. *Moreover, a conditionally sequentially weakly compact subset of* $\mathfrak{X}$ *is necessarily bounded in norm.*

J. von Neumann [3] has called attention to the fact that weak closure cannot be defined in terms of weak sequential convergence. He has constructed a Hilbert space example of a sequence having a single limit point in the weak topology and nevertheless containing no subsequence which converges weakly to this point. This implies that the neighborhoods in the weak topology do not satisfy the first countability axiom, much less the second. It should be noted, however, that the weak topology does have a convex neighborhood basis.

We also note that $\| x \|$ is not continuous in the weak topology (unless $\mathfrak{X}$ is finite dimensional). For according to Theorem 2.7.9, given a neighborhood $N = N(x_0 ; x_1^*, \cdots, x_n^* ; \epsilon)$, there exists a $y \in \mathfrak{X}$ of arbitrarily large norm such that $x_k^*(y) = 0$ for $k = 1, \cdots, n$. It follows that $x_0 + y \in N$ where $\| y \|$ is arbitrarily large.

As an immediate corollary of Theorem 2.7.5 we have

THEOREM 2.9.2. *Each strongly closed linear subspace* $\mathfrak{L}$ *of* $\mathfrak{X}$ *is necessarily weakly closed.*

The following result due to S. Mazur [1] lies somewhat deeper.

THEOREM 2.9.3. *Each strongly closed convex subset* C *of* $\mathfrak{X}$ *is necessarily weakly closed.*

PROOF. For each $x_0 \notin C$ there exists a sphere $S = [x; \| x - x_0 \| < r]$ disjoint from C. Since S is an open convex set, Theorem 2.6.3 applies and hence there exists a continuous real-linear functional $f(x)$ and a constant γ such that $f(x_0) < \gamma \leqq f(C)$. Finally $x^*(x) = f(x) - if(ix) \in \mathfrak{X}^*$ so that $N(x_0 ; x^*; \epsilon) \cap C = \emptyset$ for $\epsilon < \gamma - f(x_0)$.

COROLLARY. *If the sequence* $\{x_n\}$ *converges weakly to* x_0, *then given an integer* n *and* $\epsilon > 0$ *there exists a finite set of real numbers* $\{\alpha_i\}$, $\alpha_i \geqq 0$, $\sum_i \alpha_i = 1$, *such that* $\| x_0 - \sum_i \alpha_i x_{n+i} \| < \epsilon$.

It is clear that a sequentially weakly compact subset is necessarily weakly sequentially complete. We also have

THEOREM 2.9.4. *A weakly compact set is weakly complete.*

PROOF. Let $[x_\pi]$ be a weakly convergent directed set contained in a weakly compact subset X. We denote the weak closure of $[x_\sigma ; \sigma \geqq \pi]$ by F_π ; the F_π possess the finite intersection property and are subsets of X. As a consequence there is an $x_0 \in \bigcap F_\pi \subset X$, and it is easily shown that $\lim_\pi x^*(x_\pi) = x^*(x_0)$ for all $x^* \in \mathfrak{X}^*$.

For the sake of completeness we list two further theorems on weakly compact subsets.

Theorem 2.9.5 is due to M. Krein [1] (cf. R. S. Phillips [3]) whereas Theorem 2.9.6 is due to W. F. Eberlein [1].

THEOREM 2.9.5. *The closed convex extension of a conditionally sequentially weakly compact set is sequentially weakly compact.*

THEOREM 2.9.6. *A subset of $\mathfrak{X}$ is sequentially weakly compact if and only if it is weakly compact.*

PROBLEMS.

1. (F. Riesz [1], see Banach [2, pp. 141–143].) Show that l_p and $L_p(\mathfrak{S}, m)$, $1 \leqq p < \infty$, are weakly sequentially complete.
2. (I. Schur [1], see Banach [2, pp. 137–139].) Show that each weakly convergent sequence in l_1 is strongly convergent.

2.10. The weak* topology of $\mathfrak{X}^*$. In the adjoint space $\mathfrak{X}^*$ we can, of course, introduce the weak topology defined in the previous section, the functionals employed will in this case be elements of the second adjoint space $\mathfrak{X}^{**}$. Another very useful topology is obtained if we limit the functionals to $\mathfrak{X} \subset \mathfrak{X}^{**}$; this defines the weak* topology of $\mathfrak{X}^*$. A generic neighborhood is now determined by $x_0^* \in \mathfrak{X}^*$, $\epsilon > 0$, and a finite set, $x_1, \cdots, x_n$, of elements in $\mathfrak{X}$;

$$(2.10.1) \quad N(x_0^*; x_1, \cdots, x_n; \epsilon) = [x^*; |(x^* - x_0^*)(x_k)| < \epsilon, k = 1, \cdots, n].$$

The same reasoning as before shows that neighborhoods of the form (2.10.1) form a basis for the point x_0^* and that $\mathfrak{X}^*$ is a topological linear space in the resulting weak* topology.

THEOREM 2.10.1. *If $\mathfrak{X}$ is separable, then the sphere $S \equiv [x^*; \| x \| \leqq M]$ is sequentially compact in the weak* topology of $\mathfrak{X}^*$.*

PROOF. Let $\{x_n\}$ be a countable dense subset of $\mathfrak{X}$. Then given a sequence $\{x_k^*\} \subset S$, one proceeds by the diagonal process to obtain a subsequence $\{x_{k_i}^*\}$ such that $\lim_{i\to\infty} x_{k_i}^*(x_n)$ exists for each integer n. For an arbitrary $x \in \mathfrak{X}$ there exists a subsequence $\{x_{n_j}\}$ such that $x_{n_j} \to x$ as $j \to \infty$. Since $\lim_{j\to\infty} x_{k_i}^*(x_{n_j})$ exists uniformly with respect to i, the iterated limits theorem implies that $\lim_{i\to\infty} x_{k_i}^*(x) \equiv x_0^*(x)$ exists; the element x being arbitrary, the limit exists for each $x \in \mathfrak{X}$. It is easy to see that x_0^* is linear and that $\| x_0^* \| \leqq M$. Thus $x_0^* \in S$. Finally, it is clear from (2.10.1) that $\{x_{k_i}^*\}$ converges to x_0^* in the weak* topology of $\mathfrak{X}^*$.

For arbitrary Banach spaces we have the following result due to L. Alaoglu [1], N. Bourbaki [1], and S. Kakutani [4].

THEOREM 2.10.2. *The sphere $S \equiv [x^*; \| x^* \| \leqq M]$ is compact in the weak* topology of $\mathfrak{X}^*$.*

PROOF. With each $x \in \mathfrak{X}$ we associate a compact subset of the complex plane, namely $C_x \equiv [\alpha; |\alpha| \leqq M \| x \|]$. We then form the Cartesian product $\mathfrak{P} = \prod C_x$ of points $p(x)$ where $p(x) \in C_x$ and x ranges over $\mathfrak{X}$; $\mathfrak{P}$ is then topologized in the usual way. According to the Tychonoff theorem (see section 1.3), $\mathfrak{P}$ is a compact Hausdorff space. We now embed S in $\mathfrak{P}$ by the mapping $x^* \to p(x) \equiv x^*(x)$. It is clear from the manner in which the topologies of both $\mathfrak{P}$ and S were defined that this mapping is a homeomorphism of S onto its image $S' \subset \mathfrak{P}$. It suffices, therefore, to show that S' is a closed subset of $\mathfrak{P}$. Suppose that $p_0 \in \overline{S'}$. Then given $x, y \in \mathfrak{X}$ and $\epsilon > 0$, the set

$$N(p_0) = [p; \mid p(z) - p_0(z) \mid < \epsilon, z = x, y, \alpha x, x + y]$$

is a neighborhood of p_0 and as such contains a point $p = x^*(x)$ of S'. Thus

$$\mid x^*(x) - p_0(x) \mid < \epsilon, \qquad \mid x^*(y) - p_0(y) \mid < \epsilon,$$
$$\mid x^*(\alpha x) - p_0(\alpha x) \mid < \epsilon, \qquad \mid x^*(x + y) - p_0(x + y) \mid < \epsilon.$$

Making use of the fact that x^* is a linear functional of norm less than or equal to M, we obtain

$$\mid p_0(x + y) - p_0(x) - p_0(y) \mid < 3\epsilon,$$
$$\mid \alpha p_0(x) - p_0(\alpha x) \mid < (\mid \alpha \mid + 1)\epsilon,$$
$$\mid p_0(x) \mid \leqq M \parallel x \parallel + \epsilon.$$

Since $\epsilon > 0$ is arbitrary, it follows that $p_0(x)$ is a bounded linear functional in S; in other words $p_0 \in S'$. This concludes the proof.

We are now in a position to derive some interesting properties of reflexive spaces. The following theorem is due to S. Kakutani [4] and its corollaries are due to B. J. Pettis [2].

THEOREM 2.10.3. *A (B)-space $\mathfrak{X}$ is reflexive if and only if $S \equiv [x; \parallel x \parallel \leqq 1]$ is weakly compact.*

PROOF. If $\mathfrak{X}$ is reflexive, then the weak topology of $\mathfrak{X}$ is equivalent to the weak* topology of $\mathfrak{X}^{**} = \mathfrak{X}$. It follows from the above theorem that S is weakly compact. Conversely suppose that S (and hence MS) is weakly compact and that $x_0^{**} \in \mathfrak{X}^{**}$. Let π be a finite set of functionals in $\mathfrak{X}^*$ and order the π's by inclusion. By Theorem 2.7.8 there exists an $x_\pi \in \mathfrak{X}$ such that $x^*(x_\pi) = x_0^{**}(x^*)$ for all $x^* \in \pi$ and $\parallel x_\pi \parallel \leqq M \equiv 2 \parallel x_0^{**} \parallel$. Thus $\lim_\pi x^*(x_\pi) = x_0^{**}(x^*)$ for all $x^* \in \mathfrak{X}^*$. The weakly compact set MS is also weakly complete by Theorem 2.9.4 so that the directed set $[x_\pi]$ converges weakly to an $x_0 \in \mathfrak{X}$. Thus $x_0^{**}(x^*) = x^*(x_0)$ for all $x^* \in \mathfrak{X}^*$. It follows that the natural mapping of $\mathfrak{X}$ into $\mathfrak{X}^{**}$ is actually onto and hence that $\mathfrak{X}$ is reflexive.

COROLLARY 1. *Each closed linear subspace $\mathfrak{X}_0$ of a reflexive space $\mathfrak{X}$ is also reflexive.*

PROOF. According to Theorem 2.9.2, $\mathfrak{X}_0$ is weakly closed and hence $S \cap \mathfrak{X}_0$ is a weakly compact subset of $\mathfrak{X}$. Now each $x^* \in \mathfrak{X}^*$ contracts on $\mathfrak{X}_0$ to an element of $\mathfrak{X}_0^*$ and, by Theorem 2.7.3, each $x_0^* \in \mathfrak{X}_0^*$ can be extended on $\mathfrak{X}$ to be an element in $\mathfrak{X}^*$. It follows that $S \cap \mathfrak{X}_0$ is also weakly compact relative to the weak topology of $\mathfrak{X}_0$.

COROLLARY 2. *$\mathfrak{X}$ is reflexive if and only if $\mathfrak{X}^*$ is reflexive.*

PROOF. If $\mathfrak{X}$ is reflexive, then the weak and the weak* topology of $\mathfrak{X}^*$ are equivalent. Consequently $[x^*; \parallel x^* \parallel \leqq 1]$, which is weakly* compact by Theorem 2.10.2, is also weakly compact. Thus if $\mathfrak{X}^*$ is reflexive then so is $\mathfrak{X}^{**}$ and $\mathfrak{X}$, being a closed linear subspace of $\mathfrak{X}^{**}$ under the natural embedding, is also reflexive.

DEFINITION 2.10.1. *A linear subspace* $\Gamma \subset \mathfrak{X}^*$ *is said to be boundedly weakly* closed if it contains all weak* limit points of the bounded subsets of* Γ.

DEFINITION 2.10.2. *A linear subspace* $\Gamma \subset \mathfrak{X}^*$ *is said to be regularly closed if for each* $x_0^* \notin \Gamma$ *there exists an* $x_0 \in \mathfrak{X}$ *such that* $x_0^*(x_0) = 1$ *and* $x^*(x_0) = 0$ *for all* $x^* \in \Gamma$.

It is clear that a weakly* closed linear subspace is boundedly weakly* closed and that a regularly closed linear subspace is weakly* closed. However the basic result concerning these concepts is contained in the following lemma due to S. Banach [2, pp. 119–121].

LEMMA 2.10.1. *Let* Γ *be a boundedly weakly* closed linear subspace in* $\mathfrak{X}^*$ *and suppose* $x_0^* \notin \Gamma$. *Then for each* M *satisfying the condition* $0 < M < \inf [\| x^* - x_0^* \|; x^* \in \Gamma]$, *there exists an* $x_0 \in \mathfrak{X}$ *such that*

$$x_0^*(x_0) = 1, \quad x^*(x_0) = 0 \quad \textit{for all} \quad x^* \in \Gamma, \quad \textit{and} \quad \| x_0 \| \leqq 1/M.$$

PROOF. Choose an increasing sequence of numbers $\{M_k\}$ such that $M_1 = M$ and $M_k \to \infty$, and set $S \equiv [x; \| x \| \leqq 1]$. Then there exists a finite subset X_1 of S such that if x^* satisfies

$$\| x^* - x_0^* \| \leqq M_2 \text{ and } | x^*(x) - x_0^*(x) | \leqq M_1 \text{ for all } x \in X_1, \tag{2.10.2}$$

then $x^* \notin \Gamma$. For if this were not the case, then corresponding to each finite subset π of S there would be an $x_\pi^* \in \Gamma$ satisfying (2.10.2) for all $x \in \pi$. We order the sets π by inclusion and denote the weak* closure of $[x_\sigma^*; \sigma \geqq \pi]$ by F_π. It is clear that the F_π possess the finite intersection property. Since Γ is boundedly weakly* closed, it follows from Theorem 2.10.2 that $\Gamma_M \equiv [x^*; \| x^* \| \leqq M, x^* \in \Gamma]$ is a weakly* compact subset of $\mathfrak{X}^*$. Consequently $F_\pi \subset \Gamma_M$ for $M = M_2 + \| x_0^* \|$, and there exists an $x_1^* \in \bigcap F_\pi \subset \Gamma$. It follows that

$$| x_1^*(x) - x_0^*(x) | \leqq M_1$$

for all $x \in S$ and hence that $\| x_1^* - x_0^* \| \leqq M_1$, contrary to hypothesis. By a similar argument we successively establish the existence of a sequence of finite subsets $X_1, X_2, \cdots \subset S$ such that a functional x^* satisfying $\| x^* - x_0^* \| \leqq M_k$ and $| x^*(x) - x_0^*(x) | \leqq M_i$ for all $x \in X_i, i = 1, 2, \ldots, k-1$, can not belong to Γ. Thus, since $M_k \to \infty$, we see that $x^* \notin \Gamma$ if

$$| x^*(x) - x_0^*(x) | \leqq M \quad \text{for all} \quad x \in \frac{M}{M_i} X_i \quad \text{and} \quad i = 1, 2, \cdots. \tag{2.10.3}$$

Let $\{x_n\}$ be a sequence of elements which successively exhausts the sets $(M/M_1)X_1$, $(M/M_2)X_2, \cdots$. Then $\| x_n \| \to 0$ and $T(x^*) = \{x^*(x_n)\}$ is a bounded linear mapping of $\mathfrak{X}^*$ into c_0. It is clear from (2.10.3) that the point $p_0 = \{x_0^*(x_n)\}$ lies at a distance $\geqq M$ from the linear subspace $T(\Gamma)$. By Theorem 2.7.5 there exists a sequence $\{\alpha_n\} \in c_0^* = l_1$ such that $\sum_n \alpha_n x^*(x_n) = 0$ for all $x^* \in \Gamma$, $\sum_n \alpha_n x_0^*(x_n) = 1$, and $\sum | \alpha_n | \leqq 1/M$. The element $x_0 = \sum_n \alpha_n x_n$ clearly satisfies all of the conditions of the lemma.

We summarize the above results in the

THEOREM 2.10.4. *For a linear subspace* Γ *of* $\mathfrak{X}^*$ *the following are equivalent*: (1) Γ *is weakly* closed*; (2) Γ *is boundedly weakly* closed*; *and* (3) Γ *is regularly closed.*

PROBLEM (S. Banach [2, p. 34] and S. Kakutani [6]). A functional Lim (x) defined on m is called a *Banach limit* if it satisfies the following conditions: (1) $\operatorname{Lim}(\alpha x + \beta y) = \alpha \operatorname{Lim}(x) + \beta \operatorname{Lim}(y)$; (2) If $\alpha_n = 1$ for $n = 1, 2, \cdots$, then $\operatorname{Lim}(\{\alpha_n\}) = 1$; (3) If $\alpha_n \geqq 0$ for $n = 1, 2, \cdots$, then $\operatorname{Lim}(\{\alpha_n\}) \geqq 0$, and (4) $\operatorname{Lim}(\{\alpha_{n+1}\}) = \operatorname{Lim}(\{\alpha_n\})$. Show that there exist Banach limits in m^*.

HINT. Let $x_n^*(\{\alpha_n\}) = n^{-1}\sum_1^n \alpha_k$ and set F_n equal to the weak* closure of $[x_k^*; k \geqq n]$. Then $x_0^* \in \bigcap F_n$ is a Banach limit.

References. Alaoglu [1], Banach [**1, 2**], Bohnenblust and Sobczyk [1], Bourbaki [1], Day [1], Dieudonné [1], Dunford [**3**], Eberlein [1], Eidelheit [1], Hahn [1], Helly [**1, 2**], Hildebrandt [**2**], James [1], Kakutani [**3, 4, 6**], Kelley [1], Klee [1], Kolmogoroff [**3**], Krein [1], Krein and Milman [1], LaSalle [1], Lorch [1], Mazur [1], Milman [1], v. Neumann [**3**], Pettis [**2, 3**], Phillips [**3**], F. Riesz [**1, 2**], Schur [1], Wehausen [1].

3. Linear Transformations

2.11. General properties. Let $\mathfrak{X}$ and $\mathfrak{Y}$ be (B)-spaces over the same scalar field Φ. The transformation $y = T(x)$ is taken to be linear with domain $\mathfrak{D} \subset \mathfrak{X}$ and range $\mathfrak{R} \subset \mathfrak{Y}$, and it is assumed that $\mathfrak{D}$ is a linear subspace of $\mathfrak{X}$. The present section is concerned with basic concepts arising in the study of linear transformations. Most of the results go back to S. Banach; those dealing with bounded linear transformations are to be found in his treatise [**2**], and much of the rest consists of generalizations of his results (cf. R. S. Phillips [**12**]).

Definition 2.11.1. *Let $\mathfrak{X} \times \mathfrak{Y}$ be the product of $\mathfrak{X}$ and $\mathfrak{Y}$. The graph $\mathfrak{G}$ of a transformation $y = T(x)$ is the set of vector pairs $[(x, T(x)); x \in \mathfrak{D}] \subset \mathfrak{X} \times \mathfrak{Y}$.*

A suitable norm for $\mathfrak{X} \times \mathfrak{Y}$ is given by $\| (x, y) \| = \| x \| + \| y \|$. With this norm $\mathfrak{X} \times \mathfrak{Y}$ becomes a (B)-space.

Definition 2.11.2. *A transformation $y = T(x)$ is said to be closed if its graph is a closed subset of $\mathfrak{X} \times \mathfrak{Y}$.*

In other words, a transformation is closed if whenever $x_n \to x_0$, $\{x_n\} \subset \mathfrak{D}$, and $y_n = T(x_n) \to y_0$, it always follows that $x_0 \in \mathfrak{D}$ and $y_0 = T(x_0)$.

Definition 2.11.3. *If $T_1(x)$ and $T_2(x)$ are transformations on $\mathfrak{X}$ to $\mathfrak{Y}$ with domains $\mathfrak{D}_1$ and $\mathfrak{D}_2$ respectively, if $\mathfrak{D}_1 \subset \mathfrak{D}_2$, and if $T_1(x) = T_2(x)$ in $\mathfrak{D}_1$, then $T_2(x)$ is called an extension of $T_1(x)$, in symbols, $T_1 \subset T_2$.*

Theorem 2.11.1. *Let $y = T(x)$ be a linear transformation on $\mathfrak{D} \subset \mathfrak{X}$ to $\mathfrak{Y}$ with graph $\mathfrak{G}$. Then T has a closed extension if and only if no vector pair of the form (θ, y), $y \neq \theta$, belongs to $\bar{\mathfrak{G}}$. In this case, $\bar{\mathfrak{G}}$ is the graph of the smallest closed linear extension of T.*

Proof. Since $\mathfrak{G}$ is a linear subspace of $\mathfrak{X} \times \mathfrak{Y}$, the same is true of $\bar{\mathfrak{G}}$. Now $\bar{\mathfrak{G}}$ is the graph of some transformation, say $\bar{T}$. Suppose that (x, y_1) and $(x, y_2) \in \bar{\mathfrak{G}}$. Then $(x, y_1) - (x, y_2) = (\theta, y_1 - y_2) \in \bar{\mathfrak{G}}$ and the hypothesis asserts that $y_1 = y_2$. It follows that $\bar{T}$ is single-valued. The linearity and the closure of the transformation $\bar{T}$ is an immediate consequence of the linearity and the closure of the subspace $\bar{\mathfrak{G}}$. It is obvious that $\bar{\mathfrak{G}}$ is the graph of the smallest closed extension of T. Conversely, no well defined extension of T can take θ into a vector other than θ.

DEFINITION 2.11.4. *A linear transformation on* $\mathfrak{D} \subset \mathfrak{X}$ *to* $\mathfrak{Y}$ *is said to be bounded if* $\| T(x) \| \leqq M \| x \|$ *for all* $x \in \mathfrak{D}$. *The norm of* T *is* $\| T \| = \sup [\| T(x) \|; \| x \| \leqq 1, x \in \mathfrak{D}]$.

It is clear that a linear bounded transformation T is closed if and only if its domain $\mathfrak{D}$ is closed. If $\mathfrak{D}$ is not closed we may extend T to $\bar{\mathfrak{D}}$ as follows. Let $x_0 \in \bar{\mathfrak{D}}$. Then for any sequence $\{x_n\} \subset \mathfrak{D}$ converging to x_0, we have $\| T(x_n) - T(x_m) \| \leqq \| T \| \| x_n - x_m \|$ and hence $\{T(x_n)\}$ is a Cauchy sequence whose limit y_0 is independent of the particular choice of sequence $\{x_n\}$, $x_n \to x_0$. Defining $T(x_0) = y_0$, it readily follows that

THEOREM 2.11.2. *If* $T(x)$ *is a linear bounded transformation on* $\mathfrak{D} \subset \mathfrak{X}$ *to* $\mathfrak{Y}$ *with norm* $\| T \|$, *then* T *has a unique linear bounded extension* $\bar{T}$ *on* $\bar{\mathfrak{D}}$ *and* $\| \bar{T} \| = \| T \|$.

We denote the class of all linear bounded transformations with domain $\mathfrak{X}$ and range in $\mathfrak{Y}$ by $\mathfrak{E}(\mathfrak{X}, \mathfrak{Y})$. If $T_1, T_2 \in \mathfrak{E}(\mathfrak{X}, \mathfrak{Y})$ and $\alpha_1, \alpha_2 \in \Phi$, then we may define $(\alpha_1 T_1 + \alpha_2 T_2)(x) \equiv \alpha_1 T_1(x) + \alpha_2 T_2(x)$. It is clear that $\alpha_1 T_1 + \alpha_2 T_2 \in \mathfrak{E}(\mathfrak{X}, \mathfrak{Y})$ and further that $\mathfrak{E}(\mathfrak{X}, \mathfrak{Y})$ is, in fact, a linear system. It is also clear that $\| T \| = \sup [\|T(x) \|; \| x \| \leqq 1]$ is a suitable norm. Finally, we show that $\mathfrak{E}(\mathfrak{X}, \mathfrak{Y})$ is complete in this metric. For let $\{T_n\}$ be a Cauchy sequence and set $\delta_{nm} = \| T_n - T_m \|$. Then since $\| T_n(x) - T_m(x) \| \leqq \delta_{nm} \| x \|$ and since $\mathfrak{Y}$ is itself complete, the sequence $\{T_n(x)\}$ converges for each x to a limit $T_0(x)$ as $n \to \infty$. It is readily shown that $T_0(x)$ is linear and that $\| T_0(x) \| \leqq (\lim_n \| T_n \|) \| x \|$; hence $T_0 \in \mathfrak{E}(\mathfrak{X}, \mathfrak{Y})$. If $\delta_{nm} < \epsilon$ for $n, m > N$, we have $\| T_n(x) - T_0(x) \| \leqq \epsilon \| x \|$ and $\| T_n - T_0 \| \leqq \epsilon$. It follows that $\| T_n - T_0 \| \to 0$. This proves

THEOREM 2.11.3. *The class of all linear bounded transformations on* $\mathfrak{X}$ *to* $\mathfrak{Y}$ *forms a* (B)*-space* $\mathfrak{E}(\mathfrak{X}, \mathfrak{Y})$.

We next prove the Banach-Steinhaus theorem (cf. S. Banach [**2**, pp. 79–80]).

THEOREM 2.11.4. *Let* $\{T_n\}$ *be a sequence of linear bounded transformations in* $\mathfrak{E}(\mathfrak{X}, \mathfrak{Y})$ *such that* (i) $\| T_n \| \leqq M$ *for all* n, *and* (ii) $\lim_{n\to\infty} T_n(x)$ *exists for every* x *in a set* X *which is dense in a sphere* S. *Then* $\lim_{n\to\infty} T_n(x)$ *exists for all* x, *and the limit is a linear transformation with bound* $\leqq \liminf_{n\to\infty} \| T_n \|$.

PROOF. The first step is to prove convergence in S. Let $x_0 \in S$ and $x_n \to x_0$ where $\{x_n\} \subset X$. Then for arbitrary positive integers n, p, q we have

$$\| T_p(x_0) - T_q(x_0) \| \leqq \| T_p(x_0) - T_p(x_n) \| + \| T_p(x_n) - T_q(x_n) \| + \| T_q(x_n) - T_q(x_0) \|,$$

whence $\limsup_{p,q\to\infty} \| T_p(x_0) - T_q(x_0) \| \leqq 2M \| x_0 - x_n \|$. Since $x_n \to x_0$, the convergence of $\{T_n(x_0)\}$ is proved. If y_0 is the center of S, r its radius, and x an arbitrary point of $\mathfrak{X}$, then $y_0 + \alpha x \in S$ for $|\alpha| < r/\|x\|$, and consequently the sequence $\{T_n(y_0 + \alpha x)\}$ converges. However $T_n(y_0 + \alpha x) = T_n(y_0) + \alpha T_n(x)$ and hence $\{T_n(x)\}$ also converges; that is, $T(x) \equiv \lim_{n\to\infty} T_n(x)$ exists for all x. Since $T_n(x)$ is linear, it follows that $T(x)$ is also linear, and $\| T_n \| \leqq M$ implies that $\| T \| \leqq M$. By employing the above argument on an appropriate subsequence of the T_n's it is clear that we obtain $\| T \| \leqq \liminf_{n\to\infty} \| T_n \| + \epsilon$ for each $\epsilon > 0$. It follows that $\| T \| \leqq \liminf_{n\to\infty} \| T_n \|$.

The transformation $y = T(x)$ has an inverse $x = T^{-1}(y)$ if and only if the correspondence between domain $\mathfrak{D}$ and range $\mathfrak{R}$ is one-to-one. In this case

$$y = T[T^{-1}(y)] \quad \text{for all} \quad y \in \mathfrak{R}; \qquad x = T^{-1}[T(x)] \quad \text{for all} \quad x \in \mathfrak{D}.$$

The condition for the existence of the inverse is that $x_1 \neq x_2$ should imply $T(x_1) \neq T(x_2)$ and if T is linear this is equivalent to

$$T(x) = \theta \quad \text{if and only if} \quad x = \theta. \tag{2.11.1}$$

A simple argument shows that *the inverse of a linear transformation is also linear.*

THEOREM 2.11.5. *If a closed transformation has an inverse, then the inverse is also closed.*

PROOF. If T^{-1} exists, then the graph of T may be written $[(T^{-1}(y), y); y \in \mathfrak{R}]$. It follows that the graph of T^{-1} is closed if the graph of T is closed.

THEOREM 2.11.6. *A necessary and sufficient condition for a linear transformation $y = T(x)$ to have a bounded inverse is the existence of an $m > 0$ such that $\| T(x) \| \geqq m \| x \|$ for all x in $\mathfrak{D}$. The supremum of the admissible values of m is the reciprocal of the norm of T^{-1}.*

PROOF. If $T^{-1}(y)$ exists as a bounded transformation, then $\| x \| = \| T^{-1}(y) \| \leqq M \| y \|$, $y \in \mathfrak{R}$, and $\| T(x) \| \geqq (1/M) \| x \|$. Conversely, if $\| T(x) \| \geqq m \| x \|$, then (2.11.1) holds, $T^{-1}(y)$ exists, and $\| T^{-1}(y) \| \leqq (1/m) \| y \|$.

THEOREM 2.11.7. *If $y = T(x)$ and its inverse are bounded linear transformations, and if the domain of $T(x)$ is closed, then the range is also closed.*

PROOF. Given $y_n \to y_0$, $\{y_n\} \subset \mathfrak{R}$, and $y_n = T(x_n)$; then $\| x_n - x_m \| \leqq \| T^{-1} \| \| y_n - y_m \|$ whence it follows that $\lim x_n = x_0$ exists. Since $T(x)$ is defined and continuous at $x = x_0$, we have $T(x_0) = \lim T(x_n) = y_0$; that is, $y_0 \in \mathfrak{R}$ and $\mathfrak{R}$ is closed.

DEFINITION 2.11.5. *Let $y = T(x)$ be a linear transformation defined on a domain $\mathfrak{D}$ dense in $\mathfrak{X}$ to $\mathfrak{Y}$ and let $\mathfrak{X}^*$ and $\mathfrak{Y}^*$ be the adjoint spaces of $\mathfrak{X}$ and $\mathfrak{Y}$ respectively. The adjoint transformation T^* of T is defined as follows. Its domain $\mathfrak{D}(T^*)$ consists of the set of all $y^* \in \mathfrak{Y}^*$ for which there exists an $x^* \in \mathfrak{X}^*$ such that $y^*[T(x)] = x^*(x)$ for all $x \in \mathfrak{D}$; in this case we define $T^*(y^*) = x^*$.*

THEOREM 2.11.8. *Let $T(x)$ be a linear transformation defined on a domain $\mathfrak{D}$ dense in $\mathfrak{X}$ to $\mathfrak{Y}$. Then T^* is a closed linear transformation on $\mathfrak{D}(T^*) \subset \mathfrak{Y}^*$ to $\mathfrak{X}^*$. If, in addition, T is bounded, then $T^* \in \mathfrak{E}(\mathfrak{Y}^*, \mathfrak{X}^*)$ and $\| T^* \| = \| T \|$.*

PROOF. Since $\mathfrak{D}$ is dense in $\mathfrak{X}$ we see by Theorem 2.11.2 that $F(x) = y^*[T(x)]$ has a unique bounded extension on $\mathfrak{X}$ whenever it is bounded on $\mathfrak{D}$. Thus T^* is well defined; it is clearly linear. Suppose $\{y_n^*\} \subset \mathfrak{D}(T^*)$, $y_n^* \to y_0^*$, and $T^*(y_n^*) = x_n^* \to x_0^*$. Then $y_n^*[T(x)] = x_n^*(x)$, $x \in \mathfrak{D}$, implies that $y_0^*[T(x)] = x_0^*(x)$, $x \in \mathfrak{D}$, so that $y_0^* \in \mathfrak{D}(T^*)$ and $T^*(y_0^*) = x_0^*$. On the other hand if T is bounded, then for arbitrary $y^* \in \mathfrak{Y}^*$ we see that $F(x) = y^*[T(x)]$ defines a linear bounded functional on $\mathfrak{D}$, in fact, $| F(x) | \leqq \| y^* \| \| T \| \| x \|$. Thus $\mathfrak{D}(T^*) = \mathfrak{Y}^*$ and $\| T^* \| \leqq \| T \|$. By the definition of the norm, given $\epsilon > 0$ there exists an x_ϵ with $\| x_\epsilon \| = 1$ and $\| T(x_\epsilon) \| \geqq \| T \| - \epsilon$. If $y_\epsilon = T(x_\epsilon)$, we choose a functional $y_\epsilon^* \in Y^*$ such that $y_\epsilon^*(y_\epsilon) = \| y_\epsilon \|$, $\| y_\epsilon^* \| = 1$; such a functional exists by Theorem 2.7.4. Then $\| T^* \| \geqq \| T^*(y_\epsilon^*) \| \geqq | [T^*(y_\epsilon^*)](x_\epsilon) | = \| y_\epsilon \| \geqq \| T \| - \epsilon$. Hence $\| T^* \| = \| T \|$.

The second adjoint of a linear transformation T with $\bar{\mathfrak{D}} = \mathfrak{X}$ is not always well defined since $\mathfrak{D}(T^*)$ is not in general dense in $\mathfrak{Y}^*$. However, we do have

THEOREM 2.11.9. *Let T be a closed linear transformation defined on a domain $\mathfrak{D}$ dense in $\mathfrak{X}$ to $\mathfrak{Y}$. Then $\mathfrak{D}(T^*)$ is weakly* dense in $\mathfrak{Y}^*$ and if $\mathfrak{Y}$ is reflexive $\mathfrak{D}(T^*)$ is strongly dense in $\mathfrak{Y}^*$.*

PROOF. If $\mathfrak{D}(T^*)$ were not weakly* dense in $\mathfrak{Y}^*$, then by Theorem 2.10.4 there would exist a $y_0 \in \mathfrak{Y}$, $y_0 \neq \theta$, such that $y^*(y_0) = 0$ for all $y^* \in \mathfrak{D}(T^*)$. Now the graph $\mathfrak{G}$ of T is a closed linear subspace of $\mathfrak{X} \times \mathfrak{Y}$ and $(\theta, y_0) \notin \mathfrak{G}$. By Theorem 2.7.5 there exists an $(x_0^*, y_0^*) \in (\mathfrak{X} \times \mathfrak{Y})^* = \mathfrak{X}^* \times \mathfrak{Y}^*$ such that $x_0^*(x) + y_0^*[T(x)] = 0$ for all $x \in \mathfrak{D}$ and $x_0^*(\theta) + y_0^*(y_0) \neq 0$. It follows from the first of these assertions that $y_0^* \in \mathfrak{D}(T^*)$ and from the second that $y_0^*(y_0) \neq 0$, which is impossible in view of the way in which y_0 was chosen. Thus $\mathfrak{D}(T^*)$ is weakly* dense in $\mathfrak{Y}^*$. In case $\mathfrak{Y}$ is reflexive, we conclude that $\mathfrak{D}(T^*)$ is weakly dense and hence by Theorem 2.9.2 strongly dense in $\mathfrak{Y}^*$.

THEOREM 2.11.10. *If $T \in \mathfrak{E}(\mathfrak{X}, \mathfrak{Y})$, then its second adjoint T^{**} is an extension of T defined on $\mathfrak{X}^{**}$ to $\mathfrak{Y}^{**}$ and $\| T^{**} \| = \| T \|$. In particular, if $\mathfrak{X}$ is a reflexive space then $T^{**} = T$.*

PROOF. The second adjoint is defined as $T^{**} = (T^*)^*$. It follows from Theorem 2.11.8 that T^{**} is a linear bounded transformation on $\mathfrak{X}^{**}$ to $\mathfrak{Y}^{**}$ and that $\| T^{**} \| = \| T^* \| = \| T \|$. According to Theorem 2.8.3, $\mathfrak{X}$ is embedded in $\mathfrak{X}^{**}$ by the natural mapping $x_0 \to x_0^{**}$ defined by $x_0^{**}(x^*) = x^*(x_0)$, $x^* \in \mathfrak{X}^*$. Now

$$[T^{**}(x_0^{**})](y^*) = x_0^{**}[T^*(y^*)] = [T^*(y^*)](x_0) = y^*[T(x_0)]$$

and since this holds for all $y^* \in \mathfrak{Y}^*$, we see that $T(x_0) \to T^{**}(x_0^{**})$ under the natural mapping of $\mathfrak{Y}$ into $\mathfrak{Y}^{**}$. This shows that T^{**} coincides with T on $\mathfrak{X}$ when

$\mathfrak{X}$ and $\mathfrak{Y}$ are considered as subspaces of $\mathfrak{X}^{**}$ and $\mathfrak{Y}^{**}$, respectively, under the natural embeddings. If $\mathfrak{X}$ is reflexive, then $\mathfrak{X} = \mathfrak{X}^{**}$ and consequently $T^{**} = T$.

THEOREM 2.11.11. *If $T \in \mathfrak{E}(\mathfrak{X}, \mathfrak{Y})$, then T is continuous with respect to the weak topologies of $\mathfrak{X}$ and $\mathfrak{Y}$, and T^* is continuous with respect to the weak* topologies of $\mathfrak{Y}^*$ and $\mathfrak{X}^*$.*

PROOF. Since $y_k^*[T(x) - T(x_0)] = [T^*(y_k^*)](x - x_0)$, it follows that $T[N(x_0 ; T^*(y_1^*), \cdots, T^*(y_n^*); \epsilon] \subset N[T(x_0); y_1^*, \cdots, y_n^*; \epsilon]$, and this implies weak continuity. A similar argument shows that T^* is weak* continuous.

We turn now to the relation between a transformation, its adjoint, and their inverses. See also S. Banach [**2**, pp. 145–150].

THEOREM 2.11.12. *Let T be a linear transformation with $\bar{\mathfrak{D}} = \mathfrak{X}$. Then $(T^*)^{-1}$ exists if and only if $\bar{\mathfrak{R}} = \mathfrak{Y}$. More generally, $\bar{\mathfrak{R}}$ consists of the set of all points y such that $T^*(y^*) = \theta$ implies $y^*(y) = 0$.*

PROOF. If $T^*(y_0^*) = \theta$, then $[T^*(y_0^*)](x) = y_0^*[T(x)] = 0$ for all $x \in \mathfrak{D}$ and hence $y_0^*(\bar{\mathfrak{R}}) = 0$. In particular, $\bar{\mathfrak{R}} = \mathfrak{Y}$ implies that $y_0^* = \theta$ and hence that T^* has an inverse. On the other hand if $y_0 \notin \bar{\mathfrak{R}}$, then Theorem 2.7.5 asserts the existence of a functional $y_0^* \in \mathfrak{Y}^*$ such that $y_0^*(y_0) = 1$ and $y_0^*(\bar{\mathfrak{R}}) = 0$. Thus $y_0^*[T(x)] = 0$ for all $x \in \mathfrak{D}$; it follows that $y_0^* \in \mathfrak{D}(T^*)$ and $T^*(y_0^*) = \theta$; whereas $y_0^*(y_0) \neq 0$. In particular we see that if $\bar{\mathfrak{R}} \neq \mathfrak{Y}$, then T^* cannot have an inverse.

THEOREM 2.11.13. *Let T be a linear transformation with $\bar{\mathfrak{D}} = \mathfrak{X}$. If $\mathfrak{R}(T^*)$ is weakly* dense in $\mathfrak{X}^*$, then T has an inverse.*

PROOF. Suppose that T has no inverse; than there is an $x_0 \neq \theta$ such that $T(x_0) = \theta$. Consequently $[T^*(y^*)](x_0) = y^*[T(x_0)] = 0$ for all $y^* \in \mathfrak{D}(T^*)$ and this shows that the weak* closure of $\mathfrak{R}(T^*)$ is a proper subspace of $\mathfrak{X}^*$, contrary to assumption.

THEOREM 2.11.14. *Let T be a linear transformation with an inverse and such that $\bar{\mathfrak{D}} = \mathfrak{X}$ and $\bar{\mathfrak{R}} = \mathfrak{Y}$. Then $(T^*)^{-1} = (T^{-1})^*$; further T^{-1} is bounded if and only if $(T^*)^{-1}$ is bounded on $\mathfrak{X}^*$.*

PROOF. In the first place $(T^{-1})^*$ exists because $\mathfrak{R} = \mathfrak{D}(T^{-1})$ is dense in $\mathfrak{Y}$ and $(T^*)^{-1}$ exists by Theorem 2.11.12. If $y \in \mathfrak{R}$ and $y^* \in \mathfrak{D}(T^*)$, then $y^*(y) = y^*\{T[T^{-1}(y)]\} = [T^*(y^*)][T^{-1}(y)]$. This implies that $\mathfrak{R}(T^*) \subset \mathfrak{D}[(T^{-1})^*]$ and $(T^{-1})^*[T^*(y^*)] = y^*$ for all $y^* \in \mathfrak{D}(T^*)$. Thus $(T^{-1})^*$ is an extension of $(T^*)^{-1}$. On the other hand if $x \in \mathfrak{D}$ and $x^* \in \mathfrak{D}[(T^{-1})^*]$, then $x^*(x) = x^*\{T^{-1}[T(x)]\} = [(T^{-1})^*(x^*)][T(x)]$. This implies that $(T^{-1})^*(x^*) \in \mathfrak{D}(T^*)$ and $T^*[(T^{-1})^*(x^*)] = x^*$ for all $x^* \in \mathfrak{D}[(T^{-1})^*]$. It follows that $\mathfrak{R}(T^*) \supset \mathfrak{D}[(T^{-1})^*]$. Therefore $\mathfrak{D}[(T^{-1})^*] = \mathfrak{R}(T^*) = \mathfrak{D}[(T^*)^{-1}]$ and hence $(T^{-1})^* = (T^*)^{-1}$. If, in addition, T^{-1} is bounded, then Theorem 2.11.8 applies so that $(T^{-1})^* = (T^*)^{-1} \in \mathfrak{E}(\mathfrak{X}^*, \mathfrak{Y}^*)$. Conversely if $(T^*)^{-1}$ is bounded on $\mathfrak{X}^*$, then for all $y \in \mathfrak{R}$ and $x^* \in \mathfrak{X}^*$ we have $| x^*[T^{-1}(y)] | = | [(T^{-1})^*(x^*)](y) | \leqq \| (T^*)^{-1} \| \| x^* \| \| y \|$. It follows that T^{-1} is bounded.

A similar result which does not require $\mathfrak{R}$ to be dense in $\mathfrak{Y}$ is given by

THEOREM 2.11.15. *A linear transformation with $\bar{\mathfrak{D}} = \mathfrak{X}$ has a linear bounded inverse if and only if $\mathfrak{R}(T^*) = \mathfrak{X}^*$.*

PROOF. Suppose first that $\mathfrak{R}(T^*) = \mathfrak{X}^*$. Then T^{-1} exists by Theorem 2.11.13. Further to each $x^* \in \mathfrak{X}^*$ there corresponds a $y^* \in \mathfrak{Y}^*$ such that $x^* = T^*(y^*)$. Hence for all $y \in \mathfrak{R}$, $\| y \| \leqq 1$, we have

$$| x^*[T^{-1}(y)] | = | [T^*(y^*)][T^{-1}(y)] | = | y^*(y) | \leqq \| y^* \|.$$

Applying Theorem 2.8.6 we see that $\sup [\, \| T^{-1}(y) \|; \| y \| \leqq 1, y \in \mathfrak{R}] < \infty$. Thus T^{-1} is bounded. Conversely if T^{-1} exists and is bounded, then for each $x^* \in \mathfrak{X}^*$ the formula $x^*[T^{-1}(y)] = y^*(y)$, $y \in \mathfrak{R}$, defines a functional in $\mathfrak{Y}^*$; here y^* need not be unique. Further $y^*[T(x)] = x^*\{T^{-1}[T(x)]\} = x^*(x)$, $x \in \mathfrak{D}$, so that $y^* \in \mathfrak{D}(T^*)$ and $T^*(y^*) = x^*$. Since x^* was arbitrary, it follows that $\mathfrak{R}(T^*) = \mathfrak{X}^*$.

2.12. Closed linear transformations. The closed linear transformations include practically all of the linear operators which the analyst is likely to use. It is fortunate, therefore, that the theory of such transformations is neither meager nor trivial.

We shall denote the class of all closed linear transformations with domain in $\mathfrak{X}$ and range in $\mathfrak{Y}$ by $\mathfrak{C}(\mathfrak{X}, \mathfrak{Y})$. Unlike $\mathfrak{E}(\mathfrak{X}, \mathfrak{Y})$, the set $\mathfrak{C}(\mathfrak{X}, \mathfrak{Y})$ is in general not a linear system; that is, if T_1, T_2 are two closed linear transformations with domains $\mathfrak{D}_1$ and $\mathfrak{D}_2$ respectively, then $T_1 + T_2$ defined on $\mathfrak{D}_1 \cap \mathfrak{D}_2$ need not have a closed linear extension.

EXAMPLE. Let $\mathfrak{X} = \mathfrak{Y} = m$ and let Lim $\in m^*$ be a Banach limit (see the problem in section 2.10). We define $T_1(\{\alpha_n\}) = \{\beta_n\}$ where $\beta_1 = \text{Lim}\,\{\sum_{k=1}^{n}\alpha_k\}$, $\beta_n = \sum_{k=1}^{n-1}\alpha_k$ for $n > 1$; $T_2(\{\alpha_n\}) = \{\gamma_n\}$ where $\gamma_1 = 0$, $\gamma_n = -\beta_n$ for $n > 1$; and $\mathfrak{D}_1 = \mathfrak{D}_2 = [\{\alpha_n\}; \{\alpha_n\} \in m, \{\sum_{k=1}^{n}\alpha_k\} \in m]$. It is easy to show that T_1 and T_2 are both closed linear operators. On the other hand $(T_1 + T_2)(\{\alpha_n\}) = \{\delta_n\}$, $\delta_1 = \text{Lim}\,\{\sum_{k=1}^{n}\alpha_k\}$ and $\delta_n = 0$ for $n > 1$, does not have a closed extension. In fact, set $x_k = \{\alpha_n^k\}$ where $\alpha_n^k = 1/k$ for $n \leqq k$ and $= 0$ for $n > k$. Then $\| x_k \| \to 0$ as $k \to \infty$, whereas $\{\delta_n^k\} = (1, 0, 0, \cdots)$ for all k. It follows from Theorem 2.11.1 that $T_1 + T_2$ can not have a closed extension.

On the other hand if $T \in \mathfrak{C}(\mathfrak{X}, \mathfrak{Y})$ and B is a bounded linear transformation on $\mathfrak{D}(B) \subset \mathfrak{X}$ to $\mathfrak{Y}$ with $\mathfrak{D}(B) \supset \mathfrak{D}(T)$, then $T + B$ defined on $\mathfrak{D}(T)$ is a closed linear transformation. This remark enables one to bring a certain amount of order into $\mathfrak{C}(\mathfrak{X}, \mathfrak{Y})$ if he is content to deal only with what we shall call *closely related* transformations. Two transformations U_1, $U_2 \in \mathfrak{C}(\mathfrak{X}, \mathfrak{Y})$ are said to be closely related, in symbols $U_1 \approx U_2$, if $\mathfrak{D}(U_1) = \mathfrak{D}(U_2)$ and

$$(2.12.1) \quad d(U_1, U_2) \equiv \sup [\, \| (U_1 - U_2)(x) \|; \| x \| \leqq 1, x \in \mathfrak{D}(U_1) = \mathfrak{D}(U_2)]$$

is finite. It is clear that this relation is an equivalence and hence that $\mathfrak{C}(\mathfrak{X}, \mathfrak{Y})$ splits up into equivalence classes, the class containing the transformation T being $\mathfrak{E}(T) \equiv [U; U \approx T]$. If U_1, $U_2 \in \mathfrak{E}(T)$, then $U_1 - U_2$ is bounded on $\mathfrak{D}(T)$ and has an extension in $\mathfrak{C}[\overline{\mathfrak{D}(T)}, \mathfrak{Y}]$. Conversely, if $B \in \mathfrak{C}[\overline{\mathfrak{D}(T)}, \mathfrak{Y}]$ and $U \in \mathfrak{C}(T)$, then $U + B$, defined on $\mathfrak{D}(T)$, is closed and belongs to $\mathfrak{C}(T)$. We now introduce the metric defined by (2.12.1) in each equivalence class. It is easy to verify that this function satisfies the postulates for a distance function as given in section 1.5.

We shall now derive one of the basic theorems on closed linear transformations, a result which was first proved for bounded instead of closed operators by S. Banach [2, pp. 38–40].

THEOREM 2.12.1. *If $y = T(x)$ is a closed linear transformation whose range $\mathfrak{R}$ is of the second category in $\mathfrak{Y}$, then*

(1) $\mathfrak{R} = \mathfrak{Y}$;

(2) *there is a constant $m > 0$ such that to every $y \in \mathfrak{Y}$ there is an $x \in \mathfrak{D}$ with $y = T(x)$ and $\| x \| \leqq m \| y \|$;*

(3) *if T^{-1} exists then it is bounded.*

PROOF. Let $S = [x; \| x \| < 1]$ and set $\mathfrak{D}_1 = \mathfrak{D} \cap S$. Then $\mathfrak{D} = \bigcup_n (n\mathfrak{D}_1)$ and hence $\mathfrak{R} = \bigcup_n nT(\mathfrak{D}_1)$. Since $\mathfrak{R}$ is of the second category in $\mathfrak{Y}$, the same will be true of $n_1T(\mathfrak{D}_1)$ for some integer n_1. It follows that $n_1T(\mathfrak{D}_1)$ is dense in some sphere, say $S_0 = [y; \| y - y_0 \| < r_0]$. In particular there is an $x_1 \in n_1\mathfrak{D}_1$ such that $\| y_1 - y_0 \| < r_0/2$ where $y_1 = T(x_1)$. Thus $n_1T(\mathfrak{D}_1) - y_1$ is dense in $[y; \| y \| < r_0/2] \subset S_0 - y_1$. Finally $n_1T(\mathfrak{D}_1) - y_1 \subset n_1T(\mathfrak{D}_1) - n_1T(\mathfrak{D}_1) = 2n_1T(\mathfrak{D}_1)$ and hence

$$\overline{T(2^{-k}\mathfrak{D}_1)} \supset 2^{-k}S_1, \qquad k = 0, 1, 2, \cdots, \tag{2.12.2}$$

where $S_1 = [y; \| y \| < r_1 = (4n_1)^{-1}r_0]$.

Let y be an arbitrary point of S_1. Using (2.12.2) with $k = 0$, we pick a point $x_1 \in \mathfrak{D}_1$ such that

$$\| y - T(x_1) \| < 2^{-1}r_1, \qquad \| x_1 \| < 1.$$

Since $y - T(x_1) \in 2^{-1}S_1$, we can use (2.12.2) with $k = 1$ to choose a point $x_2 \in 2^{-1}\mathfrak{D}_1$ such that

$$\| y - T(x_1) - T(x_2) \| < 2^{-2}r_1, \qquad \| x_2 \| < 2^{-1}.$$

By employing (2.12.2) repeatedly in this manner, we obtain a sequence $\{x_n\} \subset \mathfrak{D}_1$ with

$$\| y - \sum_1^n T(x_k) \| = \| y - T(\sum_1^n x_k) \| < 2^{-n}r_1, \qquad \| x_n \| < 2^{-n+1}.$$

Upon setting $s_n = \sum_1^n x_k$, we see that (i) $s_n \in \mathfrak{D}$, (ii) $\lim_n s_n = x$ exists and $\| x \| \leqq \sum_1^\infty \| x_n \| < 2$, (iii) $\lim_n T(s_n) = y$. Since T is closed, it follows that $x \in \mathfrak{D}$ and $y = T(x)$.

Thus we have shown that $\mathfrak{R}$ contains a sphere S_1 and hence $\mathfrak{R} \supset \bigcup_n nS_1 = \mathfrak{Y}$. Statement (2) follows by taking $m = 4/r_1$; for if $y \in \mathfrak{Y}$ we have $r_1y/(2 \| y \|) \in S_1$ and hence there is an $x_0 \in \mathfrak{D}$, $\| x_0 \| < 2$, such that $T(x_0) = r_1y/(2 \| y \|)$ and thus, setting $x = (2 \| y \|/r_1)x_0$ we have $y = T(x)$ and $\| x \| < (4/r_1) \| y \|$. Statement (3) follows from (2) and Theorem 2.11.6. This completes the proof.

Corollary. *If $y = T(x)$ is a linear bounded transformation mapping $\mathfrak{X}$ in a one-to-one manner onto $\mathfrak{Y}$, then the inverse is also bounded.*

Proof. A bounded transformation defined on a closed domain is closed so that the above theorem applies.

Theorem 2.12.1 has some very interesting applications, a few of which are given below. Theorems 2.12.2 and 2.12.3 are due to S. Banach [**2**, p. 41], whereas Theorem 2.12.4 is due to N. Dunford [**3**].

Theorem 2.12.2. *If a linear system can be made into a (B)-space by two different choices of a norm, $\| x \|_1$ and $\| x \|_2$, in such a manner that $\lim_{n\to\infty} \| x_n \|_1 = 0$ always implies $\lim_{n\to\infty} \| x_n \|_2 = 0$, then the corresponding notions of convergence are equivalent and for $x \neq \theta$ we have $0 < m \leqq \| x \|_2/\| x \|_1 \leqq M < \infty$, where m and M are independent of x.*

Proof. This is an immediate consequence of the corollary of the preceding theorem if one takes for $\mathfrak{X}$ and $\mathfrak{Y}$ the (B)- spaces corresponding to the two norms $\| x \|_1$ and $\| x \|_2$ respectively and sets $T(x) = x$.

We come now to the *closed graph theorem.*

Theorem 2.12.3. *If $y = T(x)$ is a closed linear transformation with domain $\mathfrak{D}$ of the second category in $\mathfrak{X}$, then $\mathfrak{D} = \mathfrak{X}$ and T is bounded.*

Proof. We define the auxiliary transformation $U[(x, T(x))] = x$ on the (B)-space $\mathfrak{G} \subset \mathfrak{X} \times \mathfrak{Y}$ to $\mathfrak{X}$. It is clear that U is linear. Further U^{-1} exists since $U[(x, T(x))] = \theta$ implies $x = \theta$ and $T(x) = \theta$. Finally U is bounded on $\mathfrak{G}$ and *a fortiori* closed; in fact,

$$\| U[(x, T(x))] \| = \| x \| \leqq \| x \| + \| T(x) \| = \| (x, T(x)) \|.$$

It follows from Theorem 2.12.1 that the range of U, namely $\mathfrak{D}$, is actually $\mathfrak{X}$ and that U^{-1} is bounded on $\mathfrak{X}$. Thus $\| x \| + \| T(x) \| = \| U^{-1}(x) \| \leqq M \| x \|$ and hence $\| T(x) \| \leqq M \| x \|$.

Theorem 2.12.4. *If T is a linear transformation with $\bar{\mathfrak{D}} = \mathfrak{X}$ and if $\mathfrak{D}(T^*)$ contains a determining manifold Γ for $\mathfrak{Y}$, then T is bounded.*

Proof. Let T_0^* be the restriction of T^* defined on Γ to $\mathfrak{X}^*$. Since Γ is itself a complete normed space and since T_0^* is closed, it follows from the preceding theorem that T_0^* is bounded. Hence

$$| y^*[T(x)] | = | [T_0^*(y^*)](x) | \leqq \| T_0^* \| \| y^* \| \| x \| \quad \text{for all} \quad x \in \mathfrak{D} \quad \text{and} \quad y^* \in \Gamma.$$

Thus $\| T(x) \| = \sup [| y^*[T(x)] |; y^* \in \Gamma, \| y^* \| \leqq 1] \leqq \| T_0^* \| \| x \|$ so that T is bounded on $\mathfrak{D}$.

THEOREM 2.12.5. *If T is a closed linear transformation with $\bar{\mathfrak{D}} = \mathfrak{X}$, then $T(x) = y$ has a unique solution for all $y \in \mathfrak{Y}$ if and only if $T^*(y^*) = x^*$ has a unique solution for all $x^* \in \mathfrak{X}^*$. In either case both T^{-1} and $(T^*)^{-1}$ are bounded.*

PROOF. If T^{-1} exists and $\mathfrak{R} = \mathfrak{Y}$, then T^{-1} is bounded by Theorem 2.12.1 and according to Theorem 2.11.14 $(T^*)^{-1}$ will exist and be bounded on $\mathfrak{X}^*$. Conversely if $(T^*)^{-1}$ exists with $\mathfrak{R}(T^*) = \mathfrak{X}^*$, then Theorem 2.12.1 again implies that $(T^*)^{-1}$ is bounded. Theorem 2.11.13 asserts that T^{-1} exists, and Theorem 2.11.12 shows that $\bar{\mathfrak{R}} = \mathfrak{Y}$. According to Theorem 2.11.14, T^{-1} is bounded on $\mathfrak{R}$ and since it is closed by Theorem 2.11.5, it follows that $\mathfrak{R} = \mathfrak{Y}$.

2.13. Compact linear transformations. The compact linear transformations form a somewhat special but nevertheless useful class of transformations. Such transformations occur in the classical theory of integral equations and in the non-singular problems of mathematical physics (cf. R. Courant and D. Hilbert [1, Chapters III and V]); and their theory served as a model for the early work in functional analysis.

DEFINITION 2.13.1. *A compact (or weakly compact) transformation on $\mathfrak{X}$ to $\mathfrak{Y}$ takes bounded sets into conditionally compact (or weakly compact) subsets of $\mathfrak{Y}$.*

It follows from Theorems 2.9.5 and 2.9.6 that an equivalent definition for a weakly compact transformation is that it takes bounded sets into conditionally sequentially weakly compact subsets of $\mathfrak{Y}$.

If T is a compact (or weakly compact) transformation on $\mathfrak{X}_1$ to $\mathfrak{X}_2$ and U is continuous (or weakly continuous) on $\mathfrak{X}_2$ to $\mathfrak{X}_3$, then UT is compact (or weakly compact) on $\mathfrak{X}_1$ to $\mathfrak{X}_3$. Likewise if V is bounded on $\mathfrak{X}_1$ to $\mathfrak{X}_2$ and T is compact (or weakly compact) on $\mathfrak{X}_2$ to $\mathfrak{X}_3$, then TV is compact (or weakly compact) on $\mathfrak{X}_1$ to $\mathfrak{X}_3$. Further it is clear that a compact transformation is necessarily weakly compact. Since a conditionally weakly compact subset of $\mathfrak{Y}$ is bounded by Theorems 2.9.1 and 2.9.6, we have

THEOREM 2.13.1. *A compact or weakly compact transformation on $\mathfrak{X}$ to $\mathfrak{Y}$ is bounded.*

DEFINITION 2.13.2. *A transformation mapping $\mathfrak{X}$ into a finite dimensional subspace of $\mathfrak{Y}$ is said to be degenerate.*

THEOREM 2.13.2. *A bounded degenerate transformation on $\mathfrak{X}$ to $\mathfrak{Y}$ is compact.*

PROOF. This is an immediate consequence of Theorem 1.12.2.

THEOREM 2.13.3. *A degenerate bounded linear transformation T on $\mathfrak{X}$ to $\mathfrak{Y}$ has a representation of the form*

$$T(x) = \sum_{k=1}^{n} x_k^*(x) y_k \tag{2.13.1}$$

where $[y_k\,;k = 1, \cdots, n]$ *and* $[x_k^*\,;k = 1, \cdots, n]$ *are sets of linearly independent elements in* $\mathfrak{Y}$ *and* $\mathfrak{X}^*$ *respectively.*

PROOF. Suppose that $\mathfrak{R}$ is n-dimensional and choose $[y_k\,;k = 1, \cdots, n]$ to form a basis for $\mathfrak{R}$. By Theorem 2.7.5 there exist functionals $[y_k^*\,;k = 1, \cdots, n] \subset \mathfrak{Y}^*$ such that $y_i^*(y_k) = 0$ or 1 according as $i \neq k$ or $i = k$. By hypothesis $T(x) \in \mathfrak{R}$ and hence is of the form $T(x) = \sum_1^n F_k(x)y_k$ so that $F_k(x) = y_k^*[T(x)] = x_k^*(x)$, where $x_k^* = T^*(y_k^*)$. Finally we show that the $[x_k^*\,;k = 1, \cdots, n]$ are linearly independent. Otherwise there would exist $[z_j^*\,;j = 1, \cdots, m]$, $m < n$, such that $x_k^* = \sum_{j=1}^m \alpha_{kj}z_j^*$. In this case $T(x) = \sum_1^m z_j^*(x)w_j$ where $w_j = \sum_{k=1}^n \alpha_{kj}y_k$; but this would imply that the dimension of $\mathfrak{R}$ is $m < n$ which is contrary to our assumption.

REMARK. Because of the simple form of the degenerate linear bounded transformation, many problems for such transformations reduce to their matrix analogues; the same is true for the limit of a sequence of such transformations. As the following theorem shows, the limit of a sequence of degenerate linear bounded transformations is compact. The question naturally arises as to whether all compact linear transformations can be approximated in norm by degenerate linear bounded transformations. An affirmative answer has been given for certain domain and range spaces (cf. R. S. Phillips [2]); however the general problem is still open.

THEOREM 2.13.4. *If* $\{T_n\}$ *is a sequence of compact linear transformations and* $\lim_{n\to\infty} \| T_n - T \| = 0$, *then* T *is also compact.*

PROOF. Let $S_1 = [x;\, \| x \| < 1]$. It is sufficient to show that $T(S_1)$ is totally bounded in $\mathfrak{Y}$, that is, given $\epsilon > 0$ there exist a finite set of ϵ-spheres of the form $S(y';\epsilon) = [y;\, \| y - y' \| < \epsilon]$ which cover $T(S_1)$. To this end we choose n_0 so that $\| T_{n_0} - T \| < \epsilon/2$. T_{n_0} being compact, we see that $T_{n_0}(S_1)$ is covered by a finite set of $(\epsilon/2)$-spheres $[S(y_k\,;\epsilon/2);\, k = 1, \cdots, p]$. For arbitrary $x \in S_1$, $T_{n_0}(x) \in S(y_k\,;\epsilon/2)$ for some k and hence

$$\| T(x) - y_k \| \leqq \| T(x) - T_{n_0}(x) \| + \| T_{n_0}(x) - y_k \| < \epsilon.$$

It follows that $T(S_1) \subset \bigcup_1^p S(y_k\,;\epsilon)$.

We come next to a theorem due to J. Schauder [1] (cf. S. Kakutani [8]).

THEOREM 2.13.5. *If* T *is a compact linear transformation on* $\mathfrak{X}$ *to* $\mathfrak{Y}$, *then* T^* *on* $\mathfrak{Y}^*$ *to* $\mathfrak{X}^*$ *is also compact.*

PROOF. As above, there exists a finite set of $(\epsilon/4)$-spheres $[S(y_k\,;\epsilon/4);\, k = 1, \cdots, n]$ covering $T(S_1)$. We now define the auxiliary transformation $U(y^*) = \{y^*(y_k)\}$ on $\mathfrak{Y}^*$ to Z_n. It is clear that U is a degenerate (and therefore compact) linear bounded transformation. Hence $U(S_1^*)$, $S_1^* = [y^*;\, \| y^* \| < 1]$, is also covered by a finite set of $(\epsilon/4)$-spheres in Z_n with centers at, say, $U(y_j^*)$, $j = 1, \cdots, p$ and $y_j^* \in S_1^*$. Thus for a given $y^* \in S_1^*$, there is a y_j^* with $| y^*(y_k) - y_j^*(y_k) | < \epsilon/4$ for $k = 1, \cdots, n$; and for each $x \in S_1$ there is a y_k with $\| T(x) - y_k \| < \epsilon/4$. Consequently

$$| y^*[T(x)] - y_j^*[T(x)] | \leqq | y^*[T(x)] - y^*(y_k) | + | y^*(y_k) - y_j^*(y_k) | + | y_j^*(y_k) - y_j^*[T(x)] | < \tfrac{3}{4}\epsilon.$$

Hence $\| T^*(y^*) - T^*(y_j^*) \| = \sup [|y^* [T(x)] - y_j^*[T(x)] |; x \in S_1] < \epsilon$ so that $T^*(S_1^*) \subset \bigcup_1^p S[T^*(y_j^*); \epsilon]$. T^* is therefore compact.

The counterparts of Theorems 2.13.4 and 2.13.5 go back to V. Gantmacher [1] and R. S. Phillips [3]; the following proofs are due to J. Schwartz and S. Kakutani (unpublished).

LEMMA 2.13.1. *A linear bounded transformation T on $\mathfrak{X}$ to $\mathfrak{Y}$ is weakly compact if and only if $T^{**}(\mathfrak{X}^{**}) \subset \mathfrak{Y}$ in the sense of the natural embedding of $\mathfrak{Y}$ in $\mathfrak{Y}^{**}$.*

PROOF. Suppose that T is weakly compact and that $x_0^{**} \in \mathfrak{X}^{**}$. Let π denote a finite set of functionals in $\mathfrak{X}^*$ and order the π's by inclusion. By Theorem 2.7.8 there exists an $x_\pi \in \mathfrak{X}$ such that $y^*(x_\pi) = x_0^{**}(y^*)$ for all $y^* \in \pi$ and $\| x_\pi \| \leqq M = 2 \| x_0^{**} \|$. Then $[T^{**}(x_0^{**})](y^*) = x_0^{**}[T^*(y^*)] = \lim_\pi [T^*(y^*)](x_\pi) = \lim_\pi y^*[T(x_\pi)]$. Thus $[T(x_\pi)]$ is a weakly convergent directed set contained in a weakly compact subset of $\mathfrak{Y}$. By Theorem 2.9.4 there exists a $y_0 \in \mathfrak{Y}$ such that $[T^{**}(x_0^{**})](y^*) = y^*(y_0)$ for all $y^* \in \mathfrak{Y}^*$ and this implies $T^{**}(\mathfrak{X}^{**}) \subset \mathfrak{Y}$. Suppose on the other hand that T is given linear, bounded, and such that $T^{**}(\mathfrak{X}^{**}) \subset \mathfrak{Y}$. Since T^{**} is weakly* continuous on $\mathfrak{X}^{**}$ to $\mathfrak{Y}^{**}$, a closed bounded sphere is mapped by T^{**} onto a weakly* compact subset of $\mathfrak{Y}^{**}$ and since the image set lies in $\mathfrak{Y}$ it is also weakly compact considered as a subset of $\mathfrak{Y}$. Now T is a restriction of T^{**} by Theorem 2.11.10. Therefore T takes a closed bounded sphere into a conditionally weakly compact subset of $\mathfrak{Y}$.

THEOREM 2.13.6. *If $\{T_n\}$ is a sequence of weakly compact linear transformations and $\lim_{n\to\infty} \| T_n - T \| = 0$, then T is also weakly compact.*

PROOF. Taking second adjoints we see by Theorem 2.11.8 that $\| T_n^{**} - T^{**} \| \to 0$; in particular $T_n^{**}(x^{**}) \to T^{**}(x^{**})$. Hence $T_n^{**}(\mathfrak{X}^{**}) \subset \mathfrak{Y}$ implies that $T^{**}(\mathfrak{X}^{**}) \subset \mathfrak{Y}$ and it follows from the preceding lemma that T is weakly compact.

THEOREM 2.13.7. *If T is a weakly compact linear transformation on $\mathfrak{X}$ to $\mathfrak{Y}$, then T^* on $\mathfrak{Y}^*$ to $\mathfrak{X}^*$ is also weakly compact.*

PROOF. Given $y_0^{***} \in \mathfrak{Y}^{***}$, consider its restriction to the domain $\mathfrak{Y} \subset \mathfrak{Y}^{**}$. This restriction corresponds to some $y_0^* \in \mathfrak{Y}^*$ in the sense that $y_0^{***}(y^{**}) = y^{**}(y^*)$ for all $y^{**} \in \mathfrak{Y} \subset \mathfrak{Y}^{**}$. If T is weakly compact, then by the lemma $T^{**}(\mathfrak{X}^{**}) \subset \mathfrak{Y}$ so that

$$[T^{***}(y_0^{***})](x^{**}) = y_0^{***}[T^{**}(x^{**})] = [T^{**}(x^{**})](y_0^*) = x^{**}[T^*(y_0^*)]$$

for all $x^{**} \in \mathfrak{X}^{**}$. But this implies that $\mathfrak{R}(T^{***}) \subset \mathfrak{R}(T^*) \subset \mathfrak{X}^*$ in the sense of the natural embedding of $\mathfrak{X}^*$ into $\mathfrak{X}^{***}$; again the lemma implies that T^* is weakly compact.

We conclude this section with the following theorem due to R. S. Phillips [8] (cf. F. Rellich [1]).

THEOREM 2.13.8. *Let T be a compact (or weakly compact) linear transformation on $\mathfrak{X}$ to $\mathfrak{Y}$. If U is a bounded linear transformation on $\mathfrak{X}$ to $\mathfrak{Y}$ such that $\mathfrak{R}(U) \subset \mathfrak{R}(T)$, then U is also compact (or weakly compact).*

PROOF. Set $\mathfrak{M} = [x; T(x) = \theta]$. Then $\mathfrak{M}$ is a closed linear subspace and may be used as in Theorem 1.12.3 to define the quotient space $\mathfrak{X} \div \mathfrak{M}$ of cosets $[X]$. We define the transformation $\tilde{T}(X) = T(x)$ for $x \in X$, mapping $\mathfrak{X} \div \mathfrak{M}$ into $\mathfrak{Y}$. It is clear that $\tilde{T}$ is well defined, linear, and possesses an inverse. We also have

$$\| \tilde{T} \| = \sup [\| \tilde{T}(X) \| ; \| X \| < 1] = \sup [\| T(x) \| ; \| x \| < 1] = \| T \|,$$

so that $\tilde{T}$ is bounded and $\tilde{T}^{-1}$ is closed. In addition $\tilde{T}$ is compact (or weakly compact) since

$\tilde{T}([X; \| X \| < 1]) \equiv T([x; \| x \| < 1])$ is a conditionally compact (or weakly compact) subset of $\mathfrak{Y}$. By hypothesis $\mathfrak{R}(U) \subset \mathfrak{R}(T) = \mathfrak{R}(\tilde{T})$ so that $V = \tilde{T}^{-1}U$ is a closed linear transformation on $\mathfrak{X}$ to $\mathfrak{X} \div \mathfrak{M}$. Because of the closed graph theorem, V is actually bounded. Finally $U = \tilde{T}V$ is compact (or weakly compact) since it is the product of two bounded linear transformations, one of which is compact (or weakly compact).

References. Banach [2], Courant and Hilbert [1], Dunford [3], Gantmacher [1], Kakutani [8], Phillips [2, 3, 8, 12], Rellich [1], Schauder [1].

4. Spaces of Endomorphisms

2.14. The Banach algebra of endomorphisms. We come now to the study of linear transformations on a (B)-space $\mathfrak{X}$ to itself. In this context the terms *operator* and *operation* will be used as synonyms for transformation. The family of all linear bounded operations constitute one of the most important instances of a Banach algebra as defined in Definition 1.15.1.

Definition 2.14.1. *A linear bounded transformation on a (B)-space $\mathfrak{X}$ to itself will be called an endomorphism of $\mathfrak{X}$.*

This use of the term is sufficiently close to modern usage in abstract algebra so that no confusion is likely to arise.

The class of all linear bounded transformations on $\mathfrak{X}$ to $\mathfrak{Y}$, previously denoted by $\mathfrak{E}(\mathfrak{X}, \mathfrak{Y})$, was shown in section 2.11 to be a (B)-space. With endomorphisms we have the added operation of multiplication; here $(T_1T_2)(x) \equiv T_1[T_2(x)]$. If T_1, T_2 are endomorphisms of $\mathfrak{X}$, so are αT_1, $T_1 + T_2$, and T_1T_2. Further $\| \alpha T_1 \| = | \alpha | \| T_1 \|$, $\| T_1 + T_2 \| \leqq \| T_1 \| + \| T_2 \|$, $\| T_1T_2 \| \leqq \| T_1 \| \| T_2 \|$. Denoting the identity by I, we have $\| I \| = 1$. This establishes

Theorem 2.14.1. *The set of all endomorphisms of $\mathfrak{X}$ forms a (B)-algebra $\mathfrak{E}(\mathfrak{X})$ which has the identity transformation as unit element. $\mathfrak{E}(\mathfrak{X})$ is non-commutative if $\mathfrak{X}$ is of dimension greater than one.*

We call $\mathfrak{E}(\mathfrak{X})$ the *Banach algebra of endomorphisms* of $\mathfrak{X}$.

If T is a fixed element of $\mathfrak{E}(\mathfrak{X})$, a polynomial in T is of the form

$$\alpha_0 I + \sum_{k=1}^{n} \alpha_k T^k, \qquad T^k = TT^{k-1}, \qquad \alpha_k \in \Phi.$$

The set of all such polynomials in a fixed T is a commutative subalgebra of $\mathfrak{E}(\mathfrak{X})$ and so is its closure.

If T_1, $T_2 \in \mathfrak{E}(\mathfrak{X})$, then T_1T_2 is compact (or weakly compact) if the same is true of either T_1 or T_2. If T_1 and T_2 are both compact (or weakly compact) then so are αT_1 and $T_1 + T_2$. It follows from this and Theorems 2.13.4 and 2.13.6 that the set of all compact (or weakly compact) endomorphisms of $\mathfrak{X}$ forms a closed two-sided ideal in $\mathfrak{E}(\mathfrak{X})$.

DEFINITION 2.14.2. *An endomorphism J satisfying $J^2 = J$ is called a projection operator. J is said to project $\mathfrak{X}$ onto $\mathfrak{R}(J)$.*

The identity I and the zero Θ are projection operators. If J is a projection operator, then so is $I - J$. A vector x belongs to $\mathfrak{R}(J)$ if and only if $J(x) = x$. It follows from this that $\mathfrak{R}(J)$ is a closed linear manifold; the same is true of $\mathfrak{R}(I - J)$. Further the relations $J(I - J) = \Theta = (I - J)J$ and $I = J + (I - J)$ imply that $\mathfrak{R}(J) \cap \mathfrak{R}(I - J) = [\theta]$ and that $\mathfrak{X} = \mathfrak{R}(J) + \mathfrak{R}(I - J)$. A converse to this is given in the following theorem due to F. J. Murray [1].

THEOREM 2.14.2. *If $\mathfrak{M}$ and $\mathfrak{N}$ are closed linear manifolds such that $\mathfrak{M} \cap \mathfrak{N} = [\theta]$ and $\mathfrak{X} = \mathfrak{M} + \mathfrak{N}$, then there exists a projection operator which projects $\mathfrak{X}$ onto $\mathfrak{M}$ (or $\mathfrak{N}$).*

PROOF. Each $x \in \mathfrak{X}$ has a unique representation $x = y + z$, where $y \in \mathfrak{M}$ and $z \in \mathfrak{N}$. We define $J(x) = y$. It is clear that J is a linear operation defined for all $x \in \mathfrak{X}$, that $J^2 = J$, and that $\mathfrak{R}(J) = \mathfrak{M}$. Furthermore, J is closed. For if $x_n \to x_0$ and $J(x_n) = y_n \to y_0$, then $y_0 \in \mathfrak{M}$ and $z_n = x_n - y_n \to x_0 - y_0 = z_0 \in \mathfrak{N}$; so that $J(x_0) = y_0$. It now follows from the closed graph theorem that J is bounded and this concludes the proof.

We return now to the problem mentioned in section 1.15 of renorming a (B)-algebra with a unit element so that in the new norm the unit is of norm one. The following result is due to I. Gelfand [4].

THEOREM 2.14.3. *Let $\mathfrak{B}$ be a (B)-algebra with a unit element. There exists a (B)-algebra $\mathfrak{B}'$ with unit e' such that $\mathfrak{B} \cong \mathfrak{B}'$ and $\| e' \| = 1$.*

PROOF. Let $\mathfrak{E}(\mathfrak{B})$ be the (B)-algebra of endomorphisms of $\mathfrak{B}$ considered here as a (B)-space. The element $x \in \mathfrak{B}$ defines an endomorphism T_x where $T_x(y) = xy$. Set $\mathfrak{B}' = [T_x ; x \in \mathfrak{B}]$. Then $\mathfrak{B}'$ is a subalgebra of $\mathfrak{E}(\mathfrak{B})$. Since $x \neq \theta$ implies that $T_x(e) = xe \neq \theta$, the mapping $x \to T_x$ is an algebraic isomorphism of $\mathfrak{B}$ onto $\mathfrak{B}'$. Further $e \to T_e = I$ and $\| I \| = 1$. Now $\| T_x \| = \sup [\| xy \|; \| y \| \leqq 1, y \in \mathfrak{B}] \leqq \| x \|$ so that the mapping is continuous. Finally, $\| T_x \| \geqq \| xe \| / \| e \| = \| x \| / \| e \|$, and this shows that the mapping is bicontinuous.

2.15. The strong and the weak operator topologies. The system of all endomorphisms of $\mathfrak{X}$ was made into a Banach algebra in the preceding section by the introduction of the *normed topology*, in this connection usually referred to as the *uniform operator topology*. Following J. von Neumann [3] we may also introduce *the strong* and *the weak operator topologies*. The resulting algebraic spaces are *topological algebras* in the sense of Definition 1.14.1; they will be denoted by $\mathfrak{E}_s(\mathfrak{X})$ and $\mathfrak{E}_w(\mathfrak{X})$ respectively.

The corresponding neighborhood bases are defined in a manner analogous to the procedure used in section 2.9 for introducing the weak neighborhood topology in $\mathfrak{X}$. *A strong operator neighborhood* is any set of the form

$$N(T_0\,; x_1\,, \cdots, x_n\,; \epsilon) = [T;\ \| T(x_k) - T_0(x_k) \| < \epsilon, k = 1, \cdots, n]$$

where $x_1\,, \cdots, x_n$ are n arbitrary elements of $\mathfrak{X}$. Similarly a *weak operator neighborhood* is any set of the form

$$N(T_0\,; x_1\,, \cdots, x_n\,; x_1^*\,, \cdots, x_n^*\,; \epsilon) = [T;\ | x_k^*[T(x_k)] - x_k^*[T_0(x_k)] | < \epsilon, k = 1, \cdots, n]$$

where $x_1\,, \cdots, x_n$ and $x_1^*\,, \cdots, x_n^*$ are arbitrary elements of $\mathfrak{X}$ and $\mathfrak{X}^*$ respectively. A complete neighborhood basis is obtained by varying T_0, ϵ, n, x_k, and x_k^* over their domains of definition. It can be shown that neighborhoods of the above type, with T_0 fixed, form a neighborhood basis to T_0 itself in each of the two topologies. We leave to the reader the verification of the fact that in each case the resulting topology satisfies the postulates for an algebraic topological space as given in Definition 1.14.1.

Definition 2.15.1. (1) *The directed set* $[T_\pi]$ *of endomorphisms of* $\mathfrak{X}$ *is said to be strongly convergent if* $\lim_{\pi_1,\pi_2} \| T_{\pi_1}(x) - T_{\pi_2}(x) \| = 0$ *for all* $x \in \mathfrak{X}$; *it converges strongly to the endomorphism* T_0 *if* $\lim_\pi \| T_\pi(x) - T_0(x) \| = 0$ *for all* $x \in \mathfrak{X}$. (2) *The directed set is said to be weakly convergent if* $\lim_{\pi_1,\pi_2} | x^*[T_{\pi_1}(x)] - x^*[T_{\pi_2}(x)] | = 0$ *for all* $x \in \mathfrak{X}$ *and* $x^* \in \mathfrak{X}^*$; *it converges weakly to* T_0 *if* $\lim_\pi | x^*[T_\pi(x)] - x^*[T_0(x)] | = 0$ *for all* $x \in \mathfrak{X}$ *and* $x^* \in \mathfrak{X}^*$.

Definition 2.15.2. $\mathfrak{E}(\mathfrak{X})$ *is said to be strongly (or weakly) sequentially complete if every strongly (or weakly) convergent sequence converges strongly (or weakly) to an element of the space.*

Theorem 2.15.1. $\mathfrak{E}(\mathfrak{X})$ *is strongly sequentially complete.*

Proof. Suppose $\{T_n\}$ is a strongly convergent sequence of endomorphisms of $\mathfrak{X}$. Since $\mathfrak{X}$ is complete, $T_n(x) \to T_0(x) \in \mathfrak{X}$ for each $x \in \mathfrak{X}$. It is clear that $T_0(x)$ is linear and it follows from Theorem 2.5.5 that there is an M such that $\| T_n \| \leqq M$ for all n. Consequently $\| T_0 \| \leqq M$ so that $T_0 \in \mathfrak{E}(\mathfrak{X})$.

Theorem 2.15.2. *If* $\mathfrak{X}$ *is weakly sequentially complete, then* $\mathfrak{E}(\mathfrak{X})$ *is weakly sequentially complete.*

Proof. The argument here is much the same as in the previous theorem. In this case Theorem 2.9.1 in conjunction with Theorem 2.5.5 shows that $\| T_n \| \leqq M$ for all n.

Problem. Prove that the closed unit sphere in $\mathfrak{E}(\mathfrak{X})$ is compact relative to the weak operator topology if $\mathfrak{X}$ is a reflexive (B)-space.

2.16. Resolvents and spectra. In the present section we consider linear transformations, not necessarily bounded, with domain and range sets in the same *complex* (B)-space $\mathfrak{X}$. Here we shall make only a preliminary examination of the spectral properties of such operators; a more detailed study requiring additional background material will be made in paragraphs 5.2 and 5.3.

For a given linear operator T with both domain $\mathfrak{D}$ and range $\mathfrak{R}$ in $\mathfrak{X}$, the transformation $T_\lambda = \lambda I - T$ is well defined on $\mathfrak{D}$; here λ is an arbitrary complex number. We denote the range of T_λ by $\mathfrak{R}_\lambda$. The distribution of the values of λ for which T_λ has an inverse and the properties of the inverse when it exists are basic in the description of T.

DEFINITION 2.16.1. *The values of λ for which T_λ has a bounded inverse $R(\lambda; T)$ with domain dense in $\mathfrak{X}$ form the resolvent set $\rho(T)$ of T; $R(\lambda; T)$ is called the resolvent of T. The values of λ for which T_λ has an unbounded inverse with domain dense in $\mathfrak{X}$ form the continuous spectrum $C\sigma(T)$. The values of λ for which T_λ has an inverse whose domain is not dense in $\mathfrak{X}$ form the residual spectrum $R\sigma(T)$. The values of λ for which no inverse exists form the point spectrum $P\sigma(T)$. The union of $C\sigma(T)$, $R\sigma(T)$, and $P\sigma(T)$ is the spectrum $\sigma(T)$ of T.*

The definition gives

THEOREM 2.16.1. *The four sets $\rho(T)$, $C\sigma(T)$, $R\sigma(T)$, and $P\sigma(T)$ are mutually exclusive and their union is the complex plane.*

In practice it is often convenient to classify the spectral properties of the operator T_λ, for fixed λ, by means of the following concepts.

DEFINITION 2.16.2. *The operator T has the property*

P_1 *if T is not one-to-one that is, if there is an x_0, $x_0 \neq \theta$, such that $T(x_0) = \theta$;*

P_2 *if $\mathfrak{R}$ is non-dense in $\mathfrak{X}$;*

P_3 *if the image of the vectors of norm one in $\mathfrak{D}$ is not bounded away from θ, that is, if there exists a sequence $\{x_n\} \subset \mathfrak{D}$ such that* (i) $\| x_n \| = 1$ *and* (ii) $\| T(x_n) \| \to 0$.

In terms of these properties we can characterize the spectral subdivisions and the resolvent set as follows. A point λ belongs to

$\sigma(T)$ *if T_λ has at least one of the properties* P_ν;
$P\sigma(T)$ *if it has* P_1;
$R\sigma(T)$ *if it has* P_2 *but not* P_1;
$C\sigma(T)$ *if it has* P_3 *but neither* P_1 *nor* P_2;
$\rho(T)$ *if it has none of the properties* P_ν.

We note that in their influence on the spectral character of the point λ, the properties P_ν have a definite order of dominance:

$$P_1 > P_2 > P_3 \tag{2.16.1}$$

in the sense that P_1 forces λ into $P\sigma(T)$ even if P_2 and P_3 also hold; P_2 puts λ into $R\sigma(T)$ only in the absence of P_1 but then regardless of the presence of P_3; whereas P_3 is decisive only in the absence of P_1 and P_2, in which case $\lambda \in C\sigma(T)$.

DEFINITION 2.16.3. *If $\lambda_0 \in P\sigma(T)$, then λ_0 is called a characteristic value of T*

and if $T(x_0) = \lambda_0 x_0$, $x_0 \neq \theta$, *then* x_0 *is a characteristic element (vector) of* T. *The least linear manifold* $\mathfrak{M}(\lambda_0 ; T)$ *containing the characteristic elements corresponding to* λ_0 *is called the characteristic manifold of* T *corresponding to the value* λ_0.

In this connection we also have

DEFINITION 2.16.4. *A complex number* λ_0 *has index* ν *(a positive integer or zero) with respect to* T *in case* $T_{\lambda_0}^{\nu+1}(x) = \theta$ *implies* $T_{\lambda_0}^{\nu}(x) = \theta$ *and there is an* x_0 *such that* $T_{\lambda_0}^{\nu}(x_0) = \theta$ *and* $T_{\lambda_0}^{\nu-1}(x_0) \neq \theta$. *If* $\nu = 0$, *this is taken to mean that* T_{λ_0} *has an inverse. If no such integer* ν *exists we say that* λ_0 *is of infinite index.*

The change of the spectrum under extensions of T is important. If T_1 is an extension with domain $\mathfrak{D}_1$, then $T_{1\lambda} = \lambda I - T_1$ is an extension of T_λ with domain $\mathfrak{D}_1 \supset \mathfrak{D}$ and range $\mathfrak{R}_{1\lambda} \supset \mathfrak{R}_\lambda$. If $T_{1\lambda}$ has an inverse, so does T_λ, and $(T_{1\lambda})^{-1}$ is an extension of $(T_\lambda)^{-1}$. On the other hand, the existence of $(T_\lambda)^{-1}$ does not imply the existence of $(T_{1\lambda})^{-1}$.

THEOREM 2.16.2. *If* T_1 *is an extension of* T, *then* $P\sigma(T) \subseteqq P\sigma(T_1)$ *and* $\mathfrak{M}(\lambda_0 ; T) \subseteqq \mathfrak{M}(\lambda_0 ; T_1)$ *for every* $\lambda_0 \in P\sigma(T)$. *Further,* $R\sigma(T) \supseteqq R\sigma(T_1)$, $C\sigma(T) \subseteqq C\sigma(T_1) \cup P\sigma(T_1)$, *and* $\rho(T_1) \subseteqq \rho(T) \cup R\sigma(T)$.

PROOF. The first statement is obvious. If $\lambda \in R\sigma(T_1)$, then $(T_{1\lambda})^{-1}$ exists and $\mathfrak{R}_{1\lambda}$ is not dense in $\mathfrak{X}$. Hence $(T_\lambda)^{-1}$ also exists and $\mathfrak{R}_\lambda$ is non-dense; this proves $R\sigma(T) \supseteqq R(T_1)$. On the other hand a point in $R\sigma(T)$ may very well turn up in any one of the four sets associated with T_1. If $\lambda \in C\sigma(T)$, then $\mathfrak{R}_\lambda \subseteqq \mathfrak{R}_{1\lambda}$ and both are dense in $\mathfrak{X}$; $(T_{1\lambda})^{-1}$ either does not exist or is unbounded since $(T_\lambda)^{-1}$ is unbounded. Finally, if $\lambda \in \rho(T_1)$, then $(T_{1\lambda})^{-1}$ exists as a bounded transformation and $\mathfrak{R}_{1\lambda}$ is dense in $\mathfrak{X}$. It follows that $(T_\lambda)^{-1}$ also exists and is bounded, but its domain may be non-dense. This proves the last statement. Conversely, a point in $\rho(T)$ may turn up in any one of the sets associated with T_1 except in $R\sigma(T_1)$.

Thus an extension may enlarge the point spectrum. No characteristic value is ever lost; new characteristic values may be generated from $\rho(T)$ or by transfer from the rest of $\sigma(T)$. The continuous spectrum may suffer losses to the point spectrum but may gain from the residual spectrum or from the resolvent set of T. Finally, the residual spectrum never gains and the resolvent set can gain only from the residual spectrum.

We denote the class of all closed operators with domain and range in $\mathfrak{X}$ by $\mathfrak{O}(\mathfrak{X})$.

THEOREM 2.16.3. *If* $T \in \mathfrak{O}(\mathfrak{X})$, *then* $\lambda \in \rho(T)$ *if and only if* $(T_\lambda)^{-1}$ *exists and* $\mathfrak{R}_\lambda = \mathfrak{X}$. *In this case* $R(\lambda; T) \in \mathfrak{E}(\mathfrak{X})$.

PROOF. If T is closed and $\lambda \in \rho(T)$, then $R(\lambda; T)$ is closed and bounded with domain $\mathfrak{R}_\lambda$ dense in $\mathfrak{X}$. This implies that $\mathfrak{R}_\lambda = \mathfrak{X}$ and hence that $R(\lambda; T) \in \mathfrak{E}(\mathfrak{X})$. Conversely if T is closed with range $\mathfrak{X}$ and if $(T_\lambda)^{-1}$ exists, then, according to Theorem 2.12.1, $(T_\lambda)^{-1}$ is bounded on $\mathfrak{X}$ so that $\lambda \in \rho(T)$.

COROLLARY. *If* $T \in \mathfrak{O}(\mathfrak{X})$ *and* T_1 *is a proper extension of* T, *then* $\rho(T) \subseteqq P\sigma(T_1)$.

For a linear operator T with domain $\mathfrak{D}$, we define $\mathfrak{D}(T^n)$ inductively: $\mathfrak{D}(T) = \mathfrak{D}$ and $\mathfrak{D}(T^n) = [x; x \in \mathfrak{D}(T^{n-1}), T^{n-1}(x) \in \mathfrak{D}]$. In case $\rho(T) \neq \emptyset$ and say $\lambda_0 \in \rho(T)$, then one readily verifies that $\mathfrak{D}(T^n) = R^n(\lambda_0; T)[\mathfrak{X}]$. If $p(\lambda) = \sum_0^n \alpha_k \lambda^k$, then it is clear that $p(T) \equiv \alpha_0 I + \sum_1^n \alpha_k T^k$ is a well defined linear operator on $\mathfrak{D}(T^n)$. In this connection we have the following theorem due to A. E. Taylor [9].

THEOREM 2.16.4. *Let $T \in \mathfrak{O}(\mathfrak{X})$ and $\rho(T) \neq \emptyset$. If $p(\lambda)$ is a polynomial of degree n, then $p(T)$ is a closed linear operator with domain $\mathfrak{D}(T^n)$.*

PROOF. The result is obviously true for $n \leqq 1$. Assume the theorem to be valid for $n \leqq k - 1$ and choose $\lambda_0 \in \rho(T)$. We write $p(\lambda) = (\lambda_0 - \lambda)q(\lambda) + r$ where $q(\lambda)$ is a polynomial of degree $k - 1$ and r is a constant. Thus $p(T) \subset (\lambda_0 I - T)q(T) + rI$ and it is clearly sufficient if we show that $(\lambda_0 I - T)q(T)$ is a closed operator on $\mathfrak{D}(T^k)$. Hence suppose that $x_n \to x_0$, $\{x_n\} \subset \mathfrak{D}(T^k)$, and $(\lambda_0 I - T)q(T)(x_n) = y_n \to y_0$. Then $q(T)(x_n) = R(\lambda_0; T)(y_n) \to R(\lambda_0; T)(y_0)$. By the inductive hypothesis, $q(T)$ is closed on $\mathfrak{D}(T^{k-1})$. Hence $x_0 \in \mathfrak{D}(T^{k-1})$ and $q(T)(x_0) = R(\lambda_0; T)(y_0) \in \mathfrak{D}$ so that $(\lambda_0 I - T)[q(T)(x_0)] = y_0$. It remains only to prove that $x_0 \in \mathfrak{D}(T^k)$. Now $q(\lambda) = \sum_0^{k-1} \beta_j \lambda^j$, $\beta_{k-1} \neq 0$, and hence

$$q(T)(x_0) - \beta_0 x_0 - \cdots - \beta_{k-2}T^{k-2}(x_0) = \beta_{k-1}T^{k-1}(x_0).$$

Each term in the left member belongs to $\mathfrak{D}$. It follows that $T^{k-1}(x_0)$ also lies in $\mathfrak{D}$ and hence that $x_0 \in \mathfrak{D}(T^k)$.

COROLLARY. *Let $T \in \mathfrak{O}(\mathfrak{X})$ with $\overline{\mathfrak{D}} = \mathfrak{X}$ and $\rho(T) \neq \emptyset$. If $p(\lambda)$ is a polynomial of degree $n \geqq 1$, then $p(T)$ is bounded on $\mathfrak{D}(T^n)$ if and only if T is bounded.*

PROOF. For $\lambda_0 \in \rho(T)$, $R(\lambda_0; T)$ takes $\mathfrak{X}$ into $\mathfrak{D}$ which is dense in $\mathfrak{X}$. Hence $R(\lambda_0; T)$ takes any dense subset of $\mathfrak{X}$ into a dense set. It follows that $\mathfrak{D}(T^n)$ is dense in $\mathfrak{X}$. Now if $p(T)$ is bounded then $\mathfrak{D}(T^n) = \mathfrak{X}$ and hence $\mathfrak{D} = \mathfrak{X}$ so that T is bounded by the closed graph theorem. The converse is trivial.

The spectral properties of an operator T with dense domain are closely related to those of its adjoint T^*. The transformation T^* has its domain and range in $\mathfrak{X}^*$, $(T_\lambda)^* = \lambda I^* - T^*$, and $\mathfrak{D}[(T_\lambda)^*] = \mathfrak{D}(T^*)$.

THEOREM 2.16.5. *If T is a linear operator with $\overline{\mathfrak{D}} = \mathfrak{X}$ and $\mathfrak{R} \subset \mathfrak{X}$, then $\rho(T) = \rho(T^*)$ and $R(\lambda; T)^* = R(\lambda; T^*)$. Further, $P\sigma(T) \subseteqq P\sigma(T^*) \cup R\sigma(T^*)$, $R\sigma(T) \subseteqq P\sigma(T^*)$, and $C\sigma(T) \subseteqq R\sigma(T^*) \cup C\sigma(T^*)$.*

PROOF. If $\lambda \in \rho(T)$, then according to Theorem 2.11.14, $\lambda \in \rho(T^*)$ and $R(\lambda; T)^* = R(\lambda; T^*)$. On the other hand if $\lambda \in \rho(T^*)$, then Theorem 2.11.13 shows that T_λ has an inverse, Theorem 2.11.12 shows that $\overline{\mathfrak{R}}_\lambda = \mathfrak{X}$, Theorem 2.16.3 shows that $\mathfrak{R}_\lambda(T^*) = \mathfrak{X}^*$, and Theorem 2.11.14 then implies that $\lambda \in \rho(T)$. If $\lambda \in P\sigma(T)$, then, by Theorem 2.11.13, T^* has property P_2 so that $\lambda \in P\sigma(T^*) \cup R\sigma(T^*)$. If $\lambda \in R\sigma(T)$, then by Theorem 2.11.12, T^* has property P_1 so that $\lambda \in P\sigma(T^*)$. Finally, if $\lambda \in C\sigma(T)$, then, by Theorem 2.11.12, T^* does not have property P_1 so that $\lambda \in R\sigma(T^*) \cup C\sigma(T^*)$. It can be shown by examples that all of the indicated possibilities can actually occur.

All of the above considerations make sense only if λ is finite and in speaking of the spectrum $\sigma(T)$ of a linear operator T we shall never include the point at infinity. However, there are some contexts in which it is natural to adjoin the point at infinity to the spectrum of an unbounded operator. To avoid confusion we shall call this augmented set the *extended spectrum* and denote it by $\sigma_e(T)$. Thus $\sigma_e(T) = \sigma(T)$ for bounded T and $\sigma_e(T) = \sigma(T) \cup \{\infty\}$ for unbounded T (see section 5.11).

References. Gelfand [**4**], Murray [**1**], v. Neumann [**3**], Taylor [**9**].

CHAPTER III

VECTOR-VALUED FUNCTIONS

3.1 Orientation. We shall now prepare the frame-work for *a theory of functions having values in a Banach space* by generalizing the basic notions from real and complex variable theory. We begin by studying vector-valued functions defined on the real numbers and develop a Riemann integral for such functions. We next treat vector-valued functions on an abstract measure space, extending the theory of the Lebesgue integral. The consideration of vector-valued functions on the complex numbers leads us to an abstract theory of analytic functions; and this can be further generalized by allowing the domain to be a complex vector space. The material for this chapter has been selected with a view of providing an account of the general ideas as well as the special results which will be needed in the remaining parts of this treatise. The paragraph headings are: *Abstract Integration, Complex Function Theory,* and *Analytic Functions on Vectors to Vectors.*

1. Abstract Integration

3.2. Some properties of vector-valued functions. All of function theory is based upon limiting processes, and hence ultimately upon a notion of convergence. In a (B)-space we have a profusion of topologies at our disposal. For the sake of simplicity we shall limit our considerations to the norm and the weak topologies. Similarly, for a Banach space of linear bounded transformations we shall consider only the uniform, the strong, and the weak operator topologies. Corresponding to each such topology we have a distinct notion for the several basic concepts of analysis. Occasionally, however, a concept turns out to be the same in both the weak and the strong topologies.

We shall be concerned with a vector-valued function $x(\sigma)$ defined on some abstract set $\mathfrak{S}$ to a (B)-space $\mathfrak{X}$. Thus to each point $\sigma \in \mathfrak{S}$ there corresponds a unique vector $x(\sigma)$ belonging to $\mathfrak{X}$. In the case where the (B)-space is the space of linear bounded operators from $\mathfrak{X}$ to $\mathfrak{Y}$, that is $\mathfrak{E}(\mathfrak{X}, \mathfrak{Y})$, we shall speak of the function as an *operator-valued function* and denote it by $U(\sigma)$. We denote the real number system by E_1.

DEFINITION 3.2.1. *A vector-valued function $x(\xi)$ defined on a subset $\mathfrak{S}$ of E_1 to a (B)-space $\mathfrak{X}$ is* (1) *weakly continuous at $\xi = \xi_0$ if $\lim_{\xi \to \xi_0} | x^*[x(\xi) - x(\xi_0)] | = 0$ for each $x^* \in \mathfrak{X}^*$,* (2) *strongly continuous at $\xi = \xi_0$ if $\lim_{\xi \to \xi_0} \| x(\xi) - x(\xi_0) \| = 0$.*

DEFINITION 3.2.2. *An operator-valued function $U(\xi)$ on a subset $\mathfrak{S}$ of the real number system to $\mathfrak{E}(\mathfrak{X}, \mathfrak{Y})$ is (1) continuous in the weak operator topology at $\xi = \xi_0$ if $\lim_{\xi \to \xi_0} | y^*\{[U(\xi) - U(\xi_0)](x)\} | = 0$ for each $x \in \mathfrak{X}$, $y^* \in \mathfrak{Y}^*$, (2) continuous in the strong operator topology at $\xi = \xi_0$ if $\lim_{\xi \to \xi_0} \| [U(\xi) - U(\xi_0)](x) \| = 0$ for each $x \in \mathfrak{X}$, and (3) continuous in the uniform operator topology at $\xi = \xi_0$ if $\lim_{\xi \to \xi_0} \| U(\xi) - U(\xi_0) \| = 0$. We shall abbreviate these designations to weak, strong, and uniform continuity where this does not lead to ambiguity.*

For a weakly continuous vector-valued function $x(\xi)$ on the interval $[\alpha, \beta]$ to $\mathfrak{X}$, it is clear that $x^*[x(\xi)]$ is continuous and hence bounded on $[\alpha, \beta]$ for each $x^* \in \mathfrak{X}^*$. It follows from the uniform boundedness theorem that $\| x(\xi) \|$ is likewise bounded on $[\alpha, \beta]$. Further if $\xi_n \to \xi_0 \in [\alpha, \beta]$ then, according to Theorem 2.9.2 we see that $x(\xi_0)$ belongs to the least closed linear subspace containing the $[x(\xi_n)]$. Consequently the range of $x(\xi)$ for $\xi \in [\alpha, \beta]$ is contained in the *separable* closed subspace spanned by $[x(\xi), \xi$ rational in $[\alpha, \beta]]$. Similarly if $U(\xi)$ is an operator-valued function of $[\alpha, \beta]$ to $\mathfrak{E}(\mathfrak{X}, \mathfrak{Y})$, continuous in the weak operator topology, we see that $\| U(\xi) \|$ will be bounded on $[\alpha, \beta]$. However the range of $U(\xi)$ for $\xi \in [\alpha, \beta]$ need not be contained in a separable subspace of $\mathfrak{E}(\mathfrak{X}, \mathfrak{Y})$.

DEFINITION 3.2.3. *A vector-valued function $x(\xi)$ on the interval (α, β) to the (B)-space $\mathfrak{X}$ is weakly (strongly) differentiable at $\xi = \xi_0$ if there is an element $x'(\xi_0) \in \mathfrak{X}$ such that the difference quotient $\delta^{-1}[x(\xi_0 + \delta) - x(\xi_0)]$ tends weakly (strongly) to $x'(\xi_0)$ when $\delta \to 0$. We call $x'(\xi_0)$ the weak (strong) derivative of $x(\xi)$ at ξ_0.*

We note that $x(\xi)$ weakly differentiable at ξ_0 implies that $x^*[x(\xi)]$ is a differentiable numerical-valued function at $\xi = \xi_0$ for all $x^* \in \mathfrak{X}^*$; however the converse is in general not true. The above definition extends to partial derivatives of functions of several variables in an obvious fashion. In the case of operator-valued functions we again have three possibilities depending on whether the incremental ratio converges to the derivative in the weak, strong, or uniform operator topology. *A weakly (strongly, uniformly) differentiable function is evidently weakly (strongly, uniformly) continuous.*

THEOREM 3.2.1. *If the weak derivative of $x(\xi)$ equals θ everywhere in (α, β), then $x(\xi)$ is a constant.*

PROOF. The assumption implies $dx^*[x(\xi)]/d\xi = 0$ and hence that $x^*[x(\xi)] = x^*[x(\xi_0)]$ for all x^* with ξ_0 fixed in (α, β). By Theorem 2.8.2, $x(\xi) = x(\xi_0)$ for all $\xi \in (\alpha, \beta)$.

DEFINITION 3.2.4. *A vector-valued function $x(\xi)$ on the interval $[\alpha, \beta]$ to the (B)-space $\mathfrak{X}$ is of (1) weak bounded variation in $[\alpha, \beta]$ if $x^*[x(\xi)]$ is of bounded variation for every $x^* \in \mathfrak{X}^*$, (2) bounded variation if $\sup \| \sum_i [x(\beta_i) - x(\alpha_i)] \| < \infty$ over every choice of a finite number of non-overlapping intervals (α_i, β_i) in $[\alpha, \beta]$ and (3) strong bounded variation if $\sup \sum_i \| x(\alpha_i) - x(\alpha_{i-1}) \| < \infty$ where all possible finite partitions of $[\alpha, \beta]$ are allowed. The two suprema are known as the total and the strong total variations respectively.*

It is easy to verify that strong bounded variation implies bounded variation and that bounded variation implies weak bounded variation. Just as in the numerical-valued case, a vector-valued function of strong bounded variation on $[\alpha, \beta]$ to a (B)-space can have only a denumerable number of points of discontinuity and the one-sided limits exist at each point in the interval $[\alpha, \beta]$. The next theorem is due to N. Dunford [**3**] and I. Gelfand [**2**].

THEOREM 3.2.2. *A function of weak bounded variation is of bounded variation (but not necessarily of strong bounded variation).*

PROOF. Let Var $\{x^*[x(\xi)]\}$ be the total variation of $x^*[x(\xi)]$ on the interval $[\alpha, \beta]$. Then for every choice of a finite number of non-overlapping intervals (α_i, β_i) in $[\alpha, \beta]$ we have

$$| x^*\{\sum_i [x(\beta_i) - x(\alpha_i)]\} | \leqq \sum_i | x^*[x(\beta_i)] - x^*[x(\alpha_i)] | \leqq \text{Var } \{x^*[x(\xi)]\}.$$

By the uniform boundedness theorem (Theorem 2.8.6) there exists an $M > 0$ such that $\| \sum_i [x(\beta_i) - x(\alpha_i)] \| \leqq M$ for all choices of a finite number of non-overlapping intervals in $[\alpha, \beta]$.

The second part of the theorem is demonstrated by means of a counter-example. Let $\mathfrak{X}$ be the space of all bounded complex-valued functions $f(\tau)$ on the interval $[0, 1]$ with $\| f \| = \sup [\, | f(\tau) |; 0 \leqq \tau \leqq 1]$. We next define the vector function $x(\xi) = f_\xi(\cdot)$ on $[0, 1]$ to $\mathfrak{X}$ so that $f_\xi(\tau) = 1$ for $0 \leqq \tau \leqq \xi$ and $= 0$ for $\xi < \tau \leqq 1$; $f_1(\tau) \equiv 1$. For any choice of non-overlapping intervals (α_i, β_i) in $[0, 1]$ we have $\| \sum_i [x(\beta_i) - x(\alpha_i)] \| \leqq 1$. On the other hand $\| x(\tau) - x(\sigma) \| = 1$ for any choice of $\sigma, \tau \in [0, 1]$ with $\sigma \neq \tau$; it is apparent that $x(\xi)$ is not of strong bounded variation.

Weak and strong forms of a function theoretic property are never independent. If the definitions are properly made, a stronger form always implies a weaker one. It sometimes happens that the converse is also true so that the two forms are actually equivalent. An example of this is afforded by Theorem 3.2.2 above and by Theorem 3.2.3 below. Still another example will be found in Theorem 3.10.1 which asserts that all notions of holomorphism coincide for operator functions. It may happen in certain types of (B)-spaces that a strong and a weak function theoretic property are equivalent. Such a possibility is illustrated by Theorem 3.5.3 according to which weak and strong measurability coincide in separable spaces.

We list some conventions concerning infinite series with terms in a (B)-space $\mathfrak{X}$. *Such a series will be said to converge strongly (weakly) to the sum $s \in \mathfrak{X}$ if the partial sums converge strongly (weakly) to s. The series $\sum u_n$ is said to be absolutely convergent if $\sum \| u_n \|$ converges*; an absolutely convergent series is obviously strongly convergent. If $u_n = u_n(\sigma)$, $\sigma \in \mathfrak{S}$, we have the possibility that the strong or weak convergence may be uniform with respect to σ; formal definitions are left to the reader. In the case of a series of operator functions $\sum U_n(\sigma)$ care must be taken to distinguish between various types of uniform convergence. Such a series may be convergent in the uniform operator topology for fixed σ or it may converge, in one operator topology or another, uniformly with respect to σ.

Following W. Orlicz [2], we introduce still another notion of convergence for infinite sums in a (B)-space, namely *unconditional convergence*. This type of convergence has very useful applications, especially in integration theory. The notion is of little value in the numerical case where unconditional convergence coincides with absolute convergence.

DEFINITION 3.2.5. *An infinite series $\sum u_n$ is said to be strongly (weakly) unconditionally convergent if and only if every subseries is strongly (weakly) convergent.*

The following lemma and theorem are essentially due to W. Orlicz [1] who proved them for weakly complete spaces. Our proof is due to N. Dunford [3] (cf. B. J. Pettis [1]).

LEMMA 3.2.1. *If $\sum u_n$ is weakly unconditionally convergent, then*

$$\lim_{N\to\infty} \sum_{n=N}^{\infty} | x^*(u_n) | = 0 \tag{3.2.1}$$

uniformly on the unit sphere $\mathfrak{S}_1^*$ of $\mathfrak{X}^*$.

PROOF. By assumption given any subsequence of integers, $\pi = (n_1, n_2, \cdots)$, there exists an $x_\pi \in \mathfrak{X}$ such that

$$\sum_{n\in\pi} x^*(u_n) = x^*(x_\pi)$$

for all $x^* \in \mathfrak{X}^*$. It follows that $\sum_{n=1}^{\infty} | x^*(u_n) | < \infty$ for each $x^* \in \mathfrak{X}^*$. Hence we may define $V(x^*) = \{x^*(u_n)\}$ on $\mathfrak{X}^*$ to l; V is clearly linear and closed. By Theorem 2.12.3 V must be bounded. We now show that V is also compact. Let $\mathfrak{Y}$ be the least closed linear subspace containing $\{u_n\}$. Since x_π is the weak limit of elements in $\mathfrak{Y}$, it follows from Theorem 2.9.2 that x_π likewise belongs to $\mathfrak{Y}$. Let $\{x_k^*\}$ be a sequence in $\mathfrak{S}_1^*$, and let y_k^* be the restriction of x_k^* on $\mathfrak{Y}$. Since $\mathfrak{Y}$ is separable, we may extract a subsequence $y_{k_j}^*$ (by Theorem 2.10.1) which converges in the weak* topology of $\mathfrak{Y}^*$ to a functional $y_0^* \in \mathfrak{Y}^*$. Finally let x_0^* be a linear extension of y_0^* over $\mathfrak{X}$ (see Theorem 2.7.3). We wish to show that $V(x_{k_j}^*)$ converges weakly to $V(x_0^*)$ in l. In view of the fact that V is bounded, it is sufficient to show that $\lim_{j\to\infty} f[V(x_{k_j}^*)] = f[V(x_0^*)]$ for each bounded linear functional f of a fundamental set in $m = l^*$. The set of characteristic functions of the sets π of integers form such a fundamental set. Let f_π be such a functional in m. Then

$$f_\pi[V(x_{k_j}^*)] = \sum_{n\in\pi} x_{k_j}^*(u_n) = x_{k_j}^*(x_\pi) = y_{k_j}^*(x_\pi)$$

$$\to y_0^*(x_\pi) = x_0^*(x_\pi) = \sum_{n\in\pi} x_0^*(u_n) = f_\pi[V(x_0^*)].$$

Now by Problem 2, section 2.9, weak convergence in l implies strong convergence.

Hence $V(x_{k_j}^*)$ converges in norm to $V(x_0^*)$. This shows that $V(\mathfrak{S}_1^*)$ is a compact subset of l so that (3.2.1) is a direct consequence of Problem 1 of section 1.12.

THEOREM 3.2.3. *A weakly unconditionally convergent series is necessarily strongly unconditionally convergent.*

PROOF. Again let π be a subsequence of integers. Then by hypothesis there exists an $x_\pi \in \mathfrak{X}$ such that $\sum_{n \in \pi} x^*(u_n) = x^*(x_\pi)$ for each $x^* \in \mathfrak{X}^*$. Clearly

$$\left\| \sum_{\substack{n \in \pi \\ n \leqq N}} u_n - x_\pi \right\| \leqq \sup_{x^* \in \mathfrak{S}_1^*} \left[\sum_{\substack{n \in \pi \\ n > N}} | x^*(u_n) | \right]$$

and by the lemma the right side goes to zero as $N \to \infty$.

3.3. Riemann-Stieltjes integrals. The problem of developing an abstract integration theory has attracted many authors during recent years. The Riemann integral was extended by L. M. Graves (1927); the Lebesgue integral by T. H. Hildebrandt (1927), S. Bochner (1933), G. Birkhoff (1935–1937; two definitions B_0, B_1), N. Dunford (1935–1938; two definitions D_0 and D_1), I. Gelfand (1936–1938), B. J. Pettis (1938), and R. S. Phillips (1940). G. B. Price [1] and C. E. Rickart [1] have formulated still more general concepts of an integral. The Hildebrandt, the Bochner, and the D_0 integrals are equivalent and are the most restrictive of the Lebesgue integrals. The other integrals have been shown by Phillips [1] to be different aspects of the B_1 integral defined on a convex linear topological space; the integrals differing only in respect to the topology employed. Thus the B_0 integral makes use of the norm topology, the Pettis integral employs the weak topology, and the Gelfand and D_1 integrals are defined for an adjoint (B)-space with the weak* topology. We shall give a fairly complete discussion of the Bochner integral in sections 3.7 and 3.8. However a knowledge of the abstract Lebesgue integral is not required for a first reading of this book; the abstract Riemann-Stieltjes integral will be found adequate for this purpose.

The Riemann-Stieltjes integral can be extended to vector-valued functions in two ways: *either the integrand or the integrator may be vector-valued.* Let $x(\xi)$ be a vector-valued function on $[\alpha, \beta]$ to $\mathfrak{X}$ and let $g(\xi)$ be a numerical-valued function on the same interval. We denote the subdivision $(\sigma_0 = \alpha \leqq \sigma_1 \leqq \cdots \leqq \sigma_n = \beta)$ together with the points τ_i $(\sigma_{i-1} \leqq \tau_i \leqq \sigma_i)$ by π and set $| \pi | = \max_i | \sigma_i - \sigma_{i-1} |$.

DEFINITION 3.3.1. *Let*

$$S_\pi(x, g) = \sum_{i=1}^{n} x(\tau_i)[g(\sigma_i) - g(\sigma_{i-1})]. \tag{3.3.1}$$

Then if $\lim_{|\pi| \to 0} S_\pi$ *exists in a given topology, we define this limit to be the integral*

$$\int_\alpha^\beta x(\xi)\, dg(\xi) \tag{3.3.2}$$

relative to this topology.

DEFINITION 3.3.2. *Let*

$$(3.3.3) \qquad s_\pi(x, g) = \sum_{i=1}^{n} g(\tau_i)[x(\sigma_i) - x(\sigma_{i-1})].$$

Then if $\lim_{|\pi| \to 0} s_\pi$ *exists in a given topology, we define this limit to be the integral*

$$(3.3.4) \qquad \int_\alpha^\beta g(\xi)\, dx(\xi)$$

relative to this topology.

THEOREM 3.3.1. *If either of the integrals* (3.3.2) *or* (3.3.4) *exist in a given topology, then both will exist in this topology and*

$$(3.3.5) \qquad \int_\alpha^\beta x(\xi)\, dg(\xi) = x(\xi)g(\xi)\Big|_\alpha^\beta - \int_\alpha^\beta g(\xi)\, dx(\xi).$$

PROOF. This is an immediate consequence of the following identity and its dual (obtained by interchanging the roles of $x(\xi)$ and $g(\xi)$):

$$\sum_{i=1}^{n} x(\tau_i)[g(\sigma_i) - g(\sigma_{i-1})] = x(\beta)g(\beta) - x(\alpha)g(\alpha) - \sum_{i=0}^{n} g(\sigma_i)[x(\tau_{i+1}) - x(\tau_i)]$$

where we have set $\tau_0 = \alpha$ and $\tau_{n+1} = \beta$. For it is clear that $(\tau_0 = \alpha \leqq \tau_1 \leqq \cdots \leqq \tau_{n+1} = \beta)$ is likewise a subdivision of $[\alpha, \beta]$, that $\tau_i \leqq \sigma_i \leqq \tau_{i+1}$, and also that $\max_i |\tau_{i+1} - \tau_i| \leqq 2|\pi|$.

We shall therefore concern ourselves only with the existence of one or the other of the two integrals (3.3.2) and (3.3.4).

THEOREM 3.3.2. *Suppose that either* (1) $x(\xi)$ *is a strongly continuous vector-valued function on* $[\alpha, \beta]$ *to* $\mathfrak{X}$ *and* $g(\xi)$ *is a numerically-valued function of bounded variation on* $[\alpha, \beta]$, *or* (2) $x(\xi)$ *is a vector-valued function on* $[\alpha, \beta]$ *to* $\mathfrak{X}$ *of bounded variation in the sense of Definition* 3.2.4 *and* $g(\xi)$ *is a continuous numerically-valued function on* $[\alpha, \beta]$. *Then the integrals*

$$\int_\alpha^\beta x(\xi)\, dg(\xi) \qquad \text{and} \qquad \int_\alpha^\beta g(\xi)\, dx(\xi)$$

exist in the norm topology. Further, if T *is a closed linear operator on* $\mathfrak{X}$ *to* $\mathfrak{Y}$, *if* $x(\xi) \in \mathfrak{D}(T)$, *and if* $T[x(\xi)]$ *is strongly continuous in case* (1) *or of bounded variation in case* (2), *then*

$$(3.3.6) \qquad T\left\{\int_\alpha^\beta x(\xi)\, dg(\xi)\right\} = \int_\alpha^\beta T[x(\xi)]\, dg(\xi)$$

and

$$(3.3.7) \qquad T\left\{\int_\alpha^\beta g(\xi)\, dx(\xi)\right\} = \int_\alpha^\beta g(\xi)\, dT[x(\xi)].$$

PROOF. For case (1) it is clear that $x(\xi)$ is uniformly continuous in the norm topology on $[\alpha, \beta]$. Hence given $\epsilon > 0$ there exists a $\delta > 0$ such that $\| x(\xi') - x(\xi'') \| < \epsilon$ if only $| \xi' - \xi'' | < \delta$. Consequently if $|\pi_1|, |\pi_2| < \delta/2$ a simple calculation shows that

$$\| S_{\pi_1} - S_{\pi_2} \| \leqq 2\epsilon \operatorname{Var}[g(\xi)].$$

This proves the existence of the strong integral for case (1). For case (2), $g(\xi)$ is uniformly continuous on $[\alpha, \beta]$ and choosing ϵ and δ in the same manner as above we obtain

$$| x^*(S_{\pi_1} - S_{\pi_2}) | \leqq 2\epsilon \operatorname{Var}\{x^*[x(\xi)]\}$$

for $|\pi_1|, |\pi_2| < \delta/2$. Now

$$\begin{aligned} \operatorname{Var}\{x^*[x(\xi)]\} &\leqq \operatorname{Var}[\mathfrak{R}\{x^*[x(\xi)]\}] + \operatorname{Var}[\mathfrak{J}\{x^*[x(\xi)]\}] \\ &\leqq 4 \sup | x^*\{\sum_i [x(\beta_i) - x(\alpha_i)]\} | \end{aligned}$$

taken over all finite sets of non-overlapping intervals (α_i, β_i) in $[\alpha, \beta]$. Hence by Definition 3.2.4 there will exist an $M > 0$ such that $\operatorname{Var}\{x^*[x(\xi)]\} \leqq M \| x^* \|$. As a consequence

$$\| S_{\pi_1} - S_{\pi_2} \| = \sup_{\|x^*\|=1} | x^*(S_{\pi_1} - S_{\pi_2}) | \leqq 2M\epsilon.$$

This establishes the existence of the strong integral for case (2). Because of Theorem 3.3.1 both integrals will exist in each case. We turn now to the second part of the theorem. For any π we have $T[S_\pi(x, g)] = S_\pi[T(x), g]$ because of the linearity of T. Further we have just shown that

$$\lim_{|\pi|\to 0} S_\pi(x, g) = \int_\alpha^\beta x(\xi)\, dg(\xi),$$

and applying the above result to $T[x(\xi)]$ instead of to $x(\xi)$ we see that

$$\lim_{|\pi|\to 0} T[S_\pi(x, g)] = \lim_{|\pi|\to 0} S_\pi[T(x), g] = \int_\alpha^\beta T[x(\xi)]\, dg(\xi).$$

Since T is closed, it follows that $\int_\alpha^\beta x(\xi)\, dg(\xi) \in \mathfrak{D}(T)$ and that (3.3.6) holds. Replacing S_π by s_π in the above argument we obtain (3.3.7).

If T is a bounded linear operator on $\mathfrak{X}$ to $\mathfrak{Y}$ and if $x(\xi)$ is strongly continuous (or of bounded variation), then $T[x(\xi)]$ is automatically strongly continuous (or of bounded variation). Consequently (3.3.6) and (3.3.7) are valid in this case.

Corollary 1. *For function pairs of the type considered in Theorem* 3.3.2 *we have*

(i) $$\int_\alpha^\beta [\gamma_1 x_1(\xi) + \gamma_2 x_2(\xi)]\, dg(\xi) = \gamma_1 \int_\alpha^\beta x_1(\xi)\, dg(\xi) + \gamma_2 \int_\alpha^\beta x_2(\xi)\, dg(\xi);$$

(ii) $$\int_\alpha^\beta x(\xi)\, d[\gamma_1 g_1(\xi) + \gamma_2 g_2(\xi)] = \gamma_1 \int_\alpha^\beta x(\xi)\, dg_1(\xi) + \gamma_2 \int_\alpha^\beta x(\xi)\, dg_2(\xi);$$

(iii)
$$\int_\alpha^\beta x(\xi)\, dg(\xi) = \int_\alpha^\gamma x(\xi)\, dg(\xi) + \int_\gamma^\beta x(\xi)\, dg(\xi)$$

where $\alpha < \gamma < \beta$; *for functions of type* (1) *we also have*

(iv)
$$\left\| \int_\alpha^\beta x(\xi)\, dg(\xi) \right\| \leqq \left\{ \sup_{\alpha \leqq \xi \leqq \beta} \| x(\xi) \| \right\} \text{Var}\,[g(\xi)];$$

(v) *If* $x_n(\xi) \to x(\xi)$ *uniformly in* $[\alpha, \beta]$, *then*

$$\lim_{n\to\infty} \int_\alpha^\beta x_n(\xi)\, dg(\xi) = \int_\alpha^\beta x(\xi)\, dg(\xi).$$

PROOF. It is clear that Theorem 3.3.2 applies in particular if T is replaced by a bounded linear functional, $x^* \in \mathfrak{X}^*$. Thus

$$x^*\left\{\int_\alpha^\beta x(\xi)\, dg(\xi)\right\} = \int_\alpha^\beta x^*[x(\xi)]\, dg(\xi) = \int_\alpha^\gamma x^*[x(\xi)]\, dg(\xi) + \int_\gamma^\beta x^*x(\xi)]\, dg(\xi)$$
$$= x^*\left\{\int_\alpha^\gamma x(\xi)\, dg(\xi) + \int_\gamma^\beta x(\xi)\, dg(\xi)\right\}.$$

Since $\mathfrak{X}^*$ is total on $\mathfrak{X}$ we see that (iii) follows. Likewise (i) and (ii) are obtained directly from the corresponding properties of the Riemann-Stieltjes integral for numerically-valued functions. Property (iv) follows from the simple estimate $\| S_\pi(x, g) \| \leqq [\sup_{\alpha \leqq \xi \leqq \beta} \| x(\xi) \|]\, \text{Var}[g(\xi)]$. Finally (v) is an obvious consequence of (iv).

COROLLARY 2. *Let* $\{g_n(\xi)\}$ *be a sequence of numerically-valued functions of bounded variation on* $[\alpha, \beta]$ *and suppose that* $g_n(\cdot) \to g(\cdot)$ *in the weak* topology of* $C^*[\alpha, \beta]$. *If* $x(\xi)$ *is a strongly continuous vector-valued function on* $[\alpha, \beta]$ *to* $\mathfrak{X}$, *then*

$$\lim_{n\to\infty} \int_\alpha^\beta x(\xi)\, dg_n(\xi) = \int_\alpha^\beta x(\xi)\, dg(\xi)$$

in the norm topology.

PROOF. According to the uniform boundedness theorem there exists an $M > 0$ such that $\text{Var}[g_n(\xi)] \leqq M$, $n = 1, 2, \cdots$. If the assertion were false, we could find a subsequence $\{g_{n_k}(\xi)\}$ and an $\epsilon > 0$ such that $\| \int_\alpha^\beta x(\xi)\, dg_{n_k}(\xi) - \int_\alpha^\beta x(\xi)\, dg(\xi) \| \geqq \epsilon$ for all $k \geqq 1$ and $\lim_{k\to\infty} g_{n_k}(\xi) \equiv g_0(\xi)$ exists for all $\xi \in [\alpha, \beta]$ (by the Helly theorem). As in the proof of Theorem 3.3.2, $\lim_{|\pi|\to 0} S_\pi(x, g_{n_k}) = \int_\alpha^\beta x(\xi)\, dg_{n_k}(\xi)$ uniformly with respect to k. Further it is clear that

$$\lim_{k\to\infty} S_\pi(x, g_{n_k}) = S_\pi(x, g_0)$$

for each subdivision π. The iterated limits theorem therefore applies and we have $\lim_{k\to\infty} \int_\alpha^\beta x(\xi)\, dg_{n_k}(\xi) = \int_\alpha^\beta x(\xi)\, dg_0(\xi)$. By assumption $\int_\alpha^\beta x(\xi)\, dg_n(\xi) \to \int_\alpha^\beta x(\xi)\, dg(\xi)$ in the weak topology. Hence it follows from Theorem 2.8.2 that $\int_\alpha^\beta x(\xi)\, dg_0(\xi) = \int_\alpha^\beta x(\xi)\, dg(\xi)$, which is impossible.

THEOREM 3.3.3. *If $U(\xi)$ is an operator-valued function on $[\alpha, \beta]$ to $\mathfrak{E}(\mathfrak{X}, \mathfrak{Y})$ continuous in the uniform operator topology and if $g(\xi)$ is a numerically-valued function of bounded variation on $[\alpha, \beta]$, then*

$$\left\{\int_\alpha^\beta U(\xi)\, dg(\xi)\right\}^* = \int_\alpha^\beta U^*(\xi)\, dg(\xi). \tag{3.3.8}$$

Here both integrals exist in the uniform operator topology, the left side integral in $\mathfrak{E}(\mathfrak{X}, \mathfrak{Y})$ and the right side integral in $\mathfrak{E}(\mathfrak{Y}^*, \mathfrak{X}^*)$.

PROOF. The mapping $U \to U^*$ of $\mathfrak{E}(\mathfrak{X}, \mathfrak{Y})$ into $\mathfrak{E}(\mathfrak{Y}^*, \mathfrak{X}^*)$ is an isometry. Hence $U^*(\xi)$ is continuous on $[\alpha, \beta]$ in the uniform operator topology of $\mathfrak{E}(\mathfrak{Y}^*, \mathfrak{X}^*)$. Thus the existence of the integrals follows from Theorem 3.3.2. Now $[S_\pi(U, g)]^* = S_\pi(U^*, g)$ and consequently

$$[\lim_{|\pi|\to 0} S_\pi(U, g)]^* = \lim_{|\pi|\to 0}[S_\pi(U, g)]^* = \lim_{|\pi|\to 0} S_\pi(U^*, g);$$

this is a restatement of (3.3.8).

THEOREM 3.3.4. *If $U(\xi)$ is an operator-valued function on $[\alpha, \beta]$ to $\mathfrak{E}(\mathfrak{X}, \mathfrak{Y})$ continuous in the strong operator topology and if $g(\xi)$ is a numerically-valued function of bounded variation on $[\alpha, \beta]$, then the integrals*

$$\int_\alpha^\beta U(\xi)\, dg(\xi) \qquad \textit{and} \qquad \int_\alpha^\beta g(\xi)\, dU(\xi)$$

exist in the strong operator topology. Further

$$\left\{\int_\alpha^\beta U(\xi)\, dg(\xi)\right\}[x] = \int_\alpha^\beta U(\xi)[x]\, dg(\xi) \qquad \textit{and}$$

$$\left\{\int_\alpha^\beta g(\xi)\, dU(\xi)\right\}[x] = \int_\alpha^\beta g(\xi)\, dU(\xi)[x].$$

PROOF. It is clear from Theorem 3.3.2 that $\int_\alpha^\beta U(\xi)(x)\, dg(\xi)$ exists and is equal to an element, $V(x)$, of $\mathfrak{Y}$ for each $x \in \mathfrak{X}$. It follows from Corollary 1 to Theorem 3.3.2 that $V(x)$ is linear and that

$$\| V(x) \| \leqq \left[\sup_{\alpha \leqq \xi \leqq \beta} \| U(\xi)(x) \|\right] \mathrm{Var}\,[g(\xi)] \leqq M \| x \| \,\mathrm{Var}\,[g(\xi)]$$

for some $M > 0$. Thus $V \in \mathfrak{E}(\mathfrak{X}, \mathfrak{Y})$. Further $S_\pi(U, g)(x) \to \int_\alpha^\beta U(\xi)(x)\, dg(\xi) = V(x)$ for each $x \in \mathfrak{X}$ and hence $S_\pi(U, g) \to V$ in the strong operator topology.

If the underlying space is a Banach algebra, then the Riemann-Stieltjes integral admits of still another type of extension. Let $y = f(x)$ be a function on $\mathfrak{B}$ to itself, continuous in the norm topology. Let Γ be a rectifiable arc in $\mathfrak{B}$. By this we mean that Γ is given by an equation $x = x(\xi)$, $0 \leqq \xi \leqq 1$, where $x(\xi)$ is continuous and of strong bounded variation in the sense of Definition 3.2.4. We then define

$$\int_\Gamma f(x)\cdot dx = \lim_{|\pi|\to 0} \sum_{i=1}^n f(x(\tau_i))[x(\sigma_i) - x(\sigma_{i-1})]. \tag{3.3.9}$$

The existence of the integral is established in the usual manner and it has the properties of linearity and boundedness which are to be expected. In particular

$$\left\| \int_\Gamma f(x) \cdot dx \right\| \leqq \max \| f(x) \| \, l(\Gamma), \tag{3.3.10}$$

where $l(\Gamma)$ is the length of Γ, that is, the strong total variation of $x(\xi)$ in $[0, 1]$. Since $\mathfrak{B}$ is non-commutative, there is also an integral $\int_\Gamma dx \cdot f(x)$ which is ordinarily distinct from $\int_\Gamma f(x) \cdot dx$. This concept of the integral was introduced by Bochner and Taylor [1].

3.4. The calculus. We now have at our disposal the principal tools of the calculus, namely, differentiation and integration. It is to be expected that many of the results of classical analysis carry over into this general setting. For instance if $x(\xi)$ is a vector-valued function on $[\alpha, \beta]$ to $\mathfrak{X}$ continuous in the norm topology, then it is easy to see that the indefinite integral $\int_\alpha^\xi x(\tau)\, d\tau$ is strongly differentiable and that the derivative is again $x(\xi)$. As a further illustration we shall prove a basic existence theorem for differential equations.

THEOREM 3.4.1. *Let $y = f(\xi, x)$ on $E_1 \times \mathfrak{X}$ to $\mathfrak{X}$ be defined and continuous in each variable separately for $| \xi - \xi_0 | \leqq \alpha$, $\| x - x_0 \| \leqq \beta$ and satisfy*

$$\| f(\xi, x) \| \leqq \mu, \quad \| f(\xi, x_1) - f(\xi, x_2) \| \leqq \gamma \| x_1 - x_2 \| \tag{3.4.1}$$

for ξ, x, x_1, x_2 in the indicated regions. Here α, β, γ, μ are fixed positive numbers and $\alpha\mu \leqq \beta$. Then there exists one and only one strongly continuously differentiable function $x(\xi)$ such that

$$\frac{dx(\xi)}{d\xi} = f[\xi, x(\xi)] \tag{3.4.2}$$

in $| \xi - \xi_0 | \leqq \alpha$ and $x(\xi_0) = x_0$.

PROOF. It is clear that continuity in each variable separately plus the Lipschitz condition (3.4.1) implies that $f(\xi, x)$ is actually continuous in both variables. The classical method of successive approximations can now be applied. This leads to a sequence of functions $x_n(\xi)$ defined for $| \xi - \xi_0 | \leqq \alpha$ by

$$x_0(\xi) = x_0, \qquad x_n(\xi) = x_0 + \int_{\xi_0}^{\xi} f[\tau, x_{n-1}(\tau)]\, d\tau,$$

the integral being taken relative to the strong topology. We see by induction that $x_n(\xi)$ is strongly continuous and that $\| x_n(\xi) - x_0 \| \leqq \beta$ for $| \xi - \xi_0 | \leqq \alpha$. The classical proof may be followed step by step, replacing absolute values by norms throughout. One shows that $\| x_n(\xi) - x_{n-1}(\xi) \| \leqq \mu\gamma^{n-1} | \xi - \xi_0 |^n/n!$ and hence that the $x_n(\xi)$ converge uniformly in $| \xi - \xi_0 | \leqq \alpha$ to a strongly continuous function $x(\xi)$. Thus

$$\| f[\xi, x(\xi)] - f[\xi, x_n(\xi)] \| \leqq \gamma \| x(\xi) - x_n(\xi) \| \to 0$$

uniformly in $|\xi - \xi_0| \leqq \alpha$. Applying Corollary 1 to Theorem 3.3.2 we obtain

$$x(\xi) = x_0 + \int_{\xi_0}^{\xi} f[\tau, x(\tau)]\, d\tau.$$

It is obvious that $x(\xi)$ is continuously differentiable, that $x'(\xi) = f[\xi, x(\xi)]$, and that $x(\xi_0) = x_0$. The uniqueness is proved in the usual manner.

If $\mathfrak{X}$ is a Banach algebra, then $f(\xi, x) = ax$, with $a, x \in \mathfrak{X}$, satisfies the hypothesis of the above theorem (here $\gamma = \| a \|$). The differential equation

$$\frac{dx}{d\xi} = ax, \qquad x(0) = e, \tag{3.4.3}$$

is of interest in the theory of semi-groups. Here the method of successive approximations leads to the unique solution

$$x(\xi) = e + \sum_{n=1}^{\infty} \frac{\xi^n a^n}{n!} \equiv \exp(\xi a). \tag{3.4.4}$$

We take this series as the definition of the *exponential function*, $\exp(\xi a)$. It obviously reduces to the classical exponential function when $\mathfrak{X}$ is the algebra of complex numbers.

It frequently happens that the (B)-space, $\mathfrak{X}$, is a space of numerically-valued functions or of classes of functions defined on some abstract set $\mathfrak{S}$. In this case one would expect the differential equation (3.4.2) to be in some way connected with the partial differential equation

$$\frac{\partial \varphi(\xi, \sigma)}{\partial \xi} = f[\xi, \varphi(\xi, \cdot)](\sigma), \qquad | \xi - \xi_0 | \leqq \alpha, \tag{3.4.5}$$

with the initial condition $\varphi(\xi_0, \sigma) = \varphi_0(\sigma)$. The precise connection depends on the individual space $\mathfrak{X}$.

If $\mathfrak{X} = M(\mathfrak{S})$, the space of all bounded functions of σ with norm $\| \varphi \| = \sup_{\sigma \in \mathfrak{S}} | \varphi(\sigma) |$, then the connection is fairly obvious. In fact, let $x(\xi)$ be the solution to (3.4.2) obtained by means of Theorem 3.4.1. For fixed ξ, $x(\xi)$ has a unique representation $\varphi(\xi, \sigma)$ in $M(\mathfrak{S})$. Since $x(\xi)$ is strongly continuously differentiable, we see that $\partial\varphi(\xi, \sigma)/\partial\xi$ exists uniformly with respect to σ and satisfies (3.4.5) for all $| \xi - \xi_0 | \leqq \alpha$ and $\sigma \in \mathfrak{S}$.

The situation is far less obvious if each $x \in \mathfrak{X}$ corresponds to a *class* of numerically-valued functions. The solution obtained in Theorem 3.4.1 gives us no hint about how to choose a representative function for each vector $x(\xi)$ (ξ fixed); this is clearly needed in order to give meaning to (3.4.5). The familiar spaces of this kind are subsumed under what we shall call *spaces of type L*, defined as follows.

Let $\mathfrak{S}$ be an abstract set, let $\mathfrak{E}$ be a σ-ring of subsets of $\mathfrak{S}$, and let $m(E)$ be a σ-finite measure on $\mathfrak{E}$ (see P. R. Halmos [1]). Two numerically-valued functions are said to be equivalent, in symbols $\varphi_1(\sigma) \approx \varphi_2(\sigma)$, if they differ at most on a set of measure zero. We say that a (B)-space $\mathfrak{X}$ is of type L if it consists of equivalence classes of numerically-valued functions and if it has the following two properties:

(1). If $x(\xi)$ is a strongly continuous vector-valued function defined on the interval $I = [\alpha, \beta]$, then there exists a function $\varphi(\xi, \sigma)$ measurable on the product set $I \times \mathfrak{S}$ such that $x(\xi) = \varphi(\xi, \cdot)$ for each $\xi \in I$.

(2). Let $x(\xi)$ be strongly continuous on the interval $I = [\alpha, \beta]$ and suppose that $\varphi(\xi, \sigma)$ is measurable on $I \times \mathfrak{S}$ and that $x(\xi) = \varphi(\xi, \cdot)$ for each $\xi \in I$. Then

$$\left[\int_\alpha^\beta x(\xi)\, d\xi\right](\sigma) \approx \int_\alpha^\beta \varphi(\xi, \sigma)\, d\xi, \tag{3.4.6}$$

where the integral on the left is the abstract Riemann integral and the integral on the right is the ordinary Lebesgue integral for numerically-valued functions.

It is easy to show that $L_p(\mathfrak{S}, m)$ for $1 \leqq p \leqq \infty$ is of type L. In fact, let $x(\xi)$ be a vector-valued function strongly continuous on $I = [\alpha, \beta]$. We now arbitrarily choose a definite representation $\varphi_0(\xi, \sigma)$ of $x(\xi)$ for each $\xi \in I$. Let $\xi_0 = \alpha < \xi_1 < \cdots < \xi_n = \beta$ divide $[\alpha, \beta]$ into n equal parts and set

$$\varphi_n(\xi, \sigma) = \varphi_0(\xi_{k-1}, \sigma) \quad \text{for} \quad \xi_{k-1} \leqq \xi < \xi_k; \qquad k = 1, 2, \cdots, n,$$
$$\varphi_n(\beta, \sigma) = \varphi_0(\beta, \sigma).$$

It is clear that $\varphi_n(\xi, \sigma)$ is measurable on the product set $I \times \mathfrak{S}$ and that $\| \varphi_n(\xi, \cdot) - x(\xi) \| \to 0$ uniformly for $\xi \in I$. Suppose first that $1 \leqq p < \infty$. Then

$$\lim_{k,n\to\infty} \int_\alpha^\beta \int_{\mathfrak{S}} | \varphi_n(\xi, \sigma) - \varphi_k(\xi, \sigma) |^p\, dm\, d\xi = 0$$

and hence, by the usual completeness argument for an L_p space, there exists a function $\varphi(\xi, \sigma)$ measurable on $I \times \mathfrak{S}$ such that

$$\lim_{n\to\infty} \int_\alpha^\beta \int_{\mathfrak{S}} | \varphi_n(\xi, \sigma) - \varphi(\xi, \sigma) |^p\, dm\, d\xi = 0.$$

Keeping n fixed we have by the Fubini theorem,

$$\int_{\mathfrak{S}} | \varphi_n(\xi, \sigma) - \varphi(\xi, \sigma) |^p\, dm < \infty$$

almost everywhere in I and consequently $\varphi(\xi, \cdot) \in L_p(\mathfrak{S}, m)$ almost everywhere in I. Finally

$$\int_\alpha^\beta \| x(\xi) - \varphi(\xi, \cdot) \|^p\, d\xi \leqq \int_\alpha^\beta \| x(\xi) - \varphi_n(\xi, \cdot) \|^p\, d\xi + \int_\alpha^\beta \| \varphi_n(\xi, \cdot) - \varphi(\xi, \cdot) \|^p\, d\xi.$$

Both terms on the right go to zero as $n \to \infty$ and therefore $x(\xi) = \varphi(\xi, \cdot)$ almost everywhere in I. We now redefine $\varphi(\xi, \sigma)$ on a ξ-set of measure zero so that $x(\xi) = \varphi(\xi, \cdot)$ for all $\xi \in I$. The so redefined $\varphi(\xi, \sigma)$ is clearly still measurable on $I \times \mathfrak{S}$. This is property (1) for the case $1 \leqq p < \infty$. Next suppose that $\mathfrak{X} = L_\infty(\mathfrak{S}; m)$ and set

$$F_{nm} = [(\xi, \sigma), | \varphi_n(\xi, \sigma) - \varphi_m(\xi, \sigma) | > \| \varphi_n(\xi, \cdot) - \varphi_m(\xi, \cdot) \|].$$

Then F_{nm} is certainly measurable on $I \times \mathfrak{S}$ and of measure zero for each $\xi \in I$; F_{nm} is consequently of measure zero in $I \times \mathfrak{S}$. Hence $F_0 \equiv \bigcup F_{nm}$ is of measure zero both for each $\xi \in I$ and in $I \times \mathfrak{S}$. Now $\lim_{m,\, n\to\infty} | \varphi_n(\xi, \sigma) - \varphi_m(\xi, \sigma) | = 0$ uniformly on $I \times \mathfrak{S} \ominus F_0$. Define $\varphi(\xi, \sigma)$ equal to the limiting function on $I \times \mathfrak{S} \ominus F_0$ and equal to zero on F_0. Then $\varphi(\xi, \sigma)$ is the uniform limit almost everywhere in $I \times \mathfrak{S}$ of measurable functions and is therefore itself measurable. It is also clear that $x(\xi) = \varphi(\xi, \cdot)$ for all $\xi \in I$.

In order to prove property (2) we notice that $\mathfrak{X}^*$ contains all characteristic functions, $\psi(\sigma)$, of sets of finite m-measure. For $x^* = \psi(\cdot)$ we have

$$\int_{\mathfrak{S}} \psi(\sigma) \left[\int_\alpha^\beta x(\xi)\, d\xi\right] (\sigma)\, dm = x^* \left[\int_\alpha^\beta x(\xi)\, d\xi\right] = \int_\alpha^\beta x^*[x(\xi)]\, d\xi.$$

By the Fubini theorem

$$\int_\alpha^\beta x^*[x(\xi)]\, d\xi = \int_\alpha^\beta \left[\int_{\mathfrak{S}} \psi(\sigma)\varphi(\xi, \sigma)\, dm\right] d\xi = \int_{\mathfrak{S}} \psi(\sigma) \left[\int_\alpha^\beta \varphi(\xi, \sigma)\, d\xi\right] dm.$$

Hence

$$\int_{\mathfrak{S}} \psi(\sigma) \left[\int_\alpha^\beta x(\xi)\, d\xi\right] (\sigma)\, dm = \int_{\mathfrak{S}} \psi(\sigma) \left[\int_\alpha^\beta \varphi(\xi, \sigma)\, d\xi\right] dm$$

for all characteristic functions, $\psi(\sigma)$, and this implies (3.4.6).

THEOREM 3.4.2. *Let $\mathfrak{X}$ be a (B)-space of type L. If $x(\xi)$ is a vector-valued function on $I = [\alpha, \beta]$ to $\mathfrak{X}$, n-times continuously differentiable, then there exists a numerically-valued function $\varphi(\xi, \sigma)$ measurable on $I \times \mathfrak{S}$ such that for $0 \leqq k \leqq n - 1$, $\partial^k\varphi(\xi, \sigma)/\partial\xi^k$ is absolutely continuous for each $\sigma \in \mathfrak{S}$, and $\partial^k\varphi(\xi, \cdot)/\partial\xi^k = x^{(k)}(\xi)$ for each $\xi \in I$; further $\partial^n\varphi(\xi, \sigma)/\partial\xi^n$ exists almost everywhere in $I \times \mathfrak{S}$ and $\partial^n\varphi(\xi, \cdot)/\partial\xi^n = x^{(n)}(\xi)$ for almost all $\xi \in I$.*

PROOF. Let $y(\xi) = x^{(n)}(\xi)$. By property (1) there exists a numerically-valued function $\psi_0(\xi, \sigma)$ measurable on $I \times \mathfrak{S}$ such that $\psi_0(\xi, \cdot) = y(\xi)$ for all $\xi \in I$. We now set

$$\psi_1(\xi, \sigma) = \int_\alpha^\xi \psi_0(\tau, \sigma)\, d\tau.$$

It follows from property (2) that

$$\psi_1(\xi, \cdot) = \int_\alpha^\xi x^{(n)}(\tau)\, d\tau = x^{(n-1)}(\xi) - x^{(n-1)}(\alpha).$$

The integral $\int_\alpha^\beta \psi_0(\tau, \sigma)\, d\tau$ may not exist for a set of σ's of measure zero. In this case we redefine $\psi_0(\xi, \sigma)$ to vanish identically for such σ. It is clear that the so redefined $\psi_0(\xi, \sigma)$ can be used in the above just as well as the original. In this case $\psi_1(\xi, \sigma)$ will be absolutely continuous in ξ for all $\sigma \in \mathfrak{S}$. Further $\psi_1(\xi, \sigma)$ will be measurable on $I \times \mathfrak{S}$ because the indefinite integral of a function measurable on $I \times \mathfrak{S}$ is again measurable on $I \times \mathfrak{S}$. It follows that $\limsup_{\delta\to 0} [\psi_1(\xi + \delta, \sigma) - \psi_1(\xi, \sigma)]/\delta$ is measurable on $I \times \mathfrak{S}$. Let F be the subset of $I \times \mathfrak{S}$ where the lim sup of the incremental ratio and $\psi_0(\xi, \sigma)$ differ. Then F is a measurable subset of $I \times \mathfrak{S}$. Since the limit of the incremental ratio is equal to $\psi_0(\xi, \sigma)$ almost everywhere in ξ for each $\sigma \in \mathfrak{S}$, it follows that every σ-section of F is of measure zero and hence F is of measure zero in the product set. The argument applies equally well to the lim inf of the incremental ratio. It follows that $\partial\psi_1(\xi, \sigma)/\partial\xi = \psi_0(\xi, \sigma)$ almost everywhere in $I \times \mathfrak{S}$; hence the equality holds almost everywhere in σ for almost all $\xi \in I$; that is $\partial\psi_1(\xi, \cdot)/\partial\xi = y(\xi)$ for almost all $\xi \in I$. We next apply property (2) to $\psi_1(\xi, \sigma)$ and obtain

$$\psi_2(\xi, \sigma) = \int_\alpha^\xi \psi_1(\tau, \sigma)\, d\tau$$

where now

$$\begin{aligned}\psi_2(\xi, \cdot) &= \int_\alpha^\xi [x^{(n-1)}(\tau) - x^{(n-1)}(\alpha)]\, d\tau \\ &= x^{(n-2)}(\xi) - x^{(n-2)}(\alpha) - (\xi - \alpha)x^{(n-1)}(\alpha).\end{aligned}$$

Again $\psi_2(\xi, \sigma)$ is measurable on $I \times \mathfrak{S}$. Since $\psi_1(\xi, \sigma)$ is absolutely continuous in ξ for each $\sigma \in \mathfrak{S}$, it is clear that $\partial\psi_2(\xi, \sigma)/\partial\xi = \psi_1(\xi, \sigma)$ at all points of $I \times \mathfrak{S}$. Proceeding in this way we finally obtain

$$\psi_n(\xi, \sigma) = \int_\alpha^\xi \psi_{n-1}(\tau, \sigma)\, d\tau$$

where

$$\psi_n(\xi, \cdot) = x(\xi) - \sum_{k=0}^{n-1} \frac{(\xi - \alpha)^k}{k!} x^{(k)}(\alpha).$$

If we now substitute in this formula any realization whatever for the $x^{(k)}(\alpha)$ we arrive at the desired realization of $x(\xi)$, namely,

$$\varphi(\xi, \sigma) = \psi_n(\xi, \sigma) + \sum_{k=0}^{n-1} \frac{(\xi - \alpha)^k}{k!} x^{(k)}(\alpha)(\sigma).$$

REMARK. The above result may be extended in the following manner. In the definition of a space of type L we require that the consequents remain valid for functions $x(\xi)$ merely Bochner integrable (see section 3.7). Again $L_p(\mathfrak{S}, m)$ $(1 \leqq p \leqq \infty)$ can be shown to be of type L. Theorem 3.4.2 is now valid for functions $x(\xi)$ having nth derivatives which are (B)-integrable. The same reasoning applies, *mutatis mutandis*. Related considerations occur in D. C. J. Burgess [1].

3.5. Measurable functions. Let $\mathfrak{S}$ be an abstract set, let $\mathfrak{E}$ be a σ-ring of subsets of $\mathfrak{S}$, and let $m(E)$ defined on $\mathfrak{E}$ be a σ-finite complete measure function (see Halmos [1]). In this section we intend to study the notion of *measurability* for vector-valued functions on $\mathfrak{S}$, relative to the measure function $m(E)$. There are several such notions just as there were several notions of continuity for vector-valued functions. A number of definitions are required at the outset.

DEFINITION 3.5.1. *Let $x(\sigma)$ and $\{x_n(\sigma)\}$ be functions on $\mathfrak{S}$ to $\mathfrak{X}$. The sequence $\{x_n(\sigma)\}$ converges to $x(\sigma)$ in $\mathfrak{S}$*

(1) *almost uniformly if to every $\epsilon > 0$ there is a set $E_\epsilon \in \mathfrak{E}$ with $m(E_\epsilon) < \epsilon$ and*

to every $\delta > 0$ *there is an integer* $n(\delta, \epsilon)$ *such that* $\| x(\sigma) - x_n(\sigma) \| < \delta$ *for* $\sigma \in \mathfrak{S} \ominus E_\epsilon$ *and* $n \geqq n(\delta, \epsilon)$;

(2) *almost everywhere if there exists a null set* $E_0 \in \mathfrak{E}$ *such that* $\lim_{n\to\infty} \| x(\sigma) - x_n(\sigma) \| = 0$ *for each* $\sigma \in \mathfrak{S} \ominus E_0$;

(3) *in measure if for every* $\epsilon > 0$ *the outer measure of the subset of* $\mathfrak{S}$ *where* $\| x(\sigma) - x_n(\sigma) \| > \epsilon$ *tends to zero when* $n \to \infty$.

THEOREM 3.5.1. *The three types of convergence in the preceding definitions are related as follows*: (1) *implies* (2) *and* (3); *if* $\| x(\sigma) - x_n(\sigma) \|$ *is measurable and if* $m(\mathfrak{S}) < \infty$ *then* (2) *implies* (1) *and* (3); (3) *does not imply convergence anywhere. However if* (3) *holds then one can always find a subsequence of* $\{x_n(\sigma)\}$ *which converges almost uniformly to* $x(\sigma)$.

The proof paraphrases its counterpart for the numerically-valued case and is omitted here.

DEFINITION 3.5.2. (1) $x(\sigma)$ *is said to be finitely-valued if it is constant on each of a finite number of disjoint measurable sets* E_j *and equal to* θ *on* $\mathfrak{S} \ominus \bigcup E_j$. (2) *It is a simple function if it is finitely-valued and if the set for which* $\| x(\sigma) \| > 0$ *is of finite measure.* (3) $x(\sigma)$ *is countably-valued if it assumes at most a countable set of values in* $\mathfrak{X}$, *assuming each value different from* θ *on a measurable subset.*

DEFINITION 3.5.3. $x(\sigma)$ *is said to be separably-valued if its range,* $x(\mathfrak{S})$, *is separable. It is almost separably-valued if there exists a null set* $E_0 \in \mathfrak{E}$ *such that* $x(\mathfrak{S} \ominus E_0)$ *is separable.*

DEFINITION 3.5.4. (1) $x(\sigma)$ *is said to be weakly measurable in* $\mathfrak{S}$ *if the numerically-valued functions* $x^*[x(\sigma)]$ *are measurable for each* $x^* \in \mathfrak{X}^*$. (2) $x(\sigma)$ *is strongly measurable if there exists a sequence of countably-valued functions converging almost everywhere in* $\mathfrak{S}$ *to* $x(\sigma)$.

If $m(\mathfrak{S}) < \infty$ then it is easy to see that we may replace "countably-valued" in part (2) of the last definition by "simple".

THEOREM 3.5.2. *If* $x(\sigma)$ *is weakly measurable and if there exists a denumerable determining set* Λ *for* $\mathfrak{X}$, *then the numerically-valued function* $\| x(\sigma) \|$ *is measurable.*

PROOF. Let $\Lambda = \{x_n^*\}$. Then $\| x(\sigma) \| = \sup_n | x_n^*[x(\sigma)] |$. By hypothesis $x_n^*[x(\sigma)]$ is measurable in $\mathfrak{S}$, so that $| x_n^*[x(\sigma)] |$ and $\sup_n | x_n^*[x(\sigma)] |$ have the same property.

If $\mathfrak{X}$ is separable or if $\mathfrak{X}$ is the adjoint space to a separable (B)-space, then by Theorem 2.8.5 there will exist a denumerable determining set Λ for $\mathfrak{X}$. Hence Theorem 3.5.2 is applicable to these two situations. The two notions of measurability for vector-valued functions are connected by the following theorem due to B. J. Pettis [1] (cf. I. Gelfand [2]).

THEOREM 3.5.3. *A vector-valued function is strongly measurable if and only if it is weakly measurable and almost separably-valued.*

PROOF. We start with the necessity. If $x(\sigma)$ is strongly measurable, then there exists a null set $E_0 \in \mathfrak{E}$ and a sequence $\{x_n(\sigma)\}$ of countably-valued functions such that $\| x(\sigma) - x_n(\sigma) \| \to 0$ as $n \to \infty$ for each $\sigma \in \mathfrak{S} \ominus E_0$. If $x^* \in \mathfrak{X}^*$ we have *a fortiori* that $| x^*[x(\sigma) - x_n(\sigma)] | \to 0$ for each $\sigma \in \mathfrak{S} \ominus E_0$. It is clear that the numerically-valued function $x^*[x_n(\sigma)]$ is measurable. Since $x^*[x(\sigma)]$ is the limit almost everywhere of a sequence of measurable functions, it is also measurable, and hence $x(\sigma)$ is weakly measurable. The values taken on by the functions $\{x_n(\sigma)\}$ form a countable set and the least closed linear subspace containing this set is consequently separable. It is clear that $x(\mathfrak{S} \ominus E_0)$ is contained in this separable subspace and hence is separable. Thus $x(\sigma)$ is almost separably-valued.

We now prove the converse statement. Without loss of generality we may assume that $x(\sigma)$ is actually separably-valued. Replacing $\mathfrak{X}$ by the least closed linear subspace containing $x(\mathfrak{S})$, it is clear that we lose no generality in supposing that $\mathfrak{X}$ itself is separable. Thus Theorem 3.5.2 applies and $\| x(\sigma) \|$ is measurable. Let $S_0 = [\sigma, \| x(\sigma) \| > 0]$. Then $S_0 \in \mathfrak{E}$ and $x(\sigma) - x_0$ is weakly measurable on S_0 for each $x_0 \in \mathfrak{X}$. Again by Theorem 3.5.2 $\| x(\sigma) - x_0 \|$ is measurable in S_0. Now $x(\mathfrak{S})$ is separable so that there will exist a sequence $\{x_n\}$ dense in $x(\mathfrak{S})$. Given $\epsilon > 0$ we now define

$$E_n = [\sigma; \| x(\sigma) - x_n \| < \epsilon, \sigma \in S_0].$$

Then $E_n \in \mathfrak{E}$ and $\bigcup E_n = S_0$ since $\{x_n\}$ is dense in $x(\mathfrak{S})$. Setting $F_n = E_n \ominus \bigcup_{k<n} E_k$ we see that the sets F_n are measurable, disjoint, and that $\bigcup F_n = S_0$. We now define

$$\begin{aligned} x_\epsilon(\sigma) &= x_n \quad \text{for} \quad \sigma \in F_n, \qquad n = 1, 2, \cdots, \\ &= \theta \quad \text{on} \quad \mathfrak{S} \ominus S_0. \end{aligned}$$

It is clear that $x_\epsilon(\sigma)$ is a countably-valued function and that $\| x(\sigma) - x_\epsilon(\sigma) \| < \epsilon$ for all $\sigma \in \mathfrak{S}$. Hence $x(\sigma)$ is the uniform limit of countably-valued functions and consequently strongly measurable. This completes the proof.

We have actually proved a somewhat stronger result than the statement of the theorem would indicate.

COROLLARY 1. *A function $x(\sigma)$ is strongly measurable if and only if it is the uniform limit almost everywhere of a sequence of countably-valued functions.*

COROLLARY 2. *If $\mathfrak{X}$ is separable, then strong and weak measurability are equivalent notions.*

We note that a vector-valued function $x(\xi)$ defined on the interval $[\alpha, \beta]$ and weakly continuous, say on the right, is strongly measurable. It is clearly weakly measurable. Further by Theorem 2.9.2, $x([\alpha, \beta])$ is contained in the least closed linear subspace spanned by $[x(\xi);\ \xi$ rational and in $[\alpha, \beta]]$. Thus $x(\xi)$ is also separably-valued and therefore strongly measurable.

Strongly measurable vector-valued functions have properties analogous to those of measurable numerically-valued functions.

THEOREM 3.5.4. (1) *If $x(\sigma)$ and $y(\sigma)$ are strongly measurable in $\mathfrak{S}$ and γ_1, γ_2 are constants, then $\gamma_1 x(\sigma) + \gamma_2 y(\sigma)$ is strongly measurable.* (2) *If $f(\sigma)$ is a finite numerically-valued function which is measurable, then $f(\sigma)x(\sigma)$ is strongly measurable if $x(\sigma)$ has this property.* (3) *If $x(\sigma)$ is the limit almost everywhere of a sequence of strongly measurable functions, then $x(\sigma)$ is strongly measurable.* (4) *The same conclusion is valid if in* (3) *the word "limit" (that is, strong limit) is replaced by "weak limit".* (5) *The conclusion is also valid if "limit almost everywhere" is replaced by "limit in measure".*

PROOF. (1) follows directly from Definition 3.5.4. (2) follows in the same manner if one keeps in mind that $f(\sigma)$ is the limit almost everywhere of countably-valued numerical functions. (3) If $x(\sigma)$ is the limit almost everywhere of the sequence $\{x_n(\sigma)\}$ of strongly measurable functions, then aside from a set E_0 of measure zero, $x_n(\sigma) \to x(\sigma)$. By Theorem 3.5.3 there exist sets E_n of measure zero, such that $x_n(\mathfrak{S} \ominus E_n)$ is separable. Clearly $F = \bigcup E_k$ is of measure zero. Now the least closed linear subspace $\mathfrak{X}_0$ containing $[x_n(\mathfrak{S} \ominus F); n = 1, 2, \cdots]$ is separable and contains $x(\mathfrak{S} \ominus F)$. Further $x^*[x_n(\sigma)]$ is measurable and $x^*[x_n(\sigma)] \to x^*[x(\sigma)]$ on $\mathfrak{S} \ominus F$ for each $x^* \in \mathfrak{X}^*$. Thus $x(\sigma)$ is weakly measurable and almost separably-valued. (4) The argument of (3) is equally valid for (4). In this case $x(\mathfrak{S} \ominus F)$ belongs to $\mathfrak{X}_0$ because of Theorem 2.9.2. (5) If $x_n(\sigma)$ converges in measure to $x(\sigma)$, then there exists a subsequence which converges strongly to $x(\sigma)$ almost everywhere. The result then follows from (3).

Ordinarily $x(\sigma)y(\sigma)$ does not have a meaning in (B)-spaces, but if $\mathfrak{X}$ is a Banach algebra and not merely a Banach space, then the product is well defined and is strongly measurable whenever the factors have this property.

The above considerations also apply to the case in which $x(\sigma)$ is an operator-valued function. However, in this case a new set of conventions is more appropriate for the applications.

DEFINITION 3.5.5. (1) *The operator-valued function $U(\sigma)$ is said to be uniformly measurable in $\mathfrak{S}$ if there exists a sequence of countably-valued operator functions in $\mathfrak{E}(\mathfrak{X}, \mathfrak{Y})$ converging almost everywhere to $U(\sigma)$ in the uniform operator topology.* (2) *$U(\sigma)$ is strongly measurable in $\mathfrak{S}$ if the vector-valued function $U(\sigma)[x]$ is strongly measurable in the sense of Definition* 3.5.4 (2) *for all $x \in \mathfrak{X}$.* (3) *$U(\sigma)$ is weakly measurable in $\mathfrak{S}$ if $y^*\{U(\sigma)[x]\}$ is measurable for all $x \in \mathfrak{X}$, $y^* \in \mathfrak{Y}^*$.*

It is clear that uniform measurability of the operator-valued function $U(\sigma)$ is the same as strong measurability of $U(\sigma)$ considered as a vector-valued function in the (B)-space $\mathfrak{E}(\mathfrak{X}, \mathfrak{Y})$. The connection between the three different types of measurability for operator functions is given by the following theorem due to N. Dunford **[4]**.

THEOREM 3.5.5. *A necessary and sufficient condition that $U(\sigma)$ be* (1) *strongly measurable is that $U(\sigma)$ be weakly measurable and that $U(\sigma)[x]$ be almost separably-*

valued in $\mathfrak{Y}$ *for each* $x \in \mathfrak{X}$; (2) *uniformly measurable is that* $U(\sigma)$ *be weakly measurable and almost separably-valued in* $\mathfrak{E}(\mathfrak{X}, \mathfrak{Y})$.

PROOF. Part (1) is an immediate consequence of Theorem 3.5.3 and Definition 3.5.5. The proof of part (2) follows the same lines as the proof of Theorem 3.5.3 and it is only in the proof of the measurability of $\| U(\sigma) \|$ that any modifications are necessary. It should be noted first that if $U(\sigma)$ is almost separably-valued in $\mathfrak{E}(\mathfrak{X}, \mathfrak{Y})$, then $U(\sigma)[x]$ is almost separably-valued in $\mathfrak{Y}$ for each x. Thus the first conclusion from the assumptions of part (2) is that $U(\sigma)$ is strongly measurable. In order to prove that $\| U(\sigma) \|$ is measurable, we argue as follows. Without restricting the generality we may assume that $U(\mathfrak{S})$ is separable. There is then a countable set $\{U_n\} \subset \mathfrak{E}(\mathfrak{X}, \mathfrak{Y})$ and dense in $U(\mathfrak{S})$. To each n we can find a sequence $\{x_{mn}\}$ such that (i) $\| x_{mn} \| = 1$ and (ii) $\| U_n(x_{mn}) \| \geqq \| U_n \| - 1/m$. All the numerically-valued functions $\| U(\sigma)[x_{mn}] \|$ are measurable since $U(\sigma)$ is strongly measurable. Hence $F(\sigma) \equiv \sup_{m,n} \| U(\sigma)[x_{mn}] \|$ is also measurable. It is clear that $F(\sigma) \leqq \| U(\sigma) \|$. Actually equality holds. For given $\sigma \in \mathfrak{S}$ and m there exists an n depending on σ and m such that $\| U(\sigma) - U_n \| \leqq 1/m$. Hence

$$\begin{aligned} F(\sigma) &\geqq \| U(\sigma)[x_{mn}] \| \geqq \| U_n[x_{mn}] \| - \| \{U(\sigma) - U_n\}[x_{mn}] \| \\ &\geqq \| U_n \| - 2/m \geqq \| U(\sigma) \| - 3/m \end{aligned}$$

for every m. Hence $F(\sigma) = \| U(\sigma) \|$ for all σ and $\| U(\sigma) \|$ is measurable. The proof is now completed as in Theorem 3.5.3.

3.6. Countably additive set functions. The study of countably additive set functions forms an important part of measure theory. In this section we shall consider countably additive set functions defined on a σ-ring, $\mathfrak{E}$, to a Banach space $\mathfrak{X}$. Again we shall be interested in the relation between the weak and strong generalizations of familiar notions.

DEFINITION 3.6.1. *A set function* $x(E)$ *on* $\mathfrak{E}$ *to* $\mathfrak{X}$ *is said to be strongly (weakly) countably additive if for every denumerable sequence* $\{E_n\}$ *of disjoint subsets in* $\mathfrak{E}$,

$$x(\cup_n E_n) = \sum_{n=1}^{\infty} x(E_n), \tag{3.6.1}$$

where the sum converges in the norm (weak) topology.

THEOREM 3.6.1. *If* $x(E)$ *is a weakly countably additive set function on* $\mathfrak{E}$ *to* $\mathfrak{X}$, *then* $\| x(E) \|$ *is bounded on* $\mathfrak{E}$.

PROOF. The theorem is known to be true in the numerically-valued case. Hence $| x^*[x(E)] |$ is bounded on $\mathfrak{E}$ for each $x^* \in \mathfrak{X}^*$. The result now follows from the uniform boundedness theorem.

THEOREM 3.6.2. *If* $x(E)$ *is a weakly countably additive set function on* $\mathfrak{E}$ *to* $\mathfrak{X}$, *then* $x(E)$ *is necessarily strongly countably additive on* $\mathfrak{E}$ *to* $\mathfrak{X}$.

PROOF. Let $\{E_n\}$ be a sequence of disjoint sets in $\mathfrak{E}$ and let $\pi = (n_1, n_2, \ldots)$ be an arbitrary subsequence of integers. Then for a weakly countably additive set function $x(E)$, it is clear that

$$x^*\left[x\left(\bigcup_{n\in\pi} E_n\right)\right] = \sum_{n\in\pi} x^*[x(E_n)]$$

for all $x^* \in \mathfrak{X}^*$. Thus $\sum_n x(E_n)$ is weakly unconditionally convergent and hence by Theorem 3.2.3 the sum is strongly convergent.

DEFINITION 3.6.2. *A set function $x(E)$ on $\mathfrak{E}$ to $\mathfrak{X}$ is said to be* (1) *absolutely continuous relative to a σ-finite measure, $m(E)$, if for each $\epsilon > 0$ there exists a $\delta > 0$ such that $\| x(E) \| < \epsilon$ whenever $m(E) < \delta$, and* (2) *strongly absolutely continuous relative to $m(E)$ if for each $\epsilon > 0$ there exists a $\delta > 0$ such that $\sum_n \| x(E_n) \| < \epsilon$ for each sequence of disjoint sets $\{E_n\} \subset \mathfrak{E}$ such that $\sum_n m(E_n) < \delta$.*

THEOREM 3.6.3. *If $x(E)$ is a weakly countably additive set function on $\mathfrak{E}$ to $\mathfrak{X}$ and if $m(E) = 0$ implies $x(E) = \theta$, then $x(E)$ is absolutely continuous relative to $m(E)$.*

PROOF. We note that the theorem is known to be true in the numerically-valued case and hence for all set functions of the form $x^*[x(E)]$ where $x^* \in \mathfrak{X}^*$. Now suppose that the theorem were false. Then there would be an $\epsilon > 0$ and a sequence of sets $\{E_k\}$ belonging to $\mathfrak{E}$ such that $\| x(E_k) \| > 2\epsilon$ and $m(E_k) < 2^{-k}$. Clearly $\lim_{n\to\infty} m(\bigcup_{k\geqq n} E_k) = 0$. According to Theorem 2.7.4 we can find an $x_1^* \in \mathfrak{X}^*$, $\| x_1^* \| = 1$, for which $| x_1^*[x(E_1)] | > 2\epsilon$. Since $x_1^*[x(E)]$ is absolutely continuous relative to $m(E)$ there exists an n_1 such that $| x_1^*[x(E)] | < \epsilon$ for all $E \subset \bigcup_{k\geqq n_1} E_k$. Setting $F_1 = E_1 \ominus \bigcup_{k\geqq n_1} E_k$ we see that $\| x(F_1) \| \geqq | x_1^*[x(F_1)] | > \epsilon$ and that F_1 is disjoint from E_k for $k \geqq n_1$. We now treat E_{n_1} in a similar fashion. This leads to an $n_2 > n_1$ and a set $F_2 = E_{n_1} \ominus \bigcup_{k\geqq n_2} E_k$ such that $\| x(F_2) \| > \epsilon$; F_2 is clearly disjoint from F_1 and from E_k for $k \geqq n_2$. In this way we obtain a sequence of disjoint sets $\{F_k\}$ in $\mathfrak{E}$ with $\| x(F_k) \| > \epsilon$. However by Theorem 3.6.2, $\sum_k x(F_k)$ converges in norm to $x(\bigcup_k F_k)$ and this implies that $\| x(F_k) \| \to 0$. The assumption that the theorem is false has thus led to a contradiction.

3.7. Lebesgue integrals. The Lebesgue integral has been generalized for vector-valued functions in two distinct ways. The first such generalization was developed by Bochner [2] and may be described as follows. One begins with the simple functions, identifying any pair which differ only on a null set. The classes of such functions then form a linear normed space with norm $\| x(\cdot) \| = \int \| x(\sigma) \| \, dm$. Defining the integral $\int x(\sigma) \, dm$ in the obvious way it is clear that $\| \int x(\sigma) \, dm \| \leqq \| x(\cdot) \|$. If one now completes this space one can extend the integral to all Cauchy sequences and obtain the Bochner integral (see Dunford [1]). For the other type of generalization one begins with a given topology on $\mathfrak{X}$ and defines the integral relative to this topology as follows. Let $\{E_n\}$ be a denumerable subdivision of the set $S_0 \equiv [\sigma; \| x(\sigma) \| > 0]$. A meaning can be assigned to the sum $\sum x(\sigma_n) m(E_n)$ (here $\sigma_n \in E_n$) and if these sums converge in the given topology with further refinements of the subdivision, then this limit is defined to be the value of the integral $\int x(\sigma)\, dm$ (see Phillips [1]). We shall give an example of each of these types. Since the Bochner integral is most suitable for our purposes, we shall devote most attention to it. However we shall begin with a brief discussion of the Pettis integral which can be defined as an integral of the second type relative to the weak topology on $\mathfrak{X}$. We have found it convenient to use more direct definitions for both the Bochner and the Pettis integrals than the ones indicated above.

Our first result was found independently by I. Gelfand [2] and N. Dunford [3].

THEOREM 3.7.1. *If $x(\sigma)$ is weakly measurable and if $x^*[x(\sigma)] \in L(\mathfrak{S}, m)$ for each $x^* \in \mathfrak{X}^*$, then there exists an $x^{**} \in \mathfrak{X}^{**}$ such that*

$$x^{**}(x^*) = \int_{\mathfrak{S}} x^*[x(\sigma)]\, dm$$

for all $x^* \in \mathfrak{X}^*$.

PROOF. Set

$$F(x^*) = \int_{\mathfrak{S}} x^*[x(\sigma)]\, dm.$$

It is clear that F is defined and linear on $\mathfrak{X}^*$. It remains only to show that F is bounded. To this end we define the linear operator W on $\mathfrak{X}^*$ to $L(\mathfrak{S}, m)$ as $W(x^*) = x^*[x(\sigma)]$. It is easy to verify that W is closed and hence by Theorem 2.12.3 W is also bounded. Therefore

$$| F(x^*) | \leqq \int_{\mathfrak{S}} | x^*[x(\sigma)] | \, dm \leqq \| W \| \| x^* \| ,$$

which is the desired result.

As a consequence of the above theorem we may set $x^{**} = \int_{\mathfrak{S}} x(\sigma)\, dm$. In general x^{**} can not be replaced by an element of $\mathfrak{X}$; when such a replacement can be made the integral is called a Pettis integral. More precisely

DEFINITION 3.7.1. *The function $x(\sigma)$ on $\mathfrak{S}$ to $\mathfrak{X}$ is integrable (Pettis) if and only if there is an element x_E of $\mathfrak{X}$ corresponding to each $E \in \mathfrak{E}$ such that*

$$x^*(x_E) = \int_E x^*[x(\sigma)]\, dm$$

for all $x^* \in \mathfrak{X}^*$, where the integral on the right is supposed to exist in the sense of Lebesgue. By definition

$$(\mathrm{P}) \int_E x(\sigma)\, dm = x_E .$$

A useful characterization of Pettis integrable functions is not known. However if $\mathfrak{X}$ is reflexive then it is clear from Theorem 3.7.1 that $x(\sigma)$ is Pettis integrable if and only if $x^*[x(\sigma)] \in L(\mathfrak{S}, m)$ for each $x^* \in \mathfrak{X}^*$.

The following are some immediate consequences of the definition. (1) *An integrable function is weakly measurable (but not necessarily strongly measurable).* (2) *The integral is uniquely defined.* (3) *If $x_1(\sigma)$ and $x_2(\sigma)$ are Pettis integrable, then so is $\gamma_1 x_1(\sigma) + \gamma_2 x_2(\sigma)$ and $(\gamma_1 x_1 + \gamma_2 x_2)_E = \gamma_1 x_{1E} + \gamma_2 x_{2E}$.* (4) *A simple function in the sense of Definition* 3.5.2 (2) *is integrable and*

$$(\mathrm{P}) \int_E x(\sigma)\, dm = \sum_k x_k m(E_k \cap E),$$

where $x(\sigma) = x_k$ on E_k. (5) *If $\mathfrak{X}$ is the space of complex numbers, the definition coincides with the Lebesgue integral.*

THEOREM 3.7.2. *If $x(\sigma)$ is Pettis integrable, the set function $x_E = (\mathrm{P})\int_E x(\sigma)\, dm$ is strongly countably additive and absolutely continuous relative to $m(E)$.*

PROOF. Let $\{E_n\}$ be a denumerable sequence of disjoint sets in $\mathfrak{E}$. Then

$$x^*(x_{\cup E_n}) = \int_{\cup E_n} x^*[x(\sigma)]\, dm = \sum_{n=1}^{\infty} \int_{E_n} x^*[x(\sigma)]\, dm = \sum_{n=1}^{\infty} x^*(x_{E_n}).$$

Thus x_E is weakly countably additive and hence by Theorem 3.6.2 x_E is strongly countably additive. It now follows from Theorem 3.6.3 that x_E is also absolutely continuous relative to $m(E)$.

A fundamental property of the integral is contained in

THEOREM 3.7.3. *If T is a linear bounded transformation on the (B)-space $\mathfrak{X}$ to the (B)-space $\mathfrak{Y}$ with the same scalar field and if $x(\sigma) \in \mathfrak{X}$ is integrable (Pettis), so is $T[x(\sigma)]$ and*

$$(\mathrm{P}) \int_E T[x(\sigma)]\, dm = T(x_E). \tag{3.7.1}$$

PROOF. This follows from the properties of the adjoint transformation T^*. It is required to show that $y^*\{T[x(\sigma)]\}$ is integrable (Lebesgue) for every $y^* \in \mathfrak{Y}^*$ and that the value of the integral over E is $y^*[T(x_E)]$. Now to a given $y^* \in \mathfrak{Y}^*$ there corresponds a unique $x^* \in \mathfrak{X}^*$, $x^* = T^*(y^*)$, defined by $y^*[T(x)] = x^*(x)$, and

$$\int_E y^*\{T[x(\sigma)]\}\, dm = \int_E x^*[x(\sigma)]\, dm = x^*(x_E) = y^*[T(x_E)]$$

for every $y^* \in \mathfrak{Y}^*$ as asserted.

We next introduce the Bochner integral which is much more easily applied than the Pettis integral although it lacks somewhat the generality of the Pettis integral.

DEFINITION 3.7.2. *A countably-valued function $x(\sigma)$ on $\mathfrak{S}$ to $\mathfrak{X}$ is integrable (Bochner) if and only if $\| x(\sigma) \|$ is integrable (Lebesgue). By definition*

$$(B) \int_E x(\sigma)\, dm = \sum_{k=1}^{\infty} x_k m(E_k \cap E)$$

where $x(\sigma) = x_k$ *on* $E_k \in \mathfrak{E}$ $(k = 1, 2, \cdots)$.

The integral is well defined for all $E \in \mathfrak{E}$ and for $\mathfrak{S}$ itself. This follows from the fact that the series is absolutely convergent since

$$\sum_{k=1}^{\infty} \| x_k \| m(E_k \cap E) = \int_E \| x(\sigma) \| \, dm.$$

Consequently

$$\left\| (B) \int_E x(\sigma)\, dm \right\| \leqq \int_E \| x(\sigma) \| \, dm \tag{3.7.2}$$

for countably-valued functions. Further

$$x^*\left[\int_E x(\sigma)\, dm\right] = \sum_{k=1}^{\infty} x^*(x_k) m(E_k \cap E) = \int_E x^*[x(\sigma)]\, dm$$

for every $x^* \in \mathfrak{X}^*$, the series being absolutely convergent. It follows that the (B)- and (P)-integrals of such functions coincide.

Definition 3.7.3. *A function* $x(\sigma)$ *on* $\mathfrak{S}$ *to* $\mathfrak{X}$ *is integrable (Bochner) if and only if there exists a sequence of countably-valued integrable functions* $\{x_n(\sigma)\}$ *converging almost everywhere to* $x(\sigma)$ *and such that*

$$\lim_{n\to\infty} \int_{\mathfrak{S}} \| x(\sigma) - x_n(\sigma) \| \, dm = 0. \tag{3.7.3}$$

By definition

$$(B) \int_E x(\sigma)\, dm = \lim_{n\to\infty} (B) \int_E x_n(\sigma)\, dm \tag{3.7.4}$$

for each $E \in \mathfrak{E}$ *and* $E = \mathfrak{S}$.

We have to verify that (3.7.3) is meaningful and that the limit in (3.7.4) exists and is unique. In the first place $x(\sigma)$ is by definition strongly measurable so that $\| x(\sigma) - x_n(\sigma) \|$ is measurable and consequently (3.7.3) is meaningful. The existence of the limit in (3.7.4) follows from the fact that the integrals in the right members form a Cauchy sequence in $\mathfrak{X}$, as can be seen from

$$\left\| \int_E x_n(\sigma)\, dm - \int_E x_m(\sigma)\, dm \right\| = \left\| \int_E [x_n(\sigma) - x_m(\sigma)]\, dm \right\|$$

$$\leqq \int_E \| x_n(\sigma) - x_m(\sigma) \| \, dm \leqq \int_{\mathfrak{S}} \| x_n(\sigma) - x(\sigma) \| \, dm + \int_{\mathfrak{S}} \| x(\sigma) - x_m(\sigma) \| \, dm.$$

Finally it is clear that the limit is independent of the defining sequence since any two such sequences can be combined into a single sequence by alternating the terms between the two original sequences.

Suppose that $x(\sigma)$ is (B)-integrable and that the Bochner integral of $x(\sigma)$ is defined by the sequence $\{x_n(\sigma)\}$ of countably-valued integrable functions. Then $\{x^*[x_n(\sigma)]\}$ converges almost everywhere and in the mean of order one to $x^*[x(\sigma)]$ for each $x^* \in \mathfrak{X}^*$. Thus

$$x^*\left[\int_E x_n(\sigma)\, dm\right] = \int_E x^*[x_n(\sigma)]\, dm \to \int_E x^*[x(\sigma)]\, dm.$$

On the other hand, since x^* is a functional continuous in the norm topology, we have

$$x^*\left[\int_E x_n(\sigma)\, dm\right] \to x^*\left[\int_E x(\sigma)\, dm\right].$$

Hence

$$x^*\left[\int_E x(\sigma)\, dm\right] = \int_E x^*[x(\sigma)]\, dm \tag{3.7.5}$$

for all $x^* \in \mathfrak{X}^*$. In other words, *every (B)-integrable function is also (P)-integrable and the integrals have the same value.* We may therefore conclude from Theorem 3.7.2 that the set function $x_E = (B) \int_E x(\sigma)\, dm$ is strongly completely additive and absolutely continuous relative to $m(E)$.

The great virtue of the (B)-integral is that the class of (B)-integrable functions is easily characterized. This is a consequence of the following theorem.

THEOREM 3.7.4. *A necessary and sufficient condition that $x(\sigma)$ on $\mathfrak{S}$ to $\mathfrak{X}$ be integrable (Bochner) is that $x(\sigma)$ be strongly measurable and that $\int_{\mathfrak{S}} \| x(\sigma) \|\, dm < \infty$.*

PROOF. If $x(\sigma)$ is (B)-integrable then it is necessarily strongly measurable and *a fortiori* $\| x(\sigma) \|$ is measurable. Finally for an approximating sequence of countably-valued integrable functions, we have

$$\int_{\mathfrak{S}} \| x(\sigma) \|\, dm \leqq \int_{\mathfrak{S}} \| x(\sigma) - x_n(\sigma) \|\, dm + \int_{\mathfrak{S}} \| x_n(\sigma) \|\, dm < \infty$$

for each integer n.

Conversely let $x(\sigma)$ be strongly measurable and let $\| x(\sigma) \|$ be summable. Set $S_0 \equiv [\sigma;\, \| x(\sigma) \| > 0]$. Then $S_0 \in \mathfrak{E}$ and hence there exists a subdivision of S_0 into disjoint sets $\{S_n\} \subset \mathfrak{E}$ such that $S_0 = \bigcup_n S_n$ and $0 < m(S_n) < \infty$ for each $n \geqq 1$. Given $\epsilon > 0$, then according to Corollary 1 to Theorem 3.5.3 we can find a countably-valued function $x_{\epsilon,n}(\sigma)$ such that

$$\| x_{\epsilon,n}(\sigma) - x(\sigma) \| < 2^{-n}\epsilon/m(S_n) \quad \text{on} \quad S_n. \tag{3.7.6}$$

Define

$$x_\epsilon(\sigma) = x_{\epsilon,n}(\sigma) \quad \text{for} \quad \sigma \in S_n \qquad (n = 1, 2, \cdots)$$
$$= \theta \qquad \text{for} \quad \sigma \in \mathfrak{S} \ominus S_0 .$$

Then clearly $x_\epsilon(\sigma)$ is countably-valued and

$$\int_{\mathfrak{S}} \| x(\sigma) - x_\epsilon(\sigma) \| \, dm < \sum_{n=1}^{\infty} 2^{-n}[\epsilon/m(S_n)]m(S_n) = \epsilon.$$

Further

$$\int_{\mathfrak{S}} \| x_\epsilon(\sigma) \| \, dm \leqq \int_{\mathfrak{S}} \| x_\epsilon(\sigma) - x(\sigma) \| \, dm + \int_{\mathfrak{S}} \| x(\sigma) \| \, dm < \infty.$$

Thus corresponding to a sequence of ϵ_n's converging to zero, we can find a sequence of countably-valued integrable functions approximating $x(\sigma)$ in the sense of Definition 3.7.3. This implies the integrability of $x(\sigma)$.

Our construction of $x_\epsilon(\sigma)$ in the above proof actually leads to a somewhat stronger result.

COROLLARY. *Suppose $x(\sigma)$ is (B)-integrable and set $S_0 \equiv [\sigma; \| x(\sigma) \| > 0]$. Then given $\epsilon > 0$, there exists a subdivision of S_0 into disjoint sets $\{E_k\} \subset \mathfrak{E}$ such that for arbitrary $\sigma_k \in E_k$, the function*

$$x_\epsilon(\sigma) = x(\sigma_k) \quad \textit{for} \quad \sigma \in E_k \qquad (k = 1, 2, \cdots),$$
$$= \theta \qquad \textit{for} \quad \sigma \in \mathfrak{S} \ominus S_0$$

is countably-valued, integrable, and satisfies the relation

$$\int_{\mathfrak{S}} \| x(\sigma) - x_\epsilon(\sigma) \| \, dm < \epsilon.$$

Furthermore this remains valid for all refinements of the above subdivision.

PROOF. Let the $\{E_k\}$ be the totality of sets of constancy for the functions $\{x_{\epsilon,n}\}$ in the above proof. Then redefining $x_{\epsilon,n}(\sigma)$ on the E_k's in the manner suggested by the corollary can only increase the right member of (3.7.6) by a factor of two.

We denote the class of functions on $\mathfrak{S}$ to $\mathfrak{X}$, which are Bochner integrable relative to $m(E)$, by $B(\mathfrak{S}; \mathfrak{X}; m)$. As we shall see, $B(\mathfrak{S}; \mathfrak{X}; m)$ becomes a Banach space if the norm of the element $x(\cdot)$ is defined to be

$$\| x(\cdot) \| = \int_{\mathfrak{S}} \| x(\sigma) \| \, dm. \tag{3.7.7}$$

THEOREM 3.7.5. *If $x_1(\sigma)$ and $x_2(\sigma) \in B(\mathfrak{S}; \mathfrak{X}; m)$ and γ_1, γ_2 are constants, then $\gamma_1 x_1(\sigma) + \gamma_2 x_2(\sigma) \in B(\mathfrak{S}; \mathfrak{X}; m)$ and*

$$\int_E [\gamma_1 x_1(\sigma) + \gamma_2 x_2(\sigma)] \, dm = \gamma_1 \int_E x_1(\sigma) \, dm + \gamma_2 \int_E x_2(\sigma) \, dm.$$

PROOF. The theorem is obviously true for countably-valued integrable functions and follows for the general case by a limiting process.

We prove similarly

THEOREM 3.7.6. *If* $x(\sigma) \in B(\mathfrak{S}; \mathfrak{X}; m)$, *then*

$$\left\| \int_E x(\sigma)\, dm \right\| \leqq \int_E \| x(\sigma) \|\, dm. \tag{3.7.8}$$

THEOREM 3.7.7. *If* $x_n(\sigma) \in B(\mathfrak{S}; \mathfrak{X}; m)$ *for all* n *and*

$$\lim_{m,n\to\infty} \int_{\mathfrak{S}} \| x_m(\sigma) - x_n(\sigma) \|\, dm = 0,$$

then there exists an element $x(\sigma) \in B(\mathfrak{S}; \mathfrak{X}; m)$ *such that*

$$\lim_{n\to\infty} \int_{\mathfrak{S}} \| x(\sigma) - x_n(\sigma) \|\, dm = 0. \tag{3.7.9}$$

If $y(\sigma)$ *has the same property, then* $x(\sigma) = y(\sigma)$ *almost everywhere. Finally*

$$\lim_{n\to\infty} \int_E x_n(\sigma)\, dm = \int_E x(\sigma)\, dm. \tag{3.7.10}$$

PROOF. Brief indications will suffice since the argument closely follows the classical proof. We select a subsequence $\{x_{n_j}(\sigma)\}$ subject to the condition

$$\int_{\mathfrak{S}} \| x_{n_j}(\sigma) - x_n(\sigma) \|\, dm < 2^{-j} \qquad \text{for } n > n_j.$$

The series

$$x_{n_1}(\sigma) + \sum_{j=2}^{\infty} [x_{n_j}(\sigma) - x_{n_{j-1}}(\sigma)]$$

converges for almost all σ since the integral of the sum of the norms converges. The sum $x(\sigma)$ is strongly measurable by Theorem 3.5.4 (3) and $\| x(\sigma) \|$ is integrable. Hence $x(\sigma) \in B(\mathfrak{S}; \mathfrak{X}; m)$. For fixed n the function $\| x_{n_j}(\sigma) - x_n(\sigma) \|$ is dominated by a fixed integrable function for all j and $\| x_{n_j}(\sigma) - x_n(\sigma)\| \to \| x(\sigma) - x_n(\sigma) \|$ for almost all σ as $j \to \infty$. Hence

$$\int_{\mathfrak{S}} \| x(\sigma) - x_n(\sigma) \|\, dm \leqq 2^{-j} \qquad \text{for } n > n_j.$$

The uniqueness almost everywhere of the limit is proved by the usual argument. Finally (3.7.10) follows directly from (3.7.8) and (3.7.9).

This result establishes the completeness of $B(\mathfrak{S}; \mathfrak{X}; m)$.

THEOREM 3.7.8. *The set of functions* $B(\mathfrak{S}; \mathfrak{X}; m)$ *becomes a Banach space if we identify functions which differ only on sets of measure zero.*

The classical theorem of Lebesgue on passage to the limit under the sign of integration holds for (B)-integrals.

THEOREM 3.7.9. *If* $\{x_n(\sigma)\} \subset B(\mathfrak{S}; \mathfrak{X}; m)$ *converges almost everywhere to a limit function* $x(\sigma)$ *and if there exists a fixed function* $F(\sigma) \in L(\mathfrak{S}; m)$ *such that* $\| x_n(\sigma) \| \leqq F(\sigma)$ *for all* n *and* σ, *then* $x(\sigma) \in B(\mathfrak{S}; \mathfrak{X}; m)$ *and*

$$\lim_{n\to\infty} \int_E x_n(\sigma)\, dm = \int_E x(\sigma)\, dm.$$

The proof may be left to the reader. We see in particular that the conclusion is valid if $x_n(\sigma)$ converges boundedly to $x(\sigma)$ and $m(\mathfrak{S}) < \infty$.

For the (B)-integral we can improve upon the results of Theorems 3.7.2 and 3.7.3.

THEOREM 3.7.10. *Let* $\{E_n\}$ *be a denumerable sequence of disjoint sets in* $\mathfrak{E}$. *Then for* $x(\sigma) \in B(\mathfrak{S}; \mathfrak{X}; m)$

$$\int_{\cup E_n} x(\sigma)\, dm = \sum_{n=1}^{\infty} \int_{E_n} x(\sigma)\, dm,$$

where the sum on the right is absolutely convergent.

PROOF. Set $x_E = \int_E x(\sigma)\, dm$. Then as in the proof of Theorem 3.7.2, we see that $x^*[x_{\cup E_n}] = \sum_{n=1}^{\infty} x^*(x_{E_n})$ for each $x^* \in \mathfrak{X}^*$. However this sum is also absolutely convergent as is readily seen from (3.7.8). Hence $x^*[x_{\cup E_n}] = x^*[\sum_{n=1}^{\infty} x_{E_n}]$ and this implies the desired result.

THEOREM 3.7.11. *Let* $x(\sigma) \in B(\mathfrak{S}; \mathfrak{X}; m)$. *Then the set function* $x_E = \int_E x(\sigma)\, dm$ *is strongly absolutely continuous.*

PROOF. This result follows from (3.7.8) and the absolute continuity of the integral $\int_E \| x(\sigma) \|\, dm$.

Generalizing Theorem 3.3.2, E. Hille [19] has proved

THEOREM 3.7.12. *Let* T *be a closed linear transformation on* $\mathfrak{X}$ *to* $\mathfrak{Y}$. *If* $x(\sigma) \in B(\mathfrak{S}; \mathfrak{X}; m)$ *and* $T[x(\sigma)] \in B(\mathfrak{S}; \mathfrak{Y}; m)$, *then*

$$T\left[\int_E x(\sigma)\, dm\right] = \int_E T[x(\sigma)]\, dm \tag{3.7.11}$$

for all $E \in \mathfrak{E}$ *and* $E = \mathfrak{S}$.

PROOF. We avail ourselves of the Corollary to Theorem 3.7.4 and obtain two subdivisions of $S_0 \equiv [\sigma; \| x(\sigma) \| > 0]$, one furnishing an ϵ-approximation for $x(\sigma)$ and the other an ϵ-approximation for $T[x(\sigma)]$. Let $\{E_n\}$ be a common refinement of these two subdivisions and let $\sigma_n \in E_n$. We then set

$$\begin{aligned} x_\epsilon(\sigma) &= x(\sigma_n) \quad \text{for} \quad \sigma \in E_n \qquad (n = 1, 2, \cdots), \\ &= \theta \quad \text{for} \quad \sigma \in \mathfrak{S} \ominus S_0 . \end{aligned}$$

Then $\int_{\mathfrak{S}} \| x(\sigma) - x_\epsilon(\sigma) \| dm < \epsilon$ and $\int_{\mathfrak{S}} \| T[x(\sigma)] - T[x_\epsilon(\sigma)] \| dm < \epsilon$. Now

$$\int_E x_\epsilon(\sigma)\, dm = \sum_{n=1}^{\infty} x(\sigma_n)m(E_n \cap E) = \lim_{N\to\infty} \sum_{n=1}^{N} x(\sigma_n)m(E_n \cap E),$$

$$\int_E T[x_\epsilon(\sigma)]\, dm = \sum_{n=1}^{\infty} T[x(\sigma_n)]m(E_n \cap E) = \lim_{N\to\infty} T\left[\sum_{n=1}^{N} x(\sigma_n)m(E_n \cap E)\right].$$

Since T is closed it follows that $\int_E x_\epsilon(\sigma)\, dm \in \mathfrak{D}(T)$ and that $T(\int_E x_\epsilon(\sigma)\, dm] = \int_E T[x_\epsilon(\sigma)]\, dm$. If we now choose a sequence of ϵ_n's converging to zero, we have by Theorem 3.7.7

$$\int_E x_{\epsilon_n}(\sigma)\, dm \to \int_E x(\sigma)\, dm,$$

$$T\left[\int_E x_{\epsilon_n}(\sigma)\, dm\right] = \int_E T[x_{\epsilon_n}(\sigma)]\, dm \to \int_E T[x(\sigma)]\, dm.$$

Again making use of the closure of T we obtain (3.7.11).

If, in particular, T is a linear bounded transformation on $\mathfrak{X}$ to $\mathfrak{Y}$, then the theorem applies if only $x(\sigma) \in B(\mathfrak{S}; \mathfrak{X}; m)$. Indeed if $\{x_n(\sigma)\}$ is a sequence of countably-valued integrable functions approximating $x(\sigma)$ in the sense of Definition 3.7.3, then $T[x_n(\sigma)]$ is countably-valued, integrable, converges almost everywhere to $T[x(\sigma)]$, and $\int_{\mathfrak{S}} \| T[x(\sigma)] - T[x_n(\sigma)] \| dm \leqq \| T \| \int_{\mathfrak{S}} \| x(\sigma) - x_n(\sigma) \| dm \to 0$. Hence $T[x(\sigma)] \in B(\mathfrak{S}; \mathfrak{Y}; m)$.

We also have an analogue of the Fubini theorem for (B)-integrals. Suppose that $\mathfrak{S}$ and $\mathfrak{T}$ are abstract sets possessing σ-rings of subsets, $\mathfrak{E}$ and $\mathfrak{F}$, with σ-finite measures $m(E)$ and $n(F)$ defined on $\mathfrak{E}$ and $\mathfrak{F}$, respectively. We denote by $\mathfrak{E} \times \mathfrak{F}$ the σ-ring of subsets of $\mathfrak{S} \times \mathfrak{T}$ generated by the class of all rectangular sets of the form $E \times F$, where $E \in \mathfrak{E}$ and $F \in \mathfrak{F}$. Finally we denote the product measure by $m \times n$.

THEOREM 3.7.13. *If $x(\sigma, \tau)$ is (B)-integrable on $\mathfrak{S} \times \mathfrak{T}$, then the functions $y(\sigma) = \int_{\mathfrak{T}} x(\sigma, \tau)\, dn$ and $z(\tau) = \int_{\mathfrak{S}} x(\sigma, \tau) dm$ are defined almost everywhere in $\mathfrak{S}$ and $\mathfrak{T}$ respectively and*

$$\int_{\mathfrak{S}\times\mathfrak{T}} x(\sigma, \tau)\, d(m \times n) = \int_{\mathfrak{S}} y(\sigma)\, dm = \int_{\mathfrak{T}} z(\tau)\, dn. \tag{3.7.12}$$

PROOF. We may assume without loss of generality that $x(\sigma, \tau)$ is separably-valued. Then for each $x^* \in \mathfrak{X}^*$, we have $x^*[x(\sigma, \tau)]$ measurable in τ for each σ and hence $x(\sigma, \tau)$ is strongly measurable in τ for each σ. Likewise $x(\sigma, \tau)$ is strongly measurable in σ for each τ. The Fubini theorem applied to $\| x(\sigma, \tau) \|$ implies that $\int_{\mathfrak{T}} \| x(\sigma, \tau) \| dn$ and $\int_{\mathfrak{S}} \| x(\sigma, \tau) \| dm$ are finite for almost all σ and almost all τ, respectively. Hence the integrals defining $y(\sigma)$ and $z(\tau)$ exist almost everywhere in $\mathfrak{S}$ and $\mathfrak{T}$ respectively. Further $x^*[y(\sigma)] = \int_{\mathfrak{T}} x^*[x(\sigma, \tau)]\, dn$ so that $x^*[y(\sigma)]$ is measurable. Since $y(\sigma)$ belongs to the least closed linear subspace containing $[x(\sigma, \tau), (\sigma, \tau) \in \mathfrak{S} \times \mathfrak{T}]$, $y(\sigma)$ is separably-valued and hence strongly measurable. In addition $\int_{\mathfrak{S}} \| y(\sigma) \| \, dm \leqq \int_{\mathfrak{S}\times\mathfrak{T}} \| x(\sigma, \tau) \| \, d(m \times n)$ so that $y(\sigma)$ is (B)-integrable. Likewise $z(\tau)$ is (B)-integrable. If we now apply the Fubini theorem to $x^*[x(\sigma, \tau)]$ we obtain

$$\int_{\mathfrak{S}\times\mathfrak{T}} x^*[x(\sigma, \tau)] \, d(m \times n) = \int_{\mathfrak{S}} x^*[y(\sigma)] \, dm = \int_{\mathfrak{T}} x^*[z(\tau)] \, dn$$

for each $x^* \in \mathfrak{X}^*$ and this implies (3.7.12).

3.8. Further properties of the (B)-integral. We now consider a few aspects of the (B)-integral which will be of special interest for our later work. We start with operator-valued functions on $\mathfrak{S}$ to $\mathfrak{E}(\mathfrak{X}, \mathfrak{Y})$. Here we must distinguish between the uniform (B)-integral and the strong (B)-integral. If $U(\sigma)$ is uniformly measurable and if $\int_{\mathfrak{S}} \| U(\sigma) \| \, dm < \infty$ then $U(\sigma) \in B(\mathfrak{S}; \mathfrak{E}(\mathfrak{X}, \mathfrak{Y}); m)$ and the previous theory applies directly. In this case $\int_{\mathfrak{S}} U(\sigma) \, dm \in \mathfrak{E}(\mathfrak{X}, \mathfrak{Y})$ and is the limit in the uniform operator topology of the approximating integrals. On the other hand if $U(\sigma)(x) \in B(\mathfrak{S}; \mathfrak{Y}; m)$ for each $x \in \mathfrak{X}$, then the previous theory merely asserts that $\int_{\mathfrak{S}} U(\sigma)(x) \, dm = V(x)$ is an element of $\mathfrak{Y}$. It requires additional argument to show that $V \in \mathfrak{E}(\mathfrak{X}, \mathfrak{Y})$.

THEOREM 3.8.1. *If* $U(\sigma) \in B(\mathfrak{S}; \mathfrak{E}(\mathfrak{X}, \mathfrak{Y}); m)$, *then* $U^*(\sigma) \in B(\mathfrak{S}; \mathfrak{E}(\mathfrak{Y}^*, \mathfrak{X}^*); m)$ *and*

$$\left[\int_{\mathfrak{S}} U(\sigma) \, dm\right]^* = \int_{\mathfrak{S}} U^*(\sigma) \, dm. \tag{3.8.1}$$

PROOF. Let $\{U_n(\sigma)\}$ be a sequence of countably-valued integrable functions which approximate $U(\sigma)$ in the sense of Definition 3.7.3. Now $U \to U^*$ is an isometric mapping of $\mathfrak{E}(\mathfrak{X}, \mathfrak{Y})$ into $\mathfrak{E}(\mathfrak{Y}^*, \mathfrak{X}^*)$. Hence the $\{U_n^*(\sigma)\}$ approximate $U^*(\sigma)$ in the sense of Definition 3.7.3 and $U^*(\sigma) \in B(\mathfrak{S}; \mathfrak{E}(\mathfrak{Y}^*, \mathfrak{X}^*); m)$. Likewise if $\sum V_n$ is an absolutely convergent sum, then

$$\left(\sum_{n=1}^{\infty} V_n\right)^* = \lim_{N\to\infty} \left(\sum_{n=1}^{N} V_n\right)^* = \lim_{N\to\infty} \sum_{n=1}^{N} V_n^* = \sum_{n=1}^{\infty} V_n^* .$$

Consequently (3.8.1) holds for each of the countably-valued functions, $U_n(\sigma)$. Finally passing to the limit with n, we see that (3.8.1) also holds for $U(\sigma)$.

THEOREM 3.8.2. *If* $U(\sigma)(x) \in B(\mathfrak{S}; \mathfrak{Y}; m)$ *for each* $x \in \mathfrak{X}$, *then*

$$V(x) = \int_{\mathfrak{S}} U(\sigma)(x) \, dm$$

defines a linear bounded operator on $\mathfrak{X}$ *to* $\mathfrak{Y}$.

PROOF. It is clear that V is well defined and linear on $\mathfrak{X}$. In order to show that V is bounded we consider an auxiliary transformation W on $\mathfrak{X}$ to $B(\mathfrak{S}; \mathfrak{Y}; m)$, defined by $W(x) = U(\sigma)(x)$. One sees directly that W is linear and closed. Hence by Theorem 2.12.3, it follows that W is also bounded. Therefore

$$\| V(x) \| \leqq \int_{\mathfrak{S}} \| U(\sigma)(x) \| \, dm \leqq \| W \| \, \| x \| . \tag{3.8.2}$$

There are some results in integration theory which depend on the topological and group properties of the underlying space, $\mathfrak{S}$. We therefore specialize, taking $\mathfrak{S}$ to be a Lebesgue measurable subset of the k-dimensional Euclidean space, E_k. We shall further restrict $m(E)$ to be the Lebesgue measure function and we denote the corresponding family of (B)-integrable vector-functions by $B(\mathfrak{S}; \mathfrak{X})$. Finally we denote the k-tuples of E_k by bold face type ($\boldsymbol{\sigma} = (\sigma_1, \sigma_2, \cdots, \sigma_k)$).

Definition 3.7.3 shows that *countably-valued functions are dense in* $B(\mathfrak{S}; \mathfrak{X})$ and this in turn implies that *simple functions are also dense in* $B(\mathfrak{S}; \mathfrak{X})$. This means that *the two-valued simple functions* ($x(\boldsymbol{\sigma}) = a$ on E_1, θ on $\mathfrak{S} \ominus E_1$) *form a fundamental set in* $B(\mathfrak{S}; \mathfrak{X})$. This set may be further reduced, however, provided $\mathfrak{S}$ is a connected convex set as we may assume without restricting the generality. A classical argument shows that it is sufficient to limit E_1 to be a k-dimensional rectangle, I: $(\alpha_1 < \sigma_1 < \beta_1, \cdots, \alpha_k < \sigma_k < \beta_k)$. The corresponding functions $x(\boldsymbol{\sigma}) = a$ on I, θ on $\mathfrak{S} \ominus I$ form a fundamental set. Such a step function can obviously be approximated in the mean of order one by continuous functions. *Hence the continuous functions which differ from* θ *only on bounded sets are also dense in* $B(\mathfrak{S}; \mathfrak{X})$. To simplify the formulation of the following theorems we take $\mathfrak{S} = E_k$.

THEOREM 3.8.3. *If* $x(\boldsymbol{\sigma}) \in B(E_k; \mathfrak{X})$, *then*

$$\lim_{\boldsymbol{\alpha} \to 0} \int_{E_k} \| x(\boldsymbol{\sigma} + \boldsymbol{\alpha}) - x(\boldsymbol{\sigma}) \| \, d\boldsymbol{\sigma} = 0.$$

Here $x(\boldsymbol{\sigma} + \boldsymbol{\alpha}) = x(\sigma_1 + \alpha_1, \cdots, \sigma_k + \alpha_k)$.

PROOF. The theorem is clearly true for continuous functions which differ from θ only on bounded sets. Given $x(\boldsymbol{\sigma}) \in B(E_k; \mathfrak{X})$ and $\epsilon > 0$, there exists a continuous function, $x_\epsilon(\boldsymbol{\sigma})$, of this kind such that $\| x(\cdot) - x_\epsilon(\cdot) \| < \epsilon$. It follows that $\int_{E_k} \| x(\boldsymbol{\sigma} + \boldsymbol{\alpha}) - x_\epsilon(\boldsymbol{\sigma} + \boldsymbol{\alpha}) \| \, d\boldsymbol{\sigma} < \epsilon$. Consequently

$$\limsup_{\boldsymbol{\alpha} \to 0} \int_{E_k} \| x(\boldsymbol{\sigma} + \boldsymbol{\alpha}) - x(\boldsymbol{\sigma}) \| \, d\boldsymbol{\sigma}$$
$$\leqq 2\epsilon + \limsup_{\boldsymbol{\alpha} \to 0} \int_{E_k} \| x_\epsilon(\boldsymbol{\sigma} + \boldsymbol{\alpha}) - x_\epsilon(\boldsymbol{\sigma}) \| \, d\boldsymbol{\sigma} \leqq 2\epsilon.$$

THEOREM 3.8.4. *If* $x(\boldsymbol{\sigma}) \in B(E_k; \mathfrak{X})$ *and* $f(\boldsymbol{\sigma})$ *is a bounded numerically-valued measurable function, then*

$$y(\boldsymbol{\xi}) = \int_{E_k} f(\boldsymbol{\sigma}) x(\boldsymbol{\sigma} + \boldsymbol{\xi}) \, d\boldsymbol{\sigma}$$

is a continuous function of $\boldsymbol{\xi}$.

PROOF. The integral obviously exists and defines an element of $\mathfrak{X}$. If $|f(\sigma)| \leqq M$ we have

$$\| y(\xi + \alpha) - y(\xi) \| \leqq \int_{E_k} |f(\sigma)| \, \| x(\sigma + \xi + \alpha) - x(\sigma + \xi) \| \, d\sigma$$

$$\leqq M \int_{E_k} \| x(\tau + \alpha) - x(\tau) \| \, d\tau \to 0$$

with α.

We next consider the question of the differentiability of the indefinite integral. Let $C(\xi, \gamma)$ be the cube

$$\xi_1 - \gamma < \sigma_1 < \xi_1 + \gamma, \cdots, \xi_k - \gamma < \sigma_k < \xi_k + \gamma.$$

THEOREM 3.8.5. *Let $x(\sigma) \in B(E_k; \mathfrak{X})$. Then for almost all ξ*

$$\lim_{\gamma \to 0} (2\gamma)^{-k} \int_{C(\xi,\gamma)} \| x(\sigma) - x(\xi) \| \, d\sigma = 0. \tag{3.8.3}$$

PROOF. We may suppose without loss of generality that $x(\sigma)$ is separably-valued. Let $\{x_n\}$ be a denumerable set dense in $x(E_k)$. Then by the classical theorem of Lebesgue

$$\lim_{\gamma \to 0} (2\gamma)^{-k} \int_{C(\xi,\gamma)} \| x(\sigma) - x_n \| \, d\sigma = \| x(\xi) - x_n \| \tag{3.8.4}$$

for almost all ξ and each n, and consequently for almost all ξ and all n. If ξ is a point at which (3.8.4) is valid for all n, then for a given $\epsilon > 0$ choose x_n so that $\| x(\xi) - x_n \| < \epsilon$. We then have

$$\limsup_{\gamma \to 0} (2\gamma)^{-k} \int_{C(\xi,\gamma)} \| x(\sigma) - x(\xi) \| \, d\sigma$$

$$\leqq \limsup_{\gamma \to 0} (2\gamma)^{-k} \int_{C(\xi,\gamma)} [\| x(\sigma) - x_n \| + \| x_n - x(\xi) \|] \, d\sigma < 2\epsilon.$$

Since ϵ is arbitrary, this implies (3.8.3).

COROLLARY 1. *Let $x(\sigma) \in B(E_k; \mathfrak{X})$. Then for almost all ξ*

$$\lim_{\gamma \to 0} (2\gamma)^{-k} \int_{C(\xi,\gamma)} x(\sigma) \, d\sigma = x(\xi).$$

PROOF. Since

$$\left\| (2\gamma)^{-k} \int_{C(\xi,\gamma)} x(\sigma) \, d(\sigma) - x(\xi) \right\| \leq (2\gamma)^{-k} \int_{C(\xi,\gamma)} \| x(\sigma) - x(\xi) \| \, d\sigma,$$

the result is an immediate consequence of the above theorem.

In Theorem 3.8.5 we may replace the cube $C(\xi, \gamma)$ by certain other measurable point sets $S(\xi, \gamma)$ which shrink to the point ξ as $\gamma \to 0$. The quantity $(2\gamma)^k$ is then to be replaced by $m[S(\xi, \gamma)]$. If $k = 1$, we may in particular take either of the intervals $(\xi - \gamma, \xi)$ or $(\xi, \xi + \gamma)$. This leads to the following

COROLLARY 2. *Let $x(\sigma) \in B(E_1 ; \mathfrak{X})$. Then for almost all ξ*

$$\lim_{\gamma \to 0} \frac{1}{\gamma} \int_\xi^{\xi+\gamma} x(\sigma)\, d\sigma = x(\xi) \tag{3.8.4}$$

and

$$\lim_{\gamma \to 0} \frac{1}{\gamma} \int_\xi^{\xi+\gamma} \| x(\sigma) - x(\xi) \|\, d\sigma = 0. \tag{3.8.5}$$

It follows from Theorem 3.7.11 and the above corollary that the indefinite integral, $y(\xi) \equiv \int_\alpha^\xi x(\sigma)\, d\sigma$, of a Bochner integrable function is strongly absolutely continuous and possesses a strong derivative almost everywhere which is equal to the integrand $x(\xi)$. On the other hand a strongly absolutely continuous function need not be differentiable anywhere. An example of this is furnished by $y(\xi) \equiv \varphi(\xi, \cdot)$ on $0 \leqq \xi \leqq 1$ to $\mathfrak{X} = L_1(E_1)$ and defined so that $\varphi(\xi, \sigma) = 1$ for $0 \leqq \sigma \leqq \xi$ and $= 0$ elsewhere. However we do have

THEOREM 3.8.6. *If $y(\xi)$ is of strong bounded variation on E_1 to $\mathfrak{X}$ and almost everywhere weakly differentiable with derivative $x(\xi)$, then $x(\xi) \in B(E_1 ; \mathfrak{X})$. If $y(\xi)$ is also weakly absolutely continuous then it can be expressed as the indefinite integral of $x(\xi)$.*

PROOF. We note first of all that $y(\xi)$ can have only a denumerable number of points of discontinuity and hence is contained in a closed separable subspace $\mathfrak{X}_0$ of $\mathfrak{X}$. Since weak limits of elements in $\mathfrak{X}_0$ are contained in $\mathfrak{X}_0$, almost all values of $x(\xi)$ will lie in $\mathfrak{X}_0$. Thus $x(\xi)$ is weakly measurable, almost separably-valued, and hence strongly measurable. We now define

$$x_n(\xi) = 2^n[y(k2^{-n}) - y((k - 1)2^{-n})]$$

for $(k - 1)2^{-n} \leqq \xi < k2^{-n}$, $k = 0, \pm 1, \pm 2, \cdots$. Then clearly $\int \| x_n(\sigma) \|\, d\sigma \leqq \text{Var}\, [y(\xi)]$. Further $\{x_n(\sigma)\}$ converges weakly almost everywhere to $x(\sigma)$ and hence almost everywhere $\| x(\sigma) \| \leqq \liminf_n \| x_n(\sigma) \|$. Fatou's lemma then asserts that $\int \| x(\sigma) \|\, d\sigma \leqq \text{Var}\, [y(\xi)]$ so that $x(\xi) \in B(E_1 ; \mathfrak{X})$. If, in addition, $y(\xi)$ is weakly absolutely continuous, then for each $x^* \in \mathfrak{X}^*$ we have

$$x^*[y(\xi)]\, |_\alpha^\beta = \int_\alpha^\beta x^*[x(\sigma)]\, d\sigma = x^* \left[\int_\alpha^\beta x(\sigma)\, d\sigma \right],$$

and hence $y(\xi)\, |_\alpha^\beta = \int_\alpha^\beta x(\sigma)\, d\sigma$.

In addition to the class $B(\mathfrak{S}; \mathfrak{X}; m) = B_1(\mathfrak{S}; \mathfrak{X}; m)$, mention should be made of the classes $B_p(\mathfrak{S}; \mathfrak{X}; m)$, $1<p<\infty$. A function $x(\sigma)$ on $\mathfrak{S}$ to $\mathfrak{X}$ belongs to $B_p(\mathfrak{S}; \mathfrak{X}; m)$ if $x(\sigma)$ is strongly measurable in $\mathfrak{S}$ and $\int_\mathfrak{S} \| x(\sigma) \|^p dm < \infty$.

Similarly $x(\sigma) \in B_\infty(\mathfrak{S}; \mathfrak{X}; m)$ if $x(\sigma)$ is strongly measurable in $\mathfrak{S}$ and $\| x(\sigma) \|$ is bounded except in a null set. The class $B_p(\mathfrak{S}; \mathfrak{X}; m)$ becomes a (B)-space under the norm

$$\| x(\cdot) \|_p = \left\{ \int_{\mathfrak{S}} \| x(\sigma) \|^p \, dm \right\}^{1/p}, \qquad \| x(\cdot) \|_\infty = \text{ess sup} \, \| x(\sigma) \|.$$

When $\mathfrak{X}$ is reflexive and $1 < p < \infty$, Phillips [3] has shown that the adjoint space to $B_p(\mathfrak{S}; \mathfrak{X}; m)$ is simply $B_{p'}(\mathfrak{S}; \mathfrak{X}^*; m)$ where $1/p + 1/p' = 1$ (cf. M. M. Day [2] and J. Dieudonné [2]). The obvious generalizations of Theorems 3.8.3 and 3.8.4 are also valid. In the last theorem we have now to suppose that $f(\sigma) \in L_{p'}(E_k)$.

3.9. Singular integrals. A considerable body of the classical theory of singular integrals carries over to the case of vector-valued functions. We shall state several theorems concerning such integrals and give brief indications of their proofs, which follow standard patterns. Throughout the discussion the *kernel* $K(\xi, \sigma; \omega)$ is a numerically-valued function defined for $-\infty < \xi, \sigma < \infty, \omega > 0$, measurable in (ξ, σ) as well as in ξ and σ separately for all values of the other variable. We set

$$x(\xi; \omega) = \int_{E_1} K(\xi, \sigma; \omega) x(\sigma) \, d\sigma, \tag{3.9.1}$$

where $x(\sigma)$ is vector-valued and the existence of the integral will be ensured by further assumptions.

THEOREM 3.9.1. *Let $K(\xi, \sigma; \omega)$ satisfy the conditions:*

(1) *$K(\xi, \sigma; \omega) \in L_1(E_1)$ as a function of ξ for all σ and as a function of σ for all ξ when ω is fixed but arbitrary;*

(2) *$\int_{E_1} | K(\xi, \sigma; \omega) | \, d\xi < A$ for all σ and ω;*

(3) *$\lim_{\omega \to \infty} \int_{E_1 \ominus I} | K(\xi, \sigma; \omega) | \, d\xi = 0$ for every open interval I containing σ;*

(4) *$\lim_{\omega \to \infty} \int_I K(\xi, \sigma; \omega) \, d\sigma = 1$ for every open interval I containing ξ;*

(5) *To every open interval I there is a measurable function $M(\xi, I)$ such that $| \int_I K(\xi, \sigma; \omega) \, d\sigma | \leqq M(\xi, I)$ for all ω and $\int_I M(\xi, I) \, d\xi < \infty$.*

If $x(\sigma) \in B_1(E_1; \mathfrak{X})$, then

(i) *$x(\xi; \omega)$ exists for almost all ξ and belongs to $B_1(E_1; \mathfrak{X})$;*

(ii) *$\| x(\cdot; \omega) \|_1 \leqq A \| x(\cdot) \|_1$;*

(iii) *$\lim_{\omega \to \infty} \| x(\cdot) - x(\cdot; \omega) \|_1 = 0$.*

PROOF. The measurability assumptions ensure that the integrand in (3.9.1) is a strongly measurable function of σ for all ξ and ω. That $| K(\xi, \sigma; \omega) | \, \| x(\sigma) \|$ is integrable for almost all ξ follows from the Fubini theorem by virtue of the inequality

$$\int_{E_1} \left\{ \int_{E_1} |K(\xi, \sigma; \omega) | \, \| x(\sigma) \| \, d\xi \right\} d\sigma \leqq A \, \| x(\cdot) \|_1$$

which is implied by (1) and (2). This proves (i) and (ii). If $x(\sigma)$ is the step function $x_1(\sigma) = a$ in I and θ outside, $m(I) < \infty$, then

$$\| x_1(\cdot) - x_1(\cdot\,; \omega) \|_1 = \| a \| \int_I \left| \int_I K(\xi, \sigma; \omega)\, d\sigma - 1 \right| d\xi$$
$$+ \| a \| \int_{E_1 \ominus I} \left| \int_I K(\xi, \sigma; \omega)\, d\sigma \right| d\xi.$$

Both terms on the right tend to zero as $\omega \to \infty$, the first by virtue of conditions (4) and (5), and the second by virtue of conditions (2) and (3). Now such step functions form a fundamental set in $B_1(E_1\,; \mathfrak{X})$ and the operation which takes $x(\cdot)$ into $x(\cdot\,\omega)$ is linear and bounded uniformly with respect to ω. The Banach-Steinhaus theorem 2.11.4 then shows that (iii) holds for every $x(\cdot) \in B_1(E_1\,; \mathfrak{X})$.

The conditions of Theorem 3.9.1 are not suitable for a discussion of point-wise convergence. The choice of additional conditions is simplest in the case of convergence at points of continuity.

THEOREM 3.9.2. *Let $K(\xi, \sigma; \omega)$ satisfy conditions* (1) *and* (4) *of the preceding theorem and in addition:*

(6) *to every ξ there is a finite $M_1(\xi, \omega)$ such that $| K(\xi, \sigma; \omega) | \leqq M_1(\xi, \omega)$ for all σ;*

(7) *to every ξ and every $\epsilon > 0$ there is a finite $M_2(\xi, \epsilon)$ such that $| K(\xi, \sigma; \omega) | < M_2(\xi, \epsilon)$ for all ω when σ is outside of $(\xi - \epsilon, \xi + \epsilon)$;*

(8) *to every ξ there is a finite $M(\xi)$ such that $\int_{\xi-1}^{\xi+1} | K(\xi, \sigma; \omega) | \, d\sigma \leqq M(\xi)$ for all ω.*

If $x(\sigma) \in B_1(E_1\,; \mathfrak{X})$ then $x(\xi; \omega)$ exists for all ξ and

$$\lim_{\omega \to \infty} x(\xi; \omega) = x(\xi)$$

at all points of continuity of $x(\sigma)$.

PROOF. The existence of $x(\xi; \omega)$ follows from the assumptions of measurability together with (6). We note next that (4) implies that

$$\lim_{\omega \to \infty} \int_I K(\xi, \sigma; \omega)\, d\sigma = 0$$

for every closed interval I which does not contain the point ξ.

Let ξ be a point of continuity of $x(\sigma)$ and break up the integral defining $x(\xi; \omega)$ into three parts, J_1, J_2, and J_3, using $\sigma = \xi - \epsilon$ and $\xi + \epsilon$ as partition points. Here $\epsilon = \epsilon(\delta)$ is chosen so small that $\| x(\sigma) - x(\xi) \| \leqq \delta$ when $| \sigma - \xi | \leqq \epsilon$. Then

$$J_2 - x(\xi) = \left\{ \int_{\xi-\epsilon}^{\xi+\epsilon} K(\xi, \sigma; \omega)\, d\sigma - 1 \right\} x(\xi)$$
$$+ \int_{\xi-\epsilon}^{\xi+\epsilon} K(\xi, \sigma; \omega)[x(\sigma) - x(\xi)]\, d\sigma.$$

The first term on the right tends to zero as $\omega \to \infty$ by (4) and the norm of the second term is less than $\delta M(\xi)$ by (8). Thus $\limsup_{\omega\to\infty} \| J_2 - x(\xi) \| \leqq \delta M(\xi)$.

If $x(\sigma)$ is a step function, we have

$$J_1 + J_3 = \sum_{j=1}^{n} a_j \int_{I_j} K(\xi, \sigma; \omega)\, d\sigma,$$

and, since ξ lies outside of all the intervals I_j, each integral tends to zero when $\omega \to \infty$. In the general case we may approximate $x(\sigma)$ in the mean of order one by a step function $x_0(\sigma)$ such that $\| x(\cdot) - x_0(\cdot) \|_1 < \eta$. We have then, with obvious notation, $\| J_1 + J_3 - J_{10} - J_{30} \| < \eta M_2(\xi, \epsilon)$ by (7), and consequently

$$\limsup_{\omega\to\infty} \| x(\xi) - x(\xi; \omega) \| \leqq \delta M(\xi) + \eta M_2(\xi, \epsilon(\delta)).$$

Here η is arbitrary so that $\limsup_{\omega\to\infty} \| x(\xi) - x(\xi; \omega) \| \leqq \delta M(\xi)$. In this expression δ is arbitrary and can be replaced by zero. This completes the proof.

REMARK. *Suppose that condition* (4) *is satisfied in the following stronger form:* (4′) *for every* $\delta > 0$ *and all* ξ

$$\lim_{\omega\to\infty} \int_{\xi-\delta}^{\xi} K(\xi, \sigma; \omega)\, d\sigma = \mu_1, \qquad \lim_{\omega\to\infty} \int_{\xi}^{\xi+\delta} K(\xi, \sigma; \omega)\, d\sigma = \mu_2,$$

where $\mu_1 + \mu_2 = 1$ *and* μ_1, μ_2 *are independent of* ξ.

Then, under the assumptions (1), (4′), (6), (7), *and* (8),

$$\lim_{\omega\to\infty} x(\xi; \omega) = \mu_1 x(\xi - 0) + \mu_2 x(\xi + 0)$$

at all points where $x(\sigma)$ *has left- and right-hand limits.*

This is proved by the same methods as the preceding theorem.

Assumptions of a more special nature have to be made to attain convergence almost everywhere.

THEOREM 3.9.3. *Let* $K(\xi, \sigma; \omega)$ *satisfy conditions* (1), (4), (7) *and* (9) *there exists a non-negative function* $P(\beta, \omega)$ *such that*

(i) $| K(\xi, \sigma; \omega) | \leqq P(| \sigma - \xi |; \omega)$ *for all* ω *and all* σ, ξ *with* $| \sigma - \xi | \leqq 1$;

(ii) $P(\beta; \omega)$ *is a bounded decreasing function of* β *for fixed* ω;

(iii) $\int_0^1 P(\beta; \omega)\, d\beta \leqq M$ *for all* $\omega > 0$.

If $x(\sigma) \in B_1(E_1; \mathfrak{X})$ *then* $x(\xi; \omega)$ *exists for all* ξ *and* $\lim_{\omega\to\infty} x(\xi; \omega) = x(\xi)$ *almost everywhere, in particular, in the Lebesgue set of* $x(\sigma)$ *where formula* (3.8.5) *is valid.*

PROOF. The existence of $x(\xi; \omega)$ for all ξ follows from the measurability together with conditions (7) and (9ii). Now let ξ be a point where (3.8.5) is valid and break up the integral defining $x(\xi; \omega)$ as in the preceding proof. The discussion of $J_1 + J_3$ which is based on (4) and (7) goes as before. In the discussion of J_2 we note that

$$\left\| \int_{\xi-\epsilon}^{\xi+\epsilon} K(\xi, \sigma; \omega)[x(\sigma) - x(\xi)]\, d\sigma \right\|$$

$$\leqq \int_{-\epsilon}^{\epsilon} P(| \beta |; \omega) \, \| x(\xi + \beta) - x(\xi) \| \, d\beta \equiv J_4.$$

If now

$$X(\beta; \xi) = \int_0^\beta \| x(\xi + \tau) - x(\xi) \| \, d\tau,$$

then by (3.8.5) $| X(\beta; \xi) | < \delta | \beta |$ for $| \beta | \leqq \epsilon = \epsilon(\delta)$. An integration by parts gives

$$J_4 = [X(\epsilon; \xi) - X(-\epsilon; \xi)]P(\epsilon; \omega) - \int_{-\epsilon}^{\epsilon} X(\beta; \xi) \, d_\beta P(|\beta|; \omega)$$

$$< 2\delta \left[\epsilon P(\epsilon; \omega) - \int_0^\epsilon \beta \, d_\beta P(\beta; \omega) \right] = 2\delta \int_0^\epsilon P(\beta; \omega) \, d\beta \leqq 2\delta M,$$

if $\epsilon < 1$ as we may assume. The proof is then completed as before.

These three theorems are typical for the theory and the reader will have no difficulty in proving analogous theorems for other classes of functions.

References. G. Birkhoff [**1, 2**], Bochner [**2**], Bochner and Taylor [**1**], Burgess [**1**], Day [**2**], Dieudonné [**2**], Dunford [**1, 3, 4**], Gelfand [**1, 2**], Graves [**1**], Halmos [**1**], Hildebrandt [**1, 3**], Hille [**19**], Orlicz [**1, 2**], Pettis [**1**], Phillips [**1, 3**], Price [**1**], Rickart [**1**].

2. Complex Function Theory

3.10. Holomorphic functions. The theory of analytic functions on the complex plane to a linear vector space goes back to D. Hilbert [**1**] and F. Riesz [**2**]. For the case of a general (B)-space the basic extensions are due to Norbert Wiener [**1**]. In recent years N. Dunford, L. Fantappiè, I. Gelfand, and A. E. Taylor have done much to broaden our knowledge in this field.

The basic concepts are those of a *holomorphic vector function* and a *holomorphic operator function*. Let D be a domain of the complex ζ-plane, $\zeta = \xi + i\eta$, and let $x(\zeta)$ be a function on D to a complex (B)-space $\mathfrak{X}$, $U(\zeta)$ a function on D to $\mathfrak{E}(\mathfrak{X}, \mathfrak{Y})$, the (B)-space of linear bounded operators on one complex (B)-space, $\mathfrak{X}$, to another, $\mathfrak{Y}$. In the classical function theory a function $f(\zeta)$ is holomorphic in the domain D if it is single-valued and differentiable. The first of these notions carries over to abstract functions without ambiguity, but the second has two different meanings for vector functions and three for operator functions depending upon the topology employed. Nevertheless we arrive at a unique concept of holomorphism.

Definition 3.10.1. *Let $\Gamma_1 \subset \mathfrak{X}^*$ and $\Gamma_2 \subset \mathfrak{Y}^*$ be determining manifolds for $\mathfrak{X}$ and $\mathfrak{Y}$ respectively. Then $x(\zeta)$ and $U(\zeta)$ are said to be holomorphic in D if* $x^*[x(\zeta)]$

and $y^*\{U(\zeta)[x]\}$ *are holomorphic in Cauchy's sense for every choice of* $x^* \in \Gamma_1$ *and* $x \in \mathfrak{X}$, $y^* \in \Gamma_2$ *respectively.*

Even with $\Gamma_1 = \mathfrak{X}^*$ (or $\Gamma_2 = \mathfrak{Y}^*$) this assumption is weaker than weak differentiability, and still it implies the strongest possible conclusion. For vector functions this was first proved by N. Dunford [**3**, p. 354] and for operator functions by E. Hille [**7**, p. 6]. See also A. E. Taylor [**5**, p. 576] and [**7**, p. 653]). The proofs are beautiful applications of the principle of uniform boundedness.

THEOREM 3.10.1. (1) *If* $x(\zeta)$ *is holomorphic in* D, *then* $x(\zeta)$ *is strongly continuous and strongly differentiable in* D, *uniformly with respect to* ζ *in any compact subset of* D. (2) *If* $U(\zeta)$ *is holomorphic in* D, *then* $U(\zeta)$ *is continuous and differentiable in the uniform operator topology for* ζ *in* D, *again uniformly with respect to* ζ *in any compact subset of* D.

PROOF. It is enough to prove (2) which exhibits the method. We base the proof upon the fact that the difference quotient of a numerically-valued function $f(\zeta)$ which is holomorphic in D tends to its limit, the derivative, uniformly with respect to ζ in any compact subset of D. Expressed as an inequality for difference quotients we may formulate this observation as follows:

LEMMA 3.10.1. *To any function* $f(\zeta)$ *holomorphic in the domain* D *and any compact subset* S, *there is a finite quantity* $M(f; S)$ *such that for every choice of* ζ, $\zeta + \alpha$, *and* $\zeta + \beta$ *in* S

$$\left| \frac{1}{\alpha - \beta} \left\{ \frac{1}{\alpha} [f(\zeta + \alpha) - f(\zeta)] - \frac{1}{\beta} [f(\zeta + \beta) - f(\zeta)] \right\} \right| \leqq M(f; S).$$

This follows from the fact that the function inside the absolute value signs on the left is represented by the Cauchy integral

$$\frac{1}{2\pi i} \int_C \frac{f(\tau)\, d\tau}{(\tau - \zeta)(\tau - \zeta - \alpha)(\tau - \zeta - \beta)},$$

where C consists of a finite number of closed simple rectifiable curves in D having a positive minimal distance both from S and from the boundary of D. The conclusion is immediate.

We apply this lemma to the function $y^*\{U(\zeta)[x]\}$ where x and y^* are arbitrary elements of $\mathfrak{X}$ and Γ_2 respectively. To simplify the notation somewhat we write

$$\frac{1}{\alpha - \beta} \left\{ \frac{1}{\alpha} [U(\zeta + \alpha) - U(\zeta)] - \frac{1}{\beta} [U(\zeta + \beta) - U(\zeta)] \right\} = U(\zeta; \alpha, \beta).$$

The lemma then asserts that

$$| y^*\{U(\zeta; \alpha, \beta)[x]\} | \leqq M(y^*, x, U; S)$$

for every choice of ζ, $\zeta + \alpha$ and $\zeta + \beta$ in S. By Theorem 2.8.6 this implies the existence of a finite $M(x, U; S)$ such that

$$\| U(\zeta; \alpha, \beta)[x] \| \leqq M(x, U; S).$$

But now Theorem 2.5.5 applies and ensures the existence of a finite $M(U; S)$ such that

$$\| U(\zeta; \alpha, \beta) \| \leqq M(U; S). \tag{3.10.1}$$

Here we let $\alpha, \beta \to 0$ and use the fact that $\mathfrak{E}(\mathfrak{X}, \mathfrak{Y})$ is complete. Let $U'(\zeta)$ denote the limit of the difference quotient; $U'(\zeta)$ is an operator function on D to $\mathfrak{E}(\mathfrak{X}, \mathfrak{Y})$. If we take the limit as $\beta \to 0$ in (3.10.1) we obtain

$$\left\| \frac{1}{\alpha} [U(\zeta + \alpha) - U(\zeta)] - U'(\zeta) \right\| \leqq | \alpha | M(U; S)$$

for all ζ and $\zeta + \alpha$ in S. This of course implies continuity in the uniform operator topology as well as differentiability and continuity in the weak and strong topologies.

A simple consequence of Theorem 3.10.1 is that if $U(\zeta)$ is holomorphic in D as an operator function then it is also holomorphic in D as a vector function. This follows from the fact that the uniform differentiability of $U(\zeta)$ implies the differentiability of $F[U(\zeta)]$ for all linear functionals F in $\mathfrak{E}(\mathfrak{X}, \mathfrak{Y})^*$. In the remainder of this paragraph we shall not distinguish between operator-valued and vector-valued functions.

3.11. Cauchy's integral. Let C be a rectifiable curve in the complex ζ-plane given by the equation $\zeta = \zeta(\alpha)$, $0 \leqq \alpha \leqq \alpha_0$, where $\zeta(\alpha)$ is continuous and of bounded variation in $[0, \alpha_0]$. If $x(\zeta)$ is any strongly continuous function on C to a (B)-space $\mathfrak{X}$, then the integral

$$\int_0^{\alpha_0} x(\zeta(\alpha))\, d\zeta(\alpha) \equiv \int_C x(\zeta)\, d\zeta$$

exists by Theorem 3.3.2 and for any linear bounded functional $x^* \in \mathfrak{X}^*$

$$x^* \left[\int_C x(\zeta)\, d\zeta \right] = \int_C x^*[x(\zeta)]\, d\zeta. \tag{3.11.1}$$

The restrictive assumption that $x(\zeta)$ be strongly continuous is a matter of convenience and simplicity and may be replaced by (B)-integrability with respect to the arc length measure function.

So far the function $x(\zeta)$ was arbitrary except for the continuity assumption. Suppose now that $x(\zeta)$ is holomorphic in a domain D of the ζ-plane and that C is a simple closed rectifiable curve in D, the interior of C being also in D. It follows from (3.11.1) that

$$x^* \left[\int_C x(\zeta)\, d\zeta \right] = \int_C x^*[x(\zeta)]\, d\zeta = 0$$

for every $x^* \in \mathfrak{X}^*$. By Theorem 2.7.4 this requires that $\int_C x(\zeta)\, d\zeta = \theta$. We have consequently proved the *analogue of Cauchy's theorem for vector-valued functions.*

THEOREM 3.11.1. *If $x(\zeta)$ is a holomorphic vector function on the domain D to the (B)-space $\mathfrak{X}$, then*

$$\int_C x(\zeta)\, d\zeta = \theta$$

for every simple closed rectifiable contour C in D such that the interior of C belongs to D.

The fact that linear operations commute with integration provides us with one of the most powerful tools in extending classical function theory to vector-valued functions. The following very useful theorem is typical for the procedure.

THEOREM 3.11.2. *Let C be a rectifiable curve in the complex plane, let $x(\tau)$ be strongly continuous on C to $\mathfrak{X}$, and let $K(\zeta, \tau)$ be a numerically-valued bounded function with the following additional properties: There exists a domain D in the ζ-plane such that $K(\zeta, \tau)$ is a holomorphic function of ζ in D for every fixed τ on C and $K(\zeta, \tau)$ is a continuous function of τ on C for every fixed ζ in D. Then*

$$y(\zeta) = \int_C K(\zeta, \tau)x(\tau)\, d\tau$$

is a holomorphic function on D to $\mathfrak{X}$.

REMARK. The assumption that C is of finite length and that $x(\tau)$ and $K(\zeta, \tau)$ are continuous in τ may be weakened in an obvious manner. Thus it is sufficient to assume that $K(\zeta, \tau)$ is a bounded function of (ζ, τ), measurable in τ, and that $x(\tau)$ is (B)-integrable over C.

PROOF. The existence of $y(\zeta)$ as a function on D to $\mathfrak{X}$ follows from Theorem 3.3.2. Since

$$x^*[y(\zeta)] = \int_C K(\zeta, \tau)x^*[x(\tau)]\, d\tau$$

is holomorphic in D in the sense of Cauchy for every $x^* \in \mathfrak{X}^*$, $y(\zeta)$ satisfies the conditions of Definition 3.10.1 and is therefore holomorphic in D.

Taking $K(\zeta, \tau) = 1/(\tau - \zeta)$, we obtain integrals of the Cauchy type, which consequently define holomorphic functions of ζ in each of the domains into which C divides the ζ-plane. In particular, this observation gives the Cauchy integral representations of a holomorphic vector function and its derivatives. We have merely to note that if $x(\zeta)$ is holomorphic then

$$\frac{d}{d\zeta}\, x^*[x(\zeta)] = x^*\left[\frac{d}{d\zeta}\, x(\zeta)\right].$$

as is seen from

$$\frac{1}{\alpha}\{x^*[x(\zeta+\alpha)] - x^*[x(\zeta)]\} = x^*\left\{\frac{1}{\alpha}[x(\zeta+\alpha) - x(\zeta)]\right\}$$

by letting $\alpha \to 0$, remembering that the expression inside the brackets on the right tends in norm to $x'(\zeta)$.

THEOREM 3.11.3. *Let $x(\zeta)$ be a holomorphic function on the domain D to the (B)-space $\mathfrak{X}$. Let C be a simple closed rectifiable curve in D, the interior of which is contained in D, and let ζ be such that* arg $(\tau - \zeta)$ *increases by 2π when τ describes C (positive orientation). Then*

$$x^{(n)}(\zeta) = \frac{n!}{2\pi i}\int_C \frac{x(\tau)\,d\tau}{(\tau-\zeta)^{n+1}}, \qquad n = 0, 1, 2, \cdots. \tag{3.11.2}$$

It follows from Theorem 3.11.1 that the integral $\int x(\zeta)\,d\zeta$ is independent of the path and depends only upon initial and terminal points if D is simply-connected, and in the multiply-connected case we may deform paths in the customary manner. Thus we have the usual freedom of choice of paths to suit our needs.

The Cauchy-Hadamard theorem also has its analogue in this abstract setting.

THEOREM 3.11.4. *Given the power series*

$$\sum_{n=0}^{\infty} a_n(\zeta-\zeta_0)^n, \qquad a_n \in \mathfrak{X};$$

set

$$1/\rho = \limsup_{n\to\infty} \| a_n \|^{1/n}.$$

Then the series is absolutely convergent for $|\zeta - \zeta_0| < \rho$ and divergent for $|\zeta-\zeta_0| > \rho$. The series converges to a holomorphic function on $|\zeta-\zeta_0| < \rho$ to $\mathfrak{X}$, the convergence being uniform in every concentric circle of radius less than ρ.

PROOF. The classical argument for the convergence of the series applies if we replace absolute values by norms throughout. It remains to show that the function $x(\zeta) \equiv \sum_n a_n(\zeta-\zeta_0)^n$ is holomorphic for $|\zeta-\zeta_0| < \rho$. Since the series converges in norm in this circle, we have

$$x^*[x(\zeta)] = \sum_{n=0}^{\infty} x^*(a_n)(\zeta-\zeta_0)^n$$

for each $x^* \in \mathfrak{X}^*$ and all $|\zeta-\zeta_0| < \rho$. The series on the right converges to a numerically-valued function holomorphic in $|\zeta-\zeta_0| < \rho$ so that $x(\zeta)$ is itself holomorphic in this circle by Definition 3.10.1.

If $x(\zeta)$ is holomorphic in $|\zeta-\zeta_0| < R$ and $\| x(\zeta) \| \leq M$ for such values of ζ, then taking C to be the circle $|\zeta - \zeta_0| = r < R$ in Theorem 3.11.3 we obtain the *Cauchy estimates*

$$\| x^{(n)}(\zeta_0) \| \leqq MR^{-n}n!.$$

From these estimates we conclude the validity of the *Taylor expansion*

$$x(\zeta) = \sum_{n=0}^{\infty} \frac{x^{(n)}(\zeta_0)}{n!} (\zeta - \zeta_0)^n \qquad \text{for} \qquad |\zeta - \zeta_0| < R.$$

Indeed, the series on the right converges in norm to a limit for such values of ζ and since

$$x^*[x^{(n)}(\zeta)] = \frac{d^n}{d\zeta^n} x^*[x(\zeta)]$$

for every $x^* \in \mathfrak{X}^*$, the series converges weakly to $x(\zeta)$. Its strong limit must then also be $x(\zeta)$. The same type of argument may be used to prove *Laurent's expansion*:

If $x(\zeta)$ is holomorphic in $0 \leqq R_1 < |\zeta - \zeta_0| < R_2 \leqq \infty$, then

$$x(\zeta) = \sum_{n=-\infty}^{\infty} a_n(\zeta - \zeta_0)^n, \qquad a_n = \frac{1}{2\pi i}\int_C x(\tau)(\tau - \zeta_0)^{-n-1}\, d\tau$$

where C, for instance, is the circle $|\tau - \zeta_0| = r$, $R_1 < r < R_2$.

It is clear that the following estimates hold

$$\text{(3.11.3)} \qquad \| a_n \| \leqq M(r; x)r^{-n}, \qquad M(r; x) = \max_{|\zeta - \zeta_0| = r} \| x(\zeta) \|.$$

If $R_1 = 0$ and there is actually at least one $a_n \neq \theta$ with a negative subscript, then $\zeta = \zeta_0$ is a *singular point* of $x(\zeta)$, namely a *pole of order* m if $a_{-m} \neq \theta$ but $a_n = \theta$ for $n < -m$ and otherwise an (isolated) *essential singularity*. In the case of a pole of order m we have

$$M_1 |\zeta - \zeta_0|^{-m} \leqq \| x(\zeta) \| \leqq M_2 |\zeta - \zeta_0|^{-m}$$

for all small values of $|\zeta - \zeta_0|$.

From the Taylor expansion the *uniqueness theorem* is concluded in the usual manner:

THEOREM 3.11.5. *If $x(\zeta)$ and $y(\zeta)$ are holomorphic in D and if $x(\zeta_n) = y(\zeta_n)$, $n = 1, 2, 3, \cdots$, the points $\{\zeta_n\}$ having a limit point in D, then $x(\zeta) \equiv y(\zeta)$ in D.*

The property of being holomorphic is strongly adherent and is preserved by various convergence processes. The next theorem contains the most elementary result in this direction; the theorem of Vitali is given in a later section (Theorem 3.14.1).

THEOREM 3.11.6. *Let $\{x_n(\zeta)\}$ be a sequence of holomorphic functions on D to $\mathfrak{X}$ which converge uniformly with respect to ζ on a simple closed rectifiable curve C, the* *interior of which, D_0 say, is also in D. Then $\{x_n(\zeta)\}$ converges to a holomorphic function $x(\zeta)$ in D_0 and, moreover, $x_n^{(k)}(\zeta) \to x^{(k)}(\zeta)$ in D_0 for every k. The convergence is uniform with respect to ζ, in D_0 when $k = 0$, in any compact subset of D_0 when $k > 0$.*

The classical proof applies, *mutatis mutandis*, to the vector case.

The material in this section lends itself very nicely to the following problem suggested by the work of H. Müntz [1]. Given a separable (B)-space $\mathfrak{X}$, find a sequence of elements such that every infinite subsequence is fundamental in $\mathfrak{X}$. Since $\mathfrak{X}$ is separable there will exist a sequence $\{x_n\}$ dense in $\mathfrak{X}$. Set $a_n = x_n/\| x_n \|$ and let $x(\zeta) \equiv \sum_{n=1}^{\infty} a_n\zeta^n$. Then $x(\zeta)$ is clearly holomorphic in $|\zeta| < 1$. Suppose that $\zeta_n \to \zeta_0$ where $|\zeta_n| < 1, n = 0, 1, 2, \cdots$, then $x(\zeta_n)$ is a sequence of elements in $\mathfrak{X}$ of the desired kind. For if $x(\zeta_{n_k})$ were not fundamental in $\mathfrak{X}$, then by Theorem 2.7.6 there would exist a non-zero $x_0^* \in \mathfrak{X}^*$ whose null space contains the linear subspace spanned by $\{x(\zeta_{n_k})\}$. Thus $x_0^*[x(\zeta_{n_k})] = 0$ for all k and hence $x_0^*[x(\zeta)] = \sum_{n=1}^{\infty} x_0^*(a_n)\zeta^n \equiv 0$ for $|\zeta| < 1$. This implies that $x_0^*(a_n) = 0$ and hence that $x_0^*(x_n) = 0$ for all n, which is clearly impossible since $\{x_n\}$ is dense in $\mathfrak{X}$.

3.12 Analytic continuation. Theorem 3.11.5 has a number of important consequences, the basic one being that *the Weierstrass principle of analytic continuation applies to vector-valued functions.*

Starting with an element of the function, a power series in $(\zeta - \zeta_0)$ with coefficients in $\mathfrak{X}$ and strongly convergent for $|\zeta - \zeta_0| < \rho(\zeta_0)$, we obtain the totality of *regular elements* by the usual method of continuation. To this set we add the *algebraic elements* corresponding to the algebraic singularities including ordinary poles. *The set of all regular and algebraic elements constitutes the vector-valued analytic function $x(\zeta)$.*

We note that the *theorem of monodromy* is valid for vector-valued analytic functions, that is, *if a regular element of $x(\zeta)$ can be continued analytically along every path in a simply-connected domain D, then the resulting regular elements form a holomorphic function in D.*

Suppose for the sake of simplicity that the analytic function $x(\zeta)$ is single-valued. Then the *set of points which are centers of regular elements is the domain of holomorphism of $x(\zeta)$*. Adding the set of poles, we get the *domain of existence* of $x(\zeta)$ which in this case coincides with the *domain of meromorphism. Every boundary point of the domain of holomorphism is a singular point of $x(\zeta)$*. The accessible boundary points may be characterized in the usual manner by the fact that the radius of convergence of a regular element tends to zero when the center approaches the boundary point in question. We note that *a singular point of $x(\zeta)$ is necessarily a singular point of at least one of the scalar functions $x^*[x(\zeta)]$*. If Σ is the union of the singular points of all functions $x^*[x(\zeta)]$, then the boundary of Σ is the boundary of the domain of holomorphism of $x(\zeta)$, that is, the set of singular points of $x(\zeta)$. A similar situation holds for holomorphic operator functions. A singular point of $U(\zeta)$ is necessarily a singular point of at least one scalar function $y^*\{U(\zeta)[x]\}$ and the boundary of the union of the singular points of all functions $y^*\{U(\zeta)[x]\}$ is the set of singularities of $U(\zeta)$.

The *law of permanency of functional equations* has only limited scope for vector-valued functions since multiplication is not defined in ordinary (B)-spaces. However, it does hold for linear functional equations in the following sense:

Let $T[x; \zeta]$ be a linear transformation on $\mathfrak{X}$ to $\mathfrak{Y}$, not necessarily bounded. Suppose there exists a fixed domain D in the ζ-plane such that if $x(\zeta)$ is holomorphic in any domain $D_0 \subset D$, then $T[x(\zeta); \zeta]$ exists and is holomorphic in D_0. Now let $x(\zeta)$ be a solution of the functional equation

$$T[x; \zeta] = \theta$$

holomorphic in $D_0 \subset D$. If $x(\zeta)$ can be continued analytically along a path C in D, then $T[x(\zeta); \zeta]$ can be continued along the same path and $T[x(\zeta); \zeta] \equiv \theta$ on C.

The first half of the assertion follows from the initial assumption of T, the second half is then a consequence of Theorem 3.11.5.

We shall require the following theorem in Chapter XVII.

THEOREM 3.12.1. *Let $x(\zeta)$ be a vector-valued function, defined and holomorphic in a domain D to $\mathfrak{X}$. Let $\zeta(t, \tau)$ be a complex-valued function, defined on the unit square $0 \leqq t, \tau \leqq 1$, continuous in (t, τ) and such that $\zeta(0, \tau) \in D$ for $0 \leqq \tau \leqq 1$. Suppose for each $\tau \in [0, 1]$ that $x(\zeta)$ can be continued analytically along the path $\zeta(t, \tau)$, $0 \leqq t \leqq 1$. Then for any continuous map $\tau(t)$ of $[0, 1]$ into itself, $x(\zeta)$ can be continued analytically along the path $\zeta[t, \tau(t)]$, $0 \leqq t \leqq 1$. Further, $x(\zeta)$ will attain the same element at the point $\zeta[1, \tau(1)]$ along the two paths $\zeta[t, \tau(1)]$ and $\zeta[t, \tau(t)]$, $0 \leqq t \leqq 1$.*

PROOF. Given (t, τ), there exists by assumption a function element $x[\zeta; t, \tau]$ defined in a largest circle about $\zeta(t, \tau)$, say of radius $\rho(t, \tau)$, and numbers $\delta(\tau)$ such that $x[\zeta; t_1, \tau]$ coincides with $x[\zeta; t_2, \tau]$ on their common domain of existence if $| t_1 - t_2 | < \delta(\tau)$. In order to prove the theorem it clearly suffices to show for each (t_0, τ_0) that $x[\zeta; t, \tau]$ coincides with $x[\zeta; t_0, \tau_0]$ on their common domain of existence whenever $(| t - t_0 | + | \tau - \tau_0 |)$ is sufficiently small. Let $r(\tau) \equiv \inf [\rho(t, \tau); 0 \leqq t \leqq 1]$. Then the extension of $x(\zeta)$ along the path $\zeta(t, \tau_1)$ is completely determined by its extension along $\zeta(t, \tau_0)$ if

$$d(\tau_1, \tau_0) \equiv \sup [| \zeta(t, \tau_1) - \zeta(t, \tau_0) | ; 0 \leqq t \leqq 1] < \tfrac{1}{2}r(\tau_0).$$

This assertion can be established by the following argument. Choose $\eta \equiv \eta(\tau_0)$ so that

$$\sup [| \zeta(t_1, \tau) - \zeta(t_2, \tau) | ; 0 \leqq \tau \leqq 1] < \tfrac{1}{8}r(\tau_0) \equiv \sigma$$

whenever $| t_1 - t_2 | < \eta$. Then the circle C_k of radius σ about $\zeta(k\eta, \tau_1)$ is contained in the domain of existence of both $x[\zeta; k\eta, \tau_0]$ and $x[\zeta; (k + 1)\eta, \tau_0]$. Assuming $x[\zeta; k\eta, \tau_1]$ to coincide with $x[\zeta; k\eta, \tau_0]$ in C_k (as is the case for $k = 0$), we see that $x[\zeta; k\eta, \tau_1]$ also coincides with $x[\zeta; (k + 1)\eta, \tau_0]$ in C_k. On the other hand, $x[\zeta; (k + 1)\eta, \tau_1]$ also provides the analytic continuation of $x(\zeta)$ from C_k into C_{k+1}. The assertion now follows by induction since the overlapping circles $\{C_k\}$ cover the path $\zeta(t, \tau_1)$, $0 \leqq t \leqq 1$. Finally given any t_0, $k\eta \leqq t_0 \leqq (k + 1)\eta$, then for $(k - 1)\eta < t_1 < (k + 2)\eta$ we see that the domain of $x[\zeta; t_0, \tau_0]$ contains $\zeta(k\eta, \tau_0)$, $\zeta((k + 1)\eta, \tau_0)$, C_k, C_{k+1}, and consequently $\zeta(t_1, \tau_1)$. Since the successive pairs of functions $x[\zeta; t_0, \tau_0]$, $x[\zeta; k\eta, \tau_0]$ (or $x[\zeta; (k + 1)\eta, \tau_0]$), $x[\zeta; k\eta, \tau_1]$ (or $x[\zeta; (k + 1)\eta, \tau_1]$), $x[\zeta; t_1, \tau_1]$ coincide on their common domains, it follows that the same is true of $x[\zeta; t_0, \tau_0]$ and $x[\zeta; t_1, \tau_1]$. This holds therefore if $d(\tau_1, \tau_0) < \tfrac{1}{2}r(\tau_0)$ and $| t_1 - t_0 | < \eta(\tau_0)$.

3.13. The principle of the maximum. We return to the Cauchy integral representation of $x(\zeta)$. If C is taken to be the circle $| \zeta - \zeta_0 | = r$, the formula may be written

$$x(\zeta_0) = \frac{1}{2\pi i} \int_0^{2\pi} x(\zeta_0 + re^{i\theta})\, d\theta.$$

Consequently

$$\| x(\zeta_0) \| \leqq \frac{1}{2\pi} \int_0^{2\pi} \| x(\zeta_0 + re^{i\theta}) \|\, d\theta.$$

This formula has important implications.

THEOREM 3.13.1. *If $x(\zeta)$ is a holomorphic function on D to $\mathfrak{X}$, then $\| x(\zeta) \|$ is a subharmonic function of ζ in D. It follows that $\| x(\zeta) \|$ can have no maximum in D unless $\| x(\zeta) \|$ is of constant value throughout D.*

REMARK. Unlike the classical case, $\| x(\zeta) \|$ may have a minimum other than zero in D as the following example shows. Let Z_2 be the (B)-space of complex number pairs $x = (z_1, z_2)$ where $\| x \| = \max [| z_1 |, | z_2 |]$. Set $a_1 = (1, 0)$ and $a_2 = (0, 1)$. Then $x(\zeta) = a_1 + a_2\zeta$ is a non-constant entire function. However $\| x(\zeta) \| = 1$ for $| \zeta | \leqq 1$ and $\| x(\zeta) \| = | \zeta |$ for $| \zeta | > 1$.

From Theorem 3.13.1, we obtain the *principle of the maximum*:

Let $x(\zeta)$ be defined in a domain D of the extended plane and on its boundary C, holomorphic in D and strongly continuous in $D \cup C$. If $\sup [\| x(\zeta) \|; \zeta \in C] = M$, *then either $\| x(\zeta) \| \equiv M$ or $\| x(\zeta) \| < M$ in D.*

There is a large group of theorems in classical function theory which is more or less closely attached to the principle of the maximum. The common characteristic of these theorems is that fairly meager information concerning the absolute value of a holomorphic function permits far-reaching conclusions concerning the properties of the function. Such theorems as a rule may be carried over to vector-valued functions, simply replacing statements concerning $| f(\zeta) |$ by the corresponding statement concerning $\| x(\zeta) \|$. Often the classical proofs carry over directly; if not, the assumptions show that the classical theorem applies to the numerically-valued functions $x^*[x(\zeta)]$ and the desired conclusion for $x(\zeta)$ is reached with the aid of the principle of uniform boundedness or from the fact that the vanishing of all functionals of an element in $\mathfrak{X}$ implies that the element is θ. The following theorems belong to this group. The first is the *extended theorem of Liouville*. We recall that a *function which is holomorphic in the finite ζ-plane is said to be entire.*

THEOREM 3.13.2. *An entire function $x(\zeta)$ such that $\| x(re^{i\theta}) \| \leqq Mr^{\alpha}$, $0 \leqq \theta \leqq 2\pi$, α fixed $\geqq 0$, for all large r, is a polynomial in ζ of degree $\leqq \alpha$. It is a constant if $\alpha < 1$.*

THEOREM 3.13.3. *If $x(\zeta)$ is holomorphic in $0 < | \zeta - \zeta_0 | < R$ and if moreover $\| x(\zeta_0 + re^{i\theta}) \| \leqq Mr^{-\alpha}$, $0 \leqq \theta \leqq 2\pi$, α fixed $\geqq 0$, for all small values of r, then $\zeta = \zeta_0$ is a pole of order $\leqq \alpha$.*

Both of these theorems follow from the estimates (3.11.3), the first on letting $r \to \infty$ and the second on letting $r \to 0$. Another result of this kind is the extended *Schwarz's lemma.*

THEOREM 3.13.4. *Let $x(\zeta)$ be holomorphic in $| \zeta | < 1$, $\| x(\zeta) \| \leqq M$, and $x(0) = \theta$. Then $\| x(\zeta) \| \leqq M | \zeta |$ for $| \zeta | < 1$.*

PROOF. The classical form of the lemma shows that $| x^*[x(\zeta)] | \leqq M | \zeta |$ for $| \zeta | < 1$ and every functional $x^* \in \mathfrak{X}^*$ of norm one. Theorem 2.7.4 asserts that for each fixed ζ there exists an x^* such that $\| x^* \| = 1$ and $| x^*[x(\zeta)] | = \| x(\zeta) \|$. It follows that $\| x(\zeta) \| \leqq M | \zeta |$ for $| \zeta | < 1$.

LEMMA 3.13.1. *Suppose that $x(\zeta)$ is an entire function and that there exists a numerical trigonometric polynomial $P(\theta)$ such that*

$$(3.13.1) \qquad | P(\theta) | \, \| x(re^{i\theta}) \| \leqq Mr^{\alpha} \text{ for all } r \geqq r_0 ,$$

where M, r_0, α are positive constants. Then $x(\zeta)$ is a polynomial of degree $\leqq \alpha$.

REMARK. The lemma is of value only if $P(\theta)$ is allowed the value zero: otherwise the result follows from Theorem 3.13.2. The lemma generalizes a result due to M. H. Stone [6] in which $P(\theta) = [\cos \theta]^n$.

PROOF. We can express the functions $x(\zeta)$ and $P(\theta)$ as

$$x(\zeta) = \sum_{\nu=0}^{\infty} a_\nu \zeta^\nu, \qquad a_\nu \in \mathfrak{X},$$

and

$$P(\theta) = \sum_{\mu=k}^{m} c_\mu e^{\mu i\theta},$$

where $c_k \neq 0$. For each positive integer p, we then form the integral

$$J(r, p) = r^{-p} \frac{1}{2\pi} \int_0^{2\pi} P(\theta) x(re^{i\theta}) e^{-(k+p)i\theta} \, d\theta.$$

By assumption, $J(r, p) \to \theta$ as $r \to \infty$ if $p > \alpha$. On the other hand if we substitute the series expansions for $x(re^{i\theta})$ and $P(\theta)$ into the integrand and then integrate termwise, we obtain

$$J(r, p) = \sum c_\mu a_\nu r^{\nu - p}$$

where $\mu + \nu = p + k$, $k \leqq \mu \leqq m$, $\nu \geqq 0$. Here every power of r has a negative exponent except one ($\nu = p$). Consequently $\lim_{r\to\infty} J(r, p) = c_k a_p$. Combining this with the earlier limit, we obtain $a_p = \theta$ for all $p > \alpha$, as was to be proved.

If $\alpha = n$, a positive integer, then setting $p = n$ in the above argument, we see that

$$| c_k | \, \| a_n \| \leqq M.$$

From this relation we obtain the

COROLLARY. *If the inequality* (3.13.1) *is replaced by*

$$\lim_{r\to\infty} r^{-n} | P(\theta) | \, \| x(re^{i\theta}) \| = 0$$

uniformly in θ, then $x(\zeta)$ is a polynomial of degree $< n$.

LEMMA 3.13.2. *If $0 \leqq \alpha < 1$, then $n^\alpha n! \leqq 2\Gamma(n + 1 + \alpha)$ for $n \geqq 1$.*

PROOF. Set

$$J(n, \alpha) = \frac{1}{n^\alpha n!} \int_0^\infty e^{-\sigma} \sigma^{n+\alpha} \, d\sigma.$$

Then for $0 \leqq \alpha < 1$ we have

$$J(n, \alpha) = \left(\frac{n-1}{n}\right)^{\alpha} \frac{n+\alpha}{n} J(n-1, \alpha) > \left(1 - \frac{1}{n^2}\right) J(n-1, \alpha)$$

$$> \left\{\prod_{j=2}^{n} \left(1 - \frac{1}{j^2}\right)\right\} J(1, \alpha) > \tfrac{1}{2} J(1, \alpha) \geqq \tfrac{1}{2}.$$

THEOREM 3.13.5. *If $x(\zeta)$ is an entire function of $1/(\zeta - 1)$ and if in the expansions*

$$x(\zeta) = \sum_{n=0}^{\infty} a_n \zeta^n, \qquad | \zeta | < 1,$$

$$x(\zeta) = \sum_{n=0}^{\infty} b_n \zeta^{-n}, \qquad | \zeta | > 1,$$

both $\| a_n \|$ and $\| b_n \|$ are $o(n^{\alpha})$, $\alpha > 0$, then $\zeta = 1$ is a pole of order $< \alpha + 1$, that is, $x(\zeta)$ is a polynomial in $1/(\zeta - 1)$ of degree $< \alpha + 1$.

REMARK. The classical analogue of Theorem 3.13.5 is due to G. Pólya [3]. The argument was extended to the vector-valued case by E. Hille [11] and later simplified by M. H. Stone [6].

PROOF. The hypothesis together with Lemma 3.13.2 implies that for any given $\epsilon > 0$ there exists an N_ϵ such that

$$\| a_n \|, \| b_n \| \leqq \epsilon \frac{\Gamma(\alpha + n + 1)}{\Gamma(\alpha + 1)\Gamma(n + 1)} = \epsilon(-1)^n \binom{-\alpha - 1}{n}$$

for $n > N_\epsilon$. For $n \leqq N_\epsilon$ we have a similar inequality with ϵ replaced by a suitable quantity M. For $| \zeta | = r < 1$ this implies

$$\| x(\zeta) \| \leqq M(1 - r)^{-1-\alpha},$$

and there exists an r'_ϵ, $0 < r'_\epsilon < 1$, such that

$$\| x(\zeta) \| \leqq 2\epsilon(1 - r)^{-1-\alpha}$$

for $r'_\epsilon < r < 1$. Similarly we get

$$\| x(\zeta) \| \leqq 2\epsilon(1 - 1/r)^{-1-\alpha}$$

for $1 < r < r''_\epsilon$. Putting

$$1 - \zeta = (\omega + \tfrac{1}{2})^{-1},$$

we define

$$y(\omega) = x(\zeta).$$

By assumption $y(\omega)$ is an entire function of ω. Now

$$r = | \zeta | = | (\omega - \tfrac{1}{2})/(\omega + \tfrac{1}{2}) |.$$

Hence if we set $\omega = u + iv = Re^{i\theta}$, then $0 \leqq r < 1$ is equivalent to $-\pi/2 < \theta < \pi/2$ and $1 < r \leqq \infty$ is equivalent to $\pi/2 < \theta < 3\pi/2$. For $R \geqq 1$ we see that $|\omega + \frac{1}{2}|$, $|\omega - \frac{1}{2}| \leqq 3R/2$. Hence for $-\pi/2 < \theta < \pi/2$ (i.e. $0 \leqq r < 1$) and $R \geqq 1$, we have

$$(1 - r)^{-1} = \frac{|\omega + \frac{1}{2}|}{|\omega + \frac{1}{2}| - |\omega - \frac{1}{2}|} = \frac{|\omega + \frac{1}{2}|\,[|\omega + \frac{1}{2}| + |\omega - \frac{1}{2}|]}{|\omega + \frac{1}{2}|^2 - |\omega - \frac{1}{2}|^2}$$

$$= \frac{1}{2u}|\omega + \tfrac{1}{2}|\,[|\omega + \tfrac{1}{2}| + |\omega - \tfrac{1}{2}|] \leqq \frac{9}{4}\frac{R}{\cos\theta}.$$

Similarly for $\pi/2 < \theta < 3\pi/2$ (i.e. $1 < r \leqq \infty$) and $R \geqq 1$, we have

$$(1 - 1/r)^{-1} \leqq \frac{9}{4}\frac{R}{\cos\theta}.$$

Since $\zeta \to 1$ as $R \to \infty$, it follows that

$$R^{-\alpha-1}\,|\cos\theta|^{\alpha+1}\,\|y(Re^{i\theta})\| \to 0$$

as $R \to \infty$, uniformly in θ. If m is the least integer $\geqq \alpha + 1$, we have *a fortiori*

$$R^{-\alpha-1}\,|\cos\theta|^{m}\,\|y(Re^{i\theta})\| \to 0.$$

Therefore the above corollary applies to $y(\omega)$ and shows that this function is a polynomial in ω of degree less than $\alpha + 1$. But this means that $x(\zeta)$ is a polynomial in $1/(\zeta - 1)$ of degree $<\alpha + 1$. This is the required result.

The results of the classical investigations centering around the *Phragmén-Lindelöf theorem* also admit of extensions to vector-valued functions. Some of these theorems are listed below without proofs. The reader will find proofs of the classical prototypes and indications of the extensions in the following references. The classical case of Theorem 3.13.6 is due to F. and R. Nevanlinna [1] (extension I. Gelfand [5]); Theorem 3.13.7 is due to F. Carlson [1] for numerically-valued functions (extension E. Hille [7]); Theorem 3.13.8 goes back to G. Pólya [3] (see Szegö [1] for proof).

THEOREM 3.13.6. *Let $x(\zeta)$ be holomorphic in $\Re(\zeta) > 0$. For every finite point on the imaginary axis and for every $\epsilon > 0$ there shall exist a (semi-circular) neighborhood in which $\|x(\zeta)\| \leqq 1 + \epsilon$. Then either $\|x(\zeta)\| \leqq 1$ for all ζ in $\Re(\zeta) > 0$ or*

$$\liminf_{r\to\infty} \frac{1}{r}\int_{-\pi/2}^{\pi/2} \log^{+}\|x(re^{i\theta})\|\cos\theta\,d\theta > 0.$$

Here $\log^{+}\alpha = \max(\log\alpha, 0)$.

THEOREM 3.13.7. *Let $x(\zeta)$ be holomorphic in $\Re(\zeta) > 0$ and strongly continuous in $\Re(\zeta) \geqq 0$. Suppose that*

(i) $\|x(\pm ir)\| \leqq Ce^{\pi r}$,

(ii) $\| x(re^{i\theta}) \| \leqq C \exp [\lambda(\theta) r]$, $-\pi/2 \leqq \theta \leqq \pi/2$, *where* $\lambda(\theta) \leqq M$, $\lambda(-\theta) = \lambda(\theta)$, *and*

$$\limsup_{\delta \to 0} \frac{1}{\delta} \left\{ \pi - \lambda \left(\frac{\pi}{2} - \delta \right) \right\} = +\infty,$$

and

(iii) $x(n) = \theta$, $n = 1, 2, 3, \cdots$.
Then $x(\zeta) \equiv \theta$.

THEOREM 3.13.8. *If* $x(\zeta)$ *is an entire function of order one and minimal type and if* $\| x(\pm n) \| = o(n^{\alpha})$, $\alpha > 0$, *then* $x(\zeta)$ *is a polynomial in* ζ *of degree* $<\alpha$. *In particular,* $x(\zeta)$ *is a constant if* $\| x(\pm n) \|$ *is bounded with respect to* n.

For vector-valued entire functions $x(\zeta)$ the notions of order and type are defined by obvious modifications of the classical concepts Thus $x(\zeta)$ *is of order* ρ *if*

$$\rho = \limsup_{r \to \infty} [\log \log M(r; x)]/[\log r]$$

where $M(r; x) = \max \| x(re^{i\theta}) \|$. *A function of order* ρ *is of type* α *if*

$$\alpha = \limsup_{r \to \infty} r^{-\rho} \log M(r; x).$$

The function is of minimal type if $\alpha = 0$, *normal type if* $0 < \alpha < \infty$, *maximal type if* $\alpha = +\infty$.

3.14. The theorem of Vitali. This name covers two distinct but closely related propositions dealing with functions which are holomorphic and uniformly bounded in a fixed domain D. The first asserts that a sequence of such functions which converges in a point set having a limit point in the domain converges uniformly in any compact subset of D; whereas the second asserts that any family of such functions contains a sequence, uniformly convergent in any compact subset of D. The first of these propositions carries over directly to the vector-valued case without modification; the second requires an additional compactness assumption.

THEOREM 3.14.1. *Let* $\{x_n(\zeta)\}$ *be holomorphic in a fixed domain* D *and* $\| x_n(\zeta) \| \leqq M$ *for all* n *and all* ζ *in* D. *Let there be a set* $\{\zeta_k\}$ *in* D *having a limit point* ζ_0 *in* D *such that* $\lim_{n\to\infty} x_n(\zeta_k)$ *exists for each* $k = 1, 2, \cdots$. *Then* $\lim_{n\to\infty} x_n(\zeta)$ *exists everywhere in* D, *the convergence is uniform with respect to* ζ *in any compact subset of* D, *and the limit function* $x(\zeta)$ *is holomorphic in* D.

PROOF. Here we follow the argument due to E. Lindelöf [3] in the classical case. We first prove the theorem for the special case in which D is the interior of the unit circle, $\zeta_0 = 0$, and $M = 1$. We then have

$$x_n(\zeta) = \sum_{m=0}^{\infty} a_{nm} \zeta^m, \qquad | \zeta | < 1,$$

and $\| a_{nm} \| \leqq 1$ for all n and m. For $| \zeta | < \frac{1}{2}$, we have

$$\| x_n(\zeta) - a_{n0} \| \leqq \sum_{m=1}^{\infty} \| a_{nm} \| \, | \zeta |^m < 2 | \zeta |$$

for every n. Hence for a fixed k with $| \zeta_k | < \frac{1}{2}$

$$\begin{aligned} \| a_{n0} - a_{p0} \| &\leqq \| x_n(\zeta_k) - a_{n0} \| + \| x_n(\zeta_k) - x_p(\zeta_k) \| + \| x_p(\zeta_k) - a_{p0} \| \\ &\leqq 4 | \zeta_k | + \| x_n(\zeta_k) - x_p(\zeta_k) \| \end{aligned}$$

and

$$\limsup_{n, p \to \infty} \| a_{n0} - a_{p0} \| \leqq 4 | \zeta_k | .$$

Since $\zeta_k \to 0$ when $k \to \infty$, it follows that $\{a_{n0}\}$ is a Cauchy sequence and has a strong limit, say a_0, and $\| a_0 \| \leqq 1$.

We now put $x_{n,1}(\zeta) = [x_n(\zeta) - a_{n0}]/\zeta$. Again for $| \zeta | < \frac{1}{2}$ we have

$$\| x_{n,1}(\zeta) - a_{n1} \| \leqq \sum_{m=2}^{\infty} \| a_{nm} \| \, | \zeta |^{m-1} < 2 | \zeta | .$$

Furthermore $\lim_{n \to \infty} x_{n,1}(\zeta_k)$ exists for each k. Using the preceding argument we see that $\lim_{n \to \infty} a_{n1} = a_1$ exists and $\| a_1 \| \leqq 1$. By complete induction we get the existence of $\lim_{n \to \infty} a_{nm} = a_m$ for all m and $\| a_m \| \leqq 1$. Now form

$$x(\zeta) = \sum_{m=0}^{\infty} a_m \zeta^m, \qquad | \zeta | < 1.$$

Then for $| \zeta | \leqq r < 1$

$$\| x_n(\zeta) - x(\zeta) \| \leqq \sum_{m=0}^{k} \| a_{nm} - a_m \| r^m + 2r^{k+1}/(1 - r),$$

whence we conclude that $\| x_n(\zeta) - x(\zeta) \| \to 0$ as $n \to \infty$, uniformly with respect to ζ for $| \zeta | \leqq r < 1$. This completes the proof of the special case.

The extension to the general case is now routine analysis. Each point ζ' of D can be joined to ζ_0 by a broken line in D; the broken line can be covered by a finite number of circles, the first with its center at ζ_0 and each successive circle having its center in the preceding circle. The above argument proves convergence for the first circle, whence it spreads to the consecutive circles and finally to the circle about ζ'. The convergence will be uniform in some smaller circle about ζ'. We now proceed to cover each point of a given compact subset S by a circle in which we have uniform convergence. Since S is compact, a finite subcovering exists and the convergence will be uniform on this finite covering.

THEOREM 3.14.2. *Let $[x(\zeta)]$ be a family of functions, holomorphic and uniformly bounded in a fixed domain D. Suppose further that all of the ranges of the $x(\zeta)$ lie in a compact (weakly compact) subset of $\mathfrak{X}$. Then there exists a sequence of functions in this family which converges strongly (weakly) at each $\zeta \in D$ to a function holomorphic in D, the convergence being uniform on each compact subset of D.*

Proof. Choose a sequence of points $\{\zeta_k\}$ in D having a limit point ζ_0 in D. For each ζ_k, the elements $[x(\zeta_k)]$ lie in a compact (weakly compact) subset of $\mathfrak{X}$ and therefore a subsequence of the $[x(\zeta_k)]$ will converge strongly (weakly) to an element of $\mathfrak{X}$. By the diagonal process we may extract a subsequence $\{x_n(\zeta)\}$ which converges strongly (weakly) for all of the points $\{\zeta_k\}$. For the case of strong compactness, the result now follows from the previous theorem. For the case of weak compactness, it follows that $x^*[x_n(\zeta)]$ converges to a numerically-valued holomorphic function $f(\zeta; x^*)$ for each $x^* \in \mathfrak{X}^*$, the convergence being uniform on each compact subset of D. Now for fixed $\zeta' \in D$, the $\{x_n(\zeta')\}$ lie in a weakly compact subset so that a subsequence $\{x_{n_k}(\zeta')\}$ converges weakly to some $y(\zeta') \in \mathfrak{X}$. It follows that

$$x^*[y(\zeta')] = \lim_{k\to\infty} x^*[x_{n_k}(\zeta')] = f(\zeta'; x^*) = \lim_{n\to\infty} x^*[x_n(\zeta')]$$

for each $x^* \in \mathfrak{X}^*$. Thus $\{x_n(\zeta)\}$ converges weakly to $y(\zeta)$ and the convergence is uniform in each compact subset of D for each x^*. Finally we see that $y(\zeta)$ is holomorphic in D since $x^*[y(\zeta)] = f(\zeta; x^*)$ for all $x^* \in \mathfrak{X}^*$.

We conclude this section with an application of the Vitali theorem to a theorem of the Phragmén-Lindelöf type.

Theorem 3.14.3. *Let $x(\zeta)$ be holomorphic and bounded in the sector*

$$\alpha < \arg \zeta < \beta, 0 < R < |\zeta|,$$

and suppose that

$$\lim_{r\to\infty} x(re^{i\theta_0}) = x_0$$

for some $\theta_0 \in (\alpha, \beta)$. Then $\lim_{r\to\infty} x(re^{i\theta}) = x_0$ for all $\theta \in (\alpha, \beta)$, the convergence being uniform in $\alpha + \epsilon \leqq \theta \leqq \beta - \epsilon$ for each $\epsilon > 0$.

Proof. We consider the sequence of functions $x_n(\zeta) \equiv x(n\zeta)$ defined on the domain D: $\alpha < \arg \zeta < \beta$, $R < |\zeta| < 2R$. The functions $\{x_n(\zeta)\}$ are holomorphic and uniformly bounded on D and $x_n(\zeta) \to x_0$ for all $\zeta = re^{i\theta_0}$, $R < r < 2R$. By Theorem 3.14.1 there exists a function $y(\zeta)$ holomorphic in D such that $x_n(\zeta) \to y(\zeta)$ for each $\zeta \in D$, the convergence being uniform in $\alpha + \epsilon \leqq \arg \zeta \leqq \beta - \epsilon$, $5R/4 \leqq |\zeta| \leqq 7R/4$, for each $\epsilon > 0$. By assumption $y(re^{i\theta_0}) = x_0$ for $R < r < 2R$ and hence $y(\zeta) \equiv x_0$ in D. Now the intervals $[5nR/4, 7nR/4]$ and $[5(n+1)R/4, 7(n+1)R/4]$ overlap for $n \geqq 3$. Consequently if we translate the above information on the sequence $\{x_n(\zeta)\}$ into behavior of $x(\zeta)$ we obtain the desired result.

3.15. Vector-valued functions of several complex variables. A considerable portion of the theory developed in the preceding sections may be extended to the case of functions of several complex variables. In the present section we give only a couple of theorems, due to Max Zorn, which are basic for the discussion in Chapter XXVI.

Consider a function $f(\boldsymbol{\zeta}) = f(\zeta_1, \cdots, \zeta_n)$ on Z_n to a (B)-space $\mathfrak{X}$, defined in some domain Δ of Z_n. The partial derivative of $f(\zeta_1, \cdots, \zeta_n)$ with respect to ζ_k is defined by the usual convention

$$\frac{\partial f}{\partial \zeta_k} = \lim_{\alpha_k \to 0} \alpha_k^{-1}[f(\zeta_1, \cdots, \zeta_k + \alpha_k, \cdots, \zeta_n) - f(\zeta_1, \cdots, \zeta_k, \cdots, \zeta_n)],$$

the limit, which is taken in the sense of strong convergence in $\mathfrak{X}$, being independent of the manner in which α_k tends to 0.

THEOREM 3.15.1. *Let $f(\zeta_1, \cdots, \zeta_n)$ on Z_n to $\mathfrak{X}$ have first order partials with respect to each ζ_k when $\boldsymbol{\zeta} = (\zeta_1, \cdots, \zeta_n)$ lies in some domain Δ containing the origin. Then:*

(i) *$f(\zeta_1, \cdots, \zeta_n)$ has partial derivatives of all orders and the mixed partials are independent of the order of differentiation;*

(ii) *$f(\boldsymbol{\zeta})$ is continuous and bounded on every bounded closed subset of Δ;*

(iii) *if $\| f(\boldsymbol{\zeta}) \| \leqq M$ for $\| \boldsymbol{\zeta} \| \leqq \rho$ and $\sigma < \rho$, then for $\| \boldsymbol{\zeta} \| \leqq \sigma$*

$$\left\| f(\boldsymbol{\zeta}) - f(0) - \sum_{k=1}^{n} \left(\frac{\partial f}{\partial \zeta_k}\right)_0 \zeta_k \right\| \leqq [M/\rho(\rho - \sigma)] \, \| \boldsymbol{\zeta} \|^2;$$

(iv) *$f(\zeta, \zeta, \cdots, \zeta)$ is differentiable with respect to ζ.*

PROOF. We may assume that this theorem is known for the special case in which $\mathfrak{X} = Z_1$ and we reduce the general case to this particular instance with the aid of the bounded linear functionals $x^* \in \mathfrak{X}^*$.

(i) The numerical function $x^*[f(\zeta_1, \cdots, \zeta_n)]$ is partially differentiable since strong differentiability implies the weak kind. Hence

$$\frac{\partial}{\partial \zeta_j} x^*[f(\zeta_1, \cdots, \zeta_n)] = x^* \left[\frac{\partial}{\partial \zeta_j} f(\zeta_1, \cdots, \zeta_n)\right]$$

exists and is partially differentiable with respect to ζ_k, $k = 1, 2, \cdots, n$, for every $x^* \in \mathfrak{X}^*$. This implies the existence of second order partials of $f(\zeta_1, \cdots, \zeta_n)$ itself (in the sense of strong differentiability) and proves the first part of assertion (i) since the higher derivatives can be handled by an induction argument. The relation

$$x^* \left\{\frac{\partial^2 f}{\partial \zeta_j \partial \zeta_k}\right\} = \frac{\partial^2}{\partial \zeta_j \partial \zeta_k} x^*[f] = \frac{\partial^2}{\partial \zeta_k \partial \zeta_j} x^*[f] = x^* \left\{\frac{\partial^2 f}{\partial \zeta_k \partial \zeta_j}\right\},$$

which is valid for every $x^* \in \mathfrak{X}^*$, shows that the order of differentiation is immaterial.

(ii) This part as well as (iv) follows from (iii); let us only establish the boundedness as it will be necessary for the proof of (iii).

We consider the numerically-valued functions $x^*[f(\boldsymbol{\zeta})]$ on an arbitrary closed bounded subset E_0 of Δ. For every fixed x^* the set $| x^*[f(E_0)] |$ is bounded; by the theory of uniform boundedness it follows that $\| f(E_0) \|$ is also bounded.

(iii) Now consider a ρ such that $\| \boldsymbol{\zeta} \| \leqq \rho$ lies in Δ. This set may therefore serve as an E_0; let M be the corresponding bound. For the numerical functions $x^*[f(\boldsymbol{\zeta})]$ this gives

$$| x^*[f(\boldsymbol{\zeta})] | \leqq M \, \| x^* \| \text{ in } E_0 .$$

Hence, by the numerical case,

$$\left| x^*[f(\zeta)] - x^*[f(0)] - \sum_{k=1}^{n} \left\{ \frac{\partial}{\partial \zeta_k} x^*[f] \right\}_0 \zeta_k \right|$$

$$= \left| x^* \left\{ f(\zeta) - f(0) - \sum_{k=1}^{n} \left(\frac{\partial f}{\partial \zeta_k} \right)_0 \zeta_k \right\} \right|$$

$$\leqq \| x^* \| \, [M/\rho(\rho - \sigma)] \, \| \zeta \|^2$$

for $\| \zeta \| \leqq \sigma < \rho$. An application of Theorem 2.7.4 yields (iii). As was observed above, (iii) implies the continuity of $f(\zeta)$ and part (iv).

THEOREM 3.15.2. *Let $\sum_{k=0}^{\infty} P_k(\zeta_1 , \zeta_2)$ be a series of homogeneous polynomials*

$$P_k(\zeta_1 , \zeta_2) = \sum_{n=0}^{k} a_{kn} \zeta_1^n \zeta_2^{k-n},$$

a_{kn} constants in $\mathfrak{X}$, and let Δ be an open set in the space Z_2. The following facts are then equivalent:

(i) *the series converges at each point of Δ;*

(ii) *the terms P_k are uniformly bounded in any bounded neighborhood Δ_0 of any point $(\zeta_{10} , \zeta_{20})$ in Δ such that $\bar{\Delta}_0 \subset \Delta$;*

(iii) *every point of Δ has a neighborhood Δ_1 such that $\sum_{k=0}^{\infty} \sup_{\Delta_1} \| P_k(\zeta_1 , \zeta_2) \|$ is convergent.*

PROOF. It is easy to show that (ii) implies (iii). For let $\| P_k \| \leqq M$ in Δ_0; then for sufficiently small ϵ, $0 < \epsilon < 1$, the set $(1 - \epsilon)\Delta_0$ will still be a neighborhood of $(\zeta_{10} , \zeta_{20})$ and in it we have $\| P_k \| \leqq M(1 - \epsilon)^k$ by virtue of the homogeneity of P_k. That (iii) implies (i) is trivial, but the implication (i) implies (ii) is not. Here we have recourse to the numerical case for which the theorem has been proved by F. Hartogs [1]. He states that the series converges locally uniformly in Δ; this implies uniform convergence on bounded closed subsets and hence (ii) for the numerical case.

If $\sum P_k(\zeta_1 , \zeta_2)$ is convergent in Δ, then

$$x^* \left\{ \sum_{k=0}^{\infty} P_k(\zeta_1, \zeta_2) \right\} = \sum_{k=0}^{\infty} x^*[P_k(\zeta_1 , \zeta_2)]$$

will be a series of numerical polynomials, convergent in Δ and therefore the terms $x^*[P_k(\zeta_1 , \zeta_2)]$ are uniformly bounded in any Δ_0 (with $\bar{\Delta}_0 \subset \Delta$) for each $x^* \in \mathfrak{X}^*$. Applying the uniform boundedness theorem to the set

$$[P_k(\zeta_1, \zeta_2); \qquad (\zeta_1 , \zeta_2) \in \Delta_0 , k = 1, 2, \cdots]$$

we see that it is bounded, which was to be proved.

References. F. Carlson [1], Dunford [3], Fantappiè [3], Gelfand [5], Hartogs [1], Hilbert [1], Hille [7, 11], Lindelöf [3], Müntz [1], F. and R. Nevanlinna [1], Pólya [3], F. Riesz [2]. M. H. Stone [6], Szegö [1], A. E. Taylor [5, 7], and N. Wiener [1].

3. Analytic Functions on Vectors to Vectors

3.16. (G)-differentiability. The theory of analytic functions on vectors to vectors is of recent date, but the Volterra theory of functions of composition can be regarded as a forerunner, and most of the basic concepts can be traced back to the founders of functional analysis. The fundamental notions of abstract differentials, polynomials, and power series were introduced by M. Fréchet around 1909. A few years later R. Gâteaux applied the ideas of Fréchet to concrete problems (analytic functionals and functions of infinitely many unknowns) and much of the later development is based on the brilliant work of Gâteaux which was published posthumously.

The modern theory began with the work of A. D. Michal and his pupils, in particular I. E. Highberg and R. S. Martin, who developed a theory of abstract power series. Soon thereafter L. M. Graves introduced a theory of analytic functions based upon the Gâteaux differential; these results were greatly augmented by the work of A. E. Taylor. S. Banach also seems to have possessed a theory of analytic operations, but of this only some fragments dealing with polynomial operations were published. J. S. Silva has recently explored certain aspects of this subject by means of an approach which goes back to L. Fantappiè. E. R. Lorch has developed a somewhat different notion of analyticity for functions with domain and range in a commutative Banach algebra. See A. Alexiewicz and W. Orlicz for real analytic functions and also G. Suchumlinov for the case of analytic functionals.

The crux of the theory of vector-valued functions of vectors is the question of differentiability. In this general situation the differentials of Fréchet and of Gâteaux appear to be the most appropriate concepts. In the present paragraph we shall develop a few of the basic ideas in this theory, confining ourselves to those results which are required in later applications. A more complete development of the subject will be found in Chapter XXVI.

We start with a few geometrical notions which are essential to the theory. As we shall see, the finite topology discussed in section 1.10 provides a suitable setting for the Gâteaux differential whereas the norm topology is appropriate for the Fréchet differential.

Definition 3.16.1. *A set $\mathfrak{S}^*(x_0)$ is called a c-star about x_0 if $\mathfrak{S}^*(x_0) = x_0 + H$, where $h \in H$, $|\zeta| \leqq 1$ implies that $\zeta h \in H$.*

We note that if $\mathfrak{D}$ is finitely open and contains x_0, then $\mathfrak{D}$ contains a finitely open *c*-star about x_0 formed by all points $x_0 + h$ for which $|\zeta| \leqq 1$ implies $x_0 + \zeta h \in \mathfrak{D}$. We then speak of the *c*-star in $\mathfrak{D}$ about x_0.

Let us denote by $\rho(x, h)$ the supremum of all numbers ρ such that $|\zeta| \leqq \rho$ implies that $x + \zeta h \in \mathfrak{D}$.

Definition 3.16.2. *Let $y = f(x)$ on $\mathfrak{X}$ to $\mathfrak{Y}$ be defined in the finitely open set $\mathfrak{D}$ and suppose that for every $x \in \mathfrak{D}$ and $h \in \mathfrak{X}$ the quotient $[f(x + \zeta h) - f(x)]/\zeta$,* *which is defined for* $0 < |\zeta| < \rho(x, h)$, *tends to a unique limit as* $\zeta \to 0$. *We then say that*

(i) $f(x)$ *is (G)-differentiable in* $\mathfrak{D}$;

(ii) $$\delta f(x; h) = \delta_x^h f = \lim_{\zeta \to 0} \frac{1}{\zeta} [f(x + \zeta h) - f(x)]$$

is the first variation of $f(x)$ *with increment* h;

(iii) $f(x)$ *possesses a Gâteaux differential in* $\mathfrak{D}$.

We note that if f and g are (G)-differentiable, then so is $\alpha f + \beta g$ and

$$\delta_x^h[\alpha f + \beta g] = \alpha \delta_x^h f + \beta \delta_x^h g.$$

It is also a simple matter to show that

$$\delta f(x; \alpha h) = \alpha \delta f(x; h)$$

so that $\delta f(x; h)$ is *homogeneous of degree one in* h.

DEFINITION 3.16.3. *A function* $f(x)$ *defined on* $\mathfrak{X}$ *to* $\mathfrak{Y}$ *is said to be homogeneous of degree* n *if*

$$f(\alpha x) = \alpha^n f(x).$$

Such a function is called bounded if there exists an $M > 0$ *such that*

$$\| f(x) \| \leq M \| x \|^n, \qquad \text{for all } x \in \mathfrak{X}.$$

DEFINITION 3.16.4. *Let* $y = f(x)$ *on* $\mathfrak{X}$ *to* $\mathfrak{Y}$ *be defined in the open set* $\mathfrak{D}$. *We say that* $f(x)$ *is (F)-differentiable and possesses a total or a Fréchet differential in* $\mathfrak{D}$ *if*

(i) $\delta f(x; h)$ *exists as a bounded homogeneous function of degree one in* h;

(ii) $$\lim_{\|h\| \to 0} \frac{1}{\| h \|} \| f(x + h) - f(x) - \delta f(x; h) \| = 0$$

for all $x \in \mathfrak{D}$.

Actually Definition 3.16.4 contains a redundance: Zorn **[3]** has shown that condition (ii) is implied by (i).

THEOREM 3.16.1. $f(x)$ *is (G)-differentiable in the finitely open set* $\mathfrak{D}$ *if and only if for every* $x \in \mathfrak{D}$ *and* $h \in \mathfrak{X}$, $f(x + \zeta h)$ *is a holomorphic function of* ζ *when* $|\zeta| < \rho(x, h)$.

PROOF. This follows from the observation

$$\left\{\frac{d}{d\zeta} f[x + (\zeta_0 + \zeta)h]\right\}_{\zeta=0} = \lim_{\zeta \to 0} \frac{1}{\zeta} \{f[(x + \zeta_0 h) + \zeta h] - f(x + \zeta_0 h)\}$$
$$= \delta f(x + \zeta_0 h; h).$$

If $f(x)$ is (G)-differentiable in the finitely open set $\mathfrak{D}$ and if $x \in \mathfrak{D}$ we may now define the *nth variation* $\delta^n f(x; h)$ *of* $f(x)$ *with increment* h *as*

$$\delta^n f(x; h) = \left[\frac{d^n}{d\zeta^n} f(x + \zeta h)\right]_{\zeta=0}. \tag{3.16.1}$$

It is clear that $\delta^n f(x; h)$ is homogeneous of degree n in h. Actually much more can be said of the nth variation; we shall show in Chapter XXVI that $\delta^n f(x; h)$ is a homogeneous polynomial of degree n in h. In particular $\delta f(x; h)$ is linear in h. Zorn calls this function the *derivative* of $f(x)$ and denotes it by $f'(x)$ so that $\delta f(x; h) = f'(x)[h]$ in his notation; $f'(x) \in \mathfrak{E}(\mathfrak{X}, \mathfrak{Y})$ if $f(x)$ is (F)-differentiable.

Let us now look at the Taylor development of the function $f(x + \zeta h)$ which must be valid for $|\zeta| < \rho(x, h)$:

$$f(x + \zeta h) = \sum_{n=0}^{\infty} \left[\frac{d^n}{d\alpha^n} f(x + \alpha h)\right]_{\alpha=0} \frac{\zeta^n}{n!}.$$

Making use of (3.16.1) we have

$$f(x + \zeta h) = \sum_{n=0}^{\infty} \delta^n f(x; h) \frac{\zeta^n}{n!} = \sum_{n=0}^{\infty} \frac{1}{n!} \delta^n f(x; \zeta h).$$

Here we may replace ζh by h; reinterpreting the condition on h geometrically we arrive at

THEOREM 3.16.2. *Let* $f(x)$ *be (G)-differentiable in the finitely open set* $\mathfrak{D}$. *Then for* $x \in \mathfrak{D}$ *and* $x + h$ *in the c-star about* x *in* $\mathfrak{D}$, *we have*

$$f(x + h) = \sum_{n=0}^{\infty} \frac{1}{n!} \delta^n f(x; h). \tag{3.16.2}$$

Another way of stating this condition is to say that *the series* (3.16 2) *converges for* $\rho(x, h) > 1$.

From Theorem 3.11.3 it follows that

$$\delta^n f(x; h) = \frac{n!}{2\pi i} \int_C f(x + \zeta h) \zeta^{-n-1} \, d\zeta, \tag{3.16.3}$$

where C is any circle $|\zeta| = \rho'$ with $\rho' < \rho(x, h)$. In particular, if $x + h$ belongs to a c-star in $\mathfrak{D}$ about x, we may choose $\rho' = 1$. This leads to an analogue of the classical estimates of Cauchy:

THEOREM 3.16.3. *Let* $f(x)$ *be (G)-differentiable in the finitely open set* $\mathfrak{D}$. *In any c-star* $\mathfrak{C}^*(a)$ *about* $x = a$ *in which* $\| f(x) \| \leqq M$, *we have*

$$\| \delta^n f(a; h) \| \leqq Mn!$$

for $a + h \in \mathfrak{C}^* (a)$.

THEOREM 3.16.4. *Let* $f(x)$ *be (G)-differentiable in the open domain* $\mathfrak{D}$ *If* $f(x) = \theta$ *throughout some sphere, then* $f(x) \equiv \theta$ *in* $\mathfrak{D}$.

PROOF. Suppose that $f(x) = \theta$ throughout the sphere $\mathfrak{S}_a$: $\| x - a \| < r$. Then formula (3.16.3) shows that $\delta^n f(a; x - a) = \theta$ for all x. It follows from Theorem 3.16.2 that $f(x) \equiv \theta$ in the largest sphere $\mathfrak{S}$ with center at $x = a$ and contained in $\mathfrak{D}$. Now if b is any point in $\mathfrak{D}$, we may join the points a and b by a broken line and hence by a finite chain of open spheres in $\mathfrak{D}$, $\mathfrak{S}_0 = \mathfrak{S}, \mathfrak{S}_1, \cdots, \mathfrak{S}_n$, such that $\mathfrak{S}_k$ contains the center of $\mathfrak{S}_{k+1}$. Since $f(x) \equiv \theta$ in $\mathfrak{S}_0$, the preceding argument shows that $f(x) \equiv \theta$ in $\mathfrak{S}_1$, and hence by induction, $f(b) = \theta$. Thus $f(x) \equiv \theta$ in $\mathfrak{D}$.

It should be observed that the conclusion is not necessarily valid if it is known merely that $f(x)$ vanishes in a point set having a limit point in $\mathfrak{D}$. For example a linear bounded functional on $\mathfrak{X}$ is (G)-differentiable (and continuous) and vanishes on a linear subspace without vanishing identically in $\mathfrak{X}$.

3.17. (F)-Differentiability. In passing from the (G)- to the (F)-theory, finitely open sets are replaced by open sets, which are required for (F)-differentiability. Definition 3.16.4 shows that an (F)-differentiable function is continuous and (G)-differentiable. The converse is also true; in fact, continuity may be replaced by local boundedness or even continuity in the sense of Baire. The sharpest of these results will be proved in section 26.7, but for our present purposes local boundedness provides a convenient working condition.

DEFINITION 3.17.1. *A function $f(x)$ on $\mathfrak{X}$ to $\mathfrak{Y}$ defined in the open set $\mathfrak{D}$ is said to be locally bounded if for every point $a \in \mathfrak{D}$ there is a sphere $\mathfrak{S}_a$, $\| x - a \| \leqq r_a$, and a finite $M(a)$ such that $\| f(x) \| \leqq M(a)$ when $x \in \mathfrak{S}_a$.*

DEFINITION 3.17.2. *A function on $\mathfrak{X}$ to $\mathfrak{Y}$ defined in the domain $\mathfrak{D}$ is said to be analytic in $\mathfrak{D}$ if it is single-valued, locally bounded, and (G)-differentiable in $\mathfrak{D}$.*

THEOREM 3.17.1. *If $f(x)$ is analytic in the domain $\mathfrak{D}$, then it is continuous and (F)-differentiable in $\mathfrak{D}$. For $x \in \mathfrak{D}$, the nth variation, $\delta^n f(x; h)$, is a bounded homogeneous function of degree n in h and a locally bounded function of x. To each $a \in \mathfrak{D}$, there corresponds an r'_a such that the Taylor expansion*

$$f(x + h) = \sum_{n=0}^{\infty} \frac{1}{n!} \delta^n f(x; h)$$

converges uniformly for $\| x - a \| \leqq r'_a$ and $\| h \| \leqq r'_a$.

PROOF. Since $f(x)$ is (G)-differentiable, it can be expanded in a Taylor's series about each point $x \in \mathfrak{D}$. The assumption that $f(x)$ is locally bounded enables us to apply Theorem 3.16.3 with $\mathfrak{C}^*(a)$ replaced by $\mathfrak{S}_a$. This gives

$$\| \delta^n f(x; h) \| \leqq n!\, M(a) \quad \text{if} \quad \| x - a \| \leqq \tfrac{1}{2} r_a \quad \text{and} \quad \| h \| \leqq \tfrac{1}{2} r_a,$$

so that $x + h \in \mathfrak{S}_a$. Now $\delta^n f(x; h)$ is homogeneous of degree n in h and hence for all h

$$\| \delta^n f(x; h) \| \leqq n!\, M(a) \left(\frac{2}{r_a}\right)^n \| h \|^n \quad \text{if} \quad \| x - a \| \leqq \tfrac{1}{2} r_a .$$

This proves that $\delta^n f(x; h)$ is a bounded homogeneous function of degree n in h, locally bounded with respect to x in $\mathfrak{D}$ for fixed h. Combining these estimates with the Taylor expansion, we obtain

$$\| f(x + h) - f(x) \| \leqq \frac{2M(a) \| h \|}{r_a - 2\| h \|},$$

$$\| f(x + h) - f(x) - \delta f(x; h) \| \leqq \frac{4M(a) \| h \|^2}{r_a(r_a - 2 \| h \|)}$$

for $\| x - a \| \leqq \frac{1}{2}r_a$, $\| h \| < \frac{1}{2}r_a$, so that $f(x)$ is continuous and (F)-differentiable in $\mathfrak{D}$. It is also clear from the above estimates that the series converges uniformly for $\| x - a \| \leqq r_a'$, $\| h \| \leqq r_a'$ provided $r_a' < \frac{1}{2}r_a$.

3.18. Properties of analytic functions. We shall prove some theorems about vector-valued analytic functions of vectors which are obvious analogues of classical theorems. We note first of all that Theorem 3.16.4 applies in particular to analytic functions; this makes it possible to carry out the classical process of analytic continuation, with obvious modifications, for the present class of analytic functions. Likewise the convergence theorems of classical function theory extend, *mutatis mutandis*, to vector-valued analytic functions of vectors. We restrict ourselves to the analogues of well known theorems of Vitali and Osgood. Both the assumptions and the conclusions differ somewhat from the classical prototypes, however; in particular, no information is obtained relating to uniform convergence for the given sequence.

THEOREM 3.18.1. *Let* $\{f_k(x)\}$ *be a sequence of functions analytic and locally uniformly bounded in a fixed domain* $\mathfrak{D}$. *If* $\lim_{k\to\infty} f_k(x)$ *exists in a sphere* $\mathfrak{S}$ *in* $\mathfrak{D}$, *then the limit* $f(x)$ *exists everywhere in* $\mathfrak{D}$ *and is analytic there. Further*

$$\delta^n f(x; h) = \lim_{k\to\infty} \delta^n f_k(x; h)$$

for every n *and all* x *in* $\mathfrak{D}$.

PROOF. The assumption of local uniform boundedness is understood in the following sense: to each point $x_0 \in \mathfrak{D}$ there is a sphere $\mathfrak{S}(x_0)$ and a finite positive quantity $M(x_0)$ such that $\| f_k(x) \| \leqq M(x_0)$ for all $x \in \mathfrak{S}(x_0)$ and $k = 1, 2, \cdots$. The essential part of the proof consists in showing that $\lim_{k\to\infty} f_k(x)$ exists everywhere in $\mathfrak{D}$. Let a be the center of $\mathfrak{S}$ and let b be any other point of $\mathfrak{D}$. We join a to b by a polygonal line P in $\mathfrak{D}$. Each point x of P is the center of a sphere $\mathfrak{S}(x) \subset \mathfrak{D}$ of uniform boundedness and since P is compact it follows that a finite subset of these spheres will cover P. Let $\mathfrak{C}$ denote this finite covering of P. It is clear that the $f_k(x)$ will be uniformly bounded in $\mathfrak{C}$. Let ρ be the minimum distance from P to the boundary of $\mathfrak{C}$; again P compact implies that $\rho > 0$. We now construct a chain of interlacing spheres, $\mathfrak{S}_1, \mathfrak{S}_2, \cdots, \mathfrak{S}_N$, of radius ρ and with centers a distance $\rho/2$ apart along P, starting at a and ending at b. Let a_ν denote the center of $\mathfrak{S}_\nu$; here $a_1 = a$ and $a_N = b$. By assumption $\lim_{k\to\infty} f_k(x)$ exists in some sphere $\mathfrak{S}$ concentric with $\mathfrak{S}_1$ about a_1. For arbitrary $h \in \mathfrak{X}$, the functions

$f_k(a_1 + \zeta h)$ are holomorphic and uniformly bounded for $|\zeta| < \rho/\|h\|$. Further the sequence converges for $|\zeta|\,\|h\| <$ radius of $\mathfrak{S}$. By the Vitali Theorem 3.14.1, the $\lim_{k\to\infty} f_k(a_1 + \zeta h)$ exists for all $|\zeta| < \rho/\|h\|$ and this limit $f(a_1 + \zeta h)$ is holomorphic in ζ. Since h was arbitrary, we see that $\lim_{k\to\infty} f_k(x) = f(x)$ for all $x \in \mathfrak{S}_1$ and that $f(x)$ is (G)-differentiable at $x = a_1$. Now $\mathfrak{S}_1$ contains a sphere of radius $\rho/2$ about a_2 so that we may apply the same argument to the $f_k(x)$ in $\mathfrak{S}_2$. By induction we have $\lim_{k\to\infty} f_k(x) = f(x)$ for all $x \in \mathfrak{S}_N$ and $f(x)$ is (G)-differentiable at $x = b$. Since b was an arbitrary point of $\mathfrak{D}$, it follows that $\lim_{k\to\infty} f_k(x) = f(x)$ for all $x \in \mathfrak{D}$ and that $f(x)$ is (G)-differentiable in $\mathfrak{D}$. Now $\|f_k(x)\| \leqq M(x_0)$ for $x \in \mathfrak{S}(x_0)$ so that $\|f(x)\| \leqq M(x_0)$ for $x \in \mathfrak{S}(x_0)$. Thus $f(x)$ is locally bounded and hence analytic. Finally the bounded convergence of $f_k(x_0 + \zeta h)$ to $f(x_0 + \zeta h)$ for $|\zeta|\,\|h\| < r(x_0) =$ radius of $\mathfrak{S}(x_0)$ implies uniform convergence in ζ for each h, $|\zeta|\,\|h\| < r(x_0)/2$, by the Vitali Theorem 3.14.1. By Theorem 3.11.6, this in turn implies the convergence of the derivatives for $\zeta = 0$, whence it follows that

$$\lim_{k\to\infty} \delta^n f_k(x_0; h) = \delta^n f(x_0; h)$$

for all n and all $x_0 \in \mathfrak{D}$. This completes the proof.

THEOREM 3.18.2. *If the functions $\{f_k(x)\}$ are analytic in a fixed domain $\mathfrak{D}$ and if $\lim_{k\to\infty} f_k(x) = f(x)$ exists for every $x \in \mathfrak{D}$, then every open subset of $\mathfrak{D}$ contains a domain in which $f(x)$ is analytic.*

REMARK. In other words, the domains of analyticity of $f(x)$ are everywhere dense in $\mathfrak{D}$. There exists at least one such domain; there may be infinitely many, the number being limited only by the cardinality of the maximum number of non-overlapping domains possible for the space $\mathfrak{X}$. The analytic functions defined in the different domains are not necessarily related by analytic continuation. Theorem 3.18.2 is the only result of this paragraph which requires the domain space $\mathfrak{X}$ to be complete.

PROOF. Let $\mathfrak{G}$ be an open subset of $\mathfrak{D}$ and let $\mathfrak{S}$ be a closed sphere interior to $\mathfrak{G}$. Then $\mathfrak{S}$ is clearly of the second category and applying Theorem 1.4.2 to the continuous functionals $\{\|f_k(x)\|\}$, we see that there exists a sphere $\mathfrak{S}'$ interior to $\mathfrak{S}$ and a constant M such that $\|f_k(x)\| \leqq M$ for all $x \in \mathfrak{S}'$ and $k \geqq 1$. The functions $f_k(x)$ being analytic and uniformly bounded in $\mathfrak{S}'$, the existence of $\lim_{k\to\infty} f_k(x) = f(x)$ in $\mathfrak{S}'$ implies that $f(x)$ is analytic in $\mathfrak{S}'$ by virtue of the preceding theorem.

We conclude the discussion by extending the principle of the maximum.

THEOREM 3.18.3. *Let $f(x)$ be analytic in the domain $\mathfrak{D}$. Then $\|f(x)\|$ can have no maximum in $\mathfrak{D}$ unless $\|f(x)\|$ is of constant value throughout $\mathfrak{D}$.*

PROOF. Suppose $a \in \mathfrak{D}$ is such that $\|f(x)\| \leqq \|f(a)\| = M$ for all $x \in \mathfrak{D}$. Let $h \in \mathfrak{X}$ be fixed and consider the manifold $x = a + \zeta h$ generated as ζ ranges over the complex numbers. It intersects $\mathfrak{D}$ in a relatively open set corresponding to an open set Δ in the complex plane. Let Δ_0 be the component of Δ which contains $\zeta = 0$. Then $f(a + \zeta h)$ is a holomorphic function of ζ in Δ_0. By Theorem 3.13.1, this implies that $\| f(a + \zeta h) \| = M$ for all $\zeta \in \Delta_0$. In this argument h was arbitrary. Consequently $\| f(x) \| = M$ for all x in $\mathfrak{D}$ which can be joined to a by a straight line segment. This in turn implies $\| f(x) \| = M$ for every $x \in \mathfrak{D}$ which can be joined to a by a polygonal line in $\mathfrak{D}$. Since all points in $\mathfrak{D}$ have this property, we conclude that $\| f(x) \| \equiv M$ in $\mathfrak{D}$.

THEOREM 3.18.4. *Let $f(x)$ be analytic in a bounded domain $\mathfrak{D}$ and continuous in its closure $\bar{\mathfrak{D}}$. If* $\sup \| f(x) \| = M$ *for $x \in \bar{\mathfrak{D}} \ominus \mathfrak{D}$, then either $\| f(x) \| < M$ in $\mathfrak{D}$ or else $\| f(x) \| \equiv M$ in $\mathfrak{D}$.*

REMARK. The value $M = \infty$ is not precluded in Theorem 3.18.4 as the following example shows: Let $\mathfrak{X} = l$, $\mathfrak{Y}$ = space of complex numbers, and $\mathfrak{D}$ = unit sphere in

$$l \equiv [\{\xi_n\};\ \sum_{n=1}^{\infty} | \xi_n | < 1].$$

We define

$$f(\{\xi_n\}) = \sum_{n=1}^{\infty} \frac{\xi_n}{\xi_n - 1 - 1/n}.$$

Then it is easy to show that $f(x)$ is analytic in $\bar{\mathfrak{D}}$. However if $\xi_n^k = 0$ for $n \neq k$, $= 1$ for $n = k$, then $\| f(\{\xi_n^k\}) \| = k$ and hence $\sup_{x \in \bar{\mathfrak{D}}} \| f(x) \| = \infty$.

PROOF. Suppose $a \in \mathfrak{D}$ and let $h \in \mathfrak{X}$. As in the proof of the previous theorem, we consider the manifold $[x = a + \zeta h]$ and its intersection with $\mathfrak{D}$. This corresponds to an open set Δ in the complex plane; again let Δ_0 be the component of Δ which contains $\zeta = 0$. In this case $f(a + \zeta h)$ is holomorphic in Δ_0 and continuous in $\bar{\Delta}_0$. The maximum principle of section 3.13 asserts that $\| f(a) \| \leqq \sup \| f(a + \zeta h) \|$ for $\zeta \in \bar{\Delta}_0 \ominus \Delta_0$ and hence that $\| f(a) \| \leqq M$. Since a was an arbitrary point of $\mathfrak{D}$, $\| f(x) \| \leqq M$ for all $x \in \mathfrak{D}$. If $\| f(a) \| = M$, then the previous theorem applies and we have $\| f(x) \| \equiv M$ in $\mathfrak{D}$.

3.19. (L)-analyticity. The theory of analytic functions on vectors to vectors can be developed in still another direction if the underlying space is a commutative Banach algebra. For this case E. R. Lorch [3] has introduced the following definition of differentiability and analyticity.

DEFINITION 3.19.1. *Let $\mathfrak{B}$ be a commutative complex (B)-algebra with a unit element. Let $f(z)$ be a function whose domain $\mathfrak{D}$ is an open connected subset of $\mathfrak{B}$ and whose range is also contained in $\mathfrak{B}$. Then $f(z)$ is said to have an (L)-derivative $f'(z_0)$ at $z = z_0$ if for each $\epsilon > 0$ a $\delta > 0$ can be found such that for all h in $\mathfrak{B}$ with $\| h \| < \delta$*

$$\| f(z_0 + h) - f(z_0) - hf'(z_0) \| < \epsilon \| h \|. \tag{3.19.1}$$

If $f(z)$ has a derivative at each point of $\mathfrak{D}$, then it is (L)-analytic in $\mathfrak{D}$.

A function analytic according to this definition is clearly continuous and (F)-differentiable and hence analytic in the sense of Definition 3.17.2. However, *not every (F)-differentiable function on a commutative (B)-algebra to itself is (L)-analytic*. This is illustrated by the following example: Let Z_2 be the space of ordered pairs of complex numbers $z = (\zeta_1, \zeta_2)$ and define

$$(\alpha_1, \alpha_2) + (\beta_1, \beta_2) = (\alpha_1 + \beta_2, \alpha_2 + \beta_2), \qquad \alpha(\beta_1, \beta_2) = (\alpha\beta_1, \alpha\beta_2),$$

$$(\alpha_1, \alpha_2)\cdot(\beta_1, \beta_2) = (\alpha_1\beta_1, \alpha_2\beta_2), \qquad \| (\alpha_1, \alpha_2) \| = \max [|\alpha_1|, |\alpha_2|].$$

Under these conventions, Z_2 becomes a commutative (B)-algebra with a unit element. The function $f(z) = (\zeta_2, \zeta_1)$ is an (F)-differentiable function of z which is not (L)-analytic.

THEOREM 3.19.1. *Let $\mathfrak{B}$ be a commutative complex (B)-algebra with a unit element. Given the power series*

$$\sum_{n=0}^{\infty} a_n(z - a)^n \equiv f(z), \qquad a_n \in \mathfrak{B}; \tag{3.19.2}$$

set

$$1/\rho = \limsup_{n\to\infty} \| a_n \|^{1/n}.$$

Then the series converges absolutely for $\| z - a \| < \rho$ and every larger sphere about a contains points at which the series diverges. The series converges to an (L)-analytic function in $\| z - a \| < \rho$, the convergence being uniform in each sphere about a of radius less than ρ.

PROOF. Again the classical argument for the convergence of a power series applies for $\| z - a \| < \rho$. On the other hand according to Theorem 3.11.4 the series will diverge for all $z = a + \zeta e$ with $| \zeta | > \rho$. In order to show that $f(z)$ is (L)-differentiable at z_0, $\| z_0 - a \| < \rho$, we set $\sigma = \rho - \| z_0 - a \|$ and consider points z such that $\| z - z_0 \| < \sigma$. Then the series $\sum_n \| a_n \| (\| z - z_0 \| + \| z_0 - a \|)^n$ converges and hence we may expand and rearrange the terms in powers of $\| z - z_0 \|$. It follows that

$$f(z) = \sum_{k=0}^{\infty} b_k(z - z_0)^k,$$

where

$$b_k = \sum_{n=k}^{\infty} \binom{n}{k} a_n(z_0 - a)^{n-k};$$

the series converging for all $\| z - z_0 \| < \sigma$. As above this implies that $\limsup_{n\to\infty} \| b_n \|^{1/n} \leqq 1/\sigma$. Consequently

$$\| f(z_0 + h) - f(z_0) - hb_1 \| \leqq \sum_{n=2}^{\infty} \| b_n \| \| h \|^n = o(\| h \|)$$

as $\| h \| \to 0$, which was to be proved.

A considerable portion of the classical theory carries over to (L)-analytic functions (see E. R. Lorch [3]). However, the above material is sufficient for our purposes. We shall return to this subject in Chapter XXVI where it will be shown that a function (L)-analytic in $\mathfrak{D}$ may be expanded in a convergent Taylor series of the form (3.19.2) about each point of $\mathfrak{D}$.

References. Alexiewicz and Orlicz [1], Banach [**3, 4**], Evans [1], Fantappiè [**2, 3**], Fréchet [**1, 2, 3, 5**], Gâteaux [**1, 2**], Graves [**1, 2, 3**], Highberg [**1, 2**], Hildebrandt and Graves [1], Hyers [1], Kerner [1], Lorch [3], Martin [1], Mazur and Orlicz [1], Michal [1], Michal and Clifford [1], Michal and Martin [1], Silva [1], Suchumlinov [1], Taylor [**1, 3, 4, 6, 7, 8**], and Zorn [**1, 2, 3**].

CHAPTER IV

BANACH ALGEBRAS

4.1. Orientation. This chapter is devoted to a study of Banach algebras, at present one of the most active fields in functional analysis. Banach algebras were unknown before 1935; nevertheless by 1940 their importance had already been established due principally to the independent efforts of M. H. Stone and I. Gelfand. Stone was motivated in this direction by the Hilbert space spectral theory, whereas Gelfand seems to have been motivated by the underlying algebraic nature of Wiener's Tauberian theorems.

There are six paragraphs: *General Properties*, *Spectral Theory*, *Ideal Theory for Commutative Banach Algebras*, *The Banach Algebra* $S(\varphi)$, *Commutative* (A^*)-*Algebras*, and *Commutative* (B^*)-*Algebras*. The first paragraph is of an introductory nature and is mainly concerned with regular and singular elements of the algebra. The second deals with various aspects of the resolvent, that is, the inverse of $\lambda e - x$, treated as a function of λ. The third paragraph is the core of the chapter. Here we present the Gelfand representation theory for commutative complex Banach algebras with unit elements. This theory is then applied to the algebra of set functions, $S(\varphi)$, which plays a significant part in the study of semi-groups of linear operators. Paragraph five deals with (A*)-algebras and contains a proof of a vector-valued variant of Wiener's Tauberian theorem. In the final paragraph a representation theory for commutative (B*)-algebras is developed and then applied to commutative self-adjoint operator algebras on a Hilbert space. References are given at the end of each paragraph.

1. General Properties

4.2. The unit element. Throughout this and the following chapter the symbol $\mathfrak{B}$ denotes a complex Banach algebra having a unit element e. Multiplication is non-commutative unless the contrary is explicitly stated. All statements regarding limits and convergence are to be understood in terms of the normed metric of the space $\mathfrak{B}$.

It should be remarked that it is possible to carry through much of this development without assuming a unit. Such an approach is useful in ideal theory and will be discussed in Chapter XXIV. However the existence of a unit is a genuine convenience and actually does not restrict the generality in any essential way. This follows from the fact that it is always possible to adjoin a unit.

Suppose, in fact, that $\mathfrak{A}$ is a (B)-algebra without a unit, and let $\mathfrak{B}$ be the set of all pairs (x, α) where $x \in \mathfrak{A}$ and α is a complex number. The sum, scalar product, vector product, and norm are then defined as follows:

$$\begin{aligned} (x, \alpha) + (y, \beta) &= (x + y, \alpha + \beta), \\ \beta(x, \alpha) &= (\beta x, \beta\alpha), \\ (x, \alpha)(y, \beta) &= (xy + \alpha y + \beta x, \alpha\beta), \\ \| (x, \alpha) \| &= \| x \| + | \alpha | . \end{aligned}$$

It is now an easy matter to show that $\mathfrak{B}$ is a (B)-algebra with unit $e = (\theta, 1)$ and that the mapping $x \to (x, 0)$ maps $\mathfrak{A}$ isomorphically and isometrically into $\mathfrak{B}$.

4.3 Regular elements. The distribution of regular elements in $\mathfrak{B}$ is an important question on which the two following theorems shed some light.

THEOREM 4.3.1. *Every element in the open sphere $\| x - e \| < 1$ is regular and for such an x*

$$x^{-1} = e + \sum_{n=1}^{\infty} (e - x)^n. \tag{4.3.1}$$

PROOF. The series is absolutely convergent and therefore defines an element of $\mathfrak{B}$. Multiplication by $x = e - (e - x)$ on the right or left gives e.

THEOREM 4.3.2. *The regular elements form an open set in $\mathfrak{B}$.*

PROOF. We denote the set of regular elements by $\mathfrak{G}$. We shall show for $x_0 \in \mathfrak{G}$ that all of the elements in the open sphere $\| x - x_0 \| < \| x_0^{-1} \|^{-1}$ also belong to $\mathfrak{G}$. For such an x we have $\| e - xx_0^{-1} \| = \| (x_0 - x)x_0^{-1} \| < 1$. By the previous theorem

$$(xx_0^{-1})^{-1} = e + \sum_{n=1}^{\infty} (e - xx_0^{-1})^n = e + \sum_{n=1}^{\infty} [(x_0 - x)x_0^{-1}]^n.$$

It follows that

$$\begin{aligned} x^{-1} &= [(xx_0^{-1})x_0]^{-1} = x_0^{-1}(xx_0^{-1})^{-1} \\ &= x_0^{-1} + x_0^{-1} \sum_{n=1}^{\infty} [(x_0 - x)x_0^{-1}]^n. \end{aligned} \tag{4.3.2}$$

Moreover the series shows that

$$\| x^{-1} - x_0^{-1} \| \leqq \| x_0^{-1} \|^2 \| x - x_0 \| [1 - \| x - x_0 \| \| x_0^{-1} \|]^{-1}. \tag{4.3.3}$$

Hence we have

THEOREM 4.3.3. *The inverse x^{-1} is a continuous function of x in $\mathfrak{G}$.*

THEOREM 4.3.4. *The inverse of x is an analytic function of x in the sense of Definition* 3.17.2 *in each of the components of* $\mathfrak{G}$.

This is an immediate consequence of formula (4.3.2) and Definition 3.17.2.

It should be noted that the property of being regular or of being singular is not invariant under extensions and contractions of the algebra. Extensions preserve regularity but contractions may not; for singular elements the situation is reversed. We shall see later (section 4.11) that some elements may remain singular under all possible extensions of the algebra.

4.4. The kernel. We return to the set $\mathfrak{G}$ of regular elements of $\mathfrak{B}$. If $x, y \in \mathfrak{G}$, then $(xy)^{-1} = y^{-1}x^{-1}$ so that $xy \in \mathfrak{G}$. Further if $x \in \mathfrak{G}$, then $x^{-1} \in \mathfrak{G}$. Since $e \in \mathfrak{G}$ and since the associative law holds in $\mathfrak{B}$, we see that $\mathfrak{G}$ is a group. We call $\mathfrak{G}$ the *maximal group* of $\mathfrak{B}$ since every other group in $\mathfrak{B}$ having e as its unit element must be a subgroup of $\mathfrak{G}$.

According to Theorem 4.3.2, the maximal group is an open subset of $\mathfrak{B}$. The algebra $\mathfrak{B}$ being a locally connected space, it follows that $\mathfrak{G}$ is the union of disjoint maximal open connected sets, that is, $\mathfrak{G} = \bigcup_\alpha \mathfrak{G}_\alpha$ where the $\mathfrak{G}_\alpha$'s denote the *components* of $\mathfrak{G}$. The component $\mathfrak{G}_1$ containing e is called the *principal component of* $\mathfrak{G}$ or the *kernel of* $\mathfrak{B}$. It will be shown in Theorem 9.5.5 that in the commutative case $\mathfrak{G}$ has either a single component or else infinitely many components. In the latter case the number of components need not be countably infinite.

THEOREM 4.4.1. *$\mathfrak{G}_1$ is an invariant subgroup of $\mathfrak{G}$ as well as the maximal connected subgroup of $\mathfrak{G}$.*

PROOF. Let a be a fixed element of $\mathfrak{G}$ and consider the mapping $x \to ax$. This is a continuous map of $\mathfrak{B}$ into itself (actually a homeomorphism of $\mathfrak{B}$ onto $\mathfrak{B}$) and, in particular, $\mathfrak{G}$ is mapped into itself. Since connected sets go into connected sets, each component of $\mathfrak{G}$ is mapped into a component of $\mathfrak{G}$. Thus if $a \in \mathfrak{G}_1$, then $a\mathfrak{G}_1 \subset \mathfrak{G}_1$ since the transformation maps $e \in \mathfrak{G}_1$ on $a \in \mathfrak{G}_1$. It follows that $ab \in \mathfrak{G}_1$ if a and $b \in \mathfrak{G}_1$. The inverse transformation $(x \to a^{-1}x)$ is of this same type and maps $a \in \mathfrak{G}_1$ on $e \in \mathfrak{G}_1$. Thus if $a \in \mathfrak{G}_1$, then $a^{-1}\mathfrak{G}_1 \subset \mathfrak{G}_1$ so that $a^{-1} = a^{-1}e \in \mathfrak{G}_1$. Hence $\mathfrak{G}_1$ is a subgroup of $\mathfrak{G}$.

Finally, if a is any element of $\mathfrak{G}$, then the mapping $x \to a^{-1}xa$ is again continuous, taking components of $\mathfrak{G}$ into components of $\mathfrak{G}$. In particular, it leaves $\mathfrak{G}_1$ invariant since e goes into e. Hence $\mathfrak{G}_1$ is an invariant or normal subgroup of $\mathfrak{G}$. It is also a maximal connected subgroup of $\mathfrak{G}$ since any larger subgroup must be disconnected. This completes the proof.

We observe that the components of $\mathfrak{G}$ are the elements of the *quotient group (factor group)* $\mathfrak{G}/\mathfrak{G}_1$. The results of the above theorem can be somewhat strengthened as follows.

THEOREM 4.4.2. *$\mathfrak{G}_1$ is the only open connected subgroup of $\mathfrak{G}$.*

PROOF. Suppose that $\mathfrak{H}$ is an open connected subgroup of $\mathfrak{G}$. It is consequently a subgroup of $\mathfrak{G}_1$. If a is any element of $\mathfrak{G}_1$, then $a\mathfrak{H}$ is a left coset of $\mathfrak{H}$ with respect to $\mathfrak{G}_1$. As the homeomorphic image of an open set, $a\mathfrak{H}$ is also open. It follows that the union of any system of left (or right) cosets of $\mathfrak{H}$ is open. Since the complement of $\mathfrak{H}$ with respect to $\mathfrak{G}_1$ is the union of left cosets of $\mathfrak{H}$, this complement will be open. Thus $\mathfrak{H}$ is both open and closed in $\mathfrak{G}_1$ and therefore $\mathfrak{H} = \mathfrak{G}_1$.

Suppose next that $\mathfrak{S}$ is a subalgebra of $\mathfrak{B}$ containing the unit element and let $\mathfrak{Q} = \mathfrak{G} \cap \mathfrak{S}$. In this case $\mathfrak{Q}$ need not be a group since the inverse of an element of $\mathfrak{Q}$ need not lie in $\mathfrak{S}$. However $\mathfrak{Q}$ does have many properties in common with the maximal group $\mathfrak{G}$. It is clear that $e \in \mathfrak{Q}$. Further, $\mathfrak{G}$ being open in $\mathfrak{B}$, it follows that $\mathfrak{Q}$ is open in $\mathfrak{S}$. Since $\mathfrak{Q}$ is an open subset of $\mathfrak{S}$ and since $\mathfrak{S}$ is locally connected, we see that $\mathfrak{Q}$ is the union of maximal connected open sets in $\mathfrak{S}$. Let $\mathfrak{Q}_1 = \mathfrak{K}(\mathfrak{S})$ be the component of $\mathfrak{Q}$ containing the unit element. We call $\mathfrak{K}(\mathfrak{S})$ the *kernel* of $\mathfrak{S}$ with respect to $\mathfrak{B}$. The following theorem is due to M. Nagumo [**1**]. For a generalization see C. E. Rickart [**2**, Theorem 2.11].

THEOREM 4.4.3. *If $\mathfrak{S}$ is a closed subalgebra of $\mathfrak{B}$, containing the unit element, then $\mathfrak{K}(\mathfrak{S})$ is a group, actually a subgroup of $\mathfrak{G}_1$, and $\mathfrak{K}(\mathfrak{S})$ is the maximal connected group in $\mathfrak{S}$ having e as unit element.*

PROOF. It is clear that $\mathfrak{Q} \subset \mathfrak{G}$ and $\mathfrak{Q}_1 \subset \mathfrak{G}_1$. Further $\mathfrak{Q}_1$ is connected by definition. The argument used in Theorem 4.4.1 shows that if x and y belong to $\mathfrak{Q}_1$ so does xy, and if $x \in \mathfrak{Q}_1$ while $x^{-1} \in \mathfrak{S}$ then $x^{-1} \in \mathfrak{Q}_1$. It is therefore sufficient to show that $x \in \mathfrak{Q}_1$ implies that $x^{-1} \in \mathfrak{S}$.

Let $\mathfrak{Q}'$ consist of those elements of $\mathfrak{Q}_1$ which possess an inverse in $\mathfrak{S}$. It is obvious that $e \in \mathfrak{Q}'$ and we shall show that $\mathfrak{Q}'$ is open as well as closed in $\mathfrak{Q}_1$. If $x_0 \in \mathfrak{Q}'$ and $x \in \mathfrak{Q}_1$ lies in the sphere of radius $\| x_0^{-1} \|^{-1}$ about x_0, then the terms of the expansion (4.3.2) all lie in $\mathfrak{S}$ and hence $x^{-1} \in \mathfrak{S}$, that is, $\mathfrak{Q}'$ is open in $\mathfrak{Q}_1$. Suppose next that a sequence $\{x_n\} \in \mathfrak{Q}'$ converges to $x_0 \in \mathfrak{Q}_1$. By Theorem 4.3.3 we have $x_n^{-1} \rightarrow x_0^{-1}$ and since $x_n^{-1} \in \mathfrak{S}$ and $\mathfrak{S}$ is closed, it follows that $x_0^{-1} \in \mathfrak{S}$, that is, $x_0 \in \mathfrak{Q}'$. Thus $\mathfrak{Q}'$ is both open and closed in the connected set $\mathfrak{Q}_1$ whence $\mathfrak{Q}' = \mathfrak{Q}_1$.

This argument shows that $\mathfrak{Q}_1$ is a group. Being contained in $\mathfrak{G}_1$, it is obviously a subgroup of $\mathfrak{G}_1$. It is further clear that $\mathfrak{Q}_1$ is the maximal connected group in $\mathfrak{S}$ having e as unit element.

THEOREM 4.4.4. *If $\mathfrak{Q}$ is a group, that is, if ring contraction from $\mathfrak{B}$ to $\mathfrak{S}$ preserves regularity, then $\mathfrak{K}(\mathfrak{S})$ is a normal subgroup of $\mathfrak{Q}$.*

The proof is the same as in Theorem 4.4.1. In fact, the theorem is valid in any topological group.

4.5. Singular elements. The set $\mathfrak{J}$ of singular elements in $\mathfrak{B}$ is closed and contains at least the zero element. If $\mathfrak{B}$ is a field, θ is the only singular element.

We note further that $\mathfrak{F}$ is connected because, if x is singular so is αx, and therefore x can be joined with θ by a line segment of elements in $\mathfrak{F}$.

We note at this juncture some special types of elements which may occur in a (B)-algebra, all of which are singular and remain singular under arbitrary extensions of the algebra.

An element $x \neq \theta$ is said to be a *divisor of zero* if there is a $y \neq \theta$ such that either $xy = \theta$ or $yx = \theta$. It is clear that y is also a divisor of zero. This notion has been generalized by G. Šilov [1] who calls x a *generalized divisor of zero* if there is a sequence $\{y_n\} \subset \mathfrak{B}$ with $\| y_n \| = 1$ such that either $xy_n \to \theta$ or $y_n x \to \theta$. It is clear that such an element is singular.

An element $x \neq \theta$ is said to be *nilpotent* if some power of x equals θ. Such an element is clearly a divisor of zero. This notion has been generalized by I. Gelfand [4] (the term *quasi-nilpotent* is due to E. R. Lorch [3]); x is quasi-nilpotent if $\lim_{n\to\infty} \| x^n \|^{1/n} = 0$. For regular x, $e = x^n(x^{-1})^n$ so that $\| e \| \leqq \| x^n \| \| x^{-1} \|^n$ and hence $\| x^n \|^{1/n} \geqq \| e \|^{1/n}/\| x^{-1} \|^{-1}$: It follows that a quasi-nilpotent element is singular. Finally we have the *idempotent* elements: j is idempotent if $j^2 = j$. This condition is of course satisfied by the zero and the unit elements, but there may be other idempotents. Thus if $\mathfrak{B} = \mathfrak{E}(\mathfrak{X})$ and J is a projection operator, then we have $J^2 = J$. Any idempotent $j \neq e$ is singular.

4.6. Functions on scalars to the algebra. The theory of functions on scalars to a Banach algebra differs from the corresponding theory of functions on scalars to a (B)-space only in the presence of ring multiplication. For this reason we restrict ourselves to pointing out some of the new features introduced by this fact. We also stick rather faithfully to the normed metric and all notions are based on convergence in this metric. Thus we have for the most part only one form of each function theoretic concept to contend with.

In agreement with this convention, we say that a *function $x(\alpha)$ on a measurable set $\mathfrak{S}$ to $\mathfrak{B}$ is measurable if it is the limit almost everywhere of a sequence of countably-valued functions.* Theorem 3.5.4 holds for such functions and in addition *the product of two measurable functions is measurable.*

Continuity and differentiability are defined in the obvious manner in terms of the normed metric. The classical formulas of elementary differential calculus are largely valid for differentiable functions with values in $\mathfrak{B}$, but owing to the noncommutative character of multiplication, they may take unconventional forms. Thus if $x(\xi)$ and $y(\xi)$ are differentiable functions in an interval (ξ_1, ξ_2), then $x(\xi)y(\xi)$ is also differentiable and

$$[x(\xi)y(\xi)]' = x(\xi)y'(\xi) + x'(\xi)y(\xi), \tag{4.6.1}$$

where the order of the factors is essential. It is evident from (4.3.2) that

$$[x^{-1}(\xi)]' = -x^{-1}(\xi)x'(\xi)x^{-1}(\xi) \tag{4.6.2}$$

is valid if and only if $x(\xi)$ is a regular element of $\mathfrak{B}$ and $x(\xi)$ is differentiable.

The methods of §3.2 apply to holomorphic functions on a domain D of the complex plane to a Banach algebra $\mathfrak{B}$. In particular, a power series in $\zeta - \zeta_0$ with coefficients in $\mathfrak{B}$

$$\sum_{n=0}^{\infty} a_n(\zeta - \zeta_0)^n \equiv x(\zeta)$$

defines a holomorphic function on $|\zeta - \zeta_0| < \rho$ to $\mathfrak{B}$ where

$$1/\rho = \limsup_{n\to\infty} \| a_n \|^{1/n}. \tag{4.6.3}$$

The series diverges outside of this circle (see Theorem 3.11.4).

In the case of a Banach algebra, however, we may multiply absolutely convergent power series and the product is an absolutely convergent power series whose coefficients are given by the Cauchy formula so that

$$\sum_{n=0}^{\infty} a_n\zeta^n \sum_{n=0}^{\infty} b_n\zeta^n = \sum_{n=0}^{\infty} \left[\sum_{k=0}^{n} a_k b_{n-k}\right] \zeta^n.$$

The order of the factors is of course essential. Further, if the constant term a_0 of $x(\zeta)$ is a regular element of $\mathfrak{B}$ so that a_0^{-1} exists, then $x(\zeta)$ also has a holomorphic inverse for sufficiently small values of $|\zeta - \zeta_0|$. This follows from Theorem 4.3.2 and formula (4.3.2).

The formula

$$\exp(\zeta a) = e + \sum_{n=1}^{\infty} \frac{a^n\zeta^n}{n!}, \qquad a \in \mathfrak{B}, \tag{4.6.4}$$

clearly defines an entire function of ζ and we have

$$\frac{d}{d\zeta} \exp(\zeta a) = a \exp(\zeta a) = \exp(\zeta a)a. \tag{4.6.5}$$

A simple calculation shows that if a_1 and a_2 commute then

$$\exp(\zeta_1 a_1)\exp(\zeta_2 a_2) = \exp(\zeta_1 a_1 + \zeta_2 a_2) \tag{4.6.6}$$

and in particular,

$$\exp(\zeta a)\exp(-\zeta a) = e \tag{4.6.7}$$

for all ζ. Hence $\exp(\zeta a)$ has an inverse for all ζ and a. These formulas serve as further justification for the name exponential function assigned to the series in (3.4.4).

In conclusion let us remark that in dealing with analytic functions to a Banach algebra the analyst has to be prepared for functions with somewhat unconventional behavior; the similarity with the classical case is not merely suggestive but also deceptive. Thus, for instance, if q is a quasi-nilpotent element, the function $\log(e - \zeta q)$, defined by the obvious power series, is an entire function of ζ.

Similarly the function $(e - \zeta a)^{-1}$ is ordinarily not a rational function of ζ. We shall make a thorough study of this function, or rather of an equivalent function, in the next paragraph and in Chapter V.

References. Gelfand [**4**], Lorch [**3**], Nagumo [**1**], Rickart [**2**], Šilov [**1**].

2. Spectral Theory

4.7. The resolvent. We shall consider the inverse of $\lambda e - a$ as a function of λ for a fixed $a \in \mathfrak{B}$. The resulting theory is closely related to the discussion in section 2.16, where we considered the resolvent of a linear transformation T on a (B)-space to itself, and has direct and important applications to the theory of $R(\lambda; T)$. Our present point of view is somewhat different, however, inasmuch as we are interested only in what happens in the given algebra and not in any underlying space. Thus the classification of the spectrum into point spectrum, continuous spectrum, and residual spectrum has no longer any sense and we are merely concerned with the resolvent set and its complement, the spectrum.

If we so desire, however, we may consider the element a of $\mathfrak{B}$ as an operator defining the transformation $y = ax$ on $\mathfrak{B}$ to itself and to this operator we can apply the discussion of section 2.16. Thus we have to distinguish between the spectral properties of a as an element of $\mathfrak{B}$ and as an operator on $\mathfrak{B}$ to itself. It is properties of the first kind which concern us here.

Definition 4.7.1. *According as $\lambda e - a$ is regular or singular in $\mathfrak{B}$, we say that λ belongs to the resolvent set $\rho(a)$ or the spectrum $\sigma(a)$ of a. For λ in $\rho(a)$ the inverse of $\lambda e - a$ exists; it is denoted by $R(\lambda; a)$ and is called the resolvent of a.*

We have from the definition

$$R(\lambda; a)(\lambda e - a) = (\lambda e - a)R(\lambda; a) = e, \qquad \lambda \in \rho(a). \tag{4.7.1}$$

Theorem 4.7.1. *The resolvent set is open. In each of its components $R(\lambda; a)$ is a holomorphic function of λ.*

Proof. Suppose that $\lambda_0 \in \rho(a)$ and substitute $x = \lambda e - a$, $x_0 = \lambda_0 e - a$ in formula (4.3.2). Since $x_0 - x = (\lambda_0 - \lambda)e$, the formal result is

$$R(\lambda; a) = R(\lambda_0\,; a)\left\{e + \sum_{n=1}^{\infty} (\lambda_0 - \lambda)^n [R(\lambda_0\,; a)]^n\right\}. \tag{4.7.2}$$

This series is absolutely convergent at least when

$$|\lambda - \lambda_0| < \| R(\lambda_0\,; a) \|^{-1}$$

and within this circle it defines a holomorphic function of λ. Multiplication by $\lambda e - a = (\lambda - \lambda_0)e + (\lambda_0 e - a)$ on left or right gives e, so the series actually represents the resolvent. This shows that a circular neighborhood of λ_0 also belongs to $\rho(a)$ and that $R(\lambda; a)$ is holomorphic in this neighborhood. This proves the theorem.

As an open set, $\rho(a)$ is the union of a finite or countably infinite number of disjoint connected open sets, the components of $\rho(a)$. We shall see in a moment that *there is at least one component,* $\rho_1(a)$, *known as the principal component and containing the point at infinity.* But often there are infinitely many components. Thus $R(\lambda; a)$ *is in general not an analytic function in the sense of Weierstrass, but rather an analytic expression defining distinct analytic functions in distinct components.* We shall express this fact briefly by saying that $R(\lambda; a)$ is *locally analytic* or, since $R(\lambda; a)$ is always single-valued, *locally holomorphic.*

In studying the principal component we shall need the following:

LEMMA 4.7.1. *Let* $\{\alpha_n\}$ *be a sequence of real numbers such that* $\alpha_{m+n} \leqq \alpha_m + \alpha_n$ *for all integers m and n. Then* $\alpha = \lim_{n\to\infty}(\alpha_n/n)$ *exists and* $-\infty \leqq \alpha < \infty$.

For a simple proof we refer the reader to G. Pólya and G. Szegö [1, p. 171, problem 98]. Actually α is also equal to inf (α_n/n). The sequence $\{\alpha_n\}$ is subadditive in the subscript and the result is closely related to Theorem 7.6.1.

THEOREM 4.7.2. *We have*

$$R(\lambda; a) = \lambda^{-1}e + \sum_{n=1}^{\infty} \lambda^{-n-1}a^n \tag{4.7.3}$$

for $|\lambda| > \gamma$ *where*

$$\gamma = \lim_{n\to\infty} \| a^n \|^{1/n}.$$

The right side of (4.7.3) *diverges for* $|\lambda| < \gamma$.

PROOF. Putting

$$\alpha_n = \log \| a^n \|,$$

it is a simple matter to verify that $\alpha_{m+n} \leqq \alpha_m + \alpha_n$. Hence

$$\lim_{n\to\infty} \frac{\alpha_n}{n} = \lim_{n\to\infty} \log \| a^n \|^{1/n} \equiv \alpha$$

exists. We put $\gamma = \exp(\alpha)$. That the series converges absolutely for $|\lambda| > \gamma$ and diverges for $|\lambda| < \gamma$ follows from Theorem 3.11.4. Multiplication of the series by $(\lambda e - a)$ on either side gives e. Hence the series represents $R(\lambda; a)$ for $|\lambda| > \gamma$.

It follows that the exterior of the circle $|\lambda| = \gamma$ belongs to $\rho(a)$ and by definition it belongs to the principal component.

DEFINITION 4.7.2. *We define the spectral radius for an element a of* $\mathfrak{B}$, *denoted by* $r(a)$, *to be*

$$(4.7.4)\qquad r(a) = \sup[\,|\lambda|\,;\lambda \in \sigma(a)].$$

The next result is due to I. Gelfand [**4**].

THEOREM 4.7.3. *The spectral radius satisfies the following properties:*

$$(4.7.5)\qquad \begin{gathered} r(a) = \lim_{n\to\infty} \|\, a^n \,\|^{1/n}, \\ r(a^k) = [r(a)]^k, \qquad r(\alpha a) = |\,\alpha\,|\, r(a), \qquad r(a) \leqq \|\, a\,\|. \end{gathered}$$

PROOF. By the previous theorem any λ such that $|\lambda| > \gamma = \lim_{n\to\infty} \|\, a^n \,\|^{1/n}$ belongs to $\rho(a)$ and hence $r(a) \leqq \gamma$. On the other hand it follows from Theorem 4.7.1 that $R(\lambda; a)$ is holomorphic for $|\lambda| > r(a)$. Thus the Laurent expansion, given in this case by (4.7.3), converges for $|\lambda| > r(a)$. Again by the previous theorem this implies that $r(a) \geqq \gamma$. This proves the first assertion and the others readily follow from this.

THEOREM 4.7.4. *The spectrum of a is a closed bounded non-vacuous point set.*

PROOF. The spectrum, being the complement of the open resolvent set, is necessarily closed and bounded. If $\sigma(a)$ were vacuous, then $\rho(a)$ would be the whole complex plane and $R(\lambda; a)$ would be an entire function which is bounded in norm for large $|\lambda|$, as can be seen directly from (4.7.3). By the extended Liouville theorem (Theorem 3.13.2), $R(\lambda; a)$ must be a constant and, since $R(\lambda; a) \to \theta$ when $|\lambda| \to \infty$, $R(\lambda; a) \equiv \theta$ which clearly contradicts formula (4.7.1).

COROLLARY. *Let* $\mathfrak{A}$ *be a complex normed algebra with a unit element. Then the spectrum of each* $a \in \mathfrak{A}$ *is non-vacuous.*

PROOF. As we have already remarked in Chapter I, the complex normed algebra $\mathfrak{A}$ can be embedded in a complete normed algebra $\mathfrak{B}$ with a unit element. It is clear that if a is regular in $\mathfrak{A}$ then it will certainly be regular in $\mathfrak{B}$. Hence if the spectrum of a relative to $\mathfrak{A}$ is empty, then the spectrum of a relative to $\mathfrak{B}$ will likewise be empty. However this is impossible according to the above theorem.

Any closed bounded point set in the λ-plane may be the spectrum of an element in a suitably chosen Banach algebra. Thus if $x(\xi)$ is a bounded complex-valued function defined on $[0,1]$ and $\|\, x(\cdot)\,\| = \sup |\, x(\xi)\,|$, then the set of all such functions with the arithmetical operations defined in the ordinary manner is a Banach algebra. Here the spectrum of $x(\cdot)$ is clearly the closure of its range which may be a perfectly arbitrary closed bounded point set.

That $R(\lambda; a)$ is locally analytic was proved above with the aid of the series (4.7.2). A more direct proof could be had from formula (4.6.2) and the remark in section 4.6 according to which the inverse of a holomorphic function is holomorphic wherever the inverse exists. Applied to the holomorphic function $\lambda e - a$, it shows that $R(\lambda; a)$ is holomorphic wherever it exists.

For future reference we note the formula for the derivatives of $R(\lambda; a)$ which may be read off from (4.7.2):

$$R^{(k)}(\lambda; a) = (-1)^k k![R(\lambda; a)]^{k+1}. \tag{4.7.6}$$

4.8. The resolvent equations. The resolvent satisfies two very important relations, the first of which is given in the following theorem.

THEOREM 4.8.1. *If λ and μ belong to $\rho(a)$, then*

$$R(\lambda; a) - R(\mu; a) = -(\lambda - \mu)R(\lambda; a)R(\mu; a). \tag{4.8.1}$$

PROOF. The formula (4.8.1), called the *first resolvent equation*, is a consequence of

$$\begin{aligned} R(\lambda; a) &= R(\lambda; a)(\mu e - a)R(\mu; a) \\ &= R(\lambda; a)[(\mu - \lambda)e + (\lambda e - a)]R(\mu; a) \\ &= (\mu - \lambda)R(\lambda; a)R(\mu; a) + R(\mu; a). \end{aligned}$$

Incidentally, it follows from (4.8.1) that $R(\lambda; a)$ and $R(\mu; a)$ commute. It should be observed that it is not necessary for λ and μ to belong to the same component of $\rho(a)$.

The functional equation

$$R(\lambda) - R(\mu) = -(\lambda - \mu)R(\lambda)R(\mu), \qquad R(\lambda), R(\mu) \in \mathfrak{B}, \tag{4.8.2}$$

is of great importance in analysis. The equation imposes quite severe restrictions on its solutions. In fact, it is clear from (4.8.2) that a solution which is bounded is necessarily analytic. All resolvents are solutions; however, as we shall see in paragraph 5.2, there are other solutions as well. Any function satisfying (4.8.2) will be called a *pseudo-resolvent*.

One further remark about the first resolvent equation is in order. Let $a \in \mathfrak{B}$. Then the resolvent $R(\lambda; a)$ defined on $\rho(a)$ has no analytic extension. For it is clear that any such extension, say $R(\lambda)$, would continue to satisfy (4.8.2), at least for $\mu \in \rho(a)$. Consequently

$$\begin{aligned} (\lambda e - a)R(\lambda) &= (\lambda - \mu)R(\lambda) + (\mu e - a)[R(\mu; a) - (\lambda - \mu)R(\mu; a)R(\lambda)] \\ &= e. \end{aligned}$$

Similarly, $R(\lambda)(\lambda e - a) = e$ and hence $\lambda \in \rho(a)$.

We may also study the resolvent $R(\lambda; x) = (\lambda e - x)^{-1}$ as a function of x. This leads to the *second resolvent equation*, namely (4.8.3) below.

THEOREM 4.8.2. *Let x and y belong to a complex Banach algebra $\mathfrak{B}$, having a unit* *element e, and suppose that $\lambda \in \rho(x) \cap \rho(y)$ so that $R(\lambda; x)$ and $R(\lambda; y)$ exist. Then*

$$R(\lambda; x) - R(\lambda; y) = R(\lambda; x)(x - y)R(\lambda; y). \tag{4.8.3}$$

PROOF. Formula (4.8.3) is obtained from the identity

$$(\lambda e - x)[R(\lambda; x) - R(\lambda; y)](\lambda e - y) = x - y$$

upon multiplication by $R(\lambda; x)$ on the left and $R(\lambda; y)$ on the right.

THEOREM 4.8.3. *For fixed λ_0, the set S of all $x \in \mathfrak{B}$ such that $\lambda_0 \in \rho(x)$ is open. Further $R(\lambda_0; x)$ is an analytic function of x in each of the components of S.*

PROOF. Suppose that $\lambda_0 \in \rho(a)$ and substitute $x = \lambda_0 e - y$, $x_0 = \lambda_0 e - a$ in the formula (4.3.2). For $\| y - a \| = \| x - x_0 \| < \| x_0^{-1} \|^{-1} = \| R(\lambda_0; a) \|^{-1}$ we obtain

$$R(\lambda_0; y) = R(\lambda_0; a) \left\{ e + \sum_{n=1}^{\infty} [(y - a)R(\lambda_0; a)]^n \right\}. \tag{4.8.4}$$

Hence $y \in S$ for $\| y - a \| < \| R(\lambda_0; a) \|^{-1}$. Since the expansion (4.8.4) converges uniformly in any sphere about a of radius less than $\| R(\lambda_0; a) \|^{-1}$, it is easy to see that $R(\lambda_0; y)$ is locally bounded, (G)-differentiable, and therefore analytic in each of the components of S.

4.9. Normed fields. We shall first apply the spectral theory to the simplest of all Banach algebras, namely, normed fields. Here the basic theorem is due to S. Mazur [2] who extended a classical theorem of G. Frobenius.

THEOREM 4.9.1. *A complex normed algebra which is a skew-field is isomorphic to the complex field.*

This theorem is a consequence of the following:

THEOREM 4.9.2. *If $\mathfrak{A}$ is an algebra with unit element e over the field Φ and if $\mathfrak{A}$ has the property that to every $x \in \mathfrak{A}$ there is at least one $\mu \in \Phi$ such that $\mu e - x$ does not have a left (right) inverse, then $\mathfrak{A}$ is a skew-field if and only if $\mathfrak{A} = \Phi e$.*

PROOF. We know that $\mathfrak{A}$ contains the subalgebra Φe. But if $\mathfrak{A}$ is a skew-field and if $x \in \mathfrak{A}$ and $\mu e - x$ does not have a left inverse, then $\mu e - x = \theta$; that is, every x is of the form μe and $\mathfrak{A} = \Phi e$. The converse is obvious.

The assumptions of this theorem are satisfied in the case of a complex normed algebra with unit element by virtue of the corollary to Theorem 4.7.4. This observation proves Theorem 4.9.1.

In this connection we also note the following theorem which is due to E. R. Lorch [3]. See also S. Mazur [2].

THEOREM 4.9.3. *If $\mathfrak{B}$ is a complex (B)-algebra with a unit element and if the norm satisfies the condition $\| xy \| = \| x \| \| y \|$ for all x, y, then $\mathfrak{B}$ is isomorphic to the complex field.*

PROOF. Since $R(\lambda; x)(\lambda e - x) = e$, the product property of the norm shows that $\| R(\lambda; x) \| = \| \lambda e - x \|^{-1}$ for $\lambda \in \rho(x)$. Referring back to formula (4.7.2), we see that if $R(\lambda_0; x)$ exists, then $R(\lambda; x)$ exists for $|\lambda - \lambda_0| < \| \lambda_0 e - x \|$. Hence we may continue $R(\lambda; x)$ analytically along any path starting at $\lambda = \lambda_0$ on which $\lambda e - x \neq \theta$. There can be at most one value of λ for which $\lambda e - x$ vanishes; on the other hand, there must be at least one such value since, by Theorem 4.7.4, $R(\lambda; x)$ cannot exist for all finite values of λ. Hence there exists a complex number α such that $x = \alpha e$ and the theorem is proved.

This result is likewise valid for a complex normed algebra, $\mathfrak{A}$, with a unit element if the norm satisfies $\| xy \| = \| x \| \| y \|$. For by completing the algebra we obtain a (B)-algebra, $\mathfrak{B}$, satisfying the postulates of the theorem. Thus $\mathfrak{B} = Ce$ where C denotes the complex field. However, $Ce \subset \mathfrak{A} \subset \mathfrak{B}$ and hence $\mathfrak{A} = Ce$.

4.10. Quasi-nilpotent elements. For a quasi-nilpotent element, q, we have by definition $\lim_{n\to\infty} \| q^n \|^{1/n} = 0$. According to Theorem 4.7.3 this means that $r(q) = 0$ and hence that $\sigma(q)$ consists only of the number 0. It follows from Theorem 4.7.2 that the resolvent $R(\lambda; q)$ is an entire function of $1/\lambda$.

THEOREM 4.10.1. *Let $\mathfrak{B}$ be a complex Banach algebra with a unit element. If q is a quasi-nilpotent element, then a necessary and sufficient condition that $q^N = \theta$ for some $N \geqq 1$ is that $x = e + q$ have the property $\| x^{\pm n} \| = o(n^N)$ for $n = 1, 2, 3, \cdots$.*

The case $N = 1$ with $o(n)$ replaced by $O(1)$ was first treated by I. Gelfand [5]. The theorem in its present generality is due to E. Hille [11]. A simplified proof is to be found in a paper by M. H. Stone [6].

PROOF. Since q is quasi-nilpotent, $\lambda e - (e + q) = (\lambda - 1)e - q$ is regular except at $\lambda = 1$. It follows that

$$R(\lambda; e + q) = R(\lambda - 1; q) = \sum_{n=0}^{\infty} q^n(\lambda - 1)^{-n-1} \tag{4.10.1}$$

exists for $\lambda \neq 1$ and is an entire function of $1/(\lambda - 1)$. This function admits of the following two Laurent expansions:

$$R(\lambda; e + q) = -\sum_{n=0}^{\infty} (e + q)^{-n-1}\lambda^n, \qquad |\lambda| < 1, \tag{4.10.2}$$

$$R(\lambda; e + q) = \sum_{n=0}^{\infty} (e + q)^n\lambda^{-n-1}, \qquad |\lambda| > 1. \tag{4.10.3}$$

Supposing the coefficients of these two series are $o(n^N)$, by Theorem 3.13.5 this implies that $R(\lambda; e + q)$ is a polynomial in $1/(\lambda - 1)$ of degree not greater than N. A comparison with (4.10.1) then gives $q^N = \theta$.

On the other hand if $q^N = \theta$, then for $n \geqq N$ we have

$$\| x^n \| = \left\| \sum_{k=0}^{N-1} \frac{n!}{k!(n-k)!} q^k \right\| = O(n^{N-1}) = o(n^N).$$

Likewise for $q^N = \theta$, we have $x^{-1} = e + \sum_{j=1}^{N-1}(-q)^j = e + p$ where $p^N = \theta$. Hence for $n \geqq N$

$$\| x^{-n} \| = \left\| \sum_{k=0}^{N-1} \frac{n!}{k!(n-k)!} p^k \right\| = o(n^N).$$

4.11. Permanent spectral singularities. It will be recalled that the regular or singular character of an element depends on the algebra in which it is embedded. If $a \in \mathfrak{B}_1 \subset \mathfrak{B}_2$ and a is regular in $\mathfrak{B}_1$, then it is also regular in $\mathfrak{B}_2$ provided that the extension $\mathfrak{B}_2$ has the same unit element as $\mathfrak{B}_1$. However, an element may be regular in $\mathfrak{B}_2$ but not in $\mathfrak{B}_1$; in other words, an element a has, as a rule, two different spectra $\sigma(a; \mathfrak{B}_1)$ and $\sigma(a; \mathfrak{B}_2)$ according as it is considered an element in $\mathfrak{B}_1$ or $\mathfrak{B}_2$ respectively, with $\sigma(a; \mathfrak{B}_1) \supset \sigma(a; \mathfrak{B}_2)$. The first half of Theorem 4.11.1 is due to S. Bochner and R. S. Phillips [1] and the second half to C. E. Rickart [2]; the first half of Theorem 4.11.2 is due to E. R. Lorch [3] and the second half to G. Šilov [1]; and Corollary 3 to Theorem 4.11.2 is due to E. R. Lorch [3].

Definition 4.11.1. *A point $\lambda \in \sigma(a; \mathfrak{B}_1)$ is a removable spectral singularity of a in $\mathfrak{B}_1$ if $\lambda \notin \sigma(a; \mathfrak{B}_2)$ for a suitable choice of $\mathfrak{B}_2 \supset \mathfrak{B}_1$. It is a permanent spectral singularity of a if $\lambda \in \sigma(a; \mathfrak{B}_2)$ for every possible extension $\mathfrak{B}_2$ of $\mathfrak{B}_1$.*

Theorem 4.11.1. *If $\lim_{n\to\infty} a_n = a_0$ and if a_n^{-1} exists for each $n > 0$, then either $\limsup_{n\to\infty} \| a_n^{-1} \| < \infty$, in which case a_0^{-1} exists, or a_0 is a generalized divisor of zero.*

Proof. First suppose that $\| a_n^{-1} \| \leqq M < \infty$, for all n. Then

$$\| a_m^{-1} - a_n^{-1} \| = \| a_m^{-1}(a_n - a_m)a_n^{-1} \| \leqq M^2 \| a_n - a_m \| \to 0.$$

Thus there exists a b_0 such that $a_n^{-1} \to b_0$. Moreover

$$a_0 b_0 = (a_0 - a_n)b_0 + a_n(b_0 - a_n^{-1}) + a_n a_n^{-1}.$$

Since $(a_0 - a_n)b_0 \to \theta$, $a_n(b_0 - a_n^{-1}) \to \theta$, and $a_n a_n^{-1} = e$, we have $a_0 b_0 = e$. Similarly $b_0 a_0 = e$, so $a_0^{-1} = b_0$ exists. On the other hand suppose that $\limsup_{n\to\infty} \| a_n^{-1} \| = \infty$, and we may clearly assume that this holds with lim sup replaced by lim. Now define $b_n = a_n^{-1}/\|a_n^{-1}\|$ so that $\| b_n \| = 1$. Then

$$a_0 b_n = (a_0 - a_n)b_n + a_n b_n,$$

where the two summands on the right tend to θ when $n \to \infty$. Therefore $a_0 b_n \to \theta$; that is, a_0 is a generalized divisor of zero.

Corollary. *If $\mathfrak{G}$ is the maximal group in $\mathfrak{B}$, then the elements on the boundary of $\mathfrak{G}$ are generalized divisors of zero or θ.*

Another consequence is:

THEOREM 4.11.2. *All points belonging to the boundary of the spectrum of an element a in $\mathfrak{B}$ are permanent spectral singularities of a. If λ_0 is such a boundary point, then $\lambda_0 e - a$ is either θ or a generalized divisor of zero.*

PROOF. The second half of the theorem follows directly from the preceding theorem and this shows that $\lambda_0 e - a$ is a singular element of $\mathfrak{B}$ as well as of any extension of $\mathfrak{B}$.

The only (B)-algebra without generalized divisors of zero is the complex field. For clearly each $a \in \mathfrak{B}$ must have a non-empty spectrum by Theorem 4.7.4 and consequently there will exist at least one boundary point λ_0 of $\sigma(a)$. If there are no generalized divisors of zero, then $\lambda_0 e - a = \theta$ and hence $\mathfrak{B} = Ce$. This argument furnishes us with another proof of Theorem 4.9.3.

Theorem 4.11.2 has other interesting consequences.

COROLLARY 1. *The components of $\rho(a; \mathfrak{B}_1)$ are also components of $\rho(a; \mathfrak{B}_2)$ if $\mathfrak{B}_1 \subset \mathfrak{B}_2$.*

COROLLARY 2. *The spectrum $\sigma(a; \mathfrak{B})$ remains invariant under* (i) *extension of $\mathfrak{B}$ if $\sigma(a; \mathfrak{B})$ is nowhere dense,* (ii) *contraction of $\mathfrak{B}$ if $\rho(a; \mathfrak{B})$ is connected.*

COROLLARY 3. *If the spectrum $\sigma(a; \mathfrak{B})$ is real, then $\sigma(a; \mathfrak{B})$ remains invariant under a contraction of $\mathfrak{B}$.*

PROOF. Let $\mathfrak{B}_0$ be a subalgebra of $\mathfrak{B}$ which contains a. Assume first of all that $\sigma(a; \mathfrak{B}_0)$ is not real. Then clearly $\sigma(a; \mathfrak{B}_0)$ will contain boundary points which are not real. Now according to the above theorem this would imply that $\sigma(a; \mathfrak{B})$ also contained non-real points contrary to assumption. Thus $\sigma(a; \mathfrak{B}_0)$ is real and consequently nowhere dense; it now follows from Corollary 2 that $\sigma(a; \mathfrak{B}) = \sigma(a; \mathfrak{B}_0)$.

LEMMA 4.11.1. *If $\mathfrak{B}$ is a (*)-algebra and x is regular, then $(x^*)^{-1} = (x^{-1})^*$ exists. As a consequence $\sigma(x^*) = \overline{\sigma(x)}$ for all $x \in \mathfrak{B}$.*

PROOF. The first part of the lemma follows immediately from $e = (xx^{-1})^* = (x^{-1})^*x^*$ and $e = (x^{-1}x)^* = x^*(x^{-1})^*$. Since $(\lambda e - x)^* = \bar{\lambda}e - x^*$, it is clear that $\lambda e - x$ and $\bar{\lambda}e - x^*$ will be regular together, which is the second assertion of the lemma.

The next theorem is somewhat deeper and is due to C. E. Rickart [**2**].

THEOREM 4.11.3. *Let $\mathfrak{B}$ be an (A^*)-algebra. If $\mathfrak{B}_0$ is a (*)-subalgebra of $\mathfrak{B}$, then $\mathfrak{B}_0$ contains inverses (i.e. if $x \in \mathfrak{B}_0$ and if x^{-1} exists in $\mathfrak{B}$, then $x^{-1} \in \mathfrak{B}_0$).*

PROOF. Suppose that $x \in \mathfrak{B}_0$, x^{-1} exists in $\mathfrak{B}$, but that x does not have, say, a left inverse in $\mathfrak{B}_0$. Then x^*x likewise can not have a left inverse in $\mathfrak{B}_0$. On the other hand x^*x is self-adjoint and hence its spectrum relative to $\mathfrak{B}$ is real. By Corollary 3 above, $\sigma(x^*x; \mathfrak{B}_0) = \sigma(x^*x; \mathfrak{B})$. Since $0 \in \sigma(x^*x; \mathfrak{B}_0)$, it follows that x^*x is also singular in $\mathfrak{B}$. However this is impossible as both x and therefore x^* (Lemma 4.11.1) are regular. Consequently x^{-1} must belong to $\mathfrak{B}_0$.

4.12. Some properties of (B*)-algebras. As another application of spectral theory we shall derive a few properties of (B*)-algebras.

THEOREM 4.12.1. *If x is a normal element of a (B^*)-algebra, then $\| x \|^2 = \| x^2 \|$ and $r(x) = \| x \| = \| x^* \|$.*

PROOF. Making use of the norm properties for elements of a (B*)-algebra we have

$$\| x \|^2 \| x^* \|^2 = \| xx^* \|^2 = \| (xx^*)(xx^*)^* \| = \| x^2(x^*)^2 \|$$
$$\leqq \| x^2 \| \| (x^*)^2 \| \leqq \| x \|^2 \| x^* \|^2.$$

Therefore $\| x^2 \| \| (x^*)^2 \| = \| x \|^2 \| x^* \|^2$. Now $\| x^2 \| \leqq \| x \|^2$ and $\| (x^*)^2 \| \leqq \| x^* \|^2$. Hence $\| x^2 \| = \| x \|^2$. By induction we see that $\| x^{2^n} \| = \| x \|^{2^n}$. It follows from Theorem 4.7.3 that

$$r(x) = \lim_{n\to\infty} \| x^{2^n} \|^{2^{-n}} = \| x \|.$$

According to Lemma 4.11.1, $\sigma(x^*) = \overline{\sigma(x)}$. Consequently $r(x) = r(x^*)$ and therefore $\| x \| = \| x^* \|$.

In particular the unit element e is normal and $\sigma(e) = \{1\}$. Hence $\| e \| = r(e) = 1$. The next result is due to I. Gelfand and M. Neumark [1]; however the proof which we give is due to R. F. Arens [1].

THEOREM 4.12.2. *If x is a self-adjoint element of a (B^*)-algebra, then the spectrum of x is real.*

PROOF. Suppose first that $i \in \sigma(x)$ and let γ be any real number. Since $ie - x = (1 + \gamma)ie - (x + \gamma ie)$, it follows that $(1 + \gamma)i \in \sigma(x + \gamma ie)$. By Lemma 4.11.1 we have $-i \in \sigma(x^*)$ so that $-(1 + \gamma)i \in \sigma(x^* - \gamma ie)$. Therefore $(1 + \gamma)^2 \leqq r(x + \gamma ie)r(x^* - \gamma ie) = \| x + \gamma ie \| \| x^* - \gamma ie \|$ and, by the norm property for (B*)-algebras,

$$(1 + \gamma)^2 \leqq \| (x + \gamma ie)(x + \gamma ie)^* \| = \| x^2 + \gamma^2 e \| \leqq \| x^2 \| + \gamma^2.$$

Thus $1 + 2\gamma \leqq \| x^2 \|$ and this must be true for all γ. Therefore the assumption that $i \in \sigma(x)$ has lead to a contradiction. More generally, let us suppose that y is self-adjoint and that $\alpha + i\beta \in \sigma(y)$, where α and β are real and $\beta \neq 0$. Then $x = (y - \alpha e)/\beta$ is clearly self-adjoint. Further $ie - x = \beta^{-1}[(\alpha + i\beta)e - y]$ so that $i \in \sigma(x)$; again this is impossible.

COROLLARY. *A (B^*)-algebra satisfies the postulates for an (A^*)-algebra.*

References. Arens [1], Bochner and Phillips [1], Gelfand [4, 5], Gelfand and Neumark [1], Hille [11], Lorch [3], Mazur [2], Rickart [2], Šilov [1], Stone [6].

3. Ideal Theory for Commutative Banach Algebras

4.13. Ideals. In the remainder of this chapter we deal only with commutative Banach algebras with unit elements. This will be entirely adequate for our subsequent study of semi-groups of linear operators and we postpone until Chapter XXIV a parallel development for the general (B)-algebra. It is fair to say that the commutative theory had its germ in the work of N. Wiener [**3**] on Tauberian theorems. However it was I. Gelfand [**4**] who saw the underlying algebraic nature of Wiener's work and who is responsible for the theory which we now present.

Definition 4.13.1. *Let $\mathfrak{B}$ be a commutative complex Banach algebra with a unit element.*

(1) *A subset $\mathfrak{i}$ of $\mathfrak{B}$ is called an ideal if $x - y \in \mathfrak{i}$ whenever $x, y \in \mathfrak{i}$ and if $zx \in \mathfrak{i}$ whenever $z \in \mathfrak{B}$ and $x \in \mathfrak{i}$.*

(2) *An ideal $\mathfrak{i}$ is said to be proper if $\mathfrak{i} \neq \{\theta\}, \mathfrak{B}$.*

(3) *An ideal $\mathfrak{m}$ is said to be maximal if it is not equal to the whole algebra and is not properly contained in any other ideal.*

If an ideal contains a regular element x_0, then $\mathfrak{B}x_0 = \mathfrak{B}$ so that such an ideal is actually the algebra $\mathfrak{B}$ itself. In other words, any proper ideal is necessarily disjoint from the maximal group $\mathfrak{G}$.

Theorem 4.13.1. *The closure of a proper ideal is again a proper ideal.*

Proof. Let $\mathfrak{i}$ be a proper ideal with closure $\mathfrak{c}$. Then for $x, y \in \mathfrak{c}$ there exist sequences $\{x_n\}$, $\{y_n\}$ belonging to $\mathfrak{i}$ such that $x_n \to x$ and $y_n \to y$. It is clear that $x_n - y_n \to x - y$ and for any $z \in \mathfrak{B}$ that $zx_n \to zx$. Since the approximating elements belong to $\mathfrak{i}$, it follows that $x - y$ and zx belong to $\mathfrak{c}$ and hence that $\mathfrak{c}$ is an ideal. On the other hand $\mathfrak{G}$ is open and since $\mathfrak{i}$ and $\mathfrak{G}$ are disjoint the same is true of $\mathfrak{c}$ and $\mathfrak{G}$; therefore $\mathfrak{c}$ is a proper ideal.

Corollary. *A maximal ideal is necessarily closed.*

Theorem 4.13.2. *Each ideal other than the unit ideal is contained in a maximal ideal.*

Proof. The proof requires the use of the maximal principle. Let $\mathfrak{i}_0$ be the given non-unit ideal and let $\mathfrak{J}$ be the class of all proper ideals containing $\mathfrak{i}_0$. We then introduce an order relation in $\mathfrak{J}$, denoting by $\mathfrak{i}_\alpha < \mathfrak{i}_\beta$ the fact that all of the elements of $\mathfrak{i}_\alpha$ are contained in $\mathfrak{i}_\beta$. Suppose that $[\mathfrak{i}_\alpha]$ is a simply ordered subset of $\mathfrak{J}$ and set $\mathfrak{i}_\gamma = \bigcup \mathfrak{i}_\alpha$. We now show that $\mathfrak{i}_\gamma$ is an upper bound in $\mathfrak{J}$ for $[\mathfrak{i}_\alpha]$. In fact if, $x, y \in \mathfrak{i}_\gamma$, then there exist ideals $\mathfrak{i}_{\alpha_1}$ and $\mathfrak{i}_{\alpha_2}$ such that $x \in \mathfrak{i}_{\alpha_1}$ and $y \in \mathfrak{i}_{\alpha_2}$. Since $[\mathfrak{i}_\alpha]$ is simply ordered, we may suppose that $\mathfrak{i}_{\alpha_1} < \mathfrak{i}_{\alpha_2}$ so that both x and y belong to $\mathfrak{i}_{\alpha_2}$. Hence $x - y \in \mathfrak{i}_{\alpha_2} \subset \mathfrak{i}_\gamma$ and for any $z \in \mathfrak{B}$, $zx \in \mathfrak{i}_{\alpha_2} \subset \mathfrak{i}_\gamma$. Thus $\mathfrak{i}_\gamma$ is an ideal. Since each $\mathfrak{i}_\alpha$ and $\mathfrak{G}$ are disjoint, the same is true of $\mathfrak{i}_\gamma$ and $\mathfrak{G}$ and hence $\mathfrak{i}_\gamma$ is not the unit ideal. Finally it is clear that $\mathfrak{i}_\gamma > \mathfrak{i}_\alpha$.

It now follows from the maximal principle that there exists at least one maximal ideal which contains $\mathfrak{i}_0$.

THEOREM 4.13.3. *An element is regular if and only if it does not belong to any maximal ideal.*

PROOF. As we have already observed, a regular element can belong to neither a proper ideal nor to the zero ideal. Hence it can not belong to any maximal ideal. Conversely suppose that x_0 is singular. Let $\mathfrak{i} = \mathfrak{B}x_0$ and suppose that $x_1, x_2 \in \mathfrak{i}$. Then there exist $z_1, z_2 \in \mathfrak{B}$ such that $x_i = z_i x_0$ and hence $x_1 - x_2 = (z_1 - z_2)x_0 \in \mathfrak{i}$ and $zx_1 = (zz_1)x_0 \in \mathfrak{i}$ for all $z \in \mathfrak{B}$. Thus $\mathfrak{i}$ is an ideal. Since x_0 is singular, $\mathfrak{i}$ cannot contain the unit element and therefore $\mathfrak{i}$ is not the unit ideal. By the previous theorem there exists a maximal ideal containing $\mathfrak{i}$ and hence x_0 $(=ex_0)$.

THEOREM 4.13.4. *If $\mathfrak{B}$ contains no proper ideals, then $\mathfrak{B}$ is isomorphic to the complex field.*

PROOF. In particular $\mathfrak{B}$ can not contain a maximal ideal other than the zero ideal. Hence by the previous theorem each non-zero element is regular. In other words $\mathfrak{B}$ is a field and therefore, according to Theorem 4.9.1, $\mathfrak{B}$ is isomorphic to the complex field.

DEFINITION 4.13.2. *The radical $\mathfrak{p}$ of a commutative (B)-algebra with unit element is defined as the intersection of all maximal ideals. If $\mathfrak{p}$ is simply the zero element, $\mathfrak{B}$ is said to be without radical or to be semi-simple.*

It is clear that the intersection of any set of ideals is again an ideal. Hence the radical is an ideal containing at least the zero element.

4.14. Residue-class algebras. Let $\mathfrak{i}$ be an ideal of the commutative algebra $\mathfrak{A}$. In particular, $\mathfrak{i}$ is a linear subset of $\mathfrak{A}$ and we can define a relation of *congruence modulo* $\mathfrak{i}$ as in section 1.12. Under this relation the algebra $\mathfrak{A}$ breaks up into a system of mutually exclusive residue classes $X, Y, \cdots$. If $x_1 - x_2 \in \mathfrak{i}$ and $y_1 - y_2 \in \mathfrak{i}$, then

$$\begin{aligned}(\alpha x_1 + \beta y_1) - (\alpha x_2 + \beta y_2) &= \alpha(x_1 - x_2) + \beta(y_1 - y_2) \in \mathfrak{i},\\ x_1 y_1 - x_2 y_2 &= x_1(y_1 - y_2) + (x_1 - x_2)y_2 \in \mathfrak{i}.\end{aligned} \tag{4.14.1}$$

We may therefore define the operations of addition, scalar multiplication, and multiplication for the residue-classes by the conventions: $X + Y$, αX, XY are the residue-classes containing $x + y$, αx, and xy respectively, where x and y are representative elements of the respective classes X and Y. In this manner the system of residue-classes becomes an algebra which we denote by $\mathfrak{A}/\mathfrak{i}$ and refer to as a *residue-class algebra* or a *quotient algebra* or a *difference algebra*. The zero element of $\mathfrak{A}/\mathfrak{i}$ is $\Theta = \mathfrak{i}$, whereas the unit element E of $\mathfrak{A}/\mathfrak{i}$ is the residue-class determined by the unit element of $\mathfrak{A}$. Finally the mapping which takes

x of $\mathfrak{A}$ into the residue-class containing x is clearly a homomorphism of $\mathfrak{A}$ onto $\mathfrak{A}/\mathfrak{i}$. We summarize these remarks in the following theorem.

THEOREM 4.14.1. *If $\mathfrak{A}$ is a commutative algebra with a unit element over Φ and if $\mathfrak{i}$ is an ideal, then $\mathfrak{A}/\mathfrak{i}$ is likewise a commutative algebra with a unit element over Φ and $\mathfrak{A} \sim \mathfrak{A}/\mathfrak{i}$.*

We note that $\mathfrak{A}/\mathfrak{p}$, where $\mathfrak{p}$ is the radical of $\mathfrak{A}$, is semi-simple since the image under the homomorphism $\mathfrak{A} \sim \mathfrak{A}/\mathfrak{p}$ of any maximal ideal in $\mathfrak{A}$ is a maximal ideal in $\mathfrak{A}/\mathfrak{p}$.

THEOREM 4.14.2. *If $\mathfrak{B}$ is a commutative complex (B)-algebra with a unit element and if $\mathfrak{i}$ is a closed ideal, then $\mathfrak{B}/\mathfrak{i}$ becomes a commutative complex (B)-algebra with a unit element under the norm*

$$\| X \| \equiv \inf [\, \| x \|;\, x \in X]. \tag{4.14.2}$$

PROOF. It follows from Theorem 1.12.3 that $\mathfrak{B}/\mathfrak{i}$ is a complex (B)-space and from Theorem 4.14.1 that $\mathfrak{B}/\mathfrak{i}$ is a commutative complex algebra with a unit element. It remains only to verify the additional norm property introduced by the algebra.

$$\begin{aligned} \| XY \| &\equiv \inf_{z \in XY} \| z \| \leqq \inf_{x \in X, y \in Y} \| xy \| \\ &\leqq \inf_{x \in X, y \in Y} \| x \| \, \| y \| = \Big[\inf_{x \in X} \| x \| \Big]\Big[\inf_{y \in Y} \| y \|\Big] \\ &= \| X \| \, \| Y \| . \end{aligned}$$

We note that if $\mathfrak{i} \neq \mathfrak{B}$, then $E \neq \Theta$. In this case $\| E \| = \| E^2 \| \leqq \| E \| \, \| E \|$ implies that $\| E \| \geqq 1$. If, in addition, $\| e \| = 1$, then $e \in E$ implies that $\| E \| \leqq 1$ so that $\| E \| = 1$.

THEOREM 4.14.3. *If $\mathfrak{m}$ is a maximal ideal contained in a commutative complex (B)-algebra $\mathfrak{B}$ with a unit element, then $\mathfrak{B}/\mathfrak{m}$ is isomorphic to the complex field. For each $x \in \mathfrak{B}$, the homomorphism $\mathfrak{B} \sim \mathfrak{B}/\mathfrak{m}$ maps x into a complex number, $x(\mathfrak{m})$ defined by*

$$x \equiv x(\mathfrak{m})e \pmod{\mathfrak{m}}. \tag{4.14.3}$$

This correspondence has the following properties:

(1) $(x_1 + x_2)(\mathfrak{m}) = x_1(\mathfrak{m}) + x_2(\mathfrak{m})$;
(2) $(\alpha x)(\mathfrak{m}) = \alpha x(\mathfrak{m})$;
(3) $(x_1x_2)(\mathfrak{m}) = x_1(\mathfrak{m})x_2(\mathfrak{m})$;
(4) $e(\mathfrak{m}) = 1$;
(5) $x(\mathfrak{m}) = 0$ *if and only if* $x \in \mathfrak{m}$;
(6) $| x(\mathfrak{m}) | \leqq \| x \|$.

PROOF. If $\mathfrak{m}$ is the trivial ideal $\{\theta\}$, then $\mathfrak{B}$ can contain no proper ideal and hence by Theorem 4.13.4, $\mathfrak{B} \cong \mathfrak{B}/\mathfrak{m}$ is isomorphic to the complex field. On the other hand if $\mathfrak{m}$ is proper as well as maximal, then, by the corollary to Theorem 4.13.1, $\mathfrak{m}$ will be closed. By the previous theorem $\mathfrak{B}/\mathfrak{m}$ is a commutative complex (B)-algebra with a unit element. Further $\mathfrak{B}/\mathfrak{m}$ can contain no proper ideal. For suppose that $\mathfrak{i}'$ were such an ideal. It is then easy to see that the set of all elements of $\mathfrak{B}$ which map into $\mathfrak{i}'$ under the homomorphism $\mathfrak{B} \sim \mathfrak{B}/\mathfrak{m}$, themselves constitute a proper ideal which contains $\mathfrak{m}$. This is clearly impossible if $\mathfrak{m}$ is maximal. Again by Theorem 4.13.4 we see that $\mathfrak{B}/\mathfrak{m}$ is isomorphic to the complex field, C, or equivalently $\mathfrak{B}/\mathfrak{m} = CE$. The homomorphism $\mathfrak{B} \sim \mathfrak{B}/\mathfrak{m}$ therefore takes x into $X = x(\mathfrak{m})E$. Hence $x(\mathfrak{m})e$ likewise belongs to X so that $x \equiv x(\mathfrak{m})e \pmod{\mathfrak{m}}$. The properties (1) through (5) follow directly from the nature of the homomorphism $\mathfrak{B} \sim \mathfrak{B}/\mathfrak{m}$. Finally we have $|x(\mathfrak{m})| \leqq \| X \| = \inf_{y \in X} \| y \| \leqq \| x \|$.

Definition 4.14.1. *A complex-valued function $\mu(x)$ not identically zero is said to be a linear multiplicative functional on $\mathfrak{B}$ if* (i) $\mu(\alpha x + \beta y) = \alpha\mu(x) + \beta\mu(y)$ *and* (ii) $\mu(xy) = \mu(x)\mu(y)$ *for all $x, y \in \mathfrak{B}$.*

If $\mathfrak{B}$ is the complex field, then $\mu(x) = x$ is the only linear multiplicative functional on $\mathfrak{B}$. For the general commutative complex (B)-algebra with a unit element, we see from the previous theorem that each maximal ideal, $\mathfrak{m}$, defines a linear multiplicative functional, namely $\mu(x) = x(\mathfrak{m})$. As we shall now show, this exhausts the class of linear multiplicative functionals for such algebras.

Theorem 4.14.4. *Let $\mathfrak{B}$ be a commutative complex (B)-algebra with a unit element. If $\mu(x)$ is a linear multiplicative functional on $\mathfrak{B}$, then $\mathfrak{m} \equiv [x; \mu(x) = 0]$ is a maximal ideal and $\mu(x) = x(\mathfrak{m})$ for all $x \in \mathfrak{B}$.*

Proof. Suppose first of all that $x_1, x_2 \in \mathfrak{m}$. Then $\mu(x_1 - x_2) = \mu(x_1) - \mu(x_2) = 0$ and for all $z \in \mathfrak{B}$, $\mu(zx_1) = \mu(z)\mu(x_1) = 0$. Hence $\mathfrak{m}$ is an ideal. Since $\mu(x) \not\equiv 0$, $\mathfrak{m}$ is not the unit ideal. Now $x = ex$ so that $\mu(x) = \mu(e)\mu(x)$ for all x. Again since $\mu(x) \not\equiv 0$, it follows that $\mu(e) = 1$. Suppose now that $\mathfrak{m}$ were not maximal; that is, suppose that $\mathfrak{m}$ is contained in a proper ideal, $\mathfrak{i}$. There would then exist an element $x_0 \in \mathfrak{i} \ominus \mathfrak{m}$, that is, $\mu(x_0) \neq 0$. We have

$$\mu[x_0 - \mu(x_0)e] = \mu(x_0) - \mu(x_0)\mu(e) = 0.$$

Hence $x_0 - \mu(x_0)e \in \mathfrak{m}$ and therefore $\mu(x_0)e = x_0 - [x_0 - \mu(x_0)e] \in \mathfrak{i}$. This is contrary to $\mathfrak{i}$ being proper. Consequently $\mathfrak{m}$ is a maximal ideal and defines the homomorphic mapping $x \to x(\mathfrak{m})$ where $x - x(\mathfrak{m})e \in \mathfrak{m}$. Thus $0 = \mu[x - x(\mathfrak{m})e] = \mu(x) - x(\mathfrak{m})$ which concludes the proof of the theorem.

Corollary. *If $\mu(x)$ is a linear multiplicative functional, then μ is bounded and hence belongs to the adjoint space $\mathfrak{B}^*$ of $\mathfrak{B}$ considered as a (B)-space;* $\| \mu \| \leqq 1$.

Proof. It is only necessary to show that μ is bounded and this follows from $|\mu(x)| = |x(\mathfrak{m})| \leqq \| x \|$ (see Theorem 4.13.3). Thus $\| \mu \| \leqq 1$ and since $\mu(e) = 1$, we have $\| \mu \| = 1$ when $\| e \| = 1$.

Combining the results of Theorems 4.14.3 and 4.14.4, we see that *there is a one-to-one correspondence between the linear multiplicative functionals and the maximal ideals of an algebra* $\mathfrak{B}$. This remark is exceedingly useful since we can now find the maximal ideals of $\mathfrak{B}$ by determining the linear multiplicative functionals. In this way we replace an algebraic problem by an analytic problem.

Before concluding this section we shall illustrate the theory by proving a theorem due to N. Wiener [3] on absolutely convergent Fourier series. The theorem states that if $f(\xi)$ defined on $[0, 2\pi)$ has an absolutely convergent Fourier series and if $f(\xi) \neq 0$, then $1/f(\xi)$ likewise has an absolutely convergent Fourier series.

We take for our (B)-algebra, $\mathfrak{B}$, the set of all functions defined on $[0, 2\pi)$ with absolutely convergent Fourier series. Such a function $f(\xi)$ has the unique representation

$$f(\xi) = \sum_{-\infty}^{\infty} a_n e^{in\xi}.$$

We define $(f + g)(\xi) = f(\xi) + g(\xi)$, $(\alpha f)(\xi) = \alpha f(\xi)$, $(fg)(\xi) = f(\xi)g(\xi)$, and $\| f \| = \sum_{-\infty}^{\infty} | a_n |$. With these conventions it is not difficult to verify that $\mathfrak{B}$ is in fact a commutative complex (B)-algebra with unit element, $e(\xi) \equiv 1$. For a fixed $\xi_0 \in [0, 2\pi)$ it is clear that $\mu(f) = f(\xi_0)$ defines a linear multiplicative functional on $\mathfrak{B}$. We now show that this exhausts the linear multiplicative functionals on $\mathfrak{B}$. It is clear that the element $u(\xi) = e^{i\xi}$ is regular and that $[u(\xi)]^k = e^{ik\xi}$ for all $k = 0, \pm 1, \pm 2, \cdots$. For an arbitrary linear multiplicative functional μ, we have $| \mu(u^k) | \leqq \| u^k \| = 1$ and $\mu(u)\mu(u^{-1}) = \mu(e) = 1$. Thus $| \mu(u) | = 1$ so that $\mu(u) = e^{i\xi_0}$ for some $\xi_0 \in [0, 2\pi)$. Consequently $\mu(u^k) = e^{ik\xi_0}$. For any $f(\xi) = \sum_{-\infty}^{\infty} a_n e^{in\xi} \in \mathfrak{B}$, the series $\sum_{-\infty}^{\infty} a_n u^n$ converges in norm to f. Thus $\mu(f) = \mu(\sum_{-\infty}^{\infty} a_n u^n) = \sum_{-\infty}^{\infty} a_n \mu(u^n) = f(\xi_0)$. We are now in a position to prove the Wiener theorem. According to Theorem 4.13.3, an element f is regular if and only if it does not belong to any maximal ideal. This condition is equivalent to $\mu(f) \neq 0$ for all μ and, as we have just seen, this in turn means that $f(\xi) \neq 0$ on $[0, 2\pi)$. Thus $f^{-1}(\xi) = 1/f(\xi)$ is contained in $\mathfrak{B}$ if and only if $f(\xi) \neq 0$ on $[0, 2\pi)$. This is the Wiener theorem. We remark that the only element of $\mathfrak{B}$ common to all maximal ideals is the zero element. Hence $\mathfrak{B}$ is semi-simple.

4.15. Representation theory. We turn now to the problem of representing the general commutative complex Banach algebra with a unit element by means of a simple model. We shall present the exceedingly elegant solution to this problem obtained by I. Gelfand [4]. In the previous section it was found that for a fixed maximal ideal, $\mathfrak{m}$, the homomorphism $\mathfrak{B} \sim \mathfrak{B}/\mathfrak{m}$ could be represented by the mapping $x(\mathfrak{m})$ on $\mathfrak{B}$ to the complex field. We now change our point of view and, keeping x fixed, consider $x(\mathfrak{m})$ as a complex-valued function on the set of all maximal ideals. Gelfand's main result can be formulated as follows:

THEOREM 4.15.1. *Let* $\mathfrak{B}$ *be a commutative complex Banach algebra with a unit element e. Let* $\mathfrak{M} = [\mathfrak{m}]$ *be the set of all maximal ideals in* $\mathfrak{B}$. *For a given maximal ideal,* $\mathfrak{m}$, *the homomorphism* $\mathfrak{B} \sim \mathfrak{B}/\mathfrak{m}$ *assigns to each* $x \in \mathfrak{B}$ *a complex number,* $x(\mathfrak{m})$, *defined by* $x \equiv x(\mathfrak{m})e \pmod{\mathfrak{m}}$. *These mappings define a class of complex-valued functions on* $\mathfrak{M}$ *having the following properties:*

(1) $(x_1 + x_2)(\mathfrak{m}) = x_1(\mathfrak{m}) + x_2(\mathfrak{m})$;

(2) $(\alpha x)(\mathfrak{m}) = \alpha x(\mathfrak{m})$;

(3) $(x_1x_2)(\mathfrak{m}) = x_1(\mathfrak{m})x_2(\mathfrak{m})$;

(4) $e(\mathfrak{m}) \equiv 1$;
(5) $x(\mathfrak{m}) = 0$ *if and only if* $x \in \mathfrak{m}$;
(6) $\sigma(x) \equiv x(\mathfrak{M})$;
(7) *the spectral radius* $r(x) = \sup_{\mathfrak{m}\in\mathfrak{M}} | x(\mathfrak{m}) | \leqq \| x \|$;
(8) $q(\mathfrak{m}) \equiv 0$ *if and only if* q *is quasi-nilpotent*;
(9) *an element* x *is regular if and only if* $x(\mathfrak{m}) \neq 0$; *for regular* x, $x^{-1}(\mathfrak{m}) = [x(\mathfrak{m})]^{-1}$;
(10) *if* $\mathfrak{m}_1 \neq \mathfrak{m}_2$ *then there is an* $x \in \mathfrak{B}$ *such that* $x(\mathfrak{m}_1) \neq x(\mathfrak{m}_2)$.

PROOF. Properties (1) through (5) paraphrase their counterparts in Theorem 4.14.3. By Theorem 4.13.3, an element x is regular if and only if it does not belong to any maximal ideal and hence, according to (5), if and only if $x(\mathfrak{m}) \neq 0$. Thus $\lambda - x(\mathfrak{m}) = (\lambda e - x)(\mathfrak{m}) \neq 0$ if and only if $\lambda \in \rho(x)$. In other words, $\sigma(x) \equiv [x(\mathfrak{m});\ \mathfrak{m} \in \mathfrak{M}]$, and this establishes (6). It now follows from the definition of spectral radius that $r(x) = \sup_{\mathfrak{m}\in\mathfrak{M}} | x(\mathfrak{m}) |$ and from (6) of Theorem 4.14.3 that $| x(\mathfrak{m}) | \leqq \| x \|$, which together give (7). As we have already noted in section 4.10, q is quasi-nilpotent if and only if $r(q) = 0$ and therefore if and only if $q(\mathfrak{m}) \equiv 0$. If x is a regular element then $x^{-1}(\mathfrak{m})x(\mathfrak{m}) = e(\mathfrak{m}) \equiv 1$, and this together with the above remarks proves (9). Finally if $\mathfrak{m}_1 \neq \mathfrak{m}_2$, then there is at least one element x in $\mathfrak{m}_1$ which does not lie in $\mathfrak{m}_2$. Hence $x(\mathfrak{m}_1) = 0 \neq x(\mathfrak{m}_2)$, which proves (10).

COROLLARY 1. *The mapping which takes* $x \in \mathfrak{B}$ *into the complex-valued function* $x(\mathfrak{m})$ *on* $\mathfrak{M}$ *is a homomorphism.*

COROLLARY 2. *The radical of* $\mathfrak{B}$ *consists of the set of all quasi-nilpotent elements.*

COROLLARY 3. *The mapping which takes* $x \in \mathfrak{B}$ *into the complex-valued function* $x(\mathfrak{m})$ *on* $\mathfrak{M}$ *is an isomorphism if and only if* $\mathfrak{B}$ *is semi-simple.*

We remark that the spectral radius $r(x)$ has the properties of a *pseudo-norm* in a commutative (B)-algebra. For as is readily seen from property (7) of Theorem 4.15.1,

$$(4.15.1) \qquad r(x) \geqq 0, \qquad r(\alpha x) = | \alpha | r(x), \qquad r(x + y) \leqq r(x) + r(y).$$

Since $r(q) = 0$ for quasi-nilpotent elements it is in general not a true norm. The spectral radius corresponds to the usual norm in $M(\mathfrak{M})$, the space of bounded functions on $\mathfrak{M}$ with $\| f \| = \sup_{\mathfrak{m}\in\mathfrak{M}} | f(\mathfrak{m}) |$.

The final touch to the Gelfand theory is achieved by introducing a topology in $\mathfrak{M}$. A generic neighborhood for a given maximal ideal $\mathfrak{m}_0$ is defined by means of an arbitrary $\epsilon > 0$ and a finite set of elements $x_1, x_2, \cdots, x_n \in \mathfrak{B}$ as

$$(4.15.2) \quad N(\mathfrak{m}_0; x_1, x_2, \cdots, x_n; \epsilon) \equiv [\mathfrak{m};\ | x_k(\mathfrak{m}) - x_k(\mathfrak{m}_0) | < \epsilon, \quad k = 1, 2, \cdots, n].$$

THEOREM 4.15.2. *In the above topology, the functions* $x(\mathfrak{m})$ *are continuous in* $\mathfrak{m}$ *for each* $x \in \mathfrak{B}$ *and* $\mathfrak{M}$ *is a compact Hausdorff space.*

PROOF. It is clear from the definition of neighborhood that $x(\mathfrak{m})$ is a continuous function of $\mathfrak{m}$ for each $x \in \mathfrak{B}$. Further if we paraphrase (4.15.2) using the corresponding linear multiplicative functionals, then $N(\mu_0 ; x_1 , x_2 , \cdots , x_n ; \epsilon)$ is the set of all linear multiplicative functionals μ such that $| \mu(x_i) - \mu_0(x_i) | < \epsilon$ for $i = 1, 2, \cdots , n$. Moreover the linear multiplicative functionals form a subset of the unit sphere $\mathfrak{S}_1^*$ in $\mathfrak{B}^*$, the adjoint space to $\mathfrak{B}$ (see the corollary to Theorem 4.14.4), and the above neighborhoods are precisely the weak* neighborhoods for $\mathfrak{B}^*$. We have already proved in Theorem 2.10.2 that $\mathfrak{S}_1^*$ is a compact Hausdorff space in this topology; consequently the same will be true of $\mathfrak{M}$ if it can be shown that $\mathfrak{M}$ is a closed subset of $\mathfrak{S}_1^*$. Suppose, then, that x^* belongs to the closure of $\mathfrak{M}$ relative to $\mathfrak{S}_1^*$. Given $x, y \in \mathfrak{B}$ and $\epsilon > 0$, the set $N(x^*; x, y, xy, e; \epsilon)$ is a neighborhood of x^* and must therefore contain a point μ of $\mathfrak{M}$. Thus

$$| \mu(x) - x^*(x) | < \epsilon, \qquad | \mu(y) - x^*(y) | < \epsilon,$$
$$| \mu(xy) - x^*(xy) | < \epsilon, \qquad | \mu(e) - x^*(e) | < \epsilon.$$

Making use of the properties of μ, namely $\mu(xy) = \mu(x)\mu(y)$ and $\mu(e) = 1$, we obtain

$$| x^*(xy) - x^*(x)x^*(y) | < (1 + \| x \| + \| y \|)\epsilon,$$
$$| x^*(e) - 1 | < \epsilon.$$

Since $\epsilon > 0$ is arbitrary, it follows that x^* is a linear multiplicative functional. In other words $x^* \in \mathfrak{M}$. This concludes the proof of the theorem.

Let $C(\mathfrak{M})$ be the Banach algebra of all continuous complex-valued functions on the compact Hausdorff space $\mathfrak{M}$, the basic operations being defined as: $(\alpha f)(\mathfrak{m}) = \alpha f(\mathfrak{m})$, $(f + g)(\mathfrak{m}) = f(\mathfrak{m}) + g(\mathfrak{m})$, $(fg)(\mathfrak{m}) = f(\mathfrak{m})g(\mathfrak{m})$, $e(\mathfrak{m}) \equiv 1$, $\| f \| = \sup_{\mathfrak{m} \in \mathfrak{M}} | f(\mathfrak{m}) |$. As an immediate consequence of Theorems 4.15.1 and 4.15.2 we have

COROLLARY. *The mapping $x \to x(\mathfrak{m})$ is a continuous homomorphism of $\mathfrak{B}$ into the (B)-algebra $C(\mathfrak{M})$; it is an algebraic isomorphism if and only if $\mathfrak{B}$ is semi-simple.*

THEOREM 4.15.3. *The topology on $\mathfrak{M}$ is uniquely determined by the fact that $\mathfrak{M}$ is a compact Hausdorff space on which the functions $[x(\mathfrak{m}); x \in \mathfrak{B}]$ are continuous in $\mathfrak{m}$.*

PROOF. It is clear from Theorem 4.15.1 (10) that the family of functions $\mathfrak{F} \equiv [x(\mathfrak{m}); x \in \mathfrak{B}]$ separate the points of $\mathfrak{M}$. Since the topology defined by the neighborhoods (4.15.2) is just the topology induced by $\mathfrak{F}$, the assertion of the theorem is an immediate consequence of Theorem 1.4.1.

These results can be illustrated by the algebra $\mathfrak{B}$ of complex-valued functions on $[0, 2\pi)$ with absolutely convergent Fourier series which was treated at the end of the previous section. There it was found that each maximal ideal (or equivalently, each multiplicative linear functional μ) corresponds to a unique $\xi_0 \in [0, 2\pi)$ and $\mu(f) = f(\xi_0)$ for all $f \in \mathfrak{B}$. If we identify 0 and 2π, the interval $[0, 2\pi]$ becomes a compact Hausdorff space in the usual euclidean topology and $f(\xi)$ becomes a continuous function of ξ. By Theorem 4.15.3, this topology is equivalent to the $\mathfrak{M}$ topology introduced above.

PROBLEM (G. ŠILOV [2]). Suppose that $\mathfrak{B}$ is generated by the elements a and e. Show that the mapping $\zeta \equiv a(\mathfrak{m})$ of $\mathfrak{M}$ onto a subset $\mathfrak{M}'$ of the complex-plane is a homeomorphism. Then prove that the complement of $\mathfrak{M}'$ is connected.

HINT. Let $p(\zeta)$ be a polynomial in ζ. If ζ_0 belongs to a bounded component of the complement of $\mathfrak{M}'$, then

$$| p(\zeta_0) | \leqq \sup_{\zeta \in \mathfrak{M}'} | p(\zeta) | = \sup_{\mathfrak{m} \in \mathfrak{M}} | p[a(\mathfrak{m})] | \leqq \| p(a) \|.$$

Thus $\mu[p(a)] = p(\zeta_0)$ defines a multiplicative linear functional on $\mathfrak{B}$.

One of the most interesting properties of $\mathfrak{M}$ is the existence of a unique minimal closed subset on which each of the functions $| x(\mathfrak{m}) |$ achieves its maximum. This set is often referred to as the *Šilov boundary* after its discoverer (cf. Gelfand, Raikov, and Šilov [1]).

THEOREM 4.15.4. *There exists a closed subset S of $\mathfrak{M}$ with the following properties:*
(i) $\sup_{\mathfrak{m} \in S} | x(\mathfrak{m}) | = r(x)$ *for all* $x \in \mathfrak{B}$;
(ii) *if F is any closed subset of $\mathfrak{M}$ such that* $\sup_{\mathfrak{m} \in F} | x(\mathfrak{m}) | = r(x)$ *for all* $x \in \mathfrak{B}$, *then* $S \subset F$.

We illustrate this notion with two examples. Consider the algebra of functions holomorphic in $| \zeta | < 1$ and continuous in $| \zeta | \leqq 1$ with the usual arithmetic operations and $\| f(\cdot) \| = \sup | f(\zeta) |$. In this case $\mathfrak{M}$ is homeomorphic with $|\zeta| \leqq 1$ and under this correspondence the Šilov boundary maps onto the circle $| \zeta | = 1$. For the algebra $C(\mathfrak{M})$, the Šilov boundary is $\mathfrak{M}$ itself (this is a special instance of Theorem 4.19.5).

PROOF. Let $\mathfrak{F}$ be the family of all closed subsets of $\mathfrak{M}$ satisfying the property (i). It is clear that $\mathfrak{M}$ itself belongs to $\mathfrak{F}$ so that $\mathfrak{F}$ is not empty. We order the sets of $\mathfrak{F}$ by inclusion. Let $[F_\alpha]$ be a linearly ordered collection of sets in $\mathfrak{F}$ and set $H(x) \equiv [\mathfrak{m}; | x(\mathfrak{m}) | = r(x)]$. Then $H(x)$ is closed, $F_\alpha \cap H(x) \neq \emptyset$, and the sets $F_\alpha \cap H(x)$ are linearly ordered. Consequently $(\bigcap_\alpha F_\alpha) \cap H(x) = \bigcap_\alpha (F_\alpha \cap H(x)) \neq \emptyset$. In other words $\bigcap_\alpha F_\alpha$ is a lower bound in $\mathfrak{F}$ for $[F_\alpha]$. It therefore follows from the maximal principle that there exists a minimal set S in $\mathfrak{F}$. It remains to show that S satisfies (ii).

Suppose that $F \in \mathfrak{F}$ and let $\mathfrak{m}_0$ be a point of S. We shall show, for an arbitrary neighborhood $N = N(\mathfrak{m}_0 ; x_1 , x_2 , \cdots , x_n ; \epsilon)$ of $\mathfrak{m}_0$, that $N \cap F \neq \emptyset$. Since F is closed this will imply that $\mathfrak{m}_0 \in F$ and hence that $S \subset F$. Now S is a minimal set in $\mathfrak{F}$ so that $S \ominus N$ does not satisfy property (i). Therefore there exists an element y in $\mathfrak{B}$ such that $| y(\mathfrak{m}) |$ achieves its maximum $r(y)$ within $N \cap S$ and $| y(\mathfrak{m}) | < r(y)$ on $S \ominus N$. By taking a suitable power of $y/r(y)$, it is clear that we will obtain a $z \in \mathfrak{B}$ such that $| z(\mathfrak{m}) | = r(z) = 1$ for some point of

$N \cap S$ and $|z(\mathfrak{m})| < \epsilon/(2 \sup_i \|x_i\|)$ on $S \ominus N$. Setting $w_i = x_i z - x_i(\mathfrak{m}_0)z$, we clearly have $|w_i(\mathfrak{m})| \leqq 2\|x_i\| |z(\mathfrak{m})| < \epsilon$ on $S \ominus N$ and by the definition of N, $|w_i(\mathfrak{m})| \leqq |x_i(\mathfrak{m}) - x_i(\mathfrak{m}_0)| r(z) < \epsilon$ on N. Since $S \in \mathfrak{F}$, we see that $r(w_i) < \epsilon$. On the other hand $F \in \mathfrak{F}$ so that there exists an $\mathfrak{m}_1 \in F$ such that $|z(\mathfrak{m}_1)| = r(z) = 1$. Therefore $|x_i(\mathfrak{m}_1) - x_i(\mathfrak{m}_0)| |z(\mathfrak{m}_1)| = |w_i(\mathfrak{m}_1)| \leqq r(w_i) < \epsilon$ implies $|x_i(\mathfrak{m}_1) - x_i(\mathfrak{m}_0)| < \epsilon$ for $i = 1, 2, \cdots, n$. It follows that $\mathfrak{m}_1 \in N$ and this concludes the proof.

THEOREM 4.15.5. *If $\mathfrak{B}$ maps into a dense subalgebra of $C(\mathfrak{M})$ under the mapping $x \to x(\mathfrak{m})$, then the Šilov boundary is $\mathfrak{M}$.*

PROOF. Let $\mathfrak{m}_0 \in \mathfrak{M}$ and let N be a neighborhood of $\mathfrak{m}_0$. Since $\mathfrak{M}$ is in particular normal, there exists a continuous function $f(\mathfrak{m})$ on $\mathfrak{M}$ such that $f(\mathfrak{m}_0) = 1$ and $f(\mathfrak{m}) = 0$ on $\mathfrak{M} \ominus N$. By assumption there exists an $x \in \mathfrak{B}$ such that $\sup_{\mathfrak{m} \in \mathfrak{M}} |x(\mathfrak{m}) - f(\mathfrak{m})| < \frac{1}{8}$. Thus $|x(\mathfrak{m}_0)| > \frac{2}{3}$ and $|x(\mathfrak{m})| < \frac{1}{3}$ in $\mathfrak{M} \ominus N$. Thus the maximum of $|x(\mathfrak{m})|$ cannot occur in $\mathfrak{M} \ominus N$ and we see that N contains a point of the Šilov boundary S. Since N was an arbitrary neighborhood of $\mathfrak{m}_0$ and since S is closed, it follows that $\mathfrak{m}_0 \in S$.

PROBLEM (R. F. ARENS AND I. M. SINGER [1]). Let $\mathfrak{B}$ be a commutative complex (B)-algebra with Šilov boundary S. Corresponding to each maximal ideal $\mathfrak{m}_0$ there is a measure $m_0(E)$, $m_0(S) = 1$, on the Baire subsets of S such that $x(\mathfrak{m}_0) = \int_S x(\mathfrak{m})\, dm_0$ for each $x \in \mathfrak{B}$.

Let $\mathfrak{B}$ and $\mathfrak{B}'$ be two commutative complex (B)-algebras with unit elements e and e' respectively. Suppose that Ψ is a homomorphism of $\mathfrak{B}$ into $\mathfrak{B}'$ such that $\Psi(e) = e'$. If μ' is a multiplicative linear functional on $\mathfrak{B}'$, then it is clear that $\mu'[\Psi(x)]$ defines a multiplicative linear functional μ on $\mathfrak{B}$. Thus Ψ induces an adjoint mapping Ψ^* on the maximal ideals $\mathfrak{M}'$ of $\mathfrak{B}'$ into the maximal ideals $\mathfrak{M}$ of $\mathfrak{B}$. It is easy to see that Ψ^* is a continuous mapping (even when Ψ is not continuous). For let $\mathfrak{m}_0 = \Psi^*(\mathfrak{m}_0')$ and let $N = N(\mathfrak{m}_0; x_1, x_2, \cdots, x_n; \epsilon)$ be a generic neighborhood of $\mathfrak{m}_0$. Then the Ψ^* image of $N'(\mathfrak{m}_0'; \Psi(x_1), \Psi(x_2), \cdots, \Psi(x_n); \epsilon)$ is contained in N. As a consequence $\Psi^*(\mathfrak{M}')$ is a closed subset of $\mathfrak{M}$.

THEOREM 4.15.6. (ŠILOV). *Let $\mathfrak{B}$ and $\mathfrak{B}'$ be two commutative complex (B)-algebras such that $\mathfrak{B} \subset \mathfrak{B}'$ and $e = e'$. Then each maximal ideal in the Šilov boundary of $\mathfrak{B}$ is contained in a maximal ideal of $\mathfrak{B}'$.*

PROOF. The correspondence $\Psi(x) = x$ defines a homomorphism of $\mathfrak{B}$ into $\mathfrak{B}'$ such that $\Psi(e) = e'$. If Ψ^* is the induced map of $\mathfrak{M}'$ into $\mathfrak{M}$, then $[\Psi(x)](\mathfrak{m}') = x[\Psi^*(\mathfrak{m}')]$ for all $x \in \mathfrak{B}$. The theorem asserts that $\Psi^*(\mathfrak{M}')$ contains the Šilov boundary of $\mathfrak{B}$. Now for $x \in \mathfrak{B}$, $r(x) = \lim_{n\to\infty} \|x^n\|^{1/n} = r[\Psi(x)]$. Hence

$$r(x) = \sup_{\mathfrak{m}' \in \mathfrak{M}'} |[\Psi(x)](\mathfrak{m}')| = \sup_{\mathfrak{m}' \in \mathfrak{M}'} |x[\Psi^*(\mathfrak{m}')]|.$$

Since $\Psi^*(\mathfrak{M}')$ is closed, it follows from Theorem 4.15.4 that $\Psi^*(\mathfrak{M}')$ contains the Šilov boundary of $\mathfrak{B}$.

References. Arens and Singer [1], Gelfand [4], Gelfand, Raikov, and Šilov [1], Loomis [1], Šilov [2], Wiener [3].

4. The Banach Algebra $S(\varphi)$

4.16. The basic properties of $S(\varphi)$. We shall now apply the foregoing theory to a Banach algebra which will play an important role in the operational calculus (Chapter XV) and in the spectral theory (Chapter XVI) for semi-groups of linear transformations. This algebra differs from the familiar algebra of functions of bounded variation on $[0, \infty)$ with a convolution product only in the use of a weight factor in defining the norm. Both A. Beurling [1] and I. Gelfand [6] have considered related algebras (functions on $(-\infty, \infty)$ and continuous weight factors). Because of its importance in semi-group theory we shall treat this algebra in some detail; here we follow R. S. Phillips [6]. The reader is referred to P. R. Halmos [1] for the required properties of set functions.

We begin by introducing a weight factor $\varphi(\xi)$ with the following properties:

(W) (i) $\varphi(\xi)$ is a real-valued Borel measurable function defined on $[0, \infty)$;

(ii) $\varphi(\xi)$ is submultiplicative: $0 < \varphi(\xi_1 + \xi_2) \leqq \varphi(\xi_1)\varphi(\xi_2)$ for all $\xi_1, \xi_2 \geqq 0$;

(iii) $\varphi(0) = 1$.

We see that $f(\xi) \equiv \log \varphi(\xi)$ is a finite-valued measurable subadditive function. Thus it follows from Theorem 7.4.1 that both $\varphi(\xi)$ and $[\varphi(\xi)]^{-1}$ are bounded in each interval of the form $(\epsilon, 1/\epsilon)$, $\epsilon > 0$, and from Theorem 7.4.2 that $\lim\inf_{\xi \to 0+} \varphi(\xi) \geqq 1$. Thus for each $k > 0$ there exists a constant $m_k > 0$ such that $\varphi(\xi) \geqq m_k$ for $\xi \in [0, k]$. According to Theorem 7.6.1 we have

$$\omega_0 \equiv \inf_{\xi > 0} \xi^{-1} \log \varphi(\xi) = \lim_{\xi \to \infty} \xi^{-1} \log \varphi(\xi) \geqq -\infty. \tag{4.16.1}$$

Let $\mathfrak{E}$ be the Borel subsets of $[0, \infty)$. A set is said to be bounded if it is contained in a finite subinterval of $[0, \infty)$. We shall be concerned with complex-valued set functions $a(E)$ defined and countably additive on all bounded Borel sets. The *total variation* $|a|(E)$ of such a set function is defined on bounded Borel sets as

$$|a|(E) = \sup \sum_i | a(E_i) |, \tag{4.16.2}$$

the supremum being taken over all disjoint denumerable subdivisions of E. A familiar argument shows that $|a|(E)$ is a countably additive set function on the bounded Borel sets; further $|a|(E)$ can be extended to all sets of $\mathfrak{E}$ in a unique way so as to be a σ-finite measure function. It is clear that $| a(E) | \leqq |a|(E)$ for all bounded Borel sets. We now define $S(\varphi)$ to be the class of all set functions $a(E)$ such that $\int_0^\infty \varphi(\xi)\, d\, | a | < \infty$. The norm is given by

$$\| a \| = \int_0^\infty \varphi(\xi)\, d\, | a |. \tag{4.16.3}$$

It is clear from (4.16.2) that $|\alpha a|(E) = |\alpha|\, |a|(E)$ and that $| a + b |(E) \leqq | a | (E) + | b | (E)$. Consequently $S(\varphi)$ is a linear space and the norm given by (4.16.3) satisfies the postulates for a norm.

In order to establish completeness, let $\{a_n\} \subset S(\varphi)$ be a Cauchy sequence. For $E \subset [0, k]$, we have

$$(4.16.4) \qquad | a_n(E) - a_m(E) | \leqq |a_n - a_m| (E) \leqq m_k^{-1} \int_E \varphi(\xi) \, d\, |a_n - a_m|$$
$$\leqq m_k^{-1} \| a_n - a_m \|.$$

Hence $\lim_{n \to \infty} a_n(E) = a(E)$ exists uniformly for all $E \subset [0, k]$. If $\{E_i\}$ is a subdivision of a bounded set E into disjoint sets, $E_i \in \mathfrak{E}$ and $\bigcup_i E_i = E$, then

$$\lim_{n \to \infty} \sum_{i=1}^{m} a_n(E_i) = \sum_{i=1}^{m} a(E_i)$$

exists uniformly in m and $\lim_{m \to \infty} \sum_{i=1}^{m} a_n(E_i) = a_n(E)$ exists for each n. Hence the double limit exists and therefore $a(E)$ is countably additive on bounded sets. For a disjoint subdivision $\{E_i\}$ of $E \subset [0, k]$, it is clear from (4.16.4) that

$$\sum_{i=1}^{m} | a_n(E_i) - a(E_i) | \leqq m_k^{-1} \sup_{j > n} \int_E \varphi(\xi) \, d\, | a_n - a_j |$$
$$\leqq m_k^{-1} \sup_{j > n} \| a_n - a_j \|.$$

Hence $| a_n - a |(E) \leqq m_k^{-1} \sup_{j > n} \| a_n - a_j \|$ so that $| \, | a_n - a |(E) - | a_n - a_j |(E) \, | \leqq | a - a_j | (E) \to 0$. Recalling that $\varphi(\xi)$ is bounded in $(\epsilon, 1/\epsilon)$, $\epsilon > 0$, we see that

$$\int_\epsilon^{1/\epsilon} \varphi(\xi) \, d\, | a_n - a | = \lim_{j \to \infty} \int_\epsilon^{1/\epsilon} \varphi(\xi) \, d\, | a_n - a_j | \leqq \sup_{j > n} \| a_n - a_j \|$$

and therefore that $\| a_n - a \| \leqq \sup_{j > n} \| a_n - a_j \|$. From this it follows that $a = a_n - (a_n - a)$ is an element of $S(\varphi)$ and that $\lim_{n \to \infty} \| a_n - a \| = 0$. Thus $S(\varphi)$ is complete.

Each set function $a(E)$ in $S(\varphi)$ has the usual decomposition $a(E) = a_1(E) + a_2(E) + a_3(E)$ where a_1, a_2, a_3 are respectively the absolutely continuous (relative to Lebesgue measure), the atomic, and the non-atomic singular parts of a. In fact corresponding to each set function, $a(E)$, there exists a subdivision of $[0, \infty)$ into three mutually exclusive Borel sets E_1, E_2, E_3 such that $a_i(E) = a(E \cap E_i)$; here E_2 is denumerable, E_3 is of Lebesgue measure zero, and $a_3(E)$ vanishes on all denumerable sets. Since the set functions $a_i(E)$ are non-vanishing on mutually exclusive sets, it follows that $\| a \| = \| a_1 \| + \| a_2 \| + \| a_3 \|$. Moreover given any denumerable collection of set functions $\{a^n\}$ there will exist a common subdivision of $[0, \infty)$ into mutually exclusive Borel sets E_1, E_2, E_3 with the above properties such that $a_i^n(E) = a^n(E \cap E_i)$. As a consequence we see that $S(\varphi)$ is the direct sum of the three closed linear subspaces consisting of the absolutely continuous set functions $L(\varphi)$, the atomic set functions $A(\varphi)$, and the non-atomic singular set functions $N(\varphi)$, respectively.

We next define the product $c = a * b$ of two elements in $S(\varphi)$. For two σ-finite measures, a and $b \in S(\varphi)$, the *product set function* $\bar{c} = a \times b$ is a well defined σ-finite measure on the smallest sigma-ring $\mathfrak{E}_2$ generated by the class of all rectangular sets $E \times F$, where $E, F \in \mathfrak{E}$. We define $c = a * b$ on Borel sets by

$$(4.16.5) \qquad c(E) = \bar{c}[(\eta, \zeta); \quad \eta + \zeta \in E, \quad \eta, \zeta \geqq 0].$$

For measure functions, $c(E) = | c | (E) = (| a | * | b |)(E) = (a * b)(E)$. Hence making use of the Fubini theorem we see that

$$\| c \| = \int_0^\infty \varphi(\xi)\, d\,|\,c\,| \leqq \int_0^\infty \varphi(\xi)\, d(|\,a\,| * |\,b\,|)$$

$$(4.16.6) \qquad = \int_0^\infty \int_0^\infty \varphi(\eta + \zeta)\, d_\eta\,|\,a\,|\, d_\zeta\,|\,b\,|$$

$$\leqq \int_0^\infty \int_0^\infty \varphi(\eta)\varphi(\zeta)\, d_\eta\,|\,a\,|\, d_\zeta\,|\,b\,| = \|\,a\,\|\,\|\,b\,\|\,.$$

Thus $c(E)$ is finite valued on bounded Borel sets. It follows from (4.16.5) that $c(E)$ is countably additive on bounded Borel sets and from (4.16.6) that $c \in S(\varphi)$. In the general case we can decompose a into the sum of four σ-finite measures: positive and negative parts of the real and imaginary parts of a. Thus $a(E) = \sum_{k=1}^{4} i^k a_k(E)$ where the a_k are σ-finite measures belonging to $S(\varphi)$. Using a similar decomposition for b we obtain $\bar{c} = \sum_{j,k=1}^{4} i^{j+k} a_j \times b_k$. The values of $\bar{c}$ on rectangular sets are clearly independent of the particular decomposition of a and b employed; hence the same is true of the values of $\bar{c}$ on $\mathfrak{E}_2$, the smallest sigma-ring extension of the rectangular sets. We now define $c = a * b$ on bounded Borel sets as in (4.16.5). It follows from the way in which $\bar{c}$ was defined, that $c(E)$ is finite-valued and countably additive on bounded Borel sets and that $c \in S(\varphi)$. Further the product is clearly commutative; with the help of the Fubini theorem we obtain

$$(4.16.7) \qquad c(E) = \int_0^\infty a(E - \xi)\, d_\xi b = \int_0^\infty b(E - \xi)\, d_\xi a.$$

Thus $|\,c(E)\,| \leqq \int_0^\infty |a\,|\,(E - \xi)\, d_\xi\,|\,b\,| = (\,|\,a\,| * |\,b\,|\,)(E)$ and hence we have $|\,c\,|(E) \leqq (\,|\,a\,| * |\,b\,|\,)(E)$. It follows that (4.16.6) remains valid. The distributive laws are easily verified and the associativity can be established by means of the Fubini theorem. Finally we set

$$(4.16.8) \qquad e_\xi(E) = \begin{cases} 1 & \text{if } \xi \in E, \\ 0 & \text{otherwise.} \end{cases}$$

Then e_0 is the unit element of our algebra and $(e_\xi * a)(E) = a(E - \xi)$ is a translate of a (ξ units to the right); $\|\,e_\xi\,\| = \varphi(\xi)$. We have now shown that $S(\varphi)$ is a commutative complex (B)-algebra with a unit element.

If $a \in L(\varphi)$, then $a(E) = 0$ for all Borel sets of Lebesgue measure zero; it is evident from (4.16.7) that the same is true of $c = a * b$ for any $b \in S(\varphi)$. Thus $c \in L(\varphi)$ so that $L(\varphi)$ is a proper ideal of the algebra $S(\varphi)$. Further if $a(E) = 0$ on all denumerable sets, the same will be true of $c = a * b$ for any $b \in S(\varphi)$. Again it follows that $L(\varphi) + N(\varphi)$ is a proper ideal of $S(\varphi)$.

If $a \in L(\varphi)$, then by the Radon-Nikodym theorem there exists a Borel measurable density function $f(\xi)$ such that $a(E) = \int_E f(\xi)\, d\xi$. Hence $|a|(E) = \int_E |\,f(\xi)\,|\, d\xi$ and $\|\,a\,\| = \int_0^\infty \varphi(\xi)\,|\,f(\xi)\,|\, d\xi$. It is now easy to show that the translates, $e_\xi * a$, of $a \in L(\varphi)$ define a strongly continuous vector-valued function of ξ for $\xi > 0$. Since $\varphi(\xi)$ is bounded in each interval of the form $(\epsilon, 1/\epsilon)$, the result is clearly valid for set functions having continuous and ultimately vanishing densities. The set of all such functions is dense in $L(\varphi)$. Hence for an arbitrary $a \in L(\varphi)$ there exists a sequence $\{a_n\} \subset L(\varphi)$ of functions with continuous translates for $\xi > 0$ such that $\|\,a - a_n\,\| \to 0$. Now

$$(4.16.9) \qquad \|\,e_\xi * a - e_\tau * a\,\| \leqq \|\,e_\xi * (a - a_n)\,\| + \|\,e_\xi * a_n - e_\tau * a_n\,\| + \|\,e_\tau * (a_n - a)\,\|$$

$$\leqq [\varphi(\xi) + \varphi(\tau)]\,\|\,a - a_n\,\| + \|\,e_\xi * a_n - e_\tau * a_n\,\|.$$

For fixed $\xi > 0$ both of these terms can be made arbitrarily small by first choosing n sufficiently large and by then choosing $|\xi - \tau|$ sufficiently small. Since $\| e_\xi * a \| \leqq \| a \| \varphi(\xi)$, it now follows that the abstract Bochner integral of $\int_0^\infty (e_\xi * a)\, d_\xi b$ converges to an element $c \in L(\varphi)$ for each $b \in S(\varphi)$. Actually, in order to apply the theory of section 3.7 one should first decompose b into a sum $b(E) = \sum_1^4 i^k b_k(E)$, where the b_k are σ-finite measures belonging to $S(\varphi)$. Then $\int (e_\xi * a)\, d_\xi b \equiv \sum_1^4 i^k \int (e_\xi * a)\, d_\xi b_k$ and we see that this is independent of the particular decomposition of b employed on reducing the consideration to the numerical case by applying a linear bounded functional to the sum on the right. In fact, for each bounded set $E \subset [0, k]$, $x^*(a) = a(E)$ defines a linear functional in $S(\varphi)^*$; clearly $\| x^* \| \leqq m_k^{-1}$. Since x^* commutes with the operation of integration we have

$$c(E) = x^*(c) = \int_0^\infty x^*(e_\xi * a)\, d_\xi b = \int_0 a(E - \xi)\, d_\xi b$$

and hence $c = a * b$. Thus for $a \in \mathfrak{L}(\varphi)$ we have

$$a * b = \int_0^\infty (e_\xi * a)\, d_\xi b, \tag{4.16.10}$$

where the integral is to be understood as a Bochner integral in $S(\varphi)$.

We shall later have occasion to consider an algebra $S_0(\varphi)$ of countably additive set functions defined on the Borel subsets of $(-\infty, \infty)$. In this case $\varphi(\xi)$ is real-valued, non-negative, Borel measurable, and submultiplicative on $(-\infty, \infty)$, again with $\varphi(0) = 1$. As a consequence $\varphi(\xi)$ and $[\varphi(\xi)]^{-1}$ are bounded in each finite subinterval of $(-\infty, \infty)$ (see Theorem 7.4.1). Our entire discussion can be carried over almost verbatim. As a result $S_0(\varphi)$ is a commutative Banach algebra with a unit element. One difference, however, may be worth mentioning. As can be seen from (4.16.9), the translates $e_\xi * a$ of $a \in L_0(\varphi)$ are now strongly continuous at all $\xi \in (-\infty, \infty)$.

4.17. Semi-group representations. As a prelude to a general study of linear multiplicative functionals on $S(\varphi)$ we shall first investigate their behavior on the shift elements, $\mathfrak{S} = [e_\xi]$. Since $e_\xi * e_\eta = e_{\xi+\eta}$ we see that the shift elements form a semi-group of elements relative to the product operation. Let μ be a linear multiplicative functional on $S(\varphi)$ and set $\nu(\xi) = \mu(e_\xi)$. Then

$$\nu(\xi + \eta) = \nu(\xi)\nu(\eta) \qquad \text{for } \xi, \eta \geqq 0, \tag{4.17.1}$$
$$\nu(0) = 1,$$

so that $\nu(\xi)$ is a complex-valued representation of the semi-group $\mathfrak{S}$. If we consider $\nu(\xi)$ as an operator on the complex field, $\alpha \to \nu(\xi)\alpha$, then a function $\nu(\xi)$ satisfying (4.17.1) is the simplest example of a semi-group of linear transformations. It will therefore be worth while to study such functions with care.

THEOREM 4.17.1. *If $\nu(\xi)$ satisfies (4.17.1) and if $\nu(\xi_0) = 0$ for some $\xi_0 > 0$, then $\nu(\xi) = 0$ for all $\xi > 0$.*

PROOF. For $\xi > \xi_0$ we have $\nu(\xi) = \nu(\xi - \xi_0)\nu(\xi_0) = 0$. On the other hand for arbitrary $\xi > 0$, there exists an integer n such that $n\xi > \xi_0$. Thus $[\nu(\xi)]^n = \nu(n\xi) = 0$ and hence $\nu(\xi) = 0$.

We shall say that $\nu(\xi)$ satisfies (4.17.1) trivially if $\nu(\xi) \equiv 0$ for all $\xi > 0$.

THEOREM 4.17.2. *If $\nu(\xi)$ satisfies (4.17.1) non-trivially and if $|\nu(\xi)|$ is bounded in some interval $[\tau_1, \tau_2]$, then $|\nu(\xi)| = \exp(\alpha\xi)$ for some real number α.*

PROOF. Set $f(\xi) = \log |\nu(\xi)|$. We see by the previous theorem that $f(\xi)$ is well defined on $[0, \infty)$. Further $f(\xi + \eta) = f(\xi) + f(\eta)$, $f(0) = 0$, and $f(\xi)$ is bounded from above on $[\tau_1, \tau_2]$. We first show that $f(\xi)$ is bounded in a neighborhood of the origin. If this were not so there would exist a sequence $\xi_n \to 0+$ such that $|f(\xi_n)| \to \infty$. Hence $|f(\tau_1 + \xi_n)| = |f(\tau_1) + f(\xi_n)| \to \infty$. Since $f(\xi)$ is bounded above in $[\tau_1, \tau_2]$, this means that $f(\xi_n) \to -\infty$ and hence that $f(\tau_2 - \xi_n) = f(\tau_2) - f(\xi_n) \to \infty$ which is impossible. Next we show that $f(\xi) \to 0$ as $\xi \to 0+$. If the contrary were true, then there would exist a sequence $\xi_n \to 0+$ such that $f(\xi_n) \geqq \epsilon > 0$ (or $f(\xi_n) \leqq -\epsilon < 0$) for some ϵ. But then $f(\sum_{i=k}^{k+n} \xi_i) \geqq n\epsilon$ and $\lim_{k\to\infty} \sum_{i=k}^{k+n} \xi_i = 0$ for arbitrary n, which is again impossible. Not only is $f(\xi)$ right continuous at the origin, but because of the additivity it is clearly right continuous for all $\xi \geqq 0$. Finally for any rational number, r, a familiar argument gives $f(r) = rf(1)$. Making use of the right continuity we finally obtain $f(\xi) = \xi f(1)$ for all $\xi \geqq 0$. This is the desired result with $\alpha = f(1)$.

For $\nu(\xi) = \mu(e_\xi)$ we have $|\nu(\xi)| \leqq \| e_\xi \| = \varphi(\xi)$ which is bounded in every interval of the form $[\epsilon, 1/\epsilon]$. Hence in this case $|\nu(\xi)| = \exp(\alpha\xi)$ and $\alpha \leqq \inf_{\xi>0} \xi^{-1} \log \varphi(\xi) \equiv \omega_0$. However even for the general non-trivial $\nu(\xi)$ satisfying (4.17.1) we may set

$$\chi(\xi) = \nu(\xi)/|\nu(\xi)|. \tag{4.17.2}$$

It is clear that

$$\chi(\xi + \eta) = \chi(\xi)\chi(\eta) \quad \text{and} \quad |\chi(\xi)| \equiv 1. \tag{4.17.3}$$

For negative ξ we may set $\chi(\xi) = [\chi(-\xi)]^{-1}$ and (4.17.3) will remain valid for all real ξ. Such a function is called a *character* of the real line.

THEOREM 4.17.3. *If $\chi(\xi)$ is a measurable character of the real line, then $\chi(\xi) = \exp(i\beta\xi)$ for some real β.*

PROOF. For $|\delta| > 0$ and $\gamma > 0$, we have

$$[\chi(\xi + \delta) - \chi(\xi)]\gamma = \int_0^\gamma [\chi(\xi + \delta - \eta) - \chi(\xi - \eta)]\chi(\eta)\, d\eta.$$

Hence

$$|\chi(\xi + \delta) - \chi(\xi)| \leqq \gamma^{-1} \int_0^\gamma |\chi(\xi + \delta - \eta) - \chi(\xi - \eta)|\, d\eta$$

which goes to zero with $|\delta|$. Thus $\chi(\xi)$ is continuous. Further

$$\frac{\chi(\delta) - 1}{\delta} \int_0^\gamma \chi(\xi)\, d\xi = \delta^{-1} \int_0^\gamma [\chi(\xi + \delta) - \chi(\xi)]\, d\xi = \delta^{-1} \int_\gamma^{\gamma+\delta} \chi(\xi)\, d\xi - \delta^{-1} \int_0^\delta \chi(\xi)\, d\xi.$$

Choose γ so that $\int_0^\gamma \chi(\xi)\, d\xi \neq 0$. Since the limit as $|\delta| \to 0$ exists for the terms on the right side of this equation, it follows that the derivative of $\chi(\xi)$ exists at $\xi = 0$. Set $d\chi(\xi)/d\xi |_{\xi=0} = i\beta$. Then making use of $\chi(\xi + \eta) = \chi(\xi)\chi(\eta)$, we see that $d\chi(\xi)/d\xi = i\beta\chi(\xi)$ for all $\xi \in (-\infty, \infty)$. Hence $\chi(\xi) = C \exp(i\beta\xi)$. Finally $\chi(0) = 1$ implies that $C = 1$ and $|\chi(\xi)| \equiv 1$ implies that β is real.

COROLLARY. *If $\nu(\xi)$ is measurable and satisfies (4.17.1) non-trivially, then*

$$\nu(\xi) = \exp[(\alpha + i\beta)\xi]$$

for some real numbers α and β.

PROOF. In the first place $f(\xi) = \log | \nu(\xi) |$ is finite-valued, measurable, and additive on $[0, \infty]$. Hence, by Theorem 7.4.1, $f(\xi)$ will be bounded in each interval of the form $[\epsilon, 1/\epsilon]$. Making use of Theorem 4.17.2, we see that $| \nu(\xi) | = \exp(\alpha\xi)$ for some real α. Further $\chi(\xi) = \nu(\xi)/|\nu(\xi)|$ is measurable. The result now follows from the previous theorem.

We shall need the following theorem in Chapter XVI. It furnishes us with a criterion for distinguishing between measurable and non-measurable characters of the real line. It is evident from this result that the non-measurable characters are extremely pathological.

THEOREM 4.17.4. *Let $\chi(\xi)$ be a character of the real line and set $F_\tau \equiv [\chi(\xi); 0 \leqq \xi \leqq \tau]$ and $F = \bigcap_\tau \overline{F}_\tau$. There are two alternatives: either $F = \{1\}$ in which case $\chi(\xi)$ is a continuous character and $\chi(\xi) = \exp(i\beta\xi)$ or else F contains the unit circle and $\chi(\xi)$ is non-measurable.*

PROOF. The sets F_T are monotone decreasing with τ. Hence a necessary and sufficient condition that a closed subset of the unit circle have points in common with F is that it have points in common with each of the sets $\overline{F_T}$. If F consists only of the point 1, then given $\epsilon > 0$ there exists a $\delta > 0$ such that $| \chi(\xi) - 1 | < \epsilon$ for all $\xi \in [0, \delta]$. Otherwise the closed subset of the unit circle defined by $| \exp(i\theta) - 1 | \geqq \epsilon$ would contain points of $\chi(\xi)$ for arbitrary small ξ. Thus all of the sets F_τ would have points in common with this subset and so would F itself. Therefore $\chi(\xi)$ is right continuous at the origin. Since $\chi(\xi + \delta) = \chi(\xi)\chi(\delta)$, we see that $\chi(\xi)$ is right continuous at all points and hence by the previous theorem $\chi(\xi) = \exp(i\beta\xi)$ for some real β.

On the other hand suppose that $\exp(i\theta_1) \in F$ for $0 < \theta_1 < 2\pi$. This can happen if and only if there exists a sequence $\xi_n \to 0+$ such that $\chi(\xi_n) \to \exp(i\theta_1)$. It follows that $\chi(k\xi_n) \to \exp(ik\theta_1)$ for any integer k and hence that $\exp(ik\theta_1) \in F$. Thus if F is not dense on the unit circle it must contain a point of smallest positive angle, say $\theta_0 > 0$, in which case F consists precisely of the points $[\exp(ik\theta_0); k = 1, 2, \cdots, 2\pi/\theta_0]$. Again there exists a sequence $\xi_n \to 0+$ such that $\chi(\xi_n) \to \exp(i\theta_0)$. In this case $\chi(\xi_n/k)$ will contain a subsequence which converges to one of the kth roots of $\exp(i\theta_0)$. For $k = 2\pi/\theta_0$, none of these roots lies in the set $[\exp(ik\theta_0); k = 1, 2, \cdots, 2\pi/\theta_0]$, yet one of them must belong to F. Since this is impossible it follows that F is dense on the unit circle. As F is closed it must consist of all the points on the unit circle.

It should be remarked that there is nothing special about $\xi = 0$ in the above theorem and that a similar result holds for any real ξ.

As we shall see the Kronecker theorem is useful in the study of algebras such as $S(\varphi)$. For the sake of completeness we shall include a proof of this theorem.

THEOREM 4.17.5 (KRONECKER). *Let $(\tau_1, \tau_2, \cdots, \tau_n)$ be a set of n rationally independent numbers. Then given $\epsilon > 0$ and an arbitrary set of n numbers $(\delta_1, \delta_2, \cdots, \delta_n)$ there exists a ξ and a set of integers $(m_1, m_2, \cdots, m_n)$ such that*

$$| \delta_i - \tau_i\xi - m_i | < \epsilon, \qquad i = 1, 2, \cdots, n. \tag{4.17.4}$$

PROOF. Let U denote the unit cube in a real n-dimensional euclidean space: $0 \leqq \eta_i < 1$, $i = 1, 2, \cdots, n$. We define the following point transformation on U to itself,

$$T_\xi(\{\eta_i\}) = \{(\eta_i + \tau_i\xi) \pmod 1\}.$$

We now consider the cube $C_\epsilon \equiv [0 < \eta_i < \epsilon/2; i \equiv 1, 2, \cdots, n]$ and the set of all its translates, namely $E = \bigcup_\xi T_\xi(C_\epsilon)$. E is the union of open sets and is therefore measurable; in fact $\text{meas}(E) \geqq \text{meas}(C_\epsilon) > 0$. It will be sufficient to show that E is of measure one; for in this case each point in U (and in particular $(\delta_1, \delta_2, \cdots, \delta_n) \pmod 1)$) lies within $\epsilon/2$ of some point of E and hence within ϵ of some translate of the origin. This is the assertion (4.17.4). Let $\chi(\{\eta_i\})$ be the characteristic function of E and expand χ in a multiple Fourier series: $\chi \sim \sum a_{k_1, k_2, \cdots, k_n} \exp(2\pi i \sum_1^n k_j\eta_j)$. Since E is clearly invariant under T_ξ, the same is true of χ and hence of the Fourier series of χ. This means that

$$a_{k_1, k_2, \cdots, k_n} = a_{k_1, k_2, \cdots, k_n} \exp\left(2\pi i \sum_1^n k_j \tau_j \xi\right)$$

for all ξ. If $a_{k_1, \cdots, k_n} \neq 0$, then we must have $\sum_{j=1}^n k_j \tau_j = 0$. By assumption the τ_j's are rationally independent and therefore $k_j = 0$ for all j. Hence $\chi \sim a_{0,0,\cdots,0} = 1$ and consequently meas $(E) = 1$.

The following application of the Kronecker theorem is due to I. Gelfand [6].

THEOREM 4.17.6. *Let $\chi(\xi)$ be a character of the real line. Further let $\{\alpha_n\}$ be an absolutely summable sequence of complex numbers and let $\{\xi_n\}$ be an arbitrary sequence of real numbers. Then given $\epsilon > 0$ there exists a real number β such that*

$$\sum_{n=1}^{\infty} \left| \alpha_n \chi(\xi_n) - \alpha_n \exp(2\pi i \beta \xi_n) \right| < \epsilon. \tag{4.17.5}$$

PROOF. Since $\sum |\alpha_n| < \infty$, it is clear that there exists an $N(\epsilon)$ such that $\sum_{n>N} |\alpha_n| < \epsilon/3$. Let $\tau_1, \tau_2, \cdots, \tau_M$ be a rationally independent basis for the numbers $\xi_1, \xi_2, \cdots, \xi_N$. The τ_i's can be chosen so that $\xi_j = \sum_{m=1}^M k_{jm} \tau_m$, where the k_{jm}'s are integers. Since $|\chi(\xi)| = 1$, we may set $\chi(\tau_m) = \exp(2\pi i \delta_m)$. Clearly $\chi(\xi_j) = \exp(2\pi i \sum_{m=1}^M k_{jm} \delta_m)$. Now by the Kronecker theorem, given $\eta > 0$ there exists a real β and integers $n_1, n_2, \cdots, n_M$ such that $|\beta \tau_i - \delta_i - n_i| < \eta$ for $i = 1, 2, \cdots, M$. But this means that

$$\begin{aligned} |\chi(\xi_j) - \exp(2\pi i \beta \xi_j)| &= \left| \exp\left(2\pi i \sum_{m=1}^M k_{jm} \delta_m\right) - \exp\left(2\pi i \beta \sum_{m=1}^M k_{jm} \tau_m\right) \right| \\ &\leqq 2\pi \left[\sum_{m=1}^M |k_{jm}|\right] \eta. \end{aligned}$$

Hence, setting $C = \sup \left[\sum_{m=1}^M |k_{jm}|; j = 1, 2, \cdots, N\right]$, we have

$$\sum_{n=1}^{\infty} |\alpha_n \chi(\xi_n) - \alpha_n \exp(2\pi i \beta \xi_n)| < 2\epsilon/3 + 2\pi C \eta \sum_{n=1}^{N} |\alpha_n| < \epsilon$$

for η sufficiently small.

4.18. Maximal ideals in $S(\varphi)$. We shall now apply the Gelfand theory to the (B)-algebra $S(\varphi)$. This will require a study of the maximal ideals in $S(\varphi)$, or equivalently a study of the linear multiplicative functionals on $S(\varphi)$.

We distinguish between two types of maximal ideals:

$\mathfrak{W} \equiv$ *all maximal ideals which do not contain $L(\varphi)$;*

$\mathfrak{U} \equiv \mathfrak{M} \ominus \mathfrak{W} =$ *all maximal ideals which do contain $L(\varphi)$.*

It is clear that $\mathfrak{W}$ and $\mathfrak{U}$ are disjoint and that $\mathfrak{M} = \mathfrak{W} \cup \mathfrak{U}$. This decomposition is basic in our discussion of $S(\varphi)$. As we shall see in Chapter XVI there is a corresponding decomposition for the maximal ideals in the algebra associated with a semi-group of operators.

THEOREM 4.18.1. *The set of all $a \in S(\varphi)$ such that $a([0]) = 0$ defines a maximal ideal. If μ_0 is the corresponding linear multiplicative functional, then $\mu_0(a) = a([0])$ and hence $\mu_0(e_\xi)$ satisfies (4.17.1) trivially. Further μ_0 is the only linear multiplicative functional such that $\mu(e_\xi) = 0$ for some $\xi > 0$.*

PROOF. Since $(\alpha a + \beta b)([0]) = \alpha a([0]) + \beta b([0])$, $(a * b)([0]) = a([0]) b[0])$, and $e_0([0]) = 1$, it follows that $\mu_0(a) \equiv a([0])$ defines a linear multiplicative functional on $S(\varphi)$ and hence that the kernel of μ_0 is a maximal ideal. It is obvious that $\mu_0(e_\xi) = e_\xi([0]) = 0$ for $\xi > 0$. On the other hand suppose that μ is a linear multiplicative functional such that $\mu(e_{\xi_0}) = 0$ for some $\xi_0 > 0$. By Theorem 4.17.1 we have $\mu(e_\xi) = 0$ for all $\xi > 0$. Now let $a \in S(\varphi)$ and suppose that $a([0]) = 0$. We then set $I_\delta = [0, \delta]$ and define $a_\delta(E) = a(E \ominus I_\delta \cap E)$. It is clear that $\lim_{\delta\to 0+} \| a_\delta - a \| = 0$. Further $a_\delta * e_{-\delta/2} \in S(\varphi)$ and $a_\delta = (a_\delta * e_{-\delta/2}) * e_{\delta/2}$. Hence $\mu(a_\delta) = \mu(a_\delta * e_{-\delta/2})\mu(e_{\delta/2}) = 0$ and by continuity we obtain $\mu(a) = 0$. For arbitrary $a \in S(\varphi)$, $a - a([0])e_0$ vanishes on the set $[0]$ so that $\mu(a) = \mu[a - a([0])e_0] + \mu[a([0])e_0] = a([0])$; hence $\mu = \mu_0$.

Since $\mu_0(a) = 0$ for all $a \in L(\varphi)$ we see that $\mu_0 \in \mathfrak{U}$. It will be convenient to distinguish between μ_0 and the rest of $\mathfrak{U}$. We therefore define

$\mathfrak{U}_0 \equiv$ the maximal ideal consisting of all $a \in S(\varphi)$ such that $a([0]) = 0$;

$\mathfrak{U}_1 \equiv \mathfrak{U} \ominus \mathfrak{U}_0$.

THEOREM 4.18.2. *$\mathfrak{W}$ is an open subset of $\mathfrak{M}$ homeomorphic with the subset*

$$C_0 \equiv [\lambda; \Re(\lambda) \leqq \omega_0]$$

of the complex plane. Under this correspondence $\lambda \in C_0 \to \mathfrak{m}_\lambda \in \mathfrak{W}$ and

$$a(\mathfrak{m}_\lambda) = \int_0^\infty e^{\lambda\xi}\, da. \tag{4.18.1}$$

PROOF. For $\lambda \in C_0$, we see from (4.16.1) and (4.16.3) that the integral $\int_0^\infty \exp(\lambda\xi)\, da \equiv \mu(a)$ converges absolutely. It is clear that $\mu(a)$ is linear. Further

$$\begin{aligned} \mu(a * b) &= \int_0^\infty e^{\lambda\xi}\, d(a * b) \\ &= \int_0^\infty \int_0^\infty e^{\lambda(\xi+\eta)}\, d_\xi a\, d_\eta b = \mu(a)\mu(b). \end{aligned} \tag{4.18.2}$$

Hence μ is a linear multiplicative functional on $S(\varphi)$. Now $a(E) \equiv \text{meas}\,(E \cap [\delta, 3\delta]) \in L(\varphi)$ and

$$\mu(a) = \int_\delta^{3\delta} e^{\lambda\xi}\, d\xi = \lambda^{-1}(e^{3\delta\lambda} - e^{\delta\lambda})$$

is non-vanishing for a suitable choice of $\delta > 0$; hence the kernel of μ, namely $\mathfrak{m}_\lambda$, belongs to $\mathfrak{W}$. On the other hand if $\mathfrak{m} \in \mathfrak{W}$, then there exists an $a_0 \in L(\varphi)$ such that $a_0(\mathfrak{m}) \neq 0$. Let μ be the linear multiplicative functional with $\mathfrak{m}$ as kernel; then $\mu(a) = a(\mathfrak{m})$. Now $a_0 * e_\xi$ is continuous in the norm for $\xi > 0$ and hence $\mu(e_\xi) = \mu(a_0 * e_\xi)/\mu(a_0)$ is likewise continuous for $\xi > 0$. Making use of the corollary to Theorem 4.17.3, we see that $\mu(e_\xi) = \exp(\lambda\xi)$ for some complex number λ. Again by (4.16.2)

$$\Re(\lambda) = \xi^{-1} \log |\mu(e_\xi)| \leqq \inf_{\xi>0} [\xi^{-1} \log \| e_\xi \|] = \omega_0.$$

Thus $\lambda \in C_0$. Finally for any $b \in S(\varphi)$, we have by (4.16.10)

$$\begin{aligned} \mu(a_0)\mu(b) &= \mu(a_0 * b) = \mu\left[\int_0^\infty (a_0 * e_\xi)\, db\right] \\ &= \int_0^\infty \mu(a_0 * e_\xi)\, db = \int_0^\infty \mu(a_0)\mu(e_\xi)\, db. \end{aligned} \tag{4.18.3}$$

Here we have made use of the fact that $\int_0^\infty (a_0 * e_\xi)\, db$ is a Bochner integral and that the bounded linear functional μ commutes with the operation of integration. It follows that $\mu(b) = \int_0^\infty \exp(\lambda\xi)\, db$. Finally since $\exp(\lambda\xi) = \int_0^\infty \exp(\lambda\tau)\, d_\tau e_\xi$, it is clear that the correspondence between $\mathfrak{W}$ and C_0 is one-to-one.

In order to show that $\mathfrak{W}$ is open we need only to point out that $\mathfrak{W}$ is the union of the open sets $[\mathfrak{m}; |a(\mathfrak{m})| > 0, a \in L(\varphi)]$. Since $a(\mathfrak{m}_\lambda) = \int_0^\infty \exp(\lambda\xi)\, da$ is continuous in $\lambda \in C_0$ for each $a \in S(\varphi)$, it follows that the mapping $\lambda \to \mathfrak{m}_\lambda$ is continuous. If $\lambda_0 \to \mathfrak{m}_0$, we now choose $a_0 \in L(\varphi)$ such that $|a_0(\mathfrak{m}_0)| \neq 0$. We recall for $a_0 \in L(\varphi)$ that

$$\lim_{|\lambda| \to \infty} \int_0^\infty \exp(\lambda\xi)\, da_0 = 0$$

uniformly with respect to λ in the closed half-plane C_0. Hence the set of λ's in C_0 such that $|\int_0^\infty \exp(\lambda\xi)\, da_0| = |a_0(\mathfrak{m})| \geqq |a_0(\mathfrak{m}_0)|/2$ define a compact subset F of C_0 containing λ_0 as an interior point; likewise the image of F contains $\mathfrak{m}_0$ as an interior point. It follows that the mapping on F into $\mathfrak{W}$ is bicontinuous and consequently that the same is true of the extended map on C_0 to $\mathfrak{W}$.

THEOREM 4.18.3. *The linear multiplicative functional μ corresponds to a maximal ideal in $\mathfrak{U}_1$ if and only if $\mu[L(\varphi)] = 0$ and $\mu(e_\xi) = \chi(\xi) \exp(\alpha\xi)$, where $\alpha \leqq \omega_0$ and $\chi(\xi)$ is a character of the real line.*

PROOF. The maximal ideals in $\mathfrak{U}_1$ are characterized by the fact that $\mu[L(\varphi)] = 0$ and and $\mu(e_\xi) \neq 0$. Now $|\mu(e_\xi)| \leqq \varphi(\xi)$ and hence $\mu(e_\xi) \neq 0$ is found equivalent with $\mu(e_\xi) = \chi(\xi) \exp(\alpha\xi)$ where $\alpha \leqq \omega_0$ (see the discussion following Theorem 4.17.2).

One can obtain a representation for the multiplicative linear functionals in $\mathfrak{U}_1$ by methods developed by Y. A. Šreĭder [1]. We shall refrain from such a discussion since it is somewhat lengthy. However, if the maximal ideal corresponding to μ contains both $N(\varphi)$ and $L(\varphi)$, then it is clear that $\mu(a) = \sum_\xi \mu(e_\xi)a([\xi])$ for all $a \in S(\varphi)$.

THEOREM 4.18.4. *If $\omega_0 > -\infty$ then the algebra $S(\varphi)$ is semi-simple and contains no divisors of zero. If $\omega_0 = -\infty$, then there is only one maximal ideal, namely $\mathfrak{m}_0 \in \mathfrak{U}_0$, and $\mathfrak{m}_0$ is the radical of $S(\varphi)$.*

PROOF. First suppose that $\omega_0 > -\infty$. Then if q is quasi-nilpotent, we have $0 = q(\mathfrak{m}_\lambda) = \int_0^\infty \exp(\lambda\xi)\, dq$ for all $\lambda \in C_0$ and C_0 consists of a half-plane. According to the uniqueness theorem for Laplace-Stieltjes integrals this means that $q = \theta$. Hence by Corollary 2 of Theorem 4.15.1, $S(\varphi)$ is semi-simple. Further if $a * b = \theta$, then by (4.18.2) $a(\mathfrak{m}_\lambda)b(\mathfrak{m}_\lambda) = 0$ for all $\lambda \in C_0$. However both $a(\mathfrak{m}_\lambda)$ and $b(\mathfrak{m}_\lambda)$ are holomorphic functions of λ in the interior of C_0. Since at least one of these functions must vanish denumerably often on any given denumerable subset of C_0 with limit point interior to C_0, it follows that one of these functions, say $a(\mathfrak{m}_\lambda)$, vanishes identically on C_0. By the uniqueness theorem $a = \theta$ and hence $S(\varphi)$ contains no divisors of zero. Suppose next that $\omega_0 = -\infty$. Then according to Theorems 4.18.2 and 4.18.3 both $\mathfrak{W}$ and $\mathfrak{U}_1$ must be empty. Thus the only maximal ideal is $\mathfrak{m}_0 \in \mathfrak{U}_0$, and since it is clearly the intersection of all maximal ideals $\mathfrak{m}_0$ is also the radical of $S(\varphi)$.

We conclude this paragraph with two theorems of a type associated with the names of N. Wiener and H. R. Pitt [1].

THEOREM 4.18.5. *Let $a \in S(\varphi)$ have the decomposition $a = a_1 + a_2 + a_3$ into its absolutely continuous, atomic, and non-atomic singular parts, respectively. In case $\omega_0 > -\infty$, a will be regular if $\int_0^\infty \exp(\lambda\xi)\, da \neq 0$ for all $\lambda \in C_0$ and if*

$$\inf_{\lambda \in C_0} \left| \sum \exp(\lambda\xi) a([\xi]) \right| > \| a_3 \|. \tag{4.18.4}$$

If $\omega_0 = -\infty$, then a is regular if and only if $a(\{0\}) \neq 0$.

PROOF. Suppose first that $\omega_0 > -\infty$. Then according to Theorem 4.18.2, the first condition implies that a does not belong to a maximal ideal of $\mathfrak{B}$. Taking the limit as $\lambda \to -\infty$ in (4.18.4), we see that $a([0]) \neq 0$ and hence that a does not belong to the ideal in $\mathfrak{U}_0$. Now for $\mu \in \mathfrak{U}_1$, $\mu(a) = \mu(a_2) + \mu(a_3)$. It follows from Theorem 4.18.3 that $\mu(e_\xi) = \chi(\xi) \exp(\alpha\xi)$, where $\alpha \leqq \omega_0$ and $\chi(\xi)$ is a character of the real line. Since a_2 is atomic it is determined by its values on a denumerable set of points $\{\xi_n\}$. Hence $\mu(a_2) = \sum_n \chi(\xi_n) \exp(\alpha\xi_n)\, a([\xi_n])$; further from the definition of $S(\varphi)$ we have $\sum_n \exp(\alpha\xi_n)\, |a([\xi_n])| < \infty$. By Theorem 4.17.6, given $\epsilon > 0$ there exists a real β such that

$$| \mu(a_2) - \textstyle\sum_n \exp[(\alpha + i\beta)\xi_n]\, a([\xi_n]) \,| < \epsilon.$$

It follows that $| \mu(a_2) | \geqq \inf_{\lambda \in C_0} | \sum_\xi \exp(\lambda\xi)\, a([\xi]) |$. Condition (4.18.4) therefore implies that $| \mu(a_2) | > \| a_3 \| \geqq | \mu(a_3) |$ and hence that $| \mu(a) | > 0$. The result for $\omega_0 > -\infty$ now follows from Theorem 4.15.1 (9). Finally if $\omega_0 = -\infty$, then there is only one maximal ideal, namely $\mathfrak{m}_0 \in \mathfrak{U}_0$. An element a will be regular if and only if $a(\mathfrak{m}_0) = a([0]) \neq 0$.

In case $\omega_0 > -\infty$, the condition (4.18.4) requires that $| a([0]) | > \| a_3 \|$ as can be seen by letting $\lambda \to -\infty$. On the other hand (4.18.4) will be satisfied if $| a([0]) | > \frac{1}{2}(\| a_2 \| + \| a_3 \|)$. This follows from the relation

$$\begin{aligned} | \sum_\xi a([\xi]) \exp(\lambda\xi) | &\geqq | a([0]) | - | \sum_{\xi > 0} a([\xi]) \exp(\lambda\xi) | \\ &\geqq | a([0]) | - (\| a_2 \| - | a([0]) |). \end{aligned}$$

THEOREM 4.18.6. *Let $\omega_0 > -\infty$. An element a belonging to $L(\varphi) + A(\varphi)$ is regular in $S(\varphi)$ if and only if*

$$\inf_{\lambda \in C_0} \left| \int_0^\infty e^{\lambda\xi}\, da \right| > 0. \tag{4.18.5}$$

PROOF. If a is regular in $S(\varphi)$, then $a(\mathfrak{m})$ is continuous and non-vanishing on $\mathfrak{M}$ and therefore $| a(\mathfrak{m}) |$ is bounded away from zero. This implies (4.18.5) by Theorem 4.18.2. For the sufficiency argument it suffices to establish (4.18.4) with $\| a_3 \| = 0$. Let $\lambda_0 = \alpha_0 + i\beta_0$ be given, $\alpha_0 \leqq \omega_0$. The function $f(\beta) \equiv \sum_\xi \exp[(\alpha_0 + i\beta)\xi] a([\xi])$ is clearly almost periodic in β so that for $\epsilon > 0$ there exists a sequence $\beta_n \to \infty$ such that $| f(\beta_0) - f(\beta_n) | < \epsilon$ for all n. Since $a \in L(\varphi) + A(\varphi)$, the Riemann-Lebesgue theorem implies that

$$\lim_{\beta \to \infty} \left| \int_0^\infty \exp[(\alpha_0 + i\beta)\xi]\, da - f(\beta) \right| = 0.$$

It follows that $f(\beta_0)$ lies in the closure of the set $[\int_0^\infty \exp[(\alpha_0 + i\beta)\xi]\, da; -\infty < \beta < \infty]$ and hence that (4.18.5) implies (4.18.4).

It should be noticed that if a is regular and belongs to $L(\varphi) + A(\varphi)$ (again for the case $\omega_0 > -\infty$), then $b = a^{-1}$ likewise belongs to $L(\varphi) + A(\varphi)$. In fact if we decompose a and b into their absolutely continuous, atomic, and non-atomic singular parts, we obtain

$$e_0 = a * b = (a_1 * b_1 + a_1 * b_2 + a_1 * b_3 + a_2 * b_1) + (a_2 * b_2) + (a_2 * b_3),$$

where we have grouped the terms of e_0 according to its decomposition. If follows that $a_2 * b_2 = e_0$ and that $a_2 * b_3 = \theta$. Since $S(\varphi)$ contains no divisors of zero (Theorem 4.18.4) and since $a_2 \neq \theta$, we see that $b_3 = \theta$.

References. Beurling [1], Gelfand [6], Halmos [1], Phillips [6], Šreĭder [1], Wiener and Pitt [1].

5. Commutative (A*)-Algebras

4.19. Representation theory for commutative (A*)-algebras. We now restrict our algebra somewhat further, assuming $\mathfrak{B}$ to be a commutative complex (A*)-algebra with a unit element. The structure of the Gelfand representation theory is further enriched by the added algebraic properties.

Theorem 4.19.1. *For a commutative complex (A*)-algebra with a unit element we have*

$$x^*(\mathfrak{m}) = \overline{x(\mathfrak{m})} \tag{4.19.1}$$

for all $x \in \mathfrak{B}$ *and* $\mathfrak{m} \in \mathfrak{M}$.

Proof. If x is self-adjoint then the points of $\sigma(x)$ are real and hence by Theorem 4.15.1 (property (6)), $x(\mathfrak{m})$ is real valued on $\mathfrak{M}$. Now an arbitrary element x can be represented as

$$x = x_1 + ix_2 \tag{4.19.2}$$

where both $x_1 = (x + x^*)/2$ and $x_2 = (x - x^*)/(2i)$ are self-adjoint. Therefore $x^* = x_1 - ix_2$ and $x^*(\mathfrak{m}) = x_1(\mathfrak{m}) - ix_2(\mathfrak{m}) = \overline{x(\mathfrak{m})}$.

Definition 4.19.1. *We denote by* $R(\mathfrak{M})$ *the real algebra of all continuous real-valued functions over the compact Hausdorff space* $\mathfrak{M}$. *The basic operations are defined by*: $(\alpha f)(\mathfrak{m}) = \alpha f(\mathfrak{m})$, $(f + g)(\mathfrak{m}) = f(\mathfrak{m}) + g(\mathfrak{m})$, $(fg)(\mathfrak{m}) = f(\mathfrak{m})g(\mathfrak{m})$, $e(\mathfrak{m}) \equiv 1$, *and* $\|f\| = \sup_{\mathfrak{m}\in\mathfrak{M}} |f(\mathfrak{m})|$.

Theorem 4.19.2. *Let* $A(\mathfrak{M})$ *be a closed subalgebra of* $R(\mathfrak{M})$. *Then* $A(\mathfrak{M})$ *is a linear lattice under the convention*: $f < g$ *denotes that* $f(\mathfrak{m}) \leqq g(\mathfrak{m})$ *for all* $\mathfrak{m} \in \mathfrak{M}$.

Proof. It is clear that $A(\mathfrak{M})$ is a partially ordered linear system in the sense of Definition 1.11.1. In order to show that it is a lattice it suffices to prove that $f \vee \theta \in A(\mathfrak{M})$ for all $f \in A(\mathfrak{M})$. Since $2(f \vee \theta)(\mathfrak{m}) = |f(\mathfrak{m})| + f(\mathfrak{m})$, it is enough to show that $|f(\mathfrak{m})|$ belongs to $A(\mathfrak{M})$ whenever $f(\mathfrak{m})$ does. Now the function $\psi(\tau) = |\tau|$ may be approximated uniformly on $-\|f\| \leqq \tau \leqq \|f\|$ by a sequence of polynomials without constant terms $\{p_n(\tau)\}$. As $A(\mathfrak{M})$ is an algebra we see that $p_n(f) \in A(\mathfrak{M})$. Further

$$\| p_n(f) - p_m(f) \| \leqq \sup_{|\tau| \leqq \|f\|} | p_n(\tau) - p_m(\tau) |.$$

Hence the $\{p_n(f)\}$ form a Cauchy sequence in $A(\mathfrak{M})$ and since $A(\mathfrak{M})$ is closed the limit element, g, belongs to $A(\mathfrak{M})$. For each $\mathfrak{m} \in \mathfrak{M}$, we have $p_n(f)(\mathfrak{m}) = p_n[f(\mathfrak{m})] \rightarrow \psi[f(\mathfrak{m})] = |f(\mathfrak{m})|$. Therefore $g(\mathfrak{m}) = |f(\mathfrak{m})| \in A(\mathfrak{M})$.

The next theorem is due to M. H. Stone [4] and is often referred to as the Stone-Weierstrass theorem.

Theorem 4.19.3. ***Let*** $A(\mathfrak{M})$ ***be a closed subalgebra of*** $R(\mathfrak{M})$ ***containing the unit element. If*** $A(\mathfrak{M})$ ***separates points of*** $\mathfrak{M}$ ***then*** $A(\mathfrak{M}) = R(\mathfrak{M})$.

PROOF. By assumption $A(\mathfrak{M})$ is closed. Hence for arbitrary $h \in R(\mathfrak{M})$ and $\epsilon > 0$ it will be sufficient to exhibit an $f \in A(\mathfrak{M})$ such that $\| h - f \| < \epsilon$. Now let $\mathfrak{m}_0$ be a fixed point of $\mathfrak{M}$. We shall first show that for any point $\mathfrak{m}_1 \in \mathfrak{M}$ there exists a function $f(\mathfrak{m}; \mathfrak{m}_0, \mathfrak{m}_1) \in A(\mathfrak{M})$ which assumes the same values as $h(\mathfrak{m})$ at the two points $\mathfrak{m}_0$ and $\mathfrak{m}_1$. For $\mathfrak{m}_1 = \mathfrak{m}_0$, we may take $f(\mathfrak{m}; \mathfrak{m}_0, \mathfrak{m}_1) = h(\mathfrak{m}_0)e(\mathfrak{m})$. For $\mathfrak{m}_1 \neq \mathfrak{m}_0$ there exists by assumption a function $g \in A(\mathfrak{M})$ such that $g(\mathfrak{m}_1) \neq g(\mathfrak{m}_0)$. In this case we take

$$f(\cdot; \mathfrak{m}_0, \mathfrak{m}_1) = \left[\frac{h(\mathfrak{m}_0) - h(\mathfrak{m}_1)}{g(\mathfrak{m}_0) - g(\mathfrak{m}_1)}\right][g(\cdot) - g(\mathfrak{m}_1)e(\cdot)] + h(\mathfrak{m}_1)e(\cdot).$$

Since the function $[f(\mathfrak{m}; \mathfrak{m}_0, \mathfrak{m}_1) - h(\mathfrak{m})]$ is continuous in $\mathfrak{m}$, there exists a neighborhood $N(\mathfrak{m}_1)$ of $\mathfrak{m}_1$ such that $f(\mathfrak{m}; \mathfrak{m}_0, \mathfrak{m}_1) > h(\mathfrak{m}) - \epsilon$ for all $\mathfrak{m} \in N(\mathfrak{m}_1)$. Varying $\mathfrak{m}_1$, the set of all such neighborhoods, $N(\mathfrak{m})$, covers $\mathfrak{M}$ and hence a finite subset $[N(\mathfrak{m}_i), i = 1, 2, \cdots, n]$ will likewise cover $\mathfrak{M}$. We now set

$$f'(\cdot; \mathfrak{m}_0) = \bigvee_{i=1}^{n} f(\cdot; \mathfrak{m}_0, \mathfrak{m}_i).$$

Then clearly $f'(\mathfrak{m}; \mathfrak{m}_0) > h(\mathfrak{m}) - \epsilon$ for all $\mathfrak{m} \in \mathfrak{M}$ and $f'(\mathfrak{m}_0; \mathfrak{m}_0) = h(\mathfrak{m}_0)$. In addition $f'(\cdot; \mathfrak{m}_0)$ belongs to $A(\mathfrak{M})$ by the previous theorem. Again there exists a neighborhood $N'(\mathfrak{m}_0)$ of $\mathfrak{m}_0$ such that $f'(\mathfrak{m}; \mathfrak{m}_0) < h(\mathfrak{m}) + \epsilon$ for all $\mathfrak{m}$ belonging to $N'(\mathfrak{m}_0)$. Varying $\mathfrak{m}_0$, the set of all such neighborhoods, $N'(\mathfrak{m})$, covers $\mathfrak{M}$ and so will a finite subset $[N'(\mathfrak{m}_i), i = 1, 2, \cdots, p]$. Finally we set

$$f = \bigwedge_{i=1}^{p} f'(\cdot; \mathfrak{m}_i).$$

Then $f \in A(\mathfrak{M})$ and $h(\mathfrak{m}) + \epsilon > f(\mathfrak{m}) > h(\mathfrak{m}) - \epsilon$; this is the desired result.

As an application of the Stone-Weierstrass theorem we obtain the following theorem due to I. Gelfand and G. Šilov [1].

THEOREM 4.19.4. *If $\mathfrak{B}$ is a commutative complex (A^*)-algebra with a unit element, then the image of $\mathfrak{B}$ in $C(\mathfrak{M})$, under the homomorphism $x \to x(\mathfrak{m})$, is dense in $C(\mathfrak{M})$.*

PROOF. For self-adjoint elements x, y and real α, β, the elements $\alpha x + \beta y$ and xy are again self-adjoint elements. Hence the set of all self-adjoint elements in $\mathfrak{B}$, which we denote by $\mathfrak{A}_0$, forms a commutative real (B)-algebra with a unit element. According to Theorem 4.19.1, $\mathfrak{A}_0$ maps into a subalgebra $A_0(\mathfrak{M})$ of $R(\mathfrak{M})$ under the mapping $x \to x(\mathfrak{m})$. Further $A_0(\mathfrak{M})$ separates points of $\mathfrak{M}$. For if $\mathfrak{m}_1$ and $\mathfrak{m}_2$ are distinct maximal ideals, then $\mathfrak{B}$ contains an element y belonging to $\mathfrak{m}_1$ which does not belong to $\mathfrak{m}_2$; thus $y(\mathfrak{m}_1) = 0 \neq y(\mathfrak{m}_2)$. Let $x = yy^*$. Then $x \in \mathfrak{A}_0$ and by Theorem 4.19.1 $x(\mathfrak{m}_1) = 0 \neq x(\mathfrak{m}_2) = | y(\mathfrak{m}_2) |^2$. It is clear that the closure of $A_0(\mathfrak{M})$ in $R(\mathfrak{M})$ is again a subalgebra of $R(\mathfrak{M})$ containing the unit element. Hence by the Stone-Weierstrass theorem $\overline{A_0(\mathfrak{M})} = R(\mathfrak{M})$. Now $\mathfrak{B} =$

$\mathfrak{A}_0 + i\mathfrak{A}_0$. If we denote the image of $\mathfrak{B}$ by $B(\mathfrak{M})$, then

$$\overline{B(\mathfrak{M})} = \overline{A_0(\mathfrak{M}) + iA_0(\mathfrak{M})} \equiv \overline{A_0(\mathfrak{M})} + i\overline{A_0(\mathfrak{M})} = C(\mathfrak{M}).$$

COROLLARY. *If $\mathfrak{B}$ is a commutative complex (A^*)-algebra with a unit element, then the Šilov boundary is $\mathfrak{M}$.*

PROOF. This follows directly from Theorems 4.15.5 and 4.19.4.

An example of an (A*)-algebra is furnished by the set of all functions defined on $[0, 2\pi)$ with absolutely convergent Fourier series. We refer the reader back to the concluding paragraph of section 4.14 for a discussion of this algebra. To each linear multiplicative functional, μ, there corresponds a $\xi_0 \in [0, 2\pi)$ such that $\mu(f) = f(\xi_0)$. For $f(\xi) = \sum_{-\infty}^{\infty} a_n \exp(in\xi)$, we set $f^*(\xi) = \sum_{-\infty}^{\infty} \bar{a}_{-n} \exp(in\xi)$. It is clear that this is a proper adjoint operation and that $\mu(f^*) = \overline{f(\xi_0)} = \overline{\mu(f)}$. It follows that for self-adjoint elements, $\mu(f)$ is real and hence that $\sigma(f)$ contains only real numbers. The algebra therefore satisfies the postulates of an (A*)-algebra.

4.20. The algebra $LA(-\infty, \infty)$. One of the most interesting examples of an (A*)-algebra is given by a subalgebra of $S_0(\varphi)$, namely $L_0(\varphi) + A_0(\varphi)$ for the case $\varphi(\xi) \equiv 1$ on $(-\infty, \infty)$. We shall denote this algebra by $LA(-\infty, \infty)$. The larger algebra $S_0(\varphi)$ is not itself an (A*)-algebra (see Y. A. Šreĭder [1]). The results of this section are essentially due to I. Gelfand [6].

The algebra $LA(-\infty, \infty)$ consists, then, of all countably additive complex-valued set functions on the Borel subsets of $(-\infty, \infty)$ with vanishing non-atomic singular parts. The discussion of section 4.16 applies and we see that $LA(-\infty, \infty)$ is a commutative complex (B)-algebra with unit element. The only point which might be in doubt is whether or not the product of two elements in $LA(-\infty, \infty)$ is again in this algebra. If we decompose a and b into their absolutely continuous and atomic parts, then

$$a * b = (a_1 * b_1 + a_1 * b_2 + a_2 * b_1) + (a_2 * b_2) \tag{4.20.1}$$

is the corresponding decomposition for $a * b$. In particular if $a = a_1 + \alpha e_0$ is regular and if $b = a^{-1}$, then $a_2 * b_2 = \alpha b_2 = e_0$; hence $a^{-1} = b_1 + \alpha^{-1}e_0$. Thus $L(-\infty, \infty) + \{\alpha e_0\}$ contains its own inverses. Finally we set

$$a^*(E) = \overline{a(-E)}. \tag{4.20.2}$$

$LA(-\infty, \infty)$ becomes a (*)-algebra under this convention.

The set of shift elements, $\mathfrak{S} = [e_\xi]$, now form a group with respect to the product operation. If μ is a linear multiplicative functional, then $\chi(\xi) = \mu(e_\xi)$ is a complex-valued representation of the group $\mathfrak{S}$. Further $|\chi(\xi)| \leqq \|e_\xi\| = 1$ and $|\chi(\xi)\chi(-\xi)| = \chi(0) = 1$; hence $|\chi(\xi)| \equiv 1$. Consequently $\chi(\xi)$ is a *character* of the real line. According to Theorem 4.17.3 either $\chi(\xi)$ is measurable in which case $\chi(\xi) = \exp(i\beta\xi)$ for some real β, or else $\chi(\xi)$ is non-measurable.

We shall distinguish between two types of maximal ideals:

$\mathfrak{W}$ = all maximal ideals which do not contain $L(-\infty, \infty)$;

$\mathfrak{U}$ = all maximal ideals which do contain $L(-\infty, \infty)$.

It is clear that $\mathfrak{W}$ and $\mathfrak{U}$ are disjoint and that $\mathfrak{M} = \mathfrak{W} \cup \mathfrak{U}$.

THEOREM 4.20.1. *$\mathfrak{W}$ is an open subset of $\mathfrak{M}$, homeomorphic with the real numbers R. Under this correspondence $\beta \in R \to \mathfrak{m}_\beta \in \mathfrak{W}$ and*

$$a(\mathfrak{m}_\beta) = \int_{-\infty}^{\infty} \exp(i\beta\xi)\, da. \tag{4.20.3}$$

PROOF. We omit the proof since a paraphrase of the argument given in Theorem 4.18.2 suffices to prove this theorem.

THEOREM 4.20.2. *If the linear multiplicative functional, μ, corresponds to a maximal ideal in $\mathfrak{U}$, then*

$$\mu(a) = \sum_{\xi} \chi(\xi)a([\xi]) \tag{4.20.4}$$

where $\chi(\xi)$ is a character of the real line.

PROOF. As we have already remarked $\chi(\xi) = \mu(e_\xi)$ is always a character of the real line. We note that $\chi(\xi)$ may or may not be continuous. Further $\mu[L(-\infty, \infty)] = 0$. Thus if $a = a_1 + a_2$ is a decomposition of a into its absolutely continuous and atomic parts, then

$$\mu(a) = \mu(a_2) = \sum_{\xi} \chi(\xi)a([\xi]).$$

THEOREM 4.20.3. *$LA(-\infty, \infty)$ is a semi-simple (A^*)-algebra.*

PROOF. If q is quasi-nilpotent, then by Theorem 4.15.1 $\int_{-\infty}^{\infty} \exp(i\beta\xi)\, dq = 0$ for all real β. It follows from the uniqueness theorem for Fourier-Stieltjes integrals that $q = \theta$. In order to prove that $LA(-\infty, \infty)$ is an (A^*)-algebra it will be sufficient to show that $\mu(a^*) = \overline{\mu(a)}$ for all linear multiplicative functionals. For $\mu \in \mathfrak{W}$ we combine (4.20.2) and (4.20.3) to obtain

$$\mu(a^*) = \int_{-\infty}^{\infty} \exp\,(i\beta\xi)\, da^* = \int_{-\infty}^{\infty} \exp\,(-i\beta\xi)\, d\bar{a} = \overline{\mu(a)}.$$

For $\mu \in \mathfrak{U}$ we obtain a similar result from (4.20.2) and (4.20.4).

It is clear that $L(-\infty, \infty) + [\alpha e_0]$ is a (*)-subalgebra of $LA(-\infty, \infty)$. Thus Theorem 4.11.3 gives us another proof of the fact that $L(-\infty, \infty) + [\alpha e_0]$ contains inverses. If x is self-adjoint and belongs to the subalgebra, then x has a real spectrum relative to both the algebra and the subalgebra by Corollary 3 to Theorem 4.11.2. Consequently $L(-\infty, \infty) + [\alpha e_0]$ is itself an (A^*)-algebra. It follows from Theorem 4.15.6 and the corollary to Theorem 4.19.4 that all of the multiplicative linear functionals on $L(-\infty, \infty) + [\alpha e_0]$ are restrictions of functionals in $\mathfrak{W}$ and $\mathfrak{U}$. The functionals in $\mathfrak{W}$ remain distinct whereas the functionals in $\mathfrak{U}$ restrict to the single functional $\mu_0(a_1 + \alpha e_0) = \alpha$.

THEOREM 4.20.4. *$\mathfrak{W}$ is dense in $\mathfrak{M}$.*

PROOF. Let $\mu_0 \in \mathfrak{U}$ and let $N(\mu_0\,;\, a_1, a_2, \cdots, a_m\,;\, \epsilon)$ be an arbitrary neighborhood of μ_0. We shall show that N contains an element of $\mathfrak{W}$. Now only a denumerable subset of the number pairs $(a_k([\xi]), \xi)$ with arbitrary ξ and $k = 1, 2, \cdots, m$ correspond to non-vanishing $a_k([\xi])$. These we reorder as (α_n, ξ_n). Note that the ξ_n's need not be distinct. According to Theorem 4.17.6 there will exist a real β_0 such that $\sum_n |\alpha_n\chi(\xi_n) - \alpha_n \exp(i\beta_0\xi_n)| < \epsilon/3$. This clearly implies that

$$|\mu_0(a_k) - \sum_{\xi} \exp(i\beta_0\xi)a_k([\xi])| < \epsilon/3 \qquad \text{for } k = 1, 2, \cdots, m.$$

Set $f_k(\beta) = \sum_\xi \exp\,(i\beta\xi)a_k([\xi])$. Then the function $f(\beta) = \sum_{k=1}^{m} |f_k(\beta) - f_k(\beta_0)|$ is almost periodic and hence there exists a sequence $\beta_n \to \infty$ such that $|f(\beta_n) - f(\beta_0)| < \epsilon/3$ for all n. It follows that

$$|f_k(\beta_n) - f_k(\beta_0)| < \epsilon/3$$

for all n and $k = 1, 2, \cdots, m$. Finally by the Riemann-Lebesgue theorem

$$\lim_{\beta\to\infty} \left| \int_{-\infty}^{\infty} \exp\,(i\beta)\, da_k - f_k(\beta) \right| = 0,$$

again for $k = 1, 2, \cdots, m$. Combining we obtain $| \mu_0(a_k) - \int_{-\infty}^{\infty} \exp(i\beta_n\xi)\, da_k | < \epsilon$ for n sufficiently large and $k = 1, 2, \cdots, m$. In other words $\mathfrak{m}_{\beta_n} \in N(\mu_0; a_1, a_2, \cdots, a_m; \epsilon)$ for n sufficiently large.

COROLLARY. *An element a in $LA(-\infty, \infty)$ is regular if and only if*

$$\inf_{\beta \in R} \left| \int_{-\infty}^{\infty} \exp(i\beta\xi)\, da \right| > 0. \tag{4.20.5}$$

PROOF. The inequality (4.20.5) means that $a(\mathfrak{m})$ is bounded away from zero on $\mathfrak{W}$. Since $\mathfrak{W}$ is dense in $\mathfrak{M}$ and since $a(\mathfrak{m})$ is continuous in $\mathfrak{m}$, the inequality (4.20.5) is equivalent to $a(\mathfrak{m}) \neq 0$ on $\mathfrak{M}$. Finally a is regular if and only if $a(\mathfrak{m}) \neq 0$ on $\mathfrak{M}$ by Theorem 4.15.1.

For elements of $L(-\infty, \infty) + [\alpha e_0]$, the proof of the corollary can be considerably simplified. In this case the Riemann-Lebesgue theorem can be used in place of Theorem 4.20.4.

Results of the type obtained in this section have been generalized in several directions. D. A. Raikov [1] and I. E. Segal [1] have extended the theory to include complex-valued functions on a locally compact commutative group, integrable with respect to the Haar measure. S. Bochner and R. S. Phillips [1] have treated convolution algebras in which the function values lie in a non-commutative complex (B)-algebra with a unit element.

4.21. Some Tauberian theorems. We shall apply the foregoing theory to prove a vector-valued variant of Wiener's general Tauberian theorem [3].

It is convenient at this point to consider an element of $L(-\infty, \infty)$ sometimes as a set function $a(E)$ and other times as the corresponding density function, $a(\xi)$. These two representations of a are related by $a(E) = \int_E a(\xi)\, d\xi$; the argument will always indicate which of of these two forms we are using. We shall denote the Fourier transforms of $a, b, \cdots \in LA(-\infty, \infty)$ by $A(\beta), B(\beta), \cdots$ respectively. Thus

$$A(\beta) = \int_{-\infty}^{\infty} \exp(i\beta\xi)\, da(E). \tag{4.21.1}$$

DEFINITION 4.21.1. *A complex-valued functions $f(\xi) \in W(-\infty, \infty)$ if $f(\xi) \in L(-\infty, \infty)$ and $F(\beta) \neq 0$ for all real β. Similarly $g(\eta) \in W(0, \infty)$ if $g(\eta) \in L(0, \infty)$ and*

$$\int_0^{\infty} \eta^{i\beta} g(\eta)\, d\eta \neq 0$$

for all real β.

Finally we introduce the two functions

$$k_n(\xi) = (2/\pi) \frac{\sin^2(n\xi/2)}{n\xi^2}$$

and

$$h_n(\xi) = (2/\pi) \frac{\sin(3n\xi/2)\sin(n\xi/2)}{n\xi^2}.$$

Both functions belong to $L(-\infty, \infty)$. The first is the familiar Fejér kernel and for any $a \in L(-\infty, \infty)$ we have $\lim_{n\to\infty} \| a * k_n - a \| = 0$. Further $K_n(\beta) = 0$ for $|\beta| > n$; whereas $H_n(\beta) = 1$, $|\beta| < n$, $H_n(\beta) = 0$ for $|\beta| > 2n$, and $H_n(\beta) \geqq 0$ elsewhere. Thus the Fourier transform of $(e_0 - h_n) * k_n$ vanishes identically; and by the uniqueness theorem for Fourier-Stieltjes integrals we have $(e_0 - h_n) * k_n = \theta$.

THEOREM 4.21.1 (WIENER). *If $f \in W(-\infty, \infty)$, then the translates of f, namely $[f * e_\xi]$, are fundamental in $L(-\infty, \infty)$.*

PROOF. For any $n > 0$, $(1 - H_n(\beta)) + |F(\beta)|^2$ is bounded away from zero on $(-\infty, \infty)$. Hence by the corollary to Theorem 4.20.3, $e_0 - h_n + f * f^*$ has an inverse which we denote by b_n. Now for any $a_0 \in L(-\infty, \infty)$ we have

$$\begin{aligned} a_0 = e_0 * a_0 &= [(e_0 - h_n + f * f^*) * b_n] * a_0 \\ &= (e_0 - h_n) * b_n * a_0 + f * f^* * b_n * a_0 . \end{aligned}$$

Hence

$$\begin{aligned} a_0 * k_n &= [(e_0 - h_n) * k_n] * b_n * a_0 + f * f^* * b_n * a_0 * k_n \\ &= \theta + f * c_n \end{aligned} \tag{4.21.2}$$

where $c_n = f^* * b_n * a_0 * k_n \in L(-\infty, \infty)$. Making use of the properties of the Fejér kernel we see that for any $\epsilon > 0$ there exists an $N(\epsilon)$ such that $\| a_0 * k_N - a_0 \| < \epsilon$. Further by (4.16.11) we have $f * c_N = \int_{-\infty}^{\infty} (f * e_\xi)\, dc_N(E)$. The integral here is an abstract Bochner integral and can be approximated by finite sums: $\| \sum_j (f * e_{\xi_j}) c_N(E_j) - f * c_N \| < \epsilon$. Combining the above two inequalities with (4.21.2) we obtain

$$\| a_0 - \sum_j (f * e_{\xi_j}) c_N(E_j) \| < 2\epsilon.$$

Since the coefficients, $c_N(E_j)$, are merely numbers, it follows that a_0 can be approximated in the norm by linear combinations of the translates of f.

THEOREM 4.21.2. *If $f(\xi) \in W(-\infty, \infty)$, $h(\xi) \in L(-\infty, \infty)$, if $x(\xi)$ is a bounded strongly measurable function on $(-\infty, \infty)$ to a (B)-space $\mathfrak{X}$, and if*

$$\lim_{\tau \to \infty} \int_{-\infty}^{\infty} f(\xi - \tau) x(\xi)\, d\xi = x_0 \int_{-\infty}^{\infty} f(\xi)\, d\xi, \tag{4.21.3}$$

then

$$\lim_{\tau \to \infty} \int_{-\infty}^{\infty} h(\xi - \tau) x(\xi)\, d\xi = x_0 \int_{-\infty}^{\infty} h(\xi)\, d\xi. \tag{4.21.4}$$

PROOF. This is the vector form of Wiener's Tauberian theorem and we can adapt his argument for this case. In order to avoid confusion we shall denote the norm of elements in $\mathfrak{X}$ with a subscript $\mathfrak{X}$. It is clear that (4.21.3) implies that

$$\lim_{\tau \to \infty} \int_{-\infty}^{\infty} \left\{ \sum_{\nu=1}^{n} \alpha_\nu f(\xi + \lambda_\nu - \tau) \right\} x(\xi)\, d\xi = x_0 \int_{-\infty}^{\infty} \left\{ \sum_{\nu=1}^{n} \alpha_\nu f(\xi + \lambda_\nu) \right\} d\xi$$

for any choice of the complex constants α_ν and the real constants λ_ν. By the previous theorem we can choose these constants so that

$$\left\| h(\xi) - \sum_{\nu=1}^{n} \alpha_\nu f(\xi + \lambda_\nu) \right\| < \epsilon.$$

However if $\| h - k \| < \epsilon$, then for all real τ

$$\left\| \int_{-\infty}^{\infty} [h(\xi - \tau) - k(\xi - \tau)] x(\xi)\, d\xi \right\|_{\mathfrak{X}} < \epsilon \operatorname*{ess\,sup}_{-\infty < \xi < \infty} \| x(\xi) \|_{\mathfrak{X}} .$$

The desired conclusion is an immediate consequence of these estimates.

For the applications which we have in mind in Chapter XVIII, it is convenient to reformulate the theorem for the interval $(0, \infty)$.

THEOREM 4.21.3. *If* $g(\eta) \in W(0, \infty)$, $k(\eta) \in L(0, \infty)$, *if* $y(\eta)$ *is a bounded strongly measurable function on* $(0, \infty)$ *to a* (B)*-space* $\mathfrak{X}$, *and if*

$$\lim_{\sigma \to \infty} \frac{1}{\sigma} \int_0^\infty g(\eta/\sigma) y(\eta)\, d\eta = y_0 \int_0^\infty g(\eta)\, d\eta, \tag{4.21.5}$$

then

$$\lim_{\sigma \to \infty} \frac{1}{\sigma} \int_0^\infty k(\eta/\sigma) y(\eta)\, d\eta = y_0 \int_0^\infty k(\eta)\, d\eta. \tag{4.21.6}$$

PROOF. We set up an isometry from $L(-\infty, \infty)$ to $L(0, \infty)$, namely $f(\xi) \in L(-\infty, \infty) \to g(\eta) = \eta^{-1} f(\log \eta) \in L(0, \infty)$. Substituting $\xi = \log \eta$ in the following integrals we obtain

$$\int_{-\infty}^\infty |f(\xi)|\, d\xi = \int_0^\infty |g(\eta)|\, d\eta,$$

$$\int_{-\infty}^\infty \exp(i\beta\xi) f(\xi)\, d\xi = \int_0^\infty \eta^{i\beta} g(\eta)\, d\eta,$$

and

$$\int_{-\infty}^\infty f(\xi - \tau) x(\xi)\, d\xi = \frac{1}{\sigma} \int_0^\infty g(\eta/\sigma) y(\eta)\, d\eta,$$

where $\tau = \log \sigma$ and $y(\eta) = x(\log \eta)$. It is now clear that Theorem 4.21.3 is simply a reformulation of Theorem 4.21.2.

References. Bochner and Phillips [1], Gelfand [6], Gelfand and Šilov [1], Raikov [1], Segal [1], Šreĭder [1], Stone [4], Wiener [3].

6. COMMUTATIVE (B*)-ALGEBRAS

4.22. Representation theory for commutative (B*)-algebras. In this paragraph we suppose that $\mathfrak{B}$ is a commutative complex (B*)-algebra with a unit element. The representation theory for such algebras reaches the ultimate in simplicity as is evident from the following theorem due to I. Gelfand and M. Neumark [1].

THEOREM 4.22.1. *Let* $\mathfrak{B}$ *be a commutative complex* (B^*)*-algebra with a unit element. Then* $\mathfrak{B}$ *is isomorphic and isometric to* $C(\mathfrak{M})$, *the algebra of all continuous functions on the compact Hausdorff space of maximal ideals* $\mathfrak{M}$. *Under this mapping the* $(^*)$*-operation goes into the conjugation operation.*

PROOF. In a commutative algebra each element is normal. Hence Theorem 4.12.1 applies to each $x \in \mathfrak{B}$ and we obtain $r(x) = \| x \|$. Since $r(x) = 0$ implies $x = \theta$, we see that θ is the only quasi-nilpotent element in $\mathfrak{B}$. Consequently $\mathfrak{B}$ is semi-simple and the mapping $x \to x(\mathfrak{m}) \in C(\mathfrak{M})$ is an isomorphism (see the corollary to Theorem 4.15.2). However $r(x) = \| x \|$ also implies that the mapping is an isometry. By Theorem 4.12.2, $\mathfrak{B}$ is an (A*)-algebra. Making use of Theorem 4.19.1 we see that $x^*(\mathfrak{m}) = \overline{x(\mathfrak{m})}$. Further, according to Theorem 4.19.4 the image of $\mathfrak{B}$ is dense in $C(\mathfrak{M})$. Since the mapping is an isometry and since $\mathfrak{B}$ is complete, it follows that the image of $\mathfrak{B}$ must be all of $C(\mathfrak{M})$.

PROBLEM (M. H. STONE [4]). For each closed ideal $\mathfrak{i}$ in $C(\mathfrak{M})$ there exists a closed subset F of $\mathfrak{M}$ such that

$$\mathfrak{i} \equiv [x; x(\mathfrak{m}) = 0 \quad \text{for all} \quad \mathfrak{m} \in F]$$

and conversely.

An element x is said to be *real* if the range of $x(\mathfrak{m})$ on $\mathfrak{M}$ is real valued. If x is self-adjoint then $x(\mathfrak{m}) = x^*(\mathfrak{m}) = \overline{x(\mathfrak{m})}$ and hence x is real. Conversely, if x is real then these same relations hold so that both x and x^* map into $x(\mathfrak{m})$. Since the mapping is one-to-one it follows that $x = x^*$. Consequently the set of all self-adjoint elements, $\mathfrak{A}$, constitutes the class of real elements of $\mathfrak{B}$. Further $\mathfrak{A}$ maps isomorphically and isometrically onto $R(\mathfrak{M})$. We may now introduce a partial ordering into $\mathfrak{A}$.

DEFINITION 4.22.1. *For $x, y \in \mathfrak{A}$ we say that $x > y$ if $x(\mathfrak{m}) - y(\mathfrak{m}) \geqq 0$ for all $\mathfrak{m} \in \mathfrak{M}$.*

It is clear that $R(\mathfrak{M})$ and hence $\mathfrak{A}$ becomes a lattice under this ordering.

The idempotent elements of $\mathfrak{B}$ are of special interest; they are often referred to as *projections*. In particular θ and e are projections. More generally if $x^2 = x$, then $[x(\mathfrak{m})]^2 = x(\mathfrak{m})$ and hence $x(\mathfrak{m})$ can assume only the values zero or one; clearly x is then self-adjoint. Further since $x(\mathfrak{m})$ is continuous on $\mathfrak{M}$, it must assume the value one on a subset of $\mathfrak{M}$ which is both closed and open, that is clopen. Conversely, let F be a clopen subset of $\mathfrak{M}$. Then $f(m) = 1$ on F and $= 0$ elsewhere defines a continuous function on $\mathfrak{M}$. Hence there exists an $x \in \mathfrak{B}$ which maps into $f(\mathfrak{m})$ and since $[f(\mathfrak{m})]^2 = f(\mathfrak{m})$ it follows that $x^2 = x$. Thus there is a one-to-one correspondence between the projections in $\mathfrak{B}$ and the clopen subsets of $\mathfrak{M}$. The following theorem which is due to M. H. Stone [7] relates the lattice properties of $R(\mathfrak{M})$ with the clopen subsets of $\mathfrak{M}$.

THEOREM 4.22.2. *Let $R(\mathfrak{M})$ be the set of all real-valued continuous functions on a compact Hausdorff space $\mathfrak{M}$. If $R(\mathfrak{M})$ is a complete lattice under the ordering*

$$(f \vee \theta)(\mathfrak{m}) = \max[f(\mathfrak{m}), 0],$$

then the closure of each open set is clopen. More generally, to each Borel set E there corresponds a unique clopen set $\gamma(E)$ which differs from E only in a set of the first category. This correspondence is a set-algebraic homomorphism in the following *sense*:

$$\gamma(\bigcup_{n=1}^{\infty} E_n) = \overline{\bigcup_{n=1}^{\infty} \gamma(E_n)}. \tag{4.22.1}$$

PROOF. Let G be an open subset of $\mathfrak{M}$ and let $\mathfrak{p} \in G$. Since $\mathfrak{M}$ is normal there exists a continuous function $f(\mathfrak{p}, G; \mathfrak{m})$ defined on $\mathfrak{M}$ such that $0 \leqq f(\mathfrak{p}, G; \mathfrak{m}) \leqq 1$, $f(\mathfrak{p}, G; \mathfrak{p}) = 1$, and $f(\mathfrak{p}, G; \mathfrak{m}) = 0$ on $\mathfrak{M} \ominus G$. We define

$$f = \vee_{\mathfrak{p} \in G} f(\mathfrak{p}, G; \cdot).$$

It is clear from the properties of $f(\mathfrak{p}, G; \mathfrak{m})$ that $0 \leqq f(\mathfrak{m}) \leqq 1$ and that $f(\mathfrak{m}) \equiv 1$ on G. Now for $\mathfrak{p} \in G$ and $\mathfrak{q} \in \mathfrak{M} \ominus \bar{G}$ we see that $f(\mathfrak{p}, G; \mathfrak{m}) \leqq 1 - f(\mathfrak{q}, \mathfrak{M} \ominus \bar{G}; \mathfrak{m})$ and consequently $f(\mathfrak{m}) \leqq 1 - f(\mathfrak{q}, \mathfrak{M} \ominus \bar{G}; \mathfrak{m})$; thus $f(\mathfrak{q}) = 0$. Since $f \in R(\mathfrak{M})$ is continuous, it follows that f is the characteristic function of $\bar{G}$ and hence that $\bar{G}$ is clopen. We note that $\bar{G} - G$ is nowhere dense and therefore is a set of the first category.

Let $\mathfrak{R}$ denote the class of all subsets of $\mathfrak{M}$ which differ from clopen sets by sets of the first category. It is clear that if $E \in \mathfrak{R}$ then $\mathfrak{M} \ominus E$ likewise belongs to $\mathfrak{R}$. Suppose further that the sets $\{E_n ; n = 1, 2, \cdots\}$ belong to $\mathfrak{R}$. Then there exist clopen sets F_n such that E_n differs from F_n only on a set of the first category. It follows that $\bigcup_n E_n$ can differ from $\bigcup_n F_n$ only on a set of the first category. Now $\bigcup_n F_n$ is open and by the previous paragraph, $F = \overline{\bigcup_n F_n}$ is clopen and differs from $\bigcup_n F_n$ only on a set of the first category. Thus $\bigcup_n E_n$ differs from F only on a set of the first category and therefore $\bigcup_n E_n$ belongs to $\mathfrak{R}$. Thus $\mathfrak{R}$ is a σ-algebra of sets containing all of the open sets. It follows that $\mathfrak{R}$ contains the σ-algebra generated by the open sets, namely the Borel sets. In order to establish the uniqueness of the correspondence γ defined in the statement of the theorem, we suppose that F_1 and F_2 are two clopen sets which differ from a given Borel set only by sets of the first category. Then F_1 and F_2 can differ from each other only on a set of the first category. On the other hand, the sets $F_1 \ominus F_2$ and $F_2 \ominus F_1$ are both open. Now a non-vacuous open subset of a compact Hausdorff space is necessarily of the second category; consequently $F_1 = F_2$.

It is clear that $\gamma(\emptyset) = \emptyset$ and that $\gamma(\mathfrak{M}) = \mathfrak{M}$. Further if E_1 and E_2 are disjoint Borel sets, $\gamma(E_1)$ and $\gamma(E_2)$ can have only a set of the first category in common. However the common part of two clopen sets is open and this implies that $\gamma(E_1)$ and $\gamma(E_2)$ are themselves disjoint. Finally, as we have already shown in the previous paragraph, the relation (4.22.1) is valid. This concludes the proof of the theorem.

For a given Borel set E, we now set $p(\mathfrak{m}; E)$ equal to the characteristic function of the set $\gamma(E)$. It is clear that $p(E) \equiv p(\cdot\,; E)$ belongs to $R(\mathfrak{M})$ and is idempotent. Thus $p(E)$ is a projection. As a function of the Borel subsets of $\mathfrak{M}$, $p(E)$ satisfies the following properties:

(4.22.2)

(a) $p(\emptyset) = \theta, \quad p(\mathfrak{M}) = e$;

(b) If $E_1 \cap E_2 = \emptyset$, then $p(E_1)p(E_2) = \theta$;

(c) Given $\{E_n\}$, then $p(\bigcup_n E_n) = \bigvee_n p(E_n)$.

In the usual terminology, $[p(E)]$ *is a resolution of the identity* relative to the Borel subsets of $\mathfrak{M}$.

It is now an easy matter to obtain the usual integral representation for elements of the complete lattice $R(\mathfrak{M})$. Given $f \in R(\mathfrak{M})$, let $\alpha = \min f(\mathfrak{m})$ and $\beta = \max f(\mathfrak{m})$. Further set $E_\lambda = [\mathfrak{m}; f(\mathfrak{m}) \leqq \lambda]$ and define $p(\lambda) = p(E_\lambda)$. It is clear from (4.22.2) that

$$\begin{aligned} &(\alpha)\ p(\lambda) = \theta \quad \text{for} \quad \lambda < \alpha, \qquad p(\lambda) = e \quad \text{for} \quad \lambda \geqq \beta; \\ &(\beta)\ p(\lambda)p(\mu) = p(\lambda) \quad \text{for} \quad \lambda \leqq \mu; \\ &(\gamma)\ p(\lambda) = \bigwedge_{\mu > \lambda} p(\mu). \end{aligned} \tag{4.22.3}$$

A one-parameter family of projections with these properties is also called a resolution of the identity. Finally let π represent a subdivision of the interval

$$[\alpha - \delta, \beta]\colon \lambda_0 = \alpha - \delta < \lambda_1 < \lambda_2 < \cdots < \lambda_n = \beta,$$

for fixed $\delta > 0$, together with numbers $\{\lambda_i'\}$, $\lambda_{i-1} \leqq \lambda_i' \leqq \lambda_i$; and set $|\pi| = \max_i |\lambda_i - \lambda_{i-1}|$. The function

$$g_\pi(\mathfrak{m}) = \lambda_i' \quad \text{on} \quad E_{\lambda_i} \ominus E_{\lambda_{i-1}}, \qquad i = 1, 2, \cdots, n,$$

clearly differs from

$$h_\pi = \sum_{i=1}^{n} \lambda_i'[p(\lambda_i) - p(\lambda_{i-1})]$$

on at most a set of the first category. On the other hand

$$\sup_{\mathfrak{m} \in \mathfrak{M}} |f(\mathfrak{m}) - g_\pi(\mathfrak{m})| \leqq |\pi| .$$

Hence except for a set of the first category $|f(\mathfrak{m}) - h_\pi(\mathfrak{m})| \leqq |\pi|$. However both f and h_π are continuous so that the set $[\mathfrak{m}; |f(\mathfrak{m}) - h_\pi(\mathfrak{m})| > |\pi|]$ is both open and of the first category. It follows that this set is vacuous and hence that

$$\| f - h_\pi \| \leqq |\pi| .$$

Passing to the limit as $|\pi| \to 0$, we obtain $f = \int_{\alpha-\delta}^{\beta} \lambda \, dp(\lambda)$, where the integral is a Riemann-Stieltjes integral which converges in the norm topology (cf. Theorem 3.3.2). Finally it is evident from condition (α) of (4.22.3) that the result is independent of $\delta > 0$ so that we may write

$$f = \int_{\alpha-}^{\beta} \lambda \, dp(\lambda). \tag{4.22.4}$$

4.23. Operator algebras on Hilbert space. One of the most fruitful directions of research in (B)-algebras, especially for noncommutative algebras, deals with certain operator algebras on a Hilbert space. We shall limit ourselves to just a fragment of this field, developing only what will be needed for Chapter XXII. Adequate background material may be found in sections 1.12 and 2.8.

The Hilbert space $\mathfrak{H}$ is its own adjoint space and consequently the adjoint of a linear bounded operator is again an operator on $\mathfrak{H}$. Since the general linear bounded functional on $\mathfrak{H}$ can be expressed in terms of the inner product, the adjoint operator A^* to the operator A satisfies and is determined by the relation

$$(Ax, y) = (x, A^*y), \qquad x, y \in \mathfrak{H}. \tag{4.23.1}$$

It is easy to see that this defines a suitable (*)-operation. In fact $(Ax, y) = (x, A^*y) = ((A^*)^*x, y)$ implies that $A = A^{**}$;

$$\begin{aligned}((\alpha A + \beta B)x, y) &= \alpha(Ax, y) + \beta(Bx, y) = (x, \bar{\alpha}A^*y) + (x, \bar{\beta}B^*y) \\ &= (x, (\bar{\alpha}A^* + \bar{\beta}B^*)y)\end{aligned}$$

implies that $(\alpha A + \beta B)^* = \bar{\alpha}A^* + \bar{\beta}B^*$; and

$$(ABx, y) = (Bx, A^*y) = (x, B^*A^*y)$$

implies that $(AB)^* = B^*A^*$.

THEOREM 4.23.1. *If A is a self-adjoint linear bounded operator, then*

$$\sup_{\|x\|=1} |(Ax, x)| = \|A\|. \tag{4.23.2}$$

PROOF. Let $\alpha = \sup_{\|x\|=1} |(Ax, x)|$. By the Schwarz inequality we have $|(Ax, x)| \leqq \|Ax\| \|x\| \leqq \|A\| \|x\|^2$ so that $\alpha \leqq \|A\|$. In order to show that $\|A\| \leqq \alpha$, we make use of the relation

$$2(Ax, y) + 2(Ay, x) = (A(x + y), (x + y)) - (A(x - y), (x - y)).$$

Hence

$$4 |\mathfrak{R}[(Ax, y)]| \leqq \alpha(\|x + y\|^2 + \|x - y\|^2) = 2\alpha(\|x\|^2 + \|y\|^2).$$

Setting

$$y = Ax/\|Ax\|$$

we finally obtain

$$\|A\| = \sup_{\|x\|=1} \|Ax\| = \sup_{\|x\|=1} \left(Ax, \frac{Ax}{\|Ax\|}\right) \leqq \sup_{\|x\|=1} \alpha/2\,(\|x\|^2 + 1) = \alpha.$$

We say that an *algebra* is *self-adjoint* if it is closed relative to the adjoint operation.

DEFINITION 4.23.1. *A (C^*)-algebra (or (W^*)-algebra) is a self-adjoint complex algebra of bounded linear operators on a Hilbert space, closed in the uniform (or weak) operator topology.*

Since closure in the weak operator topology implies closure in the uniform operator topology, it is clear that a (W*)-algebra is also a (C*)-algebra.

THEOREM 4.23.2. *A (C^*)-algebra satisfies the postulates of a (B^*)-algebra.*

PROOF. We have already shown that a self-adjoint algebra is a (*)-algebra. It remains only to prove that $\|AA^*\| = \|A\| \|A^*\|$. Now AA^* is a self-adjoint operator. Hence by the previous theorem

$$\|AA^*\| = \sup_{\|x\|=1} |(AA^*x, x)| = \sup_{\|x\|=1} |(A^*x, A^*x)| = \|A^*\|^2.$$

However for operators in general we have $\|A\| = \|A^*\|$ (see Theorem 2.11.8).

We shall again denote by $\mathfrak{A}$ the set of self-adjoint elements in a given algebra of operators on $\mathfrak{H}$. There is a natural way of ordering the elements of $\mathfrak{A}$, namely $A > B$ is taken to mean that $(Ax, x) \geqq (Bx, x)$ for all $x \in \mathfrak{H}$. On the other hand when $\mathfrak{B}$ is a commutative (C*)-algebra with a unit element, the representation theory of the preceding section also furnished us with an ordering for the elements of $\mathfrak{A}$. We shall show that these two orderings of $\mathfrak{A}$ are equivalent. Theorems 4.23.3 and 4.23.4 are due to M. H. Stone [5, 7].

THEOREM 4.23.3. *Let $\mathfrak{B}$ be a commutative (C*)-algebra with a unit element. Then $\mathfrak{B}$ is isomorphic and isometric with $C(\mathfrak{M})$. Under this mapping the (*)-operation goes over into the conjugation operation and $A > \Theta$ is equivalent with $A(\mathfrak{m}) \geqq 0$ for all $\mathfrak{m} \in \mathfrak{M}$. The self-adjoint elements of $\mathfrak{B}$ form a lattice.*

PROOF. Most of the assertion follows directly from Theorems 4.22.1 and 4.23.2. It is however necessary to show that the two ordering relations are equivalent. First suppose that $A(\mathfrak{m}) \geqq 0$. Then $f(\mathfrak{m}) \equiv [A(\mathfrak{m})]^{1/2}$ belongs to $R(\mathfrak{M})$ so that there exists a $B \in \mathfrak{A}$ such that $B \to f(\mathfrak{m})$. It follows that $B^2 \to A(\mathfrak{m})$ and hence that $B^2 = A$. But then $(Ax, x) = (B^2x, x) = (Bx, Bx) \geqq 0$, which shows that $A > \Theta$. On the other hand suppose that $A > \Theta$ and that $A(\mathfrak{m})$ takes on negative values. Now $0 \leqq (Ax, x) \leqq \| A \| \| x \|^2$ so that

$$-\tfrac{1}{2} \| A \| \| x \|^2 \leqq (Ax, x) - \tfrac{1}{2} \| A \| \| x \|^2 \leqq \tfrac{1}{2} \| A \| \| x \|^2.$$

Since $(Ax, x) - \frac{1}{2} \| A \| \| x \|^2 = ([A - \frac{1}{2} \| A \| I]x, x)$, Theorem 4.23.1 implies that

$$\| [A - \tfrac{1}{2} \| A \| I] \| \leqq \tfrac{1}{2} \| A \| .$$

However $A - \frac{1}{2} \| A \| I \to A(\mathfrak{m}) - \frac{1}{2} \| A \|$ and if $A(\mathfrak{m})$ takes on negative values we have

$$\sup_{\mathfrak{m} \in \mathfrak{M}} | A(\mathfrak{m}) - \tfrac{1}{2} \| A \| | > \tfrac{1}{2} \| A \| .$$

This contradicts the norm preserving property of the mapping. We conclude, therefore, that $A(\mathfrak{m}) \geqq 0$ if and only if $A > \Theta$. Finally since $R(\mathfrak{M})$ is clearly a lattice the same is true of $\mathfrak{A}$.

We shall require the following lemma, the proof of which goes back to F. Riesz [4].

LEMMA 4.23.1. *Let $\mathfrak{B}$ be a commutative (C*)-algebra and let $[\pi]$ be a directed set. Suppose that $[B_\pi] \subset \mathfrak{B}$ is a non-increasing function of π with $B_\pi > \Theta$ for all π. Then $\lim_\pi B_\pi x$ exists for each $x \in \mathfrak{H}$.*

PROOF. If $\pi_1 < \pi_2$, then $B_{\pi_1} > B_{\pi_2} > \Theta$ and it is clear from the above representation theory that $(B_{\pi_1} - B_{\pi_2})B_\pi > \Theta$ for arbitrary π. Consequently for $\pi_0 < \pi_1, \pi_2 < \pi_{12}$ and $x \in \mathfrak{H}$ we get

$$(B^2_{\pi_0}x, x) \geqq (B_{\pi_1}B_{\pi_2}x, x) \geqq (B^2_{\pi_{12}}x, x) \geqq 0.$$

Thus $(B^2_\pi x, x)$ is non-increasing in π and bounded from below so that the limit exists; further we see that $\lim_\pi(B^2_\pi x, x) = \lim_{\pi_1, \pi_2}(B_{\pi_1}B_{\pi_2}x, x)$. It follows that

$$\| (B_{\pi_1} - B_{\pi_2})x \|^2 = ((B_{\pi_1} - B_{\pi_2})^2 x, x)$$
$$= (B^2_{\pi_1} x, x) - 2(B_{\pi_1} B_{\pi_2} x, x) + (B^2_{\pi_2} x, x) \to 0,$$

which is the assertion of the lemma.

THEOREM 4.23.4. *The self-adjoint elements of a commutative* (W^*)*-algebra with identity form a complete lattice.*

PROOF. According to the preceding theorem, the set of self-adjoint elements in $\mathfrak{B}$, which we denote by $\mathfrak{A}$, form a lattice. Suppose that a subset $[A_\rho] \subset \mathfrak{A}$ is bounded above by $B \in \mathfrak{A}$. We now show that $[A_\rho]$ has a least upper bound in $\mathfrak{A}$. To this end let π denote a finite subset of the ρ's and let $\pi_1 < \pi_2$ mean set inclusion. Under this convention the π's form a directed set. We define

$$A_\pi = \bigvee_{\rho \in \pi} A_\rho$$

and set $B_\pi = B - A_\pi$. Then for $\pi_1 < \pi_2$ we have $A_{\pi_1} < A_{\pi_2} < B$ so that $B_{\pi_1} > B_{\pi_2} > \Theta$. Hence the previous lemma applies to $[B_\pi]$. Consequently $B_\pi x$, and therefore $A_\pi x$, converges to a limit for each $x \in \mathfrak{H}$. Set $\lim_\pi A_\pi x \equiv A_0 x$. It is clear that A_0 is self-adjoint and further that $A_0 \in \mathfrak{B}$ since by hypothesis $\mathfrak{B}$ is weakly and *a fortiori* strongly closed. Finally we show that $A_0 = \bigvee A_\rho$. In fact if $\rho \in \pi$, then $(A_\rho x, x) \leqq (A_\pi x, x) \leqq (A_0 x, x)$ so that $A_\rho < A_0$ for all ρ. Further $(A_0 x, x) = \lim_\pi (A_\pi x, x) \leqq (Bx, x)$ and therefore $A_0 < B$. Since B was any upper bound in $\mathfrak{A}$ for the set $[A_\rho]$, it follows that A_0 is actually the least upper bound of this set in $\mathfrak{A}$.

As an application of the foregoing theory we shall obtain the integral representation for a self-adjoint operator A. Let $\mathfrak{B}'$ be the algebra of all polynomials in A and let $\mathfrak{B}$ denote the weak closure of $\mathfrak{B}'$. Then $\mathfrak{B}$ is a weakly closed commutative subalgebra of $\mathfrak{E}(\mathfrak{H})$. Since $\mathfrak{B}'$ is self-adjoint, we see from the relation $(x, (A - A_1)y) = ((A^* - A_1^*)x, y)$ together with the definition of a neighborhood in the weak operator topology that $\mathfrak{B}$ contains B whenever it contains B. Hence $\mathfrak{B}$ is a commutative (W*)-algebra with identity and the set of self-adjoint elements in $\mathfrak{B}$ maps onto $R(\mathfrak{M})$. By Theorems 4.23.3 and 4.23.4 it follows that $R(\mathfrak{M})$ is a complete lattice. Let $A \to A(\mathfrak{m})$ and let $\alpha = \inf_{\mathfrak{m} \in \mathfrak{M}} A(\mathfrak{m})$, $\beta = \sup_{\mathfrak{m} \in \mathfrak{M}} A(\mathfrak{m})$. Again we set $E_\lambda = [\mathfrak{m}; A(\mathfrak{m}) \leqq \lambda]$. If $P(\lambda)$ is chosen so that $P(\lambda)(\mathfrak{m})$ is the characteristic function of the set $\gamma(E_\lambda)$, then $P(\lambda)$ is a projection operator and the one-parameter family of operators $[P(\lambda)]$ is a resolution of the identity satisfying the condition (4.22.3). Finally the discussion leading up to (4.22.4) shows that

$$A = \int_{\alpha-}^{\beta} \lambda \, d\, P(\lambda). \tag{4.23.3}$$

References. Gelfand and Neumark [1], F. Riesz [4], Stone [4, 5, 7].

CHAPTER V

ANALYSIS IN A BANACH ALGEBRA

5.1. Orientation. In the present chapter we continue our study of Banach algebras. Whereas the previous chapter was mainly concerned with algebraic aspects, the emphasis here is on the analysis, both with regard to problems and method.

The central idea of the chapter is the operational calculus defined for any (B)-algebra, $\mathfrak{B}$. The operational calculus may be described as follows: Let Δ be an open subset of the complex plane and let $\mathfrak{G}(\Delta)$ be the set of all $x \in \mathfrak{B}$ such that $\sigma(x) \subset \Delta$. Further let $H(\Delta)$ be the algebra of all functions $f(\lambda)$, locally holomorphic in Δ. Then there exists a unique isomorphic mapping $f(\lambda) \to f(x)$ from $H(\Delta)$ onto an algebra of functions locally analytic on $\mathfrak{G}(\Delta)$ to $\mathfrak{B}$ for which $1 \to e$, $\lambda \to x$, and which is continuous in a certain sense. An indication of the intimate relation existing between x and $f(x)$ is given by the spectral mapping theorem which asserts that $\sigma[f(x)] = f[\sigma(x)]$.

The operational calculus has important applications to the study of idempotents and to the spectral resolution of elements in $\mathfrak{B}$. These results apply in particular to the elements of the (B)-algebra of endomorphisms $\mathfrak{E}(\mathfrak{X})$. However they do not apply to the closed linear operators belonging to $\mathfrak{O}(\mathfrak{X})$. In order to extend the calculus and its applications so that they apply to closed linear operators, it is first necessary to make a thorough study of the resolvents for such operators. The desired extensions then require little more than a reformulation of the previous theory.

The paragraph headings are: *The Operational Calculus*, *Pseudo-Resolvents*, and *The Extended Operational Calculus*. References are to be found at the end of each paragraph.

1. The Operational Calculus

5.2. Extension of holomorphic scalar functions. Among the analytic functions on a complex Banach algebra $\mathfrak{B}$ to itself an important class is formed by *the functions which become analytic in the classical sense on the subalgebra of complex numbers*. This class of functions has been studied from different points of view by N. Dunford, I. Gelfand, E. R. Lorch, and A. E. Taylor. We shall attempt to fit the theory of these functions into the general theory of analytic functions on

$\mathfrak{B}$ to $\mathfrak{B}$. Throughout this paragraph $\mathfrak{B}$ will denote a complex Banach algebra with a unit element e.

The subject may be looked upon as an *extension problem*. The algebra C of complex numbers is extended to a complex Banach algebra $\mathfrak{B}$ so that C becomes embedded in $\mathfrak{B}$. Is it possible to simultaneously extend holomorphic functions $f(\lambda)$ on C to analytic functions $f(x)$ on $\mathfrak{B}$ in such a manner that $f(\lambda e) = f(\lambda)e$? In this formulation the question does not have a unique solution inasmuch as an analytic function $f(x)$ on $\mathfrak{B}$ to $\mathfrak{B}$ may vanish for all x in Ce without vanishing identically.

As an illustration, let $\mathfrak{B}$ be a complex matrix algebra, A a matrix such that $A^k = \Theta$ but $A^{k-1} \neq \Theta$, then any nontrivial product of k factors A and n factors X is a homogeneous polynomial in X of degree n which vanishes when $X = \lambda E$ without vanishing identically and the extension problem becomes indeterminate.

It is possible, however, to define a unique *principal extension* in a perfectly natural manner and the resulting class of analytic functions has simple properties. Actually there are several different procedures which lead to the same result. We shall start with the local point of view which is less *ad hoc* than the methods in the large.

Definition 5.2.1. *If $f(\lambda)$ is holomorphic in the circle $|\lambda - \lambda_0| < \rho$, then*

$$f(x) = \sum_{n=0}^{\infty} \frac{f^{(n)}(\lambda_0)}{n!} (x - \lambda_0 e)^n \tag{5.2.1}$$

is by definition the principal extension of $f(\lambda)$ in the sphere $\| x - \lambda_0 e \| < \rho$.

Theorem 5.2.1. *If $f(\lambda)$ is locally holomorphic in an open set Δ of the complex plane and if $\mathfrak{D}_\Delta$ is the union of all spheres $\| x - \lambda_0 e \| < \rho$ such that the circle $|\lambda - \lambda_0| < \rho$ is in Δ, then the principal extension of $f(\lambda)$ is uniquely determined by Definition 5.2.1 in the domain $\mathfrak{D}_\Delta$ and is a locally analytic function of x in $\mathfrak{D}_\Delta$.*

Proof. We denote the power series in formula (5.2.1) by $f(x; \lambda_0 e)$. It is clear that the series converges for $\| x - \lambda_0 e \| < \rho$ and that the terms are analytic functions of x. Hence by Theorem 3.18.1 it follows that $f(x; \lambda_0 e)$ is analytic in its sphere of definition. Thus we have merely to verify that the various power series $f(x; \alpha e)$, $\alpha \in \Delta$, define the same analytic function $f(x)$. Suppose that $x_0 \in \mathfrak{D}_\Delta$ lies in the spheres of definition of $f(x; \lambda_1 e)$ as well as of $f(x; \lambda_2 e)$ so that $\| x_0 - \lambda_1 e \| < \rho_1$ and $\| x_0 - \lambda_2 e \| < \rho_2$. Then $|\lambda_1 - \lambda_2| < \rho_1 + \rho_2$ and consequently the two circles $|\lambda - \lambda_1| < \rho_1$ and $|\lambda - \lambda_2| < \rho_2$ overlap. Let λ_0 be a point common to both of these circles. The classical argument based on the rearrangement of power series now shows that $f(x; \lambda_1 e) \equiv f(x; \lambda_0 e)$ in the sphere $\| x - \lambda_0 e \| < \rho_1 - |\lambda_1 - \lambda_0|$ and $f(x; \lambda_2 e) \equiv f(x; \lambda_0 e)$ in a concentric sphere. Thus $f(x; \lambda_1 e) \equiv f(x; \lambda_2 e)$ in some sphere about $\lambda_0 e$ and hence by Theorem 3.16.4, these two functions are identical in the domain common to their two spheres of definition and in particular $f(x_0; \lambda_1 e) = f(x_0; \lambda_2 e)$.

THEOREM 5.2.2. *Let $x \in \mathfrak{D}_\Delta$ and choose a $\lambda_0 \in \Delta$ such that the circle $\Gamma_x : |\lambda - \lambda_0| = \| x - \lambda_0 e \| + \epsilon$, $\epsilon > 0$, lies in Δ. Then*

$$(5.2.2) \qquad f(x) = \frac{1}{2\pi i} \int_{\Gamma_x} f(\lambda) R(\lambda; x)\, d\lambda,$$

where $f(x)$ is the principal extension of $f(\lambda)$.

PROOF. The definition of $\mathfrak{D}_\Delta$ shows that it is always possible to find such a circle Γ_x. We apply formula (4.8.4), choosing $a = \lambda_0 e$. Since

$$R(\lambda; \lambda_0 e) = (\lambda - \lambda_0)^{-1} e,$$

the formula simplifies to

$$(5.2.3) \quad R(\lambda; x) = \sum_{n=0}^{\infty} (x - \lambda_0 e)^n (\lambda - \lambda_0)^{-n-1}, \qquad \| x - \lambda_0 e \| < |\lambda - \lambda_0|.$$

Substituting this expression into (5.2.2), we see that the integral equals $f(x; \lambda_0 e) = f(x)$ and the theorem is proved.

Formula (5.2.3) shows that $\sigma(x)$, the spectrum of x, lies in the circle $|\lambda - \lambda_0| \leqq \| x - \lambda_0 e \|$; that is, $\sigma(x)$ is interior to Δ when $x \in \mathfrak{D}_\Delta$. But for the existence of the integral in (5.2.2) it is not essential that we integrate along the circle Γ_x; any closed contour in Δ surrounding $\sigma(x)$ will do. This indicates that the integral has a meaning as long as $\sigma(x)$ lies in Δ and this may very well happen for x outside of $\mathfrak{D}_\Delta$. Before we can proceed to a study of the resolvent integral, a few observations of mixed function-theoretic and topological nature are required.

Let Φ be a compact subset of the complex λ-plane and let $f(\lambda)$ be a function locally holomorphic in an open set Δ containing Φ. Then we can find an open set Ω with the following properties: (i) $\Phi \subset \Omega \subset \Delta$, (ii) Ω has a finite number of components Ω_μ, (iii) each Ω_μ is bounded by a finite number of simple closed rectifiable curves $\Gamma_{\mu\nu}$, and (iv) Ω has a positive distance from the boundary of Δ. In fact let δ be the minimum distance from Φ to the boundary of Δ; clearly $\delta > 0$. Following a suggestion due to W. Gustin, we cover the plane with an hexagonal mesh, each hexagon being of diameter less than $\delta/2$. It is now easy to show that the interior of the union of all (closed) hexagons containing points of Φ has the requisite properties of Ω. We assign a positive orientation to each $\Gamma_{\mu\nu}$ in the usual manner and let $\Gamma = \bigcup \Gamma_{\mu\nu}$ be the boundary of Ω, the orientation of Γ being induced by that of the $\Gamma_{\mu\nu}$. We call Γ *an oriented envelope of Φ with respect to* $f(\lambda)$. For $\lambda \in \Omega$ we have

$$f(\lambda) = \frac{1}{2\pi i} \int_\Gamma \frac{f(\zeta)}{\zeta - \lambda}\, d\zeta.$$

If $\lambda \in \Phi$, the integral is independent of the choice of Γ. The proof can be reduced to the classical elementary case by a somewhat laborious discussion of the different topological possibilities.

We shall also need the following theorem concerning spectra.

THEOREM 5.2.3. *The spectrum of x is an upper semi-continuous function of x.*

REMARK. We say that $\sigma(x)$ is upper semi-continuous at $x = a$ if, given any open set Ω containing $\sigma(a)$, there exists an $\epsilon > 0$ such that $\sigma(x) \subset \Omega$ for all $\| x - a \| < \epsilon$.

PROOF. Let a be given and let Ω be an open set containing $\sigma(a)$. Set $\tilde{\Omega}$ equal to the complement of Ω relative to the extended complex plane. According to Theorems 4.7.1 and 4.7.2, $R(\lambda; a)$ is locally holomorphic and hence continuous in $\tilde{\Omega}$. Since $\tilde{\Omega}$ is compact in the topology of the extended plane, there exists an $M(a, \Omega)$ such that $\| R(\lambda; a) \| \leqq M$ in $\tilde{\Omega}$. By formula (4.8.4), $R(\lambda; x)$ will exist for every finite $\lambda \in \tilde{\Omega}$ provided that $\| x - a \| < 1/M = \epsilon$. It follows that for any such x we have $\rho(x) \supset \tilde{\Omega}$ and hence $\sigma(x) \subset \Omega$. This completes the proof.

We come now to the main extension theorem.

THEOREM 5.2.4. *Let $f(\lambda)$ be locally holomorphic in the open set Δ. Let $\mathfrak{G}(\Delta)$ be the open set of points $x \in \mathfrak{B}$ such that $\sigma(x) \subset \Delta$. For $x \in \mathfrak{G}(\Delta)$ define*

$$f(x) = \frac{1}{2\pi i} \int_{\Gamma_x} f(\zeta) R(\zeta; x)\, d\zeta, \tag{5.2.4}$$

where Γ_x is any oriented envelope of $\sigma(x)$ with respect to $f(\lambda)$. Then $f(x)$ is locally analytic in $\mathfrak{G}(\Delta)$ and coincides with the principal extension of $f(\lambda)$ in $\mathfrak{D}_\Delta$.

PROOF. It follows from the preceding theorem that $\mathfrak{G}(\Delta)$ is an open subset of $\mathfrak{B}$. Likewise it follows from Theorem 5.2.2 and the subsequent remarks that $\mathfrak{G}(\Delta) \supset \mathfrak{D}_\Delta$ and that the integral represents the principal extension of $f(\lambda)$ in $\mathfrak{D}_\Delta$. It remains to show that $f(x)$ is locally analytic in $\mathfrak{G}(\Delta)$. For this purpose, let $a \in \mathfrak{G}(\Delta)$ and choose an open sphere $\mathfrak{S}(a)$: $\| x - a \| < \rho$ so small that Φ, the closure of $\bigcup_x \sigma(x)$, $x \in \mathfrak{S}(a)$, is contained in Δ. For each $x \in \mathfrak{S}(a)$ we may replace Γ_x by Γ, a fixed oriented envelope of Φ with respect to $f(\lambda)$. Since $R(\lambda; a)$ is holomorphic on Γ, there exists a finite quantity $M(a)$ with $\| R(\lambda; a) \| \leqq M(a)$ everywhere on Γ. Formula (4.8.4) shows that for $\lambda \in \Gamma$

$$R(\lambda; x) = R(\lambda; a) \sum_{n=0}^{\infty} [(x - a)R(\lambda; a)]^n,$$

the series being absolutely convergent for $\| x - a \| \leqq \rho_0 < \min \{\rho, 1/M(a)\}$, the convergence being uniform with respect to x in this sphere as well as with respect to λ on Γ. Termwise integration gives

$$f(x) = \sum_{n=0}^{\infty} \frac{1}{2\pi i} \int_{\Gamma} f(\zeta) R(\zeta; a)[(x - a)R(\zeta; a)]^n\, d\zeta. \tag{5.2.5}$$

This is an abstract power series in $(x - a)$ which is uniformly convergent when $\| x - a \| \leqq \rho_0$. Since the individual terms are clearly analytic, it follows from Theorem 3.18.1 that $f(x)$ is itself analytic in this sphere and consequently it is locally analytic in $\mathfrak{G}(\Delta)$.

Let $\mathfrak{D}(\Delta)$ be the union of all components of $\mathfrak{G}(\Delta)$ containing points of Δe.

Formula (5.2.4) *for* $x \in \mathfrak{D}(\Delta)$ *is by definition the principal extension of* $f(\lambda)$. Since each component of $\mathfrak{D}(\Delta)$ contains an open subset of $\mathfrak{D}_\Delta$, it is clear from Theorems 3.16.4 and 5.2.4 that the principal extension of $f(\lambda)$ is uniquely determined by its values on $\mathfrak{D}_\Delta$.

The use of formula (5.2.4) in matrix theory is of old standing. It was anticipated by G. Frobenius (1896) who referred to $f(x)$ as the residue of $f(\lambda)R(\lambda; x)$ with respect to the roots of the characteristic equation of the matrix x. It first appeared in its present form during 1928 in the writings of G. Giorgi [1] and L. Fantappiè [1] as the basis for a matrix calculus. In one form or another (5.2.4) is basic in L. Fantappiè's theory of analytic functionals as well as in recent investigations of commutative (B)-algebras by I. Gelfand and E. R. Lorch, and of spectral theory by N. Dunford and A. E. Taylor.

Formula (5.2.4) presupposes that the Banach algebra has a unit element. It is possible, however, to give an alternative definition of $f(x)$ (when $f(0) = 0$) which is equivalent to (5.2.4) when the latter applies, but which holds in any complex Banach algebra. See further Chapter XXIV, especially section 24.5.

The interpretation of the symbol $f(x)$ when $f(\lambda)$ is holomorphic in the sense of Cauchy and x is an element of a (B)-algebra (usually an algebra of endomorphisms $\mathfrak{E}(\mathfrak{X})$) differs in the several recent investigations. Lorch has emphasized with justice that analyticity does not reside in the symbol $f(\cdot)$ and that the term "analytic function of x" should be reserved for differentiable functions of x. The functions defined by formula (5.2.4) satisfy this requirement, if one is satisfied with the existence of the Fréchet differentials and analyticity in the sense of Definition 3.17.2. The more restrictive definition of analyticity due to Lorch (Definition 3.19.1) works well in the commutative case for which it was constructed, but loses its significance in a non-commutative algebra.

Formula (5.2.4) may be regarded as a natural generalization of the well known formula of Cauchy. The formulas for the derivatives also admit of generalizations. If in (5.2.5) one sets $x = a + \lambda h$, then, making use of formula (3.16.3), it is evident that

$$\delta^n f(a; h) = \frac{n!}{2\pi i} \int_{\Gamma_a} f(\zeta)R(\zeta; a)[hR(\zeta; a)]^n \, d\zeta. \tag{5.2.6}$$

The correspondence between the functions $f(\lambda)$, locally holomorphic in an open set Δ, and the function $f(x)$, defined on $\mathfrak{G}(\Delta)$ by formula (5.2.4), was first investigated by I. Gelfand [4]. Such a correspondence amounts to an *operational calculus* for the elements of the Banach algebra. We shall now prove

THEOREM 5.2.5. *Let* Δ *be an open subset of the complex plane. Let* $\mathfrak{G}(\Delta)$ *be the open set of points* $x \in \mathfrak{B}$ *such that* $\sigma(x) \subset \Delta$. *Let* $H(\Delta)$ *be the complex algebra of all functions* $f(\lambda)$, *locally holomorphic in* Δ, *with the ordinary definitions of the arithmetic operations, and with a sequence topology*: $f_n \to f$ *denoting that* $f_n(\lambda)$ *converges pointwise to* $f(\lambda)$, *the convergence being uniform in each compact subset of* Δ. *Further let* $\mathfrak{B}(\Delta)$ *be the complex algebra of functions* $F(x)$, *locally analytic in* $\mathfrak{G}(\Delta)$ *and having values in* $\mathfrak{B}$, *the arithmetic operations being defined as in* $\mathfrak{B}$.

There exists an isomorphic mapping: $f(\lambda) \to f(x)$ *of* $H(\Delta)$ *on a subalgebra* $\mathfrak{B}_0(\Delta)$ *of* $\mathfrak{B}(\Delta)$ *such that* (i) $1 \to e$, (ii) $\lambda \to x$, *and* (iii) $f_n \to f$ *implies that* $\| f_n(x) - f(x) \| \to 0$ *locally uniformly in* $\mathfrak{G}(\Delta)$. *This mapping is unique and is defined by* (5.2.4).

PROOF. We start by proving that the mapping defined by (5.2.4) has the required properties. Theorem 5.2.4 assures us that $f(x)$ belongs to $\mathfrak{B}(\Delta)$. Since both the functions 1 and λ are entire, a suitable oriented envelope of $\sigma(x)$ with respect to either of these functions would be $\Gamma_x : |\lambda| = r > \|x\|$. Making use in (5.2.4) of the representation $R(\lambda; x) = \sum_0^\infty \lambda^{-n-1}x^n$ and the above choice for Γ_x, it is readily seen that $1 \to e$ and $\lambda \to x$. It is further clear that the correspondence is linear so that in order to prove the mapping to be a homomorphism it is sufficient to show that products go into products. To this end let $f(\lambda)$, $g(\lambda)$, $h(\lambda) \in H(\Delta)$ with $h(\lambda) = f(\lambda)g(\lambda)$ and suppose that Γ, Γ' are two oriented envelopes of $\sigma(x)$ with respect $f(\lambda)$, $g(\lambda)$, and $h(\lambda)$ such that the open set bounded by Γ contains Γ'. Then

$$f(x)g(x) = \left(\frac{1}{2\pi i}\right)^2 \int_{\Gamma'} \int_{\Gamma} f(\zeta)g(\mu)R(\zeta; x)R(\mu; x)\, d\mu\, d\zeta$$

$$(5.2.7) \qquad = \frac{1}{2\pi i} \int_{\Gamma'} f(\zeta)R(\zeta; x) \left(\frac{1}{2\pi i} \int_{\Gamma} \frac{g(\mu)}{\mu - \zeta}\, d\mu\right) d\zeta$$

$$- \frac{1}{2\pi i} \int_{\Gamma} g(\mu)R(\mu; x) \left(\frac{1}{2\pi i} \int_{\Gamma'} \frac{f(\zeta)}{\mu - \zeta}\, d\zeta\right) d\mu,$$

where we have employed the first resolvent equation. Since $\zeta \in \Gamma'$ is interior to Γ whereas $\mu \in \Gamma$ is exterior to Γ', the two integrals involving $(\mu - \zeta)^{-1}$ are equal to $g(\zeta)$ and 0 respectively. Hence

$$f(x)g(x) = \frac{1}{2\pi i} \int_{\Gamma} f(\zeta)g(\zeta)R(\zeta; x)\, d\zeta = h(x),$$

and the mapping is a homomorphism.

In order to prove that the homomorphism is actually an isomorphism, we have to show that the correspondence is one-to-one or, equivalently, that $f(x) \equiv \theta$ implies $f(\lambda) \equiv 0$ on Δ. Suppose that for a particular choice of $f(\lambda)$ in $H(\Delta)$ we have $f(x) = \theta$ for all x in $\mathfrak{G}(\Delta)$. But if $\lambda \in \Delta$ then $\lambda e \in \mathfrak{G}(\Delta)$ and $f(\lambda e) = f(\lambda)e$ so that in particular $f(\lambda) \equiv 0$ on Δ. Thus the mapping is an isomorphism.

Suppose that $f(\lambda) \in H(\Delta)$ and is different from zero in some domain $\Delta_0 \subset \Delta$. Let $H(\Delta_0)$ and $\mathfrak{G}(\Delta_0)$ bear the same relation to Δ_0 as $H(\Delta)$ and $\mathfrak{G}(\Delta)$ have to Δ. Then $f(\lambda)$ and $[f(\lambda)]^{-1}$ belong to $H(\Delta_0)$ so that the corresponding functions $f(x)$ and $[f(x)]^{-1}$ exist for $x \in \mathfrak{G}(\Delta_0)$. Since $f(x)$ has an inverse for such values of x, it follows, in particular, that $f(x) \neq \theta$. Necessary and sufficient conditions on x in order that $f(x) = \theta$ have been given by N. Dunford [7, p. 644, *the minimal equation theorem*].

To prove the continuity, we proceed as in the proof of Theorem 5.2.4. Given $a \in \mathfrak{G}(\Delta)$, we choose a sphere $\mathfrak{S}(a)$: $\|x - a\| < \rho$ so small that Φ, the closure of $\mathsf{U}_x\sigma(x)$, $x \in \mathfrak{S}(a)$, is in Δ. In the representation of the functions $\{f_n(x)\}$ corresponding to the given Cauchy sequence $\{f_n(\lambda)\}$, we may replace Γ_x by a fixed contour Γ which is an oriented envelope of Φ with respect to all of the functions $f_n(\lambda)$. Again we have $\| R(\lambda; a) \| \leqq M(a)$ for all $\lambda \in \Gamma$ and from this we obtain $\| R(\lambda; x) \| \leqq K(a)$ for $\| x - a \| \leqq \rho_0 < \min [\rho, 1/M(a)]$ and for all $\lambda \in \Gamma$, where $K(a) = M(a)[1 - \rho_0 M(a)]^{-1}$. It follows that

$$\| f_n(x) - f(x) \| \leqq (2\pi)^{-1}K(a)l(\Gamma) \sup_{\lambda\in\Gamma} | f_n(\lambda) - f(\lambda) |$$

for all $\| x - a \| \leqq \rho_0$, where $l(\Gamma)$ is the length of Γ. Now $f_n \to f$ implies $f_n(\lambda) \to f(\lambda)$ uniformly on Γ so that $f_n(x)$ converges locally uniformly to $f(x)$. We have now shown that the mapping defined by (5.2.4) has all of the desired properties.

It remains to show that the mapping is unique. In order to prove this we observe first that if $\mathfrak{J}$ is any isomorphism with the stated properties, then $\mathfrak{J}$ must map the zero and the unit elements of $H(\Delta)$ upon the zero and the unit elements of $\mathfrak{B}_0(\Delta)$, that is, $0 \to \theta$ and $1 \to e$. Further, $\mathfrak{J}$ maps the polynomial $P(\lambda) = \sum_0^n \beta_k\lambda^k$ upon $P(x) = \beta_0 e + \sum_1^n \beta_k x^k$. Moreover if α is not in Δ, then $(\alpha - \lambda)^{-1} \in H(\Delta)$ and for $x \in \mathfrak{G}(\Delta)$, $\alpha e - x$ is regular so that $R(\alpha; x)$ exists. Since products go into products, $\mathfrak{J}$ must take $(\alpha - \lambda)^{-1}$ into $R(\alpha; x)$ and hence $(\alpha - \lambda)^{-n}$ into $[R(\alpha; x)]^n$. It follows that $\mathfrak{J}$ is uniquely defined on each rational function whose poles are outside of Δ and in particular that $\mathfrak{J}$ must agree with the mapping (5.2.4) on such functions. We can now appeal to an extension of the Runge theorem due to J. L. Walsh [1, p. 16]. Under the present assumptions on Δ, each $f(\lambda) \in H(\Delta)$ is the limit of a sequence of rational functions with poles outside of Δ, the convergence being uniform on each compact subset of Δ; and if Δ is simply connected, the rational functions may be taken as polynomials. In other words, the rational functions of $H(\Delta)$ are dense in this space. From the fact that $\mathfrak{J}$ agrees with the mapping defined by (5.2.4) in a dense set of $H(\Delta)$ together with the continuity assumptions, we conclude that $\mathfrak{J}$ is identical with (5.2.4) everywhere in $H(\Delta)$ so that the correspondence is unique.

We note that the isomorphic mapping $\mathfrak{J}$ remains unique even if condition (iii) in the statement of Theorem 5.2.5 is replaced by (iii′) $f_n \to f$ implies that $\| f_n(x) - f(x) \| \to 0$ for each $x \in \mathfrak{G}(\Delta)$. Likewise $\mathfrak{J}$ remains unique even if it is not assumed that $f(x)$ is locally analytic.

There are several remarks which should be appended to the preceding definitions and theorems. The interpretation of $f(x)$ on $\mathfrak{D}(\Delta)$ as the principal extension of $f(\lambda)$ appears to be new. Further if we limit x to $\mathfrak{D}(\Delta)$ then the conditions in Theorem 5.2.5 contain a redundance since (i) is then implied by (ii) and the local analyticity of $f(x)$. To see this, let $\lambda_0 \in \Delta$ $(\lambda_0 \neq 0)$ and set $f(\lambda) \equiv 1$, $g(\lambda) = \lambda$. Then $\lambda_0 e \in \mathfrak{D}(\Delta)$ and for x sufficiently small $\lambda_0 e - x$ is regular and likewise belongs to $\mathfrak{D}(\Delta)$. By (ii) $g(\lambda_0 e - x) = \lambda_0 e - x$ so that $f(\lambda_0 e - x)(\lambda_0 e - x) = f(\lambda_0 e - x)g(\lambda_0 e - x) = g(\lambda_0 e - x) = \lambda_0 e - x$. Multiplying through by $(\lambda_0 e - x)^{-1}$, we obtain $f(\lambda_0 e - x) = e$. Consequently $f(x) \equiv e$ in some open subset of each component of $\mathfrak{D}(\Delta)$ and, because of the analyticity, $f(x) \equiv e$ throughout $\mathfrak{D}(\Delta)$.

If $\mathfrak{B}$ is a commutative (B)-algebra, $f(x)$ on $\mathfrak{D}(\Delta)$ is not merely the principal extension of $f(\lambda)$, but it is the only extension from scalars to $\mathfrak{B}$ which is locally analytic in the more stringent sense of Lorch, Definition 3.19.1. In fact, for the commutative (B)-algebra formula (5.2.5) takes on the somewhat more familiar form

$$f(x) = \sum_{n=0}^{\infty} (x - a)^n \frac{1}{2\pi i} \int_\Gamma f(\zeta)[R(\zeta; a)]^{n+1} \, d\zeta. \tag{5.2.8}$$

This is a power series of the type occurring in formula (3.19.2) and it consequently defines a locally (L)-analytic function on $\mathfrak{G}(\Delta)$. On the other hand suppose that $F(x)$ is (L)-analytic in a domain $\mathfrak{D}$ which intersects the complex plane in a domain Δ_0 and that $F(\lambda e) = F(\lambda)e$ for $\lambda \in \Delta_0$ where $F(\lambda)$ is holomorphic in Δ_0. This requires that $F^{(n)}(\lambda e) = F^{(n)}(\lambda)e$ for every n and by Theorem 26.4.1 we see that the Taylor series expansion of $F(x)$ about the point $\lambda_0 e$ (for $\lambda_0 \in \Delta_0$) is $F(x) = \sum_{n=0}^{\infty} (x - \lambda_0 e)^n F^{(n)}(\lambda_0)/n!$. Thus Definition 5.2.1 together with Theorem 3.16.4 shows that $F(x)$ coincides with the principal extension of $F(\lambda)$. Thus *the principal extension of $f(\lambda)$, defined on Δ, is locally (L)-analytic if $\mathfrak{B}$ is a commutative (B)-algebra and it is the only extension of $f(\lambda)$ having this property.*

I. Gelfand [4], in his study of the operational calculus, restricted himself to a commutative (B)-algebra and to a simply connected domain Δ. He showed that if $\mathfrak{J}$ is a continuous isomorphic mapping, $f(\lambda) \to f(x)$, of holomorphic functions into abstract functions taking $1 \to e$ and $\lambda \to x$, then $f(x)$ is necessarily given by (5.2.4). Later N. Dunford [8] developed an operational calculus for elements of a non-commutative (B)-algebra which made use of functions $f(\lambda)$, locally holomorphic on open sets of the type considered in Theorem 5.2.5. It appears that E. R. Lorch [4] was the first to point out the analytic character of $f(x)$ as a function of x; limiting himself to the commutative case, he showed that $f(x)$ was (L)-analytic.

5.3. The spectral mapping theorem. Composite functions. We shall now determine how the spectrum of $f(x)$ is related to the spectrum of x. The following result is essentially due to I. Gelfand [4].

THEOREM 5.3.1. *If $f(\lambda) \in H(\Delta)$ and $x \in \mathfrak{G}(\Delta)$, then $\sigma[f(x)] = f[\sigma(x)]$.*

PROOF. Let $\alpha \in \sigma(x)$ and define

$$g(\lambda) = [f(\alpha) - f(\lambda)](\alpha - \lambda)^{-1}, \qquad \lambda \in \Delta,$$

so that $g(\lambda) \in H(\Delta)$ and consequently

$$(\alpha e - x)g(x) = g(x)(\alpha e - x) = f(\alpha)e - f(x). \tag{5.3.1}$$

If $f(\alpha)e-f(x)$ were regular, then (5.3.1) implies the regularity of $\alpha e - x$; hence $f(\alpha) \in \sigma[f(x)]$. Conversely, let $\mu \in \sigma[f(x)]$ and suppose that $\mu \notin f[\sigma(x)]$. Then there exists an open set Δ_0 containing $\sigma(x)$ and contained in Δ such that $f(\lambda) = \mu$ has no roots in Δ_0. Setting $h(\lambda) = \mu - f(\lambda)$, it is clear that both $h(\lambda)$ and $[h(\lambda)]^{-1}$ belong to $H(\Delta_0)$ and hence that the corresponding elements $h(x) = \mu e - f(x)$ and $[h(x)]^{-1}$ exist. But this is impossible for $\mu \in \sigma[f(x)]$.

We come now to a theorem on composite functions due to N. Dunford [8].

THEOREM 5.3.2. *If $g(\lambda) \in H(\Delta)$ and if $f(\mu) \in H(\Delta_0)$ where $g(\Delta) \subset \Delta_0$, then $f[g(\lambda)] \in H(\Delta)$, $g(x) \in \mathfrak{G}(\Delta_0)$, $f[g(x)] \in \mathfrak{B}_0(\Delta)$, and $f[g(x)] = [f(g)](x)$ for each $x \in \mathfrak{G}(\Delta)$.*

PROOF. It is clear that $f[g(\lambda)] \in H(\Delta)$ and hence by Theorem 5.2.4 it follows that $[f(g)](x) \in \mathfrak{B}_0(\Delta)$. Consequently it remains only to prove that $f[g(x)]$ exists and that $f[g(x)] = [f(g)](x)$ for each $x \in \mathfrak{G}(\Delta)$. Now for $x \in \mathfrak{G}(\Delta)$, $\sigma[g(x)] = g[\sigma(x)] \subset g(\Delta) \subset \Delta_0$ so that $g(x) \in \mathfrak{G}(\Delta_0)$ and hence $f[g(x)]$ exists. To prove the equality let Δ_x be a bounded open set containing $\sigma(x)$ and such that $\bar{\Delta}_x \subset \Delta$; then $g(\bar{\Delta}_x)$ is a compact subset of Δ_0. For $\mu \notin g(\bar{\Delta}_x)$, the function $h(\lambda) \equiv [\mu - g(\lambda)]^{-1} \in H(\Delta_x)$ so that $h(x) = R(\mu; g(x))$ exists. Hence taking Γ_0 and Γ to be oriented envelopes of $g(\bar{\Delta}_x)$ with respect to $f(\mu) \in H(\Delta_0)$ and of $\sigma(x)$ with respect to $g(\lambda) \in H(\Delta_x)$, respectively, we have

$$f[g(x)] = \frac{1}{2\pi i}\int_{\Gamma_0} f(\mu)R(\mu; g(x))\, d\mu = \left(\frac{1}{2\pi i}\right)^2 \int_{\Gamma_0}\int_{\Gamma} \frac{f(\mu)R(\zeta; x)}{\mu - g(\zeta)}\, d\zeta\, d\mu$$

$$= \frac{1}{2\pi i}\int_{\Gamma} f[g(\zeta)]R(\zeta; x)\, d\zeta = [f(g)](x).$$

5.4. The exponential function, the logarithm, and powers. It is a simple matter to obtain the exponential function either directly from formula (5.2.4) or from Theorem 5.2.5. Given x, let Δ be the circular domain $|\lambda| < r$ where $r > \|x\|$. Then $\sigma(x) \subset \Delta$ and $\exp(\lambda) \in H(\Delta)$. Since the series $\sum_0^\infty \lambda^n/n!$ converges uniformly in Δ, it follows from the properties (i), (ii), and (iii) of Theorem 5.2.5 that

$$\exp(x) = \sum_{k=0}^{\infty} \frac{x^k}{k!}. \tag{5.4.1}$$

If x and y commute, then a straightforward calculation shows

$$\exp(x)\exp(y) = \exp(x + y). \tag{5.4.2}$$

In particular one sees that $\exp(x)$ is always a regular element of $\mathfrak{B}$ having $\exp(-x)$ as its inverse. We also have the obvious

THEOREM 5.4.1. *The functional equation*

$$F(x + y) = F(x)F(y), \qquad x, F(x) \in \mathfrak{B}, \tag{5.4.3}$$

is satisfied by

$$F(x) = j\exp[P(x)], \tag{5.4.4}$$

where j is idempotent, $j^2 = j$, $P(x)$ is linear on $\mathfrak{B}$ to itself, and j, $P(x)$, $P(y)$ commute for all $x, y \in \mathfrak{B}$.

We now proceed to define the logarithm of x which is clearly going to be infinitely many-valued. We start with a domain Δ in the complex plane which excludes zero and in which $\arg\lambda$ is single-valued. Any determination of $\log\lambda$ will then be holomorphic in Δ and defines a corresponding determination of $\log x$ by

$$\log x = \frac{1}{2\pi i}\int_{\Gamma_x} \log\zeta\, R(\zeta; x)\, d\zeta \tag{5.4.5}$$

for $x \in \mathfrak{G}(\Delta)$. Here Γ_x is any oriented envelope of $\sigma(x)$ with respect to $\log \lambda$ as defined on Δ. If x is any regular element of $\mathfrak{B}$ whose spectrum does not separate $\lambda = 0$ from $\lambda = \infty$, then we may define $\log x$ by (5.4.5) because in this case we can always find a domain Δ with the required properties and such that $\sigma(x) \subset \Delta$. An alternate statement is that $\lambda = 0$ belongs to the principal component of the resolvent set of x. The various determinations of $\log x$ defined by (5.4.5) obviously differ by multiples of $2\pi ie$.

In some instances we can obtain still other determinations of $\log x$. Suppose that x is regular, that $\lambda = 0$ belongs to the principal component of the resolvent set of x, and that $\sigma(x)$ is not connected. In this case our choice of Δ need not be connected and we can choose different determinations of $\log \lambda$ in the different components of Δ. These various determinations will differ by multiples of $2\pi i$ on the different components. Now a function $f(\lambda) \equiv 1$ on one component of Δ and zero elsewhere clearly belongs to $H(\Delta)$. Since $[f(\lambda)]^2 = f(\lambda)$, it follows that $[f(x)]^2 = f(x)$ so that $f(x)$ is idempotent. Consequently two determinations of $\log x$ will differ by an expression of the form $2\pi i \sum_1^k n_\nu j_\nu$, where the n's are integers and the j's are idempotents which commute with x.

According to Theorem 5.3.2 we have

$$\exp(\log x) = x, \qquad x \in \mathfrak{G}(\Delta), \tag{5.4.6}$$

for each of the determinations of $\log x$. This is not too surprising since for any idempotent j, substitution in (5.4.1) gives $\exp(2\pi ij) = e$. Hence if x and j commute we see that

$$\exp(x + 2\pi ij) = \exp x. \tag{5.4.7}$$

For a commutative (B)-algebra, $2\pi ij$ can be thought of as a period of the exponential function. We shall return to this point in the next section. To obtain the converse relation to (5.4.6), let Δ_0 be an open subset of the λ-plane such that no two points of Δ_0 are congruent modulo $2\pi i$. The function $\mu = \exp \lambda$ maps Δ_0 onto an open set Δ with the properties stated above and in which $\log \lambda$ is piecewise holomorphic. Theorem 5.3.2 then shows that

$$\log[\exp x] = x + 2\pi i \sum_{\nu=1}^{k} n_\nu j_\nu, \qquad x \in \mathfrak{G}(\Delta_0), \tag{5.4.8}$$

where the integers n_ν and the idempotents j_ν depend on the determination of the logarithm. In particular, the n_ν are zero if we choose the proper determination of $\log \lambda$ in (5.4.5).

In the commutative case Lorch [3] has defined the logarithm by

$$\log x = \int_e^x z^{-1}\, dz, \tag{5.4.9}$$

where the integral is defined as in (3.3.9) and the path joining e to x is an arbitrary rectifiable curve in $\mathfrak{G}_1$, the principal component of the maximal group. The resulting function $\log x$ is (L)-analytic in $\mathfrak{G}_1$ which is the maximal domain of definition of the logarithm.

If $x = \exp y$ then we may define the *power*

$$x^\alpha = \exp(\alpha y) \tag{5.4.10}$$

where α is a real or complex number. Since y is not uniquely determined, neither is x^α. However for a fixed y, it is clear that $\{x^\alpha\}$ forms a continuous one-parameter group as α ranges over the real line or the complex plane. In this case we say that x is embedded in a power group $\{x^\alpha\}$. One of the important problems in the theory of groups of linear operators is to determine when a given element x is embeddable in a power group or equivalently to determine when x is expressible as an exponential. A solution to this problem will be given in section 9.5.

5.5. Idempotent elements. The idempotent elements play an important role in the theory of (B)-algebras. They are basic for the spectral resolution of an element of the algebra as well as in the study of the reducibility of the algebra itself.

If j is an idempotent, then $j^2 = j$ and hence the resolvent series (4.7.3) becomes simply

$$R(\lambda; j) = \frac{e - j}{\lambda} + \frac{j}{\lambda - 1}. \tag{5.5.1}$$

When j is different from θ and e, the spectrum consists of two points $\lambda = 0$ and 1, each of which is a simple pole of the resolvent. In case $j = \theta$ we have $\sigma(\theta) = \{0\}$, whereas for $j = e$ we have $\sigma(e) = \{1\}$. We observe that the residue at $\lambda = 1$ is j and the residue at $\lambda = 0$ is $e - j$, both of which are idempotents. This is a special instance of the following

THEOREM 5.5.1. *Let Γ_0 be the union of a finite number of simple closed rectifiable curves having no points in common. Suppose Γ_0 bounds the open set Δ_0 in such a way that Δ_0 lies to the left of Γ_0 relative to the positive orientation of Γ_0. Then if $\Gamma_0 \subset \rho(a)$ the element*

$$j = \frac{1}{2\pi i} \int_{\Gamma_0} R(\zeta; a)\, d\zeta \tag{5.5.2}$$

is an idempotent which commutes with a. The element j is different from θ and e unless $\Delta_0 \supset \sigma(a)$ in which case $j = e$ or unless $\Delta_0 \cap \sigma(a) = \emptyset$ in which case $j = \theta$.

REMARK. The expression (5.5.2) goes back to F. Riesz [2, pp. 117–121] who derived some of the properties of j.

PROOF. For fixed $a \in \mathfrak{B}$, let Δ be an open subset of the complex plane containing $\sigma(a) \cup \Delta_0$ and such that Δ_0 is both open and closed relative to Δ. Then $a \in \mathfrak{G}(\Delta)$. Now set $f(\lambda) = 1$ on Δ_0 and $= 0$ elsewhere in Δ. We see that $f(\lambda) \in H(\Delta)$ so that $f(a)$ is defined by the operational calculus. Obviously $j = f(a)$ and since $[f(\lambda)]^2 = f(\lambda)$, it follows that $j^2 = j$. Likewise $\lambda f(\lambda) = f(\lambda)\lambda$ implies $aj = ja$. According to Theorem 5.3.1, $\sigma(j) = f[\sigma(a)]$ and therefore if $\Delta_0 \cap \sigma(a) =$

$\emptyset$ then $\sigma(j) = \{0\}$ and so $j = \theta$; if $\sigma(a) \subset \Delta_0$ then $\sigma(j) = \{1\}$ and so $j = e$. Otherwise $\sigma(j) = \{0\} \cup \{1\}$ and j will be different from both θ and e.

Combining this result with our earlier remarks about the spectrum of a non-trivial idempotent, we obtain

THEOREM 5.5.2. *A necessary and sufficient condition that a Banach algebra $\mathfrak{B}$ shall contain an idempotent different from θ and e is that there is at least one element of $\mathfrak{B}$ whose spectrum is not connected.*

For a commutative (B)-algebra the Gelfand representation theory also sheds some light on the properties of idempotents. If j is an idempotent, then $[j(\mathfrak{m})]^2 = j(\mathfrak{m})$, $\mathfrak{m} \in \mathfrak{M}$, so that $j(\mathfrak{m}) = 0$ or 1. Set $F_0 = [\mathfrak{m}; j(\mathfrak{m}) = 0]$, $F_1 = [\mathfrak{m}; j(\mathfrak{m}) = 1]$. Clearly $F_0 \cup F_1 = \mathfrak{M}$ and $F_0 \cap F_1 = \emptyset$. If $j = \theta$ then $F_0 = \mathfrak{M}$, whereas if $j = e$ then $F_1 = \mathfrak{M}$. Conversely suppose that $F_0 = \mathfrak{M}$. Then $\sigma(j) = \{0\}$ and as we have already remarked this implies $j = \theta$. On the other hand if $F_1 = \mathfrak{M}$, then $\sigma(j) = \{1\}$ and hence $j = e$. It follows that j is different from θ and e if and only if neither F_0 nor F_1 is empty. Since $j(\mathfrak{m})$ is continuous on $\mathfrak{M}$, we see in this case that both F_0 and F_1 are clopen. Consequently if $\mathfrak{B}$ contains a non-trivial idempotent then $\mathfrak{M}$ is not connected. The converse statement is also valid as Šilov [**3**] has shown.

Returning to the general (B)-algebra, we note that if j is an idempotent which commutes with the element a and hence with $R(\lambda; a)$, then $jR(\lambda; a)$ is a *quasi-resolvent* in the sense that

$$(5.5.3) \qquad (\lambda j - ja)[jR(\lambda; a)] = [jR(\lambda; a)](\lambda j - ja) = j$$

for all $\lambda \in \rho(a)$. In other words $jR(\lambda; a)$ may be regarded as the resolvent of ja in the subalgebra $j\mathfrak{B}j$ in which j plays the role of the unit element. Of course $\rho(a)$ may be a proper subset of the resolvent set for ja considered as an element of $j\mathfrak{B}j$.

The above considerations shed some light on the question of the reducibility of a (B)-algebra.

DEFINITION 5.5.1. *Let $\mathfrak{B}$ be a (B)-algebra with unit element e, $\mathfrak{B}_1$ a subalgebra with unit element e_1. $\mathfrak{B}_1$ is called an invariant subalgebra if $\mathfrak{B}_1\mathfrak{B} \subset \mathfrak{B}_1$, $\mathfrak{B}\mathfrak{B}_1 \subset \mathfrak{B}_1$, and it is proper if it is neither $\{\theta\}$ nor $\mathfrak{B}$. $\mathfrak{B}$ is said to be reducible or irreducible according as it has or has not a proper invariant subalgebra with a unit element.*

Here $\mathfrak{B}_1\mathfrak{B}_2$ is the set of all products xy for $x \in \mathfrak{B}_1$ and $y \in \mathfrak{B}_2$. An invariant subalgebra is a two-sided ideal in the sense of Definition 1.13.2.

It is clear that if $\mathfrak{B}_1$ is a proper invariant subalgebra of $\mathfrak{B}$ with unit element e_1, then $e_1^2 = e_1$, and e_1 is different from θ and e. Also e_1 is in the center of $\mathfrak{B}$ since $e_1x \in \mathfrak{B}_1$ so that $e_1x = e_1xe_1$ and similarly $xe_1 = e_1xe_1$. Thus e_1 is a non-trivial idempotent which commutes with every element of $\mathfrak{B}$. Conversely, suppose that such an element e_1 exists in $\mathfrak{B}$. Then $\mathfrak{B}_1 = e_1\mathfrak{B} = \mathfrak{B}e_1$ is a two-sided ideal in $\mathfrak{B}$ and $\mathfrak{B}_1 \neq \theta$ since $e_1 \in \mathfrak{B}_1$. Moreover, $e \notin \mathfrak{B}_1$ since $e = e_1x$ would imply that $e = e_1e = e_1$. It follows that $\mathfrak{B}$ is reducible. Hence we have the

THEOREM 5.5.3. *A (B)-algebra is reducible if and only if it contains an idempotent e_1 different from θ and e in its center.*

Combining this result with that of Theorem 5.5.2 we obtain

THEOREM 5.5.4. *A necessary condition that a (B)-algebra be reducible is that it contain an element whose spectrum is not connected. If the algebra is commutative this condition is also sufficient.*

If an algebra $\mathfrak{B}$ is reducible it can be expressed as the product of two invariant subalgebras with units. For let $\mathfrak{B}_1$ be a proper invariant subalgebra with unit element e_1. Then $e_2 = e - e_1$ is also a non-trivial idempotent in the center and if we set $\mathfrak{B}_2 = e_2\mathfrak{B}$ we can easily verify that $\mathfrak{B} = \mathfrak{B}_1 \times \mathfrak{B}_2$.

In the algebras which arise in analysis, the number of idempotents may very well be infinite and even non-countable. The latter is the case, for instance, in the algebra of bounded functions $x(\xi)$ defined on an interval (α, β) with $\| x(\cdot) \| = \sup | x(\xi) |$ where all the characteristic functions of subsets of (α, β) are idempotents. On the other hand, any algebra of continuous functions on a connected Hausdorff space can possess no non-trivial idempotent and hence is irreducible. For example, the algebra $S(\varphi)$, studied in §4.4, can be faithfully represented as the algebra of continuous functions, $f_a(\lambda) = \int_0^\infty \exp(\lambda\xi)\, da$, on the half-plane $\Re(\lambda) \leqq \omega_0$ (here we assume that $\omega_0 > -\infty$) and hence $S(\varphi)$ is irreducible. The same reasoning applies to prove that $S_0(\varphi)$ is also irreducible.

Before concluding this section we return to the question of the periods of the exponential function mentioned in the previous section. The following theorem is due to E. R. Lorch [**3**].

THEOREM 5.5.5. *If $\mathfrak{B}$ is commutative and j is an idempotent element, then $2\pi ij$ is a period of* $\exp x$ *and every period is of the form $2\pi i \sum_1^k n_\nu j_\nu$ where k is finite, the n's are integers, and the j's idempotents. In particular, if θ and e are the only idempotents of $\mathfrak{B}$, then* $\exp x$ *is simply periodic.*

PROOF. We have already verified that $\exp(2\pi ij) = e$. Suppose that p is a period of $\exp x$. Then $\exp p = e$ so that $\exp(\xi p)$ is a periodic function of ξ of period one. Hence (see section 11.2)

$$(5.5.4)\quad R(\lambda; p) = \int_0^\infty e^{-\lambda\xi} \exp(\xi p)\, d\xi = [1 - e^{-\lambda}]^{-1} \int_0^1 e^{-\lambda\xi} \exp(\xi p)\, d\xi,$$

where the three members are defined for $\lambda \in \rho(p)$, $\Re(\lambda) > 0$, and $\lambda \neq 2n\pi i$ respectively. Since $\sigma(p)$ is bounded, this shows that $R(\lambda; p)$ is a meromorphic function of λ, with simple poles at a finite number of the integral multiples of $2\pi i$, say at $\lambda_\nu = 2\pi i n_\nu$, $\nu = 1, 2, \cdots, k$. Now let Γ_ν be a small circle about the point λ_ν with the usual orientation. Then $\Gamma = \bigcup_{\nu=1}^k \Gamma_\nu$ is an oriented envelope of $\sigma(p) = \{\lambda_\nu\}$ with respect to $f(\lambda) = \lambda$ and hence

$$p = \frac{1}{2\pi i} \int_\Gamma \zeta R(\zeta; p)\, d\zeta = \sum_{\nu=1}^k \lambda_\nu \frac{1}{2\pi i} \int_{\Gamma_\nu} R(\zeta; p)\, d\zeta.$$

The desired result now follows from Theorem 5.5.1. In particular, if $k = 1$ we have $j_1 = e$, $p = 2\pi i n_1 e$.

Combining Theorems 5.5.4 and 5.5.5 we obtain

COROLLARY. *In a commutative algebra $\mathfrak{B}$ the exponential function is simply periodic if and only if $\mathfrak{B}$ is irreducible.*

5.6. The spectral resolution of an element. One of the principal goals in spectral theory is to obtain a spectral resolution for the elements of a (B)-algebra. We take as our model the canonical reduction of a matrix and even though we are in general not able to achieve the detail available in the matrix case, still the results are extremely useful. Here we follow N. Dunford [**8**].

The basic notion in this discussion will be that of a *spectral set.*

DEFINITION 5.6.1. *A non-void set σ is called a spectral set of $a \in \mathfrak{B}$ if σ is a subset of $\sigma(a)$ and σ is both open and closed in $\sigma(a)$. $\Gamma(\sigma)$ is called an oriented envelope of σ if*

(i) *$\Gamma(\sigma)$ is the union of a finite number of closed simple rectifiable curves having no points in common;*

(ii) *$\Gamma(\sigma)$ lies in the resolvent set $\rho(a)$ of a;*

(iii) *$\Gamma(\sigma)$ bounds an open set $\Delta(\sigma)$ containing σ; $\Delta(\sigma)$ lies to the left of $\Gamma(\sigma)$ relative to the orientation of $\Gamma(\sigma)$;*

(iv) *$\sigma' = \sigma(a) \ominus \sigma$ belongs to the complement of $\overline{\Delta(\sigma)}$;*

(v) *$max_{\lambda \in \Gamma(\sigma)}\, d(\lambda, \sigma) < \frac{1}{3} d(\sigma, \sigma')$.*

Condition (v) ensures that $\overline{\Delta(\sigma)}$ has no points in common with $\overline{\Delta(\sigma')}$. The construction of $\Gamma(\sigma)$ can be conveniently achieved by means of a sufficiently fine hexagonal mesh.

To each resolution of $\sigma(a)$ into disjoint spectral sets there is a corresponding resolution of a and $R(\lambda; a)$ given in the following

THEOREM 5.6.1. *Let $\sigma(a) = \bigcup_1^k \sigma_\alpha$ where each σ_α is a spectral set of a, $\sigma_\alpha \cap \sigma_\beta = \emptyset$ when $\alpha \neq \beta$, and $k > 1$. Define*

$$j_\alpha = \frac{1}{2\pi i} \int_{\Gamma_\alpha} R(\zeta; a)\, d\zeta, \tag{5.6.1}$$

where Γ_α is an oriented envelope of σ_α. Then

$$\sum_1^k j_\alpha = e, \qquad j_\alpha^2 = j_\alpha, \qquad j_\alpha j_\beta = \theta \text{ for } \alpha \neq \beta, \qquad j_\alpha \neq \theta, e.$$

Setting

$$a_\alpha = j_\alpha a, \qquad R_\alpha(\lambda; a) = j_\alpha R(\lambda; a),$$

then

$$a = \sum_1^k a_\alpha , \qquad a_\alpha a_\beta = \theta \qquad \text{for } \alpha \neq \beta;$$

$$(5.6.2) \qquad R(\lambda; a) = \sum_1^k R_\alpha(\lambda; a), \qquad R_\alpha(\lambda; a) R_\beta(\lambda; a) = \theta \qquad \text{for } \alpha \neq \beta;$$

$$R_\alpha(\lambda; a) a_\beta = a_\beta R_\alpha(\lambda; a) = \theta \qquad \text{for } \alpha \neq \beta;$$

$$(\lambda j_\alpha - a_\alpha) R_\alpha(\lambda; a) = R_\alpha(\lambda; a)(\lambda j_\alpha - a_\alpha) = j_\alpha .$$

Furthermore, the spectrum of a_α *is* $\sigma_\alpha \cup [0]$ *and* $R_\alpha(\lambda;\ a)$ *can be extended to be locally holomorphic in the complement of* σ_α. *Finally*

$$(5.6.3) \qquad R_\alpha(\lambda; a) = j_\alpha \lambda^{-1} + \sum_{n=1}^{\infty} a_\alpha^n \lambda^{-n-1}$$

for $|\lambda| > \gamma_\alpha = \lim_{n\to\infty} \| a_\alpha^n \|^{1/n}$.

Proof. We again avail ourselves of the operational calculus. Let $\Delta = \bigcup_1^k \Delta(\sigma_\alpha)$ so that $\sigma(a) \subset \Delta$ and $a \in \mathfrak{S}(\Delta)$. Further set $f_\alpha(\lambda) = 1$ on $\Delta(\sigma_\alpha)$ and $= 0$ elsewhere in Δ. Then as in the proof of Theorem 5.5.1, $j_\alpha = f_\alpha(a), j_\alpha^2 = j_\alpha , j_\alpha a = a j_\alpha$, and $j_\alpha \neq \theta, e$. From $\sum_1^k f_\alpha(\lambda) \equiv 1$ on Δ and $f_\alpha(\lambda) f_\beta(\lambda) \equiv 0$ for $\alpha \neq \beta$ we obtain $\sum_1^k j_\alpha = e$ and $j_\alpha j_\beta = \theta$ for $\alpha \neq \beta$. Since j_α commutes with a, it must also commute with $R(\lambda;\ a)$. One sees by inspection that all of the relations in (5.6.2) are immediate consequences of the above properties of the j's.

The operational calculus also shows that $a_\alpha = g_\alpha(a)$ where $g_\alpha(\lambda) = \lambda f_\alpha(\lambda)$ and hence $\sigma(a_\alpha) = g_\alpha[\sigma(a)] = \sigma_\alpha \cup [0]$ by Theorem 5.3.1. Likewise for μ exterior to $\bar{\Delta}$, $R_\alpha(\mu;\ a) = j_\alpha R(\mu;\ a) = h_\alpha(a)$ where $h_\alpha(\lambda) = f_\alpha(\lambda)(\mu - \lambda)^{-1}$. Hence

$$R_\alpha(\mu; a) = \frac{1}{2\pi i} \int_{\Gamma_\alpha} \frac{R(\zeta; a)}{\mu - \zeta} \, d\zeta.$$

It is evident that $R_\alpha(\lambda;\ a)$ can be extended by this integral representation to be locally holomorphic exterior to any oriented envelope of σ_α and hence in the complement of σ_α. The extension is of course unique. Since $R_\alpha(\lambda;\ a)$, defined on $\rho(a)$, is the resolvent of a_α relative to the subalgebra $j_\alpha \mathfrak{B} j_\alpha$, the same remains true of its extension. Finally the extended $R_\alpha(\lambda;\ a)$ must have precisely the same singularities as $R(\lambda;\ a)$ in σ_α since $R(\lambda;\ a)$ has no extension outside of $\rho(a)$ and since $R(\lambda;\ a) = \sum_1^k R_\beta(\lambda;\ a)$. Thus the spectrum of a_α relative to $j_\alpha \mathfrak{B} j_\alpha$ is precisely σ_α, whereas $\sigma(a_\alpha) = \sigma_\alpha \cup [0]$.

Formula (5.6.3) is a simple consequence of the fact that the extended $R_\alpha(\lambda; a)$ is the quasi-resolvent of a_α relative to the subalgebra $j_\alpha \mathfrak{B} j_\alpha$. For by Theorem 4.7.2 we have

$$R_\alpha(\lambda; a) = j_\alpha \lambda^{-1} + \sum_{n=1}^{\infty} a_\alpha^n \lambda^{-n-1}$$

and the equality is valid for $|\lambda| > \gamma_\alpha$. This completes the proof of the theorem.

If σ_α consists of the single point, λ_α, then we can somewhat strengthen the statement of the above theorem. In addition to a_α we also introduce the element $a_\alpha^- = j_\alpha(a - \lambda_\alpha e)$. Again $R_\alpha(\lambda; a)$ is the quasi-resolvent of a_α relative to $j_\alpha \mathfrak{B} j_\alpha$. Applying Theorem 4.7.2, we obtain

$$R_\alpha(\lambda; a) = (\lambda j_\alpha - a_\alpha)^{-1} = [(\lambda - \lambda_\alpha)j_\alpha - a_\alpha^-]^{-1}$$

$$= j_\alpha(\lambda - \lambda_\alpha)^{-1} + \sum_{n=1}^{\infty} (a_\alpha^-)^n(\lambda - \lambda_\alpha)^{-n-1}.$$

Since $R_\alpha(\lambda; a)$ is holomorphic for $|\lambda - \lambda_\alpha| > 0$ the above expansion must also remain valid in this domain. This in turn implies $\| (a_\alpha^-)^n \|^{1/n} \to 0$ and hence that a_α^- is quasi-nilpotent. Thus we have the

COROLLARY. *Let the resolvent $R(\lambda; a)$ be holomorphic except for isolated singular points $\lambda_1, \lambda_2, \cdots, \lambda_k$. Then there exist k idempotents $j_1, j_2, \cdots, j_k$ defined by*

$$j_\alpha = \frac{1}{2\pi i}\int_{\Gamma_\alpha} R(\zeta; a)\, d\zeta,$$

where Γ_α is a small circle about λ_α. In this case $j_\alpha R(\lambda; a)$ can be extended to be holomorphic for $|\lambda - \lambda_\alpha| > 0$ and can be expanded as

$$j_\alpha R(\lambda; a) = j_\alpha(\lambda - \lambda_\alpha)^{-1} + \sum_{n=1}^{\infty} (a_\alpha^-)^n(\lambda - \lambda_\alpha)^{-n-1}.$$

Here $a_\alpha^- = j_\alpha(a - \lambda_\alpha e)$ is quasi-nilpotent (or nilpotent). Furthermore $a = \sum_1^k a_\alpha = \sum_1^k a_\alpha^- + \sum_1^k \lambda_\alpha j_\alpha$.

The next theorem shows that the spectral resolution of a extends to $f(a)$. Here we assume that $\sigma(a)$ can be decomposed into the disjoint spectral sets $\{\sigma_\alpha\}$ and define j_α by (5.6.1).

THEOREM 5.6.2. *Let Δ be an open subset of the complex plane which contains $\sigma(a) \cup [0]$ and suppose that $f(\lambda) \in H(\Delta)$. Then $a_\alpha = j_\alpha a \in \mathfrak{G}(\Delta)$ and*

$$f(a_\alpha) = (e - j_\alpha)f(0) + j_\alpha f(a), \qquad f(a) = \sum_{\alpha=1}^{k} f(a_\alpha) - (k - 1)f(0)e.$$

PROOF. By Theorem 5.6.1, $\sigma(a_\alpha) = \sigma_\alpha \cup [0] \subset \Delta$ and hence $a_\alpha \in \mathfrak{G}(\Delta)$. It is clear from formula (5.6.3) that $j_\alpha R(\lambda; a) = R_\alpha(\lambda; a) = R(\lambda; a_\alpha) - (e - j_\alpha)\lambda^{-1}$. Consequently

$$f(a_\alpha) = \frac{1}{2\pi i}\int_\Gamma f(\zeta)R(\zeta; a_\alpha)\, d\zeta$$

$$= \frac{1}{2\pi i} j_\alpha \int_\Gamma f(\zeta)R(\zeta; a)\, d\zeta + (e - j_\alpha)\frac{1}{2\pi i}\int_\Gamma f(\zeta)\zeta^{-1}\, d\zeta,$$

where Γ is an oriented envelope of $\sigma(a) \cup [0]$ with respect to $f(\lambda)$. It now follows that $f(a_\alpha) = j_\alpha f(a) + (e - j_\alpha)f(0)$ and this together with the fact that $e = \sum_1^k j_\alpha$ implies the second relation in the statement of the theorem.

5.7. Spectral theory for compact linear operators. All of the results derived in the preceding sections apply in particular to the elements of $\mathfrak{E}(\mathfrak{X})$, the Banach algebra of linear bounded transformations of a (B)-space $\mathfrak{X}$ to itself. For suppose $T \in \mathfrak{E}(\mathfrak{X})$ and let $R(\lambda; T)$ be its resolvent in the sense of Definition 2.16.1. Then $R(\lambda; T) \in \mathfrak{E}(\mathfrak{X})$ for λ in $\rho(T)$ but not for any λ in $\sigma(T)$. It follows that $R(\lambda; T)$ is the inverse of $\lambda I - T$ in the algebra $\mathfrak{E}(\mathfrak{X})$ and that the notions of resolvent set and spectrum given in Definition 4.7.1 coincide with the corresponding notions of Definition 2.16.1.

If we restrict ourselves to the compact linear operators of $\mathfrak{E}(\mathfrak{X})$, it is possible to extend the results of the previous section. The resulting theory, which is mainly the work of F. Riesz [3], forms a direct generalization of the Fredholm theory of integral equations and constitutes one of the earliest and most important achievements in functional analysis. Riesz's work in 1918 was complete in so far as the spectral theory for compact linear operators was concerned, but he did not investigate the interplay between the spectral properties of the operator and those of the adjoint operator. Actually the concept of an adjoint operator was not then available. The augmented theory is due to T. H. Hildebrandt [2] and J. Schauder [1] (cf. S. Banach [2, Chapter X]). Although the following development is somewhat novel, still it leans heavily on the work of F. Riesz [3]; one can even find a suggestion of the method in Riesz's classic monograph [2, pp. 117–121].

Let T be a compact linear operator belonging to $\mathfrak{E}(\mathfrak{X})$. Then T takes bounded subsets of $\mathfrak{X}$ into compact subsets of $\mathfrak{X}$. Hence (here we assume that $\mathfrak{X}$ is not finite dimensional) the image of any bounded set cannot contain the unit sphere by Theorem 1.12.2. It follows that T can not have a bounded inverse in $\mathfrak{E}(\mathfrak{X})$ and hence that $\lambda = 0$ belongs to the spectrum of T. We shall now show that aside from $\lambda = 0$, the spectrum, $\sigma(T)$, consists of isolated points, each belonging to the point spectrum of T.

THEOREM 5.7.1. *If $T \in \mathfrak{E}(\mathfrak{X})$ is compact, then $P\sigma(T)$ consists of isolated points with the possible exception of $\lambda = 0$.*

PROOF. Suppose that $\{\lambda_n\} \in P\sigma(T)$ and that $\lambda_n \to \lambda_0 \neq 0$. Then there exist non-zero characteristic vectors $x_n \in \mathfrak{X}$ such that $T(x_n) = \lambda_n x_n$. Taking the λ_n's to be distinct we now show by an inductive argument that the x_n's are linearly independent. In fact suppose that this is true for $x_1, x_2, \cdots, x_{n-1}$ but that there exist α_ν's not all zero such that $\sum_1^n \alpha_\nu x_\nu = \theta$. Then $\theta = T(\sum_1^n \alpha_\nu x_\nu) = \sum_1^n \alpha_\nu \lambda_\nu x_\nu$ and hence $\sum_1^{n-1} \alpha_\nu(\lambda_n - \lambda_\nu)x_\nu = \theta$. Since not all of the coefficients in this last expression can be zero, this means that $x_1, x_2, \cdots, x_{n-1}$ are linearly dependent, which is contrary to our inductive hypothesis.

Now let $\mathfrak{L}_n$ be the linear extension of the vectors $x_1, x_2, \cdots, x_n$. Applying Lemma 1.12.2 we obtain a sequence $\{y_n\}$ with the following properties: $y_n \in \mathfrak{L}_n$, $\| y_n \| = 1$, and $\| y_n - x \| > \frac{1}{2}$ for all $x \in \mathfrak{L}_{n-1}$. In particular $y_n = \sum_1^n \alpha_{n\nu} x_\nu$ and hence $T(y_n) - \lambda_n y_n \in \mathfrak{L}_{n-1}$. It follows that for $n > k$,

$$\| T(y_n) - T(y_k) \| = \| \lambda_n y_n + \{[T(y_n) - \lambda_n y_n] - T(y_k)\} \| > | \lambda_n |/2;$$

however this cannot be the case if the $T(y_n)$ lie in a compact subset of $\mathfrak{X}$.

COROLLARY. *If* $T \in \mathfrak{E}(\mathfrak{X})$ *is compact, then* $R\sigma(T)$ *consists of isolated points with the possible exception of* $\lambda = 0$.

PROOF. In general if $\lambda \in R\sigma(T)$, then λ belongs to the point spectrum of the adjoint operator T^*. Since $T^* \in \mathfrak{E}(\mathfrak{X}^*)$ is also compact by Theorem 2.13.5, the desired result follows directly from Theorem 5.7.1.

THEOREM 5.7.2. *If* $T \in \mathfrak{E}(\mathfrak{X})$ *is compact, then* $\sigma(T) \equiv P\sigma(T) \cup R\sigma(T) \cup [0]$.

PROOF. Suppose that $\lambda \notin P\sigma(T) \cup R\sigma(T) \cup [0]$. Then clearly $(\lambda I - T)$ is one-to-one and the range $\mathfrak{R}_\lambda$ of $(\lambda I - T)$ is dense in $\mathfrak{X}$. It is required to show that $\lambda \in \rho(T)$, in other words, that $\mathfrak{R}_\lambda \equiv \mathfrak{X}$ and that $(\lambda I - T)^{-1}$ is bounded. To this end suppose that $(\lambda I - T)(x_n) = y_n$ and that $y_n \to y_0$. It will be sufficient to prove (i) that $y_0 \in \mathfrak{R}_\lambda$ which implies that $R_\lambda \equiv \mathfrak{X}$; and (ii) that the x_n form a Cauchy sequence which implies that $(\lambda I - T)^{-1}$ is continuous. In the first place the $\| x_n \|$ must be bounded. For suppose $\lim \sup_{n\to\infty} \| x_n \| = \infty$ and we may clearly assume that this holds with lim sup replaced by lim. Then setting $z_n = x_n/\| x_n \|$, we see that $(\lambda I - T)(z_n) \to \theta$. The operator T being compact, a subsequence $\{z_{n_k}\}$ and an element z_0 exist such that $T(z_{n_k}) \to \lambda z_0$ and since $\lambda \neq 0$ we obtain $z_{n_k} \to z_0$. Thus $T(z_0) = \lambda z_0$ and this means that $\lambda \in P\sigma(T)$, contrary to hypothesis. Consequently the $\| x_n \|$ are bounded and hence there exist a subsequence $\{x_{n_k}\}$ and an element w_0 such that $T(x_{n_k}) \to w_0$. Thus $\lambda x_{n_k} = T(x_{n_k}) + y_{n_k} \to w_0 + y_0$. Finally we set $x_0 = (w_0 + y_0)/\lambda$. Then $x_{n_k} \to x_0$ and $(\lambda I - T)(x_0) = y_0$ so that $y_0 \in \mathfrak{R}_\lambda$. Moreover since $\lambda \notin P\sigma(T)$, x_0 is uniquely determined by y_0 and hence (by the same argument) any subsequence of $\{x_n\}$ must contain a subsequence which converges to x_0. From this it follows that $x_n \to x_0$.

LEMMA 5.7.1. *If* $T \in \mathfrak{E}(\mathfrak{X})$ *is compact, then*

$$S(\lambda; T) = R(\lambda; T) - \lambda^{-1}I, \qquad \lambda \in \rho(T), \tag{5.7.1}$$

is also compact.

PROOF. For $| \lambda | > \| T \|$ we obtain directly from Theorem 4.7.2 that $S(\lambda; T) = \sum_1^\infty T^n\lambda^{-n-1}$ so that $S(\lambda; T)$ is the limit in the uniform operator topology of compact operators and hence is itself compact. Substituting (5.7.1) into the first resolvent equation we get

$$S(\mu; T) = \lambda^2\mu^{-2}S(\lambda; T) + \lambda\mu^{-1}(\lambda - \mu)S(\lambda; T)S(\mu; T).$$

Choosing $| \lambda | > \| T \|$, this relation then exhibits $S(\mu; T)$ as the sum of two compact operators and this proves the lemma.

LEMMA 5.7.2. *Let* $T \in \mathfrak{E}(\mathfrak{X})$ *be compact and let* $\lambda_0 \in \sigma(T)$, $\lambda_0 \neq 0$. *Then for any* *sufficiently small circle Γ_0 about λ_0, the only spectral point of T contained in Γ_0 will be λ_0 and*

$$J_0 = \frac{1}{2\pi i} \int_{\Gamma_0} R(\zeta; T)\, d\zeta \tag{5.7.2}$$

is a degenerate (i.e. range is finite dimensional) projection operator. Further

$$J_0^* = \frac{1}{2\pi i} \int_{\Gamma_0} R(\zeta; T^*)\, d\zeta. \tag{5.7.3}$$

PROOF. Combining Theorem 5.7.1 together with its corollary and Theorem 5.7.2, we see that the points of $\sigma(T)$ are isolated with the possible exception of $\lambda = 0$. Consequently circles Γ_0 with the required properties will exist. In particular Γ_0 will not contain $\lambda = 0$ so that $\int_{\Gamma_0} \zeta^{-1}\, d\zeta = 0$. Recalling the definition of $S(\lambda; T)$ given by (5.7.1) we see that

$$J_0 = \frac{1}{2\pi i} \int_{\Gamma_0} S(\zeta; T)\, d\zeta.$$

Now $S(\lambda; T)$ is clearly holomorphic on Γ_0 and consequently the above integral is the limit in the uniform operator topology of Riemann sums. By Lemma 5.7.1, the terms of these sums and hence the sums themselves are compact operators. It follows that J_0 is also a compact operator. Since J_0 is a projection operator any bounded set in the range of J_0 must lie in a compact subset of $\mathfrak{X}$ and, according to Theorem 1.12.2, this means that the range of J_0 is finite dimensional. Finally the relation (5.7.3) is an immediate consequence of Theorem 3.3.3 and the fact that $[R(\lambda; T)]^* = R(\lambda; T^*)$ (see Theorem 2.16.5).

In the statement of the following theorem we denote the range of J_0 by $\mathfrak{R}_0$ and that of J_0^* by $\mathfrak{R}_0^*$.

THEOREM 5.7.3. *Let $T \in \mathfrak{E}(\mathfrak{X})$ be compact. Aside from $\lambda = 0$, the points of $\sigma(T)$ are isolated and belong to the point spectrum of T. For $\lambda_0 \in \sigma(T)$, $\lambda_0 \neq 0$, let $\mathfrak{M}_k \subset \mathfrak{X}$ and $\mathfrak{M}_k^* \subset \mathfrak{X}^*$ be the null spaces of the operators $(\lambda_0 I - T)^k$ and $(\lambda_0 I^* - T^*)^k$ respectively. Then for $k > 0$, $\mathfrak{M}_k$ ($\mathfrak{M}_k^*$) is always of positive finite dimension and there exists an $n_0 > 0$ such that $\mathfrak{M}_k = \mathfrak{R}_0$ ($\mathfrak{M}_k^* = \mathfrak{R}_k^*$) for $k \geqq n_0$, whereas $\mathfrak{M}_{k-1}$ ($\mathfrak{M}_{k-1}^*$) is a proper subset of $\mathfrak{M}_k$ ($\mathfrak{M}_k^*$) for $k \leqq n_0$. The dimension of $\mathfrak{M}_k^*$ is equal to that of $\mathfrak{M}_k$ for all $k > 0$.*

PROOF. Let $\lambda_0 \neq 0$ be a point in $\sigma(T)$. We show first that if $(\lambda_0 I - T)^k(x) = \theta$ then $x \in \mathfrak{R}_0$. In case $k = 1$ and $(\lambda_0 I - T)(x) = \theta$, then $R(\zeta; T)(x) = (\zeta - \lambda_0)^{-1}x$ and direct substitution in (5.7.2) gives $J_0 x = x$. Suppose next that $(\lambda_0 I - T)(x) = y \in \mathfrak{R}_0$. Then $(\lambda_0 I - T)(I - J_0)(x) = (I - J_0)(y) = \theta$ and therefore $(I - J_0)(x) \in \mathfrak{R}_0$. This means that $J_0(I - J_0)(x) = (I - J_0)(x)$ and hence that $J_0 x = x$. Finally if $(\lambda_0 I - T)^k(x) = \theta$, then $(\lambda_0 I - T)[(\lambda_0 I - T)^{k-1}(x)] \in \mathfrak{R}_0$ and by the above $(\lambda_0 I - T)^{k-1}(x) \in \mathfrak{R}_0$. Thus k iterations of this argument will yield $x \in \mathfrak{R}_0$. Consequently $\mathfrak{M}_k \subset \mathfrak{R}_0$ and similarly $\mathfrak{M}_k^* \subset \mathfrak{R}_0^*$.

We may therefore limit ourselves to the finite dimensional subspaces $\mathfrak{N}_0$ and $\mathfrak{N}_0^*$. Since $TJ_0 = J_0T$ $(T^*J_0^* = J_0^*T^*)$ the transformation T (T^*) takes $\mathfrak{N}_0$ $(\mathfrak{N}_0^*)$ into itself. The theorem can now be reduced to the corresponding result for matrices. To this end we introduce the realizations

$$J_0(x) = \sum_{i=1}^{n} y_i^*(x)y_i, \qquad J_0^*(x^*) = \sum_{i=1}^{n} x^*(y_i)y_i^*,$$

obtained in Theorem 2.13.3. Here the $\{y_i\}$ $(\{y_i^*\})$ are linearly independent and span $\mathfrak{N}_0$ $(\mathfrak{N}_0^*)$. For an $x = \sum_1^n \alpha_j y_j \in \mathfrak{N}_0$

$$T\ (x) = J_0\{T(x)\} = \sum_{i=1}^{n}\left[\sum_{j=1}^{n} \alpha_j y_i^*\{T(y_j)\}\right] y_i,$$

whereas for $x^* = \sum_1^n \beta_j y_j^* \in \mathfrak{N}_0^*$

$$T^*(x^*) = J_0^*\{T^*(x^*)\} = \sum_{i=1}^{n}\left[\sum_{j=1}^{n} \beta_j\{T^*(y_j^*)\}(y_i)\right] y_i^*.$$

It is clear that the matrix $(y_i^*\{T(y_j)\})$ is the transpose of $(\{T^*(y_j^*)\}(y_i))$ so that we have only to appeal to the corresponding results in the finite dimensional problem. In order to show that $\mathfrak{M}_k = \mathfrak{N}_0$ $(\mathfrak{M}_k^* = \mathfrak{N}_0^*)$ for k sufficiently large and that $\lambda_0 \in P\sigma(T)$ we need in addition the fact that this matrix has only one characteristic value, namely λ_0; or equivalently, that $Tx_1 = \lambda_1 x_1$ for $x_1 \in \mathfrak{N}_0$ implies that $\lambda_1 = \lambda_0$. However in this case $R(\zeta; T)x_1 = (\zeta - \lambda_1)^{-1}x_1$ and it is clear from (5.7.2) that $J_0 x_1 = x_1$ only if λ_1 is interior to Γ_0 for all sufficiently small Γ_0, that is, only if $\lambda_1 = \lambda_0$.

References. Banach [**2**], Dunford [**7, 8**], Fantappiè [**1, 3**], Frobenius [**1**], Gelfand [**4**], Giorgi [**1**], Hildebrandt [**2**], Lorch [**3, 4**], F. Riesz [**2, 3**], Schauder [**1**], Šilov [**3**], Taylor [**7**], Walsh [**1**].

2. Pseudo-Resolvents

5.8. The first resolvent equation. We have already noted that the results of the previous paragraph are applicable to the elements of the algebra of endomorphisms $\mathfrak{E}(\mathfrak{X})$. However they do not apply to unbounded linear transformations on the (B)-space $\mathfrak{X}$ to itself and it is precisely these transformations which are of most interest to the analyst. In the remainder of this chapter we shall develop a parallel theory for unbounded operators.

We assume that T is a closed linear transformation with domain $\mathfrak{D}(T)$ and range $\mathfrak{R}(T)$, both contained in $\mathfrak{X}$; that is, $T \in \mathfrak{C}(\mathfrak{X})$. Then $T_\lambda \equiv \lambda I - T$ defined on $\mathfrak{D}(T)$ is closed and when T_λ^{-1} exists it will also be closed by Theorem 2.11.5. In case $\lambda \in \rho(T)$, then $T_\lambda^{-1} \equiv R(\lambda; T)$ belongs to $\mathfrak{E}(\mathfrak{X})$. Thus for $\lambda \in \rho(T)$

$$(5.8.1) \qquad \begin{aligned} (\lambda I - T)R(\lambda; T)x &= x, & x &\in \mathfrak{X}; \\ R(\lambda; T)(\lambda I - T)x &= x, & x &\in \mathfrak{D}(T). \end{aligned}$$

THEOREM 5.8.1. *If λ and μ belong to $\rho(T)$, then*

$$(5.8.2) \qquad R(\lambda; T) - R(\mu; T) = -(\lambda - \mu)R(\lambda; T)R(\mu; T).$$

PROOF. If $\lambda, \mu \in \rho(T)$, then for all $x \in \mathfrak{X}$

$$\begin{aligned} R(\lambda; T)x &= R(\lambda; T)(\mu I - T)R(\mu; T)x \\ &= R(\lambda; T)[(\mu - \lambda)I + (\lambda I - T)]R(\mu; T)x \\ &= -(\lambda - \mu)R(\lambda; T)R(\mu; T)x + R(\mu; T)x, \end{aligned}$$

where the last step, namely $R(\lambda; T)(\lambda I - T)R(\mu; T)x = R(\mu; T)x$, is justified since $R(\mu; T)x \in \mathfrak{D}(T)$.

It follows from the above theorem that $R(\lambda; T)$ is a solution of the first resolvent equation (4.8.2) in its domain of existence, $\rho(T)$. Although $R(\lambda; T)$ is called a resolvent and belongs to $\mathfrak{E}(\mathfrak{X})$, it is not a resolvent in the sense of Definition 4.7.1 unless T itself belongs to $\mathfrak{E}(\mathfrak{X})$. In any case, whether T is bounded or not, $R(\lambda; T)$ may properly be called a *pseudo-resolvent* since it is a solution of the first resolvent equation. We now turn our attention to the study of pseudo-resolvents.

THEOREM 5.8.2. *Let $R(\lambda)$ be a single-valued function defined on an open set Δ of the complex plane to a Banach algebra $\mathfrak{B}$, satisfying*

$$(5.8.3) \qquad R(\lambda) - R(\mu) = -(\lambda - \mu)R(\lambda)R(\mu)$$

for all values of λ and μ in Δ. Then $R(\lambda)$ is locally holomorphic in Δ.

PROOF. The equation (5.8.3) shows that $R(\lambda)$ and $R(\mu)$ commute. Let $\lambda_0 \in \Delta$ and replace μ by λ_0 in the equation. We then have

$$R(\lambda)[e - (\lambda_0 - \lambda)R(\lambda_0)] = R(\lambda_0).$$

By Theorem 4.3.1 the second factor on the left has an inverse for λ such that $|\lambda - \lambda_0| \, \| R(\lambda_0) \| < 1$ and multiplying both sides by this inverse gives

$$(5.8.4) \qquad R(\lambda) = R(\lambda_0) \left\{ e + \sum_{n=1}^{\infty} (\lambda_0 - \lambda)^n [R(\lambda_0)]^n \right\}.$$

By Theorem 3.11.4, $R(\lambda)$ is holomorphic in a neighborhood of $\lambda = \lambda_0$, and, λ_0 being arbitrary, this means that $R(\lambda)$ is locally holomorphic in Δ.

The formula for the derivatives of $R(\lambda)$ can be read off from (5.8.4):

$$(5.8.5) \qquad R^{(n)}(\lambda) = (-1)^n n! [R(\lambda)]^{n+1}.$$

THEOREM 5.8.3. *Let $R(\lambda)$ be a single-valued function defined on a subset E of the complex plane to the (B)-algebra $\mathfrak{B}$ and satisfying*

$$(5.8.6)\quad R(\lambda) - R(\lambda_0) = -(\lambda - \lambda_0)R(\lambda)R(\lambda_0) = -(\lambda - \lambda_0)R(\lambda_0)R(\lambda)$$

for all values of $\lambda \in E$; here $\lambda_0 \in E$ is fixed. A necessary and sufficient condition that $R(\lambda)$ be the resolvent of some element $a \in \mathfrak{B}$ is that $R(\lambda_0)$ be regular in $\mathfrak{B}$; in this case $a = \lambda_0 e - [R(\lambda_0)]^{-1}$. If $\mathfrak{B} = \mathfrak{E}(\mathfrak{X})$, then a necessary and sufficient condition that $R(\lambda)$ be the resolvent of some closed linear operator $T \in \mathfrak{O}(\mathfrak{X})$ is that $R(\lambda_0)$ have an inverse (not necessarily bounded); in this case $T = \lambda_0 I - [R(\lambda_0)]^{-1}$.

PROOF. If $R(\lambda)$ is the resolvent of an element $a \in \mathfrak{B}$ for all $\lambda \in E$, then $R(\lambda)$ is regular for every $\lambda \in E$ (in fact, for every $\lambda \in \rho(a)$). It follows that the stated condition is necessary. But it is also sufficient. Suppose that $[R(\lambda_0)]^{-1}$ exists in $\mathfrak{B}$ and put $a = \lambda_0 e - [R(\lambda_0)]^{-1}$. Then $\lambda e - a = (\lambda - \lambda_0)e + [R(\lambda_0)]^{-1}$ and by (5.8.6)

$$(5.8.7)\quad R(\lambda)\{R(\lambda_0)^{-1} + (\lambda - \lambda_0)e\} = R(\lambda)\{e + (\lambda - \lambda_0)R(\lambda_0)\}[R(\lambda_0)]^{-1} = e.$$

A similar argument shows that $R(\lambda)$ is also the right inverse of $\lambda e - a$ and hence that $R(\lambda)$ is the resolvent of a. Since $R(\lambda)$ is regular for all λ in E, it is clear that a is actually independent of λ_0.

In case $\mathfrak{B} = \mathfrak{E}(\mathfrak{X})$ and $R(\lambda)$ is the resolvent for some $T \in \mathfrak{O}(\mathfrak{X})$ for all $\lambda \in E$, then $R(\lambda)$ will have an inverse, namely $\lambda I - T$, for every $\lambda \in \rho(T)$ and hence for every $\lambda \in E$. Conversely, suppose that $[R(\lambda_0)]^{-1}$ exists and put $T = \lambda_0 I - [R(\lambda_0)]^{-1}$. Since $[R(\lambda_0)]^{-1}$ is closed (by Theorem 2.11.5) so is T. A calculation similar to that employed in (5.8.7) shows that $R(\lambda)$ satisfies both of the relations (5.8.1) and hence that $R(\lambda) = R(\lambda; T)$ for all $\lambda \in E$. This concludes the proof.

COROLLARY. *If $R(\lambda)$ defined on E is the resolvent of an element $a \in \mathfrak{B}$ (or of a closed linear operator $T \in \mathfrak{O}(\mathfrak{X})$), then* (1) *any analytic extension, or* (2) *any extension which satisfies* (5.8.3) *is also the resolvent of a (or T).*

PROOF. In either case the extension satisfies (5.8.6) for some $\lambda_0 \in E$ and hence the assertions follow directly from Theorem 5.8.3.

In case $\mathfrak{B} = \mathfrak{E}(\mathfrak{X})$, it may happen that $R(\lambda; T)x$ has an analytic extension beyond $\rho(T)$ for some $x \in \mathfrak{X}$. In fact this extension may even be multiple-valued as the following example, suggested by S. Kakutani, shows. Let $\mathfrak{X}$ be the (B)-space of functions $f(\zeta)$ holomorphic in $|\zeta| < 1$ and continuous in $|\zeta| \leqq 1$ with $\|f(\cdot)\| = \sup |f(\zeta)|$. Define $T(f) = [f(\zeta) - f(0)]/\zeta$. It follows from the Schwarz lemma that $\|T\| \leqq 2$. It can be easily verified that for $|\lambda| > 1$

$$R(\lambda; T)f = (1 - \lambda\zeta)^{-1}[\lambda^{-1}f(\lambda^{-1}) - \zeta f(\zeta)].$$

On the other hand for each λ, $|\lambda| < 1$, $f(\zeta) = (1 - \lambda\zeta)^{-1}$ is a characteristic function of the equation $Tf = \lambda f$. Consequently $\rho(T) \equiv [\lambda, |\lambda| > 1]$. Now let $g(\zeta) = (2 - \zeta)^{1/2}$, where we choose that branch of the radical which is positive for real ζ, $|\zeta| \leqq 1$. Then

$$R(\lambda; T)g = (1 - \lambda\zeta)^{-1}[\lambda^{-1}(2 - \lambda^{-1})^{1/2} - \zeta(2 - \zeta)^{1/2}].$$

For $|\lambda| = 1$, $R(\lambda; T)g$ lies in $\mathfrak{X}$ provided both of the radicals are evaluated on the same branch. For $0 < |\lambda| < 1$, $R(\lambda; T)g$ lies in $\mathfrak{X}$ for all possible evaluations of the term $(2 - \lambda^{-1})^{1/2}$. In particular as λ winds about $\lambda = \frac{1}{2}$, staying within the unit circle, we go from one branch to the other. We cannot then recross $|\lambda| = 1$ until we have encircled the point $\lambda = \frac{1}{2}$ an even number of times.

For the terminology used in the following two theorems the reader is referred to the Gelfand representation theory developed in paragraph 4.3.

THEOREM 5.8.4. *Let $R_0(\lambda)$ be a single-valued function defined on a subset E_0 of the complex plane to $\mathfrak{B}$ and satisfying* (5.8.3) *for all λ, $\mu \in E_0$. Let $\mathfrak{B}_0$ be any closed commutative subalgebra of $\mathfrak{B}$ containing the set $[R_0(\lambda); \lambda \in E_0]$ and the unit element. Then the set of maximal ideals $\mathfrak{M}_0$ in $\mathfrak{B}_0$ splits into two disjoint sets $\mathfrak{W}_0$, $\mathfrak{U}_0$ with $\mathfrak{M}_0 = \mathfrak{W}_0 \cup \mathfrak{U}_0$ and there exists a numerically-valued function $\alpha(\mathfrak{m})$ defined and continuous on $\mathfrak{W}_0$ such that*

$$(5.8.8) \qquad \begin{aligned} R_0(\lambda)(\mathfrak{m}) &= \frac{1}{\lambda - \alpha(\mathfrak{m})}, & \mathfrak{m} \in \mathfrak{W}_0, \\ &= 0, & \mathfrak{m} \in \mathfrak{U}_0, \end{aligned}$$

for all $\lambda \in E_0$. Further if we set $\sigma_0 \equiv [\alpha(\mathfrak{m}); \mathfrak{m} \in \mathfrak{W}_0]$, then σ_0 is a closed (but not necessarily bounded) subset of the complex plane and $\sigma_0 \cap E_0 = \emptyset$.

PROOF. Since $x \to x(\mathfrak{m})$ is a homomorphism of $\mathfrak{B}_0$ onto the complex field, it is clear that $R_0(\lambda)(\mathfrak{m})$ satisfies (5.8.3) for each $\mathfrak{m} \in \mathfrak{M}_0$ and all λ, $\mu \in E_0$. Hence if $R_0(\lambda_0)(\mathfrak{m}_0) = 0$ for some $\lambda_0 \in E_0$ and $\mathfrak{m}_0 \in \mathfrak{M}_0$, then it follows from (5.8.3) that $R_0(\lambda)(\mathfrak{m}_0) \equiv 0$ for all $\lambda \in E_0$. Thus $\mathfrak{U}_0$ is simply the set of maximal ideals such that $R_0(\lambda)(\mathfrak{m}) = 0$ for some $\lambda \in E_0$ and hence for all $\lambda \in E_0$. If $\mathfrak{m} \in \mathfrak{W}_0 \equiv \mathfrak{M}_0 \ominus \mathfrak{U}_0$, then $R_0(\lambda)(\mathfrak{m}) \not\equiv 0$ and we set $\alpha(\mathfrak{m}) = \lambda_0 - [R_0(\lambda_0)(\mathfrak{m})]^{-1}$. Since $R_0(\lambda_0)(\mathfrak{m})$ is continuous on $\mathfrak{W}_0$, the same will be true of $\alpha(\mathfrak{m})$. A simple calculation, making use of (5.8.3), shows that $R_0(\lambda)(\mathfrak{m}) = [\lambda - \alpha(\mathfrak{m})]^{-1}$ for all $\lambda \in E_0$. Since the representation is finite-valued for all $\lambda \in E_0$, the set σ_0 must be disjoint from E_0. It remains to prove that σ_0 is closed. Suppose that $\alpha_0 \neq \infty$ belongs to the closure of σ_0. Then there exists a subsequence $\{\mathfrak{m}_n\}$ such that $\alpha(\mathfrak{m}_n) \to \alpha_0$. Since $\mathfrak{M}_0$ is compact there will exist at least one limit point $\mathfrak{m}_0$ of the set $\{\mathfrak{m}_n\}$ and, since the representations are continuous, we must have $\alpha(\mathfrak{m}_0) = \alpha_0$. Thus $\alpha_0 \in \sigma_0$ and this concludes the proof.

COROLLARY 1. *Let ρ_0 be the set complementary to σ_0. Then $[e - (\mu - \lambda)R_0(\mu)]^{-1}$ exists in $\mathfrak{B}_0$ for λ, $\mu \in \rho_0$.*

PROOF. We have only to consider the representation. In fact, for λ, $\mu \in \rho_0$ we have

$$\begin{aligned} [e - (\mu - \lambda)R_0(\mu)](\mathfrak{m}) &= 1 - (\mu - \lambda)[\mu - \alpha(\mathfrak{m})]^{-1} \\ &= \frac{\lambda - \alpha(\mathfrak{m})}{\mu - \alpha(\mathfrak{m})} \neq 0, \qquad \mathfrak{m} \in \mathfrak{W}_0, \\ &= 1, \qquad \mathfrak{m} \in \mathfrak{U}_0. \end{aligned}$$

The result is now an immediate consequence of Theorem 4.15.1 (9).

COROLLARY 2. *Let $a \in \mathfrak{B}$ and suppose $\lambda_0 \in \rho(a)$. Then $\lambda \in \rho(a)$ if and only if $e - (\lambda_0 - \lambda)R(\lambda_0 ; a)$ is regular in $\mathfrak{B}$. In this case*

$$R(\lambda; a) = R(\lambda_0 ; a)[e - (\lambda_0 - \lambda)R(\lambda_0 ; a)]^{-1}. \tag{5.8.9}$$

If $\mathfrak{B} = \mathfrak{E}(\mathfrak{X})$ and $T \in \mathfrak{O}(\mathfrak{X})$, then the same result, with a replaced by T, is valid.

PROOF. Choose $E_0 = \rho(a)$ and suppose that $\lambda \in \rho(a)$. Then $\rho_0 \supset \rho(a)$ and hence by the corollary $[e - (\lambda_0 - \lambda)R(\lambda_0 ; a)]^{-1}$ exists. Conversely, suppose $e - (\lambda_0 - \lambda)R(\lambda_0 ; a)$ is regular in $\mathfrak{B}$. We then set

$$R \equiv R(\lambda_0 ; a)[e - (\lambda_0 - \lambda)R(\lambda_0 ; a)]^{-1}.$$

It is clear that the factors in the right member commute with each other and with a. Hence

$$\begin{aligned} R(\lambda e - a) &= (\lambda e - a)R \\ &= [(\lambda - \lambda_0)e + (\lambda_0 e - a)]R(\lambda_0 ; a)[e - (\lambda_0 - \lambda)R(\lambda_0 ; a)]^{-1} \\ &= [(\lambda - \lambda_0)R(\lambda_0 ; a) + e][e - (\lambda_0 - \lambda)R(\lambda_0 ; a)]^{-1} = e. \end{aligned}$$

If $\mathfrak{B} = \mathfrak{E}(\mathfrak{X})$ and $T \in \mathfrak{O}(\mathfrak{X})$, the same argument shows that $I - (\lambda_0 - \lambda)R(\lambda_0 ; T)$ is regular in $\mathfrak{E}(\mathfrak{X})$ if $\lambda \in \rho(T)$ and, conversely, if this element is regular then R is a right inverse of $(\lambda I - T)$. On the other hand it is clear that

$$R \equiv R(\lambda_0 ; T)\ [I - (\lambda_0 - \lambda)R(\lambda_0 ; T)]^{-1}$$

has domain $\mathfrak{X}$ and range equal to the range of $R(\lambda_0 ; T)$, namely $\mathfrak{D}(T)$. Now a right inverse with range $\mathfrak{D}(T)$ is also a left inverse so that $R = R(\lambda; T)$. This concludes the proof.

In the notation of Theorem 5.8.4, we have

THEOREM 5.8.5. *If $R_0(\lambda) = R(\lambda; a)$, $a \in \mathfrak{B}$, and if $E_0 = \rho(a)$, then $\sigma_0 = \sigma(a)$ and $\mathfrak{U}_0$ is empty. If $R_0(\lambda) = R(\lambda; T)$, $T \in \mathfrak{O}(\mathfrak{X})$, and $E_0 = \rho(T)$ is non-void, then $\sigma_0 = \sigma(T)$; in this case $\mathfrak{U}_0$ is empty if and only if $T \in \mathfrak{E}(\mathfrak{X})$.*

PROOF. Since $E_0 \cap \sigma_0 = \emptyset$, it is clear that σ_0 is contained in $\sigma(a)$. On the other hand suppose that $\lambda \notin \sigma_0$. Then for $\lambda_0 \in \rho(a)$, we see by Corollary 1 above that $e - (\lambda_0 - \lambda)R(\lambda_0 ; a)$ is regular in $\mathfrak{B}$ and by Corollary 2 that $\lambda \in \rho(a)$. Thus $\sigma_0 = \sigma(a)$. If $T \in \mathfrak{O}(\mathfrak{X})$ has a non-vacuous resolvent set, then the same argument proves that $\sigma_0 = \sigma(T)$. Again if $R_0(\lambda) = R(\lambda; a)$, $a \in \mathfrak{B}$, then by the operational calculus

$$a = \frac{1}{2\pi i}\int_\Gamma \zeta R(\zeta; a)\, d\zeta \tag{5.8.10}$$

where Γ is an oriented envelope of $\sigma(a)$. Since the integrand lies in $\mathfrak{B}_0$ and since $\mathfrak{B}_0$ is closed, it follows that $a \in \mathfrak{B}_0$. This means that $[R(\lambda; a)]^{-1} \in \mathfrak{B}_0$ and hence by the Gelfand theory that $R(\lambda; a)(\mathfrak{m}) \neq 0$ for all $\mathfrak{m} \in \mathfrak{M}_0$. As a consequence $\mathfrak{U}_0$ is empty. This argument applies equally well for $\mathfrak{B} = \mathfrak{E}(\mathfrak{X})$ and T bounded.

On the other hand if $\mathfrak{U}_0$ is empty, then $R(\lambda; T)(\mathfrak{m}) \neq 0$ for all $\mathfrak{m} \in \mathfrak{M}_0$ and hence by Theorem 4.15.1 the operator $[R(\lambda; T)]^{-1}$ lies in $\mathfrak{B}_0 \subset \mathfrak{E}(\mathfrak{X})$. However

$$\lambda I - T = (\lambda I - T)R(\lambda; T)[R(\lambda; T)]^{-1} = [R(\lambda; T)]^{-1} \in \mathfrak{E}(\mathfrak{X})$$

and hence $T \in \mathfrak{E}(\mathfrak{X})$.

COROLLARY. *If $R_0(\lambda) = R(\lambda; a)$, $a \in \mathfrak{B}$, and if $E_0 = \rho(a)$, then $a \in \mathfrak{B}_0$ and the function $\alpha(\mathfrak{m})$ which appears in* (5.8.8) *is the Gelfand representation of a in $\mathfrak{B}_0$.*

PROOF. This is an immediate consequence of (5.8.10) since

$$a(\mathfrak{m}) = \frac{1}{2\pi i}\int_\Gamma \zeta R(\zeta; a)(\mathfrak{m})\, d\zeta = \frac{1}{2\pi i}\int_\Gamma \zeta[\zeta - \alpha(\mathfrak{m})]^{-1}\, d\zeta = \alpha(\mathfrak{m}).$$

We conclude this section with a general characterization of pseudo-resolvents.

LEMMA 5.8.1. *Suppose $r \in \mathfrak{B}$ and λ_0 are given. Denote by $E(\lambda_0)$ the set of all λ such that $[e - (\lambda_0 - \lambda)r]^{-1}$ exists in $\mathfrak{B}$. Then the function*

$$R_{\lambda_0}(\lambda) \equiv r[e - (\lambda_0 - \lambda)r]^{-1}$$

defined on $E(\lambda_0)$ satisfies (5.8.3). *Each point of the circle $|\lambda - \lambda_0| < \|r\|^{-1}$ lies in $E(\lambda_0)$.*

PROOF. It is easy to see that all of the elements of $\mathfrak{B}$ involved in the composition of $R_{\lambda_0}(\lambda)$, $\lambda \in E(\lambda_0)$, commute. Thus for $\lambda, \mu \in E(\lambda_0)$ we have

$$\begin{aligned}
[R_{\lambda_0}(\lambda) - R_{\lambda_0}(\mu)][e - (\lambda_0 - \lambda)r][e - (\lambda_0 - \mu)r]& \\
&= r\{[e - (\lambda_0 - \mu)r] - [e - (\lambda_0 - \lambda)r]\} = -(\lambda - \mu)r^2 \\
&= -(\lambda - \mu)R_{\lambda_0}(\lambda)R_{\lambda_0}(\mu)[e - (\lambda_0 - \lambda)r][e - (\lambda_0 - \mu)r].
\end{aligned}$$

Multiplying the end members by $[e - (\lambda_0 - \mu)r]^{-1}[e - (\lambda_0 - \lambda)r]^{-1}$ we obtain the desired result. It is clear by Theorem 4.3.1 that each point of the circle, $|\lambda - \lambda_0| < \|r\|^{-1}$, lies in $E(\lambda_0)$.

THEOREM 5.8.6. *Let $R_o(\lambda)$ be a single-valued function defined on a subset E_o of the complex plane to $\mathfrak{B}$ and satisfying* (5.8.3). *Then there exists a unique extension of $R_o(\lambda)$ satisfying* (5.8.3) *which contains all other extensions of $R_o(\lambda)$ satisfying* (5.8.3). *This maximal extension can be defined as in Lemma* 5.8.1 *over $E(\lambda_0)$ where $r = R_o(\lambda_0)$ and λ_0 is an arbitrary point in E_o. $E(\lambda_0)$ is an open subset of the complex plane.*

PROOF. Because of Corollary 1 to Theorem 5.8.4 we see at once that $E(\lambda_0) \supset E_o$. Thus $[e - (\lambda_0 - \lambda)R_o(\lambda_0)]^{-1}$ exists in $\mathfrak{B}$ for each $\lambda \in E_o$ and it is easily seen from the first resolvent equation that $R_o(\lambda) = R_{\lambda_0}(\lambda)$ for all $\lambda \in E_o$. Consequently $R_{\lambda_0}(\lambda)$ is actually an extension of $R_o(\lambda)$ satisfying (5.8.3). However this argument applies to any extension of $R_o(\lambda)$ as well as to $R_o(\lambda)$ itself. Hence $R_{\lambda_0}(\lambda)$ contains every extension of $R_o(\lambda)$ which satisfies (5.8.3). The resulting extension is clearly independent of $\lambda_0 \in E_o$. If we choose $\mathfrak{B}_0$ in Theorem 5.8.4 to contain the set $[R_{\lambda_0}(\lambda); \lambda \in E(\lambda_0)]$, then we see that $E(\lambda_0) \subset \rho_o$. On the other hand according to Corollary 1 to Theorem 5.8.4, each $\lambda \in \rho_o$ belongs to $E(\lambda_0)$. It follows that $E(\lambda_0) = \rho_o$. Since σ_o, the complementary set to ρ_o, is closed, we see that $E(\lambda_0)$ is open.

COROLLARY 1. *Suppose $r \in \mathfrak{B}$ and λ_0 are given. Then there exists a unique function $R(\lambda)$ satisfying* (5.8.3) *with $R(\lambda_0) = r$ which contains all other such functions.*

PROOF. The function $R_o(\lambda) = r$ on the set $E_o = \{\lambda_0\}$ satisfies (5.8.3) trivially so that the assertion is an immediate consequence of the above theorem.

COROLLARY 2. *Suppose $R(\lambda)$ defined on Δ is a maximally extended pseudo-resolvent. Then $R(\lambda)$, which is locally holomorphic in Δ, cannot be continued analytically beyond Δ.*

PROOF. It follows from the above theorem that Δ is open and from Theorem 5.8.2 that $R(\lambda)$ is locally holomorphic in Δ. Suppose that $R(\lambda)$ could be continued analytically along a path P starting at some point within Δ and extending outside of Δ. Then for the so-extended function, $\| R(\lambda) \|$ will be bounded along P and hence there will exist a point $\lambda_0 \in \Delta \cap P$ such that the circle $| \lambda - \lambda_0 | < \| R(\lambda_0) \|^{-1}$ contains points not in Δ. However this is impossible since on the one hand $E(\lambda_0)$ must contain this circle (by Lemma 5.8.1) and on the other $E(\lambda_0) = \Delta$ (by Theorem 5.8.6).

The first resolvent equation admits of several simple transformations which leave it invariant and which consequently enable us to construct new solutions. The following are worthy of notice; the verifications are left to the reader.

(1) If $R(\lambda)$ is a solution, so is $R(\lambda + \alpha)$ for any fixed α.

(2) If j is an idempotent in $\mathfrak{B}$ and j commutes with $R(\lambda)$, then $jR(\lambda)$ is also a solution.

(3) If $R(\lambda)$ is a solution, so is

$$\frac{e}{\lambda} + \frac{1}{\lambda^2} R\left(-\frac{1}{\lambda}\right).$$

5.9. Expansion theorems for pseudo-resolvents. In the previous section it was shown that a pseudo-resolvent, defined on an open subset of the complex plane to a Banach algebra, is locally holomorphic. It is natural, therefore, to study the various local expansion theorems for such functions.

THEOREM 5.9.1. *Each solution of the first resolvent equation which is defined and single-valued in some neighborhood of $\lambda = \lambda_0$ is of the form*

$$R(\lambda) = \sum_{n=0}^{\infty} (\lambda_0 - \lambda)^n x^{n+1}, \tag{5.9.1}$$

where x is an element of $\mathfrak{B}$. Conversely, each series of this type satisfies the resolvent equation in its circle of convergence. It is a resolvent in the sense of Definition 4.7.1 if and only if x is regular.

PROOF. If we set $x = R(\lambda_0)$ in the series representation for $R(\lambda)$ found in (5.8.4), we obtain (5.9.1). Conversely the series (5.9.1) represents $R_{\lambda_0}(\lambda) \equiv x[e + (\lambda - \lambda_0)x]^{-1}$ within its circle of convergence and $R_{\lambda_0}(\lambda)$ has been shown in Lemma 5.8.1 to be a pseudo-resolvent in $E(\lambda_0)$ which clearly contains the circle of convergence. Finally we see that $R(\lambda_0) = x$ and hence by Theorem 5.8.3, $R(\lambda)$ is a resolvent in the sense of Definition 4.7.1 if and only if x is regular.

We note that if $x = q$ is quasi-nilpotent, then $R(\lambda)$ is an entire function of λ which reduces to a polynomial if q is actually nilpotent. Conversely, if a solution

$R(\lambda)$ of the resolvent equation is an entire function of λ, then $R(\lambda)$ is quasi-nilpotent in $\mathfrak{B}$ for every λ.

We next consider pseudo-resolvents which are holomorphic at infinity.

THEOREM 5.9.2. *Each solution of the first resolvent equation which exists and is bounded outside a large circle is of the form*

$$R(\lambda) = z + jR(\lambda; x), \tag{5.9.2}$$

where $z^2 = \theta$, $j^2 = j$, $zj = jz = \theta$, $xj = jx = x$, *that is,* $x \in j\mathfrak{B}j$. *Such a solution is a quasi-resolvent if* $z = \theta$; *it is a resolvent in the sense of Definition* 4.7.1 *if* $j = e$.

PROOF. Since $R(\lambda)$ is holomorphic at every point where it is defined and bounded at infinity, it must be holomorphic also at infinity. Substituting $R(\lambda) = \sum_0^\infty c_n\lambda^{-n}$ in the first resolvent equation, dividing by $(\lambda - \mu)$, and comparing coefficients of like powers, we see first that all coefficients commute and obtain the following system of equations:

$$c_0^2 = \theta, \qquad c_0c_k = \theta \qquad \text{for all } k,$$

$$c_1^2 = c_1, \qquad c_n = c_{n-k}c_{k+1} \qquad \text{for } n > 1,\ k = 0, 1, \cdots, n-1.$$

The last set of equations shows that $c_n = (c_2)^{n-1}$, $n > 1$. Putting $c_0 = z$, $c_1 = j$, $c_2 = x$, we obtain the expression stated in the theorem and the relations between z, j, and x are read off from the equations. The remaining assertions then become obvious if one notes that $j = e$ implies $z = \theta$.

We come finally to the Laurent expansion of a pseudo-resolvent $R(\lambda)$. This leads to important conclusions concerning the structure of $R(\lambda)$. Since the resolvent equation is unchanged under translations, it is no restriction to assume that the center of the annulus of holomorphism of $R(\lambda)$ is at $\lambda = 0$.

THEOREM 5.9.3. *If* $R(\lambda)$ *is a solution of the first resolvent equation for* $0 \leqq \gamma_1 < |\lambda| < \gamma_2 \leqq \infty$, *then*

$$\begin{aligned} R(\lambda) &= R^-(\lambda) + R^+(\lambda), \\ R^-(\lambda)R^+(\mu) &= R^+(\mu)R^-(\lambda) = \theta, \end{aligned} \tag{5.9.3}$$

where

(i) $R^-(\lambda)$ *and* $R^+(\lambda)$ *are also solutions of the resolvent equation;*

(ii) $R^-(\lambda)$ *is holomorphic for* $|\lambda| > \gamma_1$ *and is of the form*

$$R^-(\lambda) = jR(\lambda; a^-) = j\lambda^{-1} + \sum_{n=1}^{\infty} (a^-)^n\lambda^{-n-1},$$

j being an idempotent and $a^- \in j\mathfrak{B}j$, *if* $\gamma_1 = 0$, a^- *is quasi-nilpotent;*

(iii) $R^+(\lambda)$ *is holomorphic for* $|\lambda| < \gamma_2$ *and*

$$R^+(\lambda) = \sum_{n=0}^{\infty} (-\lambda)^n a_0^{n+1},$$

a_0 belonging to the subalgebra $(e - j)\mathfrak{B}(e - j)$.

In particular, if $R(\lambda)$ is the resolvent of the element a, then

$$a = a^+ + a^-,$$

where

$$a^- = aj = ja, \qquad a^+ = a(e - j) = (e - j)a, \qquad -a_0a^+ = e - j,$$

and $R^-(\lambda)$ and $R^+(\lambda)$ are the resolvents of a^- and a^+ respectively relative to the respective subalgebras $j\mathfrak{B}j$ and $(e - j)\mathfrak{B}(e - j)$. If $R(\lambda)$ is the resolvent of the operator $A \in \mathfrak{O}(\mathfrak{X})$, then, setting $a^- = A^-$, $a_0 = A_0$, and $j = J$, we have

$$A = A^+ + A^-,$$

where

$$A^- = AJ \supset JA,\ A^+ = A(I - J) = (I - J)A,\ -AA_0 = I - J \supset -A_0A,$$

and $R^-(\lambda)$ and $R^+(\lambda)$ are the resolvents of A^- and A^+ respectively restricted to the respective subspaces $J[\mathfrak{X}]$ and $(I - J)[\mathfrak{X}]$.

Remark. This theorem is suggested by results of M. Nagumo [1] who discussed the case in which $R(\lambda) = R(\lambda; a)$ and the center of the annulus is an isolated singularity on the boundary of the principal component of $\rho(a)$. Nagumo, however, stated that the functions $R^+(\lambda)$ and $R^-(\lambda)$ are resolvents; properly speaking they are merely quasi-resolvents as will be seen.

Proof. Substituting a Laurent series in the resolvent equation, we obtain

$$\sum_{-\infty}^{\infty} c_n \frac{\lambda^n - \mu^n}{\lambda - \mu} = -\sum_{-\infty}^{\infty}\sum_{-\infty}^{\infty} c_k c_m \lambda^k \mu^m.$$

It is easily seen that all the elements c_n commute with each other. Also on the left-hand side of the above equation the coefficient of c_n is

$$\begin{aligned} &\lambda^{n-1} + \lambda^{n-2}\mu + \cdots + \mu^{n-1}, && \text{if } n \geqq 1;\\ &0, && \text{if } n = 0;\\ &-(\lambda^n\mu^{-1} + \lambda^{n+1}\mu^{-2} + \cdots + \lambda^{-1}\mu^n), && \text{if } n < 0. \end{aligned}$$

Thus all terms involving λ^n and μ^n, n being negative, are missing as well as those involving mixed products $\lambda^k\mu^m$ in which the exponents have opposite signs. This implies that every element c_k with $k \geqq 0$ is orthogonal to every c_m with $m \leqq -1$. Then defining

$$R^+(\lambda) = \sum_{n=0}^{\infty} c_n\lambda^n, \qquad R^-(\lambda) = \sum_{n=-\infty}^{-1} c_n\lambda^n,$$

one sees that relations (5.9.3) hold. Substituting this expression for $R(\lambda)$ in the resolvent equation and using the orthogonality property, it may be seen from the power series involved that both $R^+(\lambda)$ and $R^-(\lambda)$ will be solutions of the resolvent equation.

$R^-(\lambda)$ is holomorphic at infinity and vanishes there. Hence from Theorem 5.9.2, $R^-(\lambda) = jR(\lambda; a^-)$ where $a^- \in j\mathfrak{B}j$. The series defining $R(\lambda; a^-)$ converges for $|\lambda| > \gamma_1$ and if $\gamma_1 = 0$ this implies that a^- is a quasi-nilpotent element. The idempotent j cannot be θ unless $R^-(\lambda) \equiv \theta$ and it will be seen in a moment that $j \neq e$ unless $R^+(\lambda) \equiv \theta$.

Since $R^+(\lambda)$ is holomorphic at the origin, Theorem 5.9.1 shows that $R^+(\lambda) = \sum_{n=0}^{\infty}(-\lambda)^n a_0^{n+1}$. Since $a_0 = c_0$ and $j = c_{-1}$, these two elements commute and are orthogonal to each other. It follows that a_0 belongs to the subalgebra $(e - j)\mathfrak{B}(e - j)$. The assumption $j = e$ would clearly imply $a_0 = \theta$ and $R^+(\lambda) \equiv \theta$. $R^+(\lambda)$ being orthogonal to j, we conclude that

$$R^-(\lambda) = jR(\lambda) = R(\lambda)j, \qquad R^+(\lambda) = (e - j)R(\lambda) = R(\lambda)(e - j),$$

and

$$R^-(\lambda)a_0 = a_0R^-(\lambda) = \theta, \qquad R^+(\lambda)a^- = a^-R^+(\lambda) = \theta.$$

If $R(\lambda)$ is the resolvent of a,

$$(\lambda e - a)R(\lambda) = R(\lambda)(\lambda e - a) = e;$$

upon multiplication by the idempotents j and $e - j$ this identity yields

$$(j\lambda - ja)R^-(\lambda) = R^-(\lambda)(j\lambda - ja) = j,$$

$$[(e - j)\lambda - (e - j)a]R^+(\lambda) = R^+(\lambda)[(e - j)\lambda - (e - j)a] = (e - j).$$

On the other hand, from the definition of $R^-(\lambda)$ it follows that

$$(j\lambda - a^-)R^-(\lambda) = R^-(\lambda)(j\lambda - a^-) = j.$$

Since a^- and ja are both in $j\mathfrak{B}j$, the fact that they have the same quasi-resolvent in this subalgebra implies that they are equal, $a^- = ja$. Putting $a^+ = a - a^- = (e - j)a$, it is seen that

$$-a^+R^+(0) = -a^+a_0 = e - j$$

and so a^+ is the quasi-inverse of $-a_0$ in the subalgebra $(e - j)\mathfrak{B}(e - j)$.

The argument is somewhat more delicate when $R(\lambda)$ is the resolvent of an operator $A \in \mathfrak{O}(\mathfrak{X})$, In this case we use the representations

$$J = \frac{1}{2\pi i}\int_\Gamma R(\lambda; A)\,d\lambda, \qquad A^- = \frac{1}{2\pi i}\int_\Gamma \lambda R(\lambda; A)\,d\lambda,$$

where Γ is the circle: $|\lambda| = \gamma$, $\gamma_1 < \gamma < \gamma_2$. Since $AR(\lambda; A) = \lambda R(\lambda; A) - I$ is clearly continuous on Γ in the strong operator topology, Theorem 3.3.2 implies that

$$AJx = \frac{1}{2\pi i}\int_\Gamma AR(\lambda; A)x\,d\lambda = \frac{1}{2\pi i}\int_\Gamma [\lambda R(\lambda; A)x - x]\,d\lambda = A^-x + \theta = A^-x$$

for all $x \in \mathfrak{X}$. Similarly for $x \in \mathfrak{D}(A)$ one shows that $JAx = A^-x$. It readily follows from these two relations that $A^+ = A(I - J) = (I - J)A$, where $\mathfrak{D}(A^+) = \mathfrak{D}(A)$. Again

$$A_0 = \frac{1}{2\pi i} \int_\Gamma \lambda^{-1} R(\lambda; A)\, d\lambda$$

and

$$AA_0 x = \frac{1}{2\pi i} \int_\Gamma \lambda^{-1} AR(\lambda; A)x\, d\lambda = \frac{1}{2\pi i} \int_\Gamma \lambda^{-1}[\lambda R(\lambda; A)x - x]\, d\lambda = Jx - Ix$$

for all $x \in \mathfrak{X}$; whereas for $x \in \mathfrak{D}(A)$ a similar argument shows that $A_0 A x = Jx - Ix$. It is clear from the definition of $R^-(\lambda)$ that its restriction to the subspace $J[\mathfrak{X}]$ is the resolvent of the corresponding restriction of A^-. Finally one readily verifies that $R^+(\lambda)$ restricted to the subspace $(I - J)[\mathfrak{X}]$ satisfies the relations (5.8.1) for the corresponding restriction of A^+ and is therefore its resolvent relative to this subspace. This concludes the proof of Theorem 5.9.3.

For the case in which $R(\lambda) = R(\lambda; a)$, $a \in \mathfrak{B}$, the results of the above theorem are subsumed under Theorem 5.6.1; here $\sigma(a)$ has been split into two disjoint spectral sets, one contained in $|\lambda| \leqq \gamma_1$ and the other in $|\lambda| \geqq \gamma_2$. For algebras which are semi-simple in the sense of Definition 24.8.2, we have the following interesting

Corollary. *If θ is the only quasi-nilpotent element in $\mathfrak{B}$, then any isolated singularity of $R(\lambda)$ is necessarily a simple pole.*

Proof. Indeed, if the isolated singularity is at the origin, as we may assume, the principal part of $R(\lambda)$ there is $R^-(\lambda) = jR(\lambda; a^-)$ where a^- has to be quasi-nilpotent and hence equal to θ. $R^-(\lambda)$ then reduces to its first term j/λ and $R(\lambda)$ has a simple pole.

When $\mathfrak{B} = \mathfrak{E}(\mathfrak{X})$ and $R(\lambda) = R(\lambda; T)$ these theorems yield further information about the nature of the spectrum. If T is a bounded linear transformation, then we know from Theorem 4.7.2 that $\lambda = \infty$ is a first order zero for $R(\lambda; T)$. As the following theorem shows, $\lambda = \infty$ is always a singular point of $R(\lambda; T)$ if T is unbounded. See A. E. Taylor [9].

Theorem 5.9.4. *For unbounded T, $R(\lambda; T)$ is singular at infinity. If infinity is an isolated singularity of $R(\lambda; T)$, then it is an essential singularity.*

Proof. Suppose that T is unbounded and that $\lambda = \infty$ is a regular point of $R(\lambda; T)$. Then Theorem 5.9.2 applies and

$$R(\lambda; T) = Z + R(\lambda; S)J,$$

where $S \in \mathfrak{E}(\mathfrak{X})$, $Z^2 = \Theta$, $J^2 = J$, $ZJ = JZ = \Theta$, and $J \neq I$. Choose $x \neq \theta$, $x \in \mathfrak{X}$, so that $Jx = \theta$. Then $R(\lambda; T)x = Zx$ and

$$[R(\lambda; T)]^2 x = Z^2 x + R(\lambda; S)JZx = \theta$$

which is impossible since $[R(\lambda; T)]^{-1}$ exists. It follows that $\lambda = \infty$ is a singular point of $R(\lambda; T)$. Suppose $\lambda = \infty$ were a pole of order $m - 1$, $m > 1$. Then by Theorem 5.9.3

$$R(\lambda; T) = R(\lambda; T)J + \sum_{n=0}^{\infty} (-\lambda)^n T_0^{n+1},$$

where $T_0 \in \mathfrak{E}(\mathfrak{X})$, $T_0 J = JT_0 = \Theta$, and $T_0^m \neq \Theta$ but $T_0^{m+1} = \Theta$. Thus there exists a $y \in \mathfrak{X}$ such that $x = T_0^m y \neq \theta$. Then $T_0 x = \theta$ and $Jx = JT_0^m y = \theta$ so that $R(\lambda; T)x = \theta$ and this again is impossible.

THEOREM 5.9.5. *If T is a closed linear transformation on $\mathfrak{D}(T) \subset \mathfrak{X}$ to $\mathfrak{X}$ and if λ_0 is a pole of $R(\lambda; T)$, then λ_0 belongs to the point spectrum of T.*

PROOF. By Theorem 5.9.3, the canonical representation of $R(\lambda; T)$ near $\lambda = \lambda_0$ is

$$R(\lambda; T) = J \sum_{n=0}^{\infty} Q^n(\lambda - \lambda_0)^{-n-1} + R(\lambda; T)(I - J),$$

where $J^2 = J$ and $QJ = JQ = Q$. If λ_0 is a pole of order m, $m > 1$, then $Q^{m-1} \neq \Theta$ whereas $Q^m = \Theta$. Again we can find an element y in $\mathfrak{X}$ such that $x = Q^{m-1}y = JQ^{m-1}y \neq \theta$. Since $Qx = \theta$ and $Jx = x$, it follows that $R(\lambda; T)x = (\lambda - \lambda_0)^{-1}x$. If $m = 1$ the same result can be achieved simply by choosing x so that $Jx = x$. In either case $(\lambda I - T)R(\lambda; T)x = x$ implies that $Tx = \lambda_0 x$.

It is worth noting that if x is any characteristic element corresponding to the characteristic value λ_0, then $Jx = x$. This follows from the identity

$$(\lambda - \lambda_0)R(\lambda; T)x = x,$$

valid for characteristic elements, by substituting in the canonical representation (5.9.3) and comparing coefficients.

On the other hand, if $\lambda = \lambda_0$ is an essential singularity of $R(\lambda; T)$, no assertion can be made concerning the spectral character of λ_0. In the following three examples, T is a quasi-nilpotent operator so that $\lambda = 0$ is the only point in $\sigma(T)$ and $\lambda = 0$ is an essential singular point of $R(\lambda; T)$.

(i) $\mathfrak{X} = l_2$, T takes $(x_1, x_2, \cdots, x_n, \cdots)$ into $(x_2, x_3/2, \cdots, x_{n+1}/n, \cdots)$, point spectrum;

(ii) $\mathfrak{X} = C[0, 1]$, T takes $x(t)$ into $\int_0^t x(\xi)\, d\xi$, residual spectrum;

(iii) $\mathfrak{X} = C_0[0, 1]$, subspace of $C[0, 1]$ where $x(0) = 0$, T as in (ii), continuous spectrum.

The same ambiguity is encountered at non-isolated points of the spectrum. Even an interior point may belong to the point spectrum. The following example illustrates this possibility:

(iv) $\mathfrak{X} = C[0, \infty]$, T takes $x(t)$ into $x(t + h)$ where h is a fixed positive number. Here the point spectrum is the set $|\lambda| < 1$ plus $\lambda = 1$, the rest of the unit circle is the continuous spectrum. Cf. section 19.2.

If T is bounded then $\sigma(T)$ is necessarily non-empty by Theorem 4.7.4. However if T is unbounded $\sigma(T)$ may very well be empty; in this case $\lambda = \infty$ is an essential singularity of $R(\lambda; T)$ (see Theorem 5.9.4). The following example is a case in point:

(v) $\mathfrak{X} = C_0[0, 1]$, T takes $x(t)$ into $x'(t)$, $\mathfrak{D}(T)$ is the subspace of $C_0[0, 1]$ where $x'(t)$ exists and belongs to $C_0[0, 1]$. Here $R(\lambda; T)x$ is the solution of the differential equation $\lambda y - y' = x$, $y(0) = 0$, and takes $x(t)$ into $-\int_0^t e^{\lambda(t-\xi)} x(\xi)\, d\xi$; $R(\lambda; T)$ is an entire function of λ.

5.10. The second resolvent equation. We now study various generalizations of the second resolvent equation already obtained in section 4.8 for the resolvent of elements belonging to a Banach algebra $\mathfrak{B}$. We shall eventually be interested in extending our considerations to unbounded transformations on a (B)-space $\mathfrak{X}$ to $\mathfrak{X}$. First, however, we consider the functional equation

$$R(x) - R(y) = R(x)(x - y)R(y) \tag{5.10.1}$$

on $\mathfrak{B}$ to itself.

It is evident from (5.10.1) that $R(x)$ locally bounded implies continuity and that continuity implies analyticity. Actually local boundedness need not be postulated as the following theorem shows.

THEOREM 5.10.1. *Let $R(x)$ be a single-valued function on a Banach algebra $\mathfrak{B}$ to itself, satisfying* (5.10.1) *for all values of x and y in an open subset $\mathfrak{D}$ of $\mathfrak{B}$. Then $R(x)$ is locally analytic in $\mathfrak{D}$. A necessary and sufficient condition for the existence of an element c of $\mathfrak{B}$ such that $(c - x)R(x) = R(x)(c - x) = e$ for all $x \in \mathfrak{D}$ is that $R(x)$ be a regular element of $\mathfrak{B}$ for at least one value of x in $\mathfrak{D}$.*

PROOF. If $a, x \in \mathfrak{D}$ then

$$R(x)[e - (x - a)R(a)] = R(a). \tag{5.10.2}$$

By Theorem 4.3.1 the second factor on the left has an inverse if

$$\| (x - a)R(a) \| < 1.$$

Multiplication on the right of both sides by this inverse yields

$$R(x) = R(a)\left\{e + \sum_{n=1}^{\infty} [(x - a)R(a)]^n\right\}.$$

If $R(a) = \theta$, then obviously $R(x) \equiv \theta$. Excluding this trivial case, one sees that the power series is absolutely convergent in the sphere $\| x - a \| < \| R(a) \|^{-1}$ and is uniformly convergent in any concentric sphere of smaller radius. Consequently $R(x)$ is locally bounded in $\mathfrak{D}$ and it is clear from the above expansion that $R(x)$ is (G)-differentiable at a; a being arbitrary, it follows that $R(x)$ is locally analytic in $\mathfrak{D}$.

If there exists a fixed element c of $\mathfrak{B}$ such that $(c - x)R(x) = R(x)(c - x) \equiv e$ in $\mathfrak{D}$, then $R(x)$ is regular everywhere in $\mathfrak{D}$ so that the stated condition is necessary. Suppose conversely that $[R(a)]^{-1}$ exists in $\mathfrak{B}$ at $a \in \mathfrak{D}$. From (5.10.2) one infers that $R(x)\{[R(a)]^{-1} - (x - a)\} \equiv e$ in $\mathfrak{D}$ and, from

$$[e - R(a)(x - a)]R(x) = R(a),$$

one sees that $[R(a)]^{-1} - (x - a)$ is also the left inverse of $R(x)$. This completes the proof of the theorem.

We note that if $c - x$ is a regular element of $\mathfrak{B}$ for each x in some domain $\mathfrak{D} \subset \mathfrak{B}$, then $(c - x)^{-1}$ is obviously a solution of (5.10.1) in $\mathfrak{D}$. On the other hand the assumption in Theorem 5.10.1 that $[R(x)]^{-1}$ shall exist for at least one x in $\mathfrak{D}$ is not redundant; this can be seen from the following theorem.

THEOREM 5.10.2. *Given r, $a \in \mathfrak{B}$. Denote by $\mathfrak{E}(a)$ the set of all x such that $e - (x - a)r$ and $e - r(x - a)$ are regular. Then $\mathfrak{E}(a)$ is an open subset of $\mathfrak{B}$ and the function*

(5.10.3) $R(x) \equiv r[e - (x - a)r]^{-1} = [e - r(x - a)]^{-1}r$

defined on $\mathfrak{E}(a)$ satisfies (5.10.1). *In particular for $\| x - a \| < \| r \|^{-1}$*

$$R(x) = r \sum_{n=0}^{\infty} [(x - a)r]^n \tag{5.10.4}$$

and every solution of (5.10.1) *defined in a neighborhood of $x = a$ is of this form.*

PROOF. Since the regular elements $\mathfrak{G}$ form an open subset of $\mathfrak{B}$ by Theorem 4.3.2 and since both $s(x) = e - (x - a)r$ and $t(x) = e - r(x - a)$ are continuous mappings of $\mathfrak{B}$ into $\mathfrak{B}$, it is readily seen that $\mathfrak{E}(a) \equiv s^{-1}(\mathfrak{G}) \cap t^{-1}(\mathfrak{G})$ is open. Now

$$r[e - (x - a)r] = [e - r(x - a)]r$$

and consequently the second equality in (5.10.3) is valid for all $x \in \mathfrak{E}(a)$. Hence

$$[e - r(x - a)][R(x) - R(y)][e - (y - a)r]$$
$$= r[e - (y - a)r] - [e - r(x - a)]r = r(x - y)r.$$

Multiplying on the left by $[e - r(x - a)]^{-1}$ and on the right by $[e - (y - a)r]^{-1}$, we obtain (5.10.1). For $\| x - a \| < \| r \|^{-1}$ the formula (5.10.4) follows directly from (5.10.3) and Theorem 4.3.1. Finally we see from (5.10.2) that each solution of (5.10.1) in a neighborhood of $x = a$ is of the form (5.10.4).

We shall now extend the foregoing material on the second resolvent equation to $\mathfrak{C}(\mathfrak{X})$, the closed linear operators on $\mathfrak{X}$ to $\mathfrak{X}$. We see at once that (5.10.1) is not even meaningful for arbitrary pairs of operators in $\mathfrak{C}(\mathfrak{X})$. However we obtain a clue on how to make this extension from the following two theorems.

THEOREM 5.10.3. *Let S, $T \in \mathfrak{C}(\mathfrak{X})$ and suppose that $\lambda \in \rho(S) \cap \rho(T)$ so that $R(\lambda; S)$ and $R(\lambda; T)$ exist. If $\mathfrak{D}(S) \supset \mathfrak{D}(T)$, then*

$$R(\lambda; S) - R(\lambda; T) = R(\lambda; S)(S - T)R(\lambda; T). \tag{5.10.5}$$

PROOF. For each $y \in \mathfrak{X}$, $R(\lambda; T)y = x$ belongs to $\mathfrak{D}(T)$. Further it is easily seen that

$$(\lambda I - S)[R(\lambda; S) - R(\lambda; T)](\lambda I - T)x = (S - T)x.$$

Operating on both members by $R(\lambda; S)$ and replacing x by $R(\lambda; T)y$ gives (5.10.5).

THEOREM 5.10.4. *Let $T \in \mathfrak{O}(\mathfrak{X})$ and let S be a linear operator with $\mathfrak{D}(S) = \mathfrak{D}(T)$. Suppose further that $\lambda \in \rho(T)$. Then $S \in \mathfrak{O}(\mathfrak{X})$ and $\lambda \in \rho(S)$ if and only if $I - (S - T)R(\lambda; T)$ is regular in $\mathfrak{E}(\mathfrak{X})$. In this case*

$$R(\lambda; S) = R(\lambda; T)[I - (S - T)R(\lambda; T)]^{-1}. \tag{5.10.6}$$

In particular if $\| (S - T)R(\lambda; T) \| \equiv \gamma < 1$, then $S \in \mathfrak{O}(\mathfrak{X})$, $\lambda \in \rho(S)$,

$$R(\lambda; S) = R(\lambda; T)\left\{I + \sum_{n=1}^{\infty} [(S - T)R(\lambda; T)]^n\right\}, \tag{5.10.7}$$

and

$$\| R(\lambda; S) - R(\lambda; T) \| \leqq \| R(\lambda; T) \| \gamma(1 - \gamma)^{-1}. \tag{5.10.8}$$

PROOF. If S is closed, then a simple calculation shows that $SR(\lambda; T)$ is likewise closed. Since $SR(\lambda; T)$ is linear and defined on all of $\mathfrak{X}$, it follows by Theorem 2.12.3 that $SR(\lambda; T)$ is bounded. Similarly $TR(\lambda; T)$ is bounded and hence $(S - T)R(\lambda; T) \in \mathfrak{E}(\mathfrak{X})$. Suppose next that $\lambda \in \rho(S)$. Then by formula (5.10.5)

$$R(\lambda; S)[I - (S - T)R(\lambda; T)] = R(\lambda; T).$$

The range of the right member and consequently of the left member is $\mathfrak{D}(T) = \mathfrak{D}(S)$. This implies that the range of $I - (S - T)R(\lambda; T)$ is all of $\mathfrak{X}$. Furthermore this operator must be one-to-one on $\mathfrak{X}$. For suppose $[I - (S - T)R(\lambda; T)]x = \theta$, $x \neq \theta$. Choose $y \in \mathfrak{D}(T)$ so that $(\lambda I - T)y = x$. Then

$$(\lambda I - T)y = (S - T)R(\lambda; T)(\lambda I - T)y = Sy - Ty$$

and consequently $Sy = \lambda y$ which is impossible for $\lambda \in \rho(S)$. Thus $I - (S - T)R(\lambda; T)$ is one-to-one on $\mathfrak{X}$ to all of $\mathfrak{X}$ and therefore by Theorem 2.12.1 the operator is regular in $\mathfrak{E}(\mathfrak{X})$.

Conversely if $I - (S - T)R(\lambda; T)$ is regular in $\mathfrak{E}(\mathfrak{X})$, we set

$$R \equiv R(\lambda; T)[I - (S - T)R(\lambda; T)]^{-1}.$$

Then the range of R is equal to the range of $R(\lambda; T)$, namely $\mathfrak{D}(T)$. It is further clear that R is the right inverse of $(\lambda I - S)$ since

$$\begin{aligned}(\lambda I - S)R &= [(\lambda I - T) - (S - T)]R(\lambda; T)[I - (S - T)R(\lambda; T)]^{-1} \\ &= [I - (S - T)R(\lambda; T)][I - (S - T)R(\lambda; T)]^{-1} = I.\end{aligned}$$

Now a right inverse with range $\mathfrak{D}(T) = \mathfrak{D}(S)$ is also a left inverse and therefore

$R = R(\lambda; S)$. Since $R(\lambda; S)$ is closed it follows from Theorem 2.11.5 that $\lambda I - S$ and hence S is closed.

In particular if $\| (S - T)R(\lambda; T) \| = \gamma < 1$, then formula (5.10.7) defines the operator R above and hence is equal to $R(\lambda; S)$. The bound in (5.10.8) comes directly from the expansion. This completes the proof of Theorem 5.10.4.

In order to study the second resolvent equation in this larger setting, it is now clear that we must restrict ourselves to pairs S, T such that $\mathfrak{D}(S) = \mathfrak{D}(T)$. Further

$$d(S, T) = \sup [\, \| (S - T)x \| \,;\, x \in \mathfrak{D}(T) = \mathfrak{D}(S),\ \| x \| = 1] \tag{5.10.9}$$

appears to be a suitable metric function (actually, Theorem 5.10.4 suggests that a somewhat weaker metric might do equally well). As we have already remarked in section 2.12, two operators will be a finite distance apart if and only if they lie in the same equivalence class $\mathfrak{C}(T)$; here $S \approx T$ means that $\mathfrak{D}(S) = \mathfrak{D}(T)$ and that $d(S, T) < \infty$. As an immediate consequence of this and the above theorem we have the

Corollary. *For a fixed λ_0, the set of $S \in \mathfrak{C}(T)$ such that $\lambda_0 \in \rho(S)$ is an open subset of $\mathfrak{C}(T)$.*

As defined above, $\mathfrak{C}(T)$ has a point geometry. However by arbitrarily assigning the role of origin to a particular $T_0 \in \mathfrak{C}(T)$, $\mathfrak{C}(T)$ becomes isometric with the (B)-space of operators $\mathfrak{E}[\overline{\mathfrak{D}(T)}, \mathfrak{X}]$ under the mapping $S \to B$ where B is the unique extension of $S - T_0$ on $\overline{\mathfrak{D}(T)}$. This mapping enables us to take over the theory of analyticity already developed for functions on one (B)-space to another and apply it to functions defined on $\mathfrak{C}(T)$ to $\mathfrak{E}(\mathfrak{X})$. Thus, a given function $f(S)$ on $\mathfrak{C}(T)$ to $\mathfrak{E}(\mathfrak{X})$ is said to be analytic if and only if $f(T_0 + B)$ is analytic in the usual sense relative to the argument $B \in \mathfrak{E}[\overline{\mathfrak{D}(T)}, \mathfrak{X}]$. This notion is clearly independent of the particular choice of $T_0 \in \mathfrak{C}(T)$.

Theorem 5.10.5. *Let $R(S)$ be a single-valued function on $\mathfrak{C}(T)$ to $\mathfrak{E}(\mathfrak{X})$ such that the range of $R(S)$ is contained in $\mathfrak{D}(T)$ and such that*

$$R(S) - R(T) = R(S)(S - T)R(T) \tag{5.10.10}$$

for all S, T in an open subset $\mathfrak{D}$ of $\mathfrak{C}(T)$. Then $R(S)$ is locally analytic in $\mathfrak{D}$. A necessary and sufficient condition for the existence of an operator $C \in \mathfrak{C}(T)$ such that $(C - S)R(S)x = x$ in $\mathfrak{X}$ and $R(S)(C - S)x = x$ in $\mathfrak{D}(T)$ is that $R(S)$ have an inverse belonging to $\mathfrak{C}(T)$ for at least one value of S in $\mathfrak{D}$.

The proof of Theorem 5.10.5 paraphrases that of Theorem 5.10.1 and is omitted.

Before concluding this section we remark that for each $R \in \mathfrak{E}(\mathfrak{X})$ with range contained in $\mathfrak{D}(T)$ and $A \in \mathfrak{C}(T)$, the series

$$R(S) = R \sum_{n=0}^{\infty} [(S - A)R]^n \tag{5.10.11}$$

defines a solution of (5.10.10) analytic in the sphere $d(S, A) < \| R \|^{-1}$ and every solution of (5.10.10) defined in a neighborhood of $S = A \in \mathfrak{E}(T)$ is of this form.

References. Nagumo [1], Stone [3], Taylor [2, 9].

3. The Extended Operational Calculus

5.11. An operational calculus for closed operators. The operational calculus plays an important role in the theory of semi-groups of linear transformations. However what is needed is a calculus for the closed linear operators belonging to $\mathfrak{D}(\mathfrak{X})$ rather than the bounded linear operators of $\mathfrak{E}(\mathfrak{X})$. In the present paragraph we follow A. E. Taylor [9] in extending our previous results to $\mathfrak{D}(\mathfrak{X})$; for the most part this requires little more than a reformulation of the earlier concepts. Later in Chapters XV and XVI we will develop a more interesting extension of the operational calculus, applicable to those operators in $\mathfrak{D}(\mathfrak{X})$ which are generators of semi-groups of transformations.

We recall the notion of the *extended spectrum* of an operator T, denoted by $\sigma_e(T)$, which consists of the singular points of $R(\lambda; T)$ in the extended complex plane. If T is bounded, then $\sigma_e(T) = \sigma(T)$; whereas if T is unbounded, then $\sigma_e(T) = \sigma(T) \cup [\infty]$ in accordance with Theorem 5.9.4. In all cases $\sigma_e(T)$ is a closed non-void subset of the extended complex plane. For some operators, $\sigma_e(T)$ consists of the entire extended plane; however our theory will not be applicable to such operators.

Let Φ be a closed subset of the extended λ-plane and let $f(\lambda)$ be a function locally holomorphic in an open set Δ containing Φ. The notion of an *oriented envelope* Γ of Φ with respect to $f(\lambda)$ extends in the obvious fashion. We shall always assume that Γ is disjoint from $\lambda = \infty$. For λ interior to Γ, the Cauchy formula now becomes

$$f(\lambda) = \delta f(\infty) + \frac{1}{2\pi i} \int_\Gamma \frac{f(\zeta)}{\zeta - \lambda} \, d\zeta \tag{5.11.1}$$

where $\delta = 1$ or 0 according as Γ contains $\lambda = \infty$ in its interior or not. Formula (5.11.1) is basic to the extended operational calculus.

In the following we again make use of the metric topology in $\mathfrak{D}(\mathfrak{X})$ defined by (5.10.9); here $d(S, T)$ is finite if and only if S and T belong to the same class.

Theorem 5.11.1. *The extended spectrum of T is an upper semi-continuous function of T.*

The proof of Theorem 5.2.3 applies, *mutatis mutandis*. Reference should now be made to Theorems 5.8.2 and 5.10.4.

Suppose Δ is an open subset of the extended complex plane and define $\mathfrak{G}(\Delta)$ to be the set of all operators $T \in \mathfrak{O}(\mathfrak{X})$ such that $\sigma_e(T) \subset \Delta$. Then an immediate consequence of the above theorem is that $\mathfrak{G}(\Delta)$ is an open subset of $\mathfrak{O}(\mathfrak{X})$.

THEOREM 5.11.2. *Let Δ be an open subset of the extended complex plane not containing the point α_0, $\alpha_0 \neq \infty$. Let $H(\Delta)$ be the complex algebra of all functions $f(\lambda)$, locally holomorphic in Δ, with the ordinary definitions of the arithmetic operations, and with a sequence topology: $f_n \to f$ denoting that $f_n(\lambda)$ converges pointwise to $f(\lambda)$, the convergence being uniform in each compact subset of Δ. Further let $\mathfrak{B}(\Delta)$ be the complex algebra of functions $f(T)$, locally analytic in $\mathfrak{G}(\Delta)$ and having values in $\mathfrak{E}(\mathfrak{X})$, the arithmetic operations being defined as in $\mathfrak{E}(\mathfrak{X})$.*

There exists an isomorphic mapping: $f(\lambda) \to f(T)$ of $H(\Delta)$ on a subalgebra $\mathfrak{B}_0(\Delta)$ of $\mathfrak{B}(\Delta)$ such that (i) $1 \to I$, (ii) $(\alpha_0 - \lambda)^{-1} \to R(\alpha_0; T)$, *and* (iii) $f_n \to f$ *implies that* $\| f_n(T) - f(T) \| \to 0$ *locally uniformly in $\mathfrak{G}(\Delta)$. This mapping is unique and is defined by*

$$f(T) = \delta f(\infty)I + \frac{1}{2\pi i}\int_{\Gamma_T} f(\zeta)R(\zeta; T)\, d\zeta, \tag{5.11.2}$$

where Γ_T is an oriented envelope of $\sigma_e(T)$ with respect to $f(\lambda)$.

PROOF. We first show that the mapping defined by (5.11.2) has the required properties. It is of course essential that $f(T)$ be independent of the particular oriented envelope of $\sigma_e(T)$ employed. The only case where this is at all in doubt is when $\sigma_e(T)$ is bounded and the alternative exists for Γ_T to contain $\lambda = \infty$ in its interior or not. Suppose then that $\sigma_e(T)$ is bounded (and hence that T is bounded), that Γ_1, Γ_2 are oriented envelopes of $\sigma_e(T)$ with respect to $f(\lambda)$, and that Γ_1 contains $\lambda = \infty$ in its interior whereas Γ_2 does not. Then Γ_1 is equivalent to Γ_2 plus a large circle C taken in the clockwise sense about $\lambda = 0$. On the other hand $R(\mu^{-1}; T) = \sum \mu^{n+1}T^n$ and hence

$$\frac{1}{2\pi i}\int_C f(\zeta)R(\zeta; T)\, d\zeta = -\frac{1}{2\pi i}\int_c f(\mu^{-1})R(\mu^{-1}; T)\,\frac{d\mu}{\mu^2} = -f(\infty)I$$

where c is a small circle about $\mu = 0$ taken in the counterclockwise sense. It follows that the value of δ just compensates for the choice in path. For definiteness we shall assume hereafter in verifying the properties of (5.11.2) that all paths Γ_T contain $\lambda = \infty$ in their interior. This restricts the generality only if Δ does not contain $\lambda = \infty$ and this case has already been treated in Theorem 5.2.5.

If $f(\lambda) \equiv 1$, then the integral $\int_{\Gamma_T} f(\zeta)R(\zeta; T)\, d\zeta$ can be thought of as the negative of the integral about the exterior of Γ_T and here the integrand is locally holomorphic. Consequently this integral is simply Θ so that $f(T) = I$. If $\alpha_0 \notin \Delta$ and $g(\lambda) = (\alpha_0 - \lambda)^{-1}$, then $g(\lambda) \in H(\Delta)$ and $g(\infty) = 0$. Again $\int_{\Gamma_T} g(\zeta)R(\zeta; T)\, d\zeta$ is the negative of the integral about the exterior of Γ_T and here the only singularity is a simple pole at $\zeta = \alpha_0$. The residue is $-R(\alpha_0; T)$ and hence $g(T) = R(\alpha_0; T)$. This proves (i) and (ii).

It is clear that the mapping (5.11.2) is linear. In order to show that products go into products suppose that $f(\lambda)$, $g(\lambda)$, $h(\lambda) \in H(\Delta)$, where $h(\lambda) = f(\lambda)g(\lambda)$, and suppose further that Γ, Γ' are two oriented envelopes of $\sigma_e(T)$ with respect to $f(\lambda)$, $g(\lambda)$, and $h(\lambda)$ such that Γ' is interior to Γ. The double integral appearing in the product $f(T)g(T)$ may be written as

$$\left(\frac{1}{2\pi i}\right)^2 \int_{\Gamma'}\int_{\Gamma} f(\zeta)g(\mu)R(\zeta;\, T)R(\mu;\, T)\, d\mu\, d\zeta$$
$$= \frac{1}{2\pi i}\int_{\Gamma'} f(\zeta)R(\zeta;\, T)\left(\frac{1}{2\pi i}\int_{\Gamma}\frac{g(\mu)}{\mu - \zeta}\, d\mu\right) d\zeta$$
$$- \frac{1}{2\pi i}\int_{\Gamma} g(\mu)R(\mu;\, T)\left(\frac{1}{2\pi i}\int_{\Gamma'}\frac{f(\zeta)}{\mu - \zeta}\, d\zeta\right) d\mu,$$

where we have made use of the first resolvent equation. Since $\zeta \in \Gamma'$ is interior to Γ whereas $\mu \in \Gamma$ is exterior to Γ', the two integrals involving $(\mu - \zeta)^{-1}$ are equal to $g(\zeta) - g(\infty)$ and $f(\infty)$ respectively. It follows that

$$f(T)g(T) = f(\infty)g(\infty)I + \frac{1}{2\pi i}\int_{\Gamma'} f(\zeta)g(\zeta)R(\zeta;\, T)\, d\zeta = h(T).$$

The proof that the homomorphism is actually an isomorphism goes as follows: Let λ_0 be a point of Δ, $\lambda_0 \neq \infty$. Then $\lambda_0 I \in \mathfrak{G}(\Delta)$ and $R(\lambda;\, \lambda_0 I) = (\lambda - \lambda_0)^{-1}I$. A simple calculation shows that $f(\lambda_0 I) = f(\lambda_0)I$. Hence $f(T) \equiv \Theta$ on $\mathfrak{G}(\Delta)$ implies $f(\lambda) \equiv 0$ on Δ.

In order to prove that $f(T)$ is locally analytic in $\mathfrak{G}(\Delta)$ and that the continuity property (iii) is valid it suffices to paraphrase the corresponding arguments employed in Theorems 5.2.4 and 5.2.5. Replacing $a \in \mathfrak{B}$ by $A \in \mathfrak{O}(\mathfrak{X})$, one now refers to Theorem 5.10.4 instead of Theorem 4.8.3.

It remains to show that the mapping is unique. Suppose that $\mathfrak{J}$ is any isomorphism with the stated properties. Then $0 \to \Theta$ and $1 \to I$. For $\beta \notin \Delta$, $(\beta - \lambda)^{-1} \in H(\Delta)$ so that $(\beta - \lambda)^{-1} \to R_\beta(T)$; in particular for $\beta = \alpha_0$ we have by (ii) that $R_{\alpha_0}(T) = R(\alpha_0\,;\, T)$. Now $(\beta - \lambda)^{-1} - (\alpha_0 - \lambda)^{-1} = (\alpha_0 - \beta)(\beta - \lambda)^{-1}(\alpha_0 - \lambda)^{-1}$ and therefore

$$R_\beta(T) - R(\alpha_0\,;\, T) = (\alpha_0 - \beta)R_\beta(T)R(\alpha_0\,;\, T) = (\alpha_0 - \beta)R(\alpha_0\,;\, T)R_\beta(T).$$

According to Theorem 5.8.3 this implies that $R_\beta(T) = R(\beta;\, T)$. Since products go into products we also have $(\beta - \lambda)^{-n} \to [R(\beta;\, T)]^n$. If $\lambda = \infty$ does not belong to Δ, then $\mathfrak{G}(\Delta)$ can contain only bounded operators. In this case $(\alpha_0 - \lambda) \in H(\Delta)$, $(\alpha_0 - \lambda) \to [R(\alpha_0\,;\, T)]^{-1} = \alpha_0 I - T$ and hence $\lambda \to T$. Thus in all cases $\mathfrak{J}$ is uniquely defined on each rational function regular in Δ. In particular $\mathfrak{J}$ must agree with the mapping (5.11.2) on such functions. Again, according to an extension of the Runge theorem due to J. L. Walsh [1, p. 16], each function $f(\lambda) \in H(\Delta)$ is the limit of a sequence of rational functions with poles outside of Δ, the convergence being uniform on each compact subset of Δ. From the fact that $\mathfrak{J}$ agrees with (5.11.2) on a dense subset of $H(\Delta)$ together with the continuity assumption (iii), we conclude that $\mathfrak{J}$ is identical with (5.11.2) everywhere in $H(\Delta)$.

We note that the isomorphic mapping $\mathfrak{J}$ remains unique even if condition (iii) is replaced by (iii′) $f_n \to f$ implies that $\| f_n(T) - f(T) \| \to 0$ for each $T \in \mathfrak{G}(\Delta)$. In addition the uniqueness does not depend on the fact that the functions in $\mathfrak{B}(\Delta)$ are locally analytic in $\mathfrak{G}(\Delta)$.

REMARK. The basic observation that formula (5.11.2) defines for each $T \in \mathfrak{G}(\Delta)$ a homomorphic mapping on $H(\Delta)$ into $\mathfrak{E}(\mathfrak{X})$ is due to A. E. Taylor [**9**]. The rest of Theorem 5.11.2 appears to be new.

Thus far we have dealt with unbounded operators only indirectly through the associated resolvent operator. We next consider polynomial operators in T. For a polynomial $p(\lambda)$ of degree n, the operator $p(T)$ defined on $\mathfrak{D}(T^n)$ belongs to $\mathfrak{O}(\mathfrak{X})$ whenever $\rho(T) \neq \emptyset$ by Theorem 2.16.4. The following three theorems are due to A. E. Taylor [**9**].

THEOREM 5.11.3. *Let Δ be an open proper subset of the extended plane containing $\lambda = \infty$. Suppose that $f(\lambda)$ belongs to $H(\Delta)$ and that $f(\lambda)$ has a zero of order n at $\lambda = \infty$. Then for $T \in \mathfrak{G}(\Delta)$, the range of $f(T)$ is contained in $\mathfrak{D}(T^n)$ and if $p(\lambda)$ is a polynomial of degree $\leqq n$, $p(T)f(T) = [pf](T)$.*

PROOF. We choose $\alpha_0 \notin \Delta$ and set $g(\lambda) = (\alpha_0 - \lambda)^n f(\lambda)$. By assumption $f(\lambda)$ has a zero of order n at $\lambda = \infty$ so that $g(\lambda) \in H(\Delta)$. Consequently $f(T) = [R(\alpha_0 ; T)]^n g(T)$. Since $[R(\alpha_0 ; T)]^n[\mathfrak{X}] = \mathfrak{D}(T^n)$, it now follows that the range of $f(T)$ is contained in $\mathfrak{D}(T^n)$. Next suppose that $p(\lambda)$ is a given polynomial of degree $\leqq n$. Expanded about α_0, $p(\lambda)$ takes the form $p(\lambda) = \sum_{k=0}^n \beta_k(\alpha_0 - \lambda)^k$. It is clear that

$$p(T)f(T) = \{p(T)[R(\alpha_0 ; T)]^n\}g(T) = \left\{\sum_{k=0}^n \beta_k[R(\alpha_0 ; T)]^{n-k}\right\} g(T).$$

On the other hand, $[pf](\lambda) = \{\sum_{k=0}^n \beta_k(\alpha_0 - \lambda)^{k-n}\}g(\lambda)$ and, since both factors in the right member of this equation lie in $H(\Delta)$, the operational calculus gives

$$[pf](T) = \left\{\sum_{k=0}^n \beta_k[R(\alpha_0 ; T)]^{n-k}\right\} g(T).$$

Therefore $p(T)f(T) = [pf](T)$.

A result complementing the above is given by

THEOREM 5.11.4. *Let Δ be an open proper subset of the extended plane containing $\lambda = \infty$. Suppose that $f(\lambda) \in H(\Delta)$ and that $T \in \mathfrak{G}(\Delta)$. Then for $x \in \mathfrak{D}(T^n)$ and $p(\lambda)$ a polynomial of degree $\leqq n$, we have $f(T)p(T)x = p(T)f(T)x$.*

PROOF. For $x \in \mathfrak{D}(T^n)$ and $\alpha_0 \notin \Delta$ there exists a $y \in \mathfrak{X}$ such that $x = [R(\alpha_0 ; T)]^n y$. Now $g(\lambda) = (\alpha_0 - \lambda)^{-n} \in H(\Delta)$ and $g(T) = [R(\alpha_0 ; T)]^n$. It is clear that $g(\lambda)$, and hence $f(\lambda)g(\lambda)$, has a zero of order $\geqq n$ at $\lambda = \infty$. According to the previous theorem

$$p(T)[f(T)g(T)] = [pfg](T) = f(T)\{[pg](T)\} = f(T)[p(T)g(T)]$$

and applying this to y gives the desired result.

THEOREM 5.11.5. *Let Δ be an open proper subset of the extended plane containing $\lambda = \infty$. Suppose that $f(\lambda) \in H(\Delta)$ has a zero of order n at $\lambda = \infty$ but no other zero in Δ. Then for $T \in \mathfrak{G}(\Delta)$, $f(T)$ defines a one-to-one mapping of $\mathfrak{X}$ onto $\mathfrak{D}(T^n)$.*

PROOF. Choose $\alpha_0 \notin \Delta$ and set $g(\lambda) = (\alpha_0 - \lambda)^n f(\lambda)$. Then $g(\lambda) \neq 0$ in Δ and hence $g^{-1}(\lambda) \equiv [g(\lambda)]^{-1} \in H(\Delta)$. It follows that $g^{-1}(T) = [g(T)]^{-1}$ so that $g(T)$ is regular in $\mathfrak{E}(\mathfrak{X})$. According to Theorem 5.11.3, $g(T) = (\alpha_0 I - T)^n f(T)$ and hence $f(T) = [R(\alpha_0 ; T)]^n g(T)$. Since $g(T)$ is regular in $\mathfrak{E}(\mathfrak{X})$, the range of $f(T)$ is equal to that of $[R(\alpha_0 ; T)]^n$, namely $\mathfrak{D}(T^n)$. Finally since each of the factors in the above expression for $f(T)$ has an inverse, the same is true of $f(T)$.

5.12. The spectral mapping theorem. Composite functions. We next prove a variety of spectral mapping theorems. The first two of these are concerned with the mapping $T \in \mathfrak{O}(\mathfrak{X}) \rightarrow f(T) \in \mathfrak{E}(\mathfrak{X})$, defined by the operational calculus; they are prototypes for some of our later results on semi-groups of transformations. In our present situation the second is by far the better result, extending the mapping to the fine structure of the spectrum. However the proof of the first is simpler and its use of the Gelfand representation theory is somewhat novel. The third and final theorem of this set deals with the mapping $T \in \mathfrak{O}(\mathfrak{X}) \rightarrow p(T) \in \mathfrak{O}(\mathfrak{X})$, where $p(\lambda)$ is a polynomial in λ. Many of the results of this section are to be found in A. E. Taylor [9]; Theorem 5.12.2 appears to be new.

THEOREM 5.12.1. *Let Δ be an open proper subset of the extended complex plane. Suppose $f(\lambda) \in H(\Delta)$ and $T \in \mathfrak{G}(\Delta)$. Then $\sigma[f(T)] = f[\sigma_e(T)]$.*

PROOF. Given $T \in \mathfrak{G}(\Delta)$, we set $\mathfrak{S} = [R(\lambda; T), \lambda \in \rho(T)]$. Let $\mathfrak{B}_o = \mathfrak{S}^{cc}$, the commutant of the commutant of $\mathfrak{S}$ relative to $\mathfrak{E}(\mathfrak{X})$. Then according to Theorems 1.13.1 and 1.14.1, $\mathfrak{B}_o$ is a commutative (B)-algebra containing the identity as well as the elements of $\mathfrak{S}$; and the spectrum of any element of the subalgebra $\mathfrak{B}_o$ relative to $\mathfrak{B}_o$ is the same as the spectrum of this element relative to $\mathfrak{E}(\mathfrak{X})$ itself. For $f(\lambda) \in H(\Delta)$, it follows from (5.11.2) that $f(T)$ belongs to the closed linear extension of $\mathfrak{S}$ and I. Since $\mathfrak{B}_o$ is closed in $\mathfrak{E}(\mathfrak{X})$ this means that $f(T) \in \mathfrak{B}_o$ and hence that the spectrum of $f(T)$ relative to $\mathfrak{B}_o$ is precisely $\sigma[f(T)]$.

We now make use of the Gelfand representation theory for $\mathfrak{B}_o$. Since $\mathfrak{B}_o$ contains $\mathfrak{S}$, Theorems 5.8.4 and 5.8.5 are applicable. It follows that the set of maximal ideals $\mathfrak{M}_o$ splits into two disjoint sets $\mathfrak{W}_o$, $\mathfrak{U}_o$ with $\mathfrak{M}_o = \mathfrak{W}_o \cup \mathfrak{U}_o$ and there exists a numerically-valued function $\alpha(\mathfrak{m})$ on $\mathfrak{W}_o$ such that

$$
(5.12.1)\qquad \begin{aligned} R(\lambda; T)(\mathfrak{m}) &= [\lambda - \alpha(\mathfrak{m})]^{-1}, & \mathfrak{m} &\in \mathfrak{W}_o, \\ &= 0, & \mathfrak{m} &\in \mathfrak{U}_o, \end{aligned}
$$

for all $\lambda \in \rho(T)$. Further $\sigma(T) = [\alpha(\mathfrak{m}); \mathfrak{m} \in \mathfrak{W}_o]$ and $\mathfrak{U}_o$ is empty if and only if $T \in \mathfrak{E}(\mathfrak{X})$.

The multiplicative linear functional associated with any maximal ideal is continuous in the uniform operator topology. By Theorem 3.3.2 we have

$$
f(T)(\mathfrak{m}) = \delta f(\infty) + \frac{1}{2\pi i}\int_{\Gamma_T} f(\zeta)R(\zeta; T)(\mathfrak{m})\, d\zeta.
$$

Combining this with (5.12.1) we obtain

$$
(5.12.2)\qquad \begin{aligned} f(T)(\mathfrak{m}) &= \delta f(\infty) + \frac{1}{2\pi i}\int_{\Gamma_T} f(\zeta)[\zeta - \alpha(\mathfrak{m})]^{-1}\, d\zeta \\ &= f[\alpha(\mathfrak{m})], & \mathfrak{m} &\in \mathfrak{W}_o, \\ &= \delta f(\infty), & \mathfrak{m} &\in \mathfrak{U}_o. \end{aligned}
$$

If we now apply Theorem 4.15.1 (6), we see that for T bounded ($\mathfrak{U}_o = \emptyset$) we have $\sigma[f(T)] = f[\sigma(T)]$, whereas for T unbounded ($\mathfrak{U}_o \neq \emptyset$) we have $\sigma[f(T)] = f[\sigma_e(T)]$. This concludes the proof.

Since we are dealing with operators on a (B)-space we can also consider the resolution of the spectrum into the three components $P\sigma(T)$, $R\sigma(T)$, and $C\sigma(T)$. We refer to Definition 2.16.2 for the terminology used below.

THEOREM 5.12.2. *Let Δ be an open proper subset of the extended complex plane. Suppose $f(\lambda) \in H(\Delta)$ and $T \in \mathfrak{G}(\Delta)$. If $\alpha I - T$ has the property P_ν ($\nu = 1, 2,$ or 3) then so has $f(\alpha)I - f(T)$. If $\mu I - f(T)$ has the property P_ν, $\mu \neq f(\infty)$ and $f(\lambda) \not\equiv \mu$ on any component of Δ, then there exists an α with $f(\alpha) = \mu$, $\alpha \in \sigma(T)$, such that $\alpha I - T$ has this property. If T is bounded $\mu = f(\infty)$ need not be excepted.*

PROOF. For $\alpha \in \sigma(T)$, we set

$$
g(\lambda) = [f(\alpha) - f(\lambda)](\alpha - \lambda)^{-1}.
$$

It is clear that $g(\lambda) \in H(\Delta)$ and that $g(\lambda)$ has a zero of order $\geqq 1$ at $\lambda = \infty$. Consequently by Theorems 5.11.3 and 5.11.4 we have

$$
(5.12.3)\qquad \begin{aligned} [f(\alpha)I - f(T)]x &= (\alpha I - T)g(T)x, & x &\in \mathfrak{X}, \\ &= g(T)(\alpha I - T)x, & x &\in \mathfrak{D}(T). \end{aligned}
$$

This shows that the left member vanishes whenever $(\alpha I - T)x$ does; that is, property P_1 is preserved under the mapping. On the other hand, if $\alpha I - T$ maps $\mathfrak{X}$ into a non-dense subspace $\mathfrak{X}_\alpha$, then it clearly maps $g(T)[\mathfrak{X}]$ into a subspace of

$\mathfrak{X}_\alpha$ so that $[f(\alpha)I - f(T)][\mathfrak{X}]$ is also non-dense. Finally, if the image of the unit vectors in $\mathfrak{D}(T)$ under the operator $\alpha I - T$ is not bounded away from zero, then the same holds for the image under $f(\alpha)I - f(T)$. We have now proved that if $\alpha I - T$ has the spectral property P_ν then so does $f(\alpha)I - f(T)$.

In order to prove the converse statement, choose a $\mu \neq f(\infty)$ and such that $f(\lambda) \not\equiv \mu$ on any component of Δ. For a fixed $T \in \mathfrak{G}(\Delta)$, it is clear that we may replace Δ by any open subset of Δ containing $\sigma_e(T)$. Thus we may assume without loss of generality that the equation $f(\lambda) = \mu$ has only a finite number of roots $\alpha_1, \alpha_2, \cdots, \alpha_n$ in Δ; the α_k's will all be finite. Similarly we may suppose that there exists a finite $\alpha_0 \notin \Delta$. We now set

$$q_k(\lambda) = (\alpha_k - \lambda)(\alpha_0 - \lambda)^{-1};$$

then $q_k(\lambda) \in H(\Delta)$. We further define

$$h(\lambda) = [\mu - f(\lambda)] \prod_{k=1}^{n} \left(\frac{\alpha_0 - \lambda}{\alpha_k - \lambda}\right) = [\mu - f(\lambda)] \left[\prod_{k=1}^{n} q_k(\lambda)\right]^{-1}.$$

Clearly both $h(\lambda)$ and $[h(\lambda)]^{-1}$ belong to $H(\Delta)$ and hence $B = [h(T)]^{-1}$ is regular in $\mathfrak{E}(\mathfrak{X})$. Applying the operational calculus we obtain

$$\prod_{k=1}^{n} q_k(T) = B[\mu I - f(T)] = [\mu I - f(T)]B. \tag{5.12.4}$$

Here $q_k(T)$ is a bounded linear operator. Further by Theorem 5.11.3

$$q_k(T) = (\alpha_k I - T)R(\alpha_0 ; T). \tag{5.12.5}$$

Suppose now that x is annihilated by $\mu I - f(T)$; then x is also annihilated by $\prod_1^n q_k(T)$. This means that for at least one of the factors $q_k(T)$ we can find a $y \neq \theta$ such that $q_k(T)y = \theta$. Since $z = R(\alpha_0 ; T)y \neq \theta$, we have $(\alpha_k I - T)z = \theta$. Suppose next that $\mu I - f(T)$ maps $\mathfrak{X}$ into a non-dense subspace of $\mathfrak{X}$; we see from (5.12.4) that the same is true of $\prod_1^n q_k(T)$. Now if $q_k(T)[\mathfrak{X}]$ is dense in $\mathfrak{X}$, then $q_k(T)$ will map any dense subset of $\mathfrak{X}$ onto a dense subset of $\mathfrak{X}$. Consequently at least one of the factors $q_k(T)$ must take $\mathfrak{X}$ into a non-dense subspace of $\mathfrak{X}$. Since $\mathfrak{D}(T) = R(\alpha_0 ; T)[\mathfrak{X}]$, it follows that $(\alpha_k I - T)[\mathfrak{D}(T)]$ is non-dense and hence that $(\alpha_k I - T)$ has P_2. Finally if $\mu I - f(T)$ has property P_3 then at least one of the factors $q_k(T)$ must also have property P_3. Thus there exists a sequence $\{x_n\}$, $\| x_n \| = 1$, such that $q_k(T)x_n \to \theta$. Setting $y_n \equiv R(\alpha_0 ; T)x_n$ it follows from (5.12.5) that

$$(\alpha_0 - \alpha_k)y_n = (\alpha_0 I - T)y_n - (\alpha_k I - T)y_n = x_n - q_k(T)x_n .$$

Thus $\lim_{n\to\infty} q_k(T)x_n = \theta$ implies that $\lim_{n\to\infty} | \alpha_0 - \alpha_k | \, \| y_n \| = 1$ so that the $\| y_n \|$ are bounded away from zero. Finally $(\alpha_k I - T)y_n = q_k(T)x_n \to \theta$ and therefore $(\alpha_k I - T)$ has P_3. For T bounded, we may suppose in the above argument that Δ is bounded; in this case $\mu = f(\infty)$ does not enter into the argument and therefore need not be excepted.

For T unbounded we know from Theorem 5.12.1 that $f(\infty) \in \sigma[f(T)]$. To determine the fine structure, we let $\mu = f(\infty)$; then $g(\lambda) = (\alpha_0 - \lambda)[\mu - f(\lambda)] \in H(\Delta)$ so that

$$g(T)R(\alpha_0\,;T) = R(\alpha_0\,;T)g(T) = \mu I - f(T).$$

Since $R(\alpha_0\,;T)$ has property P_3 the same will be true of $\mu I - f(T)$. Further the range of $R(\alpha_0\,;T)$ is just $\mathfrak{D}(T)$; and if $\mathfrak{D}(T)$ is non-dense in $\mathfrak{X}$ then it is clear that $\mu I - f(T)$ will have property P_2. On the other hand, suppose that $f(\lambda) \not\equiv \mu$ on any component of Δ and suppose that $\mu - f(\lambda)$ has a zero of order $m > 0$ at $\lambda = \infty$. In this case we redefine $h(\lambda)$ as

$$h(\lambda) = [\mu - f(\lambda)](\alpha_0 - \lambda)^m \left[\prod_{k=1}^{n} q_k(\lambda)\right]^{-1}.$$

Again both $h(\lambda)$ and $[h(\lambda)]^{-1} \in H(\Delta)$ and $B = [h(T)]^{-1}$ is regular in $\mathfrak{E}(\mathfrak{X})$. Formula (5.12.4) now becomes

$$[R(\alpha_0\,;T)]^m \prod_{k=1}^{n} q_k(T) = B[\mu I - f(T)] = [\mu I - f(T)]B.$$

Following the same reasoning as in the above theorem we see that if $\mu I - f(T)$ has P_1 then so does one of the $\alpha_k I - T$ and if $\mu I - f(T)$ has P_2 while $\mathfrak{D}(T)$ is dense in $\mathfrak{X}$, then one of the $\alpha_k I - T$ will also have P_2. We summarize these results in the

COROLLARY. *If T is unbounded and $\mu = f(\infty)$, then $\mu I - f(T)$ will always have property P_3; it will also have P_2 whenever $\mathfrak{D}(T)$ is non-dense in $\mathfrak{X}$. Suppose further that $f(\lambda) \not\equiv \mu$ on any component of Δ. Then if $\mu I - f(T)$ has P_1 so does $\alpha I - T$ for some $\alpha \in \sigma(T)$, $f(\alpha) = \mu$; if $\mu I - f(T)$ has P_2 and $\mathfrak{D}(T)$ is dense in $\mathfrak{X}$, then there exists an $\alpha \in \sigma(T)$ with $f(\alpha) = \mu$ such that $\alpha I - T$ has P_2.*

REMARK. In the converse statement of both Theorem 5.12.2 and its corollary we have imposed the restriction that $f(\lambda) \not\equiv \mu$ on any component of Δ. Of course if $\sigma_e(T)$ is disjoint from each component of constancy for $f(\lambda)$ this restriction is not essential. On the other hand suppose $f(\lambda) \equiv \mu$ on Δ_1 and that $\sigma_e(T) \cap \Delta_1 = \sigma_1 \neq \emptyset$; here σ_1 is a spectral set. Let Γ_1 be an oriented envelope of σ_1 and define

$$J_1 = \frac{1}{2\pi i}\int_{\Gamma_1} R(\zeta;T)\,d\zeta.$$

It is clear that J_1 is a non-zero projection operator and that $J_1 f(T) = f(T)J_1 = \mu J_1$. Because of the constancy of $f(\lambda)$ on Δ_1 we obtain $f(T) = f(T)(I - J_1) + \mu J_1$. Consequently

$$\mu I - f(T) = (I - J_1)[\mu I - f(T)] = [\mu I - f(T)](I - J_1)$$

will have properties P_1 and P_2 regardless of the nature of $\alpha I - T$ for $\alpha \in \sigma_e(T)$, $f(\alpha) = \mu$.

THEOREM 5.12.3. *If $T \in \mathfrak{C}(\mathfrak{X})$ and $\rho(T) \neq \emptyset$, then for any polynomial $p(\lambda)$, $\sigma[p(T)] = p[\sigma(T)]$. If in addition $\overline{\mathfrak{D}(T)} = \mathfrak{X}$, then $\sigma_e[p(T)] = p[\sigma_e(T)]$.*

PROOF. For $\alpha \in \sigma(T)$, we set

$$q(\lambda) = [p(\alpha) - p(\lambda)](\alpha - \lambda)^{-1}.$$

If n is the degree of $p(\lambda)$, then for $x \in \mathfrak{D}(T^n)$ we have

$$[p(\alpha)I - p(T)]x = (\alpha I - T)q(T)x = q(T)(\alpha I - T)x.$$

If $\alpha I - T$ has P_1 then so has $p(\alpha)I - p(T)$. On the other hand if the range of $\alpha I - T$ is a proper subset of $\mathfrak{X}$ then the same is true of the range of $p(\alpha)I - p(T)$. Finally if $(\alpha I - T)$ has P_3 but not P_1 we prove that $p(\alpha)$ lies in $\sigma[p(T)]$. Suppose on the contrary that $p(\alpha) \in \rho[p(T)]$. Then $(\alpha I - T)q(T)R(p(\alpha), p(T)) = I$ so that the range of $(\alpha I - T)$ is all of $\mathfrak{X}$. By Theorem 2.16.3 this requires that α lie in $\rho(T)$ which is impossible. It follows that $\sigma[p(T)] \supset p[\sigma(T)]$. Conversely if $\mu \notin p[\sigma(T)]$, then there exists an open proper subset Δ of the complex plane with the properties (i) $\sigma_e(T) \subset \Delta$ and (ii) $\mu \notin p(\Delta)$. Consequently $g(\lambda) = [\mu - p(\lambda)]^{-1} \in H(\Delta)$ and by Theorem 5.11.3

$$[\mu I - p(T)]g(T) = I.$$

Further $g(\lambda)$ has no zeros on $\sigma(T)$ and therefore by Theorem 5.11.5, $g(T)$ is one-to-one on $\mathfrak{X}$ to $\mathfrak{D}(T^n) = \mathfrak{D}[p(T)]$. It follows that $g(T) = R(\mu; p(T))$ and hence that $\mu \notin \sigma[p(T)]$. This proves that $\sigma[p(T)] = p[\sigma(T)]$. Finally if $\overline{\mathfrak{D}(T)} = \mathfrak{X}$, then by the corollary to Theorem 2.16.4 $p(T)$ is bounded if and only if T is bounded. Therefore $\sigma_e[p(T)] = p[\sigma_e(T)]$.

The following composition theorem is also valid:

THEOREM 5.12.4. *Let Δ be an open proper subset of the extended complex plane. If $g(\lambda) \in H(\Delta)$ and if $f(\mu) \in H(\Delta_0)$ where $g(\Delta) \subset \Delta_0$, then $f[g(\lambda)] \in H(\Delta)$, $f[g(T)] \in \mathfrak{B}(\Delta)$, and $f[g(T)] = [f(g)](T)$ for each $T \in \mathfrak{G}(\Delta)$.*

PROOF. The proof of Theorem 5.3.2 applies with the obvious modifications. We note only the changes in the last step. We choose an open set Δ_T with the properties $\sigma(T) \subset \Delta_T$ and $\bar{\Delta}_T \subset \Delta$; then $g(\bar{\Delta}_T)$ is a bounded closed subset of Δ_0. Again we take Γ_0 and Γ to be oriented envelopes of $g(\bar{\Delta}_T)$ with respect to $f(\mu) \in H(\Delta_0)$ and of $\sigma_e(T)$ with respect to $g(\lambda) \in H(\Delta_T)$, respectively. Without loss of generality we may suppose that $\mu = \infty$ is not interior to Γ_0. We then have

$$R(\mu; g(T)) = \delta[\mu - g(\infty)]^{-1}I + \frac{1}{2\pi i}\int_\Gamma [\mu - g(\zeta)]^{-1}R(\zeta; T)\, d\zeta$$

and

$$\begin{aligned} f[g(T)] &= \frac{1}{2\pi i}\int_{\Gamma_0} f(\mu)R(\mu; g(T))\, d\mu \\ &= \delta\left(\frac{1}{2\pi i}\int_{\Gamma_0} \frac{f(\mu)}{\mu - g(\infty)}\, d\mu\right) I + \left(\frac{1}{2\pi i}\right)^2 \int_{\Gamma_0}\int_\Gamma \frac{f(\mu)R(\zeta; T)}{\mu - g(\zeta)}\, d\zeta\, d\mu \\ &= \delta f[g(\infty)]I + \frac{1}{2\pi i}\int_\Gamma f[g(\zeta)]R(\zeta; T)\, d\zeta = [f(g)](T). \end{aligned}$$

5.13. The spectral resolution of a closed operator. We now extend the results of section 5.6 to include closed linear operators; again we employ the notion of a spectral set.

THEOREM 5.13.1. *Let $\sigma_1, \sigma_2, \cdots, \sigma_k$ be bounded spectral sets of the operator $T \in \mathfrak{D}(\mathfrak{X})$ with $\sigma_\alpha \cap \sigma_\beta = \emptyset$ for $\alpha \neq \beta$. Further suppose that $\bigcup_1^k \sigma_\alpha$ is a proper subset of $\sigma_e(T)$. Define*

$$J_\alpha = \frac{1}{2\pi i} \int_{\Gamma_\alpha} R(\zeta; T)\, d\zeta, \tag{5.13.1}$$

where Γ_α is an oriented envelope of σ_α. Then

$$J_\alpha^2 = J_\alpha, \qquad J_\alpha J_\beta = \Theta \quad \text{for} \quad \alpha \neq \beta, \qquad J_\alpha \neq \Theta, I.$$

Setting

$$T_\alpha = TJ_\alpha, \qquad R_\alpha(\lambda; T) = J_\alpha R(\lambda; T),$$

then $T_\alpha = J_\alpha T J_\alpha \in \mathfrak{E}(\mathfrak{X})$ and for $\alpha \neq \beta$

$$T_\alpha T_\beta = \Theta, \qquad T_\alpha R_\beta(\lambda; T) = R_\beta(\lambda; T) T_\alpha = \Theta, \qquad R_\alpha(\lambda; T) R_\beta(\lambda; T) = \Theta,$$
$$(\lambda J_\alpha - T_\alpha) R_\alpha(\lambda; T) = R_\alpha(\lambda; T)(\lambda J_\alpha - T_\alpha) = J_\alpha .$$

Furthermore, the spectrum of T_α is $\sigma_\alpha \cup \{0\}$ and $R_\alpha(\lambda; T)$ can be extended to be locally holomorphic in the complement of σ_α. Finally

$$R_\alpha(\lambda; T) = J_\alpha \lambda^{-1} + \sum_{n=1}^{\infty} T_\alpha^n \lambda^{-n-1} \tag{5.13.2}$$

for $|\lambda| > \gamma_\alpha = \lim_{n\to\infty} \| T_\alpha^n \|^{1/n}$.

PROOF. To the spectral sets σ_α we add the "extended" spectral set $\sigma_0 = \sigma_e(T) \ominus \bigcup_1^k \sigma_\alpha$ and we denote by Γ_0 an oriented envelope of σ_0. Let $\Delta = \bigcup_0^k \Delta(\sigma_\alpha)$; then $T \in \mathfrak{G}(\Delta)$. If we now set $f_\alpha(\lambda) = 1$ on $\Delta(\sigma_\alpha)$ and $= 0$ elsewhere in Δ, then as in the proof of Theorem 5.6.1 we have: $J_\alpha = f_\alpha(T)$, $J_\alpha^2 = J_\alpha$, $J_\alpha \neq \Theta$ or I, $\sum_0^k J_\alpha = I$, and $J_\alpha J_\beta = \Theta$ for $\alpha \neq \beta$. It can be verified directly from (5.13.1) that J_α and $R(\lambda; T)$ commute. For $\alpha \neq 0$, $f_\alpha(\lambda)$ has a zero of arbitrary high order at $\lambda = \infty$. Hence Theorem 5.11.3 applies and we see that $T_\alpha = TJ_\alpha = J_\alpha T J_\alpha$ belongs to $\mathfrak{E}(\mathfrak{X})$, the range of J_α being contained in $\mathfrak{D}(T)$. The orthogonality properties of the T_α and the $R_\alpha(\lambda; T)$, $\alpha \neq 0$, now follow from those of the J_α. The quasi-resolvent relations follow in one direction by operating on both sides of $(\lambda I - T)R(\lambda; T) = I$ by $J_\alpha^2 = J_\alpha$ and in the other by applying the operator $J_\alpha R(\lambda; T)(\lambda I - T)$ to $J_\alpha x$. The rest of the proof goes through as in the proof of Theorem 5.6.1.

THEOREM 5.13.2. *Let $R(\lambda; T)$ be holomorphic save for the isolated singularities $\lambda_1, \lambda_2, \cdots, \lambda_k, \infty$. Then there exist k idempotents $J_1, J_2, \cdots, J_k$ where*

$$J_\alpha = \frac{1}{2\pi i} \int_{\Gamma_\alpha} R(\zeta, T)\, d\zeta,$$

Γ_α being a small circle about $\zeta = \lambda_\alpha$, and

$$J_\alpha^2 = J_\alpha, \qquad J_\alpha J_\beta = \Theta \quad \textit{for} \quad \alpha \neq \beta, \qquad\qquad J_\alpha \neq \Theta, I.$$

The operator $T_\alpha^- = (T - \lambda_\alpha I)J_\alpha \in \mathfrak{E}(\mathfrak{X})$ and

$$R(\lambda; T) = \sum_{\alpha=1}^{k} R_\alpha^-(\lambda; T) + R^+(\lambda; T),$$

where

$$R_\alpha^-(\lambda; T) = J_\alpha R(\lambda; T) = J_\alpha(\lambda - \lambda_\alpha)^{-1} + \sum_{n=1}^{\infty} (T_\alpha^-)^n(\lambda - \lambda_\alpha)^{-n-1},$$

$$R^+(\lambda; T) = \sum_{n=0}^{\infty} (T_\infty)^{n+1}(-\lambda)^n.$$

Further

$$R_\alpha^-(\lambda; T)R^+(\lambda; T) = R^+(\lambda; T)R_\alpha^-(\lambda; T) = \Theta, \qquad T_\alpha^- T_\infty = T_\infty T_\alpha^- = \Theta,$$

$$R_\alpha^-(\lambda; T)R_\beta^-(\lambda; T) = \Theta \quad \textit{and} \quad T_\alpha^- T_\beta^- = \Theta \quad \textit{for} \quad \alpha \neq \beta.$$

The T_α^- and T_∞ are all quasi-nilpotents.

PROOF. The preceding theorem shows that the J_α's are mutually orthogonal idempotents. Further if Γ is the circle $|\zeta| = \gamma > \max |\lambda_\alpha|$ and

$$J = \frac{1}{2\pi i} \int_\Gamma R(\zeta; T)\, d\zeta$$

then $J = \sum_1^k J_\alpha$ and $J^2 = J$. The $R_\alpha^-(\lambda; T)$ exist, are orthogonal to one another, and are quasi-resolvents of $T_\alpha = TJ_\alpha = J_\alpha T J_\alpha \in \mathfrak{E}(\mathfrak{X})$ with respect to the subalgebra $J_\alpha \mathfrak{E}(\mathfrak{X}) J_\alpha$. Applying Theorem 4.7.2, we obtain

$$\begin{aligned} R_\alpha^-(\lambda; T) &= (\lambda J_\alpha - T_\alpha)^{-1} = [(\lambda - \lambda_\alpha)J_\alpha - T_\alpha^-]^{-1} \\ &= J_\alpha(\lambda - \lambda_\alpha)^{-1} + \sum_{n=1}^{\infty} (T_\alpha^-)^n(\lambda - \lambda_\alpha)^{-n-1}. \end{aligned}$$

Since the extended $R_\alpha^-(\lambda; T)$ is holomorphic for $|\lambda - \lambda_\alpha| > 0$ the above expansion is valid in this domain and as a consequence we see that T_α^- is quasi-nilpotent. To complete the argument, one uses the Laurent expansion (see Theorem 5.9.3) of $R(\lambda; T)$ for the annulus $\gamma < |\lambda| < \infty$, obtaining

$$R(\lambda; T) = R^-(\lambda; T) + R^+(\lambda; T)$$

where

$$R^-(\lambda; T) = JR(\lambda; T) = \textstyle\sum_{\alpha=1}^k J_\alpha R(\lambda; T) = \sum_{\alpha=1}^k R_\alpha^-(\lambda; T),$$

$$R^+(\lambda; T) = (I - J)R(\lambda; T) = \textstyle\sum_{n=0}^\infty (T_\infty)^{n+1}(-\lambda)^n.$$

Note that the idempotent of Theorem 5.9.3 is precisely J since it is the residue of

$R(\lambda;\, T)$. Likewise T_∞, being the constant term in the Laurent expansion of $R(\lambda;\, T)$, is given by

$$T_\infty = \frac{1}{2\pi i} \int_\Gamma R(\zeta;\, T) \frac{d\zeta}{\zeta}.$$

The above expansion for $R^+(\lambda;\, T)$ converges for all λ and therefore T_∞ will also be quasi-nilpotent. Finally the orthogonality of $R^+(\lambda;\, T)$ and $R_\alpha^-(\lambda;\, T)$ follows from that of $I - J$ and J_α. This completes the proof.

5.14. Spectral theory for operators with compact resolvents. A spectral theory analogous to the Riesz theory for compact linear operators can be developed for closed linear operators with compact resolvents. We shall show, in fact, that such operators have a pure point spectrum and are subject to a simple elementary divisor theory. The results in this section are due to R. S. Phillips [**8**].

THEOREM 5.14.1. *Let $R(\lambda)$, defined on a subset E of the complex plane to $\mathfrak{E}(\mathfrak{X})$, satisfy the first resolvent equation. If for some $\lambda_0 \in E$, $R(\lambda_0)$ is compact (or weakly compact), then $R(\lambda)$ is compact (or weakly compact) for all $\lambda \in E$.*

PROOF. The statement of the theorem is an immediate consequence of the first resolvent equation:

$$R(\lambda) = R(\lambda_0) - (\lambda - \lambda_0)R(\lambda_0)R(\lambda).$$

For the product of a bounded linear operator and a compact (or weakly compact) operator is compact (or weakly compact) and the sum of two compact (or weakly compact) operators is again compact (or weakly compact).

We may now speak without ambiguity of an operator T possessing a compact resolvent, since by the above theorem $R(\lambda;\, T)$ is compact at one point of $\rho(T)$ if and only if it is compact for all $\lambda \in \rho(T)$. As in section 5.7, it is convenient to assume that $\mathfrak{X}$ is not finite dimensional.

THEOREM 5.14.2. *If $T \in \mathfrak{D}(\mathfrak{X})$ possesses a compact resolvent, then T is unbounded and has a pure point spectrum consisting of isolated points.*

PROOF. Let λ_0 be a fixed point in $\rho(T)$. According to Theorem 5.7.3, the operator $I + (\lambda - \lambda_0)R(\lambda_0\, ;\, T)$ is regular in $\mathfrak{E}(\mathfrak{X})$ for all but an at most denumerable set of isolated points $\{\lambda_n\}$. Clearly $\lambda_0 \notin \{\lambda_n\}$. Corresponding to each λ_n there are one or more non-zero vectors x_n such that $(\lambda_n - \lambda_0)R(\lambda_0\, ;\, T)x_n = -x_n$. Hence

$$x_n = (\lambda_0 I - T)R(\lambda_0\, ;\, T)x_n = (\lambda_0 - \lambda_n)^{-1}(\lambda_0 I - T)x_n$$

and the obvious simplification yields $Tx_n = \lambda_n x_n$. Consequently $\lambda_n \in P\sigma(T)$. On the other hand for all $\lambda \notin \{\lambda_n\}$, $I + (\lambda - \lambda_0)R(\lambda_0\, ;\, T)$ is regular in $\mathfrak{E}(\mathfrak{X})$ so that by Corollary 2 to Theorem 5.8.4 we have $\sigma(T) \subset \{\lambda_n\}$. Therefore $\sigma(T) = P\sigma(T) = \{\lambda_n\}$. Finally if T were bounded then $R(\lambda_0\, ;\, T)\ (\lambda_0 I - T) = I$ would be a compact operator. But according to Theorem 1.12.2 this is impossible in an infinite dimensional (B)-space $\mathfrak{X}$.

We next develop an elementary divisor theory for closed linear operators T with dense domain and such that $R(\lambda; T)$ is compact. Let C_0 be a circle about $\lambda_0 \in \sigma(T)$, sufficiently small so that it contains no other point of $\sigma(T)$. Then, as in Theorem 5.13.1,

$$J_0 = \frac{1}{2\pi i} \int_{C_0} R(\zeta; T)\, d\zeta \tag{5.14.1}$$

defines a projection operator different from both Θ and I. Further by Theorem 2.16.5, $[R(\lambda; T)]^* = R(\lambda; T^*)$, so that

$$J_0^* = \frac{1}{2\pi i} \int_{C_0} R(\zeta; T^*)\, d\zeta \tag{5.14.2}$$

follows from Theorem 3.3.3. We shall denote the range of J_0 by $\mathfrak{R}_0$ and that of J_0^* by $\mathfrak{R}_0^*$.

THEOREM 5.14.3. *Suppose that $T \in \mathfrak{O}(\mathfrak{X})$ has a dense domain and that the resolvent of T is compact. For $\lambda_0 \in \sigma(T)$, let $\mathfrak{M}_k \subset \mathfrak{X}$ and $\mathfrak{M}_k^* \subset \mathfrak{X}^*$ be the null spaces for the respective operators $(\lambda_0 I - T)^k$ and $(\lambda_0 I^* - T^*)^k$. Then for $k > 0$, $\mathfrak{M}_k$ ($\mathfrak{M}_k^*$) is always of positive finite dimension and there exists an $n_0 > 0$ such that $\mathfrak{M}_k = \mathfrak{R}_0$ ($\mathfrak{M}_k^* = \mathfrak{R}_0^*$) for $k \geqq n_0$, whereas $\mathfrak{M}_{k-1}$ ($\mathfrak{M}_{k-1}^*$) is a proper subset of $\mathfrak{M}_k$ ($\mathfrak{M}_k^*$) for $k \leqq n_0$. The dimension of $\mathfrak{M}_k$ is equal to that of $\mathfrak{M}_k^*$ for $k > 0$.*

PROOF. It is clear that $R(\lambda; T)$ is holomorphic and *a fortiori* continuous on $C_0 \subset \rho(T)$. Consequently the integral in the right member of (5.14.1) is the limit in the uniform operator topology of Riemann sums. The terms of these sums and hence the sums themselves are compact operators. It follows that J_0 is a compact operator. However, since J_0 is also a projection operator, this together with Theorem 1.12.2 implies that $\mathfrak{R}_0$ is finite dimensional. Further, according to Theorem 5.13.1, $T_0 = TJ_0 = J_0TJ_0 \in \mathfrak{E}(\mathfrak{X})$. Hence $\mathfrak{R}_0 \subset \mathfrak{D}(T)$ and T takes $\mathfrak{R}_0$ into itself. Likewise $\mathfrak{R}_0^* \subset \mathfrak{D}(T^*)$ and T^* takes $\mathfrak{R}_0^*$ into itself. The rest of the proof follows the argument used in Theorem 5.7.3.

If $T \in \mathfrak{O}(\mathfrak{X})$ does not have a dense domain then the adjoint operator T^* is not well defined. However those of the above results pertaining to T alone remain valid. The relevant parts in the proof of Theorem 5.14.3 suffice to prove

THEOREM 5.14.4. *Let $T \in \mathfrak{O}(\mathfrak{X})$ possess a compact resolvent. For $\lambda_0 \in \sigma(T)$, let $\mathfrak{M}_k$ be the null space for $(\lambda_0 I - T)^k$. Then for $k > 0$, $\mathfrak{M}_k$ is always of positive finite dimension and there exists an $n_0 > 0$ such that $\mathfrak{M}_k = \mathfrak{R}_0$ for $k \geqq n_0$, whereas $\mathfrak{M}_{k-1}$ is a proper subset of $\mathfrak{M}_k$ for $k \leqq n_0$.*

References. Phillips [8], Taylor [9], Walsh [1].

CHAPTER VI

LAPLACE INTEGRALS AND BINOMIAL SERIES

6.1. Orientation. An important part of classical analysis is concerned with functions which are holomorphic in a half-plane. Such a function may be representable by a Cauchy or Poisson integral in terms of the boundary values on the line bounding the half-plane, or it may be representable by one of the several forms of the Laplace integral, or by a suitable interpolation series such as the binomial series, to mention just a few alternatives. The rate of growth of the function on rays or on vertical lines determines what representations are possible.

In the theory of semi-groups of linear bounded transformations, we shall be concerned with several vector-valued functions which are holomorphic in a half-plane. The resolvent $R(\lambda; A)$ of the infinitesimal generator is such a function and in some of the important cases the semi-group operator itself has this property. For an effective study of these functions we are forced to carry over the classical theory to vector-valued functions.

The present chapter is divided into three paragraphs: *Laplace Transforms*, *Functions Holomorphic in a Half-Plane*, and *Binomial Series*. In the first we develop the elements of the theory of vector-valued Laplace-Stieltjes integrals, including the theory of convergence, analytic properties, scalar multiplication, and inversion formulas. For the classical theory we refer the reader to D. V. Widder's excellent monograph [1]. The second paragraph is concerned with functions of class $H_p(\alpha; \mathfrak{X})$, certain concepts of order together with the associated growth measuring functions, and the problem of representing functions by Laplace integrals. For the classical theory the reader may consult E. Hille [3] and E. Hille and J. D. Tamarkin [6, 7] where further references are to be found. The theory of binomial series is equivalent to the theory of functions which are of exponential type in the half-plane. Here the classical theory is due essentially to F. Carlson [2] and N. E. Nörlund [2] to whose writings we refer for further details. An extension of this theory to vector-valued functions was given by E. Hille [7]; the present account, though sketchy, has certain advantages.

Paragraph one is indispensible for the subsequent theory of linear bounded transformations; however the reader may omit paragraphs two and three on a first reading.

References. Carlson [2], Caton and Hille [1], Hille [3, 7], Hille and Tamarkin [6, 7], Lindelöf [2], Nörlund [2], Phillips [9], Phragmén and Lindelöf [1], Post [1], Widder [1], Yosida [3].

1. Laplace Transforms

6.2. Laplace-Stieltjes integrals. Let $\mathfrak{X}$ be a complex (B)-space, $a(\xi)$ a function on $[0, \infty]$ to $\mathfrak{X}$, and let $a(\xi)$ be of *strong bounded variation* over every finite interval $[0, \omega]$ in the sense of Definition 3.2.4 (3). Since a function of strong bounded variation has right and left limits everywhere and is continuous except for a countable set of discontinuities of the first kind, we may normalize $a(\xi)$ by assuming that

$$a(0) = \theta, \qquad a(\xi) = \tfrac{1}{2}[a(\xi - 0) + a(\xi + 0)] \quad \text{for} \quad \xi > 0.$$

The integral

$$a(\xi; \lambda) = \int_0^\xi e^{-\lambda\alpha}\, da(\alpha)$$

exists for finite complex values of λ and finite positive values of ξ. If, for a particular λ, $\lim_{\xi\to\infty} a(\xi; \lambda)$ exists, we denote the limit by

$$f(\lambda) = \int_0^\infty e^{-\lambda\xi}\, da(\xi). \tag{6.2.1}$$

We say that the integral *converges* for this value of λ and call $f(\lambda)$ the *Laplace-Stieltjes transform* of $a(\xi)$. If $a(\xi)$ is absolutely continuous and $a(\xi) = \int_0^\xi g(\alpha)\, d\alpha$, then

$$f(\lambda) = \int_0^\infty e^{-\lambda\xi} g(\xi)\, d\xi \tag{6.2.2}$$

and $f(\lambda)$ is called the *Laplace transform* of $g(\xi)$.

Let $a_*(\xi)$ denote the *strong variation* of $a(\alpha)$ in $[0, \xi]$. We say that the Laplace-Stieltjes integral is *absolutely convergent* for $\lambda = \lambda_0$ if the numerical integral

$$\varphi(\lambda) = \int_0^\infty e^{-\lambda\xi}\, da_*(\xi) \tag{6.2.3}$$

converges for $\lambda = \sigma_0$ where $\lambda_0 = \sigma_0 + i\tau_0$. This implies ordinary convergence and the inequality

$$\| f(\sigma + i\tau) \| \leqq \varphi(\sigma) \tag{6.2.4}$$

whenever the right side exists. Necessary and sufficient conditions for the convergence of (6.2.3) may of course be read off from the classical theory and the latter also indicates the nature of the results relating to ordinary convergence in the abstract case.

Remark. Let $g(\xi) \in B[(0, \omega), \mathfrak{X}]$ and set $a(\xi) = \int_0^\xi g(\alpha)\, d\alpha$. Then $a(\xi)$ is of strong bounded variation and

$$a_*(\xi) = \int_0^\xi \| g(\alpha) \|\, d\alpha \quad \text{for} \quad \xi \in [0, \omega].$$

This is clearly true for simple functions $h(\xi)$, constant on intervals. As we have already noticed in section 3.8, given $\epsilon > 0$ there exists such an $h(\xi)$ for which

$$\int_0^\xi \| g(\alpha) - h(\alpha) \| \, d\alpha < \epsilon.$$

Setting $b(\xi) = \int_0^\xi h(\alpha) \, d\alpha$, we see by Theorem 3.7.6 that

$$(a - b)_*(\xi) \leqq \int_0^\xi \| g(\alpha) - h(\alpha) \| \, d\alpha < \epsilon.$$

Now for any two functions $c(\xi)$, $d(\xi)$ of strong bounded variation we have $(c + d)_*(\xi) \leqq c_*(\xi) + d_*(\xi)$. Hence $b_*(\xi) - (a - b)_*(\xi) \leqq a_*(\xi) \leqq b_*(\xi) + (a - b)_*(\xi)$ and therefore

$$\int_0^\xi \| g(\alpha) \| \, d\alpha - 2\epsilon \leqq \int_0^\xi \| h(\alpha) \| \, d\alpha - \epsilon \leqq a_*(\xi)$$

$$\leqq \int_0^\xi \| h(\alpha) \| \, d\alpha + \epsilon \leqq \int_0^\xi \| g(\alpha) \| \, d\alpha + 2\epsilon.$$

Since $\epsilon > 0$ is arbitrary, this concludes the proof.

The following approach to the convergence theory was developed by F. Bohnenblust in 1932 (unpublished) for the case of numerically-valued functions. Let $a(\infty)$ denote $\lim_{\xi\to\infty} a(\xi)$, whenever the latter exists as an element of $\mathfrak{X}$, otherwise θ. We define the *order* of $a(\xi)$ to be

$$\gamma = \limsup_{\xi\to\infty} \frac{1}{\xi} \log \| a(\xi) - a(\infty) \| . \tag{6.2.5}$$

The functions $a(\xi)$ of order $< \infty$ form a linear space $\mathfrak{A}$; we collect functions having the same order γ into a subspace $\mathfrak{A}(\gamma)$. Since the order of the sum of two functions of order γ is $\leqq\gamma$, the spaces $\mathfrak{A}(\gamma)$ are not linear.

According to Theorems 3.3.1 and 3.3.2 the integral defining $a(\xi; \lambda)$ can be integrated by parts; this gives

$$\begin{aligned} a(\xi; \lambda) &= e^{-\lambda\xi} a(\xi) + \lambda \int_0^\xi e^{-\lambda\alpha} a(\alpha) \, d\alpha \\ &= e^{-\lambda\xi}[a(\xi) - a(\infty)] + a(\infty) + \lambda \int_0^\xi e^{-\lambda\alpha}[a(\alpha) - a(\infty)] \, d\alpha. \end{aligned} \tag{6.2.6}$$

Conversely we have

$$\begin{aligned} a(\xi) = a(\xi; 0) &= \int_0^\xi e^{\lambda\alpha} \, d_\alpha a(\alpha; \lambda) \\ &= e^{\lambda\xi} a(\xi; \lambda) - \lambda \int_0^\xi e^{\lambda\alpha} a(\alpha; \lambda) \, d\alpha. \end{aligned} \tag{6.2.7}$$

Formula (6.2.6) defines a linear transformation

$$T(\lambda)[a(\xi)] = a(\xi; \lambda)$$

on $\mathfrak{A}$ to itself which maps the subspace $\mathfrak{A}(\gamma)$ on the subspace $\mathfrak{A}[\gamma - \Re(\lambda)]$ in a one-to-one manner. Indeed, if $a(\xi) \in \mathfrak{A}(\gamma)$, then (6.2.6) shows that the order of $T(\lambda)[a(\xi)]$ is $\leqq \gamma - \Re(\lambda)$ and (6.2.7), which defines the inverse transformation, shows that inequality is excluded and that the correspondence between the two subspaces is one-to-one. A simple computation shows that $T(\lambda)$ has the group property

$$T(\lambda_1 + \lambda_2) = T(\lambda_1)T(\lambda_2). \tag{6.2.8}$$

The functions $a(\xi)$ for which $\lim_{\xi\to\infty} a(\xi)$ exists form a certain linear subspace $\mathfrak{C}$ of $\mathfrak{A}$. A sufficient condition that $a(\xi) \in \mathfrak{C}$ is that the order of $a(\xi)$ is <0 and a necessary condition is that it is $\leqq 0$. Now *the Laplace-Stieltjes transform of $a(\xi)$ will exist for a particular λ if and only if $T(\lambda)[a(\xi)] \in \mathfrak{C}$*, that is, *if* the order of $T(\lambda)[a(\xi)]$ is <0 and *only if* it is $\leqq 0$. From this observation we get

THEOREM 6.2.1. *There exist two real numbers σ_0 and σ_a such that the integral (6.2.1) is convergent for $\Re(\lambda) > \sigma_0$, but not for any λ with $\Re(\lambda) < \sigma_0$, and it is absolutely convergent for $\Re(\lambda) > \sigma_a$, but not for any λ with $\Re(\lambda) < \sigma_a$. We have*

$$-\infty \leqq \sigma_0 \leqq \sigma_a \leqq \infty, \tag{6.2.9}$$

and

$$\sigma_0 = \limsup_{\xi\to\infty} \frac{1}{\xi} \log \| a(\infty) - a(\xi) \|, \tag{6.2.10}$$

$$\sigma_a = \limsup_{\xi\to\infty} \frac{1}{\xi} \log | a_*(\infty) - a_*(\xi) |, \tag{6.2.11}$$

where $a(\infty) = \lim_{\xi\to\infty} a(\xi)$ or θ according as the limit exists or not and $a_*(\infty)$ is defined similarly.

We refer to σ_0 and σ_a as the *abscissas of ordinary* and of *absolute convergence* respectively. A different abscissa of absolute convergence could be defined by replacing the strong variation by the total variation in (6.2.11). There is also an *abscissa of uniform convergence* which lies between σ_0 and σ_a. We do not wish to introduce these notions here, especially as they may lead to added confusion in the case in which $a(\xi)$ is an operator where we have to consider the uniform as well as the strong topology.

Formula (6.2.6) shows that for $\Re(\lambda) > \sigma_0$

$$f(\lambda) = \lambda \int_0^\infty e^{-\lambda\xi}[a(\xi) - a(\infty)]\, d\xi + a(\infty). \tag{6.2.12}$$

If $\Re(\lambda) > \max(\sigma_0, 0)$ we may omit $a(\infty)$ in this formula. We see that

$$\| f(\lambda) \| = O(|\lambda|), \qquad \Re(\lambda) > \sigma_0 + \epsilon.$$

From (6.2.4) we get that $f(\lambda)$ is uniformly bounded when $\Re(\lambda) > \sigma_a + \epsilon$. Further for each $x^* \in \mathfrak{X}^*$ we have

$$x^*[f(\lambda)] = \int_0^\infty e^{-\lambda\xi}\, dx^*[a(\xi)].$$

We conclude from the classical theory that $x^*[f(\lambda)]$ is holomorphic for $\Re(\lambda) > \sigma_0$ and hence the same is true of $f(\lambda)$ itself. In like manner we see that the derivatives of $f(\lambda)$ are given by

$$f^{(n)}(\lambda) = (-1)^n \int_0^\infty e^{-\lambda\xi}\xi^n \, da(\xi), \qquad \Re(\lambda) > \sigma_0. \tag{6.2.13}$$

Since $f(\lambda)$ is holomorphic in a half-plane and its norm cannot grow faster than $O(|\lambda|)$, it follows from the extensions of the principle of the maximum (see section 3.13) that its zeros must be distributed rather sparingly if the function is not to vanish identically. The classical instance of this observation is the *theorem of Lerch* which holds also for abstract Laplace-Stieltjes integrals:

THEOREM 6.2.2. *If $f(\lambda)$ is holomorphic in $\Re(\lambda) > \sigma_0$ and if*

$$f(\lambda) = \theta \quad \textit{for} \quad \lambda = \lambda_0 + n, \qquad n = 1, 2, 3, \cdots,$$

then $f(\lambda) \equiv \theta$.

PROOF. If $x^* \in \mathfrak{X}^*$ is a linear bounded functional, then

$$x^*[f(\lambda)] = \lambda \int_0^\infty e^{-\lambda\xi} x^*[a(\xi)] \, d\xi, \qquad \Re(\lambda) > \max(0, \sigma_0),$$

vanishes for $\lambda = \lambda_0 + n$ and, by the classical theorem of Lerch, this implies $x^*[f(\lambda)] \equiv 0$. Since this holds for every x^* we have $f(\lambda) \equiv \theta$.

We note that $x^*[a(\xi)]$ is also normalized. Hence by the classical uniqueness theorem for Laplace-Stieltjes integrals $x^*[a(\xi)] \equiv 0$ (not merely for almost all ξ) and this implies $a(\xi) \equiv \theta$. Thus the *uniqueness theorem* also extends:

THEOREM 6.2.3. *There cannot exist two different normalized representations of $f(\lambda)$ in terms of Laplace-Stieltjes integrals.*

In the classical theory it is shown that the product of two absolutely convergent Laplace-Stieltjes integrals is an integral of the same kind. Since only scalar multiplication has a sense in ordinary (B)-spaces, we have to be satisfied with a partial extension of this theorem. To simplify matters we introduce a slight restriction in the scalar factor.

THEOREM 6.2.4. *Let*

$$\gamma(\lambda) = \int_0^\infty e^{-\lambda\xi} \, d\beta(\xi), \qquad f(\lambda) = \int_0^\infty e^{-\lambda\xi} \, da(\xi)$$

be Laplace-Stieltjes integrals absolutely convergent for $\Re(\lambda) > \sigma_a$. Here $\beta(\xi)$ is to be a continuous numerically-valued function of bounded variation, $\beta(0) = 0$, and $a(\xi)$ is a normalized function of strong bounded variation on $[0, \infty)$ to $\mathfrak{X}$. Then

$$\gamma(\lambda)f(\lambda) = \int_0^\infty e^{-\lambda\xi} \, dc(\xi) \tag{6.2.14}$$

with

$$c(\xi) = \int_0^\xi a(\xi - \eta)\, d\beta(\eta) = \int_0^\xi \beta(\xi - \eta)\, da(\eta), \tag{6.2.15}$$

the integral being absolutely convergent for $\Re(\lambda) > \sigma_a$.

PROOF. We note that $c(\xi)$ is defined for all ξ since $\beta(\xi)$ is continuous, the two expressions for $c(\xi)$ being found to be equal by an integration by parts. We define $\beta(\xi) \equiv 0$ and $a(\xi) \equiv \theta$ for $\xi < 0$. We denote the variation of $\beta(\xi)$ in $(-\infty, \eta)$ by $\beta_*(\eta)$ with similar notation for the strong variations of $a(\xi)$ and $c(\xi)$. We have then

$$c(\xi) = \int_{-\infty}^{\infty} \beta(\xi - \eta)\, da(\eta)$$

and from this representation one concludes successively that (i) $c(0) = \theta$, (ii) $c(\xi)$ is continuous, (iii) $c(\xi)$ is of strong bounded variation, and (iv) $c_*(\xi) \leqq \beta_*(\xi)a_*(\xi)$ for all ξ. Without restricting the generality we may assume $\sigma_a = 0$. We have then $\log \beta_*(\omega) = o(\omega)$, $\log a_*(\omega) = o(\omega)$ so, by (iv), $\log c_*(\omega) = o(\omega)$ whence it follows that (6.2.14) is absolutely convergent for $\Re(\lambda) > 0 = \sigma_a$. To prove that the integral really represents the product $\gamma(\lambda)f(\lambda)$ it is enough to show that the linear functionals of the two sides of (6.2.14) are equal and this follows from the classical multiplication theorem for Laplace-Stieltjes integrals. This completes the proof.

The particular case $\beta(\xi) = \xi^\alpha/\Gamma(\alpha + 1)$, $\gamma(\lambda) = \lambda^{-\alpha}$, leads to the important representation

$$f(\lambda) = \lambda^\alpha \int_0^\infty e^{-\lambda\xi}\, d_\xi a_\alpha(\xi), \qquad \Re(\lambda) > \max\,(0, \sigma_0), \tag{6.2.16}$$

where

$$a_\alpha(\xi) = \frac{1}{\Gamma(\alpha)} \int_0^\xi (\xi - \eta)^{\alpha-1} a(\eta)\, d\eta \tag{6.2.17}$$

is the fractional integral of $a(\xi)$ of order α, $\Re(\alpha) > 0$. This is a generalization of (6.2.12) using integration by parts of fractional order. The main value of this formula lies in the fact that it defines a fairly effective method of summation for divergent Laplace-Stieltjes integrals which is equivalent to the arithmetic means (C, α) in the right half-plane. The effectiveness is based upon the fact that if $f(\lambda)$ is of finite order and holomorphic in a half-plane, then $\lambda^{-\alpha}f(\lambda)$ is representable by an absolutely convergent Laplace-Stieltjes integral for sufficiently large values of $\Re(\alpha)$ (cf. section 6.6).

It is important to realize that (6.2.16) is actually a convergent representation of $f(\lambda)$ for $\Re(\lambda) > \max\,(0, \sigma_0)$; this does not follow from Theorem 6.2.4 which asserts something different, namely absolute convergence for $\Re(\lambda) > \max\,(0, \sigma_a)$. However, it is an easy matter to show that the integral converges for $\Re(\lambda) > \max(0, \sigma_0)$ by estimating the rate of growth of $a_\alpha(\xi)$, that of $a(\xi)$ being known. To prove that the integral equals $\lambda^{-\alpha}f(\lambda)$ it is enough to show that the linear functionals are equal and this follows from another classical multiplication theorem for Laplace-Stieltjes integrals according to which the product integral converges if one of the factors converges absolutely, the other being merely convergent. In our case $\lambda^{-\alpha}$ has an absolutely convergent representation for $\Re(\lambda) > 0$.

6.3. Inversion formulas. For the complex inversion formulas we need some preliminaries. With a slight change of conventional notation we place

$$\text{si}(\xi) = \frac{1}{\pi}\int_\xi^\infty \frac{\sin \eta}{\eta}\, d\eta. \tag{6.3.1}$$

The following properties of si (ξ) are well known. It is a bounded function whose maximum $M = 1.0894 \cdots$ is reached for $\xi = -\pi$, while the minimum $-0.28 \cdots$ is reached at $\xi = \pi$. For small values of ξ

$$\text{si}(\xi) = \tfrac{1}{2} + O(\xi),$$

for large values

$$\text{si}(\xi) = O(1/\xi), \quad \xi \to +\infty, \quad \text{si}(\xi) = 1 + O(1/\xi), \quad \xi \to -\infty.$$

We shall now prove the following lemma for Dirichlet's integral which is well known in the numerical case:

Lemma 6.3.1. *Let $\mathfrak{B}$ be the class of functions $b(\xi)$ on $(-\infty, \infty)$ to $\mathfrak{X}$ such that $b(\xi) \to \theta$ when $\xi \to -\infty$ and $b(\xi)$ is of strong bounded variation on $(-\infty, \infty)$. Then*

$$b(\xi \mid \omega) = \lim_{\alpha\to\infty} \frac{1}{\pi}\int_{-\alpha}^{\alpha} \frac{\sin \omega\eta}{\eta}\, b(\xi + \eta)\, d\eta \tag{6.3.2}$$

exists for all real ξ and ω and is absolutely continuous in ξ on every finite interval. Further

$$b(\xi \mid \omega) = \int_{-\infty}^{\infty} \text{si}(\omega\eta)\, d_\eta b(\xi + \eta) = \int_{-\infty}^{\infty} \text{si}[\omega(\sigma - \xi)]\, db(\sigma), \tag{6.3.3}$$

$$\| b(\xi \mid \omega) \| \leqq M b_*(\infty), \tag{6.3.4}$$

$$\lim_{\omega\to\infty} b(\xi \mid \omega) = \tfrac{1}{2}[b(\xi - 0) + b(\xi + 0)] \tag{6.3.5}$$

for every ξ. The limit exists uniformly with respect to ξ in any closed interval of continuity of $b(\xi)$.

Remark. It is perhaps necessary to observe that $b(\xi \mid \omega)$ is ordinarily not a member of $\mathfrak{B}$. A counter example is given by the characteristic function of the interval (β, ∞) in the numerical case with $b(\xi \mid \omega) = \text{si}[\omega(\beta - \xi)]$. It is true, however, that $\lim_{\xi\to-\infty} b(\xi \mid \omega) = \theta$ for all elements of $\mathfrak{B}$.

Proof. Formula (6.3.3) is proved by integration by parts under the limit sign in (6.3.2), and (6.3.4) is an immediate consequence thereof. We recall that $b_*(\sigma)$

is the strong variation of $b(\xi)$ in the interval $(-\infty, \sigma]$. That $b(\xi \mid \omega)$ is absolutely continuous in ξ follows by differentiation with respect to ξ under the sign of integration in the last member of (6.3.3) which leads to an absolutely convergent integral. To prove (6.3.5) we break up the interval of integration in the second member of (6.3.3) into six parts denoting the integrals from left to right by I_1 to I_6. The partition points are taken at $\xi = -\omega^{-1/2}, -\omega^{-2}, 0, \omega^{-2}, \omega^{-1/2}$. Using the properties of si (ξ) stated above, we see that

$$\begin{aligned}
I_1 &= b(\xi - \omega^{-1/2}) + O(\omega^{-1/2}),\\
\| I_2 \| &\leqq M[b_*(\xi - \omega^{-2}) - b_*(\xi - \omega^{-1/2})],\\
I_3 &= \tfrac{1}{2}[b(\xi) - b(\xi - \omega^{-2})] + O(\omega^{-1}),\\
I_4 &= \tfrac{1}{2}[b(\xi + \omega^{-2}) - b(\xi)] + O(\omega^{-1}),\\
\| I_5 \| &\leqq M[b_*(\xi + \omega^{-1/2}) - b_*(\xi + \omega^{-2})],\\
I_6 &= O(\omega^{-1/2}),
\end{aligned}$$

where all the O-estimates hold uniformly with respect to ξ in $(-\infty, \infty)$. From these relations (6.3.5) follows. This completes the proof.

We can now prove the *complex inversion formula:*

THEOREM 6.3.1. *Let $f(\lambda)$ be defined by* (6.2.1), *convergent for $\Re(\lambda) > \sigma_0$, let $\gamma > \max(0, \sigma_0)$ and set*

$$a(\xi \mid \omega) = \frac{1}{2\pi i}\int_{\gamma - i\omega}^{\gamma + i\omega} e^{\lambda\xi} f(\lambda)\,\frac{d\lambda}{\lambda}. \tag{6.3.6}$$

Then

$$\lim_{\omega \to \infty} \alpha(\xi \mid \omega) = \begin{cases} a(\xi), & \xi > 0,\\ \tfrac{1}{2}a(0+), & \xi = 0,\\ 0, & \xi < 0.\end{cases} \tag{6.3.7}$$

The limit exists uniformly with respect to ξ in any finite interval of continuity of $a(\xi)$.

PROOF. We substitute (6.2.12) for $f(\lambda)$ in the definition of $a(\xi \mid \omega)$ and interchange the order of integration as we may do using the Fubini theorem. It follows that

$$a(\xi \mid \omega) = \frac{1}{\pi}\int_0^\infty a(\eta) e^{\gamma(\xi - \eta)}\,\frac{\sin \omega(\xi - \eta)}{\xi - \eta}\,d\eta. \tag{6.3.8}$$

For $\gamma > \delta > \max(0, \sigma_0)$, formula (6.2.10) shows that $\| a(\xi) \| \leqq Me^{\delta\xi}$ so that the above integral is absolutely convergent. Setting $b_\rho(\xi) = a(\xi)e^{-\gamma\xi}$ for $0 \leqq \xi \leqq \rho$ and $= \theta$ for $\xi < 0$, and $\xi > \rho$, we see that $b_\rho(\xi)$ satisfies the conditions of the preceding lemma. Further $b_\rho(\xi)$ is normalized in the interval $(0, \rho)$. Thus for

$b_\rho(\xi \mid \omega)$ defined as in formula (6.3.2) we see that

$$\lim_{\omega\to\infty} b_\rho(\xi \mid \omega)e^{\gamma\xi} = \begin{cases} a(\xi), & 0 < \xi < \rho, \\ \frac{1}{2}a(0+), & \xi = 0, \\ \theta, & \xi < 0, \end{cases}$$

the limit existing uniformly with respect to ξ in any finite interval of continuity contained in $(-\rho, \rho)$. On the other hand for $\xi < r < \rho$

$$\| b_\rho(\xi \mid \omega)e^{\gamma\xi} - a(\xi \mid \omega) \| \leqq \frac{1}{\pi}\int_\rho^\infty \frac{\| a(\eta) \| e^{\gamma(\xi-\eta)}}{|\xi - \eta|} d\eta$$
$$\leqq \frac{Me^{-(\gamma-\delta)\rho}e^{\gamma r}}{\pi(\rho - r)(\gamma - \delta)}.$$

Hence $\lim_{\rho\to\infty} b_\rho(\xi \mid \omega)e^{\gamma\xi} = a(\xi \mid \omega)$ uniformly in ω and uniformly for ξ in any finite interval. The desired result now follows from the familiar iterated limits theorem.

Applying this theorem to formula (6.2.16), we see that for $\Re(\alpha) > 0$, $\xi > 0$,

$$\text{(6.3.9)} \qquad \lim_{\omega\to\infty} \frac{1}{2\pi i}\int_{\gamma-i\omega}^{\gamma+i\omega} e^{\lambda\xi}f(\lambda)\frac{d\lambda}{\lambda^{1+\alpha}} = \frac{1}{\Gamma(\alpha)}\int_0^\xi (\xi - \eta)^{\alpha-1}a(\eta)\, d\eta,$$

the limit being θ for $\xi < 0$.

For the applications which we have in mind an inversion formula for Laplace integrals will also be needed.

THEOREM 6.3.2. *Let $g(\xi) \in B[(0, \omega); \mathfrak{X}]$ for every finite ω and let*

$$\text{(6.3.10)} \qquad f(\lambda) = \int_0^\infty e^{-\lambda\xi}g(\xi)\, d\xi$$

be absolutely convergent for $\Re(\lambda) > \sigma_a$. Let $\gamma > \max(0, \sigma_a)$ and set

$$\text{(6.3.11)} \qquad g_1(\xi \mid \omega) = \frac{1}{2\pi}\int_{-\omega}^{\omega}\left\{1 - \frac{|\tau|}{\omega}\right\} e^{(\gamma+i\tau)\xi}f(\gamma + i\tau)\, d\tau.$$

Then $\lim_{\omega\to\infty} g_1(\xi \mid \omega) = g(\xi)$ for almost all positive ξ and equals θ for $\xi < 0$. The limit equals $\frac{1}{2}[g(\xi + 0) + g(\xi - 0)]$ whenever this expression has a meaning. The limit exists uniformly with respect to ξ in any finite interval of continuity of $g(\xi)$.

PROOF. We substitute (6.3.10) for $f(\lambda)$ in the definition of $g_1(\xi \mid \omega)$ and interchange the order of integration, obtaining after some simplification

$$g_1(\xi \mid \omega) = \frac{2}{\pi\omega}\int_0^\infty \frac{\sin^2[\omega(\xi - \alpha)/2]}{(\xi - \alpha)^2} g(\alpha)e^{\gamma(\xi-\alpha)}\, d\alpha.$$

Put $h(\xi) = g(\xi)e^{-\gamma\xi}$ or θ according as $\xi > 0$ or < 0 with $h_1(\xi \mid \omega) = g_1(\xi \mid \omega)e^{-\gamma\xi}$.

Since $\gamma > \sigma_a$, $h(\xi) \in B[E_1; \mathfrak{X}]$ and

$$h_1(\xi \mid \omega) = \int_0^\infty F(\xi - \alpha; \omega)h(\alpha)\, d\alpha$$

with

$$F(\beta; \omega) = \frac{2}{\pi\omega} \frac{\sin^2(\omega\beta/2)}{\beta^2}.$$

It is a simple matter to verify that the Fejér kernel satisfies all the conditions of Theorems 3.9.1 to 3.9.3 and condition (4′) with $\mu_1 = \mu_2 = \frac{1}{2}$. It follows that $h_1(\xi \mid \omega)$ tends to $h(\xi)$ when $\omega \to \infty$, in the mean of order one as well as pointwise in the Lebesgue set of $h(\xi)$. Since the Lebesgue sets of $g(\xi)$ and $h(\xi)$ are identical, the conclusions of the theorem are immediate.

The theorem shows that the integral

$$\frac{1}{2\pi i} \int_{\gamma - i\infty}^{\gamma + i\infty} e^{\gamma\xi} f(\lambda)\, d\lambda$$

is summable $(C, 1)$ to $g(\xi)$ almost everywhere if $f(\lambda)$ is the Laplace transform of $g(\xi)$ and $\gamma > \max(0, \sigma_a)$. It is clear that other methods of summation could be used for the same purpose.

The two preceding theorems express the *determining function* $a(\xi)$ or $g(\xi)$ in terms of the values of the *generating function* $f(\lambda)$ on vertical lines in the half-plane of convergence. There are many other means at our disposal for the solution of the inversion problem. Methods involving the values of $f(\lambda)$ at the positive integers or the values of $f(\lambda)$ and its derivatives at some fixed point are available but will not be considered here. Instead we shall study two of the methods involving the values of $f(\lambda)$ for large real values of λ.

THEOREM 6.3.3. *Let $f(\lambda)$ satisfy the assumptions of Theorem 6.3.2 and set*

$$g_0(\xi \mid \omega) = e^{-\omega\xi} \sum_{n=0}^{\infty} \frac{(-1)^n (\omega^2\xi)^{n+1}}{n!(n+1)!} f^{(n)}(\omega). \tag{6.3.12}$$

Then $g_0(\xi \mid \omega)$ exists for $\omega > \sigma_a$ and $\lim_{\omega\to\infty} g_0(\xi \mid \omega) = g(\xi)$ in the Lebesgue set of $g(\xi)$. If $g(\xi)$ is continuous in some open interval, then the limit holds uniformly in each compact subinterval.

REMARK. This inversion formula, due to R. S. Phillips [9], plays a central role in our later treatment of the generation of semi-groups of linear bounded operators. It was, in fact, suggested by the proof of Theorem 12.3.1 which goes back to K. Yosida [3]. The formula is also closely related to the still earlier work of W. B. Caton and E. Hille [1].

PROOF. For $\omega > \sigma_a$ the formulas (6.2.13) apply and we may write (6.3.12) in the more compact form

$$g(\xi \mid \omega) = \int_0^\infty K(\xi, \alpha; \omega) g(\alpha)\, d\alpha$$

where

$$K(\xi, \alpha; \omega) = e^{-\omega(\xi+\alpha)} \sum_{n=0}^{\infty} \frac{(\omega^2\xi)^{n+1}\alpha^n}{n!(n+1)!}$$

$$= \omega \left(\frac{\xi}{\alpha}\right)^{1/2} e^{-\omega(\xi+\alpha)} I_1[2\omega(\xi\alpha)^{1/2}].$$

Here $I_1(v)$ is the Bessel function of first order with purely imaginary argument. The interchange of order of summation and integration is easily justified if we make use of the inequality (6.3.14a).

We now list a few of the properties of $K(\xi, \alpha; \omega)$ which we shall need.

(6.3.13a) $$K(\xi, \alpha; \omega) \geqq 0, \qquad \xi, \alpha, \omega \geqq 0,$$

(6.3.13b) $$\int_0^\infty K(\xi, \alpha; \omega)\, d\alpha = 1 - e^{-\omega\xi}.$$

Since $I_1(v)$ is bounded by $(v/2) \exp(v)$ and $Cv^{-1/2} \exp(v)$, we have the corresponding bounds for $K(\xi, \alpha; \omega)$, namely,

(6.3.14a) $$K(\xi, \alpha; \omega) \leqq \omega^2\xi \exp[-\omega(\xi^{1/2} - \alpha^{1/2})^2],$$

(6.3.14b) $$K(\xi, \alpha; \omega) \leqq C\omega^{1/2}\xi^{1/4}\alpha^{-3/4} \exp[-\omega(\xi^{1/2} - \alpha^{1/2})^2].$$

Let ξ be a point of the Lebesgue set for $g(\xi)$; then

$$G(\beta; \xi) = \int_\xi^\beta \| g(\xi) - g(\alpha) \|\, d\alpha = o(|\beta - \xi|).$$

Because of (6.3.13b) it will be sufficient to show that

$$\int_0^\infty K(\xi, \alpha; \omega) \| g(\xi) - g(\alpha) \|\, d\alpha \to 0$$

as $\omega \to \infty$. To this end we break up the range of the integral into four parts:

$$J_1 = \int_0^{\xi-\delta}, \qquad J_2 = \int_{\xi-\delta}^{\xi_0}, \qquad J_3 = \int_{\xi_0}^{\xi+\delta}, \qquad J_4 = \int_{\xi+\delta}^\infty.$$

Here ξ_0 will be specified later as a point in the interval $(\xi - \delta, \xi + \delta)$. Making use of (6.3.14a) we see that

$$J_1 \leqq \omega^2\xi \int_0^{\xi-\delta} \exp[-\omega(\xi^{1/2} - \alpha^{1/2})^2] \| g(\xi) - g(\alpha) \|\, d\alpha$$

$$\leqq \omega^2\xi \exp[-\omega\{\xi^{1/2} - (\xi - \delta)^{1/2}\}^2] \int_0^{\xi-\delta} \| g(\xi) - g(\alpha) \|\, d\alpha.$$

Thus $J_1 \to 0$ as $\omega \to \infty$. Likewise using (6.3.14a) and splitting off a factor

$\exp(-\gamma\alpha)$, $\gamma > \sigma_a$, we see that

$$J_4 \leqq \omega^2\xi \exp[\omega\gamma\xi(\omega - \gamma)^{-1}] \exp[-(\omega - \gamma)[(\xi + \delta)^{1/2} - \omega(\omega - \gamma)^{-1}\xi^{1/2}]^2]$$
$$\times \int_{\xi+\delta}^{\infty} \exp(-\gamma\alpha) \| g(\xi) - g(\alpha) \| \, d\alpha$$

and hence $J_4 \to 0$ as $\omega \to \infty$. In treating J_2 and J_3 use is made of the bound (6.3.14b). Given $\epsilon > 0$ we now choose δ so that $0 < \delta < \xi$ and $G(\beta; \xi) < \epsilon \mid \beta - \xi \mid$ for $\mid \beta - \xi \mid < \delta$. Then for ω sufficiently large and ξ fixed

$$k(\xi, \alpha; \omega) \equiv C\omega^{1/2}\xi^{1/4}\alpha^{-3/4} \exp[-\omega(\xi^{1/2} - \alpha^{1/2})^2]$$

has but one relative extremum in the interval $(\xi - \delta, \xi + \delta)$. This is a maximum of $O(\omega^{1/2})$ at $\xi_0 \sim \xi - 3/(2\omega)$. Finally

$$J_2 \leqq \int_{\xi-\delta}^{\xi_0} k(\xi, \alpha; \omega) \, dG(\alpha; \xi) = k(\xi, \alpha; \omega) \, G(\alpha; \xi) \Big|_{\xi-\delta}^{\xi_0}$$
$$- \int_{\xi-\delta}^{\xi_0} \left(\frac{\partial}{\partial\alpha} k(\xi, \alpha; \omega)\right) G(\alpha; \xi) \, d\alpha.$$
$$\leqq \left| k(\xi, \alpha; \omega) \, G(\alpha; \xi) \Big|_{\xi-\delta}^{\xi_0} \right| + \left| \epsilon k(\xi, \alpha; \omega) \mid \xi - \alpha \mid \Big|_{\xi-\delta}^{\xi_0} \right| + \epsilon \int_{\xi-\delta}^{\xi_0} k(\xi, \alpha; \omega) \, d\alpha.$$

It is clear that $k(\xi, \xi - \delta; \omega) \to 0$ exponentially for fixed δ as $\omega \to \infty$ and that $k(\xi, \xi_0; \omega) \mid \xi - \xi_0 \mid = O(\omega^{-1/2})$. A simple calculation shows that

$$\int_{\xi-\delta}^{\xi_0} k(\xi, \alpha; \omega) \, d\alpha \leqq 2C\pi^{1/2}[\xi/(\xi - \delta)]^{1/4}.$$

Consequently $\limsup_{\omega\to\infty} J_2 \leqq \epsilon \, 2C\pi^{1/2} [\xi(\xi - \delta)]^{1/4}$. A similar argument shows that the same is true of $\limsup_{\omega\to\infty} J_3$. Since ϵ is arbitrary, the first part of the theorem is proved.

If $g(\xi)$ is continuous in some interval, then it is clear from the above argument that the limit holds uniformly in every compact subinterval.

We state without proof a companion theorem for the Laplace-Stieltjes transform (see R. S. Phillips [9, Theorem 2.3]).

THEOREM 6.3.4. *Let $f(\lambda)$ satisfy the assumptions of Theorem 6.3.1 and set*

$$a_0(\xi \mid \omega) = e^{-\omega\xi} \sum_{n=0}^{\infty} \frac{(-1)^n(\omega^2\xi)^{n+1}}{n!(n+1)!} f^{(n)}(\omega).$$

Then for all $\xi \geqq 0$

$$\lim_{\omega\to\infty} \int_0^{\xi} a_0(\alpha \mid \omega) \, d\alpha = a(\xi).$$

A different class of inversion formulas has been investigated very thoroughly by D. V. Widder after preliminary work by E. L. Post. We shall consider only one of Widder's operators

$$L_{k,\xi}[f(\lambda)] = \frac{(-1)^k}{k!}\left(\frac{k}{\xi}\right)^{k+1} f^{(k)}\left(\frac{k}{\xi}\right) \tag{6.3.15}$$

which is defined for large values of k for each $\xi > 0$.

THEOREM 6.3.5. *Let $f(\lambda)$ be the Laplace transform of $g(\xi)$ satisfying the assumptions of Theorem 6.3.2. Then*

$$\lim_{k\to\infty} L_{k,\xi}[f(\lambda)] = g(\xi)$$

in the Lebesgue set of $g(\xi)$. If $g(\xi)$ is continuous in some open interval, then the limit holds uniformly in each compact subinterval. If $f(\lambda)$ is the Laplace-Stieltjes transform of $a(\xi)$ as in Theorem 6.3.1., then

$$\lim_{k\to\infty} \int_0^\xi L_{k,\tau}[f(\lambda)]\, d\tau = a(\xi) - a(0+).$$

PROOF. In the first case we are led to the representation

$$L_{k,\xi}[f(\lambda)] = \frac{1}{\xi}\int_0^\infty W_0\left(\frac{\alpha}{\xi}; k\right) g(\alpha)\, d\alpha,$$

where

$$W_0(\beta; k) = \frac{k^{k+1}}{k!}\beta^k e^{-k\beta}.$$

The substitutions $\alpha = e^\tau$, $\xi = e^\eta$ lead to the singular integral

$$L_{k,\xi}[f(\lambda)] = e^{-\eta}\int_{-\infty}^\infty W(\tau - \eta; k)\, g(e^\tau)\, e^\tau\, d\tau$$

where

$$W(\sigma; k) = \frac{k^{k+1}}{k!}\exp(k\sigma - ke^\sigma).$$

It is convenient to break up the range of this integral into three parts:

$$J_1 = \int_{-\infty}^{\eta-\delta}, \qquad J_2 = \int_{\eta-\delta}^{\eta+\delta}, \qquad J_3 = \int_{\eta+\delta}^{\infty}.$$

For $\gamma > \max(0, \sigma_a)$ we see that $g(e^\tau)\exp(\tau - \gamma e^\tau) \in B(E_1, \mathfrak{X})$. Further for fixed η and $k > \gamma e^\tau$, $W(\tau - \eta; k)\exp(\gamma e^\tau)$ has its only maximum at

$$\tau_0 \sim \eta + \gamma e^\eta/k.$$

Hence for $\gamma e^\eta/k < \delta$, the maximum in $(-\infty, \eta - \delta)$ [or $(\eta + \delta, \infty)$] occurs at

$\eta - \delta$ [or $\eta + \delta$] so that

$$\| J_1 \| \leqq \frac{k^{k+1}}{k!} \exp(-k\delta - ke^{-\delta} + \gamma e^{\eta}) \int_{-\infty}^{\eta} \| g(e^{\tau}) \| \exp(\tau - \gamma e^{\tau})\, d\tau,$$

$$\| J_3 \| \leqq \frac{k^{k+1}}{k!} \exp(k\delta - ke^{\delta} + \gamma e^{\eta+\delta}) \int_{\eta}^{\infty} \| g(e^{\tau}) \| \exp(\tau - \gamma e^{\tau})\, d\tau.$$

It follows that $\lim_{k\to\infty} \| J_1 \| = 0 = \lim_{k\to\infty} \| J_3 \|$. Finally the integral J_2 may be treated precisely as in Theorem 3.9.3; here we take $K(\eta, \tau; k) = W(\tau - \eta; k)$ and $P(\sigma; k) = k^{1/2} \exp(-k\sigma^2/3)$. Now if ξ belongs to the Lebesgue set of $g(\xi)$, then $\eta = \log \xi$ belongs to the Lebesgue set of $g(e^{\eta})e^{\eta}$, and hence $\lim_{k\to\infty} L_{k,\xi}[f(\lambda)] = e^{-\eta}[g(e^{\eta})e^{\eta}] = g(\xi)$. We also note that $K(\eta, \tau; k)$ satisfies the condition (4′) of section 3.9 with $\mu_1 = \mu_2 = \frac{1}{2}$. Thus if J_2 is treated as in Theorem 3.9.2 we have

$$\lim_{k\to\infty} L_{k,\xi}[f(\lambda)] = \tfrac{1}{2}[g(\xi + 0) + g(\xi - 0)]$$

whenever this expression has a meaning. In particular, if $g(\xi)$ is continuous in some open interval, then it is easy to see that the limit exists uniformly with respect to ξ in each compact subinterval.

For the proof of the second part of the theorem we note that

$$\begin{aligned} L_{k,\xi}[f(\lambda)] &= \frac{1}{\xi} \int_0^{\infty} W_0\left(\frac{\alpha}{\xi}; k\right) da(\alpha) = e^{-\eta} \int_{-\infty}^{\infty} W(\tau - \eta; k)\, da(e^{\tau}) \\ &= -e^{-\eta} \int_{-\infty}^{\infty} \left[\frac{\partial}{\partial\tau} W(\tau - \eta; k)\right] a(e^{\tau})\, d\tau \\ &= e^{-\eta} \int_{-\infty}^{\infty} \left[\frac{\partial}{\partial\eta} W(\tau - \eta; k)\right] a(e^{\tau})\, d\tau. \end{aligned}$$

Hence

$$\begin{aligned} \int_{\xi_1}^{\xi_2} L_{k,\xi}[f(\lambda)]\, d\xi &= \int_{\eta_1}^{\eta_2} L_{k,\xi}[f(\lambda)] e^{\eta}\, d\eta \\ &= \int_{\eta_1}^{\eta_2} \int_{-\infty}^{\infty} \left[\frac{\partial}{\partial\eta} W(\tau - \eta; k)\right] a(e^{\tau})\, d\tau\, d\eta \\ &= \int_{-\infty}^{\infty} [W(\tau - \eta; k)]_{\eta_1}^{\eta_2} a(e^{\tau})\, d\tau. \end{aligned}$$

Again making use of Theorem 3.9.2, modified as in the proof of the first part of this theorem, we see that as $k \to \infty$ this expression tends to

$$a(e^{\eta_2}) - a(e^{\eta_1}) = a(\xi_2) - a(\xi_1)$$

by virtue of the Remark appended to Theorem 3.9.2. For $\xi_2 = \xi$, $\xi_1 \to 0$, we obtain the limit $a(\xi) - a(0+)$ which is the desired result. The interchange of the limit passages, $k \to \infty$ and $\xi_1 \to 0$, can be justified and it can also be shown that $L_{k,\xi}[f(\lambda)]$ is absolutely integrable down to zero. We omit these details. An important consequence of the theorem is the following

THEOREM 6.3.6. *If $f(\lambda)$ is the Laplace transform of $g(\xi)$, then $\| g(\xi) \| \leqq M$ for almost all $\xi > 0$ if and only if for all $\lambda > 0$ and $k = 0, 1, 2, \cdots$ we have*

$$\lambda^{k+1} \| f^{(k)}(\lambda) \| \leqq Mk!. \tag{6.3.16}$$

If $f(\lambda)$ is the Laplace-Stieltjes transform of $a(\xi)$, then $a_*(\infty) \leqq M + \| f(\infty) \|$ if and only if for $k = 1, 2, 3, \cdots$

$$\int_0^\infty \lambda^{k-1} \| f^{(k)}(\lambda) \| \, d\lambda \leqq M(k-1)!. \tag{6.3.17}$$

PROOF. In the first case

$$\lambda^{k+1} \| f^{(k)}(\lambda) \| = \| \int_0^\infty e^{-\lambda\xi}(\lambda\xi)^k g(\xi) \, d(\lambda\xi) \| \leqq Mk!$$

if $\| g(\xi) \| \leqq M$ almost everywhere, so (6.3.16) is necessary. In the second case

$$\begin{aligned} \int_0^\infty \lambda^{k-1} \| f^{(k)}(\lambda) \| \, d\lambda &\leqq \int_0^\infty \int_0^\infty e^{-\lambda\xi}(\lambda\xi)^{k-1}\xi \, d\lambda \, da_*(\xi) \\ &= (k-1)! \int_{0+}^\infty da_*(\xi) \\ &= (k-1)! \, [a_*(\infty) - a_*(0+)] \leqq M(k-1)!, \end{aligned}$$

where we have made use of the fact that $a_*(0+) = \| a(0+) \| = \| f(\infty) \|$. Thus (6.3.17) is necessary. To prove the sufficiency we note that the first condition implies $\| L_{k,\xi}[f(\lambda)] \| \leqq M$ for all ξ. Since $f(\lambda)$ is assumed to be the Laplace transform of $g(\xi)$ in the first case, the preceding theorem ensures that $\| g(\xi) \| \leqq M$ for almost all ξ. On the other hand (6.3.17) implies $\int_0^\infty \| L_{k,\tau}[f(\lambda)] \| \, d\tau \leqq M$. Thus for any subdivision $0 < \xi_0 < \xi_1 < \cdots < \xi_n$ we have

$$\begin{aligned} \sum_{i=1}^n \| a(\xi_i) - a(\xi_{i-1}) \| &= \lim_{k\to\infty} \sum_{i=1}^n \left\| \int_{\xi_{i-1}}^{\xi_i} L_{k,\tau}[f(\lambda)] \, d\tau \right\| \\ &\leqq \liminf_{k\to\infty} \int_{\xi_0}^{\xi_n} \| L_{k,\tau}[f(\lambda)] \| \, d\tau \leqq M. \end{aligned}$$

It follows that $a(\xi)$ is of strong bounded variation on $[0, \infty)$ and further that $a_*(\infty) - a_*(0+) \leqq M$.

Similar conditions can be formulated for the case in which $g(\xi) \in B_p[(0, \infty); \mathfrak{X}]$, $1 \leqq p < \infty$, and $f(\lambda)$ is the Laplace transform of $g(\xi)$. Widder has shown in the numerical case that the conditions of Theorem 6.3.6 are sufficient to ensure that the function $f(\lambda)$ be a Laplace or a Laplace-Stieltjes transform. However his method of proof does not seem to apply to the abstract case. On the other hand, as we shall see in Theorems 12.3.1 and 12.4.1, these conditions do suffice for a very important class of functions, namely the resolvents of closed linear operators.

2. Functions Holomorphic in a Half-Plane

6.4. The classes $H_p(\alpha; \mathfrak{X})$. We shall investigate various classes of functions $f(\lambda)$ which are holomorphic in a fixed half-plane $\sigma > \alpha$, $\lambda = \sigma + i\tau$, and have values in a fixed complex (B)-space $\mathfrak{X}$. In addition $f(\lambda)$ will be subjected to different types of boundedness conditions. We start with the classes $H_p(\alpha; \mathfrak{X})$; for the properties of numerically-valued functions of the class $H_p(\alpha)$ which will be used in the following, we refer to the papers by E. Hille and J. D. Tamarkin where further literature is quoted.

Definition 6.4.1. $f(\lambda) \in H_p(\alpha; \mathfrak{X})$, *p fixed*, $1 \leqq p < \infty$, *if*
(i) $f(\lambda)$ *is a function on complex numbers to* $\mathfrak{X}$ *which is holomorphic for* $\sigma > \alpha$;
(ii) $\sup_{\sigma > \alpha} \{\int_{-\infty}^{\infty} \| f(\sigma + i\tau) \|^p \, d\tau\}^{1/p} \equiv \| f \|_p < \infty$;
(iii) $\lim_{\sigma \to \alpha} f(\sigma + i\tau) \equiv f(\alpha + i\tau)$ *exists for almost all values of* τ *and*

$$f(\alpha + i\tau) \in B_p[(-\infty, \infty); \mathfrak{X}].$$

For the definition of the class $B_p[S; \mathfrak{X}]$, $S = (-\infty, \infty)$, see the remarks at the close of section 3.8. Assumption (iii) may be redundant; we assume it explicitly to obviate the need of a lengthy digression. It is known that for each linear bounded functional $x^* \in \mathfrak{X}^*$, we have $x^*[f(\lambda)] \in H_p(\alpha)$, $\lim_{\sigma \to \alpha} x^*[f(\sigma + i\tau)]$ exists for almost all τ (the exceptional set may depend upon x^*), and the limit function belongs to $L_p(-\infty, \infty)$.

Theorem 6.4.1. *If* $f(\lambda) \in H_p(\alpha; \mathfrak{X})$, *then* $f(\lambda)$ *is represented by its proper Cauchy and Poisson integrals for* $\sigma > \alpha$, *that is*

$$f(\lambda) = \frac{1}{2\pi i} \int_{\alpha - i\infty}^{\alpha + i\infty} \frac{f(\mu)}{\lambda - \mu} \, d\mu, \tag{6.4.1}$$

$$f(\sigma + i\tau) = \frac{\sigma - \alpha}{\pi} \int_{-\infty}^{\infty} \frac{f(\alpha + i\beta) \, d\beta}{(\sigma - \alpha)^2 + (\tau - \beta)^2}. \tag{6.4.2}$$

Proof. The integrals exist by virtue of (iii). In order to prove that they have the value $f(\lambda)$, it is enough to observe that the functionals agree. Thus

$$x^* \left\{ \frac{1}{2\pi i} \int_{\alpha - i\infty}^{\alpha + i\infty} \frac{f(\mu)}{\lambda - \mu} \, d\mu \right\} = \frac{1}{2\pi i} \int_{\alpha - i\infty}^{\alpha + i\infty} \frac{x^*[f(\mu)]}{\lambda - \mu} \, d\mu = x^*[f(\lambda)]$$

since $x^*[f(\lambda)] \in H_p(\alpha)$. This being true for every x^*, formula (6.4.1) must hold and similarly for (6.4.2).

Theorem 6.4.2. *If* $f(\lambda) \in H_p(\alpha; \mathfrak{X})$, *then*

$$\| f(\sigma + i\tau) \| \leqq [\pi(\sigma - \alpha)]^{-1/p} \| f \|_p, \tag{6.4.3}$$

and for fixed $\delta > 0$

$$\lim_{\rho \to \infty} \| f(\alpha + \delta + \rho e^{i\psi}) \| = 0, \qquad -\tfrac{1}{2}\pi \leqq \psi \leqq \tfrac{1}{2}\pi, \tag{6.4.4}$$

uniformly in ψ.

PROOF. The inequality follows from (6.4.2). The integral of the Poisson kernel over the range $-\infty < \tau < \infty$ being identically one, a classical use of Hölder's inequality gives

$$\begin{aligned} \| f(\sigma + i\tau) \|^p &\leqq \frac{\sigma - \alpha}{\pi} \int_{-\infty}^{\infty} \frac{\| f(\alpha + i\beta) \|^p \, d\beta}{(\sigma - \alpha)^2 + (\tau - \beta)^2} \\ &\leqq [\pi(\sigma - \alpha)]^{-1} \int_{-\infty}^{\infty} \| f(\alpha + i\beta) \|^p \, d\beta \\ &= [\pi(\sigma - \alpha)]^{-1} [\, \| f \|_p]^p, \end{aligned} \tag{6.4.5}$$

which is the required estimate. From this it follows that (6.4.4) holds uniformly in every sector $-\frac{1}{2}\pi + \epsilon \leqq \psi \leqq \frac{1}{2}\pi - \epsilon$. In order to prove the sharper assertion made above, we assume $\tau > 0$, and write the integral in the second member of (6.4.5) as the sum of two integrals, one from $-\infty$ to $\frac{1}{2}\tau$ and the other from $\frac{1}{2}\tau$ to ∞. Thus

$$\begin{aligned} \| f(\sigma + i\tau) \|^p \leqq 4 \frac{\sigma - \alpha}{\pi\tau^2} \int_{-\infty}^{\tau/2} \| f(\alpha + i\beta) \|^p \, d\beta \\ + \frac{1}{\pi(\sigma - \alpha)} \int_{\tau/2}^{\infty} \| f(\alpha + i\beta) \|^p \, d\beta \\ \leqq 4 \frac{\sigma - \alpha}{\pi\tau^2} [\, \| f \|_p]^p + \frac{1}{\pi(\sigma - \alpha)} \int_{\tau/2}^{\infty} \| f(\alpha + i\beta \|^p \, d\beta \end{aligned}$$

and this tends uniformly to zero when $\lambda \to \infty$ in such a manner that $\sigma \geqq \alpha + \delta$ and $\sigma\tau^{-2} \to 0$. There is a similar inequality for negative values of τ. These inequalities together with (6.4.3) suffice to prove (6.4.4).

THEOREM 6.4.3. *If* $f(\lambda) \in H_p(\alpha; \mathfrak{X})$, *then*

(i) $\lim_{\sigma \to \alpha} \int_{-\infty}^{\infty} \| f(\sigma + i\tau) - f(\alpha + i\tau) \|^p \, d\tau = 0$,

(ii) $T(\sigma; f) \equiv \int_{-\infty}^{\infty} \| f(\sigma + i\tau) \|^p \, d\tau$ *is a continuous monotone decreasing function of* σ *for* $\sigma \geqq \alpha$. *In particular,* $T(\alpha; f) = [\, \| f \|_p]^p$ *and* $T(\infty; f) = 0$.

PROOF. As in the classical case, one proves that the integral in (i) does not exceed

$$\frac{\sigma - \alpha}{\pi} \int_{-\infty}^{\infty} \frac{d\gamma}{(\sigma - \alpha)^2 + \gamma^2} \int_{-\infty}^{\infty} \| f(\alpha + i\beta) - f(\alpha + i\gamma + i\beta) \|^p \, d\beta.$$

The inner integral is clearly bounded and, according to Theorem 3.8.3, it will converge to zero as $\gamma \to 0$. Hence by a well known property of the Poisson kernel, the repeated integral consequently tends to zero when $\sigma \to \alpha$ proving (i). The first two members of (6.4.5) give $T(\sigma; f) \leqq T(\alpha; f)$ for all $\sigma > \alpha$. But in the representation (6.4.2) we may replace α by any quantity σ_0, $\alpha < \sigma_0 < \sigma$, since every $f(\lambda)$ in $H_p(\alpha; \mathfrak{X})$ also belongs to $H_p(\sigma_0; \mathfrak{X})$. This gives $T(\sigma; f) \leqq T(\sigma_0; f)$ when $\sigma > \sigma_0$, so that $T(\sigma; f)$ is a decreasing function of σ. That $T(\sigma; f)$ is continuous for $\sigma = \alpha$ follows from (i). Since (i) also holds with α replaced by σ_0, $\alpha < \sigma_0 < \sigma$, we conclude that $T(\sigma; f) \to T(\sigma_0; f)$ when $\sigma \to \sigma_0$. This completes the proof since the statements concerning $T(\alpha; f)$ and $T(\infty; f)$ are obvious.

The class $H_p(\alpha; \mathfrak{X})$ is evidently linear; it becomes a metric space under the norm $\| f \|_p$ and it is a simple matter to prove that it is complete, so that it is a (B)-space.

6.5. Order relations. So far p was supposed to be finite. We say that $f(\lambda) \in H_\infty(\alpha; \mathfrak{X})$ if *it is holomorphic and bounded for $\sigma > \alpha$ and $\lim_{\sigma\to\alpha} f(\sigma + i\tau)$ exists for almost all τ. $H_\infty(\alpha; \mathfrak{X})$ is a (B)-space under the norm.*

$$\| f \|_\infty = \sup_{\sigma > \alpha} \| f(\sigma + i\tau) \| .$$

The representation of $f(\lambda)$ by Poisson's integral, formula (6.4.2), is valid also when $p = \infty$ while most of the other results of section 6.4 become meaningless or false.

We shall need some properties of unbounded functions whose rates of growth are properly limited.

DEFINITION 6.5.1. *Let $f(\lambda)$ be a function on complex numbers to $\mathfrak{X}$, holomorphic for $\sigma \geqq \alpha$, except at infinity. Let* $M(\rho; f) = \max \| f(\alpha + \rho e^{i\psi}) \|$ *for* $-\frac{1}{2}\pi \leqq \psi \leqq \frac{1}{2}\pi$. *We say that $f(\lambda)$ is of finite order ω in this half-plane if*

$$\limsup_{\rho\to\infty} \log M(\rho; f) \Big/ \log \rho = \omega. \tag{6.5.1}$$

It is of finite order on vertical lines in the half-plane if

$$\limsup_{|\tau|\to\infty} \log \| f(\sigma + i\tau) \| \Big/ \log |\tau| \equiv \mu(\sigma) \tag{6.5.2}$$

is finite for $\sigma \geqq \alpha$. Finally, $f(\lambda)$ is of exponential type in the half-plane if

$$\limsup_{\rho\to\infty} \rho^{-1} \log M(\rho; f) = \beta < \infty. \tag{6.5.3}$$

It is of minimal type if $\beta = 0$, normal type if $\beta > 0$.

It is clear that if $f(\lambda)$ is of finite order in $\sigma \geqq \alpha$ then it is of finite order on vertical lines and $\mu(\sigma) \leqq \omega$. The following analogue of a classical theorem of E. Lindelöf is less obvious.

THEOREM 6.5.1. *Let $f(\lambda)$ be holomorphic and of exponential minimal type in the half-plane $\sigma \geqq \alpha$ and suppose that $\mu(\alpha)$ is finite. Then $f(\lambda)$ is of finite order $\mu(\alpha)$ in the half-plane in question. Further, $\mu(\sigma)$ is monotonic non-increasing, convex, and continuous for $\sigma \geqq \alpha$.*

PROOF. Let γ be fixed, $\gamma > \mu(\alpha)$, and consider $g(\lambda) = \lambda^{-\gamma} f(\lambda)$. Without restricting the generality, we may assume that $\alpha > 0$ so that $g(\lambda)$ is holomorphic for $\sigma \geqq \alpha$ and bounded on $\sigma = \alpha$. By a classical Phragmén-Lindelöf argument or using Theorem 3.13.6 we see that $\| g(\lambda) \|$ is bounded for $\sigma \geqq \alpha$, that is, $f(\lambda)$

is of finite order $\omega \leqq \gamma$ in this half-plane. This being true for every $\gamma > \mu(\alpha)$, we have $\omega \leqq \mu(\alpha)$. On the other hand, $\| f(\alpha \pm i\rho) \| \leqq M(\rho; f)$, so that comparing (6.5.1) with (6.5.2), we find that $\mu(\alpha) \leqq \omega$ and hence $\omega = \mu(\alpha)$.

By the preceding argument $\lambda^{-\gamma} f(\lambda)$ is bounded for $\gamma > \mu(\sigma_1)$ when $\sigma \geqq \sigma_1$ and, in particular, on the line $\sigma = \sigma_2$ if $\sigma_1 < \sigma_2$. From this we get that $\mu(\sigma_2) \leqq \mu(\sigma_1) + \epsilon$ for every $\epsilon > 0$. Hence $\mu(\sigma_2) \leqq \mu(\sigma_1)$ so that $\mu(\sigma)$ is never increasing. For the convexity proof we refer to Lindelöf's article [**2**, p. 3 et seq.]. A monotonic never increasing convex function being necessarily continuous, we have completed the proof.

Other growth-measuring functions will be introduced in sections 6.7, 6.8, 17.3, and 17.6.

6.6. Representation by Laplace integrals. We shall show that every function on complex numbers to $\mathfrak{X}$ which is holomorphic and of finite order in a closed half-plane may be represented by a generalized Laplace integral. We start with functions of the class $H_p(\alpha; \mathfrak{X})$.

THEOREM 6.6.1. *Let $f(\lambda) \in H_p(\alpha; \mathfrak{X})$ where $\alpha \geqq 0$. Let $\gamma > \alpha$ and $\beta p' > 1$ where $1/p + 1/p' = 1$. Then*

$$a_\beta(\xi) = \frac{1}{2\pi i} \int_{\gamma - i\infty}^{\gamma + i\infty} e^{\mu\xi} \mu^{-\beta} f(\mu)\, d\mu \tag{6.6.1}$$

defines a continuous function on $(0, \infty)$ to $\mathfrak{X}$ and

$$f(\lambda) = \lambda^\beta \int_0^\infty e^{-\lambda\xi} a_\beta(\xi)\, d\xi, \tag{6.6.2}$$

the integral being absolutely convergent for $\sigma > \alpha$.

PROOF. That the integral defining $a_\beta(\xi)$ exists follows from Hölder's inequality. Since the integral converges uniformly with respect to ξ in every finite interval, $a_\beta(\xi)$ is continuous. The usual argument shows that the integral is independent of γ when $\gamma > \alpha$ and if $\alpha > 0$ we may even take $\gamma = \alpha$. In this case, a simple computation shows that $\| a_\beta(\xi) \| \leqq M e^{\alpha\xi}$; if $\alpha = 0$ we get instead $\| a_\beta(\xi) \| \leqq M(\epsilon) e^{\epsilon\xi}$ for every $\epsilon > 0$. It follows that the second integral is absolutely convergent for $\sigma > \alpha$. In order to find its value, we substitute the integral for $a_\beta(\xi)$, interchange the order of integration, and use (6.4.1). The details are left to the reader.

REMARK. If $p = 1$, we may take $\beta = 0$, obtaining

$$f(\lambda) = \int_0^\infty e^{-\lambda\xi} a_0(\xi)\, d\xi. \tag{6.6.3}$$

By analogy with the numerically-valued case, we would expect the same representation to hold also for $1 < p \leqq 2$ with an $a_0(\xi)$ such that $e^{-\alpha\xi} a_0(\xi) \in B_{1/p'}[(0, \infty); \mathfrak{X}]$. We do not know if this is actually so; since Bochner has shown that Bessel's inequality does not necessarily hold for functions in $B_2[S; \mathfrak{X}]$, the basis for a satisfactory Fourier transform theory seems to be lacking when $p > 1$ and this closes the usual avenue of approach to the Laplace transform theory as well.

THEOREM 6.6.2. *Let $f(\lambda)$ be holomorphic and of finite order ω in the half-plane $\sigma \geqq \alpha$. Then $a_\beta(\xi)$ exists as a continuous function of ξ for $\beta > \omega + 1$ and*

$$f(\lambda) = \lambda^\beta \int_0^\infty e^{-\lambda\xi} a_\beta(\xi)\, d\xi, \tag{6.6.4}$$

convergent for $\sigma >$ max $(0, \alpha)$.

The proof is analogous to that of the preceding theorem and is omitted.

We see, in particular, that if $f(\lambda)$ is holomorphic and of finite order in a given half-plane $\sigma \geqq \alpha \geqq 0$ and if $\sigma_1 > \alpha$, then formula (6.6.4) holds for $\sigma > \sigma_1$ provided $\beta > \mu(\sigma_1) + 1$. Increasing the value of β is a convergence preserving transformation which enables us to represent $f(\lambda)$ by Laplace integrals in the largest half-plane in which $f(\lambda)$ is holomorphic and of finite order. As applied to $a_\beta(\xi)$, this transformation amounts to a fractional integration. Indeed, we read off from formula (6.2.17) that

$$a_{\beta+\gamma}(\xi) = \frac{1}{\Gamma(\gamma)} \int_0^\xi (\xi - \tau)^{\gamma-1}\, a_\beta(\tau)\, d\tau. \tag{6.6.5}$$

3. BINOMIAL SERIES

6.7. Properties of the series. Functions which are holomorphic and of exponential type in a half-plane play an important role in classical analysis. They are equally important in vector-valued function theory. Such functions admit of representations in terms of *binomial series*, also known as *binomial coefficient series* or *Newton's interpolation series*. We shall sketch the theory of such series here since they will be used extensively in Chapter XVII.

A binomial series is an expansion of the form

$$\begin{aligned} &u_0 + u_1 \frac{\zeta}{1} + u_2 \frac{\zeta(\zeta-1)}{1\cdot 2} + \cdots + u_n \frac{\zeta(\zeta-1)\cdots(\zeta-n+1)}{1\cdot 2\cdots n} + \cdots \\ &\qquad \equiv \sum_{n=0}^\infty u_n \binom{\zeta}{n}, \end{aligned} \tag{6.7.1}$$

where in the vector-valued case the quantities u_n are elements of a complex (B)-space $\mathfrak{U}$. In the applications which we have in view, $\mathfrak{U}$ will be a space $\mathfrak{E}(\mathfrak{X})$, that is, the coefficients u_n will be linear bounded operators on a complex (B)-space $\mathfrak{X}$ to itself.

The series converges trivially when ζ is a non-negative integer since all but a finite number of terms vanish. If the series converges for $\zeta = \zeta_0$, where ζ_0 is not zero or a positive integer, then one shows with the aid of Abel's summation formula that it converges for every ζ with $\Re(\zeta) > \Re(\zeta_0)$. Moreover, the convergence is uniform with respect to ζ in every bounded closed sector, having its vertex at $\zeta = \zeta_0$ but lying otherwise in the interior of the half-plane. Thus the region of convergence is a half-plane $\xi > \sigma_0$, $\zeta = \xi + i\eta$, to which have to be added the trivial points $\zeta = 0, 1, \cdots, [\sigma_0]$, and possibly also a point set on the line of convergence $\xi = \sigma_0$. Further, the sum of the series, say $f(\zeta)$, is a function on complex numbers to $\mathfrak{U}$ which is holomorphic for $\xi > \sigma_0$ by virtue of the uniform convergence. Similarly, the region of absolute convergence is a half-plane $\xi > \sigma_a$ plus, possibly, the line of absolute convergence $\xi = \sigma_a$.

The abscissas of ordinary and of absolute convergence are given by

$$\sigma_0 = -1 + \limsup_{n\to\infty} \log \left\| \sum_{k=0}^{n} (-1)^k u_k \right\| \Big/ \log n, \tag{6.7.2}$$

$$\sigma_a = -1 + \limsup_{n\to\infty} \log \left[\sum_{k=0}^{n} \| u_k \| \right] \Big/ \log n, \tag{6.7.3}$$

whenever these limits are positive. If the first one is zero, but the series $\sum_0^\infty (-1)^k u_k$ diverges, σ_0 is still given by (6.7.2); if it converges instead, we have to replace

$$\sum_0^n (-1)^k u_k \quad \text{by} \quad \sum_{n+1}^\infty (-1)^k u_k$$

in the formula. Analogous changes have to be made in the second formula. The reader will have no difficulties in verifying these formulas; cf. N. E. Nörlund [2, pp. 111–115] where $z - 1$ should be replaced by ζ and absolute values by norms in carrying over the proof. It is obvious from the formulas that

$$0 \leqq \sigma_a - \sigma_0 \leqq 1. \tag{6.7.4}$$

The expansion of a numerical function in binomial series is not unique owing to the presence of *null series*: If m is a positive integer, the series

$$\psi_m(\zeta) = \sum_{n=m}^{\infty} (-1)^{n+m} \binom{n}{m} \binom{\zeta}{n}$$

converges for $\xi > m$ to the sum zero. It has the same sum for $\zeta = 0, 1, \cdots, m - 1$, and equals one for $\zeta = m$. This fact may be used to obtain the so-called *reduced series* for $f(\zeta)$. We shall suppose that $f(\zeta)$ is known in advance to be holomorphic in a right half-plane and to be defined at the origin and the positive integers. We can then find coefficients $\gamma_0, \gamma_1, \cdots, \gamma_p$, where $p = [\sigma_0]$, so that the function

$$f(\zeta) + \sum_{m=0}^{p} \gamma_m \psi_m(\zeta) f(m)$$

has a binomial expansion convergent for $\xi > \sigma_0$, the sum of the series being $f(m)$ for $\zeta = m$, $m = 0, 1, \cdots, p$. The resulting binomial series is known as the reduced series. In the following all binomial series will be supposed to be reduced series.

In the case of a reduced series, the coefficients u_n are uniquely determined by the values of $f(\zeta)$ at the non-negative integers. We find by substitution that

$$u_n = \sum_{k=0}^{n} (-1)^{n-k} \binom{n}{k} f(k) = \Delta^n f(0), \qquad n = 0, 1, 2, \cdots. \tag{6.7.5}$$

The function $f(\zeta)$ may also be represented by a Laplace integral of the form

$$f(\zeta) = \frac{1}{2\pi i} \int_C g(\lambda)\lambda^{\zeta} \, d\lambda, \tag{6.7.6}$$

$$g(\lambda) = \sum_{n=0}^{\infty} \Delta^n f(0)(\lambda - 1)^{-n-1} = \sum_{n=0}^{\infty} f(n)\lambda^{-n-1}, \tag{6.7.7}$$

where the first series converges for $|\lambda - 1| > 1$ and the second for $|\lambda| > 2$. Further

$$\lim_{\lambda \to 0} |\lambda|^{\sigma_0 - 1 + \epsilon} \| g(\lambda) \| = 0 \tag{6.7.8}$$

for every $\epsilon > 0$, if $\lambda \to 0$ in the sector $\frac{1}{2}\pi + \delta \leqq \arg \lambda \leqq \frac{3}{2}\pi - \delta$. Finally C is a closed rectifiable path, surrounding the circle $|\lambda - 1| = 1$ once in the positive sense, beginning and ending at the origin in the neighborhood of which the path lies in the sector just mentioned. The power has its principal determination $\lambda^{\zeta} = \exp(\zeta \log \lambda)$, where the imaginary part of the logarithm lies between $-\pi$ and π. These assertions are easily verified by direct computation; see also section 17.6 where a special case is carried through in some detail.

The function $f(\zeta)$ defined by (6.7.1) for $\xi > \sigma_0$ is of exponential type in every interior half-plane. If $\beta > \sigma_0$ and if

$$h(\varphi; f) \equiv \limsup_{r \to \infty} r^{-1} \log \| f(\beta + re^{i\varphi}) \| \tag{6.7.9}$$

denotes the *Phragmén-Lindelöf growth function* of $f(\zeta)$ defined for $-\frac{1}{2}\pi \leqq \varphi \leqq \frac{1}{2}\pi$, then

$$h(\varphi; f) \leqq l(\varphi) \tag{6.7.10}$$

where

$$l(\varphi) \equiv \varphi \sin \varphi + \cos \varphi \log (2 \cos \varphi), \tag{6.7.11}$$

and $\log 2 \leqq l(\varphi) \leqq \frac{1}{2}\pi$. More precisely, we have the estimate due to F. Carlson [2]

$$\| f(\beta + re^{i\varphi}) \| \leqq \exp [rl(\varphi)] \, r^{\beta + 1/2 + \epsilon(r)} (1 + r \cos \varphi)^{-1/2}, \tag{6.7.12}$$

where $\epsilon(r)$ tends uniformly to zero when ζ tends to infinity in the sector. The proof goes through as in the numerical case, replacing absolute values by norms.

The crux of the proof lies in showing that

$$\left| \binom{\zeta}{n} \right| < Cr^{1/2} \exp [rl(\varphi)]$$

for all n when r is large and that the terms which have the greatest influence on the order of $f(\zeta)$ are those whose subscripts lie in a certain neighborhood of the critical value $r/(2 \cos \varphi)$ for which the estimates of the binomial coefficients may be reversed. We do not insist on these details, but add the observation that $rl(\varphi)$ is the *function of support of the closed convex region*

$$e^{\xi} \leqq 2 \cos \eta, \tag{6.7.13}$$

which in its turn is the image of the circle $|\lambda - 1| \leqq 1$ under the conformal mapping $\lambda = e^{\zeta}$.

6.8. Representation and analytic continuation. Conversely every function $f(\zeta)$ which is holomorphic for $\xi \geqq \beta$ and satisfies an estimate of the above type can be represented by a convergent binomial series. This is proved as in the numerical case, that is, using Newton's interpolation formula, expressing the remainder by Cauchy's integral and estimating its norm. See Nörlund (op. cit., p. 131 et seq.). The result may be formulated as follows:

THEOREM 6.8.1. *If $f(\zeta)$ is a function on complex numbers to $\mathfrak{U}$, holomorphic in the half-plane $\xi \geqq \beta$, where it satisfies the inequality*

$$\| f(\beta + re^{i\varphi}) \| \leqq e^{rl(\varphi)}(1 + r)^{\gamma + \epsilon(r)}, \qquad -\frac{\pi}{2} \leqq \varphi \leqq \frac{\pi}{2}, \tag{6.8.1}$$

and the function $\epsilon(r)$ tends uniformly to zero when r tends to infinity, then $f(\zeta)$ may be expanded in a binomial series of type (6.7.1), *the abscissa of convergence of which does not exceed the larger of the two numbers β, $\gamma + \frac{1}{2}$.*

A numerical binomial series may be summed by the method of Cesàro which in the present case is equivalent to applying the transformation $\zeta' = \zeta + k$. The much more powerful transformation $\zeta' = \alpha\zeta$ has been used by Nörlund; it gives the analytic continuation of $f(\zeta)$ in the largest half-plane where it is holomorphic and of exponential type. This method applies also to vector-valued functions. Indeed, if $f(\zeta)$ is holomorphic in $\xi \geqq \beta$ and

$$\| f(\beta + re^{i\varphi}) \| \leqq Ae^{Br}, \qquad -\frac{\pi}{2} \leqq \varphi \leqq \frac{\pi}{2},$$

then we can clearly determine a positive α so small that $f(\alpha\zeta)$ satisfies (6.8.1) and is consequently representable by a convergent binomial series in the variable ζ. This implies a convergent binomial series in the variable ζ/α for the original function $f(\zeta)$. Since min $l(\varphi) = \log 2$, it suffices to take $\alpha \leqq (\log 2)/B$. In particular, if $f(\zeta)$ admits of a representation by a binomial series in ζ, then it is also representable by a binomial series in ζ/α for $0 < \alpha < 1$. Let $\sigma_0(\alpha)$ be the abscissa of convergence of the binomial series in ζ/α. Then $\sigma_0(\alpha)$ is a never decreasing function of α which is completely determined by the rate of growth of $\| f(\zeta) \|$ on vertical lines.

With Nörlund we introduce

$$\gamma(\xi) = \limsup_{|\eta|\to\infty} |\eta|^{-1} \log \| f(\xi + i\eta) \| , \tag{6.8.2}$$

which is clearly well defined in the largest half-plane in which $f(\zeta)$ is holomorphic and of exponential type. Let ξ_0 be so chosen that $f(\zeta)$ is holomorphic and of finite exponential type for $\xi \geqq \xi_0 + \epsilon$ when $\epsilon > 0$, but lacks at least one of these properties for $\xi > \xi_0 - \epsilon$, no matter how small ϵ is. The function $\gamma(\xi)$ is defined for $\xi > \xi_0$ and has properties analogous to those of Lindelöf's function $\mu(\sigma)$ of formula (6.5.2); *$\gamma(\xi)$ is monotonic never increasing, convex, and continuous for $\xi > \xi_0$, but, in addition, it is non-negative.* All these properties follow from the Phragmén-Lindelöf extension of the principle of the maximum.

Let (α_0, α_1) be the largest interval such that the equation

$$\gamma(\sigma) = \frac{\pi}{2\alpha}$$

has a solution, which is necessarily unique, say $\sigma = \sigma_0(\alpha)$, when $0 \leqq \alpha_0 < \alpha < \alpha_1 \leqq \infty$. This is the required abscissa of convergence. For $0 < \alpha \leqq \alpha_0$ we have $\sigma_0(\alpha) = \sigma_0(\alpha_0)$, for $\alpha_1 < \alpha$, $\sigma_0(\alpha) = +\infty$. Further $\sigma_0(\alpha)$ is monotonic, never decreasing and continuous for $\alpha_0 < \alpha < \alpha_1$. These assertions are proved as in the numerical case. For future reference we formulate the result as follows:

THEOREM 6.8.2. *Let $f(\zeta)$ be a function on complex numbers to $\mathfrak{U}$ which is holomorphic and of exponential type for $\xi \geqq \xi_0 + \epsilon$ when $\epsilon > 0$, but lacks at least one of these properties for $\xi > \xi_0 - \epsilon$, no matter how small ϵ is. If $\xi_0 > 0$, we assume in addition that $f(\zeta)$ is defined for all real non-negative values of ζ. Let $\gamma(\xi)$, $\sigma_0(\alpha)$, α_0, and α_1 have the meaning defined above. Then*

$$f(\zeta) = \sum_{n=0}^{\infty} \Delta_\alpha^n f(0) \frac{1}{n!} \zeta(\zeta - \alpha) \cdots (\zeta - (n-1)\alpha), \tag{6.8.3}$$

where the series converges for $\xi > \xi_0$ when $0 < \alpha \leqq \alpha_0$, and for $\xi > \sigma_0(\alpha)$ when $\alpha_0 < \alpha < \alpha_1$. The series fails to converge for any non-trivial value of ζ when $\alpha_1 < \alpha$.

It should be observed that the values assigned to $f(\zeta)$ on the interval $[0, \xi_0]$ are actually immaterial. Changing these values merely adds a null series to (6.8.3) and does not change the abscissas of convergence.

PART TWO

BASIC PROPERTIES OF SEMI-GROUPS

Summary. Except for two chapters, the remainder of this treatise is devoted to the theory of semi-groups with special reference to one-parameter semi-groups of endomorphisms of a (B)-space. We have divided the theory, somewhat arbitrarily, into three parts: basic, advanced, and special. The present Part Two contains the basic theory and is divided into six chapters entitled: *Subadditive Functions, Semi-Modules, Addition Theorems in a Banach Algebra, Semi-Groups in the Strong Topology, Resolvent of the Generator*, and *Generation of Semi-Groups.*

Chapters VII and VIII contain prefatory material. Subadditive functions and semi-modules are intimately connected with each other and with the theory of one-parameter semi-groups. The foundations of the theory of one-parameter semi-groups is laid in Chapters IX and X. We shall be concerned with a family of endomorphisms $[T(\alpha)]$, satisfying $T(\alpha)T(\beta) = T(\alpha + \beta)$ for all values α, β of the parameter in an open semi-module of real or complex numbers which in general has $\alpha = 0$ as limit point. $T(\alpha)$ need not tend to any limit as $\alpha \to 0$; its limitary behavior can vary considerably and is basic for the classification and properties of the semi-group. The discussion is not restricted to the functional equation of the exponential function; other addition theorems are also considered as well as the case in which the parameter set is a "positive cone" in a (B)-space. The latter includes the case of n-parameter semi-groups under addition. General parameter semi-groups are discussed in Chapter XXV on Lie semi-groups.

The infinitesimal generator of the semi-group and its resolvent are discussed in Chapter XI. The resolvent of the generator is essentially the Laplace transform of the semi-group operator. Conversely the inversion formulas for the Laplace transform give us the means of constructing the semi-group knowing the resolvent of its generator. This leads to the general problem of Chapter XII, namely how to recognize if a given operator is the infinitesimal generator of a semi-group of one class or another and how to construct the corresponding semi-group when it exists.

CHAPTER VII

SUBADDITIVE FUNCTIONS

7.1. Orientation. The present chapter is concerned with real-valued functions defined on a subset Σ of E_n and satisfying the condition

$$f(\mathbf{t}_1 + \mathbf{t}_2) \leqq f(\mathbf{t}_1) + f(\mathbf{t}_2), \qquad \mathbf{t}_1, \mathbf{t}_2 \in \Sigma. \tag{7.1.1}$$

Obviously with $\mathbf{t}_1, \mathbf{t}_2 \in \Sigma$ we must require that $\mathbf{t}_1 + \mathbf{t}_2$ also belong to Σ. Functions of this kind are called *subadditive* and their domains are called *additive semi-groups* or *semi-modules.*

We have already encountered subadditive functions in section 2.5 and in the theory of $S(\varphi)$ algebras described in section 4.16. Subadditive functions on E_1 are met with in the theory of moduli of continuity (Ch.-J. de la Vallée-Poussin [1, pp. 7–8]). Positive-homogeneous subadditive functions have important applications to the theory of convex sets (H. Minkowski [1]) and to the uniqueness theory of differential equations (E. Kamke [1], M. Hukuhara [1]).

Subadditive functions also play a basic role in the theory of semi-groups where they enter in two different connections. The first instance is in the theory of semi-modules in E_n. In the simplest and most important case, that in which the semi-module is an open set having the origin as a limit point, the boundary of the semi-module is defined by a subadditive function in E_{n-1}. The second instance is in the theory of one-parameter semi-groups $[T(\alpha)]$ of endomorphisms of a (B)-space. Here the parameter set is a semi-module of real or complex numbers and $\log \| T(\alpha) \|$ is a subadditive function of α on this set.

In view of these facts it is necessary for us to include a discussion of both subadditive functions and semi-modules; the latter are treated in Chapter VIII. As we shall see, a finite, measurable, subadditive function is bounded in any compact subset interior to its domain of definition; however it may tend to infinity at the boundary. A subadditive function need not be continuous anywhere; such properties of continuity and differentiability as it may possess are regulated by its behavior for small values of $\mathbf{t}$.

The chapter is divided into three paragraphs: *Boundedness and Growth, Continuity and Differentiability,* and *Subadditive Functions in E_n*. The first two paragraphs are essentially limited to subadditive functions defined on an open semi-module in E_1 having the origin as a limit point, in other words, one of the intervals $(-\infty, \infty)$, $(-\infty, 0)$ or $(0, \infty)$. The middle paragraph includes a discussion of various associated limit functions which are also subadditive. The final paragraph deals with the extension of some of these results to subadditive functions on semi-modules in E_n. A detailed study of the n-dimensional case can be found in the work of R. A. Rosenbaum [1].

The reader who is anxious to reach the theory of semi-groups as soon as possible can omit most of Chapters VII and VIII; however the theorems in sections 7.4 and 7.6 are indispensable for what follows.

References. Beurling [1], R. Cooper [1], Hardy, Littlewood and Pólya [1], Hille [7, pp. 13, 46–47], Hukuhara [1], Kamke [1], Minkowski [1], Phillips [9], Pólya and Szegö [1, p. 17, Ex. 98], Rosenbaum [1], de la Vallée-Poussin [1].

1. Boundedness and Growth

7.2. Elementary properties. In what follows the symbol I will refer to any one of the intervals $I_0 : (-\infty, \infty)$, $I_- : (-\infty, 0)$, and $I_+ : (0, \infty)$. Further it is desirable for the applications which we have in mind to consider solutions of (7.1.1) which assume infinite values, $+\infty$ or $-\infty$. Denoting real numbers by a, addition is defined for these symbols by the conventions: $a + \infty = \infty + \infty = \infty$, $a - \infty = -\infty - \infty = -\infty$. The symbol $\infty - \infty$ is left undefined so that if $f(t_1) = +\infty$ while $f(t_2) = -\infty$, then the value of $f(t_1 + t_2)$ is not restricted at all by the relation (7.1.1). A solution of (7.1.1) is said to be *finite* in the set E if it takes on only finite values for $t \in E$. A *finite subadditive function* is one which is finite in its interval of definition.

The following theorems are easily verified.

THEOREM 7.2.1. *A positive constant is subadditive on any interval I. If $f_1(t)$ and $f_2(t)$ are finite subadditive functions on I and if γ_1, γ_2 are positive constants, then $\gamma_1 f_1(t) + \gamma_2 f_2(t)$ is also subadditive on I.*

THEOREM 7.2.2. *Let $[f_\alpha(t)]$ be a family of subadditive functions on I. Then $p(t) = \sup_\alpha f_\alpha(t)$ is also subadditive on I.*

THEOREM 7.2.3. *If $\{f_n(t)\}$ is a pointwise convergent sequence of subadditive functions in I, then $q(t) = \lim_{n\to\infty} f_n(t)$ is also subadditive on I.*

It is well-known that the functional equation

$$F(t_1 + t_2) = F(t_1) + F(t_2) \tag{7.2.1}$$

has non-measurable solutions in addition to the continuous solutions of the form $F(t) = \alpha t$. Since any real solution of (7.2.1) also satisfies (7.1.1), it is clear that *there are non-measurable subadditive functions.* Such functions are explicitly excluded from consideration and *all subadditive functions discussed in the following are supposed to be measurable.*

In considering subadditive functions defined on the intervals I_0, I_-, and I_+ we may disregard the case $I = I_-$ for if $f(t)$ is subadditive in I_- then $f(-t)$ will be subadditive in I_+.

The inequality (7.1.1) is similar to

$$g[\tfrac{1}{2}(t_1 + t_2)] \leqq \tfrac{1}{2}[g(t_1) + g(t_2)] \tag{7.2.2}$$

which characterizes *convex functions*. The two function classes are rather remotely related. The next two theorems have a bearing on this situation; part (i) of Theorem 7.2.4 is essentially due to G. H. Hardy, J. E. Littlewood, and G. Pólya [**1**, p. 83]. Here we consider only finite-valued functions.

We refer to Hardy, Littlewood, and Pólya [1] for the properties of convex functions used below. In particular, a measurable convex function is continuous and satisfies

$$g[\alpha t_1 + (1 - \alpha)t_2] \leqq \alpha g(t_1) + (1 - \alpha)g(t_2), \qquad 0 < \alpha < 1. \tag{7.2.3}$$

The function $f(t)$ is *concave* if $-f(t)$ is convex.

THEOREM 7.2.4. (i) *If $f(t)/t$ is decreasing in I_+, then $f(t)$ is subadditive, but need not be convex or concave in I_+.* (ii) *If $f(t)$ is convex and subadditive in I_+, then $f(t)/t$ is decreasing.*

PROOF. (i) We have

$$f(t_1 + t_2) = t_1 \frac{f(t_1 + t_2)}{t_1 + t_2} + t_2 \frac{f(t_1 + t_2)}{t_1 + t_2} \leqq t_1 \frac{f(t_1)}{t_1} + t_2 \frac{f(t_2)}{t_2} = f(t_1) + f(t_2).$$

The function $f(t) = t^{-1} + t^{1/2}$ satisfies the conditions of the theorem, but has a point of inflection at $t = 4$. (ii) Take $0 < a < b$ and put $t_1 = a$, $t_2 = a + b$, $\alpha = a/b$ in (7.2.3). This gives

$$f(b) \leqq \frac{a}{b} f(a) + \left(1 - \frac{a}{b}\right) f(a + b) \leqq \frac{a}{b} f(a) + \left(1 - \frac{a}{b}\right) [f(a) + f(b)]$$

which upon simplification reduces to $af(b) \leqq bf(a)$. This completes the proof.

THEOREM 7.2.5. *A necessary and sufficient condition that a measurable concave function $f(t)$ be subadditive in I_+ is that $f(0+) \geqq 0$.*

PROOF. Theorem 7.4.3 below shows that the condition is necessary. The sufficiency is proved as follows. Since $f(t)$ is concave, $-f(t)$ satisfies (7.2.3) and upon placing $t_1 = 0$, $(1 - \alpha)t_2 = a$, $t_2 = b$, we get

$$f(a) \geqq \alpha f(0) + (1 - \alpha)f(b) \geqq \frac{a}{b} f(b).$$

Hence $f(t)/t$ is decreasing so that $f(t)$ is subadditive by the preceding theorem. In the proof we have tacitly assumed $f(t)$ to be continuous to the right at $t = 0$. If this is not true, the desired inequality follows by a suitable passage to the limit. We leave this point to the reader.

7.3. Semi-modules in E_1 and infinitary solutions. Further information on subadditive functions can be obtained from a study of semi-modules in E_1. We have already remarked that such sets are suitable domains of definition for subadditive functions. However even subadditive functions $f(t)$ defined on I give rise to more general kinds of semi-modules. For instance the sets $\Sigma(+\infty; f) = [t; f(t) < \infty]$, $\Sigma(\alpha; f) = [t; f(t) < \alpha t]$, and $\Sigma(-\infty; f) = [t; f(t) = -\infty]$ are semi-modules in E_1. Other examples are the interval (a, ∞), $a \geqq 0$; the set of natural numbers; and the set of rational numbers.

Theorem 7.3.1. *If $\Sigma \subset I_+$ is an open semi-module, then for b sufficiently large $(b, \infty) \subset \Sigma$.*

Proof. For $\alpha \in \Sigma$ there is an integer n such that $\Delta \equiv [\alpha, ((n+1)/n)\alpha) \subset \Sigma$. Since Σ is a semi-module $\Delta_k \equiv n\Delta + k\alpha = [(n+k)\alpha, (n+k+1)\alpha) \subset \Sigma$ for all integers $k \geqq 0$ and hence $[n\alpha, \infty) = \bigcup_0^\infty \Delta_k \subset \Sigma$.

A stronger result of this type is

Theorem 7.3.2. *If $\Sigma \subset I_+$ is a measurable semi-module and if $m(\Sigma) > 0$, then for a sufficiently large $(a, \infty) \subset \Sigma$.*

Proof. Since the Lebesgue measure of Σ is positive, there exists an $\alpha \in \Sigma$ at which Σ is of density one. Again set $\Delta \equiv [\alpha, ((n+1)/n)\alpha)$ where n is now chosen so that $m(\Delta \cap \Sigma) > \frac{3}{4}\alpha/n$. It is clear that $E_k \equiv n[\Delta \cap \Sigma] + k\alpha \subset \Sigma$ and that $m(E_k) > \frac{3}{4}\alpha$. The sets E_k are disjoint translates of each other and $\bigcup_0^N E_k \subset (0, (n+N+1)\alpha) \cap \Sigma$. Hence

$$\tfrac{3}{4}N\alpha < m[\bigcup_0^N E_k] \leqq m[(0, (n+N+1)\alpha) \cap \Sigma].$$

A simple calculation shows that for $t > b = (3n+6)\alpha$ we have $m[(0, t) \cap \Sigma] > t/2$. Thus for $t_0 > b$, the sets $(0, t_0) \cap \Sigma$ and $t_0 - [(0, t_0) \cap \Sigma]$ must have a point in common. Consequently there exist $t_1, t_2 \in \Sigma$ such that $t_1 + t_2 = t_0$ and this implies that $t_0 \in \Sigma$.

Corollary. *Let $\Sigma \subset I_0$ be a measurable semi-module. If $I_- \cap \Sigma \neq \emptyset$ then either $m(I_+ \cap \Sigma) = 0$ or else $\Sigma = I_0$.*

Applying this corollary to the semi-module $\Sigma(\infty; f) = [t; f(t) < \infty]$ we obtain the following result first noted by Rosenbaum [1].

Theorem 7.3.3. *Let $f(t)$ be subadditive on I_0. If $f(t) < \infty$ for some $t < 0$ then either $f(t) = \infty$ for almost all $t > 0$ or else $f(t) < \infty$ for all $t \in I_0$.*

Many of the basic facts regarding infinitary subadditive functions are contained in the following theorem.

Theorem 7.3.4. *Let $f(t)$ be subadditive in I_0. If $f(a) = -\infty$ for a fixed a, $a > 0$, and $f(t)$ is finite in $(0, a)$, then $f(t) = -\infty$ for $t \geqq a$ and $f(t) = +\infty$ for $t < 0$. If, on the other hand, $f(a) = +\infty$, $a > 0$, then $f(t) = +\infty$ on a subset of $(0, a)$ of measure $\geqq a/2$ and if $f(t)$ is finite for $t > a$, then $f(t) = +\infty$ when $t < 0$.*

Proof. Suppose that $f(a) = -\infty$. If $t > a$ we may find an h, $0 \leqq h < a$, and a positive integer n such that $t = a + nh$, whence $f(t) \leqq f(a) + nf(h)$ and $f(t) = -\infty$ since $f(h) \neq +\infty$. If t is given, $t < 0$, then there is a quantity b, $a \leqq b$, such that $f(b) = -\infty$ and $0 < t + b \leqq a$. Hence $f(t) \geqq f(t+b) - f(b) = +\infty$.

Suppose instead that $f(a) = +\infty$, $a > 0$. Then if $t_1 > 0$, $t_2 > 0$ and $t_1 + t_2 = a$, we have $+\infty = f(a) \leqq f(t_1) + f(t_2)$, that is, either $f(t_1)$ or $f(t_2)$ is $+\infty$. Since $f(t)$ is measurable by assumption, it follows that $f(t) = +\infty$ on a subset of $(0, a)$ the measure of which is at least $a/2$. If $f(t)$ is finite when $t > a$ and $t_0 < 0$, then $f(t_0) \geqq f(a) - f(a - t_0) = +\infty$.

The following examples show that solutions of the type contemplated in Theorem 7.3.4 really exist:

$$f_1(t) = \begin{cases} +\infty, & t \leqq 0, \\ \cot(\pi t/a), & 0 < t < a, \\ -\infty, & a \leqq t < \infty; \end{cases}$$

$$f_2(t) = \begin{cases} +\infty, & t \leqq a/2 \text{ or } t = a, \\ 0, & \text{elsewhere.} \end{cases}$$

We also note that if $f(t)$ is subadditive on I_0 and if $f(t) < +\infty$ for all $t \in I_0$, then $f(a) = -\infty$ implies $f(t) \equiv -\infty$ on I_0. This is a simple consequence of the relation $f(t) \leqq f(t - a) + f(a) = -\infty$.

7.4. Boundednesss. Finite subadditive functions have remarkable properties of boundedness which will now be investigated.

THEOREM 7.4.1. *If $f(t)$ is subadditive and different from $+\infty$ in I, then $f(t)$ is bounded above in any compact subset I^* of I. If $f(t)$ is also different from $-\infty$, then $f(t)$ is bounded in any I^*.*

PROOF. Suppose first that $I = I_+$, $a > 0$, and $f(a) = A$. For $t_1 + t_2 = a$, $t_1 > 0$, $t_2 > 0$, we have $A = f(a) \leqq f(t_1) + f(t_2)$. Hence with $E = [t; f(t) \geqq A/2, 0 < t < a]$, it follows that $(0, a) = E \cup (a - E)$ and therefore that $m(E) \geqq a/2$. If $f(t)$ were unbounded above in some interval (α, β) where $0 < \alpha < \beta < \infty$, then we could find a sequence of points $\{t_n\}$ such that $f(t_n) \geqq 2n$, $t_n \to t_0 \geqq \alpha$. As a consequence for each n the set $E_n = [t; f(t) \geqq n, 0 < t < \beta]$ would have a measure $\geqq \alpha/2$ and hence $f(t)$ would be equal to $+\infty$ on a set of measure $\geqq \alpha/2$. This is a contradiction and shows that to every $\delta > 0$ there is a finite M_δ such that $f(t) \leqq M_\delta$ for $\delta \leqq t \leqq 1/\delta$.

If $I = I_0$ we prove by the same type of argument that $f(t)$ is also bounded above in $-1/\delta \leqq t \leqq -\delta$. But $f(t) \leqq f(1 + t) + f(-1)$ and therefore $f(t)$ is bounded above in $[-\delta, \delta]$ and hence in every interval of the type $[-1/\delta, 1/\delta]$.

Suppose next that $f(t)$ is also different from $-\infty$ in I_+. If $f(t)$ is not bounded below in (α, β) we can find a sequence $\{t_n\}$ such that $f(t_n) \leqq -n$ and $t_n \to t_0$. We set $M = \sup [f(t); 2 < t < 5]$. For any $t' \in (2, 5)$ we have $f(t' + t_n) \leqq f(t') + f(t_n) \leqq M - n$. For large n the intervals $(t_n + 2, t_n + 5)$ contain the fixed interval $(t_0 + 3, t_0 + 4)$ and for every t in this interval $f(t) \leqq M - n$, that is, $f(t) = -\infty$ contrary to our assumption. Thus $f(t)$ is also bounded below in $[\delta, 1/\delta]$ and consequently bounded.

If $I = I_0$ the same argument shows that $f(t)$ is bounded below in $[-1/\delta, -\delta]$. From boundedness in $[1, 3]$ and the inequality $f(t) \geqq -f(2) + f(2 + t)$, one infers that $f(t)$ is also bounded below in $[-\delta, \delta]$ and hence bounded in $[-1/\delta, 1/\delta]$.

A similar argument proves

THEOREM 7.4.2. *If $f(t)$ is subadditive and finite in (a, ∞) where $a > 0$, then $f(t)$ is bounded above in every interval $(2a + \delta, 2a + 1/\delta)$ and bounded below in $(a, a + 1/\delta)$. If $f(t)$ is unbounded in $(a, 2a + \delta)$, then it does not admit of a finite subadditive extension for $0 < t \leqq a$.*

EXAMPLE. $f(t) = a^2/(3at - t^2 - 2a^2)$ for $a < t < 2a$ and $f(t) = 1$ for $2a \leqq t$ is subadditive in (a, ∞) and not bounded above in $(a, 2a)$ although it is finite.

THEOREM 7.4.3. *If $f(t)$ is subadditive in I, then* $\liminf_{t\to 0} f(t)$ *is either $-\infty$ or $\geqq 0$. If* $\liminf_{t\to 0} f(t) = -\infty$, *then $f(t)$ is infinitary in I. If $f(t)$ is subadditive in I_0 and* $\liminf_{t\to 0} f(t) = +\infty$, *then $f(t) = +\infty$ almost everywhere in at least one of the subintervals I_- or I_+.*

PROOF. Put $\liminf_{t\to 0} f(t) = \lambda$ and suppose first that λ is finite. There exists a sequence $\{t_n\}$ such that $t_n \to 0$ and $\lim_{n\to\infty} f(t_n) = \lambda$. For $n \geqq N_\epsilon$ we have $\lambda - \epsilon \leqq f(2t_n) \leqq 2f(t_n) \leqq 2(\lambda + \epsilon)$ so that $\lambda \leqq 2\lambda$ or $\lambda \geqq 0$. On the other hand if $\lambda = -\infty$ and $I = I_+$, then we can find a sequence $\{t_n\}$ such that $t_n \to 0$ and $f(t_n) \leqq -n$. If $f(t) \not\equiv +\infty$ in I_+ then by Theorem 7.4.1 we see that $f(t) \leqq M_\delta$ for all $t \in (\delta, 1/\delta)$. Thus for $t \in (\delta, 1/\delta)$ and n sufficiently large

$$f(t) \leqq f(t - t_n) + f(t_n) \leqq M_\delta - n;$$

consequently $f(t) \equiv -\infty$. If $I = I_0$ and $f(t) \not\equiv +\infty$ in one or both of the subintervals I_- and I_+, then the same argument implies that $f(t) \equiv -\infty$ in that subinterval. It is possible, however, to have $f(t) \equiv -\infty$ in I_-, $f(t) \equiv +\infty$ in I_+, and $f(0)$ perfectly arbitrary. Finally if $f(t)$ is subadditive in I_0 and $\lambda = +\infty$, we set $\Sigma = [t; f(t) < \infty]$. If $I_- \cap \Sigma$ and $I_+ \cap \Sigma$ were both of positive measure, then according to the corollary to Theorem 7.3.2 we would have $\Sigma = I_0$, and hence by Theorem 7.4.1 $f(t)$ would be bounded above in the neighborhood of $t = 0$. This is contrary to $\lambda = +\infty$.

It should be observed that $\lim_{t\to 0} f(t)$ does not have to exist. An example will be given in section 7.7 of a subadditive function for which $\liminf_{t\to 0} f(t) = 0$ and $\limsup_{t\to 0} f(t) = +\infty$.

A global bound for subadditive functions is given in the following theorem due to R. S. Phillips [9].

THEOREM 7.4.4. *If $f(t)$ is subadditive in I_+ and $\int_0^\infty \exp [f(t)]\, dt = M$, then $f(t) \leqq 2 \log (M/t)$ for $t > 0$ whereas for $t > Me$, $f(t) \leqq 2 - 2t/Me$.*

PROOF. If we set $g(t) = \exp [f(t)]$, then $0 \leqq g(t_1 + t_2) \leqq g(t_1)g(t_2)$ and therefore $2g(t) \leqq 2g(t - \alpha)g(\alpha) \leqq [g(t - \alpha)]^2 + [g(\alpha)]^2$. Thus $4g(t) \leqq [g(t - \alpha) + g(\alpha)]^2$ and

$$t[g(t)]^{1/2} = 2 \int_0^{t/2} [g(t)]^{1/2}\, d\alpha$$

$$\leqq \int_0^{t/2} g(t - \alpha)\, d\alpha + \int_0^{t/2} g(\alpha)\, d\alpha \leqq \int_0^t g(\alpha)\, d\alpha \leqq M;$$

this yields $\exp [f(t)] = g(t) \leqq (M/t)^2$ for all $t > 0$. Further $g(t) \leqq [g(t/n)]^n \leqq (nM/t)^{2n}$. For $nMe \leqq t < (n + 1)Me$, $n \geqq 1$, we replace t by nMe and the n in the exponent by $-1 + t/Me$ and obtain $g(t) \leqq \exp [2 - 2t/Me]$.

We shall later have occasion to speak of *submultiplicative* functions defined on I, that is, functions $g(t)$ such that

$$0 \leqq g(t_1 + t_2) \leqq g(t_1)g(t_2), \qquad t_1, t_2 \in I. \tag{7.4.1}$$

It is clear that $f(t) = \log [g(t)]$ is subadditive and hence that the first part of Theorem 7.4.4 may be restated as follows: *If $g(t)$ is submultiplicative in I_+ and $\int_0^\infty g(t)\, dt = M$, then $g(t) \leqq (M/t)^2$ for all $t > 0$.*

7.5. Negative subadditive functions. The behavior of subadditive functions taking on negative values differs in some respects from that of typical subadditive functions.

THEOREM 7.5.1. *If $f(t)$ is finite and subadditive in I_+ and if $f(a) < 0$, $a > 0$, then $f(t) < 0$ for all large positive values of t. If I_+ be replaced by I_0 in the assumption, then in addition $f(t) \geqq 0$ for all negative values of t.*

PROOF. According to Theorem 7.4.1. there exists a finite M such that $| f(t) | \leqq M$ for $a \leqq t \leqq 2a$. If now $na \leqq t < (n + 1)a$, then

$$f(t) \leqq f[(n - 1)a] + f[t - (n - 1)a] \leqq (n - 1)f(a) + M \leqq \frac{f(a)}{a} t + 3M,$$

which is negative for all large positive t. Moreover, the inequality shows that $f(t)$ is bounded above by a linear function of t for $t \geqq 2a$. Suppose that $f(t)$ is also defined for $t \leqq 0$ and is subadditive in I_0. Since $f(0)$ is finite and $f(0) \leqq 2f(0)$ we have $f(0) \geqq 0$ and from $0 \leqq f(0) \leqq f(t) + f(-t)$ we conclude that $f(t)$ and $f(-t)$ are not negative simultaneously. In particular, $f(t)$ is certainly positive for all large negative values of t. But if $f(t) < 0$ for any $t < 0$, then the argument used above shows that $f(t) < 0$ for all $t < t_0 < 0$, which is impossible. Hence $f(t) \geqq 0$ when $t \leqq 0$.

We note that if $f(t)$ is subadditive and $f(a) = 0$, then $f(na) \leqq 0$, $n = 1, 2, 3, \cdots$. Here equality may hold for all n as is shown by the example $f(t) = | \sin t |$, $a = \pi$.

The simplest of all functions which are negative and subadditive in I_+ is $f(t) = -t$. By Theorem 7.2.5 any concave decreasing function with $f(0+) = 0$ is negative and subadditive in I_+. The following theorem shows how to construct other such functions.

THEOREM 7.5.2. *Let $f_0(t)$ be negative and subadditive in I_+ and let $F(t)$ be any positive non-decreasing function defined in I_+. Then $f_0(t)F(t)$ is negative and subadditive in I_+.*

The simple verification is left to the reader. The theorem shows that it is possible to construct a subadditive function on the interval I_+ which tends faster to $-\infty$ as $t \to \infty$ than any fixed preassigned function of t. This is in marked contrast to the case in which $f(t)$ tends to $+\infty$ with t; here, as we shall see, nothing faster than a linear function of t is admissable.

7.6. Rate of growth. A finite subadditive function is bounded in any finite closed interval interior to its interval of definition. It may, however, become unbounded when t approaches either end point of I. We start with a theorem describing the behavior for large values of t. For the following compare G. Pólya and G. Szegö [1, p. 17, Ex. 98] and the papers by A. Beurling [1] and R. Cooper [1].

THEOREM 7.6.1. *If $f(t)$ is subadditive and finite in (a, ∞), $a \geqq 0$, then*

$$\lim_{t\to\infty} \frac{f(t)}{t} = \inf_{t>a} \frac{f(t)}{t} < \infty. \tag{7.6.1}$$

PROOF. Since $f(t)$ is finite-valued, $\beta = \inf_{t>a} f(t)/t$ is either finite or $-\infty$. We shall restrict ourselves to proving the existence of the limit under the assumption that β is finite; essentially the same argument also suffices for the case $\beta = -\infty$. We choose any $b > a$ such that $f(b) < (\beta + \epsilon)b$ and let $(n + 2)b \leqq t < (n + 3)b$; then

$$\beta \leqq \frac{f(t)}{t} \leqq \frac{nb}{t}\frac{f(b)}{b} + \frac{f(t - nb)}{t} < \frac{nb}{t}(\beta + \epsilon) + \frac{f(t - nb)}{t}.$$

Since $t - nb \in [2b, 3b]$, we see that $f(t - nb)$ stays bounded by Theorem 7.4.2 and hence the last member tends to $\beta + \epsilon$ as $t \to \infty$. It follows that the limit exists and equals β.

We remark that Theorem 7.6.1 remains valid if the assumption of measurability is replaced by boundedness in every compact subset of $(2a, \infty)$.

THEOREM 7.6.2. *If $f(t)$ is a finite subadditive function on I_0 and if*

$$\inf_{t>0} \frac{f(t)}{t} = \beta \quad \textit{and} \quad \sup_{t<0} \frac{f(t)}{t} = \alpha,$$

then

$$\lim_{t\to+\infty} \frac{f(t)}{t} = \beta \quad \textit{and} \quad \lim_{t\to-\infty} \frac{f(t)}{t} = \alpha, \tag{7.6.2}$$

$$-\infty < \alpha \leqq \beta < +\infty. \tag{7.6.3}$$

PROOF. That $\lim_{t\to+\infty} f(t)/t = \beta < +\infty$ follows directly from the previous theorem. Likewise replacing $f(t)$ in I_- by $f(-t)$ on I_+ we see that

$$\lim_{t\to-\infty} f(t)/t = -\inf_{t>0} f(-t)/t = \sup_{t<0} f(t)/t = \alpha > -\infty.$$

The inequality $\alpha \leqq \beta$ follows from $0 \leqq f(0) \leqq f(t) + f(-t)$ upon dividing by t and passing to the limit.

It is clear from (7.6.3) that a finite subadditive function on I_+ for which $\beta = -\infty$ does not admit of a finite subadditive extension on I_0. Even more is true.

THEOREM 7.6.3. *If $f(t)$ is subadditive on I_0, finite on I_+ with $\beta = -\infty$, then $f(t) \equiv +\infty$ for all $t < 0$.*

PROOF. According to Theorem 7.5.1, $f(t)$ must be non-negative for all $t < 0$. Suppose that $f(-b) < +\infty$, $b > 0$, so that $f(-kb) < +\infty$ for $k = 1, 2, 3, \cdots$. Any $t \leqq 0$ may be written in the form $t = t_0 - kb$ where $b \leqq t_0 < 2b$. Hence we have $0 \leqq f(t) \leqq f(t_0) + f(-kb) \leqq M + kf(-b)$. This shows that $f(t)$ is finite-valued for all $t \in I_0$. Thus (7.6.3) is applicable, contrary to $\beta = -\infty$.

Theorem 7.6.1 shows that every finite subadditive function in I_+ is dominated above by a suitably chosen linear function of t for large positive values of t; however, no such dominant need exist for $-f(t)$ which may grow arbitrarily fast.

A function which is finite and subadditive in I_+ may also become infinite when t decreases to zero. This is shown by

THEOREM 7.6.4. *Any function $G(t)$ which is positive and never increasing in I_+ is subadditive in I_+. If in addition $G(t) \to +\infty$ as $t \to 0$, then $G(t)$ does not have a finite subadditive extension on I_0. If $f(t)$ is a finite non-negative subadditive function in I_+, then so is $G(t)f(t)$.*

PROOF. That $G(t)$ and $G(t)f(t)$ are subadditive in I_+ is trivial. Suppose that $G(t)$ is defined and subadditive in I_0 and that $G(t) \to +\infty$ as $t \to 0+$. The inequality $G(h) \leqq G(a + h) + G(-a)$, which is valid for every $h > 0$, shows that $G(t) \equiv +\infty$ for $t < 0$.

The results on boundedness and rate of growth may be summarized as follows:

THEOREM 7.6.5. *A finite subadditive function is bounded in any finite closed interval interior to its interval of definition I. If $I = I_+$, then $\lim_{t\to+\infty} f(t)/t$ exists and equals $\beta = \inf_{t>0} f(t)/t < +\infty$. A linear function of t dominates $f(t)$ for large t, but if $\beta = -\infty$, $f(t)$ may tend to $-\infty$ faster than any preassigned function as $t \to +\infty$. Further, $\lambda = \liminf_{t\to0} f(t) \geqq 0$ and $\Lambda = \limsup_{t\to0} f(t) \leqq +\infty$. If $\Lambda = +\infty$, $f(t)$ may become infinite faster than any preassigned function as $t \to 0+$. For the same function $f(t)$, it may happen that $\lambda = 0$, $\Lambda = +\infty$, and $\beta = -\infty$. If either the second or the third relation holds, then $f(t)$ does not admit of a finite subadditive extension in I_0. If $I = I_0$, then $\beta > -\infty$, further $\alpha = \lim_{t\to-\infty} f(t)/t = \sup_{t<0} f(t)/t$ exists and $0 \leqq \beta - \alpha < \infty$.*

As we have already seen in paragraph 4.4, the behavior of a subadditive function for large values of t is of importance in the applications to Fourier integrals. A. Beurling [1] distinguishes between the non-analytic and the analytic cases according as $\beta - \alpha$ is zero or not. The particular case in which $\alpha = \beta = 0$ and the integral $\int_{-\infty}^{\infty} [f(t)/(1 + t^2)]dt$ converges is referred to as the quasi-analytic case. This terminology is a natural one for the problem considered by Beurling.

2. CONTINUITY AND DIFFERENTIABILITY

7.7. Composition of two-valued subadditive functions. Given the linear functional equation $F(t_1 + t_2) = F(t_1) + F(t_2)$, the assumption that $F(t)$ is finite and measurable ensures that $F(t)$ is bounded, continuous, and differentiable in every finite interval. Much less can be expected if "=" be replaced by "$\leqq$." We have seen that a finite measurable subadditive function is bounded in every finite closed interval interior to its interval of definition. It will be shown below that it need not be continuous anywhere, much less differentiable. To bring out this and related facts which will be useful in the discussion, we shall introduce a class of special subadditive functions.

Let Σ be a measurable semi-module of real numbers, that is, $\Sigma = [\alpha]$ is a measurable point set and $\alpha, \beta \in \Sigma$ implies $\alpha + \beta \in \Sigma$. Define

$$f(t; \Sigma) = \begin{cases} a, & t \in \Sigma, \\ b, & t \notin \Sigma, \end{cases} \qquad 0 \leqq a \leqq 2b. \tag{7.7.1}$$

This is obviously a subadditive function in I_0. The assumption that Σ is a semi-module may be dropped if in addition $b \leqq 2a$.

If now $\{a_n\}$ is any sequence of positive numbers with $\sum a_n$ convergent, if $\{\Sigma_n\}$ is any sequence of distinct measurable semi-modules of real numbers, and if $f(t; \Sigma_n)$ is defined by (7.7.1) with fixed a, b independent of n, then

$$F(t) = \sum_{n=1}^{\infty} a_n f(t; \Sigma_n) \tag{7.7.2}$$

is also subadditive in I_0.

Among these functions we single out the following for special consideration. Let Σ be the set of rational numbers and let $a \neq b$. Then $f(t; \Sigma)$ is a *two-valued measurable subadditive function which is discontinuous for all values of t*. With the aid of this function we can also construct counter examples for Theorem 7.6.5. We choose $a = 0$, $b = 1$ and form $f(t) = t^{-m} f(t; \Sigma) - t^n$ where $m > 0$ and $n > 1$. This is obviously a subadditive function in I_+ with $\lambda = 0$, $\Lambda = +\infty$, and $\beta = -\infty$.

Let Σ_n be the set of positive multiples of $1/n$ for $n = 1, 2, 3, \cdots$, let $b = \frac{1}{2}a > 0$, and $a_n = 2^{-n}$. Then

$$F(t) = \sum_{n=1}^{\infty} 2^{-n} f(|t|; \Sigma_n) \tag{7.7.3}$$

is a subadditive function in I_0 which is discontinuous for all rational values of t except $t = 0$ and continuous for irrational t. Thus, if n is an integer, $n \neq 0$, then $F(n) = a$ while $\lim_{t \to n} F(t) = \frac{1}{2}a$. Further $\lim_{t \to 0} F(t) = F(0) = \frac{1}{2}a$. This function will serve us as a counter example for Theorem 7.8.2 below. Further counter examples will be constructed with the aid of the same principle in later sections.

7.8. Limit functions and continuity. We now introduce the *upper* and *lower limit functions* $\bar{f}(t)$ and $\underline{f}(t)$ defined as follows:

$$\bar{f}(t) = \lim_{h \to 0} \sup_{|t-u|<h} f(u), \qquad \underline{f}(t) = \lim_{h \to 0} \inf_{|t-u|<h} f(u). \tag{7.8.1}$$

We recall that $\bar{f}(t)$ is *upper semi-continuous* and $\underline{f}(t)$ *lower semi-continuous*.

THEOREM 7.8.1. *If $f(t)$ is subadditive in I so are $\bar{f}(t)$ and $\underline{f}(t)$.*

PROOF. The case of $\underline{f}(t)$ is typical. If h and ϵ are given positive quantities and $\alpha_h = \inf f(u)$ in $(t_1 - h, t_1 + h)$ while $\beta_h = \inf f(u)$ in $(t_2 - h, t_2 + h)$, then there exists a u_1 in the first interval and a u_2 in the second such that $f(u_1) < \alpha_h + \epsilon$, $f(u_2) < \beta_h + \epsilon$. Then $u_1 + u_2$ is a point in $(t_1 + t_2 - 2h, t_1 + t_2 + 2h)$. Hence if γ_{2h} is the infimum of $f(u)$ in this interval,

$$\gamma_{2h} \leqq f(u_1 + u_2) \leqq f(u_1) + f(u_2) < \alpha_h + \beta_h + 2\epsilon.$$

On passing to the limit with h, the inequality

$$\underline{f}(t_1 + t_2) \leqq \underline{f}(t_1) + \underline{f}(t_2) + 2\epsilon$$

results. Since ϵ is arbitrary, it follows that $\underline{f}(t)$ is subadditive in I. The upper limit function $\bar{f}(t)$ is discussed in the same manner.

THEOREM 7.8.2. *If $f(t)$ is subadditive in I_0 and $\omega(t; f) = \bar{f}(t) - \underline{f}(t)$, then*

$$0 \leqq \omega(t; f) \leqq \bar{f}(0). \tag{7.8.2}$$

This inequality is the best of its kind. If $\bar{f}(0) > 0$, then $f(t)$ may be discontinuous everywhere. If $f(t)$ is continuous at $t = 0$ but $f(0) > 0$, the discontinuities of $f(t)$ may still be everywhere dense. If, however, $f(t)$ is continuous at $t = 0$ and $f(0) = 0$, then $f(t)$ is continuous everywhere.

PROOF. Given ϵ and h, $\epsilon > 0$, $h > 0$, and a point t, two points u_1 and u_2 may be found in the interval $(t - h, t + h)$ such that $f(u_1) > \bar{f}(t) - \epsilon$, $f(u_2) < \underline{f}(t) + \epsilon$. Hence

$$\begin{aligned} \omega(t; f) &= \bar{f}(t) - \underline{f}(t) < f(u_1) - f(u_2) + 2\epsilon \\ &\leqq f(u_1 - u_2) + 2\epsilon \leqq \bar{f}(0) + 3\epsilon, \end{aligned}$$

if h is sufficiently small. Since ϵ is arbitrary, (7.8.2) follows.

That this inequality is the best of its kind follows from the examples of the preceding section. For the function $f(|t|; \Sigma)$ where Σ is the set of positive rationals and $a = 0$, $b > 0$ but arbitrary, we have $\omega(t; f) \equiv b = \bar{f}(0; \Sigma)$ and this function is discontinuous everywhere. Formula (7.7.3) exhibits a subadditive function which is continuous at the origin but discontinuous at all other rational points. Here $\bar{F}(0) > 0$ and $\omega(n; F) = \bar{F}(0)$ for integral values of n. Finally, if $f(t)$ is continuous at $t = 0$ and $f(0) = 0$, then $\bar{f}(0) = 0$ and (7.8.2) shows that $f(t)$ is continuous everywhere. This completes the proof. See also Theorem 2.5.3.

Theorem 7.8.2 breaks down if $f(t)$ is defined merely in I_+ since $\bar{f}(0)$ does not exist. The obvious expedient for getting out of this difficulty is to replace $\bar{f}(0)$ by the quantity $\Lambda = \limsup_{t \to 0+} f(t) = \bar{f}_d(0)$, which is well defined. Though the inequality $\omega(t; f) \leqq \bar{f}_d(0)$ is false, the subadditive inequality will yield information concerning one-sided oscillations and limits.

We introduce the four one-sided limit functions and the corresponding oscillations

$$(7.8.3)\qquad \bar f_l(t) = \lim_{h\to 0}\sup_{t-h<u<t} f(u), \qquad \bar f_d(t) = \lim_{h\to 0}\sup_{t<u<t+h} f(u),$$

$$(7.8.4)\qquad \underline{f}_l(t) = \lim_{h\to 0}\inf_{t-h<u<t} f(u), \qquad \underline{f}_d(t) = \lim_{h\to 0}\inf_{t<u<t+h} f(u),$$

$$(7.8.5)\qquad \omega_l(t;f) = \bar f_l(t) - \underline{f}_l(t), \qquad \omega_d(t;f) = \bar f_d(t) - \underline{f}_d(t).$$

We note that if, for instance, $\omega_l(t;f) = 0$, then $\lim_{h\to 0} f(t-h) = f(t-0)$ exists. The following theorem refers to the case $I = I_0$ or I_+. If $I = I_-$ instead we have to make an obvious interchange of left and right in the wording of the theorem.

THEOREM 7.8.3. *If $f(t)$ is subadditive in $I = I_0$ or I_+ so are the one-sided limit functions. Further*

$$(7.8.6)\qquad 0 \leqq \omega_l(t;f) \leqq \bar f_d(0), \qquad 0 \leqq \omega_d(t;f) \leqq \bar f_d(0)$$

and these inequalities are the best possible. If $\bar f_d(0) > 0$, $f(t+0)$ and $f(t-0)$ need not exist for a single value of t. If $f(0+)$ exists but exceeds zero, $f(t+0)$ may not exist for any $t \neq 0$. If $\bar f_d(0) = 0$ so that $f(0+) = 0$, then $f(t+0)$ and $f(t-0)$ exist everywhere and

$$(7.8.7)\qquad f(t-0) \geqq f(t) \geqq f(t+0),$$

but $s(t;f) \equiv f(t-0) - f(t+0)$ may be different from zero in an everywhere dense set. Moreover, if $F(t)$ is any positive never-decreasing continuous function tending to ∞ with t, then there exists a subadditive function in I_+ such that $f(0+) = 0$ but $s(t;f) \geqq F(t)$ for infinitely many values of t tending to infinity.

PROOF. The asserted subadditivity is proved as in Theorem 7.8.1. The inequalities (7.8.6) are proved as (7.8.2); we have to observe that the points u_1 and u_2 should be chosen on the same side of t and that $u_1 > u_2$. This is always possible. That (7.8.6) cannot be improved upon is shown by the subadditive function $f(t)$ which is zero or one according as t is rational or not. Here $\omega_l(t;f) = \omega_d(t;f) \equiv 1 = \bar f_d(0)$. Further, if E is the interval $(0, 1)$ plus the set of rational numbers and if $f(t; E) = 1$ on E and $\frac{1}{2}$ elsewhere, then $f(t; E)$ and

$$f(t) = \sum_{n=1}^{\infty} 2^{-n} f(nt; E)$$

are subadditive in I_0. Here $f(0+)$ exists but equals 1 and $f(t+0)$ does not exist for $t \neq 0$ while $f(t-0)$ does not exist for any t.

Suppose now that $\bar f_d(0) = 0$ so that $f(0+)$ exists and is 0. It follows from (7.8.6) that $f(t+0)$ and $f(t-0)$ exist everywhere. The inequality (7.8.7) follows from $f(t) \leqq f(t-h) + f(h)$ and $f(t+h) \leqq f(t) + f(h)$ upon letting $h \to 0$. Denoting by $[t]$ the greatest integer $\leqq t$, one verifies easily that $-[t]$ and

$$f(t) = -\sum_{n=1}^{\infty} 2^{-n}[nt]$$

are subadditive; here $f(0+) = 0$ but $s(t;f) > 0$ for all rational values of t. Finally, if $F(t)$ has the stated properties, then by Theorem 7.5.2 we have that $f(t) = -F(t)[t]$ is subadditive in I_+, $f(t) = 0$ in $(0, 1)$ and $s(n;f) = F(n)$ for $n = 1, 2, 3, \cdots$. This completes the proof.

7.9 Continuity in the mean. If $f(t)$ is a finite measurable subadditive function, then $f(t)$ is integrable over any closed interval interior to I. We introduce the mean values

$$(7.9.1)\qquad \begin{aligned} f^*(t) &= \limsup_{h\to 0} \frac{1}{2h}\int_{-h}^{h} f(t+u)\,du, \\ f_*(t) &= \liminf_{h\to 0} \frac{1}{2h}\int_{-h}^{h} f(t+u)\,du, \end{aligned}$$

and set

$$\bar{\omega}(t; f) = f^*(t) - f_*(t).$$

THEOREM 7.9.1. *If $f(t)$ is a finite measurable function which is subadditive in I, then the mean values $f^*(t)$ and $f_*(t)$ are also subadditive in I. If $I = I_0$, then*

$$(7.9.2)\qquad 0 \leqq \bar{\omega}(t; f) \leqq f^*(0).$$

This inequality is the best of its kind; $\bar{\omega}(t; f) = 0$ for almost all t, but if $f^*(0) > 0$ the points where $\bar{\omega}(t; f) > 0$ may be everywhere dense. If $f^*(0) = 0$, $\bar{\omega}(t; f) \equiv 0$ and $f(t)$ is continuous in the mean everywhere.

The proof is obtained by integrating the inequality

$$f(t_1 + t_2 + (\alpha + \beta)s) \leqq f(t_1 + \alpha s) + f(t_2 + \beta s)$$

with respect to s from $-h$ to h and choosing the numbers α and β properly. We omit the details. For the counter examples we use the function $f(t; E)$ of formula (7.7.1) choosing $a = 1, b = \frac{1}{2}$ and E in such a manner that (i) E is invariant under the translation $s = t + 2$, (ii) the density of E is zero at $t = 0$, and (iii) the upper and lower densities are one and zero respectively at $t = 1$. For this function $f^*(0; E) = \frac{1}{2}$ and $\bar{\omega}(2n + 1; f) = \frac{1}{2}$ which shows that (7.9.2) cannot be improved upon. Condensing the singularities in the usual manner, we can obtain a subadditive function which is continuous in the mean at $t = 0$ but at no other rational point. Similar results hold for left- and right-handed mean values.

7.10. Moduli of continuity. Let $C_u(-\infty, \infty)$ be the class of complex-valued functions $f(\xi)$, uniformly continuous in $(-\infty, \infty)$, and form

$$(7.10.1)\qquad \mu(t; f) = \sup_{-\infty<\xi<\infty} |f(\xi + t) - f(\xi)|,$$

$$(7.10.2)\qquad \mathrm{M}(t; f) = \max_{0\leqq s\leqq t} \mu(s; f).$$

The name modulus of continuity is usually reserved for $\mathrm{M}(t; f)$, but we shall use this term generically for both types of functions and speak of $\mathrm{M}(t; f)$ as the *rectified modulus of continuity*.

It is clear that these moduli are even, continuous, non-negative functions on I_0. They are also subadditive. In fact

$$\begin{aligned} \mu(t_1 + t_2; f) &= \sup_\xi |f(\xi + t_1 + t_2) - f(\xi)| \\ &\leqq \sup_\xi |f(\xi + t_1 + t_2) - f(\xi + t_2)| + \sup_\xi |f(\xi + t_2) - f(\xi)| \\ &= \mu(t_1; f) + \mu(t_2; f). \end{aligned}$$

On the other hand for $\epsilon > 0$ and $t_1, t_2 > 0$ there exist t_1' and t_2', $0 \leqq t_i' \leqq t_i$, such that

$$\mathrm{M}(t_1 + t_2; f) - \epsilon \leqq \mu(t_1' + t_2'; f) \leqq \mu(t_1'; f) + \mu(t_2'; f) \leqq \mathrm{M}(t_1; f) + \mathrm{M}(t_2; f).$$

The case t_1 and t_2 of opposite sign readily follows from the fact that $M(t; f)$ is even and non-decreasing for $t > 0$. We also note that $M(t_0; f) = 0$ for $t_0 \neq 0$ implies that $M(t; f) \equiv 0$ and hence that $f(\xi)$ is a constant. Finally the functions $\mu(t; f)$ and $M(t; f)$ are bounded if and only if $f(\xi)$ is bounded.

THEOREM 7.10.1. *Let $f(\xi)$ be a continuous subadditive function in I_0 with $f(0) = 0$. Then $f(\xi) \in C_u(-\infty, \infty)$ and*

$$\mu(t; f) = \max [f(t), f(-t)]. \tag{7.10.3}$$

In particular, if $f(t)$ is even, then $\mu(t; f) \equiv f(t)$ and if, in addition, $f(t)$ is non-decreasing for $t > 0$ then $M(t; f) \equiv f(t)$.

PROOF. The inequality

$$-f(-t) \leqq f(\xi + t) - f(\xi) \leqq f(t)$$

shows that

$$|f(\xi + t) - f(\xi)| \leqq \max [\,|f(t)|, |f(-t)|\,]$$

and this bound is reached either for $\xi = 0$ or for $\xi = -t$. But $0 = f(0) \leqq f(t) + f(-t)$. Hence at least one of the quantities in the last member is non-negative and dominates the absolute value of the other. This proves (7.10.3) and the rest of the theorem is obvious.

COROLLARY. *If $f(t)$ is a continuous even subadditive function and $f(0) = 0$, then $f(t)$ is the (non-rectified) modulus of continuity of a function in $C_u(-\infty, \infty)$ and if, in addition, $f(t)$ is non-decreasing for $t > 0$, then $f(t)$ is a rectified modulus of continuity.*

7.11. Differentiability. In studying the question of differentiability of subadditive functions, we have as usual to start at $t = 0$. The following theorem should be compared with Theorem 7.6.2.

THEOREM 7.11.1. *If $f(t)$ is a finite subadditive function in I_0 and if*

$$\sup_{t>0} \frac{f(t)}{t} = B \quad \text{and} \quad \inf_{t<0} \frac{f(t)}{t} = A$$

are finite, then

$$\lim_{h \to 0+} \frac{f(h)}{h} = B \quad \text{and} \quad \lim_{h \to 0-} \frac{f(h)}{h} = A, \tag{7.11.1}$$

$$A \leqq B. \tag{7.11.2}$$

The same conclusion is valid for $B = +\infty$ and $A = -\infty$ provided $\lim f(h) = 0$ *or* $\liminf f(h) > 0$ *when $h \to 0$ in I_+ and I_- respectively. If I_0 is replaced by I_+, only the first limit has a sense.*

PROOF. We proceed as in Theorem 7.6.1 and discuss only the first limit in detail. It is clear that $-\infty < B$. Suppose that B is finite and choose an a such that $f(a) > (B - \epsilon)a$. Put $a = nh + \delta$, n positive integer, $0 \leqq \delta < h$. Then

$$B - \epsilon \leqq \frac{f(a)}{a} \leqq \frac{f(nh)}{a} + \frac{f(\delta)}{a} \leqq \frac{nh}{a}\frac{f(h)}{h} + \frac{f(\delta)}{a}.$$

Let $h \to 0$; then $nh/a \to 1$ and $f(\delta) \to 0$, whence

$$B - \epsilon \leqq \liminf \frac{f(h)}{h} \leqq \limsup \frac{f(h)}{h} \leqq B$$

so that (7.11.1) holds. The same type of argument holds if $B = +\infty$ and $f(\delta) \to 0$ with δ. On the other hand, if $\liminf f(h) > 0$, we have manifestly $B = +\infty$ and $\lim f(h)/h = +\infty$. The conclusion, however, is no longer valid if $\liminf f(h) = 0 < \limsup f(h)$ as is seen from the example of the subadditive function which is 0 or 1 according as t is rational or irrational. Finally, formula (7.11.2) is an immediate consequence of the inequality $0 \leqq f(t) + f(-t)$.

We shall now consider the derived numbers of $f(t)$ which will be denoted by a prefixed D with an index $+$ to denote right, $-$ to denote left, used as a superior for upper and as an inferior for lower derived numbers. Thus $D_-f(t)$ is the lower left derived number of $f(t)$.

THEOREM 7.11.2. *If $f(t)$ is finite and subadditive in I_+, then*

$$D^+f(t) \leqq B, \qquad D^-f(t) \leqq B \tag{7.11.3}$$

for all t. In I_- we have instead

$$D_+f(t) \geqq A, \qquad D_-f(t) \geqq A \tag{7.11.4}$$

and if $f(t)$ is finite and subadditive in I_0, all four inequalities hold for all values of t. In the latter case, if A and B are finite, $f(t)$ is necessarily absolutely continuous. In particular, if $A = B$ then $f(t) = At$.

PROOF. The theorem is obviously trivially true if $A = -\infty$ and $B = +\infty$. If B is finite, the two inequalities

$$f(t + h) - f(t) \leqq f(h), \qquad f(t) - f(t - h) \leqq f(h)$$

upon division by h and passage to the limit yield (7.11.3); replacing h by $-h$ and proceeding in the same manner we get (7.11.4). If all four inequalities hold, A and B being finite, the derived numbers are bounded measurable functions. By a classical theorem, due to Lebesgue, $f(t)$ is an indefinite integral of any one of its derived numbers and hence absolutely continuous. Finally if $A = B$ then $f'(t) \equiv A$ and $f(t) = At$ since $f(0) = 0$.

Thus *the conditions, $f(0) = 0$, $f'(0)$ exists and equals A, single out a unique subadditive function $f(t) = At$.*

3. Subadditive Functions in E_n

7.12. Positive-homogeneous subadditive functions. Much of the foregoing theory can be extended to real-valued subadditive functions defined on a semi-module Σ in E_n. We shall limit our considerations to subadditive functions defined on semi-modules of a comparatively simple type.

Definition 7.12.1. *A semi-module Σ in E_n is called an angular semi-module if it is an open point set having the origin as a limit point.*

The simplest example of an angular semi-module is an *open cone*, that is, an open subset $\mathfrak{K}$ of E_n such that $\mathbf{s}, \mathbf{t} \in \mathfrak{K}$ implies that $\mathbf{s} + \mathbf{t} \in \mathfrak{K}$ and $\alpha\mathbf{t} \in \mathfrak{K}$ for all $\alpha > 0$. It is clear that $\mathfrak{K}$ is a convex subset of E_n. If $\theta \in \mathfrak{K}$ then $\mathfrak{K} = E_n$; otherwise $\mathfrak{K}$ lies on one side of some hyperplane.

For a given angular semi-module Σ, let $\mathfrak{K}(\Sigma) = [\mathbf{t};\ \alpha\mathbf{t} \in \Sigma$ for some $\alpha > 0]$. It readily follows that $\mathfrak{K}(\Sigma)$ is an open cone. Actually $\mathfrak{K}(\Sigma)$ is the smallest open cone containing Σ. Further $\theta \in \mathfrak{K}(\Sigma)$ if and only if $\theta \in \Sigma$. Consequently either $\theta \in \Sigma$, in which case $\Sigma = E_n$, or else Σ lies on one side of some hyperplane.

Definition 7.12.2. *A subadditive function defined on a cone $\mathfrak{K}$ is said to be positive-homogeneous if $f(\alpha\mathbf{t}) = \alpha f(\mathbf{t})$ for all $\mathbf{t} \in \mathfrak{K}$ and $\alpha > 0$.*

A positive-homogeneous subadditive function on an open cone $\mathfrak{K}$ such that $f(\mathbf{t}) < \infty$ for all $\mathbf{t} \in \mathfrak{K}$ is either identically equal to $-\infty$ or else finite-valued. In fact, if $f(\mathbf{t}_0) = -\infty$, then since $\mathfrak{K}$ is open there exists for any $\mathbf{t} \in \mathfrak{K}$ a $\gamma > 0$ such that $\mathbf{t} - \gamma\mathbf{t}_0 \in \mathfrak{K}$; consequently

$$f(\mathbf{t}) = f[(\mathbf{t} - \gamma\mathbf{t}_0) + \gamma\mathbf{t}_0] \leqq f(\mathbf{t} - \gamma\mathbf{t}_0) + \gamma f(\mathbf{t}_0) = -\infty.$$

In what follows we shall assume that $f(\mathbf{t})$ is a finite-valued, positive-homogeneous, subadditive function on $\mathfrak{K}$. It is clear that for $\mathbf{t}_1, \mathbf{t}_2 \in \mathfrak{K}$

$$f[\alpha\mathbf{t}_1 + (1 - \alpha)\mathbf{t}_2] \leqq \alpha f(\mathbf{t}_1) + (1 - \alpha)f(\mathbf{t}_2), \qquad 0 < \alpha < 1$$

and hence that $f(\mathbf{t})$ is necessarily convex on $\mathfrak{K}$. It follows from well known properties of convex functions that $f(\mathbf{t})$ is continuous on $\mathfrak{K}$. Likewise if the sequence $\{\mathbf{t}_n\} \subset \mathfrak{K}$ converges to a boundary point $\mathbf{t}_0$ of $\mathfrak{K}$, $\mathbf{t}_0 \neq \theta$, then either $\lim_n f(\mathbf{t}_n)$ exists or else $f(\mathbf{t}_n) \to +\infty$.

Positive-homogeneous subadditive functions occur in the uniqueness theory of differential equations; E. Kamke [1] has employed them to generalize the Lipschitz condition. From this point of departure M. Hukuhara [1] has determined the structure of all positive-homogeneous subadditive functions in E_n.

Positive-homogeneous subadditive functions were initially introduced by Minkowski [1] to characterize convex sets. Given a closed convex set $\mathfrak{C} \subset E_n$, Minkowski defined

$$F(\mathbf{t}) = \sup\,[(\mathbf{x}, \mathbf{t});\ \mathbf{x} \in \mathfrak{C}], \tag{7.12.1}$$

where $(\mathbf{x}, \mathbf{t})$ is the usual inner product in E_n. It is easy to see that $F(\mathbf{t})$ is a positive-homogeneous subadditive function, finitely defined on a cone $\mathfrak{K}$. $\mathfrak{C}$ is bounded if and only if $\mathfrak{K} = E_n$ whereas $\mathfrak{C} = E_n$ if and only if $\mathfrak{K}$ is vacuous. The function $F(\mathbf{t})$ is called the *function of support* for the convex set $\mathfrak{C}$. A converse is furnished by the following theorem, the proof of which is due to Werner Fenchel.

THEOREM 7.12.1. *If $F(\mathbf{t})$ is a positive-homogeneous subadditive function defined on a cone $\mathfrak{K}$ in E_n, then there exists a closed convex set whose function of support is equal to $F(\mathbf{t})$ on $\mathfrak{K}$ and which is finitely defined only on a sub-cone of $\bar{\mathfrak{K}}$.*

PROOF. We define $\mathfrak{C}$ to be the intersection of all half-spaces $[\mathbf{x}; (\mathbf{x}, \mathbf{t}) \leqq F(\mathbf{t})]$ where $\mathbf{t} \in \mathfrak{K}$. It is clear that $\mathfrak{C}$ is convex and closed. In order to show that $F(\mathbf{t})$ is actually equal to the function of support for $\mathfrak{C}$ on $\mathfrak{K}$, it is sufficient to show for a given $\mathbf{t}^0 \in \mathfrak{K}$ that there exists an $\mathbf{x}^0 \in \mathfrak{C}$ such that $(\mathbf{x}^0, \mathbf{t}^0) = F(\mathbf{t}^0)$. To this end we consider the set $\mathfrak{M}$ of all points lying above the surface $u = F(\mathbf{t})$, $\mathbf{t} \in \mathfrak{K}$, in the linear space $E_n \times E_1$. $\mathfrak{M}$ is a convex set and given $\mathbf{t}^0 \in \mathfrak{K}$, there exists a supporting hyperplane of $\mathfrak{M}$ passing through $(\mathbf{t}^0, F(\mathbf{t}^0))$. In symbols, there exist numbers $x_1^0, x_2^0, \cdots, x_n^0$ such that $F(\mathbf{t}) \geqq F(\mathbf{t}^0) + \sum_1^n x_i^0(t_i - t_i^0)$ for all $\mathbf{t} \in \mathfrak{K}$. In particular for $\mathbf{t} = \alpha\mathbf{t}^0$, $\alpha > 0$, we obtain

$$(\alpha - 1)F(\mathbf{t}^0) \geqq (\alpha - 1) \sum_1^n x_i^0 t_i^0 .$$

Since this holds for both $\alpha < 1$ and $\alpha > 1$ we obtain $F(\mathbf{t}^0) = \sum_1^n x_i^0 t_i^0 = (\mathbf{x}^0, \mathbf{t}^0)$. Consequently $F(\mathbf{t}) \geqq (\mathbf{x}^0, \mathbf{t})$ for all $\mathbf{t} \in \mathfrak{K}$ and hence $\mathbf{x}^0 \in \mathfrak{C}$. We note that the boundary points of $\mathfrak{K}$ may possibly be included in the set on which the function of support to the above defined $\mathfrak{C}$ is finite.

7.13. Boundedness and growth. We now drop the assumption of positive-homogeneity and consider general subadditive functions defined on angular semi-modules in E_n. The study of subadditive functions in E_n was initiated by E. Hille in connection with analytic semi-groups (see Chapter XVII) and later developed in detail by R. A. Rosenbaum [1].

THEOREM 7.13.1. *Let $f(\mathbf{t})$ be a measurable subadditive function defined on an angular semi-module Σ in E_n. If $f(\mathbf{t})$ is different from $+\infty$ in Σ, then $f(\mathbf{t})$ is bounded above in any compact subset of Σ. If $f(\mathbf{t})$ is finite throughout Σ, then $f(\mathbf{t})$ is bounded in any compact subset of Σ.*

PROOF. Let $\mathbf{t}_0 \neq \theta$ be a point of Σ and choose $\rho > 0$ so that the sphere S: $\| \mathbf{t} - \mathbf{t}_0 \| < \rho$ is contained in Σ. For $3\delta < \min (\rho, \| \mathbf{t}_0 \|)$, denote the spheres $[\mathbf{t}; \| \mathbf{t} - \mathbf{t}_0 \| < \nu\delta]$, $\nu = 1, 2$, by S_ν and set $\Sigma(\delta) = [\mathbf{t}; \| \mathbf{t} \| < \delta] \cap \Sigma$. We note that $S_2 \cap \Sigma(\delta) = \emptyset$. Further if $\mathbf{s}'$ and $\mathbf{s}_1$ are arbitrary points of S_1 and $\Sigma(\delta)$ respectively, then the point $\mathbf{s}_2 = \mathbf{s}' - \mathbf{s}_1 \in \mathbf{s}' - \Sigma(\delta) \subset S_2$.

Suppose now that $\limsup_{\mathbf{t} \to \mathbf{t}_0} f(\mathbf{t}) = +\infty$ where $\mathbf{t}_0 \in \Sigma$, $\mathbf{t}_0 \neq \theta$. Then we can find a sequence $\{\mathbf{t}_n\} \subset \Sigma$ such that $\mathbf{t}_n \to \mathbf{t}_0$ and $f(\mathbf{t}_n) \geqq 2n$. We take $\mathbf{t}_0$ and construct the sets S, S_1, S_2, and $\Sigma(\delta)$ as above. For $n \geqq N_\delta$, $\mathbf{t}_n \in S_1$ and for each

$s_1 \in \Sigma(\delta)$ we can find a unique point $s_2 \in S_2$ such that $s_1 + s_2 = t_n$. It follows that

$$f(s_1) + f(s_2) \geqq f(t_n) \geqq 2n.$$

Hence either $f(s_1) \geqq n$ or $f(s_2) \geqq n$. Consequently the set $E_n = [s; f(s) \geqq n, s \in S_2 \cup \Sigma(\delta)]$ is of measure at least equal to $m[\Sigma(\delta)]$ since the subset E' of $\Sigma(\delta)$ not in E_n has a translate, namely $t_n - E'$, which is contained in $E_n \cap S_2$. Now $E_n \supset E_{n+1}$ so that $m[\bigcap E_n] \geqq m[\Sigma(\delta)]$ and $\bigcap E_n$ is not empty. On the other hand $f(t) = +\infty$ on $\bigcap E_n$ contrary to our hypothesis. It follows that $f(t)$ is bounded above on each compact subset of Σ which does not contain θ. If $\theta \in \Sigma$ then $\Sigma = E_n$ and for $\| t \| \leqq r$, $\| t_0 \| > r$ we have $f(t) \leqq f(t - t_0) + f(t_0) \leqq 2M$ where $M = \sup [f(s); \| s - t_0 \| \leqq r]$ is finite by the previous argument. Hence in this case $f(t)$ is bounded above in a neighborhood of θ and therefore in any compact subset of Σ.

Suppose next that $f(t)$ is finite-valued but not bounded below in some compact subset F of Σ. Then there exist points $\{t_n\} \subset F$ such that $t_n \to t_0 \in F$ and $f(t_n) \leqq -n$. M being an upper bound of $f(t)$ on F, we have for each $t \in F$

$$f(t_n + t) \leqq f(t_n) + f(t) \leqq -n + M \to -\infty$$

as $n \to \infty$. We may suppose that $m(F) > 0$. Setting $E_n = [s; f(s) \leqq -n, s \in F + F]$, we see that $E_n \supset E_{n+1}$ and $m(E_n) \geqq m(F)$. It follows that $f(t) = -\infty$ on $\bigcap E_n$, a subset of $F + F$ of positive measure; this is contrary to hypothesis.

Theorem 7.4.3 also has its analogue in the n-dimensional case. From the inequality $f(2t) \leqq 2f(t)$ we conclude that $\liminf_{t \to \theta} f(t)$ is either $\geqq 0$ or $-\infty$. If $f(t)$ is finite-valued and measurable then the second alternative does not occur since $f(t_n) \leqq -n$, $t_n \to \theta$, implies that $f(t_n + t_0) \leqq -n + M$ and hence that $f(t)$ is unbounded below in every open subset of Σ, contrary to the previous theorem.

The rate of growth properties of a measurable subadditive function $f(t)$ on an angular semi-module Σ are extremely interesting. Let $\mathfrak{K} = \mathfrak{K}(\Sigma)$, the open cone generated by Σ. If $t \in \mathfrak{K}$, then, according to Theorem 7.3.1, there exists an α_t such that $\alpha t \in \Sigma$ for all $\alpha > \alpha_t$. Now $f(\xi t)$ is a finite subadditive function of ξ (though not necessarily measurable in ξ) for each $t \in \mathfrak{K}$ and all $\xi > \alpha_t$. Hence by the remark following Theorem 7.6.1 we see that $F(t) \equiv \lim_{\xi \to \infty} f(\xi t)/\xi$ exists and $F(t) < +\infty$. Further

$$F(\alpha t) = \lim_{\xi \to \infty} \frac{f(\xi \alpha t)}{\xi} = \alpha \lim_{\xi \to \infty} \frac{f(\xi \alpha t)}{\xi \alpha} = \alpha F(t), \qquad \alpha > 0, \tag{7.13.1}$$

and

$$\begin{aligned} F(s + t) &= \lim_{\xi \to \infty} \frac{f(\xi s + \xi t)}{\xi} \\ &\leqq \lim_{\xi \to \infty} \left[\frac{f(\xi s)}{\xi} + \frac{f(\xi t)}{\xi} \right] = F(s) + F(t). \end{aligned} \tag{7.13.2}$$

Thus $F(\mathbf{t})$ is a positive-homogeneous subadditive function on $\mathfrak{K}$ and the results of section 7.12 are applicable. If $\mathfrak{K} = E_n$ we must have $\Sigma = E_n$ and hence $f(\xi\mathbf{t})$ is subadditive in ξ on I_0 ; Theorem 7.6.2 asserts that in this case $F(\mathbf{t})$ will be finite-valued. We summarize these remarks in

THEOREM 7.13.2. *Let $f(\mathbf{t})$ be a finite-valued measurable subadditive function defined on an angular semi-module Σ and let $\mathfrak{K}$ be the open cone generated by Σ. Then*

$$F(\mathbf{t}) \equiv \lim_{\xi \to \infty} \frac{f(\xi\mathbf{t})}{\xi} \tag{7.13.3}$$

exists for each $\mathbf{t} \in \mathfrak{K}$. The function $F(\mathbf{t})$ is positive-homogeneous and subadditive on $\mathfrak{K}$. It is either identically $-\infty$ or else finite and continuous on $\mathfrak{K}$; if $\mathfrak{K} = E_n$ the latter case necessarily holds. Further $F(\mathbf{t})$ can be extended on a subcone of $\bar{\mathfrak{K}}$ to be the function of support for a closed convex set which is bounded only when $\mathfrak{K} = E_n$.

CHAPTER VIII

SEMI-MODULES

8.1. Orientation. The present chapter serves several different purposes. It contains the basic definitions in the theory of abstract semi-groups; it gives the elements of a theory of semi-modules (= additive abelian semi-groups) with special reference to semi-modules in a euclidean space; and finally it offers a discussion of a special semi-group related to the theory of relativity. Semi-modules of real or complex numbers form the parameter manifolds of one-parameter semi-groups of linear bounded transformations, the study of which will start in the next chapter and occupy the greater part of this treatise. This fact justifies our studying the one- and two-dimensional semi-modules at some length.

There is a peculiar relationship between semi-modules and subadditive functions which was discovered by Max Zorn in 1942. The boundary of an angular semi-module (= open, additive set whose closure contains the zero element) is determined by a subadditive function whose domain of definition is an angular semi-module of next lower dimension.

The third topic listed above gives a simple and useful non-trivial example of a non-additive law of composition and serves to introduce some of the ideas which will be more fully explored in Chapter XXV on Lie Semi-Groups.

There are three paragraphs: *Semi-Groups*, *Euclidean Semi-Modules*, and *Hyperbolical Semi-Groups*. References are to be found at the end of the paragraphs.

1. Semi-Groups

8.2. Abstract semi-groups. The notion of a *semi-group* is of a much more recent origin than that of a group though it is a more primitive notion. It seems to have made its first appearance in the literature in 1904 in the treatise of J. A. de Séguier on abstract groups [1, p. 8] followed a year later by a paper by L. E. Dickson [1] devoted to the subject. See also G. Frobenius and I. Schur [1]. The algebraic theory of semi-groups really got under way in the late twenties but progressed rather slowly. We shall not be concerned with purely algebraic aspects of the theory in this book; the reader who is interested in such questions is referred to the very incomplete list of papers in the References at the end of the paragraph. Some attention will be paid to metric topological aspects of the theory of semi-groups in the present chapter, but our main concern throughout the book will be with applications to analysis where topological semi-groups and in particular one-parameter semi-groups of linear transformations come up in the most diversified connections.

We start the discussion by giving a number of definitions. Let $\mathfrak{S}$ be a set of elements $a, b, \cdots$ and let $\bigcirc$ be a binary operation in $\mathfrak{S}$, that is, a mapping $\mathfrak{S} \times \mathfrak{S}$ into $\mathfrak{S}$ associating with every ordered pair (a, b) of elements of $\mathfrak{S}$ a definite element c of $\mathfrak{S}$: $c = a \bigcirc b$. Such a set is known as a *groupoid.*

DEFINITION 8.2.1. *An associative groupoid is called an abstract semi-group.*

Thus two postulates are assumed:

(i) *to every pair of distinct or equal elements a and b of $\mathfrak{S}$, taken in this order, there is a unique element $a \bigcirc b$ of $\mathfrak{S}$, and*

(ii) $a \bigcirc (b \bigcirc c) = (a \bigcirc b) \bigcirc c$.

The reader will recognize (i) and (ii) as two of the classical postulates for abstract groups. He will note that the existence of a unit element and of inverses is not assumed. The semi-groups considered in this treatise will usually have a unit element, however, and some elements may have inverses. When the notion of a semi-group was first introduced a law of cancellation was also assumed:

(iii) *if for any three elements a, b, c either $a \bigcirc b = a \bigcirc c$ or $b \bigcirc a = c \bigcirc a$, then* $b = c$.

In the French mathematical literature an associative groupoid is called a *demi-group* and the term semi-group is still reserved for the case in which a law of cancellation holds. If the semi-group has a finite number of elements, (iii) implies that $\mathfrak{S}$ is a group. We shall not assume (iii) anywhere in this treatise. On the other hand, we shall frequently assume:

(iv) *the operation is commutative,* $a \bigcirc b = b \bigcirc a$.

DEFINITION 8.2.2. *A semi-group satisfying* (i), (ii) *and* (iv) *is said to be abelian. If the operation is that of addition, $\mathfrak{S}$ is also called a semi-module.*

DEFINITION 8.2.3. *A subset of a given semi-group $\mathfrak{S}$ which is itself a semi-group under the given operation, is called a sub-semi-group of $\mathfrak{S}$.*

The semi-groups occurring in analysis will have topological as well as algebraic structure. The following definition of a topological semi-group is sufficiently general for most purposes.

DEFINITION 8.2.4. *A semi-group $\mathfrak{S}$ is a topological semi-group if $\mathfrak{S}$ is a Hausdorff space and to every $x, y \in \mathfrak{S}$ and every neighborhood $N(x \bigcirc y)$ of $x \bigcirc y$ there are neighborhoods $N(x)$ of x and $N(y)$ of y such that*

$$x \bigcirc N(y) \subset N(x \bigcirc y) \text{ and } N(x) \bigcirc y \subset N(x \bigcirc y).$$

These assumptions will make the "product" $x \bigcirc y$ continuous in each variable separately.

The notions of *homomorphism*, *isomorphism* and *automorphism* for semi-groups are covered by Definition 1.6.1 where we have merely to take for the two algebraic systems $\mathfrak{X}$ and $\mathfrak{X}'$ two semi-groups $\mathfrak{S}$ and $\mathfrak{S}'$.

8.3. Transformation semi-groups. An abstract semi-group is usually obtained from a transformation semi-group by a process of abstraction which disregards the nature of the elements and preserves merely their mode of combining. Conversely, an abstract semi-group may be realized as a transformation semi-group. The following definitions explain the terminology.

DEFINITION 8.3.1. *A set $\mathfrak{T}$ of transformations T_α on an abstract space $\mathfrak{X}$ to itself is a transformation semi-group if $T_\alpha T_\beta \in \mathfrak{T}$ whenever T_α and $T_\beta \in \mathfrak{T}$.*

DEFINITION 8.3.2. *A transformation semi-group $\mathfrak{T}$ is a realization of the abstract semi-group $\mathfrak{S}$ if to every element a of $\mathfrak{S}$ corresponds an element $T(a)$ of $\mathfrak{T}$ in such a manner that $T(a \bigcirc b) = T(a)T(b)$. A realization is faithful if $a \neq b$ implies $T(a) \neq T(b)$.*

Thus a realization is a homomorphic mapping of the abstract semi-group onto a semi-group of transformations. Every abstract semi-group admits of a realization by means of left-translations called the *left regular realization*. Here we take for $\mathfrak{X}$ the semi-group manifold itself and define $T(a)$ to be the transformation $y = a \bigcirc x$. Then there is a unique $T(a)$ corresponding to every a in $\mathfrak{S}$ and $T(a \bigcirc b) = T(a)T(b)$. The realization is certainly faithful if $\mathfrak{S}$ has a unit element or, more generally, if $a \bigcirc x = b \bigcirc x$ for all x in $\mathfrak{S}$ implies $a = b$.

This method does not always lead to faithful realizations, however. Thus in the semi-group of matrices of the form

$$\begin{pmatrix} x_{11} & x_{12} \\ 0 & 0 \end{pmatrix}$$

two left-translations $T(a)$ and $T(b)$ always coincide if $a_{11} = b_{11}$ regardless of the values of a_{12} and b_{12}.

DEFINITION 8.3.3. *A realization $\mathfrak{T}$ of $\mathfrak{S}$ is called a representation of $\mathfrak{S}$ if $\mathfrak{X}$ is a Banach space and $\mathfrak{T} \subset \mathfrak{E}(\mathfrak{X})$. If $\mathfrak{X}$ is n-dimensional, the representation is of degree n.*

In classical algebra it is customary to restrict oneself to representations of finite degree. This would not be appropriate for the applications which we have in mind; even if representations of finite degree exist, it is those of infinite degree which are apt to be of interest to the analyst. Thus practically all the semi-groups of linear transformations considered in Parts Two and Three of this book are representations of very simple abstract semi-groups for which one can find trivial representations of degree one. The next definitions serve to introduce these semi-groups.

DEFINITION 8.3.4. $\mathfrak{T} = \mathfrak{T}(\Pi)$ *is called a parametric semi-group of linear transformations in $\mathfrak{E}(\mathfrak{X})$ with the associated parameter semi-group Π, if $\mathfrak{T} \subset \mathfrak{E}(\mathfrak{X})$, if Π*

is a topological semi-group, and if $\mathfrak{T}(\Pi)$ *is a homomorphic image of* Π. $\mathfrak{T}(\Pi)$ *is said to be n-parametric if* Π *is a subset of the euclidean space of n dimensions, real or complex, and if* Int $(\Pi) \neq \emptyset$.

DEFINITION 8.3.5. *A one-parameter semi-group of linear transformations is said to be canonical if the law of composition reads* $T(\alpha)T(\beta) = T(\alpha + \beta)$ *where* α, $\beta \in E_1$ *or* Z_1.

The greater part of this book is devoted to canonical one-parameter semi-groups; in referring to such semi-groups we normally omit the word canonical when the meaning is clear from the context. The case of n-parametric semi-groups will be touched upon in Chapters IX and X and later treated more systematically in Chapter XXV.

8.4. Examples and illustrations. The real and the complex number systems lead to a profusion of semi-groups corresponding to different binary operations. Let us first focus our attention on the set of positive real numbers $E_1^+ = [x; x > 0]$. Among the possible definitions of $\alpha \bigcirc \beta$ we mention the following:

$$\alpha + \beta, \qquad \alpha\beta, \qquad \frac{\alpha + \beta}{1 + \alpha\beta}, \qquad \alpha(1 + \beta^2)^{1/2} + \beta(1 + \alpha^2)^{1/2}. \tag{8.4.1}$$

All four definitions correspond to addition theorems of elementary functions and suggest various generalizations. In fact, if $F(x, y)$ is a real valued function of the real variables x and y such that

$$F(x, y) \in E_1^+ \quad \text{when} \quad x, y \in E_1^+, \tag{8.4.2}$$

$$F(x, F(y, z)) = F(F(x, y), z), \tag{8.4.3}$$

then

$$\alpha \bigcirc \beta = F(\alpha, \beta) \tag{8.4.4}$$

is an admissible definition, the first condition ensuring the existence of the "product", the second its associativity. There are interesting relations of homomorphisms between these different semi-groups of positive numbers. These are left to the reader.

It is obvious that some of these definitions apply to the whole real number system or to the set of real negative numbers.

Various sub-semi-groups of the additive semi-group E_1^+ will be determined in section 8.6.

Similar considerations apply to the complex number system where E_1^+ is replaced by $Z_1^+ = [z; \Re(z) > 0]$. Thus (8.4.4) gives an admissible definition of the "product" in this system provided (8.4.2) is replaced by

$$F(x, y) \in Z_1^+ \quad \text{when} \quad x, y \in Z_1^+ \tag{8.4.5}$$

and (8.4.3) holds for $x, y, z \in Z_1^+$. In particular, the first and third operations listed under (8.4.1) are admissible in Z_1^+ but not the second and the fourth. An important class of sub-semi-groups of the additive semi-group Z_1^+ is determined in section 8.7. We note in particular:

(i) the complex numbers in the sector $\theta_1 < \arg z < \theta_2$ form a semi-group under addition if $\theta_2 - \theta_1 \leqq \pi$;

(ii) the complex numbers $x + iy$ with $x > | y^\rho |$, $0 < \rho \leqq 1$, form a semi-group under addition.

It should be observed that every ring is a semi-group under addition as well as under multiplication and that every group is also a semi-group.

We turn now to transformation semi-groups. The left (right) regular realization of the additive semi-group E_1^+ is given by

$$\mathfrak{X} = E_1^+, \qquad T(\alpha): t \to t + \alpha, \qquad \alpha > 0. \tag{8.4.6}$$

It is faithful. Similarly, if E_1^+ is considered as a semi-group under an admissible operation (8.4.4), the left regular realization is given by

$$\mathfrak{X} = E_1^+, \qquad T(\alpha): t \to F(\alpha, t), \qquad \alpha > 0. \tag{8.4.7}$$

It is faithful if, for instance, $F(x, y)$ is a continuous function of y at $y = 0$ for each x and

$$F(x, 0) \equiv x. \tag{8.4.8}$$

All these realizations are of course of degree one. The realizations of infinite degree are apt to be more interesting. Starting with the additive semi-group E_1^+ we may take $\mathfrak{X}$ as the (B)-space of continuous functions $f(t)$ on the closed interval $[0, \infty]$ with the usual norm $\| f \| = \sup | f(t) |$. We define

$$\mathfrak{X} = C[0, \infty], \qquad T(\alpha)[f] = f(t + \alpha), \qquad \alpha > 0. \tag{8.4.9}$$

This is a canonical one-parameter semi-group of linear transformations in $C[0, \infty]$ with $\| T(\alpha) \| \equiv 1$. This semi-group will be studied in detail in section 19.2 below. It is a typical one-parameter canonical semi-group.

Similar representations for E_1^+ under permissible operations (8.4.4) are obtained by taking

$$\mathfrak{X} = C[0, \infty], \qquad T(\alpha)[f] = f(F(\alpha, t)), \qquad \alpha > 0. \tag{8.4.10}$$

The regular representation for the additive semi-group Z_1^+ is

$$\mathfrak{X} = Z_1^+, \qquad T(\alpha): z \to z + \alpha, \qquad \alpha \in Z_1^+. \tag{8.4.11}$$

We shall also give a couple of representations of infinite order. For the first one we consider the (B)-space $BH(Z_1^+)$ of all functions $f(z)$ holomorphic and bounded in Z_1^+ with $\| f \| = \sup | f(z) |$ and define

$$\mathfrak{X} = BH(Z_1^+), \qquad T(\alpha)[f] = f(z + \alpha), \qquad \alpha \in Z_1^+. \tag{8.4.12}$$

For the second one we consider the space $L(0, 1)$ with the usual norm and take

$$(8.4.13)\qquad \mathfrak{X} = L(0, 1), \qquad T(\alpha)[f] = \frac{1}{\Gamma(\alpha)} \int_0^t (t - s)^{\alpha-1} f(s)\, ds, \qquad \alpha \in Z_1^+ .$$

The well-known properties of fractional integration show that this definition gives a canonical one-parameter semi-group. For further details see section 23.16.

So far we have only considered one-parameter semi-groups. We close these examples by exhibiting a three-parameter euclidean semi-group Π_3 together with one of its representations. Here Π_3 is the set of all three-vectors $x = (x_1, x_2, x_3)$ with $x_1 \geqq -1$; $x_2, x_3 \geqq 0$ and $x \bigcirc y$ is the vector whose three components are

$$(8.4.14)\qquad \frac{x_1 + y_1 + x_1y_1 + x_2y_3 - x_3y_2}{1 + x_3y_2}, \quad \frac{(1 + x_1)y_2 + x_2}{1 + x_3y_2}, \quad \frac{(1 + y_1)x_3 + y_3}{1 + x_3y_2}.$$

A representation of Π_3 is given by

$$(8.4.15)\qquad \mathfrak{X} = C[0, \infty], \qquad T(\alpha_1, \alpha_2, \alpha_3)[f] = f\left(\frac{(1 + \alpha_1)t + \alpha_2}{1 + \alpha_3 t}\right).$$

Setting two parameters equal to zero one obtains important one-parameter semi-groups. The case

$$(8.4.16)\qquad T(0, 0, \alpha)[f] = f\left(\frac{t}{1 + \alpha t}\right)$$

has not figured above and is canonical.

In most of the cases considered above we can adjoin $\alpha = 0$ to the parametric semi-group and if (8.4.8) and the symmetric condition hold, $\alpha = 0$ plays the role of unit element. In the representations, $T(0)$ will then normally be I, the identity transformation.

References (algebraic theory of semi-groups).
History: Dickson [1], Frobenius and Schur [1], de Séguier [1].
Arithmetic: Arnold [1], Lorenzen [1].
Representation: Clifford [1], Suschkewitsch [2].
Structure: Dubreil [1], Rees [1], Suschkewitsch [1].

2. Euclidean Semi-Modules

8.5. The topology of addition. We shall be concerned with the simplest of all infinite semi-groups: *the semi-modules in the n-dimensional real euclidean space* E_n. For a given n the set of distinct semi-modules in E_n is non-denumerable and the classification and study of the structure of the various types is essentially a topological problem which, as far as we know, is unsolved even for $n = 1$.

In the case of modules the situation is different (see J. Nielsen [1] and V. Bergström [1]). The whole space is the only open module. If the module is a closed set, it is the direct sum of one-dimensional modules which are either equivalent to the module of real numbers or to a module generated by a single element. The general case requires further analysis.

We lay down some preliminary definitions.

DEFINITION 8.5.1. *If X and Y are arbitrary sets in E_n their vector sum is*

$$X + Y = [z; z = x + y, x \in X, y \in Y].$$

We write $X^{(2)}$ for $X + X$ and, generally, $X^{(p)} = X + X^{(p-1)}$, $p = 2, 3, \cdots$ with $X^{(1)} = X$.

DEFINITION 8.5.2. *If S is an arbitrary set in E_n, the least semi-module of E_n containing S is called the additive resultant of S or the semi-module generated by S and it is denoted by $(S)_a$.*

DEFINITION 8.5.3. *An element of a semi-module is said to be reducible if it is the sum of two non-zero elements of the semi-module, otherwise irreducible. The set of all irreducible elements is the irreducible core. If the latter is vacuous, the semi-module is said to be indefinitely reducible.*

These notions are related as follows.

THEOREM 8.5.1. *The additive resultant of S is the intersection of all semi-modules in E_n containing S and $(S)_a$ is equal to $\bigcup_1^\infty S^{(p)}$.*

The proof is immediate.

A necessary condition that a semi-module S be indefinitely reducible is that the origin be a limit point of S. It is obviously not sufficient. Thus an open semi-module having the origin as a limit point is necessarily indefinitely reducible but this need not hold for its closure. A case in point is the semi-module $x \geqq |y|^{1/2}$ in the plane; here the frontier $x = |y|^{1/2}$ is the irreducible core of the semi-module each element of which is the sum of not more than two elements of the core. This minimal representation is unique but interior points admit any number of representations involving finite or infinite sums of elements of the core.

We shall consider briefly how topological properties fare under addition. It is clear that the *vector sum of open sets is an open set*. Hence if G is open so are the sets $G^{(p)}$ and $(G)_a$. The situation is entirely different for closed sets. Thus the vector sum of the closed sets

$$X = \left\{n + \frac{1}{n}\right\}, \qquad Y = \left\{-n - \frac{1}{en}\right\}, \quad n = 1, 2, 3, \cdots,$$

is not closed and the additive resultant of the closed set

$$X = \left[0, \frac{1}{n}; n = 1, 2, 3, \cdots\right]$$

is not closed. One can show, however, that the *vector sum of a finite number of closed bounded sets is closed*, but the second example shows that the additive resultant of a closed bounded set need not be closed.

The situation is more favorable for what might be called positive sets. We say that a vector $x = (x_1, \cdots, x_n)$ in E_n is positive if its coordinates are non-negative and strictly positive if all coordinates are positive. A set $X \subset E_n$ is positive if all its elements are positive vectors; X will be called strictly positive if

$$\inf x \equiv (\inf x_1, \cdots, \inf x_n)$$

is strictly positive. The proof of the following result is left to the reader:

THEOREM 8.5.2. *The vector sum of a finite number of closed positive sets is a closed positive set. The additive resultant of a closed strictly positive set is a closed strictly positive set.*

Another topological property of interest in connection with the additive resultant is that of connectivity. Since the additive resultant of an open sphere has only a finite number of components, the additive resultant of an open connected set has the same property. The situation is somewhat simpler in the case of convex sets. In this case we have still another representation of the additive resultant. Let S_α denote the image of S under the affine transformation $x' = \alpha x$, $\alpha > 0$. Let $x \in (S)_a$. We can then find an integer ν and ν vectors $x_1, x_2, \cdots, x_\nu \in S$ such that $x = x_1 + x_2 + \cdots + x_\nu$. Since S is convex

$$y_\nu = \nu^{-1}(x_1 + x_2 + \cdots + x_\nu) \in S$$

and $x = \nu y_\nu$, that is, $x \in S_\nu$. Conversely $S_\nu \subset (S)_a$ for all ν so that

$$(S)_a = \bigcup_1^\infty S_\nu, \qquad S_1 = S \text{ convex.} \tag{8.5.1}$$

LEMMA 8.5.1. *If S is a convex open set in E_n, then $(S)_a$ is connected if and only if $S_1 \cap S_2 \neq \emptyset$ and if this condition is satisfied, $S_\nu \cap S_{\nu+1} \neq \emptyset$ for all ν.*

For a proof of this lemma and of Theorem 8.5.3, the latter due to E. G. Begle, we refer to pages 151–152 of the first edition of this treatise.

THEOREM 8.5.3. *The additive resultant of an open convex set in E_n is simply connected whenever it is connected.*

There is a considerable literature on the question of how an additive set function behaves under vector addition of sets or, more generally, under a binary law of composition of sets. W. Sierpiński has shown that the vector sum of two sets, each of measure zero, need not be measurable, and when it is, its measure may have any value from zero to infinity. It is true, however, that if X and Y are measurable sets in E_n, the nth root of the inner measure of $X + Y$ is at least equal to the sum of the nth roots of the measures of X and of Y. This is the Brunn-Minkowski-Lusternik inequality; for this and related matters, see R. Henstock and A. M. Macbeath [1].

8.6. Semi-modules on the line. We shall consider the structure problem for semi-modules on the line, that is in E_1, and to simplify matters we restrict ourselves to semi-modules which are either open or closed. The location of the semi-module with respect to the origin plays an essential role.

THEOREM 8.6.1. *If* $\mathfrak{S}$ *is a semi-module in* E_1 *and* $\mathfrak{S}$ *is open (closed) and has the origin as a limit point, then* $\mathfrak{S}$ *equals* E_1, E_1^+ *or* E_1^- (E_1, $\bar{E}_1^+$ *or* $\bar{E}_1^-$).

PROOF. If $\mathfrak{S}$ contains a positive null sequence $\{x_k\}$ then

$$[mx_k\ ;\ k,\ m\ =\ 1,\ 2,\ 3,\ \cdots] \subset \mathfrak{S}$$

so that $\mathfrak{S}$ is everywhere dense in E_1^+. Thus for $\mathfrak{S}$ closed, $\mathfrak{S} \supset \bar{E}_1^+$. On the other hand if $\mathfrak{S}$ is open and contains the interval (α, β) and if $x_k < \beta - \alpha$, then it is easy to see that the additive resultant of (α, β) and any interval containing x_k will contain (α, ∞); hence $\mathfrak{S} \supset E_1^+$ when $\mathfrak{S}$ is open. If $\mathfrak{S}$ does not contain any negative elements we have then $\mathfrak{S} = E_1^+$ in the open case and $\mathfrak{S} = \bar{E}_1^+$ in the closed one. If there are also negative elements $\mathfrak{S} = E_1$ and if there are only negative elements we see that $\mathfrak{S} = E_1^-$ or $\bar{E}_1^-$.

THEOREM 8.6.2. *If the semi-module* $\mathfrak{S}$ *is closed and contains both positive and negative elements, and if the origin is not a limit point of* $\mathfrak{S}$, *then* $\mathfrak{S}$ *is the group* αN *of integral multiples of a fixed number* α.

PROOF. If the origin is not a limit point, $\mathfrak{S}$ will contain a least positive element, α say, and hence all the positive multiples of α. Let $-\beta$ be the largest negative element of $\mathfrak{S}$. If $\beta = \alpha$, we see that $\mathfrak{S}$ contains the group αN. But if $\alpha \neq \beta$, the subset $[m\alpha - n\beta;\ m, n = 0, 1, 2, \cdots]$ is either everywhere dense or contains an element nearer to the origin than α or $-\beta$. This contradiction shows that $\beta = \alpha$. In the same manner one shows that αN cannot be a proper subset of $\mathfrak{S}$.

The cases listed in these two theorems are the trivial semi-modules on the line.

Assuming $\mathfrak{S}$ to be a given semi-module of positive numbers such that $\mathfrak{S}$ contains a least positive element α, we shall construct the least generator A of $\mathfrak{S}$, that is, a set such that $(A)_a = \mathfrak{S}$ and if $(X)_a = \mathfrak{S}$ then $A \subset X$. We start with the interval $[\alpha, 2\alpha)$ and let A_1 be the subset of $\mathfrak{S}$ in this interval. If $\mathfrak{S} = (A_1)_a$ we are through and $A = A_1$. If not, we examine the interval $[2\alpha, 4\alpha)$ and let A_2 be the subset of $\mathfrak{S}$ there which is not already in $(A_1)_a$. The union of A_1 and A_2 belongs to A. If $(A_1 \cup A_2)_a \neq \mathfrak{S}$, we proceed to the interval $[4\alpha, 8\alpha)$ etc. In this manner we obtain a point set A defined as the union of all sets A_k. It is clear that $(A)_a \subset \mathfrak{S}$, but the converse is also true; for, any point of $\mathfrak{S}$ in $[2^{p-1}\alpha, 2^p\alpha)$ is either in $(A_1 \cup A_2 \cup \cdots \cup A_{p-1})_a$ or in A_p, hence in $(A)_a$. It is also clear from the construction that if X is such that $(X)_a = \mathfrak{S}$ then $A \subset X$. We note further that A is the irreducible core of $\mathfrak{S}$ so that $\mathfrak{S}$ is the additive resultant of its irreducible core.

If $\mathfrak{S}$ is open, then by Theorem 7.3.1 there is a quantity β such that $(\beta, \infty) \subset \mathfrak{S}$. The subinterval $(\beta, 2\beta)$ may possibly belong to A but it is clear that A is a bounded set restricted to the interval $(\alpha, 2\beta)$. Thus all A_k with $2^{k-1}\alpha > 2\beta$ are void.

If $\mathfrak{S}$ is closed instead, $\mathfrak{S}$ is measurable and Theorem 7.3.2 asserts that if the measure is positive, then $\mathfrak{S}$ contains an interval (β, ∞). In this case A is again bounded, but if $\mathfrak{S}$ is of measure zero, A is of course also of measure zero but need not be bounded. Examples will be given below.

We end this discussion by listing some special results the verification of which is left to the reader.

Suppose that A is a finite set $A = (a_1, a_2, \cdots, a_m)$ and that the a's are linearly independent over the rational field. Then $\mathfrak{S}$ is countable and every $s_k \in \mathfrak{S}$ has a unique representation

$$s_k = p_{k1}a_1 + p_{k2}a_2 + \cdots + p_{km}a_m,$$

where the p's are non-negative integers. Two sets (a_j) and (b_j) having the same number of elements generate isomorphic semi-modules, but the isomorphism does not preserve the natural order between the elements unless the generators are proportional, $b_j = \lambda a_j$.

If A is countable, we have two distinct possibilities according as the derived set A' of A is void or not. In the first case the countable set $\mathfrak{S}$ is finite over every finite interval, in the latter case it is finite only up to the first point of A'. In the second case A may be bounded but not in the first.

If $A' \neq \emptyset$ we have, *a fortiori*, $\mathfrak{S}' \neq \emptyset$. Moreover every derived set $\mathfrak{S}^{(n)}$ is non-void and $\mathfrak{S}^{(n+1)}$ is a sub-semi-module of $\mathfrak{S}^{(n)}$ and $\mathfrak{S} + \mathfrak{S}^{(n)} \subset \mathfrak{S}^{(n)}$ for $n = 0, 1, 2, \cdots$.

If A is taken as the classical Cantor set placed in the interval $[1, 2]$, then $\mathfrak{S}$ equals the union of A with the interval $(2, +\infty)$.

8.7. Angular semi-modules. As a general discussion of the structure problem for semi-modules would take us too far afield, we shall impose restrictions which single out a well defined class of semi-modules with comparatively simple properties. For the following discussion see E. Hille and M. Zorn [**1**].

DEFINITION 8.7.1. *A semi-module in a topological additive group is called an angular semi-module if it is an open point set having the zero element as a limit point.*

We usually assume that the semi-module is a subset of E_n and a detailed discussion will be carried through only for $n = 2$.

We note a striking difference between modules and semi-modules: the only module in E_n satisfying the conditions of this definition is the space itself. According to Theorem 8.6.1 the only proper angular semi-modules for the case $n = 1$ are the positive axis and the negative axis which of course are equivalent under a reflection. In two dimensions we have a much greater variety, but the various types can be characterized in simple terms by means of subadditive functions defined on one-dimensional semi-modules. This characterization extends to higher dimensions: the angular semi-modules in E_n are determined by subadditive functions defined on angular semi-modules in E_{n-1}. The relationship between angular semi-modules and subadditive functions is consequently of a recursive nature.

We start the discussion by proving some results of varying degree of generality which are needed for our problem.

We observe first that *in any topological additive group, both the closure of a semi-module and the interior of a semi-module are semi-modules.*

THEOREM 8.7.1. *If Σ is a semi-module in E_n and if every sphere with center at the origin contains an element of Σ different from 0, then there exists a vector $b \neq 0$ such that the ray ρb, $\rho \geqq 0$, is in the closure of Σ.*

PROOF. We denote the length of the vector v by $\| v \|$. By assumption there is a sequence $\{a_j\}$, $a_j \in \Sigma$, $a_j \neq 0$ with $\lim a_j = 0$. The unit vectors $a_j/\| a_j \|$ must have at least one limit point in E_n and without loss of generality we may assume $\lim a_j/\| a_j \| = b$ where $\| b \| = 1$. If ρ is given, $\rho \geqq 0$, we set $n_j = [\rho/\| a_j \|] + 1$ where $[\alpha]$ is the greatest integer $\leqq \alpha$. Then we have $\lim n_j \| a_j \| = \rho$ and the relation $\rho b = (\lim n_j \| a_j \|)(\lim a_j/\| a_j \|) = \lim n_j a_j$ shows that ρb is the limit of a sequence from Σ.

THEOREM 8.7.2. *If $\mathfrak{X}$ is a topological additive group and if $\mathfrak{S}$ is an angular semi-module in $\mathfrak{X}$, then $\mathfrak{S}$ is the interior of its own closure.*

PROOF. We have to prove that if a point x of $\mathfrak{X}$ does not belong to $\mathfrak{S}$, then every neighborhood N_x of x contains an open set which is not void and which has no points in common with $\mathfrak{S}$. To prove this we take a neighborhood N_0 of the zero element such that $x - N_0$ is in N_x; in the neighborhood N_0 there will be a vector y which, together with a full neighborhood N_y, is contained in $\mathfrak{S} \cap N_0$. The non-void open set $x - N_y$ is contained in N_x but has no points in $\mathfrak{S}$; for if $x - u$ were in $\mathfrak{S}$, $x = u + (x - u)$ would be, which is not true. This proves the theorem. An important consequence is

THEOREM 8.7.3. *Under the assumptions of the preceding theorem $\mathfrak{S} + \bar{\mathfrak{S}} \subset \mathfrak{S}$.*

PROOF. It is required to show that if x and y are arbitrary elements of $\mathfrak{S}$ and $\bar{\mathfrak{S}}$ respectively, then their sum belongs to $\mathfrak{S}$. Since the closure of a semi-group is a semi-group, $\mathfrak{S} + \bar{\mathfrak{S}}$ is first of all contained in $\bar{\mathfrak{S}}$ so that $x + y$ is in $\bar{\mathfrak{S}}$. But with $x + y$ there is a full neighborhood of the form $U_x + y$ also in $\bar{\mathfrak{S}}$; in other words, every point of $\mathfrak{S} + \bar{\mathfrak{S}}$ is an interior point of $\bar{\mathfrak{S}}$ and therefore contained in $\mathfrak{S}$ itself.

THEOREM 8.7.4. *If Σ is an angular semi-module in E_n, there exists at least one vector $b \neq 0$ such that $x \in \Sigma$ implies $x + \rho b \in \Sigma$ for all $\rho \geqq 0$.*

PROOF. It suffices to choose the vector b in Σ which is furnished by Theorem 8.7.1 and then to apply Theorem 8.7.3. Actually a slightly stronger statement could be made: to every $x \in \Sigma$ there is a positive $\epsilon(x)$ such that $x + \rho b \in \Sigma$ for $\rho > -\epsilon(x)$.

All that we have said so far in this section applies in particular to an angular semi-module in E_2. We shall now concentrate on this case and determine the structure of Σ. Since the properties of Σ are invariant under rotations about the origin, we may take b as the vector (1, 0) and identify E_2 with the complex plane Z_1. All vectors are then of the form $z = x + iy$ and if $z_0 = x_0 + iy_0 \in \Sigma$, then $x + iy_0 \in \Sigma$ for $x > x_0 - \epsilon(z_0)$. Now it is obvious that the characteristic properties of an angular semi-module (additive, open point set whose closure contains the zero element) are preserved under a projection on a linear subspace. Hence we have

THEOREM 8.7.5. *The projection of* Σ *on the imaginary axis is one of the sets* $[y; y > 0]$, $[y; y < 0]$ *or* $[y; -\infty < y < \infty]$.

It is clear that the first two alternatives are equivalent under a reflection of Σ in the real axis. We may consequently restrict ourselves to the first and the third cases. We denote the projection by Π. The last two theorems show that we may introduce a function $f(y)$ defined as a real number or as $-\infty$ for $y \in \Pi$ by

$$f(y) = \inf [x; x + iy \in \Sigma]. \tag{8.7.1}$$

We have then the basic

THEOREM 8.7.6. $f(y)$ *is a subadditive, upper semi-continuous function on* Π *such that* $\liminf_{y\to 0} f(y) = 0$ *or* $-\infty$. *In the latter case,* $f(y) \equiv -\infty$ *in* Π.

PROOF. Suppose that y_1 and y_2 are two points of Π. This implies that $f(y_1) + \delta + iy_1$ and $f(y_2) + \delta + iy_2$ are in Σ for every $\delta > 0$, hence $f(y_1) + f(y_2) + 2\delta + i(y_1 + y_2)$ is in Σ so that $f(y_1 + y_2) < f(y_1) + f(y_2) + 2\delta$. Since δ is arbitrary, $f(y)$ is subadditive. The fact that Σ is open implies that to every $\epsilon > 0$ and $y_0 \in \Pi$ there is a $\delta > 0$ such that all points

$$[x + iy; y_0 - \delta < y < y_0 + \delta, x > f(y_0) + \epsilon]$$

belong to Σ. This implies that $f(y) < f(y_0) + \epsilon$ for $|y - y_0| < \delta$ so that $f(y)$ is upper semi-continuous. Finally, $z = 0$ is supposedly a limit point of Σ. This requires that $\liminf_{y\to 0} f(y) \leqq 0$. As an upper semi-continuous function, $f(y)$ is measurable so that Theorem 7.4.3 applies. This theorem gives two alternatives: either $\liminf_{y\to 0} f(y) \geqq 0$ or it is $-\infty$. In the former case we see that

$$\liminf_{y\to 0} f(y) = 0.$$

In the latter case we have $f(y) \equiv -\infty$ since $f(y) \not\equiv +\infty$ for any y in Π. This completes the proof. If $f(y) \equiv -\infty$, then obviously Σ is either the upper half-plane $y > 0$ or the whole plane according as Π is

$$[y; y. > 0] \text{ or } [y; -\infty < y < \infty].$$

Combining the last two theorems, we arrive at the following description of the angular semi-modules in the plane.

THEOREM 8.7.7. *There exists a pair of orthogonal unit vectors u, v such that the angular semi-module Σ consists of all vectors $\xi u + \eta v$ with $\xi > f(\eta)$, where*

(i) *η varies over one of the sets $[\eta;\ \eta > 0]$, $[\eta;\ \eta < 0]$, or $[\eta;\ -\infty < \eta < \infty]$,*

(ii) *the function $f(\eta)$ has real numbers or $-\infty$ as values, is subadditive and upper semi-continuous and satisfies the condition* $\liminf_{\eta\to 0} f(\eta) = 0$ *or* $-\infty$.

Conversely, every set of vectors satisfying these conditions is an angular semi-module in E_2.

We say that a function $f(\eta)$ satisfying condition (ii) is an *admissible subadditive function*. Thus, to every angular semi-module Σ in the plane corresponds a "restricted product" $[\Pi, f(\eta)]$ where Π is an angular semi-module of real numbers and $f(\eta)$ is an admissible subadditive function defined on Π. Conversely, every restricted product defines an angular semi-module of the plane. In the system of coordinates used above, the boundary of Σ is made up of the curve $\xi = f(\eta)$ together with horizontal line segments corresponding to the discontinuities of $f(\eta)$. If $\Pi = [\eta;\ \eta > 0]$, the positive ξ-axis is also part of the boundary. Since the vector $b = u$ of Theorem 8.7.1 ordinarily is not uniquely determined, the same holds for the coordinate system (u, v) so that the same angular semi-module may correspond to infinitely many restricted products. We list some of the geometric properties of angular semi-modules in the following

THEOREM 8.7.8. *An angular semi-module in the plane is a simply-connected point-set. It is either the whole plane or a subset of a half-plane. It is indefinitely reducible in the sense of Definition* 8.3.3.

PROOF. To show that Σ is arc-wise connected we exhibit for any two elements $\xi_1 u + \eta_1 v$ and $\xi_2 u + \eta_2 v$ of Σ a polygon which connects them and is contained in Σ. The construction is possible because the upper semi-continuous function $f(\eta)$ has a finite least upper bound in the interval $\eta_1 \leqq \eta \leqq \eta_2$. If μ exceeds this bound, the polygon is made up of the following three—possibly degenerate—arcs:

$$\text{(i)}\qquad \xi = \xi_1 + \vartheta(\mu - \xi_1), \qquad \eta = \eta_1,$$

$$\text{(ii)}\qquad \xi = \mu, \qquad \eta = \eta_1 + \vartheta(\eta_2 - \eta_1), \qquad 0 \leqq \vartheta \leqq 1,$$

$$\text{(iii)}\qquad \xi = \mu + \vartheta(\xi_2 - \mu), \qquad \eta = \eta_2.$$

To prove that Σ is simply-connected, let C be a simple closed curve consisting of points in Σ; we have to show that any point z inside C (in the sense of the Jordan Curve Theorem) is necessarily contained in Σ. Indeed, consider the ray $z - \rho u$, $\rho \geqq 0$. This ray will intersect C at a point $z_0 = z - \rho_0 u$ of Σ, and by Theorem 8.7.4 this implies that $z = z_0 + \rho_0 u$ is in Σ.

If $f(\eta) \equiv -\infty$ then Σ is a half-plane or the whole plane according as Π is a proper or an improper semi-module. We can dismiss these cases as trivial. Again, if Π is a proper semi-module, say $[\eta;\ \eta > 0]$, then Σ is restricted to the upper half-plane. The only case remaining in doubt is that in which $\Pi = [\eta;\ -\infty < \eta < \infty]$ and $\liminf f(\eta) = 0$. Since $f(\eta) \not\equiv +\infty$, Theorem 7.3.4 shows that

$f(\eta) \neq -\infty$, so that $f(\eta)$ is finite. Theorem 7.6.2 then gives the existence of two finite quantities α and β with $\alpha \leqq \beta$ such that $f(\eta) \geqq \beta\eta$ when $\eta > 0$ and $f(\eta) \geqq \alpha\eta$ when $\eta < 0$, so that Σ is contained in a sector with vertex at the origin and opening $\leqq \pi$.

Finally, let $z \in \Sigma$. Since Σ is an open set, the closure of which contains the origin, we can find a z_0 such that z_0 and $z - z_0$ are in Σ. This asserts that z is the sum of two elements of Σ so that z is reducible and Σ consequently indefinitely reducible. This completes the proof.

The extension of these results from two dimensions to n is a fairly simple matter. Let Σ_n be a given angular semi-module in E_n and choose a Cartesian coordinate system $(u_1, u_2, \cdots, u_n)$ in which the positive u_1-axis belongs to the closure of Σ_n. This is possible by virtue of Theorem 8.7.1. We then project Σ_n on the orthogonal $(n-1)$-dimensional space $E_{n-1} = (u_2, \cdots, u_n)$. The set of all vectors $y = (u_2, \cdots, u_n)$ such that there is a vector $x = (u_1, u_2, \cdots, u_n)$ in Σ_n is an angular semi-module in E_{n-1} as observed above. We denote the projection of Σ_n by Σ_{n-1}. We then define

$$f(y) = f(u_2, \cdots, u_n) = \inf [u_1; (u_1, u_2, \cdots, u_n) \in \Sigma_n].$$

The values of this function on vectors in E_{n-1} are real numbers or $-\infty$. That it is subadditive and upper semi-continuous is proved as above and the same type of argument also gives $\lim \inf_{y \to 0} f(y) = 0$ or $-\infty$. Thus every angular semi-module in E_n gives rise to a restricted product $[\Sigma_{n-1}, f(y)]$ and vice versa.

Thus, in order to characterize the angular semi-modules in n-dimensional euclidean space we have to determine (i) the angular semi-modules in $(n-1)$-dimensional space and (ii) the admissible subadditive functions on such a semi-module in E_{n-1}. The necessary tools for carrying through this recursive process have been given in the preceding discussion and we shall not go into further detail here.

The methods developed above can also be used for a study of closed semi-modules in E_n containing the origin. Theorem 8.7.1 obviously applies, but when we project on E_{n-1}, the projection, which is a semi-module and contains the origin, is not necessarily closed but is merely a set F_σ. It is therefore more natural to assume at the outset that the original semi-module is a set F_σ since this property is preserved under projection. We desist from further indications.

For the work in Chapter XVII we shall need a class of semi-modules of a slightly more general nature than the angular semi-modules.

Definition 8.7.2. *A semi-module in E_n is said to be spinal if it contains a ray from the origin and an open set intersected by the ray.*

Theorem 8.7.9. *If $\mathfrak{S}$ is a spinal semi-module in the complex plane, $\mathfrak{S} \neq Z_1$, then $\mathfrak{S}$ is simply-connected and there exist arcs Φ_1 and Φ_2 such that* (1) $0 < \Phi_2 - \Phi_1 \leqq \pi$, (2) *every point of $\mathfrak{S}$ lies in $\Phi_1 \leqq \arg z \leqq \Phi_2$, and* (3) *if $\epsilon > 0$ is given there exists a finite R_ϵ such that every point z with $\Phi_1 + \epsilon \leqq \arg z \leqq \Phi_2 - \epsilon$, $R_\epsilon < |z|$ lies in $\mathfrak{S}$.*

PROOF. Without restricting the generality we may assume that the ray in $\mathfrak{S}$ is the positive real axis. By assumption it intersects the open set G contained in $\mathfrak{S}$. Since $\mathfrak{S}$ is a semi-module its projection Π on the imaginary axis is also a semi-module and Π contains the origin as an interior point. Hence

$$\Pi = [y; -\infty < y < \infty].$$

Next if $z_0 \in \mathfrak{S}$ so does $z_0 + \rho$, $\rho > 0$. For every $y \in \Pi$ a function

$$\varphi(y) = \inf [x, x + iy \in \mathfrak{S}]$$

is defined and the argument given in the proof of Theorem 8.7.6 shows that $\varphi(y)$ is subadditive and never $+\infty$.

Let G_1 be a component of G having points in common with the positive real axis and let Π_1 be the projection of G_1 on the imaginary axis. Let $y_1, y_2 \in \Pi_1$, $y_1 < 0 < y_2$. Then there exist points $x_1 + iy_1$ and $x_2 + iy_2$ in $G_1 \subset \mathfrak{S}$. Since G_1 is connected, these two points may be joined by a polygonal line in G_1 and from this it follows that $\varphi(y)$ is bounded for $y_1 \leqq y \leqq y_2$ and hence in every finite interval.

We may then apply Theorems 7.6.1 and 7.6.2 and can assert the existence of

$$\lim_{y\to\infty} \frac{\varphi(y)}{y} = \inf_{y>0} \frac{\varphi(y)}{y} = \beta < \infty,$$

$$\lim_{y\to-\infty} \frac{\varphi(y)}{y} = \sup_{y<0} \frac{\varphi(y)}{y} = \alpha > -\infty,$$

where $\beta \geqq \alpha$. This implies that $\varphi(y) \geqq \beta y$ when $y > 0$ and $\varphi(y) \geqq \alpha y$ when $y < 0$. Thus $\mathfrak{S}$ is contained in the sector formed by the rays $\beta y = x$, $y > 0$ and $\alpha y = x$, $y < 0$, the sector being supposed to contain the positive real axis. We determine Φ_1 and Φ_2 from the equations

$$\cot \Phi_1 = \alpha, \qquad \cot \Phi_2 = \beta, \qquad -\pi < \Phi_1 < 0 < \Phi_2 < \pi$$

Since $\beta \geqq \alpha$ we have $0 < \Phi_2 - \Phi_1 \leqq \pi$.

Now if $\delta > 0$ is given, we have $\varphi(y) \leqq (\beta + \delta)y$ for $y > y_\delta$ and $\varphi(y) \leqq (\alpha - \delta)y$ for $y < -y_\delta$. This implies that the distant part of the boundary of $\mathfrak{S}$ is enclosed in the infinite triangles

$$\beta y \leqq x \leqq (\beta + \delta)y, \qquad y \geqq y_\delta; \qquad \alpha y \leqq x \leqq (\alpha - \delta)y, \qquad y \leqq -y_\delta.$$

Now for $-y_\delta \leqq y \leqq y_\delta$ we have $\varphi(y) \leqq M_\delta$ so that $\mathfrak{S}$ contains an infinite sector of the kind asserted in the theorem

$$\text{arc cot } (\alpha - \delta) \leqq \arg z \leqq \text{arc cot } (\beta + \delta), \qquad |z| \geqq (y_\delta^2 + M_\delta^2)^{1/2}.$$

To show that $\mathfrak{S}$ is connected as well as simply-connected one can use the same argument as in the proof of Theorem 8.7.8.

REMARK. *If $\mathfrak{S}$ is the interior of a spinal semi-module, then its boundary is given* *as above by an upper semi-continuous subadditive function.* The boundary function of a spinal semi-module, however, is ordinarily not upper semi-continuous.

We end this discussion by listing the category theorem of Max Zorn.

THEOREM 8.7.10. *Let $\mathfrak{X}$ be a topological additive group. Let $\mathfrak{S}$ be a semi-module in $\mathfrak{X}$. If $\mathfrak{S}$ is of the second category at the zero element and if $\mathfrak{S}$ satisfies the condition of Baire, then* Int $(\mathfrak{S})$ = Int $(\bar{\mathfrak{S}})$ *and* Int $(\mathfrak{S})$ *is dense in* $\mathfrak{S}$.

For a proof we refer to page 187 of the first edition of this treatise. The theorem shows that any non-pathological semi-module having the origin as a point of the second category and located in a topological additive group, differs from an angular semi-module by a non-dense frontier set.

References. Bergström [**1**], Hille [**7**, §2.5, **13**], Hille and Zorn [**1**], Henstock and Macbeath [**1**], and J. Nielsen [**1**].

3. HYPERBOLICAL SEMI-GROUPS

8.8. Semi-groups of the hyperbolic tangent. The attention in this treatise is concentrated on canonical one-parameter semi-groups to such an extent that the reader needs an occasional reminder of the fact that there are other semi-groups and that they pose interesting mathematical problems. This is the main purpose of the present paragraph where we shall study the complex number system as a semi-group under the law of composition

$$\alpha \bigcirc \beta = \frac{\alpha + \beta}{1 + \alpha\beta}. \tag{8.8.1}$$

It was observed in section 8.4 that E_1^+ and Z_1^+ form semi-groups under this law. It is sometimes desirable to adjoin infinity to the system with suitable interpretation of the operation in the extended system.

The law is patterned on the addition theorem for the hyperbolic tangent

$$\text{th}\,(w_1 + w_2) = \frac{\text{th}\,w_1 + \text{th}\,w_2}{1 + \text{th}\,w_1\,\text{th}\,w_2}, \tag{8.8.2}$$

and we shall see that this function plays a fundamental role in the study. In passing let us recall that (8.8.1) is the formula for composition of apparent velocities in the theory of relativity, the velocity of light being set equal to unity. For this reason the name "Einstein numbers" has been proposed for real numbers between zero and one when "added" according to (8.8.1). Cf. G. A. Baker, Jr. [**1**] where a second operation is defined so that the numbers form a field.

Suppose now that $\mathfrak{S}$ is a semi-group of complex numbers under the law (8.8.1). We start with

THEOREM 8.8.1. *If $\mathfrak{S}$ contains only purely imaginary numbers then $\mathfrak{S}$ is either dense on the imaginary axis or there exists a positive integer n such that $\mathfrak{S}$ is of the form $\Gamma_n \equiv [i \tan (p\pi/n);\ p = 0, 1, \cdots, n-1]$.*

This is easily verified. Zero is clearly the unit element of these groups. In addition to the groups Γ_n, attention should be paid to the two idempotents ± 1. We have clearly

$$1 \bigcirc \alpha = \alpha \bigcirc 1 = 1, \quad (-1) \bigcirc \alpha = \alpha \bigcirc (-1) = -1, \qquad (8.8.3)$$
$$0 \bigcirc \alpha = \alpha \bigcirc 0 = \alpha$$

for every α. One can prove that if G is a finite group under the operation (8.8.1), then G is either a subgroup of a Γ_n or $\{1\}$ or $\{-1\}$.

THEOREM 8.8.2. *If $\mathfrak{S}$ contains a point z_1, $\Re(z_1) > 0$, $z_1 \neq 1$, then $+1$ is a limit point of $\mathfrak{S}$. Similarly if $\Re(z_1) < 0$, $z_1 \neq -1$, then -1 is a limit point.*

PROOF. The equation

$$z = \operatorname{th} w \qquad (8.8.4)$$

is satisfied by

$$w = \tfrac{1}{2} \log \frac{1+z}{1-z} \qquad (8.8.5)$$

for any choice of the logarithm. Here the real parts of z and of w have the same sign. The nth iterate of $z_1 = \operatorname{th} w_1$ is

$$z_1^{(n)} = \operatorname{th} n w_1 \qquad (8.8.6)$$

and, as $n \to \infty$, this tends to 1 or -1 according as z_1 lies in the right or in the left half-plane.

THEOREM 8.8.3. *Suppose that iy_0, y_0 real $\neq 0$, is a limit point of $\mathfrak{S}$. If $y_0 \neq \tan (k\pi/n)$ for all integers k and n, then every point of the imaginary axis is a limit point of $\mathfrak{S}$. If $y_0 = \tan (k\pi/n)$, then every point $i \tan (p\pi/n)$, $p = 0, 1, \cdots, n-1$, is a limit point of $\mathfrak{S}$. If $\mathfrak{S}$ is dense in a neighborhood of a point of the imaginary axis (resp. dense in the right half or in the left half of such a neighborhood) then $\mathfrak{S}$ is dense in the whole plane (resp. in the right or in the left half-plane). If $\mathfrak{S}$ contains such a neighborhood (the right half or the left half of such a neighborhood) then $\mathfrak{S}$ is the whole plane with the possible exception of $z = \pm 1$ ($\mathfrak{S}$ contains the right or the left half-plane with corresponding possible omissions).*

PROOF. If $z_n \to iy_0$ then the kth iterate of z_n tends to the kth iterate of iy_0 and the assertion concerning limit points on the imaginary axis follows from Theorem 8.8.1. In particular we see that if any point of the imaginary axis is a limit point of $\mathfrak{S}$ then the origin is a limit point. It follows that if $\mathfrak{S}$ is dense in some neighborhood of iy_0 then $\mathfrak{S}$ is also dense in some neighborhood of the origin, N_0 say. Since the iterates of N_0 exhaust the z-plane punctured at ± 1, we conclude that $\mathfrak{S}$ is dense in the whole plane. A similar argument holds for partial neighborhoods and for the case in which $\mathfrak{S}$ contains neighborhoods or partial neighborhoods.

In the following we shall restrict ourselves to semi-groups in the right half-plane.

THEOREM 8.8.4. *If $\mathfrak{S}$ is located in $\mathfrak{R}(z) > 0$ and the origin is not a limit point of $\mathfrak{S}$, then there exists a $C_0 < 1$ such that $\mathfrak{S}$ is contained in the circular domain*

$$\left|\frac{z-1}{z+1}\right| < C_0 .$$

If $\mathfrak{S}$ is open it contains a circular domain

$$0 < \left|\frac{z-1}{z+1}\right| < C_1$$

regardless of the position of $\mathfrak{S}$ relative to the origin.

PROOF. In the first case, no point of the imaginary axis can be a limit point so it suffices to show that $\mathfrak{S}$ is bounded. But if $z_n \to \infty$ then $z_n \bigcirc z_n \to 0$. It follows that $\mathfrak{S}$ must be bounded and bounded away from the imaginary axis so that a C_0 exists with the desired property. The second assertion follows from the next theorem.

We shall now consider the relationship between additive semi-groups and hyperbolical ones.

THEOREM 8.8.5. *If $\mathfrak{A}$ is a semi-module in $\mathfrak{R}(w) > 0$, then the function $z = \text{th } w$ maps $\mathfrak{A}$ onto a hyperbolical semi-group $\mathfrak{S}$ in $\mathfrak{R}(z) > 0$. Conversely, if $\mathfrak{S}$ is a semi-group under* (8.8.1) *in $\mathfrak{R}(z) > 0$, then there exists at least one and in general infinitely many semi-modules $\mathfrak{A}$ in $\mathfrak{R}(w) > 0$ which are mapped onto $\mathfrak{S}$ by $z = \text{th } w$.*

PROOF. The first assertion is obvious. For the converse let W_∞ be the set of all points w defined by (8.8.5) for any z in $\mathfrak{S}$, $z \neq +1$, and any choice of the logarithm. The set W_∞ is a semi-module for if $w_1, w_2 \in W_\infty$ then there exist points $z_1, z_2 \in \mathfrak{S}$ such that $z_1 = \text{th } w_1$, $z_2 = \text{th } w_2$ and $w_1 + w_2$ is one of the possible solutions of the equation $z_1 \bigcirc z_2 = \text{th } w$ and consequently included in W_∞. Thus we have $\mathfrak{S} = \text{th } (W_\infty)$ with obvious notation and can set $\mathfrak{A} = W_\infty$.

This construction gives the maximal semi-module that will map onto $\mathfrak{S}$. It is, however, not necessary to take all determinations of the logarithm in (8.8.5). Introducing suitable cuts in the z-plane, we can restrict ourselves to a single determination, say the one whose imaginary part lies between $-\pi/2$ and $+\pi/2$. This choice gives a set W_0 such that $\mathfrak{S} = \text{th } (W_0)$. Since W_0 is ordinarily not a semi-module we replace it by its additive resultant and take $\mathfrak{A} = (W_0)_a$. We still have th $(\mathfrak{A}) = \mathfrak{S}$ but $(W_0)_a$ is usually distinct from W_∞ so we have constructed another solution of the converse problem.

Combining this result with Theorems 8.7.7 and 8.7.8 we get:

THEOREM 8.8.6. *If* $\mathfrak{S}$ *is an open semi-group under* (8.8.1) *in* $\Re(z) > 0$ *having the origin as a limit point, then there exists at least one angular semi-module* Σ *such that* $\mathfrak{S} = \text{th}(\Sigma)$. *Consequently* $\mathfrak{S} \cup \{1\}$ *is simply connected and every point of* $\mathfrak{S}$ *outside a neighborhood of the origin can be connected with* $z = +1$ *by a circular arc in* $\mathfrak{S}$ *(image of horizontal line segment in* Σ*).*

A basic concept in the theory of Lie semi-groups (see Chapter XXV) is the notion of a canonical orbit. These orbits are defined as solutions of a functional equation with an initial condition. For the hyperbolical case the conditions become

$$g(\rho + \sigma) = \frac{g(\rho) + g(\sigma)}{1 + g(\rho)g(\sigma)}, \qquad \lim_{\rho \to 0+} \frac{g(\rho)}{\rho} = e^{i\theta}, \quad -\frac{\pi}{2} < \theta < \frac{\pi}{2}. \tag{8.8.7}$$

The unique solution is

$$g(\rho) = \text{th}(\rho e^{i\theta}) \tag{8.8.8}$$

and the curve

$$\Omega_\theta : z = \text{th}(\rho e^{i\theta}), \qquad 0 < \rho < \infty, \tag{8.8.9}$$

is known as a canonical orbit with respect to the operation (8.8.1) and the corresponding semi-group Z_1^+. For $\theta = 0$ we are dealing with the line segment (0, 1), the other orbits are loxodromic spirals having $z = +1$ as limit point. Each point of Z_1^+ may be joined to the origin by an arc of a canonical orbit. If it be required that the arc be of minimal length, the value of θ is uniquely determined unless z is real and >1 when two opposite values of θ give the same length.

The canonical orbits are connected with the regular representation of Z_1^+ by the semi-group $\mathfrak{T}$ of fractional linear transformations of the form

$$T_\alpha : z_\alpha = \frac{\alpha + z}{1 + \alpha z}, \qquad \alpha \in Z_1^+. \tag{8.8.10}$$

For if α lies on the orbit Ω_θ the transformation T_α will map Ω_θ into itself. There is consequently a sub-semi-group of $\mathfrak{T}$ leaving a given Ω_θ invariant.

8.9. Suboperative functions. The functional equation

$$f\left(\frac{z_1 + z_2}{1 + z_1 z_2}\right) = f(z_1) + f(z_2) \tag{8.9.1}$$

is satisfied by

$$f(z) = \frac{C}{2} \log \frac{1 + z}{1 - z}; \tag{8.9.2}$$

C being an arbitrary constant, but also by the real and imaginary parts of this function. We call any measurable solution of (8.9.1) an *"operative function"* with respect to the operation (8.8.1). The operative functions correspond to the linear functions in the case of addition. It should be noted that an operative function need not be single-valued even if it is real, the imaginary part of (8.9.2) when C is real being a case in point.

Now let $\mathfrak{S}$ be a semi-group under (8.8.1) located in $\Re(z) > 0$ and let $f(z)$ be a real valued function defined in $\mathfrak{S}$ and satisfying

$$f\left(\frac{z_1 + z_2}{1 + z_1 z_2}\right) \leqq f(z_1) + f(z_2), \qquad z_1, z_2 \in \mathfrak{S}. \tag{8.9.3}$$

We say that $f(z)$ is *suboperative*, again with respect to the operation (8.8.1). Such suboperative functions form the natural generalization of the subadditive functions of Chapter VII. In the present case we can read off their properties by combining Theorems 8.8.5 and 8.8.6 with the results of section 7.13. We list the following result.

THEOREM 8.9.1. *Let $\mathfrak{S}$ be a semi-group under* (8.8.1) *located in* $\Re(z) > 0$ *and not containing* $z = +1$. *Let $\mathfrak{S}$ be open and admit the origin as a limit point. Let $f(z)$ be defined in $\mathfrak{S}$ as a single-valued, measurable, finite-valued suboperative function. Then $f(z)$ is bounded on compact subsets of $\mathfrak{S}$. Further there exists a quantity β, $-\infty \leqq \beta < \infty$, such that*

$$\lim_{z \to +1} f(z) \left\{\log \left|\frac{z+1}{z-1}\right|\right\}^{-1} = \beta. \tag{8.9.4}$$

PROOF. By Theorem 8.8.6 we can find an angular semi-module Σ such that $\text{th}(\Sigma) = \mathfrak{S}$ where we take $\Sigma = W_\infty$ in the notation used in the proof of Theorem 8.8.5. We have then $f(z) = f(\text{th } w) \equiv g(w)$ where $g(w)$ is defined in Σ as a finite valued measurable subadditive function. It is consequently bounded on compact subsets of Σ by Theorem 7.13.1. A compact subset S_1 of $\mathfrak{S}$ is bounded and bounded away from the imaginary axis and from $z = +1$. Its inverse image T_1 in Σ is not compact since th w is a periodic function of period πi, but it suffices for our purposes to consider that part T_{10} of T_1 in which the imaginary part of w lies between $-\pi/2$ and $\pi/2$. Here $S_1 = \text{th}(T_{10})$ and T_{10} is compact. Since $g(w)$ is bounded in T_{10} it follows that $f(z)$ is bounded in S_1.

Next we note that Σ is a semi-module invariant under the translation $w \to w + \pi i$. It follows that Σ contains a half strip $0 < R < u$, $-\pi/2 < v \leqq \pi/2$, $w = u + iv$, and this half-strip H_R is mapped on the punctured circle

$$C_R : 0 < \left|\frac{z-1}{z+1}\right| < \delta_R$$

in a one-to-one manner by $z = \text{th } w$. Thus (8.9.4) is equivalent to proving the existence of

(8.9.5) $$\lim_{u \to \infty} u^{-1} g(u + iv) = \tfrac{1}{2}\beta,$$

uniformly with respect to v in H_R. Now the limit exists for $v = 0$ since $g(w)$ is subadditive and bounded on compact sets and for $w = u + iv$ we have for fixed a, $R < a$,

$$g(u + iv) \leqq g(u - a) + g(a + iv), \qquad g(u + a) \leqq g(u + iv) + g(a - iv).$$

Dividing by u and passing to the limit, (8.9.5) results. This completes the proof.

We note that while the behavior of $f(z)$ as $z \to +1$ is restricted by (8.9.4), there are no limits on the rate of growth of $f(z)$ when z approaches other points on the boundary of $\mathfrak{S}$.

References. Baker [1].

CHAPTER IX

ADDITION THEOREMS IN A BANACH ALGEBRA

9.1. Introduction. We shall now take up the main theme of these Lectures; *the theory of one-parameter semi-groups of endomorphisms* and the various ramifications and applications of this theory. The point of departure is the following problem:

Let $\mathfrak{X}$ be a complex Banach space, $\mathfrak{E}(\mathfrak{X})$ the corresponding Banach algebra of endomorphisms of $\mathfrak{X}$. Further let Σ be a given spinal semi-module of real or complex numbers. Determine all functions $T(\zeta)$ on Σ to $\mathfrak{E}(\mathfrak{X})$ such that for all ζ_1 and ζ_2 in Σ we have

$$T(\zeta_1)T(\zeta_2) = T(\zeta_1 + \zeta_2). \tag{9.1.1}$$

If $T(\zeta)$ is a solution of this problem, we refer to $\mathfrak{S} = [T(\zeta)]$ as a *one-parameter semi-group of endomorphisms with parameter manifold* Σ (see Definition 8.3.6). We may regard $\mathfrak{S}$ as a representation of the semi-module Σ, but this interpretation is of no use for the following.

The first case which must be settled is that in which Σ is the interval $(0, \infty)$. This case is kept in the foreground throughout; while the complex module will play a basic role in the present chapter, proper complex semi-modules will not be considered until Chapter XVII where analytic semi-groups are studied.

Equation (9.1.1) is formally *the addition theorem of the exponential function.* Actually the classical exponential function is a special instance of our theory for if we take $\mathfrak{X} = Z_1$, the complex plane, and define $T(\xi)$ as the similitude $w = f(\xi)z$ where

$$f(\xi_1 + \xi_2) = f(\xi_1)f(\xi_2), \qquad 0 < \xi_1, \xi_2 < \infty, \tag{9.1.2}$$

then $\mathfrak{S} = [T(\xi)]$ is a one-parameter semi-group of linear transformations. If $f(\xi)$ is supposed to be measurable, then, as shown in section 4.17, either $f(\xi) \equiv 0$ or there exists a complex number α such that $f(\xi) = e^{\alpha\xi}$. Moreover, this semi-group may be extended to an analytical group defined for all real and complex values of the parameter since $e^{\alpha\zeta}$ is an entire function of ζ and satisfies (9.1.2) for all values of ζ.

It is then natural to expect that if care is taken to exclude non-measurable solutions of (9.1.1) as well as projections, $T(\zeta)$ will have the form $\exp(\zeta A)$ where A is an operator on $\mathfrak{X}$ to itself and the definite interpretation of the exponential function is left open for the time being. In analogy with the classical case of continuous groups, we shall refer to A as *the infinitesimal generator of* $\mathfrak{S}$.

These expectations are fulfilled to some extent at least. If $\Sigma = (0, \infty)$ and $T(\xi)$ is measurable in the uniform or in the strong sense in $(0, \infty)$, then $T(\xi)$ is also continuous in the same sense for $0 < \xi < \infty$, but $T(\xi)$ does not necessarily tend to a limit when $\xi \to 0$. If $\lim_{\xi \to 0} T(\xi)$ exists it has to be an idempotent, that is, a projection operator which, for the purposes of the present discussion, we may assume to be the identical transformation. *If* $\lim_{\xi \to 0} T(\xi) = I$, *we have two sharply differentiated cases according as the limit exists in the uniform or in the strong operator topology.* These two cases are referred to hereinafter as the *uniform* and the *strong* cases respectively. The uniform case shows close analogy with the classical situation: *an infinitesimal generator* A *exists,* $A \in \mathfrak{E}(\mathfrak{X})$, *and* $T(\xi) = \exp(\xi A)$ where the exponential function is interpreted as in Chapter V. Since $\exp(\zeta A)$ is well defined for all complex ζ and satisfies (9.1.1), we see that $\mathfrak{S}$ *may be embedded in the analytical group* $\mathfrak{G} = [\exp(\zeta A)]$.

Save for the existence of a unique infinitesimal generator A, none of this holds in the strong case. A is now an *unbounded* linear transformation whose domain is merely dense in $\mathfrak{X}$ and the symbol $\exp(\xi A)$ must be redefined. Several new interpretations will be given. The function $T(\xi)$ is strongly continuous but usually not differentiable, much less analytic; if $T(\xi)$ can be extended to the complex plane as an analytic function, the extension $T(\zeta)$ defines an analytic semi-group whose parameter set Σ is a proper complex semi-module, that is, a subset of a half-plane and never the whole plane. The case in which $T(\xi)$ does not tend to a limit when $\xi \to 0$ agrees in most respects with the strong case and may be handled with the same methods.

Both cases present themselves in the applications, but the strong case is by far the most interesting to the analyst. It offers more difficult problems and calls for more refined analysis, and as a consequence the resulting theory is also richer and shows greater variety.

9.2. Orientation. The present chapter is devoted to the uniform case and related questions. In the uniform case the underlying space $\mathfrak{X}$ plays no role and it is only the Banach algebra $\mathfrak{E}(\mathfrak{X})$ that matters. We can omit all reference to $\mathfrak{X}$ and formulate the problem for an arbitrary complex Banach algebra $\mathfrak{B}$:

PROBLEM A. *Determine all measurable functions* $f(\xi)$ *on* $(0, \infty)$ *to* $\mathfrak{B}$ *such that for all* ξ_1 *and* ξ_2 *in* $(0, \infty)$

$$f(\xi_1 + \xi_2) = f(\xi_1)f(\xi_2). \tag{9.2.1}$$

This problem, however, is capable of further generalization in several different directions. We list three such extensions.

PROBLEM B. *Determine all functions* $F(x)$ *on a complex Banach space* $\mathfrak{X}$ *to a complex Banach algebra* $\mathfrak{B}$ *which are measurable on rays and satisfy*

$$F(x + y) = F(x)F(y) \tag{9.2.2}$$

for all x *and* y *in a given cone.*

These functional equations have the nature of addition theorems and the two remaining problems generalize this feature of the question. If $G(\alpha, \beta)$ is a given analytic function of two complex variables and u, v are elements of a Banach algebra, the symbol $G(u, v)$ is to be interpreted in the sense of the operational calculus as in section 5.2.

PROBLEM C. *Determine all measurable functions $f(\xi)$ on real numbers to a Banach algebra $\mathfrak{B}$ such that*

$$f(\xi_1 + \xi_2) = G[f(\xi_1), f(\xi_2)] \tag{9.2.3}$$

for all ξ_1, ξ_2 and $\xi_1 + \xi_2$ in some interval $(0, \omega)$.

PROBLEM D. *Determine all functions $F(x)$ on a complex Banach space $\mathfrak{X}$ to a complex Banach algebra $\mathfrak{B}$ which are measurable on rays and satisfy*

$$F(x + y) = G[F(x), F(y)] \tag{9.2.4}$$

for all x, y and $x + y$ in some domain.

Here are four problems of increasing generality all of which will be partly solved in the present chapter. Problems C and D may be omitted on a first reading. It turns out that measurability with respect to a positive scalar variable ξ implies continuity for $\xi > 0$ but not the existence of a limit when $\xi \to 0$. In the present chapter we separate and determine the solutions which are continuous at the origin, $\xi = 0$ or $x = \theta$; they are holomorphic functions of ξ and analytic functions of x respectively. Problems A and C are treated in some detail; Problems B and D, which can be reduced to A and C respectively, are discussed quite briefly.

Problem A has been in the literature in one form or another since 1935 when D. S. Nathan made an attack on equation (9.1.1). He considered a group (or group germ) of linear transformations $T(\xi)$. No mention is made of the uniform topology, only strong continuity is assumed explicitly, but his additional assumption that $\| T(\xi_1 + \xi_2) - I \| \leqq \theta < 1$ for ξ_1 and ξ_2 in some interval induces uniform continuity. The question was reopened from the point of view of Banach algebras by M. Nagumo and K. Yosida in 1936.

No study seems to have been made of Problem B though Theorem 5.4.1 must have been known to anybody who gave the question a passing thought. Problems C and D have been discussed by Dunford and Hille (abstract 1944).

The chapter is divided into four paragraphs, one for each problem. References are listed below.

References. Dunford and Hille [1], Halmos, Lumer and Schäffer [1], Lorch [3], Nagumo [1], Nathan [1], v. Neumann [1, 2], Yosida [1].

1. Problem A

9.3. Measurability implies continuity. In the following we discuss the measurable solutions of Problem A. All such solutions which are defined for $\xi > 0$ are also continuous, but they separate into two classes according as they are also continuous for $\xi = 0$ or not. Solutions of the first class are analytic and are actually entire functions of ξ. Solutions of the second class are much more varied and do not differ essentially from the solutions of the strong problem mentioned in section 9.1. The further analysis of this class will therefore be postponed to Chapter X.

Theorem 9.3.1. *Let $\mathfrak{B}$ be a real or complex Banach algebra which need not have a unit element. Let $f(\xi)$ be an everywhere defined measurable function on the interval $(0, \infty)$ to $\mathfrak{B}$ such that for $0 < \xi_1, \xi_2 < \infty$*

$$f(\xi_1 + \xi_2) = f(\xi_1)f(\xi_2). \tag{9.3.1}$$

Then $f(\xi)$ is continuous for all positive values of ξ.

Remark. Measurability is taken in the strong sense. It may of course be replaced by the equivalent condition that $f(\xi)$ is weakly measurable and almost separately valued. It should be noted that if $\mathfrak{B} = \mathfrak{E}(\mathfrak{X})$, then $f(\xi)$ is assumed to be uniformly measurable in the sense of Definition 3.5.5 (1).

Proof. Since $f(\xi)$ is everywhere defined and strongly measurable, $\| f(\xi) \|$ is finite and measurable in the sense of Lebesgue. From (9.3.1) we get the basic inequality

$$\log \| f(\xi_1 + \xi_2) \| \leqq \log \| f(\xi_1) \| + \log \| f(\xi_2) \|, \tag{9.3.2}$$

so that $\log \| f(\xi) \|$ is a measurable subadditive function of ξ in $I_+ = (0, \infty)$ and $\log \| f(\xi) \| \not\equiv +\infty$. By Theorem 7.4.1, $\log \| f(\xi) \|$ is bounded above in any interval $(\epsilon, 1/\epsilon)$. Hence $\| f(\xi) \|$ is a bounded measurable function in any such interval. Choose three numbers α, β, ξ such that $0 < \alpha < \beta < \xi$. Then

$$\int_\alpha^\beta f(\xi - \eta)f(\eta)\, d\eta$$

exists as a (B)-integral since the integrand is a bounded measurable function of η. By (9.3.1) the value of the integral is simply $(\beta - \alpha)f(\xi)$. For small values of ϵ

$$(\beta - \alpha)[f(\xi + \epsilon) - f(\xi)] = \int_\alpha^\beta [f(\xi + \epsilon - \eta) - f(\xi - \eta)]f(\eta)\, d\eta,$$

whence

$$(\beta - \alpha) \| f(\xi + \epsilon) - f(\xi) \| \leqq M \int_{\xi-\beta}^{\xi-\alpha} \| f(\tau + \epsilon) - f(\tau) \| \, d\tau,$$

where $M = \sup \| f(\eta) \|$ for $\alpha \leqq \eta \leqq \beta$. By Theorem 3.8.3 the right member tends to zero with ϵ. It follows that $f(\xi)$ is continuous for every $\xi > 0$ and the theorem is proved.

The result of this theorem is essentially the best of its kind. Thus the assumption that $f(\xi)$ is continuous for $\xi > 0$ does not imply the existence of $\lim_{\xi\to 0} f(\xi)$ or that $f(\xi)$ is differentiable for $\xi > 0$ or satisfies a Lipschitz condition of prescribed order or, finally, that $f(\xi)$ is the boundary-value of a holomorphic function. All these plausible extensions are disproved by the following counter example.

We take $\mathfrak{B} = \mathfrak{E}(l_2)$, that is, the Banach algebra of endomorphisms of the space l_2 of sequences $\{\alpha_n ; n = 0, \pm 1, \pm 2, \cdots\}$, $\| \{\alpha_n\} \| = \{\sum | \alpha_n |^2\}^{\frac{1}{2}}$. Given a transformation of the type $T(\{\alpha_n\}) = \{\mu_n \alpha_n\}$, it is easily shown that $\| T \| = \sup_n | \mu_n |$. We now define the semi-group of operators

$$T(\xi)(\{\alpha_n\}) = \{\mu_n(\xi)\alpha_n\}, \qquad \xi \geqq 0, \tag{9.3.3}$$

where

$$\mu_0(\xi) \equiv 1 \quad \text{and} \quad \mu_n(\xi) = | n |^{-\xi} \exp\,[\operatorname{sgn} n \cdot e^{|n|} i\xi] \quad \text{for } | n | > 0.$$

It is clear that $T(\xi) \in \mathfrak{E}(l_2)$ and satisfies (9.3.1) for $\xi \geqq 0$. Further $\| T(\xi) \| \equiv 1$ and

$$\| T(\xi + \delta) - T(\xi) \| = \sup_{n>0} \varphi(n; \xi, \delta),$$

where $\varphi(n; \xi, \delta) = n^{-\xi} | n^{-\delta} \exp(e^n i\delta) - 1 |$. For $n \geqq N$ and $\xi, \xi + \delta \geqq \gamma > 0$ we have $\varphi(n; \xi, \delta) < 2N^{-\gamma}$. Since $\varphi(n; \xi, \delta) \to 0$ as $\delta \to 0$, it follows that $\varphi(n; \xi, \delta)$ can be made uniformly small for each $\xi > 0$ by choosing $| \delta |$ sufficiently small. Consequently $T(\xi)$ is continuous in the uniform operator topology for $\xi > 0$. On the other hand for $\delta = e^{-n}$, $\varphi(n; \xi, \delta) > \frac{1}{2}[\log(1/\delta)]^{-\xi}$ and hence

$$\| T(\xi + \delta) - T(\xi) \| \geqq \frac{1}{2}\left[\log\left(\frac{1}{\delta}\right)\right]^{-\xi}, \quad \delta = e^{-n}, \quad \xi \geqq 0.$$

This shows that $T(\xi)$ does not converge to $T(0) = I$ as $\xi \to 0$ and further that $T(\xi)$ does not satisfy a Lipschitz condition for any $\xi > 0$. In particular, $T(\xi)$ is not differentiable for $\xi > 0$.

If $T(\zeta)$ is holomorphic in a domain D, then clearly $\| T(\zeta) \| = \sup_n | \mu_n(\zeta) |$ is bounded in each compact subset of D. Conversely, if the $| \mu_n(\xi) |$ are uniformly bounded in each compact subset of D, then the exponential form of $\mu_n(\zeta)$ implies that $\sum \alpha_n \bar{\beta}_n \mu_n(\zeta)$, $\{\alpha_n\}$ and $\{\beta_n\} \in l_2$, is a holomorphic function of ζ in D; hence $T(\zeta)$ will be holomorphic in D. Now $T(\zeta)$ can be written as the sum of two transformations $T_1(\zeta)$ and $T_2(\zeta)$, where $T_1(\zeta)$ is obtained from (9.3.3) by setting $\mu_n(\zeta) = 0$ for $n \leqq 0$ and $T_2(\zeta)$ by setting $\mu_n(\zeta) = 0$ for $n > 0$. Using the above criterion it is easy to show that $T_1(\zeta)$ and $T_2(\zeta)$ are holomorphic functions of ζ for $\mathfrak{I}(\zeta) > 0$ and $\mathfrak{I}(\zeta) < 0$ respectively, continuous in the corresponding half-planes plus the ray $\zeta > 0$ and, with the exception of $\zeta = 0$, that neither is bounded elsewhere. It follows that $T(\zeta)$ cannot define the boundary values of an analytic function on any interval (α, β) of the positive real axis. For suppose $S(\zeta)$ were such a function, holomorphic in a domain D contained in, say, the upper half-plane and which has (α, β) on its boundary. Then $S(\zeta) - T_1(\zeta)$ is holomorphic in D and converges to $T_2(\zeta)$ on (α, β). But this implies that $T_2(\zeta)$ has an analytic extension in D which is impossible.

This example may be varied so that different moduli of continuity result. It is also possible to make $\lim_{\xi\to 0} \| T(\xi) \| = +\infty$. Examples of the latter type are to be found in Chapter XX (see Theorem 20.5.3).

At this point we have the choice of elaborating the theory of equation (9.3.1) without further assumptions on $f(\xi)$ or singling out solutions with special properties. The first alternative leads to a theory not essentially different from that presented in Chapters X and XI below in the so-called strong case. We have merely to interpret $f(\xi)$ as a linear operator acting on a suitably chosen (B)-space $\mathfrak{X}$. This may be taken as $\mathfrak{B}$ itself, the operation being left-hand multiplication of x by $f(\xi)$. The measurability assumptions on $f(\xi)$ as an element of $\mathfrak{B}$ will then imply strong measurability of the operator $f(\xi)$ and the results of the "strong" theory apply. The problem of constructing solutions of (9.3.1) which are uniformly continuous for $\xi > 0$ (but not for $\xi = 0$) will be considered briefly in §12.2.

9.4. The exponential solutions. The only case in which an essentially different theory results is that in which

$$\lim_{\xi\to 0+} f(\xi) = j \tag{9.4.1}$$

exists. We can either assume (9.4.1) outright or else introduce the assumption indirectly in an equivalent form. Both alternatives will be used in the following.

THEOREM 9.4.1. *If $f(\xi)$ is defined for $\xi > 0$ and satisfies* (9.3.1) *and* (9.4.1) *then j is an idempotent of $\mathfrak{B}$ and*

$$f(\xi) = jf(\xi) = f(\xi)j.$$

Further $f(\xi)$ is continuous for $\xi \geqq 0$ if $f(0) = j$ by definition.

PROOF. That j is an idempotent follows from the relation

$$j = \lim_{\xi+\eta\to 0} f(\xi + \eta) = \lim_{\xi\to 0,\ \eta\to 0} f(\xi)f(\eta) = j^2.$$

Formula (9.4.1) further implies that

$$\lim_{\eta\to 0+} f(\xi + \eta) = f(\xi)j = jf(\xi). \tag{9.4.2}$$

It follows from this that $f(\xi)$ is measurable. In fact, for each $\epsilon > 0$ and any point ξ_0 at which the oscillation of $f(\xi)$ is $\geqq \epsilon$, there is an open set abutting ξ_0 on the right in which the oscillation is $< \epsilon$. Since these open sets are disjoint, we see that the points of oscillation $\geqq \epsilon$ are at most denumerable and consequently the same is true of the points of discontinuity. If follows that $f(\xi)$ is separably-valued and weakly measurable, and hence strongly measurable by Theorem 3.5.3.

Theorem 9.3.1 now asserts that $f(\xi)$ is continuous for $\xi > 0$; (9.4.1) implies continuity at $\xi = 0$; and (9.4.2) shows that $f(\xi) = f(\xi)j = jf(\xi)$. In particular, if $j = \theta$ we see that $f(\xi) \equiv \theta$.

THEOREM 9.4.2. *Under the assumptions of the preceding theorem there exists an element $a \in \mathfrak{B}$ such that $a = ja = aj$ and*

$$(9.4.3) \qquad f(\xi) = j + \sum_{n=1}^{\infty} \frac{\xi^n}{n!} a^n.$$

The series is absolutely convergent for all real (complex) values of ξ and satisfies (9.3.1) *for all such values.*

PROOF. According to the preceding theorem, $j^2 = j$ and $f(\xi) = f(\xi)j = jf(\xi)$ for all $\xi > 0$. Let us now introduce the subalgebra $\mathfrak{B}_0 = j\mathfrak{B}j$ in which j plays the role of unit element. It is clear that $\mathfrak{B}_0$ consists of all $x \in \mathfrak{B}$ such that $jx = xj = x$ and this implies that $\mathfrak{B}_0$ is closed in $\mathfrak{B}$ and hence complete. Since $f(\xi)$ is continuous for $\xi \geqq 0$, the integral $\int_\alpha^\beta f(\xi)\, d\xi$ exists for finite values of $\beta > \alpha \geqq 0$ and is again an element of $\mathfrak{B}_0$. Further

$$\lim_{\beta\to\alpha} \frac{1}{\beta-\alpha} \int_\alpha^\beta f(\xi)\, d\xi = f(\alpha)$$

for each $\alpha \geqq 0$. In particular there is a $\delta > 0$ such that $\| \beta^{-1}\int_0^\beta f(\tau)\, d\tau - j \| < 1$ for $0 < \beta < \delta$. By Theorem 4.3.1, $\beta^{-1}\int_0^\beta f(\tau)\, d(\tau)$ is regular in $\mathfrak{B}_0$ for $\beta \in (0, \delta)$. Now

$$\int_0^\beta f(\xi+\tau)\, d\tau - \int_0^\beta f(\tau)\, d\tau = [f(\xi) - j] \int_0^\beta f(\tau)\, d\tau$$

which gives

$$(9.4.4) \qquad \frac{1}{\xi}\int_\beta^{\beta+\xi} f(\tau)\, d\tau - \frac{1}{\xi}\int_0^\xi f(\tau)\, d\tau = \frac{1}{\xi}[f(\xi) - j] \int_0^\beta f(\tau)\, d\tau.$$

For $\beta \in (0, \delta)$ we obtain

$$(9.4.5) \qquad \frac{1}{\xi}[f(\xi) - j] = \left[\frac{1}{\xi}\int_\beta^{\beta+\xi} f(\tau)\, d\tau - \frac{1}{\xi}\int_0^\xi f(\tau)\, d\tau\right]\left[\int_0^\beta f(\tau)\, d\tau\right]^{-1}.$$

When $\xi \to 0$ the right hand side tends to a limit since the first factor tends to $f(\beta) - j$. It follows that

$$(9.4.6) \qquad \lim_{\xi\to 0} \frac{1}{\xi}[f(\xi) - j] \equiv a$$

exists and is an element of $\mathfrak{B}_0$. If in (9.4.4) we again take the limit as $\xi \to 0$, we obtain

$$(9.4.7) \qquad f(\beta) - j = a \int_0^\beta f(\tau)\, d\tau,$$

valid for all real $\beta \geqq 0$. Iterated substitution leads to

$$f(\xi) = j + \frac{\xi}{1!} a + \frac{\xi^2}{2!} a^2 + \cdots + \frac{\xi^n}{n!} a^n + \frac{a^{n+1}}{n!} \int_0^\xi (\xi - \tau)^n f(\tau)\, d\tau,$$

and letting $n \to \infty$ we obtain formula (9.4.3) (cf. section 3.4).

If $\mathfrak{B}$ is a complex Banach algebra, we see that the series defines a holomorphic function; thus $f(\xi)$ is an entire function of ξ. A simple computation shows that $f(\xi)$ satisfies (9.3.1) for all complex values of ξ (cf. section 5.4).

The main points in the preceding proof are due to M. Nagumo [1]. Formula (9.4.5), which is crucial in the proof, may also be used to prove the result under apparently weaker assumptions. We now give two such theorems.

THEOREM 9.4.3. *Let $f(\xi)$ be a solution of* (9.3.1) *defined for $\xi > 0$ which is measurable and (B)-integrable in every finite interval* $(0, \omega)$. *If*

$$\lim_{\xi \to 0} \frac{1}{\xi} \int_0^\xi f(\tau)\, d\tau = j \tag{9.4.8}$$

exists, then j is an idempotent, the limit in (9.4.6) *exists, and $f(\xi)$ is given by formula* (9.4.3).

PROOF. By Theorem 9.3.1, $f(\xi)$ is continuous for $\xi > 0$. We have

$$f(\xi) \frac{1}{\eta} \int_0^\eta f(\tau)\, d\tau = \frac{1}{\eta} \int_0^\eta f(\tau + \xi)\, d\tau.$$

When $\eta \to 0$, the left side tends to $f(\xi)j$, the right to $f(\xi)$. Hence $f(\xi) = f(\xi)j$ and similarly $f(\xi) = jf(\xi)$ for $\xi > 0$. Further

$$j \frac{1}{\xi} \int_0^\xi f(\tau)\, d\tau = \frac{1}{\xi} \int_0^\xi f(\tau)\, d\tau$$

and in the limit this gives $j^2 = j$.

We consider the closed subalgebra $\mathfrak{B}_0 = j\mathfrak{B}j$ again and note that $\int_0^\beta f(\tau)\, d\tau$ is an element of $\mathfrak{B}_0$ which has an inverse for sufficiently small values of β. From this we conclude that (9.4.5) holds for sufficiently small values of β, so that the proof can be continued as above.

Assumption (9.4.1) may be introduced in a still more indirect manner. The following theorem is an example of such a procedure; it may be regarded as a generalization of the theorem of D. S. Nathan mentioned in section 9.2.

THEOREM 9.4.4. *Let $\mathfrak{B}$ be a real or complex Banach algebra of which $\mathfrak{B}_0$ is a closed subalgebra and let $\mathfrak{B}_0$ have a unit element j. Let $f(\xi)$ be a measurable function on the interval $(0, \infty)$ to $\mathfrak{B}_0$ which satisfies* (9.3.1) *for $0 < \xi_1, \xi_2 < \infty$ and let there exist a ξ_0 such that $f(\xi_0)$ has an inverse in $\mathfrak{B}_0$. Then there exists an element a of $\mathfrak{B}_0$ such that $f(\xi)$ is given by formula* (9.4.3).

REMARK. The assumption that $f(\xi_0)$ has an inverse in $\mathfrak{B}_0$ may be replaced by $\| f(\xi_0) - j \|$

< 1. By Theorem 4.3.1 this implies the existence of the inverse, however. The basic assumption of D. S. Nathan was of this character.

PROOF. It is clear from (9.3.1) that $f(\xi) = f(\xi + \xi_0)[f(\xi_0)]^{-1}$ for $\xi > 0$. Further $f(\xi)$ is continuous for $\xi > 0$ by Theorem 9.3.1. It follows that

$$\lim_{\xi \to 0+} f(\xi) = \lim_{\xi \to 0+} f(\xi + \xi_0)[f(\xi_0)]^{-1} = j.$$

The desired result is now an immediate consequence of Theorems 9.4.1 and 9.4.2.

In conclusion it should be observed that the real difficulties in discussing the functional equation (9.3.1) are associated with the interval $(0, \infty)$. If the function $f(\xi)$ is supposed to satisfy this equation for all real values of ξ, then the assumption that $f(\xi)$ is measurable in an interval $(-\delta, \delta)$ is enough to force $f(\xi)$ to be of the form given by (9.4.3). A solution defined on $(0, \infty)$ is only exceptionally of this form, however, and this is the reason why we have to impose fairly severe restrictions on $f(\xi)$ in order to single out these solutions.

9.5. Continuous power groups and the logarithm. If $\mathfrak{B}$ has a unit element e and and if $f(\xi)$ satisfies (9.3.1) and converges to e as $\xi \to 0$, then the results of the previous section show that $f(\xi)$ can be written in the form $f(\xi) = \exp(\xi a)$. Conversely, if $x = \exp(a)$ then x can be *embedded in a continuous power group*, namely, $f(\xi) = \exp(\xi a)$. Thus a necessary and sufficient condition for a given element x to be embeddable in a continuous power group is that $x = \exp y$ have a solution y in $\mathfrak{B}$. A partial solution to this problem was obtained in section 5.4; there it was shown that $y = \log x$ is a solution when the principal component of $\rho(x)$ contains $\lambda = 0$. In particular, if $\| x - e \| < 1$, then the principal extension of $\log x$ is furnished by the series

$$\log x = -\sum_{n=1}^{\infty} \frac{1}{n} (e - x)^n, \tag{9.5.1}$$

used by J. von Neumann [2] in the case of matrix algebras.

In the present section we shall find conditions in order that the equation

$$\exp y = x \tag{9.5.2}$$

shall have a solution for a given x; any such solution may then be referred to as $\log x$. We assume that the algebra $\mathfrak{B}$ has a unit element e. The following theorem is due to M. Nagumo [1].

THEOREM 9.5.1. *Given $x \in \mathfrak{B}$. A necessary and sufficient condition that the equation $\exp y = x$ have a solution $y \in \mathfrak{B}$ is that x belong to a connected abelian group with unit element e. If x belongs to such a group, $\mathfrak{A}$ say, then there is a $y \in \mathfrak{B}(\mathfrak{A})$, the closed commutative subalgebra determined by $\mathfrak{A}$, which satisfies* (9.5.2).

PROOF. The condition is necessary for if there is a $y \in \mathfrak{B}$ with $\exp y = x$, then $x \in [\exp(\xi y); -\infty < \xi < \infty]$ which is a connected abelian group in $\mathfrak{B}$ and has e as its unit element. Suppose now that the condition is fulfilled so that $x \in \mathfrak{A}$. We have then $\mathfrak{A} \subset \mathfrak{G}_1$ since $\mathfrak{G}_1$ is the maximal connected group in $\mathfrak{B}$ with e as unit element. Let $\mathfrak{A}_0$ be the subset of $\mathfrak{A}$ containing all x of $\mathfrak{A}$ for which the equation $\exp y = x$ has a solution in $\mathfrak{B}(\mathfrak{A})$. $\mathfrak{A}_0$ is certainly not vacuous since all elements of $\mathfrak{A}$ such that $\| x - e \| < 1$ belong to $\mathfrak{A}_0$ by virtue of (9.5.1). In order to prove that $\mathfrak{A}_0 = \mathfrak{A}$, it is enough to prove that $\mathfrak{A}_0$ is both open and closed in the connected set $\mathfrak{A}$. Assume first that $a \in \mathfrak{A}_0$ and $a = \exp b$ where $b \in \mathfrak{B}(\mathfrak{A})$. If $x \in \mathfrak{A}$ and x is so near to a that $\|xa^{-1} - e\| < 1$, then $\log (xa^{-1}) = z$ exists and $z \in \mathfrak{B}(\mathfrak{A})$. Hence $xa^{-1} = \exp z$ and $x = \exp (b + z)$ since b and z commute. This shows that $\mathfrak{A}_0$ is open in $\mathfrak{A}$. To show that $\mathfrak{A}_0$ is also closed in $\mathfrak{A}$, we choose a sequence $\{x_n\} \subset \mathfrak{A}_0$ with $x_n \to x_0 \in \mathfrak{A}$. Suppose that $x_n = \exp y_n$ where $y_n \in \mathfrak{B}(\mathfrak{A})$. Since x_0^{-1} exists, Theorem 4.3.3 implies that $x_n^{-1} \to x_0^{-1}$. Thus if n is sufficiently large, $\| x_0 x_n^{-1} - e \| < 1$ and $x_0 x_n^{-1} = \exp z_n$ with $z_n \in \mathfrak{B}(\mathfrak{A})$. Since y_n and z_n commute, $x_0 = \exp (y_n + z_n)$ so that $x_0 \in \mathfrak{A}_0$ and $\mathfrak{A}_0$ is also closed in $\mathfrak{A}$. This completes the proof.

COROLLARY. *The equation* $\exp y = x$ *has a solution in the subalgebra* $\mathfrak{B}(e \cup x)$ *if and only if* x *belongs to the kernel of* $\mathfrak{B}(e \cup x)$.

PROOF. By Theorem 4.4.3, the kernel of $\mathfrak{B}(e \cup x)$ is an abelian group in the subalgebra and it is the maximal connected abelian group with unit element e. It follows that the condition is necessary as well as sufficient.

THEOREM 9.5.2. *In a commutative* (*B*)*-algebra the condition* $x \in \mathfrak{G}_1 = \mathfrak{K}(\mathfrak{B})$ *is necessary and sufficient for the equation* $\exp y = x$ *to have a solution in* $\mathfrak{B}$.

PROOF. In the commutative case the kernel of the algebra is an abelian group and it is the maximal connected group, whence the theorem follows.

Thus if $\mathfrak{B}$ is commutative, the function $w = \exp z$ maps all of $\mathfrak{B}$ upon all of $\mathfrak{G}_1$. Actually we have a sharper result:

THEOREM 9.5.3. *If* $\mathfrak{B}$ *is commutative,* $w = \exp z$ *maps* $\mathfrak{G}_1$ *onto itself.*

PROOF. If $w \in \mathfrak{G}_1$ there is a $z \in \mathfrak{B}$ with $w = \exp z$. But for all sufficiently large integers n, the element $z + n2\pi ie$ is regular and belongs moreover to $\mathfrak{G}_1$. Since $\exp (z + n2\pi ie) = \exp z = w$ the proof is complete.

REMARK. Our knowledge of the exponential function and of the logarithm is rather incomplete in the non-commutative case. In particular, the range $\mathfrak{R}$ of $\exp y$ is in doubt. It is clear that $\mathfrak{R} \subset \mathfrak{G}_1$ and that $\mathfrak{R}$ contains the sphere $\| x - e \| < 1$. Further $\mathfrak{R}$ is arcwise connected: The arc $x = \exp [\alpha y_1 + (1 - \alpha)y_2]$ joins the points $\exp y_1$ and $\exp y_2$. Finally if $x \in \mathfrak{R}$ so does x^{-1}. On the other hand, P. R. Halmos, G. Lumer and J. J. Schäffer [1] have shown that $\mathfrak{R}$ is a proper subset of $\mathfrak{G}_1$ when $\mathfrak{B} = \mathfrak{E}(\mathfrak{H})$, the algebra of endomorphisms of a Hilbert space $\mathfrak{H}$. It follows from this that $\mathfrak{R}$ is not in general a group, that is, $\mathfrak{R}$ is not closed under multiplication. For suppose $\mathfrak{R}$ were a group. Since e is an interior point of $\mathfrak{R}$, the group property forces $\mathfrak{R}$ to be open. Since $\mathfrak{R}$ is an open connected subgroup of $\mathfrak{G}_1$, Theorem 4.4.2 asserts that $\mathfrak{R} = \mathfrak{G}_1$. We are indebted to J. von Neumann for calling this fact to our attention in May, 1946.

THEOREM 9.5.4. *If* $\lambda = 0$ *belongs to the principal component of the resolvent set of* x, *then* $x \in \mathfrak{G}_1$ *and the equation* $\exp y = x$ *has a solution in* $\mathfrak{B}(e \cup x)$.

PROOF. In view of Theorem 4.4.3 and the corollary to Theorem 9.5.1, it is enough to show that x belongs to the kernel of $\mathfrak{B}(e \cup x)$ and for this we have merely to prove that x may be joined to some element of the kernel by means of an arc in $\mathfrak{B}(e \cup x)$, all points of which are regular elements of $\mathfrak{B}$. Now for $|\lambda_0| > \|x\|$, $x - \lambda_0 e$ is regular and belongs to the kernel of $\mathfrak{B}(e \cup x)$. By assumption $\lambda = 0$ belongs to the principal component of $\rho(x)$. Hence we may join the points x and $x - \lambda_0 e$ by regular elements of the form $x - \lambda e \in \mathfrak{B}(e \cup x)$. Thus x belongs to the kernel of $\mathfrak{B}(e \cup x)$ and the result follows.

The condition that $\lambda = 0$ belong to the principal component of $\rho(x)$ is sufficient but not necessary for the existence of $\log x$; counter-examples have been given by E. R. Lorch [**3**, p. 423].

As an application of these results we prove a theorem due to Lorch [**3**] (cf. section 4.4).

THEOREM 9.5.5. *If* $\mathfrak{B}$ *is commutative, then* $\mathfrak{G}$ *has either one or infinitely many components.*

PROOF. Suppose $\mathfrak{G}$ has more than one component and choose a to be a regular element of $\mathfrak{B}$ not in $\mathfrak{G}_1$. Consider all integral powers of a; it is claimed that each power belongs to a separate component and hence that the number of components is infinite. In fact, if a^{n_1} and a^{n_2}, $n_1 - n_2 = m > 0$, belong to the same component, then $a^m \in \mathfrak{G}_1$. We may therefore embed a^m in a continuous power group $[f(\xi); -\infty < \xi < \infty] \subset \mathfrak{G}_1$ where $f(1) = a^m$. Set $f(1/m) = b$ and $c = ab^{-1}$. Then $b^m = a^m$ and hence $c^m = e$. By Theorem 5.3.1 this implies that the spectrum of c consists of mth roots of unity and, thus, does not separate the plane. It follows from Theorem 9.5.4 that $c \in \mathfrak{G}_1$ and therefore $a = bc \in \mathfrak{G}_1$ against the assumption. This completes the proof.

The number of components of $\mathfrak{G}$ may very well be non-denumerable. This is the case for $\mathfrak{B} = A(-\infty, \infty)$, the purely atomic subalgebra of $S_0(\omega)$, $\omega(\xi) \equiv 0$ (cf. section 4.20). Here $e_\xi \in \mathfrak{G}$ for each $\xi \in (-\infty, \infty)$. If $e_\xi \in \mathfrak{G}_1$, then by Theorem 9.5.2 there would exist a $y \in \mathfrak{B}$ such that $e_\xi = \exp y$. Corresponding to each real number β there is a maximal ideal $\mathfrak{m}_\beta$ such that

$$e^{i\xi\beta} = e_\xi(\mathfrak{m}_\beta) = \exp [y(\mathfrak{m}_\beta)] = \exp [\gamma(\beta)]$$

where $\gamma(\beta) = \int_{-\infty}^{\infty} e^{i\xi\beta}\, d_\xi y$. Continuity of $\gamma(\beta)$ implies that $\gamma(\beta) = i\xi\beta + 2\pi ki$. Since $\gamma(\beta)$ is necessarily bounded in β, it follows that $\xi = 0$. Finally if e_ξ and e_η belong to the same component of $\mathfrak{G}$, then $e_\xi * e_{-\eta} = e_{\xi-\eta} \in \mathfrak{G}_1$, which, as we have just seen requires that $\xi = \eta$. Hence the elements e_ξ belong to distinct components for distinct values of ξ.

9.6. Semi-groups of operators. All of the foregoing material is applicable to semi-groups of operators on a (B)-space $\mathfrak{X}$. In this case $\mathfrak{B} = \mathfrak{E}(\mathfrak{X})$, the (B)-algebra of endomorphisms of $\mathfrak{X}$. For completeness we shall include an operator interpretation of some of these results.

THEOREM 9.6.1. *Let* $T(\xi)$, *defined on* $(0, \infty)$ *to* $\mathfrak{E}(\mathfrak{X})$, *be such that*

$$T(\xi_1)T(\xi_2) = T(\xi_1 + \xi_2), \qquad 0 < \xi_1, \xi_2 < \infty. \tag{9.6.1}$$

If

$$\lim_{\xi \to 0} \| T(\xi) - J \| = 0, \tag{9.6.2}$$

then J is a projection operator, $T(\xi) = T(\xi)J = JT(\xi)$, and $T(\xi)$ is continuous in the uniform operator topology for $\xi \geqq 0$ where $T(0) = J$ by definition. Further there exists an operator $A \in \mathfrak{E}(\mathfrak{X})$ such that $A = AJ = JA$ and

$$T(\xi) = J \exp (\xi A). \tag{9.6.3}$$

If $J = I$ the series defines a one-parameter group of operators on $(-\infty, \infty)$ which is continuous in the uniform operator topology.

THEOREM 9.6.2. *Let $T(\xi)$ be a solution of* (9.6.1) *which is uniformly measurable and (B)-integrable in every finite interval* $(0, \omega)$. *If*

$$\lim_{\xi \to 0} \left\| \frac{1}{\xi} \int_0^\xi T(\tau)\, d\tau - J \right\| = 0, \tag{9.6.4}$$

then (9.6.2) *also holds. Hence J is a projection operator and $T(\xi)$ is given by formula* (9.6.3).

The condition (9.6.4) will be called uniform Cesàro summability at $\xi = 0$. We shall show in Chapter XVIII that Theorem 9.6.2 remains valid under the weaker hypothesis of uniform Abel summability at $\xi = 0$.

THEOREM 9.6.3. *A necessary and sufficient condition that $T_0 \in \mathfrak{E}(\mathfrak{X})$ be embeddable in a one-parameter group of operators $T(\xi)$, continuous in the uniform operator topology, is that T_0 belong to some connected abelian group with unit I. There exists a solution $[T(\xi);\ -\infty < \xi < \infty]$ contained in $\mathfrak{B}(I \cup T_0)$ if and only if T_0 lies in the kernel of $\mathfrak{B}(I \cup T_0)$.*

2. PROBLEM B

9.7. Solutions on Banach spaces. We shall extend the results of the preceding paragraph to functions on a Banach space $\mathfrak{X}$ to a Banach algebra $\mathfrak{B}$. Here it is desired to solve the functional equation

$$F(x + y) = F(x)F(y) = F(y)F(x) \tag{9.7.1}$$

imposing as few *a priori* conditions on $F(x)$ as possible. It is clearly desirable that the domain of definition $\mathfrak{D}$ of $F(x)$ should have the property of containing $x + y$ whenever x and y are contained; in other words $\mathfrak{D}$ should be a semi-module in $\mathfrak{X}$. But in a (B)-space we have also scalar multiplication and in order to apply the results previously obtained we have to bring the scalars into play. It is convenient to assume that $\alpha x \in \mathfrak{D}$ for all $\alpha > 0$ whenever $x \in \mathfrak{D}$. This leads to

DEFINITION 9.7.1. *A set $\mathfrak{K}$ of $\mathfrak{X}$ is called an open (finitely open) positive cone if* (i) *$\mathfrak{K}$ is an open set (a finitely open set in the sense of Definition* 1.10.2) *and* (ii) *$x, y \in \mathfrak{K}$, $\alpha > 0$ implies $x + y$, $\alpha x \in \mathfrak{K}$.*

If $F(x)$ is defined on $\mathfrak{K}$ and satisfies (9.7.1) there, then for fixed $x \in \mathfrak{K}$ the function $F(\xi x)$ is defined for $\xi > 0$ and satisfies (9.3.1).

THEOREM 9.7.1. *Let $F(x)$ be defined on a finitely open positive cone $\mathfrak{K}$ and satisfy* (9.7.1) *for $x, y \in \mathfrak{K}$. Suppose that for every fixed $x \in \mathfrak{K}$, $F(\xi x)$ is a measurable function of ξ on $(0, \infty)$. If $\mathfrak{X}_{(n)}$ is any finite-dimensional linear subspace of $\mathfrak{X}$, then $F(x)$ is continuous on $\mathfrak{X}_{(n)} \cap \mathfrak{K}$. In particular, if $\mathfrak{K}$ is itself finite-dimensional, then $F(x)$ is continuous on $\mathfrak{K}$.*

PROOF. The following simple proof is due to C. E. Rickart; it replaces a more elaborate argument of Hille's. Suppose that $\mathfrak{X}_{(n)}$ is an n-dimensional linear subspace of $\mathfrak{X}$ and that $x_1, \cdots, x_n$ are n linearly independent vectors in $\mathfrak{X}_{(n)} \cap \mathfrak{K}$. Then the set $\mathfrak{K}_n$ consisting of all vectors of the form $x = \xi_1 x_1 + \cdots + \xi_n x_n$ with $\xi_1 > 0, \cdots, \xi_n > 0$ is an open subset of $\mathfrak{X}_{(n)} \cap \mathfrak{K}$. For a given $x^{(0)} \in \mathfrak{X}_{(n)} \cap \mathfrak{K}$ we can choose vectors $x_1, \cdots, x_n \subset \mathfrak{X}_{(n)} \cap \mathfrak{K}$ so that $x^{(0)} = \xi_1^{(0)} x_1 + \cdots + \xi_n^{(0)} x_n$ is an interior point of $\mathfrak{K}_n$, that is, $\xi_1^{(0)} > 0, \cdots, \xi_n^{(0)} > 0$. It remains to show that $F(x)$ is continuous in $\mathfrak{K}_n$ at $x^{(0)}$. By Theorem 9.3.1, $F(\xi_i x_i)$ is a continuous function of ξ_i, $\xi_i > 0$, for each i. Observe also that if $x^{(k)} = \xi_1^{(k)} x_1 + \cdots + \xi_n^{(k)} x_n$, then $\lim_{k\to\infty} x^{(k)} = x^{(0)}$ is equivalent to $\lim_{k\to\infty} \xi_i^{(k)} = \xi_i^{(0)}$, $i = 1, \cdots, n$ (cf. Theorem 1.10.1). Moreover $\prod_{i=1}^n y_i$ is a continuous function of $(y_1, \cdots, y_n)$ for $y_i \in \mathfrak{B}$. Since

$$F(\xi_1 x_1 + \cdots + \xi_n x_n) = \prod_{i=1}^{n} F(\xi_i x_i),$$

it follows that the left member is a continuous function of $(\xi_1, \cdots, \xi_n)$ for $\xi_i > 0$ and hence that $F(x)$ is continuous in $\mathfrak{K}_n$ at $x^{(0)}$. Every point in $\mathfrak{X}_{(n)} \cap \mathfrak{K}$ being interior to a suitably chosen $\mathfrak{K}_n$, we see that $F(x)$ is continuous everywhere in $\mathfrak{X}_{(n)} \cap \mathfrak{K}$ and the theorem is proved.

This theorem does not seem capable of much improvement. In particular, we cannot expect $F(x)$ to be continuous in $\mathfrak{K}$. Thus if $\mathfrak{B}$ is a commutative algebra with a unit element and $P(x)$ is a linear function on $\mathfrak{X}$ to $\mathfrak{B}$, then $\exp[P(x)]$ satisfies (9.7.1) for all x and $\exp[P(\xi x)] = \exp[\xi P(x)]$ is an entire function of ξ, but $\exp[P(x)]$ is not necessarily a continuous function of x since $P(x)$ does not have to be continuous. It is also clear that in general $\lim_{\xi\to 0} F(\xi x)$ need not exist for any $x \neq \theta$. We can get somewhat further, however, by assuming the existence of the latter limit.

THEOREM 9.7.2. *Let $F(x)$ be defined and satisfy* (9.7.1) *for $x \in \mathfrak{K}$ where $\mathfrak{K}$ is now an open positive cone. If*

(9.7.2) $$\lim_{\xi \to 0} F(\xi x)$$

exists for all x in $\mathfrak{K}$, then the limit is an idempotent j of $\mathfrak{B}$, independent of x, and there exists a function $P(x)$ such that $P(x) = jP(x) = P(x)j$ and

(9.7.3) $$F(x) = j + \sum_{n=1}^{\infty} \frac{1}{n!} [P(x)]^n.$$

Here $P(x)$ may be defined for all x in $\mathfrak{X}$ as an additive, real-homogeneous function which is continuous if and only if the limit in (9.7.2) exists uniformly with respect to x in some sphere. Further $P(x)$ and $P(y)$ commute for all $x, y \in \mathfrak{X}$.

Proof. By virtue of Theorem 9.4.2.

$$F(\xi x) = j(x) + \sum_{n=1}^{\infty} \frac{\xi^n}{n!} [P(x)]^n, \qquad x \in \mathfrak{K},$$

where $[j(x)]^2 = j(x)$ and $P(x) = j(x)P(x) = P(x)j(x)$. We have first to show that $j(x)$ is independent of x. From Theorem 9.7.1 and

$$F(\alpha x)F(\beta y) = F(\beta y)F(\alpha x) = F(\alpha x + \beta y), \qquad \alpha, \beta > 0,$$

we get by letting $\alpha \to 0$

$$j(x)F(\beta y) = F(\beta y)j(x) = F(\beta y)$$

whence by letting $\beta \to 0$

$$j(x)j(y) = j(y)j(x) = j(y).$$

Reversing the order of the limiting processes we get $j(x)j(y) = j(x)$ so that $j(x) = j(y) = j$.

It remains to determine the properties of $P(x)$. Using the identity $F(\xi(x + y)) = F(\xi x)F(\xi y)$, substituting the power series in ξ, and equating coefficients of the first power of ξ, we get $P(x + y) = P(x) + P(y)$. We have also $P(\xi x) = \xi P(x)$ for $\xi > 0$. If $\mathfrak{K}$ is not the whole space, $P(x)$ may be extended from $\mathfrak{K}$ to $\mathfrak{X}$ as an additive, real-homogeneous function. Since $\mathfrak{K}$ is supposed to be open we can find a closed sphere $\| x - x_0 \| \leqq \rho$ which belongs to $\mathfrak{K}$. If $y \in \mathfrak{X}$ we can express y uniquely as $y = \alpha(x - x_0)$ where $\| x - x_0 \| = \rho$ and $\alpha > 0$. We then define $P(y) = \alpha[P(x) - P(x_0)]$. It may be shown that $P(\alpha y_1 + \beta y_2) = \alpha P(y_1) + \beta P(y_2)$ for all y_1, y_2 and real α, β, that the new definition of $P(y)$ agrees with the old one in $\mathfrak{K}$, and finally that the extension is independent of the choice of x_0 and ρ. Finally for $x, y \in \mathfrak{K}$, if we equate coefficients of $\xi\eta$ in the expansion of $F(\xi x)F(\eta y) = F(\eta y)F(\xi x)$, we obtain $P(x)P(y) = P(y)P(x)$ and the commutativity clearly persists for the extended $P(x)$. It is easy to see that the series (9.7.3) satisfies the functional equation for all values of x and y.

Suppose now that $P(x)$ is continuous so that $\| P(x) \| \leqq M \| x \|$. We have then $\| F(\xi x) - j \| \leqq \exp [\xi M \| x \|] - 1$ for $\xi > 0$; consequently the limit in

(9.7.2) exists uniformly with respect to x in any finite sphere. Suppose conversely that the limit exists uniformly with respect to x in the sphere S: $\| x - x_0 \| < \rho$. We can then find a fixed β such that for every $x \in S$ we have $\| F(\tau x) - j \| \leqq \frac{1}{2}$, $0 < \tau \leqq \beta$, and consequently also

$$\left\| \beta^{-1} \int_0^\beta F(\tau x)\, d\tau - j \right\| \leqq \tfrac{1}{2}.$$

We can then apply formula (9.4.5) obtaining

$$P(x) = [F(\beta x) - j] \left\{ \int_0^\beta F(\tau x)\, d\tau \right\}^{-1},$$

the norm of which does not exceed $(\frac{1}{2}\beta)[\| j \| + 1]$ in S. $P(x)$ being additive, real-homogeneous and bounded in a sphere is consequently bounded and hence continuous. This completes the proof of the theorem.

If $\mathfrak{B}$ is a real (B)-space, then $P(x)$, being additive and real-homogeneous, is actually linear, but this is no longer necessarily the case for complex (B)-spaces. In such a space there are always solutions of (9.7.1) defined and continuous for all x which are nowhere analytic in x. This occurs even in the simplest case $\mathfrak{X} = \mathfrak{B} = Z_1$ where $F(\zeta) = \exp[\alpha\xi + \beta\eta)]$, $\zeta = \xi + i\eta$, α and β arbitrary complex numbers, defines a solution of $F(\zeta_1)F(\zeta_2) = F(\zeta_1 + \zeta_2)$ which is clearly continuous but not analytic in ζ unless $\beta = \alpha i$.

3. Problem C

9.8. Addition theorems. The functional equations studied in the preceding paragraphs are addition formulas. Classical analysis presents us with a large number of such formulas and it is natural to ask if other addition theorems than that of the exponential function may be subjected to abstract analysis. This general question has been attacked by N. Dunford and E. Hille; a brief account of the main results will be given in the remainder of this chapter.

In the present account we restrict ourselves to the case of a single function and an addition formula of the form

$$f(\xi_1 + \xi_2) = G[f(\xi_1), f(\xi_2)], \tag{9.8.1}$$

where $G(\alpha, \beta)$ is an analytic function of α and β such that $G(\alpha, \beta) = G(\beta, \alpha)$. In classical analysis $G(\alpha, \beta)$ is ordinarily supposed to be a rational or an algebraic function. We shall not make this assumption as it does not lead to any simplification of the, essentially local. problem which we are considering.

Classical function theory is concerned with the existence and properties of numerically-valued solutions of functional equations of the type represented by

(9.8.1). The problem still has a meaning for vector-valued functions; we have merely to interpret the right-hand side of (9.8.1) properly and section 5.2 shows how this should be done.

Let $\mathfrak{B}$ be a complex Banach algebra with unit element e. Let $G(\alpha, \beta)$ be a symmetric analytic function of α and β which is holomorphic if both variables are in a certain domain Δ of the complex plane. Let u and v be two commuting elements of $\mathfrak{B}$ whose spectra are located in Δ; more precisely, u and v shall belong to the domain $\mathfrak{G}(\Delta)$ of Theorem 5.2.5. We then define

$$G(u, v) = \frac{1}{(2\pi i)^2} \int_{\Gamma_v} \int_{\Gamma_u} G(\alpha, \beta) R(\alpha; u) R(\beta; v)\, d\alpha\, d\beta, \tag{9.8.2}$$

where Γ_u and Γ_v are oriented envelopes of $\sigma(u)$ and $\sigma(v)$ in Δ.

Suppose now that $f(\xi)$ is a function on real numbers to $\mathfrak{B}$ such that for all ξ under consideration $f(\xi_1)f(\xi_2) = f(\xi_2)f(\xi_1)$ and $f(\xi) \in \mathfrak{G}(\Delta_0)$ where Δ_0 is an arbitrary bounded domain such that $\bar{\Delta}_0 \subset \Delta$. Then by definition

$$G[f(\xi_1), f(\xi_2)] = \frac{1}{(2\pi i)^2} \int_{\Gamma_1} \int_{\Gamma_2} G(\alpha, \beta) R[\alpha; f(\xi_1)] R[\beta; f(\xi_2)]\, d\alpha\, d\beta, \tag{9.8.3}$$

where Γ_1 and Γ_2 are fixed envelopes of $\bar{\Delta}_0$ in Δ.

We shall extend Theorem 9.3.1 to the solutions of Problem C. To this end it is convenient to assume that $G(u, v)$ is uniformly continuous on the range of $f(\xi)$. The most natural assumption is a Lipschitz condition of order one. Such a condition is automatically satisfied, for instance, if $G(\alpha, \beta)$ is an entire function and $f(\xi)$ is bounded, or if $f(\xi)$ is restricted to a closed set interior to the domain $\mathfrak{D}_\Delta$ of Theorem 5.2.1, or, finally, if the values of $f(\xi)$ belong to a compact subset of $\mathfrak{G}(\Delta)$.

THEOREM 9.8.1. *Let $f(\xi)$ be a measurable function of ξ on the open interval* $(0, \omega)$ *to $\mathfrak{B}$ satisfying* (9.8.1) *for all ξ_1 and ξ_2 in the interval and such that $f(\xi_1)$ and $f(\xi_2)$ commute. Let $\mathfrak{R}_\epsilon$ denote the range of $f(\xi)$ for $0 < \epsilon < \xi < \omega$ where $\mathfrak{R}_\epsilon$ is bounded and $\mathfrak{R}_\epsilon \subset \mathfrak{G}(\Delta_\epsilon)$, $\bar{\Delta}_\epsilon \subset \Delta$. Suppose that for every ϵ, $0 < \epsilon < \omega$, there is a finite positive M_ϵ such that for u_1, u_2, v in $\mathfrak{R}_\epsilon$*

$$\| G(u_1, v) - G(u_2, v) \| \leqq M_\epsilon \| u_1 - u_2 \|. \tag{9.8.4}$$

Then $f(\xi)$ is continuous in $(0, \omega)$.

PROOF. Let $0 < \alpha \leqq \eta \leqq \beta < \xi < \omega$ and note that

$$f(\xi) = G[f(\xi - \eta), f(\eta)],$$

whence

$$(\beta - \alpha) f(\xi) = \int_\alpha^\beta G[f(\xi - \eta), f(\eta)]\, d\eta$$

and

$$(\beta - \alpha) \| f(\xi + \epsilon) - f(\xi) \| \leqq \int_\alpha^\beta \| G[f(\xi + \epsilon - \eta), f(\eta)] - G[f(\xi - \eta), f(\eta)] \| \, d\eta.$$

If $\delta = \min (\alpha, \xi - \beta, \xi + \epsilon - \beta)$, then the right member is dominated by

$$M_\delta \int_\alpha^\beta \| f(\xi + \epsilon - \eta) - f(\xi - \eta) \| \, d\eta = M_\delta \int_{\xi-\beta}^{\xi-\alpha} \| f(\tau + \epsilon) - f(\tau) \| \, d\tau,$$

which tends to zero with ϵ since $f(\xi)$ is bounded and measurable. Hence $f(\xi)$ is actually continuous in $(0, \omega)$.

In passing let us note that the requirement that $f(\xi_1)$ and $f(\xi_2)$ shall commute is frequently implied by the functional equation and need not be assumed explicitly.

The counter example of section 9.3 shows that we cannot expect $f(\xi)$ to approach a limit when $\xi \to 0$ or to have any stronger properties of continuity in $(0, \omega)$. The situation is entirely different if $f(\xi)$ satisfies (9.8.1) in an interval containing $\xi = 0$ or is supposed to tend to a finite limit when $\xi \to 0$. We shall consider the latter case in some detail. If $\lim_{\epsilon \to 0} f(\epsilon)$ exists and equals an element a of $\mathfrak{G}(\Delta)$, then

$$\lim_{\epsilon \to 0+} f(\xi + \epsilon) = \lim_{\epsilon \to 0+} G[f(\xi), f(\epsilon)] = G[f(\xi), a]$$

exists for $0 < \xi < \omega$. From the fact that $f(\xi)$ has a right-hand limit everywhere, it follows that $f(\xi)$ is right-hand continuous (even continuous) except possibly in a countable set. Since $f(\xi)$ in particular is measurable, it has to be continuous everywhere in $(0, \omega)$.

9.9. Holomorphic solutions. Just as in the case of Problem A, continuity at the origin implies analyticity. We set

$$G_1(\alpha, \beta) = \frac{\partial}{\partial \alpha} G(\alpha, \beta),$$

$$Q(\alpha, \beta, \gamma) = \frac{G(\alpha, \gamma) - G(\beta, \gamma)}{\alpha - \beta}, \qquad \alpha, \beta, \gamma \in \Delta.$$

The latter is evidently a holomorphic function of α, β, γ in Δ. We can then define $Q(u, v, w)$ for u, v, w in $\mathfrak{G}(\Delta)$ by the obvious triple resolvent integral, assuming u, v, w to commute, and

$$Q(u, v, w)\,(u - v) = G(u, w) - G(v, w). \tag{9.9.1}$$

LEMMA 9.9.1. *If $f(\xi) \in \mathfrak{G}(\Delta)$, if $f(\xi)$ is continuous for $0 \leqq \xi \leqq \omega$, and if $0 \leqq \eta, \zeta \leqq \omega$, then uniformly with respect to ξ*

$$\lim_{\xi \to \eta} Q[f(\xi), f(\eta), f(\zeta)] = G_1[f(\eta), f(\zeta)].$$

PROOF. First note that since $f(\xi)$ is continuous in $[0, \omega]$, the range of $f(\xi)$ is a closed connected compact set $\mathfrak{R} \subset \mathfrak{S}(\Delta)$. If $\Phi = \bigcup \sigma(x)$, $x \in \mathfrak{R}$, then Φ is a closed subset of Δ and if Γ is an oriented envelope of Φ in Δ, then $R[\alpha; f(\xi)]$ is a continuous function of (α, ξ) for $\alpha \in \Gamma$, $0 \leqq \xi \leqq \omega$. There is consequently a finite positive $M = M(\Gamma)$ such that

$$\| R[\alpha; f(\xi)] \| \leqq M(\Gamma), \qquad \alpha \in \Gamma, 0 \leqq \xi \leqq \omega. \tag{9.9.2}$$

Further

$$\lim_{\xi \to \eta} R[\alpha; f(\xi)] = R[\alpha; f(\eta)]$$

uniformly with respect to α on Γ. It follows that uniformly in ζ, $0 \leqq \zeta \leqq \omega$

$$\begin{aligned}
&\lim_{\xi \to \eta} Q[f(\xi), f(\eta), f(\zeta)] \\
&\quad = -\frac{1}{8\pi^3 i} \int_\Gamma \int_\Gamma \int_\Gamma Q(\alpha, \beta, \gamma) R[\alpha; f(\eta)] R[\beta; f(\eta)] R[\gamma; f(\zeta)] \, d\alpha \, d\beta \, d\gamma \\
&\quad = \frac{1}{8\pi^3 i} \int_\Gamma \int_\Gamma \int_\Gamma Q(\alpha, \beta, \gamma) \frac{R[\alpha; f(\eta)] - R[\beta; f(\eta)]}{\alpha - \beta} R[\gamma; f(\zeta)] \, d\alpha \, d\beta \, d\gamma \\
&\quad = \frac{1}{8\pi^3 i} \int_{\Gamma_1} \int_{\Gamma_1} d\beta \, d\gamma R[\gamma; f(\zeta)] \left\{ \int_\Gamma \frac{Q(\alpha, \beta, \gamma) R[\alpha; f(\eta)]}{\alpha - \beta} d\alpha \right. \\
&\qquad \left. - R[\beta; f(\eta)] \int_\Gamma \frac{Q(\alpha, \beta, \gamma)}{\alpha - \beta} d\alpha \right\}.
\end{aligned}$$

Here Γ and Γ_1 are oriented envelopes of Φ in Δ and Γ_1 is interior to Γ. Now

$$\int_\Gamma \frac{Q(\alpha, \beta, \gamma)}{\alpha - \beta} d\alpha = 2\pi i Q(\beta, \beta, \gamma) = 2\pi i G_1(\beta, \gamma),$$

so

$$\begin{aligned}
\lim_{\xi \to \eta} Q[f(\xi), f(\eta), f(\zeta)] &= -\frac{1}{4\pi^2} \int_{\Gamma_1} \int_{\Gamma_1} G_1(\beta, \gamma) R[\beta; f(\eta)] R[\gamma; f(\zeta)] \, d\beta \, d\gamma + U \\
&= G_1[f(\eta), f(\zeta)] + U.
\end{aligned}$$

Here

$$\begin{aligned}
U &= \frac{1}{8\pi^3 i} \int_{\Gamma_1} \int_{\Gamma_1} d\beta \, d\gamma R[\gamma; f(\zeta)] \int_\Gamma \frac{Q(\alpha, \beta, \gamma)}{\alpha - \beta} R[\alpha; f(\eta)] \, d\alpha \\
&= \frac{1}{8\pi^3 i} \int_{\Gamma_1} d\gamma R[\gamma; f(\zeta)] \int_\Gamma d\alpha R[\alpha; f(\eta)] \int_{\Gamma_1} \frac{Q(\alpha, \beta, \gamma)}{\alpha - \beta} d\beta.
\end{aligned}$$

But for $\alpha \in \Gamma$ the last integral is zero so $U = \theta$. This completes the proof of the lemma.

We shall now prove that continuity of a solution at $\xi = 0$ implies the existence of derivatives of all orders. We give the proof under the added assumption that $G_1[f(0), f(0)]$ has an inverse. This assumption has the effect of cutting out solutions involving other idempotents than e. It would be sufficient, however, to assume the existence of the inverse in the subalgebra determined by $f(\xi)$. It is tacitly assumed in the following that this algebra is commutative.

THEOREM 9.9.1. *If $f(\xi)$, having values in $\mathfrak{S}(\Delta)$, is a continuous solution of* (9.8.1) *for $0 \leqq \xi \leqq \omega$ and if $G_1[f(0), f(0)]$ has an inverse in $\mathfrak{B}$, then $f(\xi)$ has derivatives of all orders and*

$$f'(\xi) = G_1[f(0), f(\xi)]f'(0). \tag{9.9.3}$$

If $g(\xi)$ is any solution of (9.8.1) *which is continuous in* $[0, \omega]$ *and commutes with $f(\xi)$ and if $g(0) = f(0)$, $g'(0) = f'(0)$, then $g(\xi) \equiv f(\xi)$.*

PROOF. From the contour integral definition of $G_1[f(0), f(\zeta)]$ we see that it is continuous in ζ and hence

$$\lim_{\alpha\to 0} \frac{1}{\alpha} \int_0^\alpha G_1[f(0), f(\zeta)]\, d\zeta = G_1[f(0), f(0)].$$

Thus we may fix $\alpha < \omega$ so the integral on the left has an inverse in $\mathfrak{B}$. From Lemma 9.9.1 we have

$$\lim_{\xi\to 0} \frac{1}{\alpha} \int_0^\alpha Q[f(\xi), f(0), f(\zeta)]\, d\zeta = \frac{1}{\alpha} \int_0^\alpha G_1[f(0), f(\zeta)]\, d\zeta.$$

Hence

$$\lim_{\xi\to 0} \left\{\frac{1}{\alpha} \int_0^\alpha Q[f(\xi), f(0), f(\zeta)]\, d\zeta\right\}^{-1} = \left\{\frac{1}{\alpha} \int_0^\alpha G_1[f(0), f(\zeta)]\, d\zeta\right\}^{-1}. \tag{9.9.4}$$

By formula (9.9.1) we have

$$[f(\xi) - f(\eta)]\, Q[f(\xi), f(\eta), f(\zeta)] = G[f(\xi), f(\zeta)] - G[f(\eta), f(\zeta)] \tag{9.9.5}$$

whence

$$\begin{aligned} &\frac{1}{\xi} [f(\xi) - f(0)] \frac{1}{\alpha} \int_0^\alpha Q[f(\xi), f(0), f(\zeta)]\, d\zeta \\ &\qquad = \frac{1}{\alpha\xi} \int_0^\alpha \{G[f(\xi), f(\zeta)] - G[f(0), f(\zeta)]\}\, d\zeta \\ &\qquad = \frac{1}{\alpha\xi} \int_0^\alpha [f(\xi + \zeta) - f(\zeta)]\, d\zeta \\ &\qquad = \frac{1}{\alpha\xi} \left\{\int_\alpha^{\alpha+\xi} f(\tau)\, d\tau - \int_0^\xi f(\tau)\, d\tau\right\} \to \frac{1}{\alpha} [f(\alpha) - f(0)] \end{aligned} \tag{9.9.6}$$

as $\xi \to 0$. Thus (9.9.4) and (9.9.6) give the existence of

$$\lim_{\xi \to 0} \frac{1}{\xi} [f(\xi) - f(0)] = \left\{ \frac{1}{\alpha} \int_0^\alpha G_1[f(0), f(\zeta)] \, d\zeta \right\}^{-1} \frac{1}{\alpha} [f(\alpha) - f(0)]$$

so $f(\xi)$ is differentiable at $\xi = 0$. Applying the lemma once more we have

$$\frac{1}{\eta} [f(\xi + \eta) - f(\xi)] = \frac{1}{\eta} \{G[f(\eta), f(\xi)] - G[f(0), f(\xi)]\}$$

$$= \frac{1}{\eta} [f(\eta) - f(0)] Q[f(\eta), f(0), f(\xi)] \to f'(0) G_1[f(0), f(\xi)]$$

uniformly for $0 \leqq \xi \leqq \omega$. This establishes formula (9.9.3). That the two factors in the derivative commute follows from the fact that the factors on the left in formula (9.9.5) commute.

Once the existence of the first derivative has been established, that of the higher derivatives follows by easy steps. It is enough to indicate the argument for the second derivative. We have

$$f'(\xi) = f'(0) \frac{1}{(2\pi i)^2} \int_\Gamma \int_\Gamma G_1(\alpha, \beta) R[\alpha; f(0)] R[\beta; f(\xi)] \, d\alpha \, d\beta. \tag{9.9.7}$$

Now when $\eta \to 0$

$$\frac{1}{\eta} \{R[\beta; f(\xi + \eta)] - R[\beta; f(\xi)]\} \to f'(\xi) \{R[\beta; f(\xi)]\}^2,$$

where we have used formula (4.8.3) and the fact that $f'(\xi)$ commutes with $R[\gamma; f(\zeta)]$ which is obvious from (9.9.7). Here the limit exists uniformly with respect to β on Γ. Hence

$$f''(\xi) = f'(0) f'(\xi) \frac{1}{(2\pi i)^2} \int_\Gamma \int_\Gamma G_1(\alpha, \beta) R[\alpha; f(0)] \{R[\beta; f(\xi)]\}^2 \, d\alpha \, d\beta.$$

Since

$$\{R[\beta; f(\xi)]\}^2 = - \frac{\partial}{\partial \beta} R[\beta; f(\xi)],$$

an integration by parts shows that this expression equals

$$f'(0) f'(\xi) G_{1,1}[f(0), f(\xi)],$$

where

$$G_{1,1}(\alpha, \beta) = \frac{\partial^2}{\partial \alpha \partial \beta} G(\alpha, \beta)$$

and $G_{1,1}(u, v)$ is defined by the usual contour integral for $u, v \in \mathfrak{G}(\Delta)$. We see in particular that

$$f''(0) = [f'(0)]^2 G_{1,1}[f(0), f(0)]$$

and hence is uniquely determined by $f(0)$ and $f'(0)$. Similarly it is shown that all higher derivatives exist and are uniquely determined by $f(0)$ and $f'(0)$.

There is consequently a formal power series in ξ associated with $f(\xi)$ determined by the initial values $f(0)$ and $f'(0)$. We shall not attempt to prove the convergence of this series by direct estimates. In the most important case considered below the convergence follows from other considerations.

9.10. Uniqueness of the solution. The uniqueness assertion of Theorem 9.9.1 still remains to be proved. Here we shall use the Lipschitz condition of Theorem 9.8.1 which is easily verified in the present situation.

Suppose that $g(\xi)$ is continuous in $[0, \omega]$, has values in $\mathfrak{G}(\Delta)$ and commutes with $f(\eta)$. From the integral representation of the function $Q[f(\xi), g(\xi), f(\eta)]$ combined with (9.9.2), we see that there is a finite $K = K(f, g)$ such that for $0 \leqq \xi, \eta \leqq \omega$

$$\| Q[f(\xi), g(\xi), f(\eta)] \| \leqq K, \qquad \| Q[g(\eta), f(\eta), g(\xi)] \| \leqq K.$$

Since

$$G[f(\xi), f(\eta)] - G[g(\xi), f(\eta)] = Q[f(\xi), g(\xi), f(\eta)] [f(\xi) - g(\xi)],$$

we have

$$\| G[f(\xi), f(\eta)] - G[g(\xi), f(\eta)] \| \leqq K \| f(\xi) - g(\xi) \| \tag{9.10.1}$$

and a similar inequality in which f, g, and ξ, η are interchanged.

We suppose now that $f(\xi)$ and $g(\xi)$ are continuous solutions of (9.8.1) and that $f(0) = g(0), f'(0) = g'(0)$. We know then that they have derivatives of all orders and that $f^{(n)}(0) = g^{(n)}(0)$ for all n. Placing $h(\xi) = f(\xi) - g(\xi)$ we have

$$\begin{aligned} h(\xi + \eta) &= f(\xi + \eta) - g(\xi + \eta) \\ &= G[f(\xi), f(\eta)] - G[g(\xi), g(\eta)] \\ &= G[f(\xi), f(\eta)] - G[g(\xi), f(\eta)] + G[g(\xi), f(\eta)] - G[g(\xi), g(\eta)], \end{aligned}$$

so from (9.10.1)

$$\begin{aligned} \| h(\xi + \eta) \| &\leqq K \{\| f(\xi) - g(\xi) \| + \| f(\eta) - g(\eta) \|\} \\ &= K \{\| h(\xi) \| + \| h(\eta) \|\} \end{aligned}$$

whence

$$\| h(2\xi) \| \leqq 2K \| h(\xi) \|.$$

By repeated use of this inequality we get

$$\| h(\xi) \| \leqq (2K)^m \| h(\xi 2^{-m}) \| , \qquad m = 1, 2, 3, \cdots . \tag{9.10.2}$$

Now consider

$$g_n(\xi) = g(\xi) - \sum_{\nu=0}^{n} g^{(\nu)}(0) \frac{\xi^\nu}{\nu!}$$

and let

$$M_n[g] = \max \| g_n^{(n)}(\xi) \| = \max \| g^{(n)}(\xi) - g^{(n)}(0) \|$$

in the interval $[0, \omega]$. Since

$$g_n(\xi) = \int_0^\xi \cdots \int_0^{\xi_{n-2}} \int_0^{\xi_{n-1}} g_n^{(n)}(\xi_n)\, d\xi_n\, d\xi_{n-1} \cdots d\xi_1$$

we have

$$\| g_n(\xi) \| \leqq \frac{\xi^n}{n!} M_n[g], \qquad 0 \leqq \xi \leqq \omega.$$

If $M_n[g]$ is replaced by $M_n[f]$, the same inequality is satisfied by

$$f_n(\xi) = f(\xi) - \sum_{\nu=0}^{n} f^{(\nu)}(0) \frac{\xi^\nu}{\nu!}.$$

In the following M_n denotes the larger of the two quantities $M_n[f]$ and $M_n[g]$. It should be noted that the polynomial in ξ is the same in $f_n(\xi)$ as in $g_n(\xi)$.

From

$$\| h(\xi) \| = \| f(\xi) - g(\xi) \| = \| f_n(\xi) - g_n(\xi) \| \leqq \| f_n(\xi) \| + \| g_n(\xi) \|,$$

we see that

$$\| h(\xi) \| \leqq 2 \frac{\xi^n}{n!} M_n. \tag{9.10.3}$$

Combination of this estimate with (9.10.2) gives

$$\| h(\xi) \| \leqq \frac{2}{n!} M_n \omega^n (2^{1-n} K)^m, \qquad 0 \leqq \xi \leqq \omega, \qquad m, n = 1, 2, 3, \cdots. \tag{9.10.4}$$

Here we fix n so large that $2^{1-n}K < 1$. Since m is independent of n and may be taken arbitrarily large, we see that $h(\xi) \equiv 0$ and so $g(\xi) \equiv f(\xi)$, $0 \leqq \xi \leqq \omega$. This completes the proof of Theorem 9.9.1.

Suppose now that $\varphi(\zeta)$ is an analytic scalar function such that (i) $\varphi(\zeta)$ is not a constant, (ii) $\varphi(\zeta) = \sum_0^\infty \alpha_n \zeta^n$ converges for $| \zeta | < \rho$, (iii) $\varphi(\zeta) \in \Delta$, and (iv)

$$\varphi(\zeta_1 + \zeta_2) = G[\varphi(\zeta_1), \varphi(\zeta_2)] \tag{9.10.5}$$

for $| \zeta_1 |, | \zeta_2 |, | \zeta_1 + \zeta_2 | < \rho$. Differentiating $\varphi(\zeta) = G[\varphi(\zeta), \varphi(0)]$ we get $\varphi'(\zeta) = \varphi'(\zeta) G_1[\varphi(\zeta), \varphi(0)]$; by (i) this implies that

$$G_1[\varphi(\zeta), \varphi(0)] \equiv 1$$

and, in particular,

$$G_1[\varphi(0), \varphi(0)] = 1. \tag{9.10.6}$$

Differentiating (9.10.5) by parts with respect to ζ_1 and putting $\zeta_1 = 0$ in the result, one gets [cf. formula (9.9.3)]

$$\varphi'(\zeta) = \varphi'(0)G_1[\varphi(0), \varphi(\zeta)]$$

whence $\varphi'(0) \neq 0$. But if $\varphi(\zeta)$ satisfies the conditions stated above, so does $\varphi(\alpha\zeta)$ for any value of α provided ρ be replaced by $\rho/|\alpha|$. We can consequently normalize $\varphi(\zeta)$ by assuming (v) that $\varphi'(0) = 1$.

Suppose now that $f(\zeta)$ is a continuous, and hence differentiable, solution of (9.8.1) such that $f(0) = \varphi(0)e, f'(0) = a$. Then

$$G_1[f(0), f(0)] = G_1[\varphi(0), \varphi(0)]e = e$$

and the existence of an inverse is trivial. On the other hand, the function $\varphi(a\zeta)$ is defined by the series

$$\varphi(a\zeta) = \sum_0^\infty \alpha_n a^n \zeta^n$$

at least for $|\zeta| < \rho/\|a\|$ and for such values of ζ we have also $\sigma[\varphi(a\zeta)] = \varphi[\zeta\sigma(a)] \subset \Delta$ since $|\sigma(a)| \leqq \|a\|$. From the construction of the series and the properties mentioned, it follows that it satisfies (9.8.1) for

$$|\zeta_1|, |\zeta_2|, |\zeta_1 + \zeta_2| < \rho/\|a\|.$$

Further $\varphi(0) = \varphi(0)e$, $\varphi'(0) = \alpha_1 a = a$. As a corollary of Theorem 9.9.1 we then obtain

THEOREM 9.10.1. *Let $\varphi(\zeta)$ satisfy conditions* (i) – (v) *listed above and let $f(\xi)$ be a continuous, and hence differentiable, solution of* (9.8.1) *such that*

$$f(0) = \varphi(0)e, \quad f'(0) = a.$$

Then

$$f(\xi) \equiv \varphi(a\xi)$$

in the common interval of definition of the two functions. Thus $f(\xi)$ admits of an extension to the complex ζ-plane, satisfying (9.8.1), *and this extension, if defined by $f(\zeta) = \varphi(a\zeta)$, is holomorphic in some neighborhood of $\zeta = 0$.*

We sum up the main results of our study of Problem C in

THEOREM 9.10.2. *If the addition formula*

$$f(\zeta_1 + \zeta_2) = G[f(\zeta_1), f(\zeta_2)]$$

is known to have an analytic scalar solution $\varphi(\zeta)$, holomorphic in some neighborhood of $\zeta = 0$, if $\varphi(\zeta)$ is not a constant and has values in a domain Δ where $G(\alpha, \beta)$ is holomorphic, then the formula has solutions having values in an arbitrary prescribed complex Banach algebra with unit element e. If $f(\xi)$ is such a solution defined and continuous for $0 \leqq \xi \leqq \omega$, if $f(\xi) \in \mathfrak{G}(\Delta)$ and $f(0) = \varphi(0)e$, then $f(\xi)$ is not merely *differentiable but actually analytic and in some neighborhood of* $\zeta = 0$ *we have* $f(\zeta) = \varphi(a\zeta)$ *where* $a = f'(0)$.

For the validity of our argument and even of the conclusions it is quite essential that $\varphi(0)$ belongs to the domain of holomorphism of $G(\alpha, \beta)$. This excludes from consideration some important addition theorems, for instance, those of $\cos \zeta$ and $\cos \operatorname{am} \zeta$, where $\varphi(0)$ belongs to a singular manifold of $G(\alpha, \beta)$. In these cases it is preferable to replace the given addition theorem by a pair of addition theorems for the function and its first derivative. A discussion of systems of addition theorems has been given by J. R. Lee (Yale dissertation, 1950).

4. Problem D

9.11. Power series solutions. We shall restrict ourselves to a fairly simple special case of Problem D which may be reduced to Problem C. We shall also require some elements of the theory of (F)- and (G)-power series to be found in Chapter XXVI.

Let $\mathfrak{X}$ be a complex Banach space, $\mathfrak{B}$ a complex Banach algebra with unit element e, and let $G(\alpha, \beta)$ be a symmetric analytic function which is holomorphic if α and β are in a domain Δ of the complex plane. We suppose that the functional equation (9.10.5) has a scalar solution $\varphi(\zeta)$ satisfying conditions (i) to (v) of section 9.10. We take $\rho = 1$.

Theorem 9.11.1. *Let* $F(x)$ *be a function on* $\mathfrak{X}$ *to* $\mathfrak{B}$ *defined for* $\| x \| < 1$ *and having values in* $\mathfrak{G}(\Delta)$. *It is supposed that* $F(\xi x)$ *is a continuous function of* ξ *for* $0 \leqq \xi < 1$ *if* x *is fixed,* $\| x \| < 1$. *Further*

$$\lim_{\xi \to 0} F(\xi x) = \varphi(0)e \tag{9.11.1}$$

for every fixed x *and*

$$F(x + y) = G[F(x), F(y)] \tag{9.11.2}$$

if $\| x \|, \| y \|, \| x + y \| < 1$. *Then there exists a linear function* $P(x)$ *on* $\mathfrak{X}$ *to* $\mathfrak{B}$ *such that*

$$F(x) = \varphi[P(x)] \equiv \sum_{0}^{\infty} \alpha_n [P(x)]^n \tag{9.11.3}$$

for $\| x \| < 1$. *The series is an (F)- or a (G)-power series according as* $P(x)$ *is bounded or not. In the former case* $F(x)$ *is analytic in* $\| x \| < 1$. *A necessary and sufficient condition that* $P(x)$ *be bounded is that the limit in* (9.11.1) *exists uniformly with respect to* x *in* $\| x \| < \delta$.

PROOF. For fixed x the function $F(\xi x) = f(\xi)$ satisfies the conditions of Theorem 9.10.1. There is consequently an $a = P(x)$ such that $F(\zeta x) = \varphi[\zeta P(x)]$ at least for $|\zeta| < 1/\|P(x)\|$. It remains merely to discuss the properties of $P(x)$. We shall prove that $F(x)$ is (G)-differentiable and

$$(9.11.4) \qquad \delta F(x; h) = P(h)G_1[F(\theta), F(x)].$$

Since $P(h)$ is the derivative of $\varphi[\zeta P(h)]$ with respect to ζ at $\zeta = 0$ and we have $G_1[F(\theta), F(\theta)] = e$, the formula is true for $x = \theta$. From Lemma 9.9.1 we get

$$\begin{aligned}\frac{1}{\zeta}[F(x + \zeta h) - F(x)] &= \frac{1}{\zeta}\{G[F(\zeta h), F(x)] - G[F(\theta), F(x)]\}\\ &= \frac{1}{\zeta}[F(\zeta h) - F(\theta)]Q[F(\zeta h), F(\theta), F(x)]\\ &\rightarrow P(h)G_1[F(\theta), F(x)]\end{aligned}$$

when $\zeta \rightarrow 0$. This proves that $F(x)$ is (G)-differentiable for $\|x\| < 1$.

By Theorem 26.3.2, $\delta F(x; h)$ and in particular $\delta F(\theta; h) = P(h)$ are linear. From $F(\zeta x) = \varphi[\zeta P(x)]$ we get, differentiating with respect to ζ and placing $\zeta = 0$ in the result,

$$\delta^n F(\theta; h) = n!\, \alpha_n[P(h)]^n$$

which is clearly a homogeneous polynomial in h of degree n. It follows that (9.11.3) is the MacLaurin series of $F(x)$ which by Theorem 3.16.2 converges in the c-star about θ in the domain of (G)-differentiability. This implies that the series converges and represents $F(x)$ in the unit sphere. If $P(x)$ is bounded, then the series is a convergent (F)-power series for $\|x\| < 1$ and hence, by Theorem 25.6.4, $F(x)$ is analytic in $\|x\| < 1$. As a consequence the limit in (9.11.1) holds uniformly for $\|x\| < \delta$. On the other hand, if this limit exists uniformly with respect to x in $\|x\| < \delta$, then $F(x)$ is continuous at $x = \theta$. From

$$F(x + h) = G[F(x), F(h)]$$

and the continuity of $G(u, v)$ with respect to u and v in $\mathfrak{S}(\Delta)$, it follows that $F(x)$ is continuous and therefore analytic everywhere in $\|x\| < 1$. Theorem 3.17.1 now asserts that $\delta F(\theta; h) = P(h)$ is bounded.

CHAPTER X

SEMI-GROUPS IN THE STRONG TOPOLOGY

10.1. Orientation. In the present chapter we deal with the "strong case" analogues of Problems A and B of the preceding chapter.

The first problem is that of a *one-parameter semi-group* $\mathfrak{S} = [T(\xi); \xi > 0]$ of linear bounded transformations on a complex (B)-space $\mathfrak{X}$ to itself with the property that

$$T(\xi_1 + \xi_2)x = T(\xi_1)[T(\xi_2)x]$$

for all $\xi_1, \xi_2 > 0$ and all $x \in \mathfrak{X}$. We assume throughout the remainder of this treatise that $T(\xi)$ is strongly continuous for $\xi > 0$. Actually this assumption is not as restrictive as it appears at first sight since strong measurability alone forces $T(\xi)$ to be strongly continuous for $\xi > 0$.

The *infinitesimal operator* A_0 of $\mathfrak{S}$ is defined as the limit in norm as $\eta \to 0+$ of

$$A_\eta x = \frac{1}{\eta}[T(\eta) - I]x$$

wherever it exists. In general A_0 is an unbounded linear operator; however the domain of A_0 is dense in the union of the range spaces of $[T(\alpha); \alpha > 0]$. The operator A_0 is in general not closed; its least closure A, when it exists, will be called the *infinitesimal generator* of $\mathfrak{S}$. The latter operator plays the basic role in the theory. In case $T(\eta)x \to x$ as $\eta \to 0+$ for all x, it is possible to represent $T(\xi)x$ by means of an "exponential formula"

$$T(\xi)x = \lim_{\eta \to 0+} \exp(\xi A_\eta)\, x,$$

valid for all $x \in \mathfrak{X}$ and all $\xi > 0$; this replaces the representation $T(\xi) = \exp(\xi A)$ of the "uniform case." Several other exponential formulas will be proved in Chapter XI which contains a detailed discussion of the properties of A, its resolvent $R(\lambda; A)$, and the relations between $T(\xi)$ and $R(\lambda; A)$ obtained by means of the Laplace transform.

The second problem considered is that of a semi-group $\mathfrak{S} = [T(x); x \in \mathfrak{R}]$ of linear bounded transformations on a complex (B)-space $\mathfrak{Y}$ to itself, the parameter manifold being an open positive cone $\mathfrak{R}$ in a (B)-space $\mathfrak{X}$. The elements of $\mathfrak{S}$ satisfy the condition

$$T(x_1 + x_2)y = T(x_1)[T(x_2)y]$$

for all $x_1, x_2 \in \mathfrak{R}$ and all $y \in \mathfrak{Y}$. In the special case in which $\mathfrak{X}$ is the real euclidean n-dimensional space, we have what is conventionally known as an n-parameter semi-group.

Strong measurability along each "ray" of $\mathfrak{R}$ implies strong continuity on rays and this may be extended to strong continuity on $\mathfrak{R}$ relative to the finite topology of $\mathfrak{X}$. Instead of a single infinitesimal operator, we now have a family of generators $[A_o(x); x \in \mathfrak{R}]$ which form a semi-module $\mathfrak{A}$ with positive multipliers. $\mathfrak{A}$ corresponds to the Lie ring in the classical theory of continuous groups. In particular, we are able to determine the structure of all strongly measurable commutative n-parameter semi-groups.

The literature on one-parameter transformation groups goes back to M. H. Stone [1, 2] who in 1930 found a representation theorem for unitary groups in Hilbert space; general groups in a (B)-space were treated in 1939 by I. Gelfand [3] and in 1940 by M. Fukamiya [1]. The theory of one-parameter transformation semi-groups appears to have had its origins in a 1936 paper by E. Hille [4] which dealt with some special semi-groups. In 1938 B. de Sz.-Nagy [2] and E. Hille [5] independently discovered the representation theorem for semi-groups of bounded linear self-adjoint operators in a Hilbert space. The subsequent developments of this subject have been extensive and we shall refer to them as we proceed.

Groups and semi-groups of linear transformations have also been studied from the point of view of partially ordered sets. In this connection we note the work of G. Birkhoff and L. Alaoglu on ergodic theorems in general semi-groups. More recently B. Vulich [1, 2] has investigated linear multiplicative operations on partially ordered vector spaces satisfying the axioms of Kantorovich. These studies seem to be rather remote from our line of approach so we shall not go beyond the mere mentioning of their existence.

There are two paragraphs in the present chapter: *One-Parameter Semi-Groups* and *Extensions*. References are listed at the end of each paragraph.

References. Alaoglu and Birkhoff [1], G. Birkhoff [4], Fukamiya [1], Gelfand [3], Hille [4, 5], Stone [1, 2], de Sz.-Nagy [2], and Vulich [1, 2].

1. One-Parameter Semi-Groups

10.2. Measurability and continuity. Let $\mathfrak{X}$ be a complex (B)-space, $\mathfrak{E}(\mathfrak{X})$ the complex (B)-algebra of bounded linear transformations on $\mathfrak{X}$ to itself, and let $T(\xi)$ be a function on positive numbers to $\mathfrak{E}(\mathfrak{X})$ such that

$$T(\xi_1 + \xi_2)x = T(\xi_1)[T(\xi_2)x], \qquad 0 < \xi_1, \xi_2 < \infty, \tag{10.2.1}$$

for all $x \in \mathfrak{X}$. Thus $\mathfrak{S} \equiv [T(\xi)]$ is a semi-group of operators in $\mathfrak{E}(\mathfrak{X})$.

There are as usual several kinds of continuity which may be considered in connection with the semi-group $\mathfrak{S}$. It turns out that uniform continuity is distinct from and implies strong continuity, whereas strong and weak continuity are equivalent. On the other hand, measurability implies continuity in both the uniform and the strong operator topology but not in the weak operator topology.

THEOREM 10.2.1. *If $T(\xi)$ is uniformly measurable or, equivalently, if $T(\xi)$ is weakly measurable and almost separably-valued in $\mathfrak{E}(\mathfrak{X})$, then $T(\xi)$ is continuous in the uniform operator topology.*

PROOF. This result follows directly from Theorem 3.5.5 and 9.3.1.

The following sufficient condition for uniform continuity is due to P. Lax (personal communication).

THEOREM 10.2.2. *If $T(\xi)$ is strongly continuous for $\xi > 0$ and if $T(\xi_0)$ is a compact linear operator for some $\xi_0 > 0$, then so is $T(\xi)$ for each $\xi > \xi_0$ and $T(\xi)$ is continuous in the uniform operator topology for $\xi > \xi_0$.*

PROOF. By assumption $T(\xi)$ is strongly continuous for $\xi > 0$. Hence given $\delta > 0$, the uniform boundedness theorem asserts that $\| T(\xi) \| \leqq M_\delta$ for $\delta \leqq \xi \leqq 1/\delta$. Further, since $T(\xi_0)$ is compact, we see that $E \equiv [T(\xi_0)x; \| x \| \leqq 1]$ is a totally bounded subset of $\mathfrak{X}$. Consequently for a given $\epsilon > 0$ there exist vectors $x_1, x_2, \cdots, x_n$ such that the ϵ-spheres $S[T(\xi_0)x_k ; \epsilon]$, $k = 1, 2, \cdots, n$, cover E. Next we choose $\Delta > 0$ so that $\| T(\xi + \eta)x_k - T(\xi)x_k \| < \epsilon$ for $k = 1, 2, \cdots, n$, $\xi_0 + \delta \leqq \xi \leqq \xi_0 + 1/\delta$, and $| \eta | < \Delta$. Now corresponding to each x, $\| x \| \leqq 1$, there is an x_k with the property that

$$\| T(\xi)x - T(\xi)x_k \| \leqq \| T(\xi - \xi_0) \| \, \| T(\xi_0)x - T(\xi_0)x_k \| \leqq M_\delta \epsilon$$

whenever $\xi_0 + \delta \leqq \xi \leqq \xi_0 + 1/\delta$. Combining these inequalities in the usual manner, we have

$$\| T(\xi + \eta)x - T(\xi)x \| < (2M_\delta + 1)\epsilon$$

for $\xi_0 + 2\delta \leqq \xi \leqq \xi_0 + 1/\delta - \delta$ and $| \eta | < \min(\Delta, \delta)$; since x was an arbitrary element of the unit sphere in $\mathfrak{X}$ this means that $\| T(\xi + \eta) - T(\xi) \| \leqq (2M_\delta + 1)\epsilon$ for ξ and η limited as above. This proves uniform continuity for $\xi > \xi_0$ and the compactness of $T(\xi)$ for $\xi > \xi_0$ follows from the fact that $T(\xi) = T(\xi - \xi_0)T(\xi_0)$.

We now prove the strong case analogue of Theorem 10.2.1. The first result in this direction goes back to N. Dunford [4] who showed that strong measurability plus boundedness of $\| T(\xi) \|$ on $(0, \beta)$ implies right-hand strong continuity. E. Hille [13] later showed that strong measurability plus boundedness in intervals of the type $[\delta, 1/\delta]$ implies strong continuity for $\xi > 0$. Finally R. S. Phillips [4] showed that the boundedness assumption was superfluous, being already implied by strong measurability.

LEMMA 10.2.1. *If $T(\xi)$ is strongly measurable, then $\| T(\xi) \|$ is bounded in each interval $[\alpha, \beta]$, $0 < \alpha < \beta < \infty$.*

PROOF. The following proof is due to I. Miyadera [1]. Making use of the uniform boundedness theorem, we see that it suffices to prove that $\| T(\xi)x \|$ is bounded in $[\alpha, \beta]$ for each $x \in \mathfrak{X}$. Suppose that this is not true for some x. Then there will exist a $\gamma \in [\alpha, \beta]$ and a sequence $\{\xi_n\} \subset [\alpha, \beta]$ such that $\xi_n \to \gamma$ and $\| T(\xi_n)x \| \geqq n$ for all n. On the other hand, $\| T(\xi)x \|$ being measurable, there will exist a constant M and a measurable set $F \subset [0, \gamma]$ with $m(F) > \gamma/2$ such that $\sup_{\xi \in F} \| T(\xi)x \| \leqq M$. Set $E_n \equiv [(\xi_n - \eta); \eta \in F \cap [0, \xi_n]]$. Then E_n is measurable and for n sufficiently large $m(E_n) \geqq \gamma/2$. Now for $\eta \in F \cap [0, \xi_n]$ we have

$$n \leqq \| T(\xi_n)x \| \leqq \| T(\xi_n - \eta) \| \, \| T(\eta)x \| \leqq \| T(\xi_n - \eta) \| M$$

and therefore $\| T(\xi) \| \geqq n/M$ for all $\xi \in E_n$. Hence denoting $\lim\sup_n E_n$ by E we see that $\| T(\xi) \| = \infty$ for all $\xi \in E$ and $m(E) \geqq \gamma/2$. However this contradicts the fact that $\| T(\xi) \|$ is finite-valued for all $\xi \in (0, \infty)$ and thus implies the statement of the lemma.

We note that weak measurability does not imply the boundedness of $\| T(\xi) \|$ in any subinterval of $(0, \infty)$ (see R. S. Phillips [4]). Nevertheless W. Feller [4] has obtained a result analogous to the above lemma assuming even less than weak measurability. He requires merely that the numerically-valued functions $x^*[T(\xi)x]$ be measurable for each $x \in \mathfrak{X}$ and each x^* in some given subspace contained in $\mathfrak{X}^*$ and invariant under the adjoint operators $[T^*(\xi)]$. By introducing an "essential norm" $\| T(\xi) \|_e$ which is defined by means of the essential upper limit of $| x^*[T(\xi + \eta)x] |$ as $\eta \to 0+$ (instead of $| x^*[T(\xi)x] |$), Feller is able to show that $\| T(\xi) \|_e$ is bounded in each interval $[\alpha, \beta]$, $0 < \alpha < \beta < \infty$. We shall not avail ourselves of this generality; instead we shall limit our considerations to strongly measurable semi-groups.

THEOREM 10.2.3. *If $T(\xi)$ is strongly measurable then $T(\xi)$ is strongly continuous for $\xi > 0$.*

PROOF. We proceed as in the proof of Theorem 9.3.1. We first choose four numbers α, β, ξ, τ such that $0 < \alpha < \tau < \beta < \xi$ and an η so small that $\beta < \xi - \eta$. Now

$$T(\xi)x = T(\tau)[T(\xi - \tau)x];$$

the right side, being independent of τ, is certainly integrable with respect to τ so that

$$(\beta - \alpha)[T(\xi \pm \eta) - T(\xi)]x = \int_\alpha^\beta T(\tau)\{[T(\xi \pm \eta - \tau) - T(\xi - \tau)]x\} \, d\tau.$$

By Lemma 10.2.1 there exists an M such that $\| T(\tau) \| \leqq M$ for $\alpha \leqq \tau \leqq \beta$. Thus the norm of the integrand does not exceed

$$M \, \| [T(\xi \pm \eta - \tau) - T(\xi - \tau)]x \|$$

which is a measurable function of τ on $[\alpha, \beta]$ and, according to the lemma, uniformly bounded for η sufficiently small. Hence

$$(\beta - \alpha) \, \| [T(\xi \pm \eta) - T(\xi)]x \| \leqq M \int_{\xi-\beta}^{\xi-\alpha} \| [T(\sigma \pm \eta) - T(\sigma)]x \| \, d\sigma.$$

By Theorem 3.8.3 the right member tends to zero with η. It follows that $T(\xi)x$ is continuous for $\xi > 0$ and the theorem is proved.

We recall that weak one-sided continuity implies strong measurability for any vector-valued function. We thus obtain, as an immediate consequence of the above theorem, the

COROLLARY. *Weak one-sided continuity of* $T(\xi)$ *on* $(0, \infty)$ *implies strong continuity.*

That strong continuity of $T(\xi)$ does not imply uniform continuity is shown by the example in section 19.2. On the other hand the strong continuity of $T(\xi)$ does imply that $\| T(\xi)x \|$ is continuous in $(0, \infty)$ for each $x \in \mathfrak{X}$. From this one infers without difficulty that $\| T(\xi) \|$ *is lower semi-continuous* and *a fortiori* measurable. Nevertheless $\| T(\xi) \|$ need not be continuous even in the case of a group of operators (cf. R. S. Phillips [4]). Finally we note that $\omega(\xi) \equiv \log \| T(\xi) \|$ is a measurable subadditive function different from $+\infty$ in $(0, \infty)$. As a consequence Theorem 7.6.1 applies and we have

$$\omega_0 \equiv \inf_{\xi>0} \frac{1}{\xi} \log \| T(\xi) \| = \lim_{\xi\to\infty} \frac{1}{\xi} \log \| T(\xi) \| . \tag{10.2.2}$$

We shall say that $T(\xi)$ is of *type* ω_0.

The type ω_0 of a semi-group plays an important role in connection with the Laplace transform of $T(\xi)x$ when this exists. In this case the abscissa of absolute convergence is given by

$$\sigma_a(x) = \limsup_{\xi\to\infty} \frac{1}{\xi} \log \| T(\xi)x \| . \tag{10.2.3}$$

Since we wish the transform to exist for all $x \in \mathfrak{X}$, the pertinent abscissa here is given by $\sup [\sigma_a(x); x \in \mathfrak{X}]$. In this connection we have

THEOREM 10.2.4. $\omega_0 = \sup [\sigma_a(x); x \in \mathfrak{X}]$.

PROOF. Set $\sigma \equiv \sup [\sigma_a(x); x \in \mathfrak{X}]$. It is clear that $\sigma_a(x) \leqq \omega_0$ for $\| x \| \leqq 1$ and since $\sigma_a(\alpha x) = \sigma_a(x)$ for $\alpha \neq 0$ it follows that $\sigma \leqq \omega_0$. On the other hand for each $\epsilon > 0$ and $x \in \mathfrak{X}$ there exists a positive constant $M(\epsilon, x)$ such that

$$\| T(\xi)x \| \exp [-(\sigma + \epsilon)\xi] \leqq M(\epsilon, x)$$

for all $\xi \geqq 1$. Hence by the uniform boundedness theorem there exists an $M(\epsilon) > 0$ such that $\| T(\xi) \| \exp [-(\sigma + \epsilon)\xi] \leqq M(\epsilon)$ for all $\xi \geqq 1$. It follows that $\omega_0 \leqq \sigma + \epsilon$ for all $\epsilon > 0$ and this proves the converse inequality.

10.3. The infinitesimal operator. We now proceed to a further study of the structure of the semi-group $\mathfrak{S} = [T(\xi)]$, assuming merely that $T(\xi)$ is strongly continuous for $\xi > 0$. The notion of the *infinitesimal operator* A_0 of $\mathfrak{S}$ is central in such a study. We define

(10.3.1) $$A_\eta = \frac{1}{\eta}\,[T(\eta) - I], \qquad \eta > 0,$$

(10.3.2) $$A_o x = \lim_{\eta \to 0+} A_\eta x$$

whenever the limit exists. The set of elements x for which $\lim_{\eta \to 0+} A_\eta x$ exists is the domain of A_o, denoted by $\mathfrak{D}(A_o)$. It is clear that $\mathfrak{D}(A_o)$ is a linear subspace of $\mathfrak{X}$ and that A_o is a linear operator.

We start by showing that $\mathfrak{D}(A_o)$ never reduces to the zero element. To this end we set

(10.3.3) $$x_{\alpha,\beta} = \int_\alpha^\beta T(\tau)y\,d\tau, \qquad y \in \mathfrak{X},\ 0 < \alpha < \beta < \infty.$$

Then

$$\begin{aligned} A_\eta x_{\alpha,\beta} &= \frac{1}{\eta}\int_\alpha^\beta [T(\eta) - I]T(\tau)y\,d\tau \\ &= \frac{1}{\eta}\int_\alpha^\beta [T(\tau+\eta) - T(\tau)]y\,d\tau \\ &= \frac{1}{\eta}\int_\beta^{\beta+\eta} T(\sigma)y\,d\sigma - \frac{1}{\eta}\int_\alpha^{\alpha+\eta} T(\sigma)y\,d\sigma \\ &\to [T(\beta) - T(\alpha)]y \end{aligned}$$

as $\eta \to 0+$. Hence every element of the type $x_{\alpha,\beta}$ belongs to $\mathfrak{D}(A_o)$. This observation is due to N. Dunford (unpublished).

Let $\mathfrak{X}_\alpha = T(\alpha)[\mathfrak{X}]$ be the range of the transformation $T(\alpha)$, $\alpha > 0$. We have clearly

(10.3.4) $$\mathfrak{X}_\alpha \supset \mathfrak{X}_\beta \qquad \text{if } \alpha < \beta.$$

We define

(10.3.5) $$\mathfrak{X}_0 = \bigcup\,[\mathfrak{X}_\alpha\,;\ \alpha > 0].$$

Thus $\mathfrak{X}_0$ *is the least linear subspace containing the range-spaces of* $\mathfrak{S}$.

THEOREM 10.3.1. *If $T(\xi)$ is strongly continuous for $\xi > 0$, then $\mathfrak{D}(A_o)$ is dense in $\mathfrak{X}_0$, the two sets have the same closure, and the range of A_o is contained in $\bar{\mathfrak{X}}_0$.*

PROOF, If $x \in \mathfrak{X}_0$ there exists an $\alpha > 0$ and a $y \in \mathfrak{X}$ such that $x = T(\alpha)y$. The element $x_{\alpha,\beta}(= T(\alpha/2)x_{\alpha/2,\beta-\alpha/2})$ defined by formula (10.3.3) belongs to $\mathfrak{D}(A_o) \cap \mathfrak{X}_0$ and $\lim_{\beta\to\alpha} [1/(\beta - \alpha)]x_{\alpha,\beta} = x$; that is, every point of $\mathfrak{X}_0$ lies in the closure of $\mathfrak{D}(A_o) \cap \mathfrak{X}_0$. Conversely, if $x \in \mathfrak{D}(A_o)$, then $\lim_{\eta\to 0+} T(\eta)x = x$ so that $x \in \bar{\mathfrak{X}}_0$. It follows that the closures of $\mathfrak{X}_0$ and $\mathfrak{D}(A_o)$ coincide. Finally, if $x \in \mathfrak{D}(A_o)$, then $A_\eta x \in \bar{\mathfrak{X}}_0$ and so does $A_o x$.

The next result shows that $\mathfrak{D}(A_o)$ is rather sparse in the non-uniform case. Strong continuity is not postulated.

THEOREM 10.3.2. *If $\mathfrak{D}(A_o)$ is of the second category in $\mathfrak{X}$, then*

$$\lim_{\eta\to 0+} \| T(\eta) - I \| = 0,$$

A_o is a bounded linear operator, and $T(\xi) = \exp(\xi A_o)$.

PROOF. By assumption, $\lim_{\eta\to 0+} A_\eta x$ exists on a set of the second category. It therefore follows from the uniform boundedness theorem that $\| A_\eta \| \leqq M$ for some $M > 0$. Consequently $\| T(\eta) - I \| \leqq M\eta \to 0$ as $\eta \to 0+$. By Theorem 9.6.1 there exists an operator $A_o \in \mathfrak{E}(\mathfrak{X})$ such that $T(\xi) = \exp(\xi A_o)$ and it is clear from this expression that A_o is the infinitesimal operator of $\mathfrak{S}$.

THEOREM 10.3.3. *If $T(\xi)$ is strongly continuous for $\xi > 0$, then for $x \in \mathfrak{D}(A_o)$*

$$\frac{d}{d\xi} T(\xi)x = A_oT(\xi)x = T(\xi)A_ox, \qquad \xi > 0. \tag{10.3.6}$$

PROOF. We have

$$\frac{1}{\eta}[T(\xi + \eta) - T(\xi)]x = A_\eta T(\xi)x = T(\xi)A_\eta x.$$

Since the last member tends to $T(\xi)A_ox$ as $\eta \to 0+$, we see that $T(\xi)x \in \mathfrak{D}(A_o)$, that $A_oT(\xi)x = T(\xi)A_ox$, and that the right-sided derivative of $T(\xi)x$ exists and satisfies (10.3.6). It is also easy to see that the left-sided derivative exists. For we have

$$-\frac{1}{\eta}[T(\xi - \eta) - T(\xi)]x = T(\xi - \eta)A_\eta x$$

and this tends to $T(\xi)A_ox$ as $\eta \to 0+$; here we have made use of the fact that $T(\xi)$ is strongly continuous at $\xi > 0$.

COROLLARY. *If $x \in \mathfrak{D}(A_o)$ and if $\int_0^1 \| T(\tau)A_ox \| d\tau < \infty$, then*

$$T(\xi)x - x = \int_0^\xi T(\tau)A_ox\, d\tau. \tag{10.3.7}$$

PROOF. Since the derivative of $T(\xi)x$ is strongly continuous for $\xi > 0$, we have

$$T(\xi)x - T(\alpha)x = \int_\alpha^\xi T(\tau)A_ox\, d\tau, \qquad 0 < \alpha < \xi.$$

For $x \in \mathfrak{D}(A_0)$ it is clear that $\lim_{\eta\to 0+} \| A_\eta x \| < \infty$ and hence that $\lim_{\eta\to 0+} T(\eta)x = x$. Passing to the limit as $\alpha \to 0+$ in the above equation we obtain (10.3.7).

We next consider the higher powers of A_o.

THEOREM 10.3.4. *If $T(\xi)$ is strongly continuous for $\xi > 0$, then $\bigcap_n \mathfrak{D}(A_o^n)$ is dense in $\mathfrak{X}_0$.*

PROOF. We use a construction due to I. Gelfand [3] in the case of one-parameter groups. Let $\mathfrak{F}$ be the class of all numerical functions $F(\tau)$ defined on $(-\infty, \infty)$ with the following properties:

(i) $F(\tau) = 0$ for all τ outside of a compact subset of $(0, \infty)$;

(ii) $F(\tau)$ has continuous derivatives of all orders in $(-\infty, \infty)$.

If $F(\tau) \in \mathfrak{F}$, then it is clear that $F^{(n)}(\tau) \in \mathfrak{F}$ for all $n = 0, 1, 2, \cdots$ and that $F(\tau)T(\tau)x \in B(E_1; \mathfrak{X})$. Thus the set $\mathfrak{X}(\mathfrak{F})$ of all elements of the form

$$y = \int_0^\infty F(\tau)T(\tau)x\, d\tau, \qquad x \in \mathfrak{X},$$

is well defined. Further $F(\tau + \delta) \in \mathfrak{F}$ for $\delta > 0$ sufficiently small. Consequently

$$y = \int_0^\infty F(\tau + \delta)T(\tau + \delta)x\, d\tau = T(\delta) \int_0^\infty F(\tau + \delta)T(\tau)x\, d\tau$$

and hence $\mathfrak{X}(\mathfrak{F}) \subset \mathfrak{X}_0$. Now

$$\begin{aligned} A_\eta y &= \frac{1}{\eta} \int_0^\infty F(\tau)[T(\tau + \eta) - T(\tau)]x\, d\tau \\ &= \int_0^\infty \frac{1}{\eta} [F(\tau - \eta) - F(\tau)]T(\tau)x\, d\tau. \end{aligned}$$

For η sufficiently small it is clear that the integrand of the right member vanishes outside of a compact subset E of $(0, \infty)$ and that $\eta^{-1}[F(\tau - \eta) - F(\tau)]T(\tau)x \to -F'(\tau)T(\tau)x$ as $\eta \to 0+$ uniformly with respect to τ in E. Consequently

$$\lim_{\eta \to 0+} A_\eta y \equiv A_o y = -\int_0^\infty F'(\tau)T(\tau)x\, d\tau$$

and $y \in \mathfrak{D}(A_o)$. Repeating the argument we see that $A_o^n y$ exists for each integer $n > 0$ and that

$$A_o^n y = (-1)^n \int_0^\infty F^{(n)}(\tau)T(\tau)x\, d\tau.$$

It follows that $\mathfrak{X}(\mathfrak{F}) \subset \bigcap_n \mathfrak{D}(A_o^n)$. Moreover $\mathfrak{X}(\mathfrak{F})$ is a linear set.

We next show that $\mathfrak{X}(\mathfrak{F})$ is dense in $\mathfrak{X}_0$. If this were not the case, then by Theorem 2.7.5 there would exist a bounded linear functional x_0^* such that $x_0^*[\mathfrak{X}(\mathfrak{F})] = 0$ but x_0^* does not vanish identically on $\mathfrak{X}_0$. This implies that

$$\int_0^\infty F(\tau)x_0^*[T(\tau)x]\, d\tau = x_0^* \left\{ \int_0^\infty F(\tau)T(\tau)x\, d\tau \right\} = 0$$

for each $F(\tau) \in \mathfrak{F}$ and $x \in \mathfrak{X}$. Now $x_0^*[T(\tau)x]$ is continuous in τ for $\tau > 0$. Suppose that $x_0^*[T(\tau)x] \neq 0$ for some $\tau > 0$ and $x \in \mathfrak{X}$. It is clear that we could then choose an $F(\tau) \in \mathfrak{F}$ so that $\int_0^\infty F(\tau)x_0^*[T(\tau)x]\, d\tau \neq 0$. Since this is impossible we see that $x_0^*(\mathfrak{X}_0) = 0$, contrary to the way in which x_0^* was chosen. This contradiction shows that $\mathfrak{X}(\mathfrak{F})$ is dense in $\mathfrak{X}_0$ and therefore proves the theorem.

Remark. If $y \in \mathfrak{X}(\mathfrak{F})$, then $T(\xi)y$ has derivative of all orders with respect to ξ. However this does not imply that $T(\xi)y$ is analytic in ξ. I. Gelfand [3] has shown in the case of a group that the set of elements y for which $T(\xi)y$ is analytic is dense in $\bar{\mathfrak{X}}_0$ $(=\mathfrak{X})$; we do not know if this is true for strongly continuous semi-groups.

It is clear from Theorem 10.3.3 that $T(\xi)$ maps $\mathfrak{D}(A_o)$ into itself. Thus for a group of operators, $T(\xi)x$ will be differentiable if and only if $x \in \mathfrak{D}(A_o)$. However for a semi-group of operators, it may very well happen that $T(\xi)x$ is differentiable for an x not in $\mathfrak{D}(A_o)$. The next theorem treats the case in which $T(\xi)x$ is differentiable for all $x \in \mathfrak{X}$, ξ being fixed and greater than zero (cf. E. Hille [16] and R. S. Phillips [6]).

THEOREM 10.3.5. *If $T(\xi)$ is strongly continuous for $\xi > 0$ and if $T(\xi_0)[\mathfrak{X}] \subset \mathfrak{D}(A_o)$ for some $\xi_0 > 0$, then $T^{(1)}(\xi) = A_oT(\xi)$ exists as a bounded linear operator for $\xi \geqq \xi_0$ and $T(\xi)$ is continuous in the uniform operator topology for $\xi > \xi_0$. Moreover, $T^{(n)}(\xi) = A_o^nT(\xi)$ also exists as a bounded operator for $\xi \geqq n\xi_0$, and $T(\xi)$ is n times continuously differentiable in the uniform operator topology for $\xi > (n + 1)\xi_0$, $n = 1, 2, \cdots$.*

PROOF. If $T(\xi_0)$ maps $\mathfrak{X}$ into $\mathfrak{D}(A_o)$ then by (10.3.4) the same will be true of each $T(\xi)$ for $\xi \geqq \xi_0$. Thus for $\xi \geqq \xi_0$ we see that $A_oT(\xi)x = \lim_{\eta\to0+} A_\eta T(\xi)x$ exists for all $x \in \mathfrak{X}$ and hence by the uniform boundedness theorem $A_oT(\xi)$ is a bounded linear operator. As a consequence if $\xi > \xi_0 + 2\delta$, $\delta > 0$, and $|\eta| < \delta$, then

$$T(\xi + \eta)x - T(\xi)x = \int_\xi^{\xi+\eta} A_oT(\tau)x \, d\tau$$

for all $x \in \mathfrak{X}$. It follows from the strong continuity of $T(\xi)$ in $(0, \infty)$ together with the uniform boundedness theorem that $\| A_oT(\tau) \| \leqq \| T(\tau - \xi_0) \| \; \| A_oT(\xi_0) \|$ is bounded for $\tau \in [\xi, \xi + \eta]$. Consequently $\| T(\xi + \eta) - T(\xi) \| = O(|\eta|)$ as $|\eta| \to 0$.

As for the latter part of the theorem, if $\xi \geqq 2\xi_0$ then for arbitrary $x \in \mathfrak{X}$, $T^{(1)}(\xi)x = A_oT(\xi)x = T(\xi/2)A_oT(\xi/2)x$ also belongs to $\mathfrak{D}(A_o)$ and

$$T^{(2)}(\xi)x = \lim_{\eta\to0+} A_\eta\{T(\xi/2)[A_oT(\xi/2)x]\} = [A_oT(\xi/2)]^2x = A_o^2T(\xi)x.$$

An inductive argument shows that

$$T^{(n)}(\xi)x = [A_0T(\xi/n)]^nx = A_o^nT(\xi)x \tag{10.3.8}$$

for all $x \in \mathfrak{X}$, $\xi \geqq n\xi_0$, $n = 1, 2, \cdots$; the middle member shows that $T^{(n)}(\xi)$ is a bounded operator. Finally, $T(\xi)$ being continuous in the uniform operator topology for $\xi > \xi_0$, the same is true of $T^{(n)}(\xi) = T(\xi - n\xi_0)[A_o^nT(n\xi_0)]$ for $\xi > (n + 1)\xi_0$. Hence

$$T^{(n-1)}(\xi) = T^{(n-1)}(\sigma) + \int_\sigma^\xi T^{(n)}(\tau) \, d\tau, \quad \delta > 0, \sigma = (n + 1)\xi_0 + \delta,$$

is differentiable in the uniform operator topology for $\xi > (n+1)\xi_0 + \delta$, and, $\delta > 0$ being arbitrary, this proves the last assertion of the theorem.

COROLLARY. *If $T(\xi)[\mathfrak{X}] \subset \mathfrak{D}(A_o)$ for each $\xi > 0$, then $T(\xi)$ is n times continuously differentiable in the uniform operator topology for all $\xi > 0$ and all integers $n > 0$.*

If the infinitesimal operator A_o is bounded, then it is clear that $T^{(1)}(\xi)$ exists for $\xi > 0$ and $\| T^{(1)}(\xi) \|$ remains bounded near the origin. E. Hille [16] has obtained the following rather striking converse to this statement.

THEOREM 10.3.6. *If $T(\xi)$ is strongly continuous for $\xi > 0$, if $T(\xi)[\mathfrak{X}] \subset \mathfrak{D}(A_o)$ for each $\xi > 0$, and if*

$$\limsup_{\eta \to 0+} \eta \, \| T^{(1)}(\eta) \| < \frac{1}{e}, \tag{10.3.9}$$

then A_o is bounded and there exists a projection operator J such that $A_o = JA_o = A_oJ$ and $T(\xi) = J \exp(\xi A_o)$.

PROOF. By the above corollary, $T(\xi)$ has derivatives of all orders so that we may use Taylor's theorem. We then obtain

$$T(\xi) = \sum_{k=0}^{n-1} \frac{(\xi - \alpha)^k}{k!} A_o^k T(\alpha) + \frac{1}{(n-1)!} \int_\alpha^\xi (\xi - \eta)^{n-1} A_o^n T(\eta) \, d\eta. \tag{10.3.10}$$

Suppose that $\limsup_{\eta \to 0+} e\eta \, \| T^{(1)}(\eta) \| < \rho < 1$. Applying (10.3.8) we see that the remainder term is majorized by

$$\left(\frac{n}{e}\right)^n \frac{\rho^n}{(n-1)!} \int_\alpha^\xi \frac{(\xi - \eta)^{n-1}}{\eta^n} \, d\eta$$

and for $0 < \alpha \leqq \xi < \alpha(1+\rho)/\rho$ this goes to zero as $n \to \infty$. It follows that the Taylor expansion converges for these values of ξ and represents $T(\xi)$. But the power series itself converges for $| \xi - \alpha | < \alpha/\rho$ and by Theorem 3.11.4 it defines a holomorphic function in this circle. Since the circle contains the origin, it follows in particular that $\lim_{\eta \to 0+} T(\eta)$ exists in the uniform operator topology. Theorem 9.6.1 therefore applies and gives the desired conclusion.

COROLLARY. *If $T^{(1)}(\xi)$ exists as a bounded operator for $\xi > 0$ and if the infinitesimal operator A_o is unbounded, then*

$$\limsup_{\eta \to 0+} \eta \, \| T^{(1)}(\eta) \| \geqq e^{-1}. \tag{10.3.11}$$

This inequality is sharp for if $\mathfrak{X} = l_2$, $T(\xi)\{x_n\} = \{e^{-n\xi} x_n\}$, then the sign of equality holds in (10.3.11).

10.4. The first exponential formula.

We now introduce the operator

$$\exp[(\xi - \alpha)A_\eta] \, T(\alpha) = \sum_{n=0}^{\infty} \frac{1}{n!} (\xi - \alpha)^n A_\eta^n T(\alpha), \tag{10.4.1}$$

which is clearly an operator in $\mathfrak{E}(\mathfrak{X})$ since A_η is bounded. Here $\alpha \geqq 0$ is fixed, ξ is arbitrary, $\xi \geqq \alpha$, and η is positive. In case $\alpha = 0$, it is assumed that $\lim_{\eta \to 0+} T(\eta)$ exists in the strong operator topology and $T(0)$ is taken to be the limit operator. The main question before us is what happens to the operator (10.4.1) as $\eta \to 0+$. That it has a limit under very general assumptions will be shown below.

The proof of the general limit theorem seems to require a fairly sophisticated argument and several such proofs have been given, two of which are due to E. Hille (see section 9.3 of [**13**], the earliest proof was sketched in [**9**]). For the paricular but important case in which $[T(\xi)]$ is a semi-group of contraction operators, strongly continuous at the origin and hence for all $\xi > 0$ by Theorem 10.5.5 below, N. Dunford has given a very simple and elegant argument which leads very quickly to the desired result (see N. Dunford and I. E. Segal [**1**]). We start with this special case.

THEOREM 10.4.1. *If* $\| T(\xi) \| \leqq 1$ *and* $T(\xi)$ *is strongly continuous for* $\xi \geqq 0$ *with* $T(0) = I$, *then for each* $x \in \mathfrak{X}$ *and each* $\xi \geqq 0$

$$\lim_{\eta \to 0+} \exp(\xi A_\eta) x = T(\xi)x, \tag{10.4.2}$$

the limit existing uniformly with respect to ξ *in every finite interval* $[0, \beta]$.

PROOF. Since

$$\exp(\xi A_\eta) = \exp(-\xi/\eta) \exp [T(\eta)\xi/\eta], \tag{10.4.3}$$

it follows that

$$\| \exp(\xi A_\eta) \| \leqq e^{-\xi/\eta} \exp [\, \| T(\eta) \| \, \xi/\eta] \leqq e^{-\xi/\eta} e^{\xi/\eta} \equiv 1. \tag{10.4.4}$$

Thus the operator $\exp(\xi A_\eta)$ is also a contraction operator for all positive values of ξ and η. Suppose first that $x \in \mathfrak{D}(A_o)$. Then $\exp [(\xi - \tau)A_\eta]\, x$ is a differentiable function of τ by virtue of (4.6.5) and Theorem 10.3.3. Hence

$$\begin{aligned} T(\xi)x - \exp(\xi A_\eta)\, x &= \int_0^\xi \frac{d}{d\tau} \{\exp [(\xi - \tau)A_\eta]\, T(\tau)x\}\, d\tau \\ &= \int_0^\xi \exp [(\xi - \tau)A_\eta]\, T(\tau)(A_o x - A_\eta x)\, d\tau, \end{aligned} \tag{10.4.5}$$

here we have made use of the fact that both A_o and A_η commute with $T(\tau)$. Because of (10.4.4) the last member of (10.4.5) is dominated in norm by $\xi \| A_o x - A_\eta x \|$ and this expression tends to zero with η uniformly with respect to ξ in $[0, \beta]$. Thus the theorem is valid for each $x \in \mathfrak{D}(A_o)$. Since $\lim_{\eta \to 0+} T(\eta)x = x$ for each x in $\mathfrak{X}$, the subspace $\mathfrak{X}_0$ is dense in $\mathfrak{X}$ and, by Theorem 10.3.1, the same is true of $\mathfrak{D}(A_o)$. Hence it follows from (10.4.4) that (10.4.2) holds for each $x \in \mathfrak{X}$, uniformly with respect to ξ in $[0, \beta]$.

REMARK. The proof of Theorem 10.4.1 applies with little change if $T(\xi)$ is merely assumed to be continuous in the strong operator topology for $\xi \geqq 0$ with $T(0) = I$. For in this case there exist constants M and ω such that $\| T(\xi) \| \leqq M \exp(\omega\xi)$ for all $\xi \geqq 0$.

For the general theorem we shall give a proof, the central idea of which is due to Marcel Riesz (personal communication). It is based upon an inequality for expectations of distribution functions which contains the well-known inequality of Tchebycheff as a special case. This proof shows that exponential formula (10.4.2) is closely related to the theory of the Poisson distribution.

The point of departure is a *distribution function* $p(\tau)$, that is, a non-decreasing function defined on $(-\infty, \infty)$ such that $\int_{-\infty}^{\infty} dp(\tau) = 1$. For a function $f(\tau)$ integrable with respect to dp, set

$$Ef(\tau) = \int_{-\infty}^{\infty} f(\tau)\, dp(\tau). \tag{10.4.6}$$

This quantity is known as *the expectation of* $f(\tau)$. We assume, in particular, that τ is integrable with respect to dp and set $E\tau = m$. Let $g(\tau)$ be a convex differentiable function defined on some interval I containing the support of $p(\tau)$ and such that its graph contains no rectilinear segment. Next define $h(\tau)$ as the difference between $g(\tau)$ and the ordinate of its tangent at $\tau = m$. Then $h(\tau) > 0$ for $\tau \neq m$, while $h(m) = 0$. We have

$$h(\tau) = g(\tau) - [g(m) + g'(m)(\tau - m)],$$

which gives

$$Eh(\tau) = Eg(\tau) - g(m) \equiv Eg(\tau) - g(E\tau).$$

We have now the inequality of M. Riesz.

LEMMA 10.4.1. *For an arbitrary function* $f(\tau)$ *integrable with respect to* dp *and for an arbitrary* $\delta > 0$, *we have*

$$| Ef(\tau) | \leqq E | f(\tau) | \leqq \sup | f(\tau) | + \left[\sup \frac{| f(\tau) |}{h(\tau)} \right] Eh(\tau) \tag{10.4.7}$$

where the first supremum is taken over the set $S_1 \equiv [\tau; \tau \in I, | \tau - m | \leqq \delta]$ *and the second refers to the set* $S_2 \equiv I \ominus S_1$.

The inequality follows immediately from

$$E | f(\tau) | = \int_{S_1} | f(\tau) | \, dp(\tau) + \int_{S_2} \frac{| f(\tau) |}{h(\tau)} h(\tau)\, dp(\tau).$$

In applications $f(\tau)$ is usually replaced by $f(\tau) - f(m)$; it will be noticed that this is also the case in the proof of the following theorem, though it is done in an implicit way. The Tchebycheff inequality is obtained from (10.4.7) by setting $f(\tau) = 0$ or 1 according as $| \tau - m |$ is $\leqq \delta$ or $> \delta$, and letting $g(\tau) = \tau^2$, which gives $h(\tau) = (\tau - m)^2$.

We now prove

THEOREM 10.4.2. *If the semi-group $[T(\xi)]$ is continuous in the strong operator topology for $\xi \geqq \alpha \geqq 0$, then for each $x \in \mathfrak{X}$ and $\xi \geqq \alpha$*

$$\lim_{\eta \to 0+} \| \exp [(\xi - \alpha)A_\eta] \, T(\alpha)x - T(\xi)x \| = 0. \tag{10.4.8}$$

If $[T(\xi)]$ is continuous in the uniform operator topology for $\xi \geqq \alpha$, then

$$\lim_{\eta \to 0+} \| \exp [(\xi - \alpha)A_\eta] \, T(\alpha) - T(\xi) \| = 0. \tag{10.4.9}$$

In both cases the limit exists uniformly with respect to ξ in each finite interval $[\alpha, \beta]$.

PROOF. Let α, β, ξ, x be given with $0 \leqq \alpha \leqq \xi \leqq \beta < \infty$. It follows from our assumption on the continuity of $T(\xi)$ that there exists a finite quantity M such that $\| T(\xi) \| \leqq M$ for $\alpha \leqq \xi \leqq \max(\alpha + 1, 2\alpha)$ and without restricting the generality we may assume that $M > 1$. A simple calculation then shows that $\| T(\xi) \| \leqq M^{1+\xi}$ for all $\xi \geqq \alpha$. Finally let $\mu(\delta, x)$ be the rectified modulus of continuity of $T(\xi)x$ in $[\alpha, \beta]$ so that $\| T(\xi_1)x - T(\xi_2)x \| \leqq \mu(\delta, x)$ whenever $\alpha \leqq \xi_1, \xi_2 \leqq \beta$ and $| \xi_1 - \xi_2 | \leqq \delta$. From (10.4.1) one obtains

$$\exp [(\xi - \alpha)A_\eta] \, T(\alpha)x - T(\xi)x = e^{-(\xi-\alpha)/\eta} \sum_{n=0}^{\infty} \frac{1}{n!} \left(\frac{\xi - \alpha}{\eta} \right)^n [T(\alpha + n\eta)x - T(\xi)x],$$

whence

$$\begin{aligned} &\| \exp [(\xi - \alpha)A_\eta] \, T(\alpha)x - T(\xi)x \| \\ &\qquad \leqq e^{-(\xi-\alpha)/\eta} \sum_{n=0}^{\infty} \frac{1}{n!} \left(\frac{\xi - \alpha}{\eta} \right)^n \| T(\alpha + n\eta)x - T(\xi)x \|, \end{aligned} \tag{10.4.10}$$

We now apply the estimate (10.4.7) in the following manner. Let $P_\eta(\tau)$ be a Poisson distribution with parameter η [cf. formula (23.13.9)]; that is, $P_\eta(\tau)$ is constant except for jumps of

$$\frac{1}{n!} \left(\frac{\xi - \alpha}{\eta} \right)^n e^{-(\xi-\alpha)/\eta} \text{ at the points } \tau = \alpha + n\eta, \qquad n = 0, 1, 2, \cdots.$$

In this case

$$m_\eta = E_\eta(\tau) = e^{-(\xi-\alpha)/\eta} \sum_{n=0}^{\infty} \frac{1}{n!} \left(\frac{\xi - \alpha}{\eta} \right)^n (\alpha + n\eta) \equiv \xi. \tag{10.4.11}$$

Further we set

$$g(\tau) = M^{1+\tau}$$

so that

$$\begin{aligned} E_\eta g(\tau) &= e^{-(\xi-\alpha)/\eta} \sum_{n=0}^{\infty} \frac{1}{n!} \left(\frac{\xi - \alpha}{\eta} \right)^n M^{1+\alpha+n\eta} \\ &= M^{1+\alpha} \exp \left[(\xi - \alpha) \frac{M^\eta - 1}{\eta} \right]. \end{aligned}$$

It follows that

$$E_\eta h(\tau) = E_\eta g(\tau) - M^{1+\xi}.$$

Since

$$M^{1+\alpha} \exp\left[(\xi - \alpha) \frac{M^\eta - 1}{\eta}\right] \to M^{1+\alpha} \exp\left[(\xi - \alpha) \log M\right] = M^{1+\xi}$$

as $\eta \to 0+$, we see that $\lim_{\eta \to 0+} E_\eta h(\tau) = 0$ uniformly with respect to ξ in $[\alpha, \beta]$. Setting $f(\tau) = \| T(\tau)x - T(\xi)x \|$ for $\tau \geqq \alpha$ and $= 0$ for $\tau < \alpha$, (10.4.10) becomes

$$\| \exp\left[(\xi - \alpha)A_\eta\right] T(\alpha)x - T(\xi)x \| \leqq E_\eta f(\tau).$$

By (10.4.7) we have

$$E_\eta(f) \leqq \mu(\delta, x) + \left[\sup_{\tau \in S_2} \frac{f(\tau)}{h(\tau)}\right] E_\eta h(\tau).$$

Now $f(\tau) \leqq [M^{1+\tau} + M^{1+\xi}] \| x \|$ by our earlier estimate on the bound of $\| T(\xi) \|$. Since

$$h(\tau) = M^{1+\tau} - [M^{1+\xi} + (M^{1+\xi} \log M)(\tau - \xi)]$$

we see that $\lim_{\tau \to \infty} f(\tau)/h(\tau) \leqq \| x \|$. Hence there exists a constant $M(\beta, \delta)$ such that

$$\sup_{\tau \in S_2} \frac{f(\tau)}{h(\tau)} \leqq M(\beta, \delta) \| x \|$$

for all $\xi \in [\alpha, \beta]$. It now follows that

(10.4.12) $$\| \exp\left[(\xi - \alpha)A_\eta\right] T(\alpha)x - T(\xi)x \| \leqq \mu(\delta, x) + M(\beta, \delta) \| x \| E_\eta h(\tau).$$

From this estimate we conclude that (10.4.8) holds uniformly in $[\alpha, \beta]$. If $[T(\xi)]$ is continuous in the uniform operator topology for $\xi \geqq \alpha$ and not merely in the strong operator topology, we may suppress x everywhere in (10.4.12), replacing $\mu(\delta, x)$ by the modulus of continuity $\mu(\delta)$ of $T(\xi)$ in $[\alpha, \beta]$, and from this we see that (10.4.9) holds uniformly with respect to ξ in $[\alpha, \beta]$.

In view of the importance of formula (10.4.8) for our theory we restate it in several different ways. The first formulation

(10.4.13) $$T(\xi)x = \lim_{\eta \to 0+} e^{-(\xi - \alpha)/\eta} \sum_{n=0}^{\infty} \frac{1}{n!} \left(\frac{\xi - \alpha}{\eta}\right)^n T(\alpha + n\eta)x$$

shows that the result belongs to the theory of Borel summability. Equivalently, using the more elaborate notation of (23.13.9), we can write

(10.4.14) $$T(\xi)x = \lim_{\eta \to 0+} \int_{-\infty}^{\infty} T(\tau)x \, d_\tau P(\xi - \alpha; \tau, \eta)$$

which brings out the relation to the Poisson distribution explicitly. Hence $T(\xi)x$ is simply the (strong) limit of the (vector-valued) η-expectation of $T(\tau)x$ as

$\eta \to 0+$, where the definition of $T(\tau)x$ for $\tau < \alpha$ is immaterial. For the third formulation we employ the usual notation of the difference calculus

$$\Delta_\eta^n T(\alpha) = \eta^{-n} \sum_{k=0}^{n} (-1)^{n-k} \binom{n}{k} T(\alpha + k\eta) \tag{10.4.15}$$

obtaining

$$T(\xi)x = \lim_{\eta \to 0+} \sum_{n=0}^{\infty} \frac{(\xi - \alpha)^n}{n!} \Delta_\eta^n T(\alpha)x. \tag{10.4.16}$$

This form shows that we are dealing with a generalization of Taylor's series. We refer to Chapter XIX for an application to semi-groups of translations in which case the connection with the classical Taylor's series is the closest.

Formula (10.4.14) suggests strongly that similar results hold for families of distribution functions other than that of Poisson. Further it is fairly obvious that the proof of such formulas may be based on the lemma of M. Riesz. We shall not lose ourselves in generalities at this juncture, but wish to call attention to the case of the binomial distribution. This case was considered recently by D. G. Kendall [1]. It leads to elegant results.

The binomial distribution function $B_n(\tau; \xi)$ is defined for $0 \leqq \xi \leqq 1$. It has the interval $I \equiv [0, 1]$ as its carrier, and is constant except for jumps of

$$\binom{n}{k} \xi^k (1 - \xi)^{n-k}$$

at the points $\tau = k/n$, $k = 0, 1, 2, \cdots, n$ (for the properties of this function, see H. Cramér [2, p. 193]). If $[T(\xi)]$ is strongly continuous for $\xi \geqq 0$ we have

$$T(\xi)x = \lim_{n \to \infty} \int_0^1 T(\tau)x \, d_\tau B_n(\tau; \xi), \tag{10.4.17}$$

or in Kendall's formulation:

THEOREM 10.4.3. *If $[T(\xi)]$ is continuous in the strong operator topology for $\xi \geqq 0$, then*

$$\lim_{n \to \infty} \left\| \left[(1 - \xi)I + \xi T\left(\frac{1}{n}\right) \right]^n x - T(\xi)x \right\| = 0 \tag{10.4.18}$$

for each $x \in \mathfrak{X}$ and each $\xi \in [0, 1]$, uniformly in ξ.

PROOF. Since Lemma 10.4.1 again applies, we shall indicate only the main steps in the proof. We have $E_n\tau \equiv \xi$ as above. Clearly $\| T(\xi) \| \leqq M$ for $0 \leqq \xi \leqq 1$. We can now choose $g(\tau) = M(1 + \tau^2)$ so that

$$E_n g(\tau) = M\left(1 + \frac{1}{n}\xi + \frac{n-1}{n}\xi^2\right)$$

and

$$E_n h(\tau) = E_n g(\tau) - g(E_n \tau) = M \frac{1}{n} (\xi - \xi^2) \to 0$$

as $n \to \infty$ uniformly in ξ, $0 \leqq \xi \leqq 1$. We then obtain an estimate similar to (10.4.12) for

$$\int_0^1 \| T(\xi)x - T(\tau)x \| \, d_\tau B_n(\tau; \xi)$$

with $M(\beta, \delta)$ replaced by a suitable quantity $M(\delta)$. A simple calculation shows that $2M \, \delta^{-2}$ will do for $M(\delta)$. The desired conclusion then follows.

We can obviously replace strong continuity by uniform continuity in hypothesis and conclusion. An application to S. Bernstein's approximation theorem will be given in section 19.2.

Other exponential formulas will be given in Chapter XI.

10.5. Continuity at the origin. Most of the semi-groups which we shall consider will be continuous in some sense at the origin. It is only for such semi-groups that we have been able to obtain interesting structure theorems. We now begin to impose conditions of this sort.

THEOREM 10.5.1. *Let $T(\xi)$ be strongly continuous for $\xi > 0$ and suppose further that*

(i) $\int_0^1 \| T(\tau)x \| \, d\tau < \infty$ *for each* $x \in \mathfrak{X}$,

(ii) $\lim_{\eta \to 0+} \frac{1}{\eta} \int_0^\eta T(\tau)x \, d\tau \equiv Jx$ *exists for each* $x \in \mathfrak{X}$.

Then J is a projection operator, $J^2 = J$, which maps all of $\mathfrak{X}$ onto the closure of $\mathfrak{X}_0 = \bigcup [T(\alpha)[\mathfrak{X}]; \alpha > 0]$, and

$$T(\xi)J = T(\xi) = JT(\xi), \qquad \xi > 0. \tag{10.5.1}$$

PROOF. It follows from Theorem 3.8.2 that $\eta^{-1} \int_0^\eta T(\tau)x \, d\tau$ defines a linear bounded operator and hence, by the uniform boundedness theorem, the same is true of J. For $\xi > 0$ we have

$$T(\xi) \left[\frac{1}{\eta} \int_0^\eta T(\tau)x \, d\tau \right] = \frac{1}{\eta} \int_0^\eta T(\xi + \tau)x \, d\tau = \frac{1}{\eta} \int_0^\eta T(\tau)[T(\xi)x] \, d\tau,$$

and passing to the limit as $\eta \to 0$ we obtain (10.5.1). As a consequence

$$J \left[\frac{1}{\eta} \int_0^\eta T(\tau)x \, d\tau \right] = \frac{1}{\eta} \int_0^\eta T(\tau)x \, d\tau$$

and in the limit this gives $J^2 = J$. It is clear from the definition of an integral that $\eta^{-1} \int_0^\eta T(\tau)x \, d\tau$ lies in the closed linear hull of $\mathfrak{X}_0$ and this implies that $Jx \in \bar{\mathfrak{X}}_0$. Conversely if $x \in \bar{\mathfrak{X}}_0$, then $Jx = x$. Indeed, if $x \in \mathfrak{X}_0$, then strong continuity for $\xi > 0$ implies that $\lim_{\eta\to 0+} T(\eta)x = x$. Since Jx is the Cesàro average of $T(\xi)x$ near $\xi = 0$, it is clear that $Jx = x$ for $x \in \mathfrak{X}_0$. Now $\mathfrak{X}_0$ is dense in $\bar{\mathfrak{X}}_0$ so that, by Theorem 2.11.4, $Jx = x$ for all $x \in \bar{\mathfrak{X}}_0$. Consequently $J[\mathfrak{X}] = \bar{\mathfrak{X}}_0$.

THEOREM 10.5.2. *If $T(\xi)$ satisfies the hypothesis of Theorem 10.5.1 and if U is a linear operator with the property that*

$$T(\xi)x - x = \int_0^\xi T(\tau)Ux\,d\tau \tag{10.5.2}$$

for each $x \in \mathfrak{D}(U)$, then $A_o \supset JU$, that is, A_o is an extension of JU.

PROOF. The proof is immediate. We have only to notice that for each $x \in \mathfrak{D}(U)$ the relation (10.5.2) together with property (ii) implies that $\lim_{\eta\to 0+} A_\eta x = JUx$.

THEOREM 10.5.3. *If $T(\xi)$ satisfies the hypothesis of Theorem 10.5.1 with (i) replaced by the stronger condition $\int_0^1 \| T(\tau) \| \,d\tau < \infty$, then A_o is closed.*

PROOF. If $x \in \mathfrak{D}(A_o)$, then by the corollary to Theorem 10.3.3 we have

$$T(\xi)x - x = \int_0^\xi T(\tau)A_o x\,d\tau. \tag{10.5.3}$$

Suppose that $x_n \in \mathfrak{D}(A_o)$, $x_n \to x$, and $A_o x_n \to y$. According to Theorem 10.3.1, $A_o x_n \in \bar{\mathfrak{X}}_0$ so that $y \in \bar{\mathfrak{X}}_0$ and $Jy = y$. On the other hand

$$\| T(\tau)A_o x_n \| \leqq \| T(\tau) \| \sup_n \| A_o x_n \|.$$

Hence replacing x by x_n in (10.5.3) and applying Theorem 3.7.9, we obtain in the limit $T(\xi)x - x = \int_0^\xi T(\tau)y\,d\tau$. It now follows that $A_\eta x \to Jy = y$; in other words, $x \in \mathfrak{D}(A_o)$ and $A_o x = y$. This proves that A_o is closed.

REMARK. Another sufficient condition in order that A_o be closed is that there exist an $M > 0$ such that $\| T(\xi) \| \leqq M$ for $0 < \xi \leqq 1$. This is proved in Theorem 10.9.3. See also Theorem 11.5.4.

THEOREM 10.5.4. *Let $T(\xi)$ satisfy the hypothesis of Theorem 10.5.1 with (i) replaced by $\int_0^1 \| T(\tau) \| \,d\tau < \infty$. Suppose that $y_n \equiv A_{\delta_n} x_0$ converges weakly to a limit y_0 for some sequence $\{\delta_n\}$, $\delta_n \to 0+$. Then $x_0 \in \mathfrak{D}(A_o)$ and $A_o x_0 = y_0$.*

PROOF. It follows from the uniform boundedness theorem that

$$\sup_n \| A_{\delta_n} x_0 \| < \infty$$

and hence that $\lim_n T(\delta_n)x_0 = x_0$ in norm. Consequently $x_0 \in \bar{\mathfrak{X}}_0$ and $Jx_0 = x_0$. Further $A_{\delta_n} x_0 \in \bar{\mathfrak{X}}_0$ and by Theorem 2.9.2 the same is true of the weak limit y_0 so that $Jy_0 = y_0$. Setting

$$x_n = \delta_n^{-1} \int_0^{\delta_n} T(\tau)x_0\,d\tau, \qquad z_n = \delta_n^{-1} \int_\xi^{\xi+\delta_n} T(\tau)x_0\,d\tau,$$

we see that $x_n \to Jx_0 = x_0$ and $z_n \to T(\xi)x_0$, both sequences converging in norm. Moreover

$$z_n - x_n = \delta_n^{-1} \int_0^\xi [T(\tau + \delta_n)x_0 - T(\tau)x_0]\, d\tau = \int_0^\xi T(\tau)A_{\delta_n}x_0\, d\tau$$

$$= \int_0^\xi T(\tau)y_n\, d\tau.$$

For each $x^* \in \mathfrak{X}^*$, we have $| x^*[T(\tau)y_n] | \leqq \| x^* \| \| T(\tau) \| \| y_n \|$. Moreover the weak convergence of y_n to y_0 implies the boundedness of $\{ \| y_n \| \}$ and therefore the dominated convergence of $x^*[T(\tau)y_n]$ to $x^*[T(\tau)y_0]$. The Lebesgue dominated convergence theorem applies and we obtain in the limit $x^*[T(\xi)x_0] - x^*(x_0) = \int_0^\xi x^*[T(\tau)y_0]\, d\tau$. Hence $T(\xi)x_0 - x_0 = \int_0^\xi T(\tau)y_0\, d\tau$, from which it follows that $\lim_{\eta\to 0+} A_\eta x_0 = Jy_0 = y_0$ so that $x_0 \in \mathfrak{D}(A_o)$ and $A_o x_0 = y_0$.

A consequence of the above theorem is that one obtains the same operator A_o whether $A_\eta x$ is required to converge weakly or strongly in the definition of the infinitesimal operator. The latter result is due to K. Yosida [**3**]. For a related, more general theorem, see L. Schwartz [**1**]. We now prove the strong case analogue of Theorem 9.6.1; again measurability need not be assumed.

THEOREM 10.5.5. *If* $T(\xi)$ *is defined for* $\xi > 0$ *and satisfies* (10.2.1) *and if*

$$\lim_{\eta\to 0+} T(\eta) \equiv J \tag{10.5.4}$$

exists in the strong sense, then J *is a projection operator mapping all of* $\mathfrak{X}$ *onto the closure of* $\mathfrak{X}_0$, *and*

$$T(\xi) = JT(\xi) = T(\xi)J. \tag{10.5.5}$$

$T(\xi)$ *is strongly continuous for* $\xi \geqq 0$ *if* $T(0) = J$ *by definition, and*

$$T(\xi)x = \lim_{\eta\to 0+} \exp[\xi A_\eta]\, Jx \tag{10.5.6}$$

for all $x \in \mathfrak{X}$, *the limit existing uniformly with respect to* ξ *in any finite interval* $[0, \beta]$.

Finally, a set of necessary and sufficient conditions that the limit in (10.5.4) *shall exist with* $J = I$ *is that* (i) $T(\xi)$ *be strongly measurable for* $\xi > 0$, (ii) *there exist a positive* M *such that* $\| T(\xi) \| \leqq M$ *for* $0 < \xi \leqq 1$, *and* (iii) $\bar{\mathfrak{X}}_0 = \mathfrak{X}$.

PROOF. If $T(0+)$ exists as a strong limit, then using the same type of argument as in Theorem 9.4.1, we prove successively that (1) $J^2 = J$; (2)

$$\lim_{\eta\to 0+} T(\xi + \eta)x = JT(\xi)x = T(\xi)Jx$$

for each $x \in \mathfrak{X}$; and (3) $T(\xi)$ is strongly measurable for $\xi > 0$ and hence by Theorem 10.2.3 together with (10.5.4) strongly continuous for $\xi \geqq 0$. This proves (10.5.5) and shows, incidentally, that (i) and (ii) are necessary for the existence of (10.5.4). Since the hypothesis of Theorem 10.5.1 is clearly satisfied, we also have $J[\mathfrak{X}] = \bar{\mathfrak{X}}_0$. It follows that (iii) is necessary in order that $J = I$.

The formula (10.5.6) follows directly from Theorem 10.4.2.

To prove the sufficiency of the conditions (i) to (iii) we note that (i) implies that $T(\xi)$ is strongly continuous for $\xi > 0$ and hence that $\lim_{\eta\to 0+} T(\eta)x = x$ for each $x \in \mathfrak{X}_0$. Since $\mathfrak{X}_0$ is dense in $\mathfrak{X}$ by (iii) and since $\| T(\eta) \| \leqq M$ for $0 < \eta \leqq 1$ by (ii), we conclude that the limit exists and equals x for all $x \in \mathfrak{X}$. This completes the proof.

We remark that if $T(\xi)$ is in addition continuous in the uniform operator topology for $\xi > 0$, then

$$\lim_{\eta\to 0+} \| \exp [\xi A_\eta] J - T(\xi) \| = 0 \tag{10.5.7}$$

uniformly with respect to ξ in $0 < \epsilon \leqq \xi \leqq 1/\epsilon$. This follows from the formula (10.4.12) on setting $\alpha = 0$ and replacing $\mu(\delta, x)$ by $\mu(\delta) \| x \|$, where $\mu(\delta)$ is now the modulus of continuity of $T(\xi)$ in the interval $[\epsilon/2, 2/\epsilon]$.

It should be observed that conditions (i) and (ii) of Theorem 10.5.5 are not sufficient to ensure the existence of $\lim_{\eta\to 0+} T(\eta)x = Jx$ for all $x \in \mathfrak{X}$. This is shown by the following example (a power semi-group in the sense of section 19.6). We take $\mathfrak{X} = C[0, 1]$ and define

$$[T(\xi)x](t) = \left[t\left(1 + \cos\frac{1}{t}\right)\right]^{\xi} x(t), \qquad \xi > 0$$

The semi-group is strongly continuous for $\xi > 0$; however $\lim_{\eta\to 0+} T(\eta)x$ exists if and only if $x(t) = 0$ at all the points t where the bracket expression equals zero, that is for $t = 0$ and $1/[(2n + 1)\pi]$, $n = 0, 1, 2, \cdots$. The functions $x(t)$ having this property form the set $\bar{\mathfrak{X}}_0$, which is a proper subset of $\mathfrak{X}$.

We also note that the existence of $\lim_{\eta\to 0+} T(\eta)$ in the weak operator topology does not imply convergence in the strong sense. This is illustrated by the following example. Let $\mathfrak{X}$ be the Hilbert space of functions $x(t)$ defined on $[0, \infty)$ with $\| x \| = [\sum_{t\geqq 0}|x(t)|^2]^{1/2}$ and define $[T(\xi)x](t) = x(t + \xi)$ for $\xi > 0$. Then $\lim_{\eta\to 0+} x^*[T(\eta)x] = 0$ for all $x \in \mathfrak{X}$ and $x^* \in \mathfrak{X}^* = \mathfrak{X}$. However $T(\eta)$ does not converge to a limit in the strong operator topology as $\eta \to 0$. This is also an example of a semi-group of operators which is weakly measurable but not weakly (and hence not strongly) continuous for $\xi > 0$. On the other hand, if $T(\eta)$ converges to the identity in the weak operator topology as $\eta \to 0+$, then it can be shown that $\lim_{\eta\to 0+} T(\eta)x = x$ in norm for all $x \in \mathfrak{X}$ (see Theorem 10.6.5). This exemplifies the decisive role played by the identity in these matters. If it is assumed to begin with that the semi-group of operators is strongly continuous for $\xi > 0$, then the existence of $\lim_{\eta\to 0+} T(\eta)$ or $\lim_{\eta\to 0+} \eta^{-1} \int_0^\eta T(\tau)x \, d\tau$ in the weak operator topology implies the existence of the corresponding limit in the strong operator topology (see corollary to Theorem 18.7.2.).

The projection operator J induces a decomposition of $\mathfrak{X}$ into the direct sum of two subspaces, one being the closure of the union of the range spaces of the individual semi-group operators, and the other being the intersection of their "null" spaces. A detailed study of this and other such decomposition theorems will be made in Chapter XVIII which is concerned with ergodicity.

10.6. The basic classes of semi-groups. A final restriction which we now introduce requires that the semi-group $[T(\xi)]$ converges in some sense to the identity as $\xi \to 0+$. The precise sense of convergence is critical and accordingly we subdivide such semi-groups into six basic classes, denoted by the symbols (C_0), $(1, C_1)$, $(0, C_1)$, $(1, A)$, $(0, A)$, and (A) respectively (cf. R. S. Phillips [**9, 11**]).

We shall always suppose that $T(\xi)$ is strongly measurable and hence strongly continuous for $\xi > 0$. As in section 10.2, we define the type for such a semi-group to be

$$\omega_0 \equiv \inf_{\xi>0} \frac{1}{\xi} \log \| T(\xi) \| = \lim_{\xi\to\infty} \frac{1}{\xi} \log \| T(\xi) \| . \tag{10.6.1}$$

We shall further assume that $\mathfrak{X}_0$ is dense in $\mathfrak{X}$. It follows from Theorem 10.3.4 that $\bigcap_n \mathfrak{D}(A_o^n)$ will be dense in $\mathfrak{X}$.

We shall consider three kinds of convergence at $\xi = 0$, namely, $(C, 0)$, $(C, 1)$, and Abel summability. The strongest of these is $(C, 0)$; $T(\xi)x$ is said to be $(C, 0)$ summable to x if $\lim_{\eta\to 0+} T(\eta)x = x$. On the other hand if $\int_0^1 \| T(\tau)x \| \, d\tau < \infty$, then

$$C(\eta)x \equiv \frac{1}{\eta} \int_0^\eta T(\tau)x \, d\tau \tag{10.6.2}$$

exists and $T(\xi)x$ is said to be $(C, 1)$ summable to x if $\lim_{\eta\to 0+} C(\eta)x = x$. It is clear that $T(\xi)x$ is $(C, 1)$ summable to x if it is $(C, 0)$ summable to x. Finally if $\int_0^1 \| T(\tau)x \| \, d\tau < \infty$, then

$$R(\lambda)x \equiv \int_0^\infty e^{-\lambda\tau} T(\tau)x \, d\tau \tag{10.6.3}$$

exists for all λ with $\Re(\lambda) > \omega_0$ and $T(\xi)x$ is said to be Abel summable to x if $\lim_{\lambda\to\infty} \lambda R(\lambda)x = x$. We note that $(C, 1)$ summability to x implies Abel summability to x. In fact, an integration by parts shows that

$$\lambda \int_0^\infty e^{-\lambda\tau} T(\tau)x \, d\tau = \lambda^2 \int_0^\infty e^{-\lambda\tau} \tau \left[\frac{1}{\tau} \int_0^\tau T(\sigma)x \, d\sigma \right] d\tau, \qquad \lambda > \omega_0 ,$$

so that

$$\lambda \int_0^\infty e^{-\lambda\tau} T(\tau)x \, d\tau - x = \lambda^2 \int_0^\infty e^{-\lambda\tau} \tau [C(\tau)x - x] \, d\tau.$$

For $\omega > \max(0, \omega_0)$, $\| C(\tau)x \| \leqq M(\omega)e^{\omega\tau}$ and hence

$$\left\| \lambda^2 \int_0^\delta e^{-\lambda\tau} \tau [C(\tau)x - x] \, d\tau \right\| \leqq \sup[\|C(\tau)x - x\|; 0 < \tau < \delta],$$

$$\left\| \lambda^2 \int_\delta^\infty e^{-\lambda\tau} \tau [C(\tau)x - x] \, d\tau \right\| \leqq \frac{2\lambda^2 M(\omega)}{(\lambda - \omega)^2} [1 + (\lambda - \omega)\delta] e^{-(\lambda-\omega)\delta}.$$

These estimates prove the assertion.

In the above class designations, the symbol C_0 stand for $(C, 0)$, C_1 for $(C, 1)$, and A for Abel summability. Thus C_0 denotes the condition

(C_0) $\lim_{\eta\to 0+} T(\eta)x = x$ *for each* $x \in \mathfrak{X}$;

C_1 denotes

$$(\mathrm{C}_1) \lim_{\eta\to 0+} C(\eta)x = x \textit{ for each } x \in \mathfrak{X};$$

and A denotes

$$(\mathrm{A}) \lim_{\lambda\to\infty} \lambda R(\lambda)x = x \textit{ for each } x \in \mathfrak{X}.$$

In the middle four symbols the number 0 in the first place indicates that

$$(\mathrm{i})_0 \int_0^1 \| T(\tau)x \| \, d\tau < \infty \textit{ for each } x \in \mathfrak{X};$$

whereas the number 1 indicates that

$$(\mathrm{i})_1 \int_0^1 \| T(\tau) \| \, d\tau < \infty .$$

For semi-groups of class (C_0) no integrability condition is required since in this case the uniform boundedness theorem implies that $\| T(\xi) \|$ is bounded in $0 < \xi \leqq 1$. Thus for the first five classes the operators $C(\eta)$, $\eta > 0$, and $R(\lambda)$, $\Re(\lambda) > \omega_0$, are well defined on all $x \in \mathfrak{X}$; in fact Theorem 3.8.2 asserts that these operators belong to $\mathfrak{E}(\mathfrak{X})$. Consequently there is no question in these cases as to the meaning to be assigned to the corresponding summability condition. However for semi-groups in class (A) we do not require that $(\mathrm{i})_0$ be satisfied so that the integral expression given in (10.6.3) is in general meaningless. In this case we give condition (A) the following sense:

(A)′ *There exists an* $\omega_1 > \omega_0$ *and for each* λ, $\Re(\lambda) > \omega_1$, *there is an operator* $R(\lambda) \in \mathfrak{E}(\mathfrak{X})$ *with the properties* (a) $R(\lambda)$ *coincides with* (10.6.3) *on* $\mathfrak{X}_0$, (b) $\| R(\lambda) \|$ *is bounded in the half-plane* $\Re(\lambda) > \omega_1$, *and* (c) $\lim_{\lambda\to\infty} \lambda R(\lambda)x = x$ *for each* $x \in \mathfrak{X}$.

Thus in all cases $R(\lambda)$, $\Re(\lambda) > \omega_1$, belongs to $\mathfrak{E}(\mathfrak{X})$ and satisfies (10.6.3) on $\mathfrak{X}_0$. Also it is clear that $R(\lambda)$ is uniquely determined by its values on the dense subset $\mathfrak{X}_0$. Finally we see that the boundedness condition (b) is also valid in general. For if $(\mathrm{i})_0$ is satisfied and $\omega_1 > \omega_0$, then $e^{-\omega_1\xi}T(\xi)x \in B(E_1 ; \mathfrak{X})$ for all $x \in \mathfrak{X}$ and hence by (3.8.2) there exists a constant $M(\omega_1)$ such that $\| R(\lambda)x \| \leqq \int_0^\infty e^{-\omega_1\xi} \| T(\xi)x \| \, d\xi \leqq M(\omega_1) \| x \|$ for $\Re(\lambda) > \omega_1$. We note that if $(\mathrm{i})_0$ is satisfied, then $R(\lambda)$ is holomorphic for $\Re(\lambda) > \omega_0$ by sections 3.10 and 6.2. In the absence of $(\mathrm{i})_0$, condition (A)′ is sufficient to make $R(\lambda)x$ holomorphic for $\Re(\lambda) > \omega_1$ and $x \in \mathfrak{X}_0$. Further since $\mathfrak{X}_0$ is dense in $\mathfrak{X}$ and $\| R(\lambda) \|$ is bounded in $\Re(\lambda) > \omega_1$, it readily follows from Theorem 3.11.6 that $R(\lambda)x$ is holomorphic in the half-plane for all $x \in \mathfrak{X}$ and hence that $R(\lambda)$ is itself holomorphic in this half-plane. We summarize this discussion in the following:

THEOREM 10.6.1. *The inclusion relations between the basic classes are given by*

$$\begin{array}{ccccc} & & (1, A) & & \\ & \nearrow & & \searrow & \\ (A) \to (0, A) & & & & (1, C_1) \to (C_0). \\ & \searrow & & \nearrow & \\ & & (0, C_1) & & \end{array}$$

For all of these classes except (A), the linear bounded operator $R(\lambda)$ is defined by (10.6.3) for all $x \in \mathfrak{X}$, $R(\lambda)$ is holomorphic in λ for $\Re(\lambda) > \omega_0$, and $\| R(\lambda) \|$ is bounded in $\Re(\lambda) > \omega_1$ for any fixed $\omega_1 > \omega_0$. For each semi-group of class (A), there is an $\omega_1 > \omega_0$ such that $R(\lambda)$ is holomorphic and bounded in norm for $\Re(\lambda) > \omega_1$.

A further subdivision of these classes will be indicated by a subscript, as for example $(0, A)_u$. The subscript u indicates that the member semi-group is continuous in the uniform operator topology for $\xi > 0$; the subscript ∞ indicates that the member semi-group admits of a strong derivative for each $\xi > 0$ and hence by Theorem 10.3.5 that it admits of uniform derivatives of all orders for $\xi > 0$.

The above description of the basic classes makes evident the kind of convergence to the identity at $\xi = 0$ which is associated with each class. However the assumption that $\mathfrak{X}_0$ is dense in $\mathfrak{X}$ is redundant in all of the subclasses of $(0, A)$ (see Lemma 10.6.1). In order to avoid this redundancy we can replace the conditions (C_0), (C_1), and (A) respectively by

$(C_0)''$ $\quad \| T(\eta) \| = O(1)$ as $\eta \to 0+$;

$(C_1)''$ $\quad \| C(\eta) \| = O(1)$ as $\eta \to 0+$;

$(A)''$ $\quad \| R(\lambda) \| = O(1/\lambda)$ as $\lambda \to \infty$.

We denote the corresponding basic classes by $(C_0)''$, $(1, C_1)''$, $\cdots$, and refer to such a class as a primed class.

THEOREM 10.6.2. *A semi-group belongs to a primed class if and only if it belongs to the corresponding unprimed class.*

PROOF. If $[T(\xi)]$ satisfies one of the conditions (C_0), (C_1), or (A) then the uniform boundedness theorem applies and hence $[T(\xi)]$ satisfies the corresponding primed condition. On the other hand suppose $[T(\xi)]$ belongs to one of the primed classes. If $x \in \mathfrak{X}_0$, then $\lim_{\eta \to 0+} T(\eta)x = x$ and *a fortiori* $\lim_{\eta \to 0+} C(\eta)x = x$ and $\lim_{\lambda \to \infty} \lambda R(\lambda)x = x$. Since $\mathfrak{X}_0$ is dense in $\mathfrak{X}$, the Banach-Steinhaus theorem asserts that any one of the primed conditions $(C_0)''$, $(C_1)''$, or $(A)''$ implies the corresponding unprimed condition.

LEMMA 10.6.1. *If $[T(\xi)]$ is strongly measurable for $\xi > 0$ and satisfies condition $(i)_0$, then the weak form of condition (C_0), (C_1), or (A) implies that $\bar{\mathfrak{X}}_0 = \mathfrak{X}$.*

PROOF. Since the weak form of (C_0) or (C_1) implies the weak form of (A), it suffices to consider only the latter case. Suppose that $\mathfrak{X}_0$ is not dense in $\mathfrak{X}$. Then by Theorem 2.7.5 there exists an $x_0^* \in \mathfrak{X}^*$ and an $x_0 \in \mathfrak{X}$ such that $x_0^*(\mathfrak{X}_0) = 0$ and $x_0^*(x_0) \neq 0$. This leads to a contradiction since

$$0 = \lambda \int_0^\infty e^{-\lambda\tau} x_0^*[T(\tau)x_0] \, d\tau \to x_0^*(x_0) \neq 0$$

as $\lambda \to \infty$. It follows that $\bar{\mathfrak{X}}_0 = \mathfrak{X}$.

We next show that the weak form of the condition (A), (C_1), or (C_0) implies the corresponding strong form. For (C_0) this result is due to E. Hille [**15**]; in this case measurability need not even be assumed.

THEOREM 10.6.3. *If $[T(\xi)]$ is strongly measurable for $\xi > 0$ and satisfies condition* $(\mathrm{i})_0$, *then the weak form of condition* (A) *or* (C_1) *implies the corresponding strong form.*

PROOF. The proof is essentially the same for both (A) and (C_1); to avoid repetition we consider only the condition (A) and show that weak convergence implies strong convergence. The assumed weak form of (A) states that

$$\lim_{\lambda\to\infty} x^*[\lambda R(\lambda)x] = x^*(x)$$

for each $x \in \mathfrak{X}$ and $x^* \in \mathfrak{X}^*$. By the uniform boundedness theorem this implies that $\| R(\lambda) \| = O(1/\lambda)$ as $\lambda \to \infty$. Further by Lemma 10.6.1 we see that $\bar{\mathfrak{X}}_0 = \mathfrak{X}$ so that $[T(\xi)]$ is of class $(0, A)''$. It now follows from Theorem 10.6.2 that $[T(\xi)]$ satisfies condition (A).

THEOREM 10.6.4. *If $[T(\xi)]$ belongs to any of the basic classes and if*

$$\| T(\xi) \| = O(1)$$

as $\xi \to 0+$, then $[T(\xi)]$ is of class (C_0).

PROOF. Since $\bar{\mathfrak{X}}_0 = \mathfrak{X}$, it is clear that the conditions (i) to (iii) of Theorem 10.5.5 are satisfied and therefore that $[T(\xi)]$ is of class (C_0).

THEOREM 10.6.5. *If $[T(\xi)]$ is defined for $\xi > 0$ and satisfies* (10.2.1), *and if $\lim_{\xi\to 0+} T(\xi) = I$ in the weak operator topology, then $[T(\xi)]$ is of class (C_0).*

PROOF. Again it suffices to show that $[T(\xi)]$ satisfies conditions (i) to (iii) of Theorem 10.5.5. Now

$$\lim_{\eta\to 0+} x^*[T(\xi + \eta)x] = \lim_{\eta\to 0+} x^*\{T(\eta)[T(\xi)x]\} = x^*[T(\xi)x]$$

so that $T(\xi)x$ is weakly continuous on the right for each $x \in \mathfrak{X}$. By the corollary to Theorem 10.2.3, $T(\xi)x$ is strongly continuous for $\xi > 0$; this establishes (i). It follows from this and the hypothesis that $x^*[T(\xi)x]$ is continuous in ξ, $\xi \geqq 0$, for each $x \in \mathfrak{X}$ and $x^* \in \mathfrak{X}^*$. By the uniform boundedness theorem,

$$\| T(\xi) \| = O(1)$$

as $\xi \to 0+$ and this proves (ii). Finally we have $\bar{\mathfrak{X}}_0 = \mathfrak{X}$ by Lemma 10.6.1, which is the statement of (iii).

Before closing this section we introduce still another basic class of semi-groups, namely, semi-groups holomorphic in a sector of the complex plane. These are the best behaved of all semi-groups with unbounded infinitesimal operators. More precisely we have

DEFINITION 10.6.1. *Let* Σ_2 *denote the open sector* $\Phi_1 < \arg \zeta < \Phi_2$ *where* $-\pi/2 \leqq \Phi_1 < 0 < \Phi_2 \leqq \pi/2$. *A one-parameter family of operators* $[T(\zeta)]$ *defined on* Σ_2 *to* $\mathfrak{E}(\mathfrak{X})$ *is called a semi-group of class* $H(\Phi_1, \Phi_2)$ *if*

(i) $T(\zeta)$ *is holomorphic with respect to* ζ *in* Σ_2;
(ii) $T(\zeta_1 + \zeta_2) = T(\zeta_1)T(\zeta_2)$ *for all* $\zeta_1, \zeta_2 \in \Sigma_2$;
(iii) $\| T(re^{i\varphi}) \| \leqq B(\varphi)$ *for* $0 < r \leqq 1$ *and* $\varphi \in (\Phi_1, \Phi_2)$;
(iv) $\mathfrak{X}_0 \equiv \bigcup [T(\zeta)[\mathfrak{X}]; \zeta \in \Sigma_2]$ *is dense in* $\mathfrak{X}$.

Without loss of generality we could have assumed that $B(\varphi)$ is bounded in each closed sub-interval of (Φ_1, Φ_2). We have, in fact,

LEMMA 10.6.2. *If* $T(\zeta)$ *is of class* $H(\Phi_1, \Phi_2)$, *then for each* $\epsilon > 0$,

$$\sup [\, \| T(re^{i\varphi}) \|; 0 < r \leqq 1, \Phi_1 + \epsilon < \varphi < \Phi_2 - \epsilon] < \infty.$$

PROOF. Each vector ζ can be uniquely represented as $\zeta = \alpha e^{i(\Phi_1+\epsilon)} + \beta e^{i(\Phi_2-\epsilon)}$. For all ζ with $0 < |\zeta| \leqq 1$ and $\Phi_1 + \epsilon < \arg \zeta < \Phi_2 - \epsilon$ there exists a constant $\gamma > 0$ such that $0 < \alpha, \beta \leqq \gamma$ and therefore

$$\| T(\zeta) \| \leqq \| T(\alpha e^{i(\Phi_1+\epsilon)}) \| \, \| T(\beta e^{i(\Phi_2-\epsilon)}) \| \leqq [B(\Phi_1 + \epsilon)B(\Phi_2 - \epsilon)]^{\gamma+1}.$$

The infinitesimal operator is defined as before:

$$A_0 x = \lim A_\eta x$$

as *real* $\eta \to 0+$. Now, however, $T(\zeta)$ is holomorphic in Σ_2 so that the *complex* derivative exists in the uniform operator topology for each $\zeta \in \Sigma_2$. Consequently for $\zeta \in \Sigma_2$ we have

$$\frac{d}{d\zeta}[T(\zeta)x] = A_0 T(\zeta)x, \qquad x \in \mathfrak{X};$$

$$\frac{d}{d\zeta}[T(\zeta)x] = T(\zeta)A_0 x, \qquad x \in \mathfrak{D}(A_0).$$

It follows that $\mathfrak{D}(A_0) \supset \mathfrak{X}_0$.

THEOREM 10.6.6. *If* $T(\zeta)$ *is of class* $H(\Phi_1, \Phi_2)$, *then for each* $x \in \mathfrak{X}$, $\lim T(\zeta)x = x$ *as* $|\zeta| \to 0$ *in any sector of the form* $\Phi_1 + \epsilon < \arg \zeta < \Phi_2 - \epsilon$, $\epsilon > 0$. *In particular,* $[T(\xi); \xi > 0]$ *is of class* (C_0).

PROOF. If $x \in \mathfrak{X}_0$, then there exists a $y \in \mathfrak{X}$ and a $\zeta_0 \in \Sigma_2$ such that $x = T(\zeta_0)y$. Since analyticity clearly implies continuity, $T(\zeta)x = T(\zeta + \zeta_0)y \to T(\zeta_0)y = x$ as $|\zeta| \to 0$. Hence, $\mathfrak{X}_0$ being dense in $\mathfrak{X}$, the first assertion follows from Lemma 10.6.2 together with the Banach-Steinhaus theorem. In particular, for ξ real we see that $T(\xi)$ defines a semi-group of class (C_0).

10.7. Approximation of the identity. The present section deals with the degree of approximation of the identity by the semi-group operator for small values of the parameter, that is, the order of magnitude of $\| T(\xi) - I \|$ and $\| [T(\xi) - I]x \|$

as a function of ξ. The results are of interest for applications to the theory of summability and singular integrals (cf. E. Hille [**4**] and P. L. Butzer [**1**]).

THEOREM 10.7.1. *If* $T(\xi)$ *is measurable in the uniform sense for* $\xi > 0$ *and if* $\| T(\xi) - I \| < 1$ *for a single* $\xi > 0$, *then* $T(\xi) = \exp(\xi A)$, *where* A *is a bounded operator, and*

$$T(\xi) - I = \xi A + O(\xi^2). \tag{10.7.1}$$

In particular, $\| T(\xi) - I \| = o(\xi)$ *as* $\xi \to 0+$ *if and only if* $A = \Theta$ *and* $T(\xi) \equiv I$. *Further* $\| [T(\xi) - I]x_0 \| = o(\xi)$ *as* $\xi \to 0+$ *if and only if* $Ax_0 = \theta$ *and* $T(\xi)x_0 \equiv x_0$.

PROOF. The first assertion follows from Theorem 9.4.4; the others are immediate consequences of the representation $T(\xi) = \exp(\xi A)$.

Thus in the uniform case the transform $T(\xi)x$ furnishes a first order approximation of x for small values of ξ and this holds uniformly with respect to bounded subsets of $\mathfrak{X}$. No higher degree of approximation can occur except for the invariant elements. In the strong case the results are more diversified.

THEOREM 10.7.2. *Let* $T(\xi)$ *be of class* $(1, C_1)$. *If*

$$\liminf_{\xi \to 0+} \frac{1}{\xi} \| [T(\xi) - I]x_0 \| = 0, \tag{10.7.2}$$

then $A_o x_0 = \theta$ *and* $T(\xi)x_0 \equiv x_0$. *For each* $x \in \mathfrak{D}(A_o)$ *we have*

$$[T(\xi) - I]x = \xi A_o x + o(\xi). \tag{10.7.3}$$

If $\mathfrak{X}$ *is reflexive and*

$$\liminf_{\xi \to 0+} \frac{1}{\xi} \| [T(\xi) - I]x_1 \| < \infty, \tag{10.7.4}$$

then $x_1 \in \mathfrak{D}(A_o)$,

PROOF. If the relation (10.7.2) holds, then there exists a sequence $\{\delta_n\}$, $\delta_n \to 0+$, such that $A_{\delta_n} x_0 \to \theta$ in norm. Theorem 10.5.4 applies and we see that $x_0 \in \mathfrak{D}(A_o)$ and $A_o x_0 = \theta$. By (10.3.7) $T(\xi)x_0 - x_0 = \int_0^\xi T(\tau) A_o x_0 \, d\tau \equiv \theta$. The formula (10.7.3) is an immediate consequence of the definition of $A_o x$. If condition (10.7.4) holds, then there exists a sequence $\{\delta_n\}$, $\delta_n \to 0+$, such that $\| A_{\delta_n} x_1 \|$ is bounded. In a reflexive (B)-space, bounded subsets are weakly conditionally compact. Hence there is a subsequence of $\{A_{\delta_n} x_1\}$ which converges weakly to a limit. It now follows from Theorem 10.5.4 that $x_1 \in \mathfrak{D}(A_o)$. This completes the proof. Formula (10.3.7) leads to the

COROLLARY. *If* $\| T(\xi) \| \leqq 1$ *for all* ξ *and* $x \in \mathfrak{D}(A_o)$, *then*

$$\| T(\xi)x - x \| \leqq \xi \| A_o x \|. \tag{10.7.5}$$

For semi-groups of class $(1, C_1)$ this theorem shows that unless x is an invariant element of the semi-group, the degree of approximation of x by $T(\xi)x$ cannot exceed the first; this degree is reached for all elements of $\mathfrak{D}(A_o)$. In general for such semi-groups we do not have $\lim_{\xi \to 0+} \| [T(\xi) - I]x \| = 0$ for all $x \in \mathfrak{X}$. Even for semi-groups of class (C_0) there will not in general exist a modulus of continuity which is the same for all x of norm one, since this would imply that $\lim_{\xi \to 0+} \| T(\xi) - I \| = 0$; this is the case treated in Theorem 10.7.1. In this connection the reader may find the semi-group defined by formula (9.3.3) instructive. It is of class (C_0) and $\lim_{\xi \to 0+} \| T(\xi) - I \| = 2$. Formula (9.3.4) shows that for each $\xi > 0$ and $\delta = e^{-n}$ there exists an element x (of the form $T(\xi)y$) such that

$$\| [T(\delta) - I]x \| \geqq \frac{1}{2}\left[\log\left(\frac{1}{\delta}\right)\right]^{-\xi}.$$

The case of the semi-group of translations on the space of continuous functions $C[0, \infty]$ is still simpler. If $x(t) \in C[0, \infty]$ and its modulus of continuity is $\omega(\delta)$, then $\| [T(\delta) - I]x \| = \omega(\delta)$. Here every modulus of continuity is possible; thus while it is true that $\| [T(\xi) - I]x \| \to 0$ as $\xi \to 0+$, nothing can be said about the rate of approach which is true for all x of norm one.

In constructing elements of $\mathfrak{D}(A_o)$ in section 10.3 we used a process of integration applied to $T(\xi)y$. This leads to elements x which admit of first order approximation. Using integration of fractional order instead, one obtains elements for which the degree of approximation lies between zero and one. The following case is typical.

THEOREM 10.7.3. *If $T(\xi)$ is strongly continuous for $\xi > 0$ and if x is of the form*

$$x = \int_\alpha^\beta (\beta - \tau)^{\gamma-1} T(\tau) y \, d\tau, \tag{10.7.6}$$

where α, β and γ are fixed $0 < \alpha < \beta < \infty, 0 < \gamma < 1$, then $\| [T(\xi) - I]x \| = O(\xi^\gamma)$.

The proof is elementary and is left to the reader.

References. Butzer [1], Cramér [2], Dunford [4], Dunford and Segal [1], Feller [4], Gelfand [3], Hille [**4, 9, 13, 15, 16**], Kendall [1], Miyadera [1], Phillips [**4, 6, 9, 11**], L. Schwartz [1], Yosida [3].

2. Extensions

10.8. Problem B in the strong topology. We now consider two complex (B)-spaces $\mathfrak{X}$ and $\mathfrak{Y}$ and a function $T(x)$ on $\mathfrak{X}$ to $\mathfrak{E}(\mathfrak{Y})$ satisfying the following conditions:

(1) *$T(x)$ is defined for $x \in \mathfrak{K}$, a finitely open positive cone in the sense of Defini-*

tion 9.7.1, *and has values in* $\mathfrak{E}(\mathfrak{Y})$, *that is,* $T(x)$ *is a bounded linear operator on* $\mathfrak{Y}$ *to itself for each* $x \in \mathfrak{K}$.

(2) *For* $x_1, x_2 \in \mathfrak{K}$, $y \in \mathfrak{Y}$

$$T(x_1 + x_2)y = T(x_1)[T(x_2)y]. \tag{10.8.1}$$

(3) $T(\xi x)y$ *is a strongly measurable function of* ξ *for* $\xi > 0$, $x \in \mathfrak{K}$, $y \in \mathfrak{Y}$.

(4) *For each* $x \in \mathfrak{K}$ *there exists a finite positive* $M(x)$ *such that* $\| T(\xi x) \| \leqq M(x)$ *for* $0 < \xi \leqq 1$.

Thus the operators $[T(x);\ x \in \mathfrak{K}]$ form a semi-group $\mathfrak{S}$ of linear bounded transformations, the parameter manifold being the cone $\mathfrak{K}$ in $\mathfrak{X}$ which is a semi-module with respect to vector addition. We assume that $\mathfrak{K} \neq \mathfrak{X}$.

THEOREM 10.8.1. $T(\xi x)y$ *is a continuous function of* ξ *for* $\xi > 0$ *and fixed* $x \in \mathfrak{K}$, $y \in \mathfrak{Y}$. *More generally, if* $\mathfrak{X}_{(n)}$ *is an n-dimensional linear subspace of* $\mathfrak{X}$, *then* $T(x)y$ *is a continuous function of* x *in* $\mathfrak{X}_{(n)} \cap \mathfrak{K}$. *In particular, if* $\mathfrak{K}$ *itself is finite-dimensional, then* $T(x)y$ *is continuous for* x *in* $\mathfrak{K}$.

PROOF. The first assertion follows from Theorem 10.2.3 by virtue of condition (3). The proof of the second assertion follows the same pattern as that of Theorem 9.7.1; the details are left to the reader. We note that condition (4) is not required in the proof of this theorem.

For $x \in \mathfrak{K}$ we now define

$$A_\eta(x)y = \frac{1}{\eta}[T(\eta x) - I]y, \tag{10.8.2}$$

$$A_o(x)y = \lim_{\eta \to 0} A_\eta(x)y \tag{10.8.3}$$

whenever the limit exists. The domain and range of $A_o(x)$ will be denoted by $\mathfrak{D}[A_o(x)]$ and $\mathfrak{R}[A_o(x)]$ respectively. We also introduce the range of $T(\alpha x)$ denoted by $\mathfrak{Y}_\alpha(x)$ and put $\mathfrak{Y}_0(x) = \bigcup_\alpha \mathfrak{Y}_\alpha(x)$ and $\mathfrak{Y}_0 = \bigcup_{x \in \mathfrak{K}} \mathfrak{Y}_0(x)$. The next two theorems will elucidate the relations between these linear subspaces of $\mathfrak{Y}$. In the first theorem comparisons are made for fixed x, in the second one for different values of x.

THEOREM 10.8.2. *For fixed* $x \in \mathfrak{K}$, $\mathfrak{D}[A_o(x)]$ *is dense in* $\mathfrak{Y}_0(x)$ *and the two spaces have the same closure.* $\mathfrak{R}[A_o(x)]$ *is also contained in the closure of* $\mathfrak{Y}_0(x)$. *If* y *belongs to the closure of* $\mathfrak{Y}_0(x)$, *then* $\lim_{\eta \to 0+} T(\eta x)y = y$. *Finally* $\overline{\mathfrak{Y}_0(x)} = \overline{\mathfrak{Y}}_0$.

PROOF. For fixed x in $\mathfrak{K}$, $T(\xi x)$ satisfies the conditions of Theorem 10.3.1 and the first three assertions follow from this theorem. By virtue of condition (4), the relation $\lim_{\eta \to 0+} T(\eta x)y = y$, which obviously holds in $\mathfrak{Y}_0(x)$, must also hold in the closure of this space. Finally if $y \in \mathfrak{Y}_0$ then there exists a $z \in \mathfrak{Y}$ and an $x_0 \in \mathfrak{K}$ such that $y = T(x_0)z$. Applying Theorem 10.8.1 we see that $T(\eta x)y = T(\eta x + x_0)z \to T(x_0)z = y$ as $\eta \to 0+$. Thus $\mathfrak{Y}_0 \subset \overline{\mathfrak{Y}_0(x)}$, from which it follows that $\overline{\mathfrak{Y}}_0 = \overline{\mathfrak{Y}_0(x)}$.

THEOREM 10.8.3. *For every* $\alpha > 0$ *we have* $\mathfrak{D}[A_o(\alpha x)] = \mathfrak{D}[A_o(x)]$. *Further* $\mathfrak{D}[A_o(x_1 + x_2)] \supset \mathfrak{D}[A_o(x_1)] \cap \mathfrak{D}[A_o(x_2)]$ *and* $\mathfrak{D}[A_o(x_1)A_o(x_2)]$ *is dense in the intersection. For* $y \in \mathfrak{D}[A_o(x_1)] \cap \mathfrak{D}[A_o(x_2)]$,

$$\lim_{\eta \to 0+} T(\eta x_1)A_o(x_2)y = A_o(x_2)y \tag{10.8.4}$$

and

$$A_o(x_1 + x_2)y = A_o(x_1)y + A_o(x_2)y. \tag{10.8.5}$$

PROOF. The first relation is a trivial consequence of

$$\lim_{\eta \to 0} \frac{1}{\eta\alpha} [T(\eta\alpha x) - I]y = \lim_{\eta \to 0} \frac{1}{\eta}[T(\eta x) - I]y \tag{10.8.6}$$

when either limit exists. Suppose now that $y \in \mathfrak{D}[A_o(x_1)] \cap \mathfrak{D}[A_o(x_2)]$. Then $T(\beta x_2)y \in \mathfrak{D}[A_o(x_1)]$, hence also $A_\beta(x_2)y$; the latter expression tends to a limit when $\beta \to 0$ since $y \in \mathfrak{D}[A_o(x_2)]$. Thus $A_o(x_2)y$ is in the closure of $\mathfrak{D}[A_o(x_1)]$ and (10.8.4) holds by the preceding theorem. We may of course interchange x_1 and x_2 in this relation.

Next if $y \in \mathfrak{D}[A_o(x_1)] \cap \mathfrak{D}[A_o(x_2)]$ we note that $\lim_{\alpha\to 0,\, \beta\to 0} T(\alpha x_1 + \beta x_2)y = y$. Forming

$$y_n = n^2 \int_\alpha^{\alpha+1/n} \int_\beta^{\beta+1/n} T(\xi_1 x_1 + \xi_2 x_2)y \, d\xi_2 \, d\xi_1,$$

one finds that $y_n \in \mathfrak{D}[A_o(x_1)] \cap \mathfrak{D}[A_o(x_2)]$,

$$A_o(x_1)A_o(x_2)y_n = n^2\left[T\left(\frac{1}{n}x_1\right) - I\right]\left[T\left(\frac{1}{n}x_2\right) - I\right]T(\alpha x_1 + \beta x_2)y,$$

and $\lim_{n\to\infty} y_n = T(\alpha x_1 + \beta x_2)y$. This implies that the domain of $A_o(x_1)A_o(x_2)$ is dense in the intersection of the domains of $A_o(x_1)$ and $A_o(x_2)$. Finally we shall show that if y belongs to this intersection then $A_o(x_1 + x_2)y$ exists. Indeed,

$$A_\eta(x_1 + x_2)y = T(\eta x_1)A_\eta(x_2)y + A_\eta(x_1)y.$$

Here the second term clearly tends to $A_o(x_1)y$ when $\eta \to 0$ and the first term equals

$$T(\eta x_1)[A_\eta(x_2)y - A_o(x_2)y] + T(\eta x_1)A_o(x_2)y.$$

In this expression the first term tends to θ when $\eta \to 0$, since $\| T(\eta x_1) \| \leqq M(x_1)$ for $0 < \eta \leqq 1$ and since the expression in square brackets tends to θ when $\eta \to 0$. It follows then from (10.8.4) that $A_o(x_1 + x_2)y$ exists and satisfies (10.8.5).

In the same manner we prove

THEOREM 10.8.4. $\mathfrak{D}[A_o(x_1)A_o(x_2) \cdots A_o(x_k)]$ *is dense in* $\bigcap_1^k \mathfrak{D}[A_o(x_m)]$. *In particular,* $\mathfrak{D}\{[A_o(x)]^k\}$ *is dense in* $\mathfrak{D}[A_o(x)]$.

From formulas (10.8.5) and (10.8.6) the following important result is obtained:

THEOREM 10.8.5. *$A_o(x)$ is an additive, positive-homogeneous function of x. More precisely expressed,* (i) *if $A_o(x_1)y$ and $A_o(x_2)y$ exist so does $A_o(x_1 + x_2)y$ and it is given by* (10.8.5), *and* (ii) *if $A_o(x)y$ exists so does $A_o(\alpha x)y$ and $A_o(\alpha x)y = \alpha A_o(x)y$, $\alpha > 0$.*

Finally we have the following analogue of Theorem 10.5.5.

THEOREM 10.8.6. *If*

$$\lim_{\eta \to 0+} T(\eta x)y \tag{10.8.7}$$

exists for all $x \in \mathfrak{K}$ and $y \in \mathfrak{Y}$, then the limit is a projection operator J independent of x, $J^2 = J$, which maps all of $\mathfrak{Y}$ upon $\overline{\mathfrak{Y}}_0$. Further

$$T(x) = JT(x) = T(x)J. \tag{10.8.8}$$

$T(x)$ is strongly continuous as a function of x in $\mathfrak{K} \cap \mathfrak{X}_{(n)}$ where $\mathfrak{X}_{(n)}$ is any finite-dimensional subspace of $\mathfrak{X}$, and

$$T(x)y = \lim_{\eta \to 0+} \exp [A_\eta(x)] Jy \tag{10.8.9}$$

for all y, the limit existing uniformly with respect to x in any compact subset of $\mathfrak{K} \cap \mathfrak{X}_{(n)}$.

Finally, for semi-groups satisfying (1) *and* (2), *a set of necessary and sufficient conditions that the limit in* (10.8.7) *shall exist with $J = I$ is that* (3)–(4) *hold and, in addition,* (5) *$\mathfrak{Y}_0$ is dense in $\mathfrak{Y}$.*

PROOF. Denoting temporarily the limit in (10.8.7) by $J(x)$ we obtain from Theorem 10.5.5 that $J(x)$ is a projection operator, $[J(x)]^2 = J(x)$, which maps all of $\mathfrak{Y}$ on the closure of $\mathfrak{Y}_0(x)$, and $T(\xi x) = J(x)T(\xi x) = T(\xi x)J(x)$ for $\xi > 0$. Further $T(\xi x)$ is a strongly continuous function of ξ for $\xi > 0$, and

$$T(\xi x)y = \lim_{\eta \to 0} \exp [\xi A_\eta(x)] J(x)y \tag{10.8.10}$$

for all y, the limit existing uniformly with respect to ξ in any finite interval $[0, \beta]$. Since the conditions (1) to (4) are satisfied by $T(x)$, Theorem 10.8.1 applies and $T(x)$ is strongly continuous for $x \in \mathfrak{K} \cap \mathfrak{X}_{(n)}$. Likewise by Theorem 10.8.2, $\overline{\mathfrak{Y}_0(x)} = \overline{\mathfrak{Y}}_0$ so that the range of $J(x)$ is clearly independent of x.

To prove that $J(x)$ itself is independent of x, we argue as in the proof of Theorem 9.7.2. From

$$T(\alpha x_1 + \beta x_2)y = T(\alpha x_1)T(\beta x_2)y = T(\beta x_2)T(\alpha x_1)y$$

we obtain, letting $\alpha \to 0+$ and using the continuity on finite-dimensional subspaces,

$$T(\beta x_2)y = J(x_1)T(\beta x_2)y = T(\beta x_2)J(x_1)y.$$

In the limit as $\beta \to 0+$ one has

$$J(x_2)y = J(x_1)J(x_2)y = J(x_2)J(x_1)y$$

and, interchanging x_1 and x_2, this gives $J(x_1)y = J(x_2)y$; whence $J(x_1) = J(x_2) = J$ for all $x_1, x_2 \in \mathfrak{K}$.

Formula (10.8.9) now follows from (10.8.10). In order to show that the limit exists uniformly with respect to x in any compact subset of $\mathfrak{K} \cap \mathfrak{X}_{(n)}$ it suffices to show that this is true in some relatively open set about each point $x_0 \in \mathfrak{K} \cap \mathfrak{X}_{(n)}$. As in the proof of Theorem 9.7.1, we can choose n linearly independent vectors $x_1, \cdots, x_n \subset \mathfrak{K} \cap \mathfrak{X}_{(n)}$ so that x_0 is interior to the set $\mathfrak{K}_n$ of all vectors of the form $x = \xi_1 x_1 + \xi_2 x_2 + \cdots + \xi_n x_n$ with $\xi_i \geqq 0$. It follows as in Theorem 10.8.1 that $T(x)$ is strongly continuous on $\mathfrak{K}_n$ if we define $T(\theta) = J$. Thus there exists a positive constant M such that $\| T(x) \| \leqq M$ for all $x \in \mathfrak{K}_n$, $\| x \| \leqq 1$. Likewise the rectified modulus of continuity $\mu(\delta; x, y)$ of $T(x)y$ for x on the set $E \equiv [x; x \in \mathfrak{K}_n, \| x \| \leqq \beta]$ is dominated by a quantity $\mu(\delta, \beta, y)$ which goes to zero with δ. Here we choose $\beta > \max(\| x_0 \|, 1)$ so that x_0 is interior to E relative to $\mathfrak{K}_n$. The argument of Theorem 10.4.2 now yields the estimate

$$\| \exp [A_\eta(x)]\, Jy - T(x)y \| \leqq \mu(\delta, \beta, y) + M(\beta, \delta)E_\eta h(\tau) \| y \|,$$

valid for all x in E. Here $h(\tau)$ has its previous significance, $E_\eta h(\tau) = O(\eta)$ as as $\eta \to 0$, and $M(\beta, \delta)$ depends only upon M, β, and δ. It is now evident that the convergence is uniform with respect to x in E.

The last assertion, regarding conditions (3)–(5) being necessary and sufficient in order that (10.8.7) shall exist with $J = I$, follows directly from Theorem 10.5.5 together with Theorem 10.8.2.

10.9. The set of generators. From the preceding discussion we see that with the semi-group $\mathfrak{S} = [T(x)]$ of linear bounded transformations on $\mathfrak{Y}$ to itself there is associated the set $\mathfrak{A} = \{A_o(x)\}$ of infinitesimal operators. Here $A_o(x)$ is also a linear transformation on $\mathfrak{Y}$ to itself, but ordinarily $A_o(x)$ is not bounded and hence not an element of $\mathfrak{E}(\mathfrak{Y})$. While the set $\mathfrak{S}$ is multiplicative, $\mathfrak{A}$ is additive and corresponds to the *Lie ring* of a continuous group. If $A_1, A_2 \in \mathfrak{A}$ so does $\alpha_1 A_1 + \alpha_2 A_2$ when $\alpha_1 \geqq 0$, $\alpha_2 \geqq 0$, that is, *$\mathfrak{A}$ is a semi-module with operators, the multipliers being positive numbers*. Thus there is a marked difference between our set $\mathfrak{A}$ and a Lie ring, the latter being a module admitting all real numbers as multipliers. This difference of course goes back to the basic difference between a group and a semi-group.

In the Lie ring there is also defined a notion of multiplication, the *commutator* $UV - VU$ being regarded as the product of U and V. In our case it is a priori plausible that the elements of $\mathfrak{A}$ commute, but owing to the unbounded character of $A_o(x)$, the proof of this fact requires a fairly elaborate argument in the course of which we shall establish several other important properties of $A_o(x)$.

THEOREM 10.9.1. *The operators $T(x_1)$ and $A_o(x_2)$ commute in the sense that $T(x_1)A_o(x_2)y = A_o(x_2)T(x_1)y$ provided $y \in \mathfrak{D}[A_o(x_2)]$.*

PROOF. We have

$$T(x_1)A_\eta(x_2)y = A_\eta(x_2)T(x_1)y.$$

If $y \in \mathfrak{D}[A_o(x_2)]$, the left side tends to $T(x_1)A_o(x_2)y$ when $\eta \to 0$. It follows that the right member has a limit which, by definition, is $A_o(x_2)T(x_1)y$ and that the limits are equal.

The next two theorems are due to I. Gelfand [3] for one-parameter groups but his proofs carry over to the present situation.

THEOREM 10.9.2. *If $y \in \mathfrak{D}[A_o(x)]$ and $\xi > 0$, then*

$$\frac{d}{d\xi} T(\xi x)y = A_o(x)T(\xi x)y = T(\xi x)A_o(x)y. \tag{10.9.1}$$

If y is in the closure of $\mathfrak{D}[A_o(x)]$ and $\alpha > 0$, then

$$A_o(x) \int_0^\alpha T(\xi x)y\, d\xi = [T(\alpha x) - I]y. \tag{10.9.2}$$

PROOF. The relation (10.9.1) follows directly from Theorem 10.3.3. Further we have

$$A_\eta(x) \int_0^\alpha T(\xi x)y\, d\xi = \frac{1}{\eta}\int_\alpha^{\alpha+\eta} T(\tau x)y\, d\tau - \frac{1}{\eta}\int_0^\eta T(\tau x)y\, d\tau.$$

When $\eta \to 0$ the first expression on the right tends to $T(\alpha x)y$ since $T(\xi x)y$ is a continuous function of ξ at $\xi = \alpha$ for any y, while the second tends to y provided y is in the closure of $\mathfrak{D}[A_o(x)]$. This completes the proof.

If $y \in \mathfrak{D}[A_o(x)]$ we may pass to the limit under the sign of integration obtaining

$$\int_0^\alpha T(\xi x)A_o(x)y\, d\xi = [T(\alpha x) - I]y. \tag{10.9.3}$$

These formulas have analogues for functions of several variables which may be obtained by induction from the one variable case. In deriving the formulas, attention should be paid to the order of the limiting passages and to successive restrictions imposed upon the element y. Thus if $y \in \mathfrak{D}[A_o(x_1) \cdots A_o(x_k)]$

$$\frac{\partial^k}{\partial \xi_1 \cdots \partial \xi_k} T(\xi_1 x_1 + \cdots + \xi_k x_k)y = T(\xi_1 x_1 + \cdots + \xi_k x_k)\, A_o(x_1)\cdots A_o(x_k)y. \tag{10.9.4}$$

If y is in the intersection of the closures of $\mathfrak{D}[A_o(x_1)], \cdots, \mathfrak{D}[A_o(x_k)]$ we have

$$\begin{aligned} A_o(x_1) \cdots A_o(x_k) \int_0^{\alpha_1} \cdots \int_0^{\alpha_k} T(\xi_1 x_1 + \cdots + \xi_k x_k)y\, d\xi_1 \cdots d\xi_k \\ = \prod_{j=1}^{k} [T(\alpha_j x_j) - I]y \end{aligned} \tag{10.9.5}$$

where evidently the order of the unbounded operators is immaterial. If in addition $y \in$

$\mathfrak{D}[A_o(x_1) \cdots A_o(x_k)]$ the operators $A_o(x_j)$ can be taken inside the integral, starting with $A_o(x_k)$ and ending with $A_o(x_1)$, so that

$$\int_0^{\alpha_1} \cdots \int_0^{\alpha_k} T(\xi_1 x_1 + \cdots + \xi_k x_k) A_o(x_1) \cdots A_o(x_k) y \, d\xi_k \cdots d\xi_1 = \prod_{j=1}^{k} [T(\alpha_j x_j) - I] y, \tag{10.9.6}$$

order being essential.

THEOREM 10.9.3. *For each $x \in \mathfrak{R}$ the operator $A_o(x)$ is closed.*

PROOF. Suppose that $\{y_n\}$ is a sequence of elements in $\mathfrak{D}[A_o(x)]$ and that $y_n \to y_0$, $A_o(x)y_n \to z_0$. It is required to prove that $A_o(x)y_0$ exists and equals z_0. Formula (10.9.3) holds for $y = y_n$ so that

$$\int_0^{\alpha} T(\xi x) A_o(x) y_n \, d\xi = [T(\alpha x) - I] y_n .$$

Passing to the limit with n and making use of condition (4) one obtains

$$\frac{1}{\alpha} \int_0^{\alpha} T(\xi x) z_0 \, d\xi = \frac{1}{\alpha} [T(\alpha x) - I] y_0 .$$

Here z_0 is in the closure of $\mathfrak{R}[A_o(x)]$ and hence in $\overline{\mathfrak{Y}_0(x)}$. Consequently Theorem 10.8.2 applies so that the left side tends to z_0 as $\alpha \to 0+$. Thus $A_o(x)y_0$ exists and equals z_0.

We come now to the main result:

THEOREM 10.9.4. *If y is in the intersection of $\mathfrak{D}[A_o(x_1)]$ and $\mathfrak{D}[A_o(x_1)A_o(x_2)]$ then it is also in $\mathfrak{D}[A_o(x_2)A_o(x_1)]$ and*

$$A_o(x_1)A_o(x_2)y = A_o(x_2)A_o(x_1)y.$$

PROOF. We have

$$\begin{aligned} A_o(x_1)A_o(x_2)y &= \lim_{\eta \to 0} A_\eta(x_1)A_o(x_2)y \\ &= \lim_{\eta \to 0} A_o(x_2)A_\eta(x_1)y = A_o(x_2)A_o(x_1)y \end{aligned}$$

where the last step follows from the fact that (i) $\lim_{\eta \to 0} A_\eta(x_1)y$ exists and equals $A_o(x_1)y$ and (ii) $A_o(x_2)$ is closed.

The assumption that $y \in \mathfrak{D}[A_o(x_1)]$ is essential in the above theorem and is not in general implied by the existence of $A_o(x_1)A_o(x_2)y$. The following counter example may be helpful in clarifying the situation. It is open to the objection that the corresponding set $\mathfrak{R}$ is not open, but this situation has to be faced in the next section anyway. We take $\mathfrak{Y} = C_0[0, \infty]$, the space of continuous functions on $[0, \infty]$ which vanish at both 0 and ∞; $\| y \| = \sup | y(t) |$. The example consists of the two-parameter semi-group defined by

$$[T(\xi_1, \xi_2)y](t) = \exp(-\xi_1 t - \xi_2/t)y(t)$$

for $\xi_1 \geqq 0$, $\xi_2 \geqq 0$, $(\xi_1, \xi_2) \neq (0, 0)$, the value of the transform at $t = 0$ being 0. The infinitesimal operators associated with the two boundary rays of $\mathfrak{R}$ are

$$[A_1 y](t) = -ty(t), \qquad [A_2 y](t) = -(1/t)y(t).$$

$\mathfrak{D}(A_1)$ consists of all functions in $C_0[0, \infty]$ for which $ty(t) \to 0$ as $t \to \infty$, whereas $\mathfrak{D}(A_2)$ consists of all functions in $C_0[0, \infty]$ for which $y(t)/t \to 0$ as $t \to 0+$. It is easily seen that $\mathfrak{D}(A_1A_2) = \mathfrak{D}(A_2)$ and $\mathfrak{D}(A_2A_1) = \mathfrak{D}(A_1)$; the operators A_1A_2 and A_2A_1 act like the identity on their respective domains. However it is clear that $\mathfrak{D}(A_1) \neq \mathfrak{D}(A_2)$.

10.10. The n-parameter semi-groups. One of the most important applications of the preceding theory is to the n-parameter semi-groups of linear bounded transformations. Such semi-groups occur in ergodic theory. Here $\mathfrak{X} = E_n$ is a real euclidean space of n dimensions with the usual definitions of arithmetic operations and metric. We write $x = (\xi_1, \cdots, \xi_n)$ and denote the unit vectors by $u_1, \cdots, u_n$ where $u_j = (\delta_{jk})$. The assumptions (1)–(4) of section 10.8 are now replaced by

(i) *$T(x)$ is defined for $x \in E_n^+$, the 2^n-ant in which $\xi_1 \geqq 0, \cdots, \xi_n \geqq 0$, excluding* $(0, \cdots, 0)$, *and has values in* $\mathfrak{E}(\mathfrak{Y})$.

(ii) *For $x_1, x_2 \in E_n^+$, $y \in \mathfrak{Y}$ the relation* (10.8.1) *is satisfied.*

(iii) *$T(x)y$ is a strongly measurable function of x in E_n^+ for each $y \in \mathfrak{Y}$.*

(iv) *There exists a finite positive M such that $\| T(\xi u_k) \| \leqq M$ for $0 < \xi \leqq 1$ and $k = 1, \cdots, n$.*

The set E_n^+ is a positive cone but not open. It is easy to show that measurability along the rays generated by the u_k, $k = 1, \cdots, n$, suffices to imply (iii). Thus (iii) is apparently much less restrictive than (3). Finally it is evident for $x = \xi_1 u_1 + \cdots + \xi_n u_n$ that $\| T(x) \| \leqq \prod_{k=1}^{n} \| T(\xi_k u_k) \| \leqq M^n$ for $0 \leqq \xi_k \leqq 1$, $k = 1, \cdots, n$. Consequently (iv) and (4) are equivalent.

THEOREM 10.10.1. *$T(x)y$ is a continuous function of x in the interior of E_n^+ for each $y \in \mathfrak{Y}$.*

PROOF. The proof follows the pattern of the proof of Theorem 10.2.3. Since the measurability of $T(x)y = T(\sum_1^n \xi_k u_k)y$ as a function of $(\xi_1, \cdots, \xi_n)$ is postulated, the integration can now be taken over a cube rather than an interval. The details are left to the reader. Actually condition (iv) is not required since boundedness of $\| T(x) \|$ in each interior rectangle can be proved from strong measurability following the argument of Lemma 10.2.1.

Because of the above result most of the results of sections 10.8 and 10.9 apply to the present case. We get a simpler and more satisfactory theory, however, by adding a fifth assumption:

(v) *$\mathfrak{Y}_0(u_k)$ is dense in $\mathfrak{Y}$ for $k = 1, 2, \cdots, n$.*

We recall that $\mathfrak{Y}_0(u_k)$ is the union of the range spaces of $T(\alpha u_k)$ for $\alpha > 0$. We can then state a stronger result:

THEOREM 10.10.2. *Let $\mathfrak{S} = [T(x)] = [T(\xi_1, \cdots, \xi_n)]$ be an n-parameter semi-group satisfying assumptions* (i)–(v). *Then*

(1) *$T(\xi_1, \cdots, \xi_n)$ is strongly continuous in E_n^+ and tends strongly to the identity when $(\xi_1, \cdots, \xi_n) \to (0, \cdots, 0)$;*

(2) *$T(\xi_1, \cdots, \xi_n)$ is the direct product of n continuous one-parameter semi-groups $\mathfrak{S}_k = [T_k(\xi_k)] = [T(\xi_k u_k)]$ so that*

$$T(\xi_1, \cdots, \xi_n) = \prod_1^n T_k(\xi_k);$$

(3) *$T_j(\xi_j)$ commutes with $T_k(\xi_k)$ and $\lim_{\eta \to 0} T_k(\eta)y = y$;*

(4) *$T_k(\xi_k)$ is generated by the infinitesimal operator $A_k = A_o(u_k)$ and all the generators of $\mathfrak{S}$ are of the form $A_o = \sum_1^n \xi_k A_k$, $\xi_k \geqq 0$.*

REMARK. We have assumed that all the parameters are essential. If this is not the case, the theorem is still valid but the basic generators A_k are no longer linearly independent.

PROOF. We know already that $T(\xi_1, \cdots, \xi_n)$ is continuous in Int (E_n^+), but it remains to prove continuity on the boundary. The first step is to prove that $\mathfrak{Y}_0(x)$ is dense in $\mathfrak{Y}$ for every $x \in$ Int (E_n^+). This follows from (iv) and (v) by the following argument. Let y_0 be a fixed element of $\mathfrak{Y}$. To any $\epsilon > 0$ we can find elements $y_1, \cdots, y_n$ in $\mathfrak{Y}$ and numbers ξ_{k0}, $0 < \xi_{k0} \leqq 1$, such that

$$\| y_{k-1} - T(\xi_{k0} u_k) y_k \| \leqq \epsilon, \qquad k = 1, \cdots, n.$$

If $\sum_1^n \xi_{k0} u_k = x_0$, then $x_0 \in$ Int (E_n^+) and we conclude that

$$\| y_0 - T(x_0) y_n \| \leqq \epsilon \frac{M^n - 1}{M - 1} = \epsilon_1 .$$

If $x \in$ Int (E_n^+) is given we can choose an α, $\alpha > 0$, so small that $x_0 - \alpha x \in$ Int (E_n^+). Hence

$$\| y_0 - T(\alpha x)[T(x_0 - \alpha x) y_n] \| \leqq \epsilon_1,$$

so that $\mathfrak{Y}_0(x)$ is dense in $\mathfrak{Y}$. From this it follows that the conditions of Theorem 10.8.6 are satisfied and $\lim_{\eta \to 0} T(\eta x)y = y$ for every $x \in$ Int (E_n^+), $y \in \mathfrak{Y}$. In particular

$$\lim_{\eta \to 0} T(\xi_k u_k + \eta x) y = T(\xi_k u_k) y. \tag{10.10.1}$$

Here $\xi_k u_k + \eta x \in$ Int (E_n^+) so that $T(\xi_k u_k + \eta x)y$ is a continuous function of ξ_k for $\xi_k > 0$ and fixed $\eta > 0$. It follows that $T(\xi_k u_k)y$, being the limit of a sequence of continuous functions of ξ_k, is measurable in ξ_k. Further $\| T(\xi_k u_k) \| \leqq M$ for $0 < \xi_k \leqq 1$. By Theorem 10.2.3, $T(\xi_k u_k)y$ is then a continuous function of ξ_k for $\xi_k > 0$, $k = 1, \cdots, n$, and, by Theorem 10.5.5, $\lim_{\eta \to 0} T(\eta u_k)y = y$. Since

$$T(x)y = T\left(\sum_1^n \xi_k u_k\right)y = \prod_1^n T(\xi_k u_k) y \equiv \prod_1^n T_k(\xi_k) y$$

we conclude that $T(x)y$ is continuous in E_n^+. This proves (1)–(3).

The set of generators of $\mathfrak{S}$ is $\mathfrak{A} = \{A_o(x)\}$. By Theorem 10.8.5

$$A_o(x) = A_o\left(\sum_1^n \xi_k u_k\right) = \sum_1^n \xi_k A_o(u_k) \equiv \sum_1^n \xi_k A_k .$$

Here A_k is clearly the infinitesimal operator of $T_k(\xi_k)$. This completes the proof. Regarding this theorem see also the paper of N. Dunford and I. E. Segal [**1**].

We shall return to the study of n-parameter semi-groups in Chapter XXV on Lie Semi-Groups. There the parameter set considered is a general n-parameter continuous semi-group.

References. Dunford and Segal [**1**], Gelfand [**3**].

CHAPTER XI

GENERATOR AND RESOLVENT

11.1. Orientation. Further study of the semi-group $\mathfrak{S} = [T(\xi)]$ of linear bounded transformations on a complex (B)-space to itself centers around the properties of the *generating operator* A and its *resolvent* $R(\lambda; A)$. The Laplace transform turns out to be the natural intermediary between $T(\xi)$ and $R(\lambda; A)$.

The following simple case will elucidate the situation. We take $\mathfrak{X} = Z_1$, the space of complex numbers with the usual metric, and consider linear transformations on Z_1 to itself. Here

$$T(\xi)\zeta = e^{\alpha\xi}\zeta, \qquad 0 < \xi < \infty,$$

defines a one-parameter semi-group of linear transformations, the generating transformation being

$$A\zeta = \alpha\zeta$$

with the resolvent

$$R(\lambda; A) = \frac{1}{\lambda - \alpha}.$$

The function $(\lambda - \alpha)^{-1}$ is the Laplace transform of $e^{\alpha\xi}$

$$\int_0^\infty e^{-\lambda\xi}e^{\alpha\xi}\,d\xi = \frac{1}{\lambda - \alpha}$$

for $\Re(\lambda) > \Re(\alpha)$. Conversely

$$\frac{1}{2\pi i}\int_C e^{\lambda\xi}\frac{d\lambda}{\lambda - \alpha} = e^{\alpha\xi}$$

for a suitable choice of the path of integration.

Using the heuristic *correspondence principle* $\alpha \to A$, then $e^{\alpha\xi} \to T(\xi)$, $(\lambda - \alpha)^{-1} \to R(\lambda; A)$ and we are led to the relations

$$\int_0^\infty e^{-\lambda\xi}T(\xi)\,d\xi = R(\lambda; A), \tag{11.1.1}$$

$$\frac{1}{2\pi i}\int e^{\lambda\xi}R(\lambda; A)\,d\lambda = T(\xi). \tag{11.1.2}$$

Thus *we would expect the resolvent of the generator to be the Laplace transform of the semi-group operator* $T(\xi)$, *conversely the latter should be obtainable from the resolvent by the inversion of the Laplace integral.*

These expectations turn out to be fully justified but there is of course the usual difference between the uniform and the strong case when it comes to the interpretation of the formulas. The fact that $R(\lambda; A)$ is holomorphic at infinity in the uniform case (and only in this case) simplifies its discussion.

The relations between $R(\lambda; A)$ and $T(\xi)$ provide the subject matter of the present chapter. There are two paragraphs corresponding to the uniform and the strong cases respectively.

References. Fukamiya [**1**], Hille [**9, 17**], Phillips [**9, 11**], Stone [**1, 2**], Yosida [**4**].

1. The Uniform Case

11.2. The resolvent. We suppose that $\mathfrak{B}$ is a complex Banach algebra with unit element e. Let $f(\xi)$ be a measurable function on $(0, \infty)$ to $\mathfrak{B}$ such that for all ξ_1, ξ_2

$$f(\xi_1 + \xi_2) = f(\xi_1)f(\xi_2).$$

It was shown in Theorem 9.3.1 that $f(\xi)$ is necessarily continuous for $\xi > 0$ but need not approach any limit when $\xi \to 0+$. Under the additional assumption that $\lim_{\xi\to 0} f(\xi) = e$, it was shown in Theorem 9.4.2 that there exists an $a \in \mathfrak{B}$ such that

$$f(\xi) = \exp(\xi a) = \sum_{n=0}^{\infty} \frac{\xi^n}{n!} a^n$$

and this function is also defined and satisfies the functional equation for complex values of the scalar variable.

We may consequently restrict ourselves to a study of

$$f(\zeta) = \exp(\zeta a), \qquad a \in \mathfrak{B}. \tag{11.2.1}$$

The assumption that $f(0) = e$ rather than an arbitrary idempotent $j \neq e$ is natural since we want to consider the resolvent of a. Otherwise we have to restrict ourselves to the subalgebra $j\mathfrak{B}j$ in which j plays the role of unit element.

It is clear that

$$\limsup_{\xi\to\infty} \frac{1}{\xi} \log \| \exp(\xi a) \| \leqq \| a \|$$

so that the Laplace integral (note that $e^{-\lambda\xi} = \exp(-\lambda\xi)$)

$$R(\lambda) = \int_0^\infty e^{-\lambda\xi} \exp(\xi a)\, d\xi \tag{11.2.2}$$

exists and is absolutely convergent for $\mathfrak{R}(\lambda) > \| a \|$. For such values of λ we compute the value of the integral by substituting the power series of exp (ξa) and integrating termwise. The result is

$$R(\lambda) = \sum_{n=0}^{\infty} a^n \lambda^{-n-1} = R(\lambda; a) = (\lambda e - a)^{-1}. \tag{11.2.3}$$

This function is of course holomorphic for $| \lambda | > \| a \|$ and the spectrum of a is located inside the circle $| \lambda | = \| a \|$. Since exp (ζa) is an entire function of order one and type $\leqq \| a \|$ in the sense of section 3.13, the integral in (11.2.2) may be taken along an arbitrary ray $\arg \zeta = \varphi$ instead of along the real axis in the ζ-plane. The integral will then converge absolutely for $\mathfrak{R}[\lambda \exp (i\varphi)] > \| a \|$ and represents $R(\lambda; a)$ in this half-plane. We have thus proved:

THEOREM 11.2.1. *If $a \in \mathfrak{B}$ is the generating element of $f(\zeta)$ so that $f(\zeta) =$ exp (ζa), then $R(\lambda; a)$, the resolvent of a, is the Laplace transform of $f(\zeta)$.*

11.3. Inversion of the resolvent. We can invert this Laplace transform using any one of the methods developed in section 6.3. Since $R(\lambda; a)$ is holomorphic at infinity, the situation is more favorable than in the general case and the inversion formulas give more. The best result is given by a formula which does not have a meaning in the general case.

THEOREM 11.3.1. *If Γ is a simple closed rectifiable curve surrounding the spectrum of a in the positive sense then*

$$\exp (\zeta a) = \frac{1}{2\pi i} \int_\Gamma e^{\lambda \zeta} R(\lambda; a)\, d\lambda. \tag{11.3.1}$$

PROOF. Γ may be deformed into a circle $| \lambda | = \rho > \| a \|$. Substituting the power series in $1/\lambda$ for $R(\lambda; a)$ and integrating term-wise, the power series for exp (ζa) is obtained. We note that (11.3.1) represents exp (ζa) for all complex values of ζ.

The information obtained from Theorem 6.3.2 is not quite so good. To start with we see that for $\gamma > \| a \|$, ζ real positive

$$\exp (\zeta a) = \frac{1}{2\pi i} (C, 1)\text{-}\int_{\gamma - i\infty}^{\gamma + i\infty} e^{\lambda \zeta} R(\lambda; a)\, d\lambda.$$

Actually, however, the integral exists in the Cauchy sense and we even have

$$\exp (\zeta a) = e + \frac{1}{2\pi i} \int_{\gamma - i\infty}^{\gamma + i\infty} e^{\lambda \zeta} \left[R(\lambda; a) - \frac{e}{\lambda} \right] d\lambda, \tag{11.3.2}$$

where the latter integral is absolutely convergent since the expression in square brackets is $O(| \lambda |^{-2})$. Here we have of course made use of the fact that the $(C, 1)$-limit coincides with the Cauchy limit when the latter exists. The formula can be made to represent exp (ζa) on the ray $\arg \zeta = \varphi$ by turning the line of integration so that it becomes perpendicular to the direction $\arg \lambda = -\varphi$. If use is made of this artifice, formula (11.3.2) has the same range of power as (11.3.1). The latter is of course a special case of the abstract analogue of the classical relation between an entire function of exponential type and its so-called *Borel transform*.

From Theorem 6.3.3 we conclude that

$$\exp(\xi a) = \lim_{\omega\to\infty} e^{-\omega\xi} \sum_{n=0}^{\infty} \frac{(-1)^n(\omega^2\xi)^{n+1}}{n!(n+1)!} R^{(n)}(\omega; a)$$

and if we make use of the relation (4.7.6) this becomes

$$\exp(\xi a) = \lim_{\omega\to\infty} e^{-\omega\xi} \sum_{n=0}^{\infty} \frac{(\omega^2\xi)^{n+1}}{(n+1)!} R(\omega; a)^{n+1}.$$

Summing we obtain

$$\exp(\xi a) = \lim_{\omega\to\infty} \exp\,[(\omega^2 R(\omega; a) - \omega e)\xi], \tag{11.3.3}$$

a relation which is directly verifiable since $\lim_{\omega\to\infty} \omega R(\omega; a) = e$ and hence $\omega^2 R(\omega; a) - \omega e = \omega R(\omega; a)a \to a$ as $\omega \to \infty$.

We obtain an expression of a more fundamental nature using Widder's operator

$$L_{k,\zeta}[R(\lambda; \alpha)] = \frac{(-1)^k}{k!}\left(\frac{k}{\zeta}\right)^{k+1} R^{(k)}\left(\frac{k}{\zeta}; a\right)$$

which reduces to

$$L_{k,\zeta}[R(\lambda; a)] = \left[\frac{k}{\zeta} R\left(\frac{k}{\zeta}; a\right)\right]^{k+1}$$

by formula (4.7.6). Since $\lambda R(\lambda; a) \to e$ when $|\lambda| \to \infty$, Theorem 6.3.5 leads to the following important relation

$$\exp(\zeta a) = \lim_{k\to\infty} \left[\frac{k}{\zeta} R\left(\frac{k}{\zeta}; a\right)\right]^k, \tag{11.3.4}$$

which may be rewritten in the more suggestive form

$$\exp(\zeta a) = \lim_{k\to\infty} \left[e - \frac{\zeta}{k} a\right]^{-k} \tag{11.3.5}$$

and this representation is valid for all complex values of ζ. In order to see this we note that

$$L_{k,\zeta}[R(\lambda; a)] = \frac{k}{\zeta k!} \int_0^\infty \exp\left(-\frac{k\tau}{\zeta}\right)\left(\frac{k\tau}{\zeta}\right)^k \exp(\tau a)\, d\tau$$
$$= \frac{1}{\zeta}\int_0^\infty W_0\left(\left|\frac{\tau}{\zeta}\right|; k\right) \exp(\tau a)\, d\tau$$

if the integral is taken along the ray $\arg \tau = \arg \zeta$ as is permitted. Since $d\tau/\zeta$

is real positive, this is Widder's singular integral and the conclusion of Theorem 6.3.5 is valid. This argument shows that one more of the classical definitions of the exponential function extends to the abstract case:

THEOREM 11.3.2. *For any element b of* $\mathfrak{B}$ *we have*

$$\exp(b) = \lim_{k\to\infty}\left[e - \frac{1}{k}b\right]^{-k}. \tag{11.3.6}$$

Actually a direct convergence proof shows that

$$\exp(b) = \lim_{k\to\infty}\left[e + \frac{1}{k}b\right]^{k} \tag{11.3.7}$$

is true. It turns out, however, that in operator theory the first of these definitions is preferable to the second inasmuch as the first one can be used to assign a meaning to exp (U) for certain classes of unbounded operators U while the second is useless for this purpose.

11.4. Analytical one-parameter groups of linear transformations. The most important instance of the preceding theory is that in which $\mathfrak{B} = \mathfrak{E}(\mathfrak{X})$, the space of linear bounded transformations on a complex (B)-space $\mathfrak{X}$ to itself, and $\mathfrak{G} = [T(\zeta)]$ is a one-parameter group of such transformations with

$$T(\zeta_1 + \zeta_2) = T(\zeta_1)T(\zeta_2), \qquad T(0) = I. \tag{11.4.1}$$

The discussion in sections 9.6, 11.2, and 11.3 leads to the following result.

THEOREM 11.4.1. *If the operator* $T(\xi) \in \mathfrak{E}(\mathfrak{X})$ *is defined for* $\xi > 0$ *and satisfies* (11.4.1) *for real positive values of the parameter and if* $\| T(\xi) - I \| \to 0$ *with* ξ, *then there exists an operator* $A \in \mathfrak{E}(\mathfrak{X})$ *such that* $T(\xi) = \exp(\xi A)$ *for* $\xi > 0$. *Defining* $T(\zeta) = \exp(\zeta A)$ *for all complex* ζ, *then* $\mathfrak{G} = [T(\zeta)]$ *is an analytical group. The resolvent of A is the Laplace transform of* $T(\zeta)$,

$$R(\lambda; A) = \int_0^\infty e^{-\lambda\zeta}T(\zeta)\,d\zeta, \tag{11.4.2}$$

where the integral may be taken along the ray $\arg\zeta = -\arg\lambda$ *and the representation is valid at least for* $|\lambda| > \|A\|$. *Further*

$$T(\zeta) = \frac{1}{2\pi i}\int_\Gamma e^{\lambda\zeta}R(\lambda; A)\,d\lambda, \tag{11.4.3}$$

Γ *surrounding the spectrum of A in the positive sense, and also*

$$T(\zeta) = \lim_{k\to\infty}\left\{\frac{k}{\zeta}R\left(\frac{k}{\zeta}; A\right)\right\}^{k}. \tag{11.4.4}$$

Conversely, each element A of $\mathfrak{E}(\mathfrak{X})$ *defines a one-parameter analytical group* $\mathfrak{G} = [\exp(\zeta A)]$ *and this group coincides with the family of operators defined in terms of the resolvent of A by formulas* (11.4.3) *and* (11.4.4).

COROLLARY. *A semi-group* $\mathfrak{S} = [T(\xi)]$, $\xi > 0$, *has a bounded infinitesimal generator if and only if* $T(\xi) \to I$ *in the uniform topology when* $\xi \to 0$. *Such a semi-group can always be embedded in an analytical one-parameter group.*

2. THE STRONG CASE

11.5. The resolvent. We now assume that $\mathfrak{X}$ is a complex (B)-space and that $\mathfrak{S} \equiv [T(\xi)]$ is a semi-group of linear bounded operators on $\mathfrak{X}$ to itself defined for $\xi > 0$. In order to obtain a simple theory further restrictions are required. Accordingly we shall suppose $\mathfrak{S}$ to be of class (A). As defined in section 10.6 this means that (i) $T(\xi)$ is strongly continuous for $\xi > 0$; (ii) $\bar{\mathfrak{X}}_0 = \mathfrak{X}$; and (iii) if $\mathfrak{S}$ is of type ω_0, then there exists an $\omega_1 > \omega_0$ such that for each λ, $\mathfrak{R}(\lambda) > \omega_1$, there is an operator $R(\lambda) \in \mathfrak{E}(\mathfrak{X})$ with the properties (a)

$$R(\lambda)x = \int_0^\infty e^{-\lambda\xi} T(\xi)x \, d\xi \qquad \text{for } x \in \mathfrak{X}_0, \tag{11.5.1}$$

(b) $\| R(\lambda) \|$ is bounded in the half-plane $\mathfrak{R}(\lambda) > \omega_1$, and (c) $\lim_{\lambda\to\infty} \lambda R(\lambda)x = x$ for each $x \in \mathfrak{X}$. For such semi-groups the infinitesimal operator A_0 need not be closed. However, as we shall see, A_0 has a smallest closed extension A and $R(\lambda) = R(\lambda; A)$ for $\mathfrak{R}(\lambda) > \omega_1$. For all classes except (A), ω_1 can be any number $> \omega_0$.

The postulate (iii-a) asserts that $R(\lambda)x$ is given by (11.5.1) for all $x \in \mathfrak{X}_0$ whenever $\mathfrak{R}(\lambda) > \omega_1$. Thus for $x \in \mathfrak{X}_0$ and $\eta > 0$,

$$R(\lambda)T(\eta)x = \int_0^\infty e^{-\lambda\xi} T(\xi + \eta)x \, d\xi = T(\eta)R(\lambda)x.$$

Since both operators are bounded and $\mathfrak{X}_0$ is dense in $\mathfrak{X}$, it follows that $R(\lambda)T(\eta)x = T(\eta)R(\lambda)x$ for all $x \in \mathfrak{X}$; that is, $R(\lambda)$ and $T(\eta)$ commute for $\mathfrak{R}(\lambda) > \omega_1$ and $\eta > 0$.

In view of postulate (iii-a) one would expect $R(\lambda)x$ to be represented by (11.5.1) whenever the integral in this expression is meaningful. We shall verify this conjecture in Corollary 2 to Theorem 11.5.3. For the present we are content to prove

LEMMA 11.5.1. *Suppose* $\mathfrak{S}$ *is of class* (A) *and let* x *be such that* $\int_0^1 \| T(\xi)x \| \, d\xi < \infty$ *and* $\lim_{\tau\to 0+} \tau^{-1} \int_0^\tau T(\xi)x \, d\xi = x$. *Then* $R(\lambda)x = \int_0^\infty e^{-\lambda\xi} T(\xi)x \, d\xi$ *for all* λ, $\mathfrak{R}(\lambda) > \omega_1$.

PROOF. For $\sigma > 0$, $T(\sigma)x \in \mathfrak{X}_0$ so that

$$R(\lambda)T(\sigma)x = \int_0^\infty e^{-\lambda\xi} T(\xi + \sigma)x \, d\xi, \qquad \mathfrak{R}(\lambda) > \omega_1.$$

Since $R(\lambda) \in \mathfrak{E}(\mathfrak{X})$, for $\mathfrak{R}(\lambda) > \omega_1$, we see that

$$R(\lambda)[\tau^{-1}\int_0^\tau T(\sigma)x\,d\sigma] = \tau^{-1}\int_0^\tau R(\lambda)T(\sigma)x\,d\sigma$$

(11.5.2)

$$= \tau^{-1}\int_0^\tau\int_0^\infty e^{-\lambda\xi}T(\xi+\sigma)x\,d\xi\,d\sigma.$$

Further it is clear that

$$\left\| \tau^{-1}\int_0^\tau\int_0^\infty e^{-\lambda\xi}T(\xi+\sigma)x\,d\xi\,d\sigma - \int_0^\infty e^{-\lambda\xi}T(\xi)x\,d\xi \right\|$$

$$\leqq \tau^{-1}\int_0^\tau\left[\int_0^\infty e^{-\omega_1\xi}\,\|T(\xi+\sigma)x - T(\xi)x\|\,d\xi\right]d\sigma.$$

By Theorem 3.8.3, the expression in brackets tends to zero as $\sigma \to 0+$. Hence the right side of this inequality converges to zero as $\tau \to 0+$. Thus passing to the limit in (11.5.2) as $\tau \to 0+$, the right member tends to $\int_0^\infty e^{-\lambda\xi}T(\xi)x\,d\xi$ and the left member tends to $R(\lambda)x$; this is the desired result.

The basic properties of $R(\lambda)$ are established in the following theorem.

THEOREM 11.5.1. *Let $\mathfrak{S}$ be of class (A). Then for each λ, $\Re(\lambda) > \omega_1$,*

(1) $(\lambda I - A_o)R(\lambda)x = x$ *for each x such that $\int_0^1 \|T(\xi)x\|\,d\xi < \infty$ and*

$$\lim_{\eta\to 0+}\eta^{-1}\int_0^\eta T(\xi)x\,d\xi = x;$$

(2) $R(\lambda)(\lambda I - A_o)x = x$ *for each $x \in \mathfrak{D}(A_o)$;*

(3) *if $R(\lambda)x = \theta$ for all sufficiently large real λ, then $x = \theta$.*

PROOF. In part (1) it is necessary to show that $R(\lambda)x \in \mathfrak{D}(A_o)$, $\Re(\lambda) > \omega_1$, for x such that $\int_0^1\|T(\xi)x\|\,d\xi < \infty$ and $\lim_{\eta\to 0+}\eta^{-1}\int_0^\eta T(\xi)x\,d\xi = x$. It is clear that Lemma 11.5.1 is applicable so that $R(\lambda)x = \int_0^\infty e^{-\lambda\xi}T(\xi)x\,d\xi$. Consequently

$$A_\eta R(\lambda)x = \frac{1}{\eta}\int_0^\infty e^{-\lambda\xi}[T(\xi+\eta)x - T(\xi)x]\,d\xi$$

$$= \frac{1}{\eta}(e^{\lambda\eta}-1)\int_0^\infty e^{-\lambda\xi}T(\xi)x\,d\xi - \frac{1}{\eta}\int_0^\eta e^{\lambda(\eta-\xi)}T(\xi)x\,d\xi.$$

Obviously the first term in the right member tends to $\lambda R(\lambda)x$ as $\eta \to 0+$. Furthermore the second term tends to $-x$ as $\eta \to 0+$ since

$$\left\|\frac{1}{\eta}\int_0^\eta e^{\lambda(\eta-\xi)}T(\xi)x\,d\xi - x\right\|$$

$$\leqq \int_0^\eta\left|\frac{e^{\lambda(\eta-\xi)}-1}{\eta}\right|\,\|T(\xi)x\|\,d\xi + \left\|\frac{1}{\eta}\int_0^\eta T(\xi)x\,d\xi - x\right\| = o(1).$$

This establishes (1). Next if $x \in \mathfrak{D}(A_o)$, then $\lim_{\eta\to 0+}T(\eta)x = x$ and *a fortiori* the hypothesis for (1) is satisfied. Again $R(\lambda)x = \int_0^\infty e^{-\lambda\xi}T(\xi)x\,d\xi$ so that $R(\lambda)A_\eta x = A_\eta R(\lambda)x$. Taking the limit on both sides as $\eta \to 0+$ and applying (1) we obtain $R(\lambda)A_o x = \lambda R(\lambda)x - x$, which proves (2). Finally if $R(\lambda)x = 0$ for all sufficiently large real λ, then condition (iii-c) asserts that $x = \lim_{\lambda\to\infty} \lambda R(\lambda)x = \theta$.

COROLLARY. *If $\mathfrak{S}$ is of class $(0, C_1)$ then $R(\lambda) = R(\lambda; A_o)$ for $\Re(\lambda) > \omega_0$ and A_o is a closed linear operator.*

PROOF. If $\mathfrak{S}$ is of class $(0, C_1)$ then $\int_0^1 \| T(\xi)x \| d\xi < \infty$ and further $\lim_{\eta\to 0+}\eta^{-1}\int_0^\eta T(\xi)x\, d\xi = x$ for each $x \in \mathfrak{X}$ so that according to (1) of the above theorem $(\lambda I - A_o)R(\lambda)x = x$ for all $x \in \mathfrak{X}$. By (2), $R(\lambda)(\lambda I - A_o)x = x$ for all $x \in \mathfrak{D}(A_o)$. Thus $R(\lambda)$ is both a right and a left inverse for $(\lambda I - A_o)$ and consequently $R(\lambda) = R(\lambda; A_o)$ for $\Re(\lambda) > \omega_0$. It follows that $(\lambda I - A_o)$ and hence A_o is closed since it is the inverse of a closed operator.

In general the infinitesimal operator of a semi-group of class (A) need not be closed. Thus R. S. Phillips [9] has constructed a semi-group of class $(1, A)$ whose infinitesimal operator is not closed. In order to proceed with our development it will be necessary to obtain the smallest closed extension of A_o which we denote by A and call the *infinitesimal generator*. The existence of such an extension is established in the following theorem.

THEOREM 11.5.2. *For a semi-group of class (A) the infinitesimal generator A exists and $R(\lambda) = R(\lambda; A)$ for $\Re(\lambda) > \omega_1$.*

PROOF. According to Theorem 2.11.1, the operator A_o has a closed linear extension if for each sequence $\{x_n\} \subset \mathfrak{D}(A_o)$ such that $x_n \to \theta$ and $A_o x_n \to y_0$, we necessarily have $y_0 = \theta$. By the previous theorem

$$\lambda R(\lambda)x_n - R(\lambda)A_o x_n = x_n \tag{11.5.3}$$

for $\lambda > \omega_1$. Passing to the limit we see that $R(\lambda)y_0 = \theta$ and hence by property (3) of the previous theorem that $y_0 = \theta$. Thus A is well defined. Further given an $x \in \mathfrak{D}(A)$ there exists a sequence $\{x_n\} \subset \mathfrak{D}(A_o)$ such that $x_n \to x$ and $A_o x_n \to Ax$. Again applying (11.5.3) we see that $R(\lambda)(\lambda I - A)x = x$. On the other hand if $y \in \mathfrak{X}$ then we can make use of the fact that $\mathfrak{D}(A_o)$ is dense in $\mathfrak{X}$ to obtain a sequence $\{y_n\} \subset \mathfrak{D}(A_o)$ such that $y_n \to y$. It is clear that $\lim_{\eta\to 0+} T(\eta)y_n = y_n$ for $y_n \in \mathfrak{D}(A_o)$ and hence that the hypothesis for part (1) of the previous theorem is satisfied. As a consequence $A_o R(\lambda)y_n = \lambda R(\lambda)y_n - y_n$. Now $x_n \equiv R(\lambda)y_n \to R(\lambda)y \equiv x$ and $A_o x_n = \lambda R(\lambda)y_n - y_n \to \lambda R(\lambda)y - y$. Hence $x \in \mathfrak{D}(A)$ and $Ax = AR(\lambda)y = \lambda R(\lambda)y - y$. Thus $R(\lambda)$ is both a left and a right inverse for $(\lambda I - A)$ so that $R(\lambda) = R(\lambda; A)$ for $\Re(\lambda) > \omega_1$.

A few remarks concerning the spectrum of A are now in order. We see by the above theorem that the resolvent of A, namely $R(\lambda; A)$, exists for all λ in the half-plane $\Re(\lambda) > \omega_1$. Thus the spectrum of A, $\sigma(A)$, is contained in the half-plane $\Re(\lambda) \leqq \omega_1$. As a matter of fact simple examples can be adduced showing that every point of the half-plane $\Re(\lambda) \leqq \omega_0$ may belong to $\sigma(A)$ (see section 19.2). Now $\sigma(\alpha A) = \alpha\sigma(A)$ and consequently αA is in general not the infinitesimal generator of a semi-group of class (A) unless α is real and positive; for $\alpha > 0$ it is easy to verify that αA is the infinitesimal generator of the semi-group $[T(\alpha\xi)]$ (see Theorem 13.6.1).

We next determine some properties of the elements in $\mathfrak{D}(A)$. To this end we prove

LEMMA 11.5.2. *Suppose U is a closed linear operator and that $R(\lambda; U)$ exists and is bounded for $\Re(\lambda) > \omega_1$. If $\gamma > \max(0, \omega_1)$ and $x \in \mathfrak{D}(U^2)$, then*

$$y(\xi; x) = \lim_{\omega\to\infty} \frac{1}{2\pi i} \int_{\gamma-i\omega}^{\gamma+i\omega} e^{\lambda\xi} R(\lambda; U)x \, d\lambda \tag{11.5.4}$$

defines a function continuous for $\xi \geqq 0$, $y(0; x) = x$, and

$$R(\lambda; U)x = \int_0^\infty e^{-\lambda\xi} y(\xi; x) \, d\xi, \qquad \Re(\lambda) > \gamma. \tag{11.5.5}$$

PROOF. For $x \in \mathfrak{D}(U^2)$ we obtain

$$R(\lambda; U)x = \frac{1}{\lambda} x + \frac{1}{\lambda^2} Ux + \frac{1}{\lambda^2} R(\lambda; U)U^2x \tag{11.5.6}$$

by a double application of (5.8.1). Substitution in (11.5.4) gives

$$y(\xi; x) = x + \xi Ux + \frac{1}{2\pi i} \int_{\gamma-i\infty}^{\gamma+i\infty} e^{\lambda\xi} R(\lambda; U)U^2x \frac{d\lambda}{\lambda^2}, \tag{11.5.7}$$

where the integral converges absolutely for $\xi \geqq 0$. In fact the integrand is majorized for all $\xi \geqq 0$ by the function $Me^{\gamma\xi}/|\lambda|^2$, whence it follows that $y(\xi; x)$ is continuous for $\xi \geqq 0$. For $\xi = 0$, a contour integration argument shows that the integral in (11.5.7) equals θ so that $y(0; x) = x$. Further it is clear that $\| y(\xi; x) \| = O(e^{\gamma\xi})$ as $\xi \to \infty$. Thus for $\Re(\lambda) > \gamma$ the integral (11.5.5) is absolutely convergent and substituting the right member of (11.5.7) for $y(\xi; x)$ we obtain

$$\int_0^\infty e^{-\lambda\xi} y(\xi; x) \, d\xi = \frac{1}{\lambda} x + \frac{1}{\lambda^2} Ux + \frac{1}{2\pi i} \int_0^\infty e^{-\lambda\xi} \int_{\gamma-i\infty}^{\gamma+i\infty} e^{\mu\xi} R(\mu; U)U^2x \frac{d\mu}{\mu^2} \, d\xi.$$

The double integral being absolutely convergent, we may interchange the order of integration, obtaining

$$\frac{1}{2\pi i} \int_{\gamma-i\infty}^{\gamma+i\infty} \frac{R(\mu; U)U^2x}{\lambda - \mu} \frac{d\mu}{\mu^2} = \frac{1}{\lambda^2} R(\lambda; U)U^2x.$$

Here the value of the integral may be found by the calculus of residues in the classical manner. Comparing the result with (11.5.6) we see that the proof is complete.

THEOREM 11.5.3. *Let $\mathfrak{S}$ be of class (A). If $x \in \mathfrak{D}(A^n)$, $n \geqq 1$, then*

$$\frac{d^n[T(\xi)x]}{d\xi^n} = A_0^n[T(\xi)x] = T(\xi)A^nx, \qquad \xi > 0; \tag{11.5.8}$$

if $x \in \mathfrak{D}(A^2)$, then $\lim_{\eta\to 0+} T(\eta)x = x$; further $\mathfrak{D}(A^3) \subset \mathfrak{D}(A_0)$.

PROOF. Suppose first that $x \in \mathfrak{D}(A)$. Then there exists an approximating sequence $\{x_n\} \subset \mathfrak{D}(A_0)$ such that $x_n \to x$ and $A_0 x_n \to Ax$. For $0 < \alpha < \xi < \infty$ we have $T(\xi)x_n - T(\alpha)x_n = \int_\alpha^\xi T(\tau)A_0 x_n \, d\tau$ and since $\| T(\tau) \|$ is bounded for τ in the interval $[\alpha, \xi]$, it follows that

$$T(\xi)x - T(\alpha)x = \int_\alpha^\xi T(\tau)Ax \, d\tau. \tag{11.5.9}$$

As a consequence $d[T(\xi)x]/d\xi = A_0 T(\xi)x = T(\xi)Ax$ for $\xi > 0$. We now proceed by induction. Suppose (11.5.8) is true for n and that $x \in \mathfrak{D}(A^{n+1})$. Then $d[T(\xi)A^n x]/d\xi = A_0[T(\xi)A^n x] = T(\xi)A^{n+1}x$ and this together with (11.5.8) implies (11.5.8) with n replaced by $n + 1$. It follows, incidentally, that $T(\xi)[\mathfrak{D}(A^n)] \subset \mathfrak{D}(A_0^n)$ for $\xi > 0$.

If $x \in \mathfrak{D}(A^2)$ then there exists a $y \in \mathfrak{X}$ such that $x = [R(\lambda; A)]^2 y$. Further, $\mathfrak{D}(A_0^2)$ being dense in $\mathfrak{X}$, there exists a sequence $\{y_n\} \subset \mathfrak{D}(A_0^2)$ such that $y_n \to y$. The argument employed in the latter part of Theorem 11.5.2 showed that $R(\lambda; A)y_n \in \mathfrak{D}(A_0)$ and that $A_0 R(\lambda; A)y_n \to AR(\lambda; A)y$. Likewise $A_0 R(\lambda; A)y_n = R(\lambda; A)A_0 y_n \in \mathfrak{D}(A_0)$ so that $R(\lambda; A)A_0 R(\lambda; A)y_n \in \mathfrak{D}(A_0)$ and

$$A_0 R(\lambda; A)A_0 R(\lambda; A)y_n \to AR(\lambda; A)AR(\lambda; A)y = A^2[R(\lambda; A)]^2 y = A^2 x.$$

Setting $x_n = [R(\lambda; A)]^2 y_n$, we see that $\{x_n\} \subset \mathfrak{D}(A_0^2)$, $x_n \to x$, $A_0 x_n \to Ax$, and $A_0^2 x_n \to A^2 x$. Now for $x_n \in \mathfrak{D}(A_0)$, Lemma 11.5.1 shows that

$$R(\lambda; A)x_n = \int_0^\infty e^{-\lambda\xi} T(\xi)x_n \, d\xi, \qquad \Re(\lambda) > \omega_1.$$

According to Theorem 6.3.2

$$T(\xi)x_n = \frac{1}{2\pi i} (C, 1)\text{-} \int_{\gamma - i\infty}^{\gamma + i\infty} e^{\lambda\xi} R(\lambda; A)x_n \, d\lambda, \qquad \gamma > \max(0, \omega_1).$$

However Lemma 11.5.2 also applies and shows that this integral can be replaced by

$$T(\xi)x_n = x_n + \xi Ax_n + \frac{1}{2\pi i} \int_{\gamma - i\infty}^{\gamma + i\infty} e^{\lambda\xi} R(\lambda; A)A^2 x_n \frac{d\lambda}{\lambda^2}.$$

Taking the limit as $n \to \infty$ we obtain

$$T(\xi)x = x + \xi Ax + \frac{1}{2\pi i} \int_{\gamma - i\infty}^{\gamma + i\infty} e^{\lambda\xi} R(\lambda; A)A^2 x \frac{d\lambda}{\lambda^2}, \tag{11.5.10}$$

and again by Lemma 11.5.2 it follows that $\lim_{\eta \to 0+} T(\eta)x = x$. Further we see that the hypothesis of part (1) of Theorem 11.5.1 is satisfied if $x \in \mathfrak{D}(A^2)$ and hence that $R(\lambda; A)x \in \mathfrak{D}(A_0)$. This proves the final assertion of the theorem, namely that $\mathfrak{D}(A^3) \subset \mathfrak{D}(A_0)$.

COROLLARY 1. *Let $\mathfrak{S}$ be of class (A). If $T(\xi)x_0 = \theta$ for all $\xi > 0$, then $x_0 = \theta$.*

PROOF. If $T(\xi)x_0 = \theta$ for all $\xi > 0$, then $T(\xi)[R(\lambda)]^2 x_0 = [R(\lambda)]^2 T(\xi)x_0 = \theta$

for all $\xi > 0$. According to Theorem 11.5.2, $R(\lambda) = R(\lambda; A)$ for $\Re(\lambda) > \omega_1$. Hence if $x_0 \neq \theta$, then $[R(\lambda)]^2 x_0 = [R(\lambda; A)]^2 x_0 \equiv y_0 \in \mathfrak{D}(A^2)$ and is different from the zero element. Now by Theorem 11.5.3, $T(\xi)y_0 \to y_0$ as $\xi \to 0+$, which is impossible in view of the fact that $T(\xi)y_0 = \theta$ for $\xi > 0$.

COROLLARY 2. *Let $\mathfrak{S}$ be of class (A). If x is such that $\int_0^1 \| T(\xi)x \| \, d\xi < \infty$, then $R(\lambda; A)x = \int_0^\infty e^{-\lambda\xi}T(\xi)x \, d\xi$ for all λ, $\Re(\lambda) > \omega_1$.*

PROOF. Since $T(\eta)x \in \mathfrak{X}_0$ for $\eta > 0$, it follows from postulate (iii-a) that

$$T(\eta)R(\lambda)x = R(\lambda)T(\eta)x = \int_0^\infty e^{-\lambda\xi}T(\xi + \eta)x \, d\xi = T(\eta) \int_0^\infty e^{-\lambda\xi}T(\xi)x \, d\xi.$$

Setting $y = R(\lambda)x - \int_0^\infty e^{-\lambda\xi}T(\xi)x \, d\xi$, we see that $T(\eta)y = \theta$ for all $\eta > 0$. The previous corollary asserts that $y = \theta$; and this together with Theorem 11.5.2 gives the desired result.

COROLLARY 3. *Let $\mathfrak{S}$ be of class (A). Then*

$$\mathfrak{Z}(A_0) = \mathfrak{Z}(A) = [x;\, T(\xi)x = x \text{ for all } \xi > 0].$$

PROOF. It is clear that $\mathfrak{Z}(A_0) \subset \mathfrak{Z}(A)$. If $z \in \mathfrak{Z}(A)$, then $z \in \mathfrak{D}(A^k)$ for all integers $k \geqq 1$ and hence by Theorem 11.5.3 we have $z \in \mathfrak{D}(A_0)$. Thus $z \in \mathfrak{Z}(A_0)$ and therefore $\mathfrak{Z}(A_0) = \mathfrak{Z}(A)$. Further by the corollary to Theorem 10.3.3, we see that

$$T(\xi)z - z = \int_0^\xi T(\tau)A_0 z \, d\tau = \theta, \qquad \xi > 0.$$

Conversely, if $T(\xi)x = x$ for all $\xi > 0$, then obviously $x \in \mathfrak{Z}(A_0)$.

THEOREM 11.5.4. *If $\mathfrak{S}$ is of class $(0, A)$ and $x \in \mathfrak{D}(A)$, then $\lim_{\eta \to 0+} T(\eta)x = x$ and $T(\xi)x - x = \int_0^\xi T(\tau)Ax \, d\tau$. Further $\mathfrak{D}(A^2) \subset \mathfrak{D}(A_0)$.*

PROOF. If $x \in \mathfrak{D}(A)$ there exists a $y \in \mathfrak{X}$ such that $x = R(\lambda; A)y$. We have already noted for $\mathfrak{S}$ of class $(0, A)$ that $R(\lambda)y$ has the representation (11.5.1). Consequently

$$\begin{aligned} T(\eta)x - x &= \int_0^\infty e^{-\lambda\xi}[T(\xi + \eta)y - T(\xi)y] \, d\xi \\ &= (e^{\lambda\eta} - 1) \int_0^\infty e^{-\lambda\xi}T(\xi)y \, d\xi - \int_0^\eta e^{\lambda(\eta - \xi)}T(\xi)y \, d\xi, \end{aligned}$$

and this tends to θ as $\eta \to 0+$. This proves the first assertion and the second now follows from (11.5.9). Further if $x \in \mathfrak{D}(A)$ then the above shows that the hypothesis of part (1) of Theorem 11.5.1 is satisfied and hence that $R(\lambda; A)x \in \mathfrak{D}(A_0)$. Therefore $\mathfrak{D}(A^2) = R(\lambda; A)[\mathfrak{D}(A)] \subset \mathfrak{D}(A_0)$.

THEOREM 11.5.5. *A semi-group of class $(0, A)$ is of class $(0, C_1)$ if and only if A_0 is closed, that is, if and only if $A_0 = A$.*

PROOF. We have already shown in the corollary to Theorem 11.5.1 that A_o is closed if $\mathfrak{S}$ is of class $(0, C_1)$. Conversely suppose that $A_o = A$. For $x \in \mathfrak{D}(A_o)$, $\lim_{\eta\to 0+} T(\eta)x = x$ and a fortiori $\lim_{\eta\to 0+} \eta^{-1} \int_0^\eta T(\xi)x\, d\xi = x$. Further according to the previous theorem $\eta^{-1} \int_0^\eta T(\xi)A_o x\, d\xi = A_\eta x \to A_o x$ as $\eta \to 0+$. Now by Theorem 11.5.2 each $\lambda > \omega_0$ belongs to the resolvent set $\rho(A_o)$ so that for any $y \in \mathfrak{X}$ there exists an $x \in \mathfrak{D}(A_o)$ such that $y = \lambda x - A_o x$, λ fixed and $> \omega_0$. Combining the above results we have

$$\frac{1}{\eta}\int_0^\eta T(\xi)y\, d\xi = \lambda \frac{1}{\eta}\int_0^\eta T(\xi)x\, d\xi - \frac{1}{\eta}\int_0^\eta T(\xi)A_o x\, d\xi \to \lambda x - A_o x = y$$

as $\eta \to 0+$. Thus $\mathfrak{S}$ is strongly $(C, 1)$ summable to the identity at $\xi = 0$ and therefore of class $(0, C_1)$.

For an unbounded operator A, the resolvent $R(\lambda; A)$ is not holomorphic at infinity so that $R(\lambda; A)$ can not be expanded in powers of $1/\lambda$ as in the case of a bounded operator. Nevertheless, the series exists in a certain asymptotic sense on certain subspaces dense in $\mathfrak{X}$. More precisely, we have the following

THEOREM 11.5.6. *If $\mathfrak{S}$ is of class (A), then for arbitrary $x \in \mathfrak{D}(A)$*

$$\lim \lambda R(\lambda; A)x = x \tag{11.5.11}$$

as $|\lambda| \to \infty$ in any sector of the form $|\arg \lambda| \leqq \Phi < \pi/2$. If $x \in \mathfrak{D}(A^n)$, $n \geqq 1$, and $\lambda \in \rho(A)$, then

$$R(\lambda; A)x = \sum_{k=0}^{n-1} A^k x \lambda^{-k-1} + \lambda^{-n} R(\lambda; A)A^n x. \tag{11.5.12}$$

PROOF. If $\mathfrak{S}$ is of class (A), then by assumption (iii) there exists a constant $\omega_1 > \omega_0$ such that $\| R(\lambda; A) \|$ is bounded in the half-plane $\Re(\lambda) > \omega_1$. Hence if $x \in \mathfrak{D}(A)$ then $\lambda R(\lambda; A)x = x + R(\lambda; A)Ax$ is bounded in norm in the sector defined by $|\arg \lambda| \leqq \pi/2 - \epsilon$ and $|\lambda| > \omega_1 (\sin \epsilon)^{-1}$ for each $\epsilon > 0$. Likewise by assumption (iii) we have $\lim_{\lambda\to\infty} \lambda R(\lambda; A)x = x$. The first assertion is now a direct consequence of Theorem 3.14.3. Finally if $x \in \mathfrak{D}(A^n)$, then the second relation in (5.8.1) gives

$$R(\lambda; A)A^k x = \frac{1}{\lambda}\{A^k x + R(\lambda; A)A^{k+1}x\}, \quad k = 0, 1, \cdots, n-1.$$

Successive substitutions give (11.5.12).

REMARK. If $\mathfrak{S}$ is of class $(0, C_1)$, then the relation (11.5.11) holds for all $x \in \mathfrak{X}$. Indeed, an integration by parts shows that

$$\lambda R(\lambda; A)x = \lambda^2 \int_0^\infty e^{-\lambda\tau}\tau \left[\tau^{-1}\int_0^\tau T(\sigma)x\, d\sigma\right] d\tau, \qquad \Re(\lambda) > \omega_0.$$

For any $\gamma > \max(0, \omega_0)$, there exists a constant $M(x, \gamma)$ such that $\| \tau^{-1} \int_0^\tau T(\sigma)x\, d\sigma \| \leqq Me^{\gamma\tau}$. Consequently

$$\| \lambda R(\lambda; A)x \| \leqq \frac{M \mid \lambda \mid^2}{[\Re(\lambda) - \gamma]^2}, \qquad \Re(\lambda) > \gamma.$$

The left member of this inequality is clearly bounded in each sector of the form $| \arg \lambda | \leqq \pi/2 - \epsilon$, $| \lambda | > \omega_1 (\sin \epsilon)^{-1}$, $\omega_1 > \gamma$. The rest of the proof follows as in Theorem 11.5.6.

We note that (11.5.12) holds for all n if $x \in \bigcap_n \mathfrak{D}(A^n)$, the latter subset being dense in $\mathfrak{X}$ by Theorem 10.3.4. Since $\lim_{\lambda \to \infty} \lambda R(\lambda; A)A^n x = A^n x$ by (11.5.11), we have the following

THEOREM 11.5.7. *If $\mathfrak{S}$ is of class (A) and if $x \in \bigcap_n \mathfrak{D}(A^n)$, then*

$$R(\lambda; A)x \sim \sum_{n=0}^{\infty} A^n x \lambda^{-n-1}, \tag{11.5.13}$$

the series being asymptotic to $R(\lambda; A)x$ in the sense of Poincaré, that is, for each $n \geqq 1$

$$\lim \lambda^{n+1} \left[R(\lambda; A)x - \sum_{k=0}^{n-1} A^k x \lambda^{-k-1} \right] = A^n x \tag{11.5.14}$$

as $| \lambda | \to \infty$ in any sector of the form $| \arg \lambda | \leqq \Phi < \pi/2$.

11.6. Inversion of the resolvent. The development given in section 11.3 is based on the fact that the resolvent of the infinitesimal generator can be obtained as the Laplace transform of the semi-group. For semi-groups of class (A) this is, in general, no longer the case and a parallel development does not seem feasible. However for semi-groups of class $(0, A)$ the resolvent $R(\lambda; A)$ of the infinitesimal generator is the strong Laplace transform of the semi-group operators $T(\xi)$. Consequently the standard inversion formulas developed in section 6.3 apply, giving us a representation of $T(\xi)$ in terms of $R(\lambda; A)$. We note in the strong case that there is no analogue of Theorem 11.3.1. Theorem 6.3.1 gives the following result:

THEOREM 11.6.1. *Suppose $\mathfrak{S}$ is of class $(0, A)$. Then for each $x \in \mathfrak{X}$, $\xi \geqq 0$, and $\gamma > \max(0, \omega_0)$*

$$\int_0^{\xi} T(\tau)x \, d\tau = \lim_{\omega \to \infty} \frac{1}{2\pi i} \int_{\gamma - i\omega}^{\gamma + i\omega} e^{\lambda \xi} R(\lambda; A)x \frac{d\lambda}{\lambda}, \tag{11.6.1}$$

the limit existing uniformly with respect to ξ in any finite interval. For each $x \in \mathfrak{D}(A)$ and $\xi > 0$, we have

$$T(\xi)x = \lim_{\omega \to \infty} \frac{1}{2\pi i} \int_{\gamma - i\omega}^{\gamma + i\omega} e^{\lambda \xi} R(\lambda; A)x \, d\lambda, \tag{11.6.2}$$

the limit existing uniformly with respect to ξ in any interval $(\epsilon, 1/\epsilon)$. For $\xi = 0$ the limit is $\frac{1}{2}x$.

PROOF. In applying the results of section 6.3 to the present problem, we set

$g(\xi) = T(\xi)x$, $a(\xi) = \int_0^\xi T(\tau)x\,d\tau$. The former is a continuous function of ξ, the latter a continuously differentiable function of ξ. Formula (11.6.1) is then an immediate consequence of Theorem 6.3.1. For the second assertion we make use of Theorem 11.5.4 according to which

$$\int_0^\xi T(\tau)Ax\,d\tau = T(\xi)x - x$$

for $x \in \mathfrak{D}(A)$. This together with the relation

$$R(\lambda; A)Ax = \lambda R(\lambda; A)x - x$$

gives on substitution in (11.6.1)

$$T(\xi)x - x = \lim_{\omega\to\infty} \frac{1}{2\pi i} \int_{\gamma-i\omega}^{\gamma+i\omega} e^{\lambda\xi}R(\lambda; A)x\,d\lambda - x \lim_{\omega\to\infty} \frac{1}{2\pi i} \int_{\gamma-i\omega}^{\gamma+i\omega} e^{\lambda\xi}\,\frac{d\lambda}{\lambda}.$$

The second limit is x if $\xi > 0$ but $\frac{1}{2}x$ if $\xi = 0$ and the limit exists uniformly with respect to ξ in $(\epsilon, 1/\epsilon)$. This completes the proof.

As an immediate consequence of Theorem 6.3.2 we obtain

THEOREM 11.6.2. *If $\mathfrak{S}$ is of class $(0, A)$, then for each $x \in \mathfrak{X}$, $\xi > 0$, and $\gamma >$* max $(0, \omega_0)$

$$T(\xi)x = \frac{1}{2\pi i}\,(C, 1)\text{-}\int_{\gamma-i\infty}^{\gamma+i\infty} e^{\lambda\xi}R(\lambda; A)x\,d\lambda, \tag{11.6.3}$$

the limit existing uniformly with respect to ξ in each interval $(\epsilon, 1/\epsilon)$. For $\xi = 0$ the limit equals $\frac{1}{2}x$.

Before leaving the complex inversion formulas we shall derive the duals of Theorems 11.5.6 and 11.5.7. The first is a form of Taylor's theorem with remainder.

THEOREM 11.6.3. *If $\mathfrak{S}$ is of class $(0, A)$, if $x \in \mathfrak{D}(A^n)$, and if $\xi > 0$, then*

$$T(\xi)x = \sum_{k=0}^{n-1} \frac{\xi^k}{k!} A^k x + \frac{1}{(n-1)!} \int_0^\xi (\xi - \tau)^{n-1} T(\tau)A^n x\,d\tau, \qquad n \geqq 1. \tag{11.6.4}$$

PROOF. We obtain this expansion by substituting (11.5.12) into (11.6.2). Formula (6.3.9) takes care of the remainder term and shows that

$$\lim_{\omega\to\infty} \frac{1}{2\pi i} \int_{\gamma-i\omega}^{\gamma+i\omega} e^{\lambda\xi}R(\lambda; A)A^n x\,\frac{d\lambda}{\lambda^n} = \frac{1}{(n-2)!} \int_0^\xi (\xi - \tau)^{n-2} \int_0^\tau T(\sigma)A^n x\,d\sigma\,d\tau$$

$$= \frac{1}{(n-1)!} \int_0^\xi (\xi - \tau)^{n-1} T(\tau)A^n x\,d\tau.$$

A more direct and elementary proof is obtained by repeated integration by parts in the formula

$$T(\xi)x - x = \int_0^\xi T(\tau)Ax\,d\tau;$$

here one makes use of Theorem 11.5.4 and the observation that

$$\frac{d^k}{d\xi^k}[T(\xi)x] = T(\xi)A^k x, \qquad \xi > 0, k = 1, 2, \cdots, n.$$

The dual of Theorem 11.5.7 reads as follows:

THEOREM 11.6.4. *If $\mathfrak{S}$ is of class $(0, A)$, if $x \in \bigcap_n \mathfrak{D}(A^n)$, and if $\xi > 0$, then*

$$T(\xi)x \sim \sum_{n=0}^{\infty} \frac{\xi^n}{n!} A^n x, \tag{11.6.5}$$

the series being asymptotic in the sense of Poincaré to $T(\xi)x$ on the positive real axis, that is, for each $n \geqq 1$

$$\lim_{\xi\to 0+} \xi^{-n}\left[T(\xi)x - \sum_{k=0}^{n-1} \frac{\xi^k}{k!} Ax^k\right] = \frac{1}{n!} A^n x. \tag{11.6.6}$$

PROOF. If $x \in \bigcap_n \mathfrak{D}(A^n)$, then in particular $A^n x \in \mathfrak{D}(A)$ so that $\lim_{\tau\to 0+} T(\tau)A^n x = A^n x$ by Theorem 11.5.4, as a consequence it is easy to verify that

$$\lim_{\xi\to 0+} n\xi^{-n} \int_0^\xi (\xi - \tau)^{n-1} T(\tau)A^n x\,d\tau = A^n x.$$

The relation (11.6.6) now follows directly from (11.6.4).

THEOREM 11.6.5. *If $\mathfrak{S}$ is of class $(0, A)$, then*

$$T(\xi)x = \lim_{\omega\to\infty} \exp\{[\omega^2 R(\omega; A) - \omega I]\xi\}\, x \tag{11.6.7}$$

for each $x \in \mathfrak{X}$, the limit existing uniformly with respect to ξ in each interval of the form $(\epsilon, 1/\epsilon)$.

PROOF. As in the case of formula (11.3.3) we find that for $\omega > \omega_0$

$$e^{-\omega\xi} \sum_{n=0}^{\infty} \frac{(-1)^n(\omega^2\xi)^{n+1}}{n!(n+1)!} R^{(n)}(\omega; A)x = \exp\{[\omega^2 R(\omega; A) - \omega I]\xi\}\, x - e^{-\omega\xi}x.$$

According to Theorem 6.3.3 this expression tends to $T(\xi)x$ as $\omega \to \infty$, uniformly with respect to ξ in $(\epsilon, 1/\epsilon)$. In order to verify (11.6.7) we have merely to note that $e^{-\omega\xi}x \to \theta$ as $\omega \to \infty$ uniformly in $(\epsilon, 1/\epsilon)$.

Formula (11.6.7) can be rewritten as

$$T(\xi)x = \lim_{\omega\to\infty} \exp(B_\omega \xi)x \tag{11.6.8}$$

where $B_\omega = \omega^2 R(\omega; A) - \omega I$. The formula becomes somewhat more apparent if we note that $\exp(B_\omega \xi)$ is a semi-group with infinitesimal generator B_ω and that for $x \in \mathfrak{D}(A)$ we have $B_\omega x = \omega R(\omega; A)Ax \to Ax$ as $\omega \to \infty$.

THEOREM 11.6.6. *If $\mathfrak{S}$ is of class $(0, A)$, then for each $x \in \mathfrak{X}$ and $\xi > 0$ we have*

$$T(\xi)x = \lim_{k\to\infty} \left[\frac{k}{\xi} R\left(\frac{k}{\xi}; A\right)\right]^k x, \tag{11.6.9}$$

an alternate form of which is

$$T(\xi)x = \lim_{k\to\infty} \left[I - \frac{\xi}{k} A\right]^{-k} x. \tag{11.6.10}$$

The limits exist uniformly with respect to ξ in each interval of the form $(\epsilon, 1/\epsilon)$.

PROOF. Just as in (11.3.4) we find that

$$g(\xi \mid k) \equiv L_{k,\xi}[R(\lambda; A)x] = \left[\frac{k}{\xi} R\left(\frac{k}{\xi}; A\right)\right]^{k+1} x$$

and by Theorem 6.3.5 this expression tends to $T(\xi)x$ as $k \to \infty$ uniformly with respect to ξ in $(\epsilon, 1/\epsilon)$. Since $T(\xi)x$ is continuous for $\xi > 0$, it follows that

$$g\left(\frac{k}{k+1}\xi \mid k\right) \to T(\xi)x$$

as $k \to \infty$, likewise uniformly with respect to ξ in $(\epsilon, 1/\epsilon)$. But this is just the assertion (11.6.9) with k replaced by $k + 1$.

Formula (11.6.10) is the analogue of (11.3.6) for the strong case. On the other hand, (11.3.7) does not have a meaning in the strong case. Even if we restrict x to belong to $\bigcap_n D(A^n)$, it does not seem likely that

$$\lim_{k\to\infty} \left[I + \frac{\xi}{k} A\right]^k x$$

would in general exist.

11.7. Special properties. The intimate connection existing between the semi-group of operators $[T(\xi); \xi > 0]$ and the resolvent operators $[R(\lambda; A); \lambda > \omega_0]$ is made especially clear by the inverse relations

$$\lambda R(\lambda; A)x = \int_0^\infty \lambda e^{-\lambda\xi} T(\xi)x \, d\xi, \qquad \lambda > \omega_0, \tag{11.7.1}$$

and

$$T(\xi)x = \lim_{\lambda\to\infty} e^{-\lambda\xi} \sum_{n=0}^{\infty} \frac{(\lambda\xi)^n}{n!} [\lambda R(\lambda; A)]^n x, \qquad \xi > 0, \tag{11.7.2}$$

which result form Theorems 11.5.2 and 11.6.5 respectively. We see that $\lambda R(\lambda; A)x$ is the weighted average of $[T(\xi)x; \xi > 0]$, the weighting factor being $\lambda e^{-\lambda\xi} > 0$

with $\int_0^\infty \lambda e^{-\lambda\xi}\, d\xi = 1$. Similarly the approximating sums in (11.7.2) consist of powers of $[\lambda R(\lambda;\, A)]$, the nth power being weighted by $e^{-\lambda\xi}(\lambda\xi)^n/n! > 0$ and $\sum_0^\infty e^{-\lambda\xi}(\lambda\xi)^n/n! = 1$. Thus the elements of the set $[\lambda R(\lambda;\, A);\, \lambda > \omega_0]$ lie in the strongly closed convex extension of the set $[T(\xi);\, \xi > 0]$, whereas the elements of the latter set lie in the strongly closed convex extension of the set $[[\lambda R(\lambda;\, A)]^n;\, \lambda > \omega_0,\, n \geqq 1]$. It is evident that certain operator properties possessed by one of these sets must also be possessed by the others. We now list a few of these properties. The proofs are immediate consequences of the formulas (11.7.1) and (11.7.2) and the verifications are left to the reader.

DEFINITION 11.7.1. *We say that a set of operators* $[T_\alpha;\, \alpha \in \mathfrak{A}]$ *leaves the element* $x_0 \in \mathfrak{X}$ *invariant if* $T_\alpha x_0 = x_0$ *for all* $\alpha \in \mathfrak{A}$. *It leaves the linear bounded functional* $x_0^* \in \mathfrak{X}^*$ *invariant if* $x_0^*[T_\alpha x] = x_0^*(x)$ *for all* $\alpha \in \mathfrak{A}$ *and all* $x \in \mathfrak{X}$, *that is to say, if the adjoint operators* $[T_\alpha^*;\, \alpha \in \mathfrak{A}]$ *leave* x_0^* *invariant.*

DEFINITION 11.7.2. *A linear bounded operator* T *is called a contraction operator if* $\| T \| \leqq 1$.

THEOREM 11.7.1. *Let* $\mathfrak{S}$ *be of class* $(0, A)$. *A necessary and sufficient condition that* $T(\xi)$ *for each* $\xi > 0$

(1) *leave the element* $x_0 \in \mathfrak{X}$ *invariant,*

(2) *leave the linear bounded functional* $x_0^* \in \mathfrak{X}^*$ *invariant,*

(3) *be a contraction operator,*

is that $\lambda R(\lambda;\, A)$ *have the corresponding property for each* $\lambda > \omega_0$.

Let $\mathfrak{X}$ be a complex (B)-space and suppose, in addition, that $\mathfrak{X}$ is a partially ordered linear space in the sense of Definition 1.11.1 when considered over the real number field. We further assume that the positive cone $\mathfrak{X}^+$ is closed in the norm topology. In such spaces we can introduce the concept of a *positive operator.*

DEFINITION 11.7.3. *A linear bounded operator* $T \in \mathfrak{E}(\mathfrak{X})$ *is said to be positive if it maps* $\mathfrak{X}^+$ *into itself.*

THEOREM 11.7.2. *If* $\mathfrak{S}$ *is of class* $(0, A)$, *then* $T(\xi)$ *is a positive operator for each* $\xi > 0$ *if and only if* $\lambda R(\lambda;\, A)$ *is a positive operator for each* $\lambda > \omega_0$.

Suppose that $\mathfrak{X}$ is further restricted so that

$$\| x_1 + x_2 \| = \| x_1 \| + \| x_2 \| \quad \text{whenever } x_1,\, x_2 \in \mathfrak{X}^+. \tag{11.7.3}$$

Ordinary $L_1(\mathfrak{S};\, m)$ spaces are examples of such spaces.

DEFINITION 11.7.4. *A positive contraction operator* $T \in \mathfrak{E}(\mathfrak{X})$ *with the property that* $\| Tx \| = \| x \|$ *for all* $x \in \mathfrak{X}^+$ *is called a transition operator.*

THEOREM 11.7.3. *If* $\mathfrak{S}$ *is of class* $(0, A)$, *then* $T(\xi)$ *is a transition operator for each* $\xi > 0$ *if and only if* $\lambda R(\lambda;\, A)$ *is a transition operator for each* $\lambda > \omega_0$,

It is evident from the relation (11.7.2) that in the above theorems the property required of $\lambda R(\lambda;\, A)$ need hold only for sufficiently large real λ.

11.8. The exponential formulas. Throughout this treatise we have been guided by the fact that the semi-group operator $T(\xi)$ is essentially an exponential function, $T(\xi) = \exp(\xi A)$, of the infinitesimal generator A. In the uniform case the interpretation is straightforward.

$$T(\xi) = \sum_{n=0}^{\infty} \frac{\xi^n}{n!} A^n,$$

and, since the generator is bounded, the series converges in the uniform operator topology of $\mathfrak{E}(\mathfrak{X})$. In the strong case, A is unbounded and the definition of the exponential function is less direct. We have listed below the ten exponential formulas proved so far which serve as justification of our use of the symbol $\exp(\xi A)\, x$ for $T(\xi)x$. They are the formulas (10.4.2), (10.4.16), and eight of the formulas of section 11.6. In the case of (E_1) and of (E_2) with $\alpha = 0$, $T(\xi)$ is assumed to be of class (C_0); in all other cases $T(\xi)$ is of class $(0, A)$. Unless otherwise stated, x is an arbitrary element of $\mathfrak{X}$ and $\gamma > \max(0, \omega_0)$. In general the limit relations hold uniformly with respect to ξ in each interval of the form $(\epsilon, 1/\epsilon)$; in the case of (E_1) and (E_5) the interval is $(0, 1/\epsilon)$; in the case of (E_2) it is $(\alpha, \alpha + 1/\epsilon)$.

$$(E_1)\quad T(\xi)x = \lim_{\eta \to 0+} \exp(\xi A_\eta)\, x;$$

$$(E_2)\quad T(\xi)x = \lim_{\eta \to 0+} \sum_{n=0}^{\infty} \frac{(\xi - \alpha)^n}{n!} \Delta_\eta^n T(\alpha)x, \qquad \alpha \geqq 0;$$

$$(E_3)\quad T(\xi)x = \sum_{k=0}^{n-1} \frac{\xi^k}{k!} A^k x + \frac{1}{(n-1)!} \int_0^{\xi} (\xi - \tau)^{n-1} T(\tau) A^n x \, d\tau, \qquad x \in \mathfrak{D}(A^n);$$

$$(E_4)\quad \lim_{\xi \to 0+} \xi^{-n} \left[T(\xi)x - \sum_{k=0}^{n-1} \frac{\xi^k}{k!} A^k x \right] = \frac{1}{n!} A^n x, \qquad n \geqq 1, x \in \textstyle\bigcap_n \mathfrak{D}(A^n);$$

$$(E_5)\quad \int_0^{\xi} T(\tau)x \, d\tau = \lim_{\omega \to \infty} \frac{1}{2\pi i} \int_{\gamma - i\omega}^{\gamma + i\omega} e^{\lambda\xi} R(\lambda; A)x \frac{d\lambda}{\lambda};$$

$$(E_6)\quad T(\xi)x = \lim_{\omega \to \infty} \frac{1}{2\pi i} \int_{\gamma - i\omega}^{\gamma + i\omega} e^{\lambda\xi} R(\lambda; A)x \, d\lambda, \qquad x \in \mathfrak{D}(A);$$

$$(E_7)\quad T(\xi)x = \frac{1}{2\pi i} (C, 1)\text{-} \int_{\gamma - i\infty}^{\gamma + i\infty} e^{\lambda\xi} R(\lambda; A)x \, d\lambda;$$

$$(E_8)\quad T(\xi)x = \lim_{\omega \to \infty} \exp\{\xi[\omega^2 R(\omega; A) - \omega I]\}\, x;$$

$$(E_9)\quad T(\xi)x = \lim_{k \to \infty} \left[\frac{k}{\xi} R\left(\frac{k}{\xi}; A\right) \right]^k x;$$

$$(E_{10})\quad T(\xi)x = \lim_{k \to \infty} \left[I - \frac{\xi}{k} A \right]^{-k} x.$$

Other exponential formulas will be encountered later, in particular in connection with analytical semi-groups. Formula (10.4.18) could also be added to this collection.

The formulas listed above separate into several distinct groups according to the device used in defining the exponential function. We shall encounter some of these devices in a more general setting in Chapter XV.

CHAPTER XII

GENERATION OF SEMI-GROUPS

12.1. Orientation. We next consider the converse problem: *What properties should an operator U possess in order that it be the infinitesimal generator of a semi-group $[T(\xi)]$ of linear bounded operators?*

In order to make this problem more precise the type of convergence at the origin as well as the type of continuity for $\xi > 0$ must be specified. If we require that $T(\xi)$ tend to I in the uniform operator topology as $\xi \to 0+$, then the solution is obvious: *U must be bounded and every bounded operator generates a semi-group holomorphic in the plane.* The interesting problems arise when we require that $[T(\xi)]$ belong to one of the basic classes defined in section 10.6; these are the problems considered in the present chapter. We shall obtain necessary and sufficient conditions that U generate a semi-group belonging to any of the classes (C_0), $(1, A)$, $(1, C_1)$, (A), or $H(\Phi_1, \Phi_2)$, and sufficient conditions that U generate a semi-group belonging to the class $(1, A)_u$ or $(A)_\infty$.

The desirability of treating semi-groups of a general nature now becomes apparent; the more general the semi-group class, the less stringent are the conditions on U and the easier it becomes to establish the fact that U generates a semi-group. The conditions which we consider are not conditions on U directly, but rather on $R(\lambda; U)$, the resolvent of U. The methods which we employ consist in showing that, suitably restricted, $R(\lambda; U)$ is the strong Laplace transform of a one-parameter family of linear bounded operators having the semi-group property.

The first solution to a problem of this kind was obtained independently by E. Hille [**13**] and K. Yosida [**3**] in 1948; this is the so-called Hille-Yosida theorem which gives sufficient conditions on U that it generate a semi-group of class (C_0).

W. Feller [**6**] has treated the converse problem in an extremely general setting. Defining the infinitesimal generator somewhat more narrowly than is customary, he has established necessary and sufficient conditions that a linear operator U be the "infinitesimal generator" of a semi-group of linear bounded operators strongly continuous for $\xi > 0$ and with $\mathfrak{X}_0$ dense in $\mathfrak{X}$. The result is somewhat pathological from our point of view in that the "resolvent" $(\lambda I - U)^{-1}$ is in general unbounded for all λ with $\Re(\lambda) > \omega_0$.

There are three paragraphs in the chapter: *Generation of Semi-Groups Strongly Continuous for $\xi > 0$*, *Generation of Semi-Groups Uniformly Continuous for $\xi > 0$*, and *Generation of Semi-Groups Holomorphic in a Sector.*

References. Feller [6], Hille [**13, 17, 20**], Miyadera [**2, 3, 4**], Phillips [**8, 9, 11**], Phragmén and Lindelöf [**1**], Widder [**1**], Yosida [**3**].

1. Generation of Semi-Groups Strongly Continuous for $\xi > 0$

12.2. Preliminaries. This section is concerned with the general setting of the problem at hand, namely the generation of semi-groups. We desire first of all that the problem be "well set" in the sense that

(a) *there exist at most one solution;*

(b) *there exist at least one solution;*

(c) *the solution be stable, that is, the semi-group be a continuous function of the infinitesimal generator.*

We shall take these requirements up in order. The fact that a semi-group of operators is uniquely determined by its infinitesimal generator is proved in Theorem 12.2.1 below; the existence of a semi-group having a given operator as its infinitesimal generator is established under various hypotheses in the body of the present chapter; the stability is treated in the following chapter.

THEOREM 12.2.1. *A closed linear operator U can be the infinitesimal generator of at most one semi-group of class (A).*

PROOF. Suppose that U is the infinitesimal generator of two semi-groups $[T_1(\xi)]$ and $[T_2(\xi)]$, each of class (A). Let $x \in \mathfrak{D}(U^2)$. Then according to Theorem 11.5.3 we have $\lim_{\xi \to 0+} T_k(\xi)x = x$, $k = 1, 2$, and hence by Lemma 11.5.1 and Theorem 11.5.2

$$R(\lambda; U)x = \int_0^\infty e^{-\lambda\xi} T_k(\xi)x \, d\xi, \qquad k = 1, 2,$$

for all sufficiently large real λ. Since $T_k(\xi)x$, $k = 1, 2$, is continuous for $\xi > 0$, the uniqueness theorem for Laplace transforms (Theorem 6.2.3) implies that $T_1(\xi)x = T_2(\xi)x$ for all $\xi > 0$ and $x \in \mathfrak{D}(U^2)$. Now bounded linear operators are uniquely determined by their values on a dense set. Since $\mathfrak{D}(U^2)$ is dense in $\mathfrak{X}$ by Theorem 10.3.4, it follows that the two semi-groups coincide.

COROLLARY. *If a closed linear operator generates two semi-groups, $[T_1(\xi)]$ of class $H(\Phi_{11}, \Phi_{12})$ and $[T_2(\xi)]$ of class $H(\Phi_{21}, \Phi_{22})$, then $T_1(\zeta) \equiv T_2(\zeta)$ in the sector: $\max(\Phi_{11}, \Phi_{21}) < \arg \zeta < \min(\Phi_{12}, \Phi_{22})$.*

PROOF. Since $T_k(\xi)$, $\xi > 0$, is of class (C_0) for $k = 1, 2$, it follows from the previous theorem that $T_1(\zeta) = T_2(\zeta)$ for all real $\zeta > 0$. By Theorem 3.11.5 this remains true for all ζ in the common part of their domains of analyticity.

We now turn our attention to the existence question. Here a slight notational advantage is achieved if we consider only semi-groups of negative type, that is, semi-groups with $\omega_0 < 0$. A simple transformation of the form $T(\xi) \to S(\xi) \equiv e^{-\omega\xi}T(\xi)$, $\omega > \omega_0$, transforms $[T(\xi)]$ into a semi-group of negative type. More generally, we have

THEOREM 12.2.2. *Suppose $[T(\xi)]$ is a semi-group continuous in the strong (uniform) operator topology for $\xi > 0$. Then $S(\xi) \equiv e^{-\omega\xi}T(\xi)$ defines a semi-group* *continuous in the strong (uniform) operator topology for* $\xi > 0$ *and*

$$\omega_0(S) = \omega_0(T) - \Re(\omega). \tag{12.2.1}$$

The range space for $[S(\xi)]$ *coincides with that for* $[T(\xi)]$. *If* A_o *and* B_o *are the infinitesimal operators of* $T(\xi)$ *and* $S(\xi)$, *respectively, then* $\mathfrak{D}(A_o) = \mathfrak{D}(B_o)$ *and* $B_o = A_o - \omega I$. *If* A_o *has a smallest closed extension* A, *then so does* B_o, *namely* $B = A - \omega I$ *where* $\mathfrak{D}(B) = \mathfrak{D}(A)$. *If* $T(\xi)$ *is strongly differentiable for* $\xi > 0$ *then so is* $S(\xi)$.

Proof. The semi-group property and the strong (uniform) continuity of $S(\xi)$ is evident. The relation (12.2.1) follows from the fact that

$$\log \| S(\xi) \| = \log \| T(\xi) \| - \Re(\omega)\xi.$$

It is also clear that the semi-groups have the same range space. If $x \in \mathfrak{D}(A_o)$, then

$$\frac{S(\eta)x - x}{\eta} = e^{-\omega\eta} \frac{T(\eta)x - x}{\eta} + \frac{e^{-\omega\eta} - 1}{\eta} x,$$

from which it follows that $x \in \mathfrak{D}(B_o)$ and that $B_o x = A_o x - \omega x$. Thus $\mathfrak{D}(B_o) \supset \mathfrak{D}(A_o)$ and since $T(\xi) = e^{\omega\xi}S(\xi)$ the same argument applies to prove that $\mathfrak{D}(A_o) \supset \mathfrak{D}(B_o)$. The assertion about the closure of B_o is a simple consequence of the fact that $A - \omega I$ with domain $\mathfrak{D}(A)$ is a closed operator. Finally $T(\xi)$ is strongly differentiable for $\xi > 0$ if and only if $\mathfrak{X}_0(T) \subset \mathfrak{D}(A_o)$. Since $\mathfrak{X}_0(S) = \mathfrak{X}_0(T)$ and $\mathfrak{D}(B_o) = \mathfrak{D}(A_o)$, it follows that $S(\xi)$ is likewise strongly differentiable for $\xi > 0$.

Referring to the properties $(\mathrm{i})_0$, $(\mathrm{i})_1$, (C_0), (C_1), (A), and (A)′ of section 10.6, we now show that $T(\xi)$ $(= e^{\omega\xi}S(\xi))$ and $S(\xi) = e^{-\omega\xi}T(\xi)$ possess the same properties. In view of the symmetry in this relationship, it is sufficient to show that $S(\xi)$ has all of the properties belonging to $T(\xi)$. This is quite obvious for the integrability conditions $(\mathrm{i})_0$ and $(\mathrm{i})_1$ and for the condition (C_0). On the other hand if $T(\xi)$ satisfies $(\mathrm{i})_0$ and (C_1), then

$$\left\| \frac{1}{\eta} \int_0^\eta S(\xi)x \, d\xi - x \right\|$$
$$\leq \int_0^\eta \left| \frac{e^{-\omega\xi} - 1}{\eta} \right| \| T(\xi)x \| \, d\xi + \left\| \frac{1}{\eta} \int_0^\eta T(\xi)x \, d\xi - x \right\| = o(1)$$

as $\eta \to 0+$ so that $S(\xi)$ also satisfies (C_1). Likewise if $T(\xi)$ satisfies $(\mathrm{i})_0$ and (A) then

$$R_S(\lambda)x = \int_0^\infty e^{-\lambda\xi}S(\xi)x \, d\xi = \int_0^\infty e^{-(\lambda+\omega)\xi}T(\xi)x \, d\xi = R_T(\lambda + \omega)x$$

and hence by Theorem 11.5.6

$$\lambda R_S(\lambda)x = \frac{\lambda}{\lambda + \omega} [(\lambda + \omega)R_T(\lambda + \omega)x] \to x$$

as $\lambda \to \infty$; thus $S(\xi)$ also satisfies (A). Finally if $[T(\xi)]$ satisfies (A)$'$, then since the range sets for $[T(\xi)]$ and $[S(\xi)]$ coincide, $R_S(\lambda)x = R_T(\lambda + \omega)x$ on $\mathfrak{X}_0$. Consequently $R_S(\lambda)$ has a linear bounded extension on $\mathfrak{X}$ for $\Re(\lambda) > \omega_1 - \Re(\omega)$, $\| R_S(\lambda) \|$ is bounded in this domain, and as above $\lim_{\lambda\to\infty} \lambda R_S(\lambda)x = x$ for each $x \in \mathfrak{X}$; thus $[S(\xi)]$ satisfies (A)$'$.

If $[T(\xi)]$ is of class $H(\Phi_1, \Phi_2)$ then one readily verifies that $S(\zeta) \equiv e^{-\omega\zeta}T(\zeta)$ satisfies the conditions (i) to (iv) of Definition 10.6.1. Consequently $[S(\zeta)]$ is likewise of class $H(\Phi_1, \Phi_2)$. We have now proved

THEOREM 12.2.3. *The semi-groups $[T(\xi)]$ and $[S(\xi) = e^{-\omega\xi}T(\xi)]$ belong to the same class.*

The transformation $T(\xi) \to S(\xi) = e^{-\omega\xi}T(\xi)$ on semi-groups thus induces the mapping $A \to B - \omega I$ on the infinitesimal generators and conversely. Thus if U is the infinitesimal generator of a semi-group $[T(\xi)]$ belonging to a certain class, then $U - \omega I$ is the infinitesimal generator of the semi-group $[S(\xi) = e^{-\omega\xi}T(\xi)]$ belonging to the same class. In particular, ω can be chosen so that $[S(\xi)]$ is of negative type. In terms of the resolvent this amounts to working with $R(\lambda; U - \omega I) \equiv R(\lambda + \omega; U)$ instead of $R(\lambda; U)$. Thus without loss of generality we can limit ourselves to semi-groups of negative type.

It is evident from section 10.6 that $\mathfrak{D}(A)$ is dense in $\mathfrak{X}$ and $\| R(\lambda; A) \| = O(1/\lambda)$ as $\lambda \to \infty$ for any semi-group belonging to one of the basic classes. Accordingly we shall consider as possible infinitesimal generators only closed linear operators having these properties. Nevertheless the theory remains valid even when $\mathfrak{D}(A)$ is not dense in $\mathfrak{X}$ if we replace $\mathfrak{X}$ by $\mathfrak{Y} = \overline{\mathfrak{D}(A)}$ and A by its restriction to $\mathfrak{Y}$. This is a direct consequence of Theorem 12.2.4 below.

LEMMA 12.2.1. *Let U be a closed linear operator such that $\| R(\lambda; U) \| = O(1/\lambda)$ as $\lambda \to \infty$ and set $\mathfrak{Y} = \overline{\mathfrak{D}(U)}$. Then $\lim_{\lambda\to\infty} \lambda R(\lambda; U)y = y$ for each $y \in \mathfrak{Y}$.*

PROOF. If $x \in \mathfrak{D}(U)$, then $\| \lambda R(\lambda: U)x - x \| = \| R(\lambda; U)Ux \| = O(1/\lambda)$ as $\lambda \to \infty$. The Banach-Steinhaus theorem now implies that $\lim_{\lambda\to\infty} \lambda R(\lambda; U)y = y$ for all $y \in \overline{\mathfrak{D}(U)} = \mathfrak{Y}$.

THEOREM 12.2.4. *Let U be a closed linear operator such that $\| R(\lambda; U) \| = O(1/\lambda)$ as $\lambda \to \infty$ and set $\mathfrak{Y} = \overline{\mathfrak{D}(U)}$. Define U_0 to be the restriction of U with domain $\mathfrak{D}(U_0) = [y; y$ and $Uy \in \mathfrak{Y}]$. Then U_0 is a closed linear operator with $\overline{\mathfrak{D}(U_0)} = \mathfrak{Y}$ and $R(\lambda; U_0) = R(\lambda; U)_0$, the restriction of $R(\lambda; U)$ to $\mathfrak{Y}$, for all $\lambda \in \rho(U)$.*

PROOF. It is clear that U_0 has a closed extension, namely U. The smallest closed extension will still have domain and range in $\mathfrak{Y}$ and will also be a restriction of U. Since U_0 was chosen as the maximal restriction of this sort, we see that U_0 is itself closed. Suppose next that $\lambda \in \rho(U)$ and define $R(\lambda; U)_0y = R(\lambda; U)y$ for all $y \in \mathfrak{Y}$. Then if $y \in \mathfrak{D}(U_0)$, $(\lambda I - U_0)y \in \mathfrak{Y}$ and $R(\lambda; U)_0(\lambda I - U_0)y = y$. On the other hand if $y \in \mathfrak{Y}$, then $R(\lambda; U)_0y \in \mathfrak{D}(U)$ which is contained in $\mathfrak{Y}$ and $(\lambda I - U)R(\lambda; U)_0y = y$. Thus $UR(\lambda; U)_0y = \lambda R(\lambda; U)_0y - y \in \mathfrak{Y}$. It follows that $R(\lambda; U)_0 y \in \mathfrak{D}(U_0)$ and hence that $(\lambda I - U_0)R(\lambda; U)_0 y = y$. Therefore $R(\lambda; U_0) = R(\lambda; U)_0$. Finally we show that $\mathfrak{D}(U_0)$ is dense in $\mathfrak{Y}$. Let y be an arbitrary element of $\mathfrak{Y}$; then as above $\lambda R(\lambda; U)_0 y \in \mathfrak{D}(U_0)$. Lemma 12.2.1 now asserts that $\lim_{\lambda\to\infty} \lambda R(\lambda; U)_0 y = y$ so that $y \in \overline{\mathfrak{D}(U_0)}$.

12.3. Semi-groups of class (C_0). Our first theorem on the generation of semi-groups is for semi-groups of class (C_0). A semi-group $[T(\xi)]$ of this class, normalized to be of negative type, will be bounded in norm for all $\xi > 0$. Consequently Theorem 6.3.6 applies and the Laplace transform of $T(\xi)$, namely $R(\lambda; A)$, satisfies the inequality

$$\lambda^{n+1} \| R^{(n)}(\lambda; A) \| \leqq Mn! \tag{12.3.1}$$

for $\lambda > 0$ and $n = 0, 1, 2, \cdots$. In the numerical case this condition is sufficient to imply that the function in question is a Laplace integral and, in essence, Theorem 12.3.1 shows that this remains true for the resolvent function $R(\lambda; A)$. In content this result is very close to the original Hille-Yosida theorem which now appears as a corollary. Credit for its discovery is shared by W. Feller [6], I. Miyadera [2], and R. S. Phillips [8].

THEOREM 12.3.1. *A necessary and sufficient condition that a closed linear operator U generate a semi-group $[T(\xi)]$ of class (C_0) such that $\| T(\xi) \| \leqq M$ is that $\mathfrak{D}(U)$ be dense in $\mathfrak{X}$ and*

$$\| [R(\lambda; U)]^n \| \leqq M\lambda^{-n} \tag{12.3.2}$$

for $\lambda > 0$ and $n = 1, 2, 3, \cdots$.

PROOF. We note that the two inequalities (12.3.1) and (12.3.2) are equivalent by virtue of the formula

$$R^{(n)}(\lambda; U) = (-1)^n n! \, [R(\lambda; U)]^{n+1} \tag{12.3.3}$$

which was obtained in (5.8.6). The necessity argument is now immediate. For if $[T(\xi)]$ is of class (C_0) with infinitesimal generator A, then $\mathfrak{D}(A)$ is dense in $\mathfrak{X}$ by Theorem 10.3.1. Further for $\lambda > 0$

$$R(\lambda; A)x = \int_0^\infty e^{-\lambda\xi} T(\xi)x \, d\xi$$

and hence

$$R^{(n)}(\lambda; A)x = \int_0^\infty e^{-\lambda\xi}(-\xi)^n T(\xi)x \, d\xi.$$

Since $\| T(\xi) \| \leqq M$ we obtain

$$\| R^{(n)}(\lambda; A) \| \leqq \int_0^\infty e^{-\lambda\xi}\xi^n M \, d\xi = Mn!/\lambda^{n+1}, \quad n = 0, 1, 2, \cdots,$$

which is (12.3.1).

The following sufficiency argument is modelled after Yosida's proof [3] of the Hille-Yosida theorem. The proof is based on the exponential formula (11.6.8) which in turn depends on the Laplace transform inversion formula given in Theorem 6.3.3.

We first set

$$B_\lambda \equiv \lambda^2 R(\lambda; U) - \lambda I, \qquad \lambda > 0.$$

It is clear that the B_λ's commute. For $x \in \mathfrak{D}(U)$, $B_\lambda x = \lambda R(\lambda; U)Ux$ and by Lemma 12.2.1 we see that $\lim_{\lambda\to\infty} B_\lambda x = Ux$. This suggests the use of

$$S_\lambda(\xi) \equiv \exp(B_\lambda \xi) = e^{-\lambda\xi} \sum_{n=0}^{\infty} \frac{(\lambda^2\xi)^n}{n!} [R(\lambda; U)]^n$$

as approximating semi-groups. Now the estimates (12.3.2) give

$$\| S_\lambda(\xi) \| \leqq e^{-\lambda\xi} \sum_{n=0} \frac{(\lambda^2\xi)^n}{n!} (M/\lambda^n) \equiv M \tag{12.3.4}$$

for $\lambda > 0$. Further we have

$$\begin{aligned} S_\lambda(\xi)x - S_\mu(\xi)x &= \int_0^\xi \frac{d}{d\sigma} [S_\mu(\xi - \sigma)S_\lambda(\sigma)x]\, d\sigma \\ &= \int_0^\xi S_\mu(\xi - \sigma)S_\lambda(\sigma)\, [B_\lambda x - B_\mu x]\, d\sigma. \end{aligned}$$

Applying (12.3.4) we obtain

$$\| S_\lambda(\xi)x - S_\mu(\xi)x \| \leqq M^2\xi \, \| B_\lambda x - B_\mu x \| .$$

Now if $x \in \mathfrak{D}(U)$, then $\lim_{\lambda,\mu\to\infty} \| B_\lambda x - B_\mu x \| = 0$ and hence $S_\lambda(\xi)x$ converges strongly to a limit as $\lambda \to \infty$, the convergence being uniform with respect to ξ in any finite interval $0 \leqq \xi \leqq \beta < \infty$. We denote this limit by $T(\xi)x$. Since $\mathfrak{D}(U)$ is dense in $\mathfrak{X}$, the Banach-Steinhaus theorem (Theorem 2.11.4) applies so that $\lim_{\lambda\to\infty} S_\lambda(\xi)x = T(\xi)x$ for each $x \in \mathfrak{X}$ and (12.3.4) implies that this convergence is again uniform with respect to ξ in $[0, \beta]$. It follows that $T(\xi)$ is continuous in the strong operator topology for $\xi \geqq 0$, that $\| T(\xi) \| \leqq M$, and that $T(0) = I$. Now $[S_\lambda(\xi); \xi \geqq 0]$ is obviously a semi-group, that is,

$$S_\lambda(\xi_1 + \xi_2)x = S_\lambda(\xi_1)S_\lambda(\xi_2)x \qquad \text{for} \qquad \xi_1, \xi_2 \geqq 0.$$

Taking the limit as $\lambda \to \infty$ we see that $T(\xi_1 + \xi_2)x = T(\xi_1)T(\xi_2)x$ for $\xi_1, \xi_2 \geqq 0$. We have now proved that $[T(\xi)]$ is a semi-group of class (C_0) with $\| T(\xi) \| \leqq M$.

It remains to show that the infinitesimal generator A of the semi-group $[T(\xi)]$ is actually the operator U. Here we make use of the relation

$$S_\lambda(\xi)x - x = \int_0^\xi S_\lambda(\sigma)B_\lambda x\, d\sigma.$$

For $x \in \mathfrak{D}(U)$, the integrand converges boundedly to $T(\sigma)Ux$ and therefore we have

$$T(\xi)x - x = \int_0^\xi T(\sigma)Ux\,d\sigma, \qquad x \in \mathfrak{D}(U).$$

From this it follows that $A_\eta x = \eta^{-1}\int_0^\eta T(\sigma)Ux\,d\sigma \to Ux$ as $\eta \to 0+$ so that $Ax = Ux$ for all $x \in \mathfrak{D}(U)$. In other words $A \supset U$. On the other hand both $R(\lambda; U)$ and $R(\lambda; A)$ exist for $\lambda > 0$, the first by assumption and the second because of Theorem 11.5.3. Thus $(\lambda I - A)$ maps both $\mathfrak{D}(U)$ and $\mathfrak{D}(A)$ onto $\mathfrak{X}$ in a one-to-one fashion; this is possible only if $\mathfrak{D}(U) = \mathfrak{D}(A)$, that is, only if $A = U$. This concludes the proof of Theorem 12.3.1.

Because of the importance of this result we also present Hille's [**13, 17**] method of proof for the sufficiency assertion of Theorem 12.3.1. Here the argument is based on the exponential formula (11.6.9) which was derived by means of the Post-Widder inversion formula (Theorem 6.3.5).

We now define

$$T(\xi; n) \equiv \left\{\frac{n}{\xi} R\left(\frac{n}{\xi}; U\right)\right\}^n = \left(I - \frac{\xi}{n} U\right)^{-n}.$$

By (12.3.2) we have

$$\| T(\xi; n) \| \leqq M, \tag{12.3.5}$$

and (12.3.2) together with Lemma 12.2.1 implies that $\lim_{\eta\to 0+} T(\eta; n)x = x$ for each $x \in \mathfrak{X}$. Further, $R(\lambda; U)$ being holomorphic for $\lambda > 0$, we see that $T(\xi; n)$ is holomorphic in ξ for $\xi > 0$. Thus for $x \in \mathfrak{D}(U)$ we have

$$\frac{d}{d\xi}[T(\xi; n)x] = \left(I - \frac{\xi}{n} U\right)^{-n-1} Ux.$$

We now follow a suggestion due to T. Kato and write

$$\begin{aligned} T(\xi; n)x - T(\xi; m)x &= \int_0^\xi \frac{d}{d\sigma}[T(\xi - \sigma; m)T(\sigma; n)x]\,d\sigma \\ &= \int_0^\xi \left(\frac{\sigma}{n} - \frac{\xi - \sigma}{m}\right)\left(I - \frac{\xi - \sigma}{m} U\right)^{-m-1}\left(I - \frac{\sigma}{n} U\right)^{-n-1} U^2x\,d\sigma \end{aligned}$$

for $x \in \mathfrak{D}(U^2)$. Since by (12.3.2) $\| (I - \tau U)^{-k} \| \leqq M$ for $\tau > 0$ and positive integers k, we obtain

$$\| T(\xi; n)x - T(\xi; m)x \| \leqq M^2 \frac{\xi^2}{2}\left(\frac{1}{n} + \frac{1}{m}\right) \| U^2x \|, \quad x \in \mathfrak{D}(U^2). \tag{12.3.6}$$

Thus

$$\lim_{n\to\infty} T(\xi; n)x \equiv T(\xi)x \tag{12.3.7}$$

exists for each $x \in \mathfrak{D}(U^2)$ and the convergence is uniform with respect to ξ in each finite interval $[0, \beta]$. Now $\mathfrak{D}(U) = R(\lambda; U)[\mathfrak{X}]$ is by assumption dense in $\mathfrak{X}$.

Consequently $R(\lambda; U)$ maps dense subsets of $\mathfrak{X}$ onto dense subsets of $\mathfrak{X}$ and hence $\mathfrak{D}(U^2) = R(\lambda; U)[\mathfrak{D}(U)]$ is dense in $\mathfrak{X}$. Since $\mathfrak{D}(U^2)$ is dense in $\mathfrak{X}$ and since (12.3.5) holds, the Banach-Steinhaus theorem applies and we see that (12.3.7) is valid for each $x \in \mathfrak{X}$, the limit being uniform with respect to ξ in $[0, \beta]$. Thus $T(\xi)$ is strongly continuous for $\xi > 0$, $\| T(\xi) \| \leqq M$, and $T(0) = I$.

Next we observe that

$$T(\xi; mn)x = \left[T\left(\frac{\xi}{m}; n\right)\right]^m x$$

and taking the limit as $n \to \infty$ we obtain

$$T(\xi)x = \left[T\left(\frac{\xi}{m}\right)\right]^m x, \tag{12.3.8}$$

valid for all $x \in \mathfrak{X}$ and positive integers m. The semi-group property of $[T(\xi)]$ is then an immediate consequence of (12.3.8) together with the strong continuity of $T(\xi)$.

In order to prove that U is the infinitesmimal generator of $[T(\xi)]$, we write

$$T(\xi)x = T(\xi; 1)x + [T(\xi)x - T(\xi; 1)x].$$

For $x \in \mathfrak{D}(U^2)$, the inequality (12.3.6) implies that

$$\| T(\xi)x - T(\xi; 1)x \| \leqq M^2\xi^2 \| U^2x \|$$

while a double application of formula (5.8.1) gives

$$T(\xi; 1)x = \frac{1}{\xi} R\left(\frac{1}{\xi}; U\right)x = x + \xi Ux + \xi^2\left\{\frac{1}{\xi} R\left(\frac{1}{\xi}; U\right) U^2x\right\}.$$

As a consequence

$$T(\xi)x = x + \xi Ux + O(\xi^2), \qquad x \in \mathfrak{D}(U^2),$$

so that

$$Ax \equiv \lim_{\eta \to 0+} A_\eta x = Ux, \qquad x \in \mathfrak{D}(U^2).$$

Thus $(\lambda I - A)R(\lambda; U)x = x$ for all $x \in \mathfrak{D}(U)$; $\mathfrak{D}(U)$ being dense in $\mathfrak{X}$ and A being closed, it follows that this relation holds for all $x \in \mathfrak{X}$. Thus $Ax = Ux$ for all $x \in \mathfrak{D}(U)$, and since both $R(\lambda; A)$ and $R(\lambda; U)$ exist for $\lambda > 0$, this requires that $A = U$ as in the previous proof.

It should be noted that Theorem 12.3.1 remains valid if the relation (12.3.2) holds merely for some sequence $\{\lambda_n\}$ with $\lambda_n \to \infty$. In fact, it is readily verified that this is all that is actually needed in the first version of the sufficiency argument. Further if $M = 1$ and if (12.3.2) holds for $n = 1$, then it automatically holds for all $n \geqq 1$; this proves the Hille-Yosida theorem, namely,

COROLLARY. *If U is a closed linear operator with dense domain, if $R(\lambda; U)$ exists for $\lambda > 0$, and if*

$$\lambda \| R(\lambda; U) \| \leqq 1, \qquad \lambda > 0, \tag{12.3.9}$$

then U is the infinitesimal generator of a semi-group $[T(\xi)]$ of class (C_0) such that $\| T(\xi) \| \leqq 1$ for $\xi \geqq 0$.

REMARK. Given a semi-group $[T(\xi)]$ of class (C_0) and negative type, it is always possible to find an equivalent norm $\| x \|_1$ for $\mathfrak{X}$ such that $\| T(\xi) \|_1 \leqq 1$ for $\xi \geqq 0$ (see W. Feller [6]). In fact $\| x \|_1 \equiv \sup [\, \| T(\xi)x \|; \xi \geqq 0]$ will serve this purpose for it is clear that $\| x \| \leqq \| x \|_1 \leqq M \| x \|$.

THEOREM 12.3.2. *A necessary and sufficient condition that a closed linear operator U generate a group of linear bounded operators defined and strongly continuous on $(-\infty, \infty)$ is that $\mathfrak{D}(U)$ be dense in $\mathfrak{X}$ and that there exist constants $M > 0$, $\omega \geqq 0$ such that*

$$\| [R(\lambda; U)]^n \| \leqq M(|\lambda| - \omega)^{-n} \tag{12.3.10}$$

for all real λ, $|\lambda| > \omega$, and all positive integers n.

PROOF. The necessity follows as in the proof of Theorem 12.3.1. We have only to note that in this case $T(\xi)$ satisfies the bound $\| T(\xi) \| \leqq M \exp(\omega |\xi|)$ on $(-\infty, \infty)$. On the other hand, since $R(\lambda; -U) = -R(-\lambda; U)$, it follows from Theorem 12.3.1 that both U and $-U$ generate semi-groups of class (C_0), say $[T_+(\xi)]$ and $[T_-(\xi)]$ respectively. If $x \in \mathfrak{D}(U) = \mathfrak{D}(-U)$, then $T_-(\xi)x \in \mathfrak{D}(U)$ so that $S(\xi)x \equiv T_+(\xi)T_-(\xi)x$ is strongly continuously differentiable for $\xi \geqq 0$ and we have

$$\frac{d}{d\xi} S(\xi)x = [T_+(\xi)U]T_-(\xi)x + T_+(\xi)[-UT_-(\xi)x] = \theta.$$

It follows that $S(\xi)x \equiv x$ for each $x \in \mathfrak{D}(U)$ and since $\mathfrak{D}(U)$ is dense in $\mathfrak{X}$ the same is true for each $x \in \mathfrak{X}$. Thus $T_+(\xi)T_-(\xi) = I$; a similar argument shows that $T_-(\xi)T_+(\xi) = I$ and hence $T_-(\xi) = [T_+(\xi)]^{-1}$. Setting $T(\xi) = T_+(\xi)$ for $\xi \geqq 0$ and $T_-(-\xi)$ for $\xi < 0$, we see that $[T(\xi)]$ determines a group of linear bounded operators, defined and continuous in the strong operator topology on $(-\infty, \infty)$, and with infinitesimal generator U.

12.4. Semi-groups of class $(1, A)$. We next consider the generation of semi-groups of class $(1, A)$. A semi-group of this class, normalized to be of negative type, need no longer be bounded in norm; however we do have $\int_0^\infty \| T(\xi) \| d\xi = M < \infty$. The strong Laplace transform of $T(\xi)$ exists for all λ with $\mathfrak{R}(\lambda) > 0$ and is equal to $R(\lambda; A)$, the resolvent of the infinitesimal generator. As in the second part of Theorem 6.3.6, it is easy to show that

$$\int_0^\infty \lambda^{n-1} \| R^{(n)}(\lambda; A) \| d\lambda \leqq M(n-1)!$$

for $n = 1, 2, 3, \cdots$. Roughly speaking, the theorem below establishes the converse proposition. This is somewhat more than one would expect, reasoning by analogy from the numerical-valued case [see Widder [**1**, p. 306]); however it must be remembered that the resolvent has many special properties. The results in this section are due to R. S. Phillips [**9**] (cf. I. Miyadera [**3**]).

THEOREM 12.4.1. *A necessary and sufficient condition that a closed linear operator U be the infinitesimal generator of a semi-group $[T(\xi)]$ of class $(1, A)$ such that $\int_0^\infty \| T(\xi) \| \, d\xi < \infty$, is that*

(1) $\mathfrak{D}(U)$ *is dense in* $\mathfrak{X}$;

(2) $\| R(\lambda; U) \| = O(1/\lambda)$ *as* $\lambda \to \infty$;

(3) $R(\lambda; U)$ *satisfies any one of the following conditions:*

(a) *There exists a non-negative measurable function* $\varphi(\xi)$ *such that*

$$\int_0^\infty \varphi(\xi)\, d\xi < \infty \qquad \text{and} \qquad \| R^{(n)}(\lambda; U) \| \leqq \int_0^\infty e^{-\lambda\xi}\xi^n \varphi(\xi)\, d\xi$$

for all real $\lambda > 0$ *and integers* $n \geqq 0$.

(b) *There exists an* $M > 0$ *such that* $\int_0^\infty \lambda^{n-1} \| R^{(n)}(\lambda; U) \| \, d\lambda \leqq M(n-1)!$ *for all integers* $n \geqq 1$.

(c) *There exists a bounded non-decreasing function* $a(\xi)$ *such that*

$$\| R^{(n)}(\lambda; U) \| \leqq \int_0^\infty e^{-\lambda\xi}\xi^n \, da(\xi)$$

for all real $\lambda > 0$ *and integers* $n \geqq 0$.

(d) *Set* $S_\lambda(\xi) = \exp(B_\lambda \xi)$ *where* $B_\lambda \equiv \lambda^2 R(\lambda; U) - \lambda I$. *Then there exists an* $M > 0$ *such that*

$$\int_0^\infty \| S_\lambda(\xi) \| \, d\xi \leqq M$$

for all real $\lambda > 1$.

If (3-a) *is verified, then* $\| T(\xi) \| \leqq \varphi(\xi)$ *almost everywhere and if* $\varphi(\xi)$ *is upper semi-continuous then* $\| T(\xi) \| \leqq \varphi(\xi)$ *for all* $\xi > 0$. *Finally if* $\varphi(\xi)$ *is bounded, then* $[T(\xi)]$ *is of class* (C_0).

PROOF. Suppose first that $[T(\xi)]$ is of class $(1, A)$ and that $\int_0^\infty \| T(\xi) \| \, d\xi < \infty$. Then (1) is valid by Theorem 10.3.1 and (2) follows from the strong Abelian summability at $\xi = 0$ together with the uniform boundedness theorem. Theorem 11.5.2 implies that

$$R^{(n)}(\lambda; U)x = \int_0^\infty e^{-\lambda\xi}(-\xi)^n T(\xi)x \, d\xi$$

for $\Re(\lambda) > 0$ and each $x \in \mathfrak{X}$. Hence setting $\varphi(\xi) = \| T(\xi) \|$, we obtain (3-a).

The sufficiency argument is again based on the exponential formula (11.6.8). The logical structure for the rest of the proof is as follows: (3-a) $\to$ (3-b); (2) $\cup$ (3-b) $\to$ (3-c); (3-c) $\to$ (3-d); (1) $\cup$ (2) $\cup$ (3-d) $\to$ U is the infinitesimal generator of a semi-group $[T(\xi)]$ of class $(1, A)$ such that $\int_0^\infty \| T(\xi) \| \, d\xi < \infty$.

Starting with (3-a) we see that

$$\int_0^\infty \lambda^{n-1} \| R^{(n)}(\lambda; U) \| \, d\lambda \leqq \int_0^\infty \lambda^{n-1} \left\{ \int_0^\infty e^{-\lambda\xi}\xi^n \varphi(\xi) \, d\xi \right\} d\lambda$$

$$= \int_0^\infty \varphi(\xi) \left\{ \int_0^\infty e^{-\lambda\xi}(\lambda\xi)^{n-1}\xi \, d\lambda \right\} d\xi$$

$$= (n-1)! \int_0^\infty \varphi(\xi) \, d\xi.$$

This establishes (3-b). It is obvious that (3-c) is a special case of (3-a). However it is necessary for us to obtain (3-c) from (2) and (3-b) and this requires a rather elaborate argument. Let $F \in \mathfrak{E}(\mathfrak{X})^*$. Then (3-b) clearly implies

$$\int_0^\infty \lambda^{n-1} | F[R^{(n)}(\lambda; U)] | \, d\lambda \leqq M \| F \| (n-1)!$$

for all integers $n \geqq 1$. We now apply a theorem due to Widder [**1**, p. 306], according to which there exists a normalized function of bounded variation $\alpha_F(\xi)$ on $[0, \infty)$, such that

$$F[R(\lambda; U)] = \int_0^\infty e^{-\lambda\xi} \, d\alpha_F(\xi) \tag{12.4.1}$$

for real $\lambda > 0$. By a related result of Widder's [**1**, p. 299]

$$\operatorname{Var} \alpha_F \Big|_0^\xi = \lim_{n\to\infty} \int_{n/\xi}^\infty | F[R^{(n)}(\lambda; U)] | \frac{\lambda^{n-1}}{(n-1)!} \, d\lambda;$$

here we have made use of (2) by setting $F[R(\infty; U)] = 0$. We now make use of the Banach limit functional Lim $\in m^*$ (see the problem at the end of section 2.10). By means of this functional we can define

$$\alpha(\xi) = \operatorname{Lim} \int_{n/\xi}^\infty \| R^{(n)}(\lambda; U) \| \frac{\lambda^{n-1}}{(n-1)!} \, d\lambda$$

for each $\xi > 0$; $\alpha(0) = 0$. For $\xi_1 < \xi_2$ we have

$$\alpha(\xi_2) - \alpha(\xi_1) = \operatorname{Lim} \int_{n/\xi_2}^{n/\xi_1} \| R^{(n)}(\lambda; U) \| \frac{\lambda^{n-1}}{(n-1)!} \, d\lambda$$

$$\geqq \| F \|^{-1} \operatorname{Lim} \left\{ \int_{n/\xi_2}^{n/\xi_1} | F[R^{(n)}(\lambda; U)] | \frac{\lambda^{n-1}}{(n-1)!} \, d\lambda \right\}$$

$$= \| F \|^{-1} \operatorname{Var} \alpha_F \Big|_{\xi_1}^{\xi_2} \geqq 0.$$

Thus $\alpha(\xi)$ is non-decreasing, $\alpha(0) = 0$, $\alpha(\infty) \leqq M$, and $\| F \| \alpha(\xi)$ majorizes $\alpha_F(\xi)$ in the sense of a measure. It now follows from (12.4.1) that

$$| F[R^{(n)}(\lambda; U)] | \leqq \int_0^\infty e^{-\lambda\xi}\xi^n \, | d\alpha_F(\xi) | \leqq \| F \| \int_0^\infty e^{-\lambda\xi}\xi^n \, d\alpha(\xi)$$

for $\lambda > 0$ so that (3-c) holds. Condition (3-d) is now an easy consequence of (3-c). In fact

$$\| S_\lambda(\xi) \| \leqq e^{-\lambda\xi} \sum_{n=0}^{\infty} \frac{(\lambda^2\xi)^n}{n!} \| [R(\lambda; U)]^n \|, \tag{12.4.2}$$

and making use of the relation (12.3.3) together with the bound (3-c) we obtain

$$\begin{aligned}\int_0^\infty \| S_\lambda(\xi) \| \, d\xi &\leqq \int_0^\infty \left\{ e^{-\lambda\xi} \left[1 + \sum_{n=0}^{\infty} \frac{(\lambda^2\xi)^{n+1}}{n!(n+1)!} \int_0^\infty e^{-\lambda\sigma} \sigma^n \, d\alpha(\sigma) \right] \right\} d\xi \\ &= \lambda^{-1} + \int_0^\infty e^{-\lambda\sigma} \left[\sum_{n=0}^{\infty} \frac{(\lambda\sigma)^n}{n!(n+1)!} \int_0^\infty e^{-\lambda\xi} (\lambda\xi)^{n+1} \lambda \, d\xi \right] d\alpha(\sigma) \\ &= \lambda^{-1} + \int_0^\infty d\alpha(\sigma) \leqq M\end{aligned}$$

for $\lambda > 1$.

It remains to show that (1), (2), and (3-d) imply that U is the infinitesimal generator of a semi-group $[T(\xi)]$ of class $(1, A)$ such that $\int_0^\infty \| T(\xi) \| \, d\xi \leqq M$. It is clear that $[S_\lambda(\xi)]$ is a semi-group with infinitesimal generator $B_\lambda \in \mathfrak{E}(\mathfrak{X})$ and that the B_λ's commute. Setting $\varphi_\lambda(\xi) = \| S_\lambda(\xi) \|$, we see that $\varphi_\lambda(\xi)$ is sub-multiplicative and that $\int_0^\infty \varphi_\lambda(\xi) \, d\xi \leqq M$ for $\lambda > 1$. By Theorem 7.4.4,

$$\varphi_\lambda(\xi) \leqq (M/\xi)^2 \qquad \text{for} \qquad \xi > 0, \qquad \lambda > 1.$$

It follows that

$$\begin{aligned}&\int_0^\xi \varphi_\lambda(\xi - \sigma)\varphi_\mu(\sigma) \, d\sigma \\ &\qquad = \int_0^{\xi/2} [\varphi_\lambda(\xi - \sigma)\varphi_\mu(\sigma) + \varphi_\lambda(\sigma)\varphi_\mu(\xi - \sigma)] \, d\sigma \\ &\qquad \leqq M^2(2/\xi)^2 \int_0^{\xi/2} [\varphi_\mu(\sigma) + \varphi_\lambda(\sigma)] \, d\sigma \leqq 8M^3/\xi^2,\end{aligned} \tag{12.4.3}$$

again for $\xi > 0$ and $\lambda, \mu > 1$. We now write

$$\begin{aligned}S_\lambda(\xi)x - S_\mu(\xi)x &= \int_0^\xi \frac{d}{d\sigma} [S_\mu(\xi - \sigma)S_\lambda(\sigma)x] \, d\sigma \\ &= \int_0^\xi S_\mu(\xi - \sigma)S_\lambda(\sigma)[B_\lambda x - B_\mu x] \, d\sigma.\end{aligned}$$

Applying the estimate (12.4.3) we obtain

$$\begin{aligned}\| S_\lambda(\xi)x - S_\mu(\xi)x \| &\leqq \left[\int_0^\xi \varphi_\mu(\xi - \sigma)\varphi_\lambda(\sigma) \, d\sigma \right] \| B_\lambda x - B_\mu x \| \\ &\leqq 8M^3\xi^{-2} \| B_\lambda x - B_\mu x \|\end{aligned}$$

for $\lambda, \mu > 1$. According to Lemma 12.2.1, $\lim_{\lambda \to \infty} \lambda R(\lambda; U)x = x$ for each $x \in \mathfrak{X}$

and in particular for $x \in \mathfrak{D}(U)$ we have $B_\lambda x = \lambda R(\lambda; U)Ux \to Ux$ as $\lambda \to \infty$. It follows that

$$\lim_{\lambda\to\infty} S_\lambda(\xi)x \equiv T(\xi)x \tag{12.4.4}$$

exists for each $x \in \mathfrak{D}(U)$, the convergence being uniform with respect to ξ in each interval of the form $0 < \alpha \leqq \xi$. Since $\mathfrak{D}(U)$ is dense in $\mathfrak{X}$ and since

$$\| S_\lambda(\xi) \| \leqq (M/\xi)^2,$$

the limit in (12.4.4) exists for all $x \in \mathfrak{X}$, uniformly with respect to ξ in each interval $0 < \alpha \leqq \xi$. Again, if we use the fact that $[S_\lambda(\xi)]$ is a semi-group of linear bounded operators strongly continuous for $\xi > 0$, then it is easy to show that the same is true of $[T(\xi)]$. Further (12.4.4) implies that

$$\| T(\xi) \| \leqq \liminf_{\lambda\to\infty} \| S_\lambda(\xi) \|$$

and applying Fatou's lemma we get

$$\int_0^\infty \| T(\xi) \| \, d\xi \leqq \liminf_{\lambda\to\infty} \int_0^\infty \| S_\lambda(\xi) \| \, d\xi \leqq M.$$

We next prove that $R(\lambda; U) = R(\lambda)$ for $\lambda > 0$, where by definition $R(\lambda)x = \int_0^\infty e^{-\lambda\xi}T(\xi)x \, d\xi$. It is clear that $R(\lambda)$ is a linear bounded operator for $\lambda > 0$. Now if $x \in \mathfrak{D}(U)$, then $S_\lambda(\xi)x - S_\lambda(\alpha)x = \int_\alpha^\xi S_\lambda(\sigma)B_\lambda x \, d\sigma$ and $S_\lambda(\sigma)B_\lambda x \to T(\sigma)Ux$ boundedly in every interval of the form $0 < \alpha \leqq \sigma \leqq \xi < \infty$. Passing to the limit we obtain

$$T(\xi)x - T(\alpha)x = \int_\alpha^\xi T(\sigma)Ux \, d\sigma, \qquad 0 < \alpha < \xi. \tag{12.4.5}$$

Suppose next that $x \in \mathfrak{D}(U^2)$. Then $B_\lambda^2 x = \lambda^2[R(\lambda; U)]^2U^2x \to U^2x$. Integrating $d^2[S_\lambda(\xi)x]/d\xi^2 = S_\lambda(\xi)B_\lambda^2 x$, we obtain

$$S_\lambda(\xi)x = x + \xi B_\lambda x + \int_0^\xi \int_0^\sigma S_\lambda(\tau)B_\lambda^2 x \, d\tau \, d\sigma.$$

As a consequence

$$\| S_\lambda(\xi)x - x - \xi B_\lambda x \| \leqq \| B_\lambda^2 x \| \int_0^\xi \int_0^\sigma \varphi_\lambda(\tau) \, d\tau \, d\sigma \leqq \| B_\lambda^2 x \| M\xi,$$

and taking the limit as $\lambda \to \infty$ we get

$$\| T(\xi)x - x - \xi Ux \| \leqq \| U^2 x \| M\xi, \qquad x \in \mathfrak{D}(U^2).$$

Thus for $x \in \mathfrak{D}(U^2)$ we see that $\lim_{\eta\to 0+} T(\eta)x = x$ and by (12.4.5) that

$$d[T(\xi)x]/d\xi = T(\xi)Ux \qquad \text{for} \qquad \xi > 0.$$

It follows that for $\lambda > 0$ and $x \in \mathfrak{D}(U^2)$

$$R(\lambda)Ux = \int_0^\infty e^{-\lambda\xi}T(\xi)Ux\,d\xi = \int_0^\infty e^{-\lambda\xi}\frac{d}{d\xi}[T(\xi)x]\,d\xi$$
$$= e^{-\lambda\xi}T(\xi)x\Big|_0^\infty + \lambda\int_0^\infty e^{-\lambda\xi}T(\xi)x\,d\xi$$
$$= -x + \lambda R(\lambda)x,$$

that is, $R(\lambda)(\lambda I - U)x = x$ for all $x \in \mathfrak{D}(U^2)$ and $\lambda > 0$. Now for $\lambda > 0$, $(\lambda I - U)[\mathfrak{D}(U^2)] = \mathfrak{D}(U)$. Hence $R(\lambda)x = R(\lambda; U)x$ on the set $\mathfrak{D}(U)$, dense in $\mathfrak{X}$, and therefore $R(\lambda) = R(\lambda; U)$, again for $\lambda > 0$. As we have already observed, $\lim_{\lambda\to\infty}\lambda R(\lambda; U)x = x$ for each $x \in \mathfrak{X}$; it follows that $T(\xi)$ is strongly Abel summable to the identity at $\xi = 0$, and hence of class $(1, A)$. According to Theorem 11.5.2, $R(\lambda)$ is the resolvent of the infinitesimal generator, A, for the semi-group $[T(\xi)]$. Thus $R(\lambda; A) = R(\lambda; U)$ for $\lambda > 0$ and therefore $U = A$.

The final assertions of the theorem are now readily established. Suppose that (3-a) has been verified. Substituting these estimates in (12.4.2) we get

$$\| S_\lambda(\xi) \| \leqq e^{-\lambda\xi}\left[1 + \sum_{n=0}^\infty \frac{(\lambda^2\xi)^{n+1}}{n!(n+1)!}\int_0^\infty e^{-\lambda\sigma}\sigma^n\varphi(\sigma)\,d\sigma\right].$$

In the notation of Theorem 6.3.3 this becomes

$$\| S_\lambda(\xi) \| \leqq e^{-\lambda\xi} + \int_0^\infty K(\xi, \sigma; \lambda)\varphi(\sigma)\,d\sigma$$

and hence

$$\| T(\xi) \| \leqq \liminf_{\lambda\to\infty} \| S_\lambda(\xi) \| \leqq \varphi(\xi)$$

almost everywhere. On the other hand if $\varphi(\xi)$ is upper semi-continuous, a further examination of the proof of Theorem 6.3.3 shows that

$$\limsup_{\lambda\to\infty}\int_0^\infty K(\xi, \sigma; \lambda)\varphi(\sigma)\,d\sigma \leqq \varphi(\xi)$$

for all $\xi > 0$ so that in this case $\| T(\xi) \| \leqq \varphi(\xi)$ for all $\xi > 0$. Finally if $\varphi(\xi) \leqq M$, then, since $\| T(\xi) \|$ is lower semi-continuous, we see that $\| T(\xi) \| \leqq M$ for all $\xi > 0$. It now follows from Theorem 10.6.3 that $[T(\xi)]$ is of class (C_0). This concludes the proof of Theorem 12.4.1.

The following corollary generalizes a theorem due to E. Hille [**20**].

Corollary. *A necessary and sufficient condition that a closed linear operator U be the infinitesimal generator of a semi-group of class $(1, A)$ and of negative type is that*

(1) *$\mathfrak{D}(U)$ is dense in $\mathfrak{X}$;*
(2) *$\| R(\lambda; U) \| = O(1/\lambda)$ as $\lambda \to \infty$;*
(3) *There exists a family $[T(\xi); \xi > 0]$ of linear bounded operators strongly con-*

tinuous for $\xi > 0$ *with* $\int_0^\infty \| T(\xi) \| d\xi < \infty$ *and such that*

$$R(\lambda; U)x = \int_0^\infty e^{-\lambda\xi} T(\xi)x \, d\xi \tag{12.4.6}$$

holds for all $\lambda > 0$ *and all* $x \in \mathfrak{X}$.

PROOF. The above conditions are clearly necessary. Since conditions (1) and (2) of both theorem and corollary are the same, it suffices for the proof of the converse to show that condition (3) of the corollary implies (3-a) of the theorem with $\varphi(\xi) = \| T(\xi) \|$. However this follows directly from the relation

$$R^{(n)}(\lambda; U)x = \int_0^\infty e^{-\lambda\xi}(-\lambda)^n T(\xi)x \, d\xi, \qquad \lambda > 0,$$

obtained by differentiating (12.4.6); for it is now obvious that

$$\| R^{(n)}(\lambda; U) \| \leqq \int_0^\infty e^{-\lambda\xi}\xi^n \| T(\xi) \| d\xi \quad \text{for} \quad \lambda > 0.$$

The fact that the semi-group generated by U coincides with the given family of operators $[T(\xi)]$ is a consequence of the uniqueness theorem for the Laplace transform (Theorem 6.2.3).

The above corollary may also be proved without the aid of Theorem 12.4.1, following an argument developed by E. Hille [**13,** p. 241].

A few remarks are now in order. To begin with, Theorem 12.4.1 can be somewhat sharpened. This follows from the fact that in addition to the conditions (1) and (2), the sufficiency argument requires merely that (3-a), (3-c), or (3-d) be satisfied for a sequence of λ_n's such that $\lambda_n \to \infty$.

Further it is easy to see that Theorem 12.3.1 is subsumed under Theorem 12.4.1. We consider only the non-trivial part of this assertion, namely the sufficiency argument. Suppose that $R(\lambda; U)$ satisfies (12.3.2) or equivalently that

$$\lambda^{n+1} \| R^{(n)}(\lambda; U) \| \leqq Mn! \quad \text{for} \quad \lambda > 0, \quad n \geqq 0.$$

Then $\| R(\lambda; U) \| = O(1/\lambda)$ and for fixed $\omega > 0$

$$\| R^{(n)}(\lambda + \omega; U) \| \leqq Mn!(\lambda + \omega)^{-n-1}, \qquad \lambda > 0, n \geqq 0.$$

Thus condition (2) is satisfied and (3-a) can be verified with $\varphi(\xi) = Me^{-\omega\xi}$. Hence if $\mathfrak{D}(U)$ is dense in $\mathfrak{X}$, then $U - \omega I$ is the infinitesimal generator of a semi-group $[S(\xi)]$ of class (C_0) with $\| S(\xi) \| \leqq Me^{-\omega\xi}$ so that U is the infinitesimal generator of $[T(\xi) \equiv e^{\omega\xi}S(\xi)]$, likewise of class (C_0) and with $\| T(\xi) \| \leqq M$.

The conditions (1), (2), and (3) of Theorem 12.4.1 are independent. Condition (1) forces $T(\xi)x$ to be strongly continuous for all $x \in \mathfrak{X}$; (1) and (2) together are equivalent with strong Abel summability to the identity at $\xi = 0$ (see Lemma 12.2.1); whereas (3) implies $\int_0^\infty \| T(\xi) \| d\xi < \infty$. It is possible to devise closed linear operators which satisfy any two of these conditions without satisfying the third. In fact, if $[T(\xi)]$ is a semi-group of class (C_0) with infinitesimal generator A, then $R(\lambda; A^*) = R^*(\lambda; A)$ satisfies (12.3.2) and hence conditions (2) and (3), but in general $\mathfrak{D}(A^*)$ is not dense in $\mathfrak{X}^*$ (see Chapter XIV). Example 1 below defines a closed linear operator satisfying (1) and (3) but not (2). Finally Example 2 defines a closed linear operator satisfying (1) and (2) but not (3).

EXAMPLE 1. A closed linear operator satisfying conditions (1) and (3) of Theorem 12.4.1 but not condition (2).

Define $\mathfrak{X}$ to be the set of all sequence pairs $\{(\chi_n, \eta_n)\}$ with finite norm $\| \{(\chi_n, \eta_n)\} \| = \sum_{n=1}^{\infty} [| \chi_n |^2 + n | \eta_n |^2]^{1/2}$. Set $\{(\chi_n', \eta_n')\} = T(\xi)\{(\chi_n, \eta_n)\}$ where

$$\chi_n' = e^{-(n+1)\xi}(\chi_n \cos n\xi - \eta_n \sin n\xi),$$
$$\eta_n' = e^{-(n+1)\xi}(\chi_n \sin n\xi + \eta_n \cos n\xi).$$

$T(\xi)$ defines a linear bounded operator on $\mathfrak{X}$; in fact $\| T(\xi) \| \leqq \sup_n [n^{1/2} e^{-(n+1)\xi}] \leqq (2e\xi)^{-1/2} e^{-\xi}$. It is clear that the operators $[T(\xi)]$ form a semi-group, strongly continuous for $\xi > 0$. Both $\mathfrak{X}_0$ and $\mathfrak{D}(A)_0$ contain all of the ultimately zero vectors and are therefore dense in $\mathfrak{X}$. Further $R(\lambda)x \equiv \int_0^\infty e^{-\lambda\xi} T(\xi)x \, d\xi$ exists and defines a linear bounded operator for $\Re(\lambda) > 0$. Theorem 11.5.1 goes through as before except for part (3) which now follows from Theorem 6.2.3. Likewise Theorem 11.5.2 remains valid. Consequently $R(\lambda) = R(\lambda; A)$ for $\Re(\lambda) > 0$. It is obvious that condition (3-a) holds with $\varphi(\xi) = \| T(\xi) \|$. Finally we show that $\lambda \| R(\lambda; A) \| \to \infty$ as $\lambda \to \infty$. Componentwise $R(\lambda; A)$ is given by

$$\chi_n' = \alpha_n(\lambda)\chi_n - \beta_n(\lambda)\eta_n, \qquad (12.4.7)$$
$$\eta_n' = \beta_n(\lambda)\chi_n + \alpha_n(\lambda)\eta_n,$$

where

$$\alpha_n(\lambda) = \frac{\lambda + n + 1}{(\lambda + n + 1)^2 + n^2}, \qquad \beta_n(\lambda) = \frac{n}{(\lambda + n + 1)^2 + n^2}.$$

It follows that

$$\| R(\lambda; A) \| \geqq \sup_n \left\{ n^{1/2} \frac{n}{(\lambda + n + 1)^2 + n^2} \right\} \geqq \frac{1}{10(\lambda + 1)^{1/2}}$$

for $\lambda > 0$, which proves the assertion.

EXAMPLE 2. A closed linear operator U with dense domain having the property that $\| R(\sigma + \tau i; U) \| \leqq M/\sigma$ for $\sigma > 1$, which is not the infinitesimal generator of a semi-group of class (A).

Define $\mathfrak{X}$ to be the set of all sequence pairs $\{(\chi_n, \eta_n)\}$ such that $\lim_{n \to \infty} \chi_n = 0$ and $\sum_{n=1}^{\infty} | \eta_n |^2 < \infty$, where the norm for an element $x = \{(\chi_n, \eta_n)\}$ is given by

$$\| x \| = \sup_n | \chi_n | + \left[\sum_{n=1}^{\infty} | \eta_n |^2 \right]^{1/2}.$$

Set

$$\mathfrak{D}(U) = \left[x; x \in \mathfrak{X}, \lim_{n \to \infty} | ie^n \chi_n - n\eta_n | = 0, \quad \sum_{n=1}^{\infty} | n\chi_n + ie^n \eta_n |^2 < \infty \right]$$

and define U on $\mathfrak{D}(U)$ by its behavior on the component spaces, namely,

$$\chi_n' = ie^n \chi_n - n\eta_n,$$
$$\eta_n' = n\chi_n + ie^n \eta_n.$$

It is easy to verify that $\mathfrak{D}(U)$ is a linear set dense in $\mathfrak{X}$ and that U is a closed linear operator. We next define an operator $R(\lambda)$ componentwise by (12.4.7) where now

$$\begin{pmatrix} \alpha_n(\lambda) & -\beta_n(\lambda) \\ \beta_n(\lambda) & \alpha_n(\lambda) \end{pmatrix} \equiv \begin{pmatrix} \lambda - ie^n & n \\ -n & \lambda - ie^n \end{pmatrix}^{-1} \tag{12.4.8}$$

so that

$$\alpha_n(\lambda) = \frac{\lambda - ie^n}{(\lambda - ie^n)^2 + n^2}, \qquad \beta_n(\lambda) = \frac{n}{(\lambda - ie^n)^2 + n^2}.$$

It can be shown that $R(\lambda)$ defines a bounded linear operator on $\mathfrak{X}$ for $\Re(\lambda) > 1$ and that $\| R(\sigma + i\tau) \| \leqq M/\sigma$ for $\sigma > 1$ (see R. S. Phillips [9]). It is clear from (12.4.8) that $R(\lambda)$ behaves like the resolvent of U on the component spaces and hence on the ultimately zero vectors. Since the ultimately zero vectors are dense in $\mathfrak{X}$, since $R(\lambda)$ is bounded, and since U is closed, it follows that $(\lambda I - U)R(\lambda) = I$. Further if $x \in \mathfrak{D}(U)$, then it can be verified directly that $R(\lambda)(\lambda I - U)x = x$. Consequently $R(\lambda) = R(\lambda; U)$ for $\Re(\lambda) > 1$.

On the other hand, if U is the infinitesimal generator of a semi-group $[T(\xi)]$ of class (A) and if $x \in \mathfrak{D}(U^2)$ then we have

$$T(\xi)x = \frac{1}{2\pi i}(C, 1)\text{-}\int_{\gamma - i\infty}^{\gamma + i\infty} e^{\lambda\xi} R(\lambda; U)x \, d\lambda, \qquad \gamma > 1,$$

as in the proof of Theorem 11.5.3. Now the ultimately zero vectors belong to $\mathfrak{D}(U^2)$ so that componentwise $T(\xi)$ can be expressed as

$$\chi_n' = \exp(ie^n\xi)[\chi_n \cos n\xi - \eta_n \sin n\xi],$$

$$\eta_n' = \exp(ie^n\xi)[\chi_n \sin n\xi + \eta_n \cos n\xi].$$

We now have a contradiction; for if ξ_0 is not commensurate with 2π, then $T(\xi_0)$, as defined above, does not have a linear bounded extension. In fact, for such a ξ_0 there exists an infinite sequence $\{n_k\}$ such that $|\sin n_k\xi_0| \geqq \frac{1}{2}$, $k = 1, 2, 3, \cdots$. Choosing $\chi_n = k^{-1/2}$ for $n = n_k$, $\chi_n = 0$ otherwise, and $\eta_n = 0$ for all n, we see that

$$\| T(\xi_0)x \| \geqq \frac{1}{2}\left[\sum_{n=1}^{\infty} |\chi_n|^2\right]^{1/2} = \frac{1}{2}\left[\sum_{k=1}^{\infty} \frac{1}{k}\right]^{1/2} = \infty.$$

Thus the operators $T(\xi)$ are not all bounded and hence the semi-group is not of class (A). The operator U satisfies the conditions (1) and (2) of Theorem 12.4.1; however, it cannot satisfy (3) because if this were true then $[T(\xi)]$ would be of class $(1, A)$ and *a fortiori* of class (A).

In order to formulate a theorem on the generation of semi-groups of class $(1, C_1)$ it is necessary to have a condition on the infinitesimal generator which implies strong $(C, 1)$ summability to the identity at $\xi = 0$. Theorem 11.5.5 does not furnish us with a condition of this kind since it deals with the infinitesimal operator rather than its closure. The following theorem is intended to fill this need.

THEOREM 12.4.2. *A necessary and sufficient condition that a semi-group of class* $(0, A)$ *be of class* $(0, C_1)$ *is that there exist real numbers* $M > 0$ *and* ω *such that*

$$\left\| \sum_{j=1}^{n} \lambda^j [R(\lambda + \omega; A)]^j \right\| \leqq Mn \tag{12.4.9}$$

for all $\lambda > 0$ *and all integers* $n \geqq 1$.

Since we shall not have occasion to refer to this result again, we omit the proof (see R. S. Phillips [**9**, Theorem 4.2] and Miyadera [**3**]). Theorems 12.4.1 and 12.4.2 together give necessary and sufficient conditions that a closed linear operator be the infinitesimal generator of a semi-group of class $(1, C_1)$. I. Miyadera [**4**] has also obtained generation theorems for the semi-group classes $(0, A)$ and $(0, C_1)$.

12.5. Semi-groups of class (A). We conclude this paragraph with a generation theorem for semi-groups of class (A) due to R. S. Phillips [**11**]. It is convenient to normalize the member semi-groups of this class so that $\omega_0 < \omega_1 \leqq 0$; here ω_0 is the semi-group type whereas ω_1 is defined as in condition (A)′ of section 10.6. We recall that a function $\varphi(\xi) \geqq 0$ is said to be of type $\omega \equiv \limsup_{\xi \to \infty} \xi^{-1} \log \varphi(\xi)$.

THEOREM 12.5.1. *A necessary and sufficient condition that a closed linear operator U be the infinitesimal generator of a semi-group $[T(\xi)]$ of class (A) with $\omega_0 < \omega_1 \leqq 0$ is that*

(1) *$\mathfrak{D}(U)$ is dense in $\mathfrak{X}$;*

(2) *$\| R(\lambda; U) \| = O(1/\lambda)$ as $\lambda \to \infty$;*

(3) *$\| R(\lambda; U) \|$ is bounded for $\Re(\lambda) > 0$;*

(4) *There exists a non-negative non-increasing function $\varphi(\xi)$ of negative type which satisfies either of the following:*

(a) *For each $x \in \mathfrak{D}(U^2)$ there is a non-negative measurable function $\varphi(\xi; x) \leqq \varphi(\xi) \| x \|$ such that $\int_0^\infty \varphi(\xi; x)\, d\xi < \infty$ and*

$$\| R^{(n)}(\lambda; U)x \| \leqq \int_0^\infty e^{-\lambda\xi}\xi^n \varphi(\xi; x)\, d\xi$$

for all real $\lambda \geqq 0$ and integers $n \geqq 0$;

(b) *For each $x \in \mathfrak{D}(U^2)$, each δ', δ with $0 < \delta' < \delta$, and each $\epsilon > 0$, there exists a $\lambda_0 \equiv \lambda_0(x; \epsilon, \delta', \delta)$ such that if $\lambda > \lambda_0$ and $n \geqq \lambda\delta$, then*

$$\| \lambda^n[R(\lambda; U)]^n x \| \leqq \{\varphi(\delta') + \epsilon\} \| x \| .$$

Moreover $\| T(\xi) \| \leqq \varphi(\xi)$ for all $\xi > 0$.

REMARK. Condition (4-b) was first used in connection with generation theorems by W. Feller [**6**].

PROOF. Suppose first that $[T(\xi)]$ is of class (A) with $\omega_0 < \omega_1 \leqq 0$. Then conditions (1), (2), and (3) are satisfied by Theorem 11.5.2. Since $[T(\xi)]$ is of negative type, the function

$$\varphi(\xi) \equiv \sup_{\sigma \geqq 0} \| T(\xi + \sigma) \|, \qquad \xi > 0,$$

will be finite-valued and of negative type; in addition $\varphi(\xi)$ is obviously non-negative and non-increasing. If $x \in \mathfrak{D}(U^2)$, then $\lim_{\xi \to 0+} T(\xi)x = x$. Hence Lemma 11.5.1 together with Theorem 11.5.2 implies that

$$R(\lambda; U)x = \int_0^\infty e^{-\lambda\xi} T(\xi)x\, d\xi, \qquad \Re(\lambda) > 0,\ x \in \mathfrak{D}(U^2). \tag{12.5.1}$$

From this we obtain

$$R^{(n)}(\lambda; U)x = \int_0^\infty e^{-\lambda\xi}(-\xi)^n T(\xi)x\, d\xi, \qquad \Re(\lambda) > 0,\ x \in \mathfrak{D}(U^2). \tag{12.5.2}$$

Finally setting $\varphi(\xi; x) = \| T(\xi)x \|$, we see that (4-a) follows directly. In order to establish (4-b) we make use of an argument due to W. Feller [6]. Combining (12.3.3) and (12.5.2) we obtain

$$[R(\lambda; U)]^n x = \frac{1}{(n-1)!}\int_0^\infty e^{-\lambda\xi}\xi^{n-1} T(\xi)x\, d\xi, \qquad \Re(\lambda) > 0,\ x \in \mathfrak{D}(U^2). \tag{12.5.3}$$

Now

$$\frac{1}{(n-1)!}\int_{\delta'}^\infty e^{-\lambda\xi}\xi^{n-1} \| T(\xi)x \|\, d\xi \leqq \lambda^{-n}\varphi(\delta') \| x \|. \tag{12.5.4}$$

Further if $x \in \mathfrak{D}(U^2)$, then $N(x) \equiv \sup\, [\, \| T(\xi)x \|;\ \xi > 0] < \infty$ by Theorem 11.5.3. Consequently

$$\frac{1}{(n-1)!}\int_0^{\delta'} e^{-\lambda\xi}\xi^{n-1} \| T(\xi)x \|\, d\xi \leqq \frac{N(x)}{(n-1)!}\lambda^{-n}\int_0^{\delta'\lambda} e^{-s}s^{n-1}\, ds. \tag{12.5.5}$$

If we set $q = \delta'/\delta < 1$ and let $n \geqq \lambda\delta$, then

$$\frac{1}{(n-1)!}\int_0^{\delta'\lambda} e^{-s}s^{n-1}\, ds \leqq e^{-qn}\sum_{k=n}^\infty \frac{(qn)^k}{k!} \leqq \frac{q}{n(1-q)^2}; \tag{12.5.6}$$

here we have made use of the estimate

$$e^{-w}\sum{}' \frac{w^k}{k!} \leqq N^{-2}w, \tag{12.5.7}$$

where the summation extends only over those values of k for which $| k - w | > N$. (The estimate (12.5.7) is the Tchebycheff inequality for a Poisson distribution; see (10.4.7).) Combining the relations (12.5.3) through (12.5.6) gives (4-b).

In order to prove the converse proposition, we first extend $R(\lambda; U)$ into the left half-plane. According to Theorems 5.8.3 and 5.9.1,

$$R(\lambda + \zeta; U) = \sum_{n=0}^\infty (-\zeta)^n [R(\lambda; U)]^{n+1}, \qquad \Re(\lambda) > 0,$$

within the circle of convergence of this series. By condition (2) $\| R(\lambda; U) \| \leqq M$ for all λ with $\Re(\lambda) > 0$ so that the series is majorized by $\sum_{n=0}^\infty | \zeta |^n M^{n+1}$. Consequently $R(\lambda; U)$ exists and is bounded in norm in each half-plane $\Re(\lambda) \geqq \gamma > -M^{-1}$.

Suppose next that $x \in \mathfrak{D}(U^2)$ is fixed. Then it is clear from Lemma 11.5.2 that the integral

$$T(\xi)x \equiv \lim_{\omega\to\infty} \frac{1}{2\pi i}\int_{\gamma-i\omega}^{\gamma+i\omega} e^{\lambda\xi}R(\lambda; U)x\, d\lambda, \qquad \gamma > -M^{-1}, \tag{12.5.8}$$

exists and defines a strongly continuous function of ξ for $\xi > 0$ with $T(0)x = x$. It also follows from the proof of this lemma that for $-M^{-1} < \gamma < 0$,

$$T(\xi)x = \frac{1}{2\pi i}\int_{\gamma-i\infty}^{\gamma+i\infty} e^{\lambda\xi}R(\lambda; U)U^2x\frac{d\lambda}{\lambda^2}, \tag{12.5.9}$$

from which we see that $N(x) \equiv \sup [\, \| T(\xi)x \|; \xi > 0] < \infty$. Likewise Lemma 11.5.2 shows that

$$R(\lambda: U)x = \int_0^\infty e^{-\lambda\xi}T(\xi)x\, d\xi, \qquad \Re(\lambda) > 0. \tag{12.5.10}$$

As a consequence the relation (12.5.3) is again valid and we obtain

$$\| [R(\lambda; U)]^n x \| \leqq \lambda^{-n}N(x), \qquad \lambda > 0. \tag{12.5.11}$$

Moreover Theorem 6.3.3 applies so that

$$T(\xi)x = \lim_{\lambda\to\infty} e^{-\lambda\xi}\sum_{n=0}^{\infty}\frac{(-1)^n(\lambda^2\xi)^{n+1}}{n!(n+1)!}R^{(n)}(\lambda; U)x. \tag{12.5.12}$$

Assuming condition (4-a), we have

$$\| T(\xi)x \| \leqq \limsup_{\lambda\to\infty} e^{-\lambda\xi}\sum_{n=0}^{\infty}\frac{(-1)^n(\lambda^2\xi)^{n+1}}{n!(n+1)!}\int_0^\infty e^{-\lambda\xi}(-\xi)^n\varphi(\xi; x)\, d\xi.$$

Again applying Theorem 6.3.3 with $f(\lambda) = \int_0^\infty e^{-\lambda\xi}\varphi(\xi; x)\, d\xi$ we see that the right member is equal to $\varphi(\xi; x)$ for almost all $\xi > 0$. It follows that for each $x \in \mathfrak{D}(U^2)$ that $\| T(\xi)x \| \leqq \varphi(\xi) \| x \|$ for almost all $\xi > 0$. We can also obtain this result from (4-b). In fact, choosing $0 < q' < q < 1$ arbitrarily and setting $N = [\lambda q\xi]$, the largest integer less than or equal to $\lambda q\xi$, we see that for $\lambda > \lambda_0(x; \epsilon, q'\xi, q\xi)$

$$\left\| e^{-\lambda\xi}\sum_{n=N+1}^{\infty}\frac{(\lambda\xi)^n}{n!}\lambda^n[R(\lambda; U)]^n x\right\| \leqq \{\varphi(q'\xi) + \epsilon\} \| x \|. \tag{12.5.13}$$

By (12.5.11) we have for all $\lambda > 0$

$$\left\| e^{-\lambda\xi}\sum_{n=1}^{N}\frac{(\lambda\xi)^n}{n!}\lambda^n[R(\lambda; U)]^n x\right\| \leqq N(x)e^{-\lambda\xi}\sum_{n=1}^{N}\frac{(\lambda\xi)^n}{n!} \leqq \frac{N(x)}{\lambda\xi(1-q)^2}, \tag{12.5.14}$$

where again we have made use of the estimate (12.5.7). Finally setting $R^{(n)}(\lambda; U) = (-1)^n(n+1)![R(\lambda; U)]^{n+1}$ in the relation (12.5.12) and employing the inequalities (12.5.13) and (12.5.14), we arrive at the inequality $\| T(\xi)x \| \leqq \varphi(\xi') \| x \|$ for each $\xi' < \xi$. Since $\varphi(\xi)$ is non-increasing, $\| T(\xi)x \| \leqq \varphi(\xi) \| x \|$ at all points of continuity for $\varphi(\xi)$ and hence for almost all $\xi > 0$; this is the same conclusion reached from (4-a). On the other hand, $\| T(\xi)x \|$ is continuous in ξ for $x \in \mathfrak{D}(U^2)$ and this together with the fact that $\varphi(\xi)$ is non-increasing in ξ implies that $\| T(\xi)x \| \leqq \varphi(\xi) \| x \|$ for all $\xi > 0$, $x \in \mathfrak{D}(U^2)$.

It is now clear that (12.5.8) defines a linear operator $T(\xi)$ on $\mathfrak{D}(U^2)$ with bound $\varphi(\xi)$. Assumption (1) asserts that $\mathfrak{D}(U)$ and hence $\mathfrak{D}(U^2)$ is dense in $\mathfrak{X}$. Consequently $T(\xi)$ has a unique linear bounded extension on $\mathfrak{X}$ (which we denote again by $T(\xi)$) with norm $\| T(\xi) \| \leqq \varphi(\xi)$. Since $\varphi(\xi)$ is bounded on each interval of the form $[\alpha, \infty)$, $\alpha > 0$, it is easy to see that $T(\xi)x$ is strongly continuous in ξ, $\xi > 0$, for each $x \in \mathfrak{X}$. Finally $\varphi(\xi)$ being of negative type implies that $T(\xi)$ is also of negative type, that is, $\omega_0 < 0$.

We next show that $[T(\xi)]$ defines a semi-group of operators. To this end let $x \in \mathfrak{D}(U^4)$. Then $U^2[\lambda^{-2}e^{\lambda\xi}R(\lambda;\, U)U^2x] = \lambda^{-2}e^{\lambda\xi}R(\lambda;\, U)U^4x$ is integrable along the line: $\Re(\lambda) = \gamma$, $-M^{-1} < \gamma < 0$. Hence applying Theorem 3.7.12 to the right member in (12.5.9), we see that $T(\xi)x \in \mathfrak{D}(U^2)$. Thus for $x \in \mathfrak{D}(U^4)$ and $-M^{-1} < \gamma_1 < \gamma_2 < 0$, we can write

$$T(\sigma)[T(\xi)x] = \left(\frac{1}{2\pi i}\right)^2 \int_{\gamma_2-i\infty}^{\gamma_2+i\infty} \int_{\gamma_1-i\infty}^{\gamma_1+i\infty} e^{\zeta\sigma} e^{\lambda\xi} R(\zeta;\, U)R(\lambda;\, U)U^4x \frac{d\lambda}{\lambda^2}\frac{d\zeta}{\zeta^2}.$$

Making use of the first resolvent equation, this becomes

$$T(\sigma)[T(\xi)x] = \frac{1}{2\pi i}\int_{\gamma_1-i\infty}^{\gamma_1+i\infty} e^{\lambda\xi}R(\lambda;\, U)U^4x\left\{\frac{1}{2\pi i}\int_{\gamma_2-i\infty}^{\gamma_2+i\infty}\frac{e^{\zeta\sigma}}{\zeta-\lambda}\frac{d\zeta}{\zeta^2}\right\}\frac{d\lambda}{\lambda^2}$$

$$+ \frac{1}{2\pi i}\int_{\gamma_2-i\infty}^{\gamma_2+i\infty} e^{\zeta\sigma}R(\zeta;\, U)U^4x\left\{\frac{1}{2\pi i}\int_{\gamma_1-i\infty}^{\gamma_1+i\infty}\frac{e^{\lambda\xi}}{\lambda-\zeta}\frac{d\lambda}{\lambda^2}\right\}\frac{d\zeta}{\zeta^2}.$$

The inner integral in the first term on the right is equal to $\lambda^{-2}e^{\lambda\sigma}$ and that of the second term vanishes. Thus

$$T(\sigma)[T(\xi)x] = \frac{1}{2\pi i}\int_{\gamma_1-i\infty}^{\gamma_1+i\infty} e^{\lambda(\sigma+\xi)}R(\lambda;\, U)U^4x\frac{d\lambda}{\lambda^4}.$$

However, $R(\lambda;\, U)U^2x = \lambda^{-1}U^2x + \lambda^{-2}U^3x + \lambda^{-2}R(\lambda;\, U)U^4x$ and, as is easily verified, $\int_{\gamma_1-i\infty}^{\gamma_1+i\infty}\lambda^{-k}\exp[\lambda(\sigma+\xi)]\,d\lambda = 0$ for $k \geqq 1$. It follows that

$$T(\sigma)[T(\xi)x] = \frac{1}{2\pi i}\int_{\gamma_1-i\infty}^{\gamma_1+i\infty} e^{\lambda(\sigma+\xi)}R(\lambda;\, U)U^2x\frac{d\lambda}{\lambda^2} = T(\sigma+\xi)x.$$

Since $\mathfrak{D}(U^4)$ is dense in $\mathfrak{X}$ and since the operators $T(\xi)$ are bounded, we see that $T(\sigma)[T(\xi)x] = T(\sigma+\xi)x$ for all $\sigma, \xi > 0$ and $x \in \mathfrak{X}$; $[T(\xi)]$ therefore defines a semi-group.

The semi-group $[T(\xi)]$ will be of class (A) with $\omega_1 \leqq 0$ if it can be shown that $\mathfrak{X}_0$ is dense in $\mathfrak{X}$ and that the relation (12.5.10) holds for each $x \in \mathfrak{X}_0$. In this case Theorem 11.5.2 asserts that U is the infinitesimal generator of the semi-group. Now if $\mathfrak{X}_0$ were not dense in $\mathfrak{X}$, then by Theorem 2.7.5 there exists an $x_0^* \in \mathfrak{X}^*$, $x_0^* \neq \theta$, such that $x_0^*[\mathfrak{X}_0] = 0$. Applying x_0^* to the right member of (12.5.10), we see that $x_0^*[R(\lambda;\, U)x] = 0$ for all $x \in \mathfrak{D}(U^2)$. This means that $x_0^*[\mathfrak{D}(U^3)] = 0$ which is impossible as $\mathfrak{D}(U^3)$ is dense in $\mathfrak{X}$. Consequently $\bar{\mathfrak{X}}_0 = \mathfrak{X}$. In order to establish (12.5.10) for a given $x \in \mathfrak{X}_0$, we first note that by definition x can be written as $x = T(\eta)y$ for some $y \in \mathfrak{X}$ and $\eta > 0$. Moreover, $\mathfrak{D}(U^4)$ being dense in $\mathfrak{X}$, there exists a Cauchy sequence $\{y_n\} \subset \mathfrak{D}(U^4)$ with $y_n \to y$. As above $T(\eta)y_n \in \mathfrak{D}(U^2)$ and by (12.5.10) we have

$$R(\lambda; U)T(\eta)y_n = \int_0^\infty e^{-\lambda\xi}T(\xi + \eta)y_n \, d\xi, \qquad \Re(\lambda) > 0.$$

The left member converges to $R(\lambda; U)T(\eta)y = R(\lambda; U)x$ as $n \to \infty$. On the other hand, $\| T(\xi + \eta)y_n \| \leqq \varphi(\eta) \| y_n \|$ for all $\xi \geqq 0$ and hence by the dominated convergence theorem (Theorem 3.7.9)

$$\lim_{n\to\infty} \int_0^\infty e^{-\lambda\xi}T(\xi + \eta)y_n \, d\xi = \int_0^\infty e^{-\lambda\xi}T(\xi + \eta)y \, d\xi = \int_0^\infty e^{-\lambda\xi}T(\xi)x \, d\xi, \quad \Re(\lambda) > 0.$$

This proves (12.5.10) for each $x \in \mathfrak{X}_0$ and concludes the proof of the theorem.

2. Generation of Semi-Groups Uniformly Continuous for $\xi > 0$

12.6. Semi-groups of class $(1, A)_u$. In this paragraph we shall investigate two different sets of assumptions on the operator U, both of which lead to semi-groups continuous in the uniform operator topology for $\xi > 0$. The resolvent of the infinitesimal generator for any semi-group belonging to one of our basic classes is necessarily holomorphic in a half-plane; as we have seen, this half-plane may be taken as $\Re(\lambda) > 0$. We now impose conditions on the behavior of $R(\sigma + i\tau; U)$ as a function of τ for $\sigma > 0$. The two classes of semi-groups which we consider overlap but are distinct: Semi-groups of the first class are integrable in norm but need not be strongly differentiable for $\xi > 0$; semi-groups of the second class are strongly differentiable for $\xi > 0$ but need not be integrable in norm. The methods employed in this paragraph go back to E. Hille [**13**]. The following theorem, due to R. S. Phillips [**9**], generalizes one of Hille's earlier results [**13,** Theorem 12.4.1].

Theorem 12.6.1. *Let U be a closed linear operator with domain dense in $\mathfrak{X}$ and suppose that the spectrum of U is located in $\Re(\lambda) \leqq 0$. Suppose further that*

(1) *$\| R(\sigma; U) \| = O(1/\sigma)$ as $\sigma \to \infty$.*

(2) *There exists an integer $k \geqq 1$ and an index α $(0 \leqq \alpha < 1)$ such that*

$$\sigma^{k-1-\alpha} \int_{-\infty}^{\infty} \| [R(\sigma + i\tau; U)]^k \| \, d\tau \leqq B \text{ for all } \sigma > 0.$$

Then U is the infinitesimal generator of a semi-group of class $(1, A)_u$. If $\alpha = 0$, or if (1) and (2) are replaced by

(3) *There exists an integer $k \geqq 1$ such that*

$$\sigma^{k-1} \int_{-\infty}^{\infty} \| R(\sigma + i\tau; U) \|^k \, d\tau \leqq B \text{ for all } \sigma > 0,$$

then U is the infinitesimal generator of a semi-group of class $(C_0)_u$.

PROOF. We first show that (3) implies (1). Thus assuming (3), Theorem 6.4.1 applies and if $\Re(\lambda) > \sigma > 0$ we have

$$R(\lambda; U) = \frac{1}{2\pi}\int_{-\infty}^{\infty} R(\sigma + i\tau; U)\frac{d\tau}{\lambda - \sigma - i\tau}.$$

Hence for $l = k/(k-1)$, $k > 1$, and real $\lambda > \sigma > 0$,

$$\| R(\lambda; U) \| \leqq \frac{1}{2\pi}\left\{\int_{-\infty}^{\infty} \| R(\sigma + i\tau; U) \|^k d\tau\right\}^{1/k}\left\{\int_{-\infty}^{\infty} [(\lambda - \sigma)^2 + \tau^2]^{-l/2} d\tau\right\}^{1/l}$$
$$\leqq \frac{B^{1/k}}{2\pi}\left\{\int_{-\infty}^{\infty} (1 + \eta^2)^{-l/2} d\eta\right\}^{1/l} \sigma^{1/k-1}(\lambda - \sigma)^{1/l-1}.$$

In particular for $\sigma = \beta\lambda$ $(0 < \beta < 1)$, we obtain

$$\| R(\lambda; U) \| \leqq M/\lambda$$

for all $\lambda > 0$. The case $k = 1$ can be treated similarly. It is further clear that (3) implies (2) with $\alpha = 0$. Consequently it suffices to prove the theorem assuming the conditions (1) and (2).

Let α be fixed, $0 \leqq \alpha < 1$. It follows from (1) that there exist numbers $M_1 > 0$ and $\omega > 0$ such that

$$\| [R(\lambda; U)]^n \| \leqq M_1 n^{-\alpha}\lambda^{\alpha-n}$$

for $n = 1, 2, \cdots, k-1$ and $\lambda > \omega$. Applying (12.3.3) this becomes

$$\| R^{(n-1)}(\lambda; U) \| \leqq M_1(n-1)!n^{-\alpha}\lambda^{\alpha-n}, \tag{12.6.1}$$

$n = 1, 2, \cdots, k-1$ and $\lambda > \omega$. In order to obtain a similar estimate for $n \geqq k$ we make use of (2). Theorem 6.4.1 then implies that

$$\frac{d^j}{d\lambda^j}\{[R(\lambda; U)]^k\} = (-1)^j j!\frac{1}{2\pi}\int_{-\infty}^{\infty} [R(\sigma + i\tau; U)]^k(\lambda - \sigma - i\tau)^{-j-1} d\tau$$

for $\Re(\lambda) > \sigma > 0$. Again applying (12.3.3) we obtain

$$R^{(n-1)}(\lambda; U) = \frac{d^{n-k}}{d\lambda^{n-k}}[R^{(k-1)}(\lambda; U)] = (-1)^{k-1}(k-1)!\frac{d^{n-k}}{d\lambda^{n-k}}[R(\lambda; U)]^k$$
$$= (-1)^{n-1}(k-1)!(n-k)!\frac{1}{2\pi}\int_{-\infty}^{\infty} [R(\sigma + i\tau; U)]^k(\lambda - \sigma - i\tau)^{k-n-1} d\tau.$$

Hence for real $\lambda > \sigma > 0$

$$\| R^{(n-1)}(\lambda; U) \| \leqq \frac{1}{2\pi}(k-1)!(n-k)!(\lambda - \sigma)^{k-n-1}\int_{-\infty}^{\infty} \| [R(\sigma + i\tau; U)]^k \| d\tau$$
$$\leqq M_2(n-k)!(\lambda - \sigma)^{k-n-1}\sigma^{1+\alpha-k}.$$

We now set $\sigma = \lambda/n$ and the estimate becomes

$$\| R^{(n-1)}(\lambda; U) \| \leqq M_3(n-1)!n^{-\alpha}\lambda^{\alpha-n} \tag{12.6.2}$$

for $n \geqq k$ and $\lambda > 0$. Combining (12.6.1) and (12.6.2) we have

$$\| R^{(n)}(\lambda; U - \omega I) \| = \| R^{(n)}(\lambda + \omega; U) \| \leqq Kn!(n + 1)^{-\alpha}(\lambda + \omega)^{\alpha-n-1}$$

for all integers $n \geqq 0$ and real $\lambda > 0$. According to Lemma 3.13.2

$$(n + 1)^{-\alpha}n! \leqq 4 \int_0^\infty e^{-\sigma}\sigma^{n-\alpha}\, d\sigma$$

for $n \geqq 2$, and an easy calculation shows that this estimate also holds for $n = 0, 1$. Hence if we set $\varphi(\xi) \equiv 4Ke^{-\omega\xi}\xi^{-\alpha}$, then we see that condition (3-a) of Theorem 12.4.1 is satisfied by the operator $U - \omega I$. As a consequence $U - \omega I$ satisfies the hypothesis of Theorem 12.4.1 and is therefore the infinitesimal generator of a semi-group $[S(\xi)]$ of class $(1, A)$ with $\| S(\xi) \| \leqq Me^{-\omega\xi}\xi^{-\alpha}$ for $\xi > 0$. Thus U itself is the infinitesimal generator of a semi-group $[T(\xi) \equiv e^{\omega\xi}S(\xi)]$, likewise of class $(1, A)$ and with $\| T(\xi) \| \leqq M\xi^{-\alpha}$. If $\alpha = 0$, then $T(\xi)$ is bounded in norm and therefore of class (C_0).

It remains to prove that $T(\xi)$ is continuous in the uniform operator topology for $\xi > 0$. If $x \in \mathfrak{D}(U)$ and $\gamma > \max(0, \omega_0)$, then by Theorem 11.6.1 we have

$$T(\xi)x = \lim_{\omega\to\infty} \frac{1}{2\pi i} \int_{\gamma-i\omega}^{\gamma+i\omega} e^{\lambda\xi}R(\lambda; U)x\, d\lambda.$$

Now $R(\lambda; U)$ is bounded for $\mathfrak{R}(\lambda) \geqq \gamma$; thus if $x \in \mathfrak{D}(U)$ then

$$\| R(\lambda; U)x \| = | \lambda |^{-1} \| R(\lambda; U)Ux + x \| = O(1/| \lambda |)$$

and

$$\| R^{(n)}(\lambda; U)x \| \leqq n! \| [R(\lambda; U)]^n \| \| R(\lambda; U)x \| = O(1/| \lambda |).$$

Hence integration by parts $(k - 1)$ times yields

$$\begin{aligned} T(\xi)x &= \frac{1}{2\pi i} (-1)^{k-1}\xi^{1-k} \int_{\gamma-i\infty}^{\gamma+i\infty} e^{\lambda\xi}R^{(k-1)}(\lambda; U)x\, d\lambda \\ &= \frac{1}{2\pi i} (k - 1)!\, \xi^{1-k} \int_{\gamma-i\infty}^{\gamma+i\infty} e^{\lambda\xi}[R(\lambda; U)]^k x\, d\lambda, \end{aligned}$$

again for $x \in \mathfrak{D}(U)$. Now by assumption $\int_{-\infty}^{\infty} \| R(\gamma + i\tau; U)]^k \| d\tau < \infty$. The integral in the last member therefore converges in the uniform operator topology to a bounded operator which coincides with $T(\xi)$ on the dense set $\mathfrak{D}(U)$ and hence on all of $\mathfrak{X}$. Since the convergence is uniform with respect to ξ on each finite interval, it follows that $T(\xi)$ is continuous in the uniform operator topology for $\xi > 0$. This concludes the proof of Theorem 12.6.1.

Remark. Conditions (1) and (2) of Theorem 12.6.1 are satisfied if there exists a $\beta > \frac{1}{2}$ such that $\| R(\sigma + i\tau; U) \| \leqq M(\sigma + | \tau |^\beta)^{-1}$ for all $\sigma > 0$. In fact, (1) is trivially satisfied and (2) follows from

$$\int_{-\infty}^{\infty} \| [R(\sigma + i\tau; U)]^2 \| d\tau \leqq M^2 \int_{-\infty}^{\infty} (\sigma + | \tau |^\beta)^{-2}\, d\tau \leqq M_1\sigma^{-2+1/\beta}.$$

EXAMPLE. A semi-group of class $(C_0)_u$, not strongly differentiable for $\xi > 0$, whose infinitesimal generator satisfies the conditions of Theorem 12.6.1.

Set $\mathfrak{X} = l_2$. We recall that the norm for a linear operator of the form $T(\{\alpha_n\}) = \{\tau_n\alpha_n\}$ is given by $\| T \| = \sup_n | \tau_n |$. We now define $T(\xi)(\{\alpha_n\}) = \{\tau_n(\xi)\alpha_n\}$ where the multiplier $\tau_n(\xi) \equiv \exp[-(n^2 + 1 + ie^{n^4})\xi]$. It is easy to show that $[T(\xi)]$ is of class (C_0). Further $R(\lambda; A)(\{\alpha_n\}) = \{\rho_n(\lambda)\alpha_n\}$ where $\rho_n(\sigma + i\tau) = [(\sigma + 1 + n^2) + i(\tau + e^{n^4})]^{-1}$. Consequently $\| R(\sigma; A) \| < 1/(\sigma + 1)$ and

$$\int_{-\infty}^{\infty} \| R(\sigma + i\tau; A) \|^2 d\tau = \int_{-\infty}^{\infty} \sup_n | \rho_n(\sigma + i\tau) |^2 d\tau \leqq \int_{-\infty}^{\infty} \sum_{n=-\infty}^{\infty} | \rho_n(\sigma + i\tau) |^2 d\tau$$

$$= \pi \sum_{n=-\infty}^{\infty} (\sigma + 1 + n^2)^{-1} \leqq 10(\sigma + 1)^{-1/2}$$

for $\sigma > 0$. Thus A satisfies conditions (1) and (2) of Theorem 12.6.1 (with $\alpha = \frac{1}{2}$). Finally we note that $AT(\xi)(\{\alpha_n\}) = \{-[n^2 + 1 + ie^{n^4}]\tau_n(\xi)\alpha_n\}$ is unbounded and hence, by Theorem 10.3.5, $T(\xi)$ is not strongly differentiable for $\xi > 0$.

12.7. Semi-groups of class $(A)_\infty$. In the previous section it was required that $\| R(\sigma + i\tau; U) \|$ be small on the average as a function of τ. We now require that this function approach zero in a more regular fashion as $| \tau | \to \infty$. The precise limitation is formulated in terms of

DEFINITION 12.7.1. *Let Ψ be the class of all real-valued functions $\psi(\tau)$ defined on $(-\infty, \infty)$ with the following properties:*

(i) *$\psi(\tau)$ is positive, continuously differentiable, and non-decreasing when $| \tau |$ increases;*

(ii) *$\psi(\tau) \to \infty$ as $| \tau | \to \infty$;*

(iii) *$\psi'(\tau)$ is bounded;*

(iv) *$\int_{-\infty}^{\infty} e^{-\xi\psi(\tau)} d\tau < \infty$ for each $\xi > 0$.*

Condition (iv) is satisfied, for instance, if $\psi(\tau)/[\log | \tau |] \to \infty$ as $| \tau | \to \infty$. We now prove

THEOREM 12.7.1. *Let U be a closed linear operator with domain dense in $\mathfrak{X}$ and suppose the spectrum of U is contained in $\mathfrak{R}(\lambda) < 0$. Suppose further that*

(1) *$\| R(\lambda; U) \|$ is bounded in λ for $\mathfrak{R}(\lambda) \geqq 0$;*

(2) *There exists a $\psi(\tau) \in \Psi$ and an $M > 0$ such that*

$$\| R(i\tau; U) \| \leqq M/\psi(\tau) \qquad \text{for } -\infty < \tau < \infty;$$

(3) *$\| R(\sigma; U) \| \leqq M/\sigma$ for $\sigma > 0$.*

Then U is the infinitesimal generator of a semi-group $[T(\xi)]$ of class $(A)_\infty$. If, in addition, $\int_{|\tau|>1} | \tau\psi(\tau) |^{-1} d\tau < \infty$, then $\lim_{\eta\to 0+} T(\eta)x = x$ for each $x \in \mathfrak{D}(U)$.

REMARK. Condition (1) is implied by (2) and (3) together with very mild restrictions on the rate of growth of $\| R(\lambda; U) \|$. In fact, setting

$$M(\rho) = \sup [\| R(\rho e^{i\varphi}; U) \|; -\pi/2 < \varphi < \pi/2],$$

then (1) follows if

(1)′ $$\limsup_{\rho\to\infty} \rho^{-2} \log M(\rho) = 0$$

and $R(\lambda; U)$ is bounded in norm along the positive real axis and the imaginary axis. This is a direct consequence of the Phragmén-Lindelöf theorem [1] applied to vector-valued functions (see section 3.13).

PROOF. We define

(12.7.1) $$T(\xi; \omega) = \frac{1}{2\pi i}\int_{\gamma-i\omega}^{\gamma+i\omega} e^{\lambda\xi}R(\lambda; U)\, d\lambda$$

where $\gamma > 0$ is fixed. The desired semi-group is obtained as the $\lim_{\omega\to\infty} T(\xi; \omega)$; however, in order to show that this limit exists we first extend $R(\lambda; U)$ into the left half-plane. Here we make use of Theorems 5.8.3 and 5.9.1, according to which

$$R(\sigma + i\tau; U) = \sum_{n=0}^{\infty} (-\sigma)^n[R(i\tau; U)]^{n+1}$$

within the circle of convergence of this series. By condition (2) this series is majorized by $\sum [M/\psi(\tau)]^{n+1}|\sigma|^n$ and hence

(12.7.2) $$\| R(\sigma + i\tau; U) \| \leqq \frac{2M}{\psi(\tau)}$$

for $|\sigma| \leqq \psi(\tau)/(2M)$. We next consider the integral

$$\frac{1}{2\pi i}\int_{C_\alpha(\omega)} e^{\lambda\xi}R(\lambda; U)\, d\lambda = \Theta.$$

Here $C_\alpha(\omega)$ is the closed contour $ABCDA$ where

$$A = \gamma + i\omega, \qquad B = -\alpha\psi(\omega) + i\omega, \qquad C = -\alpha\psi(-\omega) - i\omega, \qquad D = \gamma - i\omega,$$

and $0 < \alpha < 1/(2M)$. BC is an arc of the curve Γ_α: $\sigma = -\alpha\psi(\tau)$, and the other portions of $C_\alpha(\omega)$ are straight line segments. The integral along DA is precisely $T(\xi; \omega)$. It is clear from (12.7.2) that the contributions to the integral of the horizontal portions of $C_\alpha(\omega)$ do not exceed in norm

$$\frac{2M}{\psi(\pm\omega)}\int_{-\infty}^{\gamma} e^{\sigma\xi}\, d\sigma = \frac{2M}{\xi\psi(\pm\omega)}\, e^{\gamma\xi};$$

and this tends to zero as $\omega \to \infty$, uniformly with respect to ξ in each interval of the form $(\epsilon, 1/\epsilon)$, $\epsilon > 0$. The integral along CB tends to

$$\frac{1}{2\pi}\int_{-\infty}^{\infty} \exp\,[(-\alpha\psi(\tau) + i\tau)\xi]\, R(-\alpha\psi(\tau) + i\tau; U)(-\alpha\psi'(\tau) + i)\, d\tau;$$

this integral is dominated by a constant multiple of

(12.7.3) $$\int_{-\infty}^{\infty} \exp\,(-\alpha\psi(\tau)\xi)\, [\psi(\tau)]^{-1}\, d\tau.$$

which converges by (iv) uniformly with respect to ξ in each interval of the form $(\epsilon, 1/\epsilon)$, $\epsilon > 0$. It follows that $\lim_{\omega\to\infty} T(\xi; \omega) \equiv T(\xi)$ exists in the uniform operator topology and that $T(\xi)$ may be represented by the absolutely convergent integral

$$T(\xi) = \frac{1}{2\pi i}\int_{\Gamma_\alpha} e^{\lambda\xi} R(\lambda; U)\, d\lambda, \qquad 0 < \alpha < \frac{1}{2M}. \tag{12.7.4}$$

We may also differentiate under the integral sign, obtaining

$$T'(\xi) = \frac{1}{2\pi i}\int_{\Gamma_\alpha} e^{\lambda\xi}\lambda R(\lambda; U)\, d\lambda, \qquad \xi > 0. \tag{12.7.5}$$

To see that this integral is absolutely convergent we note that its norm is dominated by a constant multiple of

$$\int_{-\infty}^{\infty} \exp\,(-\alpha\psi(\tau)\xi)\,|\,\tau\,|\,d\tau < K\left(\frac{\alpha\xi}{2}\right)\int_{-\infty}^{\infty}\exp\left(-\frac{\alpha}{2}\psi(\tau)\xi\right)d\tau$$

where $K(\eta) = \max_\tau\,[\,|\,\tau\,|\,\exp\,(-\eta\psi(\tau))]$. It is clear that $K(\eta)$ is finite since

$$\int_{\beta/2}^{\beta}\exp\,(-\eta\psi(\tau))\,d\tau > \frac{\beta}{2}\exp\,(-\eta\psi(\beta))$$

and the left side tends to zero as $\beta \to \infty$. A similar estimate holds for negative values of β.

In order to prove that $[T(\xi)]$ is a semi-group of class (A) with infinitesimal generator U, it suffices to verify condition (4-a) of Theorem 12.5.1. If $x \in \mathfrak{D}(U^2)$ then Lemma 11.5.2 shows that $T(\xi)x = \lim_{\omega\to\infty} T(\xi; \omega)x$ exists, that $T(\xi)x$ is strongly continuous in ξ for $\xi > 0$, and that

$$R(\lambda; U)x = \int_0^\infty e^{-\lambda\xi}T(\xi)x\, d\xi, \qquad \mathfrak{R}(\lambda) > 0.$$

Now if $2m = \inf_\tau\,[\alpha\psi(\tau)]$, then $m > 0$ and the estimate (12.7.3) implies that

$$\|\,T(\xi)\,\| \leqq Me^{-m\xi}\int_{-\infty}^{\infty}\exp\left[-\frac{\alpha}{2}\psi(\tau)\xi\right]d\tau.$$

It follows that $\varphi(\xi) \equiv \sup_{\sigma>0}\|\,T(\xi+\sigma)\,\|$ is of negative type and that $\varphi(\xi)$ together with $\varphi(\xi; x) \equiv \|\,T(\xi)x\,\|$ satisfies the condition (4-a) of Theorem 12.5.1. Finally we see that $T(\xi)$ is differentiable in the strong operator topology for $\xi > 0$ by (12.7.5), and therefore infinitely differentiable for $\xi > 0$ by Theorem 10.3.5. Accordingly, $[T(\xi)]$ is actually of class $(A)_\infty$.

To prove the last assertion of the theorem we note that if $x \in \mathfrak{D}(U)$, then $R(\lambda; U)x = \lambda^{-1}[R(\lambda; U)Ux + x]$. Substituting this in (12.7.4) we have

$$\begin{aligned} T(\xi)x &= \frac{1}{2\pi i}\int_{\Gamma_\alpha} e^{\lambda\xi}x\frac{d\lambda}{\lambda} + \frac{1}{2\pi i}\int_{\Gamma_\alpha} e^{\lambda\xi}R(\lambda; U)Ux\frac{d\lambda}{\lambda} \\ &= \frac{1}{2\pi i}\int_{\Gamma_\alpha} e^{\lambda\xi}R(\lambda; U)Ux\frac{d\lambda}{\lambda}; \end{aligned}$$

here the first integral in the middle member vanishes because Γ_α runs to the left of $\lambda = 0$. Assuming $\int_{|\tau|>1} |\tau\psi(\tau)|^{-1} d\tau < \infty$, we obtain the estimate

$$\| T(\xi)x \| \leqq \left[\frac{1}{2\pi}\int_{-\infty}^{\infty} \frac{2M \mid -\alpha\psi'(\tau) + i \mid}{\psi(\tau) \mid -\alpha\psi(\tau) + i\tau \mid} d\tau\right] \| Ux \| = K \| Ux \|, \quad \xi > 0.$$

Now $x_n \equiv nR(n; U)x \in \mathfrak{D}(U^2)$ and $Ux_n = nR(n; U)Ux$. Further

$$\| T(\eta)x - x \| \leqq \| T(\eta)x - T(\eta)x_n \| + \| T(\eta)x_n - x_n \| + \| x_n - x \|$$
$$\leqq K \| Ux - Ux_n \| + \| T(\eta)x_n - x_n \| + \| x_n - x \|.$$

Since $x_n \to x$ and $Ux_n \to Ux$ by Lemma 12.2.1, and since $\lim_{\eta\to 0+} T(\eta)x_n = x_n$ for each n by Theorem 11.5.3, it now follows that $\lim_{\eta\to 0+} T(\eta)x = x$. This concludes the proof of Theorem 12.7.1.

Remark. Let U be a closed linear operator with dense domain and suppose that

$$\| R(\sigma + i\tau; U) \| \leqq \frac{M}{\sigma + |\tau|^\beta}, \quad \sigma > 0.$$

Then for $0 < \beta < 1$, Theorem 12.7.1 applies and hence U generates a semi-group $[T(\xi)]$ of class$(A)_\infty$. If $\beta > \frac{1}{2}$, then Theorem 12.6.1 also applies and in this case $[T(\xi)]$ is of class $(1, A)$. Finally if $\beta = 0$, Example 2 of section 12.4 shows that U need not generate a semi-group even of class (A). However if $\beta = 0$ and $M = 1$, then the Hille-Yosida theorem applies and U generates a semi-group of class (C_0).

We conclude this section with an

Example. A semi-group of class (C_0) whose infinitesimal generator satisfies the conditions of Theorem 12.7.1 but not those of Theorem 12.6.1, and which is not holomorphic for $\xi > 0$.

Set $\mathfrak{X} = l_2$. We define $T(\xi)(\{\alpha_n\}) = \{\tau_n(\xi)\alpha_n\}$ where the multiplier is now chosen to be $\tau_n(\xi) = \exp[-(|n|^{1/2} + 1 + in)\xi]$. Here it is easy to prove that $[T(\xi)]$ is of class (C_0). A componentwise calculation shows that $R(\lambda; A)(\{\alpha_n\}) = \{\rho_n(\lambda)\alpha_n\}$ where $\rho_n(\lambda) = [\lambda + |n|^{1/2} + 1 + in]^{-1}$. Now $\| [R(\lambda; A)]^k \| = \sup_n |\rho_n(\lambda)|^k =$ the kth power of the reciprocal of the distance from λ to the spectral set $\{-|n|^{1/2} - 1 - in\}$. It follows that there exist constants m and M, $0 < m < 1 < M < \infty$, such that

$$m^k[\sigma + 1 + |\tau|^{1/2}]^{-k} \leqq \| [R(\sigma + i\tau; A)]^k \| \leqq M^k[\sigma + 1 + |\tau|^{1/2}]^{-k}$$

for $\sigma > 0$. From this one readily verifies that the conditions of Theorem 12.7.1 are satisfied whereas those of Theorem 12.6.1 are not. Finally if $T(\xi)$ had an analytic extension, then componentwise the operator would be defined by $\tau_n(\xi + i\eta)$. However, this is impossible for then

$$\| T(\xi + i\eta) \| = \sup_n \exp[-(|n|^{1/2} + 1)\xi + n\eta] = \infty \quad \text{if} \quad \eta \neq 0.$$

12.8. Semi-groups of class $H(\Phi_1, \Phi_2)$. We close our discussion of the converse problem by considering semi-groups of class $H(\Phi_1, \Phi_2)$. Since this class is contained in all of the other basic classes, it follows that the conditions on the operator U will now be more severe than in the previous cases. The following theorem is due to E. Hille [**13**, Theorem 13.5.3].

Theorem 12.8.1. *A necessary and sufficient condition that a closed linear operator U be the infinitesimal generator of a semi-group of class $H(\Phi_1, \Phi_2)$ is that*

$\mathfrak{D}(U)$ be dense in $\mathfrak{X}$ and that for each $\epsilon > 0$ there exist a sector Δ_ϵ and a constant $M_\epsilon > 0$ such that for λ outside Δ_ϵ and at a distance $d_\epsilon(\lambda)$ from Δ_ϵ we have

$$\| R(\lambda; U) \| \leqq \frac{M_\epsilon}{d_\epsilon(\lambda)}. \tag{12.8.1}$$

Here Δ_ϵ^-, the conjugate set to Δ_ϵ, is a sector of the form $\Phi_2 - \epsilon + \pi/2 \leqq \arg(\lambda - \lambda_\epsilon) \leqq \Phi_1 + \epsilon + 3\pi/2$ with vertex λ_ϵ.

PROOF. Suppose first that $[T(\zeta)]$ is of class $H(\Phi_1, \Phi_2)$. $\mathfrak{D}(A)$ is of course dense in $\mathfrak{X}$. Further, given $\epsilon > 0$, then by Lemma 10.6.2

$$\sup [\| T(re^{i\varphi}) \|; 0 < r \leqq 1, \Phi_1 + \epsilon < \varphi < \Phi_2 - \epsilon] \equiv M_\epsilon < \infty;$$

and, it follows that

$$\| T(re^{i\varphi}) \| \leqq M_\epsilon e^{\omega_\epsilon r}, \qquad \Phi_1 + \epsilon < \varphi < \Phi_2 - \epsilon, \tag{12.8.2}$$

where $\omega_\epsilon = \log M_\epsilon \geqq 0$. Since $[T(\xi), \xi > 0]$ is of class (C_0), we see that

$$R(\lambda; A) = \int_0^\infty e^{-\lambda\xi} T(\xi)\, d\xi, \qquad \Re(\lambda) > \omega_0; \tag{12.8.3}$$

here the integral actually exists in the uniform operator topology.

For semi-groups holomorphic in the sector (Φ_1, Φ_2), the path of integration can be deformed from the positive real axis to the ray: $\arg \zeta = \varphi$, $\varphi \in (\Phi_1, \Phi_2)$, so that

$$R(\lambda; A) = e^{i\varphi} \int_0^\infty \exp(-\lambda re^{i\varphi}) T(re^{i\varphi})\, dr, \tag{12.8.4}$$

valid for $\Re(\lambda e^{i\varphi}) > \omega(\varphi) \equiv \lim_{r\to\infty} r^{-1} \log \| T(re^{i\varphi}) \|$. In justifying this assertion, we may suppose that $\varphi \in (\Phi_1 + \epsilon, \Phi_2 - \epsilon)$. The usual contour integration argument shows that the integrals in (12.8.3) and (12.8.4) are equal provided $\| \int_0^\varphi \exp(-\lambda re^{i\theta})\, T(re^{i\theta}) r\, d\theta \|$ tends to zero as $r \to \infty$. For real λ, $\lambda \cos \phi > \omega_\epsilon$, the estimate (12.8.2) implies

$$\left\| \int_0^\varphi \exp(-\lambda re^{i\theta})\, T(re^{i\theta}) r\, d\theta \right\| \leqq M_\epsilon r \left| \int_0^\varphi \exp[(\omega_\epsilon - \lambda \cos\theta) r]\, d\theta \right|,$$

and the right hand member tends to zero as $r \to \infty$. The full statement for all λ with $\Re(\lambda e^{i\varphi}) > \omega(\varphi)$ now follows by analytic continuation together with the corollary to Theorem 5.8.3.

Substituting (12.8.2) into (12.8.4) we obtain

$$\| R(\lambda; A) \| \leqq \frac{M_\epsilon}{\Re(\lambda e^{i\varphi}) - \omega_\epsilon}, \qquad \Phi_1 + \epsilon \leqq \varphi \leqq \Phi_2 - \epsilon. \tag{12.8.5}$$

This result has a simple geometric interpretation. In fact, $\Re(\lambda e^{i\varphi})$ is equal to the projection of $\bar{\lambda}$ on the unit vector $e^{i\varphi}$ and hence it is readily seen that

$$\sup [\Re(\lambda e^{i\varphi}) - \omega_\epsilon; \Phi_1 + \epsilon \leqq \varphi \leqq \Phi_2 - \epsilon]$$

is greater than or equal to the distance from $\bar{\lambda}$ to the sector

$$\Delta_\epsilon^-: \Phi_2 - \epsilon + \pi/2 < \arg (\lambda - \lambda_\epsilon) < \Phi_1 + \epsilon + 3\pi/2,$$

where the vertex λ_ϵ is chosen so that the boundary rays of Δ_ϵ^- are a distance ω_ϵ from the origin. Thus (12.8.1) follows directly from (12.8.5).

The sufficiency argument is similar to the proof of Theorem 12.7.1. Because of the corollary to Theorem 12.2.1, it suffices to show that U generates a semi-group of class $H(\Phi_1 + \epsilon, \Phi_2 - \epsilon)$ for arbitrary $\epsilon > 0$. Further by Theorem 12.2.3 we may replace $R(\lambda; U)$ by $R(\lambda + \lambda_\epsilon ; U)$ and hence assume that Δ_ϵ^- has its vertex at $\lambda = 0$.

We now define

$$T(\xi; \omega) = \frac{1}{2\pi i} \int_{\gamma - i\omega}^{\gamma + i\omega} e^{\lambda\xi} R(\lambda; U)\, d\lambda, \qquad \xi > 0, \tag{12.8.6}$$

for $\gamma > 0$. Let $\Gamma_{\alpha\beta}$, α and $\beta > 0$, be the path:

$$\sigma = \beta + \tau \cdot \tan (-\Phi_2 + \epsilon + \alpha - \pi/2), \qquad \tau \leqq 0,$$

$$\sigma = \beta + \tau \tan (-\Phi_1 - \epsilon - \alpha + \pi/2), \qquad \tau \geqq 0.$$

Thus $\Gamma_{\alpha\beta}$ lies exterior to Δ_ϵ, its rays make an angle of α with the boundary rays of Δ_ϵ, and its vertex is at $\lambda = \beta$. Again, as in Theorem 12.7.1, it can be shown that

$$\lim_{\omega \to \infty} T(\xi; \omega) \equiv T(\xi) \tag{12.8.7}$$

exists in the uniform operator topology and can be represented by

$$T(\zeta) = \frac{1}{2\pi i} \int_{\Gamma_{\alpha\beta}} e^{\lambda\zeta} R(\lambda; U)\, d\lambda. \tag{12.8.8}$$

Now for $\lambda \in \Gamma_{\alpha\beta}$ and $\Phi_1 + \epsilon + 2\alpha < \arg \zeta < \Phi_2 - \epsilon - 2\alpha$ we have

$$\begin{aligned} &\Re(\lambda\zeta) < -\,|\,\lambda - \beta\,|\,|\,\zeta\,| \sin \alpha + \beta\,|\,\zeta\,|, \\ &\|\, R(\lambda; U)\,\| \leqq \frac{C(\alpha)}{\beta + \alpha\,|\,\lambda - \beta\,|}. \end{aligned} \tag{12.8.9}$$

Hence for such ζ the integral in (12.8.8) is majorized by

$$\frac{2C(\alpha)}{\beta} e^{\beta|\zeta|} \int_0^\infty e^{-r|\zeta| \sin \alpha}\, dr.$$

The integral therefore converges uniformly in the sector $\Phi_1 + \epsilon + 2\alpha < \arg \zeta < \Phi_2 - \epsilon - 2\alpha$, $0 < \delta < |\,\zeta\,| < 1/\delta$. Since both α and δ can be chosen arbitrarily small, it follows that $[T(\xi); \xi > 0]$ can be extended to be holomorphic in the sector $\Phi_1 + \epsilon < \arg \zeta < \Phi_2 - \epsilon$ and that this extension is given by (12.8.8). Further it is clear that Theorem 12.7.1 applies so that $[T(\xi)]$, defined by (12.8.7) for $\xi > 0$, is a semi-group of class (A). Hence for fixed real $\xi > 0$, the expression $T(\zeta)T(\xi) - T(\zeta + \xi)$ is holomorphic in ζ throughout the sector (Φ_1, Φ_2), equals

Θ for real $\zeta > 0$, and therefore equals Θ for all ζ in the sector by Theorem 3.11.5. Thus for fixed ζ in the sector the above expression is holomorphic in ξ throughout the sector, equal to Θ for real $\xi > 0$, and consequently equal to Θ for all ξ in the sector. Thus $T(\zeta_1)T(\zeta_1) = T(\zeta_1 + \zeta_2)$ for all ζ_1, ζ_2 in the sector. Conditions (i) and (ii) of Definition 10.6.1 are now verified.

In order to verify condition (iii), we integrate (12.8.8) by parts, obtaining

$$T(\zeta) = \frac{1}{2\pi i}\zeta^{-1}e^{\lambda\zeta}R(\lambda; U)\Big| - \frac{1}{2\pi i}\zeta^{-1}\int_{\Gamma_{\alpha\beta}} e^{\lambda\zeta}R'(\lambda; U)\, d\lambda.$$

The integrated part vanishes for $\Phi_1 + \epsilon + \alpha < \arg\zeta < \Phi_2 - \epsilon - \alpha$. Thus making use of the relation $R'(\lambda; U) = -[R(\lambda; U)]^2$, we have

$$T(\zeta) = \frac{1}{2\pi i}\zeta^{-1}\int_{\Gamma_{\alpha\beta}} e^{\lambda\zeta}[R(\lambda; U)]^2\, d\lambda.$$

Setting $\beta = 1/|\zeta|$, it now follows from the estimates (12.8.9) that

$$\| T(re^{i\varphi}) \| \leqq \frac{1}{\pi}\frac{e^{\beta|\zeta|}}{|\zeta|}\int_0^\infty \frac{[C(\alpha)]^2}{(\beta + \alpha r)^2}\, dr = \frac{e}{\pi\alpha}[C(\alpha)]^2.$$

Thus $\| T(\zeta) \|$ is bounded in each sector $\Phi_1 + \epsilon + \alpha < \arg\zeta < \Phi_2 - \epsilon - \alpha$, and this proves (iii).

It remains to verify (iv) and to show that $U = A$. Since $[T(\xi);\ \xi > 0]$ is of class (A) the union of the range spaces for $\xi > 0$ is dense in $\mathfrak{X}$; this implies (iv). Moreover, Theorem 12.7.1 asserts that $U = A$. We note that the infinitesimal operator A_0 is itself closed because of (iii) so that $U = A = A_0$. This concludes the proof of Theorem 12.8.1.

PART THREE

ADVANCED ANALYTICAL THEORY OF SEMI-GROUPS

Summary. In Part Three a number of new ideas and tools are gradually introduced to round out the theory of semi-groups. Among these added concepts we note: operators "closely related" to the infinitesimal generator, restricted adjoint spaces, a commutative semi-group algebra and its Gelfand representation. There are six chapters entitled *Perturbation Theory*, *Adjoint Theory*, *Operational Calculus*, *Spectral Theory*, *Holomorphic Semi-Groups*, and *Ergodic Theory*.

Chapter XIII is devoted to the problem of admissible perturbations of the infinitesimal generator and leads to the concept of "closely related operators" which generate semi-groups of similar properties. While the adjoint operators of a semi-group of operators form a semi-group on the adjoint space, the continuity properties are usually lost in passing over to the adjoint. A more satisfactory theory is obtained by restricting the adjoint space to a subspace in which the domain of the adjoint to the infinitesimal generator is dense. These concepts are studied in Chapter XIV. The operational calculus of Chapter XV is defined by means of a mapping of the (B)-algebra $S(\varphi)$ of §4.4 into the algebra of operator-valued functions of generators A with $\| T(\xi; A) \| \leqq \varphi(\xi)$. Each such operator-valued function can be represented by means of an abstract Lebesgue-Stieltjes integral of $T(\xi; A)$. The corresponding spectral mapping theorems are obtained in Chapter XVI by employing the Gelfand representation theorem together with certain more classical tools of Fourier analysis.

A semi-group $[T(\xi)]$ may admit an analytic extension from the real axis to the complex plane. If the extension exists, it is unique and defines a semi-group in the interior of a spinal (or angular) semi-module. Various analytic representations are obtained and it is shown that the convex hull of the spectrum of the infinitesimal generator is determined by the growth properties of $\| T(\zeta) \|$ on rays. This discussion occupies Chapter XVII. Finally, in Chapter XVIII we discuss ergodicity, that is, the question of the existence of generalized limits of $T(\xi)$ as $\xi \to 0$ or ∞ in the weak, strong, and uniform operator topologies.

CHAPTER XIII

PERTURBATION THEORY

13.1. Orientation. Perturbation theory has long been a useful tool in the hands of the analyst. It is used to determine the state of a system which is in a certain sense close to a known system. In our case the known system is a semi-group $T(\xi; A_0)$ of linear bounded operators with infinitesimal generator A_0 (not to be confused with the infinitesimal operator A_o of Chapter X) and we wish to ascertain that nearby operators A likewise generate semi-groups. Moreover it is desirable that the semi-group $T(\xi; A)$ vary continuously with A, for in this case the problem of generation discussed in Chapter XII will be "well set."

We shall call a semi-group property *stable* if it holds for all semi-groups $T(\xi; A)$ with A sufficiently close to A_0 whenever it holds for $T(\xi; A_0)$. Of course, in order to make this notion precise we shall have to define a topology for the set of infinitesimal generators. In applications one would expect that the infinitesimal generator itself is known only to within certain limits of error and hence that a physical meaning could be attached only to the stable properties of the associated semi-groups. Mathematically one would expect that the stable properties of a semi-group are more basic than the others and that the significant theorems in the subject would evolve about these properties.

In the first paragraph of this chapter we shall present a detailed perturbation theory for semi-groups of class $(1, A)$. It will be shown that the set of infinitesimal generators for all semi-groups of class $(1, A)$ breaks up into equivalence classes of "closely related" operators. There are certain basic semi-group properties which are passed on by an infinitesimal generator to the semi-groups generated by all closely related operators. Further, given any two closely related generators A_0 and A, it is possible to compute $T(\xi; A)$ from the semi-group $T(\xi; A_0)$ as a power series in $(A - A_0)$. Finally, by introducing a metric into such a class of operators it can be shown that

$$\lim_{A \to A_0} \| T(\xi; A) - T(\xi; A_0) \| = 0$$

uniformly with respect to ξ in each interval of the form $(\epsilon, 1/\epsilon)$, $0 < \epsilon < 1$.

A second paragraph is devoted to two rather special results in perturbation theory. The theory for semi-groups of class $H(\Phi_1, \Phi_2)$ permits a more general kind of perturbing operator than we have treated in the first paragraph, whereas the theory for semi-groups of class $(A)_\infty$ allows only bounded perturbing operators. Both of these results are of a local character.

The two paragraph headings are: *Perturbation Theory for Semi-Groups of Class* $(1, A)$, and *Special Results in Perturbation Theory.*

References. Dye and Phillips [1], Phillips [8, 9, 11], Widder [1].

1. Perturbation Theory for Semi-Groups of Class (1, A)

13.2. Bounded perturbing operators. In order to motivate the subsequent developments, we first consider a perturbation theory for semi-groups of class (C_0) in which the infinitesimal generator is perturbed by a linear bounded operator. In this case the theory is straightforward and yet the results are not without interest.

Throughout this paragraph the given infinitesimal generator A_0 and the perturbed infinitesimal generator A will have a common domain. However the decisive assumption of this section is that $B_0 \equiv A - A_0$ is bounded on $\mathfrak{D}(B_0) = \mathfrak{D}(A_0) = \mathfrak{D}(A)$. In this case, according to Theorem 2.11.2, the operator B_0 has a unique bounded linear extension B on $\overline{\mathfrak{D}(B_0)} = \mathfrak{X}$. We now establish the converse proposition, namely,

Theorem 13.2.1. *If A is the infinitesimal generator of a semi-group of class (C_0) and if $B \in \mathfrak{E}(\mathfrak{X})$, then $A + B$ defined on $\mathfrak{D}(A)$ is likewise the infinitesimal generator of a semi-group of class (C_0).*

Proof. Since A is the infinitesimal generator of a semi-group of class (C_0), Theorem 12.3.1 applies and there exist real constants $M > 0$ and ω such that

$$\| [R(\lambda; A)]^n \| \leqq M(\lambda - \omega)^{-n} \tag{13.2.1}$$

for all $\lambda > \omega$ and integers $n \geqq 1$. Hence for $\lambda > \omega_1 \equiv \omega + M \| B \|$ we have $\| BR(\lambda; A) \| \leqq M \| B \| (\lambda - \omega)^{-1} < 1$. It follows from Theorem 5.10.4 that $R(\lambda; A + B)$ exists and is given by

$$R(\lambda; A + B) = \sum_{j=0}^{\infty} R(\lambda; A)[BR(\lambda; A)]^j \tag{13.2.2}$$

for $\lambda > \omega_1$. Again applying Theorem 12.3.1, we see that the assertion will be established if it can be shown that

$$\begin{aligned} \| [R(\lambda; A + B)]^n \| &= \left\| \left\{ \sum_{j=0}^{\infty} R(\lambda; A)[BR(\lambda; A)]^j \right\}^n \right\| \\ &\leqq M(\lambda - \omega_1)^{-n} \end{aligned} \tag{13.2.3}$$

for all $\lambda > \omega_1$ and integers $n \geqq 1$. With this in mind, we regroup the terms of $\{\sum_{j=0}^{\infty} R(\lambda; A)[BR(\lambda; A)]^j\}^n$ according to powers of B. Each term containing k of the B's must also contain $n + k$ of the $R(\lambda; A)$'s since each B introduces an additional $R(\lambda; A)$. Further the $R(\lambda; A)$'s will be grouped in $k + 1$ nonempty sets separated from each other by the k B's. In other words, the typical term containing k B's will be of the form

$$[R(\lambda; A)]^{r_1}B[R(\lambda; A)]^{r_2}B \cdots [R(\lambda; A)]^{r_k}B[R(\lambda; A)]^{r_{k+1}},$$

where $\sum_{i=1}^{k+1} r_i = n + k$ and $r_i > 0$. If we make use of the estimates (13.2.1) it is clear that this term will be bounded in norm by

$$M(\lambda - \omega)^{-r_1} \| B \| M(\lambda - \omega)^{-r_2} \| B \| \cdots M(\lambda - \omega)^{-r_k} \| B \| M(\lambda - \omega)^{-r_{k+1}} = M^{k+1} \| B \|^k (\lambda - \omega)^{-(n+k)}.$$

The number of terms containing k of the B's is precisely the coefficient of x^k in $(1 - x)^{-n} = \sum_{k=0}^{\infty} C_k^n x^k$. Hence

$$\begin{aligned} \| [R(\lambda; A + B)]^n \| &\leqq \sum_{k=0}^{\infty} C_k^n M^{k+1} \| B \|^k (\lambda - \omega)^{-(n+k)} \\ &= M(\lambda - \omega)^{-n}[1 - M \| B \| (\lambda - \omega)^{-1}]^{-n} \\ &= M(\lambda - \omega_1)^{-n} \end{aligned}$$

for $\lambda > \omega_1$. This concludes the proof.

Corollary. *Let $T(\xi; A)$ be a semi-group of class (C_0) and let $B \in \mathfrak{E}(\mathfrak{X})$. If $\| T(\xi; A) \| \leqq \exp(\omega\xi)$, then $T(\xi; A + B)$ is likewise a semi-group of class (C_0) and $\| T(\xi; A + B) \| \leqq \exp(\omega_1\xi)$ where $\omega_1 = \omega + \| B \|$.*

Proof. If $\| T(\xi; A) \| \leqq \exp(\omega\xi)$, then

$$\| R(\lambda; A) \| \leqq \int_0^\infty e^{-\lambda\xi} e^{\omega\xi} \, d\xi = (\lambda - \omega)^{-1}, \qquad \lambda > \omega,$$

so that the relations (13.2.1) are satisfied with $M = 1$. The corollary to Theorem 12.3.1 together with (13.2.3) now imply that $A + B$ generates a semi-group of class (C_0) such that $\| T(\xi; A + B) \| \leqq \exp(\omega_1\xi)$.

Theorem 13.2.2. *If A is the infinitesimal generator of a group of linear bounded operators defined and strongly continuous on $(-\infty, \infty)$ and if $B \in \mathfrak{E}(\mathfrak{X})$, then $A + B$ defined on $\mathfrak{D}(A)$ is likewise the infinitesimal generator of a group of linear bounded operators defined and strongly continuous on $(-\infty, \infty)$.*

Proof. The proof follows the same argument as that of Theorem 13.2.1, depending now on Theorem 12.3.2.

The relation (13.2.2) suggests an expansion theorem for $T(\xi; A + B)$ in powers of B. For in (13.2.2) we have the Laplace transform of $T(\xi; A + B)$, that is $R(\lambda; A + B)$, represented as the sum $\sum_{j=0}^{\infty} R(\lambda; A)[BR(\lambda; A)]^j$, where again $R(\lambda; A)$ is the Laplace transform of $T(\xi; A)$. We recall that the product of the Laplace transforms of two functions is the Laplace transform of their convolution. Thus one would expect to have

$$T(\xi; A + B) = \sum_{n=0}^{\infty} S_n(\xi) \tag{13.2.4}$$

where

$$S_0(\xi) = T(\xi; A) \quad \text{and} \quad S_n(\xi)x = \int_0^\xi T(\xi - \sigma; A)BS_{n-1}(\sigma)x \, d\sigma.$$

It is not difficult to prove that this expansion is valid; however we postpone the proof until section 13.4 where the result will be established in a more general setting. For the present we note that $\| T(\xi; A) \| \leqq Me^{\omega\xi}$ implies that

$$\| S_n(\xi) \| \leqq M(M \| B \|)^n \xi^n e^{\omega\xi}/n! \tag{13.2.5}$$

so that

$$\left\| T(\xi; A + B) - \sum_{k=0}^{n} S_k(\xi) \right\| \leqq M(M \| B \|)^{n+1}\xi^{n+1}e^{\omega_1\xi}/(n + 1)! \tag{13.2.6}$$

where $\omega_1 = \omega + M \| B \|$.

The terms of the expansion (13.2.4) are rather suggestive. The zero-th term is, of course, the unperturbed semi-group. The first term,

$$S_1(\xi) = \int_0^{\xi} T(\xi - \sigma; A)BT(\sigma; A)\, d\sigma,$$

can be thought of as resulting from a perturbation of the unperturbed semi-group after a time σ, followed by the action of the unperturbed semi-group over the remaining $\xi - \sigma$ time, and this averaged over all σ, $0 < \sigma < \xi$. The second term consists of the unperturbed semi-group perturbed at times σ and τ, and this averaged over all σ, with $0 < \sigma < \tau < \xi$. And so on.

13.3. General perturbing operators. We now begin a more systematic study of perturbation theory for semi-groups of linear bounded operators. We again assume that the given infinitesimal generator A_0 and the perturbed infinitesimal generator A have a common domain. As before, the linear operator $B \equiv A - A_0$ with domain $\mathfrak{D}(B) = \mathfrak{D}(A_0) = \mathfrak{D}(A)$ will in general not be closed; however now B need not even possess a closed extension. On the other hand, both $A_0R(\lambda; A_0)$ and $AR(\lambda; A_0)$ are closed linear operators defined on all $\mathfrak{X}$ and therefore bounded by the closed graph theorem. It follows that $BR(\lambda; A_0) = AR(\lambda; A_0) - A_0R(\lambda; A_0)$ is also bounded on $\mathfrak{X}$.

The above considerations lead us to consider the following class of perturbing operators.

Definition 13.3.1. *Let A be the infinitesimal generator of a semi-group of class (A). A linear operator B is said to belong to the class $\mathfrak{J}(A)$ if $\mathfrak{D}(B) = \mathfrak{D}(A)$ and $BR(\lambda_0; A) \in \mathfrak{E}(\mathfrak{X})$ for some $\lambda_0 \in \rho(A)$.*

We note that $BR(\lambda; A)$ is bounded for every $\lambda \in \rho(A)$ if $B \in \mathfrak{J}(A)$. This is an immediate consequence of the first resolvent equation since

$$BR(\lambda; A) = BR(\lambda_0; A) + (\lambda_0 - \lambda)BR(\lambda_0; A)R(\lambda; A).$$

Actually it is more convenient to deal with an extension of $B \in \mathfrak{J}(A)$ rather than B itself. This is due to the fact that B is in general not closed. We therefore introduce

DEFINITION 13.3.2. *Let A be the infinitesimal generator of a semi-group of class (A). A linear operator B is said to belong to the class $\mathfrak{J}^{\sim}(A)$ if*

(i) $\mathfrak{D}(B) \supset \mathfrak{D}(A)$;

(ii) $BR(\lambda_0 ; A) \in \mathfrak{E}(\mathfrak{X})$ *for some* $\lambda_0 \in \rho(A)$;

(iii) *An element x belongs to $\mathfrak{D}(B)$ if and only if*

$$\lim_{\lambda \to \infty} \lambda BR(\lambda; A)x \equiv y \tag{13.3.1}$$

exists, in which case $Bx = y$.

For purposes of clarity we also insert

DEFINITION 13.3.3. *Let C and A be linear operators and suppose that $\mathfrak{D}(C) \supset \mathfrak{D}(A)$. We shall use the notation*

$$\| C \|_A \equiv \sup [\| Cx \| ; \| x \| \leqq 1, x \in \mathfrak{D}(A)].$$

THEOREM 13.3.1. *Let $B \in \mathfrak{J}(A)$. Then B has a unique extension $\tilde{B}$ in $\mathfrak{J}^{\sim}(A)$. If B is the restriction of a closed operator B_1, then $B \subset \tilde{B} \subset B_1$. If B is bounded, then $\tilde{B} \in \mathfrak{E}(\mathfrak{X})$ and $\| \tilde{B} \| = \| B \|_A$.*

PROOF. It is obviously necessary that $\mathfrak{D}(\tilde{B})$ consist of the set of all x such that $\lim_{\lambda \to \infty} \lambda BR(\lambda; A)x \equiv y$ exist and that $\tilde{B}x = y$. The operator so defined is clearly linear and conditions (i), (ii), and (iii) follow if it can be shown that $\tilde{B}$ is in fact an extension of B. Suppose, then, that $x \in \mathfrak{D}(B) = \mathfrak{D}(A)$. For a fixed $\mu \in \rho(A)$ there exists a y with $x = R(\mu; A)y$. Now $BR(\mu; A)$ is bounded and

$$\lim_{\lambda \to \infty} \lambda R(\lambda; A)y = y.$$

Hence

$$\lambda BR(\lambda; A)x = [BR(\mu; A)][\lambda R(\lambda; A)y] \to BR(\mu; A)y = Bx$$

so that we have by definition that $\tilde{B}x = Bx$. Thus $\tilde{B} \supset B$. Finally it is clear from condition (iii) that this extension is unique.

If B_1 is a closed extension of B, then $\lim_{\lambda \to \infty} \lambda R(\lambda; A)x = x$ and

$$\lim_{\lambda \to \infty} \lambda BR(\lambda; A)x = y$$

imply that $x \in \mathfrak{D}(B_1)$ and that $B_1 x = \tilde{B}x$. It follows that $B_1 \supset \tilde{B} \supset B$. Finally suppose that B is bounded on $\mathfrak{D}(A)$. Then $\| BR(\lambda; A) \| = O(1/\lambda)$ as $\lambda \to \infty$; also, as we have shown above, $\lim_{\lambda \to \infty} \lambda BR(\lambda; A)x$ exists for each x belonging to the dense set $\mathfrak{D}(B)$. Thus the Banach-Steinhaus theorem applies and we see that the limit exists for all $x \in \mathfrak{X}$, the limit Bx defining a bounded linear operator on $\mathfrak{X}$. According to Theorem 2.11.2, B has but one bounded linear extension to $\mathfrak{X}$ and the norm of this extension is just $\| B \|_A$.

DEFINITION 13.3.4. *Let A be the infinitesimal generator of a semi-group of class (A) and $B \in \mathfrak{J}(A)$. The extension of B in $\mathfrak{J}^{\sim}(A)$ is called the A-extension of B and is denoted by $\tilde{B}$.*

The notation for $\tilde{B}$ is somewhat misleading since $\tilde{B}$ depends on the infinitesimal generator A as well as B. In situations where confusion may arise from this fact, we shall take care to distinguish between the various possible extensions of B.

We next prove three lemmas which will be useful in the subsequent discussion.

LEMMA 13.3.1. *Let $B \in \mathfrak{Z}^{\sim}(A)$ and suppose $S = \sum_{n=1}^{\infty} S_n$ where S, $\{S_n\}$, $\{BS_n\} \subset \mathfrak{E}(\mathfrak{X})$ and both $\sum \| S_n \|$ and $\sum \| BS_n \|$ converge. If there exists a summable sequence $\{M_n\}$ and a constant $\omega > \omega_0$ such that $\| \lambda BR(\lambda; A)S_n \| \leqq M_n$ for all $\lambda > \omega$, then $\mathfrak{D}(B) \supset \mathfrak{R}(S)$, $BS \in \mathfrak{E}(\mathfrak{X})$, and $BS = \sum_{n=1}^{\infty} BS_n$.*

PROOF. The operator $BR(\lambda; A)$ is bounded so that

$$\lambda BR(\lambda; A)Sx = \sum_{n=1}^{\infty} \lambda BR(\lambda; A)S_n x, \qquad x \in \mathfrak{X}. \tag{13.3.2}$$

By assumption $\lim_{\lambda\to\infty} \lambda BR(\lambda; A)S_n x = BS_n x$ and $\| \lambda BR(\lambda; A)S_n \| \leqq M_n$ for $\lambda > \omega$. Hence, by the majorized convergence theorem, the right member of (13.3.2) converges to the limit $\sum_{n=1}^{\infty} BS_n x$. It now follows from Definition 13.3.2 that $Sx \in \mathfrak{D}(B)$ and $BSx = \sum_{n=1}^{\infty} BS_n x$. It is clear that

$$\| BS \| \leqq \sum_{n=1}^{\infty} \| BS_n \|.$$

LEMMA 13.3.2. *Let $B \in \mathfrak{Z}^{\sim}(A)$ and suppose that $f(\xi)$ and $Bf(\xi)$ are both strongly measurable and Bochner integrable functions on $(0, \infty)$ to $\mathfrak{X}$. If there exists a summable function $M(\xi)$ and a constant $\omega > \omega_0$ such that $\| \lambda BR(\lambda; A)f(\xi) \| \leqq M(\xi)$ on $(0, \infty)$ for all $\lambda > \omega$, then*

$$B\left[\int_0^\infty f(\xi)\, d\xi\right] = \int_0^\infty Bf(\xi)\, d\xi. \tag{13.3.3}$$

PROOF. The proof is similar to that of the previous lemma. Again, $BR(\lambda; A)$ being bounded, we have

$$\lambda BR(\lambda; A)\left[\int_0^\infty f(\xi)\, d\xi\right] = \int_0^\infty \lambda\, BR(\lambda; A)f(\xi)\, d\xi. \tag{13.3.4}$$

By assumption $\lim_{\lambda\to\infty} \lambda BR(\lambda; A)f(\xi) = Bf(\xi)$ and $\| \lambda BR(\lambda; A)f(\xi) \| \leqq M(\xi)$ for $\lambda > \omega$. The majorized convergence theorem applies and we see that the right member of (13.3.4) converges to $\int_0^\infty Bf(\xi)\, d\xi$. According to Definition 13.3.2, $\int_0^\infty f(\xi)\, d\xi \in \mathfrak{D}(B)$ and (13.3.3) holds.

LEMMA 13.3.3. *Let $B \in \mathfrak{Z}^{\sim}(A)$ and suppose that $BT(\xi_0; A)$ is bounded on $\mathfrak{D}(A)$ for some $\xi_0 > 0$. Then for all $\xi \geqq \xi_0$, $\mathfrak{D}(B) \supset T(\xi; A)[\mathfrak{X}]$, $BT(\xi; A) \in \mathfrak{E}(\mathfrak{X})$, and $\| BT(\xi; A) \| = \| BT(\xi; A) \|_A$. Moreover $BT(\xi; A)$ is continuous in the strong operator topology for $\xi > \xi_0$ and*

$$\limsup_{\xi\to\infty} \xi^{-1} \log \| BT(\xi; A) \| \leqq \omega_0. \tag{13.3.5}$$

PROOF. We first consider the operator $BT(\xi_0; A)$ which is bounded on $\mathfrak{D}(A)$.

By Theorem 2.11.2, $BT(\xi_0; A)$ has a unique linear bounded extension on $\overline{\mathfrak{D}(A)} = \mathfrak{X}$. Denoting this extension by $C(\xi_0)$, we have

$$\| C(\xi_0) \| = \| BT(\xi_0 ; A) \|_A .$$

Since

$$\lambda BR(\lambda; A)[T(\xi_0 ; A)x] = [BT(\xi_0 ; A)][\lambda R(\lambda; A)x] = C(\xi_0)[\lambda R(\lambda; A)x],$$

we see that $\lim_{\lambda \to \infty} \lambda BR(\lambda; A)[T(\xi_0 ; A)x]$ exists and is equal to $C(\xi_0)x$. Again it follows from Definition 13.3.2 that $T(\xi_0 ; A)x \in \mathfrak{D}(B)$ and that $BT(\xi_0 ; A)x = C(\xi_0)x$. This proves the first assertion for $\xi = \xi_0$. If $\xi > \xi_0$, then $BT(\xi; A) = [BT(\xi_0 ; A)]T(\xi - \xi_0 ; A)$ and hence $\mathfrak{D}(B) \supset T(\xi; A)[\mathfrak{X}]$ and $BT(\xi; A) \in \mathfrak{E}(\mathfrak{X})$; the norm condition follows from Theorem 2.11.2. Further, the relation

$$BT(\xi; A)x = [BT(\xi_0 ; A)]T(\xi - \xi_0 ; A)x$$

shows that $BT(\xi; A)$ is strongly continuous for $\xi > \xi_0$. Finally this same relation implies that $\log \| BT(\xi; A) \| \leqq \log \| BT(\xi_0 ; A) \| + \log \| T(\xi - \xi_0 ; A) \|$, from which (13.3.5) follows.

The class $\mathfrak{Z}(A_0)$ includes all linear operators $B = A - A_0$ where A is any closed linear operator with domain $\mathfrak{D}(A_0)$. One cannot expect all operators of this kind to be suitable perturbing operators. Indeed, the operator $-2A_0 \in \mathfrak{Z}(A_0)$ and yet $A_0 + (-2A_0) = -A_0$ is not in general the infinitesimal generator of a semi-group (see the remark following Theorem 11.5.2). It is clear that we must restrict our attention to the perturbing operators in $\mathfrak{Z}(A_0)$. Actually, our methods are not sufficiently general to deal with all of the perturbing operators in $\mathfrak{Z}(A_0)$; instead we shall limit ourselves to the following class of operators.

Definition 13.3.5. *Let A be the infinitesimal generator of a semi-group of class $(1, A)$. A linear operator B is said to be of class $\mathfrak{P}(A)$ if*

(i) $B \in \mathfrak{Z}(A)$;
(ii) $BT(\xi; A)$ *defined on* $\mathfrak{D}(A)$ *is bounded for all* $\xi > 0$;
(iii) $\int_0^1 \| BT(\xi; A) \|_A \, d\xi < \infty$.

It is clear that $\mathfrak{P}(A)$ is a linear system of operators. Conditions (ii) and (iii) can be formulated in many different ways as we shall show in section 13.5. However the above formulation is the simplest for our purposes.

As an immediate consequence of Lemma 13.3.3, we have

Theorem 13.3.2. *Let $T(\xi; A)$ be of class $(1, A)$ and $B \in \mathfrak{P}(A)$. Then $\mathfrak{D}(\tilde{B}) \supset \mathfrak{X}_0$, $\tilde{B}T(\xi; A) \in \mathfrak{E}(\mathfrak{X})$, and $\| \tilde{B}T(\xi; A) \| = \| BT(\xi; A) \|_A$ for all $\xi > 0$. Moreover, $\tilde{B}T(\xi; A)$ is continuous in the strong operator topology for $\xi > 0$ and*

$$\limsup_{\xi \to \infty} \xi^{-1} \log \| \tilde{B}T(\xi; A) \| \leqq \omega_0 .$$

Remark. If $A \in \mathfrak{P}(A)$, then A is bounded. In fact, for $x \in \mathfrak{D}(A)$ we have by Theorem 11.5.4 that $T(\xi; A)x - x = \int_0^\xi AT(\sigma; A)x \, d\sigma$. Thus

$$\| T(\xi; A) - I \| = \| T(\xi; A) - I \|_A \leqq \int_0^\xi \| AT(\sigma; A) \|_A \, d\sigma$$

which tends to zero with ξ. It now follows from Theorem 9.6.1 that $A \in \mathfrak{G}(\mathfrak{X})$.

The following lemmas will be needed in the proof of the expansion theorem.

LEMMA 13.3.4. *If $T(\xi; A)$ is of class $(1, A)$ and $B \in \mathfrak{P}(A)$, then*

$$\tilde{B}R^{(n)}(\lambda; A)x = \int_0^\infty e^{-\lambda\xi}(-\xi)^n \tilde{B}T(\xi; A)x \, d\xi \tag{13.3.6}$$

for all λ with $\Re(\lambda) > \omega_0$ and all integers $n \geqq 0$.

PROOF. For $\sigma \equiv \Re(\lambda) > \omega_0$, condition (iii) of Definition 13.3.5 together with Theorem 13.3.2 shows that $e^{-\lambda\xi}(-\xi)^n \tilde{B}T(\xi; A)x$ is a strongly-continuous Bochner-integrable function on $(0, \infty)$ and that $\int_0^\infty e^{-\sigma\xi}\xi^n \| \tilde{B}T(\xi; A) \| \, d\xi < \infty$. Further, given $\omega > \omega_0$ there exists a constant $M_\omega > 0$ such that $\| \mu R(\mu; A) \| \leqq M_\omega$ for all $\mu > \omega$. Thus

$$\| \mu\tilde{B}R(\mu; A)T(\xi; A)x \| = \| [\tilde{B}T(\xi; A)][\mu R(\mu; A)x] \| \leqq \| \tilde{B}T(\xi; A) \| M_\omega \| x \|$$

for all $\mu > \omega$. The assertion now follows from Lemma 13.3.2.

LEMMA 13.3.5. *Let $T(\xi; A)$ be of class $(1, A)$ and $B \in \mathfrak{P}(A)$. Suppose further that $f(\xi)$ is strongly continuous for $\xi > 0$ and $\int_0^1 \| f(\sigma) \| \, d\sigma < \infty$. Set $g(\xi) = \int_0^\xi T(\xi - \sigma; A)f(\sigma) \, d\sigma$. Then $g(\xi) \in \mathfrak{D}(\tilde{B})$,*

$$\tilde{B}g(\xi) = \int_0^\xi \tilde{B}T(\xi - \sigma; A)f(\sigma) \, d\sigma, \tag{13.3.7}$$

and $\tilde{B}g(\xi)$ is strongly continuous for $\xi > 0$.

REMARK. Since one possible choice for $\tilde{B}$ is the identity, the existence and strong continuity of $g(\xi)$ for $\xi > 0$ also follows from the lemma.

PROOF. Set

$$m(\alpha, \beta) = \sup [\, \| f(\xi) \| \, ; \, \alpha \leqq \xi \leqq \beta]$$

and

$$M(\alpha, \beta) = \sup [\, \| \tilde{B}T(\xi; A) \| \, ; \, \alpha \leqq \xi \leqq \beta] .$$

These constants are both finite for $0 < \alpha < \beta < \infty$ because of the strong continuity of $f(\xi)$ and $\tilde{B}T(\xi; A)$ for $\xi > 0$. It is clear that $\tilde{B}T(\xi - \sigma; A)f(\sigma)$ is strongly continuous in σ for $0 < \sigma < \xi$, and that

$$\| \tilde{B}T(\xi - \sigma; A)f(\sigma) \| \leqq \begin{cases} M(\xi/2, \xi) \, \| f(\sigma) \|, & 0 < \sigma \leqq \xi/2, \\ m(\xi/2, \xi) \, \| \tilde{B}T(\xi - \sigma; A) \|, & \xi/2 \leqq \sigma < \xi. \end{cases} \tag{13.3.8}$$

Now given $\omega > \omega_0$ there exists a constant M_ω such that $\| \lambda R(\lambda; A) \| \leqq M_\omega$ for all $\lambda > \omega$. Consequently

$$\| \lambda \tilde{B} R(\lambda; A) T(\xi - \sigma; A) f(\sigma) \| = \| [\tilde{B} T(\xi - \sigma; A)][\lambda R(\lambda; A) f(\sigma)] \|$$

$$\leqq \begin{cases} M_\omega M(\xi/2, \xi) \| f(\sigma) \|, & 0 < \sigma \leqq \xi/2, \\ M_\omega m(\xi/2, \xi) \| \tilde{B} T(\xi - \sigma; A) \|, & \xi/2 \leqq \sigma < \xi \end{cases} \tag{13.3.9}$$

for all $\lambda > \omega$. The estimates (13.3.8) and (13.3.9) show that the hypothesis for Lemma 13.3.2 is satisfied and consequently that $g(\xi) \in \mathfrak{D}(\tilde{B})$ and (13.3.7) holds.

We now prove the strong continuity of $\tilde{B}g(\xi)$ for $\xi > 0$. Given $\xi_0 > 0$, choose $\delta > 0$ so that $0 < \xi_0 - 2\delta$. For ξ_1, ξ_2 with $\xi_0 - \delta \leqq \xi_1, \xi_2 \leqq \xi_0 + \delta$, we have

$$\| \tilde{B}g(\xi_2) - \tilde{B}g(\xi_1) \| \leqq \int_0^{\xi_0 - 2\delta} \| [\tilde{B}T(\xi_2 - \sigma; A) - \tilde{B}T(\xi_1 - \sigma; A)] f(\sigma) \| \, d\sigma$$

$$+ \sum_{i=1}^{2} \int_{\xi_0 - 2\delta}^{\xi_i} \| \tilde{B}T(\xi_i - \sigma; A) f(\sigma) \| \, d\sigma.$$

For $0 < \sigma < \xi_0 - 2\delta$, we have

$$\| [\tilde{B}T(\xi_2 - \sigma; A) - \tilde{B}T(\xi_1 - \sigma; A)] f(\sigma) \| \leqq 2M(\delta, \xi_0 + \delta) \| f(\sigma) \| .$$

Consequently the first integral in the right member converges to zero as $\xi_1, \xi_2 \to \xi_0$ by Theorem 3.7.9. On the other hand for $\xi_0 - 2\delta < \sigma < \xi_i$, we have

$$\| \tilde{B}T(\xi_i - \sigma; A) f(\sigma) \| \leqq m(\xi_0 - 2\delta, \xi_0 + \delta) \| \tilde{B}T(\xi_i - \sigma; A) \| .$$

Hence

$$\limsup_{\xi_1, \xi_2 \to \xi_0} \| \tilde{B}g(\xi_2) - \tilde{B}g(\xi_1) \| \leqq 2m(\xi_0 - 2\delta, \xi_0 + \delta) \int_0^{3\delta} \| \tilde{B}T(\sigma; A) \| \, d\sigma,$$

which can be made arbitrarily small with δ. This concludes the proof.

LEMMA 13.3.6. *Let $T(\xi; A)$ be of class $(1, A)$ and $B \in \mathfrak{P}(A)$. Suppose further that $f(\xi)$ is strongly continuous for $\xi > 0$ and that $\int_0^\infty e^{-\omega_1 \xi} \| f(\xi) \| \, d\xi < \infty$ for some real ω_1. Set $F(\lambda) = \int_0^\infty e^{-\lambda\xi} f(\xi) \, d\xi$. Then*

$$\tilde{B}R(\lambda; A)F(\lambda) = \int_0^\infty e^{-\lambda\xi} \tilde{B} \int_0^\xi T(\xi - \sigma; A) f(\sigma) \, d\sigma \, d\xi \tag{13.3.10}$$

for $\Re(\lambda) > \max(\omega_0, \omega_1)$.

PROOF. According to Lemma 13.3.5, the right member in (13.3.10) is equal to $\int_0^\infty e^{-\lambda\xi} \int_0^\xi \tilde{B}T(\xi - \sigma; A) f(\sigma) \, d\sigma \, d\xi$. Now, for $\gamma > \max(\omega_0, \omega_1)$,

$$\int_0^\infty e^{-\gamma\xi} \int_0^\xi \| \tilde{B}T(\xi - \sigma; A) \| \| f(\sigma) \| \, d\sigma \, d\xi$$

$$\leqq \int_0^\infty e^{-\gamma\sigma} \| \tilde{B}T(\sigma; A) \| \, d\sigma \int_0^\infty e^{-\gamma\sigma} \| f(\sigma) \| \, d\sigma.$$

Hence the abstract Fubini theorem (Theorem 3.7.13) applies and

$$\int_0^\infty e^{-\lambda\xi}\int_0^\xi \tilde{B}T(\xi-\sigma;A)f(\sigma)\,d\sigma\,d\xi = \int_0^\infty e^{-\lambda\sigma}\int_\sigma^\infty e^{-\lambda(\xi-\sigma)}\tilde{B}T(\xi-\sigma;A)f(\sigma)\,d\xi\,d\sigma$$

for $\Re(\lambda) > \max(\omega_0, \omega_1)$. Finally, making use of Lemma 13.3.4 we obtain

$$\int_0^\infty e^{-\lambda\sigma}\int_\sigma^\infty e^{-\lambda(\xi-\sigma)}\tilde{B}T(\xi-\sigma;A)f(\sigma)\,d\xi\,d\sigma = \int_0^\infty e^{-\lambda\sigma}\,\tilde{B}R(\lambda;A)f(\sigma)\,d\sigma$$

$$= \tilde{B}R(\lambda;A)\int_0^\infty e^{-\lambda\sigma}f(\sigma)\,d\sigma = \tilde{B}R(\lambda;A)F(\lambda).$$

The combined set of identities now yields (13.3.10).

13.4. The perturbed semi-group. We now make use of the foregoing results to generalize the perturbation theorem of section 13.2 to semi-groups of class $(1, A)$ and to perturbing operators of class $\mathfrak{P}(A)$. Our development will be based on the series expansion (13.2.4).

Suppose, then, that $T(\xi; A)$ is a semi-group of class $(1, A)$ and $B \in \mathfrak{P}(A)$. Throughout this section we shall use the notation

$$(13.4.1) \qquad \varphi(\xi) \equiv \| T(\xi;A) \| \quad \text{and} \quad \psi(\xi) \equiv \| \tilde{B}T(\xi;A) \| .$$

We shall also denote the n-fold convolution of the function $\psi(\xi)$ with itself by $\psi^{n*}(\xi)$. It is clear that $\varphi(\xi)$ and $\psi(\xi)$ are both non-negative, measurable, and that

$$(13.4.2) \qquad \int_0^1 [\varphi(\xi) + \psi(\xi)]\,d\xi < \infty.$$

The function $\varphi(\xi)$ is submultiplicative whereas

$$(13.4.3) \qquad \psi(\xi+\eta) \leqq \psi(\xi)\varphi(\eta), \qquad \xi, \eta > 0.$$

Finally we recall that $\lim_{\xi\to\infty} \xi^{-1}\log\varphi(\xi) = \omega_0$, from which we obtain

$$(13.4.4) \qquad \limsup_{\xi\to\infty} \xi^{-1}\log\psi(\xi) \leqq \omega_0 .$$

Before establishing the expansion theorem we shall prove three preliminary lemmas, the first of which is very similar to Theorem 7.4.4.

LEMMA 13.4.1. *If* $\omega > \omega_0$, *then* $\int_0^\infty e^{-\omega\xi}[\varphi(\xi) + \psi(\xi)]\,d\xi \equiv M_\omega < \infty$ *and*

$$(13.4.5) \qquad \psi(\xi) \leqq M_\omega^2\xi^{-2}e^{\omega\xi}, \qquad \xi > 0.$$

PROOF. Any finite-valued measurable submultiplicative function $\varphi(\xi)$ is necessarily bounded in each interval of the form $(\epsilon, 1/\epsilon)$, $0 < \epsilon < 1$, by Theorem 7.4.1; the relation (13.4.3) shows that the same is true of $\psi(\xi)$. Thus the existence of the integral for each $\omega > \omega_0$ is an immediate consequence of (13.4.2) and (13.4.4). Further it is clear from (13.4.3) that

$$0 \leqq 2e^{-\omega\xi}\psi(\xi) \leqq 2[e^{-\omega(\xi-\sigma)}\psi(\xi-\sigma)][e^{-\omega\sigma}\varphi(\sigma)] \leqq [e^{-\omega(\xi-\sigma)}\psi(\xi-\sigma)]^2 + [e^{-\omega\sigma}\varphi(\sigma)]^2,$$

and hence that $4\, e^{-\omega\xi}\psi(\xi) \leqq [e^{-\omega(\xi-\sigma)}\psi(\xi - \sigma) + e^{-\omega\sigma}\varphi(\sigma)]^2$. Thus

$$\xi[e^{-\omega\xi}\psi(\xi)]^{1/2} = 2\int_0^{\xi/2} [e^{-\omega\xi}\psi(\xi)]^{1/2}\, d\sigma$$
$$\leqq \int_0^{\xi/2} e^{-\omega(\xi-\sigma)}\psi(\xi - \sigma)\, d\sigma + \int_0^{\xi/2} e^{-\omega\sigma}\varphi(\sigma)\, d\sigma \leqq M_\omega\,;$$

and this implies (13.4.5).

LEMMA 13.4.2. *Let $g_1(\xi)$, $g_2(\xi)$ be two non-negative measurable functions such that $\int_0^\infty e^{-\omega\xi}g_i(\xi)\, d\xi \leqq M_i$ and $g_i(\xi) \leqq M_i\xi^{-k}e^{\omega\xi}$ $(i = 1, 2)$ for some real constant ω and integer $k \geqq 0$. Then*

$$g_3(\xi) = (g_1 * g_2)(\xi) \equiv \int_0^\xi g_1(\xi - \sigma)g_2(\sigma)\, d\sigma$$

also satisfies these conditions with $M_3 = 2^{k+1}M_1M_2$.

PROOF. An easy calculation shows that

$$\int_0^\infty e^{-\omega\xi}g_3(\xi)\, d\xi = \int_0^\infty e^{-\omega\xi}g_1(\xi)\, d\xi \int_0^\infty e^{-\omega\xi}g_2(\xi)\, d\xi \leqq M_1M_2\,.$$

Further

$$g_3(\xi) = e^{\omega\xi}\left\{\int_0^{\xi/2} + \int_{\xi/2}^{\xi} [e^{-\omega(\xi-\sigma)}g_1(\xi - \sigma)][e^{-\omega\sigma}g_2(\sigma)]\, d\sigma\right\}$$
$$\leqq e^{\omega\xi}\left\{M_1 2^k\xi^{-k}\int_0^{\xi/2} e^{-\omega\sigma}g_2(\sigma)\, d\sigma + M_2 2^k\xi^{-k}\int_0^{\xi/2} e^{-\omega\sigma}g_1(\sigma)\, d\sigma\right\}$$
$$\leqq 2^{k+1}M_1M_2\xi^{-k}e^{\omega\xi}.$$

LEMMA 13.4.3. *Let $\psi_0(\xi)$ and $\psi(\xi)$ be two non-negative measurable functions satisfying the conditions* (13.4.2) *and* (13.4.3). *Then*

$$\theta(\xi) \equiv \sum_{n=0}^{\infty} (\psi_0 * \psi^{n*})(\xi) \tag{13.4.6}$$

converges uniformly with respect to ξ in each interval of the form $(\epsilon, 1/\epsilon)$, $0 < \epsilon < 1$. Further if $\omega > \omega_0$ is such that $\int_0^\infty e^{-\omega\xi}\psi(\xi)\, d\xi < 1$, then $\int_0^\infty e^{-\omega\xi}\theta(\xi)\, d\xi < \infty$.

PROOF. We first choose $\omega_1 > \omega_0$ so that

$$\int_0^\infty e^{-\omega_1\xi}[\varphi(\xi) + \psi_0(\xi)]\, d\xi \leqq 1 \qquad \text{and} \qquad \int_0^\infty e^{-\omega_1\xi}[\varphi(\xi) + \psi(\xi)]\, d\xi \leqq \tfrac{1}{16}.$$

It then follows from Lemma 13.4.1 that $\psi_0(\xi) \leqq \xi^{-2}e^{\omega_1\xi}$ and $\psi(\xi) \leqq (\frac{1}{16})\xi^{-2}e^{\omega_1\xi}$. Repeated application of Lemma 13.4.2 $(k = 2)$ gives

$$(\psi_0 * \psi^{n*})(\xi) \leqq 2^{-n}\xi^{-2}e^{\omega_1\xi}. \tag{13.4.7}$$

Thus the series $\sum (\psi_0 * \psi^{n*})(\xi)$ is majorized by the series $\sum 2^{-n}\xi^{-2}e^{\omega_1 \xi}$ which obviously converges uniformly with respect to ξ in each interval of the form $(\epsilon, 1/\epsilon)$. Finally we note that

$$\int_0^\infty e^{-\omega\xi}(\psi_0 * \psi^{n*})(\xi)\, d\xi = \left[\int_0^\infty e^{-\omega\xi}\psi_0(\xi)\, d\xi\right]\left[\int_0^\infty e^{-\omega\xi}\psi(\xi)\, d\xi\right]^n.$$

Hence for any choice of $\omega > \omega_0$ such that $\int_0^\infty e^{-\omega\xi}\psi(\xi)\, d\xi < 1$, we have

$$\int_0^\infty e^{-\omega\xi}\theta(\xi)\, d\xi = \left[\int_0^\infty e^{-\omega\xi}\psi_0(\xi)\, d\xi\right]\left[1 - \int_0^\infty e^{-\omega\xi}\psi(\xi)\, d\xi\right]^{-1} < \infty.$$

We come now to the central theorem of this paragraph.

Theorem 13.4.1. *Let $[T(\xi; A)]$ be a semi-group of class $(1, A)$ and $B \in \mathfrak{P}(A)$. Then $A + B$ defined on $\mathfrak{D}(A)$ generates a semi-group of class $(1, A)$ and*

$$T(\xi; A + B) = \sum_{n=0}^{\infty} S_n(\xi) \tag{13.4.8}$$

where $S_0(\xi) = T(\xi; A)$ and $S_n(\xi)x = \int_0^\xi T(\xi - \sigma; A)\tilde{B}S_{n-1}(\sigma)x\, d\sigma$; the series converges absolutely, uniformly with respect to ξ in each interval of the form $(\epsilon, 1/\epsilon)$, $0 < \epsilon < 1$. The terms $S_n(\xi)$ are continuous in the strong operator topology for $\xi > 0$ and

$$\| S_n(\xi) \| \leqq (\varphi * \psi^{n*})(\xi), \qquad n \geqq 0. \tag{13.4.9}$$

Proof. We begin by showing that $S_n(\xi)$ and $\tilde{B}S_n(\xi)$ are both continuous in the strong operator topology for $\xi > 0$ and that $\| S_n(\xi) \| \leqq (\varphi * \psi^{n*})(\xi)$ and $\| \tilde{B}S_n(\xi) \| \leqq \psi^{(n+1)*}(\xi)$. It is clear from Theorem 13.3.2 that this is true for the case $n = 0$. Suppose that it is likewise true for $n = k$. It then follows from Lemma 13.3.5 that $S_{k+1}(\xi)$ and $\tilde{B}S_{k+1}(\xi)$ are strongly continuous. Moreover, we have

$$\| S_{k+1}(\xi) \| \leqq \int_0^\xi \| T(\xi - \sigma; A) \|\, \| \tilde{B}S_k(\sigma) \|\, d\sigma \leqq (\varphi * \psi^{(k+1)*})(\xi)$$

and

$$\| \tilde{B}S_{k+1}(\xi) \| \leqq \int_0^\xi \| \tilde{B}T(\xi - \sigma; A) \|\, \| \tilde{B}S_k(\sigma) \|\, d\sigma \leqq \psi^{(k+2)*}(\xi).$$

This completes the induction argument. Thus the series $\sum_{n=0}^\infty \| S_n(\xi) \|$ is majorized by the series $\theta(\xi) \equiv \sum_{n=0}^\infty (\varphi * \psi^{n*})(\xi)$ and therefore by Lemma 13.4.3 the series $\sum_{n=0}^\infty S_n(\xi) \equiv S(\xi)$ converges absolutely, uniformly with respect to ξ in each interval of the form $(\epsilon, 1/\epsilon)$. Since each of the terms is strongly continuous for $\xi > 0$, the uniform convergence of the sum implies that $S(\xi)$ is also strongly continuous for $\xi > 0$.

We next consider the Laplace transform of $S(\xi)x$. To this end we choose $\omega > \omega_0$ such that $\int_0^\infty e^{-\omega\xi}\psi(\xi)\, d\xi \equiv \gamma < 1$. Again by Lemma 13.4.3 we have

$\int_0^\infty e^{-\omega\xi}\theta(\xi)\,d\xi < \infty$ and applying the majorized convergence theorem (Theorem 3.7.9) we obtain

$$\int_0^\infty e^{-\lambda\xi}S(\xi)x\,d\xi = \sum_{n=0}^{\infty}\int_0^\infty e^{-\lambda\xi}S_n(\xi)x\,d\xi, \qquad \Re(\lambda) > \omega.$$

Repeated application of Lemma 13.3.6 gives

$$\begin{aligned}\int_0^\infty e^{-\lambda\xi}S_n(\xi)x\,d\xi &= R(\lambda; A)\int_0^\infty e^{-\lambda\xi}\tilde{B}S_{n-1}(\xi)x\,d\xi \\ &= R(\lambda; A)[\tilde{B}R(\lambda; A)]\int_0^\infty e^{-\lambda\xi}\tilde{B}S_{n-2}(\xi)x\,d\xi \\ &= \cdots = R(\lambda; A)[\tilde{B}R(\lambda; A)]^n x.\end{aligned}$$

Consequently

$$\int_0^\infty e^{-\lambda\xi}S(\xi)x\,d\xi = \sum_{n=0}^{\infty} R(\lambda; A)[\tilde{B}R(\lambda; A)]^n x, \qquad \Re(\lambda) > \omega. \tag{13.4.10}$$

Now according to Lemma 13.3.4, $\| \tilde{B}R(\lambda; A) \| \leqq \int_0^\infty e^{-\omega\xi}\psi(\xi)\,d\xi = \gamma < 1$ for $\Re(\lambda) > \omega$. It now follows from Theorem 5.10.4 that the right member in (13.4.10) is $R(\lambda; A + B)$ so that

$$R(\lambda; A + B)x = \int_0^\infty e^{-\lambda\xi}S(\xi)x\,d\xi, \qquad \Re(\lambda) > \omega.$$

Thus we have: (1) $\mathfrak{D}(A + B) = \mathfrak{D}(A)$ is dense in $\mathfrak{X}$; (2) $\| R(\lambda; A + B) \| \leqq \| R(\lambda; A) \| (1 - \gamma)^{-1} = O(1/\lambda)$ as $\lambda \to \infty$; and (3) $R(\lambda; A + B)$ is the Laplace transform of a strongly continuous family of linear bounded operators $S(\xi)$ with $\int_0^\infty e^{-\lambda\xi} \| S(\xi) \| \,d\xi < \infty$. The corollary to Theorem 12.4.1 now asserts that $S(\xi)$ is a semi-group of class $(1, A)$ with infinitesimal generator $A + B$ defined on $\mathfrak{D}(A)$. This concludes the proof of Theorem 13.4.1.

COROLLARY 1. *Let $T(\xi; A)$ be a semi-group of class (C_0) and let $B \in \mathfrak{P}(A)$. Then $A + B$ defined on $\mathfrak{D}(A)$ generates a semi-group $T(\xi; A + B)$ of class (C_0) and the series* (13.4.8) *converges absolutely, uniformly with respect to ξ in each interval of the form $(0, \beta)$, $0 < \beta < \infty$.*

PROOF. Set $\theta_0(\xi) \equiv \sum_{n=1}^\infty \psi^{n*}(\xi)$. Making use of the estimates (13.4.9) we see that

$$\sum_{n=1}^{\infty} \| S_n(\xi) \| \leqq \sum_{n=1}^{\infty} (\varphi*\psi^{n*})(\xi) = (\varphi*\theta_0)(\xi).$$

Now according to Lemma 13.4.3, the series $\sum_{n=1}^\infty (\varphi*\psi^{n*})(\xi)$ converges uniformly with respect to ξ in each interval $(\epsilon, 1/\epsilon)$ and the function $\theta_0(\xi) \in L(0, \beta)$ for each $\beta < \infty$. On the other hand, $T(\xi; A)$ being of class (C_0), there exists a constant $M_\beta > 0$ such that $\varphi(\xi) \leqq M_\beta$ for $0 < \xi < \beta$. Thus for $\xi \in (0, \beta)$

$$(\varphi * \theta_0)(\xi) = \int_0^\xi \varphi(\xi - \sigma)\theta_0(\sigma)\, d\sigma \leqq M_\beta \int_0^\xi \theta_0(\sigma)\, d\sigma,$$

which tends to zero as $\xi \to 0+$. As a consequence $\sum_{n=1}^\infty (\varphi * \psi^{n*})(\xi)$ and hence $\sum_{n=1}^\infty \| S_n(\xi) \|$ converges uniformly with respect to ξ in each interval of the form $(0, \beta)$. Finally we see that

$$\| T(\xi; A + B) \| \leqq \varphi(\xi) + (\varphi * \theta_0)(\xi) \leqq M_\beta \left[1 + \int_0^\xi \theta_0(\sigma)\, d\sigma \right], \quad 0 < \xi < \beta,$$

so that $\| T(\xi; A + B) \|$ is bounded for $\xi \in (0, 1)$. Theorem 10.6.4 now asserts that $T(\xi; A + B)$ is of class (C_0).

REMARK. If $T(\xi; A)$ is a group of linear bounded operators defined and strongly continuous on $(-\infty, \infty)$, then each $B \in \mathfrak{P}(A)$ is bounded on $\mathfrak{D}(A)$ and $\bar{B} \in \mathfrak{E}(\mathfrak{X})$. For, the assertion $\| \bar{B}T(\xi; A) \| < \infty$ for $\xi > 0$ implies that

$$\| \bar{B} \| = \| [\bar{B}T(\xi; A)]T(-\xi; A) \| \leqq \| \bar{B}T(\xi; A) \| \| T(-\xi; A) \| < \infty.$$

Thus as far as groups are concerned, Theorem 13.4.1 provides no more generality than Theorem 13.2.2.

COROLLARY 2. *Let $T(\xi; A)$ be a semi-group of class $(1, C_1)$ and let $B \in \mathfrak{P}(A)$. Then $A + B$ defined on $\mathfrak{D}(A)$ is the infinitesimal generator of a semi-group likewise of class $(1, C_1)$.*

PROOF. It is clear from the expansion formula (13.4.8) that

$$(13.4.11) \quad \left\| \tau^{-1} \int_0^\tau T(\xi; A + B)x\, d\xi - x \right\| \leqq \left\| \tau^{-1} \int_0^\tau T(\xi; A)x\, d\xi - x \right\| + \sum_{n=1}^\infty \left\| \tau^{-1} \int_0^\tau S_n(\xi)x\, d\xi \right\|.$$

By assumption, $T(\xi; A)$ is of class $(1, C_1)$ and hence

$$\lim_{\tau \to 0+} \left\| \tau^{-1} \int_0^\tau T(\xi; A)x\, d\xi - x \right\| = 0$$

for each $x \in \mathfrak{X}$. This being so, the uniform boundedness theorem implies that there exists an $M > 0$ such that

$$(13.4.12) \quad \left\| \tau^{-1} \int_0^\tau T(\xi; A)\, d\xi \right\| \leqq M$$

for $0 < \tau < 1$. Now

$$\int_0^\tau S_n(\xi)x\, d\xi = \int_0^\tau \int_0^\xi T(\xi - \sigma; A)\bar{B}S_{n-1}(\sigma)x\, d\sigma\, d\xi$$
$$= \int_0^\tau \left[\int_\sigma^\tau T(\xi - \sigma; A)\, d\xi \right] \bar{B}S_{n-1}(\sigma)x\, d\sigma,$$

and making use of (13.4.12) we obtain the estimate

$$\left\| \tau^{-1} \int_0^\tau S_n(\xi)x\, d\xi \right\| \leqq \int_0^\tau \frac{\tau - \sigma}{\tau} M\psi^{n*}(\sigma) \| x \| \, d\sigma \leqq M \| x \| \int_0^\tau \psi^{n*}(\sigma)\, d\sigma$$

for $0 < \tau < 1$. Again by Lemma 13.4.3, $\theta_0(\xi) \equiv \sum_{n=1}^\infty \psi^{n*}(\xi)$ is integrable on $(0, 1)$ so that

$$\sum_{n=1}^\infty \left\| \tau^{-1} \int_0^\tau S_n(\xi)x\, d\xi \right\| \leqq M \| x \| \sum_{n=1}^\infty \int_0^\tau \psi^{n*}(\sigma)\, d\sigma = M \| x \| \int_0^\tau \theta_0(\sigma)\, d\sigma,$$

which tends to zero as $\tau \to 0+$. It now follows from formula (13.4.11) that $\lim_{\tau\to 0+} \tau^{-1}\int_0^\tau T(\xi; A + B)x\, d\xi = x$ for each $x \in \mathfrak{X}$; and consequently the semi-group $T(\xi; A + B)$ is of class $(1, C_1)$.

Corollary 3. *Let $T(\xi; A)$ be a semi-group of class $(1, A)_u$ and let $B \in \mathfrak{P}(A)$. Then $A + B$ defined on $\mathfrak{D}(A)$ is the infinitesimal generator of a semi-group likewise of class $(1, A)_u$.*

Proof. Since the series expansion for $T(\xi; A + B)$, namely (13.4.8), converges in the uniform operator topology, uniformly with respect to ξ in each interval of the form $(\epsilon, 1/\epsilon)$, $0 < \epsilon < 1$, it suffices to show that each term $S_n(\xi)$ is continuous in the uniform operator topology for $\xi > 0$. This is true by assumption for $S_0(\xi) = T(\xi; A)$. Further the relation $\tilde{B}T(\xi; A) = [\tilde{B}T(\delta; A)]T(\xi - \delta; A)$ shows that $\tilde{B}S_0(\xi)$ is also continuous in the uniform operator topology for $\xi > \delta > 0$ and hence for all $\xi > 0$. Suppose next that both $S_k(\xi)$ and $\tilde{B}S_k(\xi)$ are continuous in the uniform operator topology for $\xi > 0$. Then for given $\xi_0 > 0$, choose $\delta > 0$ so that $0 < \xi_0 - 2\delta$. For ξ_1, ξ_2 such that $\xi_0 - \delta < \xi_1, \xi_2 < \xi_0 + \delta$, we have

$$\| \tilde{B}S_{k+1}(\xi_2) - \tilde{B}S_{k+1}(\xi_1) \|$$
$$\leqq \int_0^{\xi_0 - 2\delta} \| \tilde{B}T(\xi_2 - \sigma; A) - \tilde{B}T(\xi_1 - \sigma; A) \| \, \| \tilde{B}S_k(\sigma) \| \, d\sigma$$
$$+ \sum_{i=1}^2 \int_{\xi_0 - 2\delta}^{\xi_i} \| \tilde{B}T(\xi_i - \sigma; A) \| \, \| \tilde{B}S_k(\sigma) \| \, d\sigma.$$

The proof now follows the argument used in the latter part of Lemma 13.3.5, the only difference being that now the integrand in the first term of the right member converges pointwise to zero because of the fact that $\tilde{B}T(\xi; A)$ is continuous in the uniform operator topology for $\xi > 0$.

Remark. It is of interest to note that when $T(\xi; A)$ is continuous in the uniform operator topology merely for ξ sufficiently large, then $T(\xi; A + B)$ need not share this property (cf. R. S. Phillips [8]). Also if $T(\xi; A)$ is of type $\omega = -\infty$, it is possible to find operators $B \in \mathfrak{P}(A)$ bounded and of arbitrarily small norm such that $T(\xi; A + B)$ is of finite type.

Referring to the terminology used in section 11.7, we have

Corollary 4. *Let $\mathfrak{X}$ be a partially ordered linear space and suppose that the* *semi-group $T(\xi; A)$ of class $(1, A)$ consists entirely of positive operators. If $B \in \mathfrak{P}(A)$ is such that $B + \alpha I$ is a positive operator on $\mathfrak{D}(A)$ for some real α, then $T(\xi; A + B)$ likewise consists only of positive operators.*

PROOF. It is clear that $e^{-\alpha\xi}T(\xi; A) = T(\xi; A - \alpha I)$ is again a semi-group of class $(1, A)$ consisting only of positive operators. Hence by Theorem 11.7.2, $R(\lambda; A - \alpha I)$ is a positive operator for all $\lambda > \omega_0 - \alpha$. Further it is easy to see that $B + \alpha I \in \mathfrak{P}(A - \alpha I)$. It follows from Lemma 13.3.4 that there exists a real constant $\omega > \omega_0$ such that

$$\| [B + \alpha I]R(\lambda; A - \alpha I) \| \leqq \int_0^\infty e^{-\omega\xi} \| [\tilde{B} + \alpha I]T(\xi; A - \alpha I) \| \, d\xi < 1.$$

By Theorem 5.10.4

$$\begin{aligned} R(\lambda; A + B) &= R(\lambda; (A - \alpha I) + (B + \alpha I)) \\ &= \sum_{n=0}^{\infty} R(\lambda; A - \alpha I)\{[B + \alpha I]R(\lambda; A - \alpha I)\}^n. \end{aligned}$$

Since all of the factors in the last member are positive operators we see that $R(\lambda; A + B)$ is a positive operator for $\lambda > \omega$. Again applying Theorem 11.7.2 we see that $T(\xi; A + B)$ consists only of positive operators.

We summarize the preceding results in

THEOREM 13.4.2. *If $T(\xi; A)$ is a semi-group of class $(1, A)$ and $B \in \mathfrak{P}(A)$, then the following properties are inherited by $T(\xi; A + B)$:*

(1) *Being a strongly continuous group on $(-\infty, \infty)$;*

(2) *Being a semi-group of class (C_0);*

(3) *Being a semi-group of class $(1, C_1)$;*

(4) *Being a semi-group of class $(1, A)$;*

(5) *Being a semi-group of class $(1, A)_u$;*

(6) *Being a semi-group consisting of positive operators when $B + \alpha I$ is positive for some real α.*

We note in passing that if A is the infinitesimal generator of a semi-group of class $(1, A)$ and if $R(\lambda; A)$ is a compact (or weakly compact) operator, then for each $B \in \mathfrak{P}(A)$ the resolvent $R(\lambda; A + B)$ is also compact (or weakly compact). This is a direct consequence of Theorem 2.13.8 since $R(\lambda; A)$ and $R(\lambda; A + B)$ have the same range, namely $\mathfrak{D}(A)$.

The final theorem of this section deals with a holomorphic family of perturbations.

THEOREM 13.4.3. *Let $T(\xi; A)$ be a semi-group of class $(1, A)$ and suppose that $B(\zeta) \in \mathfrak{P}(A)$ for each ζ of a domain D of the complex plane. Suppose further that $\widetilde{B(\zeta)}T(\xi; A)$ is holomorphic in $\zeta \in D$ for each $\xi > 0$ and that $\int_0^1 \psi_D(\xi)\, d\xi < \infty$ where*

$$\psi_D(\xi) = \sup [\, \| \widetilde{B(\zeta)} T(\xi; A) \|; \zeta \in D].$$

Then $T(\xi; A + B(\zeta))$ *is holomorphic in* $\zeta \in D$ *for each* $\xi > 0$.

PROOF. It is clear that $\psi_D(\xi)$ could just as well have been defined as the $\sup [\, \| \widetilde{B(\zeta_n)} T(\xi; A) \|; n = 1, 2, \cdots]$ with $\{\zeta_n\}$ dense in D. It follows that $\psi_D(\xi)$ is measurable. It is also clear that $\psi_D(\xi + \eta) \leqq \psi_D(\xi)\varphi(\eta)$ and hence that all of our previous estimates go through if we replace $\psi(\xi)$ by $\psi_D(\xi)$. In particular if we set $S_0(\xi, \zeta) = T(\xi; A)$ and $S_n(\xi, \zeta)x = \int_0^\xi T(\xi - \sigma; A)\widetilde{B(\zeta)} S_{n-1}(\sigma, \zeta)x \, d\sigma$, then the series

$$T(\xi; A + B(\zeta)) = \sum_{n=0}^{\infty} S_n(\xi, \zeta)$$

converges in the uniform operator topology, uniformly with respect to ξ in $(\epsilon, 1/\epsilon)$ and $\zeta \in D$. Consequently it suffices to show that each of the terms $S_n(\xi, \zeta)$ is holomorphic in $\zeta \in D$ for each $\xi > 0$. Again we proceed by induction. According to our hypothesis both $S_0(\xi, \zeta) = T(\xi; A)$ and $\widetilde{B(\zeta)} S_0(\xi, \sigma)$ are holomorphic in $\zeta \in D$. Suppose that $S_k(\xi, \zeta)$ and $\widetilde{B(\zeta)} S_k(\xi, \zeta)$ are likewise holomorphic in $\zeta \in D$ for each $\xi > 0$. In this case $T(\xi - \sigma; A)\widetilde{B(\zeta)} S_k(\sigma, \zeta)x$ and $\widetilde{B(\zeta)} T(\xi - \sigma; A)\widetilde{B(\zeta)} S_k(\sigma, \zeta)x$ will be holomorphic in $\zeta \in D$ for each $\xi > 0$ and strongly continuous in $\sigma \in (0, \xi)$ for each $\zeta \in D$. Further

$$\| T(\xi - \sigma; A)\widetilde{B(\zeta)} S_k(\sigma, \zeta) \| \leqq \varphi(\xi - \sigma)\psi_D^{(k+1)*}(\sigma)$$

and

$$\| \widetilde{B(\zeta)} T(\xi - \sigma; A)\widetilde{B(\zeta)} S_k(\sigma, \zeta) \| \leqq \psi_D(\xi - \sigma)\psi_D^{(k+1)*}(\sigma)$$

for $0 < \sigma < \xi$ and $\zeta \in D$. It now follows as in Theorem 3.11.2 that

$$\int_\delta^{\xi-\delta} T(\xi - \sigma; A)\widetilde{B(\zeta)} S_k(\sigma, \zeta)x \, d\sigma$$

and

$$\int_\delta^{\xi-\delta} \widetilde{B(\zeta)} T(\xi - \sigma; A) \widetilde{B(\zeta)} S_k(\sigma, \zeta)x \, d\sigma$$

are holomorphic in $\zeta \in D$ for $0 < \delta < \xi/2$. Finally, since these integrals converge to the respective limits $S_{k+1}(\xi, \zeta)x$ and $\widetilde{B(\zeta)} S_{k+1}(\xi, \zeta)x$ as $\delta \to 0+$ uniformly with respect to $\zeta \in D$, we see by Theorem 3.11.6 that $S_{k+1}(\xi, \zeta)$ and $\widetilde{B(\zeta)} S_{k+1}(\xi, \zeta)$ are holomorphic in $\zeta \in D$ in the strong (and hence in the uniform) operator topology.

13.5. Classes of infinitesimal generators. So far we have considered only the effect of a perturbation on a given semi-group $T(\xi; A_0)$ of class $(1, A)$. We now study the set of all semi-groups which can be obtained from $T(\xi; A_0)$ by perturbations.

We begin by giving several equivalent formulations of the basic condition $\int_0^1 \| BT(\xi; A) \| \, d\xi < \infty$ where $B \in \mathfrak{J}^{\sim}(A)$. The argument which we employ is similar to that of Theorem 12.4.1.

THEOREM 13.5.1. *Let $T(\xi; A)$ be a semi-group of class $(1, A)$ and of negative type. For a given operator $B \in \mathfrak{J}^{\sim}(A)$ and constant $M > 0$ the following conditions are equivalent:*

(1) $\mathfrak{D}(B) \supset \mathfrak{X}_0$, $BT(\xi; A) \in \mathfrak{E}(\mathfrak{X})$ *for all* $\xi > 0$, *and*

$$\int_0^\infty \| BT(\xi; A) \| \, d\xi \leqq M;$$

(2) *There exists a non-negative measurable function $\psi(\xi)$ defined on $(0, \infty)$ such that $\int_0^\infty \psi(\xi) \, d\xi \leqq M$ and*

$$\| BR^{(n)}(\lambda; A) \| \leqq \int_0^\infty e^{-\lambda\xi}\xi^n\psi(\xi) \, d\xi$$

for all real $\lambda > 0$ and integers $n \geqq 0$;

(3) *For all integers $n \geqq 1$*

$$\int_0^\infty \lambda^{n-1} \| BR^{(n)}(\lambda; A) \| \, d\lambda \leqq M(n-1)!;$$

(4) *There exists a bounded non-decreasing function $b(\xi)$ defined on $(0, \infty)$ such that $b(0) = 0$, $b(\infty) \leqq M$, and*

$$\| BR^{(n)}(\lambda; A) \| \leqq \int_0^\infty e^{-\lambda\xi}\xi^n \, db(\xi)$$

for all real $\lambda > 0$ and integers $n \geqq 0$;

(5) *Set $S_\lambda(\xi) \equiv \exp(A_\lambda\xi)$ where $A_\lambda \equiv \lambda^2 R(\lambda; A) - \lambda I$. Then $B[S_\lambda(\xi) - e^{-\lambda\xi}I] \in \mathfrak{E}(\mathfrak{X})$ for each $\xi > 0$ and*

$$\int_0^\infty \| B[S_\lambda(\xi) - e^{-\lambda\xi}I] \| \, d\xi \leqq M$$

for all real $\lambda > 0$.

If condition (2) is verified then $\| BT(\xi; A) \| \leqq \psi(\xi)$ for almost all $\xi > 0$. Further if $\psi(\xi)$ is bounded, then $B \in \mathfrak{E}(\mathfrak{X})$.

REMARK. The above formulation of condition (1) involves the infinite interval $(0, \infty)$ rather than the interval $(0, 1)$ used in Definition 13.3.5. For semi-groups of class $(1, A)$ and of negative type these conditions are equivalent since in this case

$$\int_1^\infty \| BT(\xi; A) \| \, d\xi \leqq \| BT(1; A) \| \int_0^\infty \| T(\xi; A) \| \, d\xi < \infty.$$

PROOF. We shall prove successively that $(1) \to (2) \to (3) \to (4) \to (5) \to (1)$. Assuming condition (1), we see that B is the A-extension of an operator in $\mathfrak{P}(A)$ and hence by Lemma 13.3.4 that

$$BR^{(n)}(\lambda; A)x = \int_0^\infty e^{-\lambda\xi}(-\xi)^n BT(\xi; A)x\, d\xi \tag{13.5.1}$$

for $\Re(\lambda) > \omega_0$ and $n \geqq 0$. If we now set $\psi(\xi) = \| BT(\xi; A) \|$, then (2) is obtained directly from (13.5.1). Condition (2) in turn yields

$$\int_0^\infty \lambda^{n-1} \| BR^{(n)}(\lambda; A) \| \, d\lambda \leqq (n-1)! \int_0^\infty \psi(\xi)\, d\xi \leqq M(n-1)!,$$

thus establishing (3).

In order to prove (4) we require the relation

$$[BR(\lambda; A)]^{(n)} = BR^{(n)}(\lambda; A). \tag{13.5.2}$$

By Corollary 2 to Theorem 5.8.4

$$BR(\lambda; A) = BR(\lambda_0\,; A)[I - (\lambda_0 - \lambda)R(\lambda_0\,; A)]^{-1}$$

and successively differentiating this with respect to λ gives (13.5.2). We now take $F \in \mathfrak{E}(\mathfrak{X})^*$. Condition (3) together with (13.5.2) implies that

$$\int_0^\infty \lambda^n \,|\, \{F[BR(\lambda; A)]\}^{(n)} \,|\, d\lambda \leqq M \| F \| (n-1)!$$

for $n \geqq 1$. We may therefore apply a result due to D. V. Widder [**1**, pp. 299 and 306] according to which there exists a normalized function of bounded variation $b_F(\xi)$ on $[0, \infty)$ such that

$$F[BR(\lambda; A)] = \int_0^\infty e^{-\lambda\xi}\, db_F(\xi) \qquad \lambda > 0, \tag{13.5.3}$$

and

$$\operatorname{Var} b_F \Big|_{0+}^{\xi} = \lim_{n\to\infty} \int_{n/\xi}^\infty | F[BR^{(n)}(\lambda; A)] | \frac{\lambda^{n-1}}{(n-1)!}\, d\lambda.$$

It follows from (13.5.3) that $\lim_{\lambda\to\infty} F[BR(\lambda; A)]$ exists and hence by the uniform boundedness theorem that

$$\limsup_{\lambda\to\infty} \| BR(\lambda; A) \| \equiv \gamma < \infty. \tag{13.5.4}$$

As a consequence $| b_F(0+) - b_F(0) | \leqq \gamma \| F \|$. As in the proof of Theorem 12.4.1, we now make use of a Banach limit functional, $\text{Lim} \in m^*$, defining

$$b(\xi) = \text{Lim}\left\{\int_{n/\xi}^\infty \| BR^{(n)}(\lambda; A) \| \frac{\lambda^{n-1}}{(n-1)!}\, d\lambda + \gamma\right\}$$

for each $\xi > 0$; $b(0) = 0$. It is easily seen that $b(\xi)$ is non-decreasing, $b(\infty) - b(0+) \leqq M$, and that $\| F \| b(\xi)$ majorizes $b_F(\xi)$ in the sense of a measure. It now follows from (13.5.2) and (13.5.3) that

$$| F[BR^{(n)}(\lambda; A)] | \leqq \int_0^\infty e^{-\lambda\xi}\xi^n \,|\, db_F(\xi) \,|\, \leqq \| F \| \int_0^\infty e^{-\lambda\xi}\xi^n\, db(\xi)$$

for $\lambda > 0$. This proves condition (4) with M replaced by $M' = M + \gamma$. We shall continue with the proof, establishing (5) and then (1) for M' instead of M. It will then be shown that $\gamma = \lim_{\lambda\to\infty} \| BR(\lambda; A) \| = 0$ and therefore that $M' = M$.

In proving (5) from (4) we first notice that we can write

$$S_\lambda(\xi) - e^{-\lambda\xi}I = e^{-\lambda\xi} \sum_{n=0}^{\infty} \frac{(-1)^n(\lambda^2\xi)^{n+1}}{n!(n+1)!} R^{(n)}(\lambda; A),$$

by virtue of the relation $R^{(n)}(\lambda; A) = (-1)^n n![R(\lambda; A)]^{n+1}$. This series is absolutely convergent. Next, making use of (4), we obtain the following estimate

$$\begin{aligned} &e^{-\lambda\xi} \sum_{n=0}^{\infty} \frac{(\lambda^2\xi)^{n+1}}{n!(n+1)!} \| BR^{(n)}(\lambda; A) \| \\ &\leqq e^{-\lambda\xi} \sum_{n=0}^{\infty} \frac{(\lambda^2\xi)^{n+1}}{n!(n+1)!} \int_0^\infty e^{-\lambda\sigma}\sigma^n \, db(\sigma) = \int_0^\infty K(\xi, \sigma; \lambda) \, db(\sigma) < \infty; \end{aligned} \tag{13.5.5}$$

here we have used the notation and bounds given in the proof of Theorem 6.3.3. For $\omega > \omega_0$ there exists a constant $M_\omega > 0$ such that $\| \mu R(\mu; A) \| \leqq M_\omega$ for all $\mu > \omega$. Consequently

$$\begin{aligned} \| \mu BR(\mu; A)R^{(n)}(\lambda; A) \| &= \| [BR^{(n)}(\lambda; A)][\mu R(\mu; A)] \| \\ &\leqq M_\omega \| BR^{(n)}(\lambda; A) \|, \qquad \mu > \omega. \end{aligned}$$

It now follows from Lemma 13.3.1 that $B[S_\lambda(\xi) - e^{-\lambda\xi}I] \in \mathfrak{E}(\mathfrak{X})$ and that

$$B[S_\lambda(\xi) - e^{-\lambda\xi}I] = e^{-\lambda\xi} \sum_{n=0}^{\infty} \frac{(-1)^n(\lambda^2\xi)^{n+1}}{n!(n+1)!} BR^{(n)}(\lambda; A).$$

Thus, (13.5.5) gives

$$\int_0^\infty \| B[S_\lambda(\xi) - e^{-\lambda\xi}I] \| \, d\xi \leqq \int_0^\infty \int_0^\infty K(\xi, \sigma; \lambda) \, db(\sigma) \, d\xi = \int_0^\infty db(\sigma) \leqq M'$$

for $\lambda > 0$; and this proves (5).

Next we prove (1) from (5). According to Theorem 11.6.5, $\lim_{\lambda\to\infty} S_\lambda(\xi)x = T(\xi; A)x$ for $\xi > 0$ and each $x \in \mathfrak{X}$. Consequently as $\lambda \to \infty$ we have

$$\begin{aligned} B[S_\lambda(\xi) - e^{-\lambda\xi}I]R(\mu; A)x &= BR(\mu; A)[S_\lambda(\xi) - e^{-\lambda\xi}I]x \\ &\to BR(\mu; A)T(\xi; A)x = BT(\xi; A)R(\mu; A)x \end{aligned}$$

for $\xi > 0$ and each $x \in \mathfrak{X}$. Thus

$$\| BT(\xi; A) \|_A \leqq \liminf_{\lambda\to\infty} \| B[S_\lambda(\xi) - e^{-\lambda\xi}I] \|, \qquad \xi > 0,$$

where the left side may be infinite. In any case, Fatou's lemma applies and we obtain

$$\int_0^\infty \| BT(\xi; A) \|_A \, d\xi \leqq \liminf_{\lambda\to\infty} \int_0^\infty \| B[S_\lambda(\xi) - e^{-\lambda\xi}I] \| \, d\xi \leqq M'.$$

It follows that $\| BT(\xi; A) \|_{\Delta}$ is finite for almost all $\xi > 0$. Hence by Lemma 13.3.3 we have $\mathfrak{D}(B) \supset \mathfrak{X}_0$, $BT(\xi; A) \in \mathfrak{E}(\mathfrak{X})$ for all $\xi > 0$, and $\int_0^\infty \| BT(\xi; A) \| \, d\xi \leqq M'$. Finally we note that the relation (13.5.1) for $n = 0$ implies that $\lim_{\lambda \to \infty} \| BR(\lambda; A) \| = 0$ and hence by (13.5.4) that $\gamma = 0$. Consequently $M' = M$ and the equivalence between (1), (2), (3), (4), and (5) is proved.

The final assertions of the theorem are now easily established. If condition (2) holds, then (13.5.5) shows that

$$\| B[S_\lambda(\xi) - e^{-\lambda\xi}I] \| \leqq \int_0^\infty K(\xi, \sigma; \lambda)\psi(\sigma) \, d\sigma,$$

and applying Theorem 6.3.3 we see that

$$\| BT(\xi; A) \| \leqq \liminf_{\lambda \to \infty} \| B[S_\lambda(\xi) - e^{-\lambda\xi}I] \| \leqq \psi(\xi)$$

for almost all $\xi > 0$. Moreover if $\psi(\xi) \leqq N$, then, as we have just seen, $\| BT(\xi; A) \| \leqq N$ and (13.5.1) shows that $\| \lambda BR(\lambda; A) \| \leqq N$ for $\lambda > 0$. Since $\lim_{\lambda \to \infty} \lambda BR(\lambda; A)x = Bx$ for all $x \in \mathfrak{D}(B)$, the Banach-Steinhaus theorem applies and $\lim_{\lambda \to \infty} \lambda BR(\lambda; A)x \equiv Bx$ exists for all $x \in \overline{\mathfrak{D}(\mathfrak{B})} = \mathfrak{X}$ and $\| B \| \leqq N$. This concludes the proof of Theorem 13.5.1.

Corollary. *Let $T(\xi; A)$ be a semi-group of class $(1, A)$ and of negative type. Suppose further that $B \in \mathfrak{J}(A)$ and that there exists an integer $k \geqq 1$ and an index α $(0 \leqq \alpha < 1)$ such that*

$$\sigma^{k-1-\alpha} \int_{-\infty}^{\infty} \| B[R(\sigma + i\tau; A)]^k \| \, d\tau \leqq M \qquad \textit{for all } \sigma > 0.$$

Then $B \in \mathfrak{P}(A)$.

Proof. Proceding as in the proof of Theorem 12.6.1 we obtain the estimate

$$\| BR^{(n)}(\lambda + \omega; A) \| \leqq M_1 \int_0^\infty e^{-(\lambda+\omega)\xi}\xi^{n-\alpha} \, d\xi, \qquad \omega > 0,$$

for $n \geqq k - 1$ and $\lambda > 0$. The first resolvent equation shows that $BR(\lambda; A) = BR(\lambda_0; A) - (\lambda - \lambda_0)BR(\lambda_0; A)R(\lambda; A)$ and this together with the fact that $\| \lambda R(\lambda; A) \|$ is bounded for $\lambda > 0$ implies that $\| BR(\lambda; A) \|$ is bounded for $\lambda > 0$. Hence

$$\begin{aligned} \| BR^{(n)}(\lambda + \omega; A) \| &= n! \, \| B[R(\lambda + \omega; A)]^{n+1} \| \\ &\leqq n! \, \| BR(\lambda + \omega; A) \| \, \| R(\lambda + \omega; A) \|^n \\ &\leqq M_2(\lambda + \omega)^{-n}, \qquad \omega > 0, \end{aligned}$$

for $n < k - 1$ and $\lambda > 0$. Substituting these estimates in the left member of (13.5.5), it readily follows that condition (5) of Theorem 13.5.1 (with B replaced by $\tilde{B}$) is satisfied and hence that $B \in \mathfrak{P}(A - \omega I) = \mathfrak{P}(A)$.

In considering the class of semi-groups obtained from a given semi-group by perturbations, two questions are especially pertinent. Can the given semi-group $T(\xi; A_0)$ be recovered from the perturbed semi-group $T(\xi; A_0 + B)$ by means of a further perturbation? Can a semi-group obtained by a succession of perturbations, starting with $T(\xi; A_0)$, be obtained from $T(\xi; A_0)$ by a single perturbation? Both of these questions will be answered in the affirmative. To simplify the discussion we introduce the following notation.

DEFINITION 13.5.1. *Let A_1 and A_2 be infinitesimal generators of semi-groups of class $(1, A)$. Then A_1 and A_2 are said to be closely related, in symbols $A_1 \sim A_2$, if $A_2 = A_1 + B$ where $B \in \mathfrak{P}(A_1)$. We denote the class of all infinitesimal generators closely related to A by $\mathfrak{C}(A)$.*

LEMMA 13.5.1. *Let $T(\xi; A)$ be a semi-group of class $(1, A)$ and suppose that B_0 and $B \in \mathfrak{P}(A)$. Then $\tilde{B}_0T(\xi; A + B) \in \mathfrak{E}(\mathfrak{X})$ for all $\xi > 0$ and*

$$\tilde{B}_0T(\xi; A + B) = \sum_{n=0}^{\infty} \tilde{B}_0S_n(\xi) \tag{13.5.6}$$

where $S_0(\xi) = T(\xi; A)$ and $S_n(\xi)x = \int_0^\xi T(\xi - \sigma; A)\tilde{B}S_{n-1}(\sigma)x\, d\sigma$. Further $\int_0^1 \| \tilde{B}_0T(\xi; A + B) \| \, d\xi < \infty$.

PROOF. According to Lemma 13.3.5

$$\tilde{B}_0S_n(\xi)x = \int_0^\xi \tilde{B}_0T(\xi - \sigma; A)\tilde{B}S_{n-1}(\sigma)x\, d\sigma.$$

Hence setting $\psi_0(\xi) = \| \tilde{B}_0T(\xi; A) \|$ and $\psi(\xi) = \| \tilde{B}T(\xi; A) \|$ we see that

$$\| \tilde{B}_0S_n(\xi) \| \leqq (\psi_0 * \psi^{n*})(\xi).$$

Now

$$\theta(\xi) \equiv \sum_{n=0}^{\infty} (\psi_0 * \psi^{n*})(\xi)$$

converges by Lemma 13.4.3 to a finite limit for each $\xi > 0$. Further, given $\omega > \omega_0$ there exists an $M_\omega > 0$ such that $\| \lambda R(\lambda; A) \| \leqq M_\omega$ for all $\lambda > \omega$. Consequently

$$\begin{aligned} \| \lambda\tilde{B}_0R(\lambda; A)S_n(\xi) \| &\leqq \int_0^\xi \| [\tilde{B}_0T(\xi - \sigma; A)][\lambda R(\lambda; A)][\tilde{B}S_{n-1}(\sigma)] \| \, d\sigma \\ &\leqq M_\omega(\psi_0 * \psi^{n*})(\xi), \qquad \lambda > \omega. \end{aligned}$$

The expansion (13.5.6) now follows from Lemma 13.3.1. The final assertion of the lemma is an obvious consequence of the fact that $\int_0^1 \theta(\xi)\, d\xi < \infty$, which is in turn implied by Lemma 13.4.3.

THEOREM 13.5.2. *If $A_1 \sim A_2$, then $A_2 \sim A_1$, $\mathfrak{F}(A_1) = \mathfrak{F}(A_2)$, and $\mathfrak{P}(A_1) = \mathfrak{P}(A_2)$.*

PROOF. If $A_1 \sim A_2$, then by definition $B \equiv A_2 - A_1 \in \mathfrak{P}(A_1)$ and $\mathfrak{D}(A_2) =$

$\mathfrak{D}(A_1)$. It is clear from Lemma 13.3.4 that there exists a constant $\omega > \omega_0$ such that $\| BR(\lambda; A_1) \| \leqq \gamma < 1$ for all $\lambda > \omega$. For such λ

$$R(\lambda; A_2) = \sum_{n=0}^{\infty} R(\lambda; A_1)[BR(\lambda; A_1)]^n$$

by Theorem 5.10.4. Let $B_0 \in \mathfrak{J}(A_1)$. Lemma 13.3.1 can now be employed in the usual manner to show that $B_0R(\lambda; A_2) \in \mathfrak{E}(\mathfrak{X})$ and that

$$B_0R(\lambda; A_2) = \sum_{n=0}^{\infty} B_0R(\lambda; A_1)[BR(\lambda; A_1)]^n, \qquad \lambda > \omega.$$

Since $B_0R(\lambda; A_2) \in \mathfrak{E}(\mathfrak{X})$, it follows that $B_0 \in \mathfrak{J}(A_2)$. Moreover if $B_0 \in \mathfrak{P}(A_1)$, then Lemma 13.5.1 implies that $\int_0^1 \| B_0T(\xi; A_2) \|_{A_2} d\xi < \infty$ so that $B_0 \in \mathfrak{P}(A_2)$. In particular we have $B_0 = -B \in \mathfrak{P}(A_2)$ and hence $A_2 \sim A_1 = A_2 + (-B)$. Thus the relation $\sim$ is symmetric. Using this fact together with the above proved relations, namely $\mathfrak{J}(A_2) \supset \mathfrak{J}(A_1)$ and $\mathfrak{P}(A_2) \supset \mathfrak{P}(A_1)$, it follows that $\mathfrak{J}(A_1) = \mathfrak{J}(A_2)$ and $\mathfrak{P}(A_1) = \mathfrak{P}(A_2)$.

THEOREM 13.5.3. *If $A_1 \sim A_2$ and $A_2 \sim A_3$, then $A_1 \sim A_3$.*

PROOF. Since $A_1 \sim A_2$ and $A_2 \sim A_3$, we have by definition that

$$B_1 \equiv A_2 - A_1 \in \mathfrak{P}(A_1) \qquad \text{and} \qquad B_2 \equiv A_3 - A_2 \in \mathfrak{P}(A_2).$$

According to Theorem 13.5.2, we have $\mathfrak{P}(A_1) = \mathfrak{P}(A_2)$. Thus B_1 and $B_2 \in \mathfrak{P}(A_1)$ and since $\mathfrak{P}(A_1)$ is linear we see that $B_1 + B_2 \in \mathfrak{P}(A_1)$. However

$$A_3 = A_1 + (B_1 + B_2)$$

and therefore $A_1 \sim A_3$.

We further note that $A \sim A$ since the restriction of Θ to $\mathfrak{D}(A)$ obviously belongs to $\mathfrak{P}(A)$. These results are summarized in the statement:

The $\sim$ relation of Definition 13.5.1 *is an equivalence, that is,*

(i) $A \sim A$ *(reflexive)*,

(ii) $A_1 \sim A_2$ *implies* $A_2 \sim A_1$ *(symmetric)*,

(iii) $A_1 \sim A_2$ *and* $A_2 \sim A_3$ *implies* $A_1 \sim A_3$ *(transitive)*.

It follows that the set $\mathfrak{A}_1$ of all infinitesimal generators of semi-groups of class $(1, A)$ is divided into mutually exclusive equivalence classes $\mathfrak{C}_\alpha$, $\mathfrak{A}_1 = \bigcup_\alpha \mathfrak{C}_\alpha$. In each class we choose a particular member operator, say A_α in $\mathfrak{C}_\alpha = \mathfrak{C}(A_\alpha)$, and define a metric function in this class as follows.

DEFINITION 13.5.2. *Let $\mathfrak{C}_\alpha$ be a given equivalence class with A_α the chosen member operator in $\mathfrak{C}_\alpha$. We define*

$$(13.5.7) \qquad d_\alpha(A_1, A_2) \equiv \int_0^1 \| (A_1 - A_2)T(\xi; A_\alpha) \|_{A_\alpha} d\xi$$

for all $A_1, A_2 \in \mathfrak{C}_\alpha$.

THEOREM 13.5.4. *The function $d_\alpha(A_1, A_2)$ is a well defined metric function on $\mathfrak{C}_\alpha$.*

PROOF. It is easy to see that $d_\alpha(A_1, A_2)$ is well defined. For if $A_1, A_2 \in \mathfrak{C}_\alpha$, then $A_1 - A_\alpha$ and $A_2 - A_\alpha \in \mathfrak{P}(A_\alpha)$, and hence

$$A_1 - A_2 = (A_1 - A_\alpha) - (A_2 - A_\alpha) \in \mathfrak{P}(A_\alpha).$$

Consequently $\int_0^1 \| (A_1 - A_2)T(\xi; A_\alpha) \|_{A_\alpha} d\xi < \infty$. It remains to verify the postulates D_1 and D_2 of section 1.5. Suppose first that $d_\alpha(A_1, A_2) = 0$. We set $B = A_1 - A_2$ and as usual we denote the A_α-extension of B by $\tilde{B}$. Then $d_\alpha(A_1, A_2)$ being zero, Lemma 13.3.3 implies that $\| \tilde{B}T(\xi; A_\alpha) \| = 0$ on a set E dense in $(0, 1)$. Since $\| \tilde{B}T(\xi; A_\alpha) \| \leqq \| \tilde{B}T(\eta; A_\alpha) \| \, \| T(\xi - \eta; A_\alpha) \| = 0$ for each $\eta \in E$, it follows that $\| \tilde{B}T(\xi; A_\alpha) \| = 0$ for all $\xi > 0$. Lemma 13.3.4 now shows that $\tilde{B}R(\lambda; A_\alpha) = \Theta$. Since $\mathfrak{D}(B) = \mathfrak{R}[R(\lambda; A_\alpha)]$, this means that $A_1 = A_2$. We have therefore proved D_1. In order to prove D_2, suppose that $A_1, A_2, A_3 \in \mathfrak{C}_\alpha$. Then

$$\| (A_1 - A_3)T(\xi; A_\alpha) \|_{A_\alpha} \leqq \| (A_2 - A_1)T(\xi; A_\alpha) \|_{A_\alpha} + \| (A_2 - A_3)T(\xi; A_\alpha) \|_{A_\alpha},$$

from which it follows that $d_\alpha(A_1, A_3) \leqq d_\alpha(A_2, A_1) + d_\alpha(A_2, A_3)$.

REMARK. The metric function $d_\alpha(A_1, A_2)$ will in general introduce a weaker topology in $\mathfrak{C}_\alpha$ than the metric function $d(A_1, A_2)$ defined in section 2.12. In fact if $B = A_1 - A_2$ is bounded on $\mathfrak{D}(A_\alpha)$, then

$$d_\alpha(A_1, A_2) = \int_0^1 \| BT(\xi; A_\alpha) \|_{A_\alpha} d\xi$$

$$\leqq \| B \|_{A_\alpha} \int_0^1 \| T(\xi; A_\alpha) \| \, d\xi = \left[\int_0^1 \| T(\xi; A_\alpha) \| \, d\xi \right] d(A_1, A_2).$$

It is possible to define the metric function $d_\alpha(A_1, A_2)$ on $\mathfrak{C}_\alpha$ directly from $R(\lambda; A_\alpha)$.

THEOREM 13.5.5. *Let A_α be the defining operator in $\mathfrak{C}_\alpha$ and suppose that $T(\xi; A_\alpha)$ is of type ω_0. For a fixed $\omega > \omega_0$ set*

$$\delta_\alpha(A_1, A_2) \equiv \sup_{n \geqq 1} \left\{ \int_0^\infty \| (A_1 - A_2)R^{(n)}(\lambda + \omega; A_\alpha) \| \frac{\lambda^{n-1}}{(n-1)!} d\lambda \right\}. \tag{13.5.8}$$

Then

$$\delta_\alpha(A_1, A_2) = \int_0^\infty e^{-\omega\xi} \| (A_1 - A_2)T(\xi; A_2) \|_{A_\alpha} d\xi. \tag{13.5.9}$$

Further $\delta_\alpha(A_1, A_2)$ is a metric function equivalent to $d_\alpha(A_1, A_2)$. In fact there exist constants m_α and M_α such that

$$m_\alpha \delta_\alpha(A_1, A_2) \leqq d_\alpha(A_1, A_2) \leqq M_\alpha \delta_\alpha(A_1, A_2) \tag{13.5.10}$$

for all $A_1, A_2 \in \mathfrak{C}_\alpha$.

PROOF. It is clear from (13.5.8) that $\delta_\alpha(A_1, A_3) \leqq \delta_\alpha(A_2, A_1) + \delta_\alpha(A_2, A_3)$

for $A_1, A_2, A_3 \in \mathfrak{C}_\alpha$. This is the metric postulate D_2. The postulate D_1 and the equivalence of the metrics will follow from (13.5.10), which we now proceed to prove. Again the operator $B = A_1 - A_2 \in \mathfrak{P}(A_\alpha)$. We set

$$\psi(\xi) = \| BT(\xi; A_\alpha) \|_{A_\alpha} = \| \bar{B}T(\xi; A_\alpha) \|$$

where $\bar{B}$ is the A_α-extension of B. By definition

$$d_\alpha(A_1, A_2) = \int_0^1 \psi(\xi)\, d\xi. \tag{13.5.11}$$

On the other hand, Theorem 13.5.1 implies that

$$\delta_\alpha(A_1, A_2) = \int_0^\infty e^{-\omega\xi}\psi(\xi)\, d\xi,$$

which is the same as (13.5.9). The second inequality in (13.5.10) is now immediate since

$$\int_0^1 \psi(\xi)\, d\xi \leqq e^\beta \int_0^\infty e^{-\omega\xi}\psi(\xi)\, d\xi,$$

where $\beta = \max(\omega, 0)$. To obtain the first inequality, we make use of (13.4.3) from which we obtain the estimate

$$\psi(\xi) = \int_0^1 \psi(\xi)\, d\sigma \leqq \int_0^1 \psi(\sigma)\varphi(\xi - \sigma)\, d\sigma, \qquad \xi \geqq 1.$$

Consequently

$$\begin{aligned}\int_1^\infty e^{-\omega\xi}\psi(\xi)\, d\xi &\leqq \int_1^\infty e^{-\omega\xi}\left[\int_0^1 \psi(\sigma)\varphi(\xi - \sigma)\, d\sigma\right] d\xi \\ &= \int_0^1 e^{-\omega\sigma}\psi(\sigma)\left[\int_1^\infty e^{-\omega(\xi-\sigma)}\varphi(\xi - \sigma)\, d\xi\right] d\sigma \\ &\leqq \left[\int_0^\infty e^{-\omega\xi}\varphi(\xi)\, d\xi\right]\left[e^\gamma \int_0^1 \psi(\sigma)\, d\sigma\right],\end{aligned}$$

where $\gamma = \max(-\omega, 0)$. Finally

$$\begin{aligned}\int_0^\infty e^{-\omega\xi}\psi(\xi)\, d\xi &\leqq e^\gamma \int_0^1 \psi(\xi)\, d\xi + \int_1^\infty e^{-\omega\xi}\psi(\xi)\, d\xi \\ &\leqq e^\gamma\left[1 + \int_0^\infty e^{-\omega\xi}\varphi(\xi)\, d\xi\right]\left[\int_0^1 \psi(\xi)\, d\xi\right].\end{aligned} \tag{13.5.12}$$

Thus (13.5.10) holds with

$$m_\alpha = \left\{e^\gamma\left[1 + \int_0^\infty e^{-\omega\xi}\varphi(\xi)\, d\xi\right]\right\}^{-1} \qquad \text{and} \qquad M_\alpha = e^\beta.$$

THEOREM 13.5.6. *$\mathfrak{C}_\alpha$ is complete.*

PROOF. Suppose that $\{A_n\} \subset \mathfrak{C}_\alpha$ forms a Cauchy sequence, that is, suppose that $\lim_{m,n\to\infty} d_\alpha(A_m, A_n) = 0$. Again $B_n \equiv A_n - A_\alpha \in \mathfrak{P}(A_\alpha)$. We denote the A_α-extension of B_n by $\tilde{B}_n$. According to Lemma 13.3.4,

$$B_n R(\lambda; A_\alpha)x = \int_0^\infty e^{-\lambda\xi}\tilde{B}_n T(\xi; A_\alpha)x\, d\xi \qquad \text{for} \qquad \Re(\lambda) > \omega_0 .$$

Choosing $\omega > \max(0, \omega_0)$ and employing the estimate (13.5.12) we see that

$$\| B_m R(\omega; A_\alpha) - B_n R(\omega; A_\alpha) \| \leqq \int_0^\infty e^{-\omega\xi} \| (A_m - A_n)T(\xi; A_\alpha) \|_{A_\alpha} d\xi$$

$$\leqq \left[1 + \int_0^\infty e^{-\omega\xi}\varphi(\xi)\, d\xi \right] d_\alpha(A_m, A_n).$$

It follows that the $\{B_n R(\omega; A_\alpha)\}$ form a Cauchy sequence in $\mathfrak{E}(\mathfrak{X})$. Consequently $\lim_{n\to\infty} B_n x \equiv Bx$ exists for each $x \in \mathfrak{D}(A_\alpha)$ and $BR(\omega; A_\alpha) \in \mathfrak{E}(\mathfrak{X})$. Thus $B \in \mathfrak{I}(A_\alpha)$. Next we show that $B \in \mathfrak{P}(A_\alpha)$ and hence that $A = A_\alpha + B \in \mathfrak{C}_\alpha$. It follows from Theorem 13.5.5 that

$$\delta_\alpha(A_m, A_n) \equiv \sup_{k\geqq 1} \left\{ \int_0^\infty \| (B_m - B_n)R^{(k)}(\lambda + \omega; A_\alpha) \| \frac{\lambda^{k-1}}{(k-1)!} d\lambda \right\}$$

converges to zero as $m, n \to 0$. Hence Theorem 3.7.7 applies and we see that

$$\int_0^\infty \| BR^{(k)}(\lambda; A_\alpha) \| \frac{\lambda^{k-1}}{(k-1)!} d\lambda \leqq \lim_{n\to\infty} \delta_\alpha(A_n, A_\alpha), \qquad k \geqq 1,$$

and

$$\lim_{n\to\infty} \delta_\alpha(A_n, A) = 0.$$

The first of these assertions proves that $B \in \mathfrak{P}(A_\alpha)$ by Theorem 13.5.1 and the second shows that $A_n \to A$ by Theorem 13.5.5.

It is clearly desirable that the topology in $\mathfrak{C}_\alpha$ be independent of the choice of the operator A_α used to define the metric function. The following theorem establishes this as a fact. Here we use the notation

$$d_A(A_1, A_2) = \int_0^1 \| (A_1 - A_2)T(\xi; A) \|_A \, d\xi. \tag{13.5.13}$$

THEOREM 13.5.7. *If $A_\alpha \sim A$, then there exist positive constants m' and M' such that*

$$m' d_{A_\alpha}(A_1, A_2) \leqq d_A(A_1, A_2) \leqq M' d_{A_\alpha}(A_1, A_2) \tag{13.5.14}$$

for all $A_1, A_2 \in \mathfrak{C}(A_\alpha) = \mathfrak{C}(A)$.

PROOF. The operators $B_0 = A_1 - A_2$ and $B = A - A_\alpha$ both belong to $\mathfrak{P}(A_\alpha) = \mathfrak{P}(A)$. Let $\tilde{B}_0$ and $\tilde{B}$ denote the A_α-extension of B_0 and B respectively. Then by Lemma 13.5.1.

$$\tilde{B}_0 T(\xi; A) = \sum_{n=0}^{\infty} \tilde{B}_0 S_n(\xi) \tag{13.5.15}$$

where $S_0(\xi) = T(\xi; A_\alpha)$ and $S_n(\xi)x = \int_0^\xi T(\xi - \sigma; A_\alpha)\tilde{B}S_{n-1}(\sigma)x\, d\sigma$. Setting $\psi_0(\xi) = \| \tilde{B}_0 T(\xi; A_\alpha) \|$ and $\psi(\xi) = \| \tilde{B}T(\xi; A_\alpha) \|$, we see from (13.5.15) that

$$\| B_0 T(\xi; A) \|_A = \| \tilde{B}_0 T(\xi; A) \| \leqq \sum_{n=0}^{\infty} (\psi_0 * \psi^{n*})(\xi) = \psi_0(\xi) + (\psi_0 * \theta_0)(\xi)$$

where $\theta_0(\xi) \equiv \sum_{n=1}^{\infty} \psi^{n*}(\xi)$. The function $\theta_0(\xi)$ is summable on (0, 1) by Lemma 13.4.3 so that

$$\int_0^1 (\psi_0 * \theta_0)(\xi)\, d\xi = \int_0^1 \int_0^\xi \psi_0(\xi - \sigma)\theta_0(\sigma)\, d\sigma\, d\xi$$

$$= \int_0^1 \theta_0(\sigma) \left[\int_\sigma^1 \psi_0(\xi - \sigma)\, d\xi \right] d\sigma \leqq \left[\int_0^1 \theta_0(\sigma)\, d\sigma \right] \left[\int_0^1 \psi_0(\sigma)\, d\sigma \right].$$

Recalling that $d_{A_\alpha}(A_1, A_2) = \int_0^1 \psi_0(\sigma)\, d\sigma$, we now have

$$d_A(A_1, A_2) = \int_0^1 \| B_0 T(\xi; A) \|_A\, d\xi \leqq \left[1 + \int_0^1 \theta_0(\sigma)\, d\sigma \right] d_{A_\alpha}(A_1, A_2).$$

This proves the second of the inequalities in (13.5.14) and the first follows by symmetry.

Before concluding this section we shall prove that the generation problem for semi-groups of class (1, A) is stable. For this purpose we need the following

LEMMA 13.5.2. *Suppose that* $\omega > \omega_0$ *and let* $M_\omega \equiv \int_0^\infty e^{-\omega\xi}\varphi(\xi)\, d\xi$. *If* $\psi(\xi)$ *satisfies* (13.4.3), *then*

$$\psi(\xi) \leqq 8M_\omega^{\,2} e^{\omega\xi}\xi^{-3} \int_0^\infty e^{-\omega\sigma}\psi(\sigma)\, d\sigma, \qquad \xi > 0. \tag{13.5.16}$$

PROOF. It follows from the inequality (13.4.3) that

$$\frac{\xi}{2}\psi(\xi) = \int_0^{\xi/2} \psi(\xi)\, d\sigma \leqq e^{\omega\xi} \int_0^{\xi/2} e^{-\omega\sigma}\psi(\sigma)e^{-\omega(\xi-\sigma)}\varphi(\xi - \sigma)\, d\sigma.$$

According to Theorem 7.4.4, $e^{-\omega(\xi-\sigma)}\varphi(\xi - \sigma) \leqq 4\, M_\omega^{\,2}\xi^{-2}$ for $0 < \sigma < \xi/2$. Hence

$$\frac{\xi}{2}\psi(\xi) \leqq 4\, M_\omega^{\,2}\xi^{-2}e^{\omega\xi} \int_0^{\xi/2} e^{-\omega\sigma}\psi(\sigma)\, d\sigma$$

and this implies (13.5.16).

THEOREM 13.5.8. *For semi-groups of class* (1, A), $T(\xi; A)$ *is a continuous function of* A *in the sense that*

$$\lim_{A \to A_0} \| T(\xi; A) - T(\xi; A_0) \| = 0$$

uniformly with respect to ξ *in each interval of the form* $(\epsilon, 1/\epsilon)$, $0 < \epsilon < 1$.

REMARK. The meaning to be attached to the symbol $\lim_{A \to A_0}$ is that all of the operators considered belong to an equivalence class $\mathfrak{C}_\alpha$ and the limit is to be taken as $d_\alpha(A, A_0) \to 0$.

PROOF. Theorem 13.5.7 permits us to choose A_0 as the defining operator for the metric function in the class $\mathfrak{C}_\alpha = \mathfrak{C}(A_0)$, that is; for $A_1, A_2 \in \mathfrak{C}(A_0)$ we may take $d_\alpha(A_1, A_2) = d_{A_0}(A_1, A_2)$. Again $B = A - A_0 \in \mathfrak{P}(A_0)$. By Theorem 13.4.1 we have

$$(13.5.17) \qquad \| T(\xi; A) - T(\xi; A_0) \| \leqq \sum_{n=1}^{\infty} \| S_n(\xi) \| \leqq \sum_{n=1}^{\infty} (\varphi * \psi^{n*})(\xi),$$

where $\varphi(\xi) = \| T(\xi; A_0) \|$ and $\psi(\xi) = \| \tilde{B}T(\xi; A_0) \|$. We now choose $\omega > \max(0, \omega_0)$ and set $M = \int_0^\infty e^{-\omega\xi}\varphi(\xi)\, d\xi$. The estimate (13.5.12) gives us

$$\int_0^\infty e^{-\omega\xi}\psi(\xi)\, d\xi \leqq (M + 1)\, d_\alpha(A, A_0),$$

where, as above, $d_\alpha(A, A_0) = d_{A_0}(A, A_0) = \int_0^1 \psi(\xi)\, d\xi$. If we substitute this bound in (13.5.16) we get

$$\varphi(\xi) \leqq 8(M + 1)^3 \xi^{-3} e^{\omega\xi} \qquad \text{and} \qquad \psi(\xi) \leqq 8(M + 1)^3\, d_\alpha(A, A_0) \xi^{-3} e^{\omega\xi}.$$

Repeated application of Lemma 13.4.2 yields

$$(\varphi * \psi^{n*})(\xi) \leqq N[N\, d_\alpha(A, A_0)]^n \xi^{-3} e^{\omega\xi}$$

where $N = 2^7(M + 1)^3$. Inserting this estimate in (13.5.17) we finally obtain

$$\| T(\xi; A) - T(\xi; A_0) \| \leqq \frac{N^2}{1 - N\, d_\alpha(A, A_0)} \xi^{-3} e^{\omega\xi}\, d_\alpha(A, A_0)$$

for $N\, d_\alpha(A, A_0) < 1$. The assertion of the theorem is now obvious.

13.6. Cones of infinitesimal generators. We now consider the following problem: *When is* $\sum_{i=1}^n \alpha_i A_i$, $\alpha_i \geqq 0$, *the infinitesimal generator of a semi-group given that the* A_i*'s are infinitesimal generators?* The perturbation theory developed in this paragraph provides a partial solution to this problem.

The simplest problem of this kind deals with positive scalar multiples of a single infinitesimal generator, namely αA where $\alpha > 0$. We therefore start with a single semi-group of linear bounded operators $[T(\xi)]$ continuous in the strong (or uniform) operator topology for $\xi > 0$. Setting $T_\alpha(\xi) \equiv T(\alpha\xi)$, we see that $[T_\alpha(\xi)]$ is also a semi-group continuous in the strong (or uniform) operator topology for $\xi > 0$, and that the range space of $[T_\alpha(\xi)]$, namely $\mathfrak{X}_0(\alpha)$, coincides with $\mathfrak{X}_0$. Further

$$\omega_0(\alpha) \equiv \lim_{\xi \to \infty} \xi^{-1} \log \| T_\alpha(\xi) \| = \alpha \lim_{\alpha\xi \to \infty} (\alpha\xi)^{-1} \log \| T(\alpha\xi) \| = \alpha\omega_0 .$$

Finally,

$$\frac{T_\alpha(\eta)x - x}{\eta} = \alpha \frac{T(\alpha\eta)x - x}{\eta} \to \alpha A_0 x$$

as $\eta \to 0+$ if and only if $x \in \mathfrak{D}(A_o)$; thus $A_{o\alpha} = \alpha A_o$. Now $T(\xi)$ is strongly differentiable for all $\xi > 0$ if and only if $\mathfrak{X}_0 \subset \mathfrak{D}(A_o)$ and hence since $\mathfrak{X}_0 = \mathfrak{X}_0(\alpha)$ and $\mathfrak{D}(A_o) = \mathfrak{D}(A_{o\alpha})$, $T(\xi)$ is strongly differentiable for all $\xi > 0$ if and only if the same is true of $T_\alpha(\xi)$

Referring to the conditions stated in section 10.6, it is easy to see that $[T_\alpha(\xi)]$ satisfies $(\text{i})_0$, $(\text{i})_1$, (C_0), or (C_1) if and only if $[T(\xi)]$ satisfies the corresponding condition. If $[T(\xi)]$ satisfies the condition (A)′ then there exists an $\omega_1 > \omega_0$ for each λ, $\Re(\lambda) > \omega_1$, there is an operator $R(\lambda) \in \mathfrak{E}(\mathfrak{X})$ with the properties (a) $R(\lambda)x = \int_0^\infty e^{-\lambda\xi}T(\xi)x\,d\xi$ for each $x \in \mathfrak{X}_0$, (b) $\| R(\lambda) \|$ is bounded in $\Re(\lambda) > \omega_1$, and (c) $\lim_{\lambda\to\infty} \lambda R(\lambda)x = x$ for each $x \in \mathfrak{X}$. Now for $x \in \mathfrak{X}_0(\alpha) = \mathfrak{X}_0$

$$\int_0^\infty e^{-\lambda\xi}T_\alpha(\xi)x\,d\xi = \alpha^{-1}\int_0^\infty e^{-(\lambda/\alpha)(\alpha\xi)}T(\alpha\xi)x\,d(\alpha\xi) = \alpha^{-1}R(\lambda/\alpha)x.$$

Hence setting $R_\alpha(\lambda) \equiv \alpha^{-1}R(\lambda/\alpha)$, we see that $[T_\alpha(\xi)]$ satisfies (A)′ with the operator $R_\alpha(\lambda)$ and $\omega_1(\alpha) = \alpha\omega_1$. Further, if $[T(\xi)]$ is of class (A), then by Theorem 11.5.2 the operator $R(\lambda) = R(\lambda; A)$ where A is the infinitesimal generator for $T(\xi)$. Now

$$R(\lambda; A_\alpha) = R_\alpha(\lambda) = \alpha^{-1}R\left(\frac{\lambda}{\alpha}\right) = \alpha^{-1}R\left(\frac{\lambda}{\alpha}; A\right) = R(\lambda; \alpha A)$$

so that $A_\alpha = \alpha A$. We have therefore proved

THEOREM 13.6.1. *Let $T(\xi; A)$ be a semi-group of class (A). Then αA is the infinitesimal generator of the semi-group $T(\xi; \alpha A) = T(\alpha\xi; A)$. Moreover the semi-groups $T(\xi; A)$ and $T(\xi; \alpha A)$ belong to the same basic class.*

DEFINITION 13.6.1. *A subset $\mathfrak{K}$ of a linear system $\mathfrak{X}$ is called a cone if $x, y \in \mathfrak{K}$ implies that $x + y \in \mathfrak{K}$ and $\alpha x \in \mathfrak{K}$ for all $\alpha > 0$. Given a subset $S \subset \mathfrak{X}$, the smallest cone containing S is called the conical extension of S and is denoted by $\mathfrak{K}(S)$.*

It is easy to show that $\mathfrak{K}(S)$ consists of all elements of the form $\sum_{i=1}^n \alpha_i x_i$ where $x_i \in S$, $\alpha_i \geqq 0$, and $\sum_{i=1}^n \alpha_i > 0$. Denoting the positive reals by E_1^+ and the set of all infinitesimal generators of semi-groups of the class $(1, A)$ by $\mathfrak{A}_1$, we now have

THEOREM 13.6.2. *Let A_0 be the infinitesimal generator of a semi-group of class $(1, A)$ and set $\mathfrak{C}(A_0) \equiv [A; A \sim A_0]$. Then*

$$\mathfrak{K}[\mathfrak{C}(A_0)] = E_1^+\mathfrak{C}(A_0) \subset \mathfrak{A}_1.$$

PROOF. Let $U = \sum_{i=1}^n \alpha_i A_i$ where $A_i \in \mathfrak{C}(A_0)$, $\alpha_i \geqq 0$, and $\gamma \equiv \sum_{i=1}^n \alpha_i > 0$. Then

$$\gamma^{-1}U = A_0 + \sum_{i=1}^n \frac{\alpha_i}{\gamma}(A_i - A_0).$$

It is clear that $A_i - A_0 \in \mathfrak{P}(A_0)$. Further, $\mathfrak{P}(A_0)$ is a linear system and therefore

$\gamma^{-1}U \in \mathfrak{C}(A_0)$. It follows that $U \in E_1^+\mathfrak{C}(A_0)$. Theorem 13.6.1 implies that $E_1^+\mathfrak{C}(A_0) \subset \mathfrak{A}_1$.

REMARK. Theorem 13.6.2 leaves open the question of whether $A_1 + A_2$ is the infinitesimal generator of a semi-group of class $(1, A)$, given that A_1 and A_2 are infinitesimal generators of arbitrary semi-groups of class $(1, A)$. In this general form the conjecture is false. In fact, H. A. Dye and R. S. Phillips [1] have given an example of two infinitesimal generators A_1, A_2 of groups of class (C_0) where no extension of $A_1 + A_2$ (defined on $\mathfrak{D}(A_1) \cap \mathfrak{D}(A_2)$) is the infinitesimal generator of a semi-group. However it may be that the assertion is correct under the added assumption $\mathfrak{D}(A_1) = \mathfrak{D}(A_2)$.

2. Special Results in Perturbation Theory

13.7. Semi-groups of class $H(\Phi_1, \Phi_2)$. One would expect a semi-group to become more stable as further properties are imposed. In this paragraph we shall treat the two extremes, semi-groups of class $H(\Phi_1, \Phi_2)$ and semi-groups of class $(A)_\infty$. We are able to perturb semi-groups of the former class with operators more general than those used in the previous paragraph; whereas we are able to perturb semi-groups of the latter class with nothing more general than a bounded operator.

We begin with a lemma dealing with semi-groups of class $H(\Phi_1, \Phi_2)$.

LEMMA 13.7.1. *Let U be a closed linear operator with $\mathfrak{D}(U)$ dense in $\mathfrak{X}$ and let Δ be a sector in the complex plane whose conjugate set is given by*

$$\Phi_2 + \pi/2 \leqq \arg(\lambda - \omega) \leqq \Phi_1 + 3\pi/2,$$

where $-\pi/2 \leqq \Phi_1 < 0 < \Phi_2 \leqq \pi/2$ and ω is real. Suppose there exists a constant $M > 0$ such that for λ outside of Δ and at a distance $d(\lambda)$ from Δ we have

$$\| R(\lambda; U) \| \leqq \frac{M}{d(\lambda)}.$$

Further, suppose that for $B \in \mathfrak{Z}(U)$ there exist constants $\epsilon > 0$ and $\omega' > \omega$ such that

$$\| BR(\omega'; U) \| < [1 + M (\sin \epsilon)^{-1}]^{-1}.$$

Then $U + B$ defined on $\mathfrak{D}(U)$ generates a semi-group of class $H(\Phi_1 + \epsilon, \Phi_2 - \epsilon)$.

PROOF. Let Δ' be the sector whose conjugate set is

$$\Phi_2 - \epsilon + \pi/2 \leqq \arg(\lambda - \omega') \leqq \Phi_1 + \epsilon + 3\pi/2.$$

Then Δ' contains Δ and if λ is outside of Δ' at a distance $d'(\lambda)$ from Δ' we have $d'(\lambda) \leqq d(\lambda)$ and $| \lambda - \omega' | \leqq d(\lambda) (\sin \epsilon)^{-1}$; and in particular

$$| \lambda - \omega' | \, \| R(\lambda; U) \| \leqq M (\sin \epsilon)^{-1}.$$

Making use of the first resolvent equation we see that

$$BR(\lambda; U) = BR(\omega'; U)\,[I - (\lambda - \omega')R(\lambda; U)]$$

and for λ outside of Δ' this yields the estimate

$$\| BR(\lambda; U) \| \leqq \| BR(\omega'; U) \| \,[1 + M\,(\sin \epsilon)^{-1}] \equiv \gamma < 1.$$

According to Theorem 5.10.4, $U + B \in \mathfrak{C}(\mathfrak{X})$, $\lambda \in \rho(U + B)$, and

$$\| R(\lambda; U + B) \| \leqq \| R(\lambda; U) \| \,(1 - \gamma)^{-1} \leqq \frac{M'}{d'(\lambda)}.$$

where $M' = M(1 - \gamma)^{-1}$. It now follows from Theorem 12.8.1 that $U + B$ is the infinitesimal generator of a semi-group of class $H(\Phi_1 + \epsilon, \Phi_2 - \epsilon)$.

THEOREM 13.7.1. *Let $T(\zeta; A)$ be a semi-group of class $H(\Phi_1, \Phi_2)$ and let $B \in \mathfrak{Z}(A)$. Then given $\epsilon > 0$ there exists a $\delta_\epsilon(B) > 0$ such that $A + \eta B$ defined on $\mathfrak{D}(A)$ generates a semi-group of class $H(\Phi_1 + 2\epsilon, \Phi_2 - 2\epsilon)$ for each η with $|\eta| < \delta_\epsilon(B)$.*

PROOF. According to Theorem 12.8.1, there exists a sector Δ_ϵ and a constant $M_\epsilon > 0$ such that for λ outside Δ_ϵ and at a distance $d_\epsilon(\lambda)$ from Δ_ϵ we have $\| R(\lambda; A)\| \leqq M_\epsilon/d_\epsilon(\lambda)$. Here the sector conjugate to Δ_ϵ is of the form

$$\Phi_2 - \epsilon + \pi/2 \leqq \arg\,(\lambda - \omega_\epsilon) \leqq \Phi_1 + \epsilon + 3\pi/2$$

and we may assume ω_ϵ is real. We choose $\omega' > \omega_\epsilon$ and set

$$\delta_\epsilon(B) \equiv \{\| BR(\omega'; A) \| \,[1 + M_\epsilon\,(\sin \epsilon)^{-1}]\}^{-1}.$$

The hypothesis of Lemma 13.7.1 is now satisfied by ηB if $|\eta| < \delta_\epsilon(B)$ so that the theorem follows directly from the lemma.

The above result is necessarily of a local character since $B = -2A \in \mathfrak{Z}(A)$ whereas $A + (-2A) = -A$ does not in general generate a semi-group. A "global" perturbation theory can be obtained by further restricting B. In this direction we have

THEOREM 13.7.2. *Let $T(\zeta; A)$ be a semi-group of class $H(\Phi_1, \Phi_2)$ and let $B \subset \mathfrak{P}(A)$. Then $A + B$ defined on $\mathfrak{D}(A)$ generates a semi-group of class $H(\Phi_1, \Phi_2)$. Further*

$$T(\zeta; A + B) = \sum_{n=0}^{\infty} S_n(\zeta) \tag{13.7.1}$$

where $S_0(\zeta) = T(\zeta; A)$ and $S_n(re^{i\varphi})x = \int_0^r T[(r - \sigma)e^{i\varphi}; A]\tilde{B}S_{n-1}(\sigma e^{i\varphi})x\,d\sigma$. The series converges absolutely, uniformly with respect to ζ in each sector of the form $\Phi_1 + \epsilon \leqq \arg\,(\zeta) \leqq \Phi_2 - \epsilon$, $|\zeta| \leqq 1/\epsilon$ where $0 < \epsilon < 1$.

PROOF. In order to prove that $A + B$ generates a semi- group of class

$$H(\Phi_1 + 2\epsilon, \Phi_2 - 2\epsilon)$$

we proceed as in the previous theorem; in this case we choose ω' so that

$$\| BR(\omega'; A) \| \leqq \int_0^\infty e^{-\omega'\xi} \| BT(\xi; A) \| \, d\xi < [1 + M_\epsilon \, (\sin \epsilon)^{-1}]^{-1}.$$

Since any two semi-groups generated by $A + B$ coincide on the sector common to their domains (corollary to Theorem 12.2.1) and since $\epsilon > 0$ can be chosen arbitrarily small, it follows that $A + B$ generates a semi-group of class $H(\Phi_1, \Phi_2)$. The expansion (13.7.1) is given by Corollary 1 to Theorem 13.4.1 applied to the semi-group $T(re^{i\varphi}; A)$ of class (C_0), φ fixed and $\Phi_1 < \varphi < \Phi_2$. In order to show that the convergence is uniform with respect to ζ in each sector

$$S_\epsilon \equiv [\zeta; \Phi_1 + \epsilon \leqq \arg (\zeta) \leqq \Phi_2 - \epsilon, | \zeta | \leqq 1/\epsilon]$$

we set

$$\varphi_\epsilon(r) \equiv \sup [\| T(re^{i\varphi}; A) \| ; \quad \Phi_1 + \epsilon \leqq \varphi \leqq \Phi_2 - \epsilon],$$

$$\psi_\epsilon(r) \equiv \sup [\| BT(re^{i\varphi}; A) \| ; \Phi_1 + \epsilon \leqq \varphi \leqq \Phi_2 - \epsilon].$$

It is clear that these functions are measurable since $\| T(\zeta; A) \|$ and $\| BT(\zeta; A) \|$ are both continuous in ζ. According to Lemma 10.6.2 there exists a constant $M(\epsilon) > 0$ such that $\varphi_\epsilon(r) \leqq M(\epsilon)$ for $0 < r < 1$. Further it is easy to verify that

$$\varphi_\epsilon(r + s) \leqq \varphi_\epsilon(r)\varphi_\epsilon(s) \qquad \text{and} \qquad \psi_\epsilon(r + s) \leqq \psi_\epsilon(r)\varphi_\epsilon(s)$$

for r, $s > 0$. Finally we show that $\int_0^1 \psi_\epsilon(r) \, dr < \infty$. To this end let $\delta_\epsilon > 0$ be fixed and set $\zeta(\varphi) = e^{i\varphi} - \delta_\epsilon$. Then

$$\| BT(re^{i\varphi}; A) \| \leqq \| BT(\delta_\epsilon r; A) \| \, \| T[\zeta(\varphi)r; A] \| .$$

For δ_ϵ sufficiently small, we will have $\Phi_1 + \epsilon/2 \leqq \arg [\zeta(\varphi)] \leqq \Phi_2 - \epsilon/2$ and $| \zeta(\varphi) | < 1$ for all $\varphi \in [\Phi_1 + \epsilon, \Phi_2 - \epsilon]$, and hence

$$\psi_\epsilon(r) \leqq M(\epsilon/2) \, \| BT(\delta_\epsilon r; A) \| .$$

It follows that

$$\begin{aligned} \int_0^1 \psi_\epsilon(r) \, dr &\leqq M(\epsilon/2) \int_0^1 \| BT(\delta_\epsilon r; A) \| \, dr \\ &\leqq M(\epsilon/2)\delta_\epsilon^{-1} \int_0^1 \| BT(\xi; A) \| \, d\xi. \end{aligned} \tag{13.7.2}$$

If we now argue as in Corollary 1 to Theorem 13.4.1, replacing $\varphi(\xi)$ and $\psi(\xi)$ by $\varphi_\epsilon(r)$ and $\psi_\epsilon(r)$ respectively, then it is clear that $\sum_{n=0}^\infty \| S_n(\zeta) \|$ converges uniformly with respect to ζ in each sector of the form S_ϵ. This concludes the proof.

We see by Theorem 13.7.2 that if A_0 is the infinitesimal generator of a semi-group of class $H(\Phi_1, \Phi_2)$, then the same is true of all the operators belonging to the class $\mathfrak{C}(A_0)$. Referring to the metric of Definition 13.5.2, we have

THEOREM 13.7.3. *If A_0 is the infinitesimal generator of a semi-group of class $H(\Phi_1, \Phi_2)$, then*

$$\lim_{A \to A_0} \| T(\zeta; A) - T(\zeta; A_0) \| = 0$$

uniformly with respect to ζ in each compact subset of the sector

$$\Phi_1 < \arg(\zeta) < \Phi_2, \qquad 0 < |\zeta|.$$

PROOF. The proof follows the argument of Theorem 13.5.8 where again we replace $\varphi(\xi)$ and $\psi(\xi)$ by $\varphi_\epsilon(r)$ and $\psi_\epsilon(r)$ respectively. The only new estimate that we need is given by (13.7.2), namely

$$\int_0^1 \psi_\epsilon(r)\, dr \leqq M(\epsilon/2)\delta_\epsilon^{-1}\, d_{A_0}(A, A_0).$$

13.8. Semi-groups of class $(A)_\infty$. In the present section we present a somewhat limited perturbation theory for semi-groups of class $(A)_\infty$. The development suffers from the following two shortcomings: (1) all of the perturbing operators are bounded, and (2) the theory does not apply to all semi-groups of class $(A)_\infty$. The first of these limitations is reasonable in view of the generality of the class $(A)_\infty$. The second reflects the corresponding deficiency in Chapter XII where we have given sufficient but not necessary conditions that an operator generate a semi-group of class $(A)_\infty$.

THEOREM 13.8.1. *Let U be a closed linear operator satisfying the conditions of Theorem 12.7.1. Then there exists a constant $\delta > 0$ such that $U + B$ defined on $\mathfrak{D}(U)$ generates a semi-group of class $(A)_\infty$ for each $B \in \mathfrak{E}(\mathfrak{X})$ with $\| B \| < \delta$.*

PROOF. Let $M \equiv \sup [\| R(\lambda; U) \|, \mathfrak{R}(\lambda) \geqq 0]$ and set $\delta = M^{-1}$. If $\| B \| < \delta$, then $\| BR(\lambda; U) \| \leqq \gamma < 1$ for all λ with $\mathfrak{R}(\lambda) \geqq 0$ and by Theorem 5.10.4, the operator $U + B$ with domain $\mathfrak{D}(U)$ is closed, $\lambda \in \rho(U + B)$, and

$$\| R(\lambda; U + B) \| \leqq (1 - \gamma)^{-1} \| R(\lambda; U) \|, \qquad \mathfrak{R}(\lambda) \geqq 0.$$

It follows from this estimate that the conditions of Theorem 12.7.1 are also satisfied by $U + B$ and hence that $U + B$ generates a semi-group of class $(A)_\infty$.

Applying the methods of Theorem 12.7.1, it can be shown that

$$\lim_{\| B \| \to 0} \| T(\xi; U + B) - T(\xi; U) \| = 0,$$

uniformly with respect to ξ in each interval of the form $(\epsilon, 1/\epsilon)$ where $0 < \epsilon < 1$.

CHAPTER XIV

ADJOINT THEORY

14.1. Orientation. The concepts of adjoint space and adjoint operator have already been introduced in Chapter II. From a given (B)-space $\mathfrak{X}$, there ensues a hierarchy of adjoint spaces $\mathfrak{X}^*$, $\mathfrak{X}^{**}$, $\cdots$, each space being embeddable in its second adjoint space under the natural mapping. Likewise if $U \in \mathfrak{E}(\mathfrak{X})$, then $U^* \in \mathfrak{E}(\mathfrak{X}^*)$, $U^{**} \in \mathfrak{E}(\mathfrak{X}^{**})$, $\cdots$, each operator being a restriction of its second adjoint in the sense of the above embedding. For closed linear operators $U \in \mathfrak{C}(\mathfrak{X})$ with $\mathfrak{D}(U)$ dense in $\mathfrak{X}$, U^* is again an operator in $\mathfrak{C}(\mathfrak{X}^*)$. However $\mathfrak{D}(U^*)$ need not be dense in $\mathfrak{X}^*$ and because of this fact the hierarchy of adjoint operators corresponding to U will in general extend no further than U^*. By way of avoiding this situation, one may limit the adjoint space to $\overline{\mathfrak{D}(U^*)}$, which we denote by $\mathfrak{X}^\odot$, and take $U^\odot$ to be the restriction of U^* having $\mathfrak{D}(U^\odot) = [x^*; x^* \in \mathfrak{D}(U^*), U^*x^* \in \mathfrak{X}^\odot]$. Then, if $\mathfrak{D}(U^\odot)$ is dense in $\mathfrak{X}^\odot$ one may proceed to define a second "adjoint" operator $U^{\odot\odot}$ on the subspace $\mathfrak{X}^{\odot\odot} = \overline{\mathfrak{D}[(U^\odot)^*]}$ of $(\mathfrak{X}^\odot)^*$. If this process can be continued indefinitely, one obtains for the operator U an adjoint theory which is analogous to the ordinary adjoint theory. The purpose of the present chapter is to exhibit a family of operators for which this is the case and to develop the corresponding adjoint theory.

There are two rather natural ways in which the adjoint theory can be applied to semi-groups. Given a strongly continuous semi-group of operators $[T(\xi)] \subset \mathfrak{E}(\mathfrak{X})$ with infinitesimal generator A, we can either take the adjoint semi-group to be the semi-group generated by $A^\odot$ having the domain $\mathfrak{X}^\odot = \overline{\mathfrak{D}(A^*)}$, or we can take it to be the semi-group of adjoint operators $[T^*(\xi)]$ with domain $\mathfrak{X}^*$. When $\mathfrak{X}^\odot$ is a proper subspace of $\mathfrak{X}^*$ these two notions are distinct. We have chosen to consider the infinitesimal generator as the primary entity, and from this point of view the proper domain of the adjoint semi-group is $\mathfrak{X}^\odot$. It turns out that the semi-group operators generated by $A^\odot$ are precisely the operators obtained by restricting the adjoint operators $T^*(\xi)$ to the domain $\mathfrak{X}^\odot$. As to the other alternative, while it is true that the adjoint operators $[T^*(\xi)]$ with domain $\mathfrak{X}^*$ form a semi-group, this semi-group will in general not be continuous in the strong topology. Consequently our theory does not apply; to pursue the matter further, one would be forced to develop a more inclusive theory and accept the concomitant loss in detail (see W. Feller [5]).

The theory of the adjoint semi-group as treated in this chapter can be employed to advantage in many applications. In the first place, the fact that A is the infinitesimal generator of a semi-group on $\mathfrak{X}$ implies that $A^\odot$ is the infinitesimal generator of a semi-group on $\mathfrak{X}^\odot$. In the second place, one often obtains a certain additional leverage by dealing with A and $A^\odot$ simultaneously.

The importance of the adjoint semi-group appears to have been first recognized by W. Feller [3] in his treatise on the parabolic differential equation. In the special problem with which he was dealing, he was able to obtain the full adjoint semi-group without employing a precise notion for the adjoint of an unbounded operator. The general treatment given below is due to R. S. Phillips [12].

There are two paragraphs: *Adjoint Operators*, and *Adjoint Spaces.*

References. Feller [3, 5], Phillips [12].

1. Adjoint Operators

14.2. The adjoint space. Most of our discussion concerns the notion of an adjoint space relative to a given operator U. The operator U will be subject to certain conditions which in particular are satisfied by the infinitesimal generator of any semi-group of class (A).

Definition 14.2.1. *We shall call $U \in \mathfrak{O}(\mathfrak{X})$ a $(\odot)$-operator if*

(i) *$\mathfrak{D}(U)$ is dense in $\mathfrak{X}$;*

(ii) *$\| R(\lambda; U) \| = O(1/\lambda)$ as $\lambda \to \infty$.*

For a given $(\odot)$-operator U, the vector space $\overline{\mathfrak{D}(U^*)} \subset \mathfrak{X}^*$ will be called the $(\odot)$-adjoint space of $\mathfrak{X}$ relative to U; we shall denote this space by $\mathfrak{X}^{\odot}$.

Since U will be fixed throughout this discussion, no confusion will arise if we simply refer to $\mathfrak{X}^{\odot}$ as the $(\odot)$-adjoint space. When $\mathfrak{X}$ is reflexive or when U is bounded the $(\odot)$-adjoint space $\mathfrak{X}^{\odot}$ coincides with $\mathfrak{X}^*$. For arbitrary $\mathfrak{X}$ and unbounded U, the $(\odot)$-adjoint space may very well be a proper subspace of $\mathfrak{X}^*$. However in all cases $\mathfrak{X}^{\odot}$ will be weakly* dense in $\mathfrak{X}^*$ by Theorem 2.11.9. We shall denote the generic element of $\mathfrak{X}^{\odot}$ by $x^{\odot}$.

By definition

$$\| x^{\odot} \| = \sup [\, | x^{\odot}(x) |;\ \| x \| \leqq 1,\ x \in \mathfrak{X}]. \tag{14.2.1}$$

When $\mathfrak{X}^{\odot} = \mathfrak{X}^*$ we can obtain $\| x \|$ in like manner from $\mathfrak{X}^{\odot}$. However when $\mathfrak{X}^{\odot}$ is a proper subspace of $\mathfrak{X}^*$ this need not be the case. Setting

$$\| x \|' = \sup [\, | x^{\odot}(x) |;\ \| x^{\odot} \| \leqq 1,\ x^{\odot} \in \mathfrak{X}^{\odot}], \tag{14.2.2}$$

we see that $\| x \|' \leqq \| x \|$. Further it is clear that $\| x \|'$ is a pseudo-norm, that is $\| x \|' \geqq 0$ and $\| x \|'$ satisfies the postulates N_2 and N_3 of section 1.12. However, more is true.

Theorem 14.2.1. *The function $\| x \|'$ defines an equivalent topology for $\mathfrak{X}$; in fact,*

$$\| x \|' \leqq \| x \| \leqq M \| x \|', \qquad x \in \mathfrak{X}, \tag{14.2.3}$$

where $M = \liminf_{\lambda \to \infty} \| \lambda R(\lambda; U) \| < \infty$. *In particular, we see that if* $\liminf_{\lambda \to \infty} \| \lambda R(\lambda; U) \| = 1$ *then* $\| x \|' = \| x \|$.

PROOF. It suffices to establish (14.2.3) since the other assertions of the theorem follow directly from this. We have already noted that $\| x \|' \leqq \| x \|$. Further it is clear from Definition 14.2.1 that $M = \liminf_{\lambda \to \infty} \| \lambda R(\lambda; U) \| < \infty$. To prove the second inequality in (14.2.3), we let x_0 be an arbitrary element of $\mathfrak{X}$. By Theorem 2.7.4 there exists an $x_0^* \in \mathfrak{X}^*$, $\| x_0^* \| = 1$, such that $x_0^*(x_0) = \| x_0 \|$. According to Lemma 12.2.1, $[\lambda R^*(\lambda; U)x_0^*](x_0) = x_0^*[\lambda R(\lambda; U)x_0] \to x_0^*(x_0)$ as $\lambda \to \infty$. Consequently given $\epsilon > 0$ there is a λ_ϵ with $\| \lambda_\epsilon R(\lambda_\epsilon ; U) \| \leqq M + \epsilon$ and $| [\lambda_\epsilon R^*(\lambda_\epsilon ; U)x_0^*](x_0) - \| x_0 \| | < \epsilon$. Now $y_\epsilon^* = \lambda_\epsilon R^*(\lambda_\epsilon ; U)x_0^* \in \mathfrak{X}^\odot$. Making use of the fact that $\| y_\epsilon^* \| \leqq M + \epsilon$ we have

$$\frac{| y_\epsilon^*(x_0) |}{\| y_\epsilon^* \|} \geqq \frac{\| x_0 \| - \epsilon}{M + \epsilon}.$$

Since $\epsilon > 0$ is arbitrary this implies that $\| x_0 \|' \geqq \| x_0 \|/M$.

Renorming $\mathfrak{X}$ by means of the norm $\| x \|'$ has no effect on the determination of $\mathfrak{X}^\odot$ since only the continuity properties of bounded linear functionals are involved in the definition of $\mathfrak{D}(U^*)$. Actually even the norms of the elements in $\mathfrak{X}^\odot$ remain the same.

THEOREM 14.2.2. *Let* $\| x \|'$ *be defined as in* (14.2.2). *Then*

$$\| x^\odot \| = \sup [| x^\odot(x) |; \| x \|' \leqq 1, x \in \mathfrak{X}]. \tag{14.2.4}$$

PROOF. It follows from (14.2.2) that $\| x \|' \leqq \| x \|$ and this together with (14.2.1) show that

$$\| x^\odot \| \leqq \sup [| x^\odot(x) |; \| x \|' \leqq 1, x \in \mathfrak{X}].$$

The relation (14.2.2) also implies that $\| x \|' \geqq | x^\odot(x) |/\| x^\odot \|$ so that $\| x^\odot \| \geqq | x^\odot(x) |/\| x \|'$. From this we obtain

$$\| x^\odot \| \geqq \sup [| x^\odot(x) |; \| x \|' \leqq 1, x \in \mathfrak{X}].$$

The two inequalities establish (14.2.4).

For an operator $B \in \mathfrak{E}(\mathfrak{X})$, we shall write

$$\| B \|' = \sup [\| Bx \|'; \| x \|' \leqq 1, x \in \mathfrak{X}]. \tag{14.2.5}$$

It is readily verified that

$$M^{-1}\| B \|' \leqq \| B \| \leqq M \| B \|', \tag{14.2.6}$$

where again $M = \liminf_{\lambda \to \infty} \| \lambda R(\lambda; U) \|$.

14.3. Adjoint operators. We next consider the $(\odot)$-adjoint of operators on $\mathfrak{X}$, again relative to a fixed $(\odot)$-operator U.

DEFINITION 14.3.1. *Given a linear operator B with $\overline{\mathfrak{D}(B)} = \mathfrak{X}$, we denote by $(B^*)_0$ the restriction of B^* with domain $\mathfrak{D}[(B^*)_0] = [x^*; x^* \in \mathfrak{D}(B^*) \cap \mathfrak{X}^{\odot}]$; and we denote by $B^{\odot}$ the restriction of B^* with domain $\mathfrak{D}(B^{\odot}) = [x^*; x^* \in \mathfrak{D}(B^*) \cap \mathfrak{X}^{\odot}, B^*x^* \in \mathfrak{X}^{\odot}]$.*

We note that $B^{\odot}$ is an operator with domain and range in $\mathfrak{X}^{\odot}$ whereas the range of $(B^*)_0$ need not be contained in $\mathfrak{X}^{\odot}$.

THEOREM 14.3.1. *If B is a linear operator with $\overline{\mathfrak{D}(B)} = \mathfrak{X}$, then $B^{\odot} \in \mathfrak{O}(\mathfrak{X}^{\odot})$.*

PROOF. It is clear from Definition 14.3.1 that $B^{\odot}$ is a linear operator. Further $B^{\odot}$ has a closed extension in $\mathfrak{O}(\mathfrak{X}^*)$, namely B^*, and hence $B^{\odot}$ has a least closed extension. However the least closed extension of $B^{\odot}$ will again be an operator with domain and range in $\mathfrak{X}^{\odot}$ and at the same time it will be a restriction of B^*. Since $B^{\odot}$ is the maximal restriction of B^* with domain and range in $\mathfrak{X}^{\odot}$, it follows that $B^{\odot}$ is closed.

DEFINITION 14.3.2. *A linear operator B is said to commute with U if there exists a $\lambda \in \rho(U)$ such that $R(\lambda; U)Bx = BR(\lambda; U)x$ for all $x \in \mathfrak{D}(B)$. The set of all operators in $\mathfrak{E}(\mathfrak{X})$ which commute with U is called the commutant of U.*

If $B \in \mathfrak{E}(\mathfrak{X})$, then B commutes with U in the above sense if and only if $BUx = UBx$ for all $x \in \mathfrak{D}(U)$. In this case $R(\lambda; U)B = BR(\lambda; U)$ for all $\lambda \in \rho(U)$.

LEMMA 14.3.1. *If a linear operator B commutes with U and $\overline{\mathfrak{D}(B)} = \mathfrak{X}$, then B^* commutes with U^*.*

PROOF. According to Theorem 2.16.5, $R(\lambda; U^*) = R^*(\lambda; U)$. Hence for $x \in \mathfrak{D}(B)$ and $x^* \in \mathfrak{D}(B^*)$ we have

$$[R(\lambda;\ U^*)B^*x^*](x) = x^*[BR(\lambda;\ U)x] = x^*[R(\lambda;\ U)Bx] = [R(\lambda;\ U^*)x^*](Bx).$$

It now follows from the definition of B^* that $R(\lambda; U^*)x^* \in \mathfrak{D}(B^*)$ and that $B^*R(\lambda; U^*)x^* = R(\lambda; U^*)B^*x^*$.

THEOREM 14.3.2. *If $B \in \mathfrak{E}(\mathfrak{X})$ commutes with U, then $B^{\odot} = (B^*)_0 \in \mathfrak{E}(\mathfrak{X}^{\odot})$ and $\| B^{\odot} \| = \| B \|' \leqq \| B \|$.*

PROOF. By the above lemma, $B^*R(\lambda; U^*) = R(\lambda; U^*)B^*$ from which we infer that B^* maps $\mathfrak{D}(U^*)$ into $\mathfrak{D}(U^*)$. Since B^* is bounded, it follows that B^* maps $\overline{\mathfrak{D}(U^*)} = \mathfrak{X}^{\odot}$ into $\mathfrak{X}^{\odot}$. Consequently $B^{\odot} \doteq (B^*)_0$. It is clear that the restriction of a bounded operator is bounded so that $B^{\odot} \in \mathfrak{E}(\mathfrak{X}^{\odot})$ and $\| B^{\odot} \| \leqq \| B^* \| = \| B \|$. To prove that $\| B^{\odot} \| = \| B \|'$ we note that

$$\begin{aligned} \| B \|' &= \sup [\ \| Bx \|';\ \| x \|' \leqq 1] \\ &= \sup [\ | x^{\odot}(Bx) |\ ;\ \| x^{\odot} \| \leqq 1,\ \| x \|' \leqq 1] \\ &= \sup [\ | (B^{\odot}x^{\odot})(x) |;\ \| x^{\odot} \| \leqq 1,\ \| x \|' \leqq 1]. \end{aligned}$$

Applying Theorem 14.2.2, we see that

$$\| B \|' = \sup [\| B^{\odot}x^{\odot} \|; \| x^{\odot} \| \leqq 1] = \| B^{\odot} \|.$$

COROLLARY. *The commutant of U is a closed subalgebra of $\mathfrak{E}(\mathfrak{X})$ and the mapping $B \to B^{\odot}$ of the commutant of U onto a subalgebra of $\mathfrak{E}(\mathfrak{X}^{\odot})$ is an anti-isomorphism.*

PROOF. It follows from Definition 14.3.2 that the commutant of U is a closed subalgebra of $\mathfrak{E}(\mathfrak{X})$. If B, C belong to the commutant of U then it is easy to see that $B + C \to B^{\odot} + C^{\odot}$ and $BC \to C^{\odot}B^{\odot}$. Finally (14.2.6) implies that the mapping is one-to-one and bicontinuous.

THEOREM 14.3.3. *If a linear operator B commutes with U and $\overline{\mathfrak{D}(B)} = \mathfrak{X}$, then $\rho(B^{\odot}) = \rho(B)$, $R(\mu; B^{\odot}) = [R(\mu; B^*)]_0 = R^{\odot}(\mu; B)$ for each $\mu \in \rho(B)$, and $B^{\odot}$ commutes with $U^{\odot}$.*

PROOF. Suppose first that $\mu \in \rho(B)$. Then according to Theorem 2.16.5, $\mu \in \rho(B^*)$ and $R(\mu; B^*) = R^*(\mu; B)$. Thus for $x^{\odot} \in \mathfrak{D}(B^{\odot})$ we have

$$[R(\mu; B^*)]_0(\mu I - B^{\odot})x^{\odot} = R(\mu; B^*)(\mu I - B^*)x^{\odot} = x^{\odot}$$

since $(\mu I - B^{\odot})x^{\odot} \in \mathfrak{X}^{\odot}$. On the other hand, if B commutes with U then it is easy to show that $R(\mu; B)$ likewise commutes with U. Applying the previous theorem, we see that $[R(\mu; B^*)]_0 x^{\odot} \in \mathfrak{X}^{\odot}$ for all $x^{\odot} \subset \mathfrak{X}^{\odot}$. Furthermore since $(\mu I - B^*) [R(\mu; B^*)]_0 x^{\odot} = x^{\odot}$, it follows that $B^*[R(\mu; B^*)]_0 x^{\odot} \in \mathfrak{X}^{\odot}$ and hence that $[R(\mu; B^*)]_0 x^{\odot} \in \mathfrak{D}(B^{\odot})$. Consequently $(\mu I - B^{\odot}) [R(\mu; B^*)]_0 x^{\odot} = x^{\odot}$ for all $x^{\odot} \in \mathfrak{X}^{\odot}$. Thus $[R(\mu; B^*)]_0$ is both a left and a right inverse for $(\mu I - B^{\odot})$ and therefore $\mu \in \rho(B^{\odot})$ and $R(\mu; B^{\odot}) = [R(\mu; B^*)]_0$. Finally Theorems 2.16.5 and 14.3.2 imply that $[R(\mu; B^*)]_0 = [R^*(\mu; B)]_0 = R^{\odot}(\mu; B)$.

To prove the converse proposition, suppose that $\mu \in \rho(B^{\odot})$. Then $\mathfrak{R}(\mu I - B^*) \supset \mathfrak{R}(\mu I - B^{\odot}) = \mathfrak{X}^{\odot}$ and since $\mathfrak{X}^{\odot}$ is weakly* dense in $\mathfrak{X}^*$, Theorem 2.11.13 implies that $(\mu I - B)^{-1}$ exists. Suppose next that $\mathfrak{R}(\mu I - B)$ is not dense in $\mathfrak{X}$. In this case Theorem 2.11.12 asserts that $\mu \in P\sigma(B^*)$, that is, there exists an $x_0^* \in \mathfrak{D}(B^*)$, $x_0^* \neq \theta$, such that $(\mu I - B^*)x_0^* = \theta$. By Lemma 14.3.1

$$(\mu I - B^*)R(\lambda; U^*)x_0^* = R(\lambda; U^*)(\mu I - B^*)x_0^* = \theta.$$

Setting $y_0^* \equiv R(\lambda; U^*)x_0^*$, we see that $y_0^* \in \mathfrak{D}(B^*) \cap \mathfrak{X}^{\odot}$, $y_0^* \neq \theta$, and $B^*y_0^* = \mu y_0^* \in \mathfrak{X}^{\odot}$. Thus $y_0^* \in \mathfrak{D}(B^{\odot})$ and $\mu \in P\sigma(B^{\odot})$. Since this is impossible, $\mathfrak{R}(\mu I - B)$ must be dense in $\mathfrak{X}$. Theorem 2.11.14 now applies and we see that $(\mu I - B^*)^{-1} = [(\mu I - B)^{-1}]^*$. Thus

$$\begin{aligned} \| (\mu I - B)^{-1}x \| &\leqq M \| (\mu I - B)^{-1}x \|' \\ &= M \sup [| x^{\odot}[(\mu I - B)^{-1}x] |; \| x^{\odot} \| \leqq 1] \\ &= M \sup [| [R(\mu; B^{\odot})x^{\odot}](x) |; \| x^{\odot} \| \leqq 1] \\ &\leqq M \| R(\mu; B^{\odot}) \| \| x \|. \end{aligned}$$

It follows that $(\mu I - B)^{-1}$ is bounded: and since it has a dense domain, $\mu \in \rho(B)$.

We next show that $B^{\odot}$ commutes with $U^{\odot}$. According to Lemma 14.3.1, B^* commutes with U^*, that is $B^*R(\lambda; U^*)x^* = R(\lambda; U^*)B^*x^*$ for all $x^* \in \mathfrak{D}(B^*)$. Since U obviously commutes with U, the above result applies and hence $R(\lambda; U^{\odot}) = [R(\lambda; U^*)]_0$. Thus for $x^{\odot} \in \mathfrak{D}(B^{\odot})$ we have $R(\lambda; U^{\odot})B^{\odot}x^{\odot} = B^*R(\lambda; U^{\odot})x^{\odot} \in \mathfrak{X}^{\odot}$. It follows that $R(\lambda; U^{\odot})x^{\odot} \in \mathfrak{D}(B^{\odot})$ and that $R(\lambda; U^{\odot})B^{\odot}x^{\odot} = B^{\odot}R(\lambda; U^{\odot})x^{\odot}$. This concludes the proof of the theorem.

The terminology of the next corollary refers to the spectral properties P_1 and P_2 of Definition 2.16.2.

COROLLARY. *If a linear operator B commutes with U and $\overline{\mathfrak{D}(B)} = \mathfrak{X}$, then B has property P_2 if and only if $B^{\odot}$ has property P_1.*

PROOF. If B has property P_2, then by Theorem 2.11.12 there exists an $x_0^* \in \mathfrak{D}(B^*)$ such that $B^*x_0^* = \theta$, $x_0^* \neq \theta$. Since B^* commutes with U^* by Lemma 14.3.1, we have $B^*R(\lambda; U^*)x_0^* = R(\lambda; U^*)B^*x_0^* = \theta$. Setting $y_0^* \equiv R(\lambda; U^*)x_0^*$, we see that $y_0^* \in \mathfrak{D}(B^*) \cap \mathfrak{X}^{\odot}$, $y_0^* \neq \theta$, and $B^*y_0^* = \theta$. Thus $y_0^* \in \mathfrak{D}(B^{\odot})$ and $B^{\odot}$ has has property P_1. Conversely if $B^{\odot}$ has P_1 then *a fortiori* B^* has P_1 and hence B has P_2 by Theorem 2.11.12.

THEOREM 14.3.4. *If U is a $(\odot)$-operator for $\mathfrak{X}$, then $\| R(\lambda; U^{\odot}) \| \leqq \| R(\lambda; U) \|$ for all $\lambda \in \rho(U)$, $\mathfrak{D}(U^{\odot})$ is dense in $\mathfrak{X}^{\odot}$, and* $\lim_{\lambda\to\infty} \lambda R(\lambda; U^{\odot})x^{\odot} = x^{\odot}$ *for each $x^{\odot} \in \mathfrak{X}^{\odot}$.*

PROOF. Since U commutes with U in the sense of Definition 14.3.2, Theorem 14.3.3 applies. Hence $R(\lambda; U^{\odot}) = R^{\odot}(\lambda; U)$ exists for all $\lambda \in \rho(U)$, and by Theorem 14.3.2, $\| R(\lambda; U^{\odot}) \| \leqq \| R(\lambda; U) \|$. By Theorem 2.16.5, $R(\lambda; U^*) = R^*(\lambda; U)$ so that $\| R(\lambda; U^*) \| = \| R^*(\lambda; U) \| = \| R(\lambda; U) \| = O(1/\lambda)$ as $\lambda \to \infty$; here we have made use of the fact that U is a $(\odot)$-operator. By Lemma 12.2.1, $\lim_{\lambda\to\infty} \lambda R(\lambda; U^*)x^* = x^*$ for each $x^* \in \overline{\mathfrak{D}(U^*)} = \mathfrak{X}^{\odot}$. Since $R(\lambda; U^{\odot}) = [R(\lambda; U^*)]_0$ it follows that $\lim_{\lambda\to\infty} \lambda R(\lambda; U^{\odot})x^{\odot} = x^{\odot}$ for each $x^{\odot} \in \mathfrak{X}^{\odot}$ and hence that $\mathfrak{D}(U^{\odot}) = R(\lambda; U^{\odot})[\mathfrak{X}^{\odot}]$ is dense in $\mathfrak{X}^{\odot}$.

14.4. The adjoint semi-group. We are now ready to consider the adjoint semi-group. The infinitesimal generator A of a semi-group $[T(\xi)]$ of class (A) is a $(\odot)$-operator and we may therefore apply the foregoing theory to the $(\odot)$-adjoint space taken relative to A. According to Theorem 11.5.3, $T(\xi)Ax = AT(\xi)x$ for all $x \in \mathfrak{D}(A)$ and this implies that $T(\xi)$ commutes with A in the sense of Definition 14.3.2. By Theorem 14.3.2, $T^{\odot}(\xi) = [T^*(\xi)]_0 \in \mathfrak{E}(\mathfrak{X}^{\odot})$, and since the family of adjoint operators $[T^*(\xi)]$ form a semi-group the same is true of the restriction of these operators to $\mathfrak{X}^{\odot}$.

DEFINITION 14.4.1. *Given a semi-group $[T(\xi)]$ of class (A) with infinitesimal generator A, we define the adjoint semi-group to be the family of operators $[T^{\odot}(\xi)] \subset \mathfrak{E}(\mathfrak{X}^{\odot})$, where $\mathfrak{X}^{\odot}$ is the $(\odot)$-adjoint space relative to the $(\odot)$-operator A.*

THEOREM 14.4.1. *Let $[T(\xi)]$ be a semi-group of class (A) with infinitesimal* *generator A then the adjoint semi-group $[T^{\odot}(\xi)]$ is again a semi-group of class (A) and its infinitesimal generator is $A^{\odot}$.*

PROOF. The semi-group $[T(\xi)]$ being of class (A), it follows from Theorem 11.5.2 that there exists a constant $\omega_1 > \omega_0$ such that $R(\lambda; A)$ exists for all λ, $\Re(\lambda) > \omega_1$ and has the properties: (a) $R(\lambda; A)x = \int_0^\infty e^{-\lambda\xi}T(\xi)x\,d\xi$ for all $x \in \mathfrak{X}_0$, (b) $\| R(\lambda; A) \|$ is bounded in the half-plane $\Re(\lambda) > \omega_1$, and (c) $\lim_{\lambda\to\infty} \lambda R(\lambda; A)x = x$ for each $x \in \mathfrak{X}$. Condition (c) implies that $\mathfrak{D}(A)$ is dense in $\mathfrak{X}$ and that $\| R(\lambda; A) \| = O(1/\lambda)$ as $\lambda \to \infty$. Consequently A is a ($\odot$)-operator. Applying Theorem 14.3.4 we see that $R(\lambda; A^{\odot})$ satisfies conditions of the type (b) and (c), and further that $\mathfrak{D}(A^{\odot})$ is dense in $\mathfrak{X}^{\odot}$. From the fact that $R(\lambda; A^{\odot})$ maps $\mathfrak{X}^{\odot}$ onto the dense subset $\mathfrak{D}(A^{\odot})$, we conclude that $R(\lambda; A^{\odot})$ maps any dense subset onto a dense subset; in particular, then, $\mathfrak{D}[(A^{\odot})^2] = R(\lambda; A^{\odot})[\mathfrak{D}(A^{\odot})]$ is dense in $\mathfrak{X}^{\odot}$.

We now make use of the relation (11.5.10); this is

$$(14.4.1)\quad T(\xi)x = x + \xi Ax + \frac{1}{2\pi i}\int_{\gamma-i\infty}^{\gamma+i\infty} e^{\lambda\xi}R(\lambda; A)A^2x\,\frac{d\lambda}{\lambda^2}, \qquad \gamma > \max(0, \omega_1),$$

valid for all $x \in \mathfrak{D}(A^2)$. According to Theorems 2.16.5 and 14.3.3, $R(\lambda; A^{\odot}) = [R(\lambda; A^*)]_0 = [R^*(\lambda; A)]_0$ so that for $x \in \mathfrak{D}(A^2)$, $x^{\odot} \in \mathfrak{D}[(A^{\odot})^2]$ we have

$$x^{\odot}[R(\lambda; A)A^2x] = x^{\odot}[A^2R(\lambda; A)x] = [R(\lambda; A^{\odot})(A^{\odot})^2x^{\odot}](x).$$

Hence if we operate on both members in (14.4.1) by $x^{\odot} \in \mathfrak{D}[(A^{\odot})^2]$ and make use of the fact that $\mathfrak{D}(A^2)$ is dense in $\mathfrak{X}$, we obtain

$$T^{\odot}(\xi)x^{\odot} = x^{\odot} + \xi A^{\odot}x^{\odot} + \frac{1}{2\pi i}\int_{\gamma-i\infty}^{\gamma+i\infty} e^{\lambda\xi}R(\lambda; A^{\odot})(A^{\odot})^2x^{\odot}\,\frac{d\lambda}{\lambda^2}.$$

As in Lemma 11.5.2, we see that $\lim_{\xi\to 0+} T^{\odot}(\xi)x^{\odot} = x^{\odot}$ and that $T^{\odot}(\xi)x^{\odot}$ is strongly continuous for $\xi > 0$ whenever $x^{\odot} \in \mathfrak{D}[(A^{\odot})^2]$. The former assertion implies that the closure of the adjoint semi-group range space contains $\mathfrak{D}[(A^{\odot})^2]$ and hence that $(\mathfrak{X}^{\odot})_0$ is dense in $\mathfrak{X}^{\odot}$. On the other hand, $\| T^{\odot}(\xi) \| \leqq \| T(\xi) \|$ so that $\| T^{\odot}(\xi) \|$ is bounded in each interval of the form $(\epsilon, 1/\epsilon)$, $0 < \epsilon < 1$. Further, given an $x^{\odot} \in \mathfrak{X}^{\odot}$ there exists a sequence $\{x_n^{\odot}\} \subset \mathfrak{D}[(A^{\odot})^2]$ such that $x_n^{\odot} \to x^{\odot}$. Thus $T^{\odot}(\xi)x_n^{\odot}$ is strongly continuous in ξ for $\xi > 0$ and converges to $T^{\odot}(\xi)x^{\odot}$ as $n \to \infty$ uniformly with respect to ξ in $(\epsilon, 1/\epsilon)$. It follows that $T^{\odot}(\xi)$ is continuous in the strong operator topology for $\xi > 0$.

At this point in the proof it has been shown that $T^{\odot}(\xi)$ is strongly continuous for $\xi > 0$, that $(\mathfrak{X}^{\odot})_0$ is dense in $\mathfrak{X}^{\odot}$, and that $R(\lambda; A^{\odot})$ satisfies the analogues of conditions (b) and (c). If it can be shown that the analogue of (a) is satisfied by $R(\lambda; A^{\odot})$ and $T^{\odot}(\xi)$, then it will follow that $T^{\odot}(\xi)$ is a semi-group of class (A) and by Theorem 11.5.2 that $A^{\odot}$ is its infinitesimal generator. To this end, let $x^{\odot} = T^{\odot}(\delta)y^{\odot}$ for a fixed $\delta > 0$. Then $R(\lambda; A^{\odot})x^{\odot} = R^*(\lambda; A)T^*(\delta)y^{\odot} = T^*(\delta)R^*(\lambda; A)y^{\odot}$. Thus for each $x \in \mathfrak{X}$ we have

$$[R(\lambda; A^{\odot})x^{\odot}](x) = y^{\odot}[R(\lambda; A)T(\delta)x] = \int_0^{\infty} e^{-\lambda\xi} y^{\odot}[T(\xi + \delta)x]\, d\xi$$

$$= \int_0^{\infty} e^{-\lambda\xi}[T^{\odot}(\xi + \delta)y^{\odot}](x)\, d\xi$$

$$= \left[\int_0^{\infty} e^{-\lambda\xi} T^{\odot}(\xi)x^{\odot}\, d\xi\right](x);$$

here we have applied condition (a) to the element $T(\delta)x \in \mathfrak{X}_0$. Consequently $R(\lambda; A^{\odot})x^{\odot} = \int_0^{\infty} e^{-\lambda\xi}T^{\odot}(\xi)x^{\odot}\, d\xi$ for each $x^{\odot} \in \mathfrak{X}^{\odot}$ and all λ with $\Re(\lambda) > \omega_1$; this is the analogue of condition (a) and the proof of Theorem 14.4.1 is now complete.

The mapping $B \to B^{\odot}$ of the commutant of A into $\mathfrak{E}(\mathfrak{X}^{\odot})$ is linear and continuous in the uniform operator topology by Theorem 14.3.2. Hence if $[T(\xi)]$ is continuous in the uniform operator topology for $\xi > 0$, then so is $[T^{\odot}(\xi)]$. Likewise if $[T(\xi)]$ is differentiable in the uniform operator topology or holomorphic, the same is true of $[T^{\odot}(\xi)]$. Further if $\int_0^1 \| T(\xi) \|\, d\xi < \infty$ or if $\| T(\xi) \|$ is bounded on $(0, 1)$, then $\| T^{\odot}(\xi) \|$ will also have the corresponding property.

If $[T(\xi)]$ is of class $(0, A)$ then it is *a fortiori* of class (A) and by the previous theorem $\lim_{\lambda\to\infty} \lambda R(\lambda; A^{\odot})x^{\odot} = x^{\odot}$ for each $x^{\odot} \in \mathfrak{X}^{\odot}$; here $R(\lambda; A^{\odot})x^{\odot} = \int_0^{\infty} e^{-\lambda\xi}T^{\odot}(\xi)x^{\odot}\, d\xi$ at least for each $x^{\odot} \in (\mathfrak{X}^{\odot})_0$. On the other hand if $[T(\xi)]$ is of class $(0, C_1)$, then

$$C(\tau)x \equiv \tau^{-1} \int_0^{\tau} T(\xi)x\, d\xi$$

defines an operator $C(\tau) \in \mathfrak{E}(\mathfrak{X})$ which commutes with A and such that $\| C(\tau) \| = O(1)$ as $\tau \to 0+$. Consequently $C^{\odot}(\tau) \in \mathfrak{E}(\mathfrak{X}^{\odot})$ and $\| C^{\odot}(\tau) \| = O(1)$ as $\tau \to 0+$ by Theorem 14.3.2. For $x^{\odot} \in (\mathfrak{X}^{\odot})_0$ we see that

$$[C^{\odot}(\tau)x^{\odot}](x) = x^{\odot}[C(\tau)x] = \tau^{-1} \int_0^{\tau} x^{\odot}[T(\xi)x]\, d\xi$$

$$= \left[\tau^{-1} \int_0^{\tau} T^{\odot}(\xi)x^{\odot}\, d\xi\right](x)$$

and hence that $C^{\odot}(\tau)x^{\odot} = \tau^{-1}\int_0^{\tau} T^{\odot}(\xi)x^{\odot}\, d\xi$. Now for $x^{\odot} \in (\mathfrak{X}^{\odot})_0$ we also have $\lim_{\xi\to 0+} T^{\odot}(\xi)x^{\odot} = x^{\odot}$ so that $\lim_{\tau\to 0+} C^{\odot}(\tau)x^{\odot} = x^{\odot}$. Since $(\mathfrak{X}^{\odot})_0$ is dense in $\mathfrak{X}^{\odot}$, the Banach-Steinhaus theorem implies that $\lim_{\tau\to 0+} C^{\odot}(\tau)x^{\odot} = x^{\odot}$ for all $x^{\odot} \in \mathfrak{X}^{\odot}$. Thus if $[T(\xi)]$ is of class $(0, A)$ (or $(0, C_1)$) then the condition (A) (or (C_1)) is satisfied in a sense by $[T^{\odot}(\xi)]$. Still $[T^{\odot}(\xi)]$ need not be of class $(0, A)$ (or $(0, C_1)$) because it is not evident that the condition $(\mathrm{i})_0$ is satisfied, that is, that $\int_0^1 \| T^{\odot}(\xi)x^{\odot} \|\, d\xi < \infty$ for each $x^{\odot} \in \mathfrak{X}^{\odot}$. We therefore introduce modified classes.

DEFINITION 14.4.2. *A semi-group $[T(\xi)]$, is said to be of class $(0, A)^{\odot}$ (or $(0, C_1)^{\odot}$) if it is of class $(0, A)$ (or $(0, C_1)$) and if in addition, $\int_0^1 \| T^{\odot}(\xi)x^{\odot} \|\, d\xi < \infty$ for each $x^{\odot} \in \mathfrak{X}^{\odot}$.*

We now have

COROLLARY. *A semi-group $[T(\xi)]$ and its adjoint $[T^{\odot}(\xi)]$ belong to the same basic classes except when $[T(\xi)]$ is of class $(0, A)$ or $(0, C_1)$. Semi-groups $[T(\xi)]$ of class $(0, A)^{\odot}$ [or $(0, C_1)^{\odot}$] go into semi-groups $[T^{\odot}(\xi)]$ of class $(0, A)$ [or $(0, C_1)$].*

THEOREM 14.4.2. *Let $[T(\xi)]$ be a semi-group of class (A) and define*

$$\Gamma \equiv [x^*; \lim_{\xi \to 0+} T^*(\xi)x^* = x^*, x^* \in \mathfrak{X}^*].$$

Then $\mathfrak{X}^{\odot} = \bar{\Gamma}$.

PROOF. According to Theorem 11.5.3, $\mathfrak{D}[(A^{\odot})^2] \subset \Gamma$ and since $\mathfrak{D}[(A^{\odot})^2]$ is dense in $\mathfrak{X}^{\odot}$ we see that $\mathfrak{X}^{\odot} \subset \bar{\Gamma}$. On the other hand suppose that $x_0^* \in \Gamma$. Then $T^*(\xi + \eta)x_0^* = T^*(\xi)T^*(\eta)x_0^* \to T^*(\xi)x_0^*$ as $\eta \to 0+$ so that $T^*(\xi)x_0^*$ is continuous on the right for $\xi \geqq 0$. Further if $x \in \mathfrak{X}_0$, then

$$[R^*(\lambda; A)x_0^*](x) = x_0^*[R(\lambda; A)x] = \int_0^{\infty} e^{-\lambda\xi}x_0^*[T(\xi)x]\, d\xi$$

$$= \left[\int_0^{\infty} e^{-\lambda\xi}T^*(\xi)x_0^*\, d\xi\right](x),$$

and since $\mathfrak{X}_0$ is dense in $\mathfrak{X}$ we obtain

$$R(\lambda; A^*)x_0^* = R^*(\lambda; A)x_0^* = \int_0^{\infty} e^{-\lambda\xi}T^*(\xi)x_0^*\, d\xi.$$

Using the fact that $\lim_{\xi \to 0+} T^*(\xi)x_0^* = x_0^*$, a straightforward calculation shows that

$$\lim_{\lambda\to\infty} \lambda R(\lambda; A)^*x_0^* = \lim_{\lambda\to\infty} \lambda \int_0^{\infty} e^{-\lambda\xi}T^*(\xi)x_0^*\, d\xi = x_0^*.$$

It follows that $x_0^* \in \overline{\mathfrak{D}(A^*)} = \mathfrak{X}^{\odot}$ and this concludes the proof.

Theorem 14.4.2 shows that $\mathfrak{X}^{\odot}$ is the largest subspace of $\mathfrak{X}^*$ over which the semi-group of adjoint operators $[T^*(\xi)]$ has suitable continuity properties. This is further evidence that our choice of the adjoint semi-group is appropriate.

2. ADJOINT SPACES

14.5. The hierarchy of adjoint spaces. As in the usual theory of adjoint spaces, it is possible to define a $(\odot)$-adjoint space for $\mathfrak{X}^{\odot}$ and in this way to obtain a succession of $(\odot)$-adjoint spaces arising from a given $(\odot)$-operator. The present paragraph is occupied with various aspects of this problem. Again, we shall suppose throughout that U is a fixed $(\odot)$-operator.

DEFINITION 14.5.1. *Let* $\mathfrak{X}^{\odot}$ *be the* $(\odot)$*-adjoint space of* $\mathfrak{X}$ *relative to the* $(\odot)$*-operator* U. *We now define* $(\mathfrak{X}^{\odot})^{\odot} = \mathfrak{X}^{\odot\odot}$ *to be the* $(\odot)$*-adjoint space of* $\mathfrak{X}^{\odot}$ *relative to* $U^{\odot}$.

The above definition requires that $U^{\odot}$ be a $(\odot)$-operator, a fact established by Theorem 14.3.4. The higher order $(\odot)$-adjoint spaces are now obtained by repeated application of the above process. We now prove

THEOREM 14.5.1. *If the norm of* $\mathfrak{X}$ *is given by* $\| x \|'$ *then* $\mathfrak{X}$ *can be embedded in* $\mathfrak{X}^{\odot\odot}$ *by means of the natural mapping and in this sense* $U^{\odot\odot}$ *is an extension of* U.

PROOF. Each $x_0 \in \mathfrak{X}$ defines a unique bounded linear functional $F_0 \in (\mathfrak{X}^{\odot})^*$ given by $F_0(x^{\odot}) = x^{\odot}(x_0)$. Further

$$\| F_0 \| = \sup [| F_0(x^{\odot}) | = |x^{\odot}(x_0) |;\ \| x^{\odot} \| \leqq 1,\ x^{\odot} \in \mathfrak{X}^{\odot}] = \| x_0 \|'.$$

Hence the correspondence $x_0 \to F_0$ is a linear isometric mapping of $\mathfrak{X}$ onto a subspace of $(\mathfrak{X}^{\odot})^*$. Considering $\mathfrak{X}$ as embedded in $(\mathfrak{X}^{\odot})^*$, it is required to show that $\mathfrak{X} \subset (\mathfrak{X}^{\odot})^{\odot}$, otherwise stated, that $\mathfrak{X} \subset \overline{\mathfrak{D}[(U^{\odot})^*]}$. For $x_0 \in \mathfrak{D}(U)$, let $x_0 \to F_0$, and $Ux_0 \to G_0$; then for arbitrary $x^{\odot} \in \mathfrak{D}(U^{\odot})$ we have

$$G_0(x^{\odot}) = x^{\odot}(Ux_0) = (U^{\odot}x^{\odot})(x_0) = F_0(U^{\odot}x^{\odot}).$$

It follows that $F_0 \in \mathfrak{D}[(U^{\odot})^*]$ and that $(U^{\odot})^*F_0 = G_0$. Thus $\mathfrak{D}(U)$ maps into $\overline{\mathfrak{D}[(U^{\odot})^*]}$ and the same is true of its closure $\mathfrak{X}$, that is, $\mathfrak{X} \subset \overline{\mathfrak{D}[(U^{\odot})^*]} = \mathfrak{X}^{\odot\odot}$. In particular $G_0 \in \mathfrak{X}^{\odot\odot}$ so that $F_0 \in \mathfrak{D}(U^{\odot\odot})$ and $U^{\odot\odot}F_0 = G_0$ by Definition 14.3.1. Thus $U^{\odot\odot}$ is actually an extension of U in the sense of the embedding.

The usefulness of the norm $\| x \|'$ is apparent from the above theorem. However, even if the original norm $\| x \|$ is used in $\mathfrak{X}$, Theorem 14.2.1 implies that $\mathfrak{X}$ is isomorphic with its image in $\mathfrak{X}^{\odot\odot}$ under the natural mapping. It is interesting to note that a similar modification of the norm in $\mathfrak{X}^{\odot}$ is not required. In fact, we have

THEOREM 14.5.2. $\| x^{\odot} \| = \| x^{\odot} \|'$.

PROOF. According to Theorem 14.2.2,

$$\| x^{\odot} \| = \sup [\, | x^{\odot}(x) |;\ \| x \|' \leqq 1,\ x \in \mathfrak{X}].$$

Since $\mathfrak{X}$ with norm $\| x \|'$ can be embedded in $\mathfrak{X}^{\odot\odot}$ by the above theorem, it follows that

$$\| x^{\odot} \| \leqq \sup [\, | x^{\odot\odot}(x^{\odot}) |;\ \| x^{\odot\odot} \| \leqq 1,\ x^{\odot\odot} \in \mathfrak{X}^{\odot\odot}] = \| x^{\odot} \|'.$$

Finally, Theorem 14.2.1 asserts that $\| x^{\odot} \|' \leqq \| x^{\odot} \|$ and the two opposing inequalities prove that $\| x^{\odot} \| = \| x^{\odot} \|'$.

Thus it is only in the case of $\mathfrak{X}$ and $\mathfrak{X}^{\odot}$ that a non-symmetric condition between norms may arise; for all other pairs of successive $(\odot)$-adjoint spaces the norms are symmetric in the above sense.

THEOREM 14.5.3. *If* $B \in \mathfrak{E}(\mathfrak{X})$ *commutes with* U, *then* $B^{\odot\odot} \in \mathfrak{E}(\mathfrak{X}^{\odot\odot})$ *is an extension of* B *defined on* $\mathfrak{X} \subset \mathfrak{X}^{\odot\odot}$ *and* $\| B^{\odot\odot} \|_{.} = \| B \|'$.

PROOF. If $B \in \mathfrak{E}(\mathfrak{X})$ commutes with U, then by Theorem 14.3.2, $B^{\odot} \in \mathfrak{E}(\mathfrak{X}^{\odot})$ and $\| B^{\odot} \| = \| B \|'$. According to Theorem 14.3.3, $B^{\odot}$ commutes with $U^{\odot}$ so that we likewise have $B^{\odot\odot} \in \mathfrak{E}(\mathfrak{X}^{\odot\odot})$ and $\| B^{\odot\odot} \| = \| B^{\odot} \|'$. However by Theorem 14.5.2, $\| x^{\odot} \|' = \| x^{\odot} \|$; hence $\| B^{\odot} \|' = \| B^{\odot} \|$ and consequently $\| B^{\odot\odot} \| = \| B \|'$. It remains to show that $B^{\odot\odot}$ coincides with B on $\mathfrak{X}$ considered as a subspace of $\mathfrak{X}^{\odot\odot}$ under the natural embedding, $x_0 \to x_0^{\odot\odot}$, which is defined by the relation $x_0^{\odot\odot}(x^{\odot}) = x^{\odot}(x_0)$. Now

$$B^{\odot\odot}x_0^{\odot\odot}(x^{\odot}) = x_0^{\odot\odot}(B^{\odot}x^{\odot}) = (B^{\odot}x^{\odot})(x_0) = x^{\odot}B(x_0).$$

Since this holds for all $x^{\odot} \in \mathfrak{X}^{\odot}$, we see that $Bx_0 \to B^{\odot\odot}x_0^{\odot\odot}$ under the same mapping. This is the desired result.

14.6. Reflexive spaces. Of special interest is the case in which $\mathfrak{X}^{\odot\odot} = \mathfrak{X}$ under the natural embedding; here we assume that $\mathfrak{X}$ has been renormed with the norm $\| x \|'$. When this is the case we shall say that $\mathfrak{X}$ is ($\odot$)-reflexive. We shall see in particular that $\mathfrak{X}$ is ($\odot$)-reflexive whenever it is reflexive in the usual sense. Furthermore a non-reflexive space may be ($\odot$)-reflexive relative to certain ($\odot$)-operators. Precise conditions for ($\odot$)-reflexivity can be formulated in terms of the following concept of a weak topology.

DEFINITION 14.6.1. *Given a subset $\Gamma \subset \mathfrak{X}^*$, we define the (Γ)-weak topology in $\mathfrak{X}$ in the usual way by means of the generic neighborhood*

$$(14.6.1) \quad N(x_0\,;\, x_1^*, \cdots, x_n^*\,;\, \epsilon) = [x;\ | x_k^*(x - x_0) | < \epsilon,\quad k = 1, \cdots, n]$$

where $(x_1^*, \cdots, x_n^*)$ is any finite subset of Γ and ϵ is an arbitrary positive number.

If Γ distinguishes between elements of $\mathfrak{X}$, then the usual argument can be used to show that $\mathfrak{X}$ is a topological linear space in the (Γ)-weak topology. Further, it is readily proved that the neighborhoods of the form (14.6.1) form a basis for x_0 (cf. sections 2.9 and 2.10). We now have

THEOREM 14.6.1. *Let the norm of $\mathfrak{X}$ be given by $\| x \|'$. Then $\mathfrak{X}$ is ($\odot$)-reflexive if and only if $R(\lambda; U)$ is $(\mathfrak{X}^{\odot})$-weakly compact.*

REMARK. As in section 2.13, a (Γ)-weakly compact operator takes bounded sets into (Γ)-weakly compact sets. It is clear from the first resolvent equation that if $R(\lambda; U)$ is (Γ)-weakly compact for some $\lambda \in \rho(U)$ then it is (Γ)-weakly compact for all $\lambda \in \rho(U)$.

PROOF. Suppose first that $R(\lambda; U)$ is $(\mathfrak{X}^{\odot})$-weakly compact for a fixed $\lambda \in \rho(U)$. Let F_0 be an arbitrary element of $(\mathfrak{X}^{\odot})^*$. Then by Theorem 2.7.8, given a finite set of functionals $\pi \subset \mathfrak{X}^{\odot}$, there exists a $y_\pi \in \mathfrak{X}$, $\| y_\pi \| \leqq 2 \| F_0 \|$, such that $F_0(x^{\odot}) = x^{\odot}(y_\pi)$ for all $x^{\odot} \in \pi$. Ordering the π's by inclusion, it is clear that they form a directed set. Consequently

$$\begin{aligned}[R^*(\lambda; U^{\odot})F_0](x^{\odot}) &= F_0[R(\lambda; U^{\odot})x^{\odot}] = \lim_\pi [R(\lambda; U^{\odot})x^{\odot}](y_\pi) \\ &= \lim_\pi x^{\odot}[R(\lambda; U)y_\pi].\end{aligned}$$

Arguing as in Theorem 2.9.4 (with $x_\pi \equiv R(\lambda; U)y_\pi$), we see that there exists an

$x_0 \in \mathfrak{X}$ such that $\lim_\pi x^\odot[R(\lambda; U)y_\pi] = x^\odot(x_0)$ for each $x^\odot \in \mathfrak{X}^\odot$. Thus

$$[R^*(\lambda; U^\odot)F_0](x^\odot) = x^\odot(x_0)$$

for all $x^\odot \in \mathfrak{X}^\odot$ and hence $x_0 \to R^*(\lambda; U^\odot)F_0$ under the natural mapping. Since $R^*(\lambda; U^\odot) = R(\lambda; (U^\odot)^*)$ by Theorem 2.16.5, it now follows that

$$\mathfrak{X} \supset \overline{\mathfrak{D}[(U^\odot)^*]} = \mathfrak{X}^{\odot\odot}.$$

Theorem 14.5.1, on the other hand, asserts that $\mathfrak{X} \subset \mathfrak{X}^{\odot\odot}$ so that $\mathfrak{X} = \mathfrak{X}^{\odot\odot}$.

Conversely suppose that $\mathfrak{X} = \mathfrak{X}^{\odot\odot}$ under the natural embedding. Then $R^*(\lambda; U^\odot)[(\mathfrak{X}^\odot)^*] = \mathfrak{D}[(U^\odot)^*]$ is contained in the image of $\mathfrak{X}$. Now by Theorem 2.11.11, $R^*(\lambda; U^\odot)$ is continuous in the usual weak* topology of $(\mathfrak{X}^\odot)^*$. Hence the sphere $[x^{\odot *}; \| x^{\odot *} \| \leqq M, x^{\odot *} \in (\mathfrak{X}^\odot)^*]$, which is weakly* compact by Theorem 2.10.2, maps onto a weakly* compact subset of $(\mathfrak{X}^\odot)^*$. As we have already noted, this image set lies in $\mathfrak{X}^{\odot\odot} = \mathfrak{X}$. Further the weak* topology for $\mathfrak{X}$ considered as a subset of $(\mathfrak{X}^\odot)^*$ is the same as the $(\mathfrak{X}^\odot)$-weak topology for $\mathfrak{X}$. Hence $R^*(\lambda; U^\odot)$ takes bounded subsets of $\mathfrak{X}$ into $(\mathfrak{X}^\odot)$-weakly compact subsets of $\mathfrak{X}$. By Theorem 14.5.3, $R(\lambda; U)$ is a restriction of $R^{\odot\odot}(\lambda; U) = R^\odot(\lambda; U^\odot)$ and therefore $R(\lambda; U)$ is also a restriction of $R^*(\lambda; U^\odot)$. Consequently $R(\lambda; U)$ takes bounded subsets of $\mathfrak{X}$ into $(\mathfrak{X}^\odot)$-weakly compact subsets of $\mathfrak{X}$. This concludes the proof.

Corollary. *Let the norm of $\mathfrak{X}$ be given by $\| x \|'$. If $R(\lambda; U)$ is weakly compact relative to the usual weak topology of $\mathfrak{X}$, then $\mathfrak{X}$ is $(\odot)$-reflexive.*

Proof. This follows from the fact that a weakly compact subset of $\mathfrak{X}$ is also compact relative to any weaker topology such as the $(\mathfrak{X}^\odot)$-weak topology of $\mathfrak{X}$.

In particular we see that $\mathfrak{X}$ is $(\odot)$-reflexive when $\mathfrak{X}$ is reflexive in the usual sense; for in this case every bounded operator is weakly compact. A simple direct proof of this fact can also be obtained from Theorem 2.11.9.

Theorem 14.6.2. *Let the norm of $\mathfrak{X}$ be given by $\| x \|'$. Then $\mathfrak{X}$ is $(\odot)$-reflexive if and only if $\mathfrak{X}^\odot$ is $(\odot)$-reflexive.*

Proof. Suppose first that $\mathfrak{X}$ is $(\odot)$-reflexive. As in the converse argument of Theorem 14.6.1, we see that the operator $R(\lambda; U^*) = R^*(\lambda; U)$ maps the sphere $[x^*; \| x^* \| \leqq M, x^* \in \mathfrak{X}^*]$ onto a weakly* compact subset of $\mathfrak{X}^*$. This image set being contained in $\mathfrak{X}^\odot$, it follows that $R(\lambda; U^*)$ maps bounded subsets of $\mathfrak{X}^\odot$ into $(\mathfrak{X})$-weakly compact subsets of $\mathfrak{X}^\odot$. If $\mathfrak{X}$ is $(\odot)$-reflexive, then the $(\mathfrak{X})$-weak topology for $\mathfrak{X}^\odot$ and the $(\mathfrak{X}^{\odot\odot})$-weak topology for $\mathfrak{X}^\odot$ are equivalent and therefore $R(\lambda; U^\odot) = [R(\lambda; U^*)]_0$ maps bounded subsets of $\mathfrak{X}^\odot$ into $(\mathfrak{X}^{\odot\odot})$-weakly compact subsets of $\mathfrak{X}^\odot$. Consequently $\mathfrak{X}^\odot$ is $(\odot)$-reflexive by Theorem 14.6.1.

We establish the converse proposition by an indirect proof. Suppose that $\mathfrak{X}^\odot$ is $(\odot)$-reflexive but that $\mathfrak{X}$ is not $(\odot)$-reflexive. Let $\mathfrak{X}_0^{\odot\odot}$ be the image of $\mathfrak{X}$ in $\mathfrak{X}^{\odot\odot}$ under the natural mapping. Our assumption implies that $\mathfrak{X}_0^{\odot\odot}$ does not fill out $\mathfrak{X}^{\odot\odot}$. By Theorem 2.7.5 there exists an $F_0 \in (\mathfrak{X}^{\odot\odot})^*$, $F_0 \neq \theta$, such that $F_0[\mathfrak{X}_0^{\odot\odot}] = 0$.

Now $R^{\odot\odot}(\lambda;\ U) = R(\lambda;\ U^{\odot\odot})$, being an extension of $R(\lambda;\ U)$, maps $\mathfrak{X}_0^{\odot\odot}$ into itself so that

$$[R^*(\lambda;\ U^{\odot\odot})F_0][\mathfrak{X}_0^{\odot\odot}] = F_0\{R(\lambda;\ U^{\odot\odot})[\mathfrak{X}_0^{\odot\odot}]\} = 0.$$

Further $x_0^{\odot\odot\odot} \equiv R^*(\lambda;\ U^{\odot\odot})F_0 = R(\lambda; (U^{\odot\odot})^*)F_0 \in \mathfrak{X}^{\odot\odot\odot}$ and cannot be the zero element. On the other hand, $x_0^{\odot\odot\odot}$ is the image of some non-zero $x_0^{\odot} \in \mathfrak{X}^{\odot}$ under the natural mapping of $\mathfrak{X}^{\odot}$ onto $\mathfrak{X}^{\odot\odot\odot}$, since we have assumed $\mathfrak{X}^{\odot}$ to be $(\odot)$-reflexive. Thus $x_0^{\odot}[\mathfrak{X}] = \mathfrak{X}_0^{\odot\odot}(x_0^{\odot}) = x_0^{\odot\odot\odot}[\mathfrak{X}_0^{\odot\odot}] = 0$ and this is impossssible because no non-zero element of $\mathfrak{X}^{\odot}$ can annihilate $\mathfrak{X}$.

CHAPTER XV

OPERATIONAL CALCULUS

15.1. Orientation. In paragraph 5.3 an operational calculus for arbitrary closed operators was developed which can be described in the following way: Given an open set of the extended complex plane, the calculus defines an isomorphic mapping of the algebra of functions $f(\lambda)$ holomorphic in Δ into the algebra of operator-valued functions $f(A)$ with domain $\mathfrak{G}(\Delta) \equiv [A; \sigma_e(A) \subset \Delta]$ and range $\mathfrak{E}(\mathfrak{X})$.

In the present chapter we shall limit our considerations to infinitesimal generators of semi-groups of class (A) and obtain a different sort of calculus. Instead of Δ, the defining entity is now a submultiplicative function $\varphi(\xi)$ of type $\omega_0 \equiv \lim_{\xi\to\infty} \xi^{-1} \log \varphi(\xi)$; each of the functions $f(\lambda)$ is required to be the Laplace-Stieltjes transform of a set function $a \in S(\varphi)$; and the domain of $f(A)$ is $\mathfrak{A}(\varphi) \equiv [A; \| T(\xi; A) \| \leqq \varphi(\xi)]$. The function $f(\lambda)$ and the operator-valued function $f(A)$ are then related by the formulas

$$(15.1.1) \qquad f(A)x = \int_0^\infty T(\xi; A)x \, da, \qquad f(\lambda) = \int_0^\infty e^{\lambda\xi} \, da.$$

As before, the mapping $f \to f(A)$ is an algebraic isomorphism. We note that $f(\lambda)$ is in general singular at infinity in spite of the fact that infinity belongs to $\sigma_e(A)$ (except in the trivial case of A bounded); this situation cannot occur in the operational calculus of Chapter V.

The more familiar representation

$$(15.1.2) \qquad f(A) = \frac{1}{2\pi i} \int_{\gamma - i\infty}^{\gamma + i\infty} f(\lambda) R(\lambda; A) \, d\lambda + \frac{1}{2} a([0])I$$

requires a vertical strip common to the domain of analyticity of $f(\lambda)$ and that of $R(\lambda; A)$. Accordingly the domain set for $f(A)$ is limited to a subset $\mathfrak{A}_0(\varphi)$ of $\mathfrak{A}(\varphi)$ consisting of those operators A such that $R(\lambda; A)$ exists and is bounded in norm in the half-plane $\mathfrak{R}(\lambda) > \omega$ for some $\omega < \omega_0$. By employing the representation (15.1.2), it is now possible to extend the calculus to a larger class of functions. For the enlarged algebra of functions the mapping $f \to f(A)$, $A \in \mathfrak{A}_0(\varphi)$, is again an isomorphism.

In studying the dependence of $f(A)$ on A we restrict our considerations to infinitesimal generators of semi-groups of class $(1, A)$. In this class of infinitesimal generators we can introduce a notion of analyticity for operator-valued functions and show that $f(A)$ is locally analytic as a function of A in Int $[\mathfrak{A}(\varphi)]$.

Aside from the development in paragraph 5.3 and that of the present chapter, two other constructions for an operational calculus of unbounded operators should be mentioned: (a) the operator A is self-adjoint and acts on a Hilbert space, and (b) A is a differential operator acting on a suitable function space. For material dealing with (a) the reader is referred to B. de Sz.-Nagy **[3]** and M. H. Stone **[3]**; of the vast literature relevant to (b) we mention only N. Wiener **[2]** and E. L. Post **[1]** where further references may be found. See also N. P. Romanoff **[2]**.

There are two paragraphs: *General Theory for Semi-Groups of Class* (A), and *Analytic Dependence for Semi-Groups of Class* $(1, A)$.

References. Hille **[13]**, Hille and Tamarkin **[3, 7]**, Phillips **[6, 7, 11]**, Post **[1]**, Romanoff **[2]**, Stone **[3]**, de Sz.-Nagy **[3]**, Widder **[1]**, Wiener **[2]**.

1. General Theory for Semi-Groups of Class (A)

15.2. An elementary calculus. In the present section we develop an operational calculus based on the relation (15.1.1) which provides an operator extension for the function algebra $S(\varphi)$. The algebra $S(\varphi)$ is defined as in paragraph 4.4 in terms of a weighting function $\varphi(\xi)$ satisfying the conditions

(W) (i) *$\varphi(\xi)$ is a real-valued Borel measurable function defined on* $[0, \infty)$;

(ii) *$\varphi(\xi)$ is submultiplicative*: $0 < \varphi(\xi_1 + \xi_2) \leqq \varphi(\xi_1)\varphi(\xi_2)$ *for all* $\xi_1, \xi_2 \geqq 0$;

(iii) $\varphi(0) = 1$.

The function $\varphi(\xi)$ is said to be of type

$$\omega_0 \equiv \inf_{\xi>0} \xi^{-1} \log \varphi(\xi) = \lim_{\xi\to\infty} \xi^{-1} \log \varphi(\xi). \tag{15.2.1}$$

Definition 15.2.1. *Let A be the infinitesimal generator of a semi-group $[T(\xi; A)]$ of class (A). We say that A is majorized by $\varphi(\xi)$ if $\| T(\xi; A) \| \leqq \varphi(\xi)$ for all $\xi > 0$ and we denote this relation by $A \leqq \varphi$. We set $\mathfrak{A}(\varphi) = [A; A \leqq \varphi]$.*

Remark. When $\varphi(\xi)$ is upper semi-continuous and A is the infinitesimal generator of a semi-group of class $(1, A)$, we obtain the following criterion from Theorem 12.4.1: A necessary and sufficient condition for $A \leqq \varphi$ is that there exist a constant ω such that

$$\| R^{(n)}(\lambda; A) \| \leqq \int_0^\infty e^{-\lambda\xi}\xi^n\varphi(\xi)\, d\xi$$

for all real $\lambda > \omega$ and integers $n \geqq 0$.

If $a \in S(\varphi)$ and $A \in \mathfrak{A}(\varphi)$, then $\int_0^\infty T(\xi; A)x\, da$ exists as a Bochner integral for each $x \in \mathfrak{X}$. Strictly speaking, the theory of section 3.7 requires that we decompose a into a sum of σ-finite measures, each belonging to $S(\varphi)$. The result can be shown to be independent of the particular decomposition by considering the numerical analogue $\int_0^\infty x^*[T(\xi; A)x]\, da = x^*[\int_0^\infty T(\xi; A)x\, da]$, $x^* \in \mathfrak{X}^*$.

DEFINITION 15.2.2. *Let $a \in S(\varphi)$. We define*

$$(15.2.2) \qquad \psi(a; \lambda) = \int_0^\infty e^{\lambda\xi}\, da$$

for $\Re(\lambda) \leqq \omega_0$, and

$$(15.2.3) \qquad \Psi(a; A)x = \int_0^\infty T(\xi; A)x\, da$$

for $A \in \mathfrak{A}(\varphi)$ and $x \in \mathfrak{X}$. Here $T(0; A) = I$ by definition.

As we have already shown in section 4.18, when $\omega_0 > -\infty$ the mapping $a \to \psi(a; \lambda)$ is an isomorphism of $S(\varphi)$ into the algebra of functions holomorphic in the half-plane $\Re(\lambda) < \omega_0$, the mapping being continuous in the sense that $|\psi(a; \lambda)| \leqq \int_0^\infty \varphi(\xi)\, d\,|a| = \|a\|$. We now prove an analogous result for $\Psi(a; A)$.

THEOREM 15.2.1. *The function $\Psi(a; A)$ defines a homomorphic mapping of $S(\varphi)$ into the algebra of operator-valued functions on $\mathfrak{A}(\varphi)$ to $\mathfrak{E}(\mathfrak{X})$ which is* (i) *continuous in the sense that $\|\Psi(a; A)\| \leqq \|a\|$, and* (ii) *takes $e_\xi \to T(\xi; A)$ for all $\xi \geqq 0$. If $\omega_0 > -\infty$, then the mapping is an isomorphism.*

REMARK. The analyticity of $\Psi(a;A)$ as a function of A will be considered in paragraph 15.2.

PROOF. For fixed $a \in S(\varphi)$ and $A \in \mathfrak{A}(\varphi)$ the operator $\Psi(a; A)$ is clearly linear on $\mathfrak{X}$ and

$$(15.2.4) \quad \|\Psi(a; A)\| \leqq \int_0^\infty \|T(\xi; A)\|\, d\,|a| \leqq \int_0^\infty \varphi(\xi)\, d\,|a| = \|a\|.$$

Thus $\Psi(a; A)$ is an operator-valued function of A defined on $\mathfrak{A}(\varphi)$ to $\mathfrak{E}(\mathfrak{X})$ for each $a \in S(\varphi)$, continuous as a function of a in the sense of (15.2.4). We see directly from (15.2.3) that $e_\xi \to T(\xi; A)$.

It is obvious that $a \to \Psi(a; A)$ is a linear mapping. In order to prove that it is a homomorphism we have merely to prove that products go into products. By Theorem 3.7.12 we have

$$\Psi(b; A)\Psi(a; A)x = \int_0^\infty \int_0^\infty T(\eta; A)T(\xi; A)x\, d_\eta b\, d_\xi a,$$

and applying Theorem 3.7.13 we obtain

$$\Psi(b; A)\Psi(a; A)x = \int_0^\infty \int_0^\infty T(\xi + \eta; A)x\, d(a \times b).$$

In order to evaluate this integral it suffices to consider only sets of the form $[(\xi, \eta); \xi + \eta \in E, \xi, \eta \geqq 0]$ where $E \in \mathfrak{E}$. Since

$$(a \times b)[(\xi, \eta); \xi + \eta \in E, \xi, \eta \geqq 0] = (a*b)\,(E)$$

it follows that

$$\Psi(b; A)\Psi(a; A)x = \int_0^\infty T(\sigma)x\, d_\sigma(a*b) = \Psi(a*b; A)x.$$

Next we show that the homomorphism is actually an isomorphism when $\omega_0 > -\infty$. In fact, for a given $a \in S(\varphi)$, $a \neq 0$, the uniqueness theorem for Laplace-Stieltjes integrals implies that there exists a λ_0 with $\Re(\lambda_0) < \omega_0$ such that $\psi(a; \lambda_0) \neq 0$. It is clear that $\lambda_0 I \in \mathfrak{A}(\varphi)$ and that $\Psi(a; \lambda_0 I) = \psi(a; \lambda_0)I \neq \Theta$. Thus the mapping is one-to-one and therefore an isomorphism. This concludes the proof.

Before concluding this section, we consider the relation between the adjoint of a function extension and the function extension of the adjoint operator. It will be recalled that the infinitesimal generator A of a semi-group of class (A) is a $(\odot)$-operator determining the $(\odot)$-adjoint space $\mathfrak{X}^\odot(A) \equiv \overline{\mathfrak{D}(A^*)}$. We note that $\mathfrak{X}^\odot(A)$ can vary with A. According to Theorem 14.4.1, the adjoint semi-group $T^\odot(\xi; A)$ is again a semi-group of class (A) and $T^\odot(\xi; A) = T(\xi; A^\odot)$. In other words $\Psi^\odot(e_\xi; A) = \Psi(e_\xi; A^\odot)$. This relation can be generalized as follows:

THEOREM 15.2.2. *Let $A \in \mathfrak{A}(\varphi)$. Then $A^\odot \in \mathfrak{A}(\varphi) \subset \mathfrak{O}[\mathfrak{X}^\odot(A)]$ and $\Psi^\odot(a; A) = \Psi(a; A^\odot)$ for each $a \in S(\varphi)$.*

PROOF. According to Theorem 14.3.2, $\| T^\odot(\xi; A) \| \leqq \| T(\xi; A) \| \leqq \varphi(\xi)$ so that $A^\odot \lesssim \varphi$, that is $A^\odot \in \mathfrak{A}(\varphi) \subset \mathfrak{O}[(\mathfrak{X}^\odot(A)]$. For $x^\odot \in \mathfrak{X}^\odot(A)$ and $x \in \mathfrak{X}$ we have

$$[\Psi(a; A^\odot)x^\odot](x) = \int_0^\infty [T(\xi; A^\odot)x^\odot](x)\, da = \int_0^\infty x^\odot[T(\xi; A)x]\, da$$

$$= x^\odot[\Psi(a; A)x] = [\Psi^*(a; A)x^\odot](x).$$

Thus $\Psi(a; A^\odot)$ is the restriction of $\Psi^*(a; A)$ to $\mathfrak{X}^\odot(A)$. Since $\Psi(a; A^\odot)[\mathfrak{X}^\odot(A)] \subset \mathfrak{X}^\odot(A)$, it follows from Definition 14.3.1 that this restriction is $\Psi^\odot(a; A)$ and hence $\Psi(a; A^\odot) = \Psi^\odot(a; A)$.

15.3. The algebra $\mathcal{S}(\varphi)$. For a semi-group of class (A) such that $\| T(\xi; A) \| = \varphi(\xi)$ it would appear that the calculus developed in the previous section furnishes an operator-valued extension to the largest possible variety of functions, namely $S(\varphi)$. Unfortunately $S(\varphi)$ can be quite limited. For instance, if $\int_0^1 \varphi(\xi)\, d\xi = \infty$, set functions such as $r_\alpha(E) \equiv \int_E e^{-\alpha\xi}\, d\xi$, $\Re(\alpha) > \omega_0$, do not belong to $S(\varphi)$. On the other hand, Theorem 11.5.2 shows that $\Psi(r_\alpha; A) = R(\alpha; A)$ can be obtained from (15.2.3), at least for $x \in \mathfrak{X}_0(A)$. This suggests that we extend the calculus by making use of the defining condition (A)′ given in section 10.6 for semi-groups of class (A). As a first step in this direction, we now define a suitable extension of the algebra $S(\varphi)$.

DEFINITION 15.3.1. *Let $\varphi(\xi)$ be a submultiplicative function of type $\omega_0 > -\infty$ satisfying the conditions* (W). *Let $Q(\omega_0)$ be the family of all set functions $a \in S(e^{\omega_0\xi})$ such that $\psi(a; \omega_0 + i\tau) \in L(-\infty, \infty)$ and let α_0 be a fixed complex number with*

$\Re(\alpha_0) > \omega_0$. *We then define*

$$\mathcal{S}(\varphi) = S(\varphi) + Q(\omega_0) + S(\varphi)*r_{\alpha_0}$$

where $r_\alpha(E) = \int_E e^{-\alpha\xi}\, d\xi$.

Introducing a product in $\mathcal{S}(\varphi)$ defined by convolution, we have

THEOREM 15.3.1. $\mathcal{S}(\varphi)$ *is a subalgebra of* $S(e^{\omega_0\xi})$.

PROOF. In the first place, $\omega_0 = \inf_{\xi>0} \xi^{-1} \log \varphi(\xi)$ so that $e^{\omega_0\xi} \leqq \varphi(\xi)$; as a consequence $S(\varphi) \subset S(e^{\omega_0\xi})$. Since $Q(\omega_0) \cup r_{\alpha_0} \subset S(e^{\omega_0\xi})$ by definition, we have $\mathcal{S}(\varphi) \subset S(e^{\omega_0\xi})$. Further it is clear that $\mathcal{S}(\varphi)$ is a linear system; it therefore remains to show that $\mathcal{S}(\varphi)$ is closed under multiplication. Now if $a, b \in S(e^{\omega_0\xi})$, then

$$\psi(a*b; \lambda) = \psi(a; \lambda)\psi(b; \lambda)$$

for all λ with $\Re(\lambda) \leqq \omega_0$. Hence if $a \in Q(\omega_0)$ and $b \in S(e^{\omega_0\xi})$, then

$$|\psi(a*b; \omega_0 + i\tau)|$$

is majorized by $\| b \| \, | \psi(a; \omega_0 + i\tau) |$ and therefore belongs to $L(-\infty, \infty)$. It follows that $Q(\omega_0)$ is an ideal in $S(e^{\omega_0\xi})$ and in particular that

$$Q(\omega_0)*\mathcal{S}(\varphi) \subset Q(\omega_0) \subset \mathcal{S}(\varphi).$$

Moreover $S(\varphi)$ is itself a subalgebra of $S(e^{\omega_0\xi})$ so that

$$S(\varphi)*S(\varphi) \subset S(\varphi)$$

and therefore $S(\varphi)*\mathcal{S}(\varphi) \subset S(\varphi) + S(\varphi)*Q(\omega_0) + S(\varphi)*r_{\alpha_0} \subset \mathcal{S}(\varphi)$. Finally we note that $\psi(r_{\alpha_0}*r_{\alpha_0}; \omega_0 + i\tau) = (\alpha_0 - \omega_0 - i\tau)^{-2} \in L(-\infty, \infty)$. Thus $r_{\alpha_0}*r_{\alpha_0} \in Q(\omega_0)$ and hence

$$\begin{aligned} r_{\alpha_0}*\mathcal{S}(\varphi) &\subset S(\varphi)*r_{\alpha_0} + Q(\omega_0)*r_{\alpha_0} + S(\varphi)*r_{\alpha_0}*r_{\alpha_0} \\ &\subset S(\varphi)*r_{\alpha_0} + Q(\omega_0) \subset \mathcal{S}(\varphi). \end{aligned}$$

This concludes the proof.

COROLLARY. $[r_\alpha; \Re(\alpha) > \omega_0] \subset \mathcal{S}(\varphi)$.

PROOF. A straightforward calculation shows that

$$r_\alpha = r_{\alpha_0} + (\alpha_0 - \alpha) r_\alpha * r_{\alpha_0}. \tag{15.3.1}$$

Since $\psi(r_\alpha*r_{\alpha_0}; \omega_0 + i\tau) = [(\alpha - \omega_0 - i\tau)(\alpha_0 - \omega_0 - i\tau)]^{-1} \in L(-\infty, \infty)$, it follows that $r_\alpha*r_{\alpha_0} \in Q(\omega_0)$ and hence by (15.3.1) that $r_\alpha \in \mathcal{S}(\varphi)$ for $\Re(\alpha) > \omega_0$.

Since $\mathcal{S}(\varphi)$ is a subalgebra of $S(e^{\omega_0\xi})$, we can topologize $\mathcal{S}(\varphi)$ by means of the relative topology. With the norm

$$\| a \| \equiv \int_0^\infty e^{\omega_0\xi}\, d\, | a |, \tag{15.3.2}$$

$\mathcal{S}(\varphi)$ becomes a normed algebra, in general not complete. However, the $S(e^{\omega_0\xi})$ weak* topology is more useful for our purposes and we proceed to describe this topology.

We define

$$C(\omega_0) \equiv [f(\xi);\, f(\xi) \text{ continuous on } [0, \infty),\, \lim_{\xi\to\infty} e^{-\omega_0\xi}f(\xi) = 0]$$

with $\| f \| = \sup_{\xi \geqq 0} e^{-\omega_0\xi} | f(\xi) |$. It is clear that $f(\xi) \to e^{-\omega_0\xi}f(\xi)$ maps $C(\omega_0)$ isometrically onto $C(0)$, the space of functions continuous on $[0, \infty)$ and vanishing at infinity. Now $[C(0)]^* = S(1)$, the space of countably additive set functions defined on the Borel subsets of $[0, \infty)$. The adjoint transformation taking $[C(0)]^*$ onto $[C(\omega_0)]^*$ is given by the isometry $\alpha(E) \to \int_E e^{-\omega_0\xi}\, d\alpha$ which clearly maps $S(1)$ onto $S(e^{\omega_0\xi})$. It follows that $S(e^{\omega_0\xi}) = [C(\omega_0)]^*$. We may therefore speak of the weak* topology of $S(e^{\omega_0\xi})$.

THEOREM 15.3.2. *The following statements are equivalent:*

(i) $\{\alpha_n\}$ *converges to* α_0 *in the weak* topology of* $S(e^{\omega_0\xi})$;

(ii) $\{\alpha_n\}$ *is a bounded sequence and* $\lim_{n\to\infty} \psi(\alpha_n ; \lambda) = \psi(\alpha_0 ; \lambda)$ *uniformly in each compact subset of the half-plane* $\Re(\lambda) < \omega_0$;

(iii) $\{\alpha_n\}$ *is a bounded sequence and* $\lim_{n\to\infty} \psi(\alpha_n ; \lambda) = \psi(\alpha_0 ; \lambda)$ *for each* λ *belonging to an open subset* G *of the half-plane* $\Re(\lambda) < \omega_0$.

PROOF. By making use of the above isometry, it is clear that we may assume without loss of generality that $\omega_0 = 0$ and hence that we are dealing with $S(1)$. If $\{\alpha_n\}$ converges in the weak* topology, then $\{\alpha_n\}$ is a bounded sequence by the uniform boundedness theorem. Let C be a compact subset of the half-plane $\Re(\lambda) < 0$. Then it is easy to see that the functions $[e^{\lambda\xi}, \lambda \in C]$ are equicontinuous and tend to zero as $\xi \to \infty$ uniformly with respect to λ in C. It follows that this set of functions forms a compact subset of $C(0)$, and hence that (i) implies that $\lim_{n\to\infty} \psi(\alpha_n ; \lambda) = \psi(\alpha_0 ; \lambda)$ uniformly with respect to λ in C. Thus (i) implies (ii); obviously (ii) implies (iii). In order to show that (iii) implies (i) we employ a contra-positive argument. We may as well assume that there exists an $f_0 \in C(0)$ and an $\epsilon > 0$ such that $| \int_0^\infty f_0(\xi)\, d\alpha_n - \int_0^\infty f_0(\xi)\, d\alpha_0 | > \epsilon$ for all n. The (B)-space $C(0)$ being separable, Theorem 2.10.1 asserts that there exists a subsequence $\{\alpha_{n_k}\}$ which converges weakly* to some $\alpha \in S(1)$. In particular

$$\lim_{k\to\infty} \int_0^\infty f_0(\xi)\, d\alpha_{n_k} = \int_0^\infty f_0(\xi)\, d\alpha$$

and this shows that $\alpha \neq \alpha_0$. However by the previous argument,

$$\lim_{k\to\infty} \psi(\alpha_{n_k} ; \lambda) = \psi(\alpha; \lambda)$$

for each λ with $\Re(\lambda) < 0$ so that $\psi(\alpha; \lambda) = \psi(\alpha_0 ; \lambda)$ for all $\lambda \in G$. The uniqueness theorem for Laplace-Stieltjes transforms then implies that $\alpha = \alpha_0$ which is impossible.

LEMMA 15.3.1. *Let* $\alpha \in S(e^{\omega_0\xi})$ *be such that* $\alpha([0]) = 0$ *and set*

$$\alpha_\alpha(E) = \int_E e^{-\omega_0\xi} \left[\int_0^\infty K(\sigma, \xi; \alpha) e^{\omega_0\sigma}\, d_\sigma\alpha \right] d\xi \tag{15.3.3}$$

where

$$K(\sigma, \xi; \alpha) = e^{-\alpha(\sigma+\xi)} \sum_{n=0}^{\infty} \frac{(\alpha^2\sigma)^{n+1}\xi^n}{n!(n+1)!}.$$

Then $\lim_{\alpha\to\infty} a_\alpha = a$ *in the weak* topology of* $S(e^{\omega_0\xi})$.

PROOF. A direct proof can be obtained by means of Theorem 6.3.4; however we prefer the following argument which amounts to a verification of condition (iii) of the previous theorem. Making use of the relation (6.3.13b) we have

$$\| a_\alpha \| \leqq \int_0^\infty \int_0^\infty K(\sigma, \xi; \alpha) e^{\omega_0\sigma} d_\sigma \mid a \mid d\xi \leqq \| a \|.$$

Further, a straightforward calculation yields

$$\begin{aligned} \psi(a_\alpha; \lambda) &= \int_0^\infty e^{\lambda\xi} da_\alpha \\ &= \int_0^\infty \left\{ \exp\left[\frac{\alpha\lambda - \omega_0\lambda + \omega_0^2}{\alpha + \omega_0 - \lambda}\sigma\right] - \exp[(\omega_0 - \alpha)\sigma]\right\} d_\sigma a. \end{aligned}$$

For $\Re(\lambda) < \omega_0$, the integrand is majorized by $e^{\omega_0\sigma}$ and converges pointwise to $e^{\lambda\sigma}$ for $\sigma > 0$ as $\alpha \to \infty$. By the Lebesgue majorized convergence theorem,

$$\lim_{\alpha\to\infty} \psi(a_\alpha; \lambda) = \psi(a; \lambda)$$

for each λ with $\Re(\lambda) < \omega_0$. This verifies (iii) and therefore establishes the lemma.

THEOREM 15.3.3. *Let* $\{\alpha_n\}$ *be any sequence of positive numbers converging to infinity. Then the algebra* S_0 *generated by* $e_0 \cup [r_{\alpha_n+\omega_0}; n = 1, 2, \cdots]$ *is dense in* $S(e^{\omega_0\xi})$ *relative to the weak* sequential topology.*

PROOF. Since $e_0 \in S_0$, it suffices to consider only elements $a \in S(e^{\omega_0\xi})$ such that $a([0]) = 0$. We define

$$a_{\alpha,N}(E) \equiv \int_E e^{-\omega_0\xi} \left\{ \int_0^\infty e^{-\alpha(\sigma+\xi)} \sum_{n=0}^{N} \frac{(\alpha^2\sigma)^{n+1}\xi^n}{n!(n+1)!} e^{\omega_0\sigma} d_\sigma a \right\} d\xi.$$

Making use of the fact that $r_\alpha^{k*}(E) = [(k-1)!]^{-1} \int_E e^{-\alpha\xi}\xi^{k-1} d\xi$, we see that

$$a_{\alpha,N} = \sum_{n=0}^{N} \left\{ \int_0^\infty e^{-(\alpha-\omega_0)\sigma} \frac{(\alpha^2\sigma)^{n+1}}{(n+1)!} d_\sigma a \right\} r_{\alpha+\omega_0}^{(n+1)*} \tag{15.3.4}$$

is a polynomial in $r_{\alpha+\omega_0}$. Thus $a_{\alpha,N} \in S_0$. On the other hand

$$\begin{aligned} \| a_{\alpha,N} - a_\alpha \| &\leqq \int_0^\infty \int_0^\infty \left\{ e^{-\alpha(\sigma+\xi)} \sum_{n=N+1}^{\infty} \frac{(\alpha^2\sigma)^{n+1}\xi^n}{n!(n+1)!} \right\} e^{\omega_0\sigma} d_\sigma \mid a \mid d\xi \\ &= \int_0^\infty \left\{ e^{-\alpha\sigma} \sum_{n=N+1}^{\infty} \frac{(\alpha\sigma)^{n+1}}{(n+1)!} \right\} e^{\omega_0\sigma} d_\sigma \mid a \mid. \end{aligned}$$

The bracket expression in the last member of this relation is clearly bounded by one and converges to zero pointwise as $N \to \infty$. The Lebesgue majorized convergence theorem implies that $\lim_{N\to\infty} \| a_{\alpha,N} - a_\alpha \| = 0$. According to Lemma 15.3.1, $\lim_{n\to\infty} a_{\alpha_n} = a$ in the weak* topology. Hence if we choose N_n so that $\| a_{\alpha_n,N_n} - a_{\alpha_n} \| < 1/n$, then it readily follows that $a_{\alpha_n,N_n} \to a$, again in the weak* topology of $S(e^{\omega_0\xi})$. This concludes the proof.

We now list a few properties belonging to elements of the various subalgebras of $S(e^{\omega_0\xi})$. If $a \in L(e^{\omega_0\xi})$, then $\lim_{\lambda\to\infty} \psi(a; \lambda) = 0$ uniformly with respect to λ in the closed half-plane $\Re(\lambda) \leqq \omega_0$. This is not true in general for $a \in S(e^{\omega_0\xi})$; here the best we can assert is that $\lim_{\lambda\to\infty} \psi(a; \lambda) = a([0])$ uniformly with respect to λ in each sector $\frac{1}{2}\pi + \epsilon \leqq \arg \lambda \leqq \frac{3}{2}\pi - \epsilon$, $\epsilon > 0$; however $\psi(a; \lambda)$ need not tend to a limit along vertical lines. Sufficient conditions that $f(\lambda)$ be the Laplace transform of an element of $L(e^{\omega_0\xi})$ can be found in E. Hille and J. D. Tamarkin [7], necessary and sufficient conditions are given in D. V. Widder [1].

If $f(\lambda)$ is a rational function, vanishing at infinity and having its poles in $\Re(\lambda) > \omega_0$, then a partial fraction decomposition shows that $f(\lambda)$ is a polynomial in $[r_\alpha; \Re(\alpha) > \omega_0]$ and hence by the corollary to Theorem 15.3.1 that it is the Laplace transform of an element in $\mathcal{S}(\varphi)$. Several criteria are available for determining when a set function belongs to $Q(\omega_0)$ (see E. Hille and J. D. Tamarkin [3]). We note in particular that $a(E) = \int_E g(\xi)\, d\xi \in Q(\omega_0)$ if (a) $g(\xi)$ is absolutely continuous on $[0, \infty)$ with $g(0) = 0$, (b) $e^{\omega_0\xi} g(\xi) \in L_1(0, \infty)$, and (c) $e^{\omega_0\xi} g'(\xi) \in L_p(0, \infty)$ for some p, $1 < p \leqq 2$. A necessary condition that $a \in Q(\omega_0)$ is that $a(E)$ be an absolutely continuous set function.

Each function $\psi(a; \lambda)$, $a \in S(e^{\omega_0\xi})$, can be represented in its half-plane of holomorphism by a suitable extension of Cauchy's integral, namely,

$$\psi(a, \lambda) = \lim_{\beta\to\infty} \frac{1}{2\pi i} \int_{\gamma-i\beta}^{\gamma+i\beta} \frac{\psi(a; \zeta)}{\zeta - \lambda}\, d\zeta + \tfrac{1}{2}\, a([0]), \qquad \Re(\lambda) < \gamma \leqq \omega_0. \tag{15.3.5}$$

This is proved as follows. For $\Re(\lambda) < \gamma$ we form

$$\begin{aligned}\psi_\beta(a; \lambda) &\equiv \frac{1}{2\pi i} \int_{\gamma-i\beta}^{\gamma+i\beta} \frac{\psi(a; \zeta)}{\zeta - \lambda}\, d\zeta + \tfrac{1}{2}\, a([0]) \\ &= \frac{1}{2\pi i} \int_{\gamma-i\beta}^{\gamma+i\beta} \frac{d\zeta}{\zeta - \lambda} \int_0^\infty e^{\zeta\xi}\, d_\xi a + \tfrac{1}{2}\, a([0]) \\ &= \int_0^\infty \left\{ \frac{1}{2\pi i} \int_{\gamma-i\beta}^{\gamma+i\beta} \frac{e^{\zeta\xi}}{\zeta - \lambda}\, d\zeta \right\} d_\xi a + \tfrac{1}{2}\, a([0]) \to \int_0^\infty e^{\lambda\xi}\, d_\xi a = \psi(a; \lambda).\end{aligned}$$

Here the limiting process is justified by the following observations. The integral in brackets tends to $e^{\lambda\xi}$ as $\beta \to \infty$ provided $\xi > 0$ and to $\frac{1}{2}$ when $\xi = 0$. Further this integral is bounded by a fixed multiple (depending upon λ) of $e^{\gamma\xi}$ for all values of $\xi \geqq 0$ and $\beta > 0$. Consequently the Lebesgue majorized convergence theorem permits us to pass to the limit under the sign of integration.

For $\Re(\lambda) > \gamma \leqq \omega_0$ the same type of argument shows that

$$0 = \lim_{\beta\to\infty} \frac{1}{2\pi i} \int_{\gamma-i\beta}^{\gamma+i\beta} \frac{\psi(a; \zeta)}{\zeta - \lambda}\, d\zeta + \tfrac{1}{2}\, a([0]). \tag{15.3.6}$$

Subtracting (15.3.6) with $\lambda = 2\gamma - \sigma + i\tau$ from (15.3.5) with $\lambda = \sigma + i\tau$, we obtain

$$\psi(a; \sigma + i\tau) = \frac{\gamma - \sigma}{\pi} \int_{-\infty}^{\infty} \frac{\psi(a; \gamma + i\eta)}{(\gamma - \sigma)^2 + (\eta - \tau)^2} d\eta, \qquad \sigma < \gamma. \tag{15.3.7}$$

Now the integral of the Poisson kernel over the range $(-\infty, \infty)$ is identically one. Hence if $a \in Q(\omega_0)$, then setting $\gamma = \omega_0$ in (15.3.7) we obtain

$$\int_{-\infty}^{\infty} | \psi(a; \sigma + i\tau) | \, d\tau \leqq \int_{-\infty}^{\infty} | \psi(a; \omega_0 + i\tau) | \, d\tau$$

and therefore $a \in Q(\sigma)$ for all $\sigma \leqq \omega_0$.

15.4. The amended calculus. The operational calculus developed in section 15.2 has several shortcomings, chief among these is the limited nature of the basic function algebra $S(\varphi)$ when $\int_0^1 \varphi(\xi) \, d\xi \doteq \infty$. Moreover in maximizing the domain set $\mathfrak{A}(\varphi)$ we have made this calculus very unwieldy. Both of these deficiencies have been corrected in the calculus which we treat in the present section. The basic function algebra is now $\mathcal{S}(\varphi)$ which has been designed to include an adequate class of functions, and the domain set has been suitably restricted. The resulting calculus conforms very well with the calculi of Chapter V.

DEFINITION 15.4.1. *Let A be the infinitesimal generator of a semi-group $[T(\xi; A)]$ of class (A) and of type $\omega_0(A)$. Let $\omega_1(A) = \inf [\omega; R(\lambda; A)$ is bounded for $\Re(\lambda) > \omega > \omega_0(A)]$. We say that A is strictly majorized by $\varphi(\xi)$ if $A \lesssim \varphi$ and $\omega_1(A) < \omega_0$; we denote this relation by $A \prec \varphi$. Finally we set $\mathfrak{A}_0(\varphi) = [A; A \prec \varphi]$.*

Because of the condition $(A)'$ of section 10.6 we see that $\omega_1(A) < \infty$ for any semi-group $[T(\xi; A)]$ of class (A). If, moreover, $[T(\xi; A)]$ is of class $(0, A)$, then Theorem 10.6.1 implies that $\omega_1(A) = \omega_0(A)$.

DEFINITION 15.4.2. *Let $\mathfrak{X}_1(A) = [x; \lim_{\xi \to 0+} T(\xi; A)x = x]$. For $a \in \mathcal{S}(\varphi)$, $A \in \mathfrak{A}_0(\varphi)$, and $x \in \mathfrak{X}_1(A)$ we define*

$$\Psi(a; A)x = \int_0^{\infty} T(\xi; A)x \, da \tag{15.4.1}$$

or, alternately, for $\omega_1(A) < \gamma \leqq \omega_0$

$$\Psi(a; A)x = \frac{1}{2\pi i} (C, 1)\text{-}\int_{\gamma - i\infty}^{\gamma + i\infty} \psi(a; \zeta) R(\zeta; A)x \, d\zeta + \tfrac{1}{2}a([0])x. \tag{15.4.2}$$

In order to establish the equivalence between these two definitions we require

LEMMA 15.4.1. *Let $[T(\xi; A)]$ be a semi-group of class (A). Then*

$$R(\lambda; A)x = \int_0^{\infty} e^{-\lambda\xi} T(\xi; A)x \, d\xi, \tag{15.4.3}$$

for each $x \in \mathfrak{X}_1(A)$ and λ with $\Re(\lambda) > \omega_1(A)$.

PROOF. It follows from the way in which $\omega_1(A)$ and $\mathfrak{X}_1(A)$ were defined that both members of (15.4.3) exist and are holomorphic in the half-plane $\Re(\lambda) > \omega_1(A)$. On the other hand, Lemma 11.5.1 and Theorem 11.5.2 imply that the relation (15.4.3) is valid for all λ with $\Re(\lambda)$ sufficiently large. We therefore see by Theorem 3.11.5 that the relation holds for all λ with $\Re(\lambda) > \omega_1(A)$.

THEOREM 15.4.1. *The definitions* (15.4.1) *and* (15.4.2) *are equivalent.*

PROOF. The proof of this equivalence is much like the proof of formula (15.3.5). In accordance with (15.4.2) we form (15.4.4)

$$\Psi_\beta(a; A)x \equiv \frac{1}{2\pi}\int_{-\beta}^{\beta}\left\{1 - \frac{|\eta|}{\beta}\right\}\psi(a; \gamma + i\eta)\, R(\gamma + i\eta; A)x\, d\eta + \tfrac{1}{2}a([0])x, \qquad x \in \mathfrak{X}_1(A), \tag{15.4.4}$$

and let $\beta \to \infty$. In this integral we substitute the expressions for $\psi(a; \gamma + i\eta)$ and $R(\gamma + i\eta; A)x$ as Laplace-Stieltjes integrals and in the resulting triple integral we carry out the integration with respect to η over the range $(-\beta, \beta)$. The result may be written as

$$\int_0^\infty \left\{\int_0^\infty T(\xi; A)x\, e^{\gamma(\sigma-\xi)}\, \frac{2\sin^2 \frac{1}{2}\beta(\sigma - \xi)}{\pi\beta(\sigma - \xi)^2}\, d\xi\right\} d_\sigma a + \tfrac{1}{2}a([0])x.$$

By Theorem 6.3.2, the inner integral tends to $T(\sigma; A)x$ as $\beta \to \infty$ for $\sigma > 0$ and to $\frac{1}{2}x$ for $\sigma = 0$. We have chosen γ so that $\omega_0(A) \leqq \omega_1(A) < \gamma$ and hence

$$\| T(\xi; A)x \|\, e^{-\gamma\xi}$$

is bounded in $\xi \geqq 0$. It follows that the inner integral is majorized by a suitable multiple of $e^{\gamma\sigma}$ for $\sigma \geqq 0$ and $\beta > 0$. Finally since $\gamma \leqq \omega_0$ we see that Theorem 3.7.9 applies so that we may pass to the limit as $\beta \to \infty$ under the outer integral, obtaining

$$\lim_{\beta\to\infty} \Psi_\beta(a; A)x = \int_0^\infty T(\sigma; A)x\, da = \Psi(a; A)x$$

as asserted.

As defined in Definition 15.4.2, $\Psi(a; A)$ is an operator with domain $\mathfrak{X}_1(A)$. It is clear that $\mathfrak{X}_1(A)$ is a linear subspace of $\mathfrak{X}$, and since it contains $\mathfrak{X}_0(A)$ we see that $\mathfrak{X}_1(A)$ is dense in $\mathfrak{X}$. It is also clear that $\Psi(a; A)$ is a linear operator on $\mathfrak{X}_1(A)$.

LEMMA 15.4.2. *The mapping* $a \to \Psi(a; A)$ *is an isomorphism of* $\mathcal{S}(\varphi)$ *into the algebra of functions defined on* $\mathfrak{A}_0(\varphi)$ *to linear operators on* $\mathfrak{X}_1(A)$.

PROOF. Suppose that $a \in \mathcal{S}(\varphi)$ and $A \in \mathfrak{A}_0(\varphi)$. We first prove the inclusion $\Psi(a; A)[\mathfrak{X}_1(A)] \subset \mathfrak{X}_1(A)$. Now

$$T(\eta)[\Psi(a; A)x] = \int_0^\infty T(\xi + \eta; A)x\, da, \qquad x \in \mathfrak{X}_1(A).$$

Since $\omega_0 > \omega_1(A)$, there exists for each $x \in \mathfrak{X}_1(A)$ a constant $M(x; A) < \infty$ such that $\| T(\xi; A)x \| \leqq M(x; A)e^{\omega_0 \xi}$. Consequently the integrand is dominated by a constant multiple of $e^{\omega_0 \xi}$ as $\eta \to 0+$. It follows from Theorem 3.7.9 that

$$\lim_{\eta \to 0+} T(\eta)[\Psi(a; A)x] = \int_0^\infty T(\xi; A)x \, da = \Psi(a; A)x$$

so that $\Psi(a; A)x \in \mathfrak{X}_1(A)$. Thus for $a, b \in \mathcal{S}(\varphi)$ the operator $\Psi(b; A)\Psi(a; A)$ is well defined on the domain $\mathfrak{X}_1(A)$. The remainder of the proof paraphrases that of Theorem 15.2.1 and is left to the reader.

LEMMA 15.4.3. *Let $a \in \mathcal{S}(\varphi)$ and $A \in \mathfrak{A}_0(\varphi)$. Then the linear operator $\Psi(a; A)$ with domain $\mathfrak{X}_1(A)$ has a unique bounded linear extension on $\mathfrak{X}$.*

PROOF. The subspace $\mathfrak{X}_1(A)$ being dense in $\mathfrak{X}$, any bounded linear operator on $\mathfrak{X}_1(A)$ will have a unique bounded linear extension on $\mathfrak{X}$. It therefore suffices to show that $\Psi(a; A)$ is bounded on $\mathfrak{X}_1(A)$ for each $a \in \mathcal{S}(\varphi)$ or simply for each a in the component parts of $\mathcal{S}(\varphi)$. This has already been established for each $a \in S(\varphi)$ in Theorem 15.2.1. Suppose next that $a \in Q(\omega_0)$. Then $a \in Q(\gamma)$ for each $\gamma \leqq \omega_0$ as we have already noted. On the other hand $\| R(\gamma + i\eta; A) \|$ is bounded in $\eta \in (-\infty, \infty)$ for $\gamma > \omega_1(A)$. Consequently the right member in (15.4.2) converges in the uniform operator topology to an operator in $\mathfrak{E}(\mathfrak{X})$ and hence $\Psi(a; A)$ is bounded on $\mathfrak{X}_1(A)$. If $a = r_\alpha$, $\Re(\alpha) > \omega_0$, then, according to Lemma 15.4.1, $\Psi(r_\alpha; A)$ is the restriction of $R(\alpha; A) \in \mathfrak{E}(\mathfrak{X})$ to $\mathfrak{X}_1(A)$ and is therefore also bounded. Finally for $a \in S(\varphi)$ we have by Lemma 15.4.2 that $\Psi(a * r_\alpha; A) = \Psi(a; A)\Psi(r_\alpha; A)$; that is $\Psi(a * r_\alpha; A)$ is the product of two bounded operators and hence is itself bounded. This concludes the proof.

From this point on $\Psi(a; A)$ will denote the bounded linear extension on $\mathfrak{X}$ of the operator $\Psi(a; A)$ as defined in Definition 15.4.1 with domain $\mathfrak{X}_1(A)$. It is clear that Lemma 15.4.2 applies equally well to the so extended $\Psi(a; A)$. Moreover we have

THEOREM 15.4.2. *Let $\varphi(\xi)$ be a submultiplicative function of type $\omega_0 > -\infty$ and suppose α_0 is such that $\Re(\alpha_0) > \omega_0$. Let $\mathcal{S}(\varphi)$ be the algebra of set functions described in Definition 15.3.1 with a sequence topology: $a_n \to a_0$ denoting that the sequence $\{a_n\}$ converges to a_0 in the weak* topology of $S(e^{\omega_0 \xi})$. Further let $\mathfrak{B}(\varphi)$ be the complex algebra of functions $f(A)$ defined on $\mathfrak{A}_0(\varphi)$ and having values in $\mathfrak{E}(\mathfrak{X})$, the arithmetic operations being defined as in $\mathfrak{E}(\mathfrak{X})$.*

There exists an isomorphic mapping: $a \to \Psi(a; A)$ of $\mathcal{S}(\varphi)$ onto a subalgebra $\mathfrak{B}_0(\varphi)$ of $\mathfrak{B}(\varphi)$ such that (i) $e_0 \to I$; (ii) $r_{\alpha_0} \to R(\alpha_0; A)$; *and* (iii) $a_n \to a$ *implies that $\Psi(a_n; A)x \to \Psi(a; A)x$ for each $x \in \mathfrak{X}_1(A)$. This mapping is unique and is defined by* (15.4.1) *or, alternately, by* (15.4.2) *for each $x \in \mathfrak{X}_1(A)$.*

PROOF. It is clear from Lemma 15.4.3 that the function $\Psi(a; A)$ defined in

Definition 15.4.1 is an operator-valued function on $\mathfrak{A}_0(\varphi)$ to $\mathfrak{E}(\mathfrak{X})$. It follows from this and Lemma 15.4.2 that the mapping $a \to \Psi(a; A)$ is an isomorphism. We see from (15.4.1) that $e_0 \to I$, which proves (i). In addition Lemma 15.4.1 implies (ii). Suppose next that $a_n \to a_0$ in the weak* topology of $S(e^{\omega_0 \xi})$, that $A \in \mathfrak{A}_0(\varphi)$, and that $x \in \mathfrak{X}_1(A)$. Then for each γ with $\omega_1(A) < \gamma < \omega_0$, there exists a constant $M_\gamma(x; A) < \infty$ such that $\| T(\xi; A)x \| \leqq M_\gamma(x; A)e^{\gamma\xi}$ for all $\xi \geqq 0$. Consequently

$$\left\| \int_\beta^\infty T(\xi; A)x\, da \right\| \leqq M_\gamma(x; A) \int_\beta^\infty e^{\gamma\xi}\, d\,|\,a\,| \leqq M_\gamma(x; A)e^{(\gamma-\omega_0)\beta}\,\|\, a \,\|.$$

Since $a_n \to a_0$ in the weak* topology of $S(e^{\omega_0\xi})$, the sequence $\{a_n\}$ will be bounded and hence $\lim_{\beta\to\infty} \int_\beta^\infty T(\xi; A)x\, da_n = \theta$ uniformly with respect to n. On the other hand according to Corollary 2 of Theorem 3.3.2, $\lim_{n\to\infty} \int_0^\beta T(\xi; A)x\, da_n = \int_0^\beta T(\xi; A)x\, da_0$ for all $\beta > 0$. Hence together these two facts imply that $\lim_{n\to\infty} \Psi(a_n; A)x = \Psi(a_0; A)x$ for each $x \in \mathfrak{X}_1(A)$, which proves (iii).

We next establish the uniqueness of the mapping. Suppose that $a \to \Phi(a; A)$ is any isomorphism with the stated properties. For $\Re(\alpha) > \omega_0$ the corollary to Theorem 15.3.1 shows that $r_\alpha \in \mathcal{S}(\varphi)$. As in (15.3.1), $r_\alpha - r_{\alpha_0} = (\alpha_0 - \alpha)r_\alpha * r_{\alpha_0} = (\alpha_0 - \alpha)r_{\alpha_0} * r_\alpha$ and therefore

$$\begin{aligned}\Phi(r_\alpha; A) - \Phi(r_{\alpha_0}; A) &= (\alpha_0 - \alpha)\Phi(r_\alpha; A)\Phi(r_{\alpha_0}; A)\\ &= (\alpha_0 - \alpha)\Phi(r_{\alpha_0}; A)\Phi(r_\alpha; A).\end{aligned}$$

By (ii), $\Phi(r_{\alpha_0}; A) = R(\alpha_0; A)$; hence by Theorem 5.8.3 we see that $\Phi(r_\alpha; A) = R(\alpha; A)$ for all α with $\Re(\alpha) > \omega_0$. As a consequence, if a belongs to the algebra S_0 generated by $e_0 \cup [r_\alpha; \alpha > \omega_0]$, then $\Phi(a; A)$ is uniquely determined; that is, $\Phi(a; A) = \Psi(a; A)$. Now given any $a_0 \in \mathcal{S}(\varphi)$ there exists, according to Theorem 15.3.3, a sequence $\{a_n\} \subset S_0$ such that $a_n \to a_0$ in the weak* topology of $S(e^{\omega_0\xi})$. It follows from (iii) that $\Phi(a_n; A)x \to \Phi(a_0; A)x$ for each $x \in \mathfrak{X}_1(A)$; and since $\Psi(a_n; A)x \to \Psi(a_0; A)x$, $x \in \mathfrak{X}_1(A)$, we see that $\Phi(a_0; A) = \Psi(a_0; A)$. This concludes the proof.

The analogue of Theorem 15.2.2 holds for the amended calculus.

THEOREM 15.4.3. *Let $A \in \mathfrak{A}_0(\varphi)$ be given. Then $A^\odot \in \mathfrak{A}_0(\varphi) \subset \mathfrak{D}[\mathfrak{X}^\odot(A)]$ and $\Psi^\odot(a; A) = \Psi(a; A^\odot)$ for each $a \in \mathcal{S}(\varphi)$.*

PROOF. Theorem 14.3.4 shows that $\| R(\lambda; A^\odot) \| \leqq \| R(\lambda; A) \|$ for all $\lambda \in \rho(A)$ so that $\omega_1(A^\odot) \leqq \omega_1(A) < \omega_0$. Since $A^\odot \lesssim \varphi$ by Theorem 15.2.2, it now follows that $A^\odot \prec \varphi$; that is, $A^\odot \in \mathfrak{A}_0(\varphi) \subset \mathfrak{D}[\mathfrak{X}^\odot(A)]$. The second assertion of the theorem has already been established for $a \in S(\varphi)$ in Theorem 15.2.2. According to Theorem 14.3.3, $R^\odot(\alpha_0; A) = R(\alpha_0; A^\odot)$ for $\Re(\alpha_0) > \omega_0$, and this together with the corollary to Theorem 14.3.2 shows that

$$\Psi(a*r_{\alpha_0}; A^\odot) = \Psi(a; A^\odot)\Psi(r_{\alpha_0}; A^\odot) = \Psi^\odot(a; A)\Psi^\odot(r_{\alpha_0}; A) = \Psi^\odot(a*r_{\alpha_0}; A)$$

for $a \in S(\varphi)$. Finally for $a \in Q(\omega_0)$, $x^{\odot} \in \mathfrak{X}^{\odot}(A)$, and $x \in \mathfrak{X}$ we have

$$\begin{aligned}[\Psi(a; A^{\odot})x^{\odot}](x) &= \frac{1}{2\pi i}\int_{\gamma-i\infty}^{\gamma+i\infty} \psi(a; \zeta)[R(\zeta; A^{\odot})x^{\odot}](x)\, d\zeta \\ &= \frac{1}{2\pi i}\int_{\gamma-i\infty}^{\gamma+i\infty} \psi(a; \zeta)x^{\odot}[R(\zeta; A)x]\, d\zeta \\ &= x^{\odot}[\Psi(a; A)x] = [\Psi^*(a; A)x^{\odot}](x).\end{aligned}$$

Thus $\Psi(a; A^{\odot})$ is the restriction of $\Psi^*(a; A)$ to $\mathfrak{X}^{\odot}(A)$. Since $\Psi(a; A^{\odot}) \in \mathfrak{E}[\mathfrak{X}^{\odot}(A)]$, it follows that this restriction is precisely $\Psi^{\odot}(a; A)$ and hence that $\Psi(a; A^{\odot}) = \Psi^{\odot}(a; A)$. Any $a \in \mathcal{S}(\varphi)$ is the sum of terms of the above kind and therefore the assertion is valid for each $a \in \mathcal{S}(\varphi)$.

REMARK. The extension of the operational calculus to unbounded operator functions appears to be a very promising direction for further research. For such an extension the function algebra must be radically extended beyond $S(\varphi)$. We can look upon $S(\varphi)$ as the subalgebra of $\mathfrak{E}[L(\varphi)]$ which commutes with translations, that is with $[e_\xi; \xi > 0]$. A natural generalization would be to consider the subset of $\mathfrak{D}[L(\varphi)]$ which commutes with translations as the basic function algebra for the extended calculus. The first results in this direction are to be found in a paper by R. S. Phillips [7]; a more systematic treatment has been given by A. V. Balakrishnan (University of Southern California thesis, 1954). A somewhat different approach to this subject employing the theory of distributions has also been developed by L. Schwartz (unpublished).

2. Analytic Dependence for Semi-Groups of Class (1, A)

15.5. Analyticity. The analogy between Theorem 15.4.2 and Theorems 5.2.5 and 5.11.2 is very close. The principal difference arises from the fact that in Theorem 15.4.2 the operator-valued functions in $\mathfrak{B}_0(\varphi)$ were not shown to be locally analytic in $\mathfrak{A}_0(\varphi)$. Limiting our considerations to infinitesimal generators of semi-groups of class $(1, A)$, we now establish the analyticity of this dependence.

Let $\mathfrak{A}_1$ denote the set of all infinitesimal generators of semi-groups of class $(1, A)$. We recall that the relation $\sim$, defined as in Definition 13.5.1, is an equivalence under which $\mathfrak{A}_1$ splits up into mutually exclusive equivalence classes $\mathfrak{C}_\alpha$, $\mathfrak{A}_1 = \bigcup_\alpha \mathfrak{C}_\alpha$. In each such class we choose a particular member operator, say A_α in $\mathfrak{C}_\alpha$. Then $\mathfrak{C}_\alpha = A_\alpha + \mathfrak{P}(A_\alpha)$, where $\mathfrak{P}(A_\alpha)$ is the class of linear operators described in Definition 13.3.5. Introducing a metric in $\mathfrak{C}_\alpha$ given by

$$d_\alpha(A_1, A_2) \equiv \int_0^1 \| (A_1 - A_2)T(\xi; A_\alpha) \|_{A_\alpha}\, d\xi, \tag{15.5.1}$$

$\mathfrak{C}_\alpha$ becomes a complete metric space (Theorems 13.5.4 and 13.5.6).

The geometry of $\mathfrak{C}_\alpha$ can be brought into a more familiar setting if we arbitrarily assign the role of origin to A_α and consider difference vectors rather than points. That is, we map $A \in \mathfrak{C}_\alpha \rightarrow B \equiv A - A_\alpha \in \mathfrak{P}(A_\alpha)$; this mapping defines a one-to-one correspondence between $\mathfrak{C}_\alpha$ and $\mathfrak{P}(A_\alpha)$. Now $\mathfrak{P}(A_\alpha)$ is a complex linear system and the above mapping becomes an isometry when we define the norm

$$\| B \| = \int_0^1 \| BT(\xi; A_\alpha) \|_{A_\alpha} d\xi \tag{15.5.2}$$

for $\mathfrak{P}(A_\alpha)$. Consequently $\mathfrak{P}(A_\alpha)$ is a (B)-space with this norm. We may now take over the theory of analyticity already developed in paragraph 3.3 for functions on one (B)-space to another. *A function $f(A)$ on $\mathfrak{C}_\alpha$ to $\mathfrak{E}(\mathfrak{X})$ will be said to be analytic if and only if $f(A_\alpha + B)$ is analytic in the usual sense relative to the argument $B \in \mathfrak{P}(A_\alpha)$.* A different choice of origin, say $A'_\alpha \in \mathfrak{C}_\alpha$, has the following effect. We now have $A \rightarrow B' \in \mathfrak{P}(A'_\alpha)$ where $B' = A - A'_\alpha = (A - A_\alpha) - (A'_\alpha - A_\alpha) = B - B_0$, B and $B_0 \in \mathfrak{P}(A_\alpha)$. Since $\mathfrak{P}(A_\alpha) = \mathfrak{P}(A'_\alpha)$ (Theorem 13.5.2), this amounts to a translation in $\mathfrak{P}(A_\alpha)$ and leaves invariant the notion of (G)-differentiability. On the other hand a new origin also introduces a new norm in $P(A_\alpha)$, namely

$$\| B \|' = \int_0^1 \| B'T(\xi; A'_\alpha) \|_{A'_\alpha} d\xi.$$

However Theorem 13.5.7 shows that the two norms define equivalent topologies in $\mathfrak{P}(A_\alpha)$ so that the notion of local boundedness is also unaffected by the change. It follows that the concept of analyticity is independent of our particular choice of origin in $\mathfrak{C}_\alpha$.

THEOREM 15.5.1. *$T(\xi; A)$ is locally analytic in $\mathfrak{A}_1$ for each $\xi \geqq 0$.*

PROOF. The assertion is obviously valid for $\xi = 0$. According to Theorem 13.5.8, $T(\xi; A)$ is continuous and *a fortiori* locally bounded as a function of A for each $\xi > 0$. To prove (G)-differentiability we write the expansion (13.4.8) as

$$T(\xi; A + \zeta B) = \sum_{n=0}^{\infty} \zeta^n S_n(\xi), \qquad A \in \mathfrak{C}_\alpha,\ B \in \mathfrak{P}(A_\alpha),$$

where $S_0(\xi) = T(\xi; A)$ and $S_n(\xi)x = \int_0^\xi T(\xi - \sigma; A)\tilde{B}S_{n-1}(\sigma)x\, d\sigma$. It follows from Theorem 13.4.1 that this series converges absolutely for all ζ. Consequently $T(\xi; A + \zeta B)$ is holomorphic in ζ by Theorem 3.11.4 and therefore (G)-differentiable in $\mathfrak{C}_\alpha$. Local boundedness and (G)-differentiability suffice to establish analyticity in the sense of Definition 3.17.2.

We next prove analyticity for the operator-valued extension of an arbitrary element of $S(\varphi)$. As before, we limit the domain set to operators majorized by φ; however, we also require the domain to be an open subset of $\mathfrak{A}_1$.

DEFINITION 15.5.1. *We define* $\mathfrak{A}_1(\varphi) = \text{Int}\,[\mathfrak{A}(\varphi) \cap \mathfrak{A}_1]$.

THEOREM 15.5.2. *Let $a \in S(\varphi)$. Then $\Psi(a; A)$ is locally analytic in $\mathfrak{A}_1(\varphi)$.*

PROOF. If $A \in \mathfrak{A}_1(\varphi)$ then $A \leqq \varphi$ and hence $\| \Psi(a; A) \| \leqq \int_0^\infty \varphi(\xi)\, d\,|\,a\,|$.

Thus $\Psi(a; A)$ is certainly locally bounded in $\mathfrak{A}_1(\varphi)$. On the other hand, let $A_0 \in \mathfrak{A}_1(\varphi)$ and $B \in \mathfrak{P}(A_0)$. $\mathfrak{A}_1(\varphi)$ being open, there exists a $\delta(B) > 0$ such that $A_0 + \zeta B \in \mathfrak{A}_1(\varphi)$ for all $|\zeta| < \delta(B)$. Thus we have $\| T(\xi; A_0 + \zeta B) \| \leqq \varphi(\xi)$ and $T(\xi; A_0 + \zeta B)$ is holomorphic in ζ, $|\zeta| < \delta(B)$, for each $\xi \geqq 0$. Since $\Psi(a; A_0 + \zeta B)x = \int_0^\infty T(\xi; A_0 + \zeta B)x \, da$, it now follows as in Theorem 3.11.2 that $\Psi(a; A_0 + \zeta B)x$ is holomorphic in ζ for $|\zeta| < \delta(B)$. Thus $\Psi(a; A)$ is (G)-differentiable at A_0. This together with local boundedness proves that $\Psi(a; A)$ is locally analytic in $\mathfrak{A}_1(\varphi)$.

In order to extend the previous result to the function algebra $\mathcal{S}(\varphi)$, we prove the following three lemmas.

LEMMA 15.5.1. *$\omega_0(A)$ is upper semi-continuous as a function of A in $\mathfrak{A}_1$.*

PROOF. Let $A_0 \in \mathfrak{A}_1$ be fixed. We may assume without loss of generality that A_0 is the defining operator for the metric in the class $\mathfrak{C}_\alpha = \mathfrak{C}(A_0)$; in the notation of (13.5.13) we take $d_\alpha(A_1, A_2) = d_{A_0}(A_1, A_2)$. For any fixed $\omega > \omega_0(A_0)$, an equivalent metric is given by

$$\delta_{A_0}(A_1, A_2) = \int_0^\infty e^{-\omega\xi} \| (A_1 - A_2)T(\xi; A_0) \|_{A_0} d\xi, \tag{15.5.3}$$

according to Theorem 13.5.5. Now the expansion (13.4.8) is valid for each $B \in \mathfrak{P}(A_0)$ and Lemma 13.4.3 implies that $\int_0^\infty e^{-\omega\xi} \| T(\xi; A_0 + B) \| d\xi < \infty$ when $\delta_{A_0}(A_0, A_0 + B) < 1$. Since $\| T(\xi; A_0 + B) \|$ is submultiplicative in ξ, it follows that $\omega_0(A_0 + B) < \omega$ for all B with $\delta_{A_0}(A_0, A_0 + B) < 1$; and since $\omega > \omega_0(A_0)$ is arbitrary, this establishes the upper semicontinuity of $\omega_0(A)$.

LEMMA 15.5.2. *$\omega_0(A) < \omega_0$ for each $A \in \mathfrak{A}_1(\varphi)$.*

PROOF. If $A_0 \in \mathfrak{A}_1$, then it is clear that the restriction of I to $\mathfrak{D}(A_0)$ belongs to $\mathfrak{P}(A_0)$ and hence $A_0 + \beta I \in \mathfrak{C}_\alpha = \mathfrak{C}(A_0)$. Now $d_\alpha(A_0, A_0 + \beta I) = |\beta| \int_0^1 \| T(\xi; A_\alpha) \| d\xi$. Thus if $A_0 \in \mathfrak{A}_1(\varphi)$, then $A_0 + \beta I \in \mathfrak{A}_1(\varphi)$ for $|\beta|$ sufficiently small since $\mathfrak{A}_1(\varphi)$ is open. On the other hand, $T(\xi; A_0 + \beta I) = e^{\beta\xi}T(\xi; A_0)$. Consequently $\omega_0(A_0 + \beta I) = \omega_0(A_0) + \Re(\beta) \leqq \omega_0$ for $|\beta|$ sufficiently small and therefore $\omega_0(A_0) < \omega_0$.

LEMMA 15.5.3. *Let $A_0 \in \mathfrak{A}_1$. To each $\gamma > \omega_0(A_0)$ there is a neighborhood N of A_0 such that* (i) $\sup [\omega_0(A); A \in N] < \gamma$; (ii) $\sup [\| R(\lambda; A) \|; A \in N, \Re(\lambda) \geqq \gamma] < \infty$; *and* (iii) *$R(\lambda; A)$ is analytic in $A \in N$ for all λ with $\Re(\lambda) \geqq \gamma$.*

PROOF. It follows directly from Lemma 15.5.1 that there exists a neighborhood N_1 of A_0 such that $\sup [\omega_0(A); A \in N_1] < \gamma$. In proving (ii) we may assume without loss of generality that the metric in $\mathfrak{C}_\alpha \equiv \mathfrak{C}(A_0)$ is given by (15.5.3) with $\omega = \gamma$. For $B \in \mathfrak{P}(A_0)$, Lemma 13.3.4 clearly implies that $\| BR(\lambda; A_0) \| \leqq \delta_{A_0}(A_0, A_0 + B)$ for all λ with $\Re(\lambda) \geqq \gamma$. Thus for $\delta_{A_0}(A_0, A_0 + B) < \frac{1}{2}$ and $\Re(\lambda) \geqq \gamma$, Theorem 5.10.4 gives

$$R(\lambda; A_0 + B) = \sum_{n=0}^\infty R(\lambda; A_0)[BR(\lambda; A_0)]^n \tag{15.5.4}$$

so that

$$\| R(\lambda; A_0 + B) \| \leqq 2 \| R(\lambda; A_0) \| \leqq 2 \int_0^\infty e^{-\gamma\xi} \| T(\xi; A_0) \| \, d\xi.$$

This establishes (ii) in the neighborhood $N_2 = [A; \delta_{A_0}(A, A_0) < \frac{1}{2}]$. We now set $N = N_1 \cap N_2$. For an arbitrary operator $A' \in N$, the above argument shows that there exists a neighborhood N' of A' in which the expansion (15.5.4) with A_0 replaced by A' is valid. It follows from this that $R(\lambda; A)$ is (G)-differentiable at A' for $\mathfrak{R}(\lambda) \geqq \gamma$. This together with the boundedness in N implies (iii).

We now have

THEOREM 15.5.3. *Let $\varphi(\xi)$ be a submultiplicative function of type $\omega_0 > -\infty$ and suppose α_0 is such that $\mathfrak{R}(\alpha_0) > \omega_0$. Let $\mathcal{S}(\varphi)$ be the algebra of set functions described in Definition* 15.3.1 *with a sequence topology: $a_n \to a$ denoting that the sequence $\{a_n\}$ converges to a in the weak* topology of $S(e^{\omega_0\xi})$. Further let $\mathfrak{B}_1(\varphi)$ be the complex algebra of functions $f(A)$ locally analytic on $\mathfrak{A}_1(\varphi)$ and having values in $\mathfrak{E}(\mathfrak{X})$, the arithmetic operations being defined as in $\mathfrak{E}(\mathfrak{X})$.*

There exists an isomorphic mapping: $a \to \Psi(a; A)$ of $\mathcal{S}(\varphi)$ onto a subalgebra of $\mathfrak{B}_1(\varphi)$ such that (i) $e_0 \to I$; (ii) $r_{\alpha_0} \to R(\alpha_0; A)$; *and* (iii) *$a_n \to a$ implies that $\psi(a_n; A)x \to \Psi(a; A)x$ for each $x \in \mathfrak{X}_1(A)$. This mapping is unique and is defined by* (15.4.1) *or, alternately, by* (15.4.2) *for each $x \in \mathfrak{X}_1(A)$.*

PROOF. We have already remarked that $\omega_1(A) = \omega_0(A)$ for all $A \in \mathfrak{A}_1$. Lemma 15.5.2 therefore implies that $\mathfrak{A}_1(\varphi) \subset \mathfrak{A}_0(\varphi)$; and incidentally also that $\mathfrak{A}_1(\varphi) = \text{Int}\,[\mathfrak{A}_0(\varphi) \cap \mathfrak{A}_1]$. It will hence be easily seen that Theorem 15.4.2 remains valid if we replace $\mathfrak{A}_0(\varphi)$ by $\mathfrak{A}_1(\varphi)$. Indeed, the only part of this assertion which does not follow directly from Theorem 15.4.2 is the fact that the mapping $a \to \Psi(a; A)$ is an isomorphism rather than a homomorphism, and for this the argument given in the proof of Theorem 15.2.1 suffices. It remains to prove that $\Psi(a; A)$ is locally analytic in $\mathfrak{A}_1(\varphi)$ for each $a \in \mathcal{S}(\varphi)$. This result has already been verified for $a \in S(\varphi)$ in Theorem 15.5.2 and for r_{α_0}, $\mathfrak{R}(\alpha_0) > \omega_0$, in Lemma 15.5.3. Further for $a \in S(\varphi)$, $\Psi(a{*}r_{\alpha_0}; A) = \Psi(a; A)\Psi(r_{\alpha_0}; A)$ is likewise locally analytic in $\mathfrak{A}_1(\varphi)$ since it is the product of two operator functions with this property. Finally suppose $a \in Q(\omega_0)$. We then have the representation

$$\Psi(a; A) = \frac{1}{2\pi i} \int_{\gamma - i\infty}^{\gamma + i\infty} \psi(a; \zeta) R(\zeta; A) \, d\zeta, \qquad \omega_0(A) < \gamma \leqq \omega_0. \tag{15.5.5}$$

Given $A_0 \in \mathfrak{A}_1(\varphi)$, let γ be chosen so that $\omega_0(A_0) < \gamma \leqq \omega_0$. Applying Lemma 15.5.3, we see that there is a neighborhood N of A_0 where (i) $\sup\,[\omega_0(A); A \in N] < \gamma$; (ii) $\sup\,[\, \| R(\lambda; A) \|; A \in N, \mathfrak{R}(\lambda) \geqq \gamma] < \infty$, and finally (iii) $R(\lambda; A)$ is analytic in $A \in N$ for all λ with $\mathfrak{R}(\lambda) \geqq \gamma$. It follows from (i) that $\Psi(a; A)$ can be represented by (15.5.5) with fixed γ for all $A \in N$. Condition (ii) shows that $\| \Psi(a; A) \|$ is bounded in N, and (iii) implies that $R(\lambda; A_0 + \zeta B)$ is holomorphic in ζ for each $B \in \mathfrak{P}(A_0)$ and $|\zeta|$ sufficiently small. Making use of the representation (15.5.5) and arguing as in Theorem 3.11.2 we see that $\Psi(a; A)$

is (G)-differentiable at A_0. This together with the local boundedness proves that $\Psi(a; A)$ is locally analytic in $\mathfrak{A}_1(\varphi)$. Since the sum of locally analytic operator functions is again locally analytic and since each $a \in \mathcal{S}(\varphi)$ is the sum of terms of the above kind, the proof is now complete.

15.6. The operator class $\mathfrak{A}_1(\varphi)$. In applying the previous theory it is desirable to know when a given operator $A \in \mathfrak{A}_1$ belongs to a class $\mathfrak{A}_1(\varphi)$. This is the subject of the present section.

THEOREM 15.6.1. *If A_0 is the infinitesimal generator of a semi-group of class (C_0) and*

$$\| T(\xi; A_0) \| \leqq (1 - \epsilon)e^{-\epsilon\xi}\varphi(\xi), \qquad \xi > 0, \tag{15.6.1}$$

for some $\epsilon > 0$, then $A_0 \in \mathfrak{A}_1(\varphi)$.

PROOF. We may suppose without loss of generality that the metric in $\mathfrak{C}_a \equiv \mathfrak{C}(A_0)$ is given by (15.5.3). Then for $B \in \mathfrak{P}(A_0)$, $T(\xi; A_0 + B)$ can be represented by the expansion (13.4.8) and

$$\| T(\xi; A_0 + B) \| \leqq \sum_{n=0}^{\infty} (\varphi_0 * \psi^{n*})(\xi)$$

where $\varphi_0(\xi) \equiv \| T(\xi; A_0) \|$ and $\psi(\xi) \equiv \| BT(\xi; A_0) \|_{A_0}$. But, by assumption, $\| T(\xi; A_0) \| \leqq e^{-\epsilon\xi}\varphi(\xi)$ so that $\omega_0(A_0) < \omega_0$. Further since $\| T(\xi; A_0) \|$ is bounded near $\xi = 0$, there exists an $M > 0$ such that $\varphi_0(\xi) \leq Me^{\omega_0\xi}$. A straightforward calculation shows that $(\varphi_0 * \psi^{n*})(\xi) \leq Me^{\omega_0\xi}\delta^n$, where $\delta \equiv \delta_{A_0}(A_0, A_0 + B)$. Consequently

$$\sum_{n=1}^{\infty} (\varphi_0 * \psi^{n*})(\xi) \leq M\delta(1 - \delta)^{-1}e^{\omega_0\xi}, \qquad \delta < 1.$$

Now $e^{\omega_0\xi} \leq \varphi(\xi)$ and by assumption $\varphi_0(\xi) \leq (1 - \epsilon)\varphi(\xi)$. It follows that

$$\| T(\xi; A_0 + B) \| \leqq (1 - \epsilon)\varphi(\xi) + \epsilon\varphi(\xi) = \varphi(\xi), \qquad \xi > 0,$$

if $\delta_{A_0}(A_0, A_0 + B) \leqq \epsilon(M + \epsilon)^{-1}$. Since all of the operators belonging to this sphere are contained in $\mathfrak{A}(\varphi) \cap \mathfrak{A}_1$, we see that $A_0 \in \text{Int}\,[\mathfrak{A}(\varphi) \cap \mathfrak{A}_1] \equiv \mathfrak{A}_1(\varphi)$.

The above theorem is reasonably sharp; for, as the proof of Lemma 15.5.2 shows, corresponding to each $A \in \mathfrak{A}_1(\varphi)$ there is an $\epsilon > 0$ such that $\| T(\xi; A) \| \leqq e^{-\epsilon\xi}\varphi(\xi)$. Our result for the general operator $A \in \mathfrak{A}_1$ requires the following lemmas which are of some interest in themselves.

LEMMA 15.6.1. *Suppose $\varphi_0(\xi)$ and $\psi(\xi)$ are non-negative measurable functions with the properties:*

(i) *$\varphi_0(\xi)$ is submultiplicative: $\varphi_0(\xi + \eta) \leqq \varphi_0(\xi)\varphi_0(\eta)$ for $\xi, \eta > 0$;*

(ii) *$\psi(\xi + \eta) \leqq \varphi_0(\xi)\psi(\eta)$ for $\xi, \eta > 0$;*

(iii) *$\int_0^1 [\varphi_0(\xi) + \psi(\xi)]\, d\xi < \infty$.*

Then

$$\theta(\xi;\psi) \equiv \sum_{n=0}^{\infty} (\varphi_0 * \psi^{n*})(\xi) \tag{15.6.2}$$

is non-negative, measurable, and submultiplicative.

PROOF. It is clear that $\theta(\xi;\psi)$ is non-negative and measurable. In order to show that $\theta(\xi;\psi)$ is submultiplicative we set $\varphi_n(\xi) = (\varphi_0 * \psi^{n*})(\xi)$ and show that

$$\varphi_n(\xi + \eta) \leqq \sum_{k=0}^{n} \varphi_k(\xi)\varphi_{n-k}(\eta), \qquad \xi, \eta > 0. \tag{15.6.3}$$

This inequality is obviously valid for $n = 0$ by (i). We proceed by induction, assuming (15.6.3) to hold for the case n. Then

$$\begin{aligned}
\varphi_{n+1}(\xi + \eta) &= \int_0^{\xi+\eta} \psi(\xi + \eta - \sigma)\varphi_n(\sigma)\, d\sigma \\
&= \int_0^{\xi} \psi(\xi + \eta - \sigma)\varphi_n(\sigma)\, d\sigma + \int_0^{\eta} \psi(\eta - \sigma)\varphi_n(\sigma + \xi)\, d\sigma \\
&\leqq \varphi_0(\eta) \int_0^{\xi} \psi(\xi - \sigma)\varphi_n(\sigma)\, d\sigma + \int_0^{\eta} \psi(\eta - \sigma) \left[\sum_{k=0}^{n} \varphi_k(\sigma)\varphi_{n-k}(\xi)\right] d\sigma \\
&= \varphi_0(\eta)\varphi_{n+1}(\xi) + \sum_{k=0}^{n} \varphi_{k+1}(\eta)\varphi_{n-k}(\xi),
\end{aligned}$$

which is (15.6.3) with n replaced by $n + 1$. The submultiplicativity of $\theta(\xi;\psi)$ is now immediate. According to Lemma 13.4.3, the series (15.6.2) converges absolutely for each $\xi > 0$. Hence applying (15.6.3) and interchanging orders of summation, we obtain

$$\begin{aligned}
\theta(\xi + \eta;\psi) &= \sum_{n=0}^{\infty} \varphi_n(\xi + \eta) \leqq \sum_{n=0}^{\infty} \left[\sum_{k=0}^{n} \varphi_k(\xi)\varphi_{n-k}(\eta)\right] \\
&= \left[\sum_{n=0}^{\infty} \varphi_n(\xi)\right]\left[\sum_{n=0}^{\infty} \varphi_n(\eta)\right] = \theta(\xi;\psi)\theta(\eta;\psi).
\end{aligned}$$

LEMMA 15.6.2. *Suppose $\varphi_0(\xi)$ and $\psi(\xi)$ satisfy the conditions of Lemma 15.6.1 and choose $\omega > \inf_{\xi>0} \xi^{-1} \log \varphi_0(\xi)$. Then*

$$\theta_\alpha(\xi) \equiv \sup\left[\theta(\xi;\psi);\ \int_0^{\infty} e^{-\omega\xi}\psi(\xi)\, d\xi \leqq \alpha\right] \tag{15.6.4}$$

is finite-valued, measurable, and submultiplicative on $(0, \infty)$ for each α with $0 < \alpha < 1$.

PROOF. It is clear that

$$\int_0^{\infty} e^{-\omega\xi}\theta(\xi;\psi)\, d\xi = \sum_{n=0}^{\infty} \int_0^{\infty} e^{-\omega\xi}\varphi_0(\xi)\, d\xi \left[\int_0^{\infty} e^{-\omega\xi}\psi(\xi)\, d\xi\right]^n.$$

Hence for each α, $0 < \alpha < 1$, there exists a positive constant M_α such that $\int_0^\infty e^{-\omega\xi}\theta(\xi; \psi)\, d\xi \leqq M_\alpha$ when $\int_0^\infty e^{-\omega\xi}\psi(\xi)\, d\xi \leqq \alpha$. Since $\theta(\xi; \psi)$ is submultiplicative, Theorem 7.4.4 applies and therefore $\theta(\xi; \psi) \leqq (M_\alpha/\xi)^2 e^{\omega\xi}$. It follows that $\theta_\alpha(\xi)$ is finite-valued. The function space $L(e^{-\omega\xi})$ being separable, we can choose a dense denumerable subset $\{\psi_n\}$ of the set $[\psi; \int_0^\infty e^{-\omega\xi}\psi(\xi)\, d\xi \leqq \alpha]$. But then if $\lim_{k\to\infty} \psi_{n_k} = \psi_0$ in the norm of $L(e^{-\omega\xi})$, it is easily seen that $\lim_{k\to\infty} \psi_{n_k}^{m*} = \psi_0^{m*}$ in norm. Now

$$(\varphi_0 * \psi_{n_k}^{m*})(\xi) = \int_0^{\xi/2} \varphi_0(\xi - \sigma)\psi_{n_k}^{m*}(\sigma)\, d\sigma + \int_0^{\xi/2} \psi_{n_k}^{m*}(\xi - \sigma)\varphi_0(\sigma)\, d\sigma.$$

The estimate of Theorem 7.4.4 implies that the first integrand on the right converges in mean as $k \to \infty$, whereas the estimate of Lemma 13.4.2 implies that the same is true of the second integrand. Therefore $\lim_{k\to\infty} (\varphi_0 * \psi_{n_k}^{m*})(\xi) = (\varphi_0 * \psi_0^{m*})(\xi)$ for each $m \geqq 0$ and $\xi > 0$. Further it can be shown as in Lemma 13.4.3 that the series $\sum_{m=0}^\infty (\varphi_0 * \psi_{n_k}^{m*})(\xi)$ converges uniformly in k for each $\xi > 0$ (the ω_1 of Lemma 13.4.3 can be chosen so that $\int_0^\infty e^{-\omega_1\xi}[\varphi_0(\xi) + \psi_{n_k}(\xi)]\, d\xi \leqq \frac{1}{16}$ for all k). Consequently $\lim_{k\to\infty}\theta(\xi; \psi_{n_k}) = \theta(\xi; \psi_0)$ for each $\xi > 0$. It follows that

$$\theta_\alpha(\xi) = \sup_n [\theta(\xi; \psi_n)],$$

so that $\theta_\alpha(\xi)$ is measurable. Finally the submultiplicativity of $\theta_\alpha(\xi)$ is implied by Theorem 7.2.2.

THEOREM 15.6.2. *Let* $A_0 \in \mathfrak{A}_1$, *set* $\varphi_0(\xi) = \| T(\xi; A_0) \|$, *and define* θ_α *as in Lemma* 15.6.2. *Then* $A_0 \in \mathfrak{A}_1(\varphi)$ *if* $\theta_\alpha(\xi) \leqq \varphi(\xi)$ *for some* α, $0 < \alpha < 1$, *and all* $\xi > 0$.

PROOF. Again, without loss of generality we may assume that the metric in $\mathfrak{C}(A_0)$ is given by equation (15.5.3). It is clear that $A_0 \in \mathfrak{A}_1(\varphi)$ if the sphere $S_\alpha \equiv [A; \delta_{A_0}(A, A_0) < \alpha]$ is contained in $\mathfrak{A}(\varphi) \cap \mathfrak{A}_1$. Now $S_\alpha \subset \mathfrak{C}(A_0) \subset \mathfrak{A}_1$. Hence if $A \in S_\alpha$, then $B = A - A_0 \in \mathfrak{P}(A_0)$ and $T(\xi; A) = T(\xi; A_0 + B)$ can be represented by the expansion (13.4.8). Consequently

$$\| T(\xi; A) \| \leqq \sum_{n=0}^\infty (\varphi_0 * \psi^{n*})(\xi) \equiv \theta(\xi; \psi),$$

where again $\varphi_0(\xi) \equiv \| T(\xi; A_0) \|$ and $\psi(\xi) = \| BT(\xi; A_0) \|_{A_0}$. Also, $\delta_{A_0}(A_0, A) = \int_0^\infty e^{-\omega\xi}\psi(\xi)\, d\xi < \alpha$. Thus when $\theta_\alpha(\xi) \leqq \varphi(\xi)$, we see that $\| T(\xi; A) \| \leqq \theta(\xi; \psi) \leqq \theta_\alpha(\xi) \leqq \varphi(\xi)$ for all $A \in S_\alpha$. It follows that $S_\alpha \subset \mathfrak{A}(\varphi) \cap \mathfrak{A}_1$.

We have, incidentally, given a method for constructing a φ such that $A_0 \in \mathfrak{A}_1(\varphi)$. Such a φ is obtained by setting $\varphi(\xi) \equiv \theta_\alpha(\xi)$ for some α, $0 < \alpha < 1$.

CHAPTER XVI

SPECTRAL THEORY

16.1. Orientation. Having developed an operational calculus for semi-groups of linear operators, we now consider the relation between the spectrum of $\Psi(a; A)$ and that of A, where either $a \in S(\varphi)$ and $A \leqq \varphi$ or $a \in \mathcal{S}(\varphi)$ and $A \prec \varphi$.

In this general setting the following mapping theorem holds: $\overline{\psi(a; \sigma(A))} \subset \sigma[\Psi(a; A)]$, where the inclusion can be proper. If the set function of the infinitesimal generator is sufficiently specialized a more precise mapping theorem can be proved. For instance, if $a \in L(\varphi)$ and $A \leqq \varphi$ is otherwise arbitrary, then $\psi(a; \sigma(A)) \cup 0 = \sigma[\Psi(a; A)]$. Likewise if $a \in S(\varphi)$ and $A \leqq \varphi$ is the infinitesimal generator of a semi-group continuous in the uniform operator topology for $\xi \geqq \gamma > 0$, then $\psi(a; \sigma(A)) \cup a([0]) = \sigma[\Psi(a; A)]$. In these cases if we further assume that $A \prec \varphi$, then even the fine structure of the spectrum is preserved under the mapping. When $a = e_\xi$ so that $\Psi(a; A) = T(\xi; A)$, and A is the infinitesimal generator of an arbitrary semi-group of class (A), then the correspondence for the point and residual spectra is good. However the continuous spectrum of $T(\xi; A)$ comes in part from the continuous spectrum of A and in part from the limit points of the set $\exp [\xi\sigma(A)]$ which are not otherwise accounted for; it may also contain points which have no relation with the points of $\sigma(A)$.

There are two paragraphs: *Spectral Mapping Theorems*, and *Fine Structure Theorems*. The development of the first paragraph is based on the Gelfand representation theory and follows the work of R. S. Phillips [6]. The methods of the second paragraph are for the most part due to E. Hille [10, 13].

References. Hille [10, 13], Phillips [6, 11].

1. Spectral Mapping Theorems

16.2. The commutant of the commutant. Throughout this chapter we shall deal with a fixed semi-group $\mathfrak{S} \equiv [T(\xi; A)]$ of class (A). This is a departure from the point of view taken in the previous chapter, where we dealt with a class of infinitesimal generators majorized by a fixed submultiplicative function φ and with the function algebra determined by φ. We shall now feel free to vary φ and with it the function algebra to suit our convenience, always with the proviso that φ

majorizes A. Moreover, A being fixed, we need no longer indicate the dependence on A in our notation; thus we shall write $T(\xi)$ for $T(\xi; A)$ and $\Psi(a)$ for $\Psi(a; A)$.

Our first objective is to obtain a subalgebra of $\mathfrak{E}(\mathfrak{X})$ suitable for our purposes. To this end we set

$$(16.2.1) \qquad \mathfrak{S} \equiv [T(\xi); \xi > 0] \quad \text{and} \quad \mathfrak{R} \equiv [R(\lambda; A); \lambda \in \rho(A)].$$

As in Definition 1.13.4, we denote the commutant of a set of operators $\mathfrak{E} \subset \mathfrak{E}(\mathfrak{X})$ by $\mathfrak{E}^c$. According to Theorem 1.13.1, the commutant $\mathfrak{E}^c$ is a subalgebra of $\mathfrak{E}(\mathfrak{X})$ containing the unit I. An operator $B \in \mathfrak{E}^c$ is regular in $\mathfrak{E}^c$ if and only if it is regular in $\mathfrak{E}(\mathfrak{X})$. Consequently the spectrum of an operator $B \in \mathfrak{E}^c$ relative to $\mathfrak{E}^c$ coincides with the spectrum of B relative to $\mathfrak{E}(\mathfrak{X})$, namely $\sigma(B)$. Moreover, $\mathfrak{E}(\mathfrak{X})$ being a topological algebra relative to the strong (and weak) operator topology (see section 2.15), it follows from Theorem 1.14.1 that $\mathfrak{E}^c$ is strongly (and weakly) closed in $\mathfrak{E}(\mathfrak{X})$. Finally we recall that $\mathfrak{E}$ and $\mathfrak{E}^{cc}$ are abelian together.

THEOREM 16.2.1. *$\mathfrak{S}^c = \mathfrak{R}^c$ and $\mathfrak{S}^{cc} = \mathfrak{R}^{cc}$.*

PROOF. It suffices to show that $\mathfrak{S}^c = \mathfrak{R}^c$. Suppose, therefore, that $B \in \mathfrak{S}^c$ and let $x \in \mathfrak{X}_1 = [x; \lim_{\xi\to 0+} T(\xi)x = x]$. Then $T(\xi)Bx = BT(\xi)x \to Bx$ as $\xi \to 0+$ so that $Bx \in \mathfrak{X}_1$. On the other hand, if $x \in \mathfrak{X}_1$ and $\mathfrak{R}(\lambda) > \omega_1(A)$, then we have by Lemma 15.4.1 that $R(\lambda; A)x = \int_0^\infty e^{-\lambda\xi}T(\xi)x\,d\xi$. Consequently for $x \in \mathfrak{X}_1$,

$$\begin{aligned} R(\lambda; A)Bx &= \int_0^\infty e^{-\lambda\xi}T(\xi)Bx\,d\xi = B\left[\int_0^\infty e^{-\lambda\xi}T(\xi)x\,d\xi\right] \\ &= BR(\lambda; A)x, \qquad\qquad \mathfrak{R}(\lambda) > \omega_1(A). \end{aligned}$$

Finally, since $\mathfrak{X}_1$ is dense in $\mathfrak{X}$, we see that $R(\lambda; A)$ and B commute for $\mathfrak{R}(\lambda) > \omega_1(A)$. It is now easy to show that $ABx = BAx$ for all $x \in \mathfrak{D}(A)$, from which it follows that $BR(\lambda; A) = R(\lambda; A)B$ for all $\lambda \in \rho(A)$; that is, $B \in \mathfrak{R}^c$.

Conversely, suppose $B \in \mathfrak{R}^c$ and let $x \in \mathfrak{D}(A^2)$. Such an x will lie in $\mathfrak{X}_1$ by Theorem 11.5.3. Further $B[R(\lambda; A)]^2 = [R(\lambda; A)]^2B$ implies that $B[\mathfrak{D}(A^2)] \subset \mathfrak{D}(A^2)$ so that $Bx \in \mathfrak{X}_1$ if $x \in \mathfrak{D}(A^2)$. Applying the inversion formula of Theorem 6.3.2 (with $\gamma > \omega_1(A)$), we see that for $x \in \mathfrak{D}(A^2)$

$$\begin{aligned} T(\xi)Bx &= \frac{1}{2\pi i}(C, 1)\text{-}\int_{\gamma-i\infty}^{\gamma+i\infty} e^{\zeta\xi}R(\zeta; A)Bx\,d\zeta \\ &= B\left[\frac{1}{2\pi i}(C, 1)\text{-}\int_{\gamma-i\infty}^{\gamma+i\infty} e^{\zeta\xi}R(\zeta; A)x\,d\zeta\right] = BT(\xi)x, \qquad \xi > 0. \end{aligned}$$

Since $\mathfrak{D}(A^2)$ is dense in $\mathfrak{X}$, it follows that B commutes with $T(\xi)$ for each $\xi > 0$, that is $B \in \mathfrak{S}^c$.

In the course of the above proof we have incidentally established the

COROLLARY. *Suppose $B \in \mathfrak{S}^c$, then $Bx \in \mathfrak{X}_1$ if $x \in \mathfrak{X}_1$ and $Bx \in \mathfrak{D}(A^2)$ if $x \in \mathfrak{D}(A^2)$.*

Hereafter we shall denote the subalgebra $\mathfrak{S}^{cc} = \mathfrak{R}^{cc}$ by $\mathfrak{B}$ and refer to $\mathfrak{B}$ as *the algebra associated with the semi-group* $\mathfrak{S}$. Summarizing the earlier remarks, we have

THEOREM 16.2.2. *$\mathfrak{B}$ is a strongly closed commutative sub-algebra of $\mathfrak{E}(\mathfrak{X})$ with unit I and the spectrum of any operator $B \in \mathfrak{B}$ relative to $\mathfrak{B}$ is $\sigma(B)$. Further $\mathfrak{B} \supset \mathfrak{S} \cup \mathfrak{R}$.*

We remark that any subalgebra of $\mathfrak{E}(\mathfrak{X})$ with the properties stated in Theorem 16.2.2 would be suitable for the purposes of this chapter.

THEOREM 16.2.3. *If either $a \in S(\varphi)$ and $A \leqq \varphi$ or $a \in \mathcal{S}(\varphi)$ and $A \prec \varphi$, then $\Psi(a) \in \mathfrak{B}$.*

PROOF. It suffices to prove that $\Psi(a)$ commutes with each $B \in \mathfrak{S}^c = \mathfrak{R}^c$. Let $x \in \mathfrak{X}_1$, then as above $Bx \in \mathfrak{X}_1$ and we obtain $B\Psi(a)x = \Psi(a)Bx$ directly from Definition 15.2.2 or Definition 15.4.2 according as $a \in S(\varphi)$ and $A \leqq \varphi$, or $a \in \mathcal{S}(\varphi)$ and $A \prec \varphi$, respectively. Since $\mathfrak{X}_1$ is dense in $\mathfrak{X}$, this implies the statement of the theorem. An alternate argument is obtained by making use of the strong closure of $\mathfrak{B}$. For if $a \in S(\varphi)$ and $A \leqq \varphi$, then (15.2.3) shows that $\Psi(a)$ is in the strong closure of $\mathfrak{B}$; if $a \in Q(\omega_0)$ and $A \prec \varphi$, then (15.4.2) shows that $\Psi(a)$ is in the uniform closure of $\mathfrak{B}$; and we have assumed that $R(\lambda; A) \in \mathfrak{B}$ for $\mathfrak{R}(\lambda) > \omega_1(A)$. Finally if $a \in \mathcal{S}(\varphi)$, then $\Psi(a)$ can be formed by sums and products of the above elements and hence $\Psi(a) \in \mathfrak{B}$.

16.3. Representation theory. Many of the spectral mapping theorems for the semi-group $\mathfrak{S}$ can now be obtained by applying the Gelfand representation theory to the algebra $\mathfrak{B}$ associated with $\mathfrak{S}$. This approach is similar to that employed in Theorem 5.12.1.

Let $\mathfrak{M} = [\mathfrak{m}]$ be the set of maximal ideals in $\mathfrak{B}$, topologized as in Theorem 4.15.2 and thereby forming a compact Hausdorff space. The mapping $B \to B(\mathfrak{m})$ is a continuous homomorphism of $\mathfrak{B}$ into $C(\mathfrak{M})$. According to Theorem 4.15.1, the spectrum of B relative to $\mathfrak{B}$ is precisely $B(\mathfrak{M}) \equiv [B(\mathfrak{m}); \mathfrak{m} \in \mathfrak{M}]$ and hence by Theorem 16.2.2 the spectrum of B relative to $\mathfrak{E}(\mathfrak{X})$, namely $\sigma(B)$, is also $B(\mathfrak{M})$.

In general the infinitesimal generator A will be unbounded and will not belong to $\mathfrak{B}$. In this case it is not possible to deal with A directly and the relevant information about $\sigma(A)$ must be obtained from $R(\lambda; A)$, which is, so to speak, the shadow of A in $\mathfrak{B}$. Since $\mathfrak{B}$ was defined to contain $\mathfrak{R}$, the following means of characterizing $\sigma(A)$ is provided by Theorems 5.8.4 and 5.8.5.

THEOREM 16.3.1. *Let $\mathfrak{B}$ be the (B)-algebra associated with the semi-group $\mathfrak{S}$ and let $\mathfrak{M}$ be the set of maximal ideals in $\mathfrak{B}$. Then $\mathfrak{M}$ splits up into two disjoint sets $\mathfrak{W}$, $\mathfrak{U}$ with $\mathfrak{M} = \mathfrak{W} \cup \mathfrak{U}$, and there exists a numerically-valued function $\alpha(\mathfrak{m})$ defined and continuous on $\mathfrak{W}$ such that*

$$(16.3.1) \qquad R(\lambda; A)(\mathfrak{m}) = [\lambda - \alpha(\mathfrak{m})]^{-1}, \qquad \mathfrak{m} \in \mathfrak{W},$$

$$= 0, \qquad \mathfrak{m} \in \mathfrak{U},$$

for all $\lambda \in \rho(A)$; $\sigma(A) = \alpha(\mathfrak{W}) \equiv [\alpha(\mathfrak{m}); \mathfrak{m} \in \mathfrak{W}]$; *and* $\mathfrak{U}$ *is empty if and only if* $A \in \mathfrak{E}(\mathfrak{X})$. *If* $A \in \mathfrak{E}(\mathfrak{X})$, *then* $A \in \mathfrak{B}$ *and* $A(\mathfrak{m}) = \alpha(\mathfrak{m})$ *on* $\mathfrak{M}$.

The characterization of $\sigma[T(\xi)]$ is somewhat more complicated than that of $\sigma(A)$. Given an $\mathfrak{m} \in \mathfrak{M}$, it is clear that $\nu(\xi) \equiv T(\xi)(\mathfrak{m})$ is a numerically-valued representation of the semi-group of non-negative real numbers under addition; that is,

$$\nu(\xi + \eta) = \nu(\xi)\nu(\eta), \qquad \xi, \eta > 0, \tag{16.3.2}$$
$$\nu(0) = 1.$$

Since $| T(\xi)(\mathfrak{m}) | \leqq \| T(\xi) \|$, it follows as in section 4.17 that $\nu(\xi)$ is one of three types: (a) $\nu(\xi)$ is measurable and non-vanishing, in which case there exists a complex number α such that $\nu(\xi) = e^{\alpha\xi}$; (b) $\nu(\xi_0) = 0$ for some $\xi_0 > 0$, in which case $\nu(\xi) = 0$ for all $\xi > 0$; and (c) $\nu(\xi)$ is non-measurable, in which case there exists a real constant σ and a non-measurable character $\chi(\xi)$ such that $\nu(\xi) = e^{\sigma\xi}\chi(\xi)$. Incidentally, σ and $\Re(\alpha) = \xi^{-1} \log | \nu(\xi) | \leqq \inf_{\xi>0} \xi^{-1} \log \| T(\xi) \| = \omega_0(A)$. In view of the fact that $R(\lambda; A)$ is essentially the Laplace transform of $T(\xi)$, one might expect $\mathfrak{W}$ to coincide with the maximal ideals in group (a). However this reasoning is not valid since the linear multiplicative functionals do not commute with the operation of integration in the strong operator topology. Nevertheless the conjecture is roughly correct. More precisely, $\mathfrak{W}$ is contained in the maximal ideals of group (a). To understand the matter more fully we require the following lemmas.

Lemma 16.3.1. *If* $a \in Q(\omega_0)$ *and* $A \prec \varphi$, *then* $\Psi(a)(\mathfrak{m}) = \psi(a; \alpha(\mathfrak{m}))$ *for* $\mathfrak{m} \in \mathfrak{W}$ *and* $\Psi(a)(\mathfrak{m}) = 0$ *for* $\mathfrak{m} \in \mathfrak{U}$.

Proof. Lemma 15.4.1 implies that $\sigma(A)$ is contained in the half-plane $\Re(\lambda) \leqq \omega_1(A)$ so that $\Re[\alpha(\mathfrak{m})] \leqq \omega_1(A)$ for $\mathfrak{m} \in \mathfrak{W}$. Now if $a \in Q(\omega_0)$, then $a([0]) = 0$ and (15.4.2) becomes

$$\Psi(a) = \frac{1}{2\pi i} \int_{\gamma-i\infty}^{\gamma+i\infty} \psi(a; \zeta)R(\zeta; A)\, d\zeta\,, \qquad \omega_1(A) < \gamma \leqq \omega_0\,.$$

The integral in the right member converges in the uniform operator topology, and since linear multiplicative functionals are continuous in this topology, we have

$$\Psi(a)(\mathfrak{m}) = \frac{1}{2\pi i} \int_{\gamma-i\infty}^{\gamma+i\infty} \psi(a; \zeta)R(\zeta; A)(\mathfrak{m})\, d\zeta.$$

Substituting (16.3.1) in the integrand and making use of the relation (15.3.5), we obtain

$$\Psi(a)(\mathfrak{m}) = \frac{1}{2\pi i} \int_{\gamma-i\infty}^{\gamma+i\infty} \psi(a; \zeta)(\zeta - \alpha(\mathfrak{m}))^{-1}\, d\zeta = \psi(a; \alpha(\mathfrak{m}))$$

for $\mathfrak{m} \in \mathfrak{W}$ and $\Psi(a)(\mathfrak{m}) = 0$ for $\mathfrak{m} \in \mathfrak{U}$.

LEMMA 16.3.2. $T(\xi)(\mathfrak{m}) = \exp[\xi\alpha(\mathfrak{m})]$ *for each* $\mathfrak{m} \in \mathfrak{W}$ *and* $\xi \geqq 0$.

PROOF. Let $\varphi_\delta(\xi) \equiv \max\{\| T(\xi) \|, \exp[(\omega_1(A) + \delta)\xi]\}$, $\delta > 0$. It is clear that $\varphi_\delta(\xi)$ is a submultiplicative function satisfying (W) and that $A \prec \varphi_\delta$. For a given $\mathfrak{m} \in \mathfrak{W}$, we choose $a(E) \equiv \int_E f(\xi)\, d\xi$ with the following properties: (i) $f(\xi)$ has a compact carrier interior to $(0, \infty)$, (ii) $f(\xi)$ is continuously differentiable, and (iii) $\psi(a; \alpha(\mathfrak{m})) \neq 0$. Any smooth function which "approximates" e_0 sufficiently closely will have these properties. It is clear that $a * e_\xi$ and e_ξ both belong to $\mathcal{S}(\varphi_\delta)$ for each $\xi \geqq 0$. Hence $\Psi(a * e_\xi) = \Psi(a)\Psi(e_\xi) = \Psi(a)T(\xi)$ and

$$\Psi(a * e_\xi)(\mathfrak{m}) = \Psi(a)(\mathfrak{m})T(\xi)(\mathfrak{m}).$$

Now $a * e_\xi \in Q(\omega_1(A) + \delta)$ for $\xi \geqq 0$ so that $\Psi(a * e_\xi)(\mathfrak{m}) = \psi(a * e_\xi ; \alpha(\mathfrak{m}))$ by Lemma 16.3.1. It follows that

$$T(\xi)(\mathfrak{m}) = \frac{\psi(a * e_\xi ; \alpha(\mathfrak{m}))}{\psi(a; \alpha(\mathfrak{m}))} \doteq \psi(e_\xi ; \alpha(\mathfrak{m})) = \exp[\xi\alpha(\mathfrak{m})].$$

COROLLARY 1. $\Re[\sigma(A)] \leqq \omega_0(A) = \inf_{\xi>0} \xi^{-1} \log \| T(\xi) \|$.

PROOF. According to Theorem 4.15.1, $| T(\xi)(\mathfrak{m}) | \leqq \| T(\xi) \|$ and hence $\Re[\alpha(\mathfrak{m})] = \xi^{-1} \log | T(\xi)(\mathfrak{m}) | \leqq \inf_{\xi>0} \xi^{-1} \log \| T(\xi) \| \equiv \omega_0(A)$ for each $\mathfrak{m} \in \mathfrak{W}$. The result now follows from the fact that $\sigma(A) = \alpha(\mathfrak{W})$, which was proved in Theorem 16.3.1.

COROLLARY 2. $\exp[\xi\sigma(A)] \subset \sigma[T(\xi)]$.

PROOF. The assertion is merely a restatement of known relations, namely,

$$\sigma[T(\xi)] = T(\xi)(\mathfrak{M}) \supset T(\xi)(\mathfrak{W}) = \exp[\xi\alpha(\mathfrak{W})] = \exp[\xi\sigma(A)].$$

We now consider the operator extension of the algebra $S(\varphi)$ for the infinitesimal generator A and we suppose that $A \precsim \varphi$. Theorem 15.2.1 asserts that the mapping $a \to \Psi(a)$ is a homomorphism of $S(\varphi)$ into $\mathfrak{B}$, continuous with respect to the norm topologies of $S(\varphi)$ and $\mathfrak{B}$. Consequently Ψ^* defines a linear bounded transformation on $\mathfrak{B}^*$ into $[S(\varphi)]^*$. Moreover if μ is a linear multiplicative functional on $\mathfrak{B}$, then it is easily seen that $\mu[\Psi(a)]$ defines a linear multiplicative functional on $S(\varphi)$. Thus the adjoint mapping Ψ^* takes $\mathfrak{M}$ into the maximal ideals $\mathfrak{M}' \equiv [\mathfrak{m}']$ of $S(\varphi)$. The discussion preceding Theorem 4.15.6 shows that Ψ^* is a continuous mapping of $\mathfrak{M}$ into $\mathfrak{M}'$. It was found in section 4.18 that the maximal ideals of $S(\varphi)$ also split in a natural manner into two disjoint parts $\mathfrak{W}'$, $\mathfrak{U}'$ where $\mathfrak{M}' = \mathfrak{W}' \cup \mathfrak{U}'$. To each $\mathfrak{m}' \in \mathfrak{W}'$ there corresponds a complex number $\alpha'(\mathfrak{m}')$, $\Re[\alpha'(\mathfrak{m}')] \leqq \omega_0$, such that $a(\mathfrak{m}') = \psi(a; \alpha'(\mathfrak{m}'))$; in particular

$$e_\xi(\mathfrak{m}') = \exp[\xi\alpha'(\mathfrak{m}')].$$

In view of Lemma 16.3.2, this suggests that there is a close affinity between $\mathfrak{W}$ and $\mathfrak{W}'$. We have, in fact,

THEOREM 16.3.2. *Let* $\mathfrak{M}'$ *be the maximal ideals in* $S(\varphi)$ *and set*

$$\mathfrak{W}' = [\mathfrak{m}'; \mathfrak{m}' \not\supset L(\varphi)] \qquad \text{and} \qquad \mathfrak{U}' \equiv \mathfrak{M}' \ominus \mathfrak{W}'.$$

Suppose $A \leqq \varphi$. *Then* $\mathfrak{W} = (\Psi^*)^{-1}\mathfrak{W}'$ *and* $\mathfrak{U} = (\Psi^*)^{-1}\mathfrak{U}'$. *Further* $\Psi(a)(\mathfrak{m}) = a[\Psi^*(\mathfrak{m})]$ *for each* $\mathfrak{m} \in \mathfrak{M}$ *and* $\Psi(a)(\mathfrak{m}) = \psi(a; \alpha(\mathfrak{m}))$ *for each* $\mathfrak{m} \in \mathfrak{W}$.

PROOF. Suppose $\varphi(\xi)$ is of type ω_0 and set $\varphi_\delta(\xi) \equiv \max \{\varphi(\xi), \exp [(\omega_0 + \delta)\xi]\}$ where $\delta > 0$. It is clear that φ_δ is a submultiplicative function satisfying (W) and that $A \prec \varphi_\delta$. Next choose a β with $\Re(\beta) > \omega_0 + \delta$. Then $r_\beta * e_\xi \in L(\varphi_\delta) \subset L(\varphi)$ for each $\xi > 0$. We therefore have by Theorem 15.4.2 that

$$\Psi(r_\beta * e_\xi) = \Psi(r_\beta)\Psi(e_\xi) = R(\beta; A)T(\xi).$$

As a consequence $\Psi(r_\beta * e_\xi)(\mathfrak{m}) = R(\beta; A)(\mathfrak{m})T(\xi)(\mathfrak{m})$ for $\xi > 0$. Applying Theorem 16.3.1 and Lemma 16.3.2, we see that

$$\Psi(r_\beta * e_\xi)(\mathfrak{m}) = [\beta - \alpha(\mathfrak{m})]^{-1} \exp [\xi\alpha(\mathfrak{m})] \neq 0$$

for $\mathfrak{m} \in \mathfrak{W}$ and $= 0$ for $\mathfrak{m} \in \mathfrak{U}$, again for $\xi > 0$. On the other hand, to each $\mathfrak{m}' \in \mathfrak{W}'$ there corresponds an $\alpha'(\mathfrak{m}')$, $\Re[\alpha'(\mathfrak{m}')] \leqq \omega_0$, such that $(r_\beta * e_\xi)(\mathfrak{m}') = \psi(r_\beta * e_\xi\,; \alpha'(\mathfrak{m}')) = [\beta - \alpha'(\mathfrak{m}')]^{-1} \exp [\xi\alpha'(\mathfrak{m}')]$ for $\xi > 0$; whereas if $\mathfrak{m}' \in \mathfrak{U}'$, then $\mathfrak{m}' \supset L(\varphi)$ and hence $(r_\beta * e_\xi)(\mathfrak{m}') = 0$ for $\xi > 0$. Since $\Psi(a)(\mathfrak{m}) = a[\Psi^*(\mathfrak{m})]$ it is now evident that $\Psi^*(\mathfrak{W}) \subset \mathfrak{W}'$ and $\Psi^*(\mathfrak{U}) \subset \mathfrak{U}'$. This proves the first assertion. We have incidentally shown for $\mathfrak{m} \in \mathfrak{W}$ and $\Psi^*(\mathfrak{m}) = \mathfrak{m}'$ that

$$[\beta - \alpha(\mathfrak{m})]^{-1} \exp [\xi\alpha(\mathfrak{m})] = [\beta - \alpha'(\mathfrak{m}')]^{-1} \exp [\xi\alpha'(\mathfrak{m}')]$$

for all $\xi > 0$. Passing to the limit as $\xi \to 0+$, we see that $\alpha(\mathfrak{m}) = \alpha'(\mathfrak{m}') = \alpha'(\Psi^*(\mathfrak{m}))$. Hence for $a \in S(\varphi)$ and $\mathfrak{m} \in \mathfrak{W}$, Theorem 4.18.2 implies that $\Psi(a)(\mathfrak{m}) = a[\Psi^*(\mathfrak{m})] = \psi(a; \alpha(\mathfrak{m}))$. This concludes the proof.

In studying the structure of the maximal ideals of $S(\varphi)$ in section 4.18, it was found that the set $\mathfrak{U}'$ could be further subdivided into two disjoint classes $\mathfrak{U}_0'$ and $\mathfrak{U}_1'$. The class $\mathfrak{U}_0'$ consisted of the single maximal ideal $\mathfrak{m}_0'$ defined by $a(\mathfrak{m}_0') = a([0])$. The maximal ideals in $\mathfrak{U}_1'$ are not easily dealt with; it was shown, however, that if $\mathfrak{m}' \in \mathfrak{U}_1'$, then $e_\xi(\mathfrak{m}') = e^{\sigma\xi}\chi(\xi)$, where $\sigma \leqq \omega_0$ and $\chi(\xi)$ is a character of the real line. We denote the inverse image of $\mathfrak{U}_k'$ by $\mathfrak{U}_k$; that is, $\mathfrak{U}_k = (\Psi^*)^{-1}(\mathfrak{U}_k')$, $k = 0, 1$.

THEOREM 16.3.3. *Let* $A \leqq \varphi$. *A maximal ideal* $\mathfrak{m} \in \mathfrak{M}$ *belongs to* $\mathfrak{U}_0$ *if and only if* $T(\xi)(\mathfrak{m}) = 0$ *for some* $\xi > 0$. *If* $\mathfrak{m} \in \mathfrak{U}_0$, *then* $\Psi(a)(\mathfrak{m}) = a([0])$ *for all* $a \in S(\varphi)$.

PROOF. If $\mathfrak{m} \in \mathfrak{U}_0$, then by Theorem 4,18.1 we have $\Psi(a)(\mathfrak{m}) = a[\Psi^*(\mathfrak{m})] = a([0])$ and in particular $T(\xi)(\mathfrak{m}) = e_\xi[\Psi^*(\mathfrak{m})] = 0$ for $\xi > 0$. Conversely, if $T(\xi_0)(\mathfrak{m}) = 0$ for some $\xi_0 > 0$, then $e_{\xi_0}[\Psi^*(\mathfrak{m})] = 0$ and Theorem 4.18.1 asserts that $\Psi^*(\mathfrak{m}) \in \mathfrak{U}_0'$.

We have now determined $\Psi(a)(\mathfrak{m})$ for $S(\varphi)$ and each of the component parts of $\mathcal{S}(\varphi)$. Since $\Psi(a)(\mathfrak{m})$ is linear and multiplicative in a, these results are easily combined and we obtain

THEOREM 16.3.4. *If either* $a \in S(\varphi)$ *and* $A \leqq \varphi$ *or* $a \in \mathcal{S}(\varphi)$ *and* $A \prec \varphi$, *then*

(16.3.3) $$\Psi(a)(\mathfrak{m}) = \psi(a; \alpha(\mathfrak{m})), \qquad \mathfrak{m} \in \mathfrak{W},$$

(16.3.4) $$\Psi(a)(\mathfrak{m}) = a([0]), \qquad \mathfrak{m} \in \mathfrak{U}_0 .$$

If either $a \in L(\varphi)$ *and* $A \prec \varphi$ *or* $a \in L(\varphi) + Q(\omega_0) + S(\varphi) * r_\alpha$, $\Re(\alpha) > \omega_0$, *and* $A \prec \varphi$, *then*

(16.3.5) $$\Psi(a)(\mathfrak{m}) = 0, \qquad \mathfrak{m} \in \mathfrak{U}.$$

We have already remarked that $\sigma(B) = B(\mathfrak{M})$ for each $B \in \mathfrak{B}$ and that $\mathfrak{M} = \mathfrak{W}$ if and only if A is bounded. The following spectral mapping theorem is therefore a direct consequence of Theorem 16.3.4.

THEOREM 16.3.5. *If either* $a \in S(\varphi)$ *and* $A \leqq \varphi$ *or* $a \in \mathcal{S}(\varphi)$ *and* $A \prec \varphi$, *then*

(16.3.6) $$\overline{\psi(a; \sigma(A))} \subset \sigma[\Psi(a)];$$

and in particular, if A *is bounded then*

(16.3.7) $$\psi(a; \sigma(A)) = \sigma[\Psi(a)].$$

For unbounded A *with either* $a \in L(\varphi)$ *and* $A \leqq \varphi$ *or*

$$a \in L(\varphi) + Q(\omega_0) + S(\varphi) * r_\alpha,\ \Re(\alpha) > \omega_0,\ \text{and } A \prec \varphi,$$

then

(16.3.8) $$\psi(a; \sigma(A)) \cup 0 = \sigma[\Psi(a)].$$

Finally if $0 \in \sigma[T(\xi)]$ *for some* $\xi > 0$, *and either* $a \in S(\varphi)$ *and* $A \leqq \varphi$ *or* $a \in \mathcal{S}(\varphi)$ *and* $A \prec \varphi$, *then* $a([0]) \in \sigma[\Psi(a)]$.

We note that $\overline{\psi(a; \sigma(A))}$ may be properly contained in $\sigma[\Psi(a)]$ even when $0 \notin \sigma[T(\xi)]$. An example will be given in section 23.16 of a strongly continuous group of linear bounded operators on $(-\infty, \infty)$ for which $\sigma(A) = \emptyset$. In this case $\psi(e_\xi ; \sigma(A)) = \emptyset \neq \sigma[T(\xi)]$ and $e_\xi([0]) = 0 \notin \sigma[T(\xi)]$ for any $\xi > 0$.

THEOREM 16.3.6. *A semi-group* $[T(\xi)]$ *can be embedded in a one-parameter strongly continuous group of bounded linear operators on* $(-\infty, \infty)$ *if and only if* $0 \in \rho[T(\xi)]$ *for some* $\xi > 0$.

PROOF. If $[T(\xi)]$ is embeddable in a group, then each operator $T(\xi)$ must be regular; that is, $0 \in \rho[T(\xi)]$ for all $\xi \geqq 0$. On the other hand if $0 \in \rho[T(\xi_0)]$ for some $\xi_0 > 0$, then $0 \notin T(\xi_0)(\mathfrak{M})$, and by Theorem 16.3.3, $0 \in \rho[T(\xi)]$ for all $\xi \geqq 0$. We now define $T(\xi) \equiv [T(-\xi)]^{-1}$ for $\xi < 0$. It is then readily verified that $T(\xi_1 + \xi_2) = T(\xi_1)T(\xi_2)$ for all real ξ_1, ξ_2. In order to see that $T(\xi)$ is strongly continuous on $(-\infty, \infty)$, let $\xi_0 \in (-\infty, \infty)$ be given and write $\xi = (\xi + \beta) - \beta$ where $\beta > |\xi_0|$. Then $\xi_0 + \beta > 0$ and $T(\xi)x = T(-\beta)T(\xi + \beta)x$ is obviously continuous at $\xi = \xi_0$ for each $x \in \mathfrak{X}$.

16.4. $T(\xi)$ uniformly continuous for $\xi \geqq \gamma > 0$. The present section is devoted to semi-groups which are continuous in the uniform operator topology for

$\xi \geqq \gamma > 0$. We explicitly assume that $T(\xi)$ is not continuous in the uniform operator topology at $\xi = 0$ or, equivalently, that the infinitesimal generator is unbounded.

THEOREM 16.4.1. *If* $\mathfrak{S}$ *is a semi-group of class* (A), *continuous in the uniform operator topology for* $\xi \geqq \gamma > 0$, *and with infinitesimal generator* A *unbounded, then* $\mathfrak{U} = \mathfrak{U}_0 \neq \emptyset$. *If, in addition, either* $a \in S(\varphi)$ *and* $A \leqq \varphi$ *or* $a \in \mathcal{S}(\varphi)$ *and* $A < \varphi$, *then*

$$\psi(a; \sigma(A)) \cup a([0]) = \sigma[\Psi(a)]. \tag{16.4.1}$$

PROOF. To begin with suppose that $A \leqq \varphi$ and set

$$a_\delta(E) = \delta^{-1} \operatorname{meas} [E \cap (\gamma, \gamma + \delta)], \qquad \delta > 0.$$

Then $a_\delta \in L(\varphi)$ and

$$\Psi(a_\delta) = \frac{1}{\delta} \int_\gamma^{\gamma+\delta} T(\xi)\, d\xi \to T(\gamma)$$

as $\delta \to 0+$, the convergence being relative to the uniform operator topology. As a consequence, $\lim_{\delta \to 0+} \Psi(a_\delta)(\mathfrak{m}) = T(\gamma)(\mathfrak{m})$ for each $\mathfrak{m} \in \mathfrak{M}$. Hence if

$$T(\gamma)(\mathfrak{m}) \neq 0,$$

then $\Psi(a_\delta)(\mathfrak{m}) = a_\delta[\Psi^*(\mathfrak{m})] \neq 0$ for δ sufficiently small so that $\Psi^*(\mathfrak{m}) \in \mathfrak{W}'$, that is, $\mathfrak{m} \in \mathfrak{W}$. On the other hand if $T(\gamma)(\mathfrak{m}) = 0$, then $\mathfrak{m} \in \mathfrak{U}_0$ by Theorem 16.3.3. Thus

$$\mathfrak{U} \equiv \mathfrak{M} \ominus \mathfrak{W} = \mathfrak{U}_0$$

and, as we have assumed A to be unbounded, $\mathfrak{U}_0$ is non-vacuous. Hence if either $a \in S(\varphi)$ and $A \leqq \varphi$ or $a \in \mathcal{S}(\varphi)$ and $A < \varphi$, then Theorem 16.3.4 shows that $\Psi(a)(\mathfrak{m}) = \psi(a; \alpha(\mathfrak{m}))$ for $\mathfrak{m} \in \mathfrak{W}$ and $\Psi(a)(\mathfrak{m}) = a([0])$ for $\mathfrak{m} \in \mathfrak{U} = \mathfrak{U}_0$. It follows that $\sigma[\Psi(a)] = \psi(a; \sigma(A)) \cup a([0])$.

THEOREM 16.4.2. *If* $\mathfrak{S}$ *is a semi-group of class* (A) *continuous in the uniform operator topology for* $\xi \geqq \gamma > 0$, *then* $\sigma(A)$ *lies to the left of a bounding curve* $\sigma = \theta(|\tau|) \leqq \omega_0(A)$, $\lambda = \sigma + i\tau$, *where* $\lim_{\tau \to \infty} \theta(\tau) = -\infty$.

PROOF. According to Corollary 1 of Lemma 16.3.2, $\mathfrak{R}[\sigma(A)] \leqq \omega_0(A)$. Hence it suffices to show that $\lim \sup_{|\tau| \to \infty} [\sigma; \sigma + i\tau \in \sigma(A)] = -\infty$. Now $\sigma(A) = [\alpha(\mathfrak{m}); \mathfrak{m} \in \mathfrak{W}]$ by Theorem 16.3.1. Hence if this assertion were false, then there would exist a sequence $\{\mathfrak{m}_n\} \subset \mathfrak{W}$ such that $\mathfrak{R}[\alpha(\mathfrak{m}_n)] \to \sigma_0 \leqq \omega_0(A)$ and $\mathfrak{I}[\alpha(\mathfrak{m}_n)] \to \infty$ (or $-\infty$). Suppose that $A \leqq \varphi$. Then a slight variant of the Riemann-Lebesgue theorem shows that $\lim_{n \to \infty} \psi(a; \alpha(\mathfrak{m}_n)) = 0$ for each $a \in L(\varphi)$. Since the space $\mathfrak{M}$ is compact the sequence $\{\mathfrak{m}_n\}$ will possess at least one limit point of, say $\mathfrak{m}_0$. We recall that the function $\Psi(a)(\mathfrak{m})$ is continuous on $\mathfrak{M}$ so that $\Psi(a)(\mathfrak{m}_0) = \lim_{n \to \infty} \Psi(a)(\mathfrak{m}_n)$ whenever this limit exists. In this case $\Psi(a)(\mathfrak{m}_n) = \psi(a; \alpha(\mathfrak{m}_n)) \to 0$ for each $a \in L(\varphi)$, and therefore $\Psi(a)(\mathfrak{m}_0) = 0$ for each $a \in L(\varphi)$.

Thus $\Psi^*(\mathfrak{m}_0) \in \mathfrak{U}'$. The infinitesimal generator must therefore be unbounded and by Theorem 16.4.1 we see that $\mathfrak{m}_0 \in \mathfrak{U}_0$. On the other hand for $\xi > 0$, $T(\xi)(\mathfrak{m})$ is also continuous on $\mathfrak{M}$ and as above $| T(\xi)(\mathfrak{m}_0) | = \lim_{n\to\infty} | T(\xi)(\mathfrak{m}_n) | = e^{\sigma_0 \xi} \neq 0$; this shows that $\mathfrak{m}_0 \notin \mathfrak{U}_0$. Since we have reached a contradiction, the theorem is proved.

As we impose further restrictions on the semi-group of operators, the spectrum of the infinitesimal generator becomes more and more delimited. This is illustrated in the following theorem.

THEOREM 16.4.3. *Let $\mathfrak{S}$ be a semi-group of class (A) and suppose that*

$$T(\xi_0)[\mathfrak{X}] \subset \mathfrak{D}(A)$$

for some $\xi_0 > 0$. Then $\sigma(A)$ lies to the left of a bounding curve of the form

$$\sigma = C_1 - C_2 \log | \tau |, \qquad \lambda = \sigma + i\tau.$$

PROOF. We may as well suppose A to be unbounded since the assertion is obvious for A bounded. According to Theorem 11.5.3, if $T(\xi_0)[\mathfrak{X}] \subset \mathfrak{D}(A)$ then $T(\xi)[\mathfrak{X}] \subset \mathfrak{D}(A)$ for each $\xi > \xi_0$; and by Theorem 10.3.5 this implies that $T(\xi)$ is continuously differentiable in the uniform operator topology for $\xi > 2\xi_0$. Thus Theorem 16.4.1 applies and we see that $T(\xi)(\mathfrak{m}) = \exp[\xi\alpha(\mathfrak{m})]$ for $\mathfrak{m} \in \mathfrak{W}$ and $= 0$ for $\mathfrak{m} \in \mathfrak{U} = \mathfrak{U}_0$ and $\xi > 0$. Since the derivative exists in the uniform operator topology, it follows that $AT(\xi) \in \mathfrak{B}$ for $\xi > 2\xi_0$ and

$$[AT(\xi)](\mathfrak{m}) = \alpha(\mathfrak{m}) \exp[\xi\alpha(\mathfrak{m})]$$

for $\mathfrak{m} \in \mathfrak{W}$ and $= 0$ for $\mathfrak{m} \in \mathfrak{U}$, again for $\xi > 2\xi_0$. Hence

$$\sup [| \alpha(\mathfrak{m}) \exp[\xi\alpha(\mathfrak{m})] | ; \mathfrak{m} \in \mathfrak{W}] \leqq C(\xi) \equiv \| AT(\xi) \|, \qquad \xi > 2\xi_0 .$$

In other words, $\sigma(A)$ lies to the left of the curve $(\sigma^2 + \tau^2)^{1/2} e^{\sigma\xi} = C(\xi)$, $\xi > 2\xi_0$, and *a fortiori* to the left of $| \tau | \exp[3\xi_0\sigma] = C(3\xi_0)$. The latter curve can be written as $\sigma = C_1 - C_2 \log | \tau |$ where $C_1 = (3\xi_0)^{-1} \log C(3\xi_0)$ and $C_2 = (3\xi_0)^{-1}$.

16.5. $T(\xi)$ uniformly continuous for $\xi \geqq 0$. We now consider semi-groups continuous in the uniform operator topology for $\xi \geqq 0$; these semi-groups form the most restricted of the basic classes. As we have shown in Theorem 9.6.1, there are several equivalent ways of formulating this assumption. It suffices to assume that $T(\xi) \to I$ as $\xi \to 0+$ in the uniform operator topology or, alternately, that the infinitesimal generator A belongs to $\mathfrak{E}(\mathfrak{X})$ and hence that $T(\xi) = \exp(\xi A)$.

The following theorem gives two characterizations of such a semi-group in terms of its representation in the algebra $\mathfrak{B}$ associated with $[T(\xi)]$.

THEOREM 16.5.1. *For a semi-group $[T(\xi)]$ of class (A), the following statements are equivalent:*

(i) $T(\xi) = e^{\xi A}$ *where* $A \in \mathfrak{E}(\mathfrak{X})$;

(ii) $\mathfrak{M} = \mathfrak{W}$;

(iii) *There exists a numerically-valued function* $\alpha(\mathfrak{m})$ *defined on* $\mathfrak{M}$ *such that* $T(\xi)(\mathfrak{m}) = \exp[\xi\alpha(\mathfrak{m})]$ *for all* $\mathfrak{m} \in \mathfrak{M}$ *and* $\xi \geqq 0$.

PROOF. Assuming (i), Theorem 16.3.1 shows that $\mathfrak{M} = \mathfrak{W}$. On the other hand if $\mathfrak{M} = \mathfrak{W}$, Lemma 16.3.2 implies (iii). It remains to prove that (iii) implies (i). Here we merely assume that $T(\xi)(\mathfrak{m})$ is of the form $\exp[\xi\alpha(\mathfrak{m})]$ for all $\mathfrak{m} \in \mathfrak{M}$ and $\xi \geqq 0$; we do not assume that the function $\alpha(\mathfrak{m})$ is connected with $R(\lambda; A)$ as in Theorem 16.3.1. However, implicit in (iii) is the fact that $\exp[\xi\alpha(\mathfrak{m})]$ is continuous in $\mathfrak{m}$ for each $\xi \geqq 0$ and continuous in ξ for each $\mathfrak{m} \in \mathfrak{M}$. We first show that $\alpha(\mathfrak{m})$ is itself continuous on $\mathfrak{M}$.

LEMMA 16.5.1. *Let* $\mathfrak{M}$ *be a compact Hausdorff space. If* $g(\xi; \mathfrak{m}) \equiv \exp[\xi\alpha(\mathfrak{m})]$ *is continuous in* $\mathfrak{m}$ *for each* $\xi \geqq 0$, *then the function* $\alpha(\mathfrak{m})$ *is itself continuous on* $\mathfrak{M}$.

PROOF. Since $g(1; \mathfrak{m})$ is non-vanishing and continuous on $\mathfrak{M}$, we see that $0 < \inf_{\mathfrak{m}} |g(1; \mathfrak{m})| \leqq \sup_{\mathfrak{m}} |g(1; \mathfrak{m})| < \infty$. As a consequence $\mathfrak{R}[\alpha(\mathfrak{m})]$ is bounded on $\mathfrak{M}$. Suppose that $\mathfrak{J}[\alpha(\mathfrak{m})]$ were not bounded on $\mathfrak{M}$. Then there exists a sequence $\{\mathfrak{m}_n\}$ such that $\mathfrak{R}[\alpha(\mathfrak{m}_n)] \to \alpha_0$ and $\mathfrak{J}[\alpha(\mathfrak{m}_n)] \to \infty$ (or $-\infty$). On the other hand, $|g(\xi; \mathfrak{m})|$ is bounded on $\mathfrak{M}$ for each $\xi \geqq 0$. Hence, given a denumerable dense subset D of $[0, \infty)$, we can apply the diagonal process and obtain a subsequence of our original sequence (which we renumber) such that $g(\xi; \mathfrak{m}_n)$ converges to a limit as $n \to \infty$ for each $\xi \in D$. Since the space $\mathfrak{M}$ is compact, there exists at least one limit point, say $\mathfrak{m}_0$, of the set $\{\mathfrak{m}_n\}$. It is clear that $\lim_{n\to\infty} g(\xi; \mathfrak{m}_n) = g(\xi; \mathfrak{m}_0)$ for each $\xi \in D$. One consequence of this is that $\alpha(\mathfrak{m}_0)$ is of the form $\alpha_0 + i\beta_0$; another is that $h(\xi; \mathfrak{m}_n) \equiv g(\xi; \mathfrak{m}_n)\exp[-\xi\alpha(\mathfrak{m}_0)] = \exp\{\xi[\alpha(\mathfrak{m}_n) - \alpha_0 - i\beta_0]\} \to 1$ as $n \to \infty$ for each $\xi \in D$. We now define

$$F_n \equiv [\xi; \pi/2 \leqq \arg h(\xi; \mathfrak{m}_n) \leqq 3\pi/2, 0 \leqq \xi \leqq 2\pi].$$

Each F_n is obviously closed and *a fortiori* measurable. Since

$$\mathfrak{J}[\alpha(\mathfrak{m}_n) - \alpha_0 - i\beta_0] \to \infty$$

with n, we see that $\lim_{n\to\infty} \text{meas}(F_n) = \pi$. Hence $\text{meas}[\limsup_{n\to\infty} F_n] \geqq \pi$ so that $\limsup_{n\to\infty} F_n \neq \emptyset$. For a given $\xi_0 \in \limsup_{n\to\infty} F_n$, we choose a subsequence of $\{\mathfrak{m}_n\}$ (again we renumber) such that $\lim_{n\to\infty} h(\xi_0; \mathfrak{m}_n) = e^{i\varphi_0}$ where $\pi/2 \leqq \varphi_0 \leqq 3\pi/2$. We now have

$$\lim_{n\to\infty} g(\xi; \mathfrak{m}_n) = \exp[\xi(\alpha_0 + i\beta_0)], \qquad \xi \in D, \tag{16.5.1}$$

$$\lim_{n\to\infty} g(\xi_0; \mathfrak{m}_n) = e^{i\varphi_0}\exp[\xi_0(\alpha_0 + i\beta_0)]. \tag{16.5.2}$$

The new subsequence will again have a limit point, say $\mathfrak{m}_1$. However $g(\xi; \mathfrak{m}_1)$ must be equal to the right side of (16.5.1) on the dense set D and to the right side of (16.5.2) at ξ_0. Thus $g(\xi; \mathfrak{m}_1)$ cannot be continuous in ξ at $\xi = \xi_0$ and therefore $g(\xi; \mathfrak{m}_1)$ cannot be of the form $\exp[\xi\alpha(\mathfrak{m}_1)]$. We have now reached a contradiction and this proves that $\mathfrak{I}[\alpha(m)]$ is bounded on $\mathfrak{M}$. This, in turn, implies that $\tau^{-1}[g(\tau; \mathfrak{m}) - 1]$ converges uniformly with respect to $\mathfrak{m}$ as $\tau \to 0+$. The limit function $\alpha(\mathfrak{m})$, being the uniform limit of continuous functions, is itself continuous on $\mathfrak{M}$.

We now return to the proof of Theorem 16.5.1. According to the above lemma, the function $| \alpha(\mathfrak{m}) |$ will be bounded on $\mathfrak{M}$, say by C. For $0 \leqq \xi < \pi/(2C)$, the range of $\exp [\xi\alpha(\mathfrak{m})]$ lies in the interior of the right half-plane, $\mathfrak{R}(\zeta) > 0$, and the same is true of $\sigma[T(\xi)]$. Applying Theorem 5.2.5 we see that $\xi^{-1} \log T(\xi) \equiv A_\xi$ is a well defined operator in $\mathfrak{B}$ and that $A_\xi(\mathfrak{m}) = \alpha(\mathfrak{m})$ for all

$$\mathfrak{m} \in \mathfrak{M}, \qquad 0 < \xi < \pi/(2C).$$

If $\mathfrak{B}$ has no radical, we can now assert that A_ξ is independent of ξ and hence that $T(\xi) = e^{\xi A_1}$. Allowing for the possibility of a radical, we argue as follows. We suppose for notational convenience that $2C < \pi$. Then $T(n) = e^{nA_1}$ for $n = 0, 1, 2, \cdots$. If we define

$$S(\xi) \equiv e^{-\xi A_1}T(\xi),$$

then $S(n) = I$ for $n = 0, 1, 2, \cdots$. Further since $T(\xi)$ and $e^{-\xi A_1}$ are continuous in the strong operator topology for $\xi > 0$, the same will be true of $S(\xi)$. Moreover all of the operators involved commute. It follows that $S(\xi)$ is a semi-group of operators and therefore that $S(\xi)$ is periodic of period one on $[0, \infty]$ and can be extended to be periodic of period one on $(-\infty, \infty)$. As a consequence the so extended group $[S(\xi)]$ is bounded on $(-\infty, \infty)$ so that $\sup [\| [S(\xi)]^n \| ; n = 0, \pm 1, \pm 2, \cdots] < \infty$. Finally we note that the representation of $S(\xi)$ is given by $S(\xi)(\mathfrak{m}) = \exp[-\xi\alpha(\mathfrak{m})]\, T(\xi)(\mathfrak{m}) \equiv 1$ for all $\mathfrak{m} \in \mathfrak{M}$. By Theorem 4.15.1, the operator $S(\xi)$ differs from the identity by at most a quasinilpotent element of $\mathfrak{B}$ for each $\xi > 0$ and therefore Theorem 4.10.1 implies that $S(\xi) = I$ for each $\xi > 0$. It follows that $T(\xi) = e^{\xi A_1}$, which is the property (i). This concludes the proof.

One can avoid the use of Theorem 4.10.1 in the above proof; a more direct Fourier series argument has been given by R. S. Phillips [6].

Corollary. *Let $[T(\xi)]$ be a semi-group of class (A) and let $\mathfrak{B}$ be the (B)-algebra associated with $[T(\xi)]$. If there exists a numerically-valued function $\alpha(\mathfrak{m})$ defined on the Šilov boundary S of $\mathfrak{B}$ such that $T(\xi)(\mathfrak{m}) = \exp [\xi\alpha(\mathfrak{m})]$ for each $\mathfrak{m} \in S$, then $T(\xi) = e^{\xi A}$ where $A \in \mathfrak{C}(\mathfrak{X})$.*

Proof. Since S is a closed subset of $\mathfrak{M}$, Lemma 16.5.1 applies to the function $g(\xi; \mathfrak{m}) = \exp [\xi\alpha(\mathfrak{m})]$ defined on S and therefore $| \alpha(\mathfrak{m}) |$ is bounded on S. Consequently there exists a constant $\delta > 0$ such that $| \exp [\xi\alpha(\mathfrak{m})] - 1 | < \frac{1}{2}$ for all ξ, $0 \leqq \xi \leqq \delta$, and $\mathfrak{m} \in S$. By the definition of S we have

$$\sup [| T(\xi)(\mathfrak{m}) - I(\mathfrak{m}) | ; \mathfrak{m} \in \mathfrak{M}] = \sup [| T(\xi)(\mathfrak{m}) - 1 | ; \mathfrak{m} \in S] < \tfrac{1}{2}$$

for each $\xi \in [0, \delta]$. We recall that $T(\xi)(\mathfrak{m})$ must be of one of three types: (a)

$T(\xi)(\mathfrak{m}) = e^{\alpha\xi}$; (b) $T(\xi)(\mathfrak{m}) = 0$ for all $\xi > 0$; or (c) $T(\xi)(\mathfrak{m}) = e^{\sigma\xi}\chi(\xi)$ where σ is real and $\chi(\xi)$ is a non-measurable character of the real line. Types (b) and (c) are excluded by the above inequality; (b) because $T(\xi)(\mathfrak{m}) \neq 0$ for $\xi \in [0, \delta]$ and (c) because $-\pi/2 \leqq \arg [T(\xi)(\mathfrak{m})] = \arg [\chi(\xi)] \leqq \pi/2$ for $\xi \in [0, \delta]$ which together with Theorem 4.17.4 implies that $\chi(\xi)$ is continuous. Hence for each $\mathfrak{m} \in \mathfrak{M}$ there exists an $\alpha(\mathfrak{m})$ such that $T(\xi)(\mathfrak{m}) = \exp [\xi\alpha(\mathfrak{m})]$ for all $\xi \geqq 0$. Theorem 16.5.1 now gives the desired result.

THEOREM 16.5.2. *Let $[T(\xi)]$ be a semi-group of class (A). If there exists an unbounded open connected set Δ containing the origin and if there exists numbers $0 \leqq \alpha < \beta$ such that $\sigma[T(\xi)] \cap \Delta = \emptyset$ for all $\xi \in (\alpha, \beta)$, then $T(\xi) = e^{\xi A}$ where $A \in \mathfrak{C}(\mathfrak{X})$.*

PROOF. Since $0 \notin \sigma[T(\xi)]$ for any $\xi \in (\alpha, \beta)$, we see by Theorem 16.3.3 that $\mathfrak{U}_0 = \emptyset$. Hence to each $\mathfrak{m} \in \mathfrak{M}$ there corresponds a real number σ and a character of the real line $\chi(\xi)$ such that $T(\xi)(\mathfrak{m}) = e^{\sigma\xi}\chi(\xi)$, $\xi \geqq 0$. If $\chi(\xi)$ were a non-measurable character, then the set $[\chi(\xi); \alpha < \xi < \beta]$ would be dense in the unit circle by Theorem 4.17.4; as a consequence the set $[T(\xi)(\mathfrak{m}); \alpha < \xi < \beta]$ would be dense in the ring $e^{\sigma\alpha} \leqq |\zeta| \leqq e^{\sigma\beta}$ (or $e^{\sigma\beta} \leqq |\xi| \leqq e^{\sigma\alpha}$ if $\sigma < 0$). Δ being open, we see that Δ is disjoint from this ring. We have further assumed that Δ contains points both inside and outside of this ring. It follows that the ring disconnects Δ, which is impossible. Consequently $\chi(\xi)$ is measurable and $T(\xi)(\mathfrak{m})$ is of the form $\exp [\xi\alpha(\mathfrak{m})]$. Theorem 16.5.1 now implies that $T(\xi) = e^{\xi A}$ where $A \in \mathfrak{C}(\mathfrak{X})$.

2. FINE STRUCTURE

16.6. Fine structure mapping theorems. We now examine the relationship between the fine structure of $\sigma[\Psi(a)]$ and that of $\sigma(A)$. The analogous problem for the operational calculus of general closed operators was treated in Theorem 5.12.2 and the methods of this earlier theorem can be applied to the problems of the present paragraph. The essential limitation imposed by such methods is that A be *strictly* majorized by φ.

We refer to Definition 2.16.2 for the terminology used below.

THEOREM 16.6.1. *Suppose $a \in \mathcal{S}(\varphi)$ and $A \prec \varphi$. If $\alpha \in \sigma(A)$ and if $\alpha I - A$ has the property P_ν, then $\psi(a; \alpha)I - \Psi(a)$ has the same property.*

PROOF. We first recall that $\Re(\alpha) \leqq \omega_0(A) \leqq \omega_1(A) < \omega_0 \equiv \inf_{\xi>0} \xi^{-1} \log \varphi(\xi)$. Thus choosing ω' so that $\omega_1(A) < \omega' < \omega_0$ and setting $\varphi'(\xi) = \max [\| T(\xi) \|, e^{\omega'\xi}]$, we see that $\Re(\alpha) < \omega'$, $A \prec \varphi'$, and $\mathcal{S}(\varphi) \subset \mathcal{S}(\varphi')$. For $\Re(\lambda) < \Re(\alpha)$

$$\frac{\psi(a; \alpha) - \psi(a; \lambda)}{\alpha - \lambda} = \int_0^\infty e^{\lambda\xi} f(\xi)\, d\xi, \tag{16.6.1}$$

where

$$f(\xi) = \int_0^\xi e^{-\alpha(\xi-\sigma)} \, d_\sigma[\psi(a; \alpha)e_0 - a] = \int_\xi^\infty e^{\alpha(\sigma-\xi)} \, d_\sigma a.$$

Further $b(E) = \int_E f(\xi) \, d\xi \in S(e^{\omega'\xi})$. To see this we write

$$\| b \|' \equiv \int_0^\infty e^{\omega'\xi} \mid f(\xi) \mid d\xi \leqq \int_0^\infty e^{\omega'\xi} \left[\int_\xi^\infty e^{\omega_0(\sigma-\xi)} \, d_\sigma \mid a \mid \right] d\xi$$

$$\leqq \left[\int_0^\infty e^{-(\omega_0-\omega')\xi} \, d\xi \right] \left[\int_0^\infty e^{\omega_0\sigma} \, d_\sigma \mid a \mid \right] < \infty;$$

here we have made use of the fact that $a \in S(e^{\omega_0\xi})$. This shows, incidentally, that (16.6.1) is actually valid for all λ with $\Re(\lambda) \leqq \omega'$. We now set $b = b_1 + b_2$, defining b_1 and b_2 by

$$\psi(b_1 ; \lambda) = \frac{\psi(a; \alpha) - \psi(a; \lambda)}{\alpha_0 - \lambda},$$

$$\psi(b_2 ; \lambda) = \frac{(\alpha_0 - \alpha)[\psi(a; \alpha) - \psi(a; \lambda)]}{(\alpha_0 - \lambda)(\alpha - \lambda)},$$

where $\Re(\alpha_0) > \omega'$. It is clear that $b_1 = [\psi(a; \alpha)e_0 - a] * r_{\alpha_0} \in \mathfrak{S}(\varphi')$. Moreover $b_2 = (\alpha_0 - \alpha)b * r_{\alpha_0} \in S(e^{\omega'\xi})$ and $\psi(b_2 ; \omega' + i\tau) \in L(-\infty, \infty)$ so that

$$b_2 \in Q(\omega') \subset \mathfrak{S}(\varphi').$$

Consequently $b \in \mathfrak{S}(\varphi')$, $\Psi(b) = \Psi(b_1) + \Psi(b_2)$, $\Psi(b_1) = [\psi(a; \alpha)I - \psi(a)]R(\alpha_0 ; A)$ and $\Psi(b_2) = (\alpha_0 - \alpha)\Psi(b)R(\alpha_0 ; A)$. Hence

$$\begin{aligned} \psi(a; \alpha)I - \Psi(a) &= (\alpha_0 I - A)\Psi(b_1) = (\alpha_0 I - A)[\Psi(b) - \Psi(b_2)] \\ &= (\alpha_0 I - A)[I - (\alpha_0 - \alpha)R(\alpha_0 ; A)]\Psi(b) \qquad (16.6.2) \\ &= (\alpha I - A)\Psi(b); \end{aligned}$$

and similarly for $x \in \mathfrak{D}(A)$

$$(16.6.3) \quad [\psi(a; \alpha)I - \Psi(a)]x = \Psi(b_1)(\alpha_0 I - A)x = \Psi(b)(\alpha I - A)x.$$

If $\alpha I - A$ has property P_1 (or P_3), then (16.6.3) shows that $\psi(a; \alpha)I - \Psi(a)$ likewise has P_1 (or P_3). Further if $\alpha I - A$ has P_2 then (16.6.2) implies that $\psi(a; \alpha)I - \Psi(a)$ also has P_2. This concludes the proof.

In the converse direction our results are far from exhaustive. We have

THEOREM 16.6.2. *We consider the two cases:* (i) $a \in L(\varphi) + Q(\omega_0) + S(\varphi) * r_{\alpha_0}$, $\Re(\alpha_0) > \omega_0$, *and* $A \prec \varphi$ *but* $T(\xi)$ *is otherwise unrestricted; and* (ii) $a \in \mathfrak{S}(\varphi)$ *and* $A \prec \varphi$ *with* $T(\xi)$ *continuous in the uniform operator topology for* $\xi \geqq \gamma > 0$. *In both cases if* $\mu I - \Psi(a)$, $\mu \neq a([0])$, *has property* P_ν, *then there exists an* α *with* $\psi(a; \alpha) = \mu$, $\alpha \in \sigma(A)$, *such that* $\alpha I - A$ *has the property* P_ν.

PROOF. If case (i) holds, then $a \in L(e^{\omega_0 \xi})$ and $\lim_{\lambda \to \infty} \psi(a; \lambda) = a([0]) = 0$, $\pi/2 \leqq \arg(\lambda - \omega_0) \leqq 3\pi/2$, where the limit exists uniformly with respect to λ in the half-plane $\Re(\lambda) \leqq \omega_0$. Since $\Re[\sigma(A)] \leqq \omega_0(A) < \omega_0$, we see for case (i) that the equation $\psi(a; \lambda) = \mu \neq 0$ has at most a finite number of roots belonging to $\sigma(A)$; Theorem 16.3.5 implies that there is at least one such root if

$$\mu \in \sigma[\Psi(a)].$$

On the other hand if $a \in \mathcal{S}(\varphi)$, then

$$| \psi(a; \sigma + i\tau) - a([0]) | \leqq \int_{0+}^{\infty} e^{\sigma \xi} \, d \, | a | \to 0$$

as $\sigma \to -\infty$. It follows that the equation $\psi(a; \lambda) = \mu \neq a([0])$ can have no roots if $\sigma = \Re(\lambda)$ is sufficiently small; that is, if $\sigma < \sigma_0$ where σ_0 depends on μ. Now if $T(\xi)$ is continuous in the uniform operator topology for $\xi \geqq \gamma > 0$, then Theorem 16.4.2 applies and hence $\sigma(A)$ lies to the left of a bounding curve $\sigma = \theta(| \tau |) \leqq \omega_0(A) < \omega_0$ where $\lim_{\tau \to \infty} \theta(\tau) = -\infty$. Thus in case (ii), the equation $\psi(a; \lambda) = \mu \neq a([0])$ can have roots belonging to $\sigma(A)$ only in a compact subset of $\Re(\lambda) < \omega_0$ and therefore there will be at most a finite number of such roots; Theorem 16.4.1 shows that there is at least one such root.

The rest of the proof is essentially the same for both cases. We denote the roots of $\psi(a; \lambda) = \mu \neq a([0])$ which belong to $\sigma(A)$ by $\alpha_1, \alpha_2, \cdots, \alpha_n$. Setting $q_k(\lambda) = (\alpha_k - \lambda)(\alpha_0 - \lambda)^{-1}$, where $\Re(\alpha_0) > \omega_0$, we define

$$h(\lambda) \equiv [\mu - \psi(a; \lambda)] \prod_{k=1}^{n} \left(\frac{\alpha_0 - \lambda}{\alpha_k - \lambda} \right) = [\mu - \psi(a; \lambda)] \left\{ \prod_{k=1}^{n} q_k(\lambda) \right\}^{-1}. \tag{16.6.4}$$

If we decompose $\prod_{k=1}^{n} [(\alpha_0 - \lambda)(\alpha_k - \lambda)^{-1}]$ into partial fractions, then $h(\lambda)$ becomes a sum of terms of type

$$g_0(\lambda) = \mu - \psi(a; \lambda) \qquad \text{or} \qquad g_{kj}(\lambda) = [\mu - \psi(a; \lambda)](\alpha_k - \lambda)^{-j}$$

where α_k is a root of $\psi(a; \lambda) = \mu$ of multiplicity, say m_k, and $0 < j \leqq m_k$. Now $g_0(\lambda) = \psi(a_0; \lambda)$ where $a_0 = \mu e_0 - a \in \mathcal{S}(\varphi)$. For $j = 1$ it was shown in the proof of Theorem 16.6.1 that $g_{k1}(\lambda) = \psi(a_{k1}; \lambda)$ where $a_{k1} \in \mathcal{S}(\varphi')$,

$$\varphi'(\xi) = \max [\| T_{,}(\xi) \|, e^{\omega' \xi}],$$

and $\omega_1(A) < \omega' < \omega_0$. For $j = 2$, we have $g_{k1}(\alpha_k) = 0$; hence

$$g_{k2}(\lambda) = -[g_{k1}(\alpha_k) - g_{k1}(\lambda)](\alpha_k - \lambda)^{-1}$$

and the same reasoning shows that $g_{k2}(\lambda) = \psi(a_{k2}; \lambda)$ where now $a_{k2} \in \mathcal{S}(\varphi'')$, $\varphi''(\xi) = \max [\| T(\xi) \|, e^{\omega'' \xi}]$, and $\omega_1(A) < \omega'' < \omega'$. It follows by induction that $h(\lambda) = \psi(b; \lambda)$ where $b \in \mathcal{S}(\varphi_1)$ and $\varphi_1(\xi) = \max [\| T(\xi) \|, e^{\omega_1 \xi}]$ for some ω_1 with $\omega_1(A) < \omega_1 < \omega_0$. It is clear from (16.6.4) that $\psi(b; \lambda)$ is different from zero for

$\lambda \in \sigma(A)$ and that $b([0]) = \lim_{\lambda \to -\infty} h(\lambda) = \mu - a([0]) \neq 0$. Hence $\Psi(b)(\mathfrak{m}) \neq 0$ for $\mathfrak{m} \in \mathfrak{W} \cup \mathfrak{U}_0$. Thus in case (ii) where $\mathfrak{M} = \mathfrak{W} \cup \mathfrak{U}_0$ (Theorem 16.4.1), the Gelfand theory implies that $\Psi(b)$ is regular in $\mathfrak{B}$. As for case (i), a closer examination of the proof of Theorem 16.6.1 shows that $a_{kj} \in S(\varphi_1) * r_{\alpha_0}$, $0 < j \leqq m_k$, and hence by Theorem 16.3.4, $\Psi(a_{kj})(\mathfrak{m}) = 0$ for $\mathfrak{m} \in \mathfrak{U}$ and $0 < j \leqq m_k$. Likewise by Theorem 16.3.4, if $a \in L(\varphi) + Q(\omega_0) + S(\varphi) * r_{\alpha_0}$, then $\Psi(a)(\mathfrak{m}) = 0$ for $\mathfrak{m} \in \mathfrak{U}$. Consequently $\Psi(b)(\mathfrak{m}) = \Psi(a_0)(\mathfrak{m}) = \mu \neq 0$ for $\mathfrak{m} \in \mathfrak{U}$ and again $\Psi(b)$ is regular in $\mathfrak{B}$. Let $B = [\Psi(b)]^{-1}$.

We note that

$$q_k(\lambda) = (\alpha_k - \lambda)(\alpha_0 - \lambda)^{-1} = 1 + (\alpha_k - \alpha_0)(\alpha_0 - \lambda)^{-1} = \psi(c_k ; \lambda)$$

where $c_k = e_0 + (\alpha_k - \alpha_0)r_{\alpha_0} \in \mathcal{S}(\varphi_1)$. The operational calculus therefore gives $\Psi(b) \prod_{k=1}^{n} \Psi(c_k) = \mu I - \Psi(a)$ so that

$$\prod_{k=1}^{n} \Psi(c_k) = B[\mu I - \Psi(a)] = [\mu I - \Psi(a)]B. \tag{16.6.5}$$

It is clear that $\Psi(c_k) = I + (\alpha_k - \alpha_0)R(\alpha_0 ; A) = (\alpha_k I - A)R(\alpha_0 ; A)$. Equation (16.6.5) is the analogue of equation (5.12.4) and the remainder of the proof follows precisely as in Theorem 5.12.2.

16.7. The spectrum of $T(\xi; A)$. The fine structure mapping theorems of the previous section are rather incomplete as regards the relation between $\sigma[T(\xi)]$ and $\sigma(A)$. In view of the importance of this particular case, we are fortunate that other methods are available for its study.

We summarize the pertinent results already obtained in Theorem 16.6.1 as follows:

THEOREM 16.7.1. *If $[T(\xi)]$ is a semi-group of class (A) with infinitesimal generator A and if $\alpha I - A$ has property* P$_\nu$, *then $e^{\alpha\xi}I - T(\xi)$ has the same property for each $\xi \geqq 0$. As a consequence*

$$\exp [\xi\sigma(A)] \subset \sigma[T(\xi)]. \tag{16.7.1}$$

We now proceed to a closer study of the converse proposition, beginning with the point spectrum. We shall denote the manifold annihilated by an operator T by $\mathfrak{Z}(T)$.

THEOREM 16.7.2. *Let $[T(\xi)]$ be a semi-group of class (A). Then $P\sigma[T(\xi)] = \exp [\xi P\sigma(A)]$, plus, possibly, the point $\lambda = 0$. If $\mu \in P\sigma[T(\xi)]$ for some fixed $\xi > 0$ where $\mu \neq 0$ and if $\{\alpha_n\}$ is the set of roots of $e^{\alpha\xi} = \mu$, then at least one of the points α_n lies in $P\sigma(A)$. Further $\mathfrak{Z}[\mu I - T(\xi)]$ is the closed linear extension of the linearly independent manifolds $\mathfrak{Z}[\alpha_n I - A]$, where n ranges over all $\alpha_n \in P\sigma(A)$.*

PROOF. Let $\xi > 0$ be fixed, let $\mu \in P\sigma[T(\xi)]$ where $\mu \neq 0$, and set

$$\mathfrak{M} = \mathfrak{Z}[\mu I - T(\xi)].$$

It is clear that $\mathfrak{M}$ is a non-trivial closed linear subspace of $\mathfrak{X}$. For $x \in \mathfrak{M}$ and $\tau \geqq 0$, we see that $[\mu I - T(\xi)]T(\tau)x = T(\tau)[\mu I - T(\xi)]x = \theta$ and hence

$$T(\tau)[\mathfrak{M}] \subset \mathfrak{M}.$$

Further if $x \in \mathfrak{M}$, then $T(\tau)x = \mu^{-1}T(\xi + \tau)x \to \mu^{-1}T(\xi)x = x$ as $\tau \to 0+$ so that $T(\tau)x$ is continuous for $\tau \geqq 0$. It is clear, therefore, that $S(\tau) = e^{-\alpha_0\tau}T(\tau)$, $\alpha_0 = \xi^{-1}\log\mu$, is a strongly continuous periodic semi-group of operators on $\mathfrak{M}$ with period ξ. Finally we note that the set $\{\alpha_n = \alpha_0 + 2\pi i n/\xi; n = 0, \pm 1, \pm 2, \cdots\}$ consists of all of the roots of $e^{\alpha\xi} = \mu$.

We now define

$$J_n x = \xi^{-1}\int_0^\xi e^{-2\pi i n\tau/\xi}S(\tau)x\,d\tau, \qquad x \in \mathfrak{M}. \tag{16.7.2}$$

It is clear that $J_n \in \mathfrak{E}(\mathfrak{M})$ and further that

$$\begin{aligned} J_nJ_mx &= \xi^{-2}\int_0^\xi\int_0^\xi e^{-2\pi i(n-m)\tau/\xi}e^{-2\pi i m(\tau+\sigma)/\xi}S(\tau+\sigma)x\,d\sigma\,d\tau \\ &= \delta_{nm}J_mx, \qquad x \in \mathfrak{M}. \end{aligned} \tag{16.7.3}$$

Thus the J_n form an orthogonal system of projection operators on $\mathfrak{M}$ which commute with $T(\tau)$, $\tau \geqq 0$. Since $S(\tau)x$ is a strongly continuous periodic function for $x \in \mathfrak{M}$, its Fourier series is $(C, 1)$-summable for all $\tau \geqq 0$. The proof of this assertion proceeds as in the classical numerically-valued case by means of Theorem 3.9.2. Consequently for $x \in \mathfrak{M}$, we have

$$e^{-\alpha_0\tau}T(\tau)x = (C, 1)\text{-}\sum_{n=-\infty}^{\infty} e^{2\pi i n\tau/\xi}J_nx, \tag{16.7.4}$$

and in particular for $\tau = 0$ we obtain

$$x = (C, 1)\text{-}\sum_{n=-\infty}^{\infty} J_nx. \tag{16.7.5}$$

Setting $\mathfrak{M}_n = J_n[\mathfrak{M}]$, it is clear from (16.7.3) that the non-trivial $\mathfrak{M}_n$ are linearly independent and from (16.7.5) that $\mathfrak{M}$ is the linear closed extension of $[\mathfrak{M}_n]$.

If $x \in \mathfrak{M}_n$, $x \neq \theta$, then the relation (16.7.4) implies that $e^{-\alpha_0\tau}T(\tau)x = e^{2\pi i n\tau/\xi}x$. Consequently $Ax = \lim_{\tau\to 0+}\tau^{-1}[T(\tau)x - x] = \alpha_n x$ where $\alpha_n = \alpha_0 + 2\pi i n/\xi$. Thus $x \in \mathfrak{Z}[\alpha_nI - A]$ and $\alpha_n \in P\sigma(A)$. Conversely suppose that

$$x \in \mathfrak{Z}[\alpha_nI - A], \qquad x \neq \theta.$$

Then $A^kx = \alpha_n^kx$, $k = 1, 2, \cdots$, and by Theorem 11.5.3 we have

$$\frac{d}{d\tau}T(\tau)x = T(\tau)Ax = \alpha_nT(\tau)x, \qquad \lim_{\tau\to 0+}T(\tau)x = x.$$

The above differential equation together with the initial condition has a unique solution, namely $T(\tau)x = e^{\alpha_n\tau}x$. Substituting this in (16.7.2) we see that $J_nx =$

$x \in \mathfrak{M}_n$. We have now proved that $\mathfrak{M}_n = \mathfrak{Z}[\alpha_n I - A]$. Since $\mathfrak{M}$ is non-trivial, the same will be true of at least one of the $\mathfrak{Z}[\alpha_n I - A]$ and hence at least one of the α_n lies in $P\sigma(A)$. This concludes the proof. We shall discuss the possibility of $\lambda = 0$ belonging to $P\sigma[T(\xi)]$ below in Theorem 16.7.5.

We now pass to the residual spectrum of $T(\xi)$ and prove

THEOREM 16.7.3. *Let $[T(\xi)]$ be a semi-group of class (A). If $\mu \in R\sigma[T(\xi)]$ for some fixed $\xi > 0$ where $\mu \neq 0$, then at least one of the solutions of $e^{\alpha\xi} = \mu$ lies in $R\sigma(A)$ and none can lie in $P\sigma(A)$. Moreover, $\mu \in P\sigma[T^{\odot}(\xi)]$ and hence Theorem 16.7.2 applies to the adjoint semi-group defined in Definition* 14.4.1.

PROOF. It is clear that $T(\xi)$ commutes with A in the sense of Definition 14.3.2. Applying the corollary to Theorem 14.3.3 we see that $\mu I - T^{\odot}(\xi)$ has property P_1 if $\mu I - T(\xi)$ has P_2. The previous theorem therefore applies to the adjoint semi-group so that at least one of the roots, say α_n, of $e^{\alpha\xi} = \mu$ is in $P\sigma(A^{\odot})$. The corollary to Theorem 14.3.3 now asserts that $\alpha_n I - A$ has property P_2. Moreover, no root of $e^{\alpha\xi} = \mu$ can lie in $P\sigma(A)$ since otherwise μ would belong to $P\sigma[T(\xi)]$ by Theorem 16.7.1. It follows that $\alpha_n \in R\sigma(A)$.

The situation for the continuous spectrum of $T(\xi)$ is not completely understood. Here the difficulty arises from the fact that the continuous spectrum of $T(\xi)$ cannot be altogether accounted for in terms of the spectrum of A.

THEOREM 16.7.4. *Let $[T(\xi)]$ be a semi-group of class (A). If $\mu \in C\sigma[T(\xi)]$ for some fixed $\xi > 0$ where $\mu \neq 0$, then no solution of $e^{\alpha\xi} = \mu$ lies in $P\sigma(A) \cup R\sigma(A)$. It is possible for $\mu \in C\sigma[T(\xi)]$, $\mu \neq 0$, when all of the roots of the equation $e^{\alpha\xi} = \mu$ lie in $\rho(A)$; indeed, μ need not even belong to the closure of the set* $\exp[\xi\sigma(A)]$.

PROOF. The first assertion is a direct consequence of Theorem 16.7.1. The following example proves the first part of the second assertion, and an example proving the last part of this assertion will be given in section 23.16. We take $\mathfrak{X} = l_2$ and define

$$T(\xi)\{\beta_n\} = \{e^{in\xi}\beta_n\}.$$

Then

$$\mathfrak{D}(A) = [\{\beta_n\}; \sum_{n=-\infty}^{\infty} |n\beta_n|^2 < \infty] \qquad \text{and} \qquad A\{\beta_n\} = \{in\beta_n\}.$$

It is clear that $\sigma(A) = P\sigma(A) = \{in\}$. On the other hand for $\xi = 1$, $P\sigma[T(1)] = \{e^{in}\}$ and this set is dense in the unit circle $|\lambda| = 1$ by Kronecker's approximation theorem since the numbers 1 and 2π are rationally independent. The domains $|\lambda| < 1$ and $|\lambda| > 1$ belong to the resolvent set of $T(1)$ so that $\sigma[T(1)]$ coincides with the unit circle. There is no residual spectrum and therefore every point of the circle $|\lambda| = 1$ which is not in the countable point spectrum belongs to the continuous spectrum. Consequently if $\mu \in C\sigma[T(1)]$ and $\{\alpha_n\}$ is the set of roots of $e^{\alpha\xi} = \mu$, then none of the points $\{\alpha_n\}$ lie in $\sigma(A)$.

The following theorems shed some light on the spectral character of the point $\lambda = 0$.

THEOREM 16.7.5. *Let $[T(\xi)]$ be a semi-group of class (A). The spectral classification for $T(\xi)$ of $\lambda = 0$ is the same for all $\xi > 0$. In particular, $0 \in \rho[T(\xi)]$ if and only if $[T(\xi)]$ can be embedded in a one-parameter strongly continuous group of bounded linear operations on $(-\infty, \infty)$.*

PROOF. We have already shown in Theorem 16.3.6 that $0 \in \rho[T(\xi)]$ for some $\xi > 0$ if and only if $[T(\xi)]$ can be embedded in a strongly continuous group of bounded linear operators. Thus if we are dealing with a proper semi-group, then $T(\xi)$ cannot have a bounded inverse for any $\xi > 0$; that is, $0 \in \sigma[T(\xi)]$ for each $\xi > 0$. We now show that whenever $T(\alpha)$, $\alpha > 0$, has one of the spectral properties P_ν, then $T(\xi)$ has the same property for all values of $\xi > 0$ (which is a stronger assertion than that of the theorem). Suppose first that $T(\alpha)$, $\alpha > 0$, has the property P_1 and that $T(\alpha)x_0 = \theta$, $x_0 \neq \theta$. Then clearly $T(\tau + \alpha)x_0 = T(\tau)T(\alpha)x_0 = \theta$ for all $\tau > 0$. Further $T(\alpha/2)T(\alpha/2)x_0 = T(\alpha)x_0 = \theta$ so that either $y_0 = T(\alpha/2)x_0 = \theta$ or else $T(\alpha/2)y_0 = \theta$, $y_0 \neq \theta$. In either case $T(\alpha/2)$ will have P_1 and it follows that $T(\xi)$ has P_1 for all $\xi > \alpha/2^n$, $n = 1, 2, \cdots$; that is, for all $\xi > 0$. A similar argument can be used for the properties P_2 and P_3. The details are left to the reader.

We now consider the relationship between the spectral properties of $T(\xi)$ for $\lambda = 0$ and the properties of the infinitesimal generator A.

THEOREM 16.7.6. *Let $[T(\xi)]$ be a semi-group of class $(0, A)$. A necessary and sufficient condition that $T(\xi)$ have property P_1 is that there exist an $x \in \mathfrak{X}$, $x \neq \theta$, such that $R(\lambda; A)x$ can be extended to be an entire function of λ, say $f(\lambda)$, and*

$$\| f(\sigma + i\tau) \| \leqq M \max (1, e^{-\beta\sigma}), \tag{16.7.6}$$

where M and β are positive constants depending on x. Such an x belongs to $\mathfrak{Z}[T(\xi)]$ for $\xi \geqq \beta$.

PROOF. Suppose that $T(\beta)x = \theta$, $x \neq \theta$, so that $T(\xi)x = \theta$ for $\xi \geqq \beta$. Then

$$f(\lambda) = \int_0^\beta e^{-\lambda\xi} T(\xi)x \, d\xi$$

is an entire extension of $R(\lambda; A)x$. It readily follows that (16.7.6) holds with $M = \int_0^\beta \| T(\xi)x \| \, d\xi$. Conversely, suppose that there exists an $x \neq \theta$ such that $f(\lambda)$ is an entire extension of $R(\lambda; A)x$ satisfying (16.7.6). We shall show that $T(\xi)x = \theta$ for $\xi \geqq \beta$. The proof is based on formula (6.3.9) (cf. the proof of Theorem 11.6.3). Thus for $\xi > 0$, $\gamma > 0$, we have

$$\int_0^\xi (\xi - \eta)T(\eta)x \, d\eta = \frac{1}{2\pi i} \int_{\gamma - i\infty}^{\gamma + i\infty} e^{\lambda\xi} f(\lambda) \frac{d\lambda}{\lambda^2}.$$

The integral on the right is evaluated by integrating along a rectangle with vertices at $\gamma \pm \omega i$, $-\Omega \pm \omega i$; calculating the residue at $\lambda = 0$; and letting $\omega \to \infty$, $\Omega \to \infty$ in this order, assuming $\xi > \beta$. We then obtain

$$\int_0^\xi (\xi - \eta)T(\eta)x\, d\eta = \xi f(0) + f'(0), \qquad \xi > \beta.$$

Differentiating twice with respect to ξ, we see that $T(\xi)x = \theta$ for $\xi > \beta$ and this also holds for $\xi = \beta$ because of strong continuity. The proof of Theorem 16.7.5 shows that $T(\xi)$ has P_1 for all $\xi > 0$. This completes the proof of the theorem.

THEOREM 16.7.7. *Let $[T(\xi)]$ be a semi-group of class $(0, A)^{\odot}$. A necessary and sufficient condition that $T(\xi)$ have property P_2 is that there exist an $x^* \in \mathfrak{X}^*$, $x^* \neq \theta$, such that $R^*(\lambda; A)x^*$ can be extended to be an entire function of λ, say $g(\lambda)$, and*

$$\| g(\sigma + i\tau) \| \leqq M \max (1, e^{-\beta\sigma}), \tag{16.7.7}$$

where M and β are positive constants depending on x^*. For such an x^*, the element $x^{\odot} = R^*(\lambda_0 ; A)x^*$, $\lambda_0 \in \sigma(A)$, belongs to $\mathfrak{Z}[T^{\odot}(\xi)]$ for $\xi \geqq \beta$.

PROOF. This result is readily obtained from the previous theorem by means of the adjoint theory of Chapter XIV. If $T(\xi)$ has property P_2, then $T^{\odot}(\xi)$ will have P_1 by the corollary to Theorem 14.3.3; and the previous theorem implies the existence of an x^* satisfying the stated condition. Conversely, suppose such an $x^* \in \mathfrak{X}^*$ exists and let $\lambda_0 \in \rho(A)$. Then it is clear that $x^{\odot} = R^*(\lambda_0 ; A)x^* \in \mathfrak{X}^{\odot}$ and that $x^{\odot}$ also satisfies the stated condition. Since $R(\lambda, A^{\odot})x^{\odot} = R^*(\lambda; A)x^{\odot}$, Theorem 16.7.6 shows that $x^{\odot} \in \mathfrak{Z}[T^{\odot}(\xi)]$ for $\xi \geqq \beta$. The corollary to Theorem 14.3.3 implies that $T(\xi)$ has P_2 for all $\xi \geqq \beta$ and the proof of Theorem 16.7.5 shows that $T(\xi)$ has P_2 for all $\xi > 0$.

THEOREM 16.7.8. *Let $[T(\xi)]$ be a semi-group of class (A). A necessary and sufficient condition that $0 \in \rho[T(\xi)]$ is that there exist constants $M > 0$ and $\omega \geqq 0$ such that*

$$\| [R(\lambda; A)]^n \| \leqq M(|\lambda| - \omega)^{-n} \tag{16.7.8}$$

for all real λ, $|\lambda| > \omega$, and all positive integers n.

PROOF. According to Theorem 16.7.5, $0 \in \rho[T(\xi)]$ if and only if $[T(\xi)]$ can be embedded in a strongly continuous one-parameter group of bounded linear operators on $(-\infty, \infty)$. The condition (16.7.8) was shown in Theorem 12.3.2 to be necessary and sufficient for A to generate a group.

The remaining property, namely P_3, can be characterized by default; that is, by the absence of P_1, P_2, or $0 \in \rho[T(\xi)]$.

CHAPTER XVII

HOLOMORPHIC SEMI-GROUPS

17.1. Orientation. The study of holomorphic semi-groups of linear bounded operators introduces several new aspects into semi-group theory which we now proceed to investigate.

Ordinarily a semi-group $\mathfrak{S} \equiv [T(\xi); \xi > 0]$ does not admit of an analytic extension. However, if $\mathfrak{S}$ can be extended analytically into a part of the complex plane, then it becomes pertinent to inquire as to the properties of such an extension. It turns out that the analytic extension of a semi-group of operators is a single-valued operator function which again has the semi-group property throughout its maximal domain of existence. In general this maximal domain of existence is the interior of a spinal semi-module (see section 8.7); however if $T(\xi)$ is holomorphic on some interval of the form $(0, \delta)$, $\delta > 0$, then the maximal domain of existence is an angular semi-module.

Semi-groups $\mathfrak{S} \equiv [T(\zeta)]$, defined on a spinal semi-module, S, holomorphic in the interior of S, and of class (A) on $(0, \infty)$, possess interesting rate of growth properties. If we set

$$\mathfrak{d}(\varphi) \equiv \lim_{r\to\infty} r^{-1} \log \| T(re^{i\varphi}) \| ,$$

then the function $F(re^{i\varphi}) \equiv r\mathfrak{d}(\varphi)$ is in essence the function of support of the closed convex extension of the singularities of $R(\lambda; A)$, that is, of the spectrum of the infinitesimal generator A. This is completely analogous to the corresponding relations in the numerically-valued case between the rate of growth of a determining function, holomorphic and of exponential type in a sector, and the position of the singularities of its Laplace transform (cf. G. Pólya [**2**]). Somewhat sharper results in this direction can be obtained for semi-groups of class $H(\Phi_1, \Phi_2)$.

If $T(\zeta)$ is holomorphic and of exponential type in a half-plane we can also obtain a representation of the semi-group by means of the binomial series. This technique together with a Laplace contour integral can also be used to determine sufficient conditions on a semi-group of class (A) in order that it have a holomorphic extension in a half-plane which is of exponential type. The same method can be employed to embed a given operator in a semi-group which is holomorphic in a half-plane; here the given operator is subject to certain mild restrictions.

The question of boundary values for semi-groups holomorphic in the half-plane $\mathfrak{R}(\zeta) > 0$ is also treated. Such a semi-group may possess boundary values on the imaginary axis which themselves form a strongly continuous one-parameter group of operators. Conversely, a strongly continuous one-parameter group of operators will admit of a holomorphic semi-group extension in a half-plane if the infinitesimal generator is suitably limited. One of the first results in this direction is due to H. Kober [1], who obtained the boundary group for the semi-group of fractional integrals.

There are four paragraphs: *Domains of analyticity*, *The structure of holomorphic semi-groups*, *Semi-groups and interpolation series*, and *Boundary value problems*.

References. Carlson [2], Hille [**5, 6, 7, 13**], Kober [1], Nörlund [2], Phillips [6], Phragmén and Lindelöf [1], Pólya [2].

1. Domains of Analyticity

17.2. Analytic extension of semi-groups. The purpose of the present section is to show that the semi-group property is preserved under analytic continuation and that the maximal domain of analyticity is a semi-module. The theorems in this section are both due to E. Hille [**7, 13**]. The proofs of these theorems have been considerably simplified by H. F. Bohnenblust, who is to be credited for the lemmas in this section.

Lemma 17.2.1. *Let C be a convex subset of the complex plane and suppose that $W(\zeta)$, defined on C to $\mathfrak{E}(\mathfrak{X})$, satisfies the property*

$$W(\zeta_1)W(\zeta_2) = W(\sigma_1)W(\sigma_2) \tag{17.2.1}$$

whenever $\zeta_1, \zeta_2, \sigma_1, \sigma_2 \in C$ and $\zeta_1 + \zeta_2 = \sigma_1 + \sigma_2$. Then

$$\prod_{j=1}^{m} W(\zeta_j) = \left[W\left(m^{-1}\sum_{j=1}^{m}\zeta_j\right)\right]^m \tag{17.2.2}$$

for all sets $\{\zeta_j\} \subset C$.

Proof. We proceed by an induction argument on m. The assertion (17.2.2) is obviously true for $m = 1$. Suppose that (17.2.2) holds for all $m \leqq n - 1$ and consider the case $m = n$ with $\zeta_2 = \zeta_3 = \cdots = \zeta_n$. Since

$$\sigma_1 \equiv n^{-1}[\zeta_1 + (n-1)\zeta_2] \in C, \qquad \sigma_2 \equiv n^{-1}[(n-1)\zeta_1 + \zeta_2] \in C,$$

and $\zeta_1 + \zeta_2 = \sigma_1 + \sigma_2$, we see that (17.2.1) implies that

$$W(\zeta_1)W(\zeta_2) = W(n^{-1}[\zeta_1 + (n-1)\zeta_2])W(n^{-1}[(n-1)\zeta_1 + \zeta_2]).$$

Hence

$$\begin{aligned} W(\zeta_1)[W(\zeta_2)]^{n-1} &= [W(\zeta_1)W(\zeta_2)][W(\zeta_2)]^{n-2} \\ &= W(n^{-1}[\zeta_1 + (n-1)\zeta_2])\{W(n^{-1}[(n-1)\zeta_1 + \zeta_2])[W(\zeta_2)]^{n-2}\}. \end{aligned}$$

Now

$$(n-1)^{-1}\{n^{-1}[(n-1)\zeta_1 + \zeta_2] + (n-2)\zeta_2\} = n^{-1}[\zeta_1 + (n-1)\zeta_2].$$

Consequently if we employ (17.2.2) with $m = n - 1$, we then obtain

$$W(\zeta_1)[W(\zeta_2)]^{n-1} = [W(n^{-1}[\zeta_1 + (n-1)\zeta_2])]^n. \tag{17.2.3}$$

For an arbitrary set of n points $\{\zeta_j; j = 1, 2, \cdots, n\} \subset C$, the relation (17.2.3) together with (17.2.2) for $m = n - 1$ now gives

$$\prod_{j=1}^{n} W(\zeta_j) = W(\zeta_1)\left[\prod_{j=2}^{n} W(\zeta_j)\right] = W(\zeta_1)\left\{W\left[(n-1)^{-1}\sum_{j=2}^{n}\zeta_j\right]\right\}^{n-1}$$
$$= \left[W\left(n^{-1}\sum_{j=1}^{n}\zeta_j\right)\right]^n,$$

proving the assertion (17.2.2) for the case $m = n$.

LEMMA 17.2.2. *Let C be a convex domain in the complex plane containing a convex neighborhood N such that $N + N \subset C$. Suppose $W(\zeta)$ is a holomorphic function on C to $\mathfrak{E}(\mathfrak{X})$ such that*

$$W(\zeta)W(\sigma) = W(\zeta + \sigma) \tag{17.2.4}$$

for all $\zeta, \sigma \in N$. Then

$$\prod_{j=1}^{m} W(\zeta_j) = \prod_{k=1}^{n} W(\sigma_k) \tag{17.2.5}$$

for all finite sets $\{\zeta_j\}$, $\{\sigma_k\} \subset C$ such that $\sum_{j=1}^{m}\zeta_j = \sum_{k=1}^{n}\sigma_k$.

PROOF. For all $\zeta, \sigma \in N$ we have by (17.2.4)

$$W(\zeta)W(\sigma) = W(\zeta + \sigma) = [W(\tfrac{1}{2}(\zeta + \sigma))]^2.$$

The two extreme members of this relation are holomorphic functions of ζ and σ in C. Applying Theorem 3.11.5 for fixed $\sigma \in N$, we see that

$$W(\zeta)W(\sigma) = [W(\tfrac{1}{2}(\zeta + \sigma))]^2$$

holds for all $\zeta \in C$, and a second application of Theorem 3.11.5 for fixed $\zeta \in C$ shows that the relation holds for all $\zeta, \sigma \in C$. This obviously implies (17.2.1) so that Lemma 17.2.1 can be used. Thus (17.2.5) for the case $n = m$ is a direct consequence of (17.2.2). Moreover, given two arbitrary finite sets $\{\zeta_j\}$, $\{\sigma_k\} \subset C$ such that $\sum_{j=1}^{m}\zeta_j = \tau = \sum_{k=1}^{n}\sigma_k$, Lemma 17.2.1 asserts that $\prod_{j=1}^{m} W(\zeta_j) = [W(\tau/m)]^m$ and $\prod_{k=1}^{n} W(\sigma_k) = [W(\tau/n)]^n$. Thus under our hypothesis, (17.2.5) is equivalent to the statement:

$$[W(\zeta)]^n = [W(n\zeta/m)]^m, \qquad n \geqq m, \tag{17.2.6}$$

whenever both ζ and $n\zeta/m \in C$. For the case $n = m + 1$, we choose $\zeta \in N$. Then $(m+1)\zeta/m = (m-1)\zeta/m + 2\zeta/m \in C$ and $2\zeta \in C$. Hence $[W(\zeta)]^2 = W(2\zeta)$ by (17.2.4) and therefore by Lemma 17.2.1

$$[W(\zeta)]^{m+1} = [W(\zeta)]^{m-1}W(2\zeta) = [W((m+1)\zeta/m)]^m.$$

Since the first and last members of this relation are holomorphic in the domain $C \cap mC/(m+1)$, Theorem 3.11.5 implies that (17.2.6) with $n = m + 1$ holds wherever ζ and $(m+1)\zeta/m \in C$. As a consequence, the assertion (17.2.5) also holds for $n = m + 1$. Assume next that (17.2.5) holds as well for all n with $m + 1 \leqq n \leqq m + k - 1$ and suppose that both ζ and $(m+k)\zeta/m \in C$. Then $(m + k - 1)\zeta/m \in C$ and

$$[W(\zeta)]^{m+k} = W(\zeta)[W(\zeta)]^{m+k-1} = W(\zeta)\{W[(m+k-1)\zeta/m]\}^m.$$

Hence applying (17.2.5) with $n = m + 1$, we obtain

$$[W(\zeta)]^{m+k} = \{W[(m+k)\zeta/m]\}^m,$$

which completes the induction step.

The following theorem is basic.

THEOREM 17.2.1. *Let C be a convex domain in the complex plane and let $(C)_a$ be its additive resultant in the sense of Definition* 8.5.2. *Let $\{\zeta_n\}$ be a countable subset of C such that*

(i) $\zeta_n \to \zeta_0 \in C$, (ii) $2\zeta_0 \in C$, *and* (iii) $\zeta_j + \zeta_k \in C$ *for* $j, k = 1, 2, 3, \cdots$. *Let $W(\zeta)$ be a function defined and holomorphic on C to $\mathfrak{E}(\mathfrak{X})$ such that*

$$W(\zeta_j + \zeta_k) = W(\zeta_j)W(\zeta_k) \tag{17.2.7}$$

for all j and k. Then $(C)_a$ is a simply-connected domain, $W(\zeta)$ may be continued analytically over all of $(C)_a$, and for the extended function

$$W(\zeta)W(\sigma) = W(\zeta + \sigma) \tag{17.2.8}$$

holds for all $\zeta, \sigma \in (C)_a$.

PROOF. We recall that $(C)_a$ is an open set since C is open. By assumption (ii), $C \cap 2C \neq \emptyset$. Thus Lemma 8.5.1 shows that $(C)_a$ is connected, and this together with Theorem 8.5.3 shows that $(C)_a$ is simply-connected.

We turn now to the analytic continuation of $W(\zeta)$. Since $2\zeta_0 \in C$, there exists a convex neighborhood N of ζ_0 such that $N \subset C$ and $N + N \subset C$. We may assume without loss of generality that the countable set $\{\zeta_j\} \subset N$.

The next step in the proof is to establish (17.2.8) for all $\zeta, \sigma \in N$. For fixed $\sigma = \zeta_j$ we consider the function

$$W(\zeta + \zeta_j) - W(\zeta)W(\zeta_j)$$

which is a holomorphic function of ζ in N vanishing for $\zeta = \zeta_k$, $k = 1, 2, 3, \cdots$. By Theorem 3.11.5 the difference vanishes identically in N. Next we fix ζ in N and consider the difference

$$W(\zeta + \sigma) - W(\zeta)W(\sigma).$$

This is a holomorphic function of σ in N which vanishes for $\sigma = \zeta_k$, $k =$

$1, 2, 3, \cdots$. Again applying Theorem 3.11.5 we see that (17.2.8) holds for all ζ, $\sigma \in N$.

It follows from the above argument that the hypothesis of Lemma 17.2.2 is satisfied by the function $W(\zeta)$ defined on C. We now extend $W(\zeta)$ to $(C)_a$. If $\zeta \in (C)_a$, then ζ can be represented in the form $\zeta = \sum_{j=1}^m \zeta_j$ where $\{\zeta_j\} \subset C$. This being so, we define

$$W(\zeta) \equiv \prod_{j=1}^m W(\zeta_j).$$

It is clear from Lemma 17.2.2 that the so-defined function, $W(\zeta)$, is uniquely defined on $(C)_a$. Further if $\zeta, \sigma \in (C)_a$, then

$$\zeta = \sum_{j=1}^m \zeta_j, \qquad \{\zeta_j\} \subset C; \qquad \sigma = \sum_{k=1}^n \sigma_k, \qquad \{\sigma_k\} \subset C;$$

and

$$\zeta + \sigma = \sum_{j=1}^m \zeta_j + \sum_{k=1}^n \sigma_k \in (C)_a.$$

Consequently

$$W(\zeta + \sigma) = \left[\prod_{j=1}^m W(\zeta_j)\right]\left[\prod_{k=1}^n W(\sigma_k)\right] = W(\zeta)W(\sigma),$$

which proves (17.2.8) for all $\zeta, \sigma \in (C)_a$. Finally we show that $W(\zeta)$ is a holomorphic function of ζ in $(C)_a$. Let $\zeta = \sum_{j=1}^m \zeta_j$, $\{\zeta_j\} \subset C$. Then keeping $\zeta_2, \zeta_3, \cdots, \zeta_m$ fixed, we consider ζ to be a function of ζ_1 alone. Since $W(\zeta_1)$ is holomorphic in ζ_1, the representation $W(\zeta) = W(\zeta_1)[\prod_{j=2}^m W(\zeta_j)]$ implies at once that $W(\zeta)$ is holomorphic in ζ. This concludes the proof.

We shall require the following

DEFINITION 17.2.1. *Let D be a domain in the complex plane, B its boundary, and let $x(\zeta)$ be a vector-valued function defined and holomorphic on D to $\mathfrak{X}$. We call D the maximal domain of analytic existence of $x(\zeta)$ if every accessible point of B is a singular point of $x(\zeta)$.*

LEMMA 17.2.3. *Let $x(\zeta)$ be a vector-valued function defined and holomorphic on some ray $R \equiv [\xi; \xi > \alpha_0]$ to $\mathfrak{X}$. Suppose that whenever $x(\zeta)$ can be continued analytically along some path $\zeta(t)$, $0 \leqq t \leqq 1$, starting at $\zeta(0) \in R$, then it also has an analytic extension along the paths $\zeta(t) + \beta$, $0 \leqq t \leqq 1$, for each $\beta > 0$. In this case $x(\zeta)$ is single-valued in its maximal domain of analytic existence.*

PROOF. Suppose first that $x(\zeta)$ can be continued analytically along the path $\sigma(t)$, $0 \leqq t \leqq 1$, $\sigma(0) \in R$. Choose $\gamma > \max [\Re[\sigma(t)]; 0 \leqq t \leqq 1] > \alpha_0$ and define

$$\zeta(t, \tau) = \begin{cases} \sigma(2t) + (\gamma - \alpha_0)2\tau, & 0 \leqq t \leqq \frac{1}{2}, \\ \sigma(1) + (\gamma - \alpha_0)2\tau, & \frac{1}{2} \leqq t \leqq 1. \end{cases}$$

It is clear that the paths $[\zeta(\cdot\,, \tau); 0 \leqq \tau \leqq 1]$ satisfy the hypothesis of Theorem 3.12.1. Setting

$$\tau(t) = \begin{cases} \dfrac{\gamma - \Re[\sigma(2t)]}{2(\gamma - \alpha_0)}, & 0 \leqq t \leqq \frac{1}{2}, \\ \dfrac{\gamma - \Re[\sigma(1)]}{(\gamma - \alpha_0)}(1 - t), & \frac{1}{2} \leqq t \leqq 1, \end{cases}$$

the path $\zeta(t, \tau(t))$, $0 \leqq t \leqq 1$, is seen to consist of a broken line, starting at $\zeta = \gamma$, going vertically to $\zeta = \Im[\sigma(1)] + \gamma$, and then going horizontally to $\zeta = \sigma(1)$. According to Theorem 3.12.1, $x(\zeta)$ can be continued analytically along the path $\zeta(t, \tau(t))$ and attains the same element at $\sigma(1)$ as along the path $\zeta(t, 0)$, or equivalently, as along the path $\sigma(t)$. Thus given any two paths of analytic continuation, say $\sigma_1(t)$ and $\sigma_2(t)$, from R to the same point ζ_1, we may choose $\gamma > \max\,[\Re[\sigma_1(t)], \Re[\sigma_2(t)]; 0 \leqq t \leqq 1]$ and show as above that the element attained at ζ_1 on either path is the same as that attained along a fixed broken line path. It follows that $x(\zeta)$ is continued analytically to the same value at ζ_1 along each of the two paths $\sigma_1(t)$ and $\sigma_2(t)$. This proves that $x(\zeta)$ is single-valued in its maximal domain of analytic existence.

THEOREM 17.2.2. *Let $[T(\xi); \xi > 0]$ be a semi-group of linear bounded operators on $\mathfrak{X}$ to itself. Suppose $T(\xi)$ is holomorphic for $0 \leqq \alpha < \xi < \beta$. Then there exists a holomorphic function $W(\zeta)$ on complex numbers to $\mathfrak{E}(\mathfrak{X})$ with a maximal domain of analytic existence Δ. The domain Δ is the interior of a spinal semi-module (Definition 8.7.2) and contains the interval (α, ∞) of the real axis; if $\alpha = 0$, then Δ is an angular semi-module (Definition 8.7.1). $W(\zeta)$ coincides with $T(\xi)$ on (α, ∞) and for all ζ, σ in Δ we have*

$$W(\zeta + \sigma) = W(\zeta)W(\sigma). \tag{17.2.9}$$

PROOF. By assumption there exists a convex neighborhood N and a function $W(\zeta)$ such that $W(\zeta)$ is holomorphic in N, the interval $(\alpha', \beta') \equiv N \cap (0, \infty)$ is non-empty and is contained in (α, β), and $W(\zeta)$ coincides with $T(\xi)$ on (α', β'). For $\tau \geqq 0$ let $N_\tau \equiv N + \tau$ and set $C = \bigcup_{\tau \geqq 0} N_\tau$. We define

$$W_\tau(\zeta) = W(\zeta - \tau)T(\tau), \qquad \tau \geqq 0, \tag{17.2.10}$$

for $\zeta \in N_\tau$, replacing $T(0)$ by I if $\tau = 0$. It is clear that $W_\tau(\zeta)$ is a holomorphic function of ζ in N_τ and that $W_\tau(\xi) = T(\xi)$ for $\alpha' + \tau < \xi < \beta' + \tau$. Thus if $0 < \tau_2 - \tau_1 < \beta' - \alpha'$, then $N_{\tau_1, \tau_2} \equiv N_{\tau_1} \cap N_{\tau_2} \neq \emptyset$ and contains the interval $(\alpha' + \tau_2, \beta' + \tau_1)$ of the real axis. Since $W_{\tau_1}(\zeta)$ and $W_{\tau_2}(\zeta)$ are both holomorphic in N_{τ_1, τ_2} and coincide on the interval $(\alpha' + \tau_2, \beta' + \tau_1)$ with $T(\xi)$, it follows that $W_{\tau_1}(\zeta) \equiv W_{\tau_2}(\zeta)$ in N_{τ_1, τ_2}. Consequently $W_\tau(\zeta)$ defines the analytic continuation of $W(\zeta)$ in the domain N_τ. Thus $W(\zeta)$ is defined as a holomorphic function of ζ throughout C.

We can now apply the preceding theorem with C defined as above. It follows that $W(\zeta)$ may now be defined as a holomorphic function of ζ in $(C)_a$ where it satisfies (17.2.9). As a rule $(C)_a$ will be a proper subset of the maximal domain of analytic existence, Δ. Thus suppose that $W(\zeta)$ can be continued analytically along some path $\zeta(t)$, $0 \leqq t \leqq 1$; here we may suppose without loss of generality that $\zeta(0) \in (\alpha', \infty)$. If τ is a fixed positive number, then $W(\zeta)T(\tau)$ can also be continued analytically along $\zeta(t)$, $0 \leqq t \leqq 1$. However for ζ close to the real axis, $W(\zeta)T(\tau) = W(\zeta + \tau)$ so the continuation of the left side defines that of the right side. This amounts to the analytic continuation of $W(\zeta)$ along the path $\zeta(t) + \tau$, $0 \leqq t \leqq 1$. We see therefore that Lemma 17.2.3 is applicable and hence that $W(\zeta)$ is uniquely defined in its maximal domain of analytic existence Δ. It also follows that if Δ contains a point ζ_1, then it contains all points $\zeta_1 + \tau$, $\tau > 0$. Incidentally, since $T(\xi)$ was assumed to be holomorphic on (α, β), we will have $(\alpha, \infty) \subset \Delta$. It is now easy to show that Δ is a semi-module. For if ζ and $\sigma \in \Delta$, then there exist paths $\zeta(t)$, $0 \leqq t \leqq 1$, and $\sigma(t)$, $0 \leqq t \leqq 1$, with $\zeta(0)$, $\sigma(0) \in (\alpha', \infty)$ and $\zeta(1) = \zeta$, $\sigma(1) = \sigma$. It follows as above that $W[\zeta(t)]W[\sigma(t)]$ defines an analytic continuation of $W[\zeta(t) + \sigma(t)]$ along the path $\omega(t) \equiv \zeta(t) + \sigma(t)$, $0 \leqq t \leqq 1$, starting on (α', ∞) and ending at $\zeta + \sigma$. As a consequence, $\zeta + \sigma \in \Delta$ and $W(\zeta + \sigma) = W(\zeta)W(\sigma)$. Thus Δ coincides with the interior of a spinal semi-module containing (α, ∞) and (17.2.9) holds for all $\zeta, \sigma \in \Delta$. In particular, if $\alpha = 0$ then Δ is an angular semi-module. This concludes the proof.

According to the preceding theorem, the maximal domain of analytic existence of a holomorphic semi-group of linear bounded operators is the interior of a spinal semi-module. In particular, if the semi-group is holomorphic in an interval of the form $(0, \delta)$, then the maximal domain of analytic existence is an angular semi-module. Conversely, the interior of each spinal semi-module S in the complex plane is the maximal domain of analytic existence of a suitably chosen semi-group of operators. This will be shown in Theorem 19.5.1 for the case of translation operators acting on functions defined and continuous in $\bar{S}$ and holomorphic in Int (S).

2. The Structure of Holomorphic Semi-Groups

17.3. Growth properties of the norm. Let $[T(\zeta)]$ be a semi-group of linear bounded operators, defined and holomorphic in the interior S of a spinal semi-module. The present section is concerned with the behavior of $\| T(\zeta) \|$.

Theorem 3.13.1 tells us that $\| T(\zeta) \|$ is a subharmonic function of ζ and as such can have no maximum in S. Moreover $\| T(\zeta) \|$ is now continuous and therefore bounded in each compact subset of S. However $\| T(\zeta) \|$ may grow arbitrarily fast as ζ approaches the boundary B of S or as $\zeta \to \infty$ in such a manner that its distance from B stays bounded. These possibilities are realized by trigonometric semi-groups in $L_2(-\pi, \pi)$ as will be shown in Theorem 20.9.2.

If S coincides with the whole complex plane, then $T(\zeta)$ is an entire function of the form $T(\zeta) = J \exp(\zeta A)$ where $J^2 = J \in \mathfrak{E}(\mathfrak{X})$ and $A \in \mathfrak{E}(\mathfrak{X})$. In this case

$$\| T(\zeta) \| \leqq \| J \| \exp(\| A \| \, | \zeta |). \tag{17.3.1}$$

On the other hand, if S is a proper subset of the plane, the situation is more complicated. However a similar estimate holds if we stay at a suitable distance from the boundary as is shown in the following theorem.

THEOREM 17.3.1. *Let S be the interior of a spinal semi-module contained in the right half-plane and containing the positive real axis as well as an open set intersecting this ray. Let $h(\zeta)$ denote the horizontal distance of ζ from B and let ϵ be fixed, $0 < \epsilon < 1$. If $[T(\zeta)]$ is a semi-group of linear bounded operators defined and holomorphic in S, then there exists a finite $M(\epsilon)$ such that*

$$\| T(\zeta) \| \leqq \exp [M(\epsilon) \, | \zeta |] \tag{17.3.2}$$

in that portion of S in which $h(\zeta) \geqq \epsilon | \zeta |$ and $| \zeta | \geqq 1$.

PROOF. As in the remark following Theorem 8.7.9, B is determined by an equation of the form

$$\xi = \gamma(\eta), \qquad \zeta = \xi + i\eta, \qquad -\infty < \eta < \infty,$$

where $\gamma(\eta)$ is a subadditive, upper semi-continuous function with $\limsup_{\eta \to 0} \gamma(\eta) < \infty$. At all points of discontinuity of $\gamma(\eta)$ certain horizontal segments have to be added to the graph in order to complete B. Further $\gamma(\eta) \geqq 0$ as S is contained in the right half-plane. By Theorem 7.6.2 there exist two finite numbers α, β, $\alpha \leqq 0 \leqq \beta$, such that

$$\begin{aligned} \alpha &= \sup_{\eta < 0} \frac{\gamma(\eta)}{\eta} = \lim_{\eta \to -\infty} \frac{\gamma(\eta)}{\eta}, \\ \beta &= \inf_{\eta > 0} \frac{\gamma(\eta)}{\eta} = \lim_{\eta \to \infty} \frac{\gamma(\eta)}{\eta}. \end{aligned} \tag{17.3.3}$$

Hence for a given $\epsilon > 0$, there is an $\eta_0 = \eta_0(\epsilon)$ such that

$$\beta\eta \leqq \gamma(\eta) \leqq (\beta + \epsilon/2)\eta, \qquad \eta \geqq \eta_0,$$

$$\alpha\eta \leqq \gamma(\eta) \leqq (\alpha - \epsilon/2)\eta, \qquad \eta \leqq -\eta_0.$$

We plot the rays $R_1 : \xi = (\beta + \epsilon/2)\eta$, $\eta > 0$, and $R_2 : \xi = (\alpha - \epsilon/2)\eta$, $\eta < 0$, and a circle $| \zeta | = \rho$, ρ being subject to the following conditions. The ordinates of the intersections of the circle with the rays shall exceed η_0 numerically and the smaller arc of the circle joining the two rays shall lie in S. Let V denote the infinite sector bounded by the rays and the circular arc, and let $V(\rho)$ be the portion of V in which $\rho \leqq | \zeta | \leqq 2\rho$. Since S contains all points $\zeta + \tau$, $\tau > 0$, whenever it contains ζ, it is clear that all of V lies in S.

We have now to get a rough idea of the shape of the curve $h(\zeta) = \epsilon \mid \zeta \mid$. For $\eta > 0$ we have

$$\frac{\xi}{\eta} = \frac{\gamma(\eta)}{\eta} + \epsilon \frac{\mid \zeta \mid}{\eta} \geqq \beta + \epsilon,$$

and for $\eta < 0$

$$\frac{\xi}{\eta} = \frac{\gamma(\eta)}{\eta} + \epsilon \frac{\mid \zeta \mid}{\eta} \leqq \alpha - \epsilon,$$

whence we conclude that the curve lies to the right of the rays R_1 and R_2 except for the point $\zeta = 0$. Hence if Δ_ϵ consists of the region $h(\zeta) \geqq \epsilon \mid \zeta \mid, \mid \zeta \mid \geqq 1$, that part of Δ_ϵ in which $\mid \zeta \mid \geqq \rho$ is a subset of V. Suppose now that $\log \| T(\zeta) \| \leqq M_1(\epsilon)$ in $V(\rho)$ and that $\zeta \in V, n\rho \leqq \mid \zeta \mid < (n + 1)\rho$. Then the semi-group property gives

$$\| T(\zeta) \| \leqq \exp [nM_1(\epsilon)] \leqq \exp \left[\frac{\mid \zeta \mid}{\rho} M_1(\epsilon) \right].$$

Since an inequality of the form (17.3.2) clearly holds for $\zeta \in \Delta_\epsilon, \mid \zeta \mid < \rho$, it follows that it holds everywhere in Δ_ϵ. This completes the proof.

For a closer study of the asymptotic properties of the norm we introduce the growth function of E. Phragmén and E. Lindelöf [1]. This function plays an important role in the classical theory of holomorphic functions. In fact, for an entire function $f(\zeta)$ of exponential type with Laplace transform $F(\lambda)$, G. Pólya [2] has shown that the growth function for $f(\zeta)$ is equal to the function of support for the conjugate set to the convex extension of the singularities of $F(\lambda)$. As we shall see in section 17.4, this classical analogue carries over when $[T(\zeta)]$ is a holomorphic semi-group of class (A).

In the following S will again be the interior of a spinal semi-module contained in the right half-plane, and we may assume without loss of generality that the favored ray is the positive real axis. The boundary B of S is then determined as above by a subadditive function $\xi = \gamma(\eta)$ and the asymptotic rays are given by (17.3.3). We denote the angles of these rays by Φ_1 and Φ_2, that is

$$\text{(17.3.4)} \qquad \cot \Phi_1 = \alpha, \qquad \cot \Phi_2 = \beta, \qquad -\pi/2 \leqq \Phi_1 < 0 < \Phi_2 \leqq \pi/2.$$

THEOREM 17.3.2. *Let S be the interior of a spinal semi-module contained in the right half-plane and containing the positive real axis as spine. Let $\mathfrak{S} \equiv [T(\zeta)]$ be a semi-group of linear bounded operators, defined and holomorphic in S. The indicator $\mathfrak{d}(\varphi)$ of $T(\zeta)$ is defined as*

$$\text{(17.3.5)} \qquad \mathfrak{d}(\varphi) \equiv \lim_{r \to \infty} r^{-1} \log \| T(re^{i\varphi}) \|, \qquad \Phi_1 < \varphi < \Phi_2.$$

Then either $\mathfrak{d}(\varphi)$ is a continuous function of φ in (Φ_1, Φ_2) or $\mathfrak{d}(\varphi) \equiv -\infty$. In the former case $F(re^{i\varphi}) \equiv r\mathfrak{d}(\varphi)$ is the function of support (in the sense of (7.12.1) except *for possible values along the bounding rays) of an unbounded closed convex point set* $D = D(\mathfrak{S})$, *which will be called the indicator diagram of the semi-group* $\mathfrak{S}$. *In the latter case,* $T(\zeta)$ *is a quasi-nilpotent element of* $\mathfrak{E}(\mathfrak{X})$ *for each* $\zeta \in S$ *and we set* $D = \infty$.

We shall present two proofs of this theorem, the first depending on the theory of positive-homogeneous subadditive functions (§7.3) and the second on the Gelfand representation theory for Banach algebras (§4.3).

PROOF 1. Let K be the open cone generated by S, that is, $K \equiv [\zeta; r\zeta \in S$ for some $r > 0]$. Theorem 7.6.1 shows that $\lim_{r\to\infty} r^{-1} \log \| T(r\zeta) \| < \infty$ and exists for each $\zeta \in K$. Accordingly we define

$$F(\zeta) \equiv \lim_{r\to\infty} r^{-1} \log \| T(r\zeta) \|, \qquad \zeta \in K. \tag{17.3.6}$$

It is clear that $F(re^{i\varphi}) = r\mathfrak{d}(\varphi)$. Further the relations (7.13.1) and (7.13.2) hold so that $F(\zeta)$ is a positive-homogeneous sub-additive function on K in the sense of Definition 7.12.2. Consequently the results of section 7.12 apply and we see that $F(\zeta)$ is either (i) finite and continuous on K in which case $F(\zeta)$ is essentially the function of support for an unbounded closed convex set, or (ii) $F(\zeta)$ is identically $-\infty$. The assertions of the theorem now follow directly. We note, incidentally, that $\lim_{\varphi\to\Phi_j} \mathfrak{d}(\varphi)$, $j = 1, 2$, exists and is either finite or $+\infty$.

For the second proof we require the following lemma.

LEMMA 17.3.1. *Let* S *be the interior of a spinal semi-module containing the positive real axis. Suppose that* $\tau(\zeta)$ *is a complex-valued holomorphic function of* ζ *in* S *such that*

$$\tau(\zeta + \sigma) = \tau(\zeta)\tau(\sigma), \qquad \zeta, \sigma \in S. \tag{17.3.7}$$

Then either $\tau(\zeta) \equiv 0$ *on* S *or else there exists a complex number* α *such that* $\tau(\zeta) = e^{\alpha\zeta}$ *for all* $\zeta \in S$.

PROOF. If $\tau(\zeta_0) = 0$ for some $\zeta_0 \in S$, then $\tau(\zeta_0 + \sigma) = \tau(\zeta_0)\tau(\sigma) = 0$ for all $\sigma \in S$. Since $\tau(\zeta)$ is holomorphic on S and since S is connected (Theorem 8.7.9), this implies that $\tau(\zeta) \equiv 0$. On the other hand if $\tau(\zeta) \not\equiv 0$ on S, then for a fixed $\sigma \in S$ the function

$$\nu(\xi) \equiv \frac{\tau(\xi + \sigma)}{\tau(\sigma)}, \qquad \xi \geqq 0,$$

is holomorphic on $[0, \infty)$ and satisfies (4.17.1) non-trivially. By the corollary to Theorem 4.17.3, we have $\nu(\zeta) = e^{\alpha\xi}$, $\xi \geqq 0$. Thus $\tau(\xi + \sigma) = ce^{\alpha\xi}$, $\xi \geqq 0$, where $c = \tau(\sigma)$. By analytic continuation we see that $\tau(\zeta + \sigma) = ce^{\alpha\zeta}$ for all $\zeta \in S - \sigma$. Now

$$c^2 = [\tau(\sigma)]^2 = \tau(2\sigma) = ce^{\alpha\sigma}$$

so that $c = e^{\alpha\sigma}$. It follows that $\tau(\zeta) = e^{\alpha\zeta}$ for all $\zeta \in S$.

PROOF 2. We denote the commutant of the commutant of $\mathfrak{S} \equiv [T(\zeta);\ \zeta \in S]$ by $\mathfrak{B}$. Theorems 1.13.1 and 1.14.1 imply that $\mathfrak{B}$ is a commutative closed subalgebra of $\mathfrak{E}(\mathfrak{X})$ containing the identity. We may therefore apply the Gelfand representation theory of paragraph 4.3. Let $\mathfrak{M} \equiv [\mathfrak{m}]$ be the maximal ideals in $\mathfrak{B}$. The mapping $B \to B(\mathfrak{m})$ is a continuous holomorphism of $\mathfrak{B}$ into $C(\mathfrak{M})$. By Theorem 4.15.1

$$\lim_{n\to\infty} \| B^n \|^{1/n} = r(B) = \sup [\, | B(\mathfrak{m}) |;\ \mathfrak{m} \in \mathfrak{M}]$$

and $B(\mathfrak{m}) \equiv 0$ on $\mathfrak{M}$ if and only if B is quasi-nilpotent.

Now the analytic and semi-group properties of $T(\zeta)$ are inherited by $T(\zeta)(\mathfrak{m})$. Thus $T(\zeta)(\mathfrak{m})$ satisfies the conditions of Lemma 17.3.1. It follows that for fixed $\mathfrak{m}$ the function $T(\zeta)(\mathfrak{m})$, $\zeta \in S$, is either identically zero or else there exists a complex number $\alpha(\mathfrak{m})$ such that $T(\zeta)(\mathfrak{m}) = \exp [\zeta\alpha(\mathfrak{m})]$. We now set $\mathfrak{W} \equiv [\mathfrak{m};\ T(\zeta)(\mathfrak{m}) \not\equiv 0 \text{ on } \mathfrak{M}]$ and $\mathfrak{U} \equiv [\mathfrak{m};\ T(\zeta)(\mathfrak{m}) \equiv 0 \text{ on } \mathfrak{M}]$. Then $\mathfrak{M} = \mathfrak{W} \cup \mathfrak{U}$ and $\mathfrak{W} \cap \mathfrak{U} = \emptyset$. For a given $\varphi \in (\Phi_1, \Phi_2)$ there exists an $r_0 = r_0(\varphi)$ such that $re^{i\varphi} \in S$ for $r \geqq r_0$. Hence

$$\begin{aligned} \mathfrak{d}(\varphi) &= \lim_{n\to\infty} (nr_0)^{-1} \log \| T(nr_0e^{i\varphi}) \| \\ &= r_0^{-1} \lim_{n\to\infty} \log \| [T(r_0e^{i\varphi})]^n \|^{1/n} \\ &= r_0^{-1} \sup [\log | T(r_0e^{i\varphi})(\mathfrak{m}) |;\ m \in \mathfrak{M}]. \end{aligned}$$

If $\mathfrak{M} = \mathfrak{U}$, then $T(\zeta)(\mathfrak{m}) \equiv 0$ for all $\mathfrak{m} \in \mathfrak{M}$ and $\mathfrak{d}(\varphi) \equiv -\infty$; in this case $T(\zeta)$ is quasi-nilpotent for all $\zeta \in S$. On the other hand if $\mathfrak{W} \neq \emptyset$, then

$$\mathfrak{d}(\varphi) = \sup [\mathfrak{R}[\alpha(\mathfrak{m})e^{i\varphi}];\ \mathfrak{m} \in \mathfrak{W}] > -\infty.$$

Thus $\mathfrak{d}(\varphi)$ is the supremum of the projections of $\overline{[\alpha(\mathfrak{m});\ \mathfrak{m} \in \mathfrak{W}]}$ on the unit vector $e^{i\varphi}$. Consequently if

$$D_0 \equiv \text{closed convex extension of } \overline{[\alpha(\mathfrak{m});\ \mathfrak{m} \in \mathfrak{W}]}, \tag{17.3.8}$$

then $F(re^{i\varphi})$ is equal to the function of support for D_0 for $\varphi \in (\Phi_1, \Phi_2)$. It may happen that D_0 is a proper subset of D in which case D can be obtained from D_0 by adjoining one or both of the tangent rays to D_0 which makes angles of $\Phi_1 - \pi/2$ and $\Phi_2 + \pi/2$, and then taking the convex extension of the so-augmented D_0. The continuity of $\mathfrak{d}(\varphi)$ follows directly from its geometric meaning.

We conclude this section by proving a sharper form of (17.3.5).

THEOREM 17.3.3. *Let $[T(\zeta)]$ satisfy the hypothesis of Theorem* 17.3.2. *Then given a continuous function $\omega(\varphi) > \mathfrak{d}(\varphi)$, $\varphi \in (\Phi_1, \Phi_2)$, and an $\epsilon > 0$, there exists an r_0 depending on $\omega(\varphi)$ and ϵ such that for $r > r_0$ and $\Phi_1 + \epsilon \leqq \varphi \leqq \Phi_2 - \epsilon$,*

$$\| T(re^{i\varphi}) \| \leqq \exp [r\omega(\varphi)]. \tag{17.3.9}$$

PROOF. It clearly suffices to prove this property locally with respect to φ.

Given $\varphi_0 \in (\Phi_1, \Phi_2)$, we may choose a number α with the property

$$\delta(\varphi) < \Re(\alpha e^{i\varphi}) < \omega(\varphi)$$

for $\varphi = \varphi_0$. Since $\delta(\varphi)$ and $\omega(\varphi)$ are both continuous in φ, there exists a δ, $0 < \delta < \pi/2$, such that the same inequality holds for each φ, $|\varphi - \varphi_0| < \delta$. Thus $\lim_{r\to\infty} \| T(re^{i\varphi}) \exp(-\alpha re^{i\varphi}) \| = 0$ for each φ, $|\varphi - \varphi_0| < \delta$. Let R_- denote the ray: $re^{i(\varphi_0 - \delta)}$, $r > 0$, let R_+ denote the ray: $re^{i(\varphi_0+\delta)}$, $r > 0$, and let ρ be chosen so that the infinite sector V, which is bounded by R_-, R_+, and the smaller arc of $|\zeta| = \rho$ joining the two rays, lies entirely in S. Then $\| T(\zeta)e^{-\alpha\zeta} \|$ is bounded on the boundary of V; and the proof of Theorem 17.3.1 shows that this function is of exponential growth in V. Consequently the usual Phragmén-Lindelöf argument implies that $\| T(\zeta)e^{-\alpha\zeta} \| \leqq M$ for all $\zeta \in V$. Hence

$$r^{-1} \log \| T(re^{i\varphi}) \| \leqq \Re(\alpha e^{i\varphi}) + r^{-1} \log M < \omega(\varphi)$$

for all φ, $|\varphi - \varphi_0| < \delta$, and r sufficiently large. This concludes the proof.

REMARK. Theorem 17.3.2, Theorem 17.3.3, and a variant of Theorem 17.3.1 are valid for a somewhat larger variety of domains than the interiors of spinal semi-modules. The above discussion goes through *mutatis mutandis* if S is any open connected set such that whenever $\zeta \in S$ then $\zeta + \tau \in S$ for all $\tau \geqq 0$. In this case either Φ_1 or Φ_2 may be zero.

17.4. Holomorphic semi-groups of class (A). It was shown in the previous section that the indicator function for $\mathfrak{S} = [T(\zeta)]$ determines the function of support of an unbounded closed convex point set $D(\mathfrak{S})$, called the *indicator diagram*. The full significance of $D(\mathfrak{S})$ becomes apparent if we assume that $\mathfrak{S}$ is of class (A) on the real axis as well as holomorphic in the interior of a spinal semi-module. In this case the Laplace transform of $T(\zeta)$ is equal to the resolvent $R(\lambda; A)$ of the infinitesimal generator. The singularities of $R(\lambda; A)$ coincide with the spectrum of A, namely $\sigma(A)$, and, in analogy with the classical theory of G. Pólya [2], the conjugate of $D(\mathfrak{S})$ is essentially the closed convex extension of $\sigma(A)$. This result was originally obtained by E. Hille [13, section 13.5] for semi-groups of class $H(\Phi_1, \Phi_2)$ using Laplace transform methods. For the generality required in Theorem 17.4.1, it is necessary to make use of the Banach algebra approach developed by R. S. Phillips [6].

We recall that an *extremal point* of a closed convex point set C is by definition a point on the boundary of C which is not an interior point of any line segment belonging to the boundary of C. For our purposes it is convenient to make a distinction between *ordinary* and *exceptional extremal points*. We say that an extremal point λ_0 is exceptional if and only if it is the end point of a line segment on the boundary of C and the extension of this line segment is the only line of support of C passing through λ_0. It is clear that every exceptional point is the limit of ordinary points. In the case of an ordinary extremal point λ_0, we can always separate a small neighborhood of λ_0 from the rest of C by a suitable parallel to one of the lines of support of C through λ_0 in such a manner that part of

C cut off by the parallel also lies in the small neighborhood of λ_0. This is not possible in the case of exceptional extremal points or interior points of line segments on the boundary of C. Also in the way of notation, if E is any subset of the complex plane, we shall denote the conjugate set by E^-, that is

$$E^- \equiv [\bar{\zeta}; \zeta \in E].$$

THEOREM 17.4.1. *Let* $\mathfrak{S} \equiv [T(\zeta)]$ *be a semi-group of linear bounded operators defined on a spinal semi-module* S *containing the positive real axis as spine and whose non-interior points lie on the real axis. Suppose* $\mathfrak{S}$ *is of class* (A) *on* $(0, \infty)$ *and holomorphic in the interior of* S. *Then* $R(\lambda; A)$ *exists and is holomorphic outside of* D^-, *the conjugate indicator diagram of* $\mathfrak{S}$, *and every extremal point of* D^- *is a singular point of* $R(\lambda; A)$. *If* $\mathfrak{d}(\varphi) \equiv -\infty$, *then* $R(\lambda; A)$ *is an entire function of* λ.

PROOF. We denote the commutant of the commutant of $\mathfrak{S}$ by $\mathfrak{B}$, that is $\mathfrak{B} = \mathfrak{S}^{cc}$. We wish to make use of material in Chapter XVI where $\mathfrak{B}$ was defined somewhat differently, namely as $\mathfrak{S}_0^{cc}$, where $\mathfrak{S}_0 \equiv [T(\xi); \xi > 0]$. Actually $\mathfrak{S}_0^{cc} = \mathfrak{S}^{cc}$. To see this we note that $\mathfrak{S}_0 \subset \mathfrak{S}$ so that $\mathfrak{S}_0^c \supset \mathfrak{S}^c$. On the other hand if $B \in \mathfrak{S}_0^c$, then $T(\zeta)B = BT(\zeta)$ for all real $\zeta > 0$ and Theorem 3.11.5 implies that this relation remains true for all $\zeta \in S$; thus $B \in \mathfrak{S}^c$. Consequently $\mathfrak{S}_0^c = \mathfrak{S}^c$ and therefore $\mathfrak{S}_0^{cc} = \mathfrak{S}^{cc}$.

As in Chapter XVI, the maximal ideals split into two disjoint sets $\mathfrak{W}$ and $\mathfrak{U}$ defined in Theorem 16.3.1. By Lemma 16.3.2, $T(\zeta)(\mathfrak{m}) = \exp[\alpha(\mathfrak{m})\zeta]$ for $\mathfrak{m} \in \mathfrak{W}$ and real $\zeta > 0$; however since both members are holomorphic, it follows that this relation holds for all $\zeta \in S$. Now $T(\xi)$ is certainly continuous in the uniform operator topology for ξ sufficiently large. Thus Theorem 16.4.1 implies that $\mathfrak{U} = \mathfrak{U}_0$ so that $T(\zeta)(\mathfrak{m}) \equiv 0$ for each $\mathfrak{m} \in \mathfrak{U}$ by Theorem 16.3.3. Consequently the division of $\mathfrak{M}$ given in Chapter XVI is precisely the same as that used in the second proof of Theorem 17.3.2. Thus if $\mathfrak{d}(\varphi)$ denotes the indicator function and $F(re^{i\varphi}) \equiv r\mathfrak{d}(\varphi)$, then $F(re^{i\varphi})$ is essentially the function of support of the indicator diagram D and for $\varphi \in (\Phi_1, \Phi_2)$, $F(re^{i\varphi})$ is equal to the function of support of

$$D_0 \equiv \text{closed convex extension } [\overline{\alpha(\mathfrak{m})}; \mathfrak{m} \in \mathfrak{W}].$$

It is clear that $D_0 \subset D$ and that the extremal points of D are contained in the extremal points of D_0. Now by Theorem 16.3.1, $\sigma(A) \equiv [\alpha(\mathfrak{m}); \mathfrak{m} \in \mathfrak{W}]$ and therefore $D_0 =$ closed convex extension of $\sigma(A)^-$. The resolvent $R(\lambda; A)$ is regular outside of $\sigma(A)$ and *a fortiori* outside of D^-. If $\mathfrak{d}(\varphi) \equiv -\infty$, then $\mathfrak{W} = \emptyset$ so that $\sigma(A) = \emptyset$ and $R(\lambda; A)$ is an entire function of λ. Finally, in order to show that every extremal point of D^- is a singular point of $R(\lambda; A)$, it clearly suffices to do this for the ordinary extremal points of D_0^-. However if an ordinary extremal point λ_0 were a regular point of $R(\lambda; A)$, then we could find a neighborhood of λ_0 in which $R(\lambda; A)$ is regular. Further we could find a parallel to one of the lines of support of D_0^- at λ_0 such that the part of D_0^- cut off by this parallel lies in the above neighborhood. This would imply that λ_0 is not even a member of the closed convex extension of $\sigma(A)$, contrary to our choice of λ_0. This completes the proof.

The above considerations also shed some light on the question of the maximal domain of analytic existence. In this connection we have

THEOREM 17.4.2. *Let $[T(\zeta)]$ be a semi-group of class (A) holomorphic in the sector $\Phi_1 < \arg \zeta < \Phi_2$, $-\pi/2 \leqq \Phi_1 < 0 < \Phi_2 \leqq \pi/2$. A sufficient condition that this sector be the maximal domain of existence for $T(\zeta)$ is that for all $\lambda = \sigma + i\tau$ in D^-*

$$\Re(\lambda e^{i\Phi_2}) = \sigma \cos \Phi_2 - \tau \sin \Phi_2 < \mathfrak{d}(\Phi_2 - 0),$$

$$\Re(\lambda e^{i\Phi_1}) = \sigma \cos \Phi_1 - \tau \sin \Phi_1 < \mathfrak{d}(\Phi_1 + 0).$$

REMARK. These conditions are certainly fulfilled if $\mathfrak{d}(\Phi_2 - 0)$ and $\mathfrak{d}(\Phi_1 + 0)$ are infinite. The conditions are equivalent to the requirement that the extremal points of D^- shall form an unbounded set in the upper as well as in the lower half-planes.

PROOF. If it is possible to continue $T(\zeta)$ across the ray $\arg \zeta = \Phi_2$, then by Theorem 17.2.2 the maximal domain of existence is an angular semi-module Δ and it is easy to see that the angle of the upper asymptotic ray is greater than Φ_2. According to Theorem 17.3.2 it is possible to define $\mathfrak{d}(\varphi)$ in an interval containing $\varphi = \Phi_2$ as an interior point. The extended function $\mathfrak{d}(\varphi)$ is a function of support of a closed convex set D_1 which is a proper subset of D (but which still contains D_0). The line $\sigma \cos \Phi_2 - \tau \sin \Phi_2 = \mathfrak{d}(\Phi_2 - 0)$ will then be a line of support of D_1^- and as such it must have at least one point in common with D_1^-; this contradicts the assumption that it has no point in common with D^-. The same argument applies to the ray $\arg \zeta = \Phi_1$ and shows that $T(\zeta)$ does not admit of an analytic continuation beyond the given sector.

17.5. Semi-groups of class $H(\Phi_1, \Phi_2)$. The preceding results can be improved if we limit our study to semi-groups of class $H(\Phi_1, \Phi_2)$ which are defined in Definition 10.6.1. For such semi-groups we have

THEOREM 17.5.1. *Let U be a closed linear operator on $\mathfrak{X}$ to itself, having a domain dense in $\mathfrak{X}$. The assumptions (A_1), (A_2), and (A_3) below are necessary and sufficient in order that U be the infinitesimal generator of a semi-group $\mathfrak{S} \equiv [T(\zeta)]$ having the properties (P_1) and (P_2):*

(P_1) *$\mathfrak{S}$ is a semi-group of class $H(\Phi_1, \Phi_2)$.*

(P_2) *$T(\zeta)$ has a preassigned finite indicator $\mathfrak{d}(\varphi)$ in the sector (Φ_1, Φ_2).*

(A_1) *$R(\lambda; U)$ is holomorphic outside of D^- where D is the unbounded closed convex point set having $F(re^{i\varphi}) \equiv r\mathfrak{d}(\varphi)$ as its function of support.*

(A_2) *All extremal points of D^- are singular points of $R(\lambda; U)$.*

(A_3) *For each $\delta, \epsilon > 0$, let $D(\delta, \epsilon)$ be the unbounded closed convex point set whose function of support is $F_\delta(re^{i\varphi}) \equiv r[\mathfrak{d}(\varphi) + \delta]$, $\Phi_1 + \epsilon \leqq \varphi \leqq \Phi_2 - \epsilon$. Then there exists a constant $M(\delta, \epsilon)$ such that for λ outside of $D(\delta, \epsilon)^-$ and at a distance $d_{\delta,\epsilon}(\lambda)$*

we have

$$\| R(\lambda; U) \| \leqq M(\delta, \epsilon)/d_{\delta,\epsilon}(\lambda). \tag{17.5.1}$$

PROOF. Given a semi-group $\mathfrak{S}$ having the properties (P_1) and (P_2), Theorem 17.4.1 shows that (A_1) and (A_2) hold. In order to derive the estimate (17.5.1) we consider the integral

$$\int e^{-\lambda\zeta} T(\zeta)\, d\zeta,$$

taken along the arc $|\zeta| = \rho$ from ρ to $\rho e^{i\Phi}$, $\Phi \in (\Phi_1, \Phi_2)$. We suppose that $\lambda = \sigma + i\tau$ is such that

$$\mathfrak{R}(\lambda e^{i\varphi}) = \sigma \cos\varphi - \tau \sin\varphi \geqq \max\,[\mathfrak{d}(\varphi) + 2\delta; \varphi \in [0, \Phi]], \tag{17.5.2}$$

δ being a fixed positive number. By Theorem 17.3.3,

$$\| T(\rho e^{i\varphi}) \| \leqq \exp\, \{[\mathfrak{d}(\varphi) + \delta]\rho\}$$

for all $\varphi \in [0, \Phi]$ provided that ρ is sufficiently large. Hence

$$\left\| \int_0^\Phi \exp\,(-\lambda\rho e^{i\varphi})\, T(\rho e^{i\varphi})\rho e^{i\varphi}\, d\varphi \right\| \leqq \frac{\pi}{2}\, \rho e^{-\delta\rho} \to 0$$

as $\rho \to \infty$. We can therefore deform the path of integration of the above integral from the positive real axis to the ray $\arg \zeta = \Phi$. Thus

$$R(\lambda; U) = e^{i\Phi} \int_0^\infty \exp\,(-\lambda e^{i\Phi} r) T(re^{i\Phi})\, dr \tag{17.5.3}$$

for λ satisfying the condition (17.5.2). However the left member is now holomorphic for $\mathfrak{R}(\lambda e^{i\Phi}) > \mathfrak{d}(\Phi)$. It follows that (17.5.3) defines the analytic continuation of $R(\lambda; U)$ in this half-plane. Now let $\delta, \epsilon > 0$ be given. Then according to Lemma 10.6.2 together with Theorem 17.3.3, there exists a constant $M(\delta, \epsilon)$ such that

$$\| T(re^{i\Phi}) \| \leqq M(\delta, \epsilon) \exp\, \{[\mathfrak{d}(\Phi) + \delta]r\}$$

for all $r > 0$ and $\Phi \in [\Phi_1 + \epsilon, \Phi_2 - \epsilon]$. We therefore obtain the following estimate directly from (17.5.3)

$$\| R(\lambda; U) \| \leqq M(\delta, \epsilon)[\mathfrak{R}(\lambda e^{i\Phi}) - \mathfrak{d}(\Phi) - \delta]^{-1} \tag{17.5.4}$$

where

$$\mathfrak{R}(\lambda e^{i\Phi}) > \mathfrak{d}(\Phi) + \delta.$$

We shall now interpret this result geometrically. We see that the function $F_\delta(re^{i\varphi}) \equiv r[\mathfrak{d}(\varphi) + \delta]$ is a positive-homogeneous subadditive function since this is true of $F(re^{i\varphi}) = r\mathfrak{d}(\varphi)$. Consequently $F_\delta(\zeta)$, $\Phi_1 + \epsilon \leqq \arg \zeta \leqq \Phi_2 - \epsilon$, is the function of support of a closed convex point set which we denote by $D(\delta, \epsilon)$. It is clear that $D(\delta, \epsilon) \supset D$. Suppose that λ is outside of $D(\delta, \epsilon)^-$ and at a distance

$d_{\delta,\epsilon}(\lambda)$ from $D(\delta, \epsilon)^-$. Since $D(\delta, \epsilon)^-$ is a closed convex point set, there is a unique point λ_0 on the boundary of $D(\delta, \epsilon)^-$ such that $|\lambda - \lambda_0| = d_{\delta,\epsilon}(\lambda)$. If

$$\arg(\lambda - \lambda_0) = -\Phi,$$

then $\Phi_1 + \epsilon \leqq \Phi \leqq \Phi_2 - \epsilon$ and the line through λ_0 perpendicular to the line joining λ and λ_0 is a line of support of $D(\delta, \epsilon)^-$. Further

$$\begin{aligned}\Re(\lambda e^{i\Phi}) &= \Re[(\lambda - \lambda_0)e^{i\Phi}] + \Re(\lambda_0 e^{i\Phi}) \\ &= d_{\delta,\epsilon}(\lambda) + [\mathfrak{d}(\Phi) + \delta].\end{aligned}$$

Hence (17.5.1) follows directly from (17.5.4).

To prove the converse, we note that the hypothesis of Theorem 12.8.1 is satisfied because of (A_3). Hence U is the infinitesimal generator of a semi-group $[T(\zeta)]$ of class $H(\Phi_1, \Phi_2)$. Theorem 17.4.1 shows that the indicator function for $T(\zeta)$ is uniquely determined on (Φ_1, Φ_2) by the point set $\sigma(U)$; and assumptions (A_1) and (A_2) imply that the indicator function is precisely $\mathfrak{d}(\varphi)$. This concludes the proof.

The differentiability properties of $T(\zeta)$ at $\zeta = 0$ are presented in the following theorem, which is an extension of Theorems 11.6.3 and 11.6.4 to semi-groups of class $H(\Phi_1, \Phi_2)$. The proof is left to the reader.

THEOREM 17.5.2. *Let $[T(\zeta)]$ be a semi-group of class $H(\Phi_1, \Phi_2)$ with infinitesimal generator A. If $x \in \mathfrak{D}(A^n)$, then*

$$T(\zeta)x = \sum_{k=0}^{n-1} \frac{\zeta^k}{k!} A^k x + \frac{1}{(n-1)!} \int_0^\zeta (\zeta - \tau)^{n-1} T(\tau) A^n x \, d\tau \tag{17.5.5}$$

for $\Phi_1 < \arg \zeta < \Phi_2$, and further

$$\lim \zeta^{-n} \left[T(\zeta)x - \sum_{k=0}^{n-1} \frac{\zeta^k}{k!} A^k x \right] = \frac{1}{n!} A^n x \tag{17.5.6}$$

as $|\zeta| \to 0$ in each sector of the form $\Phi_1 + \epsilon < \arg \zeta < \Phi_2 - \epsilon$, $\epsilon > 0$. If

$$x \in \bigcap_{n=1}^{\infty} \mathfrak{D}(A^n),$$

these relations hold for all n and

$$T(\zeta)x \sim \sum_{n=0}^{\infty} \frac{\zeta^n}{n!} A^n x, \tag{17.5.7}$$

where the series is asymptotic in the sense of Poincaré to $T(\zeta)x$ in each sector of the form $\Phi_1 + \epsilon < \arg \zeta < \Phi_2 - \epsilon$, $\epsilon > 0$.

There are similar extensions of Theorems 11.5.6 and 11.5.7.

THEOREM 17.5.3. *Let $[T(\zeta)]$ be a semi-group of class $H(\Phi_1, \Phi_2)$ with infinitesimal generator A. Then for each $x \in \mathfrak{X}$*

$$\lim \lambda R(\lambda; A)x = x \tag{17.5.8}$$

as $|\lambda| \to \infty$ *outside of* $D(\delta, \epsilon)^-$ *for each* $\delta, \epsilon > 0$. *If* $x \in \mathfrak{D}(A^n)$, $n \geqq 1$, *then*

$$R(\lambda; A)x = \sum_{k=0}^{n-1} A^k x \lambda^{-k-1} + \lambda^{-n} R(\lambda; A) A^n x \tag{17.5.9}$$

and

$$\lim \lambda^{n+1} \left[R(\lambda; A)x - \sum_{k=0}^{n-1} A^k x \lambda^{-k-1} \right] = A^n x \tag{17.5.10}$$

as $|\lambda| \to \infty$ *outside of* $D(\delta, \epsilon)^-$. *Finally if* $x \in \bigcap_n \mathfrak{D}(A^n)$, *then*

$$R(\lambda; A)x \sim \sum_{n=0}^{\infty} A^n x \lambda^{-n-1}, \tag{17.5.11}$$

the series being asymptotic to $R(\lambda; A)x$ *in the sense of Poincaré as* $|\lambda| \to \infty$ *outside of* $D(\delta, \epsilon)^-$.

PROOF. According to Theorem 17.5.1,

$$\| \lambda R(\lambda; A) \| \leqq M(\delta/2, \epsilon/3) \, |\lambda| \, [d_{\delta/2, \epsilon/3}(\lambda)]^{-1}$$

for λ outside of $D(\delta/2, \epsilon/3)^-$. It follows from this that $\| \lambda R(\lambda; A) \|$ remains bounded for λ outside of $D(\delta, 2\epsilon/3)^-$. On the other hand $\lim \lambda R(\lambda; A)x = x$ as $\lambda \to \infty$, λ real. Theorem 3.14.3 therefore applies and proves the assertion (17.5.8). The relation (17.5.9) is simply a restatement of (11.5.12); and (17.5.10) follows directly from (17.5.8) and (17.5.9). The assertion (17.5.11) is an obvious consequence of (17.5.10).

3. Semi-Groups and Interpolation Series

17.6. Semi-groups holomorphic in a half-plane. The holomorphic semi-groups which arise in applications are often of exponential type on a right half-plane domain of analytic existence. In this case $\mathfrak{d}(\varphi)$ is defined for $-\pi/2 < \varphi < \pi/2$ and tends to finite limits as φ approaches the end points of this interval. This implies that the set D^-, which contains the spectrum of the infinitesimal generator, is enclosed in a horizontal strip of finite width.

Applying the theory of paragraph 6.3, we see that semi-group operators which are holomorphic and of exponential type in a right half-plane may be represented by convergent binomial series. We shall first investigate what conditions are imposed on $T(\xi)$ for small values of ξ by the existence of such a representation; we then show that these conditions are sufficient in order that $T(\xi)$ admit of an analytic extension into a semi-module containing a half-plane in which the extended semi-group is of exponential type. This material consists of a condensed and simplified version of an earlier investigation by E. Hille [7].

THEOREM 17.6.1. *Let $[T(\zeta)]$ be a semi-group of linear bounded operators defined on a spinal semi-module S which contains the half-plane $\Re(\zeta) > \xi_0 \geqq 0$. Suppose further that $T(\zeta)$ is holomorphic and of exponential type in every interior half-plane and satisfies the condition (C_0) at the origin. Let α_0, α_1, and $\sigma_0(\alpha)$ have the meaning assigned to them in section 6.8. Then*

$$T(\zeta) = \sum_{n=0}^{\infty} [T(\alpha) - I]^n \binom{\zeta/\alpha}{n}, \tag{17.6.1}$$

the representation being convergent for $\Re(\zeta) > \xi_0$ if $0 < \alpha \leqq \alpha_0$ and for $\Re(\zeta) > \sigma_0(\alpha)$ if $\alpha_0 < \alpha < \alpha_1$.

This is an immediate consequence of Theorem 6.8.2. Symbolically we have

$$T(\zeta) = \{I + [T(\alpha) - I]\}^{\zeta/\alpha} = [T(\alpha)]^{\zeta/\alpha} \tag{17.6.2}$$

which is suggestive and easy to remember, but the power has no meaning *a priori* and has to be defined.

We shall now investigate the implication of the existence of the series (17.6.1). We take $\alpha < \alpha_1$ and define $\sigma_0(\alpha) = \xi_0$ for $0 < \alpha \leqq \alpha_0$. Using the expression for the abscissa of convergence, formula (6.7.2), and taking σ fixed, $\sigma > \sigma_0(\alpha)$, we see that

$$\| [T(\alpha) - I]^n \| \leqq B(\sigma, \alpha) \frac{\Gamma(\sigma/\alpha + n + 2)}{\Gamma(\sigma/\alpha + 2)\Gamma(n + 1)}. \tag{17.6.3}$$

Let us now consider the series

$$\sum_{n=0}^{\infty} [T(\alpha) - I]^n(\lambda - 1)^{-n-1} \equiv R(\lambda; T(\alpha)). \tag{17.6.4}$$

The series converges for $|\lambda - 1| > 1$ and the sum satisfies

$$\| R(\lambda; T(\alpha)) \| \leqq B(\sigma, \alpha) \left\{1 - \frac{1}{|\lambda - 1|}\right\}^{-\sigma/\alpha-2} \tag{17.6.5}$$

That the sum of the series is actually $R(\lambda; T(\alpha))$ is shown by multiplication by

$$\lambda I - T(\alpha) = (\lambda - 1)I - [T(\alpha) - I].$$

With an obvious extension of the classical definition of order of a power series on a circle, due to J. Hadamard, we can say that $R(\lambda; T(\alpha))$ is of finite order $\omega(\alpha)$ on the circle $|\lambda - 1| = 1$ and $\omega(\alpha) \leqq \sigma_0(\alpha)/\alpha + 2$. The spectrum of $T(\alpha)$ clearly lies inside or on this circle; however the only point on the circle which is of any interest to us is $\lambda = 0$. This point always belongs to the spectrum unless $\mathfrak{S}$ can be embedded in a group. The order of the resolvent at $\lambda = 0$ and the distribution of the rest of the spectrum are of fundamental importance to us. The order concepts which are needed are given in

DEFINITION 17.6.1. *Let $x(\lambda)$ be a vector-valued function of λ holomorphic in the* *sector* V: $\psi_1 \leqq \arg \lambda \leqq \psi_2$, $0 < r \leqq 1$, *where* $\psi_2 - \psi_1 = \pi/\gamma$, $\gamma > \frac{1}{2}$. *Set*

$$M(r; x(\cdot)) = \sup \left[\| x(re^{i\psi}) \| ; \psi_1 \leqq \psi \leqq \psi_2\right], \qquad 0 < r \leqq 1.$$

We say that $x(\lambda)$ *is of finite order* ω *in* V *at* $\lambda = 0$ *if*

$$\limsup_{r\to 0} \log M(r; x(\cdot))/[\log (1/r)] = \omega.$$

$x(\lambda)$ *is of sub-exponential order in* V *at* $\lambda = 0$ *if*

$$\liminf_{r\to 0} r^\gamma \log M(r; x(\cdot)) = 0.$$

We shall now prove the following

THEOREM 17.6.2. *Let* $[T(\zeta)]$ *satisfy the hypothesis of Theorem* 17.6.1. *If* $\alpha < \alpha_1$ *and* n *is a positive integer, then* $R(\lambda; T(\alpha/n))$ *exists outside the lobe of the curve* $|\lambda^n - 1| = 1$ *which contains* $\lambda = 1$. $R(\lambda; T(\alpha/n))$ *is of finite order* ω_n *at* $\lambda = 0$ *in the sector* $\pi/(2n) + \epsilon \leqq \arg \lambda \leqq (4n - 1)\pi/(2n) - \epsilon$ *and* $\omega_n \leqq n\omega_1$.

PROOF. The case $n = 1$ has already been settled and formula (17.6.5) shows that $\omega_1 \leqq \sigma_0(\alpha)/\alpha + 2$. For $n > 1$ we use the elementary identity

$$R\left(\lambda; T\left(\frac{\alpha}{n}\right)\right) = R(\lambda^n, T(\alpha)) \sum_{k=0}^{n-1} \lambda^{n-k-1} T\left(\frac{k\alpha}{n}\right) \tag{17.6.6}$$

which holds whenever the right side has a meaning. Thus the left hand side exists for any λ outside of the rose curve $|\lambda^n - 1| = 1$. This curve has n lobes, each of which surrounds an nth root of unity. In order to eliminate the extraneous lobes, we proceed step by step.

For $n = 2$ we see that $R(\lambda; T(\alpha/2))$ exists outside of the lemniscate

$$|\lambda^2 - 1| = 1.$$

Formula (17.6.5) together with (17.6.6) shows that $R(\lambda; T(\alpha/2))$ is of finite order at $\lambda = 0$ in the sector $\pi/4 + \epsilon \leqq \arg \lambda \leqq 3\pi/4 - \epsilon$ as well as in the symmetric sector in the lower half-plane. But we already know that $R(\lambda; T(\alpha/2))$ exists outside of the circle $|\lambda - 1| = 1$ and is of finite order at $\lambda = 0$ in the sector $\pi/2 + \epsilon \leqq \arg \lambda \leqq 3\pi/2 - \epsilon$. Combining these two results, we see that the resolvent $R(\lambda; T(\alpha/2))$ exists outside of the lobe of the lemniscate which contains $\lambda = 1$ and is of finite order, say ω_2, in the sector

$$\pi/4 + \epsilon \leqq \arg \lambda \leqq 7\pi/4 - \epsilon.$$

This proves the main part of the assertion for $n = 2$. Further the result holds for any $\alpha < \alpha_1$ so we may replace α by any smaller number. This shows that $R(\lambda; T(\alpha/n))$ exists outside of the lobe of $|\lambda^2 - 1| = 1$ surrounding $\lambda = 1$ and is of finite order at $\lambda = 0$ in the sector $\pi/4 + \epsilon \leqq \arg \lambda \leqq 7\pi/4 - \epsilon$ provided $n \geqq 2$. The final step, except for the estimate of the order ω_n, is then proved by an induction argument.

To complete the proof consider the function

$$x(\lambda) \equiv \lambda^{n\omega_1+\delta} R\left(\lambda;\, T\left(\frac{\alpha}{n}\right)\right), \qquad \frac{\pi}{2n} + \epsilon \leqq \arg \lambda \leqq \frac{4n-1}{2n}\pi - \epsilon,$$

$$0 < |\lambda| \leqq 1,$$

where $\delta > 0$. It is holomorphic and of finite order at $\lambda = 0$ in the sector; using (17.6.5) and (17.6.6.) we can show that it is bounded on the boundary of the sector. The classical argument of Phragmén and Lindelöf then shows that $\| x(\lambda) \|$ is bounded throughout the sector. This shows that $\omega_n \leqq n\omega_1$ and completes the proof. The converse proposition reads:

THEOREM 17.6.3. *Let* $\mathfrak{S} \equiv [T(\xi);\, \xi > 0]$ *be a semi-group of class* (A) *such that*
(1) *there exists an* $\alpha > 0$ *and a sector* $V(\psi_0)\colon \psi_0 \leqq \arg \lambda \leqq 2\pi - \psi_0$, *where* $0 < \psi_0 < \pi/2$, *such that* $R(\lambda; T(\beta))$ *exists in* $V(\psi_0)$ *for every* β, $0 < \beta \leqq \alpha$;
(2) $R(\lambda; T(\beta))$ *is of sub-exponential order in* $V(\psi_0)$ *at* $\lambda = 0$ *for* $0 < \beta < \alpha$ *and of finite order* ω *for* $\beta = \alpha$.

Then there exists a semi-group of linear bounded operators $[W(\zeta)]$ *defined and holomorphic in a half-plane* $\Re(\zeta) > \xi_0$ *in the interior of which* $W(\zeta)$ *is of exponential type. Further* $W(\xi) = T(\xi)$ *for* $\xi > \max(0, \xi_0)$. *Either* $0 \leqq \xi_0 \leqq \alpha\omega$ *or* $\xi_0 = -\infty$. *In the latter case* $W(\zeta)$ *is an entire function and* $\mathfrak{S}$ *can be embedded in an analytical group.*

PROOF. In the first step of the proof we show that $R(\lambda; T(\alpha/2))$ exists and is of finite order $\leqq 2\omega$ at $\lambda = 0$ in the sector $V(\psi_0/2)$. The proof of this fact is analogous to that of Theorem 17.6.2 except that the assumption (2) now replaces the use of the binomial series. Formula (17.6.6) with $n = 2$ shows that $R(\lambda; T(\alpha/2))$ exists for λ in the sector $\psi_0/2 \leqq \arg \lambda \leqq \pi - \psi_0/2$ and also in the symmetric sector in the lower half-plane. We already know that $R(\lambda; T(\alpha/2))$ exists in $V(\psi_0)$ and is of sub-exponential order in this sector. Since $\lambda^{2\omega+\delta}R(\lambda; T(\alpha/2))$, δ fixed positive, is bounded on the edges of the sector, the classical Phragmén-Lindelöf argument shows that it is bounded in $V(\psi_0)$. This being true for every $\delta > 0$, we see that $R(\lambda; T(\alpha/2))$ is of order $\leqq 2\omega$ in $V(\psi_0)$ and hence also in $V(\psi_0/2)$ as asserted.

We also note that if the spectrum of $T(\alpha)$ lies in the complementary sector of $V(\psi_0)$ inside the circle $|\lambda| = \rho$, then the spectrum of $T(\alpha/2)$ lies in the complement of $V(\psi_0/2)$ inside the circle $|\lambda| = \rho^{1/2}$. Iterating the above process we see that $R(\lambda; T(\alpha 2^{-n}))$ exists in $V(2^{-n}\psi_0)$ and the spectrum lies in the complementary sector inside the circle $|\lambda| = \rho^{2^{-n}}$. There is no essential restriction in assuming that all these spectra lie inside the circle $|\lambda - 1| = 1$ except for the point $\lambda = 0$.

We now form the integral [cf. formula (6.7.6)]

$$W(\zeta; \beta) = \frac{1}{2\pi i}\int_\Gamma R(\lambda; T(\beta))\lambda^{\zeta/\beta}\, d\lambda. \tag{17.6.7}$$

Here to start with, $\beta = \alpha 2^{-n}$; Γ is a closed contour starting and ending at the origin and surrounding the rest of the spectrum of $T(\beta)$ once in the positive sense.

The power has its principal determination, $\lambda^{\zeta/\beta} = \exp[\beta^{-1}\zeta \log \lambda]$, where the imaginary part of the logarithm lies between $-\pi$ and $+\pi$.

The integral is a Laplace integral of a type familiar in the theory of binomial series since the investigations of S. Pincherle, F. Carlson, and N. E. Nörlund. It also provides an interpretation and definition of the power $[T(\beta)]^{\zeta/\beta}$ which is natural from the standpoint of the operational calculus. Cf. also formula (17.6.2).

To simplify the notation still further, we put

$$W(\zeta; \alpha 2^{-n}) = W_n(\zeta). \tag{17.6.8}$$

Since $R(\lambda; T(2^{-n}\alpha))$ is holomorphic in $V(2^{-n}\psi_0)$ and of order $\leqq 2^n\omega$ at the origin, the integral exists and defines a holomorphic function of ζ in the half-plane $\mathfrak{R}(\zeta) > \alpha\omega$.

For a ζ of the form $k2^{-n}\alpha$ (k positive integer) in this half-plane, the exponent ζ/β is a positive integer and the contour may be deformed into a circle of the form $|\lambda| = r > \| T(\beta) \|$. By Cauchy's theorem

$$W_n(k2^{-n}\alpha) = [T(2^{-n}\alpha)]^k = T(k2^{-n}\alpha) \tag{17.6.9}$$

for any integer k greater than $2^n\omega$. Thus the function $W_n(\zeta)$ interpolates the sequence $\{T(k2^{-n}\alpha)\}$.

Next we show that $W_n(\zeta)$ can be expanded in a binomial series. We note that $R(\lambda; T(\beta))$, $\beta = 2^{-n}\alpha$, is holomorphic outside the circle $|\lambda - 1| = 1$ on which it has only one singular point, $\lambda = 0$, and at $\lambda = 0$ it is of finite order $\omega(\beta) \leqq 2^n\omega$. Outside the circle the resolvent is represented by the series (17.6.4) and Cauchy's formulas show that

$$[T(\beta) - I]^k = \frac{1}{2\pi i}\int_{\Gamma_k} R(\lambda; T(\beta))\,(\lambda - 1)^k\, d\lambda,$$

where we choose Γ_k to be the circle $|\lambda - 1| = 1 + k^{-1}$. Since

$$\| R(\lambda; T(\beta)) \| \leqq A(\beta, \epsilon)\,|\lambda|^{-\omega(\beta)-\epsilon}$$

along Γ_k, a simple calculation gives for $\omega(\beta) \geqq 1$

$$\| [T(\beta) - I]^k \| \leqq B(\beta, \epsilon) k^{\omega(\beta)-1+\epsilon}. \tag{17.6.10}$$

Substituting the series (17.6.4), with α replaced by β, into the integral (17.6.7) and integrating termwise along the circle $|\lambda - 2| = 2$, we obtain the series

$$W_n(\zeta) = \sum_{k=0}^{\infty} [T(\beta) - I]^k \binom{\zeta/\beta}{k}, \qquad \beta = 2^{-n}\alpha, \tag{17.6.11}$$

which by (6.7.3) and (17.6.10) converges for

$$\mathfrak{R}(\zeta) > \beta[\omega(\beta) - 1] \equiv \sigma_\beta. \tag{17.6.12}$$

The termwise integration may be justified, for sufficiently large values of $\mathfrak{R}(\zeta)$, by a straightforward estimate of the remainder of the series. Finally, since both members of (17.6.11) are holomorphic in the half-plane $\mathfrak{R}(\zeta) > \max(\alpha\omega, \sigma_\beta) = \alpha\omega$, it follows that this relation holds throughout this half-plane.

This binomial series enables us to find an estimate of $\| W_n(\zeta) \|$. By Carlson's formula (6.7.12) we have for $\sigma > \sigma_\beta$, $-\pi/2 \leqq \varphi \leqq \pi/2$,

$$\text{(17.6.13)} \qquad \| W_n(\sigma + re^{i\varphi}) \| \leqq M_n(\sigma) \exp\left[\frac{r}{\beta} l(\varphi)\right] r^{\sigma+1/2+\epsilon}.$$

We shall now prove that all the functions $W_n(\zeta)$ are identical in their common domain of analyticity which certainly includes the half-plane $\mathfrak{R}(\zeta) > \alpha\omega$. It is enough to prove that $W_n(\zeta)$ and $W_{n+1}(\zeta)$ coincide for $\mathfrak{R}(\zeta) > \alpha\omega$. Choose $\sigma > \alpha\omega$ and form the function

$$\Delta_n(\zeta) = (1 + \zeta)^{-\sigma-1}[W_n(\zeta) - W_{n+1}(\zeta)].$$

The preceding discussion shows that $\Delta_n(\zeta)$ has the following properties:

(i) $\Delta_n(\zeta)$ is holomorphic for $\mathfrak{R}(\zeta) \geqq \sigma$;
(ii) $\| \Delta_n(\sigma + re^{i\varphi}) \| \leqq M_n \exp[\alpha^{-1} r 2^{n+1} l(\varphi)]$;
(iii) $\Delta_n(k2^{-n}\alpha) = \Theta$ for all integers $k > \alpha^{-1}\sigma 2^n$;
(iv) $\delta^{-1}[\frac{1}{2}\pi - l(\frac{1}{2}\pi - \delta)] \to +\infty$ as $\delta \to 0$.

These properties together imply that the assumptions of Theorem 3.13.7 are satisfied by the function $\Delta_n[2^{-n}\alpha(\zeta + \gamma)]$, for a suitable choice of γ, with

$$\lambda(\varphi) = 2l(\varphi).$$

It follows that $\Delta_n(\zeta) \equiv \Theta$.

Thus all the functions $W_n(\zeta)$ represent the same analytic function of ζ which we now denote by $W(\zeta)$. This function is represented by the integral (17.6.7) and the binomial series (17.6.11) for $\beta = 2^{-n}\alpha$. But by Theorem 6.8.2 the series is valid for every $\beta \leqq \alpha$ and, incidentally, this implies in turn that $W(\zeta; \beta) \equiv W(\zeta)$ for all $\beta \leqq \alpha$. Let $\sigma_0(\beta)$ be the abscissa of convergence of the series; we know that $\sigma_0(\beta)$ is an increasing function of β for $0 < \beta \leqq \alpha$ which tends to a limit, say ξ_0, as $\beta \to 0$. The limiting abscissa ξ_0 is uniquely characterized by the fact that $W(\zeta)$ is holomorphic and of exponential type for $\mathfrak{R}(\zeta) > \xi_0 + \delta$ when $\delta > 0$ but lacks at least one of these properties when $\delta < 0$.

We next show that $W(\xi) = T(\xi)$ when $\xi > \max(0, \xi_0)$. By (17.6.9) we have $W(k2^{-n}\alpha) = T(k2^{-n}\alpha)$ for all positive integers k and n such that $k2^{-n}\alpha > \sigma_0(2^{-n}\alpha)$, a quantity which tends to ξ_0 as $n \to \infty$. Thus the desired relation holds for dyadic rational multiples of α. However this implies that the relation holds for all ξ in the range in question, since $T(\xi)$ is strongly continuous for $\xi > 0$ by assumption and $W(\xi)$ is analytic for $\xi > \xi_0$. Theorem 17.2.1 now shows that $W(\zeta)$ is a semi-group of operators for all ζ in the half-plane $\mathfrak{R}(\zeta) > \xi_0$.

There are three possibilities for the value of ξ_0; we may have $\xi_0 = -\infty$, zero, or positive. Indeed, if $\xi_0 < 0$, then $W(\zeta)$ is holomorphic in a neighborhood of the origin and the semi-group property implies that $W(\zeta)$ is holomorphic in the whole finite plane. In fact, $W(0)x = \lim_{\xi\to 0+} T(\xi)x = x$ for each $x \in \mathfrak{X}_0$. Since

$\mathfrak{X}_0$ is dense in $\mathfrak{X}$ we see that $W(0) = I$. Hence by Theorem 9.6.1 we have $W(\zeta) = \exp(\zeta A)$ where $A \in \mathfrak{E}(\mathfrak{X})$. Consequently $W(\zeta)$ is an entire function of order one and exponential type. As such it is representable by a binomial series in ζ/β, convergent for all ζ as soon as $\beta < (\log 2)/ \| A \|$. Thus for $\xi_0 < 0$, we are dealing with a holomorphic group; some further comments on this case are to be found below. If $\xi_0 = 0$, then we are dealing with a proper holomorphic semi-group whose domain of analytic existence is the right half-plane. In case $\xi_0 > 0$, we note the possibility of $W(\zeta)$ being continued analytically along the real axis to values of $\xi < \xi_0$. For such values of ξ, that is for $\xi < \xi_0$, the so-extended function $W(\zeta)$ need not coincide with $T(\xi)$ even though $T(\xi) = W(\xi)$ for all $\xi > \xi_0$. See section 19.4. This completes the proof of Theorem 17.6.3.

REMARK. Formula (17.6.11) implies that

$$T(\xi)x = \sum_{n=0}^{\infty} \frac{1}{n!} \xi(\xi - \beta) \cdots (\xi - (n-1)\beta) \overset{n}{\underset{\beta}{\Delta}} T(0)x \tag{17.6.14}$$

for $0 < \beta \leqq \alpha$, $\xi > \sigma_0(\beta)$. The structure of this formula is similar to that of formula (10.4.16) and suggests the following question. Let $\mathfrak{S} = [T(\xi)]$ be a given semi-group, strongly continuous for $\xi > 0$, and denote the right member of (17.6.14) by $T(\xi; \beta)x$. Then is it true that

$$\lim_{\beta \to 0+} T(\xi; \beta)x = T(\xi)x?$$

In general this question must be answered in the negative, because, as we have seen in the proof of the preceding theorem, the mere existence of $T(\xi; \beta)x$ for all small values of β and $\xi > \xi_1$ implies that $T(\xi)$ is analytic for $\xi > \xi_1$.

If $T(\zeta)$ is a holomorphic group, other interpolation series than that of Newton, that is, binomial series, may be applied. The representation by *Stirling's interpolation series* is of the form

$$\begin{aligned} T(\zeta) &= \sum_{n=0}^{\infty} A_n \zeta^2(\zeta - \alpha^2) \cdots (\zeta^2 - (n-1)^2\alpha^2)/(2n)! \\ &\quad + \sum_{n=0}^{\infty} B_n \zeta(\zeta^2 - \alpha^2) \cdots (\zeta^2 - n^2\alpha^2)/(2n+1)!, \end{aligned} \tag{17.6.15}$$

where

$$A_n = \overset{2n}{\underset{\alpha}{\Delta}} T(-n\alpha) = \alpha^{-2n} T(-n\alpha)[I - T(\alpha)]^{2n},$$

$$B_n = \nabla \overset{2n+1}{\underset{\alpha}{\Delta}} T(-(n+1)\alpha) = \tfrac{1}{2}\alpha^{-2n-1} T(-n\alpha)[I + T(-\alpha)][I - T(\alpha)]^{2n+1}.$$

This series converges for all values of ζ if α is small; it suffices that

$$\alpha \| A \| < 2 \log (1 + \sqrt{2}),$$

A being the infinitesimal generator of the group. See N. E. Nörlund [2, p. 44].

17.7. An embedding theorem. The problem of embedding a linear bounded operator in a semi-group of operators is of considerable interest. Section 9.5 was concerned with embedding an operator in a holomorphic group. We now apply the methods of the previous section to obtain an embedding theorem for semi-groups holomorphic in a half-plane.

THEOREM 17.7.1. *Let T be a given bounded linear operator on $\mathfrak{X}$ to itself. Let $R(\lambda; T)$ be holomorphic outside the sector V: $0 \leqq |\lambda| \leqq K$, $\psi_1 \leqq \arg \lambda \leqq \psi_2$, $\psi_2 - \psi_1 < 2\pi$, and suppose that $\| R(\lambda; T) \| \leqq M |\lambda|^{-\omega}$ for $|\lambda| < K$ in the complementary sector. Then there exists a semi-group operator $W(\zeta)$, holomorphic and of exponential type for $\Re(\zeta) > \omega - 1 + \delta$, $\delta > 0$, such that $W(n) = T^n$ for*

$$n > \omega - 1.$$

We may take

$$W(\zeta) = \frac{1}{2\pi i} \int_\Gamma R(\lambda; T)\lambda^\zeta \, d\lambda, \tag{17.7.1}$$

where Γ surrounds V in the positive sense, beginning and ending at the origin.

PROOF. It is clear from (17.7.1) that $W(\zeta)$ is holomorphic for $\Re(\zeta) > \omega - 1$. Further if we take Γ to consist of two rays, $\arg \lambda = \theta_1, \theta_2$, plus a circular arc, $|\lambda| = R$, joining them, then a straightforward estimate shows that $W(\zeta)$ is of exponential type in each half-plane $\Re(\zeta) > \omega - 1 + \delta$, $\delta > 0$. It is also easy to see from Cauchy's theorem that $W(n) = T^n$ for $n > \omega - 1$. The semi-group property is proved as follows. Let Γ_1 and Γ_2 be two contours, each consisting of two rays plus a circular arc and each surrounding V in the positive sense. Suppose further that Γ_2 lies in the interior of Γ_1 except for the common end point at the origin. Assuming $\xi_\nu \equiv \Re(\zeta_\nu) > \max(\omega - 1, 0)$ for $\nu = 1, 2$ and making use of the first resolvent equation, we obtain

$$\begin{aligned} W(\zeta_1)W(\zeta_2) &= \frac{1}{(2\pi i)^2} \int_{\Gamma_1} \int_{\Gamma_2} R(\lambda; T)R(\mu; T)\lambda^{\zeta_1}\mu^{\zeta_2} \, d\lambda \, d\mu \\ &= \frac{1}{(2\pi i)^2} \int_{\Gamma_1} \int_{\Gamma_2} \frac{R(\mu; T) - R(\lambda; T)}{\lambda - \mu} \lambda^{\zeta_1}\mu^{\zeta_2} \, d\lambda \, d\mu \\ &= \frac{1}{2\pi i} \int_{\Gamma_2} R(\mu; T)\mu^{\zeta_2} \left[\frac{1}{2\pi i} \int_{\Gamma_1} \frac{\lambda^{\zeta_1} \, d\lambda}{\lambda - \mu} \right] d\mu \\ &\qquad - \frac{1}{2\pi i} \int_{\Gamma_1} R(\lambda; T)\lambda^{\zeta_1} \left[\frac{1}{2\pi i} \int_{\Gamma_2} \frac{\mu^{\zeta_2} \, d\mu}{\lambda - \mu} \right] d\lambda. \end{aligned}$$

In order to justify this last step we shall show that the inner integrals are bounded. In fact, $\int \lambda^{\zeta_1}(\lambda - \mu)^{-1} \, d\lambda$ taken along the circular arc of Γ_1 is certainly bounded for $\mu \in \Gamma_2$. Taken along one of the rays of Γ_1, say $\arg \lambda = \theta$, we see that $|\lambda - \mu| \geqq c|\lambda|$ for all $\mu \in \Gamma_2$ so that for $\zeta_1 = \xi_1 + i\eta_1$

$$\left| \int \frac{\lambda^{\zeta_1}\, d\lambda}{\lambda - \mu} \right| \leqq e^{|\eta_1 \theta|} \int_0^R \frac{|\lambda|^{\xi_1}}{c|\lambda|} d|\lambda| < \infty.$$

On evaluating the above inner integrals we obtain

$$W(\zeta_1)W(\zeta_2) = \frac{1}{2\pi i} \int_{\Gamma_2} R(\mu; T)\mu^{\zeta_1+\zeta_2}\, d\mu = W(\zeta_1 + \zeta_2).$$

This concludes the proof.

We note that if $\omega < 1$, then $W(\zeta)$ defines a holomorphic group. For if $\omega < 1$, then there exists a sequence $\{\lambda_n\} \subset \rho(T)$, $\lambda_n \to 0$, such that $\| \lambda_n R(\lambda_n; T) \| \to 0$. Hence for sufficiently large n, $[I - \lambda_n R(\lambda_n; T)]^{-1}$ exists and so does $R(0; T) = R(\lambda_n; T)\, [I - \lambda_n R(\lambda_n ; T)]^{-1}$. Thus $\lambda = 0 \in \rho(T)$, $\omega = 0$, and hence $W(\zeta)$ is holomorphic at $\zeta = 0$. It follows that $W(\zeta)$ can be extended to be a holomorphic group.

The function $W(\zeta)$ defined by (17.7.1) interpolates the sequence $\{T^n\}$; however it is not the only solution of the interpolation problem which leads to a holomorphic semi-group. If k is any integer, we can clearly multiply $W(\zeta)$ by $e^{2\pi k i \zeta}$ and still have such a solution. It is not known whether or not any other semi-group operators can interpolate the sequence $\{T^n\}$. At any rate the solution $W(\zeta)$ has an extremal property relative to all other solutions. If $U(\zeta)$ is any other solution of the interpolation problem which is holomorphic in a right half-plane, then $D(\zeta) \equiv U(\zeta) - W(\zeta)$ vanishes at all integers in the half-plane of existence of $D(\zeta)$. Its norm must then satisfy the inequality

$$\limsup_{|\eta| \to \infty} \frac{1}{|\eta|} \log \| D(\xi + i\eta) \| \geqq \pi$$

unless $D(\zeta) \equiv \Theta$ (see F. Carlson [**2**, p. 33]). On the other hand, direct estimates of the integral (17.7.1) show that

$$\text{(17.7.2)} \qquad \limsup_{|\eta| \to \infty} \frac{1}{|\eta|} \log \| W(\xi + i\eta) \| \leqq \max (|\psi_1|, |\psi_2|).$$

Hence if $-\pi < \psi_1 < \psi_2 < \pi$, we can claim that $W(\zeta)$ gives the solution to the interpolation problem with the minimum rate of growth of the norm on vertical lines. These limits are precise. For example if $T = -I$, then the spectrum consists of the single point, $\lambda = -1$, and $-\psi_1 = \psi_2 = \pi$. In this case the two semi-group operator-functions

$$W_1(\zeta) = e^{\pi i \zeta} I \qquad \text{and} \qquad W_2(\zeta) = e^{-\pi i \zeta} I$$

have the property $W_1(n) = W_2(n) = T^n$ for $n \geqq 1$ and have the same rate of growth on vertical lines.

17.8. Semi-groups with real positive spectra. We now treat the topic of section 17.6 under a somewhat more specialized hypothesis. We shall suppose that the spectrum of $T(\xi)$ is real and positive for all $\xi > 0$. According to Theorem 4.7.3,

$$\log (\text{spectral radius of } T(\xi)) = \lim_{n\to\infty} \log \| [T(\xi)]^n \|^{1/n} = \omega_0 \xi .$$

Hence our assumption implies that $0 \leqq \sigma[T(\xi)] \leqq e^{\omega_0 \xi}$. Thus if we replace $T(\xi)$ by $S(\xi) \equiv e^{-\omega_0 \xi} T(\xi)$ we see that the spectra of the operators $S(\xi)$ are confined to the interval $[0, 1]$. If $T(\xi)$ is of class (A) then so is $S(\xi)$ by Theorem 12.2.3. Consequently we may suppose without loss of generality that $0 \leqq \sigma[T(\xi)] \leqq 1$. Along with the above assumption as to the spectra of $T(\xi)$ it is natural to modify the assumptions on the rate of growth of $R(\lambda; T(\xi))$. We shall prove

THEOREM 17.8.1. *Let* $\mathfrak{S} \equiv [T(\xi); \xi > 0]$ *be a semi-group of class* (A) *such that*
(1′) *the spectrum of* $T(\xi)$ *belongs to the interval* $[0, 1]$ *for each* $\xi > 0$;
(2′) *let* $d(\lambda, \xi)$ *denote the distance of* λ *from the spectrum of* $T(\xi)$; *then there exist finite positive quantities* $C(\xi)$ *and* $\omega(\xi)$, *independent of* λ, *such that for* λ *in the resolvent set of* $T(\xi)$

$$\| R(\lambda; T(\xi)) \| \leqq C(\xi)[d(\lambda, \xi)]^{-\omega(\xi)}. \tag{17.8.1}$$

Then the conclusions of Theorem 17.6.3 *hold in the following sharper form.* $W(\zeta)$ *is of finite order in every half-plane* $\Re(\zeta) \geqq \xi_0 + \delta$, $\delta > 0$, *where*

$$\xi_0 \leqq \inf_{\alpha > 0} \alpha[\omega(\alpha) - 1].$$

If $\xi > \alpha[\omega(\alpha) - 1]$, *then the Lindelöf mu-function* $\mu(\xi; W) \leqq \omega(\alpha)$. *In particular, if* $\omega(\xi)$ *is bounded for all* $\xi > 0$, *then so is* $\mu(\xi; W)$ *and* $\xi_0 = 0$ *or* $-\infty$.

PROOF. It is clear that the assumptions of Theorem 17.6.3 are satisfied so that $W(\zeta)$ exists and is given by

$$W(\zeta) = \frac{1}{2\pi i} \int_\Gamma R(\lambda; T(\alpha))\lambda^{\zeta/\alpha}\, d\lambda \tag{17.8.2}$$

for $\Re(\zeta) > \alpha[\omega(\alpha) - 1]$. Here Γ is any closed rectifiable contour which surrounds the interval $(0, 1)$ once in the positive sense, beginning and ending at the origin, where it is not tangent to the positive real axis.

It should be observed that $\omega(\xi) \geqq 1$ for all $\xi > 0$; the contrary assumption would deprive $T(\xi)$ of a spectrum which is impossible by Theorem 4.7.4. Indeed, if $\lambda_0 \in \sigma[T(\xi)]$, if $\omega(\xi) < 1$, and if λ_1 is a point on the vertical through λ_0 and such that $C(\xi) \mid \lambda_1 - \lambda_0 \mid^{1-\omega(\xi)} < 1$, then

$$R(\lambda_0 ; T(\xi)) = R(\lambda_1 ; T(\xi))[I - (\lambda_1 - \lambda_0)R(\lambda_1 ; T(\xi))]^{-1}$$

exists; this contradicts the assumption that $\lambda_0 \in \sigma[T(\xi)]$.

In order to estimate the growth of $W(\zeta)$ in the half-plane $\Re(\zeta) > \alpha[\omega(\alpha) - 1]$ we choose a particular contour Γ. Let δ be a small number to be disposed of later and let Γ consist of the two straight line segments joining the origin with the points $1 \pm \delta i$ and the semi-circle of radius δ joining these points and passing to the right of $\lambda = 1$. The rectilinear paths give contributions to the norm of

$W(\zeta)$ of the form

$$C_1(\alpha)(1+\delta^2)^{\xi/2\alpha} \exp\left[\frac{|\eta|}{\alpha} \arctan \delta\right] \{\xi - \alpha[\omega(\alpha) - 1]\}^{-1}\delta^{-\omega(\alpha)},$$

whereas the contribution of the circular arc is bounded by

$$C_2(\alpha)(1 + \delta)^{\xi/\alpha} \exp\left[\frac{|\eta|}{\alpha} \arctan \delta\right] \delta^{-\omega(\alpha)+1}.$$

If ξ is fixed, $\xi > \alpha[\omega(\alpha) - 1]$, we choose $\delta = \alpha/|\eta|$, obtaining

$$\| W(\xi + i\eta) \| \leqq C(\alpha, \xi) \,|\eta|^{\omega(\alpha)}, \tag{17.8.3}$$

while for $\zeta = c(\alpha) + re^{i\varphi}$, $-\pi/2 \leqq \varphi \leqq \pi/2$, $c(\alpha) > \alpha[\omega(\alpha) - 1]$, we take $\delta = \alpha/r$ and get

$$\| W(c(\alpha) + re^{i\varphi}) \| \leqq C_3(\alpha)r^{\omega(\alpha)}. \tag{17.8.4}$$

These estimates prove all of the remaining assertions of the theorem.

Theorem 6.6.2 now shows that $W(\zeta)$ can be represented by a Laplace integral of the form

$$W(\zeta) = \zeta^\beta \int_0^\infty e^{-\zeta\tau} a_\beta(\tau)\, d\tau. \tag{17.8.5}$$

For $\xi_1 > \inf_{\alpha>0} \alpha[\omega(\alpha) - 1] \geqq 0$, we see by (17.8.4) and Theorem 6.5.1 that $W(\zeta)$ is of order $\mu(\xi_1 ; W)$ in the half-plane $\mathfrak{R}(\zeta) \geqq \xi_1$. Consequently the above integral converges for $\mathfrak{R}(\zeta) > \xi_1$ if $\beta > \mu(\xi_1 ; W) + 1$. The function $a_\beta(\tau)$ is defined by a formula analogous to (6.6.1); the dependence of $a_\beta(\tau)$ upon the parameter β is shown by the formula (6.6.5).

4. Boundary Value Problems

17.9. Boundary values of a semi-group. Let $\mathfrak{S} \equiv [T(\zeta)]$ be a semi-group of operators holomorphic in the right half-plane $\mathfrak{R}(\zeta) > 0$. We now add the assumption that $T(\zeta)$ tends to a limit in some sense as ζ approaches the imaginary axis. We shall show that in this case the boundary values form a group of linear bounded operators.

THEOREM 17.9.1. *Let $\mathfrak{S} \equiv [T(\zeta); \mathfrak{R}(\zeta) > 0]$ be a holomorphic semi-group of linear bounded operators of class (C_0) on $(0, \infty)$. If $\| T(\xi + i\eta) \| \leqq M$ for $0 < \xi \leqq 1$ and $|\eta| \leqq 1$, then*

$$\lim_{\xi\to 0+} T(\xi + i\eta)x \equiv T(i\eta)x, \qquad -\infty < \eta < \infty, \tag{17.9.1}$$

exists for all x, $T(i\eta) \in \mathfrak{E}(\mathfrak{X})$, $\mathfrak{S}_0 \equiv [T(i\eta);\ -\infty < \eta < \infty]$ *forms a strongly continuous group*, $T(i\eta)$ *commutes with* $T(\zeta)$ *for all admissible values of* η *and* ζ, *and*

$$T(\xi + i\eta) = T(\xi)T(i\eta). \tag{17.9.2}$$

PROOF. We first show that $\| T(\xi + i\eta) \| \leqq M(\omega)$ for $0 < \xi \leqq 1$, $| \eta | \leqq \omega$. In fact, let $\xi + i\eta = re^{i\varphi}$, $n < r \leqq n + 1$. Then $\| T(\xi + i\eta) \| = \| T(re^{i\varphi}) \| \leqq \| T(ne^{i\varphi}) \| \, \| T((r - n)e^{i\varphi}) \| \leqq M^{n+1} \leqq M^{r+1}$. For $0 < \xi \leqq 1$ and $| \eta | \leqq \omega$ we have $r = | \xi + i\eta | \leqq \omega + 1$ so that $\| T(\xi + i\eta) \| \leqq M^{\omega+2} \equiv M(\omega)$.

It is clear that the limit in (17.9.1) exists for each $x \in \mathfrak{X}_0 \equiv \bigcup_{\xi>0} T(\xi)[\mathfrak{X}]$. We have assumed that $\mathfrak{X}_0$ is dense in $\mathfrak{X}$. Thus the above bound on $\| T(\xi + i\eta) \|$ together with the Banach-Steinhaus theorem implies that this limit exists for all $x \in \mathfrak{X}$ and defines a linear bounded operator $T(i\eta)$ with $\| T(i\eta) \| \leqq M(\omega)$ for $| \eta | \leqq \omega$. Now $T(i\eta)x$ is obviously a continuous function of η, $-\infty < \eta < \infty$, for each $x \in \mathfrak{X}_0$. Again, since $\mathfrak{X}_0$ is dense in $\mathfrak{X}$ and $\| T(i\eta) \|$ is bounded on each finite interval, we may conclude in the usual manner that $T(i\eta)x$ is a continuous function of η for each $x \in \mathfrak{X}$.

We have

$$\begin{aligned} T(i\eta_1)T(i\eta_2)[T(\xi)y] &= T(i\eta_1)[T(\xi + i\eta_2)y] \\ &= T(\xi + i(\eta_1 + \eta_2))y = T(i(\eta_1 + \eta_2))[T(\xi)y], \end{aligned}$$

that is, the relation

$$T(i\eta_1)T(i\eta_2)x = T(i(\eta_1 + \eta_2))x$$

holds for all $x \in \mathfrak{X}_0$. Since $\mathfrak{X}_0$ is dense in $\mathfrak{X}$, the identity holds for all $x \in \mathfrak{X}$. Thus $T(i\eta_1)$ commutes with $T(i\eta_2)$, $T(i\eta)T(-i\eta) = T(0) = I$, and the associative property evidently holds so that $\mathfrak{S}_0$ is an abelian group. Further

$$\begin{aligned} T(i\eta)T(\zeta)x &= T(\zeta + i\eta)x = \lim_{\xi \to 0+} T(\zeta + \xi + i\eta)x = \lim_{\xi \to 0+} T(\zeta)T(\xi + i\eta)x \\ &= T(\zeta)T(i\eta)x. \end{aligned}$$

The first and last members of this relation show that $T(\zeta)$ and $T(i\eta)$ commute, and the second and last members supply the factorization given in (17.9.2).

THEOREM 17.9.2. *Let* $[T(\zeta)]$ *satisfy the assumptions of Theorem* 17.9.1 *and denote the infinitesimal generator of* $[T(\xi);\ \xi > 0]$ *by* A. *Then* iA *is the infinitesimal generator of the group* $\mathfrak{S}_0$.

PROOF. The infinitesimal generator B of a group is defined the same as though the group were regarded as a semi-group on $(0, \infty)$. However in the case of a group, it can be shown by employing the argument of Theorem 10.3.3 that $x \in \mathfrak{D}(B)$ if and only if

$$Bx = \lim_{\delta \to 0} \delta^{-1}[T(i\delta) - I]x$$

The argument of Theorem 10.3.3 also shows that if $x \in \mathfrak{D}(B)$ then

$$BT(\zeta)x = T(\zeta)Bx = \frac{\partial}{\partial\eta} T(\xi + i\eta)x$$

for all ζ with $\mathfrak{R}(\zeta) \geqq 0$. From the analyticity of $T(\zeta)$ for $\mathfrak{R}(\zeta) > 0$, it follows that

$$\frac{\partial}{\partial\xi} T(\xi + i\eta)x = -i\frac{\partial}{\partial\eta} T(\xi + i\eta)x, \qquad \xi > 0,$$

for all $x \in \mathfrak{X}$. Thus if $x \in \mathfrak{D}(B)$ we have

$$iAT(\xi)x = BT(\xi)x = T(\xi)Bx \to Bx$$

as $\xi \to 0+$. Since iA is closed and $T(\xi)x \to x$ we see that $iAx = Bx$, that is, $B \subset iA$. On the other hand if $x \in \mathfrak{D}(A)$, then

$$-iBT(\xi)x = AT(\xi)x = T(\xi)Ax \to Ax$$

as $\xi \to 0+$. Again B being closed and $T(\xi)x \to x$ we see that $-iBx = Ax$ so that $A \subset -iB$. This proves that $B = iA$.

For semi-groups of the above kind it is clear that on the positive real axis $\| T(\zeta) \|$ satisfies an inequality of the form

$$\| T(\xi) \| \leqq C_1 e^{\alpha\xi}, \qquad \alpha \geqq 0, \xi > 0,$$

and on the imaginary axis

$$\| T(i\eta) \| \leqq C_2 e^{\beta|\eta|}, \qquad \beta > 0.$$

Hence by (17.9.2)

$$\| T(\xi + i\eta) \| \leqq C_1 C_2 e^{\alpha\xi + \beta|\eta|},$$

that is, $T(\zeta)$ is of exponential type in the closed right half-plane and its indicator $\mathfrak{d}(\varphi)$ satisfies

$$\mathfrak{d}(\varphi) \leqq \alpha \cos\varphi + \beta \,|\sin\varphi\,|, \qquad -\pi/2 \leqq \varphi \leqq \pi/2. \tag{17.9.3}$$

According to Theorem 17.4.1 we see that D^-, the closed convex extension of the spectrum of A (here $D^- = D_0^-$), is contained in the semi-infinite strip: $\sigma \leqq \alpha$, $-\beta \leqq \tau \leqq \beta$.

17.10. Groups admitting holomorphic semi-group extensions. We come now to the converse problem. We suppose given a one-parameter group of linear bounded operators and wish to determine when such a group forms the boundary values of a holomorphic semi-group.

THEOREM 17.10.1. *Let $\mathfrak{S}_0 \equiv [T(i\eta);\, -\infty < \eta < \infty]$ be a strongly continuous group of linear bounded operators on $\mathfrak{X}$ to itself with $T(0) = I$. Let iA be the infinitesimal generator of $\mathfrak{S}_0$. A necessary and sufficient condition for the existence of a semi-group $[T(\zeta);\, \mathfrak{R}(\zeta) > 0]$ such that* (i) *$T(\zeta)$ is holomorphic for $\mathfrak{R}(\zeta) > 0$,*

(ii) *the indicator* $\mathfrak{d}(\varphi)$ *of* $T(\zeta)$ *satisfies* (17.9.3), *and* (iii) $\lim_{\xi \to 0+} T(\xi + i\eta)x = T(i\eta)x$ *for all* $x \in \mathfrak{X}$, *is that* A *satisfies the conditions of Theorem* 17.5.1 *with* D^{-} *contained in the strip*: $\sigma \leqq \alpha,\ -\beta \leqq \tau \leqq \beta$.

PROOF. We have already shown in the previous section that the above conditions are necessary. We now prove the sufficiency. By Theorem 17.5.1 we see that A is the infinitesimal generator of a semi-group of operators, which, for the moment, we denote by $[W(\zeta);\ \mathfrak{R}(\zeta) > 0]$, and $W(\zeta)$ satisfies the conditions (i) and (ii). It remains to prove (iii). For this purpose we note that $W(\xi)$ and $T(i\eta)$ commute. This follows from the fact that $W(\xi)$ commutes with $R(\lambda; A)$ and hence also with $R(\lambda; iA) = -iR(-\lambda i; A)$ when the latter exists. However $T(i\eta)$ is obtained from $R(\lambda; iA)$ by a Laplace inversion formula, as for instance in Theorem 11.6.2, whence it follows that $W(\xi)$ also commutes with $T(i\eta)$. The operator

$$T(\zeta) \equiv W(\xi)T(i\eta), \qquad \zeta = \xi + i\eta, \tag{17.10.1}$$

is well defined and a strongly continuous function of ζ for $\xi > 0$ since the factors have this property. Further for $x \in \mathfrak{D}(A)$

$$\frac{\partial}{\partial \xi} T(\zeta)x = AT(\zeta)x, \qquad \frac{\partial}{\partial \eta} T(\zeta)x = iAT(\zeta)x.$$

Since these partials are continuous functions of ζ in the half-plane $\mathfrak{R}(\zeta) > 0$, we conclude by an obvious modification of the classical argument that $T(\zeta)x$ has a unique derivative for $x \in \mathfrak{D}(A)$. Consequently $T(\zeta)x$ is holomorphic for $\mathfrak{R}(\zeta) > 0$ and $x \in \mathfrak{D}(A)$. However $T(0) = I$, so that $T(\xi)x \equiv W(\xi)x$ for $\xi > 0$. Thus $T(\zeta)x$ and $W(\zeta)x$ are both holomorphic functions of ζ in $\mathfrak{R}(\zeta) > 0$ for $x \in \mathfrak{D}(A)$ and coincide on the positive real axis. It follows that $T(\zeta)x \equiv W(\zeta)x$ for $\mathfrak{R}(\zeta) > 0$ and each $x \in \mathfrak{D}(A)$. Since $\mathfrak{D}(A)$ is dense in $\mathfrak{X}$, we see that $T(\zeta) \equiv W(\zeta)$ for $\mathfrak{R}(\zeta) > 0$. The condition (iii) is now a direct consequence of the definition of $T(\zeta)$. This completes the proof.

CHAPTER XVIII

APPLICATIONS TO ERGODIC THEORY

18.1. Orientation. The original *ergodic hypothesis* concerned the long run average behavior of individual trajectories in the phase space $\Sigma \equiv [\sigma]$ of a dynamical system. Such trajectories define a measure preserving flow $\varphi(\sigma, \xi)$ in Σ. The first *ergodic theorems*, which were due to G. D. Birkhoff, T. Carleman, B. O. Koopman, and J. von Neumann, evolved in 1931–1932. The problem as then formulated dealt with the isometry $[T(\xi)x](\sigma) = x[\varphi(\sigma, \xi)]$ and consisted in establishing the existence of $\lim_{\xi\to\infty}\xi^{-1}\int_0^\xi T(\tau)x\,d\tau$, the individual ergodic theorem giving pointwise convergence for almost all σ, $x \in L_1(\Sigma)$, and the mean ergodic theorem giving norm convergence in $L_2(\Sigma)$. Although each of these theorems now has a large literature, we shall limit our remarks to a few of the generalizations of the mean ergodic theorem.

In 1938 the mean ergodic theorem was extended to a more general class of linear operators and to arbitrary (B)-spaces. The initial step was taken by C. Visser who established weak convergence for a general class of linear operators in $L_2(\Sigma)$; F. Riesz obtained strong convergence for the same class of operators in $L_p(\Sigma), p \geqq 1$, and, independent of Riesz, S. Kakutani and K. Yosida obtained strong convergence for a somewhat larger class of operators in an arbitrary (B)-space. The n-parameter case was considered in the following year by N. Dunford and N. Wiener. At about the same time a much more general formulation was given to the mean ergodic theorem by L. Alaoglu and G. Birkhoff who considered directed sets of means of an operator representation of a general semi-group. The ergodic theorem in this general setting has been discussed by M. M. Day and W. F. Eberlein. Although most of ergodic theory deals with the infinitary behavior of semi-groups of operators, N. Wiener has also obtained similar results for the behavior at the origin.

Since ergodic theory deals with the behavior of semi-groups of linear bounded operators at $\xi = 0$ and $\xi = \infty$, it necessarily forms an important chapter in a treatise such as this. In general the limits $\lim_{\xi\to 0+}T(\xi)$ and $\lim_{\xi\to\infty}T(\xi)$ do not exist in any of the standard operator topologies of $\mathfrak{E}(\mathfrak{X})$. However, the mean value $\xi^{-1}\int_0^\xi T(\tau)\,d\tau$ tends to a limit as $\xi \to 0+$ or ∞ under fairly general assumptions. Thus the $(C, 1)$ limits often exist. This fact naturally raises the question whether or not other generalized limits exist, in particular the Cesàro (C, α) and the Abel limits, and if the relations between these limits are governed by theorems of Abelian and Tauberian nature. The earliest work here is due to L. W. Cohen who considered the (C, α)-limit in the discrete case.

The study of the Abel limit is particularly appropriate and interesting because it is connected with the spectral properties of the operators. This is easy to see in the discrete case since

$$(A)\text{-}\lim_{n\to\infty} T^n = \lim_{r\to 1-} (1 - r) \sum_{n=0}^{\infty} r^n T^n,$$

or, setting $r = 1/\lambda$, we have

$$(A)\text{-}\lim_{n\to\infty} T^n = \lim_{\lambda\to 1+} (\lambda - 1)R(\lambda; T), \tag{18.1.1}$$

where $R(\lambda; T)$ is the resolvent of T. Similarly in the continuous case we have

$$(A)\text{-}\lim_{\xi\to\infty} T(\xi) = \lim_{\lambda\to 0+} \lambda R(\lambda; A), \tag{18.1.2}$$

where A is the infinitesimal generator of the semi-group $\mathfrak{S}$. It is clear that the spectral properties of T at $\lambda = 1$ and of A at $\lambda = 0$ are decisive for the existence of these limits. According as the limit in question exists in the uniform or the strong or the weak operator topology we say that $\mathfrak{S}$ is Abel-ergodic in the uniform or the strong or the weak sense respectively, and we use a similar terminology for other mean values.

The relations between the spectral properties of the operator T and the convergence of a sequence of operator functions $\{f_n(T)\}$ to a projection were investigated by N. Dunford in 1943; his results unified most of the known mean ergodic theorems and introduced resolvent techniques in ergodic theory. Dunford's work led E. Hille to investigate the connections with Abel summability and Tauberian theorems in 1945. Following this, Hille obtained a unified treatment of ergodic theory for one-parameter semi-groups based on the properties of the resolvent of the infinitesimal generator. Since the latter is ordinarily not a bounded operator, Dunford's methods do not as a rule apply, however his results carry over to a large extent. Further results along this line were obtained by R. S. Phillips.

There are two paragraphs: *Abelian and Tauberian Theorems* and *Ergodic Theory*. The subsequent list of references has been held down to a minimum. For a more complete bibliography on ergodic theory we refer to E. Hopf [1] and S. Kakutani [7].

References. Alaoglu and G. Birkhoff [**1, 2**], Andersen [**1**], Cohen [**1**], Day [**3, 4**], Dunford [**5, 6, 7, 8, 9**], Dunford and J. Schwartz [**1**], Eberlein [**2, 3**], Hille [**12, 13**], E. Hopf [**1**], Kakutani [**2, 7**], Phillips [**5**], Pitt [**1**], F. Riesz [**5**], Visser [**1**], Wiener [**3, 4**], Yosida [**2**].

1. Abelian and Tauberian Theorems

18.2. Averages. We start by recalling some conventions regarding generalized limits. Let $x(\xi)$ be a function on $(0, \infty)$ to a complex (B)-space $\mathfrak{X}$ such that

$$e^{-\lambda\xi}x(\xi) \in B[(0, \infty); \mathfrak{X}]$$

for every positive λ. A generalized limit for such a function may exist as $\xi \to \infty$ even though the function does not tend to a limit in the usual sense. We are primarily interested in the (C, α) and the Abel limits, the first being based on the fractional integral of $x(\xi)$ of order α, the second on the Laplace transform of $x(\xi)$.

We set

$$x_\alpha(\xi) = \frac{1}{\Gamma(\alpha)} \int_0^\xi (\xi - \tau)^{\alpha-1} x(\tau)\, d\tau, \qquad \alpha > 0.$$

This is the fractional integral of $x(\xi)$ of order α. If $1 \leqq \alpha$, then the integral exists for all values of ξ and is an absolutely continuous function of ξ. If $0 < \alpha < 1$, a familiar application of the Fubini theorem shows that the integral exists for almost all ξ and is itself integrable on each finite subinterval of $[0, \infty)$. Dividing $x_\alpha(\xi)$ by the fractional integral of unity of order α, we get the (C, α) *average*, namely

$$(18.2.1) \qquad a[\xi; x(\cdot), \alpha] \equiv \alpha\xi^{-\alpha} \int_0^\xi (\xi - \tau)^{\alpha-1} x(\tau)\, d\tau.$$

We now define

$$(18.2.2) \qquad (C, \alpha)\text{-}\lim_{\xi\to\infty} x(\xi) = \lim_{\xi\to\infty} a[\xi; x(\cdot), \alpha],$$

whenever the right member exists. For $\alpha = 0$, the $(C, 0)$-limit of $x(\xi)$ is defined to be the ordinary Cauchy limit. We say that *$x(\xi)$ is bounded (C, α) if $a[\xi; x(\cdot), \alpha]$ is a bounded function of ξ in $(0, \infty)$.*

For the Abel limit we set

$$(18.2.3) \qquad L[\lambda; x(\cdot)] \equiv \lambda \int_0^\infty e^{-\lambda\xi}\, x(\xi)\, d\xi,$$

and define

$$(18.2.4) \qquad (A)\text{-}\lim_{\xi\to\infty} x(\xi) = \lim_{\lambda\to 0+} L[\lambda; x(\cdot)],$$

whenever the right member has a sense.

Abelian and Tauberian type theorems are theorems relating different generalized limits. The class of all these limits can be partially ordered in the sense that the existence of a "weaker" limit always implies the existence of a "stronger" limit. A theorem establishing such a relation between two generalized limits is called an Abelian theorem. A Tauberian theorem, on the other hand, establishes the existence of a given generalized limit for a restricted class of functions on the assumption that the limit can be obtained by some stronger limit process. Thus Abelian theorems go from weak to strong limit methods and Tauberian theorems go in the opposite direction.

The following Abelian theorem, which is very well known in the numerical case, is basic.

THEOREM 18.2.1. *If for a fixed* α, $\alpha \geqq 0$, *we have*

$$(C, \alpha)\text{-}\lim_{\xi\to\infty} x(\xi) = a,$$

then

$$(C, \beta)\text{-}\lim_{\xi\to\infty} x(\xi) = (A)\text{-}\lim_{\xi\to\infty} x(\xi) = a, \qquad \beta > \alpha.$$

PROOF. The relation between the Cesàro averages is given by

$$a[\xi; x(\cdot), \beta] = \frac{\Gamma(\beta + 1)}{\Gamma(\alpha + 1)\Gamma(\beta - \alpha)} \xi^{-\beta} \int_0^\xi (\xi - \tau)^{\beta-\alpha-1}\tau^\alpha a[\tau; x(\cdot), \alpha]\, d\tau \tag{18.2.5}$$

which follows from the fact that $x_\beta(\xi)$ is the fractional integral of $x_\alpha(\xi)$ of order $(\beta - \alpha)$. Using formula (6.2.16) one verifies that the Abel and the Cesàro averages are related by

$$L[\lambda; x(\cdot)] = \frac{\lambda^{\alpha+1}}{\Gamma(\alpha + 1)} \int_0^\infty e^{-\lambda\tau}\tau^\alpha a[\tau; x(\cdot), \alpha]\, d\tau, \qquad \lambda > 0. \tag{18.2.6}$$

The assertions of the theorem are simple consequences of these two formulas; we leave the details to the reader.

18.3. Some Tauberian theorems. The classical Tauberian theorems are of two distinct types according as the postulated Tauberian condition is one-sided or two-sided. A two-sided condition involves only absolute values and the corresponding theorem can usually be extended to vector-valued functions, replacing absolute values by norms. One-sided conditions involve inequalities, that is, a notion of ordering or of positivity and any contemplated extension to abstract spaces would involve partial ordering as a preliminary step. Only two-sided types will be considered here; the discussion will be based on the vector-valued form of Wiener's general Tauberian theorem, established in section 4.21.

We shall need the notion of *slow oscillation* which goes back to R. Schmidt.

DEFINITION 18.3.1. *A function* $x(\xi)$ *on* $(-\infty, \infty)$ *to* $\mathfrak{X}$ *will be called slowly oscillating when* $\xi \to +\infty$ *if*

$$\lim \| x(\xi) - x(\tau) \| = 0 \qquad (\xi \to \infty,\ \xi - \tau \to 0).$$

A function $y(\xi)$ *on* $(0, \infty)$ *to* $\mathfrak{X}$ *will be called feebly oscillating when* $\xi \to +\infty$ *if*

$$\lim \| y(\xi) - y(\tau) \| = 0 \qquad (\xi \to \infty,\ \xi/\tau \to 1).$$

The following is a special case of a theorem due to H. R. Pitt [1, Theorem 9]. For the terminology we refer to Definition 4.21.1.

THEOREM 18.3.1. *If* $g(\tau) \in W(0, \infty)$, *if* $y(\tau)$ *is a bounded strongly measurable function on* $(0, \infty)$ *to* $\mathfrak{X}$ *which is also feebly oscillating, and if*

$$\lim_{\xi\to\infty} \frac{1}{\xi} \int_0^\infty g\left(\frac{\tau}{\xi}\right) y(\tau)\, d\tau = a \int_0^\infty g(\tau)\, d\tau,$$

then

$$\lim_{\xi\to\infty} y(\xi) = a.$$

PROOF. Without loss of generality we may assume that $a = \theta$. Since $y(\tau)$ is feebly oscillating, we see that to each $\epsilon > 0$ there corresponds a $\xi(\epsilon)$ and a $\delta = \delta(\epsilon)$ such that for $\xi > \xi(\epsilon)$ and $\xi \leqq \tau \leqq (1 + \delta)\xi$ we have $\| y(\tau) - y(\xi) \| \leqq \epsilon$. Define

$$k(\tau) = \begin{cases} 1/\delta, & 1 < \tau < 1 + \delta, \\ 0, & \text{elsewhere.} \end{cases}$$

Then $k(\tau) \in L(0, \infty)$ and by Theorem 4.21.3

$$\lim_{\xi\to\infty} \frac{1}{\xi\delta} \int_\xi^{\xi(1+\delta)} y(\tau)\, d\tau = a = \theta.$$

But

$$\frac{1}{\xi\delta} \int_\xi^{\xi(1+\delta)} y(\tau)\, d\tau = y(\xi) + \frac{1}{\xi\delta} \int_\xi^{\xi(1+\delta)} [y(\tau) - y(\xi)]\, d\tau$$

where the second term on the right has a norm not exceeding ϵ when $\xi > \xi(\epsilon)$; and from this it readily follows that $\lim_{\xi\to\infty} y(\xi) = \theta$. This completes the proof.

We have purposely formulated the above result for the interval $(0, \infty)$; there is of course an analogous result for $(-\infty, \infty)$ and slowly oscillating functions.

As a preparation for Theorem 18.3.3, we now prove a result concerning Cesàro averages.

THEOREM 18.3.2. *If* $x(\xi)$ *is a measurable function on* $(0, \infty)$ *to* $\mathfrak{X}$, *if* $x(\xi)$ *is integrable over every finite interval* $(0, \omega)$, *and if* $a[\xi; x(\cdot), \alpha]$ *is a bounded function of* ξ *over* $(0, \infty)$, *for some fixed* $\alpha \geqq 0$, *then* $a[\xi; x(\cdot), \beta]$ *is bounded and feebly oscillating for each* $\beta > \alpha$. *Here* $a[\xi; x(\cdot), 0] \equiv x(\xi)$ *by definition.*

PROOF. Formula (18.2.5) gives

$$\| a[\xi; x(\cdot), \beta] \| \leqq \sup_{0<\tau<\xi} \| a[\tau; x(\cdot), \alpha] \| .$$

This shows that boundedness is preserved by the Cesàro average when the index is raised. The property of feeble oscillation lies a bit deeper. We may obviously restrict ourselves to the case in which $0 < \beta - \alpha < 1$. Set $a_\beta(\xi) \equiv a[\xi; x(\cdot), \beta]$, $\gamma + 1 \equiv \beta - \alpha$, and suppose that $0 < \xi < \eta$. We consider the difference

$$a_\beta(\eta) - a_\beta(\xi) = [\mathrm{B}(\alpha+1, \gamma+1)]^{-1}\left\{\eta^{-\beta}\int_0^\eta (\eta-\tau)^\gamma\tau^\alpha a_\alpha(\tau)\,d\tau - \xi^{-\beta}\int_0^\xi (\xi-\tau)^\gamma\tau^\alpha a_\alpha(\tau)\,d\tau\right\}.$$

The expression within the braces may be written as

$$\eta^{-\beta}\int_\xi^\eta (\eta-\tau)^\gamma\tau^\alpha a_\alpha(\tau)\,d\tau - \eta^{-\beta}\int_0^\xi [(\xi-\tau)^\gamma - (\eta-\tau)^\gamma]\tau^\alpha a_\alpha(\tau)\,d\tau - (\xi^{-\beta}-\eta^{-\beta})\int_0^\xi (\xi-\tau)^\gamma\tau^\alpha a_\alpha(\tau)\,d\tau,$$

the norm of which does not exceed ess sup $\| a_\alpha(\tau) \|$ times

$$\eta^{-\beta}\int_\xi^\eta (\eta-\tau)^\gamma\tau^\alpha\,d\tau + \eta^{-\beta}\int_0^\xi [(\xi-\tau)^\gamma - (\eta-\tau)^\gamma]\tau^\alpha\,d\tau + (\xi^{-\beta}-\eta^{-\beta})\int_0^\xi (\xi-\tau)^\gamma\tau^\alpha\,d\tau = 2\eta^{-\beta}\int_\xi^\eta (\eta-\tau)^\gamma\tau^\alpha\,d\tau < \frac{2}{\gamma+1}\left(\frac{\eta-\xi}{\eta}\right)^{\gamma+1};$$

here we have made use of the relation $\eta^{-\beta}\int_0^\eta(\eta-\tau)^\gamma\tau^\alpha\,d\tau = \xi^{-\beta}\int_0^\xi(\xi-\tau)^\gamma\tau^\alpha\,d\tau$. It follows that

$$\| a[\eta; x(\cdot), \beta] - a[\xi; x(\cdot), \beta] \| \leqq C(\alpha, \beta)\left[1 - \frac{\xi}{\eta}\right]^{\beta-\alpha} \text{ess sup } \| a[\tau; x(\cdot), \alpha] \|. \tag{18.3.1}$$

Hence $a[\xi; x(\cdot), \beta]$ is feebly oscillating and the theorem is proved.

We come now to the Tauberian theorem which will serve as the basis for the applications to ergodic theory. Its prototype is a result for numerical series due to A. F. Andersen [1, p. 80] which was extended to vector-valued series by E. Hille [**12**, pp. 250–251].

THEOREM 18.3.3. *If $x(\xi)$ is a function on $(0, \infty)$ to $\mathfrak{X}$ such that $e^{-\lambda\xi}x(\xi) \in B[(0, \infty); \mathfrak{X}]$ for every $\lambda > 0$, if for some $\alpha \geqq 0$,* ess sup $\| a[\xi; x(\cdot), \alpha] \|$ *is finite, and if*

$$(A)\text{-}\lim_{\xi\to\infty} x(\xi) = a,$$

then for each $\beta > \alpha$

$$(C, \beta)\text{-}\lim_{\xi\to\infty} x(\xi) = a.$$

The conclusion also holds for $\beta = \alpha$ if it is further assumed that $a[\xi; x(\cdot), \alpha]$ is feebly oscillating.

PROOF. We use formula (18.2.6) replacing α by β and λ by $1/\xi$. The assumption may then be written as

$$\lim_{\xi\to\infty}\frac{1}{\xi}\int_0^\infty e^{-\tau/\xi}\left(\frac{\tau}{\xi}\right)^\beta a[\tau; x(\cdot), \beta]\, d\tau = a\Gamma(\beta+1).$$

Here $e^{-\tau}\tau^\beta \in W(0, \infty)$ and $\Gamma(\beta+1) = \int_0^\infty e^{-\tau}\tau^\beta\, d\tau$. According to the preceding theorem $a[\tau; x(\cdot), \beta]$ is feebly oscillating when $\tau \to \infty$ for $\beta > \alpha$; the desired conclusion then follows from Theorem 18.3.1. This completes the proof.

We can express the above result more concisely: *A function which is bounded* (C, α) *and limitable Abel, is limitable* (C, β) *with the same limit for each* $\beta > \alpha$. This is obviously a Tauberian theorem for Laplace integrals and could be formulated as such.

So far we have been concerned with the behavior of $x(\xi)$ when $\xi \to \infty$. There are of course corresponding problems for $\xi \to 0+$. We define the Cesàro and Abel averages in the same manner, but replace $\xi \to \infty$ by $\xi \to 0+$ and $\lambda \to 0+$ by $\lambda \to \infty$. A function $y(\xi)$ on $(0, \infty)$ to $\mathfrak{X}$ is said to be feebly oscillating when $\xi \to 0+$ if $\lim \| y(\xi) - y(\tau) \| = 0$ $(\xi \to 0+, \xi/\tau \to 1)$.

Theorems 18.2.1 *and* 18.3.3 *remain valid as results about limits when* $\xi \to 0+$. This is fairly obvious in the case of the former theorem but the latter requires some additional remarks. We have

$$\begin{aligned} L[\lambda; x(\cdot)] &= \frac{\lambda^{\beta+1}}{\Gamma(\beta+1)}\int_0^\infty e^{-\lambda\tau}\tau^\beta a[\tau; x(\cdot), \beta]\, d\tau \\ &= \frac{1}{\lambda}\frac{1}{\Gamma(\beta+1)}\int_0^\infty e^{-\lambda/\sigma}\left(\frac{\sigma}{\lambda}\right)^{-\beta-2} a[1/\sigma; x(\cdot), \beta]\, d\sigma. \end{aligned}$$

This expression is of the type occurring in Pitt's theorem; the kernel $g(\sigma) = e^{-1/\sigma}\sigma^{-\beta-2} \in W(0, \infty)$ and formula (18.3.1) shows that $a[1/\sigma; x(\cdot), \beta]$ is feebly oscillating for $\sigma \to \infty$ if $\| a[1/\sigma; x(\cdot), \alpha] \|$ is bounded. Thus the existence of the Abel limit implies that of the (C, β)-limit if $x(\xi)$ is bounded (C, α), $\beta > \alpha$.

2. Ergodic Theory

18.4. Ergodicity. Let $\mathfrak{S} = [T(\xi); \xi > 0]$ be a semi-group of linear bounded operators on a (B)-space $\mathfrak{X}$ to itself. Heretofore our main concern has been with semi-groups which converge in some sense to the identity as $\xi \to 0+$. However, the possibility of $T(\xi)$ converging to a projection operator other than the identity was briefly considered in Chapters IX and X. We now return to this problem and give a more systematic treatment of the behavior of $T(\xi)$ both as $\xi \to 0+$ and ∞.

We shall use the term "ergodic theorem" in referring to any proposition asserting that the semi-group operator $T(\xi)$ *has a generalized limit in one sense or another as* $\xi \to 0+$ *or* ∞. Ergodic theorems may be classified from two different points of view:

(1) *the analytic nature of the limit;*

(2) *the topological nature of the limit.*

From the first point of view we distinguish between Cesàro limits, Abel limits, and so on. Under the second heading we distinguish between weak, strong, uniform, and point-wise convergence. Thus if the (C, α)-limit is employed and the averages exist and converge in the strong operator topology, we speak of *a strong* (C, α)*-ergodic theorem* and say that $\mathfrak{S}$ or $T(\xi)$ *is strongly* (C, α)*-ergodic at* $\xi = 0$ or ∞, as the case may be.

The ergodic theorems considered in the following will involve (C, α) and Abel means exclusively. Our primary interest is with weak and strong convergence, which, incidentally, are essentially equivalent in these cases. The uniform case is also treated. However point-wise convergence is outside the scope of our methods. Exhaustive results for point-wise convergence have recently been announced by N. Dunford and J. Schwartz [1]. See also their forthcoming treatise *Spectral Theory with Applications*, Chap. VIII, §6.

In order to obtain a sufficiently general setting for our results we now introduce a new class of semi-groups.

DEFINITION 18.4.1. *A semi-group* $\mathfrak{S}$ *of type* ω_0 *will be said to be of class* (E) *if*

(i) $T(\xi)$ *is continuous in the strong operator topology for* $\xi > 0$;

(ii) $\mathfrak{X}_2 \equiv [x; \int_0^1 \| T(\xi)x \| \, d\xi < \infty]$ *is dense in* $\mathfrak{X}$;

(iii) *the linear operator*

$$R(\lambda)x \equiv \int_0^\infty e^{-\lambda\xi} T(\xi)x \, d\xi \tag{18.4.1}$$

is defined and bounded on $\mathfrak{X}_2$ *for each* λ *with* $\lambda > \omega_0$.

REMARK. If the subspace $\mathfrak{X}_2$ is of the second category in $\mathfrak{X}$ then it necessarily coincides with $\mathfrak{X}$. To see this we define the transformation $Wx = T(\xi)x$ on $\mathfrak{X}_2$ to $B[(0, 1); \mathfrak{X}]$. It is readily verified that W is linear and closed. Consequently Theorem 2.12.3 applies and hence the domain $\mathfrak{X}_2$ must be all of $\mathfrak{X}$.

We have previously introduced the subspaces $\mathfrak{X}_0 \equiv \bigcup_{\xi>0} T(\xi)[\mathfrak{X}]$ and $\mathfrak{X}_1 \equiv [x; \lim_{\xi\to 0+} T(\xi)x = x]$. It is clear that $\mathfrak{X}_2 \supset \mathfrak{X}_1 \supset \mathfrak{X}_0$ so that $\bar{\mathfrak{X}}_2 = \mathfrak{X}$ is a weaker assertion than $\bar{\mathfrak{X}}_0 = \mathfrak{X}$. Since $\mathfrak{X}_2$ is assumed to be dense in $\mathfrak{X}$, it follows that $R(\lambda)$ has a unique bounded linear extension belonging to $\mathfrak{E}(\mathfrak{X})$; we shall also denote this extension by the symbol $R(\lambda)$, distinguishing when necessary between the original operator and its extension by explicitly mentioning the domain.

The following properties of semi-groups of class (E) are essential for our purposes.

THEOREM 18.4.1. *Let* $\mathfrak{S}$ *be a semi-group of class* (E). *Then the operators*

$[R(\lambda); \lambda > \omega_0]$ *with domain* $\mathfrak{X}$ *satisfy the first resolvent equation, that is*

$$R(\lambda) - R(\mu) = (\mu - \lambda)R(\lambda)R(\mu), \qquad \lambda, \mu > \omega_0. \tag{18.4.2}$$

PROOF. We first prove the assertion for the operators $R(\lambda)$ defined by (18.4.1) on $\mathfrak{X}_2$; the extension to $R(\lambda) \in \mathfrak{E}(\mathfrak{X})$ is then immediate. For fixed $x \in \mathfrak{X}_2$, we have $T(\eta)R(\lambda)x = \int_0^\infty e^{-\lambda\xi}T(\eta + \xi)x\, d\xi$, $\lambda > \omega_0$, so that

$$\| T(\eta)R(\lambda)x - R(\lambda)x \| \leqq \int_0^\infty e^{-\lambda\xi} \| T(\eta + \xi)x - T(\xi)x \| \, d\xi.$$

According to Theorem 3.8.3 the right member converges to zero as $\eta \to 0+$ and therefore $R(\lambda)x \in \mathfrak{X}_2$. Thus $R(\mu)R(\lambda)x$ is well defined for $\lambda, \mu > \omega_0$. Now

$$R(\mu)R(\lambda)x = \int_0^\infty \int_0^\infty e^{-\mu\sigma}e^{-\lambda\xi}T(\sigma + \xi)x \, d\xi \, d\sigma$$

and applying the Fubini theorem we obtain

$$\begin{aligned} R(\mu)R(\lambda)x &= \int_0^\infty e^{-(\mu-\lambda)\sigma}\left[\int_0^\infty e^{-\lambda(\sigma+\xi)}T(\sigma + \xi)x\, d\xi\right] d\sigma \\ &= \int_0^\infty e^{-\lambda\xi}T(\xi)x\left[\int_0^\xi e^{-(\mu-\lambda)\sigma}\, d\sigma\right] d\xi \\ &= \int_0^\infty e^{-\lambda\xi}T(\xi)x\left[\frac{e^{-(\mu-\lambda)\xi}}{\lambda - \mu} - \frac{1}{\lambda - \mu}\right] d\xi \\ &= (\lambda - \mu)^{-1}[R(\mu)x - R(\lambda)x]. \end{aligned}$$

This is the required relation.

COROLLARY. *If* $\mathfrak{S}$ *is of class* (E), *then* $R(\lambda)$ *is holomorphic for* $\lambda \in (\omega_0, \infty)$.

PROOF. The previous theorem shows that $R(\lambda)$ is a pseudo-resolvent on (ω_0, ∞) and the argument employed in the proof of Theorem 5.8.2 establishes the holomorphic property (see also Theorem 5.8.6).

The infinitesimal operator A_0 is defined for semi-groups of class (E) just as in section 10.3. Again $\bigcap_n \mathfrak{D}(A_0^n)$ is dense in $\mathfrak{X}_0$. However, the theory of the infinitesimal generator as presented in section 11.5 is no longer valid. In particular, $R(\lambda)$ need not be the resolvent of the infinitesimal generator.

THEOREM 18.4.2. *If* $\mathfrak{S}$ *is a semi-group of class* (A), *then* $\mathfrak{S}$ *belongs to the class* (E) *and the bounded linear extension of* $R(\lambda)$, *defined by* (18.4.1) *on* $\mathfrak{X}_2$ *for* $\mathfrak{R}(\lambda) > \omega_0$, *is* $R(\lambda; A)$.

PROOF. Suppose $[T(\xi)]$ is of class (A). Then conditions (i) and (ii) of Definition 18.4.1 are clearly satisfied. According to Corollary 2 to Theorem 11.5.3, if $R(\lambda)$ is defined as in (iii) then it is bounded on $\mathfrak{X}_2$ for all λ with $\mathfrak{R}(\lambda) > \omega_1 > \omega_0$, having the resolvent $R(\lambda; A)$ as its extension on $\mathfrak{X}$. On the other hand, Corollary 1 to Lemma 16.3.2 asserts that $R(\lambda; A)$ exists for all λ with $\mathfrak{R}(\lambda) > \omega_0$. Hence the left member of (18.4.1) has an analytic extension to all λ with $\Re(\lambda) > \omega_0$. It is clear that the right member of (18.4.1) also has an analytic extension to all λ with $\Re(\lambda) > \omega_0$ for each $x \in \mathfrak{X}_2$. It follows that $R(\lambda)x = R(\lambda; A)x$ for each $x \in \mathfrak{X}_2$ and all λ, $\Re(\lambda) > \omega_0$, so that condition (iii) is satisfied in its entirety.

We generalize the notion of a Cesàro average as follows.

DEFINITION 18.4.2. *Let $[T(\xi)]$ be a semi-group of class (E) and suppose for a given $\alpha > 0$ and $\xi > 0$ that*

$$a[\xi; T(\cdot)x, \alpha] \equiv \alpha\xi^{-\alpha} \int_0^\xi (\xi - \tau)^{\alpha-1} T(\tau)x \, d\tau \tag{18.4.3}$$

is defined and bounded on $\mathfrak{X}_2$. We denote the unique bounded linear extension of this operator by $C(\xi; \alpha)$. We further define $C(\xi; 0) \equiv T(\xi)$.

We are now ready to formulate the basic definitions for ergodicity. We suppose α to be a fixed number, $\alpha \geqq 0$.

DEFINITION 18.4.3. *Let $[T(\xi)]$ be a semi-group of class (E) and of type $\omega_0 \leqq 0$. The semi-group is said to be weakly (strongly, or uniformly) (C, α)-ergodic at infinity if the operator $C(\xi; \alpha)$ exists for all $\xi > 0$, if $\int_0^\infty e^{-\lambda\xi} \| C(\xi, \alpha)x \| \, d\xi < \infty$ for each $x \in \mathfrak{X}$ and $\lambda > \max[0, \omega_0]$, and if*

$$(C, \alpha)\text{-}\lim_{\xi\to\infty} T(\xi) \equiv \lim_{\xi\to\infty} C(\xi; \alpha) \tag{18.4.4}$$

exists in the weak (strong, or uniform) operator topology. The semi-group is said to be weakly (strongly, or uniformly) Abel-ergodic at infinity if

$$(A)\text{-}\lim_{\xi\to\infty} T(\xi) \equiv \lim_{\lambda\to 0+} \lambda R(\lambda) \tag{18.4.5}$$

exists in the weak (strong, or uniform) operator topology.

Replacing the limiting passage $\xi \to \infty$ by $\xi \to 0+$ and $\lambda \to 0+$ by $\lambda \to \infty$, we obtain the corresponding definition of *ergodicity at zero*. Here it is no longer essential that $[T(\xi)]$ be of type $\omega_0 \leqq 0$.

We note that $C(\xi; \alpha)x$ is clearly strongly measurable for each $x \in \mathfrak{X}_2$, and, since $\mathfrak{X}_2$ is dense in $\mathfrak{X}$, the same is true for all $x \in \mathfrak{X}$. Thus the integrability assumption in Definition 18.4.3 makes sense. Moreover this assumption is automatically satisfied for each $x \in \mathfrak{X}_2$.

THEOREM 18.4.3. *Let $[T(\xi)]$ be of class (E) and of type $\omega_0 \leqq 0$. Suppose $T(\xi)$ is weakly (C, α)-ergodic at infinity (at zero) for some $\alpha \geqq 0$. Then $T(\xi)$ is weakly (C, β)-ergodic for each $\beta > \alpha$ and also weakly Abel-ergodic at infinity (at zero), the ergodic limit being the same operator in all cases. In this statement we may replace "weakly" by "strongly" in hypotheses and conclusions. If $[T(\xi)]$ is assumed to be uniformly measurable and if $\int_0^\infty e^{-\lambda\xi} \| T(\xi) \| \, d\xi < \infty$ for each $\lambda > \omega_0$, then the statement remains valid if we replace "weakly" by "uniformly" in hypotheses and conclusions.*

PROOF. In the weak or strong case when $\mathfrak{X}_2 = \mathfrak{X}$ or in the uniform case, the assertion follows directly from Theorem 18.2.1. When $\mathfrak{X}_2$ is merely dense in $\mathfrak{X}$ we argue as follows. For $\beta > \alpha$, the relation (18.2.5) becomes

$$(18.4.6) \qquad C(\xi;\beta)x = \frac{\Gamma(\beta+1)}{\Gamma(\alpha+1)\Gamma(\beta-\alpha)}\,\xi^{-\beta}\int_0^\xi (\xi-\tau)^{\beta-\alpha-1}\tau^\alpha C(\tau;\alpha)x\,d\tau$$

for each $x \in \mathfrak{X}_2$. By Theorem 3.8.2 the right member defines a linear bounded operator on all $\mathfrak{X}$. Consequently the (C, β)-averages exist in the sense of Definition 18.4.2 and are defined by (18.4.6) for each $x \in \mathfrak{X}$. Similarly $R(\lambda)x$ is given by

$$(18.4.7) \qquad R(\lambda)x = \frac{\lambda^{\alpha+1}}{\Gamma(\alpha+1)}\int_0^\infty e^{-\lambda\tau}\tau^\alpha C(\tau;\alpha)x\,d\tau, \qquad \lambda > \omega_0,$$

for all $x \in \mathfrak{X}$. The assertions for the weak and the strong operator topologies now follow directly from the formulas (18.4.6) and (18.4.7) respectively.

LEMMA 18.4.1. *Let $V(\xi)$ be an operator-valued function on (α, β) to $\mathfrak{E}(\mathfrak{X})$ such that* $\lim_{\xi\to\alpha+} x^*[V(\xi)x]$ *exists for each $x \in \mathfrak{X}$ and $x^* \in \mathfrak{X}^*$. Then there exists an ω, $\alpha < \omega < \beta$, and an $M > 0$ such that $\| V(\xi) \| \leqq M$ for $\alpha < \xi < \omega$.*

PROOF. If the assertion were false, then there would exist a sequence $\{\xi_n\}$, $\alpha < \xi_n < \beta$ and $\xi_n \to \alpha$, for which $\| V(\xi_n) \| \to \infty$ with n. On the other hand, $\sup | x^*[V(\xi_n)x] | < \infty$ for each $x \in \mathfrak{X}$ and $x^* \in \mathfrak{X}^*$ by assumption. Hence a double application of the uniform boundedness theorem shows that $\sup \| V(\xi_n) \| < \infty$, contrary to our choice of $\{\xi_n\}$. It follows that the assertion of the lemma is true.

18.5. Weak ergodic theorems. We now prove that weak and strong Abel-ergodicity are equivalent. It will be shown in section 18.7 that weak and strong (C, α)-ergodicity at zero, $\alpha \geqq 0$, are also equivalent, and that weak (C, α)-ergodicity at infinity essentially implies strong (C, β)-ergodicity at infinity for $\beta > \alpha \geqq 0$. Consequently the weak ergodic theory can for most purposes be subsumed under the strong theory.

LEMMA 18.5.1. *Let $[T(\xi)]$ be a semi-group of class (E). Suppose that*

(i) *there exists an $\omega > \omega_0$ such that $\| \lambda R(\lambda) \| \leqq M$ for $\omega < \lambda < \infty$;*

(ii) *for a given $x \in \mathfrak{X}$ there exists a sequence $\{\lambda_n\}$, $\lambda_n \to \infty$, and a $y \in \mathfrak{X}$ such that* weak $\lim_{n\to\infty} \lambda_n R(\lambda_n)x = y$.

Then $y \in \bar{\mathfrak{X}}_0$, $(x - y) \in \bigcap_{\xi>0}\mathfrak{Z}[T(\xi)]$, and strong $\lim_{\lambda\to\infty} \lambda R(\lambda)x = y$.

PROOF. If $x \in \mathfrak{X}_0 \subset \mathfrak{X}_2$, then $\lim_{\eta\to 0+} T(\eta)x = x$ in norm and *a fortiori* $\lambda R(\lambda)x = \lambda\int_0^\infty e^{-\lambda\xi}T(\xi)x\,d\xi \to x$ in norm as $\lambda \to \infty$. Since $\| \lambda R(\lambda) \| \leqq M$ for $\lambda > \omega$, the Banach-Steinhaus argument shows that $\lim_{\lambda\to\infty} \lambda R(\lambda)x = x$ in norm for all $x \in \bar{\mathfrak{X}}_0$. Likewise if $x \in \bigcap_{\xi>0} \mathfrak{Z}[T(\xi)] \subset \mathfrak{X}_2$, then

$$\lambda R(\lambda)x = \lambda\int_0^\infty e^{-\lambda\xi}T(\xi)x\,d\xi = \theta \to \theta$$

in norm as $\lambda \to \infty$. Hence the theorem will be proved if it can be shown that

$y \in \bar{\mathfrak{X}}_0$ and $x - y \subset \bigcap_{\xi>0}\mathfrak{Z}[T(\xi)]$. Now for $x \in \mathfrak{X}_2$ it is clear that $\lambda R(\lambda)x = \lambda\int_0^\infty e^{-\lambda\xi}T(\xi)x\,d\xi \in \bar{\mathfrak{X}}_0$. Since $\mathfrak{X}_2$ is dense in $\mathfrak{X}$, we see for arbitrary $x \in \mathfrak{X}$ that $\lambda R(\lambda)x$ will be the limit of elements in $\bar{\mathfrak{X}}_0$ and hence itself an element of $\bar{\mathfrak{X}}_0$. Theorem 2.9.2 now asserts that y is also an element of $\bar{\mathfrak{X}}_0$. On the other hand

$$T(\xi)(x - y) = T(\xi)x - \text{weak}\lim_{n\to\infty} T(\xi)\lambda_n R(\lambda_n)x.$$

Obviously $T(\xi)R(\lambda)x = R(\lambda)T(\xi)x$ and $T(\xi)x \in \bar{\mathfrak{X}}_0$. It follows that

$$T(\xi)(x - y) = T(\xi)x - \text{strong}\lim_{n\to\infty} \lambda_n R(\lambda_n)T(\xi)x = T(\xi)x - T(\xi)x = \theta.$$

Hence $z = x - y \in \bigcap_{\xi>0}\mathfrak{Z}[T(\xi)]$ and therefore $\lambda R(\lambda)x = \lambda R(\lambda)y + \lambda R(\lambda)z \to y$ in norm as $\lambda \to \infty$.

THEOREM 18.5.1. *Let $[T(\xi)]$ be a semi-group of class (E). Then $T(\xi)$ is weakly Abel-ergodic at zero if and only if it is strongly Abel-ergodic at zero.*

PROOF. It is clear that strong Abel-ergodicity at zero implies weak Abel-ergodicity at zero. Conversely if $T(\xi)$ is weakly Abel-ergodic at zero, then the corollary to Theorem 18.4.1 shows that $x^*[\lambda R(\lambda)x]$ is bounded in λ, $\lambda > \omega \equiv 1 + \omega_0$, for each $x \in \mathfrak{X}$ and $x^* \in \mathfrak{X}^*$. The uniform boundedness theorem now asserts that $\| \lambda R(\lambda) \|$ is also bounded for $\lambda > \omega$ and hence the previous lemma implies that $T(\xi)$ is strongly Abel-ergodic at zero.

LEMMA 18.5.2. *Let $[T(\xi)]$ be a semi-group of class (E) and of type $\omega_0 \leqq 0$. Suppose that*

(i) *there exists an $\omega > 0$ such that $\| \lambda R(\lambda) \| \leqq M$ for $0 < \lambda < \omega$;*

(ii) *for a given $x \in \mathfrak{X}$ there exists a sequence $\{\lambda_n\}$, $\lambda_n \to 0+$, and a $y \in \mathfrak{X}$ such that* weak $\lim_{n\to\infty}\lambda_n R(\lambda_n)x = y$.

Then $\lambda R(\lambda)y = y$ for all $\lambda > 0$ and strong $\lim_{\lambda\to 0+}\lambda R(\lambda)x = y$.

PROOF. The argument here is not as straightforward as that of Lemma 18.5.1; however the Yosida [2] proof of the mean ergodic theorem can be adapted to our needs. We first show that

$$\lambda R(\lambda)y = y, \qquad \lambda > 0. \tag{18.5.1}$$

For any $\lambda > 0$, we may choose n so large that $0 < \lambda_n < \lambda$. By the first resolvent equation (Theorem 18.4.1) we have

$$\lambda R(\lambda)[\lambda_n R(\lambda_n)x] = \frac{\lambda}{\lambda - \lambda_n}\lambda_n R(\lambda_n)x - \frac{\lambda_n}{\lambda - \lambda_n}\lambda R(\lambda)x.$$

Letting $n \to \infty$, both members of this equation converge in the weak operator topology and we obtain (18.5.1) in the limit.

Writing $x = y + (x - y)$, we have

$$\lambda R(\lambda)x = y + \lambda R(\lambda)(x - y).$$

The assertion will therefore be proved if it can be shown that $\lambda R(\lambda)(x - y)$

tends to θ in norm as $\lambda \to 0+$. To this end we introduce the linear bounded operator $z = Vx \equiv [I - R(1)]x$ and set $\mathfrak{Z} = V[\mathfrak{X}]$. Again using the first resolvent equation, we have for $z \in \mathfrak{Z}$

$$\lambda R(\lambda)z = \frac{\lambda}{\lambda - 1}[\lambda R(\lambda)x - R(1)x],$$

which tends to θ in norm as $\lambda \to 0+$. Since $\| \lambda R(\lambda) \| \leqq M$ for $0 < \lambda < \omega$, the same result is evidently true for all $x \in \bar{\mathfrak{Z}}$. Thus the relation, $\lambda R(\lambda)(x - y) \to \theta$ in norm, holds whenever $x - y \in \bar{\mathfrak{Z}}$.

Suppose now that $x - y$ is not in $\bar{\mathfrak{Z}}$. By Theorem 2.7.5 there exists an $x_0^* \in \mathfrak{X}^*$ such that $x_0^*(x - y) = 1$ and $x_0^*[\bar{\mathfrak{Z}}] = 0$. Since $x - R(1)x \in \mathfrak{Z}$, we have $x_0^*(x) = x_0^*[R(1)x]$ and this holds for all $x \in \mathfrak{X}$. Further

$$R(1)R(\lambda) = \frac{1}{1 - \lambda}[R(\lambda) - R(1)]$$

whence it follows that

$$x_0^*[R(\lambda)x] = x_0^*\{R(1)[R(\lambda)x]\} = \frac{1}{1 - \lambda}\{x_0^*[R(\lambda)x] - x_0^*[R(1)x]\}$$

so that

$$x_0^*[\lambda_n R(\lambda_n)x] = x_0^*[R(1)x] = x_0^*(x).$$

In the limit as $n \to \infty$ we obtain $x_0^*(y) = x_0^*(x)$ and this contradicts the assumption that $x_0^*(x - y) = 1$. It follows that $x - y \in \bar{\mathfrak{Z}}$ and hence that $\lambda R(\lambda)(x - y) \to \theta$ in norm as $\lambda \to 0+$. This completes the proof of the lemma.

THEOREM 18.5.2. *Let $[T(\xi)]$ be a semi-group of class (E) and of type $\omega_0 \leqq 0$. Then $T(\xi)$ is weakly Abel-ergodic at infinity if and only if it is strongly Abel-ergodic at infinity.*

The proof paraphrases that of Theorem 18.5.1.

18.6. Ergodic projections. This section is concerned with the character of the ergodic limit both at zero and at infinity. We shall consider only the strong Abel-limit. In view of Theorems 18.4.3, 18.5.1, and 18.5.2, it is evident that the results also apply to the weak Abel-limit as well as the weak and strong (C, α)-limits.

THEOREM 18.6.1. *Let $[T(\xi)]$ be a semi-group of class (E). If*

$$(A)\text{-}\lim_{\xi \to 0+} T(\xi)x = Jx \tag{18.6.1}$$

exists in norm for all $x \in \mathfrak{X}$, then

(1) *J is a linear bounded projection operator, $J^2 = J$;*

(2) *$JT(\xi) = T(\xi)J = T(\xi)$ for each $\xi > 0$;*

(3) $J[A_o x] = A_o[Jx] = A_o x$ *for each* $x \in \mathfrak{D}(A_o)$;
(4) $\mathfrak{R}(J) = \bar{\mathfrak{X}}_0$ *where* $\mathfrak{X}_0 \equiv \bigcup_{\xi>0}\mathfrak{R}[T(\xi)]$;
(5) $\mathfrak{Z}(J) = \bigcap_{\xi>0}\mathfrak{Z}[T(\xi)]$;
(6) $\mathfrak{R}(J)$ *and* $\mathfrak{Z}(J)$ *have only the zero element in common and* $\mathfrak{X} = \mathfrak{R}(J) \oplus \mathfrak{Z}(J)$.

PROOF. By assumption $R(\lambda) \in \mathfrak{E}(\mathfrak{X})$ for each $\lambda > \omega_0$ and

$$\lim_{\lambda\to\infty} \lambda R(\lambda)x = Jx$$

in norm for all $x \in \mathfrak{X}$. The operator J is obviously linear. The corollary to Theorem 18.4.1 implies that $\| \lambda R(\lambda)x \|$ is bounded in λ, $\lambda > 1 + \omega_0$, for each $x \in \mathfrak{X}$. Thus the uniform boundedness theorem shows that $\| R(\lambda) \| = O(1/\lambda)$ as $\lambda \to \infty$ and hence that J is a bounded operator. According to Theorem 18.4.1,

$$(\mu - \lambda)R(\lambda)R(\mu)x = (\mu - \lambda)R(\mu)R(\lambda)x = R(\lambda)x - R(\mu)x;$$

passing to the limit as $\mu \to \infty$ and multiplying through by λ, we obtain

$$\lambda R(\lambda)Jx = J[\lambda R(\lambda)x] = \lambda R(\lambda)x.$$

Letting $\lambda \to \infty$ then gives the relation $J^2x = Jx$ for all $x \in \mathfrak{X}$, which proves (1).
Now for $x \in \mathfrak{X}_2$, we have.

$$\int_0^\infty e^{-\lambda\xi}[JT(\xi)x - T(\xi)x]\, d\xi = JR(\lambda)x - R(\lambda)x = \theta, \qquad \lambda > \omega_0.$$

Since $JT(\xi)x$ and $T(\xi)x$ are continuous functions of ξ for $\xi > 0$, the uniqueness theorem for Laplace transforms implies that $JT(\xi)x = T(\xi)x$. Also for $x \in \mathfrak{X}_2$ and $\xi > 0$,

$$\lambda R(\lambda)T(\xi)x = \lambda \int_0^\infty e^{-\lambda\tau}T(\tau + \xi)x\, d\tau = T(\xi)[\lambda R(\lambda)x], \qquad \lambda > \omega_0, \tag{18.6.2}$$

and passing to the limit as $\lambda \to \infty$ we obtain

$$JT(\xi)x = T(\xi)Jx = T(\xi)x, \qquad \xi > 0.$$

Since $\mathfrak{X}_2$ is dense in $\mathfrak{X}$, this equality holds for all $x \in \mathfrak{X}$, proving (2).

The relation $JT(\xi)x = T(\xi)x$, $\xi > 0$, shows that $\mathfrak{R}(J) \supset \bar{\mathfrak{X}}_0$. On the other hand, Lemma 18.5.1 asserts that $Jx \in \bar{\mathfrak{X}}_0$ for each $x \in \mathfrak{X}$ so that $\mathfrak{R}(J) \subset \bar{\mathfrak{X}}_0$. Consequently $\mathfrak{R}(J) = \bar{\mathfrak{X}}_0$. Further if $x \in \mathfrak{Z}(J)$, then $T(\xi)x = T(\xi)Jx = \theta$ for each $\xi > 0$ and therefore $x \in \mathfrak{M} \equiv \bigcap_{\xi>0} \mathfrak{Z}[T(\xi)]$. Conversely if $x \in \mathfrak{M}$, then $x \in \mathfrak{X}_2$ and $\lambda R(\lambda)x = \lambda \int_0^\infty e^{-\lambda\xi}T(\xi)x\, d\xi = \theta$. Thus $Jx = \theta$ and hence $x \in \mathfrak{Z}(J)$. This proves (4) and (5).

Suppose next that $x \in \mathfrak{D}(A_o)$. Then $x \in \bar{\mathfrak{X}}_0$ and therefore $Jx = x$. Consequently

$$J\left[\frac{T(\eta)x - x}{\eta}\right] = \frac{T(\eta) - I}{\eta}[Jx] = \frac{T(\eta)x - x}{\eta} \to A_o x$$

as $\eta \to 0+$. It follows that $J[A_o x] = A_o[Jx] = A_o x$, proving (3).

Finally (6) is an immediate consequence of the fact that J is a projection operator. For if $x \in \mathfrak{R}(J) \cap \mathfrak{Z}(J)$, then on the one hand $Jx = x$ and on the other $Jx = \theta$, so that $x = \theta$. The decomposition

$$x = Jx + (x - Jx)$$

gives a representation of x as the sum of a vector in $\mathfrak{R}(J)$ and a vector in $\mathfrak{Z}(J)$; this representation is evidently unique. Thus (6) holds and the theorem is proved.

We have a similar theorem for the ergodic limit at infinity.

THEOREM 18.6.2. *Let $[T(\xi)]$ be a semi-group of class (E) and of type $\omega_0 \leqq 0$. If*

$$(A)\text{-}\lim_{\xi \to \infty} T(\xi)x = Px \tag{18.6.3}$$

exists in norm for all $x \in \mathfrak{X}$, then

(1) *P is a linear bounded projection operator, $P^2 = P$;*

(2) *$PT(\xi) = T(\xi)P = P$ for each $\xi > 0$;*

(3) *$A_0[Px] = \theta$ for all $x \in \mathfrak{X}$, $P[A_0 x] = \theta$ for all $x \in \mathfrak{D}(A_0)$;*

(4) *$\mathfrak{R}(P) = \mathfrak{Z}(A_0)$ is the linear manifold which is left point-wise invariant by the semi-group $\mathfrak{S}$;*

(5a) *$\mathfrak{Z}(P) \supset \overline{\mathfrak{R}(A_0)}$;*

(6a) *$\mathfrak{R}(P)$ and $\mathfrak{Z}(P)$ have only the zero element in common and*

$$\mathfrak{X} = \mathfrak{R}(P) \oplus \mathfrak{Z}(P).$$

If, in addition, $[T(\xi)]$ is of class (A) with infinitesimal generator A, then

(5b) *$\mathfrak{Z}(P) = \overline{\mathfrak{R}(A)}$;*

(6b) *$\overline{\mathfrak{R}(A)} = \overline{\mathfrak{R}(A_0)}$, $\mathfrak{Z}(A_0) = \mathfrak{Z}(A)$, and hence*

$$\mathfrak{X} = \overline{\mathfrak{R}(A)} \oplus \mathfrak{Z}(A). \tag{18.6.4}$$

PROOF. The basic assumption now is that

$$\lim_{\lambda \to 0+} \lambda R(\lambda)x = Px$$

in norm for each $x \in \mathfrak{X}$. The uniform boundedness theorem asserts that $\| R(\lambda) \| = O(1/\lambda)$ as $\lambda \to 0+$; whence it follows that P is a linear bounded operator. By Lemma 18.5.2, $\lambda R(\lambda)Px \equiv Px$ for all $\lambda > 0$, and letting $\lambda \to 0+$ we obtain $P^2x = Px$ for all $x \in \mathfrak{X}$, which proves (1).

It is clear from the relation (18.4.2) that $PR(\lambda) = R(\lambda)P$. Hence for $x \in \mathfrak{X}_2$

$$\lambda \int_0^\infty e^{-\lambda\xi}[PT(\xi)x - Px]\, d\xi = \lambda PR(\lambda)x - Px = \lambda R(\lambda)Px - Px = \theta, \quad \lambda > 0.$$

Again by the uniqueness theorem for Laplace transforms we see that $PT(\xi)x = Px$, $\xi > 0$. Also for $x \in \mathfrak{X}_2$, we see from (18.6.2) that $\lambda R(\lambda)T(\xi)x = T(\xi)[\lambda R(\lambda)x]$,

$\lambda > 0$, and passing to the limit as $\lambda \to 0+$ we obtain

$$PT(\xi)x = T(\xi)Px = Px.$$

Since $\mathfrak{X}_2$ is dense in $\mathfrak{X}$, this relation holds for all $x \in \mathfrak{X}$. This proves (2).

For any $x \in \mathfrak{X}$,

$$\frac{T(\eta) - I}{\eta}[Px] = \frac{Px - Px}{\eta} = \theta \to A_o Px$$

as $\eta \to 0+$ so that $A_o Px = \theta$. If $x \in \mathfrak{D}(A_o)$, then

$$P\left[\frac{T(\eta) - I}{\eta}x\right] = \frac{Px - Px}{\eta} = \theta \to PA_o x$$

as $\eta \to 0+$ so that $PA_o x = \theta$. This proves (3). It also shows that $\mathfrak{R}(A_o) \subset \mathfrak{Z}(P)$. Since $\mathfrak{Z}(P)$ is closed we have $\overline{\mathfrak{R}(A_o)} \subset \mathfrak{Z}(P)$ as asserted in (5a).

It follows from $A_o Px = \theta$, $x \in \mathfrak{X}$, that $\mathfrak{R}(P) \subset \mathfrak{Z}(A_o)$. Conversely suppose $x \in \mathfrak{Z}(A_o)$. Then $T(\xi)A_o x \equiv \theta$ and by the corollary to Theorem 10.3.3

$$T(\xi)x - x = \int_0^\xi T(\tau)A_o x \, d\tau = \theta, \qquad \xi > 0.$$

Consequently $x \in \mathfrak{X}_2$, $\lambda R(\lambda)x = \lambda \int_0^\infty e^{-\lambda\xi}T(\xi)x \, d\xi = x$, and $Px = x$; that is $x \in \mathfrak{R}(P)$. This proves (4) since it is obvious that $T(\xi)x \equiv x$ implies $x \in \mathfrak{Z}(A_o)$. The assertion (6a) is a consequence of the fact that P is a projection operator.

If, in addition, $\mathfrak{S}$ is of class (A), then the smallest closed extension of A_o, namely, the infinitesimal generator A, exists. It is clear that $\overline{\mathfrak{R}(A_o)} = \overline{\mathfrak{R}(A)}$. Further Corollary 3 to Theorem 11.5.3 asserts that $\mathfrak{Z}(A_0) = \mathfrak{Z}(A)$. Finally for $x \in \mathfrak{Z}(P)$ we have

$$x + AR(\lambda; A)x = \lambda R(\lambda; A)x = \lambda R(\lambda)x \to Px = \theta,$$

so that $x = \lim_{\lambda \to 0+} -AR(\lambda; A)x \in \overline{\mathfrak{R}(A)}$. This together with (5a) shows that $\mathfrak{Z}(P) = \overline{\mathfrak{R}(A)}$, proving (5b). The relation (18.6.4) now follows from (6a). Thus (6b) holds and the theorem is proved.

The following example shows that $\overline{\mathfrak{R}(A_o)}$ may be a proper subset of $\mathfrak{Z}(P)$ if $[T(\xi)]$ is merely of class (E) and Abel-ergodic at infinity. Let $\mathfrak{X} = m$, the space of sequences $[\alpha_n ; n = 1, 2, \cdots]$ with $\| \{\alpha_n\} \| = \sup | \alpha_n |$. Set

$$T(\xi)\{\alpha_n\} = \{e^{-n\xi}\alpha_n\}.$$

It is clear that $\mathfrak{X}_0 \subset c_0 \equiv [\{\alpha_n\}; \lim_{n\to\infty} \alpha_n = 0]$ and this implies that $\overline{\mathfrak{R}(A_o)} \subset c_0$. On the other hand $\| T(\xi) \| = e^{-\xi}$. Thus $\lim_{\xi\to\infty} T(\xi) = \Theta$ in the uniform operator topology and *a fortiori* (A)-$\lim_{\xi\to\infty} T(\xi)x = \Theta x = \theta$. Consequently $\mathfrak{Z}(P) = m \supset c_0 \supset \overline{\mathfrak{R}(A_o)}$, where the first inclusion is definitely proper.

The existence of the Abel-ergodic limit at infinity has the following spectral implications for the infinitesimal generator A.

THEOREM 18.6.3. *Let* $[T(\xi)]$ *be a semi-group of class* (A) *and of type* $\omega_0 \leqq 0$. *If* $T(\xi)$ *is strongly Abel-ergodic at infinity then* $\lambda = 0$ *is of index* $\nu = 0$ *or* 1 (*in the sense of Definition 2.16.4*) *with respect to the infinitesimal generator* A.

If $\nu = 0$ *then* $\lambda = 0$ *is either in the resolvent set or in the continuous spectrum of* A.

If $\nu = 1$, *then* $\lambda = 0$ *is in the point spectrum of* A; *however for* A *restricted to the subspace* $\mathfrak{Y}_1 \equiv \overline{\mathfrak{R}(A)} = (I - P)[\mathfrak{X}]$ *and considered as an element of* $\mathfrak{O}(\mathfrak{Y}_1)$, *then* $\lambda = 0$ *is of index zero and is either in the resolvent set or in the continuous spectrum.*

PROOF. We first show that $A^2x = \theta$ implies $Ax = \theta$. Indeed, if $A^2x = \theta$ then $Ax \in \mathfrak{R}(A) \cap \mathfrak{Z}(A)$ and by Theorem 18.6.2 (6b) this requires that $Ax = \theta$. In other words, $\lambda = 0$ is of index 0 or 1 for A.

Suppose next that $\nu = 0$ so that $\lambda = 0$ is not in the point spectrum of A. Thus $\mathfrak{Z}(A) = \{\theta\}$ and (18.6.4) implies that $\mathfrak{X} = \overline{\mathfrak{R}(A)}$. It follows that $\lambda = 0$ is not in the residual spectrum of A and the second assertion is proved.

If $\nu = 1$, then $\lambda = 0$ is in the point spectrum of A. Denoting the restriction of A to $\mathfrak{Y}_1$ by A_1, it is clear that $\mathfrak{Y}_1 = \overline{\mathfrak{R}(A)} \supset \mathfrak{R}(A_1)$ and hence that A_1 is an operator on $\mathfrak{Y}_1$. Further A_1 is closed because it is the restriction of a closed operator to a closed subspace. Now $\mathfrak{Y}_1 = \overline{\mathfrak{R}(A)}$ has only the zero element in common with $\mathfrak{Z}(A)$. Hence if $y \in \mathfrak{D}(A_1)$ and $A_1y = \theta$, then $y = \theta$. It follows that $\nu = 0$ for A_1 so that $\lambda = 0$ is not in the point spectrum of A_1. Moreover if $y \in \mathfrak{R}(A)$, then $y = Aw$ for some $w \in \mathfrak{D}(A)$. According to Theorem 18.6.2 (4) and (6b), $Pw \in \mathfrak{Z}(A) \subset \mathfrak{D}(A)$; thus $w_1 = (I - P)w \in \mathfrak{D}(A) \cap \overline{\mathfrak{R}(A)} = \mathfrak{D}(A_1)$. By Theorem 18.6.2 (3) we have $A_1w_1 = Aw - APw = y - \theta = y$ so that $y \in \mathfrak{R}(A_1)$. Consequently $\mathfrak{R}(A_1) = \mathfrak{R}(A)$ is dense in $\mathfrak{Y}_1$ and therefore $\lambda = 0$ is not in the residual spectrum of A_1. Thus $\lambda = 0$ is either in the resolvent set or the continuous spectrum of A_1. This concludes the proof.

18.7. Strong ergodic theorems. This section is concerned with strong ergodic theorems, that is, theorems establishing strong ergodicity for certain classes of semi-groups. We start by proving a strong Abel-ergodic theorem dealing with the behavior at zero.

THEOREM 18.7.1. *If* $[T(\xi)]$ *is a semi-group of class* (E), *then the following statements are equivalent.*

(1) $T(\xi)$ *is strongly Abel-ergodic at zero.*

(2) *For each* $x \in \mathfrak{X}$ *the set* $[\lambda R(\lambda)x;\ \omega' < \lambda < \infty]$ *is conditionally sequentially weakly compact.*

(3) $\| \lambda R(\lambda) \| \leqq M$ *for* $\omega < \lambda < \infty$ *and* $\mathfrak{X} = \overline{\mathfrak{X}_0 + \mathfrak{M}}$ *where* $\mathfrak{X}_0 = \bigcup_{\xi>0} \mathfrak{R}[T(\xi)]$ *and* $\mathfrak{M} = \bigcap_{\xi>0} \mathfrak{Z}[T(\xi)]$.

PROOF. The statement (1) together with the corollary to Theorem 18.4.1 implies that $\lambda R(\lambda)x$ is bounded in norm and conditionally compact valued on $\omega_0 < \omega < \lambda < \infty$ for each $x \in \mathfrak{X}$. Thus (1) implies (2). The implication (1) $\rightarrow$ (3) is a consequence of Theorem 18.6.1 and the uniform boundedness theorem.

In proving (1) from (2) we first note that the compactness condition implies the boundedness of $\| \lambda R(\lambda) \|$ for $\lambda > \omega$. Further, given $x \in \mathfrak{X}$, we may appeal to the weak sequential compactness of $[\lambda R(\lambda)x; \omega < \lambda < \infty]$ to obtain a sequence $\{\lambda_n\}$, $\lambda_n \to \infty$, and a $y \in \mathfrak{X}$ such that weak $\lim_{n\to\infty} \lambda_n R(\lambda_n)x = y$. Lemma 18.5.1 now shows that strong $\lim_{\lambda\to\infty} \lambda R(\lambda)x = y$, thus proving (1).

As to the passage from (3) to (1), suppose that $x = y + z$, where $y \in \mathfrak{X}_0$ and and $z \in \bigcap_{\xi>0} \mathfrak{Z}[T(\xi)]$. Then $T(\xi)x = T(\xi)y$ and

$$y = (C, 0)\text{-}\lim_{\xi\to 0+} T(\xi)x = (A)\text{-}\lim_{\xi\to 0+} T(\xi)x.$$

By assumption such elements x are dense in $\mathfrak{X}$ so that the boundedness condition guaranties the existence of the Abel limit for all $x \in \mathfrak{X}$. This completes the proof of the theorem.

Remark. In a reflexive space bounded sets are conditionally sequentially compact by Theorems 2.9.6 and 2.10.3. Thus a special case of (2) is
(2*) *$\mathfrak{X}$ is reflexive and $\| \lambda R(\lambda) \| \leqq M$ for $\omega < \lambda\ \infty$.*

With regard to (C, α)-ergodicity at zero, we have

Theorem 18.7.2. *Let $[T(\xi)]$ be a semi-group of class (E) which is strongly Abel-ergodic at zero. Given $\alpha \geqq 0$, suppose that the operator-averages $C(\xi; \alpha)$ exist and are bounded in norm for $0 < \xi < \omega$. Then* strong $\lim_{\xi\to 0+} C(\xi; \alpha)x$ *exists for each $x \in \mathfrak{X}$.*

Proof. Theorem 18.6.1 applies and we have $\mathfrak{X} = \bar{\mathfrak{X}}_0 + \mathfrak{M}$, where $\mathfrak{M} = \bigcap_{\xi>0} \mathfrak{Z}[T(\xi)]$. For $x = y + z$, $y \in \mathfrak{X}_0$ and $z \in \mathfrak{M}$, it is clear that $T(\xi)x = T(\xi)y$ and hence

$$y = (C, 0)\text{-}\lim_{\xi\to 0+} T(\xi)x = (C, \alpha)\text{-}\lim_{\xi\to 0+} T(\xi)x.$$

Since $\mathfrak{X}_0 + \mathfrak{M}$ is dense in $\mathfrak{X}$, the boundedness of $\| C(\xi; \alpha) \|$ for $0 < \xi < \omega$ suffices to establish the existence of $\lim_{\xi\to 0+} C(\xi; \alpha)x$ for all $x \in \mathfrak{X}$.

Corollary. *Let $[T(\xi)]$ be a semi-group of class (E) and suppose that $T(\xi)$ is weakly (C, α)-ergodic at zero, $\alpha \geqq 0$. Then $T(\xi)$ is strongly (C, α)-ergodic at zero.*

Proof. We see from Definition 18.4.3 that all of the conditions for (C, α)-ergodicity at zero are satisfied automatically except the existence of a strong (C, α)-limit. Theorems 18.4.1 and 18.5.1 imply that $T(\xi)$ is strongly Abel-ergodic at zero. Further $\lim_{\xi\to 0+} x^*[C(\xi; \alpha)x]$ exists for each $x \in \mathfrak{X}$ and $x^* \in \mathfrak{X}^*$. Thus by Lemma 18.4.1 there exist numbers $\omega > 0$ and $M > 0$ such that $\| C(\xi; \alpha) \| \leqq M$ for $0 < \xi < \omega$. Consequently Theorem 18.7.2 applies so that strong $\lim_{\xi\to 0+} C(\xi; \alpha)x$ exists for all $x \in \mathfrak{X}$.

The results at infinity are similar but not quite as sharp as those at zero. We have

THEOREM 18.7.3. *If* $[T(\xi)]$ *is a semi-group of class* (E) *and of type* $\omega_0 \leqq 0$, *then the following statements are equivalent.*

(1) $T(\xi)$ *is strongly Abel-ergodic at infinity.*

(2) *For each* $x \in \mathfrak{X}$ *the set* $[\lambda R(\lambda)x;\ 0 < \lambda < \omega]$ *is conditionally sequentially weakly compact.*

If, in addition, $[T(\xi)]$ *is of class* (A), *then a third equivalent statement is*

(3) $\| \lambda R(\lambda; A) \| \leqq M$ *for* $0 < \lambda < \omega$ *and* $\mathfrak{X} = \overline{\mathfrak{Z}(A) + \mathfrak{R}(A)}$.

PROOF. That (1) implies (2) is clear because of the corollary to Theorem 18.4.1. For semi-groups of class (A), we have $R(\lambda) = R(\lambda; A)$, $\lambda > \omega_0$, by Theorem 18.4.2. Hence the implication (1) $\rightarrow$ (3) follows from Theorem 18.6.2 and the uniform boundedness theorem.

We now prove that (2) $\rightarrow$ (1). The weak compactness implies (a) that $\| \lambda R(\lambda) \|$ is bounded for $0 < \lambda < \omega$, and (b) that given an $x \in \mathfrak{X}$ there exists a sequence $\{\lambda_n\}$, $\lambda_n \rightarrow 0+$, and a $y \in \mathfrak{X}$ such that weak $\lim_{n\to\infty} \lambda_n R(\lambda_n)x = y$. The strong Abel-ergodicity at zero now follows from Lemma 18.5.2.

In order to prove that (3) implies (1) we proceed as follows. If $z \in \mathfrak{Z}(A)$, then $\lambda R(\lambda; A)z = z + R(\lambda; A)Az = z$ for all $\lambda > 0$. Consequently $\lim_{\lambda\to 0+} \lambda R(\lambda; A)z = z$ in norm. On the other hand if $y \in \mathfrak{R}(A)$, then $y = Aw$ for some $w \in \mathfrak{D}(A)$ and $\lambda R(\lambda; A)y = \lambda R(\lambda; A)Aw = \lambda[\lambda R(\lambda; A)w - w]$. Making use of the boundedness of $\| \lambda R(\lambda; A) \|$, $0 < \lambda < \omega$, we see that $\lim_{\lambda\to 0+} \lambda R(\lambda; A)y = \theta$ in norm. It now follows that strong $\lim_{\lambda\to 0+} \lambda R(\lambda; A)x$ exists for all $x \in \mathfrak{R}(A) + \mathfrak{Z}(A)$ and the Banach-Steinhaus theorem shows that this is also valid for all $x \in \mathfrak{X} = \overline{\mathfrak{R}(A) + \mathfrak{Z}(A)}$. This concludes the proof.

There are several remarks which can be appended to this theorem.

A. As a special case of (2) we have

(2*) $\mathfrak{X}$ *is reflexive and* $\| \lambda R(\lambda) \| \leqq M$ *for* $0 < \lambda < \omega$.

B. The statement of Theorem 18.6.3 suggests a fourth alternative which may be formulated as follows: $[T(\xi)]$ is of class (A), $\| \lambda R(\lambda; A) \| \leqq M$ for $0 < \lambda < \omega$, and $\lambda = 0$ is not in the residual spectrum of either A or its restriction to $\mathfrak{Y}_1 = \overline{\mathfrak{R}(A)}$. Unfortunately these conditions are not sufficient to ensure Abel-ergodicity at infinity as the subsequent example shows.

Let $\mathfrak{X} = m$, the space of bounded sequences $\{\alpha_n;\ n = 0, 1, 2, \cdots\}$ with norm $\| \{\alpha_n\} \| = \sup | \alpha_n |$. Define $T(\xi)\{\gamma_n\} \equiv \{\tau_n(\xi)\}$ by

$$\tau_0(\xi) = \gamma_0 \quad \text{and} \quad \tau_n(\xi) = e^{i\xi/n}\gamma_n \qquad \text{for } n \geqq 1.$$

Clearly $\| T(\xi) \| \equiv 1$ and $\| T(\xi) - I \| \leqq | e^{i\xi} - 1 |$ for $0 < \xi < \pi$. Thus $T(\xi) \rightarrow I$ in the uniform operator topology as $\xi \rightarrow 0+$ so that $A \in \mathfrak{E}(\mathfrak{X})$ and $[T(\xi)]$ is *a fortiori* of class (C_0). Obviously $\omega_0 = 0$. Now $A\{\gamma_n\} \equiv \{\alpha_n\}$ where

$$\alpha_0 = 0 \qquad \text{and} \qquad \alpha_n = \frac{i}{n}\gamma_n \text{ for } n \geqq 1.$$

Consequently $\lambda = 0$ lies in the point spectrum of A and $\overline{\mathfrak{R}(A)} = [\{\gamma_n\}; \gamma_0 = 0, \lim_{n\to\infty}\gamma_n = 0]$. Further if A_1 is the restriction of A to $\overline{\mathfrak{R}(A)}$, then it is clear that $\overline{\mathfrak{R}(A_1)} = \overline{\mathfrak{R}(A)}$. Thus $\lambda = 0$ is not in the residual spectrum of either A or A_1. Finally setting $\lambda R(\lambda; A)\{\gamma_n\} \equiv \{\rho_n(\lambda)\}$ we have

(18.7.1) $$\rho_0(\lambda) = \gamma_0 \quad \text{and} \quad \rho_n(\lambda) = \frac{n\lambda}{n\lambda - i}\gamma_n \text{ for } n \geqq 1.$$

If $T(\xi)$ were strongly Abel-ergodic to the projection operator P, then it is clear from (18.7.1) that $P(\gamma_0, \gamma_1, \gamma_2, \cdots) = (\gamma_0, 0, 0, \cdots)$. However the strong limit does not exist for $x = (1, 1, 1, \cdots)$ since

$$\| [\lambda R(\lambda; A) - P](1, 1, 1, \cdots) \| = \left\| \left(0, \frac{\lambda}{\lambda - i}, \frac{2\lambda}{2\lambda - i}, \cdots\right) \right\| = \lim_{n\to\infty} \left| \frac{n\lambda}{n\lambda - i} \right| = 1$$

for all $\lambda > 0$.

C. Suppose $[T(\xi)]$ is a semi-group of class (E), strongly Abel-ergodic at infinity and such that $\| T(\xi) \| \leqq M(1 + \xi^k)$, $\xi > 0$. Let $F(\zeta) = \sum_{n=0}^{\infty}\alpha_n\zeta^n$ be a numerically-valued entire function. Then

$$F[T(\xi)] = \sum_{n=0} \alpha_n T(n\xi)$$

is well defined for $\xi > 0$ and

$$(A)\text{-}\lim_{\xi\to\infty} F[T(\xi)]x = F(P)x = \alpha_0 x + \left[\sum_{n=1} \alpha_n\right] Px.$$

THEOREM 18.7.4. *Let $[T(\xi)]$ be a semi-group of class (E) and of type $\omega_0 \leqq 0$, strongly Abel-ergodic at infinity. If for some $\alpha \geqq 0$, $C(\xi; \alpha)$ exists and is bounded in norm on $(0, \infty)$, then $T(\xi)$ is strongly (C, β)-ergodic at infinity for each $\beta > \alpha$. This holds as well for $\beta = \alpha$ provided $C(\xi; \alpha)x$ is feebly oscillating as $\xi \to \infty$ for each $x \in \mathfrak{X}$.*

PROOF. By assumption $\| C(\xi; \alpha) \| \leqq M$ for all $\xi > 0$. From this we see that $C(\xi; \beta)$ exists and is bounded in norm for all $\beta \geqq \alpha$ and $\xi > 0$. In particular, $\int_0^\infty e^{-\lambda\xi} \| C(\xi; \beta)x \| \, d\xi < \infty$ for each $x \in \mathfrak{X}$ and $\lambda > 0$. It remains to prove that the strong (C, β)-limits exist. However for $x \in \mathfrak{X}_2$ this follows directly from Theorem 18.3.3 and since $\mathfrak{X}_2$ is dense in $\mathfrak{X}$, the Banach-Steinhaus theorem can be employed to prove that the assertion is true for all $x \in \mathfrak{X}$.

It is clear from Theorems 18.4.3 and 18.5.2 that the statement of Theorem 18.7.4 remains valid if we replace "strong Abel-ergodicity" in the hypothesis by "weak (C, α)-ergodicity" or even by "weak Abel-ergodicity."

18.8. Uniform ergodic theorems. We turn now to a study of ergodic theorems dealing with limits in the uniform operator topology. Here we shall require the following auxiliary theorem which, for the case where $R(\lambda)$ is the resolvent of a bounded operator, is contained in a general theorem due to N. Dunford [**8**, Theorem 3.6].

THEOREM 18.8.1. *Let $R(\lambda)$ be a function on complex numbers to $\mathfrak{E}(\mathfrak{X})$ which is defined and satisfies the first resolvent equation in some domain D. Let α be a boundary point of D and $\{\alpha_n\}$ a sequence contained in D with $\alpha_n \to \alpha$. If in the sense of the uniform operator topology $\lim_{n\to\infty} (\alpha_n - \alpha)R(\alpha_n) = E$, then* (i) *$E$ is a bounded linear projection operator, $E^2 = E$;* (ii) *$E = I$ if and only if $R(\lambda) =$*

$(\lambda - \alpha)^{-1}I$; (iii) *$R(\lambda)$ has a holomorphic extension in some circle* $0 < |\lambda - \alpha| < \rho$ *which likewise satisfies the first resolvent equation*; *and* (iv) *the point* $\lambda = \alpha$ *is either a simple pole or a regular point of the extended* $R(\lambda)$.

PROOF. Setting $S(\lambda) = (\lambda - \alpha)R(\lambda)$, we see that $S(\lambda)$ is holomorphic in D and satisfies the functional equation

$$(18.8.1) \qquad (\lambda - \mu)S(\lambda)S(\mu) = (\lambda - \alpha)S(\mu) - (\mu - \alpha)S(\lambda).$$

Putting $\mu = \alpha_n$ and letting $n \to \infty$, one obtains

$$(18.8.2) \qquad S(\lambda)E = ES(\lambda) = E, \qquad \lambda \in D.$$

Putting $\lambda = \alpha_n$ and letting $n \to \infty$ gives $E^2 = E$. Since $\mathfrak{E}(\mathfrak{X})$ is complete we have $E \in \mathfrak{E}(\mathfrak{X})$. This proves (i), and (ii) follows directly from (18.8.2).

We now set $T(\lambda) = S(\lambda) - E$ so that $T(\alpha_n) \to \Theta$. Substituting $T(\lambda)$ in (18.8.2) one gets $T(\lambda)E = ET(\lambda) = \Theta$ for $\lambda \in D$. Substituting in (18.8.1) and reducing we see that $T(\lambda)$ satisfies the same equation as $S(\lambda)$. From this we find that

$$T(\lambda)[(\alpha_n - \alpha)I + (\lambda - \alpha_n)T(\alpha_n)] = (\lambda - \alpha)T(\alpha_n).$$

Since $T(\alpha_n) \to \Theta$ we can choose n so large that $\| T(\alpha_n) \| < 1$. For such an n the expression in square brackets has an inverse provided that $|\lambda - \alpha_n| < |\alpha - \alpha_n|/\| T(\alpha_n) \|$, a quantity which exceeds $|\alpha - \alpha_n|$. Hence

$$T(\lambda) = \frac{\lambda - \alpha}{\alpha_n - \alpha}\left\{I + \frac{\lambda - \alpha_n}{\alpha_n - \alpha}T(\alpha_n)\right\}^{-1} T(\alpha_n) = -\sum_{k=1}^{\infty}\left[\frac{\lambda - \alpha_n}{\alpha - \alpha_n}\right]^k [T(\alpha_n)]^k$$

has a holomorphic extension in a circle with center at α_n containing α in its interior. Thus both $T(\lambda)$ and $S(\lambda)$ have holomorphic extensions in some circle $|\lambda - \alpha| < \rho$. Consequently $R(\lambda)$ has a holomorphic extension in $0 < |\lambda - \alpha| < \rho$ and by Corollary 2 to Theorem 5.8.6 this extension satisfies the first resolvent equation. The analytic form of $R(\lambda)$ can be obtained from Theorem 5.9.3 which gives

$$(18.8.3) \qquad R(\lambda) = \frac{E}{\lambda - \alpha} + \sum_{n=0}^{\infty} (\alpha - \lambda)^n B^{n+1},$$

where $B \in \mathfrak{E}(\mathfrak{X})$ and $EB = BE = \Theta$. This proves (iii) and (iv).

The same method can be used to characterize poles of $R(\lambda)$ of order n. Let $R(\lambda)$ be defined as above and set $S_n(\lambda) \equiv (\lambda - \alpha)^n R(\lambda)$. We now assume the existence of a rectifiable curve Γ in D ending at the point $\lambda = \alpha$ and a constant M such that the length of arc on Γ from λ to α is $\leqq M |\lambda - \alpha|$ and we further suppose that $S_n^{(p)}(\lambda) \to L_{p+1}$, $p = 0, 1, 2, \cdots, n - 1$, in the uniform operator topology as $\lambda \to \alpha$ along Γ. In this case, $\lambda = \alpha$ is a pole of $R(\lambda)$ of order $\leqq n$ ($= n$ if $L_1 \neq \Theta$). The proof is somewhat lengthy and we shall content ourselves by mentioning only the important steps. Substituting $R(\lambda) = (\lambda - \alpha)^{-n} S_n(\lambda)$ in the resolvent equation and simplifying, we get

(18.8.4) $$(\lambda - \mu)S_n(\lambda)S_n(\mu) = (\lambda - \alpha)^n S_n(\mu) - (\mu - \alpha)^n S_n(\lambda).$$

Differentiating this relation p times, $p \leqq n - 1$, with respect to μ and passing to the limit as $\mu \to \alpha$ along Γ, we obtain

$$\begin{aligned}(\lambda - \alpha)S_n(\lambda)L_{p+1} - pS_n(\lambda)L_p &= (\lambda - \alpha)L_{p+1}S_n(\lambda) - pL_pS_n(\lambda)\\ &= (\lambda - \alpha)^n L_{p+1}.\end{aligned}$$

An inductive argument then gives

(18.8.5) $$\begin{aligned}S_n(\lambda)L_p = L_pS_n(\lambda) &= (\lambda - \alpha)^{n-1}L_p + (p - 1)(\lambda - \alpha)^{n-2}L_{p-1}\\ &+ (p - 1)(p - 2)(\lambda - \alpha)^{n-3}L_{p-2} + \cdots + (p - 1)!(\lambda - \alpha)^{n-p}L_1.\end{aligned}$$

Differentiating this relation m times, $m \leqq n - 1$, and passing to the limit as $\lambda \to \alpha$ along Γ then yields a set of identities in the L_p's from which we finally obtain

(18.8.6) $$\begin{aligned}L_{n-p} &= (n - p - 1)!B^p, \qquad p = 1, 2, \cdots, n - 1;\\ L_n &= (n - 1)!E, E^2 = E;\\ EB &= BE = B, B^n = \Theta.\end{aligned}$$

We now set

$$Q(\lambda) \equiv \sum_{p=0}^{n-1} \frac{(\lambda - \alpha)^p}{p!} L_{p+1} = \sum_{p=0}^{n-2} (\lambda - \alpha)^p B^{n-p-1} + (\lambda - \alpha)^{n-1}E,$$

$$U(\lambda) \equiv S_n(\lambda) - Q(\lambda).$$

Making use of the identities (18.8.6) it is easy to see that $Q(\lambda)$ itself satisfies the equation (18.8.4). Further (18.8.5) with $p = n$ gives $ES_n(\lambda) = S_n(\lambda)E = Q(\lambda)$. Since $EQ(\lambda) = Q(\lambda)E = Q(\lambda)$, we also have $ES_n(\lambda) = Q(\lambda) + EU(\lambda)$ and $S_n(\lambda)E = Q(\lambda) + U(\lambda)E$. Consequently $EU(\lambda) = \Theta = U(\lambda)E$ and therefore $Q(\lambda)U(\lambda) = \Theta = U(\lambda)Q(\lambda)$. It now follows that $U(\lambda)$ also satisfies the equation (18.8.4). Further, we see by the way in which $U(\lambda)$ was defined that $U^{(p)}(\lambda) \to \Theta$, $p = 0, 1, 2, \cdots, n - 1$, as $\lambda \to \alpha$ along Γ. We can therefore represent $U(\lambda)$ by

$$U(\lambda) = \frac{1}{(n - 2)!} \int_\alpha^\lambda (\lambda - \zeta)^{n-2} U^{(n-1)}(\zeta)\, d\zeta$$

where the path of integration is taken along Γ. From this we obtain the estimate $\| U(\lambda) \| = o(|\lambda - \alpha|^{n-1})$ as $\lambda \to \alpha$ along Γ. If we now set $T(\lambda) = (\lambda - \alpha)^{1-n}U(\lambda)$, then $T(\lambda) \to \Theta$ as $\lambda \to \alpha$ along Γ and $T(\lambda)$ satisfies the equation (18.8.1). From this point on the proof proceeds as in the case $n = 1$.

In the above characterization of a pole of $R(\lambda)$ of order n, it is necessary to assume the existence of $\lim_{\lambda \to \alpha} S_n^{(p)}(\lambda)$ for all of the integers $p = 0, 1, 2, \cdots, n - 1$. The existence of the limit for merely $p = 0, 1, 2, \cdots, n - 2$ does not exclude the possibility of $\lambda = \alpha$ being an essential singular point. Further, a function $R(\lambda)$ very well may have an essential singular point at $\lambda = \alpha$ and still $\lim_{\lambda \to \alpha} (\lambda - \alpha)^{1+\epsilon} R(\lambda)$ may exist for every positive ϵ as $\lambda \to \alpha$ in a suitable manner. Examples illustrating these phenomena are to be found in section 23.16 in connection with fractional integration.

In the preceding discussion it was tacitly assumed that α is finite. For a corresponding result at infinity we have

THEOREM 18.8.2. *Let $R(\lambda)$ be a function on complex numbers to $\mathfrak{E}(\mathfrak{X})$ which is defined and satisfies the first resolvent equation in some domain D extending to infinity. Let $\{\beta_n\}$ be a sequence contained in D with $\beta_n \to \infty$. If* $\lim_{n\to\infty} \beta_n R(\beta_n) =$ *J in the sense of the uniform operator topology, then* (i) *J is a bounded linear projection operator, $J^2 = J$;* (ii) *$J = \Theta$ if and only if $R(\lambda) \equiv \Theta$; and* (iii) *if $J \neq \Theta$, then $R(\lambda)$ has a holomorphic extension having a simple zero at infinity so that*

$$R(\lambda) = JR(\lambda; A), \tag{18.8.7}$$

where $A \in \mathfrak{E}(\mathfrak{X})$ and $A = JA = AJ$.

PROOF. Making use of the transformation mentioned at the end of section 5.8, we see that

$$R_1(\lambda) \equiv \frac{I}{\lambda} + \frac{1}{\lambda^2} R\left(-\frac{1}{\lambda}\right)$$

satisfies the first resolvent equation in the domain $D_1 \equiv [\lambda; -\lambda^{-1} \in D]$. Setting $\alpha_n = -\beta_n^{-1}$, we have $\{\alpha_n\} \subset D_1$, $\alpha_n \to 0$, and our basic assumption becomes $\lim_{n\to\infty} \alpha_n R_1(\alpha_n) = I - J = E$. The assertion of the theorem can now be read off from the previous theorem. In fact, $E \in \mathfrak{E}(\mathfrak{X})$ and $E^2 = E$ implies $J \in \mathfrak{E}(\mathfrak{X})$ and $J^2 = J$. Further $J = \Theta$ corresponds to $E = I$ so that $R_1(\lambda) = I/\lambda$ and $R(\lambda) = \Theta$; the converse statement is obvious. The extended $R_1(\lambda)$ is either regular or has a simple pole at $\lambda = 0$. Hence if $J \neq \Theta$, that is if $E \neq I$, then it is seen from (18.8.3) that $R_1(\lambda) - I/\lambda$ has a simple pole at $\lambda = 0$ so that $R(\lambda)$ will have a simple zero at $\lambda = \infty$. The relation (18.8.7) now follows directly from Theorem 5.9.2.

The situation regarding uniform ergodicity at zero is rather striking. We recall that in Theorem 9.6.2 it was shown that the existence of the uniform $(C, 1)$-$\lim_{\xi\to 0+} T(\xi)$ gives that the Cauchy limit exists and that $T(\xi) = J \exp(\xi A)$ where $A \in \mathfrak{E}(\mathfrak{X})$. We shall now prove the same type of result for the uniform Abel limit.

THEOREM 18.8.3. *Let $[T(\xi)]$ be a semi-group of class (E) and suppose that*

$$(A)\text{-}\lim_{\xi\to 0+} T(\xi) = J$$

in the uniform operator topology. Then

$$T(\xi) = J \exp(\xi A), \tag{18.8.8}$$

where $J^2 = J$, $A \in \mathfrak{E}(\mathfrak{X})$, *and* $AJ = JA = A$. *Further,* uniform $\lim_{\xi\to 0+} T(\xi) = J$.

PROOF. If $T(\xi)$ is uniformly Abel-ergodic at zero, then Theorem 18.8.2 applies and we see that $J^2 = J$ and $R(\lambda) = JR(\lambda; A)$, $\lambda > \omega_0$, where $A \in \mathfrak{E}(\mathfrak{X})$ and $AJ = JA = A$. It follows that $R(\lambda)$ has a holomorphic extension, namely $JR(\lambda; A)$, for $|\lambda| > \|A\|$. Now for $x \in \mathfrak{X}_2$ it is known that $R(\lambda)x = \int_0^\infty e^{-\lambda\xi}T(\xi)x\, d\xi$, $\lambda > \omega_0$, where the right member is holomorphic for $\Re(\lambda) > \omega_0$. Consequently for $x \in \mathfrak{X}_2$ and $\gamma > \max(\omega_0, \|A\|)$ we have $JR(\lambda; A)x = \int_0^\infty e^{-\lambda\xi}T(\xi)x\, d\xi$ for $\Re(\lambda) \geqq \gamma$ and the inversion formula of Theorem 6.3.2 gives

$$T(\xi)x = \frac{1}{2\pi i}(C, 1)\text{-}\int_{\gamma-i\infty}^{\gamma+i\infty} e^{\lambda\xi}JR(\lambda; A)x\, d\lambda$$
$$= Jx + \frac{1}{2\pi i}\int_{\gamma-i\infty}^{\gamma+i\infty} e^{\lambda\xi}\left[JR(\lambda; A)x - \frac{Jx}{\lambda}\right] d\lambda.$$

Since the integrand in the last member is $O(|\lambda|^{-2})$ as $|\lambda| \to \infty$, the path of integration can be deformed into the circle $\Gamma: |\lambda| = \gamma$. Thus

$$T(\xi)x = \frac{1}{2\pi i}\int_\Gamma e^{\lambda\xi}JR(\lambda; A)x\, d\lambda,$$

and, $\mathfrak{X}_2$ being dense in $\mathfrak{X}$, this relation holds for all $x \in \mathfrak{X}$. Theorem 11.3.1 now shows that $T(\xi) = J\exp(\xi A)$ and it follows that $\lim_{\xi\to 0+}\|T(\xi) - J\| = 0$.

The situation at infinity is more interesting. Here we have

THEOREM 18.8.4. *Let* $[T(\xi)]$ *be a semi-group of class* (A) *and of type* $\omega_0 \leqq 0$. *Then the following statements are equivalent*:

(1) $T(\xi)$ *is uniformly Abel-ergodic at infinity, that is,* $(A)\text{-}\lim_{\xi\to\infty} T(\xi) = P$ *where* $P^2 = P \in \mathfrak{E}(\mathfrak{X})$.

(2) *The point* $\lambda = 0$ *is a simple pole of* $R(\lambda; A)$ *with residue* P.

(3) $\lim_{\lambda\to 0+}\lambda^2 R(\lambda; A)x = \theta$ *for each* $x \in \mathfrak{X}$ *and* $\mathfrak{X} = \Re(A) \oplus \mathfrak{Z}(A)$ *where* $\Re(A)$ *and* $\mathfrak{Z}(A)$ *are closed and* $\Re(A) \cap \mathfrak{Z}(A) = \{\theta\}$.

(4) $\lim_{\lambda\to 0+}\lambda^2 R(\lambda; A)x = \theta$ *for each* $x \in \mathfrak{X}$ *and* $\Re(A^2)$ *is closed.*

PROOF. We shall show that (1) → (2) → (3) → (4) → (1). The first of these implications follows from Theorem 18.8.1. For $R(\lambda; A)$ is holomorphic for $\Re(\lambda) > \omega_0 \leqq 0$ by Theorem 18.4.2 and the assumption, uniform $\lim_{\lambda\to 0+}\lambda R(\lambda; A) = P$, gives

$$(18.8.9) \qquad R(\lambda; A) = \frac{P}{\lambda} + \sum_{n=0}^{\infty}(-\lambda)^n B^{n+1},\ BP = PB = \theta,$$

the series being convergent for small values of $|\lambda|$ as in (18.8.3).

Let us now show that (2) → (3). It is clear from (2) that

$$\text{uniform } \lim_{\lambda\to 0+}\lambda R(\lambda; A) = P.$$

Consequently Theorem 18.8.1 implies that P is a bounded projection operator and as such determines the decomposition

$$\mathfrak{X} = \mathfrak{R}(P) \oplus \mathfrak{Z}(P),$$

where the subspaces $\mathfrak{R}(P)$ and $\mathfrak{Z}(P)$ are closed linear manifolds having only the zero element in common. We now desire certain identities between the operators A, P, and B. To this end we consider the relations

$$R(\lambda; A)(\lambda I - A)x = x, \qquad x \in \mathfrak{D}(A),$$
$$(\lambda I - A)R(\lambda; A)x = x, \qquad x \in \mathfrak{X},$$

which are valid for $0 < |\lambda| < \rho$. Substituting the series (18.8.9) into these relations and comparing coefficients of like powers, we obtain the following identities:

$$\text{(18.8.10)} \quad \begin{aligned} PAx &= \theta,\ x \in \mathfrak{D}(A), & APx &= \theta,\ x \in \mathfrak{X}, \\ BAx &= (P - I)x,\ x \in \mathfrak{D}(A), & ABx &= (P - I)x,\ x \in \mathfrak{X}. \end{aligned}$$

The above procedure requires that A operate termwise on the series expansion for $R(\lambda; A)$. That this is the case follows from the relation

$$A\left[\frac{1}{2\pi i}\oint R(\lambda; A)\frac{d\lambda}{\lambda^k}\right] = \frac{1}{2\pi i}\oint AR(\lambda; A)\frac{d\lambda}{\lambda^k},$$

which is in turn a consequence of Theorem 3.7.12. Now if $x \in \mathfrak{R}(P)$, then $x = Px$ and $Ax = APx = \theta$ so that $\mathfrak{R}(P) \subset \mathfrak{Z}(A)$. On the other hand if $x \in \mathfrak{Z}(A)$, that is if $Ax = \theta$, then $\lambda R(\lambda; A)x = x + R(\lambda; A)Ax = x$ for all $\lambda \in \rho(A)$ and hence

$$Px - \sum_{n=0}^{\infty}(-\lambda)^{n+1}B^{n+1}x \equiv x.$$

This implies $Px = x$ so that $x \in \mathfrak{R}(P)$. Thus $\mathfrak{R}(P) = \mathfrak{Z}(A)$. Suppose next that $x \in \mathfrak{Z}(P)$. Then by (18.8.10), $x = (I - P)x = A[-Bx] \in \mathfrak{R}(A)$ and therefore $\mathfrak{Z}(P) \subset \mathfrak{R}(A)$. But if $x \in \mathfrak{R}(A)$, then $x = Ay$ and $Px = PAy = \theta$ so that $x \in \mathfrak{Z}(P)$. It follows that $\mathfrak{R}(A) = \mathfrak{Z}(P)$. Consequently $\mathfrak{R}(A)$ is closed, $\mathfrak{R}(A) \cap \mathfrak{Z}(A) = \{\theta\}$, and $\mathfrak{X} = \mathfrak{R}(A) \oplus \mathfrak{Z}(A)$. This proves the implication (2) $\rightarrow$ (3).

In proving (4) from (3) it suffices to show that $\mathfrak{R}(A^2) = \mathfrak{R}(A)$. It is obvious that $\mathfrak{R}(A^2) \subset \mathfrak{R}(A)$. Conversely, suppose $x \in \mathfrak{R}(A)$, that is, $x = Aw$ where $w \in \mathfrak{D}(A)$. Then by condition (3) we have $w = y + z$ where $y \in \mathfrak{R}(A)$ and $z \in \mathfrak{Z}(A) \subset \mathfrak{D}(A)$. Thus $y = w - z \in \mathfrak{D}(A)$ and $Ay = Aw - Az = Aw = x$. However $y \in \mathfrak{R}(A)$ so that $y = Av$ where $v \in \mathfrak{D}(A)$. Consequently $x = Ay = A^2v \in \mathfrak{R}(A^2)$. Thus $\mathfrak{R}(A^2) = \mathfrak{R}(A)$ and since $\mathfrak{R}(A)$ is assumed to be closed, this proves (4).

The passage from (4) to (1) lies considerably deeper. Let $x \in \mathfrak{D}(A^2)$ be given. Then by the relation (11.5.12) we have

$$\text{(18.8.11)} \quad \lambda^2R(\lambda; A)x - \lambda x - Ax = R(\lambda; A)A^2x = A^2R(\lambda; A)x.$$

This together with the assumed limit condition implies $\lim_{\lambda \to 0+} A^2R(\lambda; A)x = -Ax$ so that $Ax \in \mathfrak{R}(A^2)$, which is closed by hypothesis. Thus if $y \in \mathfrak{D}(A) \cap \mathfrak{R}(A)$, then there exists an $x \in \mathfrak{D}(A^2)$ such that $y = Ax$ and by the above argument $y \in \mathfrak{R}(A^2)$. In other words, $\mathfrak{D}(A) \cap \mathfrak{R}(A) \subset \mathfrak{R}(A^2)$ and since $\mathfrak{R}(A^2) \subset \mathfrak{R}(A)$, it follows that

$$\mathfrak{D}(A) \cap \mathfrak{R}(A) = \mathfrak{D}(A) \cap \mathfrak{R}(A^2).$$

The operator A is actually one-to-one on $\mathfrak{D}(A) \cap \mathfrak{R}(A^2)$. For if $y \in \mathfrak{D}(A) \cap \mathfrak{R}(A^2)$ with $Ay = \theta$, then $y \in \mathfrak{D}(A) \cap \mathfrak{R}(A)$ and there exists an $x \in \mathfrak{D}(A^2)$ such that $y = -Ax$. By (18.8.11) we have $y = -Ax = \lim_{\lambda \to 0+} R(\lambda; A)A^2x = \lim_{\lambda \to 0+} - R(\lambda; A)Ay = \theta$. We now define A_1 to be the restriction of A on $\mathfrak{R}(A^2)$. Then A_1 is closed, one-to-one, and $\mathfrak{R}(A_1) = A[\mathfrak{D}(A) \cap \mathfrak{R}(A)] = \mathfrak{R}(A^2)$. According to Theorem 2.12.1, A_1^{-1} is necessarily a linear bounded operator on $\mathfrak{R}(A^2)$ to itself.

Suppose next that $y \in \mathfrak{R}(A^2)$ is given and set $x = A_1^{-1}y$. Then $x \in \mathfrak{D}(A_1)$ and

$$R(\lambda; A)y = R(\lambda; A)Ax = -x + \lambda R(\lambda; A)x.$$

Since $\lim_{\lambda \to 0+} \lambda^2R(\lambda; A)x = \theta$ by hypothesis, we see that $\lim_{\lambda \to 0+} \lambda R(\lambda; A)y = \theta$ for all $y \in \mathfrak{R}(A^2)$. The uniform boundedness theorem applied to $\mathfrak{R}(A^2)$ asserts that $\| \lambda R(\lambda; A)y \| \leqq M \| y \|$ for $0 < \lambda < 1$ and all $y \in \mathfrak{R}(A^2)$. It follows that

$$\text{(18.8.12)} \qquad \| \lambda R(\lambda; A)y \| = \lambda \| \lambda R(\lambda; A)x - x \| \leqq \lambda(M + 1) \| A_1^{-1} \| \| y \|$$

for $0 < \lambda < 1$ and all $y \in \mathfrak{R}(A^2)$.

It is clear that $\mathfrak{Z}(A)$ is a closed linear subspace of $\mathfrak{X}$. Further for $z \in \mathfrak{Z}(A)$ we have $\lambda R(\lambda; A)z = z + R(\lambda; A)Az = z$ for all $\lambda > 0$ so that $\lim_{\lambda \to 0+} \lambda R(\lambda; A)z = z$. Since $\lim_{\lambda \to 0+} \lambda R(\lambda; A)y = \theta$ for each $y \in \mathfrak{R}(A^2)$, we see that $\mathfrak{R}(A^2) \cap \mathfrak{Z}(A) = \{\theta\}$. We next show that $\mathfrak{D}(A^2) \subset \mathfrak{R}(A^2) \oplus \mathfrak{Z}(A)$. In fact, if $x \in \mathfrak{D}(A^2)$, then $A^2x = y \in \mathfrak{R}(A^2)$. Set $x_1 = A_1^{-2}y$. Then $x_1 \in \mathfrak{D}(A^2) \cap \mathfrak{R}(A^2)$ and $A^2x = y = A_1^2x_1 = A^2x_1$. Thus $x_2 = x - x_1 \in \mathfrak{D}(A^2)$ and $A^2x_2 = \theta$. By (18.8.11) we see that $Ax_2 = \theta$ so that $x_2 \in \mathfrak{Z}(A)$. Consequently $x = x_1 + x_2 \in \mathfrak{R}(A^2) \oplus \mathfrak{Z}(A)$. This proves that $\mathfrak{D}(A^2) \subset R(A^2) \oplus \mathfrak{Z}(A)$ and since $\mathfrak{D}(A^2)$ is dense in $\mathfrak{X}$ by Theorem 10.3.4, the same will be true of $\mathfrak{R}(A^2) \oplus \mathfrak{Z}(A)$.

Actually we can prove much more; in fact, $\mathfrak{X} = \mathfrak{R}(A^2) \oplus \mathfrak{Z}(A)$. To see this, let $x \in \mathfrak{X}$ be given. As we have just shown, there exists a sequence $x_n = y_n + z_n$ such that $x_n \to x$, $\{y_n\} \subset \mathfrak{R}(A^2)$, and $\{z_n\} \subset \mathfrak{Z}(A)$. By Corollary 3 to Theorem 11.5.3, we have $T(\xi)z_n = z_n$ for all $\xi > 0$. Hence

$$\begin{aligned} [T(\xi) - I]x_n &= [T(\xi) - I]y_n = [T(\xi) - I]A^2w_n \\ &= A^2[T(\xi) - I]w_n \in \mathfrak{R}(A^2), \end{aligned}$$

where $y_n = A^2w_n$, $w_n \in \mathfrak{D}(A^2)$. Taking the limit as $n \to \infty$, we see that $[T(\xi) - I]x \in \mathfrak{R}(A^2)$ for all $x \in \mathfrak{X}$ and $\xi > 0$. Next let $F(\tau)$ be chosen from the class $\mathfrak{F}$ of Theorem 10.3.4 with $\int_0^\infty F(\tau)\, d\tau = 1$. Since $\mathfrak{R}(A^2)$ is a closed linear subspace, it is clear that $\int_0^\infty F(\tau)[T(\xi) - I]x\, d\tau \in \mathfrak{R}(A^2)$ for each $x \in \mathfrak{X}$. Moreover, as in the proof of Theorem 10.3.4 we have $\int_0^\infty F(\tau)T(\tau)x\,d\tau \in \mathfrak{D}(A^2)$ and hence $\int_0^\infty F(\tau)T(\tau)x\,d\tau \in \mathfrak{R}(A^2) \oplus \mathfrak{Z}(A)$ for each $x \in \mathfrak{X}$. Consequently

$$x = \int_0^\infty F(\tau)x\,d\tau = \int_0^\infty F(\tau)T(\tau)x\,d\tau - \int_0^\infty F(\tau)[T(\tau) - I]x\,d\tau$$

belongs to $\mathfrak{R}(A^2) \oplus \mathfrak{Z}(A)$. Thus for each $x \in \mathfrak{X}$ there exists a unique decomposition $x = y + z$ where $y \in \mathfrak{R}(A^2)$ and $z \in \mathfrak{Z}(A)$. We define the projection operator P by $Px = y$. According to Theorem 2.14.2, the operator P is a bounded projection operator with $\mathfrak{R}(P) = \mathfrak{Z}(A)$ and $\mathfrak{Z}(P) = \mathfrak{R}(A^2)$. Thus $(I - P)x \in \mathfrak{R}(A^2)$, $Px \in \mathfrak{Z}(A)$, and, as above, $\lambda R(\lambda; A)Px = Px$ for all $\lambda > 0$. The inequality (18.8.12) now implies that

$$\| \lambda R(\lambda; A)x - Px \| = \| \lambda R(\lambda; A)(I - P)x \| \leqq \lambda K \| x \|, \qquad x \in \mathfrak{X},$$

where $K = (1 + M) \| A_1^{-1} \| \| I - P \|$. It follows that

$$\text{uniform } (A)\text{-}\lim_{\xi \to \infty} T(\xi) = P$$

where $P^2 = P \in \mathfrak{E}(\mathfrak{X})$. This concludes the proof of Theorem 18.8.4.

COROLLARY. *Let $[T(\xi)]$ be a semi-group of class (A) and of type $\omega_0 \leqq 0$. If $T(\xi)$ is uniformly Abel-ergodic at infinity, then $\mathfrak{R}(A^k) = \mathfrak{R}(A)$, $k = 1, 2, 3, \cdots$.*

PROOF. We proceed by induction. The assertion being obviously valid for $k = 1$, we suppose that it has been verified for all $k \leqq n$. Then by condition (3) of Theorem 18.8.4 we have $\mathfrak{X} = \mathfrak{R}(A^n) \oplus \mathfrak{Z}(A)$. It is clear that $\mathfrak{R}(A^{n+1}) \subset \mathfrak{R}(A^n)$. On the other hand, if $x \in \mathfrak{R}(A^n)$, then $x = Aw$ where $w \in \mathfrak{D}(A)$. As above, $w = y + z$ where $y \in \mathfrak{R}(A^n)$ and $z \in \mathfrak{Z}(A) \subset \mathfrak{D}(A)$. Thus $y = w - z \in \mathfrak{D}(A)$ and $Ay = Aw = x$. However, $y \in \mathfrak{R}(A^n)$ implies that $y = A^n v$ and hence that $x = A^{n+1}v \in \mathfrak{R}(A^{n+1})$. Consequently $\mathfrak{R}(A^{n+1}) = \mathfrak{R}(A^n)$ and this completes the inductive step.

Once we have Abel-ergodicity we can prove Cesàro-ergodicity provided the assumptions of Theorem 18.3.3 apply. We state the result without proof.

THEOREM 18.8.5. *If the semi-group $[T(\xi)]$ is uniformly measurable for $\xi > 0$, if $\int_0^\infty e^{-\lambda\xi} \| T(\xi) \| d\xi < \infty$ for each $\lambda > 0$, if for some fixed $\alpha \geqq 0$ the Cesàro averages satisfy $\| C(\xi; \alpha) \| \leqq M$ for $\xi \in (0, \infty)$, and if $T(\xi)$ is uniformly Abel-ergodic at infinity, then $T(\xi)$ is also uniformly (C, β)-ergodic at infinity for each $\beta > \alpha$. This result also holds for $\beta = \alpha$ provided $C(\xi; \alpha)$ is feebly oscillating in the uniform operator topology as $\xi \to \infty$.*

PART FOUR

SPECIAL SEMI-GROUPS AND APPLICATIONS

Summary. The fourth part of this treatise is concerned with the theory of special semi-groups arising in analysis. There are five chapters: *Translations and Powers*, *Trigonometric Semi-Groups*, *Semi-Groups in* $L_p(-\infty, \infty)$, *Semi-Groups in Hilbert Space*, and *Miscellaneous Applications*.

In the main we shall be concerned with semi-groups of operators acting in one of the fundamental function spaces of analysis: Lebesgue spaces, spaces of continuous functions, functions of bounded variation, all of which are accessible to Fourier analysis. There are two basic patterns which, in one form or another, underlie most of our semi-groups, viz. translations (differentiation) and powers (exponentials). If the function space in question is invariant under one or two-sided translations, then the translations form a semi-group or a group having the operation of differentiation as its infinitesimal generator. In the associated space of Fourier transforms or Fourier coefficients there is a semi-group of powers and the elements are multiplied by an exponential function of the parameter. Such semi-groups of translations and powers are studied *per se* in Chapter XIX. The basic space of Chapter XX is $L_p(-\pi, \pi)$ and the semi-group operator is supposed to commute with translations; this leads to a power semi-group for the Fourier coefficients. In Chapter XXI the basic space is $L_p(-\infty, \infty)$ instead, but the same pattern occurs. Chapter XXII deals with semi-groups of normal operators in Hilbert space and contains integral representation theorems for such semi-groups.

Chapter XXIII contains a multitude of loosely related topics such as summability, the abstract Cauchy problem, stochastic processes, and fractional integration. The power pattern is clearly discernable in most of these applications. Considerations of space have forced us to omit a systematic discussion of the applications to partial differential equations, a field where the theory has made its most striking contributions in recent years; the paragraph on the abstract Cauchy problem at least lays the theoretical foundations for these applications of semi-group theory. The theory of stochastic processes is another field where semi-group theory has been able to make its mark; here we have done more justice to recent developments.

CHAPTER XIX

TRANSLATIONS AND POWERS

19.1. Orientation. The two special classes of semi-groups namely translations and powers, which are to be considered in the present chapter, are among the simplest concrete illustrations of the general theory that can be found. For this reason alone they would merit a place in this treatise. In addition they provide us with a number of examples and counterexamples which enable us to answer several of the existence questions raised in parts two and three.

Thus the simple semi-group of left translations on $C[0, \infty]$ shows that the spectrum of the infinitesimal generator, in this case the operator of differentiation, may fill out a half-plane. Another example, this time a power semi-group, shows that the whole spectrum may be residual and coincide with the closure of any preassigned simply-connected domain in the left half-plane. An example is given of a nilpotent semi-group which admits of an analytic continuation in a half-plane. We also prove that the interior of each proper spinal semi-module in the complex plane is the maximal domain of analytic existence for some holomorphic semi-group of translations. An example is also given of a semi-group $[T(\xi); \xi > 0]$ such that $\log \| T(\xi) \|$ is equal to an arbitrary continuous subadditive function.

There are two paragraphs: *Translations* and *Powers*.

References. S. Bernstein [1], Dahlgren [1], Dunford and Segal [1], Favard [1], Hille [10], Hille and Zorn [1], Kendall [1], Nörlund [1], Pierpont [1], Plessner [1], Szász [1].

1. Translations

19.2. Translations in $C[0, \infty]$. The semi-group of translations in $C[0, \infty]$ is perhaps the simplest of all semi-groups of linear bounded operators, but nevertheless it is quite instructive as an illustration of the general theory. For this reason we shall consider it in some detail.

The basic space being $\mathfrak{X} = C[0, \infty]$, the elements $x(t)$ are numerically-valued continuous functions of t on the closed interval $[0, \infty]$. The norm of $x(t)$ is defined as

$$\| x \| \equiv \max [\, | x(t) |; 0 \leqq t \leqq \infty].$$

The space $C[0, \infty]$ is obviously a (B)-space and the notion of strong convergence in this space coincides with the classical notion of uniform convergence with respect to t in $[0, \infty]$. Further $C[0, \infty]$ is a separable space and the functions $\{e^{-nt}; n = 0, 1, 2, \cdots\}$ form a fundamental set. Indeed, by the approximation theorem of Weierstrass the powers $\{u^n; n = 0, 1, 2, \cdots\}$ form a fundamental set in $C[0, 1]$ which is mapped on $C[0, \infty]$ by $f(u) \to g(t) = f(e^{-t})$.

We define

$$[T(\xi)x](t) = x(t + \xi), \qquad \xi \geqq 0. \tag{19.2.1}$$

Since the maximum of $|x(t + \xi)|$ cannot exceed that of $|x(t)|$, we have $\| T(\xi)x \| \leqq \| x \|$ with equality holding if $x(t)$ is constant-valued. Hence

$$\| T(\xi) \| \equiv 1. \tag{19.2.2}$$

Since $x(t)$ is uniformly continuous in $[0, \infty]$,

$$\lim_{\eta \to 0} \max [\, | x(t + \xi + \eta) - x(t + \xi) |; 0 \leqq t \leqq \infty] = 0.$$

Here we may admit negative values of η, $\eta > -\xi$, if $\xi > 0$. It follows that $T(\xi)x$ is a continuous function of ξ, $\xi \geqq 0$, for each $x \in C[0, \infty]$. On the other hand,

$$\| T(\xi + \eta) - T(\xi) \| = 2, \qquad \xi + \eta > 0, \xi > 0, \eta \neq 0. \tag{19.2.3}$$

In fact, among the elements of $C[0, \infty]$ we can always find one such that for given ξ and η satisfying the above inequalities we have $x(\xi) = 1$, $x(\xi + \eta) = -1$, and $\max | x(t) | = 1$. For such an x we have $\| [T(\xi + \eta) - T(\xi)]x \| = 2$ so that (19.2.3) holds. It follows that the operator-function $T(\xi)$ is continuous in the strong operator topology for $\xi \geqq 0$ but not continuous in the uniform operator topology.

With the usual notation we have

$$[A_\eta x](t) = \eta^{-1}[x(t + \eta) - x(t)]. \tag{19.2.4}$$

If the right member tends to a limit for a given $t = t_0$ as $\eta \to 0+$, then $x(t)$ has a right-hand derivative at t_0. In order that $x(t)$ belong to $\mathfrak{D}(A)$ it is necessary, however, that the difference quotient tend to its limit uniformly with respect to t in $[0, \infty]$. This makes the right-hand derivative continuous for all values of t and by a classical theorem (see J. Pierpont [1, pp. 508–509]) this implies that $x(t)$ has a continuous derivative. Conversely, if $x(t)$ has a derivative continuous on $[0, \infty]$, then the difference quotient tends to a limit uniformly with respect to t on $[0, \infty]$. Thus the operator A is simply ordinary differentiation and

$$\mathfrak{D}(A) = \mathfrak{D}(d/dt) \equiv [x; x'(t) \text{ exists and belongs to } C[0, \infty]].$$

We now consider the resolvent of $A = d/dt$. This amounts to finding the inverse of the operator $\lambda I - d/dt$. We are consequently led to the first order linear differential equations

$$\lambda y(t) - y'(t) = x(t), \tag{19.2.5}$$

(19.2.6) $$\lambda z(t) - z'(t) = 0.$$

The homogeneous equation has the solution $z(t) = Ce^{\lambda t}$ which belongs to $C[0, \infty]$ if $\lambda = 0$ or if $\Re(\lambda) < 0$ and for no other values of λ. It follows that $\lambda I - d/dt$ has a unique inverse when $\Re(\lambda) \geqq 0$, $\lambda \neq 0$; however this inverse may possibly be unbounded or its domain may not be dense in $C[0, \infty]$. Since $\| T(\xi) \| \equiv 1$, Theorem 11.5.1 shows that the operator $R(\lambda; d/dt)$ exists as a bounded linear operator on $C[0, \infty]$ to a dense subset of $C[0, \infty]$ for $\Re(\lambda) > 0$ and further that

(19.2.7) $$[R(\lambda; d/dt)x](t) = \int_0^\infty e^{-\lambda\xi} x(t + \xi)\, d\xi, \qquad \Re(\lambda) > 0.$$

Actually Theorem 11.5.1 requires that the right member be treated as a Bochner integral. However for each $t_0 \in [0, \infty]$ there exists a functional $x^*_{t_0} \in C[0, \infty]^*$ such that $x^*_{t_0}(x) = x(t_0)$ for all $x \in C[0, \infty]$. Applying the relation (3.7.5), we see that the right member in (19.2.7) can also be treated as an ordinary Lebesgue integral for numerically-valued functions. Consequently it is easy to verify directly that this function satisfies (19.2.5) and belongs to $C[0, \infty]$ when $\Re(\lambda) > 0$. If $\lambda = \tau i$, $\tau \neq 0$, and $y_n(t) = \exp[(\tau i - 1/n)t]$ we find that $(\tau i I - d/dt)y_n = n^{-1}y_n$ so that $(\tau i I - d/dt)^{-1}$ is necessarily unbounded. On the other hand for $w_n(t) = e^{-n}$ we have

$$(\tau i I - d/dt)w_n = (\tau i + n)w_n, \qquad n = 0, 1, 2, \cdots,$$

so that $\tau i I - d/dt$ takes the linear extension of $\{w_n\}$ onto itself. This shows that the range of $\tau i I - d/dt$ is dense in $C[0, \infty]$.

From the above discussion we see that the spectrum of d/dt, considered as an operator on $C[0, \infty]$, is precisely the half-plane $R(\lambda) \leqq 0$; the points on the imaginary axis except for $\lambda = 0$ are in the continuous spectrum and the rest of the half-plane is in the point spectrum. To each characteristic value λ corresponds essentially only one characteristic function $x(t) = e^{\lambda t}$. This semi-group therefore serves as a proof of the assertion made in section 11.5 according to which every point in the half-plane $\Re(\lambda) \leqq \omega_0$ may belong to $\sigma(A)$.

It is of interest to examine the spectral properties of $T(\xi)$ itself and to compare these with the spectral properties of A. The results will illustrate the theorems of section 16.7. Here the first question is the existence of solutions in $C[0, \infty]$ of the linear difference equations

(19.2.8) $$\lambda u(t) - u(t + \xi) = x(t),$$

(19.2.9) $$\lambda v(t) - v(t + \xi) = 0.$$

The latter is satisfied by $v(t) = \lambda^{t/\xi}$, which belongs to $C[0, \infty]$ if $\lambda = 1$ or $0 < |\lambda| < 1$. For $|\lambda| > 1$ the non-homogeneous equation has the familiar solution

(19.2.10) $$u(t) = \sum_{n=0}^{\infty} \lambda^{-n-1} x(t + n\xi),$$

where the series evidently is uniformly convergent with respect to t so that $u(t) \in C[0, \infty]$ and, moreover, is the only solution having this property (cf. N. E. Nörlund [1, pp. 296–297]). The series (19.2.10) is actually the familiar power series expression for the resolvent $[R(\lambda; T(\xi))x](t)$ obtained in Theorem 4.7.2.

For $\lambda = e^{\tau i}$, $\tau \not\equiv 0 \pmod{2\pi}$, let $y_n = [((n-1)/n)e^{\tau i}]^{t/\xi}$. Then $[e^{\tau i}I - T(\xi)]y_n = n^{-1}e^{\tau i}y_n$ so that $[e^{\tau i}I - T(\xi)]^{-1}$ is unbounded. Furthermore, $[e^{\tau i}I - T(\xi)]e^{-nt} = (e^{\tau i} - e^{-n\xi})e^{-nt}$ shows that the operator maps a fundamental set of $C[0, \infty]$ onto itself. Thus the range of $e^{\tau i}I - T(\xi)$ is dense in the space. It follows that each point $e^{\tau i}$ belongs to $C\sigma[T(\xi)]$, $\tau \not\equiv 0 \pmod{2\pi}$; $\lambda = 1$ and the points $0 < |\lambda| < 1$ are in $P\sigma[T(\xi)]$. The point $\lambda = 0$ remains; however it is clear that every function $x(t)$ in $C[0, \infty]$ which vanishes for $t \geqq \xi$ is annihilated by $T(\xi)$, so that $\lambda = 0$ is also in the point spectrum. These spectral properties of $T(\xi)$ were announced in example (iv) following Theorem 5.9.5. We note that the correspondence between the spectra of A and of $T(\xi)$ is perfect. If in the transformation $\lambda = \exp(\alpha\xi)$ we substitute for α a point in $P\sigma(A)$ or in $C\sigma(A)$, then λ will be in $P\sigma[T(\xi)]$ or $C\sigma[T(\xi)]$ respectively. Conversely, if $\lambda \in P\sigma[T(\xi)]$, $\lambda \neq 0, 1$, then all solutions α are in $P\sigma(A)$. If $\lambda = 1$, one solution is in $P\sigma(A)$ and all the rest are in $C\sigma(A)$. Finally if $\lambda \in C\sigma[T(\xi)]$, then all corresponding values of α are in $C\sigma(A)$.

Theorem 10.4.2 takes on a particularly interesting form for the semi-group of translations. Setting

$$\overset{n}{\underset{\eta}{\Delta}}\, x(t) = \eta^{-n} \sum_{k=0}^{n} (-1)^{n-k} \binom{n}{k} x(t + k\eta), \tag{19.2.11}$$

with the usual notation of the calculus of finite differences, then the theorem asserts that for each $x \in C[0, \infty]$ and each $\xi \geqq 0$

$$x(t + \xi) = \lim_{\eta \to 0+} \sum_{n=0}^{\infty} \frac{\xi^n}{n!} \overset{n}{\underset{\eta}{\Delta}}\, x(t), \tag{19.2.12}$$

where the limit exists uniformly with respect to t in $[0, \infty]$. Moreover the convergence is uniform with respect to ξ in every finite interval $[0, \omega]$. The abbreviated form of (19.2.12) is

$$x(t + \xi) = \lim_{\eta \to 0+} \exp\,[\xi \underset{\eta}{\Delta}]\, x(t), \tag{19.2.13}$$

which recalls the symbolic form of Taylor's theorem:

$$x(t + \xi) = \exp\,[\xi d/dt]\, x(t) = \sum_{n=0}^{\infty} \frac{\xi^n}{n!} x^{(n)}(t).$$

If the latter series converges for all positive values of t and ξ, then $x(t)$ must be an entire function of t. Further if $x(t)$ has derivatives of order n, then $\lim_{\eta \to 0+} \Delta_\eta^n x(t) = x^{(n)}(t)$. Thus the individual terms of the series in (19.2.12) converge to the corresponding terms of Taylor's series when $x(t)$ has derivatives of all orders. Since the limit in (19.2.12) exists without any differentiability assumptions whatsoever, the formula represents a rather far-going generalization of Taylor's theorem. See also L. Dahlgren [1], J. Favard [1], and O. Szász [1]; the latter paper deals with pointwise convergence and functions which are not necessarily continuous.

N. Dunford and I. Segal [1] have called attention to the fact that a fairly simple proof of Weierstrass's approximation theorem can be based on formula (19.2.12). Let $f(t)$ be a continuous function defined over the finite interval $[\alpha, \beta]$, $0 < \alpha < \beta < \infty$, and define an element $x(t)$ of $C[0, \infty]$ by setting $x(t) = f(\alpha)$ in $[0, \alpha)$, $x(t) = f(t)$ in $[\alpha, \beta]$, and $x(t) = f(\beta)$ in (β, ∞). We then set $t = 0$ in (19.2.12). The entire functions $[\exp(\xi\Delta_\eta)\, x](0)$ for $\eta = 1/m$, $m = 1, 2, 3, \cdots$, form a sequence converging uniformly to $f(\xi)$ in $[\alpha, \beta]$. The function $[\exp(\xi\Delta_\eta)\, x](0)$ can be approximated with any required degree of accuracy, say to within ϵ on $[\alpha, \beta]$, by a suitable partial sum of its power series and the resulting sequence of polynomials will converge to $f(\xi)$ uniformly in $[\alpha, \beta]$.

Theorem 10.4.3 also has a direct bearing on the theory of continuous functions (see D. G. Kendall [1]). When applied to the semi-group of translations in $C[0, \infty]$, the theorem shows that

$$x(t + \xi) = \lim_{n\to\infty} \sum_{k=0}^{n} \binom{n}{k} \xi^k (1 - \xi)^{n-k} x\left(t + \frac{k}{n}\right), \qquad 0 \leqq \xi \leqq 1, \tag{19.2.14}$$

where the limit exists uniformly with respect to t in $[0, \infty]$ and with respect to ξ. In particular, for $t = 0$ we obtain the approximation theorem of S. Bernstein [1].

The adjoint semi-group to the semi-group of translations is also of interest. For $\mathfrak{X} = C[0, \infty]$, the adjoint space $C[0, \infty]^*$ is the space of all countably additive set functions on the Borel subsets of $[0, \infty]$, here the point at infinity is explicitly included. The adjoint operator $T^*(\xi)$ is readily seen to be

$$[T^*(\xi)a](E) = a\{(E - \xi) \cap [0, \infty]\}. \tag{19.2.15}$$

We note that $T^*(\xi)$ is not continuous in the strong operator topology; indeed $T^*(\xi)a$ is not even continuous for atomic measures. Now the ($\odot$)-adjoint space $C[0, \infty]^{\odot}$ relative to the infinitesimal generator A is given by $\overline{\mathfrak{D}(A^*)}$, that is, the closure of the range of $R(\lambda; A^*) = R^*(\lambda; A)$. On the other hand

$$[R^*(\lambda; A)a](E) = \int_E \left\{\int_0^t e^{-\lambda(t-u)}\, d_u a\right\} dt$$

so that $\mathfrak{D}(A^*)$ is contained in the subspace of absolutely continuous (relative to Lebesgue measure) set functions, namely $L_1(0, \infty)$. Since $[T(\xi)]$ is of class (C_0), Theorem 14.4.2 implies that $C[0, \infty]^{\odot}$ can also be characterized as the set of all functions $a \in C[0, \infty]^*$ such that $T^*(\xi)a$ is continuous in ξ for $\xi \geqq 0$. The latter set obviously includes $L_1(0, \infty)$; consequently $C[0, \infty]^{\odot} = L_1(0, \infty)$. We have therefore shown that the set of all functions with continuous right translates is precisely the set of absolutely continuous set functions in $C[0, \infty]^*$. This is a classical result due to A. Plessner [1]. The adjoint semi-group operator $T^{\odot}(\xi)$ is simply the restriction of $T^*(\xi)$ to the ($\odot$)-adjoint space, $L_1(0, \infty)$.

19.3. Translations in Lebesgue spaces. The same type of discussion applies to translations on the Lebesgue space $L_p(0, \infty)$ where p is fixed, $1 \leqq p < \infty$. We now return to the usual representation of such spaces by measurable functions rather than by set functions. In this case the norm is defined as

$$\| x \|_p = \left\{ \int_0^\infty | x(t) |^p \, dt \right\}^{1/p}.$$

The system of functions $\{e^{-nt};\ n \geqq 1\}$ is a fundamental set in each of these spaces.

We define $T(\xi)$ by (19.2.1) and note that (19.2.2) and (19.2.3) hold so that the translation operator $T(\xi)$ is again continuous in the strong but not in the uniform operator topology on $L_p(0, \infty)$. If $x \in \mathfrak{D}(A)$, then the differential quotient (19.2.4) converges in the mean of order p to a function $y(t) \in L_p(0, \infty)$. Consequently

$$\frac{1}{\eta} \int_\beta^{\beta+\eta} x(s)\, ds - \frac{1}{\eta} \int_\alpha^{\alpha+\eta} x(s)\, ds = \int_\alpha^\beta \frac{x(s + \eta) - x(s)}{\eta}\, ds \to \int_\alpha^\beta y(s)\, ds$$

as $\eta \to 0+$. Now for almost all $\gamma \in [0, \infty)$ we have $\eta^{-1}\int_\gamma^{\gamma+\eta} x(s)\, ds \to x(\gamma)$. Let α be chosen from this set. Then for almost all $\beta \in [0, \infty)$, $x(\beta) - x(\alpha) = \int_\alpha^\beta y(s)\, ds$. Thus $x(t)$ can be redefined on a set of measure zero so that $x(t) \equiv \int_\alpha^t y(s)\, ds + x(\alpha)$ and the so-defined $x(t)$ is absolutely continuous having a derivative almost everywhere which is equal to $y(t)$. Interpreted in this way, $A = d/dt$. Further, if $x(t)$ is equivalent to an absolutely continuous function with derivative almost everywhere equal to $y(t) \in L_p(0, \infty)$, then it is clear that $\eta^{-1}[x(t + \eta) - x(t)] = \eta^{-1}\int_0^\eta y(t + s)\, ds$ converges in mean of order p to $y(t)$ so that $x \in \mathfrak{D}(A)$. Thus

$$\mathfrak{D}(A) \equiv [x;\ x(t) \text{ is absolutely continuous and } x'(t) \in L_p(0, \infty)].$$

The spectrum of the infinitesimal generator is the same as for translations in $C[0, \infty]$ except that the point $\lambda = 0$ is now in the continuous instead of the point spectrum. Similarly $\lambda = 1$ now belongs to $C\sigma[T(\xi)]$. The reader will have no difficulty in verifying these facts with the aid of the formulas developed in the preceding section.

The formulas (19.2.7) representing $R(\lambda; d/dt)$, $\Re(\lambda) > 0$, can again be interpreted as a Lebesgue instead of a Bochner integral. This may be justified as follows. Let $F \in L_p(0, \infty)^*$. Then by (3.7.5) together with the Fubini theorem we have

$$\begin{aligned} F[R(\lambda; d/dt)x] &= \int_0^\infty e^{-\lambda\xi} F[T(\xi)x]\, d\xi = \int_0^\infty e^{-\lambda\xi} \left[\int_0^\infty F(t) x(t + \xi)\, dt \right] d\xi \\ &= \int_0^\infty F(t) \left[\int_0^\infty e^{-\lambda\xi} x(t + \xi)\, d\xi \right] dt. \end{aligned}$$

The Fubini theorem also implies that $\int_0^\infty e^{-\lambda\xi} x(t + \xi)\, d\xi$ exists for almost all t and the above relation shows that it defines the same element in $L_p(0, \infty)$ as $R(\lambda; d/dt)x$.

Formula (19.2.12) is also valid for the space $L_p(0, \infty)$ if we replace "limit" by "limit in the mean of order p." This is a generalized Taylor theorem for the functions of $L_p(0, \infty)$.

We next consider briefly the semi-groups of left translations on a Lebesgue space where the norm is given in terms of a suitable weight factor. Let $\mu(t)$ be a fixed non-negative measurable function defined on $[0, \infty)$, which will be more carefully defined later, and let $L[(0, \infty); \mu]$ be the class of all measurable functions on $[0, \infty)$ such that

$$\| x \| = \int_0^\infty \mu(t) \mid x(t) \mid dt$$

is finite. This is also a (B)-space.

If we define $T(\xi)$ again by (19.2.1), we find that $T(\xi)x$ is not always an element of $L[(0, \infty); \mu]$ unless $\mu(t)$ satisfies certain conditions. Thus if $\mu(t) = e^{-t^2}$, the function $e^{t^2}(1 + t^2)^{-1}$ belongs to the space, but none of its transforms has this property. Roughly speaking, we cannot allow $\mu(t)$ to tend to zero faster than $e^{-\alpha t}$ as $t \to \infty$. We now make use of the fact that the norm of the translation is strongly affected by the choice of the weight factor to construct translation semi-groups in which $\| T(\xi) \|$ has a prescribed admissible rate of growth. Here the qualifier "admissible" refers to the fact that $\lim_{\xi \to \infty} \xi^{-1} \log \| T(\xi) \|$ must exist and be finite.

THEOREM 19.3.1. *Let $\alpha(t)$ be any continuous subadditive function defined on $[0, \infty)$ such that $\alpha(0) = 0$. Let $[T(\xi); \xi > 0]$ be the semi-group of left translations on $L[(0, \infty); \mu]$ where $\mu(t) = e^{-\alpha(t)}$. Then*

$$\| T(\xi) \| \equiv e^{\alpha(\xi)}, \qquad \xi > 0. \tag{19.3.1}$$

PROOF. We have

$$\| T(\xi)x \| = \int_0^\infty e^{-\alpha(t)} \mid x(t + \xi) \mid dt.$$

Since $\alpha(t)$ is subadditive, $-\alpha(t) \leqq -\alpha(t + \xi) + \alpha(\xi)$ so that

$$\| T(\xi)x \| \leqq e^{\alpha(\xi)} \int_0^\infty e^{-\alpha(t+\xi)} \mid x(t + \xi) \mid dt$$

$$= e^{\alpha(\xi)} \int_\xi^\infty e^{-\alpha(s)} \mid x(s) \mid ds \leqq e^{\alpha(\xi)} \| x \|.$$

On the other hand if

$$x_n(t) = \begin{cases} ne^{\alpha(t)}, & \xi < t < \xi + 1/n, \\ 0 \text{ elsewhere}, \end{cases}$$

then $\| x_n \| = 1$ while

$$\| T(\xi)x_n \| = n \int_0^{1/n} e^{\alpha(t+\xi)-\alpha(t)} \, dt,$$

which tends to $e^{\alpha(\xi)}$ as $n \to \infty$. This completes the proof.

REMARK. Theorem 19.3.1 serves as a partial converse of the fact, already observed in section 10.2, that $\log \| T(\xi) \|$ is a lower semi-continuous subadditive function of ξ. For this reason it would be desirable to remove the continuity assumption on $\alpha(t)$ in the hypothesis of the theorem. Unfortunately the theorem is no longer valid in this more general form. Indeed, if $\alpha(t) = 0$ for $t = 0$ and $= 1$ for $t > 0$, then for arbitrary x, $\| T(\xi)x \| = \| x \|$ for all $\xi \geqq 0$; thus $\| T(\xi) \| \equiv 1 \neq e^{\alpha(\xi)}$.

The spectrum of d/dt is appreciably affected by this change in metric. It is suggested that the reader consider the four cases: $\alpha(t) = \log (1 + t)$, $2 \log (1 + t)$, t, and $-t^2$.

19.4. A nilpotent semi-group. We now consider the semi-group $\mathfrak{S}$ of left translations on the space $C_0[0, \beta]$ consisting of continuous functions $x(t)$ which vanish at $t = \beta$, where the norm $\| x \| = \max [\, | x(t) |\, ;\ 0 \leqq t \leqq \beta]$. In this case $[T(\xi)x](t) = x(t + \xi)$ for $0 \leqq t \leqq \max (0, \beta - \xi)$ and $= 0$ for $\max (0, \beta - \xi) \leqq t \leqq \beta$. Since $T(\beta)$ annihilates each $x \in C_0[0, \beta]$, we see that every operator $T(\xi) \in \mathfrak{S}$ is nilpotent in the sense that

$$[T(\xi)]^n = \Theta \qquad \text{for } n \geqq \beta/\xi; \tag{19.4.1}$$

however the index of nilpotency is not bounded on $\mathfrak{S}$. The power series expansion for $R(\lambda; T(\xi))$ obtained in Theorem 4.8.2 shows that the resolvent for $T(\xi)$, $\xi > 0$, is a polynomial in $1/\lambda$ of degree equal to the largest integer $< 1 + \beta/\xi$. Thus $\sigma[T(\xi)] = \{0\}$ for $\xi > 0$; actually the nilpotent property implies that $0 \in P\sigma[T(\xi)]$.

We note that $T(\xi) = \Theta$ for all $\xi \geqq \beta$ so that $T(\xi)$ is holomorphic in the interval (β, ∞). It is clear that the holomorphic portion of $T(\xi)$ can be continued analytically to the entire complex plane. However the so-extended operator function does not agree with $T(\xi)$ on the interval $[0, \beta)$. The semi-group $\mathfrak{S}$ satisfies the hypothesis of Theorem 17.6.3 and furnishes an example of the case in which $\xi_0 = \beta > 0$.

The infinitesimal generator of $\mathfrak{S}$ is the operator d/dt with domain

$$\mathfrak{D}(d/dt) = [x; x'(t) \text{ exists and belongs to } C_0[0, \beta]].$$

The resolvent of d/dt is given by

$$[R(\lambda; d/dt)x](t) = \int_0^{\beta - t} e^{-\lambda\xi} x(t + \xi) \, d\xi. \tag{19.4.2}$$

Thus $R(\lambda; d/dt)$ is an entire function of λ and hence $\sigma(d/dt) = \emptyset$. This is in agreement with Theorem 16.4.1 which asserts that $\sigma[T(\xi)] = 0 \cup \exp [\xi\sigma(d/dt)] = 0$. Theorem 16.7.6 gives another proof of the fact that $0 \in P\sigma[T(\xi)]$.

19.5. Semi-modules and translations. It was shown in Theorem 17.2.2 that if a semi-group of operators $[T(\zeta)]$ is holomorphic on some interval of the positive real axis, say (α, β), then the maximal domain of analytic existence of $T(\zeta)$ is the interior of a spinal semi-module. In the present section we shall prove that the interior of every (proper) spinal semi-module is the maximal domain of analytic existence of a suitably chosen semi-group of operators. The corresponding result for angular semi-modules was first obtained by E. Hille and Max Zorn [1].

THEOREM 19.5.1. *Given a proper spinal semi-module S in the complex plane, let $\mathfrak{X} = HC(S)$ be the class of all functions $f(z)$ continuous on $\bar{S} \cup \infty$, holomorphic in z in the interior of S, and with $\| f \| = \sup [\, | f(z) | ; z \in S]$. Define $[T(\zeta)f](z) = f(z + \zeta)$ for each $\zeta \in S$. Then $T(\zeta)$ is holomorphic in the interior of S,* Int (S) *is the maximal domain of analytic existence for $T(\zeta)$, and $[T(\xi); \xi > 0]$ is of class (C_0).*

PROOF. Since a spinal semi-module contains a ray, we may assume without loss of generality that S contains the positive real axis. It is clear that $HC(S)$ is a (B)-space in which the notion of norm convergence means uniform convergence with respect to z in S. Further the interior of S is a simply-connected domain which, by assumption, is contained in the right half-plane. Now given an arbitrary circle C with center z_0, there exists a function continuous in the extended plane, holomorphic outside of C, and having C as the natural boundary of its holomorphic part. If z_0 is not contained in the closure of Int(S), a circle about z_0 can be found exterior to Int(S) and the corresponding function will belong to $HC(S)$.

It is clear that $T(\xi)$ is of class (C_0) on $(0, \infty)$. In order to show that $T(\zeta)$ is holomorphic in Int(S) it suffices, by Theorem 3.10.1, to prove that the difference quotient of $T(\zeta)$ tends strongly to a limit. This will be a consequence of the following lemma.

LEMMA 19.5.1. *Let $f(z) \in HC(S)$. If ζ and $\zeta + \epsilon$ belong to* Int(S), *we have*

$$\left\| \frac{1}{\epsilon} [f(z + \zeta + \epsilon) - f(z + \zeta)] - f'(z + \zeta) \right\| \leqq 2 \,| \epsilon |[\delta(\zeta)]^{-2} \, \| f(z) \|$$

for $| \epsilon | < \frac{1}{2}\delta(\zeta)$; here $\delta(\zeta)$ is the greatest lower bound of the distance of $z + \zeta$ from the boundary of Int(S) *as z ranges over S, ζ being a fixed point in* Int(S).

PROOF. We first note that $\delta(\zeta) > 0$ for each $\zeta \in$ Int(S). Indeed, by Theorem 8.7.9 there exists a subadditive function $\varphi(v)$ such that S consists of all vectors $z = u + iv$ with $u > \varphi(v)$ (including perhaps some with $u \geqq \varphi(v)$). Let $\zeta = \xi + i\eta$ and $z = u + iv$. Then, since $\zeta \in$ Int(S), there exist positive constants δ_1 and δ_2 such that $\xi \geqq \varphi(\eta') + \delta_1$ for $| \eta' - \eta | < \delta_2$. Consequently

$$\xi + u \geqq \varphi(\eta') + \varphi(v) + \delta_1 \geqq \varphi(\eta' + v) + \delta_1,$$

again for $| \eta' - \eta | < \delta_2$. It follows that $z + \zeta$ is at a distance from the boundary of Int(S) greater than min(δ_1, δ_2) for all $z \in S$. The estimate now follows from the representation

$$\frac{1}{\epsilon}[f(z+\zeta+\epsilon) - f(z+\zeta)] - f'(z+\zeta)$$

$$= \frac{\epsilon}{2\pi i}\int_\Gamma \frac{f(w)\,dw}{(w-z-\zeta)^2(w-z-\zeta-\epsilon)}$$

where Γ is a circle in Int(S) with center at $w = z + \zeta$ and which also contains $z + \zeta + \epsilon$.

The above estimate shows that the difference quotient of $T(\zeta)$ tends strongly to a limit in $HC(S)$ so that $T(\zeta)$ is holomorphic in Int(S) as asserted.

In order to prove that Int(S) is the maximal domain of analytic existence for $T(\zeta)$ it is sufficient to exhibit a linear bounded functional x^* defined on $HC(S)$ such that for each z_0 not in the closure of Int(S) there exists an element f of $HC(S)$ for which the numerically-valued function $x^*[T(\zeta)f]$, which is holomorphic in Int(S), cannot be continued analytically to z_0. We construct the functional as follows. Let $\{u_n\}$ be a fixed sequence of positive numbers with $\lim_{n\to\infty} u_n = 0$. Then $x^*(f) = \lim_{n\to\infty} f(u_n)$ is a bounded linear functional on $HC(S)$. It is clear that $x^*[T(\zeta)f] = f(\zeta)$ for each $f \in HC(S)$ and $\zeta \in S$. We have seen that there exist elements of $HC(S)$ having an arbitrary circle about z_0 as their natural boundary. For such a choice of f, the function $x^*[T(\zeta)f]$ cannot be continued analytically to the point z_0. This completes the proof of Theorem 19.5.1.

2. Powers

19.6. Power semi-groups. A semi-group defined on a function space and of the following general type:

$$[T(\xi)f](t) = [g(t)]^\xi f(t), \tag{19.6.1}$$

will be called a *power semi-group*. We shall not undertake a systematic study of such semi-groups. However we shall make use of their interesting spectral properties to answer some existence questions for the spectrum of the infinitesimal generator of a semi-group.

THEOREM 19.6.1. *Let Δ be any simply-connected domain in the half-plane $\mathfrak{R}(\lambda) \leqq \alpha$. Then there exists a (B)-space $\mathfrak{X}$ and a semi-group $\mathfrak{S}$ of operators on $\mathfrak{X}$, with infinitesimal generator A, such that $\mathfrak{S}$ is of class (C_0) and the spectrum of A coincides with the closure of Δ and is a pure residual spectrum, that is, $\sigma(A) = R\sigma(A)$.*

PROOF. We start with the space $\mathfrak{Y} = HB[\,|z| < 1]$ of functions $f(z)$ holomorphic and bounded in the unit circle and define $\|f\| = \sup[\,|f(z)|\,; |z| < 1]$; the desired space $\mathfrak{X}$ will be a suitably chosen subspace of $\mathfrak{Y}$. There exists a function $h(z)$ holomorphic in $|z| < 1$ which maps $|z| < 1$ conformally upon Δ. It follows that

$$\sup\,[\Re[h(z)];\, |z| < 1] \equiv \beta < \infty.$$

We now define the operator $T(\xi)$ on $\mathfrak{Y}$ as

$$[T(\xi)f](z) = e^{\xi h(z)} f(z), \qquad \xi > 0. \tag{19.6.2}$$

It is clear that $T(\xi)$ is a linear bounded operator with

$$\|T(\xi)\| = e^{\beta\xi}$$

as is seen by taking $f(z) \equiv 1$.

If $h(z)$ is bounded, that is, if Δ is a bounded point set, then we may embed $\mathfrak{S}$ in a group $\mathfrak{G} = [T(\zeta)]$ defined for all complex values of ζ by (19.6.2) with ξ replaced by ζ. $T(\zeta)$ is then an entire function of ζ and *a fortiori* continuous in the uniform operator topology. On the other hand if merely the imaginary part of $h(z)$ is bounded, then $T(\zeta)$ is defined and holomorphic in the right half-plane $\Re(\zeta) > 0$, but ordinarily $T(\xi)$ does not converge to I in the strong operator topology as $\xi \to 0+$. In the general case in which $\Re[h(z)] \leqq \beta$ but $h(z)$ is not otherwise limited, then even strong continuity fails for positive values of ξ. This is the reason for restricting the discussion to a subspace of $\mathfrak{Y}$, which we now introduce.

The function $f(z)$ of $HB[\,|z| < 1]$ belongs to $\mathfrak{X}$ if to each $\epsilon > 0$ there exists an $M = M(\epsilon; f)$ such that $|f(z)| \leqq \epsilon$ on the set $[z;\, |h(z)| \geqq M]$.

It is clear that the function class $\mathfrak{X}$ is a linear normed subspace of $\mathfrak{Y}$. Furthermore $\mathfrak{X}$ contains non-vanishing functions. In fact, if $\alpha > \beta$ and the square root is properly chosen, then the function $g_0(z) \equiv \exp\{[\alpha - h(z)]^{1/2}\} \in \mathfrak{X}$. The space $\mathfrak{X}$ is also complete. For suppose that $\{f_n\} \subset \mathfrak{X}$ and $\lim_{n\to\infty} \|f - f_n\| = 0$ where $f \in \mathfrak{Y}$. Given $\epsilon > 0$, there exists an m such that $\|f - f_m\| < \epsilon/2$. Setting $M = M(\epsilon/2; f_m)$, we have

$$|f(z)| < \frac{\epsilon}{2} + |f_m(z)| \leqq \frac{\epsilon}{2} + \frac{\epsilon}{2} = \epsilon$$

on the set $[z;\, |h(z)| \geqq M]$. Thus $f(z) \in \mathfrak{X}$ and $\mathfrak{X}$ is a (B)-space.

In order to prove that $T(\xi)$ is continuous in the strong $\mathfrak{E}(\mathfrak{X})$ operator topology for $\xi \geqq 0$, it suffices to show that $\lim_{\eta\to 0+} T(\eta)f = f$ for each $f \in \mathfrak{X}$. Let $\epsilon > 0$ be given and set $M = M(\epsilon; f)$, $E_1(M) = [z;\, |h(z)| \geqq M]$, and $E_2(M) =$ the complement of $E_1(M)$ in $|z| < 1$. Then

$$\{[T(\eta) - I]f\}(z) = [e^{\eta h(z)} - 1]f(z),$$

the absolute value of which is dominated by the quantity $[e^{\beta\eta} + 1]\epsilon$ on $E_1(M)$ and by $[e^{M\eta} - 1]\,\|f\|$ on $E_2(M)$. It follows that

$$\limsup_{\eta\to 0+} \|[T(\eta) - I]f\| \leqq 2\epsilon,$$

which can be made arbitrarily small. Hence $\lim_{\eta\to 0+} T(\eta)f = f$ for each $f \in \mathfrak{X}$.

Since

$$[A_\eta f](z) = \frac{1}{\eta}[e^{\eta h(z)} - 1]f(z) \to h(z)f(z)$$

for each fixed z in $|z| < 1$ as $\eta \to 0+$, we see that for $f \in \mathfrak{D}(A)$

$$[Af](z) = h(z)f(z) \in \mathfrak{X},$$

and a direct calculation shows that

$$\mathfrak{D}(A) = [f; h(z)f(z) \in \mathfrak{X}].$$

It is clear that A is bounded if and only if Δ is bounded.

Let us now determine the spectrum of A. If $g(z) \in \mathfrak{X}$, then the equation

$$[\lambda - h(z)]f(z) = g(z)$$

has a solution in $\mathfrak{X}$ if the first factor on the left is bounded away from zero, that is, if λ does not belong to $\bar{\Delta}$. Suppose next that $\lambda \in \Delta$ so that there is a unique z_λ with $|z_\lambda| < 1$ such that $h(z_\lambda) = \lambda$. If $f(z) \in \mathfrak{D}(A)$, then $[\lambda - h(z)]f(z) \in \mathfrak{X}$ but every such function vanishes at $z = z_\lambda$. The function $g_0(z)$ mentioned above is obviously not the limit of functions $g(z) \in \mathfrak{X}$ with $g(z_\lambda) = 0$. Thus if $\lambda \in \Delta$, the range of $\lambda I - A$ is not dense in $\mathfrak{X}$ and, since there is no function $f(z) \not\equiv 0$ with $[\lambda - h(z)]f(z) \equiv 0$, we see that $\lambda \in R\sigma(A)$. A similar argument applies if $\lambda \in \bar{\Delta} \ominus \Delta$. There is then a sequence of points $\{z_n\}$, $|z_n| < 1$, which converge to a point z_0 with $|z_0| = 1$ such that $h(z_n) \to \lambda$. Consequently $[\lambda - h(z_n)]f(z_n) \to 0$ for each $f \in \mathfrak{D}(A)$. Now $g_0(z_n) \to \exp\{[\alpha - \lambda]^{1/2}\} \neq 0$ so that $g_0(z)$ is not the limit of functions $g(z)$ such that $g(z_n) \to 0$. Further $[\lambda - h(z)]f(z) \equiv 0$ implies that $f(z) \equiv 0$. It follows that $\lambda \in R\sigma(A)$ also in the case $\lambda \in \bar{\Delta} \ominus \Delta$. This completes the proof.

The results of this theorem are of interest in connection with Theorem 17.5.1 inasmuch as it provides an example of an operator $U = A$ having the properties (A_1), (A_2), and (A_3) of the latter theorem for a given set D. We have merely to choose $\Delta = \text{Int}(D^-)$ in Theorem 19.6.1 and define $[Uf](z) = h(z)f(z)$; the operator U will then be the infinitesimal generator of a semi-group having the properties (P_1) and (P_2) of Theorem 17.5.1.

The results also throw some light on the notion of "the maximal domain of existence of a semi-group". The expression

$$e^{\zeta h(z)}f(z)$$

is an entire function of ζ for each fixed z with $|z| < 1$. Thus if $T(\xi)$ does possess an analytic continuation, say $T(\zeta)$, then $T(\zeta)$ will be given by this expression. However the expression defines a bounded linear transformation if and only if $\sup[\Re[\zeta h(z)]; |z| < 1]$ is finite, that is, if and only if the function of support $F(\zeta)$ for the closed convex extension of $\bar{\Delta}$ is finite for the argument ζ. The set of ζ for which $F(\zeta)$ is finite fills out a sector so that this sector contains the maximal domain of existence of $T(\zeta)$. On the other hand Theorem 17.5.1 (or a direct calculation) shows that $T(\zeta)$ is holomorphic in this sector. It follows that the maximal domain of analytic existence for $T(\zeta)$ is this sector.

Power semi-groups defined as above show a preference for the residual spectrum. It is possible, however, to modify the construction in such a manner that a point spectrum also appears. The following example shows how this may be done. We take $\mathfrak{X} = C_0[0, 1]$, the elements $f(t)$ being continuous in $[0, 1]$ and $f(0) = 0$. It is a (B)-space under the norm $\| f \| = \max [| f(t) | ; 0 \leqq t \leqq 1]$. Let μ_n be a given sequence of positive numbers, $\mu_n > \mu_{n+1}$, $\mu_n \to 0$, and put $\lambda_n = \log \mu_n$. Define $g(t) = \mu_n$ for $2^{1-2n} \leqq t \leqq 2^{2-2n}$, $n = 1, 2, 3, \cdots$, and equal to a linear function of t in each of the intermediary intervals. Thus $g(t) \in C_0[0, 1]$ and

$$[T(\zeta)f](t) \equiv [g(t)]^{\zeta} f(t), \qquad \Re(\zeta) > 0,$$

is a holomorphic semi-group on $C_0[0, 1]$. The infinitesimal generator is

$$[Af](t) = [\log g(t)]f(t),$$

which is an unbounded operator. Its spectrum is the interval $(-\infty, \lambda_1]$. Each point λ_n belongs to $P\sigma(A)$; the characteristic manifold of the operator $\lambda_n I - A$ is made up of the functions $f(t)$ in $C_0[0, 1]$ which vanish outside of $(2^{1-2n}, 2^{2-2n})$. All other points of $(-\infty, \lambda_1]$ belong to $R\sigma(A)$.

CHAPTER XX

TRIGONOMETRIC SEMI-GROUPS

20.1. Orientation. In the theory of summability of trigonometric Fourier series one encounters factor sequence transformations of the form

$$[T(\xi)f](t) \sim \sum_{-\infty}^{\infty} e^{\lambda_n \xi} f_n e^{nit}, \qquad f(t) \sim \sum_{-\infty}^{\infty} f_n e^{nit}.$$

One of the more familiar instances of this is the case of Abel-Poisson summability where $\lambda_n = -|n|$; here it is customary to set $e^{-\xi} = r$. For a proper choice of $\{\lambda_n\}$ these transformations form semi-groups on $L_p(-\pi, \pi)$ (cf. E. Hille [7]). Such semi-groups may be characterized by the fact that they are the only measurable semi-groups of linear bounded operators on $L_p(-\pi, \pi)$ to itself in which the member operators commute with the group of real translations on the numerical variable t, that is, if the $T(\xi)$-transform of $f(t)$ is $f(t, \xi)$, then the $T(\xi)$-transform of $f(t + \alpha)$ is $f(t + \alpha, \xi)$.

Operations which commute with translations, or, equivalently, with differentiation, have been the object of much study. We refer to the papers of F. Tricomi and T. Kitagawa for a survey of the problems and a bibliography. It is well known that in such problems the group characters of the underlying group, in this case the exponential functions $\{e^{nit}\}$, play a decisive role. This will also be the case in our problem; however our results do not seem to be covered by already existing theories.

We shall consider only the factor sequence type semi-groups on $L_p(-\pi, \pi)$ mentioned above. Our results are fairly complete for the case $p = 2$, less so for $p = 1$, and quite sketchy for general values of p. These differences merely reflect the state of our knowledge of the structure of class preserving factor sequences for different values of p.

Our results on trigonometric semi-groups, incomplete as they are, throw considerable light on the general theory of semi-groups of linear bounded operators. Thus by constructing a semi-group in $L_2(-\pi, \pi)$ with preassigned $\sigma(A)$, we show that the spectrum of the infinitesimal generator may be any closed point set in a left half-plane. The difficult problem of the rate of growth of the norm of $T(\xi)$ as $\xi \to 0+$, and the related question of the existence of a limit for $T(\xi)$ itself as $\xi \to 0+$, can be illustrated by trigonometric semi-groups in $L_1(-\pi, \pi)$. The theory of holomorphic semi-groups is also exemplified by trigonometric semi-groups. Any convex semi-module may be the maximal domain of analytic existence of a semi-group in $L_1(-\pi, \pi)$. That the rate of growth of the norm of $T(\zeta)$ is not subject to any limitations as ζ tends to a point on the boundary of the domain of analytic existence of $T(\zeta)$, or as $\zeta \to \infty$ parallel to the boundary, is also illustrated by examples in $L_2(-\pi, \pi)$.

There are two paragraphs: *Semi-Groups with Real Parameter*, and *Semi-Groups with Complex Parameter.*

References. Hille [**1, 7, 10**], Hille and Tamarkin [**2**], Kitagawa [**1**], M. Riesz [**1**], Tricomi [**1**], and Zygmund [**1**].

1. Semi-Groups with Real Parameter

20.2. Factor sequences; Lacunary series. For the following discussion we shall need some results from the classical theory of trigonometric Fourier series relating to factor sequences and lacunary series. In order to state the problem of factor sequences in a general form we require a few preliminary definitions.

A sequence of functions $\{u_n(t); n \in \mathfrak{N}\}$ defined and measurable on (α, β) is said to form an orthonormal system if

$$\int_\alpha^\beta u_j(t)\overline{u_k(t)}\, dt = \delta_{jk}, \qquad j, k \in \mathfrak{N}.$$

Such a system is said to be complete with respect to a function space $\mathfrak{X}$ if (i) $\{u_n\} \subset \mathfrak{X} \cap \mathfrak{X}^*$ and (ii) $\int_\alpha^\beta f(t)\, \overline{u_n(t)}\, dt = 0$ for all $n \in \mathfrak{N}$ implies that $f = \theta$. If $\{u_n\}$ is a complete orthonormal system relative to $\mathfrak{X}$, then each $f \in \mathfrak{X}$ has a unique representation as a formal series:

$$f \sim \sum_{n \in \mathfrak{N}} f_n u_n(t), \qquad f_n = \int_\alpha^\beta f(t)\overline{u_n(t)}\, dt.$$

Let $\{u_n; n \in \mathfrak{N}\}$ be a complete orthonormal system relative to the function spaces $\mathfrak{X}$ and $\mathfrak{Y}$. Then a sequence $\{\mu_n; n \in \mathfrak{N}\}$ is said to be a factor sequence of type $(\mathfrak{X}, \mathfrak{Y})$ if for each $f \sim \sum f_n u_n(t)$ in $\mathfrak{X}$ the formal series $\sum \mu_n f_n u_n(t)$ represents a function in $\mathfrak{Y}$.

Lemma 20.2.1. *Let $\{\mu_n; n \in \mathfrak{N}\}$ be a factor sequence of type $(\mathfrak{X}, \mathfrak{Y})$ where $\mathfrak{X}$ and $\mathfrak{Y}$ are (B)-spaces. Then the transformation U which maps $f \sim \sum f_n u_n(t) \in \mathfrak{X}$ into $Uf \sim \sum \mu_n f_n u_n(t) \in \mathfrak{Y}$ is linear and bounded.*

Proof. It is clear that U is defined and linear on $\mathfrak{X}$. According to the closed graph theorem (Theorem 2.12.3), U will be bounded if it can be shown that U is closed. To this end suppose that $f^k \to f$ in $\mathfrak{X}$ and that $Uf^k \to g$ in $\mathfrak{Y}$. Then

$$(f^k)_n = \int_\alpha^\beta f^k(t)\overline{u_n(t)}\, dt \to \int_\alpha^\beta f(t)\overline{u_n(t)}\, dt = f_n$$

and

$$\mu_n(f^k)_n = (Uf^k)_n = \int_\alpha^\beta [Uf^k](t)\overline{u_n(t)}\,dt \to \int_\alpha^\beta g(t)\overline{u_n(t)} = g_n.$$

Thus $\mu_n f_n = g_n$ so that $Uf = g$; this proves that U is closed.

In the present chapter we are concerned exclusively with the function spaces $L_p(-\pi, \pi)$, $1 \leqq p \leqq \infty$, and $C[-\pi, \pi]$, and with the trigonometric system

$$\{(2\pi)^{-1/2}e^{nit};\ n = 0, \pm 1, \pm 2, \cdots\},$$

which constitutes a complete orthonormal system for each of these spaces. The corresponding factor sequence problem for $L_p(-\infty, \infty)$ and the normalized functions of Hermite will be treated in Chapter XXI.

Necessary and sufficient conditions are known in the trigonometric case in order that a factor sequence $\{\mu_n\}$ be of type (L_1, L_1), (L_2, L_2), (L_∞, L_∞), or (C, C). Only sufficient conditions are known for the case (L_p, L_p), $1 < p < 2$; however M. Riesz [1] has shown that every sequence which is of type (L_p, L_p) is also of type (L_q, L_q) for any q with $p \leqq q \leqq p'$, where $1/p + 1/p' = 1$, and this also holds for $p = 1$ and $p' = \infty$.

In the following lemmas we use the notation

$$(20.2.1)\quad f \sim \sum_{-\infty}^{\infty} f_n e^{nit}, \qquad f_n = \frac{1}{2\pi}\int_{-\pi}^{\pi} f(t)e^{-nit}\,dt, \qquad Uf \sim \sum_{-\infty}^{\infty} \mu_n f_n e^{nit}.$$

LEMMA 20.2.2. *A necessary and sufficient condition that $\{\mu_n\}$ be of type (L_2, L_2) is that the sequence $\{\mu_n\}$ be bounded. In this case*

$$(20.2.2)\qquad \| U \| = \sup |\mu_n|.$$

The proof follows from Parseval's identity.

LEMMA 20.2.3. *A necessary and sufficient condition that $\{\mu_n\}$ be of type (L_1, L_1), (L_∞, L_∞), or (C, C) is that*

$$(20.2.3)\qquad 2\pi K(t) \equiv \mu_0 t + \sum_{-\infty}^{\infty}{}' \mu_n \frac{e^{nit} - 1}{ni}$$

define a function of bounded variation on $[-\pi, \pi]$. In this case

$$(20.2.4)\qquad [Uf](t) = \int_{-\pi}^{\pi} f(t - u)\,dK(u),$$

where the integral exists in the Lebesgue-Stieltjes sense for almost all t and

$$(20.2.5)\qquad \| U \| = \int_{-\pi}^{\pi} |dK(u)|.$$

For the proof we refer the reader to A. Zygmund [1, section 4.6]. The only point which calls for further comment is the expression for the norm in the case (L_1, L_1). Since the norm of a transformation is equal to the norm of the adjoint transformation, we have merely to consider the adjoint of the transformation (20.2.4). The adjoint transformation takes $L_\infty(-\pi, \pi)$ into itself and it is seen without difficulty that its norm is equal to the total variation of $K(u)$. See also E. Hille and J. D. Tamarkin [2, pp. 334–337].

We shall have need of a property of trigonometric lacunary series found by S. Banach. For the proof we refer to A. Zygmund [1, pp. 216–218].

LEMMA 20.2.4. *Let $\{n_k ; k = 1, 2, \cdots\}$ be a given sequence of positive integers such that $n_{k+1}/n_k \geqq 2$ for all k. In order that the lacunary series*

$$\sum_{k=1}^{\infty} c_k \exp(n_k it) \tag{20.2.6}$$

define a function $f(t) \in L_1(-\pi, \pi)$ it is necessary and sufficient that $\sum_{k=1}^{\infty} | c_k |^2$ be convergent. If this condition is satisfied, then there exist two positive absolute constants A and B such that

$$A \| f \|_2 \leqq \| f \|_1 \leqq B \| f \|_2 . \tag{20.2.7}$$

20.3. Semi-groups commuting with translations. The following theorem is basic.

THEOREM 20.3.1. *Let $\mathfrak{S} \equiv [T(\xi); \xi > 0]$ be a semi-group of linear bounded operators on $L_p(-\pi, \pi)$, $1 \leqq p \leqq \infty$, or $C[-\pi, \pi]$ to itself, and suppose that $\mathfrak{S}$ commutes with the group of real translations on the independent variable so that $[T(\xi)f(\cdot)](t) = f(t, \xi)$ implies that $[T(\xi)f(\cdot + \alpha)](t) = f(t + \alpha, \xi)$. Then $T(\xi)$ defines a factor sequence transformation*

$$T(\xi)f \sim \sum_{-\infty}^{\infty} \mu_n(\xi) f_n e^{nit} \qquad \textit{if} \qquad f \sim \sum_{-\infty}^{\infty} f_n e^{nit}, \tag{20.3.1}$$

where

$$\mu_n(\xi_1 + \xi_2) = \mu_n(\xi_1)\mu_n(\xi_2), \qquad \xi_1, \xi_2 > 0. \tag{20.3.2}$$

Suppose, in addition, that $T(\xi)$ is weakly measurable. Then if $p < \infty$, the function $T(\xi)$ is continuous in the strong operator topology for $\xi > 0$. In any case there exists a set of distinct integers $\mathfrak{N}$ and a set of complex numbers $\{\lambda_n; n \in \mathfrak{N}\}$ such that $\mu_n(\xi) = e^{\lambda_n \xi}$ or 0 according as n is in $\mathfrak{N}$ or not. If $p = 2$, then $\sup \mathfrak{R}(\lambda_n) \equiv \mu < \infty$ and $\| T(\xi) \| = e^{\mu \xi}$. If $p = 1$ or ∞ or if $\mathfrak{X} = C[-\pi, \pi]$, we form the function

$$2\pi K(t; \xi) = \sum_{n \in \mathfrak{N}} e^{\lambda_n \xi} \frac{e^{nit} - 1}{ni}, \tag{20.3.3}$$

where, if $n = 0$ is in $\mathfrak{N}$, the indeterminate fraction is to be replaced by t. A necessary condition on the set $\{\lambda_n\}$ is now that $K(t; \xi)$ be of bounded variation in $[-\pi, \pi]$ for each $\xi > 0$. In this case

(20.3.4) $$[T(\xi)f](t) = \int_{-\pi}^{\pi} f(t-u)\, d_u K(u;\xi),$$

where the integral exists in the Lebesgue-Stieltjes sense for almost all t; and $\| T(\xi) \| = \int_{-\pi}^{\pi} | d_t K(t;\xi) |$.

PROOF. Let $u_n(t) = e^{nit}$ and set $[T(\xi)u_n](t) = e_n(t, \xi)$. Since the transform of the translation is the translation of the transform and since $u_n(t+\alpha) = e^{ni\alpha}u_n(t)$, we have for each fixed α

$$e_n(t+\alpha, \xi) = e^{ni\alpha}e_n(t, \xi)$$

for almost all t. Now both members of this relation are clearly measurable in the product set Π: $-\pi \leqq t, \alpha < \pi$. Consequently the relation holds for almost all (t, α) in Π. We can therefore find a t_0 such that $e_n(t_0+\alpha, \xi) = e^{ni\alpha}e_n(t_0, \xi)$ for almost all α, that is, aside from a null set

(20.3.5) $$[T(\xi)u_n](t) = e_n(t, \xi) = \mu_n(\xi)e^{nit},$$

where we have set $\mu_n(\xi) = e_n(t_0, \xi)$. It is clear that $e_n(t, \xi)$ can be redefined so that (20.3.5) holds for all $t \in [-\pi, \pi)$. The semi-group property evidently requires that (20.3.2) holds and the linearity together with the boundedness of the transformation $T(\xi)$ shows that $T(\xi)$ has the form (20.3.1).

If $p < \infty$, then the space $L_p(-\pi, \pi)$ is separable. Thus for $L_p(-\pi, \pi)$, $1 \leqq p < \infty$, or $C[-\pi, \pi]$, weak measurability implies strong measurability by Theorem 3.5.3, Corollary 2, and this in turn implies strong continuity for $\xi > 0$ by Theorem 10.2.3. In any case, if $T(\xi)$ is weakly measurable, then $\mu_n(\xi)$ must be measurable Lebesgue for each fixed n. It now follows by the corollary to Theorem 4.17.3 that $\mu_n(\xi)$ is either identically zero for $\xi > 0$ or else an exponential function which we write as $e^{\lambda_n \xi}$. The set $\mathfrak{N}$ then consists of those integers n for which $\mu_n(\xi) \not\equiv 0$.

If $p = 2$, Lemma 20.2.2 shows that the condition $\sup | \mu_n(\xi) | \equiv M(\xi) < \infty$ is necessary and sufficient in order that $T(\xi)$ be bounded, and then $\| T(\xi) \| = M(\xi)$. The corresponding condition for the set $\{\lambda_n\}$ is simply $\sup \mathfrak{R}(\lambda_n) \equiv \mu < \infty$ which gives $\| T(\xi) \| = e^{\mu\xi}$. For the cases $p = 1$, $p = \infty$, and $C[-\pi, \pi]$, we have merely to refer to Lemma 20.2.3. We note that a factor sequence of type (L_1, L_1) is also of type (L_p, L_p) for each p such that $1 \leqq p \leqq \infty$. This concludes the proof of Theorem 20.3.1.

Suppose now that $[T(\xi)]$ is a weakly measurable semi-group of linear bounded operators which commute with translations. Then we have

(20.3.6) $$T(\xi)f \sim \sum_{\mathfrak{N}} e^{\lambda_n \xi} f_n e^{nit};$$

we shall refer to this representation of $[T(\xi)]$ as the *Dirichlet representation* of the semi-group. Formally this series is a Dirichlet series in ξ, but, since the exponents are usually widely scattered complex numbers, no domain of convergence in the complex parameter plane is to be expected when $f(t)$ is an arbitrary element of $L_p(-\pi, \pi)$.

Let $L_p(\mathfrak{N})$ denote the closed linear subspace in $L_p(-\pi, \pi)$ spanned by the system $\{e^{nit}; n \in \mathfrak{N}\}$. If $\mathfrak{X}_0 = \bigcup_{\xi>0} T(\xi)[\mathfrak{X}]$, then it is clear that $L_p(\mathfrak{N}) = \bar{\mathfrak{X}}_0$. The mapping $T(0)$, defined as

$$(20.3.7) \qquad \sum_{-\infty}^{\infty} f_n e^{nit} \to \sum_{\mathfrak{N}} f_n e^{nit},$$

is always a bounded projection operator when $p = 2$; however, it may fail to be bounded if $p \neq 2$. For $p = 1$, a necessary and sufficient condition that the projection be well defined on $L_1(-\pi, \pi)$ (and hence be bounded) follows from Lemma 20.2.3 which asserts that $\{\mu_n\}$ ($\mu_n = 1$ or 0 according as $n \in \mathfrak{N}$ or not) is a factor sequence of type (L_1, L_1) if and only if $K(t; 0)$ is of bounded variation in $[-\pi, \pi]$. This condition is also sufficient for $p \neq 1$. For the existence of subspaces of the form $L_1(\mathfrak{N})$ not having bounded projections, see section 20.5. The following theorem serves as a converse to Theorem 20.3.1.

THEOREM 20.3.2. *Let $[T(\xi); \xi > 0]$ be a family of operators defined by* (20.3.6) *on $L_p(-\pi, \pi)$, $1 \leqq p < \infty$, or $C[-\pi, \pi]$ to itself. Then $[T(\xi)]$ defines a semi-group of linear bounded operators, the elements of which commute with the group of real translations on t, and which is continuous in the strong operator topology for $\xi > 0$.*

If $f \in \mathfrak{D}(A_o)$, then $f_n = 0$ for $n \notin \mathfrak{N}$ and

$$(20.3.8) \qquad A_o f \sim \sum_{\mathfrak{N}} \lambda_n f_n e^{nit}.$$

Finally if $[T(\xi)]$ is of class (A), then $\mathfrak{N}$ consists of all the integers,

$$(20.3.9) \qquad \mathfrak{D}(A) = [f; g \sim \sum_{-\infty}^{\infty} \lambda_n f_n e^{nit} \in \mathfrak{X}],$$

and for $f \in \mathfrak{D}(A)$ we have $Af \sim \sum_{-\infty}^{\infty} \lambda_n f_n e^{nit}$.

PROOF. The semi-group property and the fact that the operators commute with translations follow directly from the form of the defining relation (20.3.6). Lemma 20.2.1 shows that $T(\xi)$ is linear and bounded. It is further clear that $T(\xi)f$ is continuous in ξ, $\xi > 0$, for each f in the linear extension $\mathfrak{L}$ of the fundamental set $\{e^{nit}\}$. Now given $f \in \mathfrak{X}$, there exists a sequence $\{f^k\} \subset \mathfrak{L}$ such that $f^k \to f$. Since $T(\xi)f^k \to T(\xi)f$ for each $\xi > 0$, we see by Theorem 3.5.4 that $T(\xi)f$ is strongly measurable. Theorem 10.2.3 now implies that $T(\xi)$ is continuous in the strong operator topology for $\xi > 0$.

If $f \in \mathfrak{D}(A_o)$ so that $\eta^{-1}[T(\eta)f - f] \to g$ in norm, then $\eta^{-1}[\mu_n(\eta) - 1]f_n \to g_n$. Consequently $f_n = 0$ if $n \notin \mathfrak{N}$ and $A_o f$ is determined by (20.3.8). Suppose next that $[T(\xi)]$ is of class (A) with infinitesimal generator A. In this case $\mathfrak{X}_0$ is dense in $\mathfrak{X}$ so that $\mathfrak{N}$ contains all of the integers. For $f \in \mathfrak{X}_0$, $\mathfrak{R}(\lambda) > \omega_1$, we have

$$[R(\lambda; A)f]_n = \int_0^\infty e^{-\lambda\xi}[T(\xi)f]_n \, d\xi = \int_0^\infty e^{-\lambda\xi}e^{\lambda_n\xi}f_n \, d\xi = (\lambda - \lambda_n)^{-1}f_n .$$

Since $\mathfrak{X}_0$ is dense in $\mathfrak{X}$, this implies that

$$[R(\lambda; A)f]_n = (\lambda - \lambda_n)^{-1}f_n \tag{20.3.10}$$

for all $f \in \mathfrak{X}$, $\Re(\lambda) > \omega_1$. Now if $f \in \mathfrak{D}(A)$ and $\lambda > \omega_1$ is fixed, then there exists a $g \in \mathfrak{X}$ such that $f = R(\lambda; A)g$. Thus

$$(Af)_n = [\lambda R(\lambda; A)g - g]_n = \lambda(\lambda - \lambda_n)^{-1}g_n - g_n = \lambda_n(\lambda - \lambda_n)^{-1}g_n = \lambda_n f_n .$$

Conversely if f is such that $h \sim \sum_{-\infty}^{\infty} \lambda_n f_n e^{nit}$ belongs to $\mathfrak{X}$, then $g = \lambda f - h \in \mathfrak{X}$ and

$$[R(\lambda; A)g]_n = (\lambda - \lambda_n)^{-1}g_n = (\lambda - \lambda_n)^{-1}(\lambda f_n - \lambda_n f_n) = f_n$$

so that $f \in \mathfrak{D}(A)$. This proves the last assertion of the theorem.

It is clear from (20.3.8) that the point spectrum of A_0 coincides with the set $\{\lambda_n\}$.

The relation between $T(\xi)$ and the projection operator $T(0)$, defined in (20.3.7), is clarified by the following two theorems.

THEOREM 20.3.3. *Let $[T(\xi); \xi > 0]$ be a semi-group of linear bounded operators on $L_p(-\pi, \pi)$, $1 \leqq p < \infty$, or $C[-\pi, \pi]$ to itself, the elements of which commute with the group of real translations on t, and suppose that $T(\xi)$ is weakly measurable. A necessary and sufficient condition that*

$$\lim_{\xi \to 0+} T(\xi)f = T(0)f, \qquad f \in \mathfrak{X}, \tag{20.3.11}$$

is that $\| T(\xi) \| \leqq M$ for $0 < \xi < 1$. In this case $T(0)$ is a bounded projection operator.

PROOF. As before weak measurability implies strong continuity for $\xi > 0$ when $\mathfrak{X}$ is separable. Hence if the relation (20.3.11) holds for all $f \in \mathfrak{X}$, then the uniform boundedness theorem shows that $\| T(\xi) \|$ is bounded on $(0, 1)$. Conversely, suppose that $\| T(\xi) \| \leqq M$ for $0 < \xi < 1$. Setting $u_n(t) = e^{nit}$, we see that $\lim_{\xi \to 0+} T(\xi)u_n = T(0)u_n = u_n$ or θ according as $n \in \mathfrak{N}$ or not. Consequently $\lim_{\xi \to 0+} T(\xi)f = T(0)f$ for all f in the linear extension $\mathfrak{L}$ of the fundamental set $\{u_n\}$. The Banach-Steinhaus theorem now implies that the limit holds for all $f \in \mathfrak{X}$ and defines a linear bounded operator, say J. It is clear that $T(0)$ and J agree on $\mathfrak{L}$; actually $T(0) = J$. For given any $f \in \mathfrak{X}$ there exists a sequence $\{f^k\} \subset \mathfrak{L}$ such that $f^k \to f$. It follows from (20.3.7) that

$$(Jf)_n = \lim_{k \to \infty} (Jf^k)_n = \lim_{k \to \infty} (T(0)f^k)_n = f_n = (T(0)f)_n .$$

Hence $T(0) = J$ so that $T(0)$ is bounded. It is clear from (20.3.7) that $T(0)$ is also a projection operator. This completes the proof.

We now consider the case in which $T(\xi)$ converges to $T(0)$ in the uniform operator topology. Here it is convenient to introduce the notation $\{\lambda_n, 0\}$ to represent the factor sequence $\{\mu_n\}$ where $\mu_n = \lambda_n$ or 0 according as $n \in \mathfrak{N}$ or not.

THEOREM 20.3.4. *Let* $[T(\xi); \xi > 0]$ *be a semi-group of linear bounded operators on* $L_p(-\pi, \pi)$, $1 \leq p \leq \infty$, *(or* $C[-\pi, \pi]$*) to itself, weakly measurable for* $\xi > 0$, *and suppose that* $T(\xi)$ *commutes with the group of real translations on* t *for each* $\xi > 0$. *Then* $\lim_{\xi\to 0+} \| T(\xi) - T(0) \| = 0$ *if and only if the sequences* $\{1, 0\}$ *and* $\{\lambda_n, 0\}$ *are both factor sequences of type* (L_p, L_p) *(or* (C, C)*).*

PROOF. If $T(\xi) \to T(0)$ in the uniform operator topology as $\xi \to 0+$, then Theorem 9.6.1 shows that $T(\xi) = T(0) \exp(\xi A)$ where $T(0)$ and A are bounded linear operators such that $A = T(0)A = AT(0)$. Now $T(0)f \sim \sum_{\mathfrak{N}} f_n e^{nit}$ so that $\{1, 0\}$ is a factor sequence of type $(\mathfrak{X}, \mathfrak{X})$. Further if $f \in T(0)[\mathfrak{X}]$, then

$$Af = \lim_{\eta\to 0+} \eta^{-1}[T(\eta) - I]f \sim \sum_{\mathfrak{N}} \lambda_n f_n e^{nit}.$$

Hence for arbitrary $f \in \mathfrak{X}$, we have $Af = AT(0)f \sim \sum_{\mathfrak{N}} \lambda_n f_n e^{nit}$. Since A is bounded, it follows that the factor sequence $\{\lambda_n, 0\}$ is also of type $(\mathfrak{X}, \mathfrak{X})$. Conversely, suppose that the sequences $\{1, 0\}$ and $\{\lambda_n, 0\}$ are both factor sequences of type $(\mathfrak{X}, \mathfrak{X})$, defining the bounded linear operators $T(0)$ and A respectively. In this case $S(\xi) \equiv T(0) \exp(\xi A)$ defines a semi-group of linear bounded operators for which $\lim_{\xi\to 0+} \| S(\xi) - T(0) \| = 0$. On the other hand

$$(S(\xi)f)_n = \sum_{k=0}^{\infty} \frac{\xi^k}{k!} (T(0)A^k f)_n = e^{\lambda_n \xi} f_n \qquad \text{or} \qquad 0$$

according as $n \in \mathfrak{N}$ or not. Thus $S(\xi) \equiv T(\xi)$ so that $T(\xi) \to T(0)$ in the uniform operator topology as $\xi \to 0+$.

20.4. The case $p = 2$. Here the results are particularly simple and easy to formulate.

THEOREM 20.4.1. *Let* $\mathfrak{S} \equiv [T(\xi); \xi > 0]$ *be a semi-group of linear bounded operators on* $L_2(-\pi, \pi)$ *to itself, the elements of which commute with the group of real translations on* t. *If* $T(\xi)$ *is weakly measurable for* $\xi > 0$, *then* (1) *formula* (20.3.6) *holds,* (2) $\| T(\xi) \| = e^{\mu\xi}$ *where* $\mu = \sup \{\Re(\lambda_n)\}$, *and* (3) $T(\xi)$ *is strongly continuous for* $\xi \geq 0$, *converging in the strong operator topology to the projection operator* $T(0)$ *as* $\xi \to 0+$. $T(\xi)$ *converges to* $T(0)$ *in the uniform operator topology if and only if the set* $\{\lambda_n\}$ *is bounded and in this case* $T(\xi) = T(0) \exp(\xi A)$ *where*

$$Af \sim \sum_{\mathfrak{N}} \lambda_n f_n e^{nit}, \qquad f \in L_2(-\pi, \pi)$$

PROOF. The assertions (1) and (2) as well as strong continuity for $\xi > 0$ was proved in Theorem 20.3.1. In connection with (3) we recall that $T(0)$ is defined by (20.3.7) and, obviously, $\| T(0) \| = 1$. Now for each $f \in L_2(-\pi, \pi)$ we have

$$\| [T(\xi) - T(0)]f \|^2 = 2\pi \sum_{\mathfrak{N}} | e^{\lambda_n \xi} - 1 |^2 | f_n |^2$$

and this clearly tends to zero with ξ. This establishes strong continuity at $\xi = 0$ and Theorem 10.5.5 now gives another proof of the fact that $T(\xi)$ is strongly continuous for $\xi \geqq 0$. On the other hand

$$\| T(\xi) - T(0) \| = \sup \{ | e^{\lambda_n \xi} - 1 | ; n \in \mathfrak{N} \}$$

and the left member tends to zero with ξ if and only if the set $\{\lambda_n\}$ is bounded. Theorem 9.6.1 then gives the desired representation of $T(\xi)$.

REMARK. The semi-group $T(\xi)$ is continuous in the uniform operator topology for $\xi > 0$ if and only if

$$\| T(\xi + \eta) - T(\xi) \| = \sup \{ | e^{\lambda_n \eta} - 1 | e^{\Re(\lambda_n)\xi} \}, \qquad \xi > 0,$$

tends to zero with η. This condition is satisfied trivially if $\{\lambda_n\}$ is bounded; a necessary and sufficient condition is that for every positive M the λ_n's satisfying $| \lambda_n | > M$ also satisfy $\Re(\lambda_n) < m$ where $m \to -\infty$ as $M \to \infty$.

Thus for the space $L_2(-\pi, \pi)$ the assumption that $T(\xi)$ is weakly measurable suffices to make $\| T(\xi) \|$ bounded on every finite interval $[0, \beta]$. This is obviously a much stronger statement than one can make in general and, as we shall see in the following section, it is no longer true in $L_1(-\pi, \pi)$.

In the present case the domain of the infinitesimal operator A_o is easy to determine. In fact

$$\mathfrak{D}(A_o) \equiv [f; \sum_{\mathfrak{N}} | \lambda_n f_n |^2 < \infty, f_n = 0 \quad \text{for} \quad n \notin \mathfrak{N}].$$

In general the spectrum of A_o fills out the plane. However if

$$\mathfrak{N} = \{n; -\infty < n < \infty\},$$

then $\sigma(A_o)$ equals the closure of the point spectrum. Indeed $[T(\xi)]$ is then of class (C_0), A_o coincides with the infinitesimal generator A, and by (20.3.10) the resolvent of A is given by

$$R(\lambda; A)f \sim \sum_{-\infty}^{\infty} (\lambda - \lambda_n)^{-1} f_n e^{nit}. \tag{20.4.1}$$

Hence if λ is at a distance $d(\lambda) > 0$ from $\{\lambda_n\}$, then $R(\lambda; A)$ exists as a linear bounded operator with $\| R(\lambda; A) \| = [d(\lambda)]^{-1}$. This fact enables us to prove

THEOREM 20.4.2. *If F is any given closed point set contained in a half-plane $\Re(\lambda) \leqq \alpha < \infty$, then there exists a semi-group $\mathfrak{S}$ of class (C_0) on $L_2(-\pi, \pi)$ whose elements commute with translations and such that $\sigma(A) = F$.*

PROOF. We have merely to choose a sequence $\{\lambda_n ; -\infty < n < \infty\}$ such that $\overline{\{\lambda_n\}} = F$ and define $T(\xi)$ by (20.3.6). The assertion now follows from Theorem 20.3.2 and the above remarks.

We see in particular that if D is any open set in the λ-plane with a finite or countably infinite number of components, one of which contains a right half-plane, then we can find a semi-group of linear bounded operators on $L_2(-\pi, \pi)$

which commute with translations and such that D is the resolvent set of the corresponding infinitesimal generator A. The resolvent $R(\lambda; A)$ is holomorphic in each of the components D_n of D, and the boundary of D_n is a natural boundary of $R(\lambda; A)$. Nevertheless, the values of $R(\lambda; A)$ in any one of these components determine its values in any other component.

20.5. The case $p = 1$. The situation for $p = 1$ is much more complex and varied than that for $p = 2$. In particular the behavior of $T(\xi)$ near $\xi = 0$ can be extremely pathological in the case $p = 1$. The following assertion can be read off from Theorem 20.3.1.

THEOREM 20.5.1. *Let $\mathfrak{S} \equiv [T(\xi); \xi > 0]$ be a semi-group of linear bounded operators on $L_1(-\pi, \pi)$ to itself, the elements of which commute with the group of real translations on t. If $T(\xi)$ is weakly measurable for $\xi > 0$, then* (1) *formula* (20.3.6) *holds,* (2) $\| T(\xi) \| = \int_{-\pi}^{\pi} | d_t K(t, \xi) |$, *and* (3) *$T(\xi)$ is strongly continuous for $\xi > 0$.*

In general for the case $p = 1$, neither the limit relation (20.3.11) exists nor is $T(0)$ bounded. In fact, even when $T(0)$ is bounded, the relation (20.3.11) need not hold. The following two examples illustrate this behavior. See E. Hille [7, pp. 45–46].

EXAMPLE 1. *There exists a semi-group $\mathfrak{S}$ satisfying the hypothesis of Theorem 20.5.1 such that $T(0) = I$ and $\lim_{\xi \to 0+} \| T(\xi) \| = \infty$.*

We let $\mathfrak{N}$ be the set of all integers so that the transformation defined by (20.3.7) is just the identity I. We take

$$\begin{aligned} \lambda_0 &= 0, \\ \lambda_n &= -\log |n| \quad \text{for } |n| \neq 2^k, \\ \lambda_n &= -2\log |n| \quad \text{for } |n| = 2^k, \end{aligned} \qquad k = 1, 2, 3 \cdots .$$

Here $K(t; \xi)$ is an absolutely continuous function of t; its derivative with respect to t may be written as

$$\begin{aligned} \pi K'(t; \xi) &= \left\{ \tfrac{1}{2} + \sum_{n=1}^{\infty} n^{-\xi} \cos nt \right\} + \sum_{k=1}^{\infty} (2^{-2k\xi} - 2^{-k\xi}) \cos (2^k t) \\ &\equiv F_1(t, \xi) + F_2(t, \xi). \end{aligned}$$

The lacunary series is of the type considered in Lemma 20.2.4. Its norm in $L_1(-\pi, \pi)$ consequently exceeds a constant multiple of the square root of

$$\sum_{k=1}^{\infty} 2^{-2k\xi}[1 - 2^{-k\xi}]^2 > \sum_{1 \leq k\xi \leq 2} 2^{-2k\xi}[1 - 2^{-k\xi}]^2 > \frac{1}{64\xi}.$$

On the other hand, a two-fold partial summation on the F_1 series gives

$$\tfrac{1}{2} + \sum_{n=1}^{\infty} n^{-\xi} \cos nt$$
$$= -(1 - 2^{-\xi})T_0 + \sum_{n=1}^{\infty} [n^{-\xi} - 2(n+1)^{-\xi} + (n+2)^{-\xi}]T_n,$$

where

$$T_n = \tfrac{1}{2}\left\{\frac{\sin (n+1)\frac{1}{2}t}{\sin \frac{1}{2}t}\right\}^2.$$

The series on the right consists of only positive terms. Integrating termwise and then applying a partial summation we obtain

$$\| F_1(t, \xi) \| \leqq \pi\left\{(1 - 2^{-\xi}) + \sum_{n=1}^{\infty} (n+1)[n^{-\xi} - 2(n+1)^{-\xi} + (n+2)^{-\xi}]\right\}$$
$$= \pi(3 - 2^{1-\xi}) < 3\pi.$$

We therefore have

$$\pi \| T(\xi) \| = \pi \int_{-\pi}^{\pi} | d_t K(t; \xi) | \geqq \| F_2(t, \xi) \| - \| F_1(t, \xi) \| > C\xi^{-1/2} - 3\pi,$$

so that $\| T(\xi) \| \to \infty$ as $\xi \to 0+$. Finally, Theorem 20.3.2 shows that $[T(\xi)]$ satisfies the hypothesis of Theorem 20.5.1.

Remark. The semi-group defined in Example 1 does not belong to class (A). In fact for $f \in \mathfrak{X}_0$ it is clear that

$$R(\lambda)f = \int_0^{\infty} e^{-\lambda\xi}\, T(\xi)f\, d\xi \sim \sum_{-\infty}^{\infty} (\lambda - \lambda_n)^{-1} f_n e^{nit}.$$

For $\lambda > 0$ this operator has a bounded linear extension on $L_1(-\pi, \pi)$. The corresponding kernel $k(t; \lambda)$ is again absolutely continuous in t and its derivative is given by

$$\pi k'(t; \lambda) = \left\{\frac{1}{2\lambda} + \sum_{n=1}^{\infty} \frac{1}{\lambda + \log n} \cos nt\right\} + \sum_{k=1}^{\infty}\left[\frac{1}{\lambda + 2k \log 2} - \frac{1}{\lambda + k \log 2}\right] \cos (2^k t)$$
$$\equiv f_1(t, \lambda) + f_2(t, \lambda).$$

Proceeding precisely as in Example 1 we obtain the following estimates:

$$\| f_2(t, \lambda) \| > c\lambda^{-1/2},$$
$$\| f_1(t, \lambda) \| \leqq \pi[3\lambda^{-1} - 2(\lambda + \log 2)^{-1}] \leqq 3\pi\lambda^{-1}.$$

Consequently

$$\pi \| R(\lambda) \| = \pi \int_{-\pi}^{\pi} | d_t k(t; \lambda) | \geqq \| f_2(t, \lambda) \| - \| f_1(t, \lambda) \| > c\lambda^{-1/2} - 3\pi\lambda^{-1}$$

so that $\| R(\lambda) \|$ is not $O(1/\lambda)$ as $\lambda \to \infty$. It follows that $\mathfrak{S}$ cannot be of class (A).

EXAMPLE 2. *There exists a semi-group* $\mathfrak{S}$ *satisfying the hypothesis of Theorem* 20.5.1 *such that* $T(0)$ *is unbounded and* $\lim\sup_{\xi\to 0+} \| T(\xi) \| = \infty$.

Let $\mathfrak{N}$ be the set of integers $\{\pm(4k+1); k = 0, 1, 2, \cdots\}$, and set $\lambda_n = -|n|$. Here

$$\pi K(t, \xi) = \sum_{k=0}^{\infty} e^{-(4k+1)\xi} \frac{\sin (4k+1)t}{4k+1}$$

is of bounded variation for $\xi > 0$. In fact

$$\pi \int_{-\pi}^{\pi} | d_t K(t; \xi) | \leqq \sum_{k=0}^{\infty} e^{-(4k+1)\xi} \int_{-\pi}^{\pi} |\cos (4k+1)t \mid dt \leqq (4 + \xi^{-1})e^{-\xi}.$$

It follows from Theorem 20.3.2 that $[T(\xi)]$ satisfies the hypothesis of Theorem 20.5.1. On the other hand, if $T(0)$ is unbounded then we see by Theorem 20.3.3 that $\lim\sup_{\xi\to 0+} \| T(\xi) \| = \infty$. Hence it suffices to show that $T(0)$ is unbounded. However

$$\pi K(t; 0) = \sum_{k=0}^{\infty} \frac{\sin (4k+1)t}{4k+1}$$

is not of bounded variation since the series diverges as the harmonic series when $t = \pi/2$. This shows that $T(0)$ is unbounded and completes the proof of Example 2.

Examples 1 and 2 adduced in connection with Theorem 20.3.3 raise the question whether or not there is any limitation on the rate of growth of $\| T(\xi) \|$ as $\xi \to 0+$. We recall that $\log \| T(\xi) \|$ is a measurable subadditive function on $(0, \infty)$ and by Theorem 7.6.4 such a function can grow as fast as any preassigned function. It turns out that this possibility is actually realized in $L_1(-\pi, \pi)$.

THEOREM 20.5.2. *Let* $\omega(\xi)$ *be a positive monotone decreasing continuous function on* $(0, \infty)$. *Then there exists a semi-group of linear bounded operators on* $L_1(-\pi, \pi)$ *to itself, continuous in the uniform operator topology for* $\xi > 0$, *and such that*

$$\| T(\xi) \| \geqq Ce^{\omega(\xi)}, \qquad 0 < \xi < \infty.$$

PROOF. For the construction we use a lacunary series. We define a kernel $K(t; \xi)$ by

$$\pi K'(t; \xi) = \sum_{k=1}^{\infty} k^{-\xi\alpha(k)} \cos (2^k t), \qquad \xi > 0,$$

where $\alpha(x)$ is a monotone increasing unbounded function to be disposed of later. The corresponding operator $T(\xi)$, which is defined by (20.3.4), has a norm which by Lemma 20.2.4 lies between constant multiples of

$$\left\{\sum_{k=1}^{\infty} k^{-2\xi\alpha(k)}\right\}^{1/2}. \tag{20.5.1}$$

Since this expression is finite for each $\xi > 0$, Theorem 20.3.2 shows that $[T(\xi)]$ is a semi-group of linear bounded operators, continuous in the strong operator topology for $\xi > 0$. Actually the semi-group is continuous in the uniform operator topology for $\xi > 0$. For $T_n(\xi)$ defined by the kernel $K_n(t; \xi)$,

$$\pi K_n'(t; \xi) = \sum_{k=1}^{n} k^{-\xi\alpha(k)} \cos (2^k t), \qquad \xi > 0,$$

is obviously continuous in the uniform operator topology for $\xi > 0$ and

$$\| T(\xi) - T_n(\xi) \| \leqq B \left\{ \sum_{k=n+1}^{\infty} k^{-2\xi\alpha(k)} \right\}^{1/2}$$

converges to zero as $n \to \infty$ uniformly with respect to ξ in each interval of the form $(\epsilon, 1/\epsilon)$, $0 < \epsilon < 1$.

In order to obtain a suitable $\alpha(x)$, let N be the least integer such that $4\xi\alpha(N) \geqq 1$. Then the square root of the sum of the first N terms in the series (20.5.1) exceeds $\frac{1}{2}N^{1/4}$. Thus

$$\| T(\xi) \| > CN^{1/4} \geqq C \left\{ \alpha^{-1} \left(\frac{1}{4\xi} \right) \right\}^{1/4}$$

and we have merely to choose $\alpha(x)$ such that

$$\alpha^{-1} \left(\frac{1}{4\xi} \right) \geqq e^{4\omega(\xi)},$$

where $\alpha^{-1}(y)$ is the inverse of $y = \alpha(x)$. If $\omega(\xi)$ is unbounded, we can satisfy this condition by choosing

$$\alpha(x) = \{4\omega^{-1}[\tfrac{1}{4} \log x]\}^{-1}.$$

This completes the proof.

20.6. Approximation of the identity. The semi-groups studied in the preceding sections provide important applications of Theorem 10.7.2. We now disregard the lacunary case so that

$$T(\xi)f \sim \sum_{-\infty}^{\infty} e^{\lambda_n \xi} f_n e^{nit}, \qquad \xi > 0 \tag{20.6.1}$$

where $\{e^{\lambda_n \xi}\}$ must be a factor sequence of type (L_p, L_p) for each $\xi > 0$. Such transformations of Fourier series are often used in the theory of summability where the problem is to determine how closely $T(\xi)f$ approximates f for small values of ξ. Here the degree of approximation may, of course, be measured locally or in the large. Our results have no bearing on the local problem, but they give fairly precise information if the approximation is taken in the sense of the metric.

THEOREM 20.6.1. *Let $\mathfrak{X}$ denote either $L_p(-\pi, \pi)$, $1 \leqq p < \infty$, or $C[-\pi, \pi]$ and suppose $T(\xi)$ is a factor sequence operator of type* (20.6.1) *such that* (i) $\| T(\xi) \|$ *is* *integrable on* (0, 1) *and* (ii) $\lim_{\eta \to 0+} \eta^{-1} \int_0^\eta T(\xi) f \, d\xi = f$ *in the norm for each* $f \in \mathfrak{X}$. *Then*

$$\liminf_{\eta \to 0+} \frac{1}{\eta} \| T(\eta) f - f \| = 0 \tag{20.6.2}$$

if and only if f *belongs to the closed linear extension of* $\{e^{nit}; \lambda_n = 0\}$. *Further*

$$T(\xi) f - f = \xi \{ A f + o(1) \} \tag{20.6.3}$$

for each $f \in \mathfrak{D}(A)$, *where* A *is the infinitesimal generator and*

$$\mathfrak{D}(A) \equiv \left[f; \sum_{-\infty}^{\infty} \lambda_n f_n e^{nit} \in \mathfrak{X} \right]. \tag{20.6.4}$$

PROOF. Theorem 20.3.2 implies that $[T(\xi); \xi > 0]$ forms a semi-group of linear bounded operators which is strongly continuous for $\xi > 0$. The assumptions (i) and (ii) show that $[T(\xi)]$ is of class $(1, C_1)$; in this case $A_o = A$. Theorem 10.7.2 therefore applies and we see that f satisfies (20.6.2) if and only if $T(\xi) f = f$ for all $\xi > 0$, that is, if and only if $(T(\xi)f)_n = e^{\lambda_n \xi} f_n$ is independent of ξ. This is equivalent to the asserted condition. The relation (20.6.3) is also implied by Theorem 10.7.2. Finally (20.6.4) is given by Theorem 20.3.2.

The case $\lambda_n = - | n |$ leads to the classical Abel-Poisson summability and deserves reformulation into more customary notation.

THEOREM 20.6.2. *Let* $f \in \mathfrak{X}$ *where* $\mathfrak{X}$ *is either* $L_p(-\pi, \pi)$, $1 \leqq p < \infty$, *or* $C[-\pi, \pi]$. *Let* $f(r, t)$ *be the harmonic function defined in the unit circle with the boundary values* $f(t)$, *viz.*

$$f(r, t) = \frac{1}{2\pi} \int_{-\pi}^{\pi} \frac{(1 - r^2) f(t - u) \, du}{1 - 2r \cos u + r^2}. \tag{20.6.5}$$

If

$$\liminf_{r \to 1-} (1 - r)^{-1} \| f(r, \cdot) - f(\cdot) \| = 0, \tag{20.6.6}$$

then $f(t)$ *is a constant. On the other hand*

$$\| f(r, \cdot) - f(\cdot) \| \leqq \log (1/r) \| \tilde{f}' \| \tag{20.6.7}$$

whenever $\tilde{f}' \sim - \sum | n | f_n e^{nit} \in \mathfrak{X}$.

PROOF. In order to obtain this theorem from the preceding one, we have merely to replace r by $e^{-\xi}$. Thus $[T(\xi)f](t) = f(e^{-\xi}, t)$ and $\| T(\xi) f \| \leqq \| f \|$ in all of the metrics involved. Theorems 20.3.2 and 20.3.3 therefore imply that $T(\xi)$ defines a semi-group of class (C_0) so that the previous theorem applies. The first assertion then follows from the fact that $\{e^{nit}; \lambda_n = 0\}$ contains only the function $f(t) \equiv 1$. The second assertion follows directly from the corollary of Theorem 10.7.2.

In the above theorem we note that Af is the conjugate of the derivative. If

$1 < p < \infty$, then $f'(t)$ and $\tilde{f}'(t)$ belong to $L_p(-\pi, \pi)$ if either one of them does. This is no longer true if $p = 1$ or if $\mathfrak{X} = C[-\pi, \pi]$.

2. Semi-Groups with Complex Parameter

20.7. Domains of existence for trigonometric semi-groups. We now consider the case in which the semi-group is a holomorphic operator-valued function of the parameter. To fix our ideas, suppose that the operator $T(\xi)$ on $L_p(-\pi, \pi)$ to itself is defined for $\xi > 0$ and commutes with translations on the real variable t so that $T(\xi)$ is given by its Dirichlet representation (20.3.6) for $\xi > 0$. Suppose further that there exists a holomorphic function $W(\zeta)$ defined in some domain containing the line segment (α, β), $0 \leqq \alpha < \beta$, and that $W(\xi) = T(\xi)$ for $\xi \in (\alpha, \beta)$. Theorem 17.2.2 then asserts that $W(\zeta)$ may be defined as a holomorphic semi-group in the interior of a spinal semi-module S where this set is the maximal domain of existence for $W(\zeta)$ and that $W(\xi) = T(\xi)$ for $\xi > \alpha$. We now show that $W(\zeta)$ inherits the Dirichlet representation from $T(\xi)$.

Theorem 20.7.1. *Let $W(\zeta)$ be a semi-group of linear bounded operators on $L_p(-\pi, \pi)$, $1 \leqq p \leqq \infty$, or $C[-\pi, \pi]$ to itself, defined and holomorphic in the interior of a spinal semi-module S, where S includes, say, the positive real axis. Suppose for each real $\xi > \alpha \geqq 0$ that $W(\xi)$ commutes with the group of real translations on t. Then $W(\zeta)$ has the representation*

$$W(\zeta)f \sim \sum_{\mathfrak{N}} e^{\lambda_n \zeta} f_n e^{nit}, \qquad \zeta \in \operatorname{Int}(S). \tag{20.7.1}$$

Proof. The argument used in the proof of Theorem 20.3.1 shows that

$$W(\xi)f \sim \sum_{-\infty}^{\infty} \mu_n(\xi) f_n e^{nit}, \qquad \xi > \alpha,$$

where

$$\mu_n(\xi_1 + \xi_2) = \mu_n(\xi_1)\mu_n(\xi_2), \qquad \xi_1, \xi_2 > \alpha.$$

It is clear that $\mu_n(\xi)$ is holomorphic for $\xi > \alpha$. It follows by Lemma 17.3.1 that $\mu_n(\xi) = e^{\lambda_n \xi}$ or 0; and we denote the set of n such that $\mu_n(\xi) \not\equiv 0$ by $\mathfrak{N}$. Now

$$(W(\zeta)f)_n = \frac{1}{2\pi} \int_{-\pi}^{\pi} [W(\zeta)f](t) e^{-nit}\, dt$$

is a numerically-valued function of ζ, holomorphic in $\operatorname{Int}(S)$, and equal to $\mu_n(\xi) f_n$ for all real $\xi > \alpha$. It must then be identically equal to $e^{\lambda_n \zeta} f_n$ or 0 according as $n \in \mathfrak{N}$ or not. This gives the representation (20.7.1).

As an analogue to Theorem 20.3.2 we have

THEOREM 20.7.2. *Let S be an open semi-module in the complex plane and suppose $[T(\zeta); \zeta \in S]$ is a family of operators defined by (20.7.1) on $L_p(-\pi, \pi)$, $1 \leqq p \leqq \infty$, or $C[-\pi, \pi]$ to itself. Then $[T(\zeta)]$ defines a semi-group of linear bounded operators, the elements of which commute with the group of real translations on t, and $T(\zeta)$ is holomorphic on $S + S$. If S is an angular semi-module or if $\| T(\zeta) \|$ is assumed to be bounded on each compact subset of S, then $T(\zeta)$ is holomorphic on all of S.*

PROOF. It is clear from (20.7.1) that $[T(\zeta)]$ defines a semi-group of operators, each of which commutes with translations. Lemma 20.2.1 shows that $T(\zeta)$ is linear and bounded. Let $\mathfrak{L}$ be the linear extension of the fundamental set $\{e^{nit}\}$. Then $T(\zeta)f$ is clearly holomorphic in S for each $f \in \mathfrak{L}$. Omitting for the moment the case $p = \infty$, we see that for each $f \in \mathfrak{X}$ there exists a sequence $\{f^k\} \subset \mathfrak{L}$ such that $f^k \to f$. Thus $T(\zeta)f^k \to T(\zeta)f$ for each $\zeta \in S$. If it is assumed that $\| T(\zeta) \|$ is bounded in each compact subset of S, then Theorem 3.14.1 applies and it follows that $T(\zeta)f$ is a holomorphic function of ζ in S. Even if local boundedness is not assumed, Theorem 3.18.2 applies. Thus for each $f \in \mathfrak{X}$ there exists a family of open sets $\{G_n\}$ dense in S such that $T(\zeta)f$ is holomorphic on each G_n. It follows that $T(\tau + \sigma)f = T(\sigma)T(\tau)f$ is a holomorphic function of τ in G_n, $n = 1, 2, 3, \cdots$, for each $\sigma \in S$. We now show that the set $\bigcup_n[\tau + \sigma; \tau \in G_n, \sigma \in S]$ covers $S + S$. In fact, given $\zeta_0 \in S + S$, there exist circles $| \tau - \tau_0 | < \rho$ and $| \sigma - \sigma_0 | < \rho$ contained in S such that $\tau_0 + \sigma_0 = \zeta_0$. Since $\{G_n\}$ is dense in S at least one of these open sets, say G_{n_0}, will have a point in common with $| \tau - \tau_0 | < \rho$. Choosing such a point $\tau \in G_{n_0}$, it is clear that $\sigma = \sigma_0 - (\tau - \tau_0)$ belongs to $| \sigma - \sigma_0 | < \rho$ and hence $T(\zeta)f$ will be holomorphic at $\zeta_0 = \tau + \sigma$. Since ζ_0 was an arbitrary point of $S + S$, we see that $T(\zeta)f$ is holomorphic throughout $S + S$ and, f being arbitrary, we see that $T(\zeta)$ is holomorphic throughout $S + S$. Finally if S is an angular semi-module, then it is easy to see that $S = S + S$. In this case, therefore, $T(\zeta)$ will be holomorphic on all of S without the local boundedness assumption. Returning to the case $p = \infty$, we note that a factor sequence of type (L_∞, L_∞) is also of type (L_1, L_1); thus $T(\zeta) = S^*(\zeta)$, where $S(\zeta)$ is a factor sequence type semi-group of operators on $L_1(-\pi, \pi)$. The above argument shows that $S(\zeta)$ is holomorphic on $S + S$, and even on S if S is an angular semi-module or if $\| T(\zeta) \| = \| S(\zeta) \|$ is locally bounded. Since the adjoint mapping of $\mathfrak{E}[L_1(-\pi, \pi)]$ into $\mathfrak{E}[L_\infty(-\pi, \pi)]$ is an isometry, it follows that $T(\zeta)$ is holomorphic as asserted.

We proceed now to a direct study of the domain of analyticity of a semi-group given *a priori* by formula (20.7.1). The problem is to determine for what values of ζ the sequence $\{e^{\lambda_n \zeta}, 0\}$ is a factor sequence of type (L_p, L_p) or (C, C). To be of type (L_2, L_2) it is necessary and sufficient that

$$\text{(20.7.2)} \qquad \| T(\zeta) \| = \sup_n e^{\Re(\lambda_n \zeta)} < \infty.$$

This condition is also necessary in order that it be of type (L_p, L_p), $1 \leqq p \leqq \infty$, or (C, C). On the other hand, if $\{e^{\lambda_n \zeta}, 0\}$ is of type (L_1, L_1) then it is also of type

(L_p, L_p), $1 \leqq p \leqq \infty$, and (C, C). The following discussion is based upon these two criteria.

Given the sequence $\{\lambda_n\}$, we shall first of all determine the point set $S(\{\lambda_n\})$ in the ζ-plane where the condition (20.7.2) is satisfied. It is clear that this set is a semi-module, that it contains the origin, and that it is convex. This permits $S(\{\lambda_n\})$ to have one of five distinct configurations: (i) the origin, (ii) a ray from the origin, (iii) a straight line through the origin, (iv) a sector with vertex at the origin and opening $\leqq \pi$, and (v) the whole plane. Here we may dismiss (i) as trivial; cases (ii) and (iii) correspond to non-holomorphic semi-groups and groups respectively; and (v) is that of a holomorphic group. The really interesting case from our present point of view is (iv).

The set $S(\{\lambda_n\})$ can be obtained by a *conjugate indicator diagram construction* analogous to that of section 17.4. Let D^- be the least convex hull of the point set $\{\bar{\lambda}_n\}$. Then

$$F(\zeta) \equiv \sup_n \Re(\lambda_n \zeta)$$

is the function of support for D^-, and $S(\{\lambda_n\})$ is precisely the set for which $F(\zeta)$ is finite. If D^- is the whole plane, then $F(\zeta)$ is finite only for $\zeta = 0$ and case (i) results. If D^- is not the whole plane this construction will in general lead to case (iv). More precisely, we get case (iv) if and only if the set $\{\lambda_n\}$ is unbounded and its distant portion may be enclosed in a sector of opening less than π. We get case (ii) when there is a single line of support, and case (iii) when there are only two lines of support. Finally, case (v) corresponds to the case in which D^- is bounded. In case (iv) a boundary ray of the sector $S(\{\lambda_n\})$ belongs to this sector if and only if an infinite line segment, perpendicular to the ray in question, lies outside of D^-. Expressed in terms of the indicator function $\mathfrak{d}(\varphi) \equiv F(e^{i\varphi})$ for D^-, we see that $S(\{\lambda_n\})$ is the sector $[re^{i\varphi}; \mathfrak{d}(\varphi) < \infty]$. Let Φ_1 and Φ_2, $\Phi_1 < \Phi_2$, denote the bounding rays of $S(\{\lambda_n\})$. The ray $\arg \zeta = \Phi_1$ (or Φ_2) will lie in $S(\{\lambda_n\})$ if and only if $\mathfrak{d}(\Phi_1 + 0)$ (or $\mathfrak{d}(\Phi_2 - 0)$) is finite. Applying Theorem 20.7.2, we now have the following

THEOREM 20.7.3. *Let* $\mathfrak{S} \equiv [T(\zeta)]$ *be a semi-group in* $L_p(-\pi, \pi)$, $1 \leqq p \leqq \infty$, *or* $C[-\pi, \pi]$ *to itself with Dirichlet representation* (20.7.1). *Let* $S(\{\lambda_n\})$ *be determined as above. Then the maximal domain of analytic existence of* $T(\zeta)$ *coincides with* Int $[S(\{\lambda_n\})]$ *if* $p = 2$ *and is a subset of* Int $[S\{\lambda_n\})]$ *otherwise.*

The above theorem gives very little information for $p \neq 2$. It could, perhaps, be complemented by a determination of the values of ζ for which the sequence $\{e^{\lambda_n \zeta}, 0\}$ is of type (L_1, L_1), but the discussion of the kernel $K(t; \zeta)$ required for this purpose is apt to be rather complicated. The following special case is of some importance and is easy to handle.

THEOREM 20.7.4. *Let* $\dot{\mathfrak{S}} \equiv [T(\zeta)]$ *be a semi-group on* $L_p(-\pi, \pi)$, $1 \leqq p \leqq \infty$, *or* $C[-\pi, \pi]$ *to itself with Dirichlet representation* (20.7.1). *Let* $S(\{\lambda_n\})$ *be determined as above. Then the maximal domain of analytic existence of* $T(\zeta)$ *coincides with* Int $[S(\{\lambda_n\})]$ *if* (1) *the distant portion of the set* $\{\lambda_n\}$ *lies in a sector of opening* $<\pi$, *and* (2) $\lim_{|n|\to\infty} |\lambda_n|/(\log|n|) = \infty$.

PROOF. The first condition ensures that $S(\{\lambda_n\})$ does not degenerate by excluding the cases (i), (ii), and (iii); the second condition excludes the case (v). Condition (2) is of a type used in the theory of Dirichlet series with complex exponents (see E. Hille [**1**, p. 266]). It forces the interior of the region of ordinary convergence of the series to coincide with the interior of the region of absolute convergence and leads to a simpler rule for the determination of the latter. However, without the use of this theory, we may argue as follows. Without loss of generality we may suppose that Int $[S(\{\lambda_n\})]$ is the sector $\Phi_1 < \arg\zeta < \Phi_2$ where $-\pi/2 \leqq \Phi_1 < 0 < \Phi_2 \leqq \pi/2$. We set $\zeta = re^{i\varphi}$, $\lambda_n = \rho_n e^{i\theta_n}$. From the construction of $S(\{\lambda_n\})$ together with condition (2) it follows that $\Phi_2 + \pi/2 - \epsilon < 2\pi - \theta_n < \Phi_1 + 3\pi/2 + \epsilon$ for $n \geqq n_1(\epsilon)$, $\epsilon > 0$. Suppose now that $\Phi_1 + 2\epsilon \leqq \varphi \leqq \Phi_2 - 2\epsilon$ and set $\sin\epsilon = \delta$. Condition (2) shows that to every given positive ω, no matter how large, we can find an $n_2(\omega)$ such that for $|n| \geqq n_2(\omega)$ we have $|\lambda_n| > \omega\log|n|$. Hence for $|n| \geqq \max[n_1(\epsilon), n_2(\omega)]$

$$e^{\Re(\lambda_n\zeta)} \leqq |n|^{-\delta r\omega}. \tag{20.7.3}$$

This estimate shows that the series

$$2\pi K'(t;\zeta) = \sum_{\mathfrak{N}} e^{\lambda_n\zeta}e^{nit}$$

converges absolutely for each ζ in Int $[S(\{\lambda_n\})]$, uniformly with respect to t. It follows that the kernel $K(t;\zeta)$ of formula (20.3.3) is of bounded variation in $[-\pi,\pi]$ for each ζ in Int $[S(\{\lambda_n\})]$. As a consequence the sequence $\{e^{\lambda_n\zeta}, 0\}$ is of type (L_1, L_1) and *a fortiori* of type (L_p, L_p), $1 \leqq p \leqq \infty$, and type (C, C). Theorem 20.7.2 now asserts that $[T(\zeta)]$ is a semi-group of linear bounded operators, holomorphic in Int $[S(\{\lambda_n\})]$. Moreover this is the maximal domain of analytic existence by the previous theorem. This completes the proof.

The following is a simple example of a semi-group which may be defined on each space $L_p(-\pi,\pi)$, $1 \leqq p \leqq \infty$, or on $C[-\pi,\pi]$, having the sector $|\arg\zeta| < \Phi \leqq \pi/2$ as its maximal domain of analytic existence. We set $\alpha = \cos\Phi$, $\beta = \sin\Phi$, $\lambda_n = ine^{i\Phi}$ for $n \geqq 0$, and $\lambda_n = ine^{-i\Phi}$ for $n < 0$. In this case

$$[T(\zeta)f](t) = \frac{1}{2\pi}\int_{-\pi}^{\pi} \frac{[1 - e^{-2\beta\zeta}]f(t-u)\,du}{1 - 2e^{-\beta\zeta}\cos(u + \alpha\zeta) + e^{-2\beta\zeta}}. \tag{20.7.4}$$

When $\Phi = \pi/2$ this becomes the classical Poisson integral after a change of variable: $r = e^{-\zeta}$.

20.8. Convex semi-module domains of existence. It follows from Theorems 20.7.3 and 20.7.4 that if we desire holomorphic semi-groups whose elements commute with translations and which have maximal domains of analytic existence which are not sectors, then we must have $p \neq 2$ and the sequence $\{\lambda_n\}$ must not grow too rapidly. The construction given below applies to $p = 1$ and ∞, and leads to a semi-group having a preassigned convex semi-module as its maximal domain of analytic existence. We shall use the notion of orthogonal semi-groups defined by

DEFINITION 20.8.1. *If $\mathfrak{T} \equiv [T(\zeta)]$ and $\mathfrak{U} \equiv [U(\zeta)]$ are two semi-groups of linear bounded operators on a complex (B)-space $\mathfrak{X}$ to itself, having a common domain of definition S in the parameter space, and if for every choice of ζ_1 and ζ_2 in S and all $x \in \mathfrak{X}$ we have*

$$T(\zeta_1)[U(\zeta_2)x] = U(\zeta_2)[T(\zeta_1)x] = \theta, \tag{20.8.1}$$

then $\mathfrak{T}$ and $\mathfrak{U}$ are said to be orthogonal on S.

While the "sum" of two semi-groups, that is, the operator family $[T(\zeta) + U(\zeta)]$, is ordinarily not a semi-group, any number of mutually orthogonal semi-groups may be "added" and the sum defines a semi-group on the common domain of definition of the components. All of the semi-groups considered in the present chapter are constructed by means of this concept of orthogonality. Thus the elementary semi-groups $[T_n(\zeta)]$ on $L_p(-\pi, \pi)$ (or $C[-\pi, \pi]$) defined by

$$[T_n(\zeta)f](t) = e^{\lambda_n \zeta} f_n e^{nit}$$

are mutually orthogonal for all ζ, and the Dirichlet representation of $T(\zeta)$ in (20.7.1) is simply the canonical representation of $T(\zeta)$ as the sum of orthogonal semi-groups:

$$T(\zeta) = \sum_{\mathfrak{N}} T_n(\zeta).$$

THEOREM 20.8.1. *Let S be a given convex open semi-module in the complex ζ-plane and suppose $\mathfrak{X} = L_1(-\pi, \pi)$, $L_\infty(-\pi, \pi)$, or $C[-\pi, \pi]$. Then there exists a holomorphic semi-group of linear bounded operators on $\mathfrak{X}$ to itself having S as its maximal domain of analytic existence.*

PROOF. If $\zeta = 0$ is an interior point of S, then S is the whole plane. In this case $T(\zeta)$ is an entire function of ζ and the construction of $T(\zeta)$ may be based on Theorem 20.3.4. We therefore define λ_n for all n in such a manner that $\{\lambda_n\}$ becomes a factor sequence of type $(\mathfrak{X}, \mathfrak{X})$ and define $T(\zeta)$ by (20.7.1). Next if $\zeta = 0$ is a boundary point of S, then S is necessarily a sector of opening $\leqq \pi$, and (20.7.4) gives the desired example when the sector is symmetric about the positive real axis as we may assume without restricting the generality.

If $\zeta = 0$ is not in the closure of S, the problem is no longer subsumed under any of the previous material. We start by constructing a semi-group of operators $T(\zeta)$ with maximal domain of analytic existence equal to the half-plane $\Re(\zeta) > 1$. Here we take $\mathfrak{N} = \{2^k; k = 1, 2, 3, \cdots\}$ and define $\lambda_n = \frac{1}{2}[\log \log n - \log \log 2]$ for each $n \in \mathfrak{N}$. Thus

(20.8.2) $$T(\zeta)f \sim \sum_{k=1}^{\infty} k^{-(\zeta/2)} f_{2^k} \exp\,(2^k it).$$

This defines a factor type transformation on $\mathfrak{X}$ to $\mathfrak{X}$ if and only if the corresponding kernel $K(t;\zeta)$ is of bounded variation. By Lemma 20.2.4 there exist positive constants A, B such that

$$A\left\{\sum_{k=1}^{\infty} k^{-\xi}\right\}^{1/2} \leqq \| T(\zeta) \| \leqq B\left\{\sum_{k=1}^{\infty} k^{-\xi}\right\}^{1/2}, \qquad \zeta = \xi + i\eta.$$

Hence $T(\zeta)$ is defined only for $\Re(\zeta) > 1$. It is also clear that $\| T(\zeta) \|$ is bounded on every compact subset of this half-plane. Theorem 20.7.2 therefore shows that $T(\zeta)$ defines a semi-group of linear bounded operators, holomorphic in the half-plane $\Re(\zeta) > 1$. Since $\|T(\xi + i\eta)\| \to \infty$ as $\xi \to 1+$, this half-plane is also the maximal domain of analytic existence for $T(\zeta)$.

It is clear that the above construction is based on the divergence of the harmonic series. Thus the same conclusion could be drawn if in (20.8.2) we summed merely over the values of k in a certain set $\mathfrak{K}$ such that $\sum_{\mathfrak{K}} (1/k)$ diverges. Also we note that if we replace ζ by (ζ/ζ_0) in (20.8.2), ζ_0 being an arbitrary complex number $\neq 0$, then the resulting semi-group exists and is holomorphic in the half-plane $\Re(\zeta/\zeta_0) > 1$ and nowhere else. Combining these two observations, we are led to a general construction.

We recall that the indicator function $\mathfrak{d}(\varphi)$ for S is defined as

$$\mathfrak{d}(\varphi) \equiv \sup\,[\Re(\bar{\zeta}e^{i\varphi});\ \zeta \in S].$$

Let (Φ_1, Φ_2) be the interior of the set on which $\mathfrak{d}(\varphi)$ is finite. Since S is non-empty, we see that $\mathfrak{d}(\varphi)$ is bounded from below, that is, $\mathfrak{d}(\varphi) \geqq -K$ on (Φ_1, Φ_2). Moreover

$$l(\varphi):\ \xi \cos\varphi + \eta \sin\varphi = \mathfrak{d}(\varphi)$$

is the equation of the line of support of S whose exterior normal at the point of support, say $\zeta(\varphi)$, makes the angle φ with the positive real axis. The foot of the perpendicular to $l(\varphi)$ from the origin is $\zeta = \mathfrak{d}(\varphi)e^{i\varphi}$. If $\zeta = 0$ does not belong to $\bar{S}$, then $\mathfrak{d}(\varphi) < 0$ for all $\varphi \in (\Phi_1, \Phi_2)$. For suppose $\mathfrak{d}(\varphi_0) \geqq 0$, $\varphi_0 \in (\Phi_1, \Phi_2)$. Then S, being a semi-module, will contain points arbitrarily close to $\{k\zeta(\varphi_0);\ k = 1, 2, \cdots\}$. It follows from this that $\mathfrak{d}(\varphi) = \infty$ for all $\varphi < \varphi_0$ or $\varphi > \varphi_0$, depending upon whether S lies below or above $l(\varphi_0)$; this contradicts the fact that $\varphi_0 \in (\Phi_1, \Phi_2)$.

Let us now choose a sequence $\{\varphi_n\}$ dense in (Φ_1, Φ_2) and set

$$\zeta_n = \mathfrak{d}(\varphi_n)e^{i\varphi_n}.$$

Let $\mathfrak{K}_n$ be the set of all positive integers whose least prime factor is the nth prime. Then $\sum_{\mathfrak{K}_n} (1/k) = \infty$. We now form

(20.8.3) $$T(\zeta)f \sim \sum_n \sum_{\mathfrak{K}_n} k^{-\zeta/(2\zeta_n)} f_{2^k} \exp\,(2^k it) \equiv \sum_n T_n(\zeta)f.$$

Here each $T_n(\zeta)$ defines a semi-group on $\mathfrak{X}$ to itself with maximal domain of analytic existence $\Re(\zeta/\zeta_n) > 1$, and the $T_n(\zeta)$ are mutually orthogonal. For ζ interior to S, say at a distance $\gamma(\zeta)$ from the boundary of S, we have

$$\Re(\zeta e^{-i\varphi_n}) = \Re(\bar{\zeta} e^{i\varphi_n}) \leqq \sigma(\varphi_n) - \gamma(\zeta)$$

so that

$$\Re(\zeta/\zeta_n) = (\sigma(\varphi_n))^{-1}\Re(\zeta e^{-i\varphi_n}) \geqq 1 + \delta$$

where $\delta \equiv \delta(\zeta) = \gamma(\zeta)/K$ does not depend on n. Applying Lemma 20.2.4 as above we see that

$$\| T(\zeta) \| \leqq B \left\{ \sum_{k=1}^{\infty} k^{-(1+\delta)} \right\}^{1/2}.$$

It is also clear from the way in which $\delta(\zeta)$ is defined that $\| T(\zeta) \|$ is bounded on each compact subset of S. Theorem 20.7.2 now shows that $T(\zeta)$ defines a semi-group of linear bounded operators, holomorphic in S. On the other hand if ζ_0 is not in $\bar{S}$, then it is separated from S by at least one line of support of the set $\{l(\varphi_n)\}$, say $l(\varphi_{n_0})$. Lemma 20.2.4 also shows that $\| T(\zeta) \| \geqq C \| T_n(\zeta) \|$ for all n. It follows that $\| T(\zeta) \|$ becomes infinite with $\| T_{n_0}(\zeta) \|$ as ζ crosses $l(\varphi_{n_0})$ along any path from S to ζ_0. Consequently S is the maximal domain of analytic existence for $T(\zeta)$. This concludes the proof.

If S happens to be a polygonal figure with a finite number of sides, $\zeta = 0 \notin \bar{S}$, then the above construction can be simplified. In this case we can select one point ζ_n on each side, extended if necessary, and omit all of the other terms in (20.8.3). The method appears to break down for other values of p. One is lead to conjecture that the maximal domain of analytic existence must be a sector when $p \neq 1$ or ∞, and that the maximal domain of existence for the case $p = 1$ and ∞ is a convex semi-module.

20.9. Properties of the norm. We now consider the norm properties of a holomorphic semi-group of operators $[T(\zeta)]$ defined on an open sector Σ_2 by the Dirichlet representation (20.7.1). In Theorem 3.13.1 it was proved that the norm of any holomorphic function is subharmonic in its domain of holomorphism. For the case $p = 2$ the norm of $T(\zeta)$ has a much stronger property. In this case $\| T(\zeta) \|$ is a convex function of ζ, that is $\| T(\zeta) \|$ is convex on every line segment in Σ_2. For the notation used below we refer to section 20.7.

THEOREM 20.9.1. *Let* $[T(\zeta)]$ *be a semi-group of operators on* $L_2(-\pi, \pi)$ *to itself, defined by the Dirichlet representation* (20.7.1) *on the sector* $\Sigma_2 \equiv \text{Int}\,[S(\{\lambda_n\})]$. *Let* $F(\zeta)$ *denote the function of support for* $D^- \equiv$ *closed convex extension of* $\{\bar{\lambda}_n\}$, *that is,*

$$F(\zeta) \equiv \sup\,[\Re(\zeta\lambda);\ \lambda \in D^-]. \tag{20.9.1}$$

Then

$$(20.9.2) \qquad \| T(\zeta) \| = \exp [F(\zeta)]$$

is a continuous convex function of ζ *in* Σ_2.

PROOF. It is easy to see that $F(\zeta) = \sup [\Re(\zeta\lambda_n); n \in \mathfrak{N}]$ and this together with (20.7.2) implies (20.9.2). Now the function of support of a convex set is a continuous convex function on its sector of definition (see section 7.12). Since e^σ is also a continuous convex function of σ, it readily follows that $\exp [F(\zeta)]$ is a continuous convex function of ζ on Σ_2.

It was proved in Theorem 17.3.1 that the norm of a semi-group of operators $[T(\zeta)]$, holomorphic in the interior of a spinal semi-module S, grows at most exponentially with respect to $|\zeta|$ if $\zeta \to \infty$ in such a manner that its distance from the boundary of S exceeds a fixed ϵ times $|\zeta|$. It was observed that no such result could be expected to hold if the distance of ζ from the boundary of S stays bounded, nor is the norm necessarily bounded as ζ approaches the boundary. The following theorem shows that these phenomena arise for trigonometric semi-groups even in the case $p = 2$.

THEOREM 20.9.2. *Let* Σ_2 *be an open sector in the complex* ζ*-plane of opening* $\leqq \pi$; *let* $\{\zeta_n\} \subset \Sigma_2$ *be a sequence of points such that either* ζ_n *approaches a point* ζ_0 *on the boundary of* Σ_2, $\zeta_0 \neq 0$, *or* $\zeta_n \to \infty$ *along a line parallel to the boundary of* Σ_2; *and let* $\{\alpha_n\}$ *be an arbitrary sequence of positive numbers. Then there exists a semi-group of operators on* $L_2(-\pi, \pi)$ *to itself, given by* (20.7.1) *and having* Σ_2 *as its maximal domain of analyticity, such that* $\| T(\zeta_n) \| \geqq \alpha_n$, $n = 1, 2, 3, \cdots$.

PROOF. It is sufficient to prove the theorem for the case in which $\Sigma_2 = \text{Int}\,[S(\{\lambda_n\})]$ is the right half-plane. The general case can be proved by the same method using two orthogonal semi-groups defined in half-planes the common part of which is the desired sector. We may therefore suppose the sequence $\{\zeta_n\}$ to be such that $\Im(\zeta_n) > 0$ and hence either $\zeta_0 = i\tau_0$, $\tau_0 > 0$, or $\Re(\zeta_n) = \gamma_0 > 0$, $\Im(\zeta_n) \to \infty$.

Set $\zeta_n = r_n e^{i\varphi_n}$. Then $\varphi_n \to \pi/2$ and we may reorder the ζ_n so that $\varphi_n < \varphi_{n+1}$. Finally set $\omega_n = \max \{(r_n)^{-1} \log \alpha_n, 0\}$. We now proceed step-wise to construct a sequence $\{\lambda_n\}$ such that the indicator function $\mathfrak{d}(\varphi)$ for the closed convex extension D^- of $\{\bar{\lambda}_n\}$ is defined on $(-\pi, \pi)$ and satisfies the condition $\mathfrak{d}(\varphi_n) \geqq \omega_n$. In this case $\text{Int}\,[S(\{\lambda_n\})]$ is the right half-plane and the relation (20.9.2) shows that $\| T(\zeta_n) \| = \exp [r_n \mathfrak{d}(\varphi_n)] \geqq \alpha_n$. Let $l(\varphi, \delta)$ denote the line

$$\xi \cos \varphi + \eta \sin \varphi = \delta, \qquad \zeta = \xi + i\eta.$$

Set $\delta_1 = \omega_1$, $\gamma_1 = 1$, and $\bar{\lambda}_1 = \delta_1 e^{i\varphi_1} + \gamma_1 i e^{i\varphi_1}$. Next choose $\delta_2 > \omega_2$ so that $\bar{\lambda}_1$ lies below the line $l(\varphi_2, \delta_2)$ and then choose $\gamma_2 > 2$ so that $\bar{\lambda}_2 \equiv \delta_2 e^{i\varphi_2} + \gamma_2 i e^{i\varphi_2}$ lies below $l(\varphi_1, \delta_1)$. In general choose $\delta_n > \omega_n$ so that the points $\{\bar{\lambda}_k; k = 1, \cdots, n - 1\}$ lie below $l(\varphi_n, \delta_n)$ and then choose $\gamma_n > n$ so that $\bar{\lambda}_n = \delta_n e^{i\varphi_n} + \gamma_n i e^{i\varphi_n}$ lies below each of the lines $\{l(\varphi_k, \delta_k); k = 1, \cdots, n - 1\}$. Further set $\lambda_n = n$ for $n \leqq 0$. Then D^- is a polygonal figure bounded on one side by the negative real axis and otherwise by the line segments $[\bar{\lambda}_{n-1}, \bar{\lambda}_n]$, $n = 1, 2, 3, \cdots$. It readily follows that Int $[S(\{\lambda_n\})]$ is the right half-plane and that $\delta(\varphi_n) = \delta_n \geqq \omega_n$. In the above we have implicitly assumed that no two φ_n's were equal. If this is not the case we may group the ζ_n's into sets, the jth set K_j consisting of all ζ_n's for which φ_n is equal to the jth largest distinct argument, say $\bar{\varphi}_j$, in the set $\{\varphi_n\}$. To K_j we order the number $\bar{\omega}_j = \max\{\omega_n ; \zeta_n \in K_j\}$. We then proceed as above, replacing φ_n by $\bar{\varphi}_n$ and ω_n by $\bar{\omega}_n$. As a result if $\zeta_n \in K_j$, then $\| T(\zeta_n) \| = \exp[r_n \delta(\varphi_n)] \geqq \exp[r_n \bar{\omega}_j] \geqq \alpha_n$.

CHAPTER XXI

SEMI-GROUPS IN $L_p(-\infty, \infty)$

21.1. Orientation. In the present chapter we shall discuss some classes of semi-groups defined in $L_p(-\infty, \infty)$. We shall treat only the cases $p = 1$ and $p = 2$ in any detail. The natural tool for such a study is Fourier analysis, which now may be of the continuous or discrete type. In the former case we use the theory of Fourier transforms, in the latter, expansions in terms of the orthogonal functions of Hermite. In either case we shall assume that the semi-group operator defines a factor transformation on the Fourier representation. In the Fourier transform case, the resulting semi-group operators have the important property of commuting with translations.

While the factor sequence problem has been solved for trigonometric Fourier series, at least for the types (L_1, L_1) and (L_2, L_2), there does not seem to be any information in the literature concerning factor functions for Fourier transforms and factor sequences for Hermitian series of type (L_1, L_1). These problems are discussed in the first paragraph, *Factor Theory*, whereas the second paragraph, *Semi-Groups of Factor Type*, gives the applications to semi-group theory.

References. Bochner [1], Hille [2, 4], Hille and Tamarkin [5], Orlicz [1], Phillips [7], M. Riesz [1], Szegö [2], Titchmarsh [1], Tricomi [1], Widder [1].

1. Factor Theory

21.2. Factor functions for Fourier transforms. In section 20.2 we considered factor sequences for Fourier series. We now consider the corresponding problem for Fourier transforms, namely,

What conditions should be satisfied by $\mu(\sigma)$, $-\infty < \sigma < \infty$, in order that $G(\sigma) = \mu(\sigma)F(\sigma)$ be the Fourier transform of a function $g(t) \in L_q(-\infty, \infty)$ whenever $F(\sigma)$ is the Fourier transform of a function $f(t) \in L_p(-\infty, \infty)$?

Such a function $\mu(\sigma)$ will be called a *factor function for Fourier transforms of type* (L_p, L_q). Very little is actually known about the theory of factor functions and we shall present only a few scattered results dealing in the main with functions of type (L_1, L_1) and (L_2, L_2). The reader is referred to E. C. Titchmarsh [1] for background material on Fourier transforms.

THEOREM 21.2.1. *If $\mu(\sigma)$ is a factor function for Fourier transforms of type*

(L_p, L_q), $1 \leqq p, q \leqq 2$, *defining the transformation* U, *then* U *is linear, bounded, and commutes with the group of real translations on* t.

PROOF. It is clear that U is defined and linear on all of $L_p(-\infty, \infty)$. It is equally clear that U commutes with translations since the Fourier transform of $f(t + \alpha)$ is just $e^{i\alpha\sigma}$ times the Fourier transform of $f(t)$. In order to prove that U is bounded, it is sufficient, according to the closed graph theorem, to show that U is closed. Hence suppose $f_n \to f$ in $L_p(-\infty, \infty)$ and $Uf_n \to g$ in $L_q(-\infty, \infty)$. We shall show that $Uf = g$. In fact, $F_n(\sigma) \to F(\sigma)$ either pointwise in case $p = 1$ or in the mean of order $p' = p/(p - 1)$ if $1 < p \leqq 2$. Likewise $\mu(\sigma)F_n(\sigma) \to G(\sigma)$ either pointwise in case $q = 1$ or in the mean of order $q' = q/(q - 1)$ in case $1 < q \leqq 2$. In any case we see that $G(\sigma) = \mu(\sigma)F(\sigma)$ for almost all σ so that $Uf = g$.

The case $p = 2 = q$ seems to be the only one which is well known:

THEOREM 21.2.2. *A necessary and sufficient condition that* $\mu(\sigma)$ *be a factor function for Fourier transforms of type* (L_2, L_2) *is that* $\mu(\sigma)$ *be an essentially bounded measurable function on* $(-\infty, \infty)$. *If this condition is satisfied, then* $\mu(\sigma)$ *defines a linear bounded operator* $g = Uf$ *on* $L_2(-\infty, \infty)$ *to itself, the norm of which is the essential supremum of* $|\mu(\sigma)|$.

We leave the proof of Theorem 21.2.2 to the reader and we go on to the case $p = 1 = q$ where the result does not seem to be so commonly known. We begin by proving two lemmas.

LEMMA 21.2.1. *If* $g(t, u)$, $-\infty < t, u < \infty$, *is a measurable function of* (t, u), *whose pth power is integrable with respect to t for each* u, $1 \leqq p < \infty$, *then* $x(u) \equiv g(\cdot, u)$ *is a strongly measurable vector-valued function on* $(-\infty, \infty)$ *to* $L_p(-\infty, \infty)$. *If, in addition,* $\| x(u) \| = \{\int_{-\infty}^{\infty} | g(t, u) |^p \, dt\}^{1/p} \leqq B$, $-\infty < u < \infty$, *and* $f(u) \in L_1(-\infty, \infty)$, *then*

$$h(t) = \int_{-\infty}^{\infty} g(t, u)f(u) \, du \tag{21.2.1}$$

exists as a Lebesgue integral for almost all t, $h(t) \in L_p(-\infty, \infty)$, $\| h \|_p \leqq B \| f \|_1$, *and*

$$h(\cdot) = (\mathrm{B}) \int_{-\infty}^{\infty} x(u)f(u) \, du, \tag{21.2.2}$$

where the integral is now a Bochner integral; moreover if $T \in \mathfrak{E}[L_p(-\infty, \infty)]$, *then*

$$Th = (\mathrm{B}) \int_{-\infty}^{\infty} T[x(u)]f(u) \, du. \tag{21.2.3}$$

PROOF. Let $x^* = \varphi(\cdot) \in L_p(-\infty, \infty)^* = L_{p'}(-\infty, \infty)$, $1/p + 1/p' = 1$. Then

$$x^*[x(u)] = \int_{-\infty}^{\infty} \varphi(t)g(t, u) \, dt$$

is a measurable function of u. Consequently $x(u)$ is weakly measurable and since $L_p(-\infty, \infty)$ is separable, it follows by Corollary 2 of Theorem 3.5.3 that $x(u)$ is actually strongly measurable. That (21.2.1) exists as a Lebesgue integral for almost all t, that $h(t) \in L_p(-\infty, \infty)$, and that $\| h \|_p \leqq B \| f \|_1$, are direct consequences of the Fubini theorem and the Hölder inequality. Further, according to the relation (3.7.5),

$$\begin{aligned} x^*\left[(\mathrm{B}) \int_{-\infty}^{\infty} x(u)f(u)\, du\right] &= \int_{-\infty}^{\infty} x^*[x(u)]f(u)\, du \\ &= \int_{-\infty}^{\infty}\int_{-\infty}^{\infty} \varphi(t)g(t, u)f(u)\, dt\, du = \int_{-\infty}^{\infty} \varphi(t)\left[\int_{-\infty}^{\infty} g(t, u)f(u)\, du\right] dt \\ &= x^*(h). \end{aligned}$$

This holds for all $x^* \in L_p(-\infty, \infty)^*$, proving (21.2.2). Finally (21.2.3) follows from Theorem 3.7.12.

LEMMA 21.2.2. *Let $f, g \in L_1(-\infty, \infty)$ and suppose that $T \in \mathfrak{E}[L_1(-\infty, \infty)]$ and commutes with the group of real translations on t. If $k = Tg$ and*

$$h(t) = \int_{-\infty}^{\infty} g(t - u)f(u)\, du,$$

then

$$[Th](t) = \int_{-\infty}^{\infty} k(t - u)f(u)\, du. \tag{21.2.4}$$

PROOF. We see that $g(t, u) \equiv g(t - u)$ satisfies all of the hypotheses of the previous lemma so that $h \in L_1(-\infty, \infty)$ and $Th = (\mathrm{B}) \int_{-\infty}^{\infty} Tg(\cdot - u)f(u)\, du$. However T commutes with translations and this means that $[Tg(\cdot - u)](t) = k(t - u)$ for each u. Now the previous lemma also applies to $k(t, u) \equiv k(t - u)$ and hence

$$[Th](t) = \left[(\mathrm{B}) \int_{-\infty}^{\infty} k(\cdot - u)f(u)\, du\right](t) = \int_{-\infty}^{\infty} k(t - u)f(u)\, du,$$

which proves (21.2.4).

THEOREM 21.2.3. *Let $T \in \mathfrak{E}[L_1(-\infty, \infty)]$ and commute with the group of real translations on t. Then there exists a function $K(t)$ which is of bounded variation on $(-\infty, \infty)$ and such that*

$$[Tf](t) = \int_{-\infty}^{\infty} f(t - u)\, dK(u), \qquad f \in L_1(-\infty, \infty), \tag{21.2.5}$$

where the integral exists in the Lebesgue-Stieltjes sense for almost all t; $\| T \| = \int_{-\infty}^{\infty} | dK(t) |$. Further the function

$$\mu(\sigma) = \int_{-\infty}^{\infty} e^{-i\sigma t}\, dK(t) \tag{21.2.6}$$

serves to define T as a factor transformation.

PROOF. Let $\{g_n\}$ be a sequence of functions in $L_1(-\infty, \infty)$ approximating the identity, for instance, let $g_n(t) = n$ for $0 < t < 1/n$, $= 0$ elsewhere. Then it is easy to show that $\lim_{n\to\infty} g_n * f = f$ in the sense of convergence in the norm for each $f \in L_1(-\infty, \infty)$. Thus if $k_n(t) = [Tg_n](t)$, then Lemma 21.2.2 implies that

$$Tf = \lim_{n\to\infty} T(g_n * f) = \lim_{n\to\infty} k_n * f \tag{21.2.7}$$

for each $f \in L_1(-\infty, \infty)$. We now define $K_n(t) = \int_0^t k_n(u)\, du$. Since $\| g_n \| = 1$, we see that

$$\text{Var}\,(K_n) = \int_{-\infty}^{\infty} | k_n(u) |\, du = \| Tg_n \| \leqq \| T \|.$$

By the Helly theorem (see Widder [1, p. 29]) there exists a subsequence $\{n_j\}$ such that $\lim_{j\to\infty} K_{n_j}(t) = K(t)$ exists for each t, $-\infty < t < \infty$, $K(t)$ is of bounded variation on $(-\infty, \infty)$ and $\text{Var}\,(K) \leqq \| T \|$. Now if $f(t)$ is continuous and has a compact support, then it is clear that

$$\int_{-\infty}^{\infty} f(t-u)\, dK(u) = \lim_{j\to\infty} \int_{-\infty}^{\infty} f(t-u)\, dK_{n_j}(u)$$

for each $t \in (-\infty, \infty)$. This together with (21.2.7) shows that the relation (21.2.5) holds for such functions f. However both members in (21.2.5) define linear bounded operators on $L_1(-\infty, \infty)$ to itself and, since the set of continuous functions with compact support is dense in $L_1(-\infty, \infty)$, it follows that (21.2.5) holds for all $f \in L_1(-\infty, \infty)$. It is clear from (21.2.5) that $\| T \| \leqq \int_{-\infty}^{\infty} | dK(t) | \leqq \text{Var}\,(K)$; here we distinguish between the variation of $K(t)$ as a point function and the variation of the set function defined by $K(t)$, the latter being denoted by $\int_{-\infty}^{\infty} | dK(t) |$. Combining this with the opposite inequality obtained above, we see that $\| T \| = \int_{-\infty}^{\infty} | dK(t) |$. Finally if $g = Tf$, then by the convolution theorem

$$\begin{aligned} G(\sigma) &= (2\pi)^{-1/2} \int_{-\infty}^{\infty} e^{-i\sigma t} g(t)\, dt = (2\pi)^{-1/2} \int_{-\infty}^{\infty} e^{-i\sigma t} \int_{-\infty}^{\infty} f(t-u)\, dK(u)\, dt \\ &= \left\{ \int_{-\infty}^{\infty} e^{-i\sigma t}\, dK(t) \right\} (2\pi)^{-1/2} \int_{-\infty}^{\infty} e^{-i\sigma t} f(t)\, dt = \mu(\sigma) F(\sigma); \end{aligned}$$

so that T is a factor transformation with factor function $\mu(\sigma)$. This concludes the proof.

THEOREM 21.2.4. *A necessary and sufficient condition that $\mu(\sigma)$ be a factor function for Fourier transforms of type (L_1, L_1) is that $(2\pi)^{-1/2}\mu(\sigma)$ be the Fourier-*

Stieltjes transform of a function $K(t)$ *which is of bounded variation on* $(-\infty, \infty)$. *The norm of the corresponding operator is equal to* $\int_{-\infty}^{\infty} | dK(t) |$.

PROOF. If $\mu(\sigma)$ is a factor function for Fourier transforms of type (L_1, L_1), then by Theorem 21.2.1 $\mu(\sigma)$ defines a linear bounded operator U which commutes with translations. By Theorem 21.2.3 there exists a function $K(t)$ of bounded variation on $(-\infty, \infty)$ such that $\| U \| = \int_{-\infty}^{\infty} | dK(t) |$; further if we define

$$\mu_1(\sigma) \equiv \int_{-\infty}^{\infty} e^{-i\sigma t}\, dK(t),$$

then U is represented by the transformation $F(\sigma) \rightarrow \mu_1(\sigma)F(\sigma)$. Thus $\mu_1(\sigma)F(\sigma) = \mu(\sigma)F(\sigma)$ for all Fourier transforms $F(\sigma)$ of functions in $L_1(-\infty, \infty)$ and hence $\mu_1(\sigma) \equiv \mu(\sigma)$.

Conversely suppose that we are given a function $K(u)$ of bounded variation on $(-\infty, \infty)$ defining $\mu(\sigma)$ as above. Then

$$g(t) = [Tf](t) = \int_{-\infty}^{\infty} f(t-u)\, dK(u)$$

defines a linear bounded operator on $L_1(-\infty, \infty)$ to itself. Taking the Fourier transform of both members we see that $G(\sigma) = \mu(\sigma)F(\sigma)$ so that $\mu(\sigma)$ is a factor function of type (L_1, L_1). This completes the proof of the theorem.

Throughout the above discussion we could have insisted that $K(t)$ be normalized, that is,

$$K(t) = \tfrac{1}{2}[K(t+0) + K(t-0)].$$

Such a normalization has no effect whatever on the set function determined by $K(t)$ and utilized in (21.2.5). It might possibly reduce the variation of $K(t)$ if the value of this function at some point t_0 lay outside of the interval $[K(t_0 - 0), K(t_0 + 0)]$; however this reduction in the variation is required anyway if $\| T \|$ is to equal Var (K). Now if $K(t)$ is normalized, then it is uniquely determined by the factor function; for in this case

$$K(t) = \frac{1}{2\pi} \lim_{\beta \to \infty} \int_{-\beta}^{\beta} \frac{e^{i\sigma t} - 1}{i\sigma}\, \mu(\sigma)\, d\sigma. \tag{21.2.8}$$

For a proof of this inversion formula see S. Bochner [1, Theorem 17, p. 66].

THEOREM 21.2.5. *Let* $\mu(\sigma)$ *be a factor function for Fourier transforms of type* (L_p, L_p), *defining the operator* U. *If* $p = 1$ *the range of* U *is dense in* $L_1(-\infty, \infty)$ *if and only if* $\mu(\sigma) \neq 0$ *for every* σ; *if* $p = 2$ *the range of* U *is dense in* $L_2(-\infty, \infty)$ *if and only if* $\mu(\sigma) \neq 0$ *except for a set of measure zero*; *finally if* $1 < p < 2$ *the latter condition is necessary in order that the range of* U *be dense in* $L_p(-\infty, \infty)$.

PROOF. Suppose first that $p = 1$. If $\mu(\sigma) = 0$ for $\sigma = \tau$, then it is clear that each $g \in \mathfrak{R}[U]$ has the property that $G(\tau) = 0$. Since $x_\tau^*(g) \equiv G(\tau)$ is a non-trivial continuous functional on $L_1(-\infty, \infty)$, it follows that $\mathfrak{R}[U]$ is not dense. On the other hand if $\mu(\sigma) \neq 0$ for all σ, then the Fourier transform of $g = Uf$ will be non-vanishing for any $f \in L_1(-\infty, \infty)$ with non-vanishing Fourier transform. However the set of translates of such a g is fundamental in $L_1(-\infty, \infty)$ by Theorem 4.21.1. Since $g(\cdot + \alpha) = Uf(\cdot + \alpha) \in \mathfrak{R}[U]$ and since there exist $f \in L_1(-\infty, \infty)$ with non-vanishing Fourier transforms, this shows that $\mathfrak{R}[U]$ is dense in $L_1(-\infty, \infty)$. If $1 < p \leqq 2$ and $\mu(\sigma) = 0$ on a set F of positive finite measure, then choose $H(\sigma)$ to be the characteristic function for the set F. Then $H(\sigma) \in L_p(-\infty, \infty)$ and its inverse Fourier transform $h(t)$ belongs to $L_q(-\infty, \infty)$, $1/p + 1/q = 1$. For each $g \in \mathfrak{R}[U]$ the Parseval formula now gives

$$0 = \int_{-\infty}^{\infty} G(\sigma)\overline{H(\sigma)}\, d\sigma = \int_{-\infty}^{\infty} g(t)\overline{h(t)}\, dt \equiv x^*(g). \tag{21.2.9}$$

Again x^* is a non-trivial linear functional on $L_p(-\infty, \infty)$ and it follows that $\mathfrak{R}[U]$ can not be dense in $L_p(-\infty, \infty)$. Finally suppose $p = 2$ and that $\mathfrak{R}[U]$ is not dense in $L_2(-\infty, \infty)$. Then since $L_2(-\infty, \infty)$ is its own adjoint, there exists a function $k \in L_2(-\infty, \infty)$ such that

$$0 = \int_{-\infty}^{\infty} [Uf](t)\overline{k(t)}\, dt = \int_{-\infty}^{\infty} \mu(\sigma)F(\sigma)\overline{K(\sigma)}\, d\sigma$$

for all $f \in L_2(-\infty, \infty)$. Now $K(\sigma)$ is non-vanishing on some set F of finite positive measure. We choose f so that $|F(\sigma)| = 1$ on F, $= 0$ elsewhere, and further $\mu(\sigma)F(\sigma)\overline{K(\sigma)} \geqq 0$ for all σ. The above relation then implies that $\mu(\sigma) = 0$ for almost all $\sigma \in F$. We have therefore shown in the case $p = 2$ that $\mu(\sigma) = 0$ on a set of positive measure if and only if $\mathfrak{R}[U]$ is not dense in $L_2(-\infty, \infty)$, and this is equivalent to the assertion of the theorem for $p = 2$.

21.3. Factor sequences for Hermitian series. We now consider the factor sequence problem for the complete orthonormal system of Hermite functions

$$\mathbf{H}_n(t) = \{\pi^{1/2}2^n n!\}^{-1/2}(-1)^n e^{t^2/2} \frac{d^n}{dt^n} e^{-t^2}. \tag{21.3.1}$$

We shall make use of various properties of these functions. For fixed t we have the estimates

$$\mathbf{H}_n(t) = O(n^{-1/4}), \qquad \int_0^t \mathbf{H}_n(u)\, du = O(n^{-3/4}); \tag{21.3.2}$$

whereas

$$\max[\,|\mathbf{H}_n(t)|;\ -\infty < t < \infty] = O(n^{-1/12}), \qquad \|\mathbf{H}_n(\cdot)\|_1 = O(n^{1/4}). \tag{21.3.3}$$

Further we have the well known expression for the Abel-Hermite kernel

$$
(21.3.4)\qquad \begin{aligned} K(t, u; r) &= \sum_{k=0}^{\infty} r^k \mathbf{H}_k(t)\mathbf{H}_k(u) \\ &= [\pi(1 - r^2)]^{-1/2} \exp\left\{- \frac{(1 + r^2)(t^2 + u^2) - 4rtu}{2(1 - r^2)}\right\}. \end{aligned}
$$

Thus $K(t, u; r) > 0$ for t, u real and $-1 < r < 1$, and

$$
(21.3.5)\qquad \int_{-\infty}^{\infty} K(t, u; r)\, dt = \left(\frac{2}{1 + r^2}\right)^{1/2} \exp\left\{- \frac{1}{2}\frac{1 - r^2}{1 + r^2} u^2\right\},
$$

the right member being bounded in t and r, $-\infty < t < \infty, 0 < r < 1$.

For the properties of Hermite functions used here see E. Hille [2] and G. Szegö [2]. Szegö's formula (8.22.8) implies the first formula under (21.3.2) (which is Hille's formula (27)), and upon being integrated yields the second formula under (21.3.2). The first formula under (21.3.3) is Szegö's formula (8.91.10). The second formula under (21.3.3) is obtained by the following argument. Since the function $\mathbf{H}_n(t)$ is normalized, the integral of $|\mathbf{H}_n(t)|$ over an interval $(-\alpha, \alpha)$ cannot exceed $(2\alpha)^{1/2}$. We choose $\alpha = (2n)^{1/2}$ and obtain a contribution $\leqq 2n^{1/4}$. The contributions from the outside intervals can be estimated with the aid of (21.3.4). Putting $t = u$, $r = 1 - n^{-1/2}$, and observing that the sum of the series exceeds the nth term, we obtain an inequality for $|\mathbf{H}_n(t)|$ which when integrated from $(2n)^{1/2}$ to ∞, gives $O(n^{1/8})$. For (21.3.4) and the Abel summability of Hermitian series which is used below, see E. Hille, op. cit.

We note that Lemma 20.2.1 applies to factor sequences for Hermite functions of type (L_p, L_q), $1 \leqq p, q \leqq \infty$. The solution of the factor sequence problem for sequences of type (L_2, L_2) is trivial as usual, but we state it for the sake of completeness.

THEOREM 21.3.1. *A necessary and sufficient condition that* $\{\mu_n\}$ *be a factor sequence of type* (L_2, L_2) *for Hermitian series is that* $\sup |\mu_n| \equiv M(\mu)$ *be finite, in which case the norm of the corresponding operator is equal to* $M(\mu)$.

The case (L_1, L_1) is more interesting:

THEOREM 21.3.2. *A necessary and sufficient condition that* $\{\mu_n\}$ *be a factor sequence of type* (L_1, L_1) *for Hermitian series is that the series*

$$
(21.3.6)\qquad K(t, u; \{\mu_n\}; r) \sim \sum_{n=0}^{\infty} \mu_n r^n \mathbf{H}_n(t)\mathbf{H}_n(u)
$$

shall define a function in $L_1(-\infty, \infty)$ *in the variable* t *for each* $u \in (-\infty, \infty)$ *and* $r \in (0, 1)$ *such that*

$$
(21.3.7)\qquad \sup_u \| K(\cdot, u; \{\mu_n\}; r) \|_1 \equiv N(\mu; r) < \infty
$$

and

$$
(21.3.8)\qquad \liminf_{r\to 1-} N(\mu; r) \equiv N(\mu) < \infty.
$$

In this case the norm of the transformation defined by the factor sequence $\{\mu_n\}$ *equals* $N(\mu)$.

PROOF. We start by proving the necessity. If $\{\mu_n\}$ is a factor sequence, then the formulas

$$(21.3.9) \qquad f \sim \sum_{n=0}^{\infty} f_n \mathbf{H}_n(t), \qquad Uf \sim \sum_{n=0}^{\infty} \mu_n f_n \mathbf{H}_n(t)$$

define a bounded linear operator on $L_1(-\infty, \infty)$ to itself by Lemma 20.2.1. In particular this implies that $|\mu_n| \leqq \|U\|$ for all n and this together with the first formula under (21.3.3) shows that the series (21.3.6) is absolutely convergent, uniformly with respect to $-\infty < t, u < \infty, 0 < r < 1 - \delta$. For fixed u and r, $K(\cdot, u; \mu_n; r)$ is the U-transform of $K(\cdot, u; r)$. The relation (21.3.5) now shows that $N(\mu; r) = [2/(1 + r^2)]^{1/2} \|U\|$ and hence that $N(\mu) \leqq \|U\|$. This proves the necessity and furnishes one of the inequalities needed for the last assertion of the theorem.

Suppose conversely that conditions (21.3.7) and (21.3.8) are satisfied by the sequence $\{\mu_n\}$. It then follows as in Lemma 21.2.1 that

$$[U_r f](t) \equiv \int_{-\infty}^{\infty} K(t, u, \{\mu_n\}; r) f(u)\, du, \qquad 0 < r < 1,$$

defines a linear bounded operator on $L_1(-\infty, \infty)$ to itself of norm $\|U_r\| \leqq N(\mu; r)$. It is clear from (21.3.6) and the Fubini theorem that

$$\int_{-\infty}^{\infty} (U_r f)(t) \mathbf{H}_k(t)\, dt = \int_{-\infty}^{\infty} f(u) \int_{-\infty}^{\infty} K(t, u; \{\mu_n\}; r) \mathbf{H}_k(t)\, dt\, du = \mu_k r^k f_k .$$

Let $\mathfrak{L}$ be the linear extension of the vector set $\{\mathbf{H}_k\}$ which is fundamental in $L_1(-\infty, \infty)$. If $f \in \mathfrak{L}$, say $f = \sum_{k=1}^{n} f_k \mathbf{H}_k$, then $U_r f = \sum_{k=1}^{n} \mu_k r^k f_k \mathbf{H}_k$ converges in the $L_1(-\infty, \infty)$ norm as $r \to 1-$. Now by assumption there exists a sequence $\{r_m\}$, $r_m \to 1-$, such that $\lim_{m\to\infty} \|U_{r_m}\| = N(\mu)$. Since $\mathfrak{L}$ is dense in $L_1(-\infty, \infty)$, the Banach-Steinhaus theorem implies that $\lim_{m\to\infty} U_{r_m} f \equiv Uf$ exists for all $f \in L_1(-\infty, \infty)$ and defines a linear bounded operator U with norm $\|U\| \leqq N(\mu)$. Finally

$$r_m^k \mu_k f_k = \int_{-\infty}^{\infty} [U_{r_m} f](t) \mathbf{H}_k(t)\, dt \to \int_{-\infty}^{\infty} [Uf](t) \mathbf{H}_k(t)\, dt$$

as $m \to \infty$ and this shows that $Uf \sim \sum_{k=0}^{\infty} \mu_k f_k \mathbf{H}_k(t)$. Thus the factor sequence $\{\mu_n\}$ defines a linear bounded operator, namely U on $L_1(-\infty, \infty)$ to itself. The two opposing inequalities for $\|U\|$ and $N(\mu)$ found above imply that $\|U\| = N(\mu)$. This completes the proof of the theorem.

It should be observed that the methods of W. Orlicz [1, pp. 22–23] and M. Riesz [1, p. 481] when applied to the assertion of Theorem 21.3.2 show that the conditions of this theorem are also necessary and sufficient in order that $\{\mu_n\}$

shall be a factor sequence of type (L_∞, L_∞) and are sufficient but not necessary in order that it shall be of type (L_p, L_p) for all p, $1 < p < \infty$.

Since our knowledge of Hermitian series defining functions in $L_1(-\infty, \infty)$ is very meager, it will be worth while to give a couple of special instances in which the conditions of Theorem 21.3.2 are satisfied.

COROLLARY 1. *The sequence* $\{\mu_n\}$ *is of type* (L_1, L_1) *for Hermitian series if*

$$\sum_{n=1}^{\infty} n^{1/6} |\mu_n| < \infty. \tag{21.3.10}$$

PROOF. It is clear that

$$\int_{-\infty}^{\infty} | K(t, u; \{\mu_n\}; r) | \, dt \leqq \sum_{n=0}^{\infty} |\mu_n| \, |\mathbf{H}_n(u)| \, \| \mathbf{H}_n \|_1$$

and hence the sufficiency of condition (21.3.10) follows from (21.3.3).

COROLLARY 2. *The sequence* $\{\mu_n\}$ *is of type* (L_1, L_1) *for Hermitian series if it is a moment sequence, that is, if there exists a function of bounded variation on* [0, 1] *such that*

$$\mu_n = \int_0^1 \tau^n \, d\varphi(\tau), \qquad n = 0, 1, 2, \cdots. \tag{21.3.11}$$

PROOF. This may be concluded from the fact that

$$\int_0^1 K(t, u; r\tau) \, d\varphi(\tau) \sim \sum_{n=0}^{\infty} \mu_n r^n \mathbf{H}_n(t) \mathbf{H}_n(u),$$

which together with (21.3.5) shows that $N(\mu; r) \leqq 2^{1/2} \operatorname{Var}(\varphi)$.

2. Semi-Groups of the Factor Type

21.4. Semi-groups commuting with translations. Our purpose in this section is to prove the analogues of Theorems 20.3.1 and 20.3.2 for the interval $(-\infty, \infty)$. Our results deal with the function spaces $L_p(-\infty, \infty)$, $1 \leqq p < \infty$, and are most complete for $L_1(-\infty, \infty)$ which is the most stringent case inasmuch as this case provides sufficiency conditions for all of the others.

THEOREM 21.4.1. *If* $\mathfrak{S} \equiv [T(\xi); \xi > 0]$ *is a semi-group of linear bounded operators on* $L_1(-\infty, \infty)$ *to itself, the elements of which commute with the group of real translations on* t, *then* $T(\xi)$ *defines a factor transformation on the Fourier transforms. In fact, there exists a function* $K(t; \xi)$ *of bounded variation in* t, $-\infty < t < \infty$, *for each* ξ *such that*

$$[T(\xi)f](t) = \int_{-\infty}^{\infty} f(t-u)\, d_u K(u;\xi), \tag{21.4.1}$$

the integral being taken in the Lebesgue-Stieltjes sense, $\| T(\xi) \| = \int_{-\infty}^{\infty} | d_t K(t;\xi) |$; *and, further,*

$$T(\xi)f \leftrightarrow \mu(\sigma;\xi)F(\sigma) \tag{21.4.2}$$

where

$$\mu(\sigma;\xi) = \int_{-\infty}^{\infty} e^{-i\sigma t}\, d_t K(t;\xi) \tag{21.4.3}$$

and

$$\mu(\sigma;\xi_1+\xi_2) = \mu(\sigma;\xi_1)\mu(\sigma;\xi_2), \qquad \xi_1, \xi_2 > 0. \tag{21.4.4}$$

If $T(\xi)$ *is weakly measurable, then* $T(\xi)$ *is continuous in the strong operator topology for* $\xi > 0$ *and* $\mu(\sigma;\xi)$ *is a continuous function of* ξ, $\xi > 0$, *for each* σ. *Further, there exists a closed set* F *(with complement* $\tilde{F}$*) on the* σ*-axis with* $\mu(\sigma;\xi) \equiv 0$ *for* $\sigma \in F$ *and* $\mu(\sigma;\xi) = \exp[\xi\lambda(\sigma)]$ *for* $\sigma \in \tilde{F}$; *here* $\lambda(\sigma)$ *is a continuous function of* σ *on* $\tilde{F}$.

PROOF. The first part of the theorem is a direct consequence of Theorem 21.2.3. We then note that $T(\xi_1+\xi_2)f = T(\xi_1)T(\xi_2)f$ implies that $\mu(\sigma;\xi_1+\xi_2)F(\sigma) = \mu(\sigma;\xi_1)\,\mu(\sigma;\xi_2)F(\sigma)$ for the Fourier transform of each $f \in L_1(-\infty, \infty)$, whence (21.4.4) follows.

If $T(\xi)$ is in addition weakly measurable, then we deduce from Theorem 10.2.3 and the separability of $L_1(-\infty, \infty)$ that $T(\xi)$ is also continuous in the strong operator topology for $\xi > 0$. Now for each real σ, $x_\sigma^*(f) \equiv F(\sigma)$ defines a linear bounded functional on $L_1(-\infty, \infty)$. Thus the strong continuity of $T(\xi)$ shows that $\mu(\sigma;\xi)F(\sigma) = x_\sigma^*[T(\xi)f]$ and hence $\mu(\sigma;\xi)$ itself is continuous in ξ for $\xi > 0$. Let F be the σ-set on which $\mu(\sigma;\xi_0)$ vanishes for some fixed $\xi_0 > 0$. It is clear from (21.4.4) that if $\mu(\sigma_0;\xi) = 0$ for some $\xi > 0$, then $\mu(\sigma_0;\xi) = 0$ for all $\xi > 0$; thus F does not depend on our choice of ξ_0. By (21.4.3) the function $\mu(\sigma;\xi_0)$ is continuous in σ so that F is a closed subset. Further it follows from the functional equation (21.4.4) together with continuity in ξ that $\mu(\sigma;\xi) = \exp[\xi\lambda(\sigma)]$ for each $\sigma \in \tilde{F}$. It remains to show that $\lambda(\sigma)$ is continuous on $\tilde{F}$. Given $\sigma_0 \in \tilde{F}$, there exists a compact closed interval $[\alpha, \beta] \subset \tilde{F}$ such that $\alpha < \sigma_0 < \beta$. The function $\exp[\xi\lambda(\sigma)]$ is continuous in σ for each $\xi \geqq 0$. Lemma 16.5.1 therefore applies and we see that $\lambda(\sigma)$ is continuous on $[\alpha, \beta]$ and hence at σ_0 as a function on the set $\tilde{F}$. However it should be noted that as σ approaches a finite end point of an interval of $\tilde{F}$, the real part of $\lambda(\sigma)$ tends to $-\infty$.

An alternate form of (21.4.1) which avoids the use of Lebesgue-Stieltjes integrals is given by

$$[T(\xi)f](t) = \frac{d}{dt}\int_{-\infty}^{\infty} f_1(t-u)K(u;\xi)\, du; \qquad f_1(t) = \int_0^t f(v)\, dv. \tag{21.4.5}$$

Using (21.4.2) we get the Dirichlet representation

$$(21.4.6)\quad [T(\xi)f](t) = (2\pi)^{-1} \int_{\tilde{F}} e^{i\sigma t} \exp [\xi\lambda(\sigma)] \left[\int_{-\infty}^{\infty} e^{-i\sigma u} f(u)\, du \right] d\sigma,$$

where the integral with respect to σ exists in the $(C, 1)$ sense.

THEOREM 21.4.2. *Let* $\mathfrak{S} \equiv [T(\xi); \xi > 0]$ *be a family of operators on* $L_p(-\infty, \infty)$ *to itself, p fixed with* $1 \leqq p \leqq 2$, *defined by*

$$(21.4.7)\quad T(\xi)f \leftrightarrow \begin{cases} 0, & \sigma \in F, \\ \exp [\xi\lambda(\sigma)]F(\sigma), & \sigma \in \tilde{F}, \end{cases} \qquad \xi > 0,$$

where F is a given measurable subset of the σ*-axis and* $\lambda(\sigma)$ *is a given measurable numerically-valued function on* $\tilde{F}$. *Then* $\mathfrak{S}$ *defines a semi-group of linear bounded operators, the elements of which commute with the group of real translations on t, and which is continuous in the strong operator topology for* $\xi > 0$. *If* $\mathfrak{X}_0$ *is dense in* $L_p(-\infty, \infty)$, *then* $F = \emptyset$ *for* $p = 1$ *and* meas $(F) = 0$ *for* $1 < p \leqq 2$.

If $f \in \mathfrak{D}(A_o)$, $f \leftrightarrow F(\sigma)$, *then* $F(\sigma)$ *vanishes on F for* $p = 1$ *and almost everywhere on F for* $1 < p \leqq 2$; *moreover*

$$(21.4.8)\quad A_o f \leftrightarrow \begin{cases} 0, & \sigma \in F, \\ \lambda(\sigma)F(\sigma), & \sigma \in \tilde{F}. \end{cases}$$

Finally if $\mathfrak{S}$ *is of class* (A), *then*

$$(21.4.9)\quad \mathfrak{D}(A) = [f;\ \lambda(\sigma)F(\sigma) \text{ is the Fourier transform of an element in } L_p(-\infty, \infty)],$$

and for $f \subset \mathfrak{D}(A)$ *we have* $Af \leftrightarrow \lambda(\sigma)F(\sigma)$.

PROOF. Theorem 21.2.1 shows that $T(\xi)$ is linear, bounded, and commutes with translations; moreover the relation (21.4.7) shows that the operators $[T(\xi)]$ have the semi-group property. Next suppose that $f \in L_p(-\infty, \infty)$ is such that $F(\sigma)$ has a compact support. Then

$$g(t, \xi) \equiv [T(\xi)f](t) = (2\pi)^{-1/2} \int_{\tilde{F}} e^{-i\sigma t} \exp [\xi\lambda(\sigma)]F(\sigma)\, d\sigma$$

is a Lebesgue measurable function of (t, ξ) since the integrand is measurable in (t, ξ, σ). Applying Lemma 21.2.1 we see that $T(\xi)f$ is strongly measurable. Since the set of functions having Fourier transforms with compact supports is dense in $L_p(-\infty, \infty)$, it follows at once that $T(\xi)f$ is strongly measurable for each $f \in L_p(-\infty, \infty)$. Theorem 10.2.3 then asserts that $T(\xi)$ is continuous in the strong operator topology for $\xi > 0$.

If $\mathfrak{X}_0$ is dense in $L_p(-\infty, \infty)$, then F is necessarily vacuous or essentially so. The argument here is essentially the same as that employed in the necessity part of Theorem 21.2.5. If $p = 1$ and $F \neq \emptyset$, then it is clear by (21.4.7) that each

$g \in \mathfrak{X}_0$ is such that $G(\sigma) = 0$ on F. Consequently $\mathfrak{X}_0$ can not be dense in $L_1(-\infty, \infty)$ when $F \neq \emptyset$. If $1 < p \leqq 2$ and meas $(F) \neq 0$ we let $H(\sigma)$ denote the characteristic function of F (or a subset of F of positive finite measure if meas $(F) = \infty$). Then $H(\sigma) \in L_p(-\infty, \infty)$ and its inverse Fourier transform $h(t) \in L_q(-\infty, \infty)$, $1/p + 1/q = 1$. For each $g \in \mathfrak{X}_0$ the Parseval formula gives (21.2.9) and therefore $\mathfrak{X}_0$ can not be dense in $L_p(-\infty, \infty)$ when meas $(F) \neq 0$.

Only the assertions about the infinitesimal operator now remain to be verified. Let $f \in \mathfrak{D}(A_0)$ so that $\eta^{-1}[T(\eta)f - f] \to g$ in norm. Then $\eta^{-1}[\mu(\sigma; \eta) - 1] F(\sigma) \to G(\sigma)$ pointwise for $p = 1$ and in mean of order q if $1 < p \leqq 2$. Now if $\sigma \in F$, then $| \eta^{-1}[\mu(\sigma; \eta) - 1]F(\sigma) | \to 0$ if $F(\sigma) = 0$ and $\to \infty$ if $F(\sigma) \neq 0$; whereas if $\sigma \in \tilde{F}$, then $\eta^{-1}[\mu(\sigma; \eta) - 1]F(\sigma) \to \lambda(\sigma)F(\sigma)$. Consequently $F(\sigma) = 0$ for all $\sigma \in F$ if $p = 1$ and for almost all $\sigma \in F$ if $1 < p \leqq 2$; and the Fourier transform of $A_0 f$ is given by (21.4.8). Suppose next that $\mathfrak{S}$ is of class (A) with infinitesimal generator A. In this case $\mathfrak{X}_0$ is dense in $L_p(-\infty, \infty)$ so that F no longer plays a role. The Fourier transform V is a bounded linear transformation mapping $L_1(-\infty, \infty)$ into $C_0[-\infty, \infty]$ and $L_p(-\infty, \infty)$ into $L_q(-\infty, \infty)$, $1 < p \leqq 2$. Consequently if $f \in \mathfrak{X}_0$, $\Re(\lambda) > \omega_1$, we have

$$V[R(\lambda; A)f] = \int_0^\infty e^{-\lambda\xi} V[T(\xi)f] \, d\xi \equiv (\mathrm{B}) \int_0^\infty e^{-\lambda\xi} \exp [\xi\lambda(\sigma)]F(\sigma) \, d\xi,$$

and Lemma 21.2.1 shows that $V[R(\lambda; A)f] = [\lambda - \lambda(\sigma)]^{-1}F(\sigma)$. Since $\mathfrak{X}_0$ is dense in $L_p(-\infty, \infty)$ we see that

$$R(\lambda; A)f \leftrightarrow [\lambda - \lambda(\sigma)]^{-1}F(\sigma), \qquad \Re(\lambda) > \omega_1, \tag{21.4.10}$$

for each $f \in L_p(-\infty, \infty)$. Given an $f \in \mathfrak{D}(A)$ and fixing $\lambda > \omega_1$, there exists a $g \in L_p(-\infty, \infty)$ such that $f = R(\lambda; A)g$. Thus

$$Af = \lambda R(\lambda; A)g - g \leftrightarrow \frac{\lambda}{\lambda - \lambda(\sigma)} G(\sigma) - G(\sigma) = \frac{\lambda(\sigma)G(\sigma)}{\lambda - \lambda(\sigma)} = \lambda(\sigma)F(\sigma).$$

Conversely, if f is such that $h \sim \lambda(\sigma)F(\sigma)$ belongs to $L_p(-\infty, \infty)$, then $g \equiv \lambda f - h \in L_p(-\infty, \infty)$ and

$$R(\lambda; A)g \leftrightarrow [\lambda - \lambda(\sigma)]^{-1}G(\sigma) = F(\sigma) \leftrightarrow f$$

so that $f \in \mathfrak{D}(A)$. This establishes (21.4.9) and completes the proof of the theorem.

THEOREM 21.4.3. *If* $\mathfrak{S} \equiv [T(\xi); \xi > 0]$ *is a family of operators on* $L_1(-\infty, \infty)$ *or* $L_2(-\infty, \infty)$ *to itself, defined by* (21.4.7), *if* $\| T(\xi) \| \leqq M$ *for* $0 < \xi < 1$, *and if* $F = \emptyset$ *for* $p = 1$ *or* meas $(F) = 0$ *for* $p = 2$, *then* $\mathfrak{S}$ *is of class* (C_0).

PROOF. By the previous theorem we see that $\mathfrak{S}$ is a semi-group of linear bounded operators, continuous in the strong operator topology for $\xi > 0$. By Theorem 21.2.5, $\mathfrak{X}_0$ is dense in $\mathfrak{X}$. The result now follows from Theorem 10.6.4.

Special choices of $\lambda(\sigma)$ lead to interesting semi-groups of operators. The following two examples of this play important roles in classical analysis.

EXAMPLE 1. *The Gauss-Weierstrass operator*, $1 \leqq p \leqq 2$.

$$\lambda(\sigma) = -\sigma^2, \qquad -\infty < \sigma < \infty. \tag{21.4.11}$$

In this case we get

$$\begin{aligned} K(t; \xi) &= \tfrac{1}{2}(\pi\xi)^{-1/2} \int_{-\infty}^{t} \exp(-u^2/4\xi)\, du, \\ [T(\xi)f](t) &= \tfrac{1}{2}(\pi\xi)^{-1/2} \int_{-\infty}^{\infty} f(t-u) \exp(-u^2/4\xi)\, du. \end{aligned} \tag{21.4.12}$$

A simple calculation shows that $\| T(\xi) \| \leqq \int_{-\infty}^{\infty} | d_t K(t; \xi) | = 1$ for all $\xi > 0$. Since $F = \emptyset$ it follows from Theorem 21.4.3 that $[T(\xi)]$ is of class (C_0) when $p = 1$ or 2. More generally, the singular integral type argument of Theorem 3.9.1 shows that $[T(\xi)]$ is of class (C_0) for all p, $1 \leqq p$. As a result $A_o = A$. If $1 \leqq p \leqq 2$, then Theorem 21.4.2 gives

$$\begin{aligned} Af &\leftrightarrow -\sigma^2 F(\sigma), \\ \mathfrak{D}(A) &= [f; g \leftrightarrow \sigma^2 F(\sigma) \text{ belongs to } L_p(-\infty, \infty)]. \end{aligned} \tag{21.4.13}$$

In the present example both A and $\mathfrak{D}(A)$ have simple representations in terms of the original function space. In order to obtain this characterization of A, we recall that $R(\lambda; A)f \leftrightarrow (\lambda + \sigma^2)^{-1}F(\sigma)$ by (21.4.10). Consequently for λ not real negative we have

$$[R(\lambda; A)f](t) = \tfrac{1}{2}\lambda^{-1/2} \int_{-\infty}^{\infty} \exp\{-\lambda^{1/2} | t - u |\} f(u)\, du. \tag{21.4.14}$$

Given any $g \in \mathfrak{D}(A)$ and fixed $\lambda > 0$, there exists an $f \in L_p(-\infty, \infty)$ such that $g = R(\lambda; A)f$. A direct calculation with $g(t)$ determined by (21.4.14), shows that $g(t)$ and $g'(t)$ are absolutely continuous and that $g''(t) = \lambda g(t) - f(t)$ for almost all t. In particular $g''(t) \in L_p(-\infty, \infty)$. Thus if U is defined by

$$\begin{aligned} [Ug](t) &= g''(t), \\ \mathfrak{D}(U) &= [g; g(t) \text{ and } g'(t) \text{ absolutely continuous}, g''(t) \in L_p(-\infty, \infty)], \end{aligned} \tag{21.4.15}$$

then $(\lambda I - U)R(\lambda; A)f = f$ for all $f \in L_p(-\infty, \infty)$ and hence $U \supset A$. On the other hand if $g \in \mathfrak{D}(U)$, then $f = \lambda g - g'' \in L_p(-\infty, \infty)$ and $h \equiv R(\lambda; A)f \in \mathfrak{D}(A) \subset \mathfrak{D}(U)$. Set $w = g - h$. Then by the above $(\lambda I - U)w = \theta$. However the only solution of

$$\lambda w(t) - w''(t) = 0, \qquad \lambda > 0,$$

in $\mathfrak{D}(U)$ is $w = \theta$ and therefore $g = h \in \mathfrak{D}(A)$. It follows that $A = U$. We note that (21.4.12) with $u(t, \xi) = [T(\xi)f](t)$ is the solution to the heat equation:

$$\begin{aligned} \frac{\partial u}{\partial \xi} &= \frac{\partial^2 u}{\partial t^2}, \\ \lim_{\xi \to 0+} \| u(\cdot, \xi) &- f(\cdot) \|_p = 0. \end{aligned} \tag{21.4.16}$$

EXAMPLE 2. *The Poisson integral for the half-plane,* $1 \leqq p \leqq 2$.

$$\lambda(\sigma) = -|\sigma|, \qquad -\infty < \sigma < \infty. \tag{21.4.17}$$

Here we get

$$\begin{aligned} K(t;\xi) &= \frac{1}{\pi} \arc\tan (t/\xi), \\ [T(\xi)f](t) &= \frac{\xi}{\pi} \int_{-\infty}^{\infty} \frac{f(t-u)\,du}{u^2+\xi^2}. \end{aligned} \tag{21.4.18}$$

It is easy to show that $\| T(\xi) \| \leqq \int_{-\infty}^{\infty} | d_t K(t;\xi) | = 1$ for all $\xi > 0$ and again $[T(\xi)]$ is of class (C_0). Further for $1 \leqq p \leqq 2$, Theorem 21.4.2 gives

$$\begin{aligned} Af &\leftrightarrow -|\sigma| F(\sigma), \\ \mathfrak{D}(A) &= [f;\, g \leftrightarrow |\sigma| F(\sigma) \text{ belongs to } L_p(-\infty, \infty)]. \end{aligned} \tag{21.4.19}$$

The operator A has the obvious representation in $L_p(-\infty, \infty)$ determined by the definition of the infinitesimal generator, namely,

$$[Af](t) = \underset{\eta \to 0+}{\text{l.i.m.}} \frac{1}{\pi} \int_{-\infty}^{\infty} \frac{f(t-u) - f(t)}{u^2 + \eta^2}\, du \tag{21.4.20}$$

whenever this limit exists; here l.i.m. denotes the limit in the mean of order p. Finally we note the following analogue to Theorem 20.6.2.

THEOREM 21.4.4. *Let $f(u, v)$ be the harmonic function defined in the upper half of the (u, v)-plane with the boundary values $f(u) \in L_p(-\infty, \infty)$, $1 \leqq p \leqq 2$, that is,*

$$f(u, v) = \frac{v}{\pi} \int_{-\infty}^{\infty} \frac{f(u-t)}{t^2+v^2}\, dt.$$

If

$$\liminf_{v \to 0+} \frac{1}{v} \| f(\cdot, v) - f(\cdot) \|_p = 0, \tag{21.4.21}$$

then $f(u) = 0$ for almost all $u \in (-\infty, \infty)$. On the other hand if $f(u) \in \mathfrak{D}(A)$ as defined in (21.4.19), *then*

$$\| f(\cdot, v) - f(\cdot) \|_p \leqq v \| Af \|_p. \tag{21.4.22}$$

PROOF. The assertion is an immediate consequence of the corollary of Theorem 10.7.2. For if f satisfies the condition (21.4.21) then $Af = \theta$ and we see by (21.4.19) that this requires $F(\sigma)$ to vanish for almost all σ. The uniqueness theorem for Fourier transforms implies that $f = \theta$. Formula (10.7.5) gives (21.4.22).

The connection between the Gauss-Weierstrass operator and the Poisson integral operator requires some comment. It appears from the relations (21.4.13) and (21.4.19) that the infinitesimal generator of the Poisson semi-group is in some sense equal to $-(-A)^{1/2}$, where A is the infinitesimal generator of the Gauss-Weierstrass semi-group. Further one can obtain the Poisson kernel directly from the Gauss-Weierstrass kernel as

$$\frac{1}{\pi}\frac{\xi}{t^2+\xi^2} = \int_0^\infty \tfrac{1}{2}(\pi\tau)^{-1/2}\exp(-t^2/4\tau)k(\tau;\xi)\,d\tau,$$

where

$$k(\tau;\xi) = \frac{\xi}{2}(\pi^{1/2}\tau^{3/2})^{-1}\exp(-\xi^2/4\tau).$$

In terms of the semi-group operators we have

$$S(\xi)f = \int_0^\infty T(\tau)f\,k(\tau;\xi)\,d\tau;$$

here $S(\xi)$ is the Poisson and $T(\xi)$ is the Gauss-Weierstrass semi-group operator. The above is a special instance of a general theory developed by R. S. Phillips [7] for obtaining from a given semi-group $[T(\xi)]$ with generator A, a second semi-group with generator $-(-A)^\alpha$, $0 < \alpha < 1$. See also A. V. Balakrishnan (University of Southern California thesis).

The Gauss-Weierstrass and the Poisson integral semi-groups were studied by E. Hille [4] in one of the earliest papers on semi-groups of operators. The formal semi-group character of the Gauss-Weierstrass operator had previously been emphasized by F. Tricomi [1].

21.5. Hermitian semi-groups. We conclude this chapter by considering semi-groups defined in terms of factor sequences for Hermitian series. The semi-groups defined in this manner will in general be distinct from those considered in the preceding section.

THEOREM 21.5.1. *Let $\mathfrak{N}$ be a set of distinct non-negative integers and let $\{\lambda_n;\ n \in \mathfrak{N}\}$ be a numerical sequence. Suppose $[T(\xi);\ \xi > 0]$ is defined by*

$$T(\xi)f \sim \sum_{\mathfrak{N}} e^{\xi\lambda_n}f_n\mathbf{H}_n(t), \qquad f \sim \sum_{n=0}^\infty f_n\mathbf{H}_n(t), \tag{21.5.1}$$

for each $f \in L_p(-\infty, \infty)$, p fixed and $1 \leqq p \leqq \infty$. Then $[T(\xi)]$ defines a semi-group of linear bounded operators, and if $1 \leqq p < \infty$, then $T(\xi)$ is continuous in the strong operator topology for $\xi > 0$. If $f \in \mathfrak{D}(A_0)$, then $f_n = 0$ for $n \notin \mathfrak{N}$ and

$$A_0 f \sim \sum_{\mathfrak{N}} \lambda_n f_n\mathbf{H}_n(t). \tag{21.5.2}$$

If $[T(\xi)]$ is of class (A), then $\mathfrak{N}$ consists of all non-negative integers,

$$\mathfrak{D}(A) = \left[f;\ g \sim \sum_{n=0}^\infty \lambda_n f_n\mathbf{H}_n(t) \in L_p(-\infty, \infty)\right], \tag{21.5.3}$$

and for $f \in \mathfrak{D}(A)$ we have $Af \sim \sum_{n=0}^\infty \lambda_n f_n\mathbf{H}_n(t)$.

If $p = 2$ a necessary and sufficient condition that $\{\lambda_n\}$ be a factor sequence defining the operators (21.5.1) is that $\sup\{\Re(\lambda_n);\ n \in N\} \equiv \mu < \infty$; in this case $\|T(\xi)\| = e^{\mu\xi}$. The corresponding condition for $p = 1$ is that for each $\xi > 0$ the series

$$K(t, u; \{\lambda_n\}; \xi; r) \sim \sum_{\mathfrak{N}} e^{\xi\lambda_n} r^n \mathbf{H}_n(t)\mathbf{H}_n(u) \tag{21.5.4}$$

defines a function in $L_1(-\infty, \infty)$ *in the variable* t *for each* $u \in (-\infty, \infty)$ *and* $r \in (0, 1)$ *such that*

$$\liminf_{r\to 1-} \sup_u \| K(\cdot, u; \{\lambda_n\}; \xi; r) \|_1 \equiv N(\xi) < \infty;$$

in this case $\| T(\xi) \| = N(\xi)$. *The same condition is necessary and sufficient for* $p = \infty$, *and sufficient for* $1 < p < \infty$.

Proof. It follows by Lemma 20.2.1 that $T(\xi)$ is a linear bounded operator, and the semi-group property is a direct consequence of the representation (21.5.1). It is clear that

$$\sum_{n\in\mathfrak{N},\, n\leqq m} e^{\xi\lambda_n} f_n r^n \mathbf{H}_n(t), \qquad 0 < r < 1,$$

is continuous and *a fortiori* strongly measurable in $\xi > 0$. Suppose now that $1 \leqq p < \infty$. Then as m tends to ∞, the above function converges in norm for each $\xi > 0$ to $\sum_{\mathfrak{N}} e^{\xi\lambda_n} f_n r^n \mathbf{H}_n(t)$ so that the latter function is likewise strongly measurable in ξ. This function in turn converges in norm for each $\xi > 0$ to $T(\xi)f$ as $r \to 1-$, and therefore $T(\xi)f$ is strongly measurable. Since this holds for each $f \in L_p(-\infty, \infty)$, Theorem 10.2.3 asserts that $T(\xi)$ is continuous in the strong operator topology for $\xi > 0$.

The statements about the infinitesimal generator are proved precisely as in Theorem 20.3.2 and the remainder of the theorem follows directly from Theorem 21.3.1 and 21.3.2. This concludes the proof.

Corollaries 1 and 2 to Theorem 21.3.2 concerning special factor sequences of type (L_1, L_1) apply. Thus if $\Re(\lambda_n)/\log n \to -\infty$ as $n \to \infty$, then $\{\lambda_n\}$ is an admissable sequence and $\{e^{\xi\lambda_n}\}$ is a factor sequence. Further, if $-\lambda_n$ defines a logarithmico-exponential function of n which tends to infinity with n but is $o(n)$, then $\{e^{\xi\lambda_n}\}$ is a moment sequence for each $\xi > 0$ and hence also a factor sequence. Cf. E. Hille and J. D. Tamarkin [**5**, p. 904, conditions C_1 and C_2].

It is clear that semi-groups defined by (21.5.1) may be used for solving construction problems in $L_1(-\infty, \infty)$ and $L_2(-\infty, \infty)$ analogous to those solved in the preceding chapter in $L_1(-\pi, \pi)$ and $L_2(-\pi, \pi)$. We shall content ourselves with a few remarks concerning the special case $\lambda_n = -n$ which is connected with the Abel summability of Hermitian series.

Here

$$K(t, u; \{\lambda_n\}; \xi; r) = K(t, u; e^{-\xi}r),$$

where the function $K(t, u; r)$ is given by (21.3.4). We conclude from (21.3.5) that

$$\| T(\xi) \| = \left(\frac{2}{1 + e^{-2\xi}}\right)^{1/2} < 2^{1/2}. \tag{21.5.5}$$

In this case we see that a kernel is available for the operator $T(\xi)$, namely,

$$[T(\xi)f](t) = \int_{-\infty}^{\infty} f(u)K(t, u; e^{-\xi})\, du. \tag{21.5.6}$$

One can prove directly from equation (21.5.6) that $\lim_{\xi\to 0+} T(\xi)f = f$ for each $f \in L_1(-\infty, \infty)$ or one can deduce this property from (21.5.5) and the fact that the function set $\{\mathbf{H}_n(t)\}$ is fundamental (cf. Theorem 20.3.3). It follows that $[T(\xi)]$ is of class (C_0). The approximation Theorem 10.7.2 implies that

$$\liminf_{\eta\to 0+} \eta^{-1} \| [T(\eta) - I]f \|_1 = 0$$

if and only if

$$f(t) = Ce^{-t^2/2};$$

and if $f \in \mathfrak{D}(A)$, then formula (10.3.7) gives

$$\| [T(\eta) - I]f \|_1 < 2^{-1/2}\eta \, \| f''(t) + (1 - t^2)f(t) \|_1 .$$

Here we have used the fact that

$$[Af](t) = \tfrac{1}{2}[f''(t) + (1 - t^2)f(t)], \tag{21.5.7}$$

which follows from the relations

$$Af \sim -\sum_{n=1}^{\infty} nf_n\mathbf{H}_n(t)$$

and

$$\mathbf{H}_n''(t) + (2n + 1 - t^2)\mathbf{H}_n(t) = 0.$$

This concludes our remarks on Hermitian semi-groups. It is evident that similar considerations apply to the other classical orthogonal series.

CHAPTER XXII

SEMI-GROUPS IN HILBERT SPACE

22.1. Orientation. The present chapter is concerned with semi-groups of bounded normal operators on a Hilbert space. Such a semi-group of operators can be made to generate a commutative (W*)-algebra and we have employed the representation theory for this associated (W*)-algebra as the basic tool in our development. The results generalize those parts of Chapters XX and XXI dealing with L_2 semi-groups.

We begin by treating the self-adjoint semi-groups since this case permits certain simplifications in our method. However the principal result is an integral representation theorem for semi-groups of bounded normal operators. This result contains the classical Stone theorem for groups of unitary operators as a special case. We show as another corollary that all holomorphic semi-groups of bounded normal operators are essentially of only one class, namely $H(\Phi_1, \Phi_2)$. The general theory is illustrated in a concluding section which deals with the unitary group of translations in $L_2(-\infty, \infty)$ and two related semi-groups.

The earliest and perhaps the deepest of these results is the above mentioned Stone theorem discovered in 1930 by M. H. Stone [**1, 2**]. This was generalized in 1936 by B. de Sz.-Nagy [**1**] to cover groups of bounded normal operators. In 1938, E. Hille [**5**] and B. de Sz.-Nagy [**2**] independently obtained the representation theorem for semi-groups of bounded self-adjoint operators. The (W*)-algebra approach to this subject which we have adopted is due to R. S. Phillips [**6**].

It should perhaps be noted that Hilbert space offers a special advantage over other (B)-spaces only if we deal with classes of operators which are peculiar to Hilbert space. In addition to semi-groups of bounded normal operators treated below, semi-groups of isometries can also be given an interesting characterization; see J. L. B. Cooper [**1**].

There is only one paragraph: *Normal Semi-Groups.*

References. Bochner [**3**], Cooper [**1**], Devinatz [**1**], Fuglede [**1**], Hille [**5, 6, 7, 17**], Phillips [**6**], Sierpiński [**1**], Stone [**1, 2, 3**], de Sz.-Nagy [**1, 2, 3**], Titchmarsh [**1**], Widder [**1**].

1. Normal Semi-Groups

22.2. Background material. In this section we present a brief summary of the Hilbert space theory which will be needed in the following. The essential parts of this material have already been established in this treatise. For additional information the reader is referred to monographs by M. H. Stone [3] and by B. de Sz.-Nagy [3].

A definition of complex Hilbert space $\mathfrak{H}$ will be found in section 1.12. It was shown in Theorem 2.8.1 that $\mathfrak{H}$ is its own adjoint space and, as was noted in section 4.23, the adjoint operation mapping $A \in \mathfrak{E}[\mathfrak{H}]$ into $A^* \in \mathfrak{E}[\mathfrak{H}]$ is a (*)-operation in the sense of Definition 1.15.2. We now define the following classes of linear bounded operators in terms of the adjoint operation.

(1) A is *self-adjoint* if $A = A^*$;

(2) N is *normal* if $NN^* = N^*N$;

(3) U is *unitary* if $UU^* = I = U^*U$;

(4) P is a *projection operator* if $P^2 = P$ and $P = P^*$.

The order relation $A > B$ introduced in the class of self-adjoint operators will be taken to mean that $(Ax, x) \geqq (Bx, x)$ for all $x \in \mathfrak{H}$. A subalgebra of $\mathfrak{E}[\mathfrak{H}]$ is said to be self-adjoint if it is closed with respect to the adjoint operation.

A self-adjoint subalgebra of $\mathfrak{E}[\mathfrak{H}]$ is called a (C*)-algebra if it is closed in the uniform operator topology, it is called a (W*)-algebra if it is closed in the weak operator topology. It is clear that a (W*)-algebra is also a (C*)-algebra and Theorem 4.23.2 shows that a (C*)-algebra is a (B*)-algebra. Hence if $\mathfrak{B}$ is a commutative (C*)-algebra with identity and if $\mathfrak{M} \equiv [\mathfrak{m}]$ is its maximal ideal space, then according to the Theorem 4.22.1 the correspondence $B \to B(\mathfrak{m})$ maps $\mathfrak{B}$ isomorphically and isometrically onto $C(\mathfrak{M})$. Further Theorem 4.23.3 asserts that under this mapping $B^* \to \overline{B(\mathfrak{m})}$ and $A > \Theta$ if and only if $A(\mathfrak{m}) \geqq 0$ for all $\mathfrak{m} \in \mathfrak{M}$. As a consequence the self-adjoint operators in $\mathfrak{B}$ form a lattice. If $\mathfrak{B}$ is a commutative (W*)-algebra with identity, then the self-adjoint operators in $\mathfrak{B}$ actually form a complete lattice by Theorem 4.23.4. In this case the closure of each open subset of $\mathfrak{M}$ is clopen and, more generally, to each Borel set E there corresponds a unique clopen set $\gamma(E)$ which differs from E at most on a set of the first category; moreover this correspondence is a set-algebraic homomorphism in the sense that

$$\gamma[\cup_1^\infty E_n] = \overline{\cup_1^\infty \gamma(E_n)}. \tag{22.2.1}$$

We shall denote the Borel sets of $\mathfrak{M}$ by $\mathfrak{E}$.

It readily follows from the above properties of $\mathfrak{M}$ that each bounded Borel measurable function $f(\mathfrak{m})$ differs from a unique continuous function on $\mathfrak{M}$, say $g(\mathfrak{m})$, on at most a set of the first category. To prove this we let π denote a finite subdivision of $\mathfrak{M}$ into disjoint Borel sets $E_1, E_2, \cdots, E_n$ together with a set of n points $\mathfrak{m}_i \in E_i$, and we set $p(\mathfrak{m}; E)$ equal to the characteristic function of the clopen set $\gamma(E)$, $E \in \mathfrak{E}$. If $\omega(\pi; f)$ is equal to the maximum of the oscillations of $f(\mathfrak{m})$ in each of the sets E_i associated with π, then

$$|f(\mathfrak{m}) - \sum_\pi f(\mathfrak{m}_i)p(\mathfrak{m}; E_i)| \leqq \omega(\pi; f)$$

except on a set $E(\pi; f)$ of the first category. It is clear that $\sum_\pi f(\mathfrak{m}_i)p(\mathfrak{m}; E_i)$ is a continuous function of $\mathfrak{m}$ so that

$$|\sum_{\pi_1} f(\mathfrak{m}_i)p(\mathfrak{m}; E_i) - \sum_{\pi_2} f(\mathfrak{m}_i)p(\mathfrak{m}; E_i)| \leqq \omega(\pi_1; f) + \omega(\pi_2; f)$$

except on an *open* set of the first category, and hence except on the null set. Choosing a sequence $\{\pi_n\}$ such that $\lim_{n\to\infty} \omega(\pi_n; f) = 0$, we see that

$$\lim_{n\to\infty} \sum_{\pi_n} f(\mathfrak{m}_i)p(\mathfrak{m}; E_i) \equiv g(\mathfrak{m})$$

exists, the convergence being uniform on $\mathfrak{M}$. Hence $g(\mathfrak{m})$ is continuous in $\mathfrak{m}$ and differs from $f(\mathfrak{m})$ at most on the set $\bigcup_n E(\pi_n; f)$ which is of the first category. Finally, the uniqueness of g follows from the fact that two such functions could differ from each other only on an open set of the first category, that is, on the null set. Now if the bounded Borel measurable functions $f_1(\mathfrak{m})$, $f_2(\mathfrak{m})$ map into $g_1, g_2 \in C(\mathfrak{M})$, respectively, under the above correspondence, then

$$\text{(22.2.2)} \quad \begin{aligned} &\text{(i) } 1 \to 1,\\ &\text{(ii) } \alpha_1 f_1 + \alpha_2 f_2 \to \alpha_1 g_1 + \alpha_2 g_2,\\ &\text{(iii) } f_1 f_2 \to g_1 g_2,\\ &\text{(iv) } \| g \| = \text{ess sup } | f(\mathfrak{m}) |. \end{aligned}$$

Incidentally, in establishing the correspondence, $f \to g$, we have actually given all of the steps needed to obtain a useful integral representation for g. In fact if we set $p(E) = p(\cdot; E)$ and introduce the obvious ordering in $[\pi]$, namely, $\pi_2 > \pi_1$ denotes that the π_2-subdivision is a refinement of the π_1-subdivision, then it is clear that $\lim_\pi \omega(\pi; f) = 0$ and hence that

$$\text{(22.2.3)} \qquad g = \lim_{\pi} \sum_{\pi} f(m_i)p(E_i) \equiv \int_{\mathfrak{M}} f(\mathfrak{m})\, d_{\mathfrak{m}} p(E)$$

where the last member is an abstract Lebesgue-Stieltjes integral converging in the norm topology of $C(\mathfrak{M})$.

Returning to the commutative (W*)-algebra $\mathfrak{B}$ with identity, suppose $P(E)$, $E \in \mathfrak{E}$, corresponds to the function $p(E) \in C(\mathfrak{M})$, that is $P(E)(\mathfrak{m}) = p(\mathfrak{m}; E)$. Then it is clear that $P(E)$ is a projection operator and

$$\text{(22.2.4)} \quad \begin{aligned} &\text{(a) } P(\emptyset) = \Theta,\ P(\mathfrak{M}) = I;\\ &\text{(b) If } E_1 \cap E_2 = \emptyset, \text{ then } P(E_1)P(E_2) = \Theta;\\ &\text{(c) Given } \{E_n\},\ P(\cup_n E_n) = \vee_n P(E_n). \end{aligned}$$

Thus $[P(E); E \in \mathfrak{E}]$ forms a *resolution of the identity* relative to the Borel subsets of $\mathfrak{M}$. Property (c) can be replaced by an equivalent condition, namely,

(c′) Given $\{E_n\}$, $E_i \cap E_j = \emptyset$ for $i \neq j$, then

$$P(\cup_n E_n)x = \lim_{n\to\infty} \sum_{k=1}^{n} P(E_k)x, \qquad x \in \mathfrak{H}.$$

The equivalence between (c) and (c′) is easy to establish. It is clear that (c) may be equivalently stated in terms of disjoint sets. From the representation we see that for a finite number of disjoint sets $\bigvee_1^n P(E_k) = P(\bigcup_1^n E_k) = \sum_1^n P(E_k)$. Hence it suffices to show for disjoint sets $\{E_k\}$ that $\bigvee_n P(\bigcup_1^n E_k)$ equals the strong $\lim_{n\to\infty} P(\bigcup_1^n E_k)$ and this follows directly from Lemma 4.23.1.

It is now easy to obtain an operational calculus mapping the bounded Borel measurable functions isomorphically and isometrically onto $\mathfrak{B}$; here we identify any two functions which differ only on a set of the first category. We have shown above that to each bounded Borel function $f(\mathfrak{m})$ there corresponds a unique function in $C(\mathfrak{M})$ and hence a unique operator $F \in \mathfrak{B}$ such that $F(\mathfrak{m})$ differs from $f(\mathfrak{m})$ on at most a set of the first category. The calculus is then defined by the map $f \to F$. It is clear from (22.2.3) that this mapping is realized by the representation of (22.2.3) in $\mathfrak{B}$, that is,

$$F = \int_{\mathfrak{M}} f(\mathfrak{m})\, d_{\mathfrak{m}} P(E), \tag{22.2.5}$$

and it follows from (22.2.2) that if $f_n \to F_n$, $n = 1, 2$, then

$$\begin{aligned} &\text{(i) } 1 \to I, \\ &\text{(ii) } \alpha_1 f_1 + \alpha_2 f_2 \to \alpha_1 F_1 + \alpha_2 F_2, \\ &\text{(iii) } f_1 f_2 \to F_1 F_2, \\ &\text{(iv) } \| F \| = \operatorname{ess\,sup} | f(\mathfrak{m}) | . \end{aligned} \tag{22.2.6}$$

The integral (22.2.5) converges in the strong as well as the uniform operator topology. Hence if $x \in \mathfrak{H}$, then

$$\begin{aligned} \| Fx \|^2 &= \lim_{\pi} (\sum_{\pi} f(\mathfrak{m}_i) P(E_i) x, \sum_{\pi} f(\mathfrak{m}_i) P(E_i) x) \\ &= \lim_{\pi} \sum_{\pi} | f(\mathfrak{m}_i) |^2 \, \| P(E_i) x \|^2 = \int_{\mathfrak{M}} | f(\mathfrak{m}) |^2 \, d_{\mathfrak{m}} \, \| P(E) x \|^2; \end{aligned} \tag{22.2.7}$$

here we have made use of the fact that $(P(E_i)x, P(E_j)x) = (P(E_j)P(E_i)x, x) = 0$ if $i \neq j$ and $= \| P(E_i)x \|^2$ if $i = j$. We note $\| P(E)x \|^2 = (P(E)x, x)$ is a countably additive set function on $\mathfrak{E}$ and hence that the integral on the extreme right is an ordinary Lebesgue-Stieltjes integral.

We next extend this calculus to arbitrary Borel measurable functions $f(\mathfrak{m})$, defining the operator F with domain

$$\mathfrak{D}(F) \equiv \left[x; \int_{\mathfrak{M}} | f(\mathfrak{m}) |^2 \, d_{\mathfrak{m}} \, \| P(E)x \|^2 < \infty \right] \tag{22.2.8}$$

by the abstract Lebesgue-Stieltjes integral

$$Fx = \int_{\mathfrak{M}} f(\mathfrak{m})\, d_{\mathfrak{m}}\, P(E)x, \tag{22.2.9}$$

which now converges in the norm of $\mathfrak{H}$, for each $x \in \mathfrak{D}(F)$. The convergence is readily established by means of (22.2.7). Again

$$\| Fx \|^2 = \int_{\mathfrak{M}} | f(\mathfrak{m}) |^2 \, d_{\mathfrak{m}} \| P(E)x \|^2. \tag{22.2.10}$$

The operator F is always closed and linear with domain $\mathfrak{D}(F)$ dense in $\mathfrak{H}$; it is unbounded if and only if the function $f(\mathfrak{m})$ is unbounded in an essential way. If λ is such that the ess inf $| \lambda - f(\mathfrak{m}) | \equiv d(\lambda) > 0$, then it can be shown that

$$R(\lambda; F) = \int_{\mathfrak{M}} (\lambda - f(\mathfrak{m}))^{-1} \, d_{\mathfrak{m}} P(E) \tag{22.2.11}$$

exists and by (22.2.6)

$$\| R(\lambda; F) \| = [d(\lambda)]^{-1}. \tag{22.2.12}$$

If $f(\mathfrak{m})$ is real valued then an equivalent representation for F is obtained as follows. Let $p(\mathfrak{m}; \lambda)$ be the characteristic function of the set $[\mathfrak{m}; f(\mathfrak{m}) \leqq \lambda]$ and let $P(\lambda)$ be the corresponding operator in $\mathfrak{B}$. Again $P(\lambda)$ is a projection operator. The family of operators $[P(\lambda); -\infty < \lambda < \infty]$ is called a *resolution of the identity* for F, having the properties

$$\begin{aligned} &(\alpha) \ \wedge P(\lambda) = \Theta, \ \vee P(\lambda) = I; \\ &(\beta) \ P(\lambda)P(\mu) = P(\lambda) \quad \text{for} \quad \lambda \leqq \mu; \\ &(\gamma) \ P(\lambda) = \wedge_{\mu > \lambda} P(\mu); \end{aligned} \tag{22.2.13}$$

and

$$Fx = \int_{-\infty}^{\infty} \lambda \, d_\lambda P(\lambda)x, \tag{22.2.14}$$

where

$$\mathfrak{D}(F) \equiv \left[x; \int_{-\infty}^{\infty} \lambda^2 \, d_\lambda \| P(\lambda)x \|^2 < \infty \right]. \tag{22.2.15}$$

The integral (22.2.14) converges in the norm of $\mathfrak{H}$ as a Cauchy integral. Property (γ) is equivalent with

$$(\gamma') \ P(\lambda)x = \lim_{\mu \to \lambda +} P(\mu)x, \qquad x \in \mathfrak{H}.$$

22.3. Self-adjoint semi-groups. It is convenient to begin the present discussion by treating semi-groups of bounded self-adjoint operators. The theory is somewhat simpler in this case and the results can be used to advantage in dealing with semi-groups of bounded normal operators. The following lemma is due to B. de Sz.-Nagy [2]; it should be noted that $\mathfrak{H}$ is not assumed to be separable.

LEMMA 22.3.1. *Let* $\mathfrak{S} \equiv [T(\xi); \xi > 0]$ *be a semi-group of bounded self-adjoint* *operators on $\mathfrak{H}$ to itself, weakly measurable for $\xi > 0$. Then $T(\xi)$ is continuous in the strong operator topology for $\xi > 0$.*

PROOF. Since $T(\xi)$ is self-adjoint, we have

$$(T(2\xi)x, x) = (T(\xi)T(\xi)x, x) = (T(\xi)x, T(\xi)x) = \| T(\xi)x \|^2 \geqq 0,$$

whence it follows that

$$(T(\alpha + \beta)x, x)^2 = (T(\alpha)x, T(\beta)x)^2$$
$$\leqq \| T(\alpha)x \|^2 \| T(\beta)x \|^2 = (T(2\alpha)x, x)(T(2\beta)x, x).$$

Setting

$$\varphi(\xi; x) = \log (T(\xi)x, x),$$

the above inequality becomes

$$\varphi(\tfrac{1}{2}(\xi_1 + \xi_2); x) \leqq \tfrac{1}{2}[\varphi(\xi_1 ; x) + \varphi(\xi_2 ; x)], \tag{22.3.1}$$

which expresses the fact that $\varphi(\xi; x)$ is a convex function of ξ in the sense of (7.2.2) for each $x \in \mathfrak{H}$.

Suppose now that $T(\xi)$ is a weakly measurable function of ξ, that is, $(T(\xi)x, y)$ is a measurable function of ξ for each $x, y \in \mathfrak{H}$. This evidently implies that $\varphi(\xi; x)$ is measurable in ξ for each $x \in \mathfrak{H}$. However a measurable convex function is necessarily continuous in the interior of its interval of definition (see W. Sierpiński [1]). Thus $\varphi(\xi; x)$ is a continuous function of ξ, $\xi > 0$, for each $x \in \mathfrak{H}$ and the same is true of $(T(\xi)x, x)$. Finally the relation

$$\| T(\xi_1)x - T(\xi_2)x \|^2 = \{(T(2\xi_1)x, x) - (T(\xi_1 + \xi_2)x, x)\}$$
$$+ \{(T(2\xi_2)x, x) - (T(\xi_1 + \xi_2)x, x)\}$$

shows that $T(\xi)$ is continuous in the strong operator topology for $\xi > 0$.

THEOREM 22.3.1. *Let $\mathfrak{S} \equiv [T(\xi); \xi > 0]$ be a semi-group of bounded self-adjoint operators on $\mathfrak{H}$ to itself, weakly measurable for $\xi > 0$. Then $\lim_{\xi \to \infty} \xi^{-1} \log \| T(\xi) \| \equiv \omega_0$ exists, $\omega_0 < \infty$, and $\| T(\xi) \| = \exp(\omega_0 \xi)$. Further*

$$\lim_{\xi \to 0+} T(\xi)x = Jx, \qquad x \in \mathfrak{H}, \tag{22.3.2}$$

where J is the projection operator with $\mathfrak{R}(J) = \overline{\mathfrak{X}}_0$ and $T(\xi)J = JT(\xi) = T(\xi)$ for all $\xi > 0$. Finally $T(\xi)$ has a holomorphic extension, $T(\zeta)$, having either the whole plane or the right half-plane as its maximal domain of analytic existence, and there exists a unique integral representation of $T(\zeta)$ of the form

$$T(\zeta)x = \int_{-\infty}^{\omega_0} e^{\lambda\zeta} \, dP(\lambda)x, \qquad x \in \mathfrak{H}. \tag{22.3.3}$$

Here $[P(\lambda)]$ is the resolution of the identity relative to $\overline{\mathfrak{X}}_0$ for the infinitesimal operator *of* $\mathfrak{S}$, *namely*,

$$(22.3.4)\qquad \begin{aligned} A_o x &= \int_{-\infty}^{\omega_0} \lambda \, dP(\lambda)x, \qquad\qquad x \in \mathfrak{D}(A_o), \\ \mathfrak{D}(A_o) &= \left[x; x \in \bar{\mathfrak{X}}_0, \int_{-\infty}^{\omega_0} \lambda^2 \, d \, \| P(\lambda)x \|^2 < \infty\right]. \end{aligned}$$

REMARK. It follows from the above that $\mathfrak{S}$ is of class (C_0) if and only if $\bar{\mathfrak{X}}_0 = \mathfrak{H}$. A necessary and sufficient condition for $\mathfrak{X}_0$ to be dense in $\mathfrak{H}$ is that $0 \notin P\sigma[T(\xi_0)]$ for some $\xi_0 > 0$. In fact, if $0 \notin P\sigma[T(\xi_0)]$, $\xi_0 > 0$, then $T^*(\xi_0) = T(\xi_0)$ has an inverse so that by Theorem 2.11.12 $\mathfrak{R}[T(\xi_0)]$ (and hence $\mathfrak{X}_0$) is dense in $\mathfrak{H}$. Conversely suppose $\bar{\mathfrak{X}}_0 = \mathfrak{H}$. Then $Jx = P(\omega_0)x = x$ and therefore $\| T(\xi)x \|^2 = \int_{-\infty}^{\omega_0} e^{2\lambda\xi} \, d \, \| P(\lambda)x \|^2 > 0$ for each $x \neq \theta$ and $\xi > 0$.

PROOF. Lemma 22.3.1 shows that $T(\xi)$ is continuous in the strong operator topology for $\xi > 0$. Thus $\| T(\xi) \|$ is lower semi-continuous and Theorem 7.6.1 can be applied to show that the type ω_0 exists and $\omega_0 < \infty$. This fact together with Theorem 4.12.1 implies that

$$(22.3.5)\qquad \| T(\xi) \| = \lim_{n\to\infty} \| T(n\xi) \|^{1/n} = e^{\omega_0\xi}.$$

Next let $\mathfrak{B}$ be a commutative (W*)-algebra with identity containing $\mathfrak{S}$; for instance $\mathfrak{B}$ could be the commutant of the commutant of $\mathfrak{S}$. Making use of the representation theory for such algebras, we see that $T(\xi)(\mathfrak{m})$ is real-valued,

$$T(\xi_1 + \xi_2)(\mathfrak{m}) = T(\xi_1)(\mathfrak{m})T(\xi_2)(\mathfrak{m}), \qquad \xi_1, \xi_2 > 0,$$

and

$$0 \leqq [T(\xi/2)(\mathfrak{m})]^2 = T(\xi)(\mathfrak{m}) \leqq e^{\omega_0\xi}.$$

According to Theorems 4.17.1 and 4.17.2 only one of two possibilities can occur; either $T(\xi)(\mathfrak{m}) \equiv 0$ for $\xi > 0$ or else there exists a real number $\alpha(\mathfrak{m})$ such that $T(\xi)(\mathfrak{m}) = \exp[\xi\alpha(\mathfrak{m})]$ for all $\xi > 0$. Let

$$(22.3.6)\qquad \begin{aligned} \mathfrak{W} &\equiv [\mathfrak{m}; T(\xi)(\mathfrak{m}) \neq 0 \quad \text{for} \quad \xi > 0], \\ \mathfrak{U} &\equiv [\mathfrak{m}; T(\xi)(\mathfrak{m}) \equiv 0 \quad \text{for} \quad \xi > 0]. \end{aligned}$$

Then $\mathfrak{M} = \mathfrak{W} \cup \mathfrak{U}$ and $\mathfrak{W} \cap \mathfrak{U} = \emptyset$. It is obvious that $\mathfrak{W}$ is an open subset of $\mathfrak{M}$ because $\mathfrak{W}$ is just the set on which the continuous function $T(\xi_0)(\mathfrak{m})$, $\xi_0 > 0$, is different from zero. Likewise $\alpha(\mathfrak{m}) = \log[T(1)(\mathfrak{m})]$ is clearly continuous on $\mathfrak{W}$. This being so, the set $E_\lambda \equiv [\mathfrak{m}; \alpha(\mathfrak{m}) \leqq \lambda]$ is a Borel set and $P(\lambda) \equiv P(E_\lambda)$ defines a resolution of the identity relative to the subspace $\mathfrak{R}[P(\mathfrak{W})]$, that is, $P(\lambda)$ satisfies all of the conditions (22.2.13) except that $\vee P(\lambda) = P(\mathfrak{W}) \equiv J$; the projection operator J may differ from I. It now follows from (22.2.5) that

$$(22.3.7)\qquad T(\xi)x = \int_{\mathfrak{M}} T(\xi)(\mathfrak{m}) \, dP(E)x = \int_{-\infty}^{\omega_0} e^{\lambda\xi} \, dP(\lambda)x.$$

Consequently $(e^{-\omega_0\xi}T(\xi)x, y) = \int_{-\infty}^{0} e^{\lambda\xi}\, d(P(\lambda + \omega_0)x, y)$ for each $x, y \in \mathfrak{H}$, and the numerical analogue of Theorem 6.2.3 implies that the right-continuous function of bounded variation $(P(\lambda)x, y)$ is uniquely determined on the half-line $-\infty < \lambda \leqq \omega_0$. It follows that the resolution of the identity $[P(\lambda)]$ is uniquely determined by $\mathfrak{S}$. The relation (22.3.2) is a further consequence of (22.3.7). For

$$\left\| \int_{-\infty}^{\beta} e^{\lambda\xi}\, dP(\lambda)x \right\|^2 = \int_{-\infty}^{\beta} e^{2\lambda\xi}\, d\, \| P(\lambda)x \|^2 \to 0$$

as $\beta \to -\infty$, uniformly with respect to $\xi > 0$. Therefore

$$\lim_{\xi\to 0+} T(\xi)x = \lim_{\beta\to-\infty} \lim_{\xi\to 0+} \int_{\beta}^{\omega_0} e^{\lambda\xi}\, dP(\lambda)x = Jx.$$

The relation (22.3.2) in turn implies that $\mathfrak{R}[J] \subset \bar{\mathfrak{X}}_0$. On the other hand $\gamma(\mathfrak{W}) = \overline{\mathfrak{W}}$ so that $p(\mathfrak{m}; \mathfrak{W})T(\xi)(\mathfrak{m}) = T(\xi)(\mathfrak{m})$ and hence $JT(\xi) = T(\xi)J = T(\xi)$. It follows that $\mathfrak{X}_0 \subset \mathfrak{R}[J]$ and this together with the above inclusion shows that $\mathfrak{R}[J] = \bar{\mathfrak{X}}_0$.

As noted above, the real-valued function $\alpha(\mathfrak{m})$ is continuous and bounded above by ω_0 on $\mathfrak{W}$. Suppose in addition that $\alpha(\mathfrak{m}) \geqq -K$. Then $T(1)(\mathfrak{m}) \geqq e^{-K}$ on $\mathfrak{W}$ and $= 0$ on $\mathfrak{U}$ so that $\mathfrak{W}$ is clopen. In this case there exist operators $T(\zeta)$ in $\mathfrak{B}$ which correspond to the continuous functions

$$T(\zeta)(\mathfrak{m}) = \begin{cases} \exp\,[\zeta\alpha(\mathfrak{m})], & \mathfrak{m} \in \mathfrak{W}, \\ 0, & \mathfrak{m} \in \mathfrak{U}. \end{cases} \tag{22.3.8}$$

It is clear from this representation that the operators $T(\zeta)$ form a group (with unit J) defined and holomorphic on the whole complex plane and that $T(\zeta)$ can be realized by (22.3.3). On the other hand if $\alpha(\mathfrak{m})$ is not bounded below, it is still possible to define $T(\zeta)$ for $\mathfrak{R}(\zeta) > 0$ by (22.3.8). In fact, $\alpha(\mathfrak{m})$ is continuous on $\mathfrak{W}$ and $| T(\zeta)(\mathfrak{m}) | = | T(\xi)(\mathfrak{m}) |$, $\zeta = \xi + i\eta$, goes to zero as $\mathfrak{m}$ tends to any point of $\mathfrak{U}$, so that it is again true that $T(\zeta)(\mathfrak{m}) \in C(\mathfrak{M})$ for $\mathfrak{R}(\zeta) > 0$. Making use of the realization (22.3.3), we see that

$$(T(\zeta)x, y) = \int_{-\infty}^{\omega_0} e^{\lambda\zeta}\, d(P(\lambda)x, y), \qquad x, y \in \mathfrak{H},$$

and therefore $T(\zeta)$ is holomorphic in ζ, $\mathfrak{R}(\zeta) > 0$. Any holomorphic extension of $T(\zeta)$ must satisfy (22.3.8). However if $\alpha(\mathfrak{m})$ is not bounded from below, the right member in (22.3.8) will not be bounded and hence not be continuous on $\mathfrak{M}$ for any ζ with $\mathfrak{R}(\zeta) < 0$. Therefore in this case the half-plane $\mathfrak{R}(\zeta) > 0$ is the maximal domain of analytic existence.

It remains to establish (22.3.4). Suppose first that $x \in \mathfrak{D}(A_o)$. Since

$$\mathfrak{D}(A_o) \subset \bar{\mathfrak{X}}_0 = J[\mathfrak{H}],$$

we see that $x = \int_{-\infty}^{\omega_0} dP(\lambda)x$. Hence

$$A_\eta x = \eta^{-1} \int_{-\infty}^{\omega_0} (e^{\lambda\eta} - 1)\, dP(\lambda)x,$$

and for $\mu < \omega_0$ we have

$$(22.3.9) \quad [J - P(\mu)]A_\eta x = A_\eta[J - P(\mu)]x = \eta^{-1} \int_{\mu+}^{\omega_0} (e^{\lambda\eta} - 1)\, dP(\lambda)x.$$

Taking the limit first as $\eta \to 0+$ and then as $\mu \to -\infty$, we obtain

$$A_o x = \int_{-\infty}^{\omega_0} \lambda\, dP(\lambda)x;$$

the existence of the second limit requires that $\int_{-\infty}^{\omega_0} \lambda^2\, d \,\| P(\lambda)x \|^2 < \infty$ so that x lies in the domain $\mathfrak{D} \equiv [x;\, x \in \bar{\mathfrak{X}}_0,\, \int_{-\infty}^{\omega_0} \lambda^2\, d\, \| P(\lambda)x \|^2 < \infty]$. Conversely suppose that a given $x \in \mathfrak{D}$. Then (22.3.9) implies that $[J - P(\mu)]x \in \mathfrak{D}(A_o)$ and that

$$A_o[J - P(\mu)]x = \int_{\mu+}^{\omega_0} \lambda\, dP(\lambda)x.$$

Now $\lim_{\mu\to-\infty} [J - P(\mu)]x = Jx = x$ and $\lim_{\mu\to-\infty} A_o[J - P(\mu)]x = \int_{-\infty}^{\omega_0} \lambda\, dP(\lambda)x$. Since A_o is closed by Theorem 10.5.3, it follows that $x \in \mathfrak{D}(A_o)$. The resolution of the identity $[P(\lambda)]$ is also uniquely determined by A_o. For a second such resolution of the identity, say $[P_1(\lambda)]$, for A_o can be used to define the semi-group $T_1(\xi)x = \int_{-\infty}^{\omega_0} e^{\lambda\xi}\, dP_1(\lambda)x$. Both $[T(\xi)]$ and $[T_1(\xi)]$ are semi-groups of class (C_0) on $\bar{\mathfrak{X}}_0$ and have the same infinitesimal generator, namely A_o. Theorem 12.2.1 asserts that $T(\xi) \equiv T_1(\xi)$ and the above uniqueness argument for $P(\lambda)$ shows that $P(\lambda) \equiv P_1(\lambda)$. This concludes the proof of Theorem 22.3.1.

B. de Sz.-Nagy [2] has shown that Theorem 22.3.1 remains valid if we replace the assumption of weak measurability by the assumption that $\| T(\xi) \|$ be bounded on every compact subset of (α, ∞) for some $\alpha > 0$. We have in fact

LEMMA 22.3.2. *Let $\mathfrak{S} \equiv [T(\xi);\, \xi > 0]$ be a semi-group of bounded self-adjoint operators on $\mathfrak{H}$ to itself such that $\| T(\xi) \|$ is bounded for*

$$0 < \alpha < \xi \leqq \max(\alpha + 1, 2\alpha).$$

Then $T(\xi)$ is continuous in the strong operator topology for $\xi > 0$.

PROOF. It is clear that the hypothesis implies that $\| T(\xi) \|$ is bounded on each compact subset of (α, ∞). In this case the type ω_0 exists, $\omega_0 < \infty$, as was remarked at the close of Theorem 7.6.1. The relation (22.3.5) is valid and we may proceed to apply the representation theory as in the proof of Theorem 22.3.1. Thus $T(\xi)(\mathfrak{m}) \doteq \exp[\xi\alpha(\mathfrak{m})]$, $\alpha(\mathfrak{m}) \leqq \omega_0$, or $\equiv 0$ according as $\mathfrak{m} \in \mathfrak{W}$ or $\mathfrak{m} \in \mathfrak{U}$. Omitting the trivial case $\omega_0 = -\infty$ and setting $S(\xi) \equiv e^{-\omega_0\xi}T(\xi)$, we see that $S(\xi)(\mathfrak{m})$ is continuous and non-increasing in ξ for each $\mathfrak{m} \in \mathfrak{M}$. Thus

$$\lim_{\xi\to\xi_0-} S(\xi)(\mathfrak{m}) = \bigwedge_{\xi<\xi_0} S(\xi)(\mathfrak{m}) = S(\xi_0)(\mathfrak{m}) = \bigvee_{\xi>\xi_0} S(\xi)(\mathfrak{m}) = \lim_{\xi\to\xi_0+} S(\xi)(\mathfrak{m}).$$

Consequently Lemma 4.23.1 applies to the right and left limits of $S(\xi)$ and shows that $S(\xi)$ is continuous in the strong operator topology at $\xi = \xi_0 > 0$.

Theorem 22.3.1 admits of generalizations in more than one direction. A. Devinatz [1] has obtained an integral representation of the form (22.3.3) for semi-groups of unbounded self-adjoint operators. On the other hand S. Bochner [3] has shown that the Bernstein-Hausdorff-Widder theorem (see D. V. Widder [1, Ch. IV]) holds in a real partially ordered linear system in which every monotone increasing sequence which is bounded above has a least upper bound. Thus any function $T(\xi)$ whose values lie in such a space and which satisfies the condition

$$(-1)^n \overset{n}{\underset{\eta}{\Delta}} T(\xi) \geqq \theta, \qquad n = 0, 1, 2, \cdots, \quad \eta > 0,$$

has an integral representation of the form (22.3.3).

Theorem 22.3.1 is obviously connected with Theorem 17.8.1; however it is not evident that the assumptions of the latter theorem are satisfied. In fact, $\mathfrak{S}$ is a semi-group of class (A) if and only if $\bar{\mathfrak{X}}_0 = \mathfrak{H}$ as was noted above. As for condition (1′), it is clear that $\sigma[T(\xi)] = [T(\xi)(\mathfrak{m});\ \mathfrak{m} \in \mathfrak{M}]$ so that $\sigma[T(\xi)]$ is contained in the interval $[0, e^{\omega_0\xi}]$. Thus (1′) is equivalent with $\mathfrak{S}$ being of type $\omega_0 \leqq 0$. On the other hand $R(\lambda;\ T(\xi)) \to [\lambda - T(\xi)(\mathfrak{m})]^{-1}$; thus $\| R(\lambda; T(\xi)) \| = [d(\lambda, \xi)]^{-1}$, where $d(\lambda, \xi)$ is the distance from λ to $\sigma[T(\xi)]$. Consequently the inequality (17.8.1) and hence condition (2′) is satisfied with $C(\xi) = 1 = \omega(\xi)$. It follows that Theorem 17.8.1 applies to the present situation provided $\bar{\mathfrak{X}}_0 = \mathfrak{H}$ and $\omega_0 \leqq 0$. The theorem then asserts that $T(\xi)$ is analytic on the positive real axis and admits of a holomorphic extension to the right half-plane where it can be represented by a Laplace integral of the type (17.8.5) with $\beta > 2$. However this is less precise information than we may obtain from formula (22.3.3).

A rather striking ergodic theorem holds for semi-groups of bounded self-adjoint operators. We have

THEOREM 22.3.2. *Let $\mathfrak{S} \equiv [T(\xi);\ \xi > 0]$ be a semi-group of bounded self-adjoint operators on $\mathfrak{H}$ to itself such that the type $\omega_0 \leqq 0$. Then*

$$\lim_{\xi\to\infty} T(\xi)x = [P(0) - P(0-)]x, \qquad x \in \mathfrak{H}, \tag{22.3.10}$$

where $[P(\lambda)]$ is the resolution of the identity relative to $\bar{\mathfrak{X}}_0$ for the infinitesimal operator A_0 of $\mathfrak{S}$.

PROOF. The condition, $\omega_0 \leqq 0$, shows that Lemma 22.3.2 applies and hence that $\mathfrak{S}$ is actually continuous in the strong operator topology for $\xi > 0$. We may therefore make use of Theorem 22.3.1 and the representation (22.3.3) according to which

$$T(\xi)x = \int_{-\infty}^{0} e^{\lambda\xi}\, dP(\lambda)x, \qquad x \in \mathfrak{H}.$$

Let $P = P(0) - P(0-)$; in the terminology of Theorem 22.3.1 we have $P = P[\mathfrak{m};\, \alpha(\mathfrak{m}) = 0]$. Then

$$T(\xi)x - Px = \int_{-\infty}^{0-} e^{\lambda\xi}\, dP(\lambda)x = \lim_{\beta\to 0-} \int_{-\infty}^{\beta} e^{\lambda\xi}\, dP(\lambda)x.$$

Now for $\beta < 0$,

$$\left\| \int_{-\infty}^{\beta} e^{\lambda\xi}\, dP(\lambda)x \right\|^2 = \int_{-\infty}^{\beta} e^{2\lambda\xi}\, d\, \| P(\lambda)x \|^2 \to 0$$

as $\xi \to \infty$ by the Lebesgue majorized convergence theorem. Hence

$$\limsup_{\xi\to\infty} \| T(\xi)x - Px \|^2 \leqq \int_{\beta}^{0-} e^{2\lambda\xi}\, d\, \| P(\lambda)x \|^2 \leqq \| [P(0-) - P(\beta)]x \|^2,$$

and since $\lim_{\beta\to 0-} \| [P(0-) - P(\beta)]x \| = 0$ by definition, we see that this proves (22.3.10).

REMARK. It should be noted that the corresponding theorem for groups of unitary operators is false. This may be seen by considering the group of translations on $L_2(-\infty, \infty)$,

$$\mathfrak{G}\colon\ [U(\xi)f](t) = f(t + \xi).$$

Let $f_0(t) = 1$ for $0 < t < 1$ and $= 0$ elsewhere. Then $\lim_{\xi\to\infty} U(\xi)f_0$ does not exist since

$$\| U(\xi_1)f_0 - U(\xi_2)f_0 \| = \sqrt{2}$$

whenever $| \xi_1 - \xi_2 | \geqq 1$. Nevertheless Theorems 18.7.3 and 18.7.4 together show that $\mathfrak{G}$ is strongly (C, α) ergodic at infinity for each $\alpha > 0$.

22.4. Normal semi-groups.

22.4. Normal semi-groups. We now consider the more general case of a semi-group of normal operators, first treated by B. de Sz.-Nagy [1, 3].

THEOREM 22.4.1. *Let $\mathfrak{S} \equiv [N(\xi);\, \xi > 0]$ be a semi-group of bounded normal operators on $\mathfrak{H}$ to itself, strongly measurable for $\xi > 0$. Then $\mathfrak{S}$ is of finite type ω_0,*

$$\| N(\xi) \| = e^{\omega_0\xi}, \tag{22.4.1}$$

and

$$\lim_{\xi\to 0+} N(\xi)x = Jx, \qquad x \in \mathfrak{H}, \tag{22.4.2}$$

where J is a projection operator such that $J[\mathfrak{H}] = \bar{\mathfrak{X}}_0$ and $JN(\xi) = N(\xi)J = N(\xi)$ for all $\xi > 0$. Further if $\mathfrak{S}^* \equiv [N^*(\xi);\, \xi > 0]$, then $\mathfrak{S} \cup \mathfrak{S}^*$ is an abelian set of operators.

PROOF. Since we have assumed $N(\xi)$ to be strongly measurable, it follows from Theorem 10.2.3 that $N(\xi)$ is continuous in the strong operator topology for $\xi > 0$. Thus $\lim_{\xi\to\infty} \xi^{-1} \log \| N(\xi) \| \equiv \omega_0$ exists ($\omega_0 < \infty$), and it follows from this together with Theorem 4.12.1 that

$$\| N(\xi) \| = \lim_{n\to\infty} \| N(n\xi) \|^{1/n} = e^{\omega_0 \xi}.$$

Thus except in the trivial case where $N(\xi) \equiv \Theta$, we see that ω_0 is finite.

Next we show that $N(\xi)N^*(\eta) = N^*(\eta)N(\xi)$ for all $\xi, \eta > 0$. If ξ and η are commensurate, say $\xi = k\sigma$ and $\eta = m\sigma$, then $N(\xi)$ and $N(\eta)$ are both powers of the normal operator $N(\sigma)$ so that the assertion is obviously true in this case. For arbitrary $\xi, \eta > 0$, let $\{\xi_n\}$ be a sequence of real numbers each commensurate with η and such that $\xi_n \to \xi$. Making use of the strong continuity we have

$$N(\xi)N^*(\eta)x = \lim_{n\to\infty} N(\xi_n)N^*(\eta)x = \lim_{n\to\infty} N^*(\eta)N(\xi_n)x = N^*(\eta)N(\xi)x$$

for each $x \in \mathfrak{H}$. This proves the assertion for general $\xi, \eta > 0$ and shows that $\mathfrak{S} \cup \mathfrak{S}^*$ is abelian. Actually the above assertion follows from a more general theorem proved by B. Fuglede [1] which states that a bounded operator which commutes with a normal operator N also commutes with N^*.

As a consequence of the above we see that

$$T(\xi) \equiv N^*(\xi/2)N(\xi/2)$$

defines a semi-group of bounded self-adjoint operators, continuous in the weak (and hence the strong) operator topology for $\xi > 0$. According to Theorem 22.3.1 there exists a projection operator J such that $\bigcup_{\xi>0} T(\xi)[\mathfrak{H}]$ is dense in $J[\mathfrak{H}]$ and $T(\xi)J = JT(\xi) = T(\xi)$ for all $\xi > 0$. Hence

$$\| N(\xi)x \|^2 = (N^*(\xi)N(\xi)x, x) = (JN^*(\xi)N(\xi)Jx, x) = \| N(\xi)Jx \|^2, \quad \xi > 0.$$

Thus if $Jx = \theta$, then $N(\xi)x = \theta$ and this shows that $N(\xi)[I - J] = \Theta$, that is $N(\xi) = N(\xi)J$. The same argument can be used to show that $N^*(\xi) = N^*(\xi)J$. Consequently $N(\xi) = [N^*(\xi)]^* = [N^*(\xi)J]^* = JN(\xi)$. Thus $\mathfrak{R}[J] \supset \mathfrak{X}_0$ and since $\bigcup_{\xi>0} T(\xi)[\mathfrak{H}] \subset \mathfrak{X}_0$ we see that $\mathfrak{R}[J] = \bar{\mathfrak{X}}_0$. Finally the usual argument shows that $\lim_{\xi\to 0+} N(\xi)x = x$ for each $x \in \bar{\mathfrak{X}}_0$ and therefore

$$\lim_{\xi\to 0+} N(\xi)x = \lim_{\xi\to 0+} N(\xi)Jx = Jx.$$

This concludes the proof.

We see from the above theorem that given a semi-group of bounded normal operators, the space $\mathfrak{H}$ splits into two orthogonal subspaces, $\mathfrak{H}_1 \equiv J[\mathfrak{H}] = \bar{\mathfrak{X}}_0$ and $\mathfrak{H}_2 \equiv (I - J)[\mathfrak{H}]$, such that $[N(\xi)]$ is of class (C_0) on $\mathfrak{H}_1$ and $N(\xi)$ annihilates $\mathfrak{H}_2$ for each $\xi > 0$. We may therefore assume in what follows without any essential loss of generality that $\bar{\mathfrak{X}}_0 = \mathfrak{H}$.

THEOREM 22.4.2. *Let* $\mathfrak{S} \equiv [N(\xi);\ \xi > 0]$ *be a semi-group of bounded normal operators of type* ω_0, *strongly measurable for* $\xi > 0$, *and such that* $\bar{\mathfrak{X}}_0 = \mathfrak{H}$. *Then* $\mathfrak{S}$ *is of class* (C_0) *and there exists a unique integral representation of* $\mathfrak{S}$ *of the form*

$$N(\xi)x = \iint_\Delta e^{\lambda\xi}\, d_\lambda Q(E)x, \qquad x \in \mathfrak{H}, \tag{22.4.3}$$

where $[Q(E)]$ *is the resolution of the identity relative to the Borel subsets of the half-plane* $\Delta \equiv [\lambda;\ \Re(\lambda) \leqq \omega_0]$ *for the infinitesimal generator*

$$\begin{aligned} Ax &= \iint_\Delta \lambda\, d_\lambda Q(E)x, \qquad x \in \mathfrak{D}(A), \\ \mathfrak{D}(A) &= \left[x;\ \iint_\Delta |\lambda|^2\, d_\lambda \|Q(E)x\|^2 < \infty\right]. \end{aligned} \tag{22.4.4}$$

PROOF. The previous theorem asserts that $\mathfrak{S}$ is of class (C_0) and that $\mathfrak{S} \cup \mathfrak{S}^*$ is an abelian set of operators. We now let $\mathfrak{B}$ denote $(\mathfrak{S} \cup \mathfrak{S}^*)^{cc}$. According to Theorems 1.13.1 and 1.14.1, the commutant of the commutant of an abelian set is a weakly closed commutative algebra with identity, and the spectrum of each operator B of $\mathfrak{B}$ taken relative to $\mathfrak{B}$ coincides with $\sigma(B)$. Since $\mathfrak{S} \subset \mathfrak{S} \cup \mathfrak{S}^*$, we see that $\mathfrak{S}^{cc} \subset \mathfrak{B}$. Further if we define $\mathfrak{R}$ as in (16.2.1), then Theorem 16.2.1 implies that $\mathfrak{S} \cup \mathfrak{R} \subset \mathfrak{S}^{cc}$. Hence by the remark following Theorem 16.2.2, $\mathfrak{B}$ can be substituted for the algebra associated with the semi-group $\mathfrak{S}$ throughout Chapter XVI. Finally we note that $\mathfrak{S} \cup \mathfrak{S}^*$ and consequently $\mathfrak{B} = (\mathfrak{S} \cup \mathfrak{S}^*)^{cc}$ is self-adjoint. Thus $\mathfrak{B}$ is a commutative (W*)-algebra with identity.

According to Theorem 16.3.1 the maximal ideals $\mathfrak{M}$ of $\mathfrak{B}$ split into two disjoint sets $\mathfrak{W}$ and $\mathfrak{U}$ with $\mathfrak{M} = \mathfrak{W} \cup \mathfrak{U}$, and there exists a numerically-valued function $\alpha(\mathfrak{m})$ defined and continuous on $\mathfrak{W}$ such that

$$R(\lambda; A)(\mathfrak{m}) = \begin{cases} (\lambda - \alpha(\mathfrak{m}))^{-1}, & \mathfrak{m} \in \mathfrak{W}, \\ 0, & \mathfrak{m} \in \mathfrak{U}. \end{cases} \tag{22.4.5}$$

Since $R(\lambda; A)(\mathfrak{m})$ is continuous on $\mathfrak{M}$, it is clear that $\mathfrak{W}$ is an open subset of $\mathfrak{M}$. We next show that $\overline{\mathfrak{W}} = \mathfrak{M}$. To begin with we know that $\overline{\mathfrak{W}}$ is clopen so that $F \equiv \mathfrak{M} \ominus \overline{\mathfrak{W}}$ is likewise clopen. If $F \neq \emptyset$, then there exists a non-zero projection operator P in $\mathfrak{B}$ such that $P(\mathfrak{m}) = 1$ on F and $= 0$ elsewhere. Let $a_n(E) \equiv n$ meas $[E \cap (0, 1/n)]$ on the Borel subsets E of $(0, \infty)$. In the terminology of Theorem 16.3.4, with $\varphi(\xi) = \|N(\xi)\| = e^{\omega_0\xi}$, we see that $A \leq \varphi$ and $a_n \in L(\varphi)$. Consequently $\Psi(a_n)(\mathfrak{m}) = 0$ on $\mathfrak{U}$. It follows that $[P\Psi(a_n)](\mathfrak{m}) = P(\mathfrak{m})\Psi(a_n)(\mathfrak{m}) \equiv 0$ on $\mathfrak{M}$ and therefore that $P\Psi(a_n) = \Theta$. Now

$$\Psi(a_n)x = n\int_0^{1/n} N(\xi)x\, d\xi \to x, \qquad x \in \mathfrak{H},$$

as $n \to \infty$. On the other hand

$$0 = (P\Psi(a_n)x, y) = (\Psi(a_n)x, Py) \to (x, Py), \qquad x, y \in \mathfrak{H},$$

and this implies that $P = \Theta$. Since this is inconsistent with our assumption on F, we see that $F = \emptyset$. Finally since $\mathfrak{W}$ is open and $\overline{\mathfrak{W}} = \mathfrak{M}$, it follows that $\mathfrak{M} \ominus \mathfrak{W}$ is of the first category and hence that $\gamma(\mathfrak{W}) = \mathfrak{M}$.

The mapping

$$\alpha: \mathfrak{m} \to \alpha(\mathfrak{m}),$$

which maps $\mathfrak{W}$ onto $\sigma(A)$ and hence into the half-plane $\Re(\lambda) \leqq \omega_0$, is continuous as was noted above. Thus α^{-1} sends open subsets of $\Re(\lambda) \leqq \omega_0$ into open subsets of $\mathfrak{W}$, closed subsets into the complements (relative to $\mathfrak{W}$) of open subsets in $\mathfrak{W}$, and, more generally, Borel subsets into Borel subsets of $\mathfrak{M}$. Setting $Q(E) = P(\alpha^{-1}(E))$, we see that $[Q(E)]$ defines a resolution of the identity relative to the Borel subsets of the half-plane $\Re(\lambda) \leqq \omega_0$.

It is now an easy matter to derive the integral representation (22.4.3). According to Lemma 16.3.2, $N(\xi)(\mathfrak{m}) = \exp[\xi\alpha(\mathfrak{m})]$ for each $\mathfrak{m} \in \mathfrak{W}$ so that $N(\xi)[\alpha^{-1}(\lambda)] \equiv e^{\lambda\xi}$ is a continuous function on $\sigma(A) \subset [\mathfrak{m};\, \Re(\lambda) \leqq \omega_0]$. Hence given $\epsilon > 0$ with $\xi > 0$ fixed, there exists a subdivision of $\Re(\lambda) \leqq \omega_0$ into disjoint Borel sets $(E_1, E_2, \cdots, E_n)$ and there exist numbers $(a_1, a_2, \cdots, a_n)$ such that $|e^{\lambda\xi} - a_i| \leqq \epsilon$ for all $\lambda \in E_i$, $i = 1, 2, \cdots, n$. Thus $|\exp[\xi\alpha(\mathfrak{m})] - a_i| \leqq \epsilon$ for all $\mathfrak{m} \in \alpha^{-1}(E_i)$ and hence for all $\mathfrak{m} \in \gamma[\alpha^{-1}(E_i)]$ except for a set of the first category. Since $\gamma(\cdot)$ is a set-algebraic homomorphism, the sets $\gamma[\alpha^{-1}(E_i)]$, $i = 1, 2, \cdots, n$, are disjoint and $\bigcup\gamma[\alpha^{-1}(E_i)] = \mathfrak{M}$. Therefore

$$\left| N(\xi)(\mathfrak{m}) - \sum_{i=1}^{n} a_i Q(E_i)(\mathfrak{m}) \right| \leqq \epsilon$$

except for a set of the first category in $\mathfrak{M}$. On the other hand both $N(\xi)(\mathfrak{m})$ and $\sum_{i=1}^{n} a_i Q(E_i)(\mathfrak{m})$ are continuous functions on $\mathfrak{M}$ and since they can differ by more than ϵ only on an open set of the first category, it follows that they can nowhere differ by more than ϵ. In other words

$$\left\| N(\xi) - \sum_{i=1}^{n} a_i Q(E_i) \right\| \leqq \epsilon.$$

Consequently

$$N(\xi) = \iint_{\Delta} e^{\lambda\xi}\, d_\lambda Q(E); \tag{22.4.6}$$

this is a Lebesgue-Stieltjes integral converging in the uniform operator topology.

We may obtain the resolvent of A at the point μ, $\Re(\mu) > \omega_0$, by taking the Laplace transform of $N(\xi)$. We have

$$(R(\mu; A)x, y) = \int_0^\infty e^{-\mu\xi} \left[\iint_{\Delta} e^{\lambda\xi}\, d_\lambda(Q(E)x, y) \right] d\xi, \qquad x, y \in \mathfrak{H},$$

and interchanging the order of integration we get

$$(R(\mu; A)x, y) = \iint_{\Delta} (\mu - \lambda)^{-1} \, d_\lambda(Q(E)x, y), \qquad x, y \in \mathfrak{H},$$

whence

$$R(\mu; A)x = \iint_{\Delta} (\mu - \lambda)^{-1} \, d_\lambda Q(E)x, \qquad x \in \mathfrak{H}. \tag{22.4.7}$$

Formula (22.4.4) now follows from (22.2.11); here the integral defining A is an abstract Lebesgue-Stieltjes integral which converges in the norm of $\mathfrak{H}$.

Neither $\mathfrak{S}$ nor its infinitesimal generator A permits more than one resolution of the identity. It is clear from the above development that the uniqueness of $[Q(E)]$ for A implies the uniqueness of $[Q(E)]$ for $\mathfrak{S}$. The converse is also true. For if A had a second resolution of the identity $[Q'(E)]$, then we could use $[Q'(E)]$ to define the semi-group

$$N'(\xi)x \equiv \iint_{\Delta} e^{\lambda\xi} \, d_\lambda Q'(E)x, \qquad x \in \mathfrak{H}.$$

Now $[N(\xi)]$ and $[N'(\xi)]$ are both of class (C_0) and both have A as their infinitesimal generator. Theorem 12.2.1 shows that $N(\xi) \equiv N'(\xi)$. Thus to prove uniqueness it suffices to prove that $[N(\xi)]$ determines $[Q(E)]$ uniquely. A direct calculation with (22.4.3) shows that

$$(N(\xi)x, N(\eta)y) = \iint_{\Delta} \exp(\lambda\xi + \bar{\lambda}\eta) \, d_\lambda(Q(E)x, y), \qquad x, y \in \mathfrak{H}.$$

Taking the double Laplace transform and interchanging the order of integration, we obtain for $\mu > \omega_0$

$$\int_0^\infty \int_0^\infty \exp[-\mu(\xi + \eta)] \, \xi^k \eta^m (N(\xi)x, N(\eta)y) \, d\xi \, d\eta$$

$$= k!m! \iint_{\Delta} (\mu - \lambda)^{-k-1}(\mu - \bar{\lambda})^{-m-1} \, d_\lambda(Q(E)x, y).$$

If there were two such resolutions of the identity, say $[Q(E)]$ and $[Q'(E)]$, then for each $x, y \in \mathfrak{H}$, $a(E; x, y) \equiv (Q(E)x, y) - (Q'(E)x, y)$ would be a countably additive set function on the Borel subsets of $\Re(\lambda) \geqq \omega_0$ such that

$$\iint_{\Delta} f(\lambda) \, d_\lambda a(E; x, y) = 0$$

for all $f(\lambda)$ in the Banach algebra $\mathfrak{C}$ generated by $\{(\mu - \lambda)^{-1}, (\mu - \bar{\lambda})^{-1}\}$, $\mu > \omega_0$ fixed, with norm $\| f \| = \sup |f(\lambda)|$. Since $(\mu - \lambda)^{-1}$ separates points of the extended half-plane $\Delta_1 \equiv \{\infty\} \cup \Delta$, we see by the Stone-Weierstrass theorem (Theorem 4.19.3) that $\mathfrak{C}$ plus the unit function, $f(\lambda) \equiv 1$, fills out $C(\Delta_1)$. Hence $\mathfrak{C}$ itself consists of the class of continuous functions on Δ_1 which vanish at infinity, that is $\mathfrak{C} = C_0(\Delta_1)$. Since $a(E; x, y) \in \mathfrak{C}^*$, it follows that $a(E; x, y) \equiv 0$. This implies that $Q(E) \equiv Q'(E)$ and so concludes the proof of Theorem 22.4.2.

Referring to the terminology of Theorem 16.3.4, we have

COROLLARY. *Let* $S \equiv [N(\xi); \xi > 0]$ *be a semi-group of bounded normal operators of class* (C_0) *and of type* ω_0. *Set* $\varphi(\xi) = \| N(\xi) \|$. *If* $a \in S(\varphi)$, $\Psi(a)x = \int_0^\infty N(\xi)x\, da$, *and* $\psi(a; \lambda) = \int_0^\infty e^{\lambda\xi}\, da$, $\Re(\lambda) \leqq \omega_0$, *then*

$$\Psi(a) = \iint_\Delta \psi(a; \lambda)\, d_\lambda Q(E), \tag{22.4.8}$$

where $[Q(E)]$ *is the resolution of the identity for* $\mathfrak{S}$.

PROOF. Employing the (W*)-algebra $\mathfrak{B} = (\mathfrak{S} \cup \mathfrak{S}^*)^{cc}$ as above, Theorem 16.3.4 asserts that $\Psi(a)(\mathfrak{m}) = \psi(a; \alpha(\mathfrak{m}))$ for each $\mathfrak{m} \in \mathfrak{M}$. The same argument used in the derivation of (22.4.6) will now establish (22.4.8).

The Stone theorem is obtained as a special case of the preceding theorem.

THEOREM 22.4.3. *Let* $\mathfrak{G} \equiv [U(\xi); -\infty < \xi < \infty]$ *be a group of unitary operators, strongly measurable on* $(-\infty, \infty)$. *Then* $U(\xi)$ *is continuous in the strong operator topology on* $(-\infty, \infty)$ *and there exists a unique representation for* $U(\xi)$ *of the form*

$$U(\xi)x = \int_{-\infty}^{\infty} e^{i\lambda\xi}\, d_\lambda Q(E)x, \qquad x \in \mathfrak{H}, \tag{22.4.9}$$

where $[Q(E)]$ *is the resolution of the identity for the infinitesimal generator*

$$\begin{aligned} Ax &= i\int_{-\infty}^{\infty} \lambda\, d_\lambda Q(E)x, \qquad x \in \mathfrak{D}(A), \\ \mathfrak{D}(A) &= \left[x;\ \int_{-\infty}^{\infty} |\lambda|^2\, d\, \| Q(E)x \|^2 < \infty\right]. \end{aligned} \tag{22.4.10}$$

PROOF. The strong continuity of $U(\xi)$ follows as usual from the strong measurability as in Theorem 10.2.3. The rest of the proof depends upon the (W*)-algebra $\mathfrak{B} = (\mathfrak{S} \cup \mathfrak{S}^*)^{cc}$. Since $U(\xi)$ is unitary, we have $U^*(\xi)U(\xi) = I$ so that $| U(\xi)(\mathfrak{m}) | \equiv 1$ on $\mathfrak{M}$. Consequently $\alpha(\mathfrak{m})$ is purely imaginary and the set $\alpha(\mathfrak{M}) = \sigma(A)$ reduces to a subset of the imaginary axis. Thus in the previous theorem $Q(E) \equiv Q(E \cap \Gamma)$ where Γ denotes the imaginary axis. Specialized in this way the previous theorem implies the above assertion, at least for $\xi > 0$. For $\xi < 0$ we have only to note that

$$U(\xi) = U^*(-\xi) = \left[\int_{-\infty}^{\infty} e^{-i\lambda\xi}\, d_\lambda Q(E)\right]^* = \int_{-\infty}^{\infty} e^{i\lambda\xi}\, d_\lambda Q(E),$$

and for $\xi = 0$ (22.4.9) follows from $Q(\Gamma) = I$.

The above proof is easily extended to establish the Stone theorem for any unitary representation of a locally compact abelian group; see R. S. Phillips [6].

THEOREM 22.4.4. *Let* $\mathfrak{S} \equiv [N(\xi);\ \xi > 0]$ *be a semi-group of bounded normal operators of class* (C_0). *Let* D^- *be the closed convex extension of the spectrum of the infinitesimal generator of* $\mathfrak{S}$ *and let* $F(\zeta)$ *be the function of support for* D. *Finally suppose that* $F(\zeta)$ *is finite in the sector* $V(\Phi_1, \Phi_2)$. *Then* V *has a non-vacuous interior, which contains the ray:* $\arg \zeta = 0$, *if and only if* $N(\xi)$ *has a holomorphic extension. The maximal extension,* $N(\zeta)$, *will be of class* $H(\Phi_1, \Phi_2)$, Int (V) *being the maximal domain of analytic existence for* $N(\zeta)$.

PROOF. As in the proof of Theorem 22.3.1, one could establish the existence of a holomorphic extension for $N(\xi)$ directly from the integral representation (22.4.3), making use of the fact that $Q(E) \equiv Q(E \cap \sigma(A))$. We prefer to obtain the result from Theorem 17.5.1. To this end let $d(\lambda)$ be the distance from a point outside of D^- to D^-. Then we see from (22.4.5) that

$$\| R(\lambda; A) \| \leqq [d(\lambda)]^{-1}.$$

Hence if $\Phi_1 < 0 < \Phi_2$, then Theorem 17.5.1 applies and we see that $N(\xi)$ has a holomorphic extension of class $H(\Phi_1, \Phi_2)$. Conversely suppose that $N(\xi)$ has a holomorphic extension with domain Δ. Then according to Lemma 16.3.2, $N(\zeta)(\mathfrak{m}) = \exp [\zeta\alpha(\mathfrak{m})]$ for $\mathfrak{m} \in \mathfrak{W}$ and all real $\zeta > 0$ and since both members are holomorphic in Δ we see that this relation holds for all $\zeta \in \Delta$. It is easy to see that $N(\xi)$ is continuous in the uniform operator topology for ξ sufficiently large. Thus Theorem 16.4.1 implies that $N(\zeta)(\mathfrak{m}) = 0$ for $\mathfrak{m} \in \mathfrak{U}$ and real $\zeta > 0$; as above this relation continues to hold for all $\zeta \in \Delta$. Here we have not assumed that the extended $N(\zeta)$ forms a semi-group on Δ. It now follows that

$$\log \| N(\zeta) \| = \sup [\Re[\zeta\alpha(\mathfrak{m})];\ \mathfrak{m} \in \mathfrak{W}] = F(\zeta) < \infty,$$

for each $\zeta \in \Delta$. As a consequence $\Delta \subset$ Int (V) and hence the holomorphic extension of class $H(\Phi_1, \Phi_2)$ found above is the maximal extension. This completes the proof.

22.5. The unitary group of translations on $L_2(-\infty, \infty)$. One of the simplest non-trivial examples of the foregoing theory is the unitary group of translations on $L_2(-\infty, \infty)$, namely,

$$\mathfrak{G}\colon\ [[U(\xi)f](t) = f(t + \xi);\ -\infty < \xi < \infty]. \tag{22.5.1}$$

As we shall see, this group is basic for many of the transformations of classical analysis.

In studying the group of translations on $L_2(-\infty, \infty)$ we shall have recourse to the Plancherel theorem, according to which the mapping

$$V\colon\ f(t) \rightarrow F(\sigma) = \underset{\beta\to\infty}{\text{l.i.m.}}\ (2\pi)^{-1/2} \int_{-\beta}^{\beta} e^{-i\sigma t} f(t)\, dt \tag{22.5.2}$$

is an isometric mapping of $L_2(-\infty, \infty)$ onto itself. It is better to think of the image space as $L_2[(-\infty, \infty)^\wedge]$, where $(-\infty, \infty)^\wedge$ denotes the character group to $(-\infty, \infty)$; it just happens in the case of $(-\infty, \infty)$ that the character group is isomorphic with the given group. We shall denote the elements of the character group by σ. The inverse transformation is given by

$$(22.5.3) \qquad V^{-1}: \; F(\sigma) \to f(t) = \operatorname*{l.i.m.}_{\beta \to \infty} (2\pi)^{-1/2} \int_{-\beta}^{\beta} e^{i\sigma t} F(\sigma)\, d\sigma.$$

Since V is an isometry, it follows that

$$(22.5.4) \qquad (f, g) = \int_{-\infty}^{\infty} f(t)\overline{g(t)}\, dt = \int_{-\infty}^{\infty} F(\sigma)\overline{G(\sigma)}\, d\sigma = (F, G).$$

A suitable reference for the Plancherel theorem is E. C. Titchmarsh [**1**, Chapter III].

The mapping V induces a corresponding isometric isomorphism

$$(22.5.5) \qquad B \to B^\wedge \equiv VBV^{-1}$$

of $\mathfrak{E}[L_2(-\infty, \infty)]$ onto $\mathfrak{E}[L_2[(-\infty, \infty)]^\wedge]$. It is clear that this mapping not only leaves invariant algebraic properties, but also the weak, strong, and uniform topological properties of the operator algebras. It therefore suffices to find a suitable integral representation for $\mathfrak{G}^\wedge \equiv V\mathfrak{G}V^{-1}$; the corresponding representation for $\mathfrak{G}$ is then immediately obtainable from that of $\mathfrak{G}^\wedge$.

Under the mapping V, the translation operator $U(\xi)$ maps into multiplication by $e^{i\sigma\xi}$, that is,

$$(22.5.6) \qquad [U^\wedge(\xi)F](\sigma) = e^{i\sigma\xi}F(\sigma).$$

Thus $U^\wedge(\xi)$ is a factor type operator. Likewise if $a(\xi) \in L_1(-\infty, \infty)$, then the operator

$$(22.5.7) \qquad \Psi(a)f \equiv (\mathrm{B}) \int_{-\infty}^{\infty} U(\xi)f\, a(\xi)\, d\xi$$

maps into a factor operator. In fact, it follows from Theorem 3.7.12 that

$$V[\Psi(a)f] = (\mathrm{B}) \int_{-\infty}^{\infty} V[U(\xi)f]a(\xi)\, d\xi = (\mathrm{B}) \int_{-\infty}^{\infty} e^{i\sigma\xi}F(\sigma)a(\xi)\, d\xi,$$

and hence by Lemma 21.2.1

$$(22.5.8) \qquad [\Psi^\wedge(a)F](\sigma) = \left[\int_{-\infty}^{\infty} e^{i\sigma\xi}a(\xi)\, d\xi\right] F(\sigma) \equiv \psi(i\sigma; a)F(\sigma).$$

Finally we note that any essentially bounded measurable function $\mu(\sigma)$ defines a normal factor operator on $L_2[(-\infty, \infty)^\wedge]$ whose norm is the ess. sup $|\mu(\sigma)|$. Such an operator is self-adjoint if $\mu(\sigma)$ is real-valued, unitary if $|\mu(\sigma)| \equiv 1$, and a projection operator if $\mu(\sigma)$ takes on only the values 0 and 1.

We note that the resolvent for the infinitesimal generator of $\mathfrak{G}$ is a special instance of (22.5.7) with $a(\xi) = e^{-\lambda\xi}$ for $\xi \geqq 0$ and $= 0$ for $\xi < 0$, $\Re(\lambda) > 0$. Hence by (22.5.8) we have

$$[R^{\wedge}(\lambda; A)F](\sigma) = (\lambda - i\sigma)^{-1}F(\sigma). \tag{22.5.9}$$

Our derivation of (22.5.9) applies only for $\Re(\lambda) > 0$. However, if we define an operator extension by the right member of (22.5.9) for each λ, $\Re(\lambda) \neq 0$, it is clear that we obtain a family of linear bounded operators which satisfy the first resolvent equation so that by the corollary of Theorem 5.8.3 the so-defined operator family is actually $R^{\wedge}(\lambda; A)$. Since $\| R^{\wedge}(\lambda; A) \| = \sup_\sigma | \lambda - i\sigma |^{-1} = | \Re(\lambda) |^{-1}$, we conclude that $\sigma(A)$ coincides with the imaginary axis.

Theorem 22.4.3 requires a resolution of the identity $[Q(E)]$ relative to the Borel subsets of $\sigma(A)$ which yields the integral representation (22.4.9). Let $\mu(\sigma; E)$ be the characteristic function for the Borel set E. It is clear that $\mu(\sigma; E)$ defines a factor projection operator, say $\mu(E)$, on $L_2[(-\infty, \infty)^{\wedge}]$ and it is easy to see that the family $[\mu(E)]$ satisfies the conditions (a), (b), (c) of (22.2.4) and hence is a resolution of the identity. Further one can verify directly that

$$U^{\wedge}(\xi) = \int_{-\infty}^{\infty} e^{i\lambda\xi}\, d_\lambda \mu(E),$$

where the integral is an abstract Lebesgue-Stieltjes integral converging in the uniform operator topology. Setting

$$Q(E) \equiv V^{-1}\mu(E)V,$$

we obtain the corresponding result for $U(\xi)$ itself:

$$U(\xi) = \int_{-\infty}^{\infty} e^{i\lambda\xi}\, d_\lambda Q(E).$$

Because of the uniqueness of this representation, it follows that $[Q(E)]$ is necessarily the resolution of the identity determined in Theorem 22.4.3.

Theorem 22.4.3 asserts that the infinitesimal generator of $\mathfrak{G}$ is given by

$$\begin{aligned} Af &= i\int_{-\infty}^{\infty} \lambda\, d_\lambda Q(E)f, \qquad f \in \mathfrak{D}(A), \\ \mathfrak{D}(A) &= \left[f;\ \int_{-\infty}^{\infty} \lambda^2\, d_\lambda \| Q(E)f \|^2 < \infty\right]. \end{aligned} \tag{22.5.10}$$

The corresponding representation for $A^{\wedge}$ is particularly simple:

$$\begin{aligned} [A^{\wedge}F](\sigma) &= \left[i\int_{-\infty}^{\infty} \lambda\, d_\lambda \mu(E)F\right](\sigma) = i\sigma F(\sigma), \qquad F \in \mathfrak{D}(A^{\wedge}), \\ \mathfrak{D}(A^{\wedge}) &= \left[F;\ \int_{-\infty}^{\infty} | \sigma F(\sigma) |^2\, d\sigma < \infty\right]. \end{aligned} \tag{22.5.11}$$

Actually A itself also has a simple representation, namely

$$[Af](t) = \frac{d}{dt} f(t), \qquad f \in \mathfrak{D}(A), \tag{22.5.12}$$
$$\mathfrak{D}(A) = [f; f(t) \text{ absolutely continuous and } f'(t) \in L_2(-\infty, \infty)].$$

The representation (22.5.12) can be verified directly as was done in section 19.3 or it can be established as follows by means of the resolvent operator. Applying Lemma 21.2.1 we see that

$$f(t) \equiv [R(\lambda; A)g](t) = \left[\int_0^\infty e^{-\lambda\xi} U(\xi) g \, d\xi\right](t) = e^{\lambda t} \int_t^\infty e^{-\lambda u} g(u) \, du.$$

A direct calculation shows that $f(t)$ is absolutely continuous and that $f'(t) = \lambda f(t) - g(t)$ for almost all t. Thus if the operator U is defined by (22.5.12) we see that $U \supset A$. On the other hand for any $f \in \mathfrak{D}(U)$ we may set $g = \lambda f - Uf$ and $h = R(\lambda; A)g$. Then $h \in \mathfrak{D}(A)$ and the above calculation shows that $(\lambda I - U)(f - h) = \theta$. However $\lambda \notin P\sigma(U)$ for $\lambda > 0$ so that $f = h \in \mathfrak{D}(A)$ and this shows that $U = A$.

We have previously considered the Gauss-Weierstrass semi-group in section 21.4, Example 1. As an illustration of Theorem 22.3.1 we now return to this semi-group, this time treating it within the context of the present chapter. The member operators are given by

$$[T_1(\zeta)f](t) = \tfrac{1}{2}(\pi\zeta)^{-1/2} \int_{-\infty}^{\infty} f(t-u) \exp(-u^2/4\zeta) \, du, \qquad \Re(\zeta) > 0. \tag{22.5.13}$$

Employing Lemma 21.2.1 we see that this operator can be written in terms of the group of translations $\mathfrak{G} \equiv [U(\xi)]$ as

$$T_1(\zeta)f = \tfrac{1}{2}(\pi\zeta)^{-1/2} \int_{-\infty}^{\infty} U(u) f \exp(-u^2/4\zeta) \, du, \qquad \Re(\zeta) > 0. \tag{22.5.14}$$

Next let

$$\tau_1(\sigma; \zeta) = \tfrac{1}{2}(\pi\zeta)^{-1/2} \int_{-\infty}^{\infty} e^{i\sigma u} \exp(-u^2/4\zeta) \, du = \exp(-\sigma^2\zeta).$$

Then the relation (22.5.8) shows that

$$[T_1^{\wedge}(\zeta)F](\sigma) = e^{-\sigma^2\zeta} F(\sigma), \qquad \Re(\zeta) > 0. \tag{22.5.15}$$

It is clear from (22.5.15) that $\mathfrak{S}_1 \equiv [T_1(\zeta); \Re(\zeta) > 0]$ defines a semi-group of linear bounded operators, holomorphic in ζ with the right half-plane as its maximal domain of analytic existence. Further if $\zeta = \xi$ is real, then $[T_1(\xi)]$ forms a semi-group of self-adjoint operators of class (C_0) and of type $\omega_0 = 0$.

By the corollary to Theorem 22.4.2 we see that $T_1(\zeta)$ can be represented as

$$T_1(\zeta) = \int_{-\infty}^{\infty} e^{-\lambda^2\zeta} \, d_\lambda Q(E),$$

where $[Q(E)]$ is the resolution of the identity for $\mathfrak{G}$. However this is not the same as (22.3.3). To obtain the latter representation, we set $\rho_1(\sigma; \lambda) = 1$ for $-\sigma^2 \leqq \lambda$ and $= 0$ otherwise, λ real. Then $\rho_1(\sigma; \lambda)$ defines a factor projection operator $\rho_1(\lambda)$ on $L_2[(-\infty, \infty)^\wedge]$ and the family $[\rho_1(\lambda)]$ forms a resolution of the identity in the sense of (22.2.13). It is easily verified that

$$T_1^\wedge(\zeta)F = \int_{-\infty}^0 e^{\lambda\zeta}\, d_\lambda \rho_1(\lambda)F.$$

Hence setting $P_1(\lambda) = V^{-1}\rho_1(\lambda)V$, it follows that $[P_1(\lambda)]$ is a resolution of the identity in $L_2(-\infty, \infty)$ and that

$$T_1(\zeta)f = \int_{-\infty}^0 e^{\lambda\zeta}\, d_\lambda P_1(\lambda)f. \tag{22.5.16}$$

Because of the uniqueness of the representation (22.3.3), we see that $[P_1(\lambda)]$ is the resolution of the identity determined by Theorem 22.3.1.

According to Theorem 22.3.1 the infinitesimal generator A_1 of $\mathfrak{S}_1$ is given by

$$\begin{aligned} A_1 f &= \int_{-\infty}^0 \lambda\, d_\lambda P_1(\lambda)f, \qquad f \in \mathfrak{D}(A_1), \\ \mathfrak{D}(A_1) &= \left[f;\ \int_{-\infty}^0 \lambda^2\, d_\lambda \| P(\lambda)f \|^2 < \infty\right]. \end{aligned} \tag{22.5.17}$$

The corresponding operator on $L_2[(-\infty, \infty)^\wedge]$ is

$$\begin{aligned} [A_1^\wedge F](\sigma) &= \left[\int_{-\infty}^0 \lambda\, d_\lambda \rho_1(\lambda)F\right](\sigma) = -\sigma^2 F(\sigma), \qquad F \in \mathfrak{D}(A_1^\wedge), \\ \mathfrak{D}(A_1^\wedge) &= \left[F;\ \int_{-\infty}^{\infty} | \sigma^2 F(\sigma) |^2\, d\sigma < \infty\right], \end{aligned} \tag{22.5.18}$$

which is precisely the relation (21.4.13) with $p = 2$. It is clear from (22.5.11) and (22.5.18) that $A_1 = A^2$, where A is the infinitesimal generator of $\mathfrak{G}$. It therefore follows from (22.5.12) that

$$\begin{aligned} [A_1 f](t) &= \frac{d^2}{dt^2} f(t), \qquad f \in \mathfrak{D}(A_1), \\ \mathfrak{D}(A_1) &= [f;\ f(t) \text{ and } f'(t) \text{ absolutely continuous,} \\ &\qquad \text{with } f'(t) \text{ and } f''(t) \in L_2(-\infty, \infty)]. \end{aligned} \tag{22.5.19}$$

This is essentially the same as (21.4.15).

Remark. It is possible to give a more explicit determination of $P_1(\lambda)$ by means of the Dirichlet transform. In fact

$$[\{I - P_1(\lambda)\}f](t) = (2\pi)^{-1/2} \int_{-\infty}^{\infty} e^{i\sigma t}[1 - \rho_1(\sigma; \lambda)]F(\sigma)\, d\sigma.$$

Since $e^{i\sigma t}F(\sigma)$ is the Fourier transform of $f(t + u)$ treated as a function of u and since the inverse Fourier transform of $(2\pi)^{-1/2}[1 - \rho_1(\sigma; \lambda)]$ is simply

$$\frac{1}{2\pi}\int_{-\sqrt{-\lambda}}^{\sqrt{-\lambda}} e^{i\sigma u}\, d\sigma = \frac{1}{\pi}\frac{\sin\sqrt{-\lambda}\, u}{u}, \qquad \lambda \leqq 0,$$

we have by (22.5.4)

$$[\{I - P_1(\lambda)\}f](t) = \frac{1}{\pi}\int_{-\infty}^{\infty} f(t + u)\,\frac{\sin\sqrt{-\lambda}\, u}{u}\, du \equiv [D(\lambda)f](t).$$

We may therefore rewrite (22.5.16) as

$$T_1(\zeta)f = -\int_{-\infty}^{0} e^{\lambda\zeta}\, d_\lambda D(\lambda)f.$$

The Poisson integral for the half-plane can be treated similarly and is to be compared with Example 2 of section 21.4. In this case

$$[T_2(\zeta)f](t) = \frac{\zeta}{\pi}\int_{-\infty}^{\infty}\frac{f(t-u)}{\zeta^2 + u^2}\, du, \qquad \Re(\zeta) > 0. \tag{22.5.20}$$

Proceeding as above we obtain

$$[T_2^{\wedge}(\zeta)F](\sigma) = e^{-|\sigma|\zeta}F(\sigma). \tag{22.5.21}$$

Again $T_2(\zeta)$ is clearly a semi-group of linear bounded operators, defined and holomorphic in the half-plane $\Re(\zeta) > 0$. For real $\zeta = \xi$, $T_2(\xi)$ is a semi-group of self-adjoint operators of class (C_0) and of type $\omega_0 = 0$. In order to obtain the integral representation (22.3.3) we now set $\rho_2(\sigma; \lambda) = 1$ for $-|\sigma| \leqq \lambda$ and $= 0$ elsewhere, λ real. Then $\rho_2(\sigma; \lambda)$ defines a factor projection operator $\rho_2(\lambda)$, the family $[\rho_2(\lambda)]$ is a resolution of the identity on $L_2[(-\infty, \infty)^{\wedge}]$, and

$$T_2^{\wedge}(\zeta)F = \int_{-\infty}^{0} e^{\lambda\zeta}\, d_\lambda\rho_2(\lambda)F.$$

It follows that $[P_2(\lambda) = V^{-1}\rho_2(\lambda)V]$ is a resolution of the identity on $L_2(-\infty, \infty)$ and that

$$T_2(\zeta)f = \int_{-\infty}^{0} e^{\lambda\zeta}\, d_\lambda P_2(\lambda)f; \tag{22.5.22}$$

this is the representation determined by Theorem 22.3.1. The infinitesimal generator A_2 is again given by (22.3.4) and consequently

$$[A_2^{\wedge}F](\sigma) = \left[\int_{-\infty}^{0}\lambda\, d_\lambda\rho_2(\lambda)F\right](\sigma) = -|\sigma|F(\sigma), \quad F \in \mathfrak{D}(A_2^{\wedge}), \tag{22.5.23}$$

$$\mathfrak{D}(A_2^{\wedge}) = \left[F;\ \int_{-\infty}^{\infty}|\sigma F(\sigma)|^2\, d\sigma < \infty\right],$$

which checks with (21.4.19). On comparing (22.5.11) and (22.5.23) it is seen that

$\mathfrak{D}(A_2) = \mathfrak{D}(A)$. The precise relation between A and A_2 is

$$A_2 = AC = CA, \tag{22.5.24}$$

where C is the conjugate operator (Hilbert transform):

$$[Cf](t) = \underset{\epsilon\to 0+}{\text{l.i.m.}}\ \frac{1}{\pi}\int_{\epsilon}^{1/\epsilon} [f(t-u) - f(t+u)]\frac{du}{u}. \tag{22.5.25}$$

This result becomes evident once we compute $C^{\wedge}$. By (22.5.8) we have

$$[C^{\wedge}F](\sigma) = \underset{\epsilon\to 0+}{\text{l.i.m.}}\ \frac{1}{\pi}\left[\int_{\epsilon}^{1/\epsilon} [e^{-i\sigma u} - e^{i\sigma u}]\frac{du}{u}\right]F(\sigma) = -i(\operatorname{sgn}\sigma)F(\sigma). \tag{22.5.26}$$

We see therefore that C defines a unitary operator which maps $\mathfrak{D}(A)$ onto itself. The relations (22.5.24) are now obvious from the $L_2[(-\infty, \infty)^{\wedge}]$ representations of the operators involved. E. Hille **[17]** has shown that (22.5.24) remains valid in all $L_p(-\infty, \infty)$ spaces, $1 < p < \infty$.

CHAPTER XXIII

MISCELLANEOUS APPLICATIONS

23.1. Orientation. In this chapter we shall try to bring out the role that the semi-group concept plays in analysis by discussing a number of problems, in classical and modern analysis, where semi-groups of linear operators enter in a significant manner. The nature of these problems is reflected by the titles of the six paragraphs: *Abel Summability*, *Hausdorff Matrices*, *The Abstract Cauchy Problem*, *Probability Matrices*, *Stochastic Processes in* E_1, and *Fractional Integration*. References are to be found at the end of each paragraph.

The subject matter is fairly heterogeneous but there are connections between the several paragraphs as well as with earlier chapters. Thus Abel summability of the classical orthogonal series is closely connected with the differential operators generating the series in question. The theory of fractional integration, both of the classical Riemann-Liouville type and the recent extensions introduced by Marcel Riesz and Lars Gårding, deals with semi-groups associated with a differential operator, possibly related to the semi-groups generated by the operator, and involve kernels that form semi-groups of the power type with ordinary multiplication replaced by convolution. The arithmetic of Hausdorff matrices appears to be related to that of distribution functions in E_1. Further every Hausdorff matrix is also a matrix of transition probabilities and the determination of a semi-group of transition probabilities is an instance of the Cauchy problem for the corresponding system of Kolmogoroff equations.

We regret not being able to include a discussion of the applications of semi-group theory to the integration of partial differential equations; the discussion in Chapter XX of the first edition is outmoded and to bring it up to date would involve a further enlargment of this treatise beyond all measure. We have included a paragraph of the abstract Cauchy problem which is the theoretical basis for the applications to partial differential equations as well as to other functional equations. For the actual applications to diffusion equations we refer to the papers of W. Feller [3], E. Hille [21] and K. Yosida (numerous papers starting with [4], references are to be found in [6]). For hyperbolic equations we refer to K. Yosida [5, 7] and R. S. Phillips [13]; an attack along totally different lines but involving semi-groups is furnished by the investigations of M. Riesz and his school which are discussed in brief in section 23.17.

References. Feller [3], Hille [21], Phillips [13], Yosida [4, 5, 6, 7].

1. Abel Summability

23.2. Semi-groups connected with Abel summability. We encounter one-parameter semi-groups of linear bounded transformations in connection with several definitions of summability. Here we shall concentrate our attention on the definition of Abel and its generalizations connected with the theory of Dirichlet series.

Let $\{\lambda_n\}$, $0 \leqq \lambda_1 < \lambda_n < \lambda_{n+1}$, $\lambda_n \to \infty$, be a given sequence and let m be the (B)-space of bounded sequences with $x = \{\alpha_n\}$ and $\| x \| = \sup | \alpha_n |$. We set $\gamma_1 = \alpha_1$, $\gamma_n = \alpha_n - \alpha_{n-1}$, $n > 1$, and define

$$T(\zeta)x = \{\beta_n\}, \quad \beta_n = \beta_n(\zeta) = \sum_{k=1}^{n} \gamma_k e^{-\lambda_k \zeta}. \tag{23.2.1}$$

Thus the nth component of $T(\zeta)x$ is the nth partial sum of the formal Dirichlet series

$$\sum_{k=1}^{\infty} \gamma_k e^{-\lambda_k \zeta}.$$

For $\Re(\zeta) > 0$ this definition gives a semi-group of linear bounded transformations on m to itself with a norm not exceeding

$$\xi^{-1} | \zeta | e^{-\lambda_1 \xi}, \text{ where } \zeta = \xi + i\eta.$$

An equivalent definition is obtained by introducing the triangular matrix $\mathfrak{T}(\zeta) \equiv (t_{jk}(\zeta))$

$$t_{jk}(\zeta) = e^{-\lambda_k \zeta} - e^{-\lambda_{k+1} \zeta}, \quad k < j, \quad t_{jj}(\zeta) = e^{-\lambda_j \zeta}, \quad t_{jk}(\zeta) = 0, \quad k > j, \tag{23.2.2}$$

letting $\{\beta_n\}$ be the one column matrix obtained by multiplying the column matrix x on the left by $\mathfrak{T}(\zeta)$. If $\lambda_1 = 0$ and $\zeta = \xi$ is real positive, $\mathfrak{T}(\zeta)$ is a transition matrix in the sense of section 23.12 below.

The interest of the transformation $x \to T(\zeta)x$ in the theory of summability is twofold. First, it is a *convergence preserving transformation*: if $\lim \alpha_n = \alpha$ exists, so does $\lim \beta_n(\zeta) \equiv \beta(\zeta)$ for every ζ with $\Re(\zeta) > 0$. Thus $T(\zeta)$ defines a semi-group also in the subspace c of convergent sequences. Further we have the Abelian theorem: $\lim_{\zeta \to 0} \beta(\zeta) = \alpha$ uniformly in the sector $| \arg \zeta | \leqq \frac{1}{2}\pi - \epsilon$, $\epsilon > 0$. Secondly, we have the possibility that $\lim_{\zeta \to 0} \beta(\zeta) \equiv \beta$ exists even if $\lim_{n \to \infty} \alpha_n$ does not, in which case we may define β as the generalized limit of α_n.

We shall not pursue these familiar ideas any further but restrict ourselves to a brief discussion of $T(\zeta)$ as a semi-group operator in c. The infinitesimal generator is the unbounded operator A taking x into $Ax = \{\delta_n\}$ where

$$\delta_n = -\sum_{k=1}^{n} \lambda_k \gamma_k = (\lambda_2 - \lambda_1)\alpha_1 + \cdots + (\lambda_n - \lambda_{n-1})\,\alpha_{n-1} - \lambda_n \alpha_n .$$

This transformation is also based upon a triangular matrix $\mathfrak{A}$

(23.2.3) $\mathfrak{A} = (a_{jk}), \quad a_{jk} = \lambda_{k+1} - \lambda_k, \quad k < j, \quad a_{jj} = -\lambda_j, \quad a_{jk} = 0, \quad k > j.$

If $\lambda_1 = 0$, $\mathfrak{A}$ is a Kolmogoroff matrix in the terminology of section 23.12. Here $\mathfrak{D}(A) = [\{\alpha_n\}; \{\alpha_n\} \in c, \{\delta_n\} \in c]$. For it is clear that $\mathfrak{D}(A)$ is a subset of the set just defined. Suppose on the other hand that $\{\alpha_n\}$ and $\{\delta_n\}$ are in c, and set $\delta_n = \delta - \rho_n$ where $\rho_n \to 0$. Consider

$$\frac{1}{\eta}[T(\eta) - I]\{\alpha_n\} - \{\delta_n\} \equiv \{\epsilon_n(\eta)\},$$

where

$$\epsilon_n(\eta) = \sum_{k=1}^{n} \gamma_k \lambda_k f(\eta\lambda_k), \qquad f(u) \equiv u^{-1}(e^{-u} - 1 + u).$$

Here $f(u)$ is positive and monotone increasing from 0 to 1 as u goes from 0 to $+\infty$. A summation by parts gives

$$\epsilon_n(\eta) = -\delta f(\eta\lambda_1) + \sum_{1}^{n-1} \rho_k[f(\eta\lambda_k) - f(\eta\lambda_{k+1})] + \rho_n f(\eta\lambda_n).$$

Suppose that $\epsilon > 0$ is given and that N is so large that $|\rho_n| \leqq \epsilon$ for $n \geqq N$ Using the listed properties of $f(u)$ and the inequality $|\rho_k| \leqq 2\,\|\{\delta_m\}\|$ we see that $|\epsilon_n(\eta)| \leqq 4\,\|\{\delta_m\}\|\, f(\eta\lambda_N) + \epsilon$, $n \geqq N$, and the same inequality holds for $n < N$, even without the ϵ. It follows that $\lim\sup_{\eta\to 0} \sup_n |\epsilon_n(\eta)| \leqq \epsilon$ and hence that $\sup_n |\epsilon_n(\eta)| \to 0$ with η. This proves that $\{\alpha_n\} \in \mathfrak{D}(A)$ and that $A\{\alpha_n\} = \{\delta_n\}$.

The spectral properties in c of A and of $T(\zeta)$, $\Re(\zeta) > 0$, are interesting. One verifies by direct computation that the point spectrum of A contains the set $\{-\lambda_n\}$ and that of $T(\zeta)$ contains the set $\{e^{-\lambda_n\zeta}\}$, a characteristic vector of $-\lambda_n$ and of $e^{-\lambda_n\zeta}$ being $\{0, \cdots, 0, 1, 1, \cdots\}$ or $n - 1$ zeros followed by ones. Actually this gives the complete spectrum of A so that A has a pure point spectrum whereas the spectrum of $T(\zeta)$ consists of $\{e^{-\lambda_n\zeta}\}$ plus the limit point 0, which can be shown to be in the continuous spectrum of $T(\zeta)$. To show that these point sets exhaust the spectra of the respective operators we shall exhibit the resolvent matrices. A simple computation shows the existence of unique triangular matrices

$$\mathfrak{B}(\lambda) = (\lambda I - \mathfrak{A})^{-1}, \qquad \lambda \notin \{-\lambda_n\}$$

and

$$\mathfrak{B}(\lambda, \zeta) = (\lambda I - \mathfrak{T}(\zeta))^{-1}, \qquad \lambda \notin \{e^{-\lambda_n\zeta}, 0\}.$$

The elements of $\mathfrak{B}(\lambda)$ for $k \leqq j$ are

$$b_{jk}(\lambda) = (\lambda_{k+1} - \lambda_k)(\lambda + \lambda_k)^{-1}(\lambda + \lambda_{k+1})^{-1}, \quad k < j, \quad b_{jj}(\lambda) = (\lambda + \lambda_j)^{-1},$$

while those of $\mathfrak{B}(\lambda, \zeta)$ are

$$b_{jk}(\lambda, \zeta) = (e^{-\lambda_k\zeta} - e^{-\lambda_{k+1}\zeta})\,(\lambda - e^{-\lambda_k\zeta})^{-1}\,(\lambda - e^{-\lambda_{k+1}\zeta})^{-1}, \qquad k < j,$$

$$b_{jj}(\lambda, \zeta) = (\lambda - e^{-\lambda_j\zeta})^{-1}.$$

Because of the nature of $\mathfrak{D}(A)$ and of $\mathfrak{D}[T(\zeta)] = c$, it suffices to prove that these matrices are bounded (in the normed topology of section 23.10 below) and take convergent sequences into convergent sequences. To prove this one has to verify that the three classical conditions for preservation of convergence (see F. Hausdorff [1]) are satisfied: *the matrix* $\mathfrak{B} = (b_{jk})$ *is convergence preserving if and only if*

(i) $\| \mathfrak{B} \| = \sup_j \sum_{k=1}^{\infty} | b_{jk} | < \infty$,

(ii) $\lim_{j\to\infty} \sum_{k=1}^{\infty} b_{jk}$ *exists,*

(iii) $\lim_{j\to\infty} b_{jk}$ *exists for each* k.

In the cases under consideration, conditions (ii) and (iii) are trivially satisfied since $\sum_k b_{jk}$ is independent of j and b_{jk} is independent of j as soon as $j > k$. To prove (i) for $\mathfrak{B}(\lambda)$, suppose that $\sigma \equiv \mathfrak{R}(\lambda) > -\lambda_m$. Then for $j > m$

$$\sum_{k=1}^{j} | b_{jk}(\lambda) | = \sum_{k=1}^{m-1} | b_{jk}(\lambda) | + \sum_{k=m}^{j} | b_{jk}(\lambda) |$$

$$< (\lambda_m - \lambda_1)[\delta(\lambda)]^{-2} + \sum_{k=m}^{j} b_{jk}(\sigma) = (\lambda_m - \lambda_1)[\delta(\lambda)]^{-2} + (\sigma + \lambda_m)^{-1}$$

where $\delta(\lambda) = \min | \lambda + \lambda_k |$. Similarly, if $d(\lambda) = \min [| \lambda - e^{-\lambda_k\zeta} |, | \lambda |]$, we have

$$\sum_{k=1}^{j} | b_{jk}(\lambda, \zeta) | < [d(\lambda)]^{-2} \sum_{1}^{\infty} | e^{-\lambda_k\zeta} - e^{-\lambda_{k+1}\zeta} | < \xi^{-1} | \zeta | [d(\lambda)]^{-2}.$$

It follows that both matrices are bounded and convergence preserving and consequently that the spectra of A and $T(\zeta)$ have the properties stated above. It is clear that for the admissible values of λ we have $\mathfrak{B}(\lambda)\cdot x = R(\lambda; A)[x]$ and $\mathfrak{B}(\lambda, \zeta)\cdot x = R(\lambda; T(\zeta))[x]$. Finally, a simple calculation shows the existence of a projection operator $E(\lambda)$, equivalently an idempotent triangular matrix $\mathfrak{E}(\lambda)$, in terms of which

$$Ax = -\int_0^{\infty} \lambda\, dE(\lambda)x,\ x \in \mathfrak{D}(A), \quad T(\zeta)x = \int_0^{\infty} e^{-\lambda\zeta}\, dE(\lambda)x,\ x \in c,$$

with similar formulas in the matrix notation.

For application of semi-groups of the Dirichlet's series type, in particular with $\lambda_n = n \log n$, to the problem of analytic continuation, see E. Lindelöf [1] and section 21.4 of the first edition of this treatise.

23.3. Some differential operators. The application of Abel's transformation with $\lambda_n = n$ to the classical orthogonal expansion leads to semi-groups whose infinitesimal generators are differential operators closely connected with the expansion in question. We recall that in the case of trigonometric Fourier series (see Theorem 20.6.2) we have

$$T_F(\zeta)\left\{\sum_{-\infty}^{\infty} f_n e^{nit}\right\} = \sum_{-\infty}^{\infty} f_n e^{-\zeta|n|} e^{nit}, \tag{23.3.1}$$

with the generator

$$A_F\left\{\sum_{-\infty}^{\infty} f_n e^{nit}\right\} \sim -\sum_{-\infty}^{\infty} |n| f_n e^{nit}, \tag{23.3.2}$$

or

$$A_F[f(t)] = \tilde{f}'(t). \tag{23.3.3}$$

Hermite, Hermite-Weber, and Laguerre series also lead to simple operators (cf. N. P. Romanoff [**2**, pp. 230–231]). We define

$$T_H(\zeta)[f] = \sum_{0}^{\infty} f_n e^{-n\zeta} H_n(t), \qquad f(t)e^{-t^2/2} \in L_2(-\infty, \infty); \tag{23.3.4}$$

$$T_{HW}(\zeta)[f] = \sum_{0}^{\infty} f_n e^{-n\zeta} \mathbf{H}_n(t), \qquad f(t) \in L_2(-\infty, \infty); \tag{23.3.5}$$

$$T_L(\zeta)[f] = \sum_{0}^{\infty} f_n e^{-n\zeta} L_n(t), \qquad f(t)e^{-t/2} \in L_2(0, \infty). \tag{23.3.6}$$

The restriction to the L_2-case is merely a matter of convenience. For the Hermite-Weber case (= expansion in terms of the orthonormal Hermite functions) we refer to section 21.5. These semi-group operators are all analytic in $\Re(\zeta) > 0$. The infinitesimal generator in each case multiplies the nth term of the orthogonal series of $f(t)$ by $-n$ and is essentially the differential operator whose characteristic functions are the orthogonal functions in question. Thus

$$A_H[f] = \tfrac{1}{2}f''(t) - tf'(t), \tag{23.3.7}$$

$$A_{HW}[f] = \tfrac{1}{2}[f''(t) + (1 - t^2)f(t)], \tag{23.3.8}$$

$$A_L[f] = -tf''(t) + (t - 1)f'(t). \tag{23.3.9}$$

These operators can provide illustrations of the theory in section 15.2. Let us take the second operator and consider the series

$$U[f] \sim \sum_{0}^{\infty} (n + \alpha)^{-\beta} f_n \mathbf{H}_n(t), \qquad \alpha > 0, \beta > 0,$$

which is of some importance in the transformation theory of Hermitian series. For $\Re(\lambda) < \alpha$ we have

$$\int_0^{\infty} e^{-(\alpha-\lambda)\xi}\xi^{\beta-1}\, d\xi = \Gamma(\beta)(\alpha - \lambda)^{-\beta} \tag{23.3.10}$$

with the principal determination of the power, so that

$$U[f] = [\Gamma(\beta)]^{-1} \int_0^\infty T_{HW}(\xi)[f] \quad e^{-\alpha\xi}\xi^{\beta-1}\, d\xi.$$

It is clear that $a(E) \equiv \int_E [\Gamma(\beta)]^{-1} e^{-\alpha\xi}\xi^{\beta-1}\, d\xi$ defines a set function $a \in S(1)$ and formula (23.3.10) shows that $\psi(a; \lambda) = (\alpha - \lambda)^{-\beta}$ for $\Re(\lambda) < \alpha$. Since $A \leqq \varphi$, it follows from Theorem 15.2.1 that

$$U[f] = \Psi(a; A_{HW})[f] = (\alpha I - A_{HW})^{-\beta}[f] = [R(\alpha; A_{HW})]^{\beta}[f].$$

In the case of the series of Legendre, the Abel transformation leads to the semi-group

(23.3.11) $$T_{Le}(\zeta)[f] = \sum_0^\infty f_n e^{-n\zeta} P_n(t), \qquad f(t) \in L_2(-1, +1),$$

for $\Re(\zeta) > 0$. While this is a fairly simple transformation, its infinitesimal generator is not so elementary as in the preceding cases. Keeping in mind that $\| P_n(t) \|_2 = [2/(2n + 1)]^{1/2}$, we see that

(23.3.12) $$A_{Le}[f] \sim -\sum_1^\infty n f_n P_n(t), \qquad \sum_1^\infty n\, |f_n|^2 < \infty.$$

Since we are dealing with the space $L_2(-1, +1)$ we can also characterize the domain of A_{Le} by function theoretical properties. To this end we observe (cf. G. Szegö [2, p. 71]) that

$$P_n'(t) = \tfrac{1}{2}(n + 1)P_{n-1}^{(1,1)}(t)$$

where the Jacobi polynomials $P_n^{(1,1)}(t)$ are orthogonal with respect to the weight function $(1 - t^2)$. This implies that the derived Legendre series of $f(t)$ may be written

$$\sum_1^\infty \tfrac{1}{2}(n + 1) f_n P_{n-1}^{(1,1)}(t)$$

and if $f(t) \in \mathfrak{D}(A_{Le})$ this series becomes the Jacobi orthogonal series of a function $g(t)$ with $(1 - t^2)^{1/2} g(t) \in L_2(-1, 1)$ since $\sum n\, |f_n|^2$ converges. The mean convergence of the series then implies that it may be integrated termwise and that $\int_a^b g(t)\, dt = f(b) - f(a)$, $-1 < a < b < 1$. It follows that

(23.3.13) $$\mathfrak{D}(A_{Le}) = [f(t);\ f(t) \text{ absolutely continuous}, \ (1 - t^2)^{1/2} f'(t) \in L_2(-1, +1)].$$

Explicit expressions for A_{Le} may be found. We start with the Legendre operator

(23.3.14) $$L[f] = (1 - t^2) f''(t) - 2t f'(t)$$

with the basic property

$$L[P_n(t)] = -n(n + 1)P_n(t).$$

Here $L - \frac{1}{4}I$ is obviously the infinitesimal generator of the contraction semi-group

$$S(\zeta)[f] = \sum_0^\infty e^{-(n+1/2)^2\zeta} f_n P_n(t). \tag{23.3.15}$$

From this one concludes the relation

$$A_{Le} - \tfrac{1}{2}I = -(\tfrac{1}{4}I - L)^{1/2}, \tag{23.3.16}$$

for instance, in the sense of R. S. Phillips [7]. Cf. corresponding discussion at the end of section 21.4. Using the formula (see G. Doetsch [1, p. 50])

$$\int_0^\infty e^{-a^2\tau} k(\tau; \xi)\, d\tau = e^{-a\xi}$$

with $k(\tau; \xi)$ defined by (21.4.23), one gets the analogue of (21.4.24)

$$T_{Le}(\xi)[f] = e^{\xi/2} \int_0^\infty S(\tau)[f]\, k(\tau; \xi)\, d\tau. \tag{23.3.17}$$

A different expression for A_{Le} is

$$A_{Le} = L[\tfrac{1}{2}I + (\tfrac{1}{4}I - L)^{1/2}]^{-1} \tag{23.3.18}$$

and here the operator $[\frac{1}{2}I + (\frac{1}{4}I - L)^{1/2}]^{-1}$ is definable by the elementary operational calculus since $[\frac{1}{2} + (\frac{1}{4} - \lambda)^{1/2}]^{-1}$ as a completely monotone function admits a representation of type (15.2.2) for $\lambda < \frac{1}{4}$ by the Bernstein-Widder theorem. It is to be observed that neither of the explicit formulas for A_{Le} gives information as precise as (23.3.13) concerning the domain of the operator although the method of deriving (23.3.18) shows that $\mathfrak{D}(A_{Le}) \supset \mathfrak{D}(L)$.

The semi-groups defined above all admit representations of the form

$$T(\zeta)[f] = \int_a^b K(t, u; e^{-\zeta}) f(u)\, du \tag{23.3.19}$$

where (a, b) is the basic interval and the kernel $K(t, u; r)$ is positive when $0 < r < 1$. In the case of the Hermite-Weber, the Laguerre, and the two Legendre cases

$$\int_a^b K(t, u; r)\, du \equiv 1. \tag{23.3.20}$$

For the Hermite case see formulas (21.3.4) and (21.3.5). The expressions for the other kernels are fairly well known, except possibly in the Legendre cases. Using the Mehler-Dirichlet integral for $P_n(\cos\theta)$, $0 < \theta < \pi$, one obtains

$$\sum_0^\infty P_n(\cos\theta) P_n(\cos\varphi) e^{-(n+1/2)^2\xi} = \frac{1}{2\pi^2} \int_0^\theta \int_0^\varphi \left[\vartheta_2\left(\frac{\alpha - \beta}{2}, e^{-\xi}\right) + \vartheta_2\left(\frac{\alpha + \beta}{2}, e^{-\xi}\right)\right] [(\cos\alpha - \cos\theta)(\cos\beta - \cos\varphi)]^{-1/2}\, d\alpha\, d\beta.$$

Since the sum of the two theta functions is always positive for the values under consideration, it follows that the kernel corresponding to $S(\xi)$ is positive; formula (23.3.17) then shows that the kernel of $T_{Le}(\xi)$ has the same property.

References. Doetsch [1], Hausdorff [1], Hille [7, 13], Lindelöf [1], Phillips [7], Romanoff [2], Szegö [2].

2. Hausdorff Matrices

23.4. Hausdorff means. The well known *logarithmic* scale of Hausdorff provides good examples of multi-parameter semi-groups of bounded linear transformations. In addition, the set of Hausdorff matrices with a suitable norm forms a normed semi-group of fairly simple structure. These facts justify a brief discussion of Hausdorff means at this juncture.

Let $\mathfrak{C}$ be the matrix (c_{mn}) where

$$c_{mn} = \begin{cases} (m+1)^{-1}, & 0 \leqq n \leqq m, \\ 0, & m < n, \end{cases} \qquad m = 0, 1, 2, \cdots,$$

which corresponds to $(C, 1)$ summability. In 1917 W. A. Hurwitz and L. L. Silverman defined a class $\mathcal{H}$ of triangular matrices $\mathfrak{H} = (h_{mn})$ by the condition that $\mathfrak{H}$ should commute with $\mathfrak{C}$:

$$\mathfrak{H}\mathfrak{C} = \mathfrak{C}\mathfrak{H}.$$

They showed that the general form of such a matrix is

$$h_{mn} = \binom{m}{n} \Delta^{m-n}\mu_n = \binom{m}{n} \sum_{k=0}^{m-n} (-1)^k \binom{m-n}{k} \mu_{n+k}, \tag{23.4.1}$$

where the sequence of the diagonal elements $h_{mm} = \mu_m$ could be given arbitrarily. Thus

$$\mathfrak{H} = \mathfrak{D}\mathfrak{M}\mathfrak{D}^{-1}, \tag{23.4.2}$$

where $\mathfrak{M}$ is a diagonal matrix with the elements $\delta_{mn}\mu_m$ and $\mathfrak{D} = (d_{mn})$ is formed by binomial coefficients

$$d_{mn} = (-1)^n \binom{m}{n}.$$

$\mathfrak{D}$ is its own inverse: $\mathfrak{D}^{-1} = \mathfrak{D}$. Hurwitz and Silverman also found special conditions under which $\mathfrak{H}$ will define a regular (= limit preserving) method of summation.

The problem was attacked *ab initio* and independently but from a slightly different and more general point of view by F. Hausdorff in 1920. He also was led to formula (23.4.2), but succeeded in showing that *a necessary and sufficient condition for $\mathfrak{H}$ to be regular is the existence of a function $q(u)$ with the properties:*

(i) *$q(u)$ is of bounded variation for $0 \leqq u \leqq 1$,*

(ii) *$q(u)$ is continuous to the right at $u = 0$ and $q(0) = 0$,*

(iii) $q(1) = 1$,

(iv) $\mu_n = \int_0^1 u^n dq(u), \quad n = 0, 1, 2, \cdots$.

Without restricting the generality, we may normalize $q(u)$ in $(0, 1)$ so that

(v) $q(u) = \frac{1}{2}[q(u - 0) + q(u + 0)], \quad 0 < u < 1$.

In this connection Hausdorff also solved the *moment problem* for a finite interval; $\{\mu_n\}$ is a *moment sequence* if and only if there exists a function $q(u)$ satisfying (i) and (iv). Incidentally, any such function defines a convergence preserving method of summability; condition (ii) ensures that null sequences go into null sequences and (iii) that all limits are preserved. In the following the symbol $q(u)$ will be used for any function satisfying all five conditions while $Q(u)$ will denote a function satisfying all conditions except possibly (iii). We say that $\mathfrak{H}$ is a *Hausdorff matrix* (normalized Hausdorff matrix) if $\{\mu_n\}$ is a moment sequence defined by a function $Q(u)$ ($q(u)$ respectively).

Let $x = \{\alpha_n\}$ be any element of the space m of bounded sequences and set

$$y = \{\beta_m\} = \mathfrak{H}x,$$

where $\mathfrak{H}$ is a Hausdorff matrix so that

$$\beta_m = \sum_{n=0}^{m} h_{mn}\alpha_n = \sum_{n=0}^{m} \binom{m}{n} \alpha_n \int_0^1 u^n(1 - u)^{m-n}\, dQ(u). \tag{23.4.3}$$

This shows that the norm of the transformation does not exceed $V[Q]$, the total variation of $Q(u)$ in $[0, 1]$. We say that the sequence $\{\alpha_n\}$ is *limitable* $[H, Q]$ to the limit β if $\lim \beta_m = \beta$.

With every Hausdorff matrix $\mathfrak{H}$ there is associated a uniquely determined "*mass function*" $Q(u)$ and a corresponding "*moment function*"

$$\mu(z) = \int_0^1 u^z\, dQ(u), \qquad \Re(z) \geqq 0, \tag{23.4.4}$$

holomorphic for $\Re(z) > 0$. Conversely, every such mass function $Q(u)$ or moment function $\mu(z)$ defines a unique Hausdorff matrix. Putting $u = e^{-t}$, $P(t) = Q(1) - Q(u)$, we obtain

$$\mu(z) = \int_0^\infty e^{-zt}\, dP(t),$$

where $P(0) = 0$, $P(t)$ is normalized and of bounded variation on $[0, \infty)$. Thus $\mu(z)$ is the Laplace-Stieltjes transform of a function in $BV[0, \infty)$.

The operations of addition, scalar multiplication, and matrix multiplication may be performed on Hausdorff matrices and lead to other such matrices:

$$\begin{aligned} \mathfrak{H}_1 + \mathfrak{H}_2 &= \mathfrak{D}(\mathfrak{M}_1 + \mathfrak{M}_2)\mathfrak{D}, \\ \alpha\mathfrak{H} &= \mathfrak{D}(\alpha\mathfrak{M})\mathfrak{D}, \\ \mathfrak{H}_1\mathfrak{H}_2 &= \mathfrak{D}(\mathfrak{M}_1\mathfrak{M}_2)\mathfrak{D}. \end{aligned} \tag{23.4.5}$$

To these operations correspond addition, scalar multiplication, and ordinary multiplication for the moment functions and addition, scalar multiplication, and convolution for the mass functions. Thus the ring product of $Q_1(u)$ and $Q_2(u)$ is the function

$$Q_3(u) = Q_1(1)Q_2(u) + \int_u^1 Q_1(u/v)\, dQ_2(v) \tag{23.4.6}$$

where the subscripts 1 and 2 may be interchanged and the integrals are taken in the Lebesque-Stieltjes sense. The corresponding function $P_3(t) = Q_3(1) - Q_3(u)$ is given by the simpler formula

$$P_3(t) = \int_0^t P_1(t - s)\, dP_2(s)$$

from which we conclude that $V[Q_3] \leqq V[Q_1]V[Q_2]$. It follows that if we define

$$\| \mathfrak{H} \| = \| \mu \| = \| Q \| = V[Q], \tag{23.4.7}$$

then the three algebras of Hausdorff matrices, moment functions, and mass functions become isomorphic and isometric (B)-algebras, since completeness is easily proved. We denote these three algebras by **H**, **M**, and **V** respectively.

We note that the normalized Hausdorff matrices form a semi-group of **H** but not a sub-algebra since neither addition nor scalar multiplication leaves the normalization invariant.

For the properties of Hausdorff matrices used in this exposition, we refer to the papers listed in the References at the end of the paragraph.

23.5. Semi-groups in M; the arithmetic of M. The set **M** contains a number of important semi-groups. The simplest example of a one-parameter semi-group in **M** is the set of functions $\{2^{-\xi z}\}$, $0 < \xi$, which corresponds to the Euler-Knopp E_ξ-methods of summability, with the mass function $q(u) = 0$ or 1 according as $u < 2^{-\xi}$ or $> 2^{-\xi}$. The infinitesimal generator of the corresponding semi-group of linear transformations on c to itself is the matrix

$$\mathfrak{A} = -\log 2\ \mathfrak{D}(\delta_{mn}m)\mathfrak{D}. \tag{23.5.1}$$

The E_ξ-methods occupy in some respects an extreme position among the Hausdorff methods of summation (rate of growth of the moment function, incommensurability with other methods); we note also that as a linear operator, E_ξ is not continuous in the uniform topology of $\mathfrak{E}(c)$.

The semi-group $\{\exp[-\zeta z^\alpha]\}$, where α is fixed, $0 < \alpha < 1$, and $|\arg \zeta| < \frac{1}{2}(1-\alpha)\pi$, shows a different behavior. These moment functions correspond to a one-parameter semi-group of regular methods of summability, incommensurable with E_ξ but including every Cesàro method. The semi-group is analytic in the sector indicated above and has as its infinitesimal generator the matrix

$$\mathfrak{A}^{(\alpha)} = -\mathfrak{D}(\delta_{mn} m^\alpha)\mathfrak{D}. \tag{23.5.2}$$

The Hölder methods correspond to the semi-group in $\mathbf{M}$ defined by $\{(z+1)^{-\zeta}\}$, $\Re(\zeta) > 0$. The generating matrix is

$$\mathfrak{A}_0 = -\mathfrak{D}(\delta_{mn} \log(m+1))\mathfrak{D}. \tag{23.5.3}$$

Hausdorff has shown how to extend the Hölder methods into a logarithmic scale of methods of summability. Let us write

$$e_1 = e, \quad e_k = \exp(e_{k-1}), \quad \log_2 x = \log\log x, \quad \log_k x = \log(\log_{k-1} x)$$

and

$$\mu(z; \zeta_1, \cdots, \zeta_n) = [\log(z+e)]^{-\zeta_1} \cdots [\log_n(z+e_n)]^{-\zeta_n}, \tag{23.5.4}$$

where the first of the parameters ζ_k which is different from zero has a positive real part and the others are arbitrary complex numbers. This is an n-parameter family of moment functions to which corresponds an n-parameter semi-group of linear operators on c which may serve as an illustration of Theorem 10.10.2. The infinitesimal generators are the n matrices

$$\mathfrak{A}_k = -\mathfrak{D}(\delta_{mp} \log_{k+1}(m+e_k))\mathfrak{D}, \qquad k = 1, \cdots, n. \tag{23.5.5}$$

The functions of the logarithmic scale have an important bearing on the arithmetic of $\mathbf{M}$ which in turn is basic in the theory of Hausdorff summability. We have arranged our definitions so that the sets $\mathbf{H}$, $\mathbf{M}$, and $\mathbf{V}$ are (B)-algebras. They are also *domains of integrity* since there are obviously no divisors of zero in $\mathbf{M}$. This means that the concepts of *divisibility*, *factor*, *multiple*, *unit*, *associate*, and *prime* can be defined in $\mathbf{M}$ and these concepts carry over to the sets $\mathbf{H}$ and $\mathbf{V}$. In the former set they reflect important properties of the corresponding Hausdorff methods of summation. Thus if $\mu(z)$ is a unit in $\mathbf{M}$, that is, has an inverse in $\mathbf{M}$, then the corresponding $[H, Q]$ is equivalent to convergence and this condition is necessary as well as sufficient. If $\mu(z)$ is a prime, then $[H, Q]$ is not equivalent to convergence but includes [= is stronger than] no Hausdorff method which is not equivalent to convergence. Two Hausdorff methods are equivalent if and only if the corresponding moment functions divide each other, that is, are associates. Further, $[H, Q_1]$ includes $[H, Q_2]$ if and only if $\mu_2(z)$ is a divisor of $\mu_1(z)$.

It follows from some results of L. L. Silverman and J. D. Tamarkin [1] that every function of the form

$$\frac{z-a}{z+b}, \qquad \Re(a) > 0, \qquad \Re(b) > 0, \tag{23.5.6}$$

is a prime in **M** but it is not known if all primes are of this form. We say that $\mu(z) \in \mathbf{M}$ is *indefinitely divisible* if $[\mu(z)]^\alpha \in \mathbf{M}$ for every $\alpha > 0$. The semi-groups listed above provide examples of such elements. Another example is $\mu(z) = 1 - e^{-cz}$, $c > 0$, and (23.5.6) for $\Re(a) = 0$ (for $\Re(a) < 0$, the function is a unit). The logarithmic scale exhibits another interesting feature of division in **M**: it is *non-Archimedean* in the sense that an element $\mu_1(z)$ may be divisible by every power of another element $\mu_2(z)$. Thus $[\log_k (z + e_k)]^{-1}$ is divisible by every power of $[\log_{k+1} (z + e_{k+1})]^{-1}$.

References. Garabedian, Hille and Wall [1], Hausdorff [1], Hille and Tamarkin [4, 5], Silverman and Tamarkin [1], Widder [1, Chapter III].

3. The Abstract Cauchy Problem

23.6. Cauchy's problem and the principle of scientific determinism. We refer to Chapter XX of the first edition of this treatise for some applications of semi-group theory to partial differential equations. The underlying abstract theory will be summarized in this paragraph.

The classical problem of Cauchy calls for the determination of a solution of a system of linear partial differential equations given the initial data on a suitable carrier. Following the usage of J. Hadamard one calls the problem *correctly set* ("*bien posé*") if it has one and only one solution. The solution does not always depend continuously upon the initial data: if it possesses this additional property, we speak of a *stable solution.* Uniqueness and stability are usually attained by imposing suitable accessory conditions serving to restrict the class of admissible solutions: conditions affecting boundary values, rate of growth, integrability, minimality or the like.

As long ago as 1903 Hadamard observed that Cauchy's problem for the wave equation leads to certain transformation groups and that the group properties imply and are implied by certain transcendental addition theorems satisfied by the elementary solutions (Riemann's function, Green's function, etc.) used in constructing the solution. See in particular his papers [3, 4] where he also called attention to related investigations of Felix Bernstein on the heat equation and the addition theorems for the functions of composition found by V. Volterra. Transformation groups of the.type here contemplated were obtained by E. Picard as early as 1895 in connection with ordinary differential equations and by Le Roux in 1903 for systems of partial differential equations.

In the cases actually studied by Hadamard, involving equations of the hyperbolic type, the phenomena have reversible character and one gets true transformation groups, but in the irreversible case, equations of parabolic type, semi-groups appear instead.

Hadamard found that the group properties are a consequence of the *principle of scientific determinism* which may be formulated as follows:

From the state of a physical system at the time t_0 we may deduce its state at a later instant t.

An important corollary of this principle is:

The state of the system at the time t may be deduced from its state at an intermediary time t_1 by first computing the state at the time t_1 and then from the latter the state at the time t, the result being the same as that obtainable by direct computation from the original state.

This formulation is due to Hadamard in his analysis of Huygens' principle and he refers to it as the *major premise of Huygens' principle* (see **[3]**, for later literature see B. B. Baker and E. T. Copson [1]).

If the state is described by a function $f(P, t)$ of position and time for an initial state at $t = 0$ given by $F(P)$, then the mapping $F(P) \to f(P, t)$ defines a transformation

$$f(P, t) = T[F(\cdot); P, t], \tag{23.6.1}$$

usually linear, and the major premises gives

$$T[F(\cdot); P, t] = T[T[F(\cdot); \cdot, t_1]; P, t - t_1] \tag{23.6.2}$$

and this is the semi-group property.

Here a word of caution seems to be necessary. The principle of scientific determinism is probably less respected in modern theoretical physics than in the days before Heisenberg's theory when Hadamard wrote. For us the main point is that the principle of determinism as well as the modern indeterminancy principle apply to physical states and not to solutions of partial differential equations or other functional equations. On the other hand even if we know that the functions describing the physical state of a system satisfy such equations, not every solution of the equation describes a possible physical state. Thus in the case of a partial differential equation corresponding to a physical system, the major premise of the principle of Huygens merely suggests that the two sides of equation (23.6.2) are equal; if we know that both sides of this equation are solutions of the same initial value problem and that this problem has a unique solution within the class of admissible functions, then and only then are we entitled to conclude the desired equality.

We shall see below, however, that there is a very close connection between on one hand the existence of a unique stable solution of Cauchy's problem within a given class of functions and corresponding to an admissible class of initial values and, on the other, the existence of a semi-group of operators which produces the solution when it is applied to the admissible initial values.

23.7. The abstract Cauchy problem; uniqueness and non-uniqueness. To gain precision we introduce an abstract Cauchy problem, denoted by ACP in the following. This problem is closely related to semi-group theory, it covers a number of problems of interest to the applications. It leaves to one side many other problems of equal or greater importance, but there seems to be some foundation for believing that the number of omitted cases will gradually be much reduced by perturbation methods or other devices. For the following see E. Hille [**18, 19, 20, 21**] and R. S. Phillips [**10**].

ACP. *Given a complex (B)-space* $\mathfrak{X}$ *and a linear operator* U *with domain* $\mathfrak{D}(U)$ *and range* $\mathfrak{R}(U)$ *in* $\mathfrak{X}$ *and given an element* $y_0 \in \mathfrak{X}$, *find a function* $y(t) = y(t; y_0)$ *such that*

(i) $y(t)$ *is strongly absolutely continuous and continuously differentiable in each finite subinterval of* $[0, \infty)$ [*of* $(0, \infty)$];

(ii) *for each* $t > 0$, $y(t) \in \mathfrak{D}(U)$ *and*

$$U[y(t)] = y'(t); \tag{23.7.1}$$

(iii) $\lim_{t \to 0+} \| y(t; y_0) - y_0 \| = 0.$

This is the formulation given by R. S. Phillips and differs somewhat from earlier versions of E. Hille in the phrasing of condition (i); it has certain advantages, in particular, it leads to a much stronger form of Theorem 23.8.3 below than was previously available. The reader will note that there are two alternatives in condition (i); the corresponding problems will be denoted by ACP_1 and ACP_2 respectively. An extension denoted by ACP^n will be studied in section 23.9.

If for a given y_0 and a given operator U the corresponding ACP has two distinct solutions, then their difference is a *nul solution* $y(t; \theta)$. The existence of such solutions is regulated by the next two theorems where a solution is said to be of normal type ω if

$$\limsup_{t \to +\infty} t^{-1} \log \| y(t) \| = \omega < \infty. \tag{23.7.2}$$

THEOREM 23.7.1. *For each* $y_0 \in \mathfrak{X}$ *the ACP has at most one solution of normal type if* U *is a closed operator whose point spectrum is not dense in any right half-plane.*

PROOF. Suppose on the contrary that there exists a nul solution $y(t)$ of normal type ω and form

$$L(\lambda; y) \equiv \int_0^\infty e^{-\lambda t} y(t)\, dt. \tag{23.7.3}$$

This integral exists in the sense of Bochner for $\mathfrak{R}(\lambda) > \omega$ and defines a holomorphic function of λ in this half-plane. For such values of λ and with $0 < \alpha < \beta < \infty$ we have

$$\int_\alpha^\beta e^{-\lambda t} y'(t)\, dt = \int_\alpha^\beta e^{-\lambda t} U[y(t)]\, dt = U\left\{\int_\alpha^\beta e^{-\lambda t} y(t)\, dt\right\}$$

by Theorem 3.7.12 since U is closed. The integral in the first member equals

$$e^{-\beta\lambda}y(\beta) - e^{-\alpha\lambda}y(\alpha) + \lambda \int_\alpha^\beta e^{-\lambda t}y(t)\, dt \to \lambda L(\lambda; y)$$

as $\alpha \to 0$, $\beta \to \infty$. Again using the closure of U we see that

$$U[L(\lambda; y)] = \lambda L(\lambda; y). \tag{23.7.4}$$

This implies a contradiction, for if $L(\lambda; y) \not\equiv \theta$ every point in the half-plane $\Re(\lambda) > \omega$, excepting at most a discrete set, must belong to the point spectrum of U. We conclude that $L(\lambda; y) \equiv \theta$ which implies that $y(t) \equiv \theta$. This proves the assertion.

REMARK. For the validity of this theorem it is not necessary that $y(t; y_0)$ be continuously differentiable. It suffices that $y(t; y_0)$ be the (B)-integral of its derivative.

THEOREM 23.7.2. *If U is closed, a necessary and sufficient condition that the ACP have a nul solution of type $\leqq \omega$ is that the characteristic equation*

$$U[x(\lambda)] = \lambda x(\lambda) \tag{23.7.5}$$

have a solution $x(\lambda) \not\equiv \theta$, bounded and holomorphic in each half-plane $\Re(\lambda) \geqq \omega + \epsilon$, $\epsilon > 0$.

PROOF. The necessity follows from the proof of the preceding theorem. To prove the sufficiency, let us suppose that (23.7.5) has a solution $x_0(\lambda) \not\equiv \theta$ which is bounded and holomorphic for $\Re(\lambda) \geqq \omega + \epsilon$, $\epsilon > 0$. Since a solution of (23.7.5) may be multiplied by any numerically valued bounded holomorphic function, we see that $x(\lambda) \equiv (\lambda - \omega + 1)^{-3}x_0(\lambda)$ is also a solution. We then form

$$y(t) = \frac{1}{2\pi i}\int_{\gamma - i\infty}^{\gamma + i\infty} e^{\lambda t}x(\lambda)\, d\lambda, \qquad \gamma > \omega. \tag{23.7.6}$$

Here $y(t)$ is a strongly continuous function of t, independent of γ, and for $t \leqq 0$ we have $y(t) = \theta$ since

$$\| y(t) \| \leqq 2\,[\max_\nu \| x_0(\gamma + i\nu) \|]\, e^{\gamma t}(\gamma - \omega + 1)^{-2}.$$

This estimate also shows that the type of $y(t)$ does not exceed ω. Using Theorem 3.7.12 again we see that

$$U[y(t)] = \frac{1}{2\pi i}\int_{\gamma - i\infty}^{\gamma + i\infty} e^{\lambda t}U[x(\lambda)]\, d\lambda = \frac{1}{2\pi i}\int_{\gamma - i\infty}^{\gamma + i\infty} e^{\lambda t}\lambda x(\lambda)\, d\lambda$$

and the third member evidently equals

$$y'(t) = \operatorname{str}\lim_{h\to 0} h^{-1}[y(t + h) - y(t)].$$

It follows that $y(t)$ is a nul solution of type $\leqq \omega$. It should be noted that $y(t)$ cannot be identically zero since its Laplace transform $x(\lambda) \not\equiv \theta$.

REMARK. The choice of the multiplier is to a large extent arbitrary, but has an important bearing on the properties of the nul solution. Thus replacing the factor $(\lambda - \omega + 1)^{-3}$ by $\exp[-(\lambda - \omega + 1)^\alpha]$, $0 < \alpha < 1$, for instance, we obtain a solution having derivatives of all orders and vanishing at $t = 0$ together with all its derivatives. Further, if $y(t)$ is a nul solution, not necessarily of normal type, then so is

$$\frac{1}{\Gamma(\beta)} \int_0^t (t - s)^{\beta-1} y(s)\, ds, \qquad \beta > 0,$$

and integration obviously improves the "smoothness" of the solution and makes it tend faster to zero with t. An example of a nul solution having derivatives of all orders, vanishing at $t = 0$, is furnished by $\int_0^t \exp[(s - t)^{-1}]\, y(s)\, ds$.

If the point spectrum of U covers the whole plane other pathological phenomena may arise such as "explosive" solutions which become infinite in norm for preassigned values of t. See Hille [**21**, pp. 97–99] and remarks at the end of section 23.9 below.

23.8. Relations to semi-group theory. It is clear from Theorems 10.3.3 and 11.5.4 that the theory of semi-groups and the ACP are related. The following theorems due to R. S. Phillips exhibit this connection in much sharper form.

THEOREM 23.8.1. *If U is the infinitesimal generator of a semi-group $[S(t; U)]$ of class $(0, A)$, then the corresponding ACP_2 for the operator U has a unique solution $y(t; y_0) \equiv S(t; U)[y_0]$ for each $y_0 \in \mathfrak{D}(U)$.*

PROOF. Theorem 11.5.4 shows that under the present assumptions the ACP_2 for the operator U has the solution $y(t; y_0) = S(t; U)[y_0]$ for each $y_0 \in \mathfrak{D}(U)$. Since U is closed and its spectrum is confined to a left half-plane, Theorem 23.7.1 shows that this is the only solution of normal type. To prove the stronger assertion that no other solution exists, we assume the existence of a nul solution $y(t)$ and, using a modification of an artifice due to N. Dunford, we form

$$S(t - s; U)[y(s)] \quad \text{for} \quad 0 < s < t.$$

Since $y(s) \in \mathfrak{D}(U)$ and is strongly differentiable, $S(t - s; U)[y(s)]$ has the same property and

$$\frac{d}{ds} \{S(t - s; U)[y(s)]\} = S(t - s; U)\{y'(s) - U[y(x)]\} = \theta.$$

Integrating between the limits α and β, $0 < \alpha < \beta < t$, gives

$$\theta = S(t - \beta; U)[y(\beta)] - S(t - \alpha; U)[y(\alpha)],$$

and, passing to the limit first as $\alpha \to 0+$ and then as $t \to \beta+$, we obtain $y(\beta) = \theta$ for all $\beta > 0$. Thus there are no nul solutions and consequently the solution of the ACP_2 for U is unique as asserted.

COROLLARY. *If U generates a semi-group of class (C_0), then the ACP_1 for U has the unique solution* $y(t; y_0) \equiv S(t; U)[y_0]$ *for* $y_0 \in \mathfrak{D}(U)$.

PROOF. To the preceding proof we have only to add the observation that for $y_0 \in \mathfrak{D}(U)$ we have $(d/dt)S(t; U)[y_0] = S(t; U)[Uy_0]$ which tends to Uy_0 as $t \to 0+$ since $S(t; U)$ is of class (C_0). This shows that $S(t; U)[y_0]$ is continuously differentiable in $[0, \infty)$ as required for the ACP_1.

REMARK. It should be observed that, $S(t; U)$ being a bounded operator, $S(t; U)[y_0]$ is well defined for every y_0 and not merely for $y_0 \in \mathfrak{D}(U)$. It does not follow, however, that the ACP_2 has a solution for every y_0 since $S(t; U)[y_0]$ need not belong to $\mathfrak{D}(U)$. Moreover, $S(t; U)[y_0]$ need not tend to y_0 when $t \to 0+$ if $y_0 \in \mathfrak{D}(U)$ and $S(t; U)$ is not of class (C_0). On the other hand, if $S(t; U)[y_0] \in \mathfrak{D}(U)$ for all $t > 0$ and if $\lim_{t \to 0+} S(t; U)[y_0] = y_0$, then $S(t; U)[y_0]$ is a solution of an ACP in a somewhat weaker sense inasmuch as this function need not be strongly absolutely continuous in $(0, 1)$, though it possesses the other properties postulated for a solution. It may very well happen that the ACP_2 has a solution for every $y_0 \in \mathfrak{X}$, a case in point being that in which $[S(t; U)]$ is of class (C_0) and $US(t; U)$ is a bounded operator for each $t > 0$, in particular, if $S(t; U)$ is analytic for $t > 0$. There is one case, however, where the admissible initial values are strictly limited to $\mathfrak{D}(U)$:

THEOREM 23.8.2. *If U generates a group $\{G(t; U); -\infty < t < +\infty\}$ strongly continuous at $t = 0$ with $G(0; U) = I$, then $G(t; U)[y_0]$ is a solution of the ACP if and only if $y_0 \in \mathfrak{D}(U)$.*

PROOF. From

$$h^{-1}[G(t + h; U) - G(t; U)][y_0] = G(t - t_0 ; U)h^{-1}[G(t_0 + h; U) - G(t_0 ; U)][y_0]$$

one sees that $G(t; U)[y_0] \in \mathfrak{D}(U)$ either for all values of t or for none.

We shall now state the converses of Theorem 23.8.1 and its corollary.

THEOREM 23.8.3. *Let U be a closed operator with dense domain and non-vacuous resolvent set and suppose that for each $y_0 \in \mathfrak{D}(U)$ there is a unique solution to ACP_1. Then U generates a semi-group of class (C_0) such that $S(t; U)[y_0] = y(t; y_0)$ for each $y_0 \in \mathfrak{D}(U)$.*

THEOREM 23.8.4. *Let U be a closed operator with dense domain and such that $R(\lambda; U) = O(\lambda^{-1})$ as $\lambda \to +\infty$. Suppose that for each $y_0 \in \mathfrak{D}(U)$ there is a unique solution to ACP_2. Then U generates a semi-group $\{S(t; U)\}$ of class $(0, A)$ such that $S(t; U)[y_0] = y(t; y_0)$ for each $y_0 \in \mathfrak{D}(U)$.*

The considerations leading to these theorems apply with suitable modifications to a more general situation to be discussed in section 23.9; for this reason the reader will find the proof of Theorem 23.8.3 after Lemma 23.9.5. For the proof of Theorem 23.8.4, however, we refer to Phillips [**10**].

As a consequence of the above theorems, we see that the assumption of a unique solution to the ACP for all initial values in $\mathfrak{D}(U)$ is in itself sufficient to force the solution to vary continuously with the initial data. Hence in this case, existence and uniqueness suffice to make the solution stable.

23.9. The abstract Cauchy problem of higher order. The ACP = ACP^1 admits of a natural generalization which we shall now discuss. For the following see Hille [**19**] and Phillips [**10**].

ACP^n. *Given a complex* (B)*-space* $\mathfrak{X}$ *and a linear operator* U *with domain and range in* $\mathfrak{X}$ *and given* n *elements* $y_0, y_1, \cdots, y_{n-1}$ *in* $\mathfrak{X}$*, find a function* $y(t) = y(t; y_0, y_1, \cdots, y_{n-1})$ *on* $[0, \infty)$ *to* $\mathfrak{X}$ *such that*

(i) $y(t)$ *is* n *times continuously differentiable in* t *on* $[0, \infty)$;

(ii) *for each* $t \geqq 0$, $y^{(k)}(t) \in \mathfrak{D}(U^{n-k})$ *where* $U^{n-k}[y^{(k)}(t)]$ *is strongly continuous in* t*, on* $[0, \infty)$ *for* $k = 0, 1, \cdots, n-1$*, while*

$$y^{(n)}(t) = U^n[y(t)], \qquad t \geqq 0; \tag{23.9.1}$$

(iii) $\lim_{t\to 0+} \| y^{(k)}(t; y_0, y_1, \cdots, y_{n-1}) - y_k \| = 0, \quad k = 0, 1, 2, \cdots, n-1.$

This formulation is more restrictive than Hille's original version, but has the advantage of leading to sharper results.

For $\sigma > 0$ we define

$$\Delta_n(\sigma) \equiv \left[\lambda; \lambda = re^{i\theta}, r > \sigma\left(\sec\frac{\theta}{n}\right)^n, |\theta| \leqq \pi\right], \tag{23.9.2}$$

where the sign of equality is admitted only for $n > 2$. If $n = 1$, the last condition should read $|\theta| < \frac{1}{2}\pi$ instead.

We have the following uniqueness theorems, analogous to Theorems 23.7.1 and 23.7.2, which we state without proof.

THEOREM 23.9.1. *If* U^n *is a closed operator whose point spectrum is not dense in any domain* $\Delta_n(\sigma)$*, then, for any choice of* $y_0, y_1, \cdots, y_{n-1}$ *in* $\mathfrak{X}$*, the ACP*^{n} *has at most one solution such that* $y^{(n-1)}(t)$ *is of normal type.*

THEOREM 23.9.2. *If* U^n *is closed, then a necessary and sufficient condition that the ACP*^{n} *have a nul solution such that* $y^{(n-1)}(t)$ *is of normal type* $\leqq \omega$ *is that the characteristic equation*

$$U^n[x(\lambda)] = \lambda^n x(\lambda) \tag{23.9.3}$$

have a solution $x(\lambda)$, $\not\equiv \theta$*, which is bounded and holomorphic in every half-plane* $\Re(\lambda) \geqq \omega + \epsilon$, $\epsilon > 0$.

The proofs go as for $n = 1$. We note that if $y^{(n-1)}(t)$ is of normal type $\leqq \omega$ so are $y^{(k)}(t)$, $k = 0, 1, \cdots, n-2$. Just as in the case $n = 1$, if nul solutions exist at all, they may be constructed so as to have derivatives of all orders vanishing at the origin.

It is obvious that the ACP^n is also related to semi-group theory when $n > 1$. But while the ACP^1 leads to the question whether U generates a semi-group, the ACP^n for $n > 1$ poses the same question for each of the n operators

$$U,\ \eta U,\ \eta^2 U,\ \cdots,\ \eta^{n-1}U, \qquad \eta = e^{2\pi i/n}.$$

Here there is a fundamental difference between $n = 2$ and $n > 2$. In the former case it is possible for both U and $-U$ to generate semi-groups of class (C_0), but

$$S(t;\ U)S(t;\ -U) = S(t;\ -U)S(t;\ U) = I, \tag{23.9.4}$$

so that U actually generates a group. Still this does not require that U be a bounded operator. But for $n > 2$ the existence of the n semi-groups $[S(t;\ \eta^k U)]$, of class (C_0) say, implies the existence and boundedness of $R(\lambda;\ U)$ in n half-planes $\Re(\lambda\eta^{-k}) > \sigma_k + \epsilon$. It follows that $R(\lambda;\ U)$ is holomorphic at infinity and hence, by Theorem 5.9.4, that U is a bounded operator. For bounded operators the ACP^n is trivially solvable:

THEOREM 23.9.3. *If U is a linear bounded operator, then the ACP^n for U has the unique solution*

$$y(t;\ y_0,\ y_1,\ \cdots,\ y_{n-1}) \equiv \sum_{k=0}^{n-1}\sum_{m=0}^{\infty} \frac{t^{mn+k}}{(mn+k)!}\,U^{mn}y_k \tag{23.9.5}$$

for any choice of the y_k's in $\mathfrak{X}$.

For $n = 2$ the corollary of Theorem 23.8.1 has a more interesting analogue:

THEOREM 23.9.4. *If U is the infinitesimal generator of a group $[G(t;\ U);\ -\infty < t < +\infty]$ of class (C_0), then the ACP^2 for U has the unique solution*

$$y(t;\ y_0,\ y_1) = \tfrac{1}{2}[G(t;\ U)(y_0 + z_1) + G(-t;\ U)(y_0 - z_1)] \tag{23.9.6}$$

for any $y_0 \in \mathfrak{D}(U^2)$, $y_1 \in \mathfrak{D}(U) \cap \Re(U)$ and z_1 such that $Uz_1 = y_1$.

It is an easy matter to verify that these formulas give solutions. We observe that the particular choice of z_1 is immaterial as long as $Uz_1 = y_1$. This is due to the fact that if $Uz_0 = \theta$ then $G(t;\ U)[z_0] \equiv z_0$ so that replacing z_1 by $z_1 + z_0$ leaves the right member of (23.9.6) unchanged. In order to prove uniqueness we need

LEMMA 23.9.1. *Let U be a linear closed operator on $\mathfrak{X}$ to $\mathfrak{X}$. Let $y(t)$ on $[0,\ \infty)$ to $\mathfrak{X}$ be continuously differentiable for $t \geqq 0$. Further suppose for $t \geqq 0$ that $y(t) \in \mathfrak{D}(U)$, $y'(t) \in \mathfrak{D}(U)$, and that $U[y'(t)]$ is strongly continuous. Then*

$$\frac{d}{dt}\,U[y(t)] = U[y'(t)], \qquad t \geqq 0. \tag{23.9.7}$$

PROOF. From

$$y(t) = y(0) + \int_0^t y'(s)\,ds$$

together with the assumptions on U it follows that

$$U[y(t)] = U[y(0)] + \int_0^t U[y'(s)]\, ds \tag{23.9.8}$$

differentiation of which yields (23.9.7).

We shall prove the uniqueness of the solution (23.9.6) by a twofold use of the Dunford device. If there exists a nul solution $y(t)$ we form

$$Y(s, t) \equiv G(t - s; U) \{y'(s) + U[y(s)]\}, \qquad 0 < s < t.$$

The partial of $Y(s, t)$ with respect to s exists and equals

$$G(t - s; U) \{- U[y'(s)] - U^2[y(s)] + y''(s) + (U[y(s)])'\}.$$

This reduces to θ by Lemma 23.9.1, the assumptions of which are satisfied since U is closed. It follows that $Y(s, t)$ is independent of s. Using closure again together with (23.9.8) we see that

$$\lim_{s \to 0+} Y(s, t) = G(t; U) \{y'(0) + U[y(0)]\} = \theta,$$

$$\lim_{s \to t-} Y(s, t) = y'(t) + U[y(t)]$$

so that the latter expression vanishes identically. On the other hand, treating

$$Z(s, t) \equiv G(s - t; U) \{y'(s) - U[y(s)]\}$$

in the same manner one concludes that $y'(t) - U[y(t)]$ vanishes identically. Adding the two expressions one sees that $y'(t)$ vanishes so that $y(t)$ is a constant which must be θ since $y(t)$ is a nul solution. This proves the uniqueness of (23.9.6).

In the case of a bounded operator U one sees that a nul solution $y(t)$ has to satisfy

$$y(t) = [(kn - 1)!]^{-1} \int_0^t (t - s)^{kn-1} U^{kn}[y(s)]\, ds$$

for $k = 1, 2, \cdots$, and this implies $y(t) \equiv \theta$.

In order to proceed to the proofs of Theorem 23.8.3 and its various generalizations we shall state and prove four lemmas. These will be based on the following two assumptions:

A_1. *V is a linear closed operator with dense domain and non-vacuous resolvent set.*

A_2. *The ACP^n for the operator V has a unique solution if the initial values satisfy*

$$y_k = V^k[y_0], \quad k = 0, 1, \cdots, n - 1, \qquad y_0 \in \mathfrak{D}(V^n). \tag{23.9.9}$$

V will later be identified with one or more of the operators $\eta^k U$,

$$\eta = \exp(2\pi i/n).$$

LEMMA 23.9.2. *The ACP_1^1 for V has a unique solution for $y_0 \in \mathfrak{D}(V^n)$ given by*

$$y(t; y_0, Vy_0, \cdots, V^{n-1} y_0) \equiv y_n(t; y_0). \tag{23.9.10}$$

PROOF. For $n = 1$ we have nothing to prove. If $n > 1$ it is clear that

$$y_n(t; y_0) = \sum_{k=0}^{n-1} \frac{t^k}{k!} V^k y_0 + \frac{1}{(n-1)!} \int_0^t (t-s)^{n-1} V^n[y_n(s; y_0)]\, ds \tag{23.9.11}$$

so that

$$y_n'(t; y_0) = \sum_{k=0}^{n-2} \frac{t^k}{k!} V^{k+1} y_0 + \frac{1}{(n-2)!} \int_0^t (t-s)^{n-2} V^n[y_n(s; y_0)]\, ds.$$

We wish to bring one of the V's out from under the integral sign and for this purpose it suffices to show that $V^{n-1}[y_n(t; y_0)]$ is strongly continuous on $[0, \infty)$. This is readily proved from the relation (11.5.12) which asserts that

$$R(\lambda; V)[y_n(t; y_0)] = \sum_{k=0}^{n-1} V^k[y_n(t; y_0)]\lambda^{-k-1} + \lambda^{-n} R(\lambda; V) V^n[y_n(t; y_0)].$$

Substituting n distinct values in $\rho(V)$ for λ we obtain n equations in the $V^k[y_n(t; y_0)]$, $k = 0, 1, \cdots, n-1$, where the coefficient determinant is a Vandermonde determinant and thus non-zero. It follows that the $V^k[y_n(t; y_0)]$ can be expressed in terms of $R(\lambda; V)\,[y_n(t; y_0)]$ and $R(\lambda; V)V^n[y_n(t; y_0)]$ and hence we see that these functions are continuous on $[0, \infty)$. Thus setting

$$w(t) = \sum_{k=0}^{n-2} \frac{t^k}{k!} V^k y_0 + \frac{1}{(n-2)!} \int_0^t (t-s)^{n-2} V^{n-1}[y_n(s; y_0)]\, ds,$$

we see that

$$y_n'(t; y_0) = V[w(t)].$$

Now $w(t)$ is clearly $(n-1)$ times continuously differentiable on $[0, \infty)$ with $w^{(n-1)}(t) = V^{n-1}[y_n(t; y_0)]$. Applying Lemma 23.9.1 with U replaced by V^{n-1} we obtain

$$w^{(n)}(t) = V^{n-1}[y_n'(t; y_0)] = V^n[w(t)],$$

and according to ACP^n (ii) applied to $y_n(t; y_0)$ this expression is continuous on $[0, \infty)$; hence ACP^n (i) is satisfied by $w(t)$. Further since $y_n'(t; y_0)$ is $(n-1)$ times continuously differentiable and since V is closed we may interchange the order of differentiation and V and therefore conclude that $y_n^{(k+1)}(t; y_0) = V[w^{(k)}(t)]$ for $0 \leqq k < n$ and $t \geqq 0$. One sees from this that the remainder of condition ACP^n (ii) is satisfied by $w(t)$. Finally one can read off from the defining relation for $w(t)$ that this function satisfies the initial conditions $y_0, Vy_0, \cdots, V^{n-1}y_0$. Applying the uniqueness assumptions A_2 we infer that $w(t) = y_n(t; y_0)$ and it follows that

$$y_n'(t; y) = V[y_n(t; y_0)]. \tag{23.9.12}$$

Now we see from (23.9.11) that $y_n(t; y_0)$ has all the properties required of a solution of ACP$_1^1$. As to the uniqueness of this solution, suppose there exists a nul solution of $y'(t) = V[y(t)]$. As was remarked at the end of section 23.7, we may assume, without loss of generality, the existence of continuous derivatives of any order, vanishing at the origin. Successive differentiations and repeated use of the closure property of V then shows that $y^{(n)}(t) = V^n[y(t)]$ so that the ACP^n for V has a nul solution, contrary to assumption A_2. This concludes the proof of the lemma.

COROLLARY. *We have for* $t \geqq 0$

$$y_n^{(k)}(t; y_0) = V^k[y_n(t; y_0)], \qquad k = 1, 2, \cdots, n. \tag{23.9.13}$$

The proof is obtained by repeated use of (23.9.12) and the fact that V^k is closed.

LEMMA 23.9.3. *If* $y_0 \in \mathfrak{D}(V^{n+1})$

$$V[y_n(t; y_0)] = y_n(t; Vy_0). \tag{23.9.14}$$

PROOF. If $y_0 \in \mathfrak{D}(V^{n+1})$, then by the preceding lemma $y_n(t; Vy_0)$ is a unique solution of the ACP$_1^1$ for V with the initial value Vy_0. As such, the function is continuously differentiable on $[0, \infty)$ and therefore $V[y_n(t; Vy_0)] = y_n'(t; Vy_0)$ is Bochner integrable in each interval $(0, \beta)$. Hence

$$y_n(t; Vy_0) = Vy_0 + \int_0^t y_n'(s; Vy_0)\,ds = V\left\{y_0 + \int_0^t y_n(s; Vy_0)\,ds\right\}.$$

Setting $z(t) = y_0 + \int_0^t y_n(s; Vy_0)\,ds$, we see that $z(t)$ is continuously differentiable for $t \geqq 0$ and that $z'(t) = y_n(t; Vy_0) = V[z(t)]$. Thus $z(t)$ is a solution of the ACP$_1^1$ with initial value y_0 and by Lemma 23.9.2 this problem has a unique solution as long as the initial value is in $\mathfrak{D}(V^n)$. Hence uniqueness implies that $z(t) = y_n(t; y_0)$ and (23.9.14) follows.

LEMMA 23.9.4. *Let* $y_0 \in \mathfrak{D}(V^n)$. *Then for each* $m \geqq 1$, *there exists a constant* C_m *such that*

$$\sum_{k=0}^{n} \| V^k[y_n(t; y_0)] \| \leqq C_m \sum_{k=0}^{n} \| V^k y_0 \|, \qquad 0 \leqq t \leqq m. \tag{23.9.15}$$

PROOF. The method of proof employed in this lemma was suggested by some work of P. Lax on the continuous dependence of solutions of linear partial differential equations on initial data. Simplifications proposed by H. G. Tillmann have been incorporated in the proof. We denote by $\mathfrak{D}(V^n)^\circ$ the set $\mathfrak{D}(V^n)$ with the norm

$$\| y \|^\circ = \sum_{k=0}^{n} \| V^k y \|.$$

The operators V^k being closed, it follows that $\mathfrak{D}(V^n)^\circ$ is a Banach space. We also introduce the auxiliary space C_{mn}°, consisting of all strongly continuous functions $x(t)$ on $[0, m]$ to $\mathfrak{D}(V^n)^\circ$. It is easy to see that C_{mn}° is a Banach space under the norm

$$\| x(\cdot) \|_m = \sup\,[\| x(t) \|^\circ; \quad 0 \leqq t \leqq m].$$

We now consider the transformation T_m on $\mathfrak{D}(V^n)^\circ$ to C°_{mn} which takes $y_0 \in \mathfrak{D}(V^n)^\circ$ into the corresponding solution $y_n(t; y_0)$ of ACP^n with $t \in [0, m]$. Assumption A_2 implies that T_m is linear on $\mathfrak{D}(V^n)^\circ$. We shall now show that T_m is closed. To this end suppose that $y_k \to y_0$ in $\mathfrak{D}(V^n)^\circ$ and that $T_m(y_k) \to v(\cdot)$ in C°_{mn}. In particular, $y_n^{(j)}(t; y_k) = V^j[y_n(t; y_k)] \to V^j[v(t)]$ in the $\mathfrak{X}$ norm for $j = 0, 1, \cdots, n$, uniformly in $[0, m]$. As a consequence $v(t)$ is n times strongly continuously differentiable on $[0, m]$, $v^{(j)}(t) = V^j[v(t)]$ for $j = 0, 1, \cdots, n$, and $t \in [0, m]$. If we now extend $v(t)$ over (m, ∞), setting

$$v(t) = y_n(t - m; v(m)), \qquad t > m,$$

then $v(t)$ becomes a solution of ACP^n with initial values $(y_0, Vy_0, \cdots, V^{n-1}y_0)$ and the uniqueness assumption implies that $v(t) = y_n(t; y_0)$. This proves that T_m is closed. It follows by the closed graph theorem that T_m is bounded and hence there is a finite C_m such that

$$\| y_n(t; y_0) \|_m \leqq C_m \| y_0 \|^\circ,$$

which proves (23.9.15).

LEMMA 23.9.5. *V is the infinitesimal generator of a semi-group $S(t; V)$ of class (C_0) and for $y_0 \in \mathfrak{D}(V^n)$*

$$S(t; V)[y_0] = y_n(t; y_0).$$

PROOF. We start by defining the operator $S(t)[y_0] = y_n(t; y_0)$ on $\mathfrak{D}(V^n)^\circ$ to itself for $t \geqq 0$. The uniqueness of the solution of the ACP^n implies that $S(t)$ is linear, and Lemma 23.9.4 implies that $S(t)$ is bounded. The corollary to Lemma 23.9.2 shows that $S(t)$ is continuous in the strong operator topology of $\mathfrak{D}(V^n)^\circ$ for $t \geqq 0$. Further, uniqueness implies that $y_n(t_1 + t_2; y_0) = y_n[t_1; y_n(t_2; y_0)]$ for each $y_0 \in \mathfrak{D}(V^n)$ and each $t_1, t_2 > 0$. In terms of the operator $S(t)$ this means that $S(t_1 + t_2) = S(t_1)S(t_2)$, so that $S(t)$ defines a semi-group of linear bounded operators on $\mathfrak{D}(V^n)^\circ$, strongly continuous for $t \geqq 0$. Next we shall prove that $S(t)$ is bounded also in the $\mathfrak{X}$ norm. To this end we choose a $\lambda_0 \in \rho(V)$. Then for each $x \in \mathfrak{D}(V^n)$ there exists a $y \in \mathfrak{D}(V^{2n})$ such that $x = (\lambda_0 I - V)^n y$. By repeated use of Lemma 23.9.3 we find that $S(t)V^k y = V^k S(t)y$, $k = 1, 2, \cdots, n$, and hence that

$$S(t)x = \sum_{k=0}^{n} (-1)^k \binom{n}{k} \lambda_0^{n-k} V^k S(t)y.$$

Consequently

$$\| S(t)x \| \leqq M_1 \| S(t)y \|^\circ \leqq M_1 \| S(t) \|^\circ \| y \|^\circ$$

where

$$M_1 = \max \left\{ \binom{n}{k} | \lambda_0 |^{n-k} \right\}.$$

Now $y = [R(\lambda_0; V)]^n x$ so that $\| y \| \leqq R^n \| x \|$ where $R = \| R(\lambda_0; V) \|$.

Further we have

$$V^k y = V^k[R(\lambda_0; V)]^n x = \sum_{p=0}^{k} (-1)^p \binom{k}{p} \lambda_0^p [R(\lambda_0 ; V)]^{n-k+p} x$$

so that

$$\| V^k y \| \leqq R^{n-k}(1 + |\lambda_0| R)^k \| x \| .$$

Combination of these estimates gives

$$\| y \|^\circ \leqq \{\textstyle\sum_k R^{n-k}(1 + |\lambda_0| R)^k\} \| x \| \equiv M_2 \| x \| ,$$

and, finally,

$$\| S(t)x \| \leqq M_1 M_2 \| S(t) \|^\circ \| x \| \tag{23.9.16}$$

for all x in $\mathfrak{D}(V^n)$. Now $\mathfrak{D}(V)$ is supposed to be dense in $\mathfrak{X}$ by A_1 and this implies that $\mathfrak{D}(V^n)$ is also dense. Hence the operator $S(t)$ can be uniquely extended to be linear and bounded on $\mathfrak{X}$ itself. The inequality (23.9.16) remains valid for the extended $S(t)$, and it is easy to show that $\{S(t)\}$ becomes a semi-group of linear bounded operators on $\mathfrak{X}$, strongly continuous for $t \geqq 0$. By definition $S(t)[y_0] = y_n(t; y_0)$ for each $y_0 \in \mathfrak{D}(V^n)$ and $t \geqq 0$. Now $\lim_{t\to 0+} \| S(t) [y_0] - y_0 \| = 0$ for each $y_0 \in \mathfrak{D}(V^n)$ and $\| S(t) \|$ is bounded with $\| S(t) \|^\circ$ near $t = 0$ by (23.9.16). It therefore follows from the Banach-Steinhaus theorem $\lim_{t\to 0+} \| S(t)[x] - x \| = 0$ for each $x \in \mathfrak{X}$. Consequently $[S(t)]$ is of class (C_0).

It remains to show that the infinitesimal generator A of $[S(t)]$ equals V. This is proved in three steps. First we note that if $y_0 \in \mathfrak{D}(V^n)$, then $S(t)[y_0] = y_n(t; y_0)$ is n times continuously differentiable for $t \geqq 0$. In particular, it follows that y_0, $Vy_0, \cdots, V^{n-1}y_0$ are in $\mathfrak{D}(A)$. In the relations

$$V^k[y_n(t; y_0)] = y_n^{(k)}(t; y_0) = A^k S(t)[y_0]$$

the first member converges to $V^k y_0$ and the third member to $A^k y_0$ as $t \to 0+$, $k = 0, 1, \cdots, n$. Since A and V are both closed we conclude that $\mathfrak{D}(V^n) \subset \mathfrak{D}(A^n)$ and $A^k y_0 = V^k y_0$ if $y_0 \in \mathfrak{D}(V^n)$. Secondly, let ω be the type of $S(t)$. Then for $\Re(\lambda) > \omega$ and $y_0 \in \mathfrak{D}(V^{n+1})$ the function

$$e^{-\lambda t} V S(t) y_0 = e^{-\lambda t} S(t) V y_0 = e^{-\lambda t} S(t) A y_0$$

is Bochner integrable over $(0, \infty)$ so that

$$VR(\lambda; A) y_0 = \int_0^\infty e^{-\lambda t} V S(t) y_0 \, dt = \int_0^\infty e^{-\lambda t} S(t) A y_0 \, dt$$

exists. Thus

$$VR(\lambda; A) y_0 = R(\lambda; A) A y_0 = AR(\lambda; A) y_0 , \qquad y_0 \in \mathfrak{D}(V^{n+1}).$$

Here both V and A are closed and $\mathfrak{D}(V^{n+1})$ is dense in $\mathfrak{X}$. It follows that the equality between the first and the third members is valid for all y_0 in $\mathfrak{X}$. But

$R(\lambda; A)[\mathfrak{X}] = \mathfrak{D}(A)$ so that $\mathfrak{D}(V) \supset \mathfrak{D}(A)$. Finally, to prove the opposite inclusion when $n > 1$, we note that if $y_k \in \mathfrak{D}(V^n) \subset \mathfrak{D}(A^n)$ so does $R(\lambda_0; V)y_k$ when $\lambda_0 \in \rho(V)$. Hence by the first step $AR(\lambda_0; V)y_k = VR(\lambda_0; V)y_k$. But if $y_k \to y_0$ in $\mathfrak{X}$

$$VR(\lambda_0; V)y_k = \lambda_0 R(\lambda_0; V)y_k - y_k \to \lambda_0 R(\lambda_0; V)y_0 - y_0,$$

$$R(\lambda_0; V)y_k \to R(\lambda_0; V)y_0.$$

A being closed, we conclude that $AR(\lambda_0; V)y_0 = VR(\lambda_0; V)y_0$ for every $y_0 \in \mathfrak{X}$. Here $R(\lambda_0; V)[\mathfrak{X}] = \mathfrak{D}(V)$ whence it follows that $Ax = Vx$ for $x \in \mathfrak{D}(V)$ so that $\mathfrak{D}(A) \supset \mathfrak{D}(V)$ and, finally, $\mathfrak{D}(A) = \mathfrak{D}(V)$. This proves that V is the infinitesimal generator of $[S(t)]$ and $S(t) = S(t; V)$, completing the proof of Lemma 23.9.5.

PROOF OF THEOREM 23.8.3. For $n = 1$, $V = U$, Lemma 23.9.5 reduces to Theorem 23.8.3.

We come now to the case $n > 1$ and start with $n = 2$.

THEOREM 23.9.5. *If $n = 2$ and assumptions* A_1 *and* A_2 *hold for $V = U$ and $V = -U$, then U is the infinitesimal generator of a group $[G(t; U); -\infty < t < +\infty]$ of class (C_0) and the solution $y(t; y_0, y_1)$ of the ACP^2 is given by formula* (23.9.6).

REMARK. It is important to observe that assuming (y_0, Uy_0) and $(y_0, -Uy_0)$ to be admissible initial values for every $y_0 \in \mathfrak{D}(U^2)$ implies both the group property and the expansion of the domain of admissible initial values to include every (y_0, y_1) with $y_0 \in \mathfrak{D}(U^2)$, $y_1 \in \mathfrak{D}(U) \cap \mathfrak{R}(U)$.

PROOF. By Lemma 23.9.5 U and $-U$ generate the semi-groups $[S(t; U)]$ and $[S(t; -U)]$ respectively, both of class (C_0). It remains to prove (23.9.4). This follows from an argument already employed in the proof of Theorem 12.3.2. Suppose that $y_0 \in \mathfrak{D}(U^2)$; then $S(t; U)[S(t; -U)y_0]$ is a strongly differentiable function of t whose derivative equals

$$S(t; U)US(t; -U)y_0 + S(t; U)(-U)S(t; -U)y_0 \equiv \theta$$

whence it follows that $S(t; U)S(t; -U)y_0 \equiv y_0$, its limit for $t \to 0+$, and, $\mathfrak{D}(U^2)$ being dense in $\mathfrak{X}$, this relation must hold for every $y_0 \in \mathfrak{X}$. In the same manner it is shown that $S(t; -U)S(t; U)y_0 \equiv y_0$. This proves (23.9.4). We can then define $G(t; U) = S(t; U)$ for $t > 0$ and $G(t; U) = S(t; -U)$ for $t < 0$ and refer to Theorem 23.9.4 to complete the proof.

THEOREM 23.9.6. *If $n > 2$ and assumptions* A_1 *and* A_2 *are satisfied by each of the operators $\eta^k U$, $\eta = \exp(2\pi i/n)$, $k = 0, 1, \cdots, n - 1$, then U is a bounded operator and the solution of the ACP^n is given by formula* (23.9.5).

PROOF. Since U satisfies A_1 and all n operators satisfy A_2, each $\eta^k U$ will generate a semi-group $\{S(t; \eta^k U)\}$ of class (C_0). This implies that $R(\lambda; \eta^k U) =$

$\eta^{-k}R(\lambda\eta^{-k}; U)$ exists and is bounded in a half-plane $\Re(\lambda) \geqq \sigma_k + \epsilon$. Hence $R(\lambda; U)$ has to exist and be bounded in n half-planes $\Re(\lambda\eta^{-k}) \geqq \sigma_k + \epsilon$. Since $n > 2$ this implies that $R(\lambda; U)$ is bounded in a neighborhood of the point at infinity, whence it follows that $R(\lambda; U)$ is holomorphic at infinity and that U is bounded. The rest follows from Theorem 23.9.3.

COROLLARY. *If $n > 3$ the conclusion of Theorem 23.9.6 holds if U satisfies A_1 and $\eta^k U$ satisfies A_2 for three values of k, $0 \leqq k_1 < k_2 < k_3 \leqq n - 1$, such that $k_2 - k_1$ and $k_3 - k_2$ are less than $\frac{1}{2}n$ while $k_3 - k_1$ exceeds $\frac{1}{2}n$.*

PROOF. The assumptions imply that $R(\lambda; U)$ is bounded in three half-planes which together cover a complete neighborhood of the point at infinity.

The last theorem and its Corollary show that if $n > 2$ and U is unbounded, the admissible initial values for the ACP^n are severely limited. The following theorem throws more light on this situation.

THEOREM 23.9.7. *If $n > 2$, if U satisfies A_1 and $\eta^{k_1}U$ and $\eta^{k_2}U$ satisfy A_2 where $0 < k_2 - k_1 < \frac{1}{2}n$, then $\eta^k U$ satisfies A_2 for $k_1 \leqq k \leqq k_2$ and there exists a semi-group $[S(t)]$, holomorphic in the sector $2k_1\pi/n < \arg t < 2k_2\pi/n$ and strongly continuous in the closed sector, such that*

$$S(s; \eta^k U) = S(s\eta^k), \qquad k_1 \leqq k \leqq k_2, \; s > 0. \tag{23.9.17}$$

Furthermore, if $k_2 - k_1 + 1 = p$ and p elements $y_0, y_1, \cdots, y_{p-1}$ of $\mathfrak{X}$ are given subject to the conditions

$$y_0 \in \mathfrak{D}(U^n), \quad y_1 \in \mathfrak{D}(U^{n-1}) \cap \Re(U), \quad \cdots, \quad y_{p-1} \in \mathfrak{D}(U^{n-p+1}) \cap \Re(U^{p-1}),$$

then there exists a unique set of p elements $x_{k_1}, x_{k_1+1}, \cdots, x_{k_2}$ in $\mathfrak{D}(U^{n-p})$ such that

$$y(t) \equiv \sum_{h=0}^{p-1} \frac{t^h}{h!} y_h + \frac{1}{(p-1)!}\int_0^t (t-s)^{p-1}\left[\sum_{k=k_1}^{k_2} \eta^{pk} S(\eta^k s; U)x_k\right] ds \tag{23.9.18}$$

satisfies conditions (i) *and* (ii) *of ACP^n as well as the first p conditions under* (iii). *Here*

$$\sum_k \eta^{(p-j)k} x_k = U^j y_{p-j}, \qquad j = 1, 2, \cdots, p. \tag{23.9.19}$$

The remaining $n - p$ conditions are satisfied if and only if

$$y_l = \sum_k \eta^{lk} U^{l-p} x_k, \qquad l = p, p+1, \cdots, n-1. \tag{23.9.20}$$

PROOF. We shall sketch the argument in the case in which p is odd. Without restricting the generality we may then assume that $\eta^{k_1} = e^{-i\alpha}$ and $\eta^{k_2} = e^{i\alpha}$ are conjugate imaginaries. The existence of $S(t; e^{\pm i\alpha}U)$ of class (C_0) implies the existence and boundedness of $R(\lambda; U)$ in a sector $|\arg(\lambda - \lambda_0)| \leqq \pi/2 + \alpha$, for sufficiently large real values of λ_0, together with the estimate

$$\| R(\lambda; U) \| \leqq M[d(\lambda) + \delta]^{-1},$$

where $d(\lambda)$ is the distance of λ from the boundary of the sector and $\delta > 0$. An application of Theorem 12.8.1 shows that U is the infinitesimal generator of a semi-group $S(t; U)$ of class $H(-\alpha, \alpha)$. Consequently, for fixed k, $k_1 \leqq k \leqq k_2$, the operators $[S(\eta^k s; U), s > 0]$ form a semi-group of class (C_0) with infinitesimal generator $\eta^k U$ so that (23.9.17) holds. By the corollary of Theorem 23.8.1 the function $S(\eta^k t; U)[y_0]$, $t \geqq 0$, is the unique solution of the ACP$_1^1$ for the operator $\eta^k U$ when $y_0 \in \mathfrak{D}(U)$. If $y_0 \in \mathfrak{D}(U^n)$, the solution has derivatives of order $\leqq n$ and the jth derivative tends to $(\eta^k U)^j y_0$ when $t \to 0+$. Thus $S(\eta^k t; U)[y_0]$, $y_0 \in \mathfrak{D}(U^n)$, gives a solution of the ACP^n for the operator $\eta^k U$ with initial values $(y_0, \eta^k U y_0, \cdots, (\eta^k U)^{n-1} y_0)$. This solution is unique since otherwise already the ACP$_1^1$ would have a nul solution. It follows that $\eta^k U$ satisfies A_2 for $k_1 \leqq k \leqq k_2$.

Let us now consider (23.9.18). It is clear that $y(t)$ has strongly continuous derivatives of order $\leqq n$. In particular,

$$y^{(n)}(t) = \sum_k S(\eta^k t; U) U^{n-p} x_k ,$$

which is meaningful for any choice of the x's in $\mathfrak{D}(U^{n-p})$. On the other hand it is obvious that $y(t)$ belongs at least to $\mathfrak{D}(U^{n-p})$. We can operate under the sign of integration with the closed operator U. Since

$$\sum_k \eta^{pk} U S(\eta^k s; U) x_k = \frac{d}{ds} \left\{ \sum_k \eta^{(p-1)k} S(\eta^k s; U) x_k \right\},$$

we can integrate by parts; using the first of the relations under (23.9.19) we can reduce $Uy(t)$ to an expression involving a polynomial in t of degree $p - 2$ and an integral of order $p - 1$. Repeating this process and using successively the relations under (23.9.19) we obtain

$$U^p y(t) = \textstyle\sum_k S(\eta^k t; U) x_k .$$

It follows that $U^{n-p}[U^p y(t)]$ exists and equals $y^{(n)}(t)$. Thus $y(t)$ is a solution of (23.9.1) and it is clear that it has the other properties stated in the theorem. Since $\det(\eta^{(p-j)k}) \neq 0$, the x's are uniquely determined by (23.9.19). Conditions (23.9.20) are the result of computing $\lim_{t \to 0+} y^{(l)}(t)$. It is not obvious that the solution is uniquely determined by the n initial conditions though Theorem 23.9.1 shows that there is no other solution of normal type.

A similar argument takes care of the case when p is even.

In Theorem 23.9.7 we have assumed that $k_2 - k_1 < \frac{1}{2}n$. If n is odd this is no restriction but if n is even it may happen that both $\eta^j U$ and $-\eta^j U$ satisfy A_2, U being unbounded. In this case $\eta^j U$ generates a group, just as in the case $n = 2$ discussed in Theorem 23.9.4. There are now two possibilities. Either $\eta^j U$ and $-\eta^j U$ are the only ones of the n operators $\eta^k U$ satisfying A_2, in which case formula (23.9.18) applies with $p = 2$ and k taking on the values j and $j + n/2$. Or else there are other admissible operators; we have then $p = \frac{1}{2}n + 1$ and formula (23.9.18) still applies.

If p is the number of operators $\eta^k U$ satisfying A_2, we call p the number of *degrees of freedom* of the ACP^n and $m = n - p$ the *deficiency*. It is clear that p gives the number of "arbitrary" initial conditions that may be prescribed. Assuming U to be unbounded, it follows from Theorems 23.9.6 and 23.9.7 that

$$(23.9.21) \qquad \begin{aligned} p &\leqq \tfrac{1}{2}(n+1) && \text{if } n \text{ is odd,} \\ p &\leqq \tfrac{1}{2}n + 1 && \text{if } n \text{ is even.} \end{aligned}$$

In concluding let us examine in a particular case what happens when non-admissible initial values are prescribed. We take $U = d/dx, \mathfrak{X} = L(-\infty, \infty)$, $n = 3$ so that the functional equation becomes

$$(23.9.22) \qquad \frac{\partial^3 y}{\partial t^3} = \frac{\partial^3 y}{\partial x^3}\,.$$

Here d/dx generates the group of translations: $G(t; d/dx)[f] = f(x + t)$, and the (continuous) spectrum of d/dx coincides with the imaginary axis. We have $p = 1$ and $\{f(x), \eta f'(x), \eta^2 f''(x)\}$, η a complex third root of unity, are non-admissible initial values. Let us try to find a solution anyway. We take

$$(23.9.23) \qquad f(x) = a_0(x - \eta)^{-1}(x + \eta)^{-1} + \sum_{k=-\infty}^{\infty}{}' a_k(x - k\eta)^{-2}, \qquad \sum_{-\infty}^{\infty} |a_k| < \infty,$$

where the dash indicates that $k = 0$ is omitted. This function is in $L(-\infty, \infty)$ and so are its derivatives of all orders. Moreover, if we vary $\{a_k\}$ in l we obtain a subset dense in $L(-\infty, \infty)$. It is easy to give a formal solution of the ACP^3 with these prohibited initial values: it is simply $f(x + \eta t)$. This function exists and is in $L(-\infty, \infty)$ together with its partials of all orders provided t is not an integer. If $a_k \neq 0$ for a particular k, then $f(x + k\eta)$ is not in $L(-\infty, \infty)$ since its norm is infinite. Thus we are dealing with a type of "explosive" solution and the initial values $f(x)$ leading to such solutions are dense in the space. Using the classical method of condensing singularities, we can evidently vary the pattern of explosions *ad lib.* We note that in this case the point spectrum of U is void and the spectrum is non-dense; for the ACP^1 explosive solutions have been encountered only when the point spectrum fills the plane.

Judging from this example it would seem as if the requirement that the solution of the ACP^n shall exist for all future time cannot be maintained when $n > 2$ unless there is proper consonance between the initial values.

References. Baker and Copson [1], F. Bernstein [1], Hadamard [**2, 3, 4**], Hille [**13, 18, 19, 20, 21**], Le Roux [1], Phillips [10], Picard [1], Volterra [1].

4. Probability Matrices

23.10. Stochastic processes; the Markoff algebra. We now consider semi-groups connected with stochastic processes of the homogeneous type depending upon a continuous time parameter t. A brief account of the terminology is necessary, but for further details the reader is referred to the authors quoted in the References.

We shall be concerned with the mathematical description of the changes with the time of a system the state of which is supposed to be described by a variable x taking values in a certain space $\mathfrak{X}$. The system defines a *stochastic process* and x is called a *stochastic variable*, if the probability of x belonging to a subset $\mathfrak{E}$ of $\mathfrak{X}$ is a known function of $\mathfrak{E}$ and t. The process is a *Markoff process* (*stochastically definite* in the terminology of Kolmogoroff) if the future development depends on the present state but not on the past history of the system. A Markoff process is characterized by its transition probabilities: a function $P(t_0, a, t, \mathfrak{E})$ which gives the probability that if $x(t_0) = a$ then $x(t) \in \mathfrak{E}$ where $t_0 < t$. These *transition probabilities* have to be consistent; under suitable assumptions concerning measurability this is expressed by the fact that $P(t_0, a, t, \mathfrak{E})$ satisfies the *Chapman-Kolmogoroff functional equation*, special instances of which will be given below. It is this equation which connects stochastic theory with the theorem of semi-groups. If $P(t_0, a, t + t_0, \mathfrak{E})$ is independent of t_0, we speak of a *temporally homogeneous Markoff process*; we restrict ourselves to this case in the following.

In the particular case in which $\mathfrak{X}$ is a discrete space with a finite or infinite set of points x_k and the process is homogeneous, we speak of a *simple Markoff chain*. Here the transition probabilities are determined by the probabilities

$$p_{jk}(t) = P(t_0, x_j, t_0 + t, x_k).$$

The probability matrix $\mathfrak{P}(t) = (p_{jk}(t))$ has the following properties:

$$\text{(23.10.1)} \qquad \begin{aligned} &\text{(i) } 0 \leqq p_{jk}(t) \leqq 1, && j, k = 1, 2, 3, \cdots, \\ &\text{(ii) } \sum_k p_{jk}(t) \equiv 1, && j = 1, 2, 3, \cdots, \\ &\text{(iii) } \mathfrak{P}(t_1 + t_2) = \mathfrak{P}(t_1)\mathfrak{P}(t_2), && t_1 > 0, t_2 > 0. \end{aligned}$$

Here the first two properties are obvious consequences of the definition of $p_{jk}(t)$ while the semi-group property is the form which the functional equation of Chapman and Kolmogoroff assumes in the present case.

With a view of applying our general theory to the present situation, we start by embedding the probability matrices in a complex ($\mathfrak{B}$)-algebra $\mathfrak{M}$, hereinafter called the *Markoff algebra*. The matrix $\mathfrak{B} = (b_{jk})$, where the b_{jk} are complex numbers, belongs to $\mathfrak{M}$ if

$$\text{(23.10.2)} \qquad \sup_j \sum_k |b_{jk}| = \|\mathfrak{B}\| < \infty.$$

We define the arithmetical operations by the usual conventions in matrix theory and note that the system is closed under these operations. In particular we find that multiplication is associative and non-commutative, that the unit matrix is the unit element, and that $\|(\delta_{jk})\| = 1$, $\|\mathfrak{B}_1\mathfrak{B}_2\| \leqq \|\mathfrak{B}_1\| \|\mathfrak{B}_2\|$. Finally it is a simple matter to prove that $\mathfrak{M}$ is complete under this norm so that $\mathfrak{M}$ is actually a (B)-algebra.

The Markoff algebra plays a role also outside the theory of probability. Thus all convergence preserving matrices of sequence to sequence transformations belong to $\mathfrak{M}$. In particular the algebra **H** of section 23.5 is (isomorphic to) a subalgebra of $\mathfrak{M}$ and the matrices $\mathfrak{T}(\zeta)$, $\mathfrak{R}(\lambda)$ and $\mathfrak{R}(\lambda;\zeta)$ of section 23.2 belong to $\mathfrak{M}$.

23.11. One-parameter semi-groups in the Markoff algebra. The problem of determining all probability matrices is evidently a special instance of determining all matrices $\mathfrak{B}(t) = (b_{jk}(t))$, defined for $t > 0$, such that $\mathfrak{B}(t) \in \mathfrak{M}$ and

$$\mathfrak{B}(t_1 + t_2) = \mathfrak{B}(t_1)\mathfrak{B}(t_2) \tag{23.11.1}$$

for all positive values of t_1 and t_2.

The metric imposed on $\mathfrak{B}$ is fairly restrictive. Thus continuity in norm of $\mathfrak{B}(t)$ requires that

$$\lim_{h\to 0} \sup_j \sum_k |b_{jk}(t+h) - b_{jk}(t)| = 0 \tag{23.11.2}$$

for fixed t. This implies not merely that each function $b_{jk}(t)$ is continuous, but that the sets $\{b_{jk}(t)\}$ and $\{\sum_k b_{jk}(t)\}$ are made up of equi-continuous functions.

We shall see below that measurability of every $b_{jk}(t)$ for $t > 0$ is (necessary and) sufficient for continuity of these functions and actually implies considerably more but, apparently, not continuity in norm of $\mathfrak{B}(t)$.

The continuity properties of $\mathfrak{B}(t)$ for $t \to 0+$ are still more complicated. For probability matrices J. L. Doob [2, p. 42] has determined the form of the possible limiting matrices.

The simplest, but also the most restrictive, case is that in which $\lim_{t\to 0+} \mathfrak{B}(t)$ exists in the sense of the metric of $\mathfrak{M}$. Assuming the limit to be the unit matrix (δ_{jk}), this case will arise if and only if

$$\lim_{t\to 0+} \sup_j \sum_k |b_{jk}(t) - \delta_{jk}| = 0. \tag{23.11.3}$$

Theorem 9.4.2 then guarantees the existence of a generating matrix $\mathfrak{A} = (a_{jk}) \in \mathfrak{M}$ such that

$$\mathfrak{B}(t) = \exp(t\mathfrak{A}), \qquad \mathfrak{A} = (b'_{jk}(0)). \tag{23.11.4}$$

Since

$$\mathfrak{B}'(t) = \mathfrak{A}\mathfrak{B}(t) = \mathfrak{B}(t)\mathfrak{A}, \tag{23.11.5}$$

we see that the functions $b_{jk}(t)$ satisfy the following two infinite systems of linear differential equations

$$b'_{jk}(t) = \sum_m a_{jm}b_{mk}(t), \tag{23.11.6}$$

$$b'_{jk}(t) = \sum_m a_{mk}b_{jm}(t), \tag{23.11.7}$$

$$j, k = 1, 2, 3, \cdots,$$

which are analogous to the differential equations of Kolmogoroff for the case of probability matrices.

In the probability case (23.11.3) requires that $p_{jj}(t) \to 1$ when $t \to 0+$ uniformly in j so that $p_{jk}(t) \to \delta_{jk}$ uniformly in (j, k). The elements of the matrix $\mathfrak{A}$ satisfy the conditions

$$(23.11.8) \qquad a_{jj} \leqq 0, \qquad a_{jk} \geqq 0, \qquad j \neq k, \qquad \text{and} \qquad \sum_k a_{jk} = 0,$$

where the a_{jj}'s form a bounded sequence. As an example illustrating this case we may take the *Poisson distribution* defined by

$$p_{jk}(t) = \frac{(at)^{k-j}}{(k-j)!} e^{-at}, \qquad k \geqq j,$$

$$p_{jk}(t) = 0, \qquad j > k.$$

The infinitesimal generator is the matrix (a_{jk}) with

$$a_{j,j+1} = a, \qquad a_{jj} = -a, \qquad a_{jk} = 0 \text{ otherwise}.$$

Cf. A. Kolmogoroff [**1**, p. 437]. The matrix (a_{jk}) evidently satisfies (23.11.8) and $(p_{jk}(t))$ satisfies (23.11.3).

23.12. Transition operators. We get results of a more general nature by letting $\mathfrak{B}(t)$ act as an operator, to be denoted by $B(t)$, on a suitably chosen (B)-space $\mathfrak{X}$. We take $\mathfrak{X} = l$, the space of absolutely convergent series, $x = \{\xi_n\}$, with $\| x \| = \sum_1^\infty | \xi_n |$, and define

$$(23.12.1) \qquad y = B(t)[x] = \{\eta_n\}, \qquad \eta_n = \sum_{m=1}^{\infty} \xi_m b_{mn}(t).$$

The norm of $B(t)$ as an operator on l to itself is equal to the norm of $\mathfrak{B}(t)$ as an element of $\mathfrak{M}$. It is obvious that $\| B(t) \| \leqq \| \mathfrak{B}(t) \|$ and, since $\eta_n = b_{kn}(t)$ if $\xi_m = 1$ for $m = k$ and 0 otherwise, we must have equality.

These transformations have statistical significance. Let $B(t) = P(t)$ and $0 \leqq \xi_m$, $\sum_1^\infty \xi_n = 1$. Then $x = \{\xi_n\}$ is the distribution function of a stochastic variable X having a countable number of possible states, the probability of X being in the nth state being ξ_n. If X is subjected to a temporally homogeneous Markoff process with transition probabilities $\mathfrak{P}(t) = (p_{jk}(t))$, then $y = P(t)[x]$ is the distribution function of X at the time t.

$P(t)$ acting in l is a *transition operator* in the sense of Definition 11.7.4. For the terminology, see G. Birkhoff [**5**, p. 137] where the term transition operator is introduced for a linear operator on a Banach lattice mapping the set of positive elements of norm one into itself.

We start by proving the following

THEOREM 23.12.1. *A necessary and sufficient condition that the semi-group* $B(t)$

be strongly continuous as a function on $(0, \infty)$ *to* $\mathfrak{E}(l)$ *is that every* $b_{jk}(t)$ *is measurable for* $t > 0$.

PROOF. The condition is obviously necessary. To prove the sufficiency we observe that if the $b_{jk}(t)$ are measurable for $t > 0$ and if $\{\alpha_k\}$ is any bounded sequence of complex numbers, then the absolutely convergent series

$$\sum_k \alpha_k \sum_j \xi_j b_{jk}(t), \qquad \sum |\xi_j| < \infty,$$

represents a measurable function of t. It follows that $B(t)$ is a weakly measurable operator. Since l is a separable space, weak measurability implies strong measurability and by Theorem 10.2.3 this makes $B(t)$ strongly continuous for $t > 0$. We have consequently proved that

$$\lim_{h\to 0} \sum_{k=1}^{\infty} \left| \sum_{j=1}^{\infty} \xi_j[b_{jk}(t + h) - b_{jk}(t)] \right| = 0 \tag{23.12.2}$$

for $t > 0$ and every choice of quantities ξ_j such that $\sum |\xi_j|$ converges. If $\xi_j = 1$ for $j = n$ and 0 otherwise, we see that

$$\lim_{h\to 0} \sum_{k=1}^{\infty} |b_{nk}(t + h) - b_{nk}(t)| = 0, \qquad n = 1, 2, 3, \cdots. \tag{23.12.3}$$

Thus, in particular, *every* $b_{nk}(t)$ *is continuous for* $t > 0$, a result first established by J. L. Doob [2] for a probability matrix. For the case of finite probability matrices we refer to the treatise of M. Fréchet [6].

On the other hand, measurability does not necessarily imply strong continuity at the origin, that is,

$$\lim_{t\to 0+} \sum_{k=1}^{\infty} \left| \sum_{j=1}^{\infty} \xi_j[b_{jk}(t) - \delta_{jk}] \right| = 0, \tag{23.12.4}$$

where we have assumed the limit matrix to be the unit matrix $\mathfrak{J}$ which is a simple but typical case. For the case of probability matrices this condition may be given a simpler form.

THEOREM 23.12.2. *A necessary and sufficient condition that a semi-group of probability matrices* $\mathfrak{P}(t) = (p_{jk}(t))$ *be strongly continuous at* $t = 0$ *as an operator on* l *is that*

$$\lim_{t\to 0+} p_{mm}(t) = 1, \qquad m = 1, 2, 3, \cdots. \tag{23.12.5}$$

PROOF. Suppose that (23.12.4) holds for all $\{\xi_n\} \in l$ where $b_{jk}(t)$ is replaced by $p_{jk}(t)$. It holds then, in particular, for $\xi_n = \delta_{mn}$, m fixed but arbitrary. In this case the double series reduces to

$$\sum_{n=1}^{\infty} |p_{mn}(t) - \delta_{mn}| = 2[1 - p_{mm}(t)].$$

Thus the condition is necessary. It is also sufficient, because if it holds for all m, then (23.12.4) holds for all finite sums, that is, the case in which all but a finite number of the ξ_n's are zero. Such elements are dense in l. Since $\| \mathfrak{P}(t) \| = 1$, the Banach-Steinhaus theorem shows that (23.12.4) holds for all $\{\xi_n\}$. It is clear that (23.12.4) implies

$$\lim_{t\to 0+} p_{mn}(t) = \delta_{mn}, \qquad m, n = 1, 2, 3, \cdots. \tag{23.12.6}$$

In the remainder of this section we restrict ourselves to semi-groups of probability matrices. If (23.12.5) holds, the general theory in Chapter X asserts the existence of a subset $\mathfrak{D} = \mathfrak{D}(A)$ dense in l such that for $x \in \mathfrak{D}$ the strong limit of $t^{-1}[P(t) - I][x]$ exists when $t \to 0+$, but we cannot conclude from this fact that the $p_{jk}(t)$ are differentiable. On the other hand, it has been shown by J. L. Doob [2, 3] and A. N. Kolmogoroff [4] that the conditions (23.12.5) imply the existence of

$$\lim_{t\to 0+} t^{-1}[p_{jj}(t) - 1] \equiv a_{jj}, \qquad \lim_{t\to 0+} t^{-1}p_{jk}(t) \equiv a_{jk}, \tag{23.12.7}$$

where $0 \leqq -a_{jj} \leqq +\infty$ and $0 \leqq a_{jk} < \infty$. Further

$$\sum_{k=1}^{\infty} a_{jk} \leqq 0, \tag{23.12.8}$$

and, if equality holds in (23.12.8), Doob showed the existence of $p'_{jk}(t)$ for $t > 0$ satisfying the equations

$$p'_{jk}(t) = \sum_{m=1}^{\infty} a_{jm}p_{mk}(t), \qquad k = 1, 2, \cdots. \tag{23.12.9}$$

Recently D. G. Austin [1] has shown the existence of $p'_{jk}(t)$ assuming only the finiteness of a_{jj}. Assuming all a_{jj} to be finite and the equality to hold in (23.12.8) for all j, Doob showed that

$$p'_{jk}(t) \geqq \sum_{m=1}^{\infty} p_{jm}(t)a_{mk}$$

where inequality is not excluded.

Equations (23.12.9) and the formally adjoint equations

$$p'_{jk}(t) = \sum_{m=1}^{\infty} p_{jm}(t)a_{mk} \tag{23.12.10}$$

are known as the *differential equations of Kolmogoroff* (see the latter's paper [1] where these equations were first derived under more restrictive assumptions). The corresponding matrix $\mathfrak{A} = (a_{jk})$ with

$$a_{jj} \leqq 0, \qquad a_{jk} \geqq 0, \qquad j \neq k, \qquad \sum_{k=1}^{\infty} a_{jk} = 0, \qquad j = 1, 2, \cdots, \tag{23.12.11}$$

is called a *Kolmogoroff matrix*. Not every semi-group of transition matrices $\mathfrak{P}(t)$ satisfies a canonical system of Kolmogoroff equations. Thus Kolmogoroff himself [4] has given examples, analyzed in more detail from the point of view of the theory of semigroups by D. G. Kendall and G. E. H. Reuter [1], examples in which either $a_{11} = -\infty$ or a_{11} is finite but $\sum a_{1k} < 0$. In both cases the differential equations take unconventional forms:

$$p'_{1k}(t) = \sum_{m=2}^{\infty} [p_{mk}(t) - p_{1k}(t)] \quad \text{and} \quad p'_{1k}(t) = \lim_{m \to \infty} p_{mk}(t) - p_{1k}(t)$$

respectively for the elements of the first row.

So far the problem of determining the infinitesimal generator of a semi-group of class (C_0), defined by transition matrices $\mathfrak{P}(t)$, has not been solved in the general case. More progress has been made with the inverse question, that of deciding when a Kolmogoroff matrix $\mathfrak{A}$ generates a semi-group $[P(t)]$ of class (C_0) in l.

Here we have first to mention the problem of integrating the Kolmogoroff equations

$$\mathfrak{Y}'(t) = \mathfrak{A}\mathfrak{Y}(t), \qquad \mathfrak{Y}(0) = \mathfrak{I}, \tag{23.12.12}$$

$$\mathfrak{Z}'(t) = \mathfrak{Z}(t)\mathfrak{A}, \qquad \mathfrak{Z}(0) = \mathfrak{I}, \tag{23.12.13}$$

where $\mathfrak{A}$ is a given matrix satisfying conditions (23.12.11) and the matrix equations are simply abbreviated notation for systems of equations of the type (23.12.9) and (23.12.10) respectively. Existence and uniqueness proofs were given by Kolmogoroff [1] under severe restrictions on $\mathfrak{A}$. The first general existence proof for these and more general functional equations of probabilistic origin is due to W. Feller [2], this was followed by a simple proof applicable to the denumerable case due to J. L. Doob [3]. An elegant argument based on a limiting passage from the finite case has been given by G. E. H. Reuter and W. Ledermann [1].

We may formulate the existence theorems as follows.

THEOREM 23.12.3. *Given a Kolmogoroff matrix $\mathfrak{A} = (a_{jk})$, there exists at least one positive continuously differentiable matrix function $\mathfrak{Y}(t) = (y_{jk}(t)) \in \mathfrak{M}$ and at least one positive continuously differentiable matrix function $\mathfrak{Z}(t) = (z_{jk}(t)) \in \mathfrak{M}$, each defined by matrices of norm $\leqq 1$, such that*

$$y'_{jk}(t) = \sum_{m=1}^{\infty} a_{jm} y_{mk}(t), \qquad \lim_{t \to 0+} y_{jk}(t) = \delta_{jk},$$

$$j, k = 1, 2, 3, \cdots,$$

$$z'_{jk}(t) = \sum_{m=1}^{\infty} z_{jm}(t) a_{mk}, \qquad \lim_{t \to 0+} z_{jk}(t) = \delta_{jk}.$$

In particular, there exists a minimal solution, a positive matrix function $\mathfrak{P}(t) =$

$(p_{jk}(t))$, *satisfying both systems and such that for any solutions* $\mathfrak{Y}(t)$ *and* $\mathfrak{Z}(t)$ *we have* $\mathfrak{Y}(t) > \mathfrak{P}(t)$, $\mathfrak{Z}(t) > \mathfrak{P}(t)$. *Here* $\mathfrak{P}(s + t) = \mathfrak{P}(s)\mathfrak{P}(t) = \mathfrak{P}(t)\mathfrak{P}(s)$.

REMARK. We recall that a matrix is said to be positive if it has non-negative elements and $\mathfrak{B}_1 > \mathfrak{B}_2$ signifies that $\mathfrak{B}_1 - \mathfrak{B}_2$ is positive.

PROOF. Using an adaptation of Doob's argument we set

(23.12.14) $\mathfrak{A} = \mathfrak{D} + \mathfrak{T}$, $\mathfrak{D} = (\delta_{jk}a_{jk})$, $\mathfrak{T} = ((1 - \delta_{jk})a_{jk})$, $\mathfrak{E}(t) = (\delta_{jk}e^{a_{jj}t})$,

and solve the symbolic integral equations

$$\mathfrak{Y}(t) = \mathfrak{E}(t) + \int_0^t \mathfrak{E}(t - s)\mathfrak{T}\mathfrak{Y}(s)\, ds, \tag{23.12.15}$$

$$\mathfrak{Z}(t) = \mathfrak{E}(t) + \int_0^t \mathfrak{Z}(s)\mathfrak{T}\mathfrak{E}(t - s)\, ds, \tag{23.12.16}$$

by the method of successive approximations. Here we are using the same type of abbreviated notation as above so that, for instance, the integral of a matrix is defined as the matrix of the integrals. We set

$$\mathfrak{Y}_0(t) = \mathfrak{E}(t), \qquad \mathfrak{Y}_n(t) = \mathfrak{E}(t) + \int_0^t \mathfrak{E}(t - s)\mathfrak{T}\mathfrak{Y}_{n-1}(s)\, ds,$$

$$\mathfrak{Z}_0(t) = \mathfrak{E}(t), \qquad \mathfrak{Z}_n(t) = \mathfrak{E}(t) + \int_0^t \mathfrak{Z}_{n-1}(s)\mathfrak{T}\mathfrak{E}(t - s)\, ds.$$

We start with the first set of equations. It is an easy matter to verify that the matrices $\mathfrak{Y}_n(t)$ exist, are positive, and form a monotone increasing sequence. Further, if $\mathfrak{Y}_n(t) = (y^n_{jk}(t))$, one proves by induction that $\sum_k y^n_{jk}(t) \leqq 1$ so that $\|\mathfrak{Y}_n(t)\| \leqq 1$. It follows that the sequence $\{y^n_{jk}(t)\}$ (j, k fixed) is monotone increasing and dominated by one. Hence $\lim_{n\to\infty} y^n_{jk}(t) \equiv p_{jk}(t)$ exists. Moreover the convergence is uniform with respect to k for j fixed and

$$\sum_{k=1}^{\infty} y^n_{jk}(t) \to \sum_{k=1}^{\infty} p_{jk}(t) \leqq 1.$$

We set $(p_{jk}(t)) = \mathfrak{P}(t)$. It is evidently a solution of (23.12.15) as well as of (23.12.12). Next we note that

$$\int_0^t \mathfrak{E}(t - s)\mathfrak{T}\mathfrak{E}(s)\, ds = \int_0^t \mathfrak{E}(s)\mathfrak{T}\mathfrak{E}(t - s)\, ds$$

so that $\mathfrak{Y}_1(t) = \mathfrak{Z}_1(t)$ and the iteration of this argument shows that $\mathfrak{Y}_n(t) = \mathfrak{Z}_n(t)$ for all n. It follows that $\mathfrak{P}(t)$ also satisfies (23.12.16) as well as (23.12.13). This implies that the two matrices $\mathfrak{A}$ and $\mathfrak{P}(t)$ commute in the following sense. Each matrix product $\mathfrak{A}\mathfrak{P}(t)$ and $\mathfrak{P}(t)\mathfrak{A}$ is a well defined matrix, the elements of which are given by absolutely convergent series, and corresponding matrix elements are identical. Normally $\mathfrak{A}\mathfrak{P}(t)$ is not in $\mathfrak{M}$, but the sum of the elements in each fixed row is an absolutely convergent series since

$$\sum_k \sum_m |a_{jm}|\, p_{mk}(t) = \sum_m |a_{jm}| \sum_k p_{mk}(t) \leqq \sum_m |a_{jm}| = 2\,|a_{jj}|.$$

If $\mathfrak{Y}(t)$ is any positive solution of (23.12.12) or, equivalently, of (23.12.15), then $\mathfrak{Y}(t) > \mathfrak{E}(t)$ and

$$\mathfrak{Y}(t) - \mathfrak{Y}_n(t) = \int_0^t \mathfrak{E}(t - s)\mathfrak{T}[\mathfrak{Y}(s) - \mathfrak{Y}_{n-1}(s)]\, ds$$

so that the left member is positive for all n and hence this is true of the limit $\mathfrak{Y}(t) - \mathfrak{P}(t)$. The same argument applies to the other system. Thus $\mathfrak{P}(t)$ is the minimal solution of both systems.

To prove the semi-group property we form

$$\mathfrak{P}(s, t) = \mathfrak{P}(t - s)\mathfrak{P}(s).$$

Representing $\mathfrak{P}(t - s)$ by (23.12.16) and $\mathfrak{P}(s)$ by (23.12.15), multiplying out, and making use of the exponential property of $\mathfrak{E}(t)$, we obtain

$$\mathfrak{P}(s, t) = \mathfrak{E}(t) + \int_0^{t-s} \mathfrak{P}(u)\mathfrak{T}\mathfrak{E}(t)\, du + \int_0^s \mathfrak{E}(t)\mathfrak{T}\mathfrak{P}(u)\, du + \int_0^{t-s}\int_0^s \mathfrak{P}(u)\mathfrak{T}\mathfrak{E}(t - u - v)\mathfrak{T}\mathfrak{P}(v)\, dv\, du;$$

the interchange of summation and integration is justified by the non-negativity of the elements of all the matrices employed. If we now differentiate with respect to s we get for almost all s in $(0, t)$

$$-\mathfrak{P}(t - s)\mathfrak{T}\mathfrak{E}(t) + \mathfrak{E}(t)\mathfrak{T}\mathfrak{P}(s) - \int_0^s \mathfrak{P}(t - s)\mathfrak{T}\mathfrak{E}(s - v)\mathfrak{T}\mathfrak{P}(v)\, dv + \int_0^{t-s} \mathfrak{P}(u)\mathfrak{T}\mathfrak{E}(t - s - u)\mathfrak{T}\mathfrak{P}(s)\, du,$$

and combining alternate terms by means of (23.12.15) and (23.12.16) we see that $\partial\mathfrak{P}(s, t)/\partial s = 0$ for almost all s. On the other hand $\lim_{s\to 0+} \mathfrak{P}(s, t) = \mathfrak{P}(t)$ so that $\mathfrak{P}(s, t) \equiv \mathfrak{P}(t)$ for all s in $(0, t)$ and the semi-group property is proved. This completes the proof of Theorem 23.12.3.

If equations (23.12.12) or (23.12.13) do not have unique solutions, there will exist nul solutions in $\mathfrak{M}$. We can then construct solutions having strong derivatives of order $\leqq n$ and tending to the zero matrix together with the first n derivatives when $t \to 0+$, all limits being taken with respect to the norm in $\mathfrak{M}$. Though Theorem 23.8.2 does not apply to the present situation, a similar result holds (see Hille [**23**] and cf. T. Kato [**1**]):

THEOREM 23.12.4. *A necessary and sufficient condition for the existence of a continuously differentiable matrix function $\mathfrak{Y}(t) \in \mathfrak{M}$ such that*

$$\mathfrak{Y}'(t) = \mathfrak{A}\mathfrak{Y}(t), \qquad \lim_{t\to 0+} \| \mathfrak{Y}(t) \| = 0, \qquad \limsup_{t\to\infty} t^{-1} \log \| \mathfrak{Y}(t) \| \leqq \omega_0,$$

is the existence of a matrix $\mathfrak{X}(\lambda) \in \mathfrak{M}$ such that $\mathfrak{X}(\lambda)$ is bounded and holomorphic in every half-plane $\mathfrak{R}(\lambda) \geqq \omega_0 + \epsilon$, $\epsilon > 0$, and

$$\mathfrak{A}\mathfrak{X}(\lambda) = \lambda\, \mathfrak{X}(\lambda). \tag{23.12.17}$$

An equivalent condition is the existence of a column vector $x^*(\lambda) \in m$ such that

$$\mathfrak{A}x^*(\lambda) = \lambda x^*(\lambda), \tag{23.12.18}$$

$x^*(\lambda)$ being bounded and holomorphic for $\mathfrak{R}(\lambda) \geqq \omega_0 + \epsilon$.

PROOF. The conditions are equivalent, for if $\mathfrak{X}(\lambda)$ exists, then each of its columns satisfies (23.12.18) and is in m, while if $x^*(\lambda)$ exists the matrix whose kth column is $2^{-k}x^*(\lambda)$ will serve for $\mathfrak{X}(\lambda)$. One verifies the first condition as in the proof of Theorem 23.8.2. If $\mathfrak{Y}(t)$ is a nul solution, then its Laplace transform $\mathfrak{X}(\lambda)$ gives the required solution of (23.12.17). Conversely, if $\mathfrak{X}(\lambda)$ exists, the inverse Laplace transform of a suitable scalar multiple of $\mathfrak{X}(\lambda)$ gives the desired nul solution $\mathfrak{Y}(t)$. In both cases the interchange of integration and matrix multiplication is justified by absolute and dominated convergence of the series involved.

In the case of the adjoint equation (23.12.13) the existence of nul solutions presents more of a problem, since the product $\mathfrak{B}\mathfrak{A}$ need not exist for an arbitrary $\mathfrak{B} \in \mathfrak{M}$. In all cases known to us, however, nul solutions of (23.12.13) coexist with solutions of $\mathfrak{W}(\lambda)\mathfrak{A} = \lambda\mathfrak{W}(\lambda)$ and the analogue of Theorem 23.12.4 holds. See Hille [22, 23].

We turn now to the problem of when a Kolmogoroff matrix $\mathfrak{A}$ generates a semi-group of transition operators in l. Here the significant results are due to Tosio Kato [1]. The matrix $\mathfrak{A}$ defines an operator A on l by the formulas

$$(23.12.19) \quad y = A[x] = \{\eta_n\}, \qquad x = \{\xi_n\}, \qquad \eta_n = \sum \xi_m a_{mn},$$

where $\sum |\xi_n| < \infty$. It is supposed that the series defining η_n are all absolutely convergent and that $\sum |\eta_n| < \infty$. The set $\mathfrak{D}(A)$ of all vectors x satisfying these conditions is by definition the domain of A. We note that all the base vectors $x_j = \{\delta_{jk}\}, j = 1, 2, \cdots$, belong to $\mathfrak{D}(A)$; these vectors span a linear subspace $\mathfrak{D}_0$ of $\mathfrak{D}(A)$. It is clear that $\mathfrak{D}_0$ is dense in l. We define A_0 as the restriction of A to $\mathfrak{D}_0$ so that $\mathfrak{D}(A_0) = \mathfrak{D}_0$. Kato's main theorem then reads:

THEOREM 23.12.5. *Given a Kolmogoroff matrix $\mathfrak{A}$ and the corresponding linear operator A_0, there exists at least one semi-group of class (C_0) acting in l the infinitesimal generator of which is an extension of A_0. In particular, there exists a minimal solution, a semi-group $[P(t)]$ of positive contraction operators, such that any semi-group $[P_1(t)]$ whose generator is an extension of A_0 satisfies $P_1(t) > P(t)$. $P(t)$ is a transition operator if and only if there is no non-trivial $x^*(\lambda) \in m$ satisfying* (23.12.18) *for some $\lambda > 0$.*

REMARK. An element $x = \{\xi_n\} \in l$ is said to be positive if $\xi_n \geqq 0$ for each n. We denote the set of positive elements by l^+. The space l is partially ordered in the sense of section 1.11 under the convention that $x < y$ means $y - x \in l^+$. We observe that a monotone increasing sequence $\{x_n\}, x_n \in l$, such that $\| x_n \| < M$ for all n is convergent. The proof of this is left to the reader. A linear operator U with domain, $\mathfrak{D}(U)$, and range in l is positive in the sense of section 11.7 if it maps $\mathfrak{D}(U) \cap l^+$ into l^+ and the inequality $U < V$ signifies that $V - U$ is positive. We observe that a monotone increasing sequence of bounded linear operators $\{U_n\}$ on l to l such that $\| U_n \| \leqq M$ for all n converges to a limit in the strong operator topology in l; this is an immediate consequence of the previous observation.

PROOF. The proof of Theorem 23.12.5 is based upon a perturbation argument just as that of Theorem 23.12.3; we shall see below that Theorem 13.5.1 does not apply, however, to the present situation. To the matrices $\mathfrak{D}$ and $\mathfrak{T}$ of (23.12.14) correspond linear operators D and T acting in l. The domain of D, $\mathfrak{D}(D)$ say, is the subset of l formed by the vectors $x = \{\xi_n\}$ such that $\sum |a_{nn}| |\xi_n| < \infty$ and for such x

$$D[x] = \{a_{nn}\xi_n\}.$$

We note that every $x_j = \{\delta_{jk}\}$ is in $\mathfrak{D}(D)$ so the latter is dense in l. Further D is closed; its resolvent is given by

$$R(\lambda; D)[x] = \left\{\frac{\xi_n}{\lambda - a_{nn}}\right\}$$

and $\lambda R(\lambda; D)$ is a positive contraction operator when $\lambda > 0$. For $x \in \mathfrak{D}(D)$ the series

$$\sum_m \sum_{n \neq m} a_{nm} |\xi_n| = \sum_n |a_{nn}| |\xi_n| = \| D[x] \|$$

converges. We can then define $T[x]$ by

$$T[x] = \{\sum_{n \neq m} a_{nm}\xi_n\}, \qquad x \in \mathfrak{D}(D),$$

so that $\| T[x] \| \leqq \| D[x] \|$, $x \in \mathfrak{D}(D)$, with equality for positive elements $x \in l^+ \cap \mathfrak{D}(D) \equiv \mathfrak{D}^+(D)$.

Next we note that $B(\lambda) \equiv TR(\lambda; D)$ is a positive linear contraction operator for $\lambda > 0$ since

$$\| TR(\lambda; D)[x] \| \leqq \sum_m \sum_{n \neq m} a_{nm} \frac{|\xi_n|}{\lambda - a_{nn}} = \sum_n \frac{|a_{nn}|}{\lambda - a_{nn}} |\xi_n| \leqq \| x \|.$$

A simple computation shows that $B(\lambda) < B(\mu)$ if $0 < \mu < \lambda$.

It is easily shown that D is the infinitesimal generator of a semi-group $[S(t; D)]$ of class (C_0) and, since $\| B(\lambda) \| \leqq 1$ for $\lambda < 0$, the operator T belongs to the class $\mathfrak{J}(D)$ in the sense of Definition 13.3.1. It does not belong to the class $\mathfrak{P}(D)$ of Definition 13.3.5, however, because $\| TS(t; D) \|_D$ is monotone decreasing and $\geqq (et)^{-1}$ when $t = |a_{nn}|^{-1}$. Hence this norm cannot be integrable over $(0, 1)$ unless D is a bounded operator. This excludes direct application of the perturbation theory of Chapter XIII and calls for a more roundabout attack.

We now introduce an auxiliary parameter r, $0 \leqq r < 1$, and form the resolvent of the operator $D + rT$ with domain $\mathfrak{D}(D)$. Since

$$(\lambda I - D - rT)^{-1} = R(\lambda; D)\, [I + rB(\lambda)]^{-1}$$

we see that $D + rT$ is closed and that

$$R(\lambda; D + rT) = R(\lambda; D) + \sum_1^\infty R(\lambda; D)[B(\lambda)]^n r^n.$$

This series is absolutely convergent for $\lambda > 0$ and defines a linear bounded positive operator on l. Now for $y \in l^+$, $x \equiv R(\lambda; D + rT)y \in \mathfrak{D}^+(D)$ and we have $\| T[x] \| = \| D[x] \|$; hence

$$\| (\lambda I - D - rT)[x] \| \geqq \| (\lambda I - D)[x] \| - r \| T[x] \|$$
$$= \lambda \| x \| + (1 - r) \| D[x] \| \geqq \lambda \| x \|.$$

This implies that $\lambda \| R(\lambda; D + rT)[y] \| \leqq \| y \|$ for all y in l^+ and inasmuch as $R(\lambda; D + rT)$ is positive the same inequality holds everywhere in l, that is, $\lambda R(\lambda; D + rT)$ is a positive contraction operator. Since $D + rT$ is closed and its domain is dense in l, Theorems 12.3.1, Corollary, and 11.7.2 show that $D + rT$ is the infinitesimal generator of a semi-group $[P(t; r)]$ of class (C_0) where $P(t; r)$ is a positive contraction operator. Obviously $R(\lambda; D + rT)$ is a positive monotone increasing operator function of r. Since

$$P(t; r)[x] = \lim_{n\to\infty} \left\{\frac{n}{t} R\left(\frac{n}{t}; D + rT\right)\right\}^n [x],$$

it follows that $P(t; r_1) < P(t; r_2)$ if $r_1 < r_2 < 1$. Since further $\| P(t; r) \| \leqq 1$, the observations made in the Remark above show that $\lim_{r\to 1-} P(t; r) \equiv P(t)$ exists in the strong operator topology. Here $P(t)$ is also a positive contraction operator, $P(t)$ is strongly continuous for $t > 0$, and $P(t)$ has the semi-group property for $P(t; r)$ has these properties and they are preserved under strong convergence. As to the continuity it suffices to observe that $P(t)$, as a strong limit of the strongly continuous operator function $P(t; r)$, is at least strongly measurable, this together with the semi-group property gives strong continuity for $t > 0$ by Theorem 10.2.3.

It remains to prove continuity at $t = 0$. We may write

$$P(t; r)[x_j] = \sum_{k=1}^{\infty} p_{jk}(t; r)x_k, \qquad 0 < r < 1,\ j = 1, 2, \cdots.$$

Here the coefficients $p_{jk}(t; r)$ are numerically valued non-negative continuous functions of t for $t \geqq 0$, $p_{jk}(0; r) = \delta_{jk}$, $\sum_{k=1}^{\infty} p_{jk}(t; r) \leqq 1$, and $p_{jk}(t; r)$ is a monotone increasing function of r for each t, j, k. As $r \to 1-$ each $p_{jk}(t; r)$ converges to a limit $p_{jk}(t)$ where $p_{jk}(t) \geqq 0$ and $\sum_k p_{jk}(t) \leqq 1$. We have naturally

$$P(t)[x_j] = \sum_{k=1}^{\infty} p_{jk}(t)x_k .$$

Given a j and ϵ, $\epsilon > 0$, we can find a δ such that $0 \leqq 1 - p_{jj}(t; 0) \leqq \epsilon$ for $0 < t < \delta$, whence it follows that the same inequalities hold for $1 - p_{jj}(t; r)$ and $1 - p_{jj}(t)$. Consequently $p_{jj}(t) \to 1$ as $t \to 0+$, for each j, and Theorem 23.12.2 shows that $P(t)$ is of class (C_0) since the conclusion of this theorem evidently is valid under the weaker assumption $\sum_k p_{jk}(t) \leqq 1$ instead of equality.

Let us now consider the infinitesimal generator G of $P(t)$. Its resolvent is given by

$$R(\lambda; G)[x] = \int_0^{\infty} e^{-\lambda t} P(t)[x]\, dt, \qquad \Re(\lambda) > 0.$$

But

$$R(\lambda; D + rT)[x] = \int_0^\infty e^{-\lambda t} P(t; r)[x] \, dt$$

where $P(t; r) < P(t)$ and $P(t; r) \to P(t)$ in the strong operator topology. We conclude that for $\lambda > 0$

$$R(\lambda; D + rT) < R(\lambda; G)$$

and, by Theorem 3.7.9 and the above Remark, that $R(\lambda; D + rT) \to R(\lambda; G)$ in the strong operator topology as $r \to 1-$. We have also

$$R_n(\lambda) \equiv R(\lambda; D) \sum_{k=0}^{n} [B(\lambda)]^k \to R(\lambda; G)$$

in the strong operator topology as $n \to \infty$. In fact, if

$$R_n(\lambda; r) \equiv R(\lambda; D) \sum_{k=0}^{n} [B(\lambda)]^k r^k, \qquad 0 < r < 1,$$

then

$$R_n(\lambda; r) < R(\lambda; D + rT) < R(\lambda; G),$$

whence it follows that $R_n(\lambda) < R_{n+1}(\lambda) < R(\lambda; G)$. Thus the sequence $\{R_n(\lambda)\}$ converges to a limit $R^\circ(\lambda)$ and $R^\circ(\lambda) < R(\lambda; G)$. On the other hand, we conclude from $R_n(\lambda; r) < R_n(\lambda)$ that

$$R(\lambda; D + rT) < R^\circ(\lambda), \; 0 < r < 1, \quad \text{and} \quad R(\lambda; G) < R^\circ(\lambda)$$

so that $R^\circ(\lambda) = R(\lambda; G)$.

We observe next that G is a closed extension of $D + T$ and, *a fortiori*, of A_0. Indeed, from the identity

$$R_n(\lambda) = R(\lambda; D) + R_{n-1}(\lambda) T R(\lambda; D),$$

assuming $x \in \mathfrak{D}(D)$, we find that

$$R_n(\lambda)(\lambda I - D)x = x + R_{n-1}(\lambda) T x.$$

Passing to the limit with n gives

$$R(\lambda; G)(\lambda I - D - T)x = x$$

so that x belongs to the range of $R(\lambda; G)$, that is, to the domain of G, and for such an x we have $(\lambda I - G)x = (\lambda I - D - T)x$. Hence G is an extension of $D + T$ and it is closed being the infinitesimal generator of a semi-group. On the other hand it may be shown that G is a restriction of A.

To prove the minimal property of $P(t)$, Kato argues as follows. Suppose that G_1 is also a closed extension of A_0 and that it generates a positive semi-group $[P_1(t)]$. First we note that G_1 is also an extension of $D + T$. In fact if $z = \sum_k \zeta_k x_k \in$

$\mathfrak{D}(D)$, then $z_n = \sum_{k=1}^{n} \zeta_k x_k \in \mathfrak{D}_0$ and $z_n \to z$, $Dz_n \to Dz$. It follows that

$$\| T(z_n - z) \| \leqq \| D(z_n - z) \| \to 0, \qquad A_0 z_n = (D + T)z_n \to (D + T)z.$$

Since G_1 is a closed extension of A_0, we have $z \in \mathfrak{D}(G_1)$ and $G_1 z = (D + T)z$, that is, G_1 is an extension of $D + T$. The resolvent $R(\lambda; G_1)$ exists for sufficiently large positive λ and is also a positive operator. By the second resolvent equation (Theorem 5.10.3) we have for $x \in l$

$$\begin{aligned} R(\lambda; G_1)x - R(\lambda; D + rT)x &= R(\lambda; G_1)(G_1 - D - rT)R(\lambda; D + rT)x \\ &= (1 - r)R(\lambda; G_1)TR(\lambda; D + rT)x. \end{aligned}$$

All the operators in the third member being positive, we conclude that $R(\lambda; G_1) > R(\lambda; D + rT)$ for $0 < r < 1$ whence it follows that $R(\lambda; G_1) > R(\lambda; G)$ and hence $P_1(t) > P(t)$ as asserted.

We turn now to the question of when the positive contraction operator $P(t)$ is actually a transition operator, that is, such that $x \in l^+$ implies $\| P(t)x \| = \| x \|$. For this we know that it is necessary and sufficient that $\lambda R(\lambda; G)$ be a transition operator for all $\lambda > 0$ and this in turn implies and is implied by $[B(\lambda)]^n \to \Theta$ in the strong operator topology. To prove the last step we note that

$$I + TR_n(\lambda) = (\lambda I - D)R_n(\lambda) + [B(\lambda)]^{n+1}$$

so that for $\lambda > 0$, $x \in l^+$,

$$\| x \| + \| TR_n(\lambda)x \| = \lambda \| R_n(\lambda)x \| + \| DR_n(\lambda)x \| + \| [B(\lambda)]^{n+1}x \|,$$

whence, cancelling the equal second terms and passing to the limit with n,

$$\| x \| = \lambda \| R(\lambda; G)x \| + \lim_{n\to\infty} \| [B(\lambda)]^{n+1}x \|.$$

Thus, the condition $[B(\lambda)]^n x \to \theta$ is necessary as well as sufficient. This condition will hold for all x if and only if it holds for each $x_j = \{\delta_{jk}\}$.

On the other hand, if $[B(\lambda)]^n x \to \theta$ for each x, we see that $(\lambda I - D - T)R_n(\lambda)x \to x$ for each x in l, that is, the range of $\lambda I - D - T$ is dense in l for each $\lambda > 0$. Thus $(\lambda I - D - T)[\mathfrak{D}(D)]$ is dense in l. Since $D + T$ is a restriction of the smallest closed extension of A_0, we see that $(\lambda I - A_0)[\mathfrak{D}_0]$ is also dense in l. Suppose now that equation (23.12.18) has a non-trivial solution $x^*(\lambda)$ in $m = l^*$ for some $\lambda > 0$. Rewriting the matrix equation as a system of linear equations, we see that the assumption implies the existence of a linear bounded functional x^* such that

$$x^*\{(\lambda I - A_0)[\mathfrak{x}_j]\} = 0, \qquad j = 1, 2, \cdots.$$

But this would imply $x^*\{(\lambda I - A_0)[\mathfrak{D}_0]\} = 0$ so that $x^*(l) = 0$ and $x^* = \theta$ whence it would follow that $x^*(\lambda) = \theta$. Hence if $P(t)$ is a transition operator, $x^*(\lambda) = 0$ is the only solution of (23.12.18) in m.

Conversely, if the latter equation has no solution except $x^*(\lambda) = \theta$ for some $\lambda > 0$, then there is no linear functional $x^*(\neq \theta)$ vanishing on $(\lambda I - A_0)[\mathfrak{D}_0]$

and, a fortiori, none that can vanish on

$$(\lambda I - D - T)[\mathfrak{D}(D)] = (\lambda I - D - T)R(\lambda; D)[l] = [I - B(\lambda)][l].$$

Hence the latter set is dense in l. We set $B_n(\lambda) \equiv (n+1)^{-1} \sum_{k=0}^{n} [B(\lambda)]^k$. Then $B_n(\lambda)$ is a positive operator, $\| B_n(\lambda) \| \leqq 1$, and for $x \in l^+$

$$\| B_n(\lambda)x \| \geqq \| [B(\lambda)]^n x \|.$$

Hence it is sufficient to prove that $\| B_n(\lambda)x \| \to 0$ as $n \to \infty$. But for every x of the form $(I - B(\lambda))y$ we have

$$\| B_n(\lambda)x \| = (n+1)^{-1} \| y - [B(\lambda)]^{n+1} y \| \to 0 \qquad \text{as } n \to \infty.$$

Such elements x being dense in l, we conclude that if $x^*(\lambda) = \theta$ is the only solution of (23.12.18), then $[B(\lambda)]^n$ tends strongly to zero and, consequently, $P(t)$ is a transition operator. This completes the proof of Theorem 23.12.5.

It remains to formulate the connections between the minimal solutions and the uniqueness questions involved in the last three theorems.

THEOREM 23.12.6. *The minimal solutions $P(t)$ and $\mathfrak{P}(t)$ are related by the formulas*

$$(23.12.20) \qquad P(t)[x_j] = \sum_{k=1}^{\infty} p_{jk}(t)x_k, \qquad \mathfrak{P}(t) = (p_{jk}(t)).$$

If $P(t)$ is a transition operator, that is, if the only solution of (23.12.18) *in m is $x^*(\lambda) = \theta$, then $[P(t)]$ is the only semi-group generated by an extension of A_0 and $\mathfrak{P}(t)$ is the only solution of* (23.12.12) *which is of normal type.*

PROOF. To avoid confusion let us temporarily denote the minimal solution of Theorem 23.12.5 by $P^\circ(t)$ and write $(p^\circ_{jk}(t)) = \mathfrak{P}^\circ(t)$ for the corresponding matrix. Since $x_j = \{\delta_{jk}\} \in \mathfrak{D}(D) \subset \mathfrak{D}(G)$, we have

$$\frac{d}{dt} P^\circ(t)[x_j] = GP^\circ(t)[x_j] = P^\circ(t)G[x_j] = P^\circ(t)(D+T)[x_j]$$

or

$$\frac{d}{dt} \sum_k p^\circ_{jk}(t)x_k = \sum_k \left[\sum_n a_{jn} p^\circ_{nk}(t) \right] x_k .$$

On the left we have the strong derivative of an element of l. We conclude that each $p^\circ_{jk}(t)$ is differentiable and that

$$\frac{d}{dt} p^\circ_{jk}(t) = \sum_{n=1}^{\infty} a_{jn} p^\circ_{nk}(t), \qquad p^\circ_{jk}(0) = \delta_{jk},$$

that is, $\mathfrak{P}^\circ(t)$ is a solution of (23.12.12). Hence $\mathfrak{P}^\circ(t) > \mathfrak{P}(t)$ by Theorem 23.12.3. On the other hand, to the minimal matrix $\mathfrak{P}(t)$ of Theorem 23.12.3 corresponds an operator $P(t)$ defined by (23.12.20) and $[P(t)]$ is a positive semi-group generated by an extension of A_0. Theorem 23.12.5 then shows that $P(t) > P^\circ(t)$ and hence also that $\mathfrak{P}(t) > \mathfrak{P}^\circ(t)$. Consequently we have $\mathfrak{P}(t) = \mathfrak{P}^\circ(t)$ and $P(t) = P^\circ(t)$. Thus the two minimal solutions correspond uniquely.

Suppose now that $P(t)$ is a transition operator. By Theorem 23.12.5 the equation (23.12.18) has $x^*(\lambda) = \theta$ as only solution in m and by Theorem 23.12.4 this implies that $\mathfrak{P}(t)$ is the only solution of (23.12.12) of normal type. From this it follows that $[P(t)]$ is the only semi-group, positive or not, which is generated by an extension of A_0.

If $P(t)$ is not a transition operator, the results are less clear cut. Doob has shown [**3**, p. 468] that if $P(t)$ is not a transition operator, then there are infinitely many positive matrices $\mathfrak{P}_1(t)$ of norm one having the semi-group property and satisfying (23.12.12) while $\mathfrak{P}_1'(t) > \mathfrak{P}_1(t)\mathfrak{A}$ and equality need not hold. In this case there are infinitely many positive contraction semi-groups $[P_1(t)]$ on l associated with the matrix $\mathfrak{A}$.

References. Austin [**1**], G. Birkhoff [**5**], Cramér [**1, 2**], Doob [**2, 3**], Feller [**2**], Fréchet [**6**], Hille [**22, 23**], Kato [**1**], Kendall and Reuter [**1**], Kolmogoroff [**1, 4**], Reuter and Ledermann [**1**].

5. Stochastic Processes in E_1

23.13. Spatially and temporally homogeneous processes. We shall consider a stochastic variable $X(t)$ depending upon the time t and taking on arbitrary real values. For a stochastically definite process we denote by $P(t_0, a; t, x)$ the transition probability that $X(t) \leqq x$ if $X(t_0) = a$, $t_0 < t$. The process is said to be *temporally homogeneous* if $P(t_0, a; t, x)$ is a function of $t - t_0$ independent of t_0; it is *spatially homogeneous* if $P(t_0, a; t, x)$ is a function of $x - a$ independent of a. In the latter case (that of a *differential stochastic process* in the terminology of Doob [**1**]) the increase of the variable $X(t)$ during any time interval is independent of the value assumed by the variable at the beginning of the interval.

If the process is spatially and temporally homogeneous, we write

$$P(0, 0; t, x) = P(t; x) \tag{23.13.1}$$

so that $P(t_0, a; t, x) = P(t - t_0; x - a)$. Here $P(t; x)$ is defined for $t > 0$, $-\infty < x < \infty$; $P(t, x)$ is a *distribution function* of x for each fixed t, that is, a never decreasing function of x such that

$$\lim_{x \to -\infty} P(t; x) = 0, \qquad \lim_{x \to \infty} P(t; x) = 1. \tag{23.13.2}$$

$P(t; x)$ satisfies the functional equation of Chapman-Kolmogoroff

(23.13.3) $$P(t_1 + t_2; x) = \int_{-\infty}^{\infty} P(t_1 ; x - y) \, d_y P(t_2; y)$$

or, symbolically,

$$P(t_1 + t_2 ; x) = P(t_1 ; x) * P(t_2 ; x).$$

All distribution functions and, in particular, $P(t; x)$ for fixed t, are elements of the space $\mathfrak{X} = BV_0(-\infty, \infty)$ of functions $F(x)$ which are of bounded variation in $(-\infty, \infty)$ with $\lim_{x \to -\infty} F(x) = 0$. This is a complex (B)-space under the norm

$$\| F \| = \int_{-\infty}^{\infty} | dF(x) | .$$

It is also a (B)-algebra if products are defined by

(23.13.4) $$F_1(x) * F_2(x) = \int_{-\infty}^{\infty} F_1(x - y) \, dF_2(y).$$

We see that the product is associative and commutative and there is a unit element, the function

$$E(x) = \begin{cases} 1, & x \geqq 0, \\ 0, & x < 0. \end{cases}$$

It should be observed that the zero and unit elements are the only idempotents of this algebra. Indeed, if $F*F = F$ and $\pi(s)$ is the Fourier-Stieltjes transform of $F(x)$, then $\pi(s)$ is continuous and $[\pi(s)]^2 = \pi(s)$. The only continuous solutions of this equation are $\pi(s) \equiv 0$ and $\equiv 1$ respectively to which correspond the zero and unit elements of our algebra. This remark is used below.

The problem of characterizing spatially and temporally homogeneous Markoff processes is consequently that of finding all semi-groups having the special form $[P(t; \cdot); 0 < t]$ in $BV_0(-\infty, \infty)$.

The notion of convergence in this space, that is, *convergence in variation*, being highly restrictive, it is not surprising that quite simple solutions of (23.13.3) turn out to be non-measurable functions of t when $P(t; \cdot)$ is considered as an abstract function on $(0, \infty)$ to $BV_0(-\infty, \infty)$. Nelson Dunford observed that the function $P(t; x) = E(x - t)$, which obviously satisfies (23.13.3) for all real values of t, is non-measurable in this sense. In fact, the obvious relation $\| P(t_1 ; \cdot) - P(t_2 ; \cdot) \| = 2$ for $t_1 \neq t_2$ prevents $P(t; \cdot)$ from being almost separably valued which is a necessary condition for measurability.

If $P(t; \cdot)$ is a measurable solution of (23.13.3), then Theorem 9.3.1 shows that it is continuous in variation for $t > 0$:

$$\lim_{h \to 0} \int_{-\infty}^{\infty} | d_x[P(t + h; x) - P(t; x)] | = 0.$$

If $P(t; \cdot) \not\equiv \theta$ and has a limit in variation when $t \to +0$, the limit must be an idempotent, and, by the remark made above, this idempotent must be $E(x)$. But if

$$\lim_{t \to 0+} \int_{-\infty}^{\infty} |d_x[P(t; x) - E(x)]| = 0, \tag{23.13.5}$$

then Theorem 9.4.2 assures the existence of an element $a(x)$ of $BV_0(-\infty, \infty)$ such that

$$\lim_{t \to 0+} \int_{-\infty}^{\infty} \left| d_x \left\{ \frac{1}{t} [P(t; x) - E(x)] - a(x) \right\} \right| = 0, \tag{23.13.6}$$

and

$$P(t; x) = \exp [ta(x)] = E(x) + \sum_{n=1}^{\infty} \frac{t^n}{n!} [a(x)]^{n*}$$

where $[a(x)]^{n*}$ is the product of n factors $a(x)$. From the definition of $a(x)$ it follows that $a(x) \geqq 0$ or $\leqq 0$ according as $x < 0$ or $\geqq 0$. Further $\lim_{x \to \pm\infty} a(x) = 0$ and $a(x)$ is never decreasing in each of the intervals $(-\infty, 0)$ and $(0, \infty)$. It is discontinuous for $x = 0$ with $a(0+) - a(0-) < 0$ unless $a(x) \equiv 0$. These facts imply that

$$a(x) = \omega(x) - \omega(\infty)E(x), \tag{23.13.7}$$

where $\omega(x)$ is never decreasing and $\omega(-\infty) = 0$. Conversely, every such function $a(x)$ generates a transition probability satisfying (23.13.5). As a simple example we may take the function

$$a(x) = \lambda[E(x - \alpha) - E(x)], \qquad \alpha \neq 0, \lambda > 0, \tag{23.13.8}$$

which generates the Poisson distributions

$$P(\lambda t; x, \alpha) = e^{-\lambda t} \sum_{n=0}^{[x/\alpha]} \frac{(\lambda t)^n}{n!}, \qquad x > 0, \alpha > 0, \tag{23.13.9}$$

with $P(\lambda t; x, \alpha) = 0$ for $x \leqq 0$ and $P(\lambda t; x, -\alpha) = 1 - P(\lambda t; -x, \alpha)$.

The abstract function $P(t; \cdot)$ may be interpreted as a transition operator on a suitably chosen space $\mathfrak{X}$. A possible choice of $\mathfrak{X}$ would be $BV_0(-\infty, \infty)$ to which all distribution functions belong. We would then define

$$P(t; \cdot)[F] = \int_{-\infty}^{\infty} F(x - y) \, d_y P(t; y)$$

for every $F(x) \in \mathfrak{X}$. The operator $P(t; \cdot)$ is continuous in variation to the right at $t = 0$ if $P(t; \cdot)[F]$ converges in variation to $F(x)$ when $t \to +0$ for every choice of $F(x)$. Taking $F(x) = E(x)$ we see, however, that this condition reduces to (23.13.5).

The case in which $\mathfrak{X}$ is chosen as $L(-\infty, \infty)$ is more favorable. We define

$$P(t; \cdot)[f] = f(t; x) = \int_{-\infty}^{\infty} f(x - y) \, d_y P(t; y), \quad f(y) \in L(-\infty, \infty). \tag{23.13.10}$$

This transformation also has statistical significance inasmuch as it takes *frequency functions* into themselves. More precisely, *if $f(x)$ is the frequency function of the stochastic variable $X(t)$ at the time $t = 0$ and if $X(t)$ is subjected to a Markoff process with transition probability $P(t; x)$, then $P(t; \cdot)[f]$ is the frequency function at the time t.* The formula obviously defines a semi-group of linear bounded transformations on $L(-\infty, \infty)$ to itself, the operations of which commute with translations, so that Theorem 21.4.1 applies.

Let us first investigate the situation with respect to measurability of $P(t; \cdot)$. If we assume that $P(t; x)$ is normalized (or right continuous or left continuous) as a function of x for each fixed t, then *the operator $P(t; \cdot)$ on $(0, \infty)$ to $L(-\infty, \infty)$ is weakly measurable if and only if $P(t; x)$ is a measurable function of t for each x.* Indeed, if $P(t; x)$ has this property, then every function of the form

$$\int_{-\infty}^{\infty} g(x)\, dx \int_{-\infty}^{\infty} f(x - y)\, d_y P(t; y),$$

where $f(x)$ and $g(x)$ are step functions, is measurable in t. This obviously implies weak measurability of $P(t; \cdot)$. Conversely, if $P(t; \cdot)$ is weakly measurable, then $P(t; x)$ is, for each fixed x, the limit of a suitably chosen sequence of measurable functions of the type used above.

Since $L(-\infty, \infty)$ is a separable space, weak measurability implies strong measurability. By Theorem 10.2.3 the operator $P(t; \cdot)$ is then strongly continuous for $t > 0$, that is, if $P(t; x)$ is a *measurable function of t for each fixed x, then*

$$\lim_{h \to 0} \int_{-\infty}^{\infty} |f(t + h; x) - f(t; x)|\, dx = 0, \qquad t > 0, \tag{23.13.11}$$

for each $f(x) \in L(-\infty, \infty)$. In general we cannot expect that either $f(t; x)$ or $P(t; x)$ shall be a continuous function of t for fixed x. A counter example is given by $P(t; x) = E(x - t)$.

Formula (23.13.11) is not necessarily true for $t = 0$, $h \to 0+$ with $f(0; x) = f(x)$, but it holds in all cases which seem to be of interest to mathematical statistics as shown by

THEOREM 23.13.1. *A necessary and sufficient condition that*

$$\lim_{t \to 0+} \int_{-\infty}^{\infty} \left| \int_{-\infty}^{\infty} f(x - y)\, d_y P(t; y) - f(x) \right| dx = 0 \tag{23.13.12}$$

for every $f(x) \in L(-\infty, \infty)$ is that for each $x \neq 0$

$$\lim_{t \to 0+} P(t; x) = E(x). \tag{23.13.13}$$

PROOF. If (23.13.12) holds for every $f(x)$ then it also holds for the characteristic function of the interval $(0, a)$ in which case the $P(t; \cdot)$ transform becomes simply $P(t; x) - P(t; x - a)$. Here the convergence in the mean gives

$$\lim_{t \to 0+} \int_0^a [1 - P(t; x) + P(t; x - a)] \, dx = 0.$$

But

$$\int_0^a [1 - P(t; x)] \, dx \geqq a[1 - P(t; a)] \geqq 0,$$

$$\int_0^a P(t; x - a) \, dx \geqq aP(t; -a) \geqq 0,$$

so (23.13.13) follows.

For the sufficiency we consider the characteristic function of the interval (a, b) with the $P(t; \cdot)$-transform $P(t; x - a) - P(t; x - b)$ which is non-negative and does not exceed one. Further

$$\int_{-\infty}^{\infty} [P(t; x - a) - P(t; x - b)] \, dx = b - a.$$

If $P(t; x) \to E(x)$ for all $x \neq 0$ when $t \to 0+$, then the integral over the interval (a, b) tends to $b - a$ so that the integrals over the intervals $(-\infty, a)$ and (b, ∞) tend to zero. But this implies that (23.13.12) holds for characteristic functions of intervals. Since linear combinations of such functions are dense in $L(-\infty, \infty)$ and $\| P(t; \cdot) \| \equiv 1$, we conclude from the Banach-Steinhaus theorem that (23.13.12) holds for every $f(x)$. This completes the proof.

23.14. Generation of transition semi-groups. Formula (23.13.8) implies that the bounded difference operator

$$\Delta_{\alpha,\lambda}[f] \equiv \lambda[f(x - \alpha) - f(x)] \tag{23.14.1}$$

generates a semi-group of transition operators in $L(-\infty, \infty)$. The two differential operators D_1 and D_2, with obvious domains, which are defined by

$$D_1[f] = f'(x), \qquad D_2[f] = f''(x), \tag{23.14.2}$$

also play a basic role in the theory of transition probabilities. The shift operator $f(x) \to f(x + mt)$ is generated by mD_1 while D_2 is the infinitesimal generator of the Gauss-Weierstrass operator defined by (21.4.12). A simple computation shows that the operator

$$\tfrac{1}{2}\sigma_0^2 D_2 - mD_1 \tag{23.14.3}$$

generates the semi-group of transition operators represented by

$$(2\sigma_0^2 \pi t)^{-1/2} \int_{-\infty}^{\infty} \exp\left\{-\frac{(x - s - mt)^2}{2\sigma_0^2 t}\right\} f(s) \, ds. \tag{23.14.4}$$

It is not difficult to verify that the operators D_2 and $D_2 + mD_1$ are "closely related" in the sense of Definition 13.5.1 and all the operators of (23.14.3) belong to the cone $\mathfrak{R}[\mathfrak{C}(D_2)]$

of Theorem 13.6.2. That the corresponding semi-groups are of class $H(-\frac{1}{2}\pi, \frac{1}{2}\pi)$ follows from Theorem 13.7.2.

We shall see that in a certain sense the operators $\Delta_{\alpha,\lambda}$, D_1 and D_2 form a base for the infinitesimal generators of all transition semi-groups. For the further study of these questions we introduce the *characteristic function*, in the sense of the theory of probability, of a transition probability, defined by

$$\pi(t; s) = \int_{-\infty}^{\infty} e^{isx} \, d_x P(t; x). \tag{23.14.5}$$

We have $|\pi(t; s)| \leqq 1$ and $\pi(t; s)$ is a continuous function of s for $-\infty < s < +\infty$ when t is fixed. Since the product operation in $BV_0(-\infty, \infty)$ corresponds to ordinary multiplication of the Fourier-Stieltjes transforms, it follows, in particular, that the product of two characteristic functions is again a characteristic function.

Condition (23.13.13) implies that

$$\lim_{t\to 0+} \pi(t; s) \equiv 1 \tag{23.14.6}$$

uniformly with respect to s in every fixed finite interval. Since (23.13.3) gives

$$\pi(t_1 ; s)\pi(t_2 ; s) = \pi(t_1 + t_2 ; s),$$

we conclude that $\pi(t; s) \neq 0$ for all s and all t provided (23.14.6) holds. Cf. Theorem 21.4.1 which shows that $\pi(t; s)$ is a continuous function of t for all s provided $P(t; x)$ is measurable in t for all x. Condition (23.14.6) implies that the set F of Theorem 21.4.1 is void.

Paul Lévy [**1**, p. 49] has shown that a sequence of characteristic functions which converges for all s and converges uniformly in an interval $[-a, a]$, converges uniformly in every finite interval and the limit is again a characteristic function. Among the numerous consequences of this theorem we note the following. If (23.14.6) holds uniformly in s in some interval $[-a, a]$, then $\lim_{t\to 0+} P(t; x) = E(x)$ for each $x \neq 0$. Further, since $[\pi(\frac{1}{2}t; s)]^2 = \pi(t; s)$, we see that the square root of $\pi(t; s)$ is definable as $\pi(\frac{1}{2}t; s)$ and is thus also a characteristic function. This result extends immediately to dyadic rational exponents and, by a limiting passage based on P. Lévy's theorem, to arbitrary positive exponents so that $[\pi(t; s)]^\alpha = \pi(\alpha t; s)$, $\alpha > 0$. In particular we see that $[\pi(1; s)]^t = \pi(t; s)$. Hence if we define $\log \pi(t; s)$ to be continuous in s and vanish for $s = 0$ we see that

$$\log \pi(t; s) = t \log \pi(1; s) \tag{23.14.7}$$

so that $t^{-1} \log \pi(t; s)$ is independent of t for all s.

A. Kolmogoroff [**2**] found necessary and sufficient conditions in order that a function $F(s)$ shall be of the form $F(s) = \log \pi(1; s)$ when $\pi(1; s)$ is the characteristic function of a transition probability $P(t; x)$ having finite second moments. See also H. Cramér [**1**, p. 91]. The general case was solved by P. Lévy [**1**, Chapter VII, especially p. 180]. Later analytical proofs of the representation theorem were given by W. Feller [1] and A. Khintchine [1]; as one of the first applications of the new semi-group theory K. Yosida [4] sketched a new mode of attack on these problems. The following discussion is largely based upon the ideas of Khintchine and Yosida. We start by proving

THEOREM 23.14.1. *Let $P(t; x)$ be a transition probability satisfying* (23.13.3) *and* (23.13.13). *Then*

$$t^{-1} \log \pi(t; s) = ims - \tfrac{1}{2}\sigma_0^2 s^2 + \int_{-\infty}^{\infty} \left[e^{isx} - 1 - \frac{isx}{1+x^2} \right] \frac{1+x^2}{x^2}\, dG_0(x), \tag{23.14.8}$$

where m is real, $\sigma_0 \geqq 0$, $G_0(x)$ is bounded, never decreasing, and continuous at $x = 0$. Conversely every such choice of m, σ_0, $G_0(x)$ defines the characteristic function of a transition probability satisfying (23.13.3) *and* (23.13.13).

PROOF. We already know that (23.13.13) implies (23.14.7). It is sufficient then to prove that the right member of (23.14.8) represents $\log \pi(1; s)$. For this purpose we introduce

$$I_h(s) \equiv \frac{1}{h} [\pi(h; s) - 1] = \frac{1}{h} \int_{-\infty}^{\infty} [e^{isx} - 1]\, d_x P(h; x)$$

and observe that

$$\log \pi(1; s) = \lim_{h \to 0+} I_h(s), \tag{23.14.9}$$

uniformly with respect to s in any fixed finite interval. We define

$$G(h; x) = \frac{1}{h} \int_0^x \frac{y^2}{1+y^2}\, d_y P(h; y) \tag{23.14.10}$$

which is obviously a bounded non-decreasing function of x for fixed $h > 0$. We base the estimates of $G(h; x)$ upon the following elementary inequality, the use of which was suggested by H. Cramér:

$$\frac{y^2}{1+y^2} \leqq a \int_0^{2/a} [1 - \cos sy]\, ds. \tag{23.14.11}$$

This holds for all real values of y, if $a = 1$, and at least for $|y| \geqq a$, if $a > 0$. In fact, the second member equals $2 - (a/y) \sin (2y/a)$ which exceeds one when $|y| \geqq a$ while the first member is always less than one. If $a = 1$, $0 < |y| < 1$, then

$$2 - \frac{1}{y} \sin 2y - y^2 = \frac{1}{3} y^2 - \frac{2^5}{5!} y^4 + \frac{2^7}{7!} y^6 - \cdots .$$

This is an alternating series and the terms decrease in absolute value so the sum is positive. Since $y^2 \geqq y^2/(1 + y^2)$, the inequality holds for all values of y as asserted.

From (23.14.11) we get immediately that the total variation of $G(h; x)$ in $(-\infty, \infty)$ satisfies the inequality

$$\begin{aligned} V[G(h; \cdot)] &= \frac{1}{h} \int_{-\infty}^{\infty} \frac{y^2}{1 + y^2} d_yP(h; y) \\ &\leqq \int_0^2 ds \frac{1}{h} \int_{-\infty}^{\infty} [1 - \cos sy] \, d_yP(h; y) = - \int_0^2 \Re[I_h(s)] \, ds, \end{aligned}$$

where the last member is uniformly bounded for $0 < h < 1$ by virtue of (23.14.9). Since $G(h; 0) = 0$, the functions $G(h; x)$ are uniformly bounded and of uniformly bounded total variation. Further

$$\begin{aligned} \int_{|x|>a} d_xG(h; x) &= \frac{1}{h} \int_{|y|>a} \frac{y^2}{1 + y^2} d_yP(h; y) \\ &\leqq a \int_0^{2/a} ds \frac{1}{h} \int_{|y|>a} [1 - \cos sy] \, d_yP(h; y) \\ &\leqq a \int_0^{2/a} ds \frac{1}{h} \int_{-\infty}^{\infty} [1 - \cos sy] \, d_yP(h; y) \\ &= -a \int_0^{2/a} \Re[I_h(s)] \, ds. \end{aligned}$$

When $h \to 0+$, the last member tends to

$$-a \int_0^{2/a} \Re[\log \pi(1; s)] \, ds.$$

This expression is small for large values of a since $\log \pi(1; s) \to 0$ with s. Given any $\epsilon > 0$, we can consequently choose an $a = a_\epsilon$ such that the total variation of $G(h; x)$ outside of $(-a, a)$ does not exceed ϵ for $0 < h < h_\epsilon$.

Since the functions $G(h; x)$ are uniformly bounded, we can find a sequence $\{h_n\}$ tending to zero such that $G(h_n; x) \to G(x)$, a bounded non-decreasing function. We put

$$m_n = \int_{-\infty}^{\infty} \frac{1}{x} \, dG(h_n ; x)$$

and find, using (23.14.9), that

$$\log \pi(1; s) = \lim_{n \to \infty} \left\{ im_n s + \int_{-\infty}^{\infty} \left[e^{isx} - 1 - \frac{isx}{1 + x^2} \right] \frac{1 + x^2}{x^2} \, dG(h_n ; x) \right\}.$$

Here the integrand is bounded and continuous; for $x = 0$ we replace the indeterminate expression by its limit $-\frac{1}{2}s^2$. Since the total variation of $G(h_n ; x)$ outside of a large interval $[-a, a]$ is uniformly small for large values of n, we conclude that the limit of the integral equals

$$\int_{-\infty}^{\infty}\left[e^{isx}-1-\frac{isx}{1+x^2}\right]\frac{1+x^2}{x^2}\,dG(x).$$

It follows that m_n also tends to a finite limit m when $n \to \infty$. If $G(x)$ is discontinuous at $x = 0$ with the saltus σ_0^2, we set $G(x) = G_0(x) + \sigma_0^2 E(x)$. This completes the proof of formula (23.14.8).

The function $G(x)$ is uniquely determined at its points of continuity. Indeed, if

$$\Delta(s) = \int_{s-1}^{s+1} \log \pi(1;\alpha)\,d\alpha - 2\log \pi(1;\,s),$$

then

$$\Delta(s) = -2\int_{-\infty}^{\infty} e^{isx}\left[1-\frac{\sin x}{x}\right]\frac{1+x^2}{x^2}\,dG(x) = \int_{-\infty}^{\infty} e^{isx}\,dK(x)$$

with

$$K(x) = -2\int_0^x\left[1-\frac{\sin y}{y}\right]\frac{1+y^2}{y^2}\,dG(y).$$

The inversion formula of P. Lévy (cf. formula (21.2.8)) then gives

$$K(x) = \frac{1}{2\pi}\lim_{\omega\to\infty}\int_{-\omega}^{\omega}\frac{1-e^{-isx}}{is}\,\Delta(s)\,ds$$

which shows that $K(x)$ and hence also $G(x)$ is uniquely determined by $\pi(1;\,s)$, that is, by $P(t;\,x)$. This also implies $\lim_{h\to 0+} G(h;\,x) = G(x)$ no matter how h tends to zero.

The converse proposition is established by the following argument. Suppose that we have shown that the integral

$$\int_{|x|>\epsilon}(e^{isx}-1)\,dG(x) \tag{23.14.12}$$

is the logarithm of a characteristic function whenever $G(x)$ is a bounded non-decreasing function and $\epsilon > 0$. The function

$$\int_{|x|>\epsilon}\left\{e^{isx}-1-\frac{isx}{1+x^2}\right\}dG(x),$$

which differs from the integral in (23.14.12) by a term of the form ism_ϵ, m_ϵ real constant, evidently has the same property. Here we may insert the factor $(1 + x^{-2})$ in front of the differential without affecting the basic property of the integral. In fact, if we set

$$G_\epsilon(x) = \begin{cases} G(x), & x \leqq -\epsilon,\\ G(-\epsilon), & -\epsilon < x < \epsilon,\\ G(x) - [G(\epsilon) - G(-\epsilon)], & \epsilon \leqq x,\end{cases}$$

we see that

$$K_\epsilon(x) \equiv \int_0^x \frac{1+y^2}{y^2}\, dG_\epsilon(y)$$

is also a bounded non-decreasing function to which the previous argument applies. It follows that

$$\int_{|x|>\epsilon} \left\{e^{isx} - 1 - \frac{isx}{1+x^2}\right\} \frac{1+x^2}{x^2}\, dG(x)$$

is the logarithm of a characteristic function. When $\epsilon \to +0$ the integral tends to a definite limit, uniformly with respect to s in every fixed finite interval $[-a, a]$. By the convergence theorem of P. Lévy, the limit is then the logarithm of a characteristic function. But this limit differs from the right member of (23.14.8) at most by an expression of the form $i\gamma s - \frac{1}{2}\sigma_0^2 s^2$ (where the second term arises from the possible saltus of $G(x)$ at $x = 0$) which is the logarithm of the characteristic function of a Gaussian distribution. Since the product of two characteristic functions is a characteristic function, we infer that the right member of (23.14.8) is the logarithm of such a function.

In order to complete the proof we have merely to verify the original hypothesis that (23.14.12) defines the logarithm of a characteristic function. Following a suggestion of H. Cramér, we observe that the function in (23.14.12) is the limit of a suitably chosen sequence of Riemann-Stieltjes sums, where the limit exists uniformly with respect to s in any fixed interval $[-a, a]$. The terms in such a sum are of the form $\alpha(e^{i\beta s} - 1)$, $\alpha > 0$, which is the logarithm of the characteristic function of a Poisson distribution. It follows that the approximating sums are logarithms of characteristic functions satisfying (23.13.3) and (23.13.13) and that the limit has the same properties. This completes the proof.

In the semi-group formed by all distribution functions under the operation of convolution, the transition probabilities $P(t; x)$ have the property of being indefinitely divisible. For such arithmetical questions we refer to P. Lévy [1] and A. Khintchine [2].

An important special case of transition probabilities $P(t; x)$ is furnished by the so called stable laws of probability discovered by P. Lévy [**1**; sections 30 and 56]. These correspond to

$$\log \pi(1; s) = -\tau \mid s \mid^\alpha [1 + i\gamma \tan(\tfrac{1}{2}\pi\alpha) \operatorname{sgn} s] \tag{23.14.13}$$

where $\tau > 0$, $-1 \leqq \gamma \leqq 1$, $0 < \alpha < 2$. For these functions one has representations of the form

$$\log \pi(1; s) = \int_{-\infty}^{\infty} (e^{isx} - 1 - isx)\mu(x) \mid x \mid^{-1-\alpha}\, dx,$$

where $\mu(x)$ is a non-negative step function having a single jump at $x = 0$ and the term $-isx$ is to be omitted if $0 < \alpha < 1$. This expression can be reduced to the form of (23.14.8). If $\alpha = 1$ we must take $\gamma = 0$; the corresponding function

$P(t; x) = \frac{1}{2} + \pi^{-1} \arctan x/\tau$ is known in the theory of probability as the Cauchy distribution, its partial derivative with respect to x is the Poisson kernel. W. Feller [4] has given various series expansions for the functions $P(t; x)$ corresponding to stable laws. See also section 23.17.

We turn now to the determination of the infinitesimal generator of the semi-group $[P(t; \cdot)]$ acting in $L(-\infty, \infty)$. We assume condition (23.13.13) so that the semi-group is of class (C_0) and $\log \pi(1; s)$ is represented by (23.14.8). We can now apply Theorem 21.4.2 according to which

$$(Af)(x) \leftrightarrow [\log \pi(1; s)]F(s), \qquad f \in \mathfrak{D}(A), \tag{23.14.14}$$

where $F(s)$ is the Fourier transform of $f(x)$. By formula (21.4.9)

$$\mathfrak{D}(A) = [f; \log \pi(1; s)\, F(s) \text{ is the Fourier transform of an element of } L(-\infty, \infty)]. \tag{23.14.15}$$

Though $\mathfrak{D}(A)$ clearly depends upon the choice of m, σ_0, and $G_0(x)$, we shall see that all domains $\mathfrak{D}(A)$ contain

$$\mathfrak{D}_2 \equiv [f; f \text{ and } f' \text{ absolutely continuous}; f, f', f'' \in L(-\infty, \infty)]. \tag{23.14.16}$$

Let $C_2[-\infty, \infty]$ be the space of functions $\varphi(s)$ such that $\psi(s) \equiv (1 + s^2)^{-1}\varphi(s) \in C[-\infty, \infty]$ and define the C_2-norm of $\varphi(s)$ as the C-norm of $\psi(s)$. We start by proving

LEMMA 23.14.1. *The functions* $\log \pi(1; s)$ *defined by formula* (23.14.8) *form a positive cone* $\mathfrak{K}$ *in* $C_2[-\infty, \infty]$, *closed in the metric of this space.*

PROOF. To simplify the notation we write $\omega(s)$ for the integral and $\omega(s; x)\, dG_0(x)$ for the integrand in formula (23.14.8). We have then the inequality

$$\begin{aligned} |\omega(s)| \leqq (s^2 + \epsilon |s|)[G_0(\epsilon) - G_0(-\epsilon)] \\ + (4\epsilon^{-2} + \epsilon^{-1} |s|)[G_0(\infty) - G_0(-\infty)] \end{aligned} \tag{23.14.17}$$

valid for any ϵ, $0 < \epsilon < 1$. To prove this we observe that

$$e^{isx} - 1 - \frac{isx}{1 + x^2} = -x^2 \int_0^s (s - \sigma)e^{\sigma ix}\, d\sigma + \frac{isx^3}{1 + x^2}$$

so that

$$\int_{-\epsilon}^{\epsilon} |\omega(s; x)|\, dG_0(x) < \tfrac{1}{2}s^2 \int_{-\epsilon}^{\epsilon} (1 + x^2)\, dG_0(x) + |s| \int_{-\epsilon}^{\epsilon} |x|\, dG_0(x),$$

while

$$\int_{\epsilon}^{\infty} |\omega(s; x)|\, dG_0(x) < \int_{\epsilon}^{\infty} |e^{isx} - 1| \frac{1 + x^2}{x^2}\, dG_0(x) + |s| \int_{\epsilon}^{\infty} x^{-1}\, dG_0(x),$$

with a similar estimate for the remaining integral over $(-\infty, -\epsilon)$. From these estimates formula (23.14.17) follows readily. Choosing $\epsilon = |s|^{-1/2}$, we see that $\omega(s) = o(s^2)$ as $|s| \to \infty$ and formula (23.14.13) shows that 2 is the best exponent obtainable.

It follows that every function $\log \pi(1; s)$ belongs to $C_2[-\infty, \infty]$. That these functions form a positive cone $\mathfrak{K}$ follows from the representation (23.14.8). Finally, if $\{\varphi_n\} \subset \mathfrak{K}$ and if the sequence converges to an element φ_0 of $C_2[-\infty, \infty]$, then the convergence is uniform with respect to s in any fixed finite interval $[-a, a]$ and by the convergence theorem of P. Lévy the sequence $\{\exp t\varphi_n\}$ converges to a characteristic function satisfying (23.13.3) and (23.13.13) so that $\varphi_0 \in \mathfrak{K}$.

THEOREM 23.14.2. *For every choice of* $\log \pi(1; s)$ *in* $\mathfrak{K}$ *the domain* $\mathfrak{D}(A)$ *of the infinitesimal generator of the corresponding semi-group* $[P(t; x)]$ *contains* $\mathfrak{D}_2$. *For* $f \in \mathfrak{D}_2$ *we have*

$$\begin{aligned}[Af](x) &= -mf'(x) + \tfrac{1}{2}\sigma_0^2 f''(x) \\ &\quad + \int_{-\infty}^{\infty} \left[f(x - y) - f(x) + \frac{f'(x)y}{1 + y^2}\right] \frac{1 + y^2}{y^2}\, dG_0(y).\end{aligned} \tag{23.14.18}$$

PROOF. Let us denote the integral by $S(x; f)$ and the integrand by

$$S(x, y; f)\, dG_0(y).$$

Proceeding as in the proof of Lemma 23.14.1 we have

$$\begin{aligned}\int_{-1}^{1} |S(x, y; f)|\, dG_0(y) < \int_{-1}^{1} \left|\int_{x-y}^{x} (u - x + y)\, |f''(u)|\, du\right| \frac{1 + y^2}{y^2}\, dG_0(y) \\ + |f'(x)| \int_{-1}^{1} |y|\, dG_0(y)\end{aligned}$$

so that

$$\begin{aligned}\int_{-\infty}^{\infty} \int_{-1}^{1} |S(x, y; f)|\, dG_0(y)\, dx < \|f''\| \int_{-1}^{1} (1 + y^2)\, dG_0(y) \\ + \|f'\| \int_{-1}^{1} |y|\, dG_0(y).\end{aligned}$$

Similarly we get

$$\begin{aligned}\int_{-\infty}^{\infty} \int_{|y|>1} |S(x, y; f)|\, dG_0(y)\, dx < 4\, \|f\| \int_{|y|>1} dG_0(y) \\ + \|f'\| \int_{|y|>1} |y|^{-1}\, dG_0(y)\end{aligned}$$

and hence

$$\|S(\cdot; f)\| < [4\,\|f\| + \|f'\| + 2\,\|f''\|][G_0(\infty) - G_0(-\infty)].$$

Since $S(\cdot\,; f) \in L(-\infty, \infty)$, it has a Fourier transform and a straightforward computation shows that

$$\int_{-\infty}^{\infty} S(x; f)e^{isx}\, dx = \omega(s) \int_{-\infty}^{\infty} f(x)e^{isx}\, dx.$$

Since the Fourier transform of $-mf'(x) + \frac{1}{2}\sigma_0^2 f''(x)$ equals $(ims - \frac{1}{2}\sigma_0^2 s^2)$ times the Fourier transform of $f(x)$, formulas (23.14.14) and (23.14.15) show that $\mathfrak{D}_2 \subset \mathfrak{D}(A)$ and that (23.14.18) holds. This completes the proof.

It may very well happen that $\mathfrak{D}_2$ is a proper subset of $\mathfrak{D}(A)$ and it may even happen that $\mathfrak{D}(A) = L(-\infty, \infty)$ so that A is a bounded operator. Now A is bounded if and only if $\log \pi(1; s)$ is the Fourier transform of a function $a(x)$ satisfying (23.13.7), that is, if and only if

$$\sigma_0 = 0, \int_{-\infty}^{\infty} [1 + x^{-2}]\, dG_0(x) < \infty, \quad \text{and} \quad m = \int_{-\infty}^{\infty} x^{-1}\, dG_0(x).$$

23.15. Transition semi-groups in E_1^+. The preceding discussion refers to distribution functions in E_1. By specialization we may obtain a theory of distribution functions in E_1^+ which, however, involves some novel features of interest. A study of such unilateral distributions and related semi-groups has been made by R. S. Phillips [7]. Some of the main results are listed below but we refer to the original paper for further details.

Suppose then that $[P(t; x)]$ is a semi-group of unilateral distribution functions defined on $(0, \infty)$ such that $P(t; 0) = 0$, $P(t; \infty) = 1$ and

(23.15.1) $$P(t_1 + t_2\, ; x) = \int_0^x P(t_1\, ; x - y)\, d_y P(t_2\, ; y).$$

We impose the continuity condition

(23.15.2) $$\lim_{t \to 0+} P(t; x) = 1, \qquad x > 0,$$

and introduce the Laplace-Stieltjes transform

(23.15.3) $$\Pi(t; z) = \int_0^{\infty} e^{zx}\, d_x P(t; x), \qquad \Re(z) \leqq 0,$$

which we shall refer to as the unilateral characteristic function of $P(t; x)$. It is related to the characteristic function of (23.14.5) by the formula

(23.15.4) $$\Pi(t; is) = \pi(t; s).$$

Just as in the bilateral case, products of characteristic functions are characteristic functions and so is $[\Pi(t; z)]^{\alpha} = \Pi(\alpha t; z)$ for any choice of $\alpha > 0$.

We have the following analogue of Theorem 23.14.1:

THEOREM 23.15.1. *If $[P(t; x)]$ is a semi-group of unilateral distribution functions* *satisfying* (23.15.2), *then*

$$t^{-1} \log \Pi(t; z) = mz + \int_0^\infty [e^{zx} - 1] \, dG(x) \tag{23.15.5}$$

where $m \geqq 0$; $G(x)$ *is non-decreasing on* $(0, \infty)$, $G(\infty) = 0$ *and* $\int_0^1 x \, dG(x) < \infty$. *The representation is valid for* $\Re(z) \leqq 0$. *Conversely, every such choice of* m *and* $G(x)$ *defines a semi-group of unilateral distribution functions satisfying* (23.15.2).

$P(t; \cdot)$ may be used as a transition operator in $L(0, \infty)$ where we define

$$[P(t; \cdot)f](x) = \int_0^x f(x - y) \, d_y P(t; y). \tag{23.15.6}$$

Condition (23.15.2) is still necessary and sufficient in order that the resulting semi-group be of class (C_0). We have the following analogue of Theorem 23.14.2:

THEOREM 23.15.2. *Let* $[P(t; \cdot)]$ *be a semi-group of unilateral transition operators satisfying* (23.15.2) *and acting on* $L(0, \infty)$. *Then the domain of the infinitesimal generator* A *of the semi-group contains*

$$\mathfrak{D}_{10} \equiv [f; f \text{ absolutely continuous}; f(0) = 0, f, f' \in L(0, \infty)] \tag{23.15.7}$$

and for $f \in \mathfrak{D}_{10}$

$$[Af](x) = -mf'(x) + \int_0^\infty [f(x - y) - f(x)] \, dG(y), \tag{23.15.8}$$

where $f(x - y) = 0$ *for* $x \leqq y$.

PROOF. Since this theorem does not occur explicitly in Phillips' paper, we shall sketch a proof; it may be read off from his Theorem 4.3, however. Suppose that $f \in \mathfrak{D}_{10}$. It is first desired to prove that the right member of (23.15.8) belongs to $L(0, \infty)$. This is obvious for the first term. Denoting the second term by $S(x; f)$ we have

$$\begin{aligned} \int_0^\infty | S(x; f) | \, dx &\leqq \int_0^\infty \int_0^1 \int_{x-y}^x | f'(u) | \, du \, dG(y) \, dx \\ &\quad + \int_0^\infty \int_1^\infty [| f(x - y) | + | f(x) |] \, dG(y) \, dx \\ &\leqq \| f' \| \int_0^1 y \, dG(y) + 2 \| f \| \int_1^\infty dG(y) \end{aligned}$$

so that the right member of (23.15.8) is actually in $L(0, \infty)$. We denote it temporarily by $[Bf](x)$ and observe that for $\Re(z) \leqq 0$

$$\int_0^\infty e^{zx}[Bf](x) \, dx = \left[mz + \int_0^\infty (e^{zy} - 1) \, dG(y) \right] \int_0^\infty e^{zx} f(x) \, dx \tag{23.15.9}$$

as is seen by replacing the lower limit 0 of the integral in (23.15.8) by $\epsilon > 0$, interchanging the order of integration in the absolutely convergent integrals and letting ϵ tend to zero.

This shows that $\log \Pi(1; z)$ is an admissible multiplier for Laplace transforms if $f(x)$ is restricted to $\mathfrak{D}_{10}$, in which case the product is the Laplace transform of a function in $L(0, \infty)$, namely Bf. In order to prove that $\mathfrak{D}_{10} \subset \mathfrak{D}(A)$ and that $Af = Bf$ for $f \in \mathfrak{D}_{10}$ we use Theorem 10.5.2. We take $\Re(z) \leqq 0$ and form the functional

$$\begin{aligned}\int_0^\infty e^{zx} \int_0^t [P(\alpha; \cdot)Bf](x)\, d\alpha\, dx &= \int_0^t \int_0^\infty e^{zx}[P(\alpha; \cdot)Bf](x)\, dx\, d\alpha \\ &= \int_0^t \int_0^\infty e^{zx} \int_0^x [Bf](y)\, d_yP(\alpha; x - y)\, dx\, d\alpha \\ &= \int_0^t \left[\int_0^\infty e^{zx}\, d_xP(\alpha; x) \int_0^\infty e^{zx}[Bf](x)\, dx\right] d\alpha,\end{aligned}$$

where we have used the convolution property of the Laplace transform. But the last member equals

$$\log \Pi(1; z) \int_0^t \Pi(\alpha; z)\, d\alpha \int_0^\infty e^{zx}f(x)\, dx = [\Pi(t; z) - 1] \int_0^\infty e^{zx}f(x)\, dx, \tag{23.15.10}$$

since $\Pi(t; z) = \exp [t \log \Pi(1; z)]$. The right member of (23.15.10) is the Laplace transform of $P(t; \cdot)f - f$ and, therefore, by the uniqueness theorem for Laplace transforms,

$$\int_0^t [P(\alpha; \cdot)Bf](x)\, d\alpha = [P(t; \cdot)f](x) - f(x).$$

Since $[P(t; \cdot)]$ is of class (C_0), Theorem 10.5.2 applies and shows that $\mathfrak{D}_{10} \subset \mathfrak{D}(A)$ and that $Af = Bf$ for $f \in \mathfrak{D}_{10}$. This completes the proof.

On the other hand, if $f \in \mathfrak{D}(A)$, formula (23.15.9) holds with Bf replaced by Af. Dividing by z and using the convolution property, one obtains, after some simplification, the relation

$$mf(x) - \int_0^x G(y)f(x - y)\, dy = - \int_0^x [Af](y)\, dy. \tag{23.15.11}$$

This may be regarded as a singular integral equation of the Volterra type for the determination of $f(x)$ in terms of its A-transform. If $m > 0$ one can use this equation to show that $f(x)$ is necessarily absolutely continuous and $f(0) = 0$. Using these facts and $\lim_{y \to 0+} yG(y) = 0$, one can differentiate (23.15.11). The resulting integral equation for $f'(x)$ has a unique solution in $L(0, \infty)$ provided

$\int_0^\infty | G(y) | \, dy < m$ (more generally, if the resolvent kernel is in $L(0, \infty)$). In this case $\mathfrak{D}(A) = \mathfrak{D}_{10}$. If the integrability conditions are not satisfied or if $m = 0$, it is possible for D_{10} to be a proper subset of $\mathfrak{D}(A)$.

Phillips has extended these results in various directions. First he replaced semi-groups of unilateral distributions by semi-groups $[\alpha(t; \cdot)]$ where $\alpha(t, \xi)$ belongs to the Banach algebra $S(\varphi)$ of §4.4 (with $\varphi(\xi)$ of type $\omega_0 > -\infty$) and such that

(i) for each $t \geqq 0$, $\alpha(t; \xi)$ is non-decreasing in ξ;
(ii) $\alpha(t_1, \cdot)*\alpha(t_2; \cdot) = \alpha(t_1 + t_2, \cdot)$, $\alpha(0, \cdot) = e_0$;
(iii) $\| \alpha(t; \cdot) \| \leqq M$ for $0 \leqq t \leqq 1$; and
(iv) $\Phi[z; \alpha(t, \cdot)] \equiv \int_0^\infty e^{z\xi} \, d_\xi\alpha(t, \xi)$ converges to one as $t \to 0+$, uniformly on any bounded subset of $\Re(z) \leqq \omega_0$.

For $t^{-1} \log \Phi[z; \alpha(t, \cdot)]$ he then obtained a representation differing from (23.15.5) in three respects: the variable z in the exponent is replaced by $z - \omega_0$, there is an additive real constant a, and $G(\xi)$ is further restricted by the condition $\int_1^\infty \varphi(\xi)e^{-\omega_0\xi} \, dG(\xi) < \infty$. A mild restriction has been imposed on $\varphi(\xi)$, namely that $\lim \sup_{\delta\to 0+} [\sup_{\xi \geqq 0} \varphi(\xi)/\varphi(\xi + \delta)] < \infty$.

Secondly Phillips discussed semi-groups of the form

$$T(t; B)[f] \equiv \int_0^\infty T(\xi; A)[f] \, d_\xi\alpha(t, \xi), \tag{23.15.12}$$

where $[T(\xi; A)]$ is a given semi-group of class (C_0) and infinitesimal generator A, while $[\alpha(t, \cdot)]$ is a semi-group in the Banach algebra $S(\varphi)$ with $\varphi(\xi) = \| T(\xi; A) \|$ satisfying the above assumptions. It turns out that the new semi-group is also of class (C_0), the domain of its infinitesimal generator B contains $\mathfrak{D}(A)$ and for $f \in \mathfrak{D}(A)$

$$B[f] = mA[f] + \int_0^\infty \{e^{-\omega_0\xi}T(\xi; A)[f] - f\} \, dG(\xi) + af. \tag{23.15.13}$$

Formula (23.15.8) may be regarded as a special case of this involving the shift operator $f(x) \to f(x - \xi)$ and its infinitesimal generator $-d/dx$. Here $f(x - \xi)$ is defined as zero for $x \leqq \xi$.

References. Cramér [1], Doob [1], Feller [1, 4], Khintchine [1, 2], Kolmogoroff [1, 2], P. Lévy [1], Phillips [7], Yosida [4].

6. Fractional Integration

23.16. Riemann-Liouville integrals. The familiar Riemann-Liouville integral of fractional order furnishes us with a very useful example of a semi-group of linear bounded operators. Here we take $\mathfrak{X} = L_p(0, 1)$, $p \geqq 1$, and define

$$[J^\zeta f](t) \equiv \frac{1}{\Gamma(\zeta)} \int_0^t (t - u)^{\zeta-1} f(u)\, du, \qquad \Re(\zeta) > 0. \tag{23.16.1}$$

In discussing this and similar kernel transformations the following classical lemma is indispensible.

LEMMA 23.16.1. *If* $K(t) \in L(-1, 1)$, *then*

$$[Uf](t) \equiv \int_0^1 K(t - u) f(u)\, du \tag{23.16.2}$$

defines a linear bounded transformation on $L_p(0, 1)$ *to itself and*

$$\| U \|_p \leqq \int_{-1}^1 | K(u) |\, du. \tag{23.16.3}$$

The proof is based on a judicious use of Hölder's inequality and may be left to the reader.

Using this lemma together with the Fubini theorem and the definition of the Eulerian integral of the first kind one establishes the fact that the operators J^ζ form a semi-group of linear bounded operators in $L_p(0, 1)$ for ζ in the half-plane $\Re(\zeta) > 0$ such that $\lim_{\xi \to 0+} \| J^\xi f - f \|_p = 0$ for all $f \in L_p(0, 1)$. Thus $[J^\xi; \xi > 0]$ is of class (C_0). Formula (23.16.3) gives the bound

$$\| J^\zeta \|_p \leqq \frac{1}{\xi \,| \Gamma(\zeta) |}, \qquad \zeta = \xi + i\eta, \xi > 0,$$

which turns out to be exact when $p = 1$ and ζ is real positive. It follows from this estimate that

$$\omega_0 = \lim_{\xi \to +\infty} \xi^{-1} \log \| J^\xi \|_p = -\infty.$$

Thus the spectrum of the infinitesimal generator A of $[J^\zeta]$ is empty and $R(\lambda; A)$, the resolvent of A, is an entire function. These operators will be determined below.

We note that J^ζ is a holomorphic function of ζ in $\Re(\zeta) > 0$. To establish this it suffices to show that $x^*[J^\zeta f]$ is holomorphic in ζ, $\Re(\zeta) > 0$, for each $f \in L_p(0, 1)$ and each x^* defined by an arbitrary element g of $L_{p'}(0, 1)$ where $1/p + 1/p' = 1$. But the analyticity of

$$\frac{1}{\Gamma(\zeta)} \int_0^1 \int_0^t g(t)(t - u)^{\zeta-1} f(u)\, du\, dt$$

is clear since the integrand and its ζ-derivative are dominated by the integrable function $| g(t) |\, (t - u)^{\delta-1}[1 + | \log (t - u) |]\, | f(u) |$ for all ζ with $\Re(\zeta) > \delta > 0$.

The operator J^ζ was defined above for $\Re(\zeta) > 0$. H. Kober [1] has shown that it is possible to extend the operator to purely imaginary values of the parameter.

The best results are obtainable when $p = 2$. We shall show that the boundary values $[J^{i\beta}]$ form a strongly continuous one parameter group of operators in $L_2(0, 1)$ with infinitesimal generator iA. Applying Theorems 17.9.1 and 17.9.2 we see that all that is required for this end is that $\| J^{\zeta} \|_2 \leqq M$ for $0 < \xi \leqq 1$, $| \eta | \leqq 1$. We prove this by means of the following lemma due to Kober and which we state without proof.

LEMMA 23.16.2. *Let* (i) $K(t, u)$ *be homogeneous of degree* -1, $0 < t, u < \infty$, (ii) $K(t, 1)t^{-1/2} \in L(0, \infty)$, *and* (iii) $f(t) \in L_2(0, \infty)$. *Then the function*

$$[Wf](t) \equiv \int_0^\infty K(t, u)f(u)\, du$$

exists for almost all $t \in (0, \infty)$ *and*

$$\| W \|_2 = \sup [\| Wf \|_2 ; \| f \|_2 \leqq 1] = \max [| k(\tau) | ; -\infty < \tau < \infty]$$

where

$$k(\tau) \equiv \int_0^\infty K(1, u) u^{-\frac{1}{2}+i\tau}\, du.$$

We apply this result to the kernel

$$K_\zeta(t, u) \equiv \begin{cases} [\Gamma(\zeta)]^{-1} t^{-\zeta} (t - u)^{\zeta - 1}, & 0 < u < t, \\ 0, & t \leqq u, \end{cases}$$

and find that

$$k_\zeta(\tau) = \frac{1}{\Gamma(\zeta)} \int_0^1 (1 - u)^{\zeta-1} u^{-\frac{1}{2}+i\tau}\, du = \frac{\Gamma(\frac{1}{2} + i\tau)}{\Gamma(\zeta + \frac{1}{2} + i\tau)}\, .$$

An application of Stirling's formula shows that

$$\sup_\tau [| k_\zeta(\tau) | ; -\infty < \tau < \infty, 0 < \xi \leqq 1, | \eta | \leqq 1] \equiv M < \infty.$$

Now J^{ζ} is simply the product of $t^{\zeta} I$ and the restriction of W_ζ to the subspace $L_2(0, 1)$ of $L_2(0, \infty)$. Consequently $\| J^{\zeta} \|_2 \leqq M$ for $0 < \xi \leqq 1$, $| \eta | \leqq 1$. We have therefore established the fact that $[J^{i\eta}]$ *forms a strongly continuous group,* $\mathfrak{G}$ *say, of linear bounded operators in* $L_2(0, 1)$ *with infinitesimal generator* Ai.

Moreover we can conclude with the help of Kober's lemma that

$$\| J^{i\eta} \|_2 = \sup_\tau | k_{i\eta}(\tau) | = \sup_\tau \left\{ \frac{\cosh (\tau + \eta)\pi}{\cosh \tau\pi} \right\}^{1/2} = e^{\frac{1}{2}\pi|\eta|}. \tag{23.16.4}$$

The group $\mathfrak{G}$ is of particular interest from the point of view of spectral theory. For $\sigma(iA) = i\sigma(A)$ is in this case empty whereas $\sigma(J^{i\eta})$ is certainly not empty, nor does it contain the point $\lambda = 0$ which is apt to be exceptional. It follows from Theorems 16.7.2 and 16.7.3 that $\sigma(J^{i\eta}) = C\sigma(J^{i\eta})$; that $P\sigma(J^{i\eta}) = \emptyset$ was proved by Kober. Further information about the spectrum will be found below. At any rate $\mathfrak{G}$ is a group of operators such that the spectra of the member operators have no relation whatever with the spectrum of the infinitesimal generator iA (cf. Theorem 16.7.4). We shall see below that $R(\lambda; A)$ is a compact operator.

We now proceed to the determination of A assuming $p \geqq 1$.

THEOREM 23.16.1. *A necessary and sufficient condition that $f \in \mathfrak{D}(A)$ is that*

$$f_*(t) \equiv \int_0^t \log (t - u) f(u)\, du \tag{23.16.5}$$

be the integral of a function in $L_p(0, 1)$. For such an f

$$A[f] = -Cf + f_*', \tag{23.16.6}$$

where C is Euler's constant. A sufficient condition that $f \in \mathfrak{D}(A)$ is that there exists an α, $0 \leqq \alpha < 1$, such that $J^\alpha f$ is the integral of a function in $L_p(0, 1)$.

PROOF. Suppose that $f \in \mathfrak{D}(A)$. We use the basic identity

$$\frac{d}{dt}[J^{\zeta+1}f](t) = [J^\zeta f](t) \qquad \text{or} \qquad [J^{\zeta+1}f](t) = \int_0^t [J^\zeta f](s)\, ds, \tag{23.16.7}$$

where $\Re(\zeta) > 0$. We differentiate the integrated form of the identity with respect to ζ; on the left we use formula (23.16.1), on the right we can differentiate under the sign of integration observing that $(\partial/\partial\zeta)J^\zeta f = J^\zeta Af$. Consequently,

$$\int_0^t [J^\zeta Af](s)\, ds$$
$$= -\psi(\zeta + 1)[J^{\zeta+1}f](t) + \frac{1}{\Gamma(\zeta + 1)} \int_0^t (t - u)^\zeta \log (t - u) f(u)\, du,$$

where $\psi(\cdot)$ is the logarithmic derivative of the gamma function. Here we let $\zeta \to 0$ obtaining

$$\int_0^t [Af](s)\, ds = -\psi(1)[Jf](t) + f_*(t).$$

Now the function on the left and the first one on the right are absolutely continuous functions having their derivatives in $L_p(0, 1)$. Hence f_* has the same property. Differentiation with respect to t gives (23.16.6). Hence the condition is necessary.

Conversely suppose that the right member of (23.16.6) exists as an element of $L_p(0, 1)$. Then an elementary but tedious calculation shows that

$$\int_\alpha^\beta J^\xi[-Cf + f_*']\, d\xi = J^\beta f - J^\alpha f, \qquad 0 < \alpha < \beta < \infty.$$

Here we let $\alpha \to 0+$. Since $[J^\xi]$ is of class (C_0) the right member tends to $J^\beta f - f$. Theorem 10.5.2 applies and shows that $f \in \mathfrak{D}(A)$ and that (23.16.6) holds, so the condition is also sufficient.

Finally suppose that $J^\alpha f = Jg$ where $0 \leqq \alpha < 1$ and $g \in L_p(0, 1)$. This implies that $f = J^{1-\alpha}g$, that is, f is in the range space of the holomorphic semi-group and therefore in $\mathfrak{D}(A)$. This completes the proof.

Using formula (11.5.1) we get

$$[R(\lambda; A)f](t) = \int_0^t E(t - u; \lambda)f(u)\, du, \tag{23.16.8}$$

where

$$E(s; \lambda) = \int_0^\infty e^{-\lambda\xi} s^{\xi-1} \frac{d\xi}{\Gamma(\xi)}\,.$$

Here $E(s; \lambda)$ is an entire function of λ for fixed $s > 0$. Since $E(s; \lambda)$ tends to infinity along the negative real axis and to zero along every other ray from the origin in the λ-plane, we conclude that this entire function is of infinite order. More precisely, in the sector $|\arg \lambda| \leqq \pi - \epsilon$ we have $|\lambda|^2 s\, |E(s; \lambda)| \leqq M_1(\epsilon)$, $0 \leqq s \leqq 1$, and $|\lambda|\, \| E(\cdot\,; \lambda) \|_1 \leqq M_2(\epsilon)$. By Lemma 23.16.1 the latter property implies that $|\lambda|\, \| R(\lambda; A) \|_p \leqq M_2(\epsilon)$ in agreement with Theorem 17.5.1.

Still assuming $p \geqq 1$ we shall now prove that $R(\lambda; A)$ is a compact operator. By Theorem 5.14.2 this implies that A has to have a pure point spectrum with isolated points; the present example shows that also the point spectrum may be missing. By Theorem 5.14.1 it is sufficient to show that $R(0; A)$ is compact. For this purpose set

$$K(u) = \begin{cases} E(u; 0), & 0 < u \leqq 1, \\ 0, & -1 \leqq u \leqq 0. \end{cases}$$

Then

$$\int_{-1}^1 |K(u)|\, du = \int_0^1 \int_0^\infty s^{\xi-1} \frac{d\xi}{\Gamma(\xi)}\, ds = \int_0^\infty \frac{d\xi}{\Gamma(\xi + 1)} < \infty.$$

We approximate $K(u)$ in mean of order one on $(-1, 1)$ by a polynomial, say $p_\epsilon(u)$, so that $\int_{-1}^1 |K(u) - p_\epsilon(u)|\, du < \epsilon$. Now

$$[P_\epsilon f](t) \equiv \int_0^1 p_\epsilon(t - u)f(u)\, du$$

maps $L_p(0, 1)$ boundedly into a finite dimensional subspace and is therefore compact. Finally by Lemma 23.16.1 we have $\| R(0; A) - P_\epsilon \| < \epsilon$ and it follows from Theorem 2.13.4 that $R(0; A)$ is also compact as well as $R(\lambda; A)$.

The resolvent of J^ζ is also of considerable interest and is, in fact, well known (see Hille and Tamarkin [1, pp. 524–525]). Since

$$R(\lambda; J^\zeta) = \sum_{n=0}^\infty \lambda^{-n-1} J^{n\zeta}$$

we have

$$(23.16.9)\qquad [R(\lambda; J^{\zeta})f](t) = \lambda^{-1}f(t) + \lambda^{-1}\int_0^t K_{\zeta}(t-u;\lambda)f(u)\,du,$$

where

$$K_{\zeta}(w;\lambda) = \frac{d}{dw}E_{\zeta}(w^{\zeta}/\lambda) \qquad \text{and} \qquad E_{\alpha}(z) = \sum_{n=0}^{\infty}\frac{z^n}{\Gamma(n\alpha+1)}$$

is the Mittag-Leffler function which is an entire function of z as long as $\Re(\alpha) > 0$.

These formulas show that $R(\lambda; J^{\zeta})$ is an entire function of $1/\lambda$ so that J^{ζ} is a quasi-nilpotent operator and its spectrum (= continuous spectrum) reduces to $\lambda = 0$. The well known properties of $E_{\alpha}(z)$ for large values of $|z|$ lead to simple estimates of the resolvent at least when ζ is real. If $0 < \xi < 2$ then for $|\lambda| < 1$

$$(23.16.10)\qquad \| R(\lambda; J^{\xi})\|_p \leqq \begin{cases} M(\xi)\,|\lambda|^{-1}, & \tfrac{1}{2}\xi\pi < \arg\lambda < (2-\tfrac{1}{2}\xi)\pi, \\ M(\xi)\,|\lambda|^{-1}\exp[|\lambda|^{-1/\xi}], & \text{elsewhere.}\end{cases}$$

These properties of the resolvent have important consequences for ergodic theory. We take $\xi = 1$ in order to obtain examples relating to Theorem 18.8.1. Here

$$(23.16.11)\qquad [R(\lambda; J)f](t) = \lambda^{-1}f(t) + \lambda^{-2}\int_0^t \exp[(t-u)\lambda^{-1}]f(u)\,du$$

and $\lambda R(\lambda; J)$ is bounded in the sector $\frac{1}{2}\pi + \epsilon \leqq \arg\lambda \leqq \frac{3}{2}\pi - \epsilon$, $\epsilon > 0$, whereas $\lambda = 0$ is an essential singular point of $R(\lambda; J)$. According to Theorem 18.8.1 the point $\lambda = \alpha$ is a simple pole of a solution $R(\lambda)$ of the first resolvent equation if $(\lambda_n - \alpha)R(\lambda_n)$ tends to a non-zero limit in the uniform topology for some choice of $\{\lambda_n\}$, $\lambda_n \to \alpha$. The example (23.16.11) shows that this condition cannot be replaced by boundedness of $(\lambda_n - \alpha)R(\lambda_n)$. Similarly, the condition for the existence of an n-tuple pole, viz. that $(\lambda - \alpha)^n R(\lambda)$ together with its derivatives of order $\leqq (n-1)$ should approach limits for a suitable approach of λ to α, is the best of its kind. In fact, $\lambda^n R(\lambda; J)$ together with its derivatives of order $\leqq (n-2)$ tends to finite limits when $\lambda \to 0$ in the sector indicated above while the $(n-1)$st derivative stays bounded. See E. Hille [**12**, §§5, 6] for other ergodic properties of the operator J^{ζ}.

Assuming $p = 2$ so that $J^{i\eta}$ is a bounded operator, we note that formula (23.16.8) no longer represents $[R(\lambda; J^{\zeta})f](t)$ when $\zeta = i\eta$. In fact, the kernel ceases to exist for $|\lambda| < \rho_{\eta} = \exp(\frac{1}{2}\pi|\eta|)$ and even when $|\lambda| > \rho_{\eta}$ the integral exists only in a generalized sense. The function $E_{\alpha}(z)$ changes character when $\alpha \to i\eta$, the limit is not an entire function, it exists only in the circle $|z| < \rho_{\eta}^{-1}$ the boundary of which is a natural boundary of $E_{i\eta}(z)$. Formula (23.16.4) shows that the two operator series

$$\sum_0^{\infty}\lambda^{-n-1}J^{ni\eta} \qquad \text{and} \qquad \sum_0^{\infty}\lambda^n J^{-ni\eta}$$

converge in the strong operator topology for $|\lambda| > \rho_\eta$ and $|\lambda| < \rho_\eta^{-1}$ respectively; they represent $R(\lambda; J^{i\eta})$ in these circular domains. Both boundaries belong to the spectrum of $J^{i\eta}$ as may be shown by computing $\int_0^1 [R(\lambda; J^{i\eta})1](t)\, dt$ with the aid of formula (23.16.13) below. The resulting power series in λ^{-1} and λ have these circles as natural boundaries. Consequently $C\sigma(J^{i\eta})$ is confined to the annulus $\rho_\eta^{-1} \leqq |\lambda| \leqq \rho_\eta$ and contains the boundary. The interior remains in doubt.

The operator J^ζ can serve to illustrate the discussion of abstract valued binomial series given in section 6.8. We restrict ourselves to $p = 1$. For the following consult E. Hille [7, section 4.4] and [12, §§5, 6]. Theorem 17.6.1 gives

$$J^\zeta = \sum_{n=0}^{\infty} (J - I)^n \binom{\zeta}{n}, \qquad J = J^1, \tag{23.16.12}$$

where a simple computation shows that

$$(-1)^n (J - I)^n [f] = f(t) - \int_0^t L_{n-1}^{(1)}(t - u) f(u)\, du$$

and $L_{n-1}^{(1)}(s)$ is a Laguerre polynomial. Cf. G. Szegö [2] for the properties of Laguerre polynomials used in the following. Since

$$\sum_{k=0}^{n} L_k^{(\alpha)}(s) = L_n^{(\alpha+1)}(s),$$

we find that

$$\sum_{k=0}^{n} (-1)^k (J - I)^k [f] = (n + 1) f(t) - \int_0^t L_{n-1}^{(2)}(t - u) f(u)\, du,$$

the norm of which does not exceed

$$\left\{ n + 1 + \int_0^1 | L_{n-1}^{(2)}(s) |\, ds \right\} \| f \| \leqq Cn \| f \|.$$

Formula (6.7.2) then gives $\sigma_0 \leqq 0$. Similarly we find that $\sigma_a \leqq \frac{1}{4}$. Here we have used the (best possible) estimates

$$\int_0^1 | L_n^{(1)}(s) |\, ds < C_1 n^{1/4}, \qquad \int_0^1 | L_{n-1}^{(2)}(s) |\, ds < C_2 n.$$

That $\sigma_0 = 0$ is fairly obvious since the assumption that $\sigma_0 = \beta < 0$ implies that $J^\zeta[f]$ exists as a holomorphic function on ζ to $L(0, 1)$ for $\Re(\zeta) > \beta$, $f(t) \in L(0, 1)$. But when $f(t) = t^\alpha$ we have

$$J^\zeta[t^\alpha] = \frac{\Gamma(\alpha + 1)}{\Gamma(\zeta + \alpha + 1)} t^{\zeta+\alpha} \tag{23.16.13}$$

which belongs to $L(0, 1)$ if and only if $\Re(\zeta) > -\Re(\alpha) - 1$ and this quantity is as near to zero as we please if $\Re(\alpha)$ is near to -1. Thus $\sigma_0 = 0$. It may be shown that

$$J^\zeta[t^\alpha] = \Gamma(\alpha + 1) \sum_{n=0}^{\infty} (-1)^n \frac{\Gamma(n + 1)}{\Gamma(n + \alpha + 1)} t^\alpha L_n^{(\alpha)}(t) \binom{\zeta}{n}$$

and if α is real, $\alpha > -1$, the series converges in norm if and only if $\Re(\zeta) > -\frac{1}{4} - \frac{1}{2}\alpha$ which tends to $\frac{1}{4}$ when $\alpha \to -1$. Hence we have proved that in the uniform topology of $\mathfrak{E}(L)$ the operator series (23.16.12) has its abscissas of ordinary and of absolute convergence equal to 0 and $\frac{1}{4}$ respectively.

We can use (23.16.11) to get a representation of the infinitesimal generator A of $[J^\zeta]$. Since

$$\frac{1}{\zeta}(J^\zeta - I)[f] = \sum_{n=1}^{\infty} \frac{1}{n!}(\zeta - 1) \cdots (\zeta - n + 1)(J - I)^n[f] \tag{23.16.14}$$

we have formally

$$A[f] = \sum_{n=1}^{\infty} (-1)^{n-1} \frac{1}{n} (J - I)^n[f]. \tag{23.16.15}$$

If $f(t)$ is a polynomial, the limiting process is easily justified, because in this case the abscissa of convergence of (23.16.14) is $\leqq -1$ and term-wise differentiation of the series at $\zeta = 0$ is permissible and leads to (23.16.15). Actually this procedure is valid under more general assumptions, for instance, if $f(t)$ is absolutely continuous.

Extension of these results concerning fractional integration is possible in various directions. We may vary the definition of the fractional integral by changing the lower limit. In this connection the reader may consult E. L. Post's discussion of generalized differentiation [1]. We may also replace the interval (0, 1) by (a, b). In the case of an infinite interval the ordinary Lebesgue metric leads to unbounded operators, but fractional integrals may be defined as bounded operators on the following (B)-space:

$$L_e(-\infty, \infty) = \left[f(t); \int_{-\infty}^{\infty} e^{-t} |f(t)| \, dt \equiv \|f\| < \infty \right].$$

Set

$$[J^\zeta f](t) = \frac{1}{\Gamma(\zeta)} \int_{-\infty}^{t} (t - u)^{\zeta-1} f(u) \, du, \qquad \Re(\zeta) > 0. \tag{23.16.16}$$

In this case it is easy to see that

$$\| J^{\xi+i\eta} \| \leqq \frac{\Gamma(\xi)}{|\Gamma(\xi + i\eta)|}, \qquad \xi > 0,$$

and that $\lim_{\xi\to 0+} J^\xi = I$ in the strong operator topology of L_e. In fact, if we map $f \in L_e(-\infty, \infty)$ isometrically into $g(t) \equiv e^{-t} f(t) \in L(-\infty, \infty)$, the transformation J^ζ goes into

$$[P^\zeta g](t) = \frac{1}{\Gamma(\zeta)} \int_{-\infty}^{t} e^{-(t-u)} (t - u)^{\zeta-1} g(u) \, du, \tag{23.16.17}$$

where $[P^\zeta]$ is readily seen to be a semi-group of transition operators on $L(-\infty, \infty)$ for ζ real and >0. Thus the results of section 23.14 apply. In particular

$$\pi(\xi, s) = (1 - is)^{-\xi} \quad \text{and} \quad \log \pi(1, s) = -\log(1 - is),$$

the power and the logarithm having their principal values. The relations (23.14.14) and (23.14.15) characterize A and $\mathfrak{D}(A)$ in terms of the Fourier transforms. By inversion one obtains two expressions for Ag, namely

$$(23.16.18)\qquad \begin{aligned} [Ag](x) &= \underset{\epsilon\to 0+}{\text{l.i.m.}} \int_\epsilon^\infty \frac{e^{-y}}{y} [g(x-y) - g(x)]\, dy, \\ \int_{-\infty}^x [Ag](t)\, dt &= - \int_0^\infty g(x-y) \int_y^\infty \frac{e^{-u}}{u}\, du\, dy, \end{aligned}$$

the first being valid whenever the limit exists while the second holds for every $g \in \mathfrak{D}(A)$. The spectrum of A coincides with the set $S_1 \equiv [-\log(1 - is); -\infty < s < \infty]$ where the logarithm has its principal value. The proof of Theorem 21.4.2 shows only that $\sigma(A) \supset S_1$. To prove the converse we note first that S_1 bounds a convex domain whose function of support is $r\delta_0(\varphi)$ where $\delta_0(\varphi) \equiv \cos\varphi \log\cos\varphi + \varphi\sin\varphi$. But $[P^\zeta]$ is of class $H[-\pi/2, \pi/2]$ so that Theorem 17.5.1 applies. The estimate of $\| J^{\xi+i\eta} \|$ given above shows that the indicator $\delta(\varphi)$ satisfies $\delta(\varphi) \leqq \delta_0(\varphi)$. We have then necessarily $\delta(\varphi) = \delta_0(\varphi)$ so that the exterior of S_1 is in $\rho(A)$. For the interior of S_1 we use Theorem 16.7.1 and the relation $\exp[\sigma(A)] \subset \sigma(P)$ where $P = P^1$. To determine $\sigma(P)$ we note that the Fourier transform of $(\lambda I - P)g$ is obtained by multiplying the Fourier transform of g by $\lambda - (1 - is)^{-1}$. It is however a simple matter to show that $[\lambda - (1 - is)^{-1}]^{-1}$ is a factor function of type (L, L) provided $|\lambda - \frac{1}{2}| \neq \frac{1}{2}$ so that $\sigma(P)$ is confined to the circle $|\lambda - \frac{1}{2}| = \frac{1}{2}$ or, in other notation, to the set $S_2 \equiv [(1 - is)^{-1}; -\infty < s < \infty]$. From $\exp[\sigma(A)] \subset S_2$ and the fact that $\sigma(A)$ belongs to the convex hull of S_1, we conclude that $\sigma(A) \subset S_1$ and finally that $\sigma(A) = S_1$. This completes the argument.

An extension in a different direction is furnished by a fractional integral introduced by H. Weyl [1] in the theory of Fourier series. Let $L_{p,0}(-\pi, \pi)$ denote the subspace of $L_p(-\pi, \pi)$ made up of functions of period 2π whose mean value over the period is zero so that $f_0 = 0$ and define

$$(23.16.19)\qquad T^\alpha[f] \sim \sum_{-\infty}^{\infty} |n|^{-\alpha} e^{\pm\pi i\alpha/2} f_n e^{nit},$$

where the ambiguous sign is opposite that of n. The formula

$$[T^\alpha f](t) = \lim_{\omega\to-\infty} \frac{1}{\Gamma(\alpha)} \int_\omega^t (t-u)^{\alpha-1} f(u)\, du, \qquad 0 < \Re(\alpha) < 1,$$

shows the connection with the classical Riemann-Liouville operator. The semi-group property of T^α is obvious and it is a holomorphic function of α in the right half-plane. The infinitesimal generator of T^α is the operator

$$(23.16.20)\qquad A[f] \sim - \sum_{-\infty}^{\infty} [\tfrac{1}{2}\pi i \operatorname{sgn} n + \log|n|] f_n e^{nit},$$

where $\mathfrak{D}(A)$ is defined as in formula (20.3.9).

23.17. The integrals of Marcel Riesz. We have seen that the Riemann-Liouville operator J^α defines an analytical semi-group associated with the operation of differentiation:

$$J^\alpha J^\beta = J^{\alpha+\beta}, \qquad \frac{d}{dt} J^{\alpha+1} = J^\alpha, \qquad \Re(\alpha) > 0, \qquad \Re(\beta) > 0.$$

Similar semi-groups may be associated with other differential operators. A classical instance is the operator of Hadamard [1] defined by

$$[H^\alpha f](t) = \frac{1}{\Gamma(\alpha)} \int_0^1 \left(\log \frac{1}{u}\right)^{\alpha-1} f(tu) \frac{du}{u} \tag{23.17.1}$$

with the properties

$$H^\alpha H^\beta = H^{\alpha+\beta}, \qquad t\frac{d}{dt} H^{\alpha+1} = H^\alpha, \qquad \Re(\alpha) > 0, \qquad \Re(\beta) > 0.$$

As a further example we may take an operator from the theory of Hermite-Weber series which is closely related to the operator U of section 21.3. We define the U^α-transform of the Hermite-Weber series by

$$U^\alpha[f] \sim \sum_{n=0}^{\infty} (2n+1)^{-\alpha} f_n \mathrm{H}_n(t), \qquad \Re(\alpha) > 0. \tag{23.17.2}$$

It may also be written in closed form as a singular integral if we use the remark at the end of section 21.3 and observe that $\{(2n+1)^{-\alpha}\}$ is a moment sequence. U^α is a bounded operator on $L_p(-\infty, \infty)$ to itself, $\lim_{\alpha\to 0+} \| [U^\alpha - I]f \| = 0$, and

$$U^\alpha U^\beta = U^{\alpha+\beta}, \qquad \left\{t^2 - \frac{d^2}{dt^2}\right\} U^{\alpha+1} = U^\alpha.$$

Incidentally, U^α is the proper analog for Hermitian series of the Hadamard operator for power series and has similar regularity preserving properties in the complex plane. Analogous operators may seemingly be constructed for all classical series expansions associated with second order linear differential equations.

We now turn to the integral operators for partial differential equations studied by Marcel Riesz and his school in penetrating investigations going back to 1933. See M. Riesz [**3, 4, 5**]. This study has led to a profusion of transformation semi-groups, mostly of a nature different from the common run of semi-groups considered in this treatise inasmuch as the operators are frequently unbounded when acting on one of the conventional (B)-spaces $\mathfrak{X}$. At best we obtain a semi-group $[T(\zeta)]$ of linear bounded operators for $\Re(\zeta) > \sigma_0 \geqq 0$. While the line $\Re(\zeta) = \sigma_0$ is the natural boundary of the analytic operator function $T(\zeta)$, it may very well happen that $T(\zeta)f$ admits of an analytic continuation for a suitable choice of f in $\mathfrak{X}$. The characterization of these elements by differentiability properties and the effective determination of the analytic continuation process are central features of the Riesz theory. For the applications to partial differential equations it is essential that for certain choices of the initial data, symbolized by f, it is possible to continue $T(\zeta)f$ analytically to $\zeta = 0$ or beyond.

In the *elliptic* case, typified by the *equation of Laplace*, Riesz introduces the operator

$$(23.17.3) \qquad I^\alpha[f] = [H_m(\alpha)]^{-1} \int f(Q) r_{PQ}^{\alpha-m}\, dQ$$

where

$$(23.17.4) \qquad H_m(\alpha) = \frac{\pi^{m/2} 2^\alpha \Gamma(\frac{1}{2}\alpha)}{\Gamma(\frac{1}{2}(m-\alpha))},$$

r_{PQ} is the euclidean distance between the points P and Q, dQ is the volume element of the m-dimensional space, and the integral is extended over the whole space. Provided $0 < \Re(\alpha) < m$, the integral will exist for all P if, for instance, $f(Q)$ is a continuous function of bounded support. Though I^α does not exist as a bounded operator on $C[E_m]$ or corresponding Lebesgue spaces, it has restricted semi-group properties and for the class of functions indicated one finds that

$$(23.17.5) \qquad I^\alpha I^\beta = I^{\alpha+\beta}, \qquad \Delta I^{\alpha+2} = -I^\alpha$$

where the real parts of α, β, and $\alpha + \beta$ lie between 0 and m and Δ is the Laplacean. Further one has

$$\lim_{\alpha \to 0+} [I^\alpha f](P) = f(P).$$

This implies that the operator I^2 becomes a right-hand inverse of the negative of the Laplacean. In general $I^2[f]$ is a Newtonian potential, but for $m = 2$ a limiting process is required and leads to a logarithmic potential. For further details we refer to Riesz's papers.

The case $m = 1$ has been investigated by W. Feller **[4]** who brought out interesting connections between the corresponding Riesz potentials, the stable distributions of Paul Lévy (see Remark in section 23.14) and the related diffusion equations studied by S. Bochner **[4]**. Feller defined a two-parameter family of operators I_δ^α by the equation

$$(23.17.6) \qquad [I_\delta^\alpha f](x) \equiv [\Gamma(\alpha) \sin \alpha\pi]^{-1} \int_{-\infty}^{\infty} f(y) \,|y - x|^{\alpha-1} \sin \alpha \left[\frac{\pi}{2} + \delta \operatorname{sgn}(y - x)\right] dy.$$

This reduces to the Riesz potential for $\delta = 0$ and gives Riemann-Liouville integrals over semi-infinite ranges for $\delta = \pm\frac{1}{2}\pi$. I_δ^α has the semi-group property in α. If $P_{\alpha\gamma}(t; x)$ is the transition probability whose characteristic function is defined by (23.14.13), then

$$(23.17.7) \qquad [T(\xi)f](x) \equiv \xi^{-1/\alpha} \int_{-\infty}^{\infty} f(y)\, d_y P_{\alpha\gamma}[1; \xi^{-1/\alpha}(x - y)]$$

defines a semi-group of linear bounded operators in $L(-\infty, \infty)$. When $\gamma = 0$ the infinitesimal generator of this semi-group is, according to Bochner, a suitable definition of

$$-\left(-\frac{d^2}{dx^2}\right)^{\alpha/2},$$

while Feller interprets it as $-I_0^{-\alpha}$. Incidentally, Bochner's interpretation is consistent with the Phillips-Balakrishnan calculus (cf. section 15.4). Feller also solves the Cauchy problem for the operator $-I_\delta^{-\alpha}$ (definable by analytic continuation) and obtains the semi-group solution in the form of series or integrals.

In the theory of partial differential equations of the *hyperbolic* type, the introduction of the semi-group operators associated with the basic differential operator has proved to be of fundamental importance. This device has led M. Riesz to a new method of solving Cauchy's problem for wave equations which has marked advantages over the classical methods of Hadamard and Volterra inasmuch as delicate limiting processes involving divergent integrals are replaced by analytic continuation with respect to the complex semi-group parameter and the profound difference between even and odd dimensions appears only in the final result and not in the intermediary analysis. In brief outline the method involves the following steps.

It is required to find a solution of the initial value problem

$$L(u) \equiv \frac{\partial^2 u}{\partial x_1^2} - \frac{\partial^2 u}{\partial x_2^2} - \cdots - \frac{\partial^2 u}{\partial x_m^2} = f(x_1, \cdots, x_m),$$

$$u(0, x_2, \cdots, x_m) = U(x_2, \cdots, x_m), \tag{23.17.8}$$

$$\frac{\partial}{\partial x_1} u(0, x_2, \cdots, x_m) = V(x_2, \cdots, x_m),$$

where f, U and V are given functions defined for all values of the variables and satisfying suitable conditions of continuity and differentiability. For the sake of simplicity we have taken the hyperplane $x_1 = 0$ as the carrier of the data of Cauchy. General surfaces as well as other differential operators are considered by Riesz. We start by defining the associated semi-group operator

$$I^\alpha[F] = \{H_m(\alpha)\}^{-1} \int_\Delta F(Q) r_{PQ}^{\alpha-m}\, dQ \tag{23.17.9}$$

where, however,

$$H_m(\alpha) = \pi^{m/2-1} 2^{\alpha-1} \Gamma(\tfrac{1}{2}\alpha) \Gamma[\tfrac{1}{2}(\alpha + 2 - m)], \tag{23.17.10}$$

r_{PQ} is the *Lorentzian* distance between the points $P = (x_1, \cdots, x_m)$ and $Q = (y_1, \cdots, y_m)$, that is

$$r_{PQ} = [(x_1 - y_1)^2 - (x_2 - y_2)^2 - \cdots - (x_m - y_m)^2]^{1/2},$$

and the integral is extended over the domain $r_{PQ}^2 > 0$, $0 < y_1 < x_1$ (= the interior of the retrograde light cone with vertex at P). Assuming $f(Q)$ to be continuous we see that the integral exists for all P provided $\Re(\alpha) > m - 2$ and for such values of α and β the operator has the properties

$$I^\alpha I^\beta = I^{\alpha+\beta}, \qquad LI^{\alpha+2} = I^\alpha. \tag{23.17.11}$$

The two functions $H_m(\alpha)$ are closely related; denoting for a moment the functions of (23.17.4) and (23.17.10) temporarily by $H_m^e(\alpha)$ and $H_m^h(\alpha)$ respectively, one verifies that $\exp(\frac{1}{2}i\pi\alpha)\ H_m^e(\alpha)$ and $H_m^h(\alpha)$ satisfy the same linear second order difference equation

$$H_m(\alpha + 2) = \alpha(\alpha + 2 - m)H_m(\alpha).$$

If $F(P)$ has continuous partial derivatives, the analytic function $I^\alpha[F]$ may be continued analytically across the line $\Re(\alpha) = m - 2$. In particular, if such partials exist of all orders $\leqq \frac{1}{2}(m - 1)$, $I^\alpha[F]$ is holomorphic for $\Re(\alpha) > 0$ and continuous at $\alpha = 0$ and

$$I^0[F] = F, \qquad LI^2[F] = F.$$

Thus I^2 is a right-hand inverse of the operator L. If now $u(P)$ is a solution of the initial value problem (23.13.5), an application of Green's theorem gives

$$\begin{aligned} I^\alpha[u(P)] &= I^{\alpha+2}[f(P)] \\ &\quad + \{H_m(\alpha + 2)\}^{-1} \int_\Sigma \left\{V(Q)r_{PQ}^{\alpha+2-m} - U(Q)\frac{\partial}{\partial x_1}(r_{PQ}^{\alpha+2-m})\right\} d\Sigma \end{aligned} \tag{23.17.12}$$

where the integral is taken over the domain

$$(x_2 - y_2)^2 + \cdots + (x_m - y_m)^2 \leqq x_1^2, \qquad y_1 = 0.$$

The value of $u(P)$ is then found by analytic continuation with respect to α and letting $\alpha \to 0+$. We cannot enter into further details here. Two methods of performing the analytic continuation have been given by N. E. Fremberg [**1**, **2**], a third is to be found in M. Riesz [**5**, pp. 49–70]. The case $m = 3$ is discussed in B. B. Baker–E. T. Copson [**1**], $m = 4$ in E. T. Copson [**1**].

In contrast to the elliptic case, the hyperbolic operator I^α is bounded, if the operand space is suitably chosen and if $\Re(\alpha) > m - 2$. Let $a > 0$ be arbitrary, but fixed and finite, and let $\mathfrak{X}$ be the space of functions defined and continuous in P if P is restricted to the strip $0 \leqq x_1 \leqq a$ with the usual metric $\| f \| = \sup | f(P) |$. Then $I^\alpha[f]$ exists as a linear bounded operator on $\mathfrak{X}$ to itself provided that $\Re(\alpha) > m - 2$. The norm of I^α turns out to be

$$\| I^{\sigma+i\tau} \| = \frac{H_m(\sigma)}{|H_m(\sigma + i\tau)|}\frac{a^\sigma}{\Gamma(\sigma + 1)}. \tag{23.17.13}$$

In fact the right member equals the supremum of the integral of $| H_m(\alpha) |^{-1} | r_{PQ}^{\alpha-m} |$

taken over the intersection of a retrograde cone with the strip and a simple consideration shows that no smaller bound will do and that it is actually reached at least when α is real. Since the right member of (23.17.13) becomes infinite when σ decreases to $m - 2$ if $\tau \neq 0$, we conclude that the line $\Re(\alpha) = m - 2$ is the natural boundary of the analytic operator function I^α as an element of $\mathfrak{E}(\mathfrak{X})$.

For $\Re(\alpha) > m - 2$ the operator I^α is quasi-nilpotent and the resolvent of I^α is consequently an entire function of $1/\lambda$ which is given by

$$(23.17.14) \qquad [R(\lambda; I^\alpha)f](P) = \lambda^{-1}f(P) + \lambda^{-1}\int_\Delta R_\alpha(r_{PQ}^\alpha \lambda^{-1})\, r_{PQ}^{-m} f(Q)\, dQ,$$

where the integral is taken over the same domain as in (23.17.9) and $R_\alpha(z)$ is the entire function

$$R_\alpha(z) = \sum_{n=1}^{\infty} [H_m(n\alpha)]^{-1} z^n.$$

This function has properties analogous to those of the Mittag-Leffler function $E_\alpha(z)$ and as a consequence $R(\lambda; I^\alpha)$ satisfies inequalities similar to (23.16.10) provided $m = 2$ or 3. In particular, if $m = 2$ and $0 < \alpha < 2$ or $m = 3$ and $1 < \alpha < 2$, then $|\lambda|\, \| R(\lambda; I^\alpha) \|$ is bounded in the sector $\frac{1}{2}\alpha\pi < \arg \lambda < (2 - \frac{1}{2}\alpha)\pi$. The infinitesimal generator A of the semi-group $[I^\alpha]$ acting in $\mathfrak{X}$ exists if and only if $m = 2$ in which case $[I^\alpha]$ is of class $H(-\frac{1}{2}\pi, \frac{1}{2}\pi)$. Its resolvent is then given by

$$(23.17.15) \qquad [R(\lambda; A)f](P) = \int_\Delta E_2(r_{PQ}\,; \lambda)\, r_{PQ}^{-2} f(Q)\, dQ,$$

where

$$E_2(s; \lambda) = \int_0^\infty e^{-\lambda\xi} s^\xi [H_2(\xi)]^{-1}\, d\xi$$

is an entire function of λ of infinite order which becomes infinite when $\lambda \to \infty$ along the negative real axis but tends to zero along every other ray $\arg \lambda = \theta$. Thus $R(\lambda; A)$ is also an entire function and the point at infinity is the only point in the extended spectrum of A. Further $|\lambda|\, \| R(\lambda; A) \|$ stays uniformly bounded in every sector $-\pi + \epsilon < \arg \lambda < \pi - \epsilon$. We see that the analogy between the Riemann-Liouville operator J^α and the hyperbolic operator I^α of M. Riesz is very close when $m = 2$ but essentially new phenomena appear for $m > 2$. In particular, for $m > 2$ the operators $[I^\alpha]$ define a semi-group whose domain of analytic existence $\Re(\alpha) > m - 2$ is a proper subset of the right half-plane. Thus we obtain a simple and natural example of a phenomenon which was first encountered in section 20.8. From the point of view of the applications, however, our emphasis on the non-continuable character of the operators $[I^\alpha]$ is misplaced; there the important fact is that $I^\alpha f$ is continuable for every function $f(P)$ having sufficiently many derivatives.

M. Riesz has extended his investigations to equations with variable coefficients (see Chapter VII of [5]) in particular the case in which L is the second order differential operator of Beltrami associated with a Riemannian space having a Lorentzian metric. A basic semi-group operator I^α is encountered having the properties (23.17.11).

Extensions of the Riesz operators have been found by L. Gårding [2, 3] in examining various classes of totally hyperbolic differential equations. In [2] the Lorentzian geometry of Riesz is replaced by the symplectic geometry of C. L. Siegel. Here the space is the set of real symmetric n by n matrices, partially ordered by defining a matrix to be positive when the associated quadratic form is positive definite. There is an associated semi-group operator I^α, acting on numerically-valued bounded continuous matrix functions and defined for $\Re(\alpha) > \frac{1}{2}(n-1)$ by an integral involving a simple kernel. There exists an associated linear differential operator D of order n such that $DI^{\alpha+1} = I^\alpha$. Similar results hold in the space of Hermitian matrices. In the general case studied in [3], it is as a rule not possible to obtain explicit expressions for the kernels defining the Riesz operators.

The situation is much simpler in the parabolic case and well worth mentioning. The following remarks are based on information kindly supplied by M. Riesz. We consider the heat equation in n dimensions

$$L(u) \equiv \frac{\partial u}{\partial t} - \sum_{k=1}^{n} \frac{\partial^2 u}{\partial x_k^2} = 0, \; L = \frac{\partial}{\partial t} - \Delta_n , \tag{23.17.16}$$

and define the parabolic integral

$$\Pi^\alpha[F] = \frac{1}{K_n(\alpha)} \int_{-\infty}^{t}\int_{-\infty}^{\infty} \cdots \int_{-\infty}^{\infty} F(\tau, Q) \exp\left[\frac{-\frac{1}{4}r_{PQ}^2}{t-\tau}\right] (t-\tau)^{\alpha-1-n/2}\, dQ\, d\tau, \tag{23.17.17}$$

where

$$P = (x_1, \cdots, x_n),\; Q = (\xi_1, \cdots, \xi_n),\; r_{PQ}^2 = \sum_{k=1}^{n} (x_k - \xi_k)^2,\; K_n(\alpha) = 2^n \pi^{n/2} \Gamma(\alpha).$$

This integral exists under fairly general assumptions on F for $\Re(\alpha) > 1$. The operator has the basic properties

$$\Pi^\alpha \Pi^\beta = \Pi^{\alpha+\beta}, \qquad L\Pi^{\alpha+1} = \Pi^\alpha. \tag{23.17.18}$$

In order to analyze the operator, let us consider the following simple case. Let $F(t, P) \equiv 0$ for $t < 0$, let $a > 0$ be fixed but arbitrary, let $S = S_a$ be the slab of E_{n+1} defined by $0 < t < a$, and let $\mathfrak{X}$ be the system of all functions $F(t, P)$ integrable over S with

$$\| F \| = \int_S | F(t, P) | \, dt\, dP.$$

Further we introduce the Gauss-Weierstrass transform

$$(23.17.19) \qquad W_{s,n}[f] = (4\pi s)^{-n/2} \int_{-\infty}^{\infty} \cdots \int_{-\infty}^{\infty} f(Q) \exp\left[-r_{PQ}^2/(4s)\right] dQ.$$

Cf. sections 21.4 and 22.5. We can now write (23.17.17) in the form

$$(23.17.20) \qquad \Pi^\alpha[F] = \frac{1}{\Gamma(\alpha)} \int_0^t W_{t-\tau,n}[F(\tau, \cdot)](t - \tau)^{\alpha-1}\, d\tau, \qquad \Re(\alpha) > 0.$$

A simple calculation shows that this is a bounded transformation on $\mathfrak{X}$ to itself. We note that $W_{s,n}[f]$ is the solution of $L(u) = 0$ which tends to $f(P)$ when $s \to 0+$. Further $W_{s,n}$ is the semi-group operator generated by Δ_n. Thus the *parabolic integral* $\Pi^\alpha[F]$, *which is associated with the operator* $L = \partial/\partial t - \Delta_n$, *is obtained from the semi-group operator generated by* Δ_n *essentially by an ordinary Riemann-Liouville integration of order* α.

From (23.17.20) we can easily obtain the further properties of $\Pi^\alpha[F]$ such as resolvent and spectra of Π^α and of its infinitesimal generator. The analogy with the Riemann-Liouville operator turns out to be perfect.

References. Baker and Copson [1], Bochner [4], Copson [1], Feller [4], Fremberg [1, 2], Gårding [2, 3], Hadamard [1], Hille [7, 12], Hille and Tamarkin [1], Kober [1], Liouville [1], Post [1], Riemann [1], M. Riesz [3, 4, 5], Szegö [2], Weyl [1].

PART FIVE

EXTENSIONS OF THE THEORY

Summary. The last part of the book is concerned with various extensions of the previously developed theory. There are three chapters entitled: *Notes on Banach Algebras*, *Lie Semi-Groups*, and *Functions on Vectors to Vectors*.

Chapter XXIV is devoted to the theory of Banach algebras with emphasis on the non-commutative case and algebras without unit element. The discussion supplements that of Chapters IV and V. The subject matter of Chapter XXV is the theory of parametric semi-groups of linear bounded transformations,

$$[T(p); p \in \Pi, T(p) \in \mathfrak{E}(\mathfrak{X})]$$

and the associated parameter semi-groups Π. Here Π is a subset of a (B)-space, in particular a euclidean n-space, and Π is closed under a product operation $p \circ q = F(p, q)$ so that $T(p)T(q) = T(p \circ q)$. The resulting theory extends the discussion given in Chapters VII to X. There are also connections with section 24.14. Finally Chapter XXVI extends the theory of functions on vectors to vectors for which the foundations were laid in §3.3.

CHAPTER XXIV

NOTES ON BANACH ALGEBRAS

24.1. Orientation. These Notes contain supplementary material having a bearing on the theory of Banach algebras developed in §1.4 and Chapters IV and V. Emphasis is now on the non-commutative case and algebras without unit element. Moreover we do not severely restrict ourselves to (B)-algebras; when the situation permits we consider also topological algebras and even algebras without topology.

There are four paragraphs: *Algebras without Unit Element, Ideal Theory, Representation of (B)-Algebras*, and *A Structure Theorem.*

Ideas and constructive suggestions for this chapter have been received from N. Jacobson, I. E. Segal, and, especially, Max Zorn.

1. Algebras without Unit Element

24.2. Reversible elements. So far our discussion of (B)-algebras has been based upon the assumed existence of a unit element. Many algebras of interest to analysis do not have this property, however, and while it is always possible to adjoin an ideal unit element, such an adjunction is often unnatural and distorts essential features of the algebra. In the present paragraph we shall develop an analytic theory of (B)-algebras without reference to a unit element and this theory will be consistent with that of Chapters IV and V when both apply.

For this purpose it will be necessary to introduce concepts capable of replacing the inverse and the resolvent which were basic in the earlier theory. Here a device used by S. Perlis (see [1] where there is a reference to Marshall Hall) is decisive. Perlis noted that if an algebra has a unit element e and if $e + x$ is regular with the inverse $e + y$, then

$$x + y + xy = x + y + yx = \theta. \tag{24.2.1}$$

In this equation there is no reference to the unit element and the relation between x and y can hold in any ring regardless of the existence of a unit element. Perlis called x *quasi-regular* and y its *quasi-inverse*; the relation between x and y is evidently symmetric. The usefulness of this concept in abstract algebra has been amply demonstrated by N. Jacobson [2], I. Kaplansky [1] and others. In applying these notions to (B)-algebras, we have found it convenient to modify the defining property and the terminology. The use of the cross product was suggested to E. Hille by N. Jacobson in July 1946; it leads to great simplification of the formulas and adds to the insight in the subject matter.

DEFINITION 24.2.1. *The cross product of two elements a and b of a ring* $\mathbf{R}$ *i defined as*

$$a \times b = a + b - ab. \tag{24.2.2}$$

If the cross product vanishes, a is said to be right reversible and to have b as a right reverse; b is left reversible and has a as a left reverse. If

$$a \times b = b \times a = \theta, \tag{24.2.3}$$

then a is reversible and has b as a reverse. The reverse of a is denoted by a^{-}.

We note that the cross product is associative

$$\begin{aligned}(a \times b) \times c &= a + b + c - (ab + ac + bc) + abc \\ &= a \times (b \times c).\end{aligned} \tag{24.2.4}$$

It is normally not commutative, $a \times b = b \times a$ if and only if $ab = ba$. The cross product is analogous to the symmetric difference $A \cup B \ominus A \cap B$ in set theory.

The uniqueness of the reverse is easily established; we leave the proof to the reader and merely state

THEOREM 24.2.1. *If a has a right reverse b as well as a left reverse c, then* $b = c$ *and a has a unique reverse.*

We use the symbols $\mathfrak{R}$, $\mathfrak{R}^r$, and $\mathfrak{R}^l$ to denote the set of reversible, right reversible, and left reversible elements of $\mathbf{R}$ respectively.

THEOREM 24.2.2. $\mathfrak{R}$ *is a group,* $\mathfrak{R}^r$ *and* $\mathfrak{R}^l$ *semi-groups, under the operation of cross multiplication. The zero element of* $\mathbf{R}$ *is the unit element of* $\mathfrak{R}$.

PROOF. It follows from (24.2.2) that $a \times \theta = \theta \times a = a$. Hence if $a \times b = \theta$, $c \times d = \theta$, then

$$(a \times c) \times (d \times b) = a \times (c \times d) \times b = a \times \theta \times b = a \times b = \theta,$$

which proves the semi-group property.

We shall now suppose that the ring is a complex (B)-algebra $\mathfrak{B}$ and proceed to give analogues of Theorems 4.3.1 and 4.3.2.

THEOREM 24.2.3. *Every element x of* $\mathfrak{B}$ *with* $\| x \| < 1$ *is in* $\mathfrak{R}$ *and*

$$x^{-} = -\sum_{n=1}^{\infty} x^n. \tag{24.2.5}$$

The proof is immediate. The condition $\| x \| < 1$ is merely sufficient for reversibility and could be replaced by $\lim_{n\to\infty} \| x^n \|^{1/n} < 1$. Thus *every quasi-nilpotent element is reversible.*

THEOREM 24.2.4. *The sets* $\mathfrak{R}$, $\mathfrak{R}^r$, $\mathfrak{R}^l$ *are open in* $\mathfrak{B}$.

PROOF. Suppose that $a \in \mathfrak{R}^r$ and $a \times b = \theta$ and form

$$u = x \times b = x - a - (x - a)b.$$

For $\| x - a \| < (1 + \| b \|)^{-1}$ we have $\| u \| < 1$ and u has a unique reverse u^-. But $(x \times b) \times u^- = \theta$ implies

$$x \times (b \times u^-) = \theta, \tag{24.2.6}$$

showing that $y_r(x; a, b) \equiv b \times u^-$ is a right reverse of x at least for $\| x - a \| < (1 + \| b \|)^{-1}$ and that $\mathfrak{R}^r$ is an open set. Similarly, if $a \in \mathfrak{R}^l$ and $c \times a = \theta$ we find that $(c \times x)^- \times c$ is a left reverse of x at least for $\| x - a \| < (1 + \| c \|)^{-1}$ so that $\mathfrak{R}^l$ is open. Finally $\mathfrak{R}$ is the intersection of the two open sets $\mathfrak{R}^r$ and $\mathfrak{R}^l$ and is therefore open.

Each of these three open sets is the union of disjoint maximal open sets called the components of the set in question. The component containing the element a is denoted by $\mathfrak{R}(a)$ and so on. The component containing the zero element is called the principal component.

If a is in $\mathfrak{R}^r$ but not in $\mathfrak{R}$, there are in general infinitely many right reverses of a. In fact, if y is any element of $\mathfrak{B}$ such that $(a \times y)^-$ exists, then $y \times (a \times y)^-$ is a right reverse of a. If $y = b + h$, $a \times b = \theta$, and $\| h - ah \| < 1$, the right reverse becomes

$$z_r(a; h) \equiv b + (ab - ba)h + (ab - ba)h \sum_{n=1}^{\infty} (h - ah)^n. \tag{24.2.7}$$

It is consequently an analytic function of h in the sense of Definition 3.17.2. As such it reduces to b identically if and only if $(ab - ba)h = \theta$ for all h in $\mathfrak{B}$. Disregarding this case, we see that the formula gives infinitely many distinct right reverses of a. A similar situation holds for left reverses. Sufficient conditions that $a \times b = \theta$ shall imply $b \times a = \theta$ have been given by N. Jacobson [**3**].

We have seen that if $a \times b = \theta$ then there is at least one right reverse of x which tends to b when $x \to a$, namely

$$\begin{aligned} &y_r(x; a, b) \\ &\quad = b - \sum_{n=1}^{\infty} [(x - a) - (x - a)b]^n + b \sum_{n=1}^{\infty} [(x - a) - (x - a)b]^n. \end{aligned} \tag{24.2.8}$$

This right reverse is clearly an analytic function of x in the sense of Definition 3.17.2. It is not obvious that $y_r(x; a, b)$ is the only right reverse of x which tends to b when $x \to a$. Leaving this question undecided we shall, however, throw some light on the character of "nearby" solutions. Consider two fixed spheres, $\mathfrak{S}_1$: $\| y \| < M$, where $M > \| b \|$, and $\mathfrak{S}_2$: $\| x - a \| < \frac{1}{2}(1 + M)^{-1}$. The latter is in $\mathfrak{R}^r$ as shown above. Take a point $x_0 \in \mathfrak{S}_2$ and suppose that there is a corresponding point $y_0 \in \mathfrak{S}_1$ such that $x_0 \times y_0 = \theta$. We can then form the function $y_r(x; x_0, y_0)$ using formula (24.2.8). This is a right reverse of x, defined and analytic in $\mathfrak{S}_2$ and reducing to y_0 for $x = x_0$. In particular it is defined and analytic at $x = a$ though we have of course no guarantee that $y_r(a; x_0, y_0) = b$. But we do see that the right reverses of x_0 in $\mathfrak{S}_2$ whose values lie in $\mathfrak{S}_1$, may be embedded in locally analytic functions of x which are right reverses of x. In this sense the right reverses of x are locally analytic functions of x in the various components of $\mathfrak{R}^r$ and the same is true for the left reverses in $\mathfrak{R}^l$. In particular, if $a \in \mathfrak{R}$ then $y_r(x; a, b)$ is a left reverse as well as a right one and hence the unique reverse of x. Consequently the latter is analytic in each component of $\mathfrak{R}$. We have therefore proved

THEOREM 24.2.5. *The reverse of x is an analytic function of x in each component of $\mathfrak{R}$. Each right (left) reverse of x is locally analytic in the components of $\mathfrak{R}^r$ ($\mathfrak{R}^l$).*

The following theorem shows that components of $\mathfrak{R}^r$ and $\mathfrak{R}^l$ are either disjoint or coincide.

THEOREM 24.2.6. *If $a \in \mathfrak{R}$, then $\mathfrak{R}(a) = \mathfrak{R}^r(a) = \mathfrak{R}^l(a)$. In particular, the principal components of $\mathfrak{R}$, $\mathfrak{R}^r$, and $\mathfrak{R}^l$ are equal.*

PROOF. It is sufficient to prove that $\mathfrak{R}$ is both open and closed in $\mathfrak{R}^r$ as well as in $\mathfrak{R}^l$. Since $\mathfrak{R}$ is the intersection of these two open sets, it is open in each of them. To prove the closure, suppose, for instance, that $a \in \bar{\mathfrak{R}} \cap \mathfrak{R}^r$. Let b be a right reverse of a and let $y_r(x; a, b)$ defined by (24.2.8) be the corresponding right reverse of x in the sphere $\mathfrak{S}$: $\| x - a \| < (1 + \| b \|)^{-1}$, where it is analytic. Since $a \in \bar{\mathfrak{R}}$, we can find a point $a_0 \in \mathfrak{R}$ and a $\rho > 0$ such that the sphere $\mathfrak{S}_0$: $\| x - a_0 \| < \rho$ satisfies $\mathfrak{S}_0 \subset \mathfrak{S} \cap \mathfrak{R}$. But for x in $\mathfrak{S}_0$, $y_r(x; a, b)$ is also a left reverse of x so that

$$x + y_r(x; a, b) = y_r(x; a, b)x, \qquad x \in \mathfrak{S}_0.$$

By Theorem 3.16.4 this identity between analytic functions must hold for all x in $\mathfrak{S}$. In particular, it holds for $x = a$, that is, a has a unique reverse and belongs to $\mathfrak{R}$. Hence $\mathfrak{R}$ is closed in $\mathfrak{R}^r$ and in the same manner we prove that $\mathfrak{R}$ is closed in $\mathfrak{R}^l$ so that $\mathfrak{R}(a) = \mathfrak{R}^r(a) = \mathfrak{R}^l(a)$. Setting $a = \theta$ we obtain the special case of the principal components. This completes the proof.

We note that the *principal component of $\mathfrak{R}$ is a normal subgroup of $\mathfrak{R}$ under cross multiplication.*

THEOREM 24.2.7. *If $\mathfrak{B}$ has a unit element e, then the transformation $y = e - x$ maps $\mathfrak{B}$ in a one-to-one manner onto itself in such a manner that products go into cross products, $\mathfrak{G}$ is mapped onto $\mathfrak{R}$, and components of $\mathfrak{G}$ onto components of $\mathfrak{R}$. In particular, x has an inverse if and only if y has a reverse.*

The proof is obvious. Combining this result with the example in section 9.5, we see that the components of $\mathfrak{R}$ may very well be non-countable, the algebra $A(-\infty, \infty)$ being a case in point.

24.3. The dissolvent. The process by means of which the resolvent is obtained from the inverse may be applied, *mutatis mutandis*, to the reverse and leads to a new concept which, lacking a better name, we shall call the *dissolvent* of x. We are also led to a new notion of the spectrum.

DEFINITION 24.3.1. *With respect to a given element x of $\mathfrak{B}$, the complex numbers fall into two complementary classes: the dissolvent set $\delta(x)$ and the spectrum $\alpha(x)$ of x. If $\lambda \neq 0$, then $\lambda \in \delta(x)$ if and only if $(1/\lambda)x$ has a unique reverse; $\lambda = 0$ belongs to $\delta(x)$ if and only if there are elements j and y of $\mathfrak{B}$ with $j^2 = j$, $x = xj = jx$, $y = yj = jy$, and $xy = yx = j$.*

That the origin requires special treatment is obvious and the particular form given to the condition will become clear later on.

DEFINITION 24.3.2. *The reverse of $(1/\lambda)x$ is denoted by $D(\lambda; x)$ and called the dissolvent of x. If $\lambda = 0$ is in $\delta(x)$ we set $D(0; x) = j$.*

It follows from the definition that

$$x + \lambda D(\lambda; x) = xD(\lambda; x) = D(\lambda; x)x \tag{24.3.1}$$

for $\lambda \in \delta(x)$ and this holds also for $\lambda = 0$.

THEOREM 24.3.1. *We have*

$$D(\lambda; x) = -\sum_{n=1}^{\infty} x^n \lambda^{-n}, \tag{24.3.2}$$

when

$$|\lambda| > \gamma = \lim_{n\to\infty} \| x^n \|^{1/n}.$$

The proof follows from Theorem 24.2.3, the existence of the limit being a consequence of Lemma 4.7.1. We note in particular that the dissolvent of a quasi-nilpotent element is an entire function of $1/\lambda$ just as the resolvent.

THEOREM 24.3.2. *The dissolvent set is open. In each of its components, $D(\lambda; x)$ is a holomorphic function of λ.*

PROOF. Suppose that $\lambda_0 \in \delta(x)$ and $\lambda_0 \neq 0$. We base the proof on formula (24.2.6) replacing a by $(1/\lambda_0)x$, b by $D(\lambda_0; x)$, and x by $(1/\lambda)x$. A simple calculation gives

$$u = \frac{\lambda - \lambda_0}{\lambda} D(\lambda_0; x)$$

and

$$D(\lambda; x) = \frac{\lambda_0}{\lambda} D(\lambda_0; x) + \frac{\lambda_0}{\lambda} D(\lambda_0; x) \sum_{n=1}^{\infty} \left\{\frac{\lambda - \lambda_0}{\lambda} D(\lambda_0; x)\right\}^n \tag{24.3.3}$$

which is clearly holomorphic in the circle

$$\delta\,|\,\lambda - \lambda_0\,| < |\,\lambda\,|, \qquad \delta = \lim_{n\to\infty} \|\,[D(\lambda_0\,;x)]^n\,\|^{1/n}. \tag{24.3.4}$$

From the way in which the series was formed, it follows that it defines a right reverse of $(1/\lambda)x$ within its circle of convergence. Since, however, x commutes with $D(\lambda_0\,; x)$, it also commutes with the sum of the series which is consequently the reverse of $(1/\lambda)x$ as implied by the notation. Thus if $\lambda_0 \in \delta(x)$, so does the circular domain in (24.3.4).

If $\lambda_0 = 0$ is in $\delta(x)$, we form the series

$$D(\lambda; x) = j + \sum_{n=1}^{\infty} y^n\lambda^n, \tag{24.3.5}$$

where j and y are the quantities postulated in Definition 24.3.1. This series converges for $|\,\lambda\,| < 1/\beta$, $\beta = \lim_{n\to\infty} \|\,y^n\,\|^{1/n}$ and satisfies (24.3.1) by virtue of the relations between j, x, and y. Incidentally, the existence of such quantities j and y is necessary and sufficient for the validity of the theorem when $\lambda = 0$ is in $\delta(x)$ and this explains why the conditions were chosen in such a peculiar manner. This argument shows that $\delta(x)$ is an open set. Since $D(\lambda; x)$ is single-valued and locally holomorphic, it is holomorphic in the large in each component of $\delta(x)$. This completes the proof.

The argument used in proving Theorem 4.7.3 also gives

THEOREM 24.3.3. *The spectral radius of x equals*

$$r(x) \equiv \sup\,[\,|\,\lambda\,|\,; \lambda \in \alpha(x)] = \lim_{n\to\infty} \|\,x^n\,\|^{1/n}. \tag{24.3.6}$$

The principal component of the dissolvent set is by definition that containing the point at infinity. We note but omit the proof of the following theorem which is the analogue of Theorem 4.7.4:

THEOREM 24.3.4. *The spectrum is a closed bounded non-vacuous point set if $x \neq 0$.*

Finally we have to compare the two concepts of the spectrum:

THEOREM 24.3.5. *If $\mathfrak{B}$ has a unit element e, then the spectra $\alpha(x)$ and $\sigma(x)$ are identical except for the point $\lambda = 0$ which may belong to $\sigma(x)$ without belonging to $\alpha(x)$. This contingency arises if and only if quantities j and y exist with the properties specified above and $j \neq e$.*

PROOF. This follows essentially from Theorem 24.2.7 which shows that $(1/\lambda)x$ has a reverse if and only if $e - (1/\lambda)x$ has an inverse. This means that a $\lambda \neq 0$ belongs to $\alpha(x)$ if and only if it belongs to $\sigma(x)$. The point $\lambda = 0$ belongs to $\sigma(x)$ if x does not have an inverse and to $\alpha(x)$ if the quantities j and y of Definition 24.3.1 do not exist. It follows that the discrepancy will arise only in the case mentioned in the theorem. Finally we note the relation

$$R(\lambda; x) = \frac{e}{\lambda} - \frac{1}{\lambda} D(\lambda; x). \tag{24.3.7}$$

24.4. The dissolvent equations. The dissolvent $D(\lambda;\ x)$ satisfies two functional equations: one involving the parameter λ and the other the abstract variable x. We refer to these equations as the *first and second dissolvent equations.* They are obvious analogs of the resolvent equations and are derived in the same manner. They read as follows:

$$\lambda D(\lambda) - \mu D(\mu) = (\lambda - \mu) D(\lambda) D(\mu), \tag{24.4.1}$$

$$x D(y) - D(x) y = D(x)(x - y) D(y). \tag{24.4.2}$$

The theory of the first of these equations can be developed along the same lines as that of the first resolvent equation so we shall restrict ourselves to a fairly brief discussion of the salient facts. An analogue of Theorems 5.82 and 5.8.3 reads:

THEOREM 24.4.1. *Let Δ be a domain in the complex plane, not containing the origin, and let $D(\lambda)$ be a function on Δ to a complex (B)-algebra, satisfying* (24.4.1) *for all λ and μ in Δ. Then $D(\lambda)$ is holomorphic in Δ. It is a dissolvent of an element of the algebra, if and only if $D(\lambda)$ is reversible for at least one value of λ in Δ.*

PROOF. Let $\lambda_0 \in \Delta$ and rewrite (24.4.1) in the form

$$\lambda D(\lambda) = \lambda_0 D(\lambda_0) + (\lambda - \lambda_0) D(\lambda_0) D(\lambda).$$

By successive substitution and passing to the limit we obtain

$$\lambda D(\lambda) = \lambda_0 D(\lambda_0) + \lambda_0 D(\lambda_0) \sum_{n=1}^{\infty} \left\{ \frac{\lambda - \lambda_0}{\lambda} D(\lambda_0) \right\}^n \tag{24.4.3}$$

which converges for

$$\delta \,|\, \lambda - \lambda_0 \,| < |\, \lambda \,|, \qquad \delta = \lim_{n\to\infty} \| D^n(\lambda_0) \|^{1/n}.$$

Cf. formula (24.3.3). Thus $D(\lambda)$ is locally holomorphic in Δ and being single-valued (see below) it is consequently holomorphic in Δ.

If $D(\lambda)$ is the dissolvent of x, then $D(\lambda)$ is the reverse of $(1/\lambda)x$ so the condition is necessary. If the condition is satisfied, we may suppose that $D(\lambda_0)$ is the reverse of $(1/\lambda_0)x$, so that $x + \lambda_0 D(\lambda_0) = x D(\lambda_0) = D(\lambda_0) x$. Elimination of $D(\lambda_0)$ between this equation and (22.4.1) with $\lambda = \lambda_0$ shows that $D(\lambda)$ is the dissolvent of x for all λ in Δ. This completes the proof.

REMARK 1. The exclusion of $\lambda = 0$ from Δ is a matter of convenience. If the origin is in Δ, it becomes necessary to impose an additional condition, for instance

$$\lim_{\lambda\to 0} \lambda D(\lambda) = \theta$$

in order to ensure that the solution be holomorphic at $\lambda = 0$. For small values of λ the solution is then given by a series of type (24.3.5).

REMARK 2. It is worth noticing that (24.4.1) is a functional equation which gives much information about its solutions by inspection. Thus the equation shows that a solution which is bounded in some neighborhood of $\lambda = \lambda_0$ is continuous there. Further continuity implies differentiability since

$$\frac{\lambda D(\lambda) - \lambda_0 D(\lambda_0)}{\lambda - \lambda_0} \longrightarrow D^2(\lambda_0).$$

Thus, every locally bounded solution is locally holomorphic. In passing we note that $D(\lambda)$ satisfies a *Riccati differential equation*

$$\lambda D'(\lambda) = D^2(\lambda) - D(\lambda).$$

It should also be observed that a locally holomorphic solution is necessarily single-valued since the equation shows that $\lambda D(\lambda)$ has to return to its initial value if it is continued analytically along a closed curve.

It is also possible to obtain analogues of the results of sections 5.6 and 5.9. We shall give some samples.

Any solution of (24.4.1) which is holomorphic in an annulus with center at the origin is of the form

$$D(\lambda) = D^+(\lambda) + D^-(\lambda) \tag{24.4.4}$$

where

$$\begin{aligned} D^+(\lambda) &= j + \sum_{n=1}^{\infty} a^n \lambda^n, \qquad aj = ja = a, \qquad j^2 = j, \\ D^-(\lambda) &= -\sum_{n=1}^{\infty} b^n \lambda^{-n}, \qquad bj = jb = \theta, \end{aligned} \tag{24.4.5}$$

so that $D^+(\lambda)$ and $D^-(\mu)$ annihilate each other.

We also have "excentric" developments of the form (24.4.4) where

$$\begin{aligned} D^+(\lambda) &= \frac{\alpha}{\lambda} \sum_{n=0}^{\infty} b^{n+1} \left(\frac{\lambda - \alpha}{\lambda}\right)^n, \\ D^-(\lambda) &= -\frac{\alpha j}{\lambda - \alpha} - \frac{\alpha}{\lambda} \sum_{n=2}^{\infty} a^{n-1} \left(\frac{\lambda}{\lambda - \alpha}\right)^n, \end{aligned} \tag{24.4.6}$$

and a, b, j are subject to the same conditions as above. We note that with each such expansion there is associated an idempotent

$$j = \pm \frac{1}{2\pi i} \int_\Gamma D(\lambda) \frac{d\lambda}{\lambda}, \tag{24.4.7}$$

where Γ surrounds either $\lambda = \alpha$ $(\neq 0)$ or $\lambda = 0$ and the sign is minus in the former and plus in the latter case.

More generally, an integral of this type defines an idempotent in $\mathfrak{B}$ if Γ is a simple closed rectifiable oriented curve in the domain of holomorphism of $D(\lambda)$ and the sign is plus if $\lambda = 0$ is interior to Γ but otherwise minus. This is verified as in section 5.6. We have the following analogue of Theorem 5.6.1 which we state without proof.

THEOREM 24.4.2. *Let $a \in \mathfrak{B}$ and let $\alpha(a) \cup \{0\} = \bigcup_0^k \alpha_m$ where $\alpha_1, \cdots, \alpha_k$ are spectral sets of a and α_0 is a spectral set containing $\lambda = 0$ or this point alone according as $\lambda = 0$ as in $\alpha(a)$ or not. Further $\alpha_m \cap \alpha_n = \emptyset$ when $m \neq n$. We define*

$$(24.4.8) \qquad j_m = \pm \frac{1}{2\pi i} \int_{\Gamma_m} D(\lambda; a) \frac{d\lambda}{\lambda},$$

where Γ_m is an oriented envelope of α_m and the sign is plus if $m = 0$, otherwise minus. Then

$$j_0 = \sum_1^k j_m, \qquad j_m j_n = \delta_{mn} j_m, \qquad m, n = 1, 2, \cdots, k.$$

Setting

$$a_m = j_m a, \qquad D_m(\lambda; a) = j_m D(\lambda; a),$$

we have

$$a_0 = \sum_1^k a_m, \qquad D_0(\lambda; a) = \sum_1^k D_m(\lambda; a),$$

$$a_m + \lambda D_m(\lambda; a) = a_m D_m(\lambda; a) = D_m(\lambda; a) a_m,$$

$$a_m a_n = \theta, \qquad a_m D_n(\lambda; a) = D_n(\lambda; a) a_m = \theta, \qquad m \neq n \neq 0.$$

$D_m(\lambda; a)$ can be extended to be holomorphic in each domain of the complex plane which has no points in common with α_m when $m \neq 0$.

In the special case in which $\alpha_0 = \{0\} \in \delta(a)$, the spectral resolution leads to an actual resolution of a and of $D(\lambda; a)$ since $a = j_0 a$ so that

$$(24.4.9) \qquad a = \sum_1^k a_m, \qquad D(\lambda; a) = \sum_1^k D_m(\lambda; a).$$

The second dissolvent equation is much less tractable than the other functional equations which we have encountered so we shall give only brief indications of what may be achieved and the reasons why the results are so restricted.

Suppose that $D(x)$ is a single-valued locally bounded solution of (24.4.2), that is, a function on $\mathfrak{B}$ to itself defined in some domain $\mathfrak{D}$ of $\mathfrak{B}$. It follows easily from the equation that $xD(x)$ is continuous in $\mathfrak{D}$ and, less easily, that $xD(x)$ is analytic in the sense of Definition 3.17.2. If $D(x)$ is continuous in $\mathfrak{D}$, then $xD(x)$ is analytic. In the first case we cannot conclude that $D(x)$ is continuous, in the second case $D(x)$ need not be analytic. Here the limitations on our knowledge are not dictated so much by defects of the method as by intrinsic properties of the functional equation. This is seen by the following example.

We take for $\mathfrak{B}$ a nilpotent algebra with the elements

$$x = \alpha z + \beta z^2, \qquad z^3 = \theta, \quad \alpha, \beta \text{ complex},$$

define the arithmetical operations in the natural manner, and set $\| x \| = \max (| \alpha |, | \beta |)$. Placing $D(x) = f(\alpha, \beta) z^2$ we find that $D(x)$ satisfies (24.4.2) no matter how $f(\alpha, \beta)$ is chosen. Here $xD(x) \equiv \theta$ is a continuous and analytic function of x, but $D(x)$ need not be continuous and if it is continuous, it need not be (G)-differentiable.

If $D(x)$ is analytic, its first variation satisfies

$$x\delta D(x; h) = D(x)h - D(x)hD(x)$$

which is the variational analogue of the Riccati equation noted above.

The situation is a different one if we assume that $D(x)$ is a dissolvent. We shall restrict ourselves to listing some properties of $D(\lambda; x)$ as a function of x which will be needed in the following.

THEOREM 24.4.3. *If* $\lambda \in \delta(a)$, $\lambda \neq 0$, *and* $D(\lambda; a)$ *is the dissolvent of* a, *then* $\lambda \in \delta(x)$ *for*

$$\| x - a \| < | \lambda | [1 + \| D(\lambda; a) \|]^{-1}$$

and

$$D(\lambda; x) = D(\lambda; a) + [U(\lambda; x, a)]^{-} - D(\lambda; a)[U(\lambda; x, a)]^{-}, \tag{24.4.10}$$

where

$$U(\lambda; x, a) = \frac{1}{\lambda}[(x - a) - (x - a)D(\lambda; a)],$$

is an analytic function of x *for such values.*

PROOF. This follows from formula (24.2.8) and Theorem 24.2.4 upon replacing x, a, and b by $(1/\lambda)x$, $(1/\lambda)a$, and $D(\lambda; a)$ respectively. Formula (24.4.10) is the result of making this substitution and clearly represents a right reverse of $(1/\lambda)x$ in the sphere indicated. But, by the proof of formula (24.2.7), $(1/\lambda)x$ has a left reverse in the same sphere, that is, there is a unique reverse and the formula represents $D(\lambda; x)$ as asserted.

In this argument it is necessary to exclude $\lambda = 0$ because the assumption that $0 \in \delta(a)$ usually does not prevent $\lambda = 0$ from belonging to $\alpha(x)$ for values of x in any neighborhood of a. This goes back to the fact that if $0 \in \delta(a)$, then $a \in j\mathfrak{B}j$ and such a subalgebra is not necessarily open in $\mathfrak{B}$.

24.5. The operational calculus. We come now to the main problem in the theory: *the mapping of locally holomorphic scalar functions into the algebra of locally analytic functions with domain and range in the complex* (B)*-algebra* $\mathfrak{B}$.

For the case of an algebra $\mathfrak{B}$ having a unit element e the basic correspondence principle of Theorem 5.2.5 gave us an isomorphic mapping. From $\lambda \to x$, $1 \to e$, we concluded step by step that scalar power series go into vector power series, the kernel $(\zeta - \lambda)^{-1}$ of Cauchy's formula goes into the resolvent $R(\zeta; x)$ while Cauchy's integral goes into the resolvent integral defining the desired mapping.

When $\mathfrak{B}$ has no unit element the correspondence principle becomes

$$\lambda \to x, \qquad 0 \to \theta,$$

with the aid of which a power series in λ (without constant term) is carried into a power series in x and

$$\lambda(\mu - \lambda)^{-1} \to -D(\mu; x).$$

There are corresponding modifications of Cauchy's integral so that

$$\frac{1}{2\pi i}\int_{\Gamma} f(\mu)\,\frac{\lambda}{\mu-\lambda}\,\frac{d\mu}{\mu} \to -\frac{1}{2\pi i}\int_{\Gamma_x} f(\mu)D(\mu;x)\,\frac{d\mu}{\mu}.$$

These heuristic considerations will now be made more precise.

DEFINITION 24.5.1. *If*

$$f(\lambda) = \sum_{n=1}^{\infty} \alpha_n \lambda^n \qquad \textit{for} \qquad |\lambda| < \rho,$$

then the principal extension of $f(\lambda)$ to $\mathfrak{B}$ is defined by

$$f(x) = \sum_{n=1}^{\infty} \alpha_n x^n \qquad \textit{for} \qquad \|x\| < \rho.$$

When there is a unit element, Theorem 5.2.3 shows that the spectrum of x is an upper semi-continuous function of x. If the unit element is missing, the theorem is modified as follows:

THEOREM 24.5.1. *The set $\alpha(x) \cup \{0\}$ is an upper semi-continuous function of x.*

PROOF. The argument used in proving Theorem 5.2.3 applies with the obvious changes replacing $\sigma(a)$ by $\alpha(a) \cup \{0\}$, $R(\lambda; a)$ by $D(\lambda; a)$, $R(\lambda; x)$ by $D(\lambda; x)$, the reference to (4.8.4) by (24.4.10), and M^{-1} by $(1 + M)^{-1}$.

Let Δ be an open subset of the complex plane containing the origin and let $\rho(\Delta)$ be the distance from 0 to the boundary of Δ. We then define $\mathfrak{D}_\Delta = [x; \|x\| < \rho(\Delta)]$. For a function $f(\lambda)$ locally holomorphic in Δ with $f(0) = 0$, the principal extension of $f(\lambda)$ is determined by Definition 24.5.1 on $\mathfrak{D}_\Delta$. We now obtain a further extension of $f(\lambda)$.

THEOREM 24.5.2. *Let $f(\lambda)$ be locally holomorphic in the open set Δ which contains the origin where $f(0) = 0$. Let $\mathfrak{G}(\Delta)$ be the set of all elements $x \in \mathfrak{B}$ such that $\alpha(x) \subset \Delta$. For $x \in \mathfrak{G}(\Delta)$ we define*

$$f(x) = -\frac{1}{2\pi i}\int_{\Gamma_x} f(\lambda)D(\lambda;x)\,\frac{d\lambda}{\lambda}, \tag{24.5.1}$$

where Γ_x is an oriented envelope of $\alpha(x)$ with respect to $f(\lambda)$. Then $f(x)$ is locally analytic in the open set $\mathfrak{G}(\Delta)$, and $f(x)$ coincides with the principal extension of $f(\lambda)$ in $\mathfrak{D}_\Delta$.

PROOF. That $\mathfrak{G}(\Delta)$ is an open set follows from the preceding theorem. The sphere $\mathfrak{D}_\Delta$ and the integral are well defined. If $\|x\| = \rho_1 < \rho_2 < \rho(\Delta)$, we take Γ_x as the circle $|\lambda| = \rho_2$, use formula (24.3.2), and integrate termwise obtaining

$$f(x) = \sum_{n=1}^{\infty} x^n \frac{1}{2\pi i}\int_{\Gamma_x} f(\lambda)\lambda^{-n-1}\,d\lambda,$$

which coincides with the principal extension of $f(\lambda)$ in $\mathfrak{D}_\Delta$.

It remains to show that $f(x)$ is analytic in the larger domain $\mathfrak{G}(\Delta)$. Let $a \in \mathfrak{G}(\Delta)$ and choose a sphere $\mathfrak{S}(a): \| x - a \| < \rho$ so small that Φ, the closure of the set $\bigcup_x \alpha(x) \cup \{0\}$, $x \in \mathfrak{S}(a)$, is in Δ. We then replace Γ_x by Γ, a fixed oriented envelope of Φ with respect to $f(\lambda)$. Since $D(\lambda; a)$ is holomorphic on each component of Γ, which is compact, there exists a finite $M(a)$ such that $\| D(\lambda; a) \| \leqq M(a)$ on Γ. Formula (24.4.10) shows that for each fixed λ on Γ the function $D(\lambda; x)$ can be written as an abstract power series in $x - a$ which is absolutely convergent for $\| x - a \| \leqq \rho_0 < \min(\rho, [1 + M]^{-1})$, the convergence being uniform with respect to x in the sphere and with respect to λ on Γ. Substitution of this series into (24.5.1) and termwise integration gives

$$f(x) = f(a) + \sum_{n=1}^{\infty} \frac{1}{n!} \delta^n f(a; x - a) \tag{24.5.2}$$

where

$$\delta^n f(a; h) = \frac{n!}{2\pi i} \int_{\Gamma} f(\lambda) \{[h - hD(\lambda; a)]^n - D(\lambda; a)[h - hD(\lambda; a)]^n\} \lambda^{-n-1} \, d\lambda. \tag{24.5.3}$$

This is an abstract power series in $x - a$ which converges absolutely and uniformly when $\| x - a \| \leqq \rho_0$. It follows that $f(x)$ is analytic in this sphere and hence everywhere in $\mathfrak{G}(\Delta)$. The nth variation of $f(x)$ is given by formula (24.5.3) as suggested by the notation. We note that if $\mathfrak{B}$ is commutative, $f(x)$ will be analytic in the sense of Lorch, Definition 3.19.1. This completes the proof.

THEOREM 24.5.3. *If $\mathfrak{B}$ has a unit element e, then the definitions for $f(x)$ given by* (5.2.4) *and* (24.5.1) *are equal when both apply.*

PROOF. If $\mathfrak{B}$ has a unit element, formula (24.3.7) holds, when $R(\lambda; x)$ and $D(\lambda; x)$ are both defined, and $\sigma(x) \subset \alpha(x) \cup \{0\}$. Formula (24.3.7) certainly holds for $|\lambda| > \| x \|$ and consequently whenever both sides have a sense. If Δ contains $\lambda = 0$, the set $\mathfrak{G}(\Delta)$ of section 5.2 coincides with the set denoted by the same symbol in the current section; if $x \in \mathfrak{G}(\Delta)$ and $f(0) = 0$, then

$$\frac{1}{2\pi i} \int_{\Gamma_x} f(\lambda) R(\lambda; x) \, d\lambda = e \frac{1}{2\pi i} \int_{\Gamma_x} \frac{f(\lambda)}{\lambda} \, d\lambda - \frac{1}{2\pi i} \int_{\Gamma_x} f(\lambda) D(\lambda; x) \frac{d\lambda}{\lambda}.$$

Here the first integral in the right member is zero since $f(\lambda)/\lambda$ is holomorphic inside Γ_x, so the two definitions of $f(x)$ coincide.

Since the two theories are consistent, we may expect the two classes of functions $f(x)$ to have further properties in common. Analogues of Theorems 5.2.5, 5.3.1 and 5.3.2 are given below; the theorems are stated in full, but the proofs are reduced to a bare minimum.

THEOREM 24.5.4. *Let Δ be an open subset of the complex plane containing the origin. Let $\mathfrak{G}(\Delta)$ be the open set of elements $x \in \mathfrak{B}$ such that $\alpha(x) \subset \Delta$. Let $H(\Delta)$ be the complex algebra of all functions $f(\lambda)$, vanishing at $\lambda = 0$, locally holomorphic* *in* Δ, *with the ordinary definitions of the arithmetic operations, and with a sequence topology*: $f_n \to f$ *denoting that* $f_n(\lambda)$ *converges pointwise to* $f(\lambda)$, *the convergence being uniform in each compact subset of* Δ. *Further let* $\mathfrak{B}(\Delta)$ *be the complex algebra of functions* $F(x)$, *locally analytic in* $\mathfrak{G}(\Delta)$ *and having values in* $\mathfrak{B}$, *the arithmetic operations being defined as in* $\mathfrak{B}$.

Then there exists a homomorphic mapping: $f(\lambda) \to f(x)$ *of* $H(\Delta)$ *on a subalgebra* $\mathfrak{B}_0(\Delta)$ *of* $\mathfrak{B}(\Delta)$ *such that* (i) $\lambda \to x$ *and* (ii) $f_n \to f$ *implies that* $\| f_n(x) - f(x) \| \to 0$ *locally uniformly in* $\mathfrak{G}(\Delta)$. *This mapping is uniquely defined by* (24.5.1).

Proof. We can carry over the argument used in proving Theorem 5.2.5 with obvious modifications replacing resolvents by dissolvents. The proof that (24.5.1) defines a continuous homomorphic mapping is left to the reader, but we shall say a few words about the uniqueness of the mapping. If $\mathfrak{H}$ is a mapping with the properties stated above, then $\mathfrak{H}$ will map the polynomial $P(\lambda) = \sum_1^n \alpha_k \lambda^k$ on the polynomial $P(x) = \sum_1^n \alpha_k x^k$. If α is not in Δ, then $\lambda(\lambda - \alpha)^{-1} \in H(\Delta)$ and if $x \in \mathfrak{G}(\Delta)$ then $D(\alpha; x)$ exists. In fact, in the identity

$$\lambda + \alpha \frac{\lambda}{\lambda - \alpha} = \lambda \frac{\lambda}{\lambda - \alpha}$$

the terms are elements of $H(\Delta)$ so that

$$x + \alpha \mathfrak{H}\left\{\frac{\lambda}{\lambda - \alpha}\right\} = x\mathfrak{H}\left\{\frac{\lambda}{\lambda - \alpha}\right\} = \mathfrak{H}\left\{\frac{\lambda}{\lambda - \alpha}\right\} x$$

and hence

$$\mathfrak{H}\left\{\frac{\lambda}{\lambda - \alpha}\right\} = D(\alpha; x).$$

Moreover, since

$$\alpha \frac{\lambda}{(\lambda - \alpha)^k} = \frac{\lambda}{\lambda - \alpha} \frac{\lambda}{(\lambda - \alpha)^{k-1}} - \frac{\lambda}{(\lambda - \alpha)^{k-1}},$$

an induction argument suffices to show that $\mathfrak{H}\{\lambda(\lambda - \alpha)^{-k}\}$ is uniquely determined for all $k \geqq 1$.

It follows that $\mathfrak{H}$ takes the rational functions of $H(\Delta)$ into their principal extensions defined by (24.5.1). But the rational functions of $H(\Delta)$ are dense in the space, by the extended Runge theorem, and $\mathfrak{H}$ is continuous, whence it follows that $\mathfrak{H}$ must be identical with the extension (24.5.1). It should be noted that we can no longer assert that the mapping is an isomorphism. Thus in the case of the nilpotent algebra considered in the preceding section we have $f(x) \equiv \theta$ if $f(0) = f'(0) = f''(0) = 0$. All functions of this algebra are of the form $f(x) = \alpha_1 x + \alpha_2 x^2$.

For the spectral mapping theorem we need a lemma to replace the familiar fact that a product cannot have an inverse unless its factors have inverses.

LEMMA 24.5.1. *If $a \times b$ has a right reverse (left reverse), then a has a right reverse (b has a left reverse).*

PROOF. If $(a \times b) \times c = \theta$, then $a \times (b \times c) = \theta$ and $b \times c$ is the right reverse of a. Left reverses are handled in the same manner.

The spectral mapping theorem requires a rather careful choice of the auxiliary functions. In order to simplify matters, we adjoin the origin to the spectrum and write $\beta(x) = \alpha(x) \cup \{0\}$ with similar notation for other elements of $\mathfrak{B}$.

THEOREM 24.5.5. *If $f(\lambda) \in H(\Delta)$ and if $x \in \mathfrak{G}(\Delta)$, then*

$$\beta[f(x)] = f[\beta(x)].$$

PROOF. Since $f(0) = 0$ we know that $\lambda = 0$ belongs both to $\beta[f(x)]$ and $f[\beta(x)]$. Suppose now that $\zeta \in \beta(x)$, $\zeta \neq 0$, $f(\zeta) \neq 0$, and form

$$g(\lambda) = \frac{\lambda f(\zeta) - \zeta f(\lambda)}{\lambda - \zeta}$$

which is an element of $H(\Delta)$. If $g(x)$ is the corresponding element of $\mathfrak{B}_0(\Delta)$, we have $xf(\zeta) - \zeta f(x) = xg(x) - \zeta g(x)$ which may be rewritten

$$\frac{1}{f(\zeta)} f(x) = \frac{1}{\zeta} x + \frac{1}{f(\zeta)} g(x) - \frac{1}{\zeta} x \frac{1}{f(\zeta)} g(x).$$

Since x and $g(x)$ commute, the assumption that the left member is reversible would imply that $(1/\zeta)x$ has a reverse, by the preceding lemma, which contradicts $\zeta \in \alpha(x)$. Hence $f(\zeta) \in \alpha[f(x)]$. Suppose conversely that $\mu \in \alpha[f(x)]$, $\mu \neq 0$, $\mu \notin f[\alpha(x)]$. We then form $h(\lambda) = f(\lambda)[f(\lambda) - \mu]^{-1}$ which is locally holomorphic in $\beta(x)$. We can therefore find an open set Δ' with $\beta(x) \subset \Delta' \subset \Delta$ such that $h(\lambda)$ is locally holomorphic in Δ'. We note that $x \in \mathfrak{G}(\Delta')$. If $h(x)$ is the corresponding function in $\mathfrak{B}_0(\Delta')$, we have $f(x) + \mu h(x) = f(x)h(x) = h(x)f(x)$, contradicting the assumption that $\mu \in \alpha[f(x)]$. This completes the proof.

THEOREM 24.5.6. *If $g(\lambda) \in H(\Delta)$, if $f(\lambda) \in H(\Delta_0)$ where $g(\Delta) \subset \Delta_0$, and if $f(0) = 0$, then $f[g(\lambda)] \in H(\Delta)$, $g(x) \in \mathfrak{G}(\Delta_0)$, $f[g(x)] \in \mathfrak{B}_0(\Delta)$, and $f[g(x)] = [f(g)](x)$ for each $x \in \mathfrak{G}(\Delta)$.*

PROOF. The basic relations are

$$\begin{aligned} f(g(x)) &= -\frac{1}{2\pi i}\int_\Gamma f(\mu)D(\mu; g(x))\,\frac{d\mu}{\mu} \\ &= \frac{1}{(2\pi i)^2}\int_\Gamma\int_{\Gamma_0}\frac{f(\mu)g(\lambda)}{g(\lambda)-\mu}\,D(\lambda; x)\,\frac{d\lambda}{\lambda}\,\frac{d\mu}{\mu} \\ &= -\frac{1}{2\pi i}\int_{\Gamma_0} f(g(\lambda))D(\lambda; x)\,\frac{d\lambda}{\lambda} = (f(g))(x), \end{aligned}$$

where Γ_0 is an oriented envelope of $\beta(x)$ with respect to $k(\lambda) = g(\lambda)[g(\lambda) - \mu]^{-1}$

and Γ is an oriented envelope of $g(\bar{\Delta}_x)$ with respect to $f(\mu)$ in the μ-plane; here Δ_x is a bounded open set, $\beta(x) \subset \Delta_x$ and $\bar{\Delta}_x \subset \Delta$. The rest of the argument goes as in the proof of Theorem 5.3.2.

References: Jacobson [2, 3], Kaplansky [1], Perlis [1], Segal [1].

2. Ideal Theory

24.6. Ideals. The concepts of right ideal, left ideal, two-sided ideal, proper and improper ideals were introduced in Definition 1.13.2 for an arbitrary algebra. An ideal theory for commutative (B)-algebras with unit element was developed in §4.3. In this and the following paragraphs we shall be concerned with ideal theory in more general situations. While emphasis will be on (B)-algebras, commutativity and existence of a unit element will not be assumed unless explicitly stated, some results will be stated for topological algebras and some for algebras which are not topologized.

We recall that a group under addition is called a module. We shall also need the concept of a regular ideal.

Definition 24.6.1. *A right ideal (left ideal) is said to be regular if there is a $j \in \mathfrak{A}$ such that $jy - y \in \mathfrak{i}$ ($yj - y \in \mathfrak{i}$) for every $y \in \mathfrak{A}$. Here j is called a left (right) unit element of $\mathfrak{A}$ modulo $\mathfrak{i}$.*

If $\mathfrak{A}$ has a unit element all ideals are regular and $j = e$.

It was observed in section 1.13 that every algebra $\mathfrak{A}$ contains the two improper ideals $\{\theta\}$ and $\mathfrak{A}$ and Theorem 4.13.4 states that if a commutative (B)-algebra with unit element has no proper ideals then $\mathfrak{A}$ is isomorphic to the complex field. The general situation is described in

Theorem 24.6.1. *If $\mathfrak{A}$ has no proper left (right) ideals, then $\mathfrak{A}$ is either a (skew) field or a zero algebra, $\mathfrak{A}^2 = \{\theta\}$, of dimension one. Conversely, a (skew) field or a zero algebra of dimension one has no proper ideals.*

Proof. We show first that if $\mathfrak{A}$ has proper zero divisors then $\mathfrak{A}$ is a zero algebra. For let $ab = \theta$, $a \neq \theta$, $b \neq \theta$. The totality $\mathfrak{z}_b$ of elements z such that $zb = \theta$ is a left ideal containing a. Hence $\mathfrak{z}_b \neq \{\theta\}$ and this forces $\mathfrak{z}_b = \mathfrak{A}$. Thus $\mathfrak{A}b = \{\theta\}$. Now consider the totality $\mathfrak{n}$ of elements v such that $\mathfrak{A}v = \{\theta\}$. This is a two-sided ideal containing b. Hence $\mathfrak{n} = \mathfrak{A}$ and $\mathfrak{A}^2 = \{\theta\}$ so that any subspace of $\mathfrak{A}$ is an ideal. Hence if $\mathfrak{A}$ has no proper left ideals it is a zero algebra of dimension one. We consider next the case $\mathfrak{A}^2 \neq \{\theta\}$. Let a be any element $\neq \theta$ in $\mathfrak{A}$ and consider the left ideal $\mathfrak{i} = \mathfrak{A}a$. Since $\mathfrak{A}$ has no zero divisors, $\mathfrak{A}a \neq \{\theta\}$ and $\mathfrak{A} = \mathfrak{A}a$. This implies that there is an element e such that $a = ea$. Then

$ea = e^2a$ and $(e^2 - e)a = \theta$ so that $e^2 = e$. Now if b is any element $\neq \theta$ in $\mathfrak{A}$, $e(eb - b) = e^2b - eb = eb - eb = \theta$. Hence $eb = b$ and similarly $be = b$ for all $b \neq \theta$. This implies that e is a unit element for $\mathfrak{A}$. Since $\mathfrak{A}a = \mathfrak{A}$ for any $a \neq \theta$, the equation $xa = e$ has a solution for every $a \neq \theta$. It follows that the non-zero elements of $\mathfrak{A}$ form a group under multiplication and $\mathfrak{A}$ is a (skew) field. The converse is obvious.

THEOREM 24.6.2. *If $\mathfrak{A}$ is a topological algebra in the sense of Definition 1.14.1, then the closure of a right (left, two-sided) ideal is an ideal of the same kind.*

PROOF. Suppose, to fix the ideas, that $\mathfrak{i}$ is a right ideal. We denote its closure by $\mathfrak{c}$ and let x, y, z denote arbitrary elements of $\mathfrak{i}$, $\mathfrak{c}$, and $\mathfrak{A}$ respectively. The neighborhood postulates satisfied by $\mathfrak{A}$ imply that the closures of $z - \mathfrak{i}$, $\alpha\mathfrak{i}$ and $\mathfrak{i}z$ are respectively $z - \mathfrak{c}$, $\alpha\mathfrak{c}$ and $\mathfrak{c}z$. Since $x - \mathfrak{i} = \mathfrak{i}$, the first of these implications gives that $x - \mathfrak{c} = \mathfrak{c}$. Hence $x - y \in x - \mathfrak{c} = \mathfrak{c}$ so that $\mathfrak{i} - y \subset \mathfrak{c}$ and, passing to the closures $\mathfrak{c} - y \subset \mathfrak{c}$. This shows that $\mathfrak{c}$ is a module. Secondly, since $\alpha\mathfrak{i} = \mathfrak{i}$, we conclude from $\alpha y \in \alpha\mathfrak{c}$ that $\alpha y \in \mathfrak{c}$ or $\mathfrak{c}$ is a linear set. Similarly $\mathfrak{i}z = \mathfrak{i}$ shows that $yz \in \mathfrak{c}z$ implies $yz \in \mathfrak{c}$ so that $\mathfrak{c}$ is a right ideal as asserted. It is clear from the proof that the theorem holds if the postulated topology is replaced by a closure topology in which the closure properties used above are valid.

The question of when the closure of a proper ideal is also proper is important. In this connection we state and prove a couple of theorems of which the second was suggested by a result due to I. E. Segal [1, Theorem 1.6]. The first is a generalization of Theorem 4.13.1.

THEOREM 24.6.3. *If $\mathfrak{A}$ is a topological algebra with unit element e and if there is a neighborhood of e all the elements of which have inverses, then the closure of any proper right (left, two-sided) ideal of $\mathfrak{A}$ is a proper ideal of the same kind.*

PROOF. Suppose that $\mathfrak{i}$ is a proper right ideal. We know that $\mathfrak{c}$, the closure of $\mathfrak{i}$, is a right ideal. Suppose that $\mathfrak{c} = \mathfrak{A}$ and let N_e be the neighborhood of e where inverses exist. It follows that there exists an $x \in \mathfrak{i} \cap N_e$, that is, $\mathfrak{i}$ contains a regular element of $\mathfrak{A}$. This leads to $\mathfrak{i} = \mathfrak{A}$ which is a contradiction.

THEOREM 24.6.4. *If $\mathfrak{A}$ is a topological algebra and if there is a neighborhood of the zero element all the elements of which have reverses, then the closure of any proper regular right (left, two-sided) ideal is a proper ideal of the same kind.*

PROOF. Suppose that $\mathfrak{i}$ is a proper regular right ideal and let j be the corresponding left unit element modulo $\mathfrak{i}$. We know that $\mathfrak{c}$, the closure of $\mathfrak{i}$, is a right ideal and it is evidently regular. If N_θ is the neighborhood of θ where reverses exist and if $\mathfrak{c} = \mathfrak{A}$, then we can find an element $j - x$ in N_θ with $x \in \mathfrak{i}$. Let z be the reverse of $j - x$. It follows that $j = x + (jz - z) - xz$ is in $\mathfrak{i}$ since it is the sum of three elements of $\mathfrak{i}$. But then $jy \in \mathfrak{i}$ for all $y \in \mathfrak{A}$ and, since $jy - y \in \mathfrak{i}$, it follows that $y \in \mathfrak{i}$ and $\mathfrak{i}$ is not proper against the assumption.

The assumptions of the last two theorems are obviously satisfied in any complex (B)-algebra.

The notions of homomorphism and isomorphism used in the following were defined in Definition 1.6.1. It follows from the definition that if $\mathfrak{A} \sim \mathfrak{A}'$ then a module (linear set, ideal) of $\mathfrak{A}$ is mapped onto a module (linear set, ideal) of $\mathfrak{A}'$.

THEOREM 24.6.5. *If $\mathfrak{A}' = H(\mathfrak{A})$ is the homomorphic image of $\mathfrak{A}$, then the elements of $\mathfrak{A}$ which are mapped on the zero element of $\mathfrak{A}'$ form a two-sided ideal of $\mathfrak{A}$ known as the kernel of the homomorphism.*

PROOF. If $H(a) = \theta'$, $H(b) = \theta'$, then $H(\alpha a + \beta b) = \theta'$ and $H(ax) = H(a)H(x) = \theta' = H(x)H(a) = H(xa)$ for all x in $\mathfrak{A}$.

Conversely, a two-sided ideal of $\mathfrak{A}$ generates a homomorphism of $\mathfrak{A}$ onto an algebra $\mathfrak{A}/\mathfrak{i}$ defined in the next section.

24.7. Residue-class algebras. The notion of a congruence relation in a linear space was introduced in section 1.12 and appeared again in section 4.14 for the case of commutative (B)-algebras with a unit element. For the moment let us dispense with the topology; let $\mathfrak{X}$ be any linear system, $L \subset \mathfrak{X}$ a linear subsystem and define $x \equiv y \pmod{L}$ if $x - y \in L$. The elements of $\mathfrak{X}$ which are congruent modulo L to a given element x form a residue-class X represented by x. We define addition and scalar multiplication for residue-classes by the convention that $X + Y$ and αX are the residue-classes determined by $x + y$ and αx respectively where $x \in X$, $y \in Y$ and $\alpha \in \Phi$. The resulting residue-class system is a linear system and will be denoted by $\mathfrak{X} \div L$. Its zero element is $\Theta = L$.

In particular these conventions apply to the case in which $\mathfrak{X} = \mathfrak{A}$ is an algebra and $L = \mathfrak{i}$ is an ideal of $\mathfrak{A}$. If $\mathfrak{i}$ is two-sided, we can go one step further and define multiplication: XY is the residue-class determined by xy. This makes sense for $(x + \mathfrak{i})(y + \mathfrak{i}) = xy + \mathfrak{i}y + x\mathfrak{i} + \mathfrak{i} = xy + \mathfrak{i}$ since the ideal is two-sided. Thus the system $\{X\}$ is also an algebra referred to as a *residue-class algebra* or *quotient algebra* or *difference algebra* and commonly denoted by $\mathfrak{A}/\mathfrak{i}$.

THEOREM 24.7.1. *If $\mathfrak{i}$ is a regular two-sided ideal and if j is the associated unit element modulo $\mathfrak{i}$, then $\mathfrak{A}/\mathfrak{i}$ has a unit element, namely the residue class J determined by j. In particular, if $\mathfrak{A}$ has the unit element e, then the residue class E determined by e is the unit element of $\mathfrak{A}/\mathfrak{i}$.*

PROOF. By assumption $xj - x \in \mathfrak{i}$ and $jx - x \in \mathfrak{i}$ for each x in $\mathfrak{A}$. Residuation modulo $\mathfrak{i}$ gives $XJ = JX = X$ for each X, which is the first assertion. If $\mathfrak{A}$ has the unit element e, then every ideal $\mathfrak{i}$ is regular with $j = e$.

THEOREM 24.7.2. *If $\mathfrak{A}$ is homomorphic to $\mathfrak{A}'$ and if $\mathfrak{i}$ is the kernel of the homomorphism, then $\mathfrak{A}' \cong \mathfrak{A}/\mathfrak{i}$.*

PROOF. We write $\mathfrak{A}' = H_1(\mathfrak{A})$ and $\mathfrak{A}/\mathfrak{i} = H_2(\mathfrak{A})$. Now $H_1(x_1) = H_1(x_2)$ and $H_2(x_1) = H_2(x_2)$ if and only if $x_1 \equiv x_2 \pmod{\mathfrak{i}}$. The correspondence $H_1(x) \leftrightarrow H_2(x)$

is defined for all x in $\mathfrak{A}$ and establishes a one-to-one mapping of $\mathfrak{A}'$ onto $\mathfrak{A}/\mathfrak{i}$. Since $H_1(\alpha x + \beta y)$ corresponds to $H_2(\alpha x + \beta y)$ and $H_1(xy)$ to $H_2(xy)$, the mapping is an isomorphism and the theorem is proved. This is the fundamental theorem on homomorphisms.

With the residue-class system $\mathfrak{R} = \mathfrak{A} \div \mathfrak{i}$ there is associated a two-sided ideal $\mathfrak{i}_0 = \mathfrak{i}:\mathfrak{A}$ known as the *quotient of* $\mathfrak{i}$ *relative to* $\mathfrak{A}$ which we now proceed to define. If $\mathfrak{i}$ is a left ideal and a is any element of $\mathfrak{A}$, then the mapping T_a which sends the residue-class $x + \mathfrak{i}$ into the residue-class $ax + \mathfrak{i}$ is an endomorphism on $\mathfrak{R}$, that is, a linear transformation of $\mathfrak{R}$ onto part of itself. Since

$$\alpha T_a = T_{\alpha a}, \qquad T_a + T_b = T_{a+b}, \qquad T_a T_b = T_{ab},$$

the transformations $\{T_a\}$ form an algebra $\mathfrak{T}$ which is a subset of the algebra of all endomorphisms on $\mathfrak{R}$, and $\mathfrak{T} = H(\mathfrak{A})$ is a homomorphic image of $\mathfrak{A}$.

Definition 24.7.1. *The quotient of* $\mathfrak{i}$ *relative to* $\mathfrak{A}$ *is the kernel of the homomorphism H just defined.*

Theorem 24.7.3. *We have* $\mathfrak{T} \cong \mathfrak{A}/(\mathfrak{i}:\mathfrak{A})$. *Further,* $(\mathfrak{i}:\mathfrak{A})\mathfrak{A} \subset \mathfrak{i}$ *and, if* $\mathfrak{A}$ *has a unit element, then* $\mathfrak{i}:\mathfrak{A}$ *is the largest two-sided ideal contained in* $\mathfrak{i}$.

Proof. The first assertion follows from the preceding theorem. An element $a \in \mathfrak{i}:\mathfrak{A}$ if and only if $ax \in \mathfrak{i}$ for all x in $\mathfrak{A}$, whence it follows that $(\mathfrak{i}:\mathfrak{A})\mathfrak{A} \subset \mathfrak{i}$. If $\mathfrak{A}$ has a unit element, then $a \in \mathfrak{i}:\mathfrak{A}$ implies that $a \in \mathfrak{i}$, that is, $\mathfrak{i}:\mathfrak{A} \subset \mathfrak{i}$, and if $\mathfrak{i}_1$ is any two-sided ideal contained in $\mathfrak{i}$, then from $a \in \mathfrak{i}_1$ we conclude that $ax \in \mathfrak{i}_1 \subset \mathfrak{i}$ for all x so that $\mathfrak{i}_1 \subset \mathfrak{i}:\mathfrak{A}$ and $\mathfrak{i}:\mathfrak{A}$ is the largest two-sided ideal contained in $\mathfrak{i}$. (For quotient ideals, cf. N. Jacobson [**2**].)

If $\mathfrak{A}$ is a (B)-algebra over the real or complex field, we use the generic notation $\mathfrak{B}$ as usual. The problem of introducing a normed topology in the residue class system $\mathfrak{B} \div \mathfrak{R}$, where $\mathfrak{R}$ is a closed subspace or an ideal, was first treated by I. Gelfand [**4**]. We have already discussed this problem for a closed linear subspace in Theorem 1.12.3 and for a closed ideal in a commutative (B)-algebra in Theorem 4.14.2. No new considerations enter in the noncommutative case; we have consequently the following

Theorem 24.7.4. *If* $\mathfrak{B}$ *is a (B)-algebra over* Φ *and if* $\mathfrak{i}$ *is a closed ideal, then* $\mathfrak{B} \div \mathfrak{i}$ *becomes a (B)-space over* Φ *under the norm* $\| X \| = \inf [\| x \| ; x \in X]$. *If* $\mathfrak{i}$ *is two-sided, then* $\mathfrak{B}/\mathfrak{i}$ *is a (B)-algebra under this norm.*

We state without proof

Theorem 24.7.5. *When* $\mathfrak{i}$ *is closed, each residue-class X, modulo* $\mathfrak{i}$, *is a closed, arcwise connected point set in* $\mathfrak{B}$.

Theorem 24.7.6. *If* $\mathfrak{i}$ *is a two-sided ideal of the algebra* $\mathfrak{A}$, *and if the residue-class X, modulo* $\mathfrak{i}$, *contains an element x regular in* $\mathfrak{A}$ *(reversible in* $\mathfrak{A}$*), then X is regular in* $\mathfrak{A}/\mathfrak{i}$ *(reversible in* $\mathfrak{A}/\mathfrak{i}$*).*

PROOF. Suppose first that $\mathfrak{A}$ has a unit element e and that E is the corresponding residue-class of $\mathfrak{A}/\mathfrak{i}$ so that E is the unit element of $\mathfrak{A}/\mathfrak{i}$. If now x has the inverse y and Y is the residue-class determined by y, then $XY = YX = E$. Similarly, if instead y is the reverse of x, the relation $x + y = xy = yx$ implies $X + Y = XY = YX$.

In the next theorem we shall consider an arbitrary algebra $\mathfrak{A}$ over the complex field. We define the notion of the spectrum of an element x of $\mathfrak{A}$ by Definition 4.7.1 or Definition 24.3.1 according as $\mathfrak{A}$ has a unit element or not. The complementary set then forms the resolvent (resp. dissolvent) set of x. These were the definitions given for (B)-algebras, but they are evidently meaningful in the absence of a topology. We then have the following

THEOREM 24.7.7. *If $\mathfrak{A}$ is an algebra over the complex field and $\mathfrak{i}$ is a two-sided ideal, then the spectrum of X, considered as an element of $\mathfrak{A}/\mathfrak{i}$, is contained in the intersection of the spectra of the elements x of X and the resolvent (dissolvent) set of X contains the union of the corresponding sets for the elements x of X.*

PROOF. Suppose first the $\mathfrak{A}$ has a unit element e. Since $\lambda e - x \in \lambda E - X$, the preceding theorem shows that $\lambda E - X$ is regular whenever $\lambda e - x$ is so. Hence $\rho(X) \supset \bigcup \rho(x)$ and $\sigma(X) \subset \bigcap \sigma(x)$. If there is no unit element, we see, again by the preceding theorem, that X/λ is reversible whenever x/λ is, $\lambda \neq 0$. Further, if there exist elements $y, j \in \mathfrak{A}$ such that $x = xj = jx, y = yj = jy, xy = yx = j$, $j^2 = j$, then the corresponding relations hold between the residue classes X, Y, J determined by x, y, j, so that $\lambda = 0$ is in the dissolvent set of X whenever $\lambda = 0$ belongs to $\delta(x)$. Hence $\delta(X) \supset \bigcup \delta(x)$ and $\alpha(X) \subset \bigcap \alpha(x)$. This completes the proof.

THEOREM 24.7.8. *Let $\mathfrak{B}$ be a complex (B)-algebra with unit element. Let $\mathfrak{i}$ be a closed two-sided ideal. If $f(\lambda)$ is locally holomorphic in an open set containing $\sigma(x)$, $x \in \mathfrak{B}$, and if x determines the residue-class X, modulo $\mathfrak{i}$, then $f(X)$ exists as an element of $\mathfrak{B}/\mathfrak{i}$, and*

$$f(X) \equiv \frac{1}{2\pi i} \int_\Gamma f(\lambda) R(\lambda; X)\, d\lambda = F(x),$$

where $F(x)$ is the residue class determined by $f(x)$, modulo $\mathfrak{i}$, and Γ is a directed envelope of $\sigma(x)$ with respect to $f(\lambda)$.

PROOF. The reader has merely to combine Theorem 5.2.4 and 24.7.7. It is obviously sufficient that $f(\lambda)$ be locally holomorphic in an open set containing $\sigma(X)$, but it is worth noticing that the existence of $f(x)$ for any x in X implies that of $f(X)$. Further, the mapping $x \to X$ being in particular a bounded linear transformation, it follows by Theorem 3.3.2 that $f(X)$ is the residue-class modulo $\mathfrak{i}$ in $\mathfrak{B}$ generated by $f(x)$. If $f(x)$ is analytic in some domain of $\mathfrak{B}$, then $f(X)$ is analytic in the corresponding domain of $\mathfrak{B}/\mathfrak{i}$ obtained by residuation modulo $\mathfrak{i}$.

A similar extension may be obtained by Theorem 24.5.2 if $f(\lambda)$ is locally holomorphic in an open set containing $\alpha(x) \cup \{0\}$ and $f(0) = 0$.

24.8. The radical. The following three sections will be devoted to the theory of the radical, maximal ideals, and their interrelations. In Definition 4.13.2 we defined the radical of a commutative (B)-algebra with unit element as the intersection of all maximal ideals. A slight modification of this definition (see Theorem 24.10.1 below) will work in the general case contemplated here, but we shall use another alternative in defining the radical as the union of the properly quasi-nilpotent elements. The resulting concept agrees with the radical of N. Jacobson [2], but is wider than the radical of I. E. Segal [1]. For the history of the development of this notion for algebras without finiteness assumptions, see also the papers of I. Gelfand [4], S. Perlis [1], R. Baer [1] and I. Kaplansky [1].

DEFINITION 24.8.1. *An element q of the algebra $\mathfrak{A}$ over the scalar field Φ is said to be quasi-nilpotent if αq has a reverse for all α in Φ. The set of all quasi-nilpotent elements is the quasi-radical $\mathfrak{q}$ of $\mathfrak{A}$.*

This definition of quasi-nilpotency agrees with that of section 4.5 for the case of a commutative (B)-algebra with unit element.

DEFINITION 24.8.2. *An element p of the algebra $\mathfrak{A}$ over the scalar field Φ is said to be properly quasi-nilpotent if $\alpha p + xp$ has a reverse for all α in Φ and all x in $\mathfrak{A}$. The set of properly quasi-nilpotent elements is the radical $\mathfrak{p}$ of $\mathfrak{A}$. If $\mathfrak{p}$ reduces to the zero element, $\mathfrak{A}$ is said to be without radical or to be semi-simple.*

The asymmetry in the condition on p is only apparent as is shown by the following

LEMMA 24.8.1. *If $\alpha p + xp$ has the reverse r, then $\alpha p + px$ has the reverse $\alpha pr + prx - \alpha p - px$.*

REMARK. It is an easy matter to verify that if ab has the reverse r then ba has the reverse $bra - ba$. If in this expression we set a formally equal to $\alpha + x$ and $b = p$ and multiply out, the result turns out to be the desired reverse.

THEOREM 24.8.1. *The radical is a two-sided ideal and it is a normal sub-group of the principal component of $\mathfrak{R}$ under the operation of cross-multiplication.*

PROOF. If $p \in \mathfrak{p}$, so do $\alpha p + ap$ and $\alpha p + pa$ for every $\alpha \in \Phi$ and every $a \in \mathfrak{A}$ by virtue of the definition of p and the preceding lemma. In order to prove that $\mathfrak{p}$ is a two-sided ideal it remains to show that $p_1 + p_2 \in \mathfrak{p}$ whenever p_1 and p_2 do so. For this purpose it is clearly enough to show that $p_1 + p_2$ is reversible. Let r be the reverse of p_1 and let us try to choose s in such a manner that $r \times s$ becomes a right reverse of $p_1 + p_2$. Using the identity

$$(a + b) \times c = a \times c + b \times c - c,$$

we obtain, suppressing some intermediary steps,

$$\begin{aligned}(p_1 + p_2) \times (r \times s) &= p_1 \times r \times s + p_2 \times r \times s - r \times s \\ &= s + p_2 - p_2(r \times s) \\ &= s + p_2 - p_2 r - (p_2 - p_2 r)s \\ &= (p_2 - p_2 r) \times s.\end{aligned}$$

This will reduce to zero if s is taken as the reverse of $p_2 - p_2 r$ which exists since $p_2 - p_2 r \in \mathfrak{p}$. Similarly one shows that $t \times r$ is a left reverse of $p_1 + p_2$ if t is the reverse of $p_2 - rp_2$. It follows that $p_1 + p_2$ is reversible and belongs to $\mathfrak{p}$ which is consequently a two-sided ideal.

We note that $\theta \in \mathfrak{p}$ and if $p \in \mathfrak{p}$, so does $p^- = -p + pp^-$. Further, if $p_1, p_2 \in \mathfrak{p}$, then $p_1 \times p_2 = p_1 + p_2 - p_1 p_2 \in \mathfrak{p}$, showing that $\mathfrak{p}$ is a group under the associative operation of cross-multiplication. It is clear that $\mathfrak{p} \subset \mathfrak{R}$ and $\mathfrak{p}$ is arc-wise connected since the elements αp, $0 \leqq \alpha \leqq 1$, connect p with θ. Hence $\mathfrak{p}$ is a subgroup of the principal component of $\mathfrak{R}$. For every $a \in \mathfrak{R}$ we have $a \times p = (a \times p \times a^-) \times a$ and here $a \times p \times a^- = p - ap - pa^- + apa^- \in \mathfrak{p}$ so that $a \times \mathfrak{p} = \mathfrak{p} \times a$ and $\mathfrak{p}$ is a normal sub-group of the principal component of $\mathfrak{R}$. This completes the proof.

THEOREM 24.8.2. *$\mathfrak{p} \subset \mathfrak{q}$ and $\mathfrak{p} = \mathfrak{q}$ if and only if $\mathfrak{q}$ is a left or right ideal. In particular, $\mathfrak{p} = \mathfrak{q}$ if $\mathfrak{A}$ is a commutative complex (B)-algebra.*

PROOF. The inclusion is obvious. Suppose that $\mathfrak{q}$ is a left ideal. Then $q \in \mathfrak{q}$ implies that $\alpha q + xq \in \mathfrak{q}$ for all $\alpha \in \Phi$ and $x \in \mathfrak{A}$ so that $q \in \mathfrak{p}$ and $\mathfrak{p} = \mathfrak{q}$. In a commutative complex (B)-algebra the series $-\sum_{n=1}^{\infty}(\alpha q + xq)^n$ converges for all α and x if $q \in \mathfrak{q}$ and the sum of the series is the reverse of $(\alpha q + xq)$ so that $\mathfrak{q} \in \mathfrak{p}$ and $\mathfrak{p} = \mathfrak{q}$.

The quasi-radical is of little interest to the algebraist, but quasi-nilpotent elements occur in so many algebras of interest to the analyst that they deserve at least a mention in passing.

THEOREM 24.8.3. *The residue-class algebra $\mathfrak{A}/\mathfrak{p}$ is semi-simple.*

PROOF. Suppose that s is in $\mathfrak{A}$ and determines the coset S in $\mathfrak{A}/\mathfrak{p}$ where S belongs to the radical of $\mathfrak{A}/\mathfrak{p}$. This means that $\alpha S + XS$ has a reverse in $\mathfrak{A}/\mathfrak{p}$ for all $\alpha \in \Phi$, $X \in \mathfrak{A}/\mathfrak{p}$. Hence $\alpha s + xs$ has a reverse modulo $\mathfrak{p}$ for every $x \in \mathfrak{A}$, that is, there are elements $y \in \mathfrak{A}$, $p \in \mathfrak{p}$ such that $y \times z = p$, $z = \alpha s + xs$. Since p^- exists, $(p^- \times y) \times z = \theta$ and z has a left reverse. The existence of a right reverse is proved in the same manner. Hence $\alpha s + xs$ has a reverse for all α and x so that $s \in \mathfrak{p}$ and $S = \Theta$. This completes the proof.

We note that the *radical cannot contain any idempotent except zero.* This follows from the fact that if $j + y = jy$ with $j^2 = j$, then $j^2 + jy = j^2 y$ and $j = \theta$. Thus the only idempotent having a right (left) reverse is the zero element.

THEOREM 24.8.4. *The radical contains every nilpotent ideal of the algebra.*

PROOF. If $\mathfrak{n}$ is a nilpotent left ideal there is an integer k with $\mathfrak{n}^k = \{\theta\}$ and if $z \in \mathfrak{n}$ then $(\alpha z + xz)^k = \theta$ for all $\alpha \in \Phi$ and $x \in \mathfrak{A}$. Since

$$(\alpha z + xz)^- = -\sum_{n=1}^{k-1}(\alpha z + xz)^n,$$

it follows that $z \in \mathfrak{p}$ and $\mathfrak{n} \subset \mathfrak{p}$.

THEOREM 24.8.5. *If $\mathfrak{A}$ is an algebra that satisfies the descending chain condition for left (right) ideals, then the radical of $\mathfrak{A}$ is nilpotent.*

PROOF. Suppose that $\mathfrak{A}$ satisfies the descending chain condition for left ideals. The integral powers of the radical are two-sided ideals of $\mathfrak{A}$ and $\mathfrak{p} \supset \mathfrak{p}^2 \supset \cdots$. By assumption there is an integer n such that $\mathfrak{p}^{n+1} = \mathfrak{p}^n$. If $\mathfrak{p}^n = \mathfrak{n}$, then $\mathfrak{n}$ is a two-sided ideal and $\mathfrak{n}^2 = \mathfrak{n}$. It is desired to prove that $\mathfrak{n} = \{\theta\}$. If $\mathfrak{n} \neq \{\theta\}$ we can use the descending chain condition to find a minimal left ideal $\mathfrak{i}$ of $\mathfrak{A}$ such that $\mathfrak{i} \subset \mathfrak{n}$ and $\mathfrak{n}\mathfrak{i} \neq \{\theta\}$. Let b be an element of $\mathfrak{i}$ such that $\mathfrak{n}b \neq \{\theta\}$. Then $\mathfrak{n}b$ is a left ideal of $\mathfrak{A}$ contained in $\mathfrak{i}$ and $\mathfrak{n}b = \mathfrak{i}$ since $\mathfrak{i}$ is minimal. Now b is in $\mathfrak{i}$, so there is an element $y \in \mathfrak{n}$ with $yb = b$. But y has a reverse so that $yb + y^-b = y^-yb$ or $b = \theta$ which contradicts the assumption that $\mathfrak{n}b \neq \{\theta\}$. It follows that $\mathfrak{n} = \{\theta\}$ and $\mathfrak{p}^n = \{\theta\}$ which proves the theorem.

In particular, we see that Definition 24.8.2 coincides with the classical definition of the radical as the union of all nilpotent ideals if the algebra satisfies the descending chain condition for left (right) ideals. The elements of the radical of an arbitrary algebra need not be nilpotent. The following example is of some interest to the analyst owing to its close relations to the kernels of the closed cycle. Let $\mathfrak{B}$ be the set of functions $f(z)$, holomorphic for $|z| < 1$ and continuous for $|z| \leqq 1$, with the usual definition of addition and scalar multiplication, but multiplication being defined as convolution so that

$$f*g = \int_0^z f(z - t)g(t)\, dt, \qquad |z| < 1.$$

Taking $\|f\| = \sup |f(z)|$ for $|z| < 1$, we see that $\mathfrak{B}$ becomes a commutative complex (B)-algebra without unit element. A fairly simple argument shows that there are no divisors of zero; thus $f(z) \equiv 0$ is the only nilpotent element of $\mathfrak{B}$. On the other hand, the series $-\sum_1^\infty[f(z)]^{n*} = [f(z)]^-$ converges in norm, for every $f(z) \in \mathfrak{B}$ so that each $f(z)$ is properly quasi-nilpotent and $\mathfrak{B}$ coincides with its own radical.

DEFINITION 24.8.3. *$\mathfrak{A}$ is called a radical algebra if $\mathfrak{A} = \mathfrak{p}$.*

Thus $\mathfrak{B}$ in the preceding example is a radical algebra. Composition of the Volterra type can be used to construct such algebras, the elements being functions or operators. An example of a radical algebra of operators is furnished by a simple modification of this example. We take $\mathfrak{X} = C[0, 1]$, let $f(t)$ and $F(t)$ be any elements of this space, and define

$$T[f; F] = \int_0^t F(t - u)f(u)\, du, \qquad 0 \leqq t \leqq 1.$$

For a fixed $F(t)$, this is a linear bounded transformation on $C[0, 1]$ to itself. If we let F run through $C[0, 1]$, the operators $T[\cdot\,; F]$ form a radical algebra. Here

$$T[\cdot\,; F_1]T[\cdot\,; F_2] = T[\cdot\,; F_3] \quad \text{with} \quad F_3(t) = \int_0^t F_1(t-u)F_2(u)\,du.$$

In this case there are nilpotent elements: if $F(t) = 0$ when $0 \leqq t \leqq 1/n$, then the nth power of $T[\cdot\,; F]$ is the zero operator. This algebra contains as a subset the fractional integration operators J^ζ, $\Re(\zeta) \geqq 1$, of section 23.16.

This example shows that operator algebras may have the extreme property $\mathfrak{p} = \mathfrak{A}$. However, for such algebras the other extreme, $\mathfrak{p} = \{\theta\}$, is more likely to occur as is shown by the next two theorems of which the first is due to N. Jacobson [2] and the second to E. Hille.

THEOREM 24.8.6. *Let $\mathfrak{X}$ be a linear system and let $\mathfrak{T} \neq \{\Theta\}$ be an irreducible algebra of endomorphisms of $\mathfrak{X}$. Then $\mathfrak{T}$ is semi-simple.*

REMARK. The undefined terms figuring in this statement are used in the following sense: $\mathfrak{T}$ is a set of linear transformations $\{T\}$ on $\mathfrak{X}$ to itself such that $T(x)$ is defined for all x in $\mathfrak{X}$, if $T_1, T_2 \in \mathfrak{T}$ so do $\alpha T_1 + \beta T_2$ and T_1T_2 for all $\alpha, \beta \in \Phi$. Further, $\mathfrak{T}$ is irreducible if there is no proper linear subspace of $\mathfrak{X}$ which is left invariant by every T.

PROOF. We note first that if y is any non-zero element of $\mathfrak{X}$ then the set of transforms $T(y)$, $T \in \mathfrak{T}$, is the whole space $\mathfrak{X}$. For this set is clearly an invariant subspace; hence the only other alternative is that $T(y) = \theta$ for all T. But then the set of vectors y with this property is not $\{\theta\}$ and since it is invariant it must be the whole space. Hence $T(y) = \theta$ for all T and all y and this means that $\mathfrak{T} = \{\Theta\}$ contrary to assumption. Now let P be any element of the radical of $\mathfrak{T}$. If $P \neq \Theta$ there is an x such that $P(x) \neq \theta$. Then the set of all elements $T[P(x)]$, $T \in \mathfrak{T}$, is the whole space. We can then find a particular transformation T such that $T[P(x)] = x$. But TP is also in the radical of $\mathfrak{T}$ so it has a reverse, R say, and

$$x = x + RTP(x) - R(x) - TP(x) = \theta$$

against our assumption. Hence $\mathfrak{T}$ is without radical.

A slightly sharper result can be proved for (B)-spaces $\mathfrak{X}$ and the associated full (B)-algebra $\mathfrak{E}(\mathfrak{X})$.

THEOREM 24.8.7. *If $\mathfrak{X}$ is a (B)-space and $\mathfrak{E}(\mathfrak{X})$ is the associated (B)-algebra of all endomorphisms of $\mathfrak{X}$, then $\mathfrak{E}(\mathfrak{X})$ is semi-simple and every element of $\mathfrak{E}(\mathfrak{X})$, except the zero element, is semi-regular in the sense that there exists a right multiple which when acted upon by the element becomes idempotent.*

PROOF. Let $T \in \mathfrak{E}(\mathfrak{X})$, $T \neq \theta$. Choose an $a \in \mathfrak{X}$ such that $T(a) \neq \theta$. If $T(a) = b$, choose a linear functional x^* on $\mathfrak{X}$ with $x^*(b) = 1$ and define $U(x) = x^*(x)a$ for all $x \in \mathfrak{X}$. This transformation U is evidently an element of $\mathfrak{E}(\mathfrak{X})$. Then

$$TU(x) = x^*(x)b, \qquad UTU(x) = x^*(x)x^*(b)a = x^*(x)a, \qquad (TU)^2(x) = x^*(x)b$$

and

$$(TU)^2 = TU \neq \theta.$$

This shows that T is semi-regular and, since no idempotent can belong to the radical, $\mathfrak{E}(\mathfrak{X})$ is without radical as asserted.

The last theorem of this section is concerned with spectral properties connected with the radical and the quasi-radical. The restrictive assumptions imposed on the algebra are largely for the sake of convenience.

THEOREM 24.8.8. *Let $\mathfrak{B}$ be a complex (B)-algebra with unit element. If $p \in \mathfrak{p}$, then $\sigma(x + p) = \sigma(x)$, $\sigma(xp) = \sigma(px) = \sigma(p) = \{\theta\}$ for all x in $\mathfrak{B}$. These relations are not necessarily true if p is replaced by an element $q \in \mathfrak{q}$. Further, if $X = x + \mathfrak{p}$ is one of the elements of $\mathfrak{B}/\mathfrak{p}$, then $\sigma(X) = \sigma(x)$.*

PROOF. Since $\lambda e - px$ and $\lambda e - xp$ have inverses for every $\lambda \neq 0$, it is clear that $\sigma(px) = \sigma(xp) = \sigma(p) = \{0\}$. The two formal identities

$$\lambda e - x - p = (\lambda e - x)[e - (\lambda e - x)^{-1}p],$$

$$\lambda e - x = (\lambda e - x - p)[e + (\lambda e - x - p)^{-1}p]$$

show that x and $x + p$ are equi-regular, whence it follows that $\sigma(x + p) = \sigma(x)$ for all x.

The last result implies that all elements of $\mathfrak{B}$ which belong to the same residue-class modulo $\mathfrak{p}$ have identical spectra. By Theorem 22.11.7 we have $\sigma(X) \subset \sigma(x)$. Suppose now that $\lambda E - X$ is a regular element of $\mathfrak{B}/\mathfrak{p}$ for some value of λ. Then there exist elements $y \in \mathfrak{B}$, $p \in \mathfrak{p}$ with $(\lambda e - x)y = e + p$. But $e + p$ is regular, hence also $\lambda e - x$. It follows that $\sigma(X) = \sigma(x)$.

It is clear that $\sigma(q) = \{0\}$, but it is not necessarily true that $\sigma(xq)$ and $\sigma(qx)$ reduce to $\{0\}$ for all x. Counter-examples may be obtained from Theorem 24.8.7. Suppose that $\mathfrak{X}$ is so chosen that $\mathfrak{E}(\mathfrak{X})$ contains a quasi-nilpotent operator Q. We can then find an operator U such that UQ is idempotent and then $\sigma(UQ) = \{0, 1\} \neq \{0\}$. Counter-examples to $\sigma(x + q) = \sigma(x)$ are harder to find, but occur in the theory of integral equations. Thus in $C[0, 1]$ the projection operator $\int_0^1$ is the sum of the quasi-nilpotent operators $\int_0^t$ and $\int_t^1$.

24.9. Maximal ideals. For our present purposes we must replace Definition 4.13.1(3) by:

DEFINITION 24.9.1. *An ideal is said to be maximal if it is not equal to the whole algebra and is not properly contained in any other ideal of the same kind.*

Maximal ideals will still be denoted by $\mathfrak{m}$. We note that the definition does not exclude the possibility that a two-sided ideal may be maximal as such but nevertheless be properly contained in a larger proper left or right ideal. Similarly, a maximal left ideal may be properly contained in a larger right ideal. We note also that the zero ideal may be a maximal ideal.

In a commutative algebra all ideals are two-sided. A sufficient condition that all ideals of an algebra be two-sided is that to every order pair of elements x, y there are elements u and v with $xy = ux = yv$. According to an observation by A. Putnam, this condition is also necessary if the algebra has a unit element. In the following such an algebra will be said to be *two-sided.*

THEOREM 24.9.1. *In an algebra satisfying the conditions of Theorems* 24.6.3 *or* 24.6.4 *every regular maximal ideal is closed.*

PROOF. Since $\bar{\mathfrak{m}}$ is an ideal of the same kind as $\mathfrak{m}$ and $\mathfrak{m}$ is maximal, the obvious inclusion $\mathfrak{m} \subset \bar{\mathfrak{m}}$ implies $\mathfrak{m} = \bar{\mathfrak{m}}$. One or the other of these conditions is obviously satisfied by any complex (B)-algebra.

THEOREM 24.9.2. *If $\mathfrak{A}$ is not a radical algebra, $\mathfrak{A} \neq \mathfrak{p}$, then $\mathfrak{A}$ contains left and right regular maximal ideals. If $\mathfrak{A} = \mathfrak{p}$, then $\mathfrak{A}$ may contain maximal ideals, but no such ideal can be regular.*

PROOF. Cf. Theorem 4.13.2 for the case of a commutative (B)-algebra with a unit element. If every element of $\mathfrak{A}$ were left reversible, then, the left reverse of a left reverse being the original element by Theorem 24.2.1, it follows that each element of $\mathfrak{A}$ would be reversible and hence that $\mathfrak{A} = \mathfrak{p}$. Thus if $\mathfrak{A} \neq \mathfrak{p}$, then $\mathfrak{A}$ contains an element a which is not left reversible. If x runs through $\mathfrak{A}$, the set $\{x - xa\} = \mathfrak{i}$ is a regular left ideal with a as the associated right unit element. It is claimed that $\mathfrak{i}$ does not contain a. If this were not so, then an element b would exist such that $ba - b = a$, that is, a would have a left reverse against our assumption. Thus $\mathfrak{i}$ does not contain a and if $\mathfrak{i}_0$ is a left ideal containing $\mathfrak{i}$ as well as a, then $\mathfrak{i}_0 = \mathfrak{A}$. Now let Γ be the collection of all the left ideals of $\mathfrak{A}$ that (i) contain $\{x - xa\}$ and (ii) do not contain a. Every such ideal is obviously regular. Relative to inclusion Γ is a partially ordered set (see section 1.2). If Σ is a simply ordered subset of Γ, the union of all the ideals contained in Σ belongs to Γ. For it is a left ideal, it contains $\{x - xa\}$, and it does not contain a since a is not contained in any of the ideals belonging to Γ. This shows that Σ has an upper bound in Γ. Consequently, by the maximal principle there exists a maximal element $\mathfrak{m}$ in Γ, that is, a left ideal in Γ not properly contained in any left ideal of Γ. From its construction $\mathfrak{m}$ contains $\{x - xa\}$, does not contain a, and is regular. We assert now that $\mathfrak{m}$ is a maximal left ideal of $\mathfrak{A}$. For if $\mathfrak{i}_0$ is a left ideal properly containing $\mathfrak{m}$, $\mathfrak{i}_0$ is not in Γ. Hence $\mathfrak{i}_0$ contains a and $\mathfrak{i}_0 = \mathfrak{A}$. Thus $\mathfrak{m}$ is maximal. The existence of right maximal regular ideals is proved in the same manner, replacing the set $\{x - xa\}$ by $\{x - bx\}$ where b does not have a right reverse.

Suppose now that $\mathfrak{A} = \mathfrak{p}$ and suppose that $\mathfrak{i}$ is a left regular ideal in $\mathfrak{A}$ with j as the associated right unit element so that $x - xj \in \mathfrak{i}$ for all x. But $j \in \mathfrak{p}$ and has a reverse j^- and $j^-j - j^- = j$ so that $j \in \mathfrak{i}$. This implies that $xj \in \mathfrak{i}$ and hence also x, that is, $\mathfrak{i} = \mathfrak{A}$. Thus a radical algebra does not have any proper regular ideals, *a fortiori*, no maximal regular ideals. This does not exclude the possibility of the existence of maximal ideals in $\mathfrak{A}$, however. Thus the nilpotent algebra $\mathfrak{B}$, considered in section 24.4, of elements of the form $\alpha z + \beta z^2$ with $z^3 = \theta$, has a maximal ideal formed by the elements βz^2. This completes the proof of the theorem which is an extension of results due to Jacobson and Segal

THEOREM 24.9.3. *A proper regular ideal is contained in some regular maximal ideal of the same kind. If $\mathfrak{A}$ has a unit element, every proper ideal is contained in some maximal ideal of the same kind.*

PROOF. Suppose that $\mathfrak{i}$ is a proper regular left ideal and that j is the corresponding right unit element modulo $\mathfrak{i}$. Then, as was shown above, $\mathfrak{i}$ does not contain j and any left ideal which contains $\mathfrak{i}$ as well as j must be $\mathfrak{A}$ itself. Further, any left ideal which contains $\mathfrak{i}$ is necessarily regular. The reasoning used in the proof of the preceding theorem then shows that we can find a maximal regular left ideal containing $\mathfrak{i}$. If $\mathfrak{A}$ has a unit element, then every ideal is regular and the proof shows that any proper ideal may be embedded in a maximal ideal of the same kind.

If $\mathfrak{A}$ does not have a unit element, ideals may exist which cannot be embedded in a maximal ideal. The following example, due to I. Kaplansky (personal communication), illustrates this. Let $\mathfrak{A} = c_0$ be the commutative (B)-algebra of sequences converging to zero with the obvious definition of arithmetic operations. Let $\mathfrak{i}$ be the set of all sequences the elements of which are ultimately zero. Then $\mathfrak{i}$ is obviously a proper ideal. Suppose $\mathfrak{i}$ were contained in a maximal ideal $\mathfrak{m}$. In this case $c_0/\mathfrak{m}$ is either a skew field or a zero algebra of dimension one by Theorem 24.6.1. If it is a skew field then $\mathfrak{m}$ is regular and hence closed; since $\mathfrak{i}$ is dense in c_0 this is impossible. On the other hand $c_0/\mathfrak{m}$ cannot be a zero algebra since every element of c_0 (and hence of $c_0/\mathfrak{m}$) has a square root.

THEOREM 24.9.4. *Let $\mathfrak{A}$ be an algebra with a unit element. If an element a of $\mathfrak{A}$ has no left (right) inverse, then there exists a maximal left (right) ideal containing a. Conversely, an element which belongs to no left (right) maximal ideals has a left (right) inverse. In a two-sided algebra (in particular, in a commutative algebra) every singular element belongs to at least one maximal ideal and an element belonging to no maximal ideal is regular.*

PROOF. If a has no left inverse, then $\mathfrak{A}a$ is a proper left ideal containing a. By the preceding theorem there is a maximal left ideal containing $\mathfrak{A}a$ and hence also a. Conversely, if a is not in any maximal left ideal, then $\mathfrak{A}a = \mathfrak{A}$ and there is an element b such that $ba = e$. The discussion of the two-sided case is immediate.

In an algebra which is not two-sided, it is by no means true that a singular element can always be embedded in a two-sided maximal ideal. This is illustrated by the situation in the algebra of all bounded linear transformations on a Hilbert space to itself. J. W. Calkin [1] has shown that in this algebra $\mathfrak{E}[\mathfrak{H}]$ there is a single two-sided maximal ideal $\mathfrak{c}$ formed by the compact transformations (see Definition 2.13.1; totally continuous in Calkin's terminology). It is clear that $\mathfrak{c}$ does not exhaust the singular elements of $\mathfrak{E}[\mathfrak{H}]$. Further properties of this algebra will be used below.

The preceding theorem has been extended to reverses by I. E. Segal [**1**].

THEOREM 24.9.5. *A necessary and sufficient condition that an element of an algebra have a left (right) reverse is that it have a left (right) reverse modulo every maximal regular left (right) ideal in the algebra.*

PROOF. The condition is obviously necessary. Now suppose that $a \in \mathfrak{A}$ and that a has a left reverse modulo every maximal regular left ideal, but has no left reverse in $\mathfrak{A}$. Let us now consider the set $\{x - xa\}$, $x \in \mathfrak{A}$. We know from the previous discussion that this set is a regular left ideal $\mathfrak{i}$ which does not contain a. It is claimed that $\mathfrak{i} = \mathfrak{A}$. If this were not so, then we could find a maximal regular left ideal $\mathfrak{m}$ containing $\mathfrak{i}$. Since a has a left reverse modulo $\mathfrak{m}$, there exists an element c such that $a + c - ca \in \mathfrak{m}$. But $c - ca \in \mathfrak{i} \subset \mathfrak{m}$ whence it follows that $a \in \mathfrak{m}$ which implies $\mathfrak{m} = \mathfrak{A}$. From this contradiction we conclude that $\mathfrak{i} = \mathfrak{A}$. But then we can find an element b with $ba - b = a$, that is, a has a left reverse.

Segal has also proved that an element of the center of the algebra has a reverse if and only if it has a reverse modulo every maximal regular two-sided ideal of the algebra.

THEOREM 24.9.6. *If $\mathfrak{B}$ is a complex (B)-algebra with unit and if $\mathfrak{m}$ is a two-sided ideal which is maximal as a left (right) ideal, then $\mathfrak{m}$ is also maximal as a right (left) ideal and $\mathfrak{B}/\mathfrak{m}$ is isomorphic to the complex field.*

PROOF. Suppose that $\mathfrak{m}$ is a maximal left ideal. It is consequently closed and being two-sided its residue-class system $\mathfrak{B}/\mathfrak{m}$ is a complex (B)-algebra by Theorem 24.7.4. If $y \in \mathfrak{B} \ominus \mathfrak{m}$, the set $\mathfrak{m} + \mathfrak{B}y$ is a left ideal containing $\mathfrak{m}$ properly and hence equals $\mathfrak{B}$. We can then find $x \in \mathfrak{B}$, $m \in \mathfrak{m}$ such that $m + xy = e$ whence $XY = E$ for the corresponding residue-classes modulo $\mathfrak{m}$. Since y is arbitrary in $\mathfrak{B} \ominus \mathfrak{m}$, this means that every element Y of $\mathfrak{B}/\mathfrak{m}$, excepting Θ, has a left inverse. Thus $\mathfrak{B}/\mathfrak{m}$ is a skew-field as well as a complex (B)-algebra; by Theorem 4.9.1 it is consequently isomorphic to the complex field. Now any right ideal containing $\mathfrak{m}$ properly would map modulo $\mathfrak{m}$ onto a proper right ideal of the complex field. Since the complex field has no proper right ideals, it follows that $\mathfrak{m}$ is also maximal as a right ideal.

The proof of this theorem obviously applies to all maximal ideals of a commutative or a two-sided algebra with unit element. It does not apply to two-sided maximal ideals in arbitrary complex (B)-algebras with unit element. Such an ideal is not necessarily maximal either as a left or as a right ideal and consequently the residue-classes need not form a field. A case in point is the algebra $\mathfrak{E}[\mathfrak{H}]$ mentioned above; Calkin has shown that the residue-class algebra $\mathfrak{E}/\mathfrak{c}$ is non-separable in the normed topology and contains a continuum of idempotents. It is consequently not a field.

THEOREM 24.9.7. *If $\mathfrak{B}$ is a two-sided complex (B)-algebra and if $\mathfrak{m}$ is a regular maximal ideal, then $\mathfrak{B}/\mathfrak{m}$ is isomorphic to the complex field.*

PROOF. Since $\mathfrak{m}$ is closed $\mathfrak{B}/\mathfrak{m}$ is again a (B)-algebra. Let j be a (two-sided) unit, modulo $\mathfrak{m}$. Then for $y \in \mathfrak{B} \ominus \mathfrak{m}$, the set $\mathfrak{m} + \mathfrak{B}y$ is a left and hence a two-sided regular ideal which contains $\mathfrak{m}$ properly since it obviously contains $y = (y - jy) + jy$. Thus $\mathfrak{m} + \mathfrak{B}y = \mathfrak{B}$ and we can find $x \in \mathfrak{B}$, $m \in \mathfrak{m}$ such that $m + xy = j$, whence $XY = J$ for the corresponding residue classes. Consequently every $Y \neq \Theta$ in $\mathfrak{B}/\mathfrak{m}$ has a left inverse; $\mathfrak{B}/\mathfrak{m}$ is a skew field and therefore isomorphic to the complex field by Theorem 4.9.1.

THEOREM 24.9.8. *Under the assumptions of Theorems* 24.9.6 *or* 24.9.7, *let* X_α *denote the residue-class of elements* $x \equiv \alpha j \pmod{\mathfrak{m}}$, *where* α *is an arbitrary complex number. Then* $\mathfrak{B}/\mathfrak{m} = \{X_\alpha\}$ *where* α *ranges over the complex field. The spectrum of* X_α *reduces to a single point,* $\sigma(X_\alpha) = \{\alpha\}$, *and this point is common to the spectra of all elements of* X_α.

The proof is obtained by combining Theorems 24.7.7 and 24.9.6 or 24.9.7. The following partial converse is important:

THEOREM 24.9.9. *If* $\mathfrak{B}$ *is a two-sided complex* (B)*-algebra, if* $a \in \mathfrak{B}$, *and if* $\lambda \in \alpha(a)$, $\lambda \neq 0$, *then there exists a homomorphic mapping of* $\mathfrak{B}$ *onto the complex field under which* a *is mapped into* λ.

PROOF. If $\lambda \in \alpha(a)$, $\lambda \neq 0$, then $\lambda^{-1}a$ does not have a reverse and by Theorem 24.9.5 there is a regular maximal ideal $\mathfrak{m}$, two-sided by assumption, such that $\lambda^{-1}a$ is not reversible modulo $\mathfrak{m}$. Since $\mathfrak{B}/\mathfrak{m}$ is isomorphic to the complex field by Theorem 24.9.7, this implies that $a \rightarrow \lambda$ under the homomorphism $\mathfrak{B} \sim \mathfrak{B}/\mathfrak{m}$; for unless $\lambda^{-1}a \rightarrow 1$ it is readily seen that $\lambda^{-1}a$ has a reverse modulo $\mathfrak{m}$.

The last two theorems show that in the case of a two-sided (in particular, a commutative) complex (B)-algebra there are available enough homomorphisms of the algebra onto the complex field to enable us to map any given element of the algebra on any preassigned complex number belonging to its spectrum.

On the other hand, if the algebra is not two-sided, none of its homomorphic images need be a field and in such a case there are no homomorphisms mapping the elements upon their spectra. The operator ring $\mathfrak{E}[\mathfrak{H}]$ considered above illustrates this point. Here there is only one maximal two-sided ideal $\mathfrak{c}$ and $\mathfrak{E}/\mathfrak{c}$ is neither a field nor a subalgebra of the complex field. If $\mathfrak{i}$ is any two-sided ideal of $\mathfrak{E}[\mathfrak{H}]$, then $\mathfrak{i} \subset \mathfrak{c}$ and $\mathfrak{E}/\mathfrak{i}$ is homomorphic to $\mathfrak{E}/\mathfrak{c}$ so that $\mathfrak{E}/\mathfrak{i}$ cannot be a field.

Finally we list a theorem which has a bearing on the quasi-radical.

THEOREM 24.9.10. *Let* $\mathfrak{B}$ *be a two-sided complex* (B)*-algebra with unit element. A necessary and sufficient condition that under all homomorphic mapping of* $\mathfrak{B}$ *onto the complex field a given element* a *is always mapped on the same complex number* α *is that* $\alpha e - a$ *be quasi-nilpotent.*

ROPOF. If $\alpha e - a = q$ is quasi-nilpotent, then $\sigma(q) = \{0\}$; hence q is always mapped onto 0 and a onto α under such homomorphisms. Conversely, if $\alpha e - a$ is not quasi-nilpotent but $\alpha \in \sigma(a)$, then $\sigma(a)$ must contain at least one other point β besides α and there is a homomorphism mapping a onto β. This completes the proof.

24.10 Maximal ideals and the radical. We shall prove some intersection theorems. It will simplify the exposition if we introduce a slightly more expressive notation. We shall denote a left ideal by $'\mathfrak{i}$, changing this to $^*\mathfrak{i}$ if the ideal is known to be regular. Similarly, $\mathfrak{m}^*$ denotes a maximal regular right ideal and so on. We start with a theorem which is suggested by the definition of the radical given by Segal.

THEOREM 24.10.1. *If* $\mathfrak{A} \neq \mathfrak{p}$, *then* $\mathfrak{p} = \bigcap_\alpha {}^*\mathfrak{m}_\alpha = \bigcap_\alpha \mathfrak{m}_\alpha^*$.

PROOF. Suppose that $\mathfrak{A} \neq \mathfrak{p}$ so that $\mathfrak{A}$ contains regular maximal ideals. If there exists an element p of $\mathfrak{p}$ and a maximal regular left ideal $^*\mathfrak{m}$ which does not contain p, then $^*\mathfrak{m} + \alpha p + \mathfrak{A}p$ is a left ideal containing $^*\mathfrak{m}$ properly and hence equals $\mathfrak{A}$. Since $^*\mathfrak{m}$ is regular, there is a right unit j such that $x - xj \in {}^*\mathfrak{m}$ for all x. We can then find $\alpha \in \Phi$, $a \in \mathfrak{A}$ such that $m + \alpha p + ap = j$ for some $m \in {}^*\mathfrak{m}$. Since $\alpha p + ap \in \mathfrak{p}$, $j - m$ has a reverse, r say, and $j = (rj - r) + m - rm \in {}^*\mathfrak{m}$ whence it follows that $^*\mathfrak{m} = \mathfrak{A}$. This contradiction shows that $p \in {}^*\mathfrak{m}$ and $\mathfrak{p} \subset \bigcap_\alpha {}^*\mathfrak{m}_\alpha$.

Suppose now that $s \in \bigcap_\alpha {}^*\mathfrak{m}_\alpha$ and s is not in $\mathfrak{p}$. This means that there is a left multiple of s of the form $t = \alpha s + ys$ which does not have a reverse. If t should have a left reverse u, then u cannot have a left reverse since it would have to coincide with t, making u the reverse of t. Hence either t or u does not a left reverse. Here $t \in \bigcap_\alpha {}^*\mathfrak{m}_\alpha$ and if u exists $u = ut - t \in \bigcap_\alpha {}^*\mathfrak{m}_\alpha$. Thus there is an element $w \in \bigcap_\alpha {}^*\mathfrak{m}_\alpha$ which does not have a left reverse. From the proofs of Theorems 24.9.2 and 24.9.3 we know that $^*\mathfrak{i} \equiv [xw - x, x \in \mathfrak{A}]$ is a regular left ideal not containing w and as such it may be embedded in a maximal ideal $^*\mathfrak{m}$ having the same properties. But w belongs to every $^*\mathfrak{m}_\alpha$ hence also to $^*\mathfrak{m}$. This contradiction shows that every multiple of s is reversible so that $s \in \mathfrak{p}$. Hence $\mathfrak{p} = \bigcap_\alpha {}^*\mathfrak{m}_\alpha$ as asserted and the other identity is proved in the same manner.

COROLLARY 1. *If* $\mathfrak{A}$ *has a unit element* $\mathfrak{p} = \bigcap_\alpha {}'\mathfrak{m}_\alpha = \bigcap_\alpha \mathfrak{m}'_\alpha$.

For all ideals are regular.

COROLLARY 2. *If* $\mathfrak{B}$ *is a* (B)*-algebra* $\mathfrak{p}$ *is closed.*

If $\mathfrak{p} = \mathfrak{B}$ the assertion is trivially valid. Otherwise, $\mathfrak{p}$ is the intersection of left regular maximal ideals which are closed by Theorem 24.9.1.

COROLLARY 3. *If* $\mathfrak{A}$ *contains left maximal ideals,* $\mathfrak{p} \supset \bigcap_\alpha {}'\mathfrak{m}_\alpha$, *if* $\mathfrak{A}$ *has right maximal ideals* $\mathfrak{p} \supset \bigcap_\alpha \mathfrak{m}'_\alpha$.

The assertion follows from Theorem 24.10.1 if $\mathfrak{p} \neq \mathfrak{A}$ and is trivial if $\mathfrak{p} = \mathfrak{A}$. Inclusions going in the opposite direction are given in Theorem 24.10.3 below.

I. E. Segal defined the radical as the intersection of all two-sided regular maximal ideals. This definition does not agree with ours or with the equivalent definition of Jacobson. Thus the operator algebra $\mathfrak{E}[\mathfrak{H}]$, which is semi-simple in our terminology, has the radical $\mathfrak{c}$ and is merely weakly semi-simple according to Segal.

THEOREM 24.10.2. *If* $'\mathfrak{m}$ *is a maximal left ideal, then the residue-class algebra* $\mathfrak{A}/('\mathfrak{m}:\mathfrak{A})$ *is semi-simple and the quotient ideal* $'\mathfrak{m}:\mathfrak{A}$ *contains the radical of* $\mathfrak{A}$. *The same result holds for maximal right ideals.*

PROOF. See Definition 24.7.1 for the notion of quotient ideals. If $'\mathfrak{m}:\mathfrak{A} = \mathfrak{A}$ there is nothing to prove. Hence suppose that $'\mathfrak{m}:\mathfrak{A} \neq \mathfrak{A}$ so that $\bar{\mathfrak{A}} = \mathfrak{A}/('\mathfrak{m}:\mathfrak{A})$ does not reduce to zero. Then $\bar{\mathfrak{A}}$ is isomorphic to an irreducible algebra $\mathfrak{T}$ of endomorphisms of the linear residue-class system $\mathfrak{L} = \mathfrak{A} \div {}'\mathfrak{m}$ (for the details see below). By Theorem 24.8.6, $\mathfrak{T}$ and hence also $\bar{\mathfrak{A}}$ are semi-simple. In order to prove that $'\mathfrak{m}:\mathfrak{A}$ contains $\mathfrak{p}$, let us consider the residue-class $\bar{P} = p + ('\mathfrak{m}:\mathfrak{A})$ where $p \in \mathfrak{p}$. It belongs to the radical of $\bar{\mathfrak{A}}$ since every element of $\bar{P}$ is properly quasi-nilpotent modulo $'\mathfrak{m}:\mathfrak{A}$. But $\bar{\mathfrak{A}}$ is without radical, so $\bar{P} = \Theta$. This means that $\mathfrak{p} \subset {}'\mathfrak{m}:\mathfrak{A}$.

In order to complete the proof we have to introduce the algebra $\mathfrak{T}$ and prove that it is irreducible relative to $\mathfrak{L}$. Consider the residue-classes

$$A = a + ('\mathfrak{m}:\mathfrak{A}), \qquad X = x + {}'\mathfrak{m}$$

which are elements of $\bar{\mathfrak{A}}$ and $\mathfrak{L}$ respectively. With each class A we associate a transformation $T(\cdot\,; A)$ defined by the convention that $Y = T(X; A)$ is the residue-class $Y = ax + {}'\mathfrak{m}$. This transformation is linear in X as well as in A. Further

$$T(\cdot\,; AB) = T[T(\cdot\,; B); A] = T(\cdot\,; A) \circ T(\cdot\,; B).$$

Thus $\mathfrak{T} = \{T(\cdot\,; A)\}$ is an algebra of endomorphisms of $\mathfrak{L}$ which is a homomorphic image of $\bar{\mathfrak{A}}$. Actually the mapping is an isomorphism. Indeed, if $T(\cdot\,; A) = T(\cdot\,; B)$, then the residue-classes $ax + {}'\mathfrak{m}$ and $bx + {}'\mathfrak{m}$ are identical for each fixed x. This implies that $(a - b)x \in {}'\mathfrak{m}$ for all x or that $a - b \in {}'\mathfrak{m}:\mathfrak{A}$ and $A = B$. Now let $\mathfrak{L}_0$ be a linear subspace of $\mathfrak{L}$ which is invariant under $\mathfrak{T}$. Let $\mathfrak{L}_0$ be made up of residue-classes $\{y + {}'\mathfrak{m}\}$. The set $'\mathfrak{i}$ of all such elements y of $\mathfrak{A}$ is linear and, since $T(\mathfrak{L}_0) \subset \mathfrak{L}_0$, we see that $ay \in {}'\mathfrak{i}$ for all a in $\mathfrak{A}$, that is, $'\mathfrak{i}$ is a left ideal containing $'\mathfrak{m}$. It is then either $'\mathfrak{m}$ or $\mathfrak{A}$ since $'\mathfrak{m}$ is maximal. It follows that $\mathfrak{L}_0$ is either the zero element or $\mathfrak{L}$ so $\mathfrak{T}$ is irreducible relative to $\mathfrak{L}$ as asserted. The same type of argument applies if $\mathfrak{m}$ is a right maximal ideal.

THEOREM 24.10.3. *If* $\mathfrak{A}$ *is an algebra that contains left (right) maximal ideals then* $\bigcap_\alpha{}'\mathfrak{m}_\alpha$ $(\bigcap_\alpha \mathfrak{m}'_\alpha)$ *is a two-sided ideal contained in* $\mathfrak{p}$ *and containing* $\mathfrak{p}\mathfrak{A}$ $(\mathfrak{A}\mathfrak{p})$.

PROOF. Let us take the left-handed case. We know already that $\bigcap_\alpha{}'\mathfrak{m}_\alpha \subset \mathfrak{p}$. By the preceding theorem we have

$$\mathfrak{p} \subset \textstyle\bigcap_\alpha('\mathfrak{m}_\alpha:\mathfrak{A}) \tag{24.10.1}$$

and, by Theorem 24.7.3, we have consequently $\mathfrak{p}\mathfrak{A} \subset \bigcap_\alpha{}'\mathfrak{m}_\alpha$. We know that $\bigcap_\alpha{}'\mathfrak{m}_\alpha$ is a left ideal and from $\bigcap_\alpha{}'\mathfrak{m}_\alpha\mathfrak{A} \subset \mathfrak{p}\mathfrak{A} \subset \bigcap_\alpha{}'\mathfrak{m}_\alpha$ we conclude that it is also a right ideal. This completes the proof.

This theorem involves four two-sided ideals: $\bigcap_\alpha{}'\mathfrak{m}_\alpha$, $\bigcap_\alpha\mathfrak{m}'_\alpha$, $\mathfrak{A}\mathfrak{p}$, and $\mathfrak{p}\mathfrak{A}$, and it is natural to ask if they are all equal. This is trivially true if $\mathfrak{A}$ has a unit element, but the general situation is rather obscure. In this connection we mention a definition of the radical proposed by Max Zorn in 1943.

DEFINITION 24.10.1. *Let* $\mathfrak{u}$ *be the set of all elements u such that* $'\mathfrak{i} + \mathfrak{A}u$ *is a proper left ideal whenever* $'\mathfrak{i}$ *is proper. Similarly, let* $\mathfrak{v}$ *be the set of all elements v such that* $\mathfrak{i}' + v\mathfrak{A}$ *is a proper right ideal whenever* $\mathfrak{i}'$ *is proper. Then* $\mathfrak{u}$ *is called the left weak radical and* $\mathfrak{v}$ *the right weak radical.*

To analyze these concepts we shall use a lemma due to N. Jacobson [**2**].

LEMMA 24.10.1. *If* $\mathfrak{A}z\mathfrak{A} \subset \mathfrak{p}$, *then* $z \in \mathfrak{p}$.

PROOF. If $\mathfrak{i}' = \Phi z + z\mathfrak{A}$ is the right ideal generated by z, $(\mathfrak{i}')^3 \subset \mathfrak{A}z\mathfrak{A} \subset \mathfrak{p}$. Hence $\mathfrak{J} \equiv (\mathfrak{i}' + \mathfrak{p})/\mathfrak{p}$ is nilpotent in the semi-simple algebra $\mathfrak{A}^\circ = \mathfrak{A}/\mathfrak{p}$. By Theorem 24.8.4, $\mathfrak{J}$ belongs to the radical of $\mathfrak{A}^\circ$. Hence $\mathfrak{J} = \Theta$, $\mathfrak{i}' \subset \mathfrak{p}$, and $z \in \mathfrak{p}$.

THEOREM 24.10.4. *In the notation of Definition* 24.10.1 $\mathfrak{u}$ *is a left ideal and* $\mathfrak{u} \subset \mathfrak{p}$ *while* $\mathfrak{v}$ *is a right ideal and* $\mathfrak{v} \subset \mathfrak{p}$. *If* $\mathfrak{A}$ *contains maximal ideals, then* $\mathfrak{A}\mathfrak{u} \subset \bigcap_\alpha {'\mathfrak{m}_\alpha}$, $\mathfrak{v}\mathfrak{A} \subset \bigcap_\alpha \mathfrak{m}'_\alpha$. *We have* $\mathfrak{u} = \mathfrak{v} = \mathfrak{p}$ *if, in addition, one of the following conditions hold,* (i) *every one-sided ideal is contained in a maximal ideal of the same kind and* $\mathfrak{A}\mathfrak{p} = \mathfrak{p}\mathfrak{A}$, *or* (ii) *every one-sided ideal can be embedded in a regular maximal ideal of the same kind.*

PROOF. The ideal properties of $\mathfrak{u}$ and $\mathfrak{v}$ are obvious from the definition. If $\mathfrak{A}$ is a radical algebra the inclusions $\mathfrak{u} \subset \mathfrak{p}$, $\mathfrak{v} \subset \mathfrak{p}$ are trivial. If $\mathfrak{A}$ is not a radical algebra, maximal ideals will certainly exist and we conclude from $'\mathfrak{m} + \mathfrak{A}\mathfrak{u} = {'\mathfrak{m}}$ that $\mathfrak{A}\mathfrak{u} \subset {'\mathfrak{m}}$ or $\mathfrak{A}\mathfrak{u} \subset \bigcap_\alpha {'\mathfrak{m}_\alpha}$. Similarly it is proved that $\mathfrak{v}\mathfrak{A} \subset \bigcap_\alpha \mathfrak{m}'_\alpha$. By Theorem 24.10.3 we have then $\mathfrak{A}\mathfrak{u} \subset \mathfrak{p}$ and $\mathfrak{v}\mathfrak{A} \subset \mathfrak{p}$ so that $\mathfrak{A}\mathfrak{u}\mathfrak{A} \subset \mathfrak{p}\mathfrak{A} \subset \mathfrak{p}$ and $\mathfrak{A}\mathfrak{v}\mathfrak{A} \subset \mathfrak{A}\mathfrak{p} \subset \mathfrak{p}$ whence, by Lemma 24.10.1, $\mathfrak{u} \subset \mathfrak{p}$, $\mathfrak{v} \subset \mathfrak{p}$ as asserted. Suppose now that (i) holds and let $p \in \mathfrak{p}$. Using Theorem 24.10.3 once more we obtain the series of inclusions

$$'\mathfrak{i} + \mathfrak{A}p \subset {'\mathfrak{m}} + \mathfrak{A}\mathfrak{p} = {'\mathfrak{m}} + \mathfrak{p}\mathfrak{A} \subset {'\mathfrak{m}} + \bigcap_\alpha {'\mathfrak{m}_\alpha} = {'\mathfrak{m}}$$

where $'\mathfrak{i}$ is an arbitrary left proper ideal of $\mathfrak{A}$ and $'\mathfrak{m}$ is the left maximal ideal in which it is contained. This shows that $p \in \mathfrak{u}$, $\mathfrak{p} \subset \mathfrak{u}$, and consequently that $\mathfrak{p} = \mathfrak{u}$. In the same manner one proves that $\mathfrak{p} = \mathfrak{v}$. If condition (ii) holds instead the inclusions follow the pattern

$$'\mathfrak{i} + \mathfrak{A}p \subset {}^*\mathfrak{m} + \mathfrak{A}\mathfrak{p} \subset {}^*\mathfrak{m} + \mathfrak{p} = {}^*\mathfrak{m} + \bigcap_\alpha {}^*\mathfrak{m}_\alpha = {}^*\mathfrak{m},$$

where we have used Theorem 24.10.1. The conclusions are the same.

THEOREM 24.10.5. *If* $\mathfrak{A}$ *contains maximal left (right) ideals, then*

$$\mathfrak{p} = \bigcap_\alpha({'\mathfrak{m}_\alpha}:\mathfrak{A}) \doteq \bigcap_\alpha(\mathfrak{m}'_\alpha:\mathfrak{A}).$$

PROOF. By formula (24.10.1), $\mathfrak{p}$ is included in the first intersection. Conversely, suppose that $y \in \bigcap_\alpha({'\mathfrak{m}_\alpha}:\mathfrak{A})$, then $y\mathfrak{A} \subset \bigcap_\alpha {'\mathfrak{m}_\alpha} \subset \mathfrak{p}$. Hence $\mathfrak{A}y\mathfrak{A} \subset \mathfrak{p}$ and this implies that $y \in \mathfrak{p}$. Thus we have equality for left ideals and similarly for right ideals.

Finally we introduce the concept of being *primitive*.

DEFINITION 24.10.2. *An algebra* $\mathfrak{A}$ *is primitive if it contains a maximal left (right) ideal* $'\mathfrak{m}$ *such that the quotient ideal* $'\mathfrak{m}:\mathfrak{A} = \{\theta\}$. *A two-sided ideal* $\mathfrak{i}$ *in* $\mathfrak{A}$ *is said to be a primitive ideal if* $\mathfrak{i} \neq \mathfrak{A}$ *and* $\mathfrak{A}/\mathfrak{i}$ *is a primitive algebra.*

THEOREM 24.10.6. *A necessary and sufficient condition that* $\mathfrak{A}$ *be a primitive algebra is that* $\mathfrak{A}$ *be isomorphic to an irreducible algebra of endomorphisms.*

PROOF. Suppose that $\mathfrak{A}$ is primitive and $'\mathfrak{m}:\mathfrak{A} = \{\theta\}$. Let $\mathfrak{L} = \mathfrak{A} \div {'\mathfrak{m}}$ with elements $X = \{x + {'\mathfrak{m}}\}$ and define $T(X; a) = \{ax + {'\mathfrak{m}}\}$ where $a, x \in \mathfrak{A}$. Then $\mathfrak{T} = \{T(\cdot; a)\}$ is an algebra of endomorphisms in $\mathfrak{L}$. The correspondence between $\mathfrak{A}$ and $\mathfrak{T}$ is clearly a homomorphism. But if $T(\cdot; a) = T(\cdot; b)$ then $(a - b)x \in {'\mathfrak{m}}$ for all x which implies that $a = b$ since $'\mathfrak{m}:\mathfrak{A} = \{\theta\}$. Thus $\mathfrak{A} \cong \mathfrak{T}$. If $\mathfrak{L}_0$ is an invariant linear subspace of $\mathfrak{L}$ so that $\mathfrak{T}(\mathfrak{L}_0) \subset \mathfrak{L}_0$, then $\mathfrak{L}_0$ is made up of residue-classes $\{y + {'\mathfrak{m}}\}$ and the corresponding set of points in $\mathfrak{A}$ is a left ideal containing $'\mathfrak{m}$. Since $'\mathfrak{m}$ is maximal, we see that $\mathfrak{L}_0 = \{\Theta\}$ or $\mathfrak{L}$, that is, $\mathfrak{T}$ is irreducible. Thus the condition is necessary.

Conversely, let $\mathfrak{A} \cong \mathfrak{T}$ where $\mathfrak{T}$ is an irreducible algebra of endomorphisms in a linear system $\mathfrak{X}$. We denote elements of $\mathfrak{A}$ by $a, b, c, \cdots$, of $\mathfrak{X}$ by $x, y, \cdots$, and let T_a denote the element of $\mathfrak{T}$ which corresponds to a in the isomorphism. Take a fixed element y, $y \neq \theta$, and consider the set of all elements b such that $T_b(y) = \theta$. This set is a left ideal $'\mathfrak{i}$ in $\mathfrak{A}$. If c is any fixed element of $\mathfrak{A}$ not in $'\mathfrak{i}$, then $T_c(y) \neq \theta$, and $\mathfrak{T}[T_c(y)]$ is an invariant linear manifold, not zero, and hence equal to $\mathfrak{X}$. If a is an arbitrary element of $\mathfrak{A}$, we can then find an element d such that $T_d[T_c(y)] = T_{dc}(y) = T_a(y)$ whence it follows that $a - dc \in {'\mathfrak{i}}$. Hence $a = (a - dc) + dc \in {'\mathfrak{i}} + \mathfrak{A}c$, that is, $\mathfrak{A} = {'\mathfrak{i}} + \mathfrak{A}c$ for each c not in $'\mathfrak{i}$. This shows that $'\mathfrak{i} = {'\mathfrak{m}}$ is maximal. If $c \in \mathfrak{A}$, $c \neq \theta$, there is a $z \in \mathfrak{X}$ such that $T_c(z) \neq \theta$ and also an a with $T_a(y) = z$. It follows that $T_{ca}(y) \neq \theta$. Thus for every $c \neq \theta$ there is a right multiple ca which is not in $'\mathfrak{m}$. But if $c \in {'\mathfrak{m}}:\mathfrak{A}$, then $cx \in {'\mathfrak{m}}$ for all x. Hence $c = \theta$, $'\mathfrak{m}:\mathfrak{A} = \{\theta\}$, and $\mathfrak{A}$ is primitive. This completes the proof.

For further details concerning primitive algebras and related topics we refer to N. Jacobson [2].

References. Baer [1], Calkin [1], Gelfand [4], Jacobson [2], Kaplansky [1], Perlis [1], Segal [1].

3. REPRESENTATION OF (B)-ALGEBRAS

24.11. Linear representation systems. The question of finding realizations and representations of an abstract algebraic system by linear transformations is a classical problem in algebra for the solution of which standardized methods are available. While these methods were originally developed with a view to applications to rings satisfying some conditions of finiteness, their power goes much further. In the following we shall examine this question for the case of (B)-algebras, referring to section 4.15 for a detailed discussion of the case in which the algebra is commutative and has a unit element.

The basic definitions are analogous to those of section 8.3 for the case of semi-groups.

DEFINITION 24.11.1. *A representation $\mathfrak{R}$ of a (B)-algebra $\mathfrak{B}$ is a continuous homomorphic mapping of $\mathfrak{B}$ onto a normed algebra $\mathfrak{T}$ of endomorphisms of a (B)-space $\mathfrak{X}$. Here $\mathfrak{B}$, $\mathfrak{T}$, and $\mathfrak{X}$ have the same scalar field. The representation is irreducible if $\mathfrak{T}$ is irreducible relative to $\mathfrak{X}$.*

Most of the terms have already been explained in connection with Theorem 24.8.6 and representations of algebras also figured in the proofs of Theorems 24.10.2 and 24.10.6, but a summary of the explanations may be useful to the reader. Thus it is required that $\mathfrak{B} \sim \mathfrak{T}$, that is, to every $a \in \mathfrak{B}$ there is a unique $T_a \in \mathfrak{T}$ and

$$T_a + T_b = T_{a+b}, \qquad \alpha T_a = T_{\alpha a}, \qquad T_a \bigcirc T_b = T_{ab}. \tag{24.11.1}$$

Here T_a is a linear bounded transformation on $\mathfrak{X}$ to itself; to every a there is a finite non-negative M_a such that $\| T_a \| \leqq M_a$. $\mathfrak{T}$ is a normed algebra, a (B)-algebra only if it is complete. The mapping is supposed to be continuous in the sense of the metric, that is, $\| a_n - a_0 \| \to 0$ implies $\| T_{a_n} - T_{a_0} \| \to 0$. Finally, $\mathfrak{T}$ is irreducible relative to $\mathfrak{X}$ if, $\mathfrak{X}_0$ being a linear subspace of $\mathfrak{X}$ such that $\mathfrak{T}(\mathfrak{X}_0) \subset \mathfrak{X}_0$, then $\mathfrak{X}_0$ is either $\{\theta\}$ or $\mathfrak{X}$.

A classical method of obtaining a representation of an algebra is based on left (right) translations in the residue-class system $\mathfrak{A} \div \mathfrak{m}$. This method is useful also for (B)-algebras. We shall carry through the details for the left representations.

We suppose that $\mathfrak{B}$ contains left maximal ideals and that $'\mathfrak{m}$ is such an ideal. With this ideal we form the residue-class system $\mathfrak{B} \div {'\mathfrak{m}} = \mathfrak{L}$. By Theorem 24.7.4, $\mathfrak{L}$ becomes a (B)-space under the norm

$$\| X \| = \inf_{x \in X} \| x \|.$$

To each a in $\mathfrak{B}$ we assign a transformation T_a defined on $\mathfrak{L}$ by

$$T_a(X) = T_a(x + {'\mathfrak{m}}) = (ax + {'\mathfrak{m}}). \tag{24.11.2}$$

This is clearly a linear transformation on $\mathfrak{L}$ to itself. Since

$$\| (ax + {'\mathfrak{m}}) \| \leqq \| a \| \, \| (x + {'\mathfrak{m}}) \|$$

or

$$\| T_a(X) \| \leqq \| a \| \, \| X \|, \tag{24.11.3}$$

we see also that T_a is a bounded transformation whose norm does not exceed $\| a \|$. From the basic properties of residue-classes together with the definition of $T_a(\cdot)$ we see that the relations (24.11.1) hold. Hence the mapping of $\mathfrak{B}$ onto $\mathfrak{T}$ is a homomorphism. Since

$$\| T_a - T_b \| = \| T_{a-b} \| \leqq \| a - b \|, \tag{24.11.4}$$

the mapping is also continuous.

It remains to show that $\mathfrak{T}$ is irreducible. Indeed, if $\mathfrak{L}_0$ is an invariant linear subspace of $\mathfrak{L}$ so that $\mathfrak{T}(\mathfrak{L}_0) \subset \mathfrak{L}_0$, then the points of $\mathfrak{B}$, which generate residue-classes belonging to $\mathfrak{L}_0$, form a left ideal containing $'\mathfrak{m}$. This ideal is consequently either $'\mathfrak{m}$ itself or $\mathfrak{B}$. In the former case $\mathfrak{L}_0 = \{\Theta\}$, in the latter $\mathfrak{L}_0 = \mathfrak{L}$ so $\mathfrak{T}$ is irreducible. Thus (24.11.2) defines a representation of $\mathfrak{B}$ which will be denoted by $\mathfrak{R}['\mathfrak{m}]$ in the following. The corresponding algebra of endomorphisms will be denoted by $\mathfrak{T}['\mathfrak{m}]$ and we write $T_a = T_a['\mathfrak{m}]$ when greater precision is desirable. The results proved so far are summed up in

THEOREM 24.11.1. *If $\mathfrak{B}$ is a (B)-algebra containing left maximal ideals, then each such ideal gives rise to an irreducible left representation $\mathfrak{R}['\mathfrak{m}]$ of $\mathfrak{B}$ defined by* (24.11.1).

If $\mathfrak{B}$ contains right maximal ideals, we have similar right (anti-) representations $[\mathfrak{m}']\mathfrak{R}$ defined by

$$(X)_a T = (x + \mathfrak{m}')_a T = (xa + \mathfrak{m}'). \tag{24.11.5}$$

These representations are also irreducible.

Some representations have interesting special properties. Suppose that Φ is the complex field and that $\alpha \neq 0$ belongs to the dissolvent spectrum of $a \in \mathfrak{B}$. More precisely, we assume that $(1/\alpha)a$ does not have a left reverse. By Theorem 24.9.5 it does not have a left reverse modulo some maximal regular left ideal. Form the set $\{xa - \alpha x\}$, $x \in \mathfrak{B}$. This is a left regular ideal; it does not contain a since $(1/\alpha)a$ does not have a left reverse. It is consequently a proper ideal and as such it is contained in a left regular maximal ideal, $^*\mathfrak{m}$ say. Consider the corresponding representation $\mathfrak{R}[^*\mathfrak{m}]$ and, in particular, the transformation $T_a[^*\mathfrak{m}]$. Let $c \in \mathfrak{C}$ be an element of the center of $\mathfrak{B}$. We have then

$$\begin{aligned} T_a(c + {}^*\mathfrak{m}) &= (ac + {}^*\mathfrak{m}) = (ca + {}^*\mathfrak{m}) \\ &= (\alpha c + {}^*\mathfrak{m}) = \alpha(c + {}^*\mathfrak{m}) \end{aligned}$$

since $ca - \alpha c \in {}^*\mathfrak{m}$. Thus on $\mathfrak{C}$ the transformation $T_a[^*\mathfrak{m}]$ reduces to a multiplication by the scalar α. We say that $\mathfrak{R}[^*\mathfrak{m}]$ *is locally spectral at* $x = a$. It is clear that $\mathfrak{R}[^*\mathfrak{m}]$ is locally spectral not merely at $x = a$ but also at all points of the subalgebra generated by a. This follows from (24.11.1).

THEOREM 24.11.2. *The representation $\mathfrak{R}['\mathfrak{m}]$ is an isomorphic mapping of $\mathfrak{B}$ onto $\mathfrak{T}['\mathfrak{m}]$ if and only if the quotient ideal $'\mathfrak{m}:\mathfrak{B} = \{\theta\}$ in which case $\mathfrak{B}$ is primitive.*

The proof follows by Theorem 24.10.6.

The properties of the quotient ideal are also decisive for the question of when the algebra $\mathfrak{T}['\mathfrak{m}]$ is simple in the sense of

DEFINITION 24.11.2. *An algebra is simple if it contains no proper two-sided ideals.*

Since $\mathfrak{T}['\mathfrak{m}]$ is irreducible, it is semi-simple by Theorem 24.8.6 but ordinarily not simple.

THEOREM 24.11.3. $\mathfrak{T}['\mathfrak{m}]$ *is simple if and only if the quotient ideal* $'\mathfrak{m}:\mathfrak{B}$ *is maximal as a two-sided ideal.*

PROOF. Suppose first that $'\mathfrak{m}:\mathfrak{B}$ is maximal. Let $\mathfrak{J}$ be a two-sided ideal of $\mathfrak{T}['\mathfrak{m}]$ and let $\mathfrak{i}$ be the set of all elements a of $\mathfrak{B}$ such that $T_a \in \mathfrak{J}$. It is clear that $\mathfrak{i}$ is a two-sided ideal in $\mathfrak{B}$. Since $\Theta \in \mathfrak{J}$ and $a \in {}'\mathfrak{m}:\mathfrak{B}$ if and only if $T_a = \Theta$, we see that $'\mathfrak{m}:\mathfrak{B} \subset \mathfrak{i}$, that is, $\mathfrak{i} = {}'\mathfrak{m}:\mathfrak{B}$ or $\mathfrak{B}$. In the former case $\mathfrak{J} = \{\Theta\}$, in the latter $\mathfrak{J} = \mathfrak{T}['\mathfrak{m}]$. Thus $\mathfrak{T}['\mathfrak{m}]$ is simple. Conversely, suppose that $\mathfrak{T}['\mathfrak{m}]$ is simple and let $\mathfrak{i}$ be a two-sided ideal of $\mathfrak{B}$ containing $'\mathfrak{m}:\mathfrak{B}$. Under the homomorphism $\mathfrak{i}$ is mapped on a set $\mathfrak{J}$ of $\mathfrak{T}['\mathfrak{m}]$ and $\mathfrak{J}$ is clearly a two-sided ideal, hence either $\{\Theta\}$ or $\mathfrak{T}['\mathfrak{m}]$. In the former case $\mathfrak{i} = {}'\mathfrak{m}:\mathfrak{B}$, in the latter $\mathfrak{i} = \mathfrak{B}$ so $'\mathfrak{m}:\mathfrak{B}$ is maximal. This completes the proof.

Use of the right representations was proposed by Max Zorn to E. Hille in 1943; the left representations were used by Gelfand and Neumark [1] in the same year.

24.12. Sufficiency of the representations. A fundamental question in any theory of representations is to decide whether or not enough irreducible representations are available to distinguish between the elements of the system. For $\mathfrak{R}['\mathfrak{m}]$ the situation is described by

THEOREM 24.12.1. *If* $a \neq b$, *then* $T_a['\mathfrak{m}] = T_b['\mathfrak{m}]$ *for all* $'\mathfrak{m}$ *if and only if* $a - b \in \mathfrak{p}$. *Thus, if* $\mathfrak{B}$ *is a radical algebra, the representations* $\mathfrak{R}['\mathfrak{m}]$ *are either trivial or fail to exist. The representations suffice (in the sense that* $a \neq b$ *implies the existence of at least one* $'\mathfrak{m}$ *with* $T_a['\mathfrak{m}] \neq T_b['\mathfrak{m}]$) *if and only if* $\mathfrak{B}$ *is semi-simple. Finally, if* $\mathfrak{B}$ *is primitive and* $'\mathfrak{m}:\mathfrak{B} = \{\theta\}$, *then* $T_a['\mathfrak{m}] \neq T_b['\mathfrak{m}]$ *whenever* $a \neq b$; *conversely, the existence of such a representation implies that* $\mathfrak{B}$ *is primitive.*

PROOF. If $T_a['\mathfrak{m}] = T_b['\mathfrak{m}]$ for a particular $'\mathfrak{m}$, then the residue-classes $(ax + {}'\mathfrak{m})$ and $(bx + {}'\mathfrak{m})$ are identical for all x. This implies, and is implied by, $(a - b)x \in {}'\mathfrak{m}$ or $a - b \in {}'\mathfrak{m}:\mathfrak{B}$. If this is to hold for every $'\mathfrak{m}$, then $a - b \in \bigcap_\alpha('\mathfrak{m}_\alpha:\mathfrak{B}) = \mathfrak{p}$. Since the reasoning may be reversed, the converse is also true and the first assertion is proved. A radical algebra does not always contain maximal ideals, so the proposed representations may fail to exist. But if a $'\mathfrak{m}$ exists, then $T_a['\mathfrak{m}] = \Theta$ for all a; the representation is then trivial and fails to differentiate between any elements of the algebra. If $\mathfrak{B}$ is semi-simple and $a \neq b$, then since $a - b$ does not lie in $\mathfrak{p}$ there exists at least one $'\mathfrak{m}$ such that

$a - b$ is not in $'\mathfrak{m}:\mathfrak{B}$ and consequently $T_a['\mathfrak{m}] \neq T_b['\mathfrak{m}]$. Conversely, if to every $a \neq \theta$ there is a $'\mathfrak{m}$ such that $T_a['\mathfrak{m}] \neq \theta$, then a does not belong to $'\mathfrak{m}:\mathfrak{B} \supset \mathfrak{p}$, that is, $\mathfrak{p} = \{\theta\}$ or $\mathfrak{B}$ is semi-simple. Finally, if $\mathfrak{B}$ is primitive and if $'\mathfrak{m}:\mathfrak{B} = \{\theta\}$, then $a - b$ cannot belong to $'\mathfrak{m}:\mathfrak{B}$ when $a \neq b$ and $T_a['\mathfrak{m}] \neq T_b['\mathfrak{m}]$. Conversely, if there is a $'\mathfrak{m}$ with $T_a['\mathfrak{m}] \neq T_b['\mathfrak{m}]$ whenever $a \neq b$, then $\mathfrak{B} \cong \mathfrak{T}['\mathfrak{m}]$ and $\mathfrak{B}$ is primitive by Theorem 24.11.2. This completes the proof.

These considerations may also be used to obtain representations of normed groups and semi-groups. With a given normed semi-group $\mathfrak{S}$ we associate a suitable *semi-group algebra* $\mathfrak{B}$ to which the representation theory applies and we obtain an induced representation of $\mathfrak{S}$. We have carried through this process in great detail in §16.1 for the case of a canonical one-parameter semi-group of linear transformations $\mathfrak{S} = [T(\xi); 0 < \xi < \infty, T(\xi) \in \mathfrak{E}(\mathfrak{X})]$. In this case the second commutant of $\mathfrak{S}$ in $\mathfrak{E}(\mathfrak{X})$ provides a commutative semi-group algebra to which the Gelfand representation theory applies. For the general case the preceding theorem obviously gives the following result:

THEOREM 24.12.2. *Let $\mathfrak{S}$ be a semi-group embedded in a (B)-algebra $\mathfrak{B}$. The representation $\mathfrak{R}['\mathfrak{m}]$ of $\mathfrak{B}$ maps $\mathfrak{S}$ upon a semi-group $\mathfrak{S}['\mathfrak{m}]$ which gives a continuous representation of $\mathfrak{S}$. A necessary and sufficient condition in order that to every pair of elements a, b of $\mathfrak{S}$ there exists at least one $'\mathfrak{m}$ such that $T_a['\mathfrak{m}] \neq T_b['\mathfrak{m}]$ is that $x - y$ is never in $\mathfrak{p}$ when $x, y \in \mathfrak{S}$.*

By Theorem 24.8.8 a sufficient (but not necessary) condition for the sufficiency of the representations of $\mathfrak{S}$ is that no two elements of $\mathfrak{S}$ have the same spectrum in $\mathfrak{B}$. It seems to be rather difficult to get a condition for the sufficiency of the representations which is expressible in terms of properties of the elements of the semi-group without reference to the semi-group algebra.

The same theorem holds for groups but here the situation is much more favorable. The basic result in the commutative case is due to I. Gelfand [5] who showed that the representations $\mathfrak{R}[\mathfrak{m}]$ suffice if the group is bounded. E. Hille has shown [11] that boundedness may be replaced by a limitation on the rate of growth of the iterates of the elements. In the following we shall extend the discussion to the non-commutative case. See also M. H. Stone [6].

THEOREM 24.12.3. *Let $\mathfrak{B}$ be a complex (B)-algebra with unit element e and let $\mathfrak{G}$ be a subset of $\mathfrak{B}$, the elements of which form a group under multiplication. If for each $x \in \mathfrak{G}$ we have $\| x^{\pm n} \| = o(n)$, where the condition need not hold uniformly with respect to x, and if $a \in \mathfrak{G}$, $b \in \mathfrak{G}$, $a \neq b$, then there exists a $'\mathfrak{m}$ of $\mathfrak{B}$ such that $T_a['\mathfrak{m}] \neq T_b['\mathfrak{m}]$. The theorem is false if $o(n)$ is replaced by $O(n)$.*

PROOF. We have to show that under the conditions of the theorem $x \in \mathfrak{G}$, $y \in \mathfrak{G}$, $x - y \in \mathfrak{p}$ implies that $x = y$. Indeed, if $x - y = p_1 \in \mathfrak{p}$, then $xy^{-1} = e - p_1x^{-1} \in \mathfrak{G}$ and $-p_1x^{-1} = p \in \mathfrak{p}$. Thus $\mathfrak{G}$ contains an element of the form $e + p$ and by assumption $\| (e + p)^{\pm n} \| = o(n)$. It follows from Theorem 4.10.1 that $p = \theta$ and $x = y$ so that the representations distinguish between the elements of $\mathfrak{G}$. We cannot replace $o(n)$ by $O(n)$ in the condition. For if $\mathfrak{B}$ is generated by e and a nilpotent element p of index 2 so that $p \neq \theta$, $p^2 = \theta$, then

$$\mathfrak{G}: (e + p)^n = e + np$$

is a group contained in $\mathfrak{B}$ and its elements are clearly $O(n)$ and not $o(n)$. In this case all representations $\mathfrak{R}['\mathfrak{m}]$ map the whole group on the unit element of $\mathfrak{T}['\mathfrak{m}]$. This completes the proof.

It is clear that the condition on the elements of $\mathfrak{G}$ is designed to exclude from $\mathfrak{G}$ all elements of the form $e + p$ where p is nilpotent of index two. Moreover, Theorem 4.10.1 shows that if we know in advance that $\mathfrak{B}$ does not contain any nilpotent element of index less than a given integer $k + 1$ (except the zero element), then we may replace $o(n)$ by $o(n^k)$ in the condition of the theorem and this is also the best condition of its kind. Finally, if we know that $\mathfrak{B}$ does not contain any nilpotent elements, then the condition may be replaced by

$$\limsup_{n\to\infty} \frac{\log \| x^{\pm n} \|}{\log n} < \infty \tag{24.12.1}$$

for all x in $\mathfrak{G}$.

In the commutative case studied by Gelfand all representations $\mathfrak{R}[\mathfrak{m}]$ are of degree one in the sense that T_x acts in the complex euclidian space of one dimension and T_x is of the form

$$T_x(\mathfrak{m}):\ z' = x(\mathfrak{m})z \tag{24.12.2}$$

where the $x(\mathfrak{m})$'s are defined in Theorem 4.15.1.

For a group $\mathfrak{G}$ embedded in $\mathfrak{B}$ and $x \in \mathfrak{G}$, we observe that $x(\mathfrak{m})$, for fixed $\mathfrak{m}$, plays the role of a *group character*.

REMARK. Let us recall that a representation of degree n of a group $\mathfrak{G}$ is a collection of non-singular n-rowed square matrices $D(x)$, where $D(x)$ is uniquely determined by $x \in \mathfrak{G}$ and $D(xy) = D(x)\,D(y)$. The trace of $D(x)$, namely $\chi(x) = \operatorname{tr} D(x)$, is independent of the particular choice of basis elements of the n-dimensional vector space. Each such function $\chi(x)$ is known as a character of the representation. Each matrix $D(x)$ defines a linear transformation T_x acting in the n-dimensional vector space. In the case of a finite group, the matrices $D(x)$ can always be assumed to be unitary. If, in addition, the group is abelian, then all irreducible representations are of degree one, so that the characters themselves give the representation: $|\chi(x)| = 1$ and $\chi(xy) = \chi(x)\chi(y)$.

THEOREM 24.12.4. *If $\mathfrak{B}$ is a commutative complex (B)-algebra with unit element, if $\mathfrak{B}$ contains the multiplicative group $\mathfrak{G}$, and if*

$$\lim_{n\to\infty} \| x^n \|^{1/n} = 1 = \lim_{n\to\infty} \| x^{-n} \|^{1/n} \tag{24.12.3}$$

for all $x \in \mathfrak{G}$, then all of the representations $x(\mathfrak{m})$ are unitary in the sense that $|x(\mathfrak{m})| = 1$ for all $x \in \mathfrak{G}$ and all $\mathfrak{m} \in \mathfrak{M}$. The condition is necessary as well as sufficient.

PROOF. By Theorem 4.7.3 the condition (24.12.3) is equivalent to $r(x) = 1 = r(x^{-1})$. Hence by properties (7) and (8) of Theorem 4.15.1, $\sup_{\mathfrak{m}\in\mathfrak{M}} | x(\mathfrak{m}) | = 1 = \sup_{\mathfrak{m}\in\mathfrak{M}} | x(\mathfrak{m}) |^{-1}$ so that $| x(\mathfrak{m}) | \equiv 1$. Conversely, if $| x(\mathfrak{m}) | \equiv 1$, then $r(x) = 1 = r(x^{-1})$.

24.13. Functions of maximal ideals. The representation theory developed in section 4.15 presupposes that the commutative (B)-algebra $\mathfrak{B}$ has a unit element. We can easily dispense with this assumption (cf. L. H. Loomis [**1**, p. 59]) as is shown by

THEOREM 24.13.1. *Let $\mathfrak{B}$ be a complex commutative (B)-algebra. Let $\mathfrak{M} \equiv [\mathfrak{m}]$ be the set of all regular maximal ideals in $\mathfrak{B}$. For a given regular maximal ideal $\mathfrak{m}$, with $j_\mathfrak{m}$ the unit element of $\mathfrak{B}$ modulo $\mathfrak{m}$, the homomorphism $\mathfrak{B} \sim \mathfrak{B}/\mathfrak{m}$ maps each $x \in \mathfrak{B}$ into a complex number $x(\mathfrak{m})$, defined by*

$$x \equiv x(\mathfrak{m})j_\mathfrak{m} \pmod{\mathfrak{m}}. \tag{24.13.1}$$

These mappings define a class of complex-valued functions on $\mathfrak{M}$ having the following properties:

(1) $(x_1 + x_2)(\mathfrak{m}) = x_1(\mathfrak{m}) + x_2(\mathfrak{m})$;
(2) $(\alpha x)(\mathfrak{m}) = \alpha x(\mathfrak{m})$;
(3) $(x_1x_2)(\mathfrak{m}) = x_1(\mathfrak{m})x_2(\mathfrak{m})$;
(4) $j_\mathfrak{m}(\mathfrak{m}) = 1$;
(5) $x(\mathfrak{m}) = 0$ *if and only if* $x \in \mathfrak{m}$;
(6) $\alpha(x) = x(\mathfrak{M})$ *plus possibly the point* $\lambda = 0$;
(7) *The spectral radius* $r(x) = \sup_{\mathfrak{m}\in\mathfrak{M}} | x(\mathfrak{m}) | \leqq \| x \|$;
(8) $q(\mathfrak{m}) \equiv 0$ *if and only if q is quasi-nilpotent*;
(9) *An element x has a reverse if and only if $x(\mathfrak{m}) \neq 1$ for all $\mathfrak{m}$ in which case* $x^{-}(\mathfrak{m}) = - x(\mathfrak{m})[1 - x(\mathfrak{m})]^{-1}$;
(10) *If $\mathfrak{m}_1 \neq \mathfrak{m}_2$, then there is an $x \in \mathfrak{B}$ such that $x(\mathfrak{m}_1) \neq x(\mathfrak{m}_2)$.*

PROOF. Properties (1) through (5) are direct consequences of Theorem 24.9.7. For the elements of the complex field Z it is clear that an element λ has a reverse if and only if $\lambda \neq 1$. Hence if $\lambda \neq 0$ it follows from Theorem 24.9.5 that x/λ has a reverse if and only if $x(\mathfrak{m})/\lambda \neq 1$, that is if and only if $x(\mathfrak{m}) \neq \lambda$; this proves (6). The first half of (9) follows similarly whereas the second half can be obtained from the definition of a reverse with the help of properties (1) and (3). From (6) and the Theorem 24.3.3 we see that $r(x) = \sup_{\mathfrak{m}\in\mathfrak{M}} | x(\mathfrak{m}) |$; the inequality $| x(\mathfrak{m}) | \leqq \| x \|$ follows from (24.3.6). According to Definition 24.8.1 an element is quasi-nilpotent if and only if its spectrum consists of the single point $\{0\}$ so that (8) is a direct consequence of (6). Finally (10) is obvious from (5).

We next proceed to topologize $\mathfrak{M}$ as in section 4.15. In the present case, however, the multiplicative linear functionals need not form a closed subset (in the weak* topology) of the unit sphere in $\mathfrak{B}^*$. But the argument of section 4.15 shows that any limit point of $\mathfrak{M}$ is again a homomorphism of $\mathfrak{B}$ into the complex field. If such a homomorphism is non-trivial then its kernel is a regular maximal ideal and hence an element of $\mathfrak{M}$. The only other possibility is that the corresponding functional be the zero element of $\mathfrak{B}^*$. In this case $\mathfrak{M}$ will be locally compact and each function $x(\mathfrak{m})$ will be continuous on $\mathfrak{M}$ and "vanish at infinity".

The above results can also be derived directly from the representation theory of §4.3. It suffices to embed $\mathfrak{B}$ in a commutative (B)-algebra $\mathfrak{B}_e$ with unit element as in section 4.2. In this case the mapping $\mathfrak{m}_e \to \mathfrak{m} \equiv \mathfrak{B} \cap \mathfrak{m}_e$ is a one-to-one correspondence between the family of all maximal ideals in $\mathfrak{B}_e$ which are not included in $\mathfrak{B}$ and the family of all regular maximal ideals in $\mathfrak{B}$.

Representation theorems of the kind given above furnish information only if there exist regular maximal one-sided ideals and a sufficient supply of such ideals which have the added property of being two-sided. In particular such a theorem is of no value in the case of a radical algebra where there are no regular maximal ideals. It is also useless in the case of the algebra of endomorphisms in a Hilbert space, the algebra $\mathfrak{E}(\mathfrak{H})$ of section 24.8, where there is only one two-sided maximal ideal $\mathfrak{c}$ and $\mathfrak{E}(\mathfrak{H})/\mathfrak{c}$ is not a field. On the other hand, an analogue of Theorem 24.13.1 holds for two-sided algebras so one can go a trifle beyond the commutative case.

References. Gelfand [5], Gelfand and Neumark [1], Hille [11], Loomis [1], Stone [6].

4. A Structure Theorem

24.14. The structure of normed groups and semi-groups. Let $\mathfrak{B}$ be a complex (B)-algebra with unit element e. We recall that the regular elements of $\mathfrak{B}$ form a group, the maximal group of $\mathfrak{B}$, which we shall now denote by $\mathfrak{G}[\mathfrak{B}]$ while $\mathfrak{G}_1[\mathfrak{B}]$ stands for its principal component. Certain subgroups of $\mathfrak{G}_1[\mathfrak{B}]$ have a fairly simple structure which may be determined. The decisive step in this connection was taken by J. v. Neumann [1, 2]. See also E. Cartan [1], G. Pólya [1], I. Schur [2], and B. de Sz.-Nagy [1].

Let $\mathfrak{B}_m$ be the algebra of m-rowed square matrices $A = (a_{jk})$ where the a_{jk} are complex numbers. This is a complex (B)-algebra under the norm of Frobenius-Wedderburn

$$\| A \| = \left\{ \sum_{j=1}^{n} \sum_{k=1}^{n} | a_{jk} |^2 \right\}^{1/2} .$$

The reader should observe that the norm of the unit element E is $\sqrt{m}$. Supposing that $\mathfrak{G}$ is a group of regular matrices in $\mathfrak{B}_m$ which is connected and closed in $\mathfrak{G}[\mathfrak{B}_m]$, v. Neumann showed that $\mathfrak{G}$ is a Lie group. More explicitly stated, there exists a set of infinitesimal generators $\mathfrak{i}$, the so-called Lie ring of $\mathfrak{G}$, with the following properties:

(a) *every element of* $\mathfrak{i}$ *belongs to* $\mathfrak{B}_m$ *and* $\mathfrak{i}$ *is closed*;

(b) *if* $U, V \in \mathfrak{i}$ *so do* $U + V$, $UV - VU$, *and* αU, α *real*;

(c) $\mathfrak{i}$ *has a finite basis*;

(d) *If* $X = \exp(U)$, $U \in \mathfrak{i}$, *then* $X \in \mathfrak{G}$ *and there is a neighborhood of* E *in which all elements of* $\mathfrak{G}$ *are of this form*;

(e) *every element of* $\mathfrak{G}$ *is of the form*

$$\exp(U_1)\exp(U_2)\cdots\exp(U_k), \qquad U_j \in \mathfrak{i}.$$

Here (c) is implied by (a): it is clear that $\mathfrak{i}$ cannot contain more than $2m^2$ elements which are linearly independent with respect to real numbers.

These results were extended by K. Yosida [**1**] to the case of a connected and locally compact group in an arbitrary complex (B)-algebra with unit element. The properties (a)–(e) hold with $\mathfrak{B}_m$ replaced by $\mathfrak{B}$. In this case (c) is no longer implied by (a). We shall consider the extension of these results to a particular class of semi-groups. Except for some special results due to I. Schur [**2**, p. 122 et seq.], such problems do not seem to have been studied for semi-groups.

DEFINITION 24.14.1. *Let* $\mathfrak{F}$ *be a set in* $\mathfrak{B}$. $\mathfrak{F}$ *is said to be differentiable at* $x = x_0$, $x_0 \in \mathfrak{F}$, *if every sequence* $\{x_n\}$ *in* $\mathfrak{F}$ *with* $x_n \to x_0$ *contains a subsequence* $\{x_{n_k}\}$ *such that there is a sequence of real numbers* $\{\epsilon_k\}$, $\epsilon_k \to 0$, *and*

$$\lim_{k\to\infty} \frac{1}{\epsilon_k}(x_{n_k} - x_0) = u \neq \theta$$

exists as an element of $\mathfrak{B}$. *We call* u *a differential coefficient of* $\mathfrak{F}$ *at* x_0.

It is clear from the definition that if u is a differential coefficient, so is αu for every real α.

THEOREM 24.14.1. *Let* $\mathfrak{S}$ *be a semi-group in* $\mathfrak{B}$ *with the following properties*:

(1) $\mathfrak{S}$ *contains at least two elements of* $\mathfrak{G}[\mathfrak{B}]$;

(2) $\mathfrak{S}$ *is closed under inversion*: *if* $x \in \mathfrak{S} \cap \mathfrak{G}[\mathfrak{B}]$, *then* $x^{-1} \in \mathfrak{S}$;

(3) $\mathfrak{G} = \mathfrak{S} \cap \mathfrak{G}[\mathfrak{B}]$ *is closed in* $\mathfrak{G}(\mathfrak{B})$, *locally compact, and connected*;

(4) $\mathfrak{S}$ *is locally a division ring*: *if* $x, y \in \mathfrak{S}$ *and* $\| x - y \| < \delta_x$ *then there exists a* $g \in \mathfrak{G}$ *such that* $x = yg$ *and* $g \to e$ *when* $y \to x$.

Then

(i) $\mathfrak{G}$ *is a Lie group having the properties* (a)–(e) [*with* $\mathfrak{B}_m$ *replaced by* $\mathfrak{B}$];

(ii) *each left coset of* $\mathfrak{G}$ *is connected, locally compact, and relatively open in* $\mathfrak{S}$. *No point of a coset can be a limit point of any other coset. There exists a discrete subset* $\mathfrak{S}_0 = \{s_\alpha\}$ *of* $\mathfrak{S}$ *such that every element of* $\mathfrak{S}$ *is of the form* sg, $s \in \mathfrak{S}_0$, $g \in \mathfrak{G}$.

The proof will be given in a number of steps.

LEMMA 24.14.1. *If* $\mathfrak{S}$ *is differentiable at a regular point, then it is differentiable everywhere.*

PROOF. Suppose that $\mathfrak{S}$ is differentiable at $x = x_0$ and $x_0 \in \mathfrak{G} \equiv \mathfrak{S} \cap \mathfrak{G}[\mathfrak{B}]$. We shall show that this implies differentiability at $x = e$. We note first that $\mathfrak{S}$ is locally perfect and contains $x = e$. If $\{g_n\} \in \mathfrak{S}$ and $g_n \to e$, we may assume that $g_n \in \mathfrak{G}$ and we know that $g_n x_0 \to x_0$. Since $\mathfrak{S}$ is differentiable at $x = x_0$, we have

$$\lim_{k\to\infty} \frac{1}{\epsilon_k} [g_{n_k} x_0 - x_0] = u \neq \theta$$

for a suitable choice of $\{n_k\}$ and $\{\epsilon_k\}$. Hence

$$\lim_{k\to\infty} \frac{1}{\epsilon_k} [g_{n_k} - e] = u x_0^{-1}$$

since x_0 is regular. This shows that $\mathfrak{S}$ is differentiable at $x = e$. Next, if y is any element of $\mathfrak{S}$ and $y_n \to y$, $y_n \in \mathfrak{S}$, then for large values of n we have $\| y - y_n \| < \delta_y$, so that, by (4), $y_n = y g_n$ where $g_n \in \mathfrak{G}$ and $g_n \to e$. Hence for a suitable choice of $\{n_k\}$ and $\{\epsilon_k\}$ we have

$$\frac{1}{\epsilon_k} [y_{n_k} - y] = y \frac{1}{\epsilon_k} [g_{n_k} - e] \to y u x_0^{-1}$$

so that $\mathfrak{S}$ is differentiable everywhere.

LEMMA 24.14.2. *If $\mathfrak{S}$ is differentiable at $x = e$, if u is one of its differential coefficients at e, and if $\{\eta_n\}$ is any null sequence of real numbers, $\eta_n \neq 0$ for all n, then there exists a sequence of elements $\{g_n\}$ in $\mathfrak{G}$ such that*

$$\lim_{n\to\infty} \frac{1}{\eta_n} [g_n - e] = u.$$

PROOF. By assumption

$$\lim_{n\to\infty} \frac{1}{\epsilon_n} [x_n - e] = u$$

for a suitable choice of $\{x_n\} \subset \mathfrak{G}$ and $\{\epsilon_n\} \subset E_1$. Choose the least integer $p = p(n)$ so that for a given n we have $| \epsilon_p | \leqq \eta_n^2$ and let $q = q(n)$ be the nearest integer to η_n/ϵ_p. It is clear that $| q(n) | \to \infty$ when $n \to \infty$. We then set $g_n = (x_p)^q$; it is an element of $\mathfrak{G}$ since $x_p \in \mathfrak{G}$. From

$$x_p = e + \epsilon_p[u + o(1)] \qquad \text{and} \qquad \epsilon_p = \frac{\eta_n}{q}\left\{1 + O\left(\frac{1}{q}\right)\right\}$$

one infers that

$$x_p = e + \frac{\eta_n}{q}[u + o(1)].$$

The binomial series shows that, uniformly with respect to q,

$$(x_p)^q = e + \eta_n[u + o(1)] + O(\eta_n^2)$$

and, for $q = q(n)$,

$$\frac{1}{\eta_n}[g_n - e] = u + o(1) + O(\eta_n) \to u$$

when $n \to \infty$. This proves the lemma.

LEMMA 24.14.3. *$\mathfrak{S}$ is differentiable at $x = e$ and hence everywhere.*

PROOF. Since $\mathfrak{G}$ is locally compact, there is an ϵ-neighborhood of e, $\epsilon < 1$, whose closure is compact in $\mathfrak{G}$. Further given a sequence of elements in $\mathfrak{S}$ converging to e we can choose a subsequence whose nth member, say x_n, belongs to $\mathfrak{G}$ and is such that

$$\frac{\epsilon}{2(n+1)} < \| x_n - e \| < \frac{\epsilon}{2n}.$$

Since

$$(x_n)^n - e = \sum_{k=1}^{n} \binom{n}{k} (x_n - e)^k,$$

a simple computation shows that for $n > 3$

$$\tfrac{1}{4}\epsilon < \| (x_n)^n - e \| < \tfrac{3}{4}\epsilon.$$

It follows that the sequence $\{(x_n)^n\}$ is bounded and bounded away from e. Since $\{(x_n)^n\}$ belongs to a compact set a subsequence converges to a limit, $e + v$ say, where $\epsilon/4 \leq \| v \| \leq 3\epsilon/4$. But $2n \| x_n - e \| < \epsilon$, so the binomial series gives

$$\begin{aligned} x_n &= [e + (x_n)^n - e]^{1/n} \\ &= e + \sum_{k=1}^{\infty} \binom{1/n}{k} [(x_n)^n - e]^k \end{aligned}$$

and

$$n(x_n - e) = \sum_{1}^{\infty} n \binom{1/n}{k} [(x_n)^n - e]^k.$$

Here the series

$$\sum_{1}^{\infty} n \binom{1/n}{k} y^k$$

converges uniformly with respect to n as well as with respect to y for $\| y \| \leq \rho < 1$. Now let $n \to \infty$ in such a manner that $(x_n)^n - e \to v$. Then

$$\lim_{n\to\infty} n(x_n - e) = \sum_{1}^{\infty} \frac{(-1)^{k-1}}{k} v^k = \log(e + v) \neq \theta.$$

This shows that $\mathfrak{S}$ is differentiable at $x = e$.

LEMMA 24.14.4. *The set of differential coefficients of* $\mathfrak{S}$ *at* $x = e$ *form a Lie ring* $\mathfrak{i}$ *having properties* (a) *to* (e).

PROOF. That $\mathfrak{i} \subset \mathfrak{B}$ follows from the completeness of the space $\mathfrak{B}$. The closure of $\mathfrak{i}$ will be proved below.

If $u \in \mathfrak{i}$, then by definition $\alpha u \in \mathfrak{i}$. If $u, v \in \mathfrak{i}$, then Lemma 24.14.2 assures the existence of sequences $\{x_n\}$ and $\{y_n\}$ in $\mathfrak{G}$ such that

$$n(x_n - e) \to u, \qquad n(y_n - e) \to v.$$

It follows that

$$n(x_n y_n - e) = n(x_n - e) + n(y_n - e) + n(x_n - e)(y_n - e) \to u + v,$$

so that $u + v \in \mathfrak{i}$. Next we consider

$$n^2(x_n y_n x_n^{-1} y_n^{-1} - e) = n^2(x_n y_n - y_n x_n) x_n^{-1} y_n^{-1}.$$

Since $x_n^{-1} y_n^{-1} \to e$ and

$$n^2(x_n y_n - y_n x_n) = n(x_n - e) n(y_n - e) - n(y_n - e) n(x_n - e) \to uv - vu,$$

we see that

$$n^2(x_n y_n x_n^{-1} y_n^{-1} - e) \to uv - vu$$

and $uv - vu \in \mathfrak{i}$. This proves (b).

Thus $\mathfrak{i}$ is an algebra over the real field with addition as in $\mathfrak{B}$ but with multiplication defined by the commutator

$$[u, v] = uv - vu.$$

This operation is neither commutative nor associative; we have

$$[u, v] = -[v, u],$$

$$[u, [v, u]] + [v, [w, u]] + [w, [u, v]] = \theta.$$

We have now to show that $\exp u \in \mathfrak{G}$ if $u \in \mathfrak{i}$. Since there is a sequence $\{x_n\}$ in $\mathfrak{G}$ such that $n(x_n - e) \to u$, we have

$$x_n = e + \frac{1}{n}[u + o(1)]$$

whence it follows that

$$(x_n)^n = \left\{e + \frac{1}{n}[u + o(1)]\right\}^n \to \exp u$$

when $n \to \infty$. Since $\mathfrak{G}$ is closed in $\mathfrak{G}(\mathfrak{B})$, $\exp u \in \mathfrak{G}$ as asserted. This proves the first part of (d).

We can now finish the proof of (a) by showing that $\mathfrak{i}$ is closed. Indeed, if $\{u_n\} \subset \mathfrak{i}$ and $u_n \to u$, then $\exp(\alpha u_n) \in \mathfrak{G}$ for every real α and $\exp(\alpha u_n) \to$

$\exp(\alpha u) \in \mathfrak{G}$. But

$$\frac{1}{\alpha}[\exp(\alpha u) - e] \to u$$

when $\alpha \to 0$ so that $u \in \mathfrak{i}$. Hence $\mathfrak{i}$ is closed.

The next assertion is that $\mathfrak{J}$ is locally compact. Let $\mathfrak{J}_0$ be the set of elements u of $\mathfrak{J}$ with $\| u \| \leqq \epsilon/2$, where the ϵ-neighborhood of e in $\mathfrak{G}$ has a compact closure. This is a closed set and so is $\mathfrak{G}_0 \equiv \exp(\mathfrak{J}_0)$ which is a subset of $\mathfrak{G}$. Since $\| \exp u - e \| \leqq \frac{3}{4}\epsilon$ when $\epsilon < 1$, it follows that $\mathfrak{G}_0$ is compact. Further, by formula (9.5.1)

$$\log(\exp u) = \sum_{k=0}^{\infty} \frac{(-1)^{k-1}}{k} (\exp u - e)^k = \sum_{k=1}^{\infty} \frac{(-1)^{k-1}}{k} \left\{ \sum_{n=1}^{\infty} \frac{u^n}{n!} \right\}^k = u.$$

Since $\log x$ is continuous for $\| x - e \| \leqq \epsilon < 1$ and the correspondence between x and $\log x$ is one-to-one, the mapping $x \to \log x$ takes compact sets onto compact sets (see section 1.4). It follows that $\mathfrak{J}_0$ is compact and, since $\mathfrak{J}$ is linear, $\mathfrak{J}$ is locally compact. But by Theorem 1.12.2 a locally compact linear subspace of a (B)-space is of finite dimension. Hence there is a finite base $u_1, u_2, \cdots, u_n$ such that every element of $\mathfrak{J}$ is of the form $\alpha_1 u_1 + \alpha_2 u_2 + \cdots + \alpha_n u_n$ with real coefficients. This proves (c).

The second part of (d) is proved by an indirect argument. Let $\{g_n\} \in \mathfrak{G}$, $\| g_n - e \| < 1$, and $g_n \to e$, so that $\log g_n$ exists as defined by formula (9.5.1). Suppose that $\log g_n$ does not belong to $\mathfrak{J}$ for any n. We shall show that this leads to a contradiction. Since $\mathfrak{J}$ is locally compact, there is a point $u_n \in \mathfrak{J}$ whose distance from $\log g_n$ is a minimum. Set $\| \log g_n - u_n \| = \epsilon_n$. It is clear that $\epsilon_n \to 0$ as $n \to \infty$. We note that $\epsilon_n \leqq \| \log g_n \| \equiv \delta_n$ and $\| u_n \| \leqq \| \log g_n \| + \epsilon_n \leqq 2\delta_n$. Further $\delta_n \to 0$ as $n \to \infty$ by (9.5.1). If we now express g_n as $\exp(\log g_n)$, then a simple calculation shows that

$$\| [g_n - \exp(u_n)] - [\log g_n - u_n] \| \leqq 2\epsilon_n[\exp(2\delta_n) - 1]. \tag{24.14.1}$$

Hence

$$\epsilon_n\{1 - 2[\exp(2\delta_n) - 1]\} \leqq \| g_n - \exp(u_n) \| \leqq 2\epsilon_n \exp(2\delta_n)$$

and since $\| g_n \exp(-u_n) - e \| \leqq \| g_n - \exp(u_n) \| \, \| \exp(-u_n) \|$ where $\| \exp(-u_n) \| \leqq \exp(2\delta_n)$, we have

$$\tfrac{1}{4}\epsilon_n \leqq \| g_n \exp(-u_n) - e \| \leqq 4\epsilon_n$$

for n sufficiently large. It follows now by Lemma 24.14.3 that there exists a $v \in \mathfrak{J}$, $v \neq \theta$, and a null sequence $\{\eta_k\}$ such that

$$\frac{1}{\eta_k}\{g_{n_k} \exp(-u_{n_k}) - e\} \to v.$$

It is clear that ϵ_{n_k}/η_k is bounded away from zero and infinity. Now $\exp(u_n) \to e$

and therefore

$$\frac{1}{\eta_k}[g_{n_k} - \exp(u_{n_k})] \to v.$$

The estimate (24.14.1) now implies that

$$\log g_{n_k} - u_{n_k} - \eta_k v = o(\eta_k).$$

Here $u_{n_k} + \eta_k v \in \mathfrak{i}$ and its distance from $\log g_{n_k}$ is $o(\eta_k) = o(\epsilon_{n_k})$ while the best approximation of $\log g_{n_k}$ by elements in $\mathfrak{i}$ is ϵ_{n_k} and is supposed to be furnished by u_{n_k}. This is a contradiction and shows that the sequence $\{\log g_n\}$ must contain infinitely many elements belonging to $\mathfrak{i}$. Actually we can draw the stronger conclusion that there exists a $\delta = \delta(\mathfrak{G})$ such that for all elements of $\mathfrak{G}$ with $\| g - e \| < \delta$ we have $\log g \in \mathfrak{i}$. Indeed, if this were false, then we could find a sequence $\{g_n\} \subset \mathfrak{G}$ converging to e such that $\log g_n$ is never in $\mathfrak{i}$ and this we have seen is impossible. This proves the second assertion under (d).

It remains to prove (e). The elements g of $\mathfrak{G}$ which can be represented in the form

$$\exp(u_1)\exp(u_2)\cdots\exp(u_k)$$

with k arbitrary and $u_1, u_2, \cdots, u_k$ in $\mathfrak{i}$, form a subset $\mathfrak{G}'$ of $\mathfrak{G}$. If $\mathfrak{G}'$ does not exhaust $\mathfrak{G}$, let its complement in $\mathfrak{G}$ be $\mathfrak{G}''$. Since $\mathfrak{G}$ is connected, one of these subsets has a limit point in the other. Let this point be x. Then $x \in \mathfrak{G}$ and every neighborhood of x contains points of $\mathfrak{G}'$ as well as of $\mathfrak{G}''$. We can choose a neighborhood N_x such that for all $y, z \in N_x \cap \mathfrak{G}$ we have $\| y^{-1}z - e \| < \delta = \delta(\mathfrak{G})$. We then choose $y \in \mathfrak{G}'$, $z \in \mathfrak{G}''$. Then $\log(y^{-1}z) = v \in \mathfrak{i}$ and $y^{-1}z = \exp v$ so that

$$z = y \exp v = \exp(u_1)\exp(u_2)\cdots\exp(u_k)\exp v$$

since $y \in \mathfrak{G}'$. But this shows that $z \in \mathfrak{G}'$, that is, $\mathfrak{G}''$ is void and $\mathfrak{G}'$ exhausts $\mathfrak{G}$. This proves property (e) and completes the proof of Lemma 24.14.4 as well as of conclusion (i) of Theorem 24.14.1.

What we have established so far constitutes a proof of the theorems of von Neumann and Yosida and the proof is essentially a paraphrase of their arguments. If $\mathfrak{S}$ is not a group, $\mathfrak{S} \neq \mathfrak{G}$, we have just one more step to take.

We consider the left cosets of $\mathfrak{G}$ in $\mathfrak{S}$, that is, the sets of $s\mathfrak{G}$ with $s \in \mathfrak{S}$. Each such set is connected and locally compact since $\mathfrak{G}$ has these properties. Suppose that x is in the left coset $\mathfrak{S}_\alpha$ and that $y \in \mathfrak{S}$ with $\| x - y \| < \delta_x$. By property (4), $y = xg$ where $g \in \mathfrak{G}$ so that $y \in \mathfrak{S}_\alpha$. Thus no point of a coset can be limit point of any other coset and each coset is open relative to $\mathfrak{S}$.

The fact that the cosets are relatively open shows that there exists a discrete subset $\mathfrak{S}_0 = \{s_\alpha\}$ of $\mathfrak{S}$, having exactly one element in common with each left coset, such that every element of $\mathfrak{S}$ is of the form sg where $s \in \mathfrak{S}_0$, $g \in \mathfrak{G}$. This completes the proof of Theorem 24.14.1.

It is not obvious that it is always possible to select the representatives of the left cosets in such a manner that $\mathfrak{S}_0$ becomes a semi-group though this would seem plausible.

It should be observed that it is perfectly feasible for a point of $\mathfrak{B}$ not in $\mathfrak{S}$ to be limit point of infinitely many left cosets of $\mathfrak{S}$. As an example we may take the semi-group

$$x^n(1 + x)^\alpha \qquad n = 0, 1, 2, \cdots ; -\infty < \alpha < \infty,$$

in the algebra of continuous functions $C[0, 1]$ with the usual definitions of norm and operations. Here each fixed value of n gives a coset and the zero-element is a limit point of each coset.

References. Cartan [1], v. Neumann [1, 2], Pólya [1], Schur [2], Sz.-Nagy [1], Yosida [1].

CHAPTER XXV

LIE SEMI-GROUPS

25.1. Orientation. The present chapter is devoted to a study of certain classes of parametric semi-groups of bounded linear transformations in the sense of Definition 8.3.4. Such a semi-group $\mathfrak{T} = [T(p); p \in \Pi, T(p) \in \mathfrak{E}(\mathfrak{X})]$ is a collection of operators $T(p)$ where the parameter p belongs to a topological semi-group Π and the composition of operators obeys the law

$$T(p)T(q) = T(p \bigcirc q);$$

here $p \bigcirc q$ is the "product operation" in Π.

The investigation falls naturally into two parts: a study of the underlying parameter semi-group Π, and a study of its representation in $\mathfrak{E}(\mathfrak{X})$ by $[T(p)]$. For the first study we shall assume that Π is embedded in a (B)-space $\mathfrak{P}$ from which Π takes its metric. It is convenient to assume that Π is a positive cone as well as closed under the product operation. Normally it is not feasible to assume that $\mathfrak{P}$ is a (B)-algebra for desirable product operations as a rule are not distributive with respect to addition and scalar multiplication. We get the sharpest results when $\mathfrak{P}$ is the real euclidean n-space E_n and

$$\Pi = [p = (p^1, p^2, \cdots, p^n); p^i \geqq 0, i = 1, 2, \cdots, n] \equiv \bar{E}_n^+$$

for short. However with only a few exceptions the results hold for an arbitrary (B)-space $\mathfrak{P}$. The main problem is to determine the canonical sub-semi-groups of Π and to characterize that part of Π which may be reached by canonical orbits from the origin.

We then pass over to $\mathfrak{T} = [T(p); p \in \Pi, T(p) \in \mathfrak{E}(\mathfrak{X})]$ where the first problem is to study the continuity properties of $T(p)$ as a function of p. Here and in the subsequent discussion we restrict ourselves to the case in which $\Pi \subset E_n$ and prove under reasonable assumptions on the product operation that strong (uniform) measurability of $T(p)$ implies strong (uniform) continuity in the interior of Π less the idempotents. A similar result holds if one assumes strong (uniform) continuity at the origin. The argument is based upon an analogue of Theorem 7.13.1: If the product operation is suitably restricted, then measurable finite-valued suboperative functions are bounded above on compact sets.

Next we consider the infinitesimal properties of $[T(p)]$. To every canonical orbit of Π there corresponds a canonical one-parameter sub-semi-group $\mathfrak{T}_a$ of $\mathfrak{T}$ generated by some $a \in \Pi$, and to each such entity is associated an infinitesimal generator $A(a)$. These infinitesimal generators form a positive cone and satisfy the analogues of the three fundamental theorems of Lie.

There are three paragraphs: *Semi-groups in a (B)-Space*, *Parametric Transformation Semi-groups*, and *The Lie Theory*.

For background material the reader is referred to papers by G. Birkhoff and P. A. Smith. Our presentation is based upon papers by E. Hille [14, 15] with various additions of unpublished material.

References. Alexandroff and Hopf [1], G. Birkhoff [3], N. Dunford [4], L. Gårding [1], I. Gelfand [3], E. Hille [14, 15], S. Lie and G. Scheffers [1], Miyadera [1], J. von Neumann [1], F. Severi and G. Scorza Dragoni [1], P. A. Smith [1, 2], A. D. Wallace [1].

1. Semi-Groups in a (B)-Space

25.2. Basic assumptions. We shall examine the structure of a topological semi-group Π embedded in a real (B)-space $\mathfrak{P}$. Various restrictive conditions will be imposed referring partly to the topological nature of Π and partly to the properties of the product operation

$$a \bigcirc b = F(a, b). \tag{25.2.1}$$

One of the following assumptions will be made concerning Π of which the second obviously is a special case of the first.

S. *$\mathfrak{P}$ is a real (B)-space. $\Pi \subset \mathfrak{P}$ is closed and* $0 \in \Pi \ominus \Pi_0$ *where* $\Pi_0 = \text{Int}(\Pi)$. *If $a \in \Pi_0$, then $-a \notin \Pi_0$. Π is closed under addition and multiplication by positive scalars.*

S_0. *$\mathfrak{P} = E_n$, the real euclidean n-space, and $\Pi = \bar{E}_n^+$.*

The norm in $\mathfrak{P}$ is denoted by $|a|$ and the metric in the product space $\mathfrak{P} \times \mathfrak{P}$ is defined by $|(a, b)| = \max(|a|, |b|)$. The zero element of $\mathfrak{P}$ is denoted by 0. The product operation will be subjected to several of the following assumptions.

P_1. *$F(a, b)$ defines a continuous mapping of $\Pi \times \Pi$ into Π.*

P_2. $F(a, 0) = F(0, a) = a$.

P_3. $F(a, F(b, c)) = F(F(a, b), c)$.

P_4. $F(a, b) = F(b, a)$.

P_5. *There exists a fixed positive constant B such that for all points a_1, a_2, and b in Π*

$$\max \{| F(a_1, b) - F(a_2, b) |, | F(b, a_1) - F(b, a_2) |\} < (1 + B | b |) | a_1 - a_2 |.$$

P_6. *There exists a positive monotone increasing continuous function $\omega(t)$, $0 < t < \infty$, tending to zero with t, such that*

$$| F(a, b) - a - b | \leqq r\omega(s), \qquad r = \min(| a |, | b |), \qquad s = | a | + | b |.$$

P_7. $t^{-1}\omega(t) \in L(0, 1)$.

P_8. *If* $\mathfrak{P} = E_n$, $\Pi \subset \bar{E}_n^+$, *the n coordinates of* $F(p, q)$ *shall have continuous first order partials with respect to the n coordinates of q.*

P_9. *To every compact set K in* Π_0 *there is a positive* $\delta \equiv \delta(K)$ *such that for* $c \in K, h \in \Pi_0, |h| < \delta$, *the equation* $F(h, b) = c$ *has a unique solution* $b = \psi(c, h)$ *in* Π_0 *which is a continuous function of* (c, h).

Further restrictions on the product operation will be introduced in §25.2.

25.3. The canonical function. Our main problem is the determination of the one-parameter canonical sub-semi-groups of Π. We shall consider only sub-semi-groups abutting the origin. These sub-semi-groups will be defined with the aid of the so-called *canonical function* $f(p)$ of Π. This function maps Π continuously into itself in such a manner that the rays

$$[\rho a;\ a \in \Pi,\ 0 \leqq \rho < \infty] \tag{25.3.1}$$

are mapped onto the *canonical orbits*

$$\Gamma_a \equiv [p = f(\rho a);\ 0 \leqq \rho < \infty]. \tag{25.3.2}$$

The elements of Π on the orbit Γ_a form a *one-parameter canonical sub-semi-group* satisfying the functional equation

$$F(f(\rho a), f(\sigma a)) = f((\rho + \sigma)a), \qquad 0 \leqq \rho, \sigma < \infty, \tag{25.3.3}$$

and the initial condition

$$\lim_{\rho \to 0+} \rho^{-1} f(\rho a) = a. \tag{25.3.4}$$

The construction of $f(p)$ is based on a device due to J. von Neumann [1] (cf. section 24.14), namely, the forming of high powers of elements close to the neutral element of the operation which, because of P_2, means close to zero in our case.

To simplify the discussion more compact notation for complicated powers will be needed. If $p \in \Pi$ and m is a positive integer, the product of m factors p will be denoted by $(p; m)$.

Let us further define a quantity τ by the convention: τ is the root of the equation

$$\omega(t) = \tfrac{1}{2} \tag{25.3.5}$$

if $\omega(t)$ can take on values $> \frac{1}{2}$, otherwise $\tau = \infty$.

The following lemmas will be basic for the discussion.

LEMMA 25.3.1. *The validity of postulates* S, P_1, P_3, *and* P_6 *is assumed. If* $p \in \Pi$, *if m is a positive integer, and if* $2m\,|\,p\,| < \tau$, *then*

$$(p; m) = mp + R_m(p), \qquad |\,R_m(p)\,| < m\,|\,p\,|\,\omega(2m\,|\,p\,|). \tag{25.3.6}$$

PROOF. The estimate is clearly true for $m = 2$ by P_6. If it is true for $m = k$

and $2(k + 1) | p | < \tau$ we have

$$(p; k + 1) = F(p, (p; k)) = p + (p; k) + G_k(p) \equiv (k + 1)p + R_{k+1}(p)$$

where by P_6

$$| G_k(p) | \leqq r_k\omega(s_k), \qquad r_k = \min (| p |, | (p; k) |), \qquad s_k = | p | + | (p; k) |.$$

The induction hypothesis implies that $| (p; k) |$ lies between

$$k | p | [1 + \omega(2k | p |)] < \tfrac{3}{2}k | p | \quad \text{and}$$

$$k | p | [1 - \omega(2k | p |)] > \tfrac{1}{2}k | p | \geqq | p|.$$

It follows that $r_k = | p |$ and $s_k < 2(k + 1) | p |$. Since $\omega(t)$ is increasing, one concludes that

$$\begin{aligned} | R_{k+1}(p) | &\leqq | R_k(p) | + | G_k(p) | \\ &\leqq k | p | \omega(2k | p |) + | p | \omega[2(k + 1) | p |] \\ &< (k + 1) | p | \omega[2(k + 1) | p |] \end{aligned}$$

and the lemma is proved.

LEMMA 25.3.2. *Assume postulates* S, P_1, P_2, P_3, *and* P_5. *If* $p \in \Pi$ *so does* $(p/\nu; \mu)$ *where* μ, ν *are arbitrary positive integers and*

$$| (p/\nu; \mu) | < B^{-1}[\exp (B | p | \mu/\nu) - 1]. \tag{25.3.7}$$

PROOF. To simplify the writing we set $(p/\nu; \mu) = p_{\mu,\nu}$. That $p_{\mu,\nu} \in \Pi$ follows from S together with P_1. By P_1, P_2, and P_5 we have

$$\begin{aligned} | p_{\mu+1,\nu} - p_{\mu,\nu} | &= | F(p_{1,\nu}, p_{\mu,\nu}) - F(p_{1,\nu}, p_{\mu-1,\nu}) | \\ &< (1 + B | p_{1,\nu} |) | p_{\mu,\nu} - p_{\mu-1,\nu} | \\ &< \cdots\cdots\cdots \\ &< (1 + B | p_{1,\nu} |)^{\mu} | p_{1,\nu} | \end{aligned}$$

so that

$$| p_{\mu+1,\nu} - p_{\mu,\nu} | < \left(1 + B \frac{| p |}{\nu}\right)^{\mu} \frac{| p |}{\nu}.$$

Addition of these inequalities gives

$$| p_{\mu,\nu} | < B^{-1}\left[\left(1 + B \frac{| p |}{\nu}\right)^{\mu} - 1\right]$$

and, *a fortiori*, (25.3.7).

Repeated squaring and estimates of differences of squares plays a fundamental role in the following. A twofold use of P_6 shows that

(25.3.8) $$| F(a, a) - F(b, b) | < [2 + B(| a | + | b |)] | a - b |.$$

A further extension of P_5 is given in

LEMMA 25.3.3. *Assume* S, P_1, P_2, P_3, *and* P_5. *If* $x, y \in \Pi$ *and if* $j \leqq k$ *are positive integers, then*

(25.3.9) $$| (x; 2^k) - (y; 2^j) | < 2^j e^{Br} | (x; 2^{k-j}) - y |$$

where $r = \max (2^k | x |, 2^j | y |)$.

PROOF. We note that $(x; 2^k)$ is the square of $(x; 2^{k-1})$ and by (25.3.7)

$$1 + B | (x; 2^{k-1}) | < \exp (2^{k-1} B | x |) < \exp (\tfrac{1}{2} Br).$$

Similar statements are true if x and k are replaced by y and j. It follows from (25.3.8) that

$$| (x; 2^k) - (y; 2^j) | < 2 \exp (\tfrac{1}{2} Br) | (x; 2^{k-1}) - (y; 2^{j-1}) | .$$

Here we may replace k and j successively by $k - 1$ and $j - 1$, $k - 2$ and $j - 2$ and so on. After j steps we have

$$\begin{aligned} | (x; 2^k) - (y; 2^j) | &\leqq 2^j \exp [\textstyle\sum_1^j 2^{-\nu} Br] | (x; 2^{k-j}) - y | \\ &< 2^j e^{Br} | (x; 2^{k-j}) - y | \end{aligned}$$

as asserted.

In particular

(25.3.10) $$| (x; 2^k) - (y; 2^k) | < 2^k e^{Br} | x - y | .$$

We can now formulate the main theorem.

THEOREM 25.3.1. *Assume postulates* S, P_1, P_2, P_3, P_5, *and* P_6. *Then the canonical function* $f(p)$ *of* Π *defined by*

(25.3.11) $$f(p) \equiv \lim_{k \to \infty} (2^{-k} p; 2^k), \qquad p \in \Pi,$$

exists and is continuous in p. *Moreover* $f(p)$ *satisfies a Lipschitz condition.*

(25.3.12) $$| f(p) - f(q) | \leqq e^{BR} | p - q |, \qquad \max (| p |, | q |) \leqq R.$$

For any choice of a, $a \in \Pi$, *the function* $g(\rho) \equiv f(\rho a)$, *satisfies functional equation*

(25.3.13) $$g(\rho + \sigma) = F(g(\rho), g(\sigma)), \qquad 0 < \rho, \sigma < \infty,$$

with the initial condition

(25.3.14) $$\lim_{\rho \to 0+} \rho^{-1} g(\rho) = a,$$

and it is the only solution in Π *of these equations. If* P_4 *holds, then for all* p, q *in* Π

(25.3.15) $$f(p + q) = F(f(p), f(q)).$$

PROOF. Let $p_k = (2^{-k}p; 2^k)$. Then Lemma 25.3.3 with $x = 2^{-k}p$, $y = 2^{-j}p$, $j < k$, shows that

$$| p_k - p_j | < e^{B|p|} | 2^j(2^{-k}p; 2^{k-j}) - p |.$$

If j is sufficiently large Lemma 25.3.1 applies and the difference in the right member is dominated by $| p | \omega(2^{1-j} | p |)$ and this quantity tends to zero as $j \to \infty$. The convergence being uniform for $| p | < R$, it follows that $f(p)$ is continuous.

To prove the Lipschitz condition we use (25.3.10) with $x = 2^{-k}p$, $y = 2^{-k}q$. Passing to the limit with k, one obtains (25.3.12).

Next we note that for any $p \in \Pi$ and any positive integer m

$$f(mp) = (f(p); m). \tag{25.3.16}$$

To prove this we consider the difference

$$(2^{-k}mp; 2^k) - ((2^{-k}p; 2^k); m) = (2^{-k}mp; 2^k) - ((2^{-k}p; m); 2^k)$$

by the associativity of the "product". The limit of the left member as $k \to \infty$ is $f(mp) - (f(p); m)$. By Lemma 25.3.2

$$| (2^{-k}p; m) | \leqq B^{-1}[\exp (m2^{-k}B | p |) - 1] < m2^{1-k} | p |$$

for sufficiently large values of k. Applying Lemmas 25.3.1 and 25.3.3, we see that the difference in question does not exceed

$$e^{2mB|p|} | mp - 2^k(2^{-k}p; m) | \leqq e^{2mB|p|}m | p |\omega(2^{1-k}m | p |)$$

for large k, and the last member tends to zero as $k \to \infty$. This proves (25.3.16).

Formula (25.3.16) evidently implies that

$$f((m_1 + m_2)p) = F(f(m_1p), f(m_2p))$$

for arbitrary positive integers m_1, m_2. Setting $p = a/n$ one sees that (25.3.13) holds for arbitrary positive rational values of ρ and σ. Since $f(p)$ is continuous, it follows that the functional equation holds for all positive reals. Lemma 25.3.1 then shows that for $2\rho | a | < \tau$

$$| (2^{-k}\rho a; 2^k) - \rho a | < \rho | a | \omega(2\rho | a |) = o(\rho) \qquad \text{as } \rho \to 0+,$$

and this implies the initial condition (25.3.14).

To prove uniqueness we first show that

$$| f(\rho a) | \leqq B^{-1}[e^{B\rho|a|} - 1]. \tag{25.3.17}$$

This inequality is a consequence of the functional equation and the initial condition; it does not depend upon the particular solution. In fact, suppose that $g(\rho)$ is a function on positive reals to Π satisfying (25.3.13) and (25.3.14). We have then by P_5

$$| g(\beta) - g(\alpha) | \leqq | g(\beta - \alpha) | [1 + B | g(\alpha) |]$$

for $0 < \alpha < \beta$. The initial condition shows that $g(\rho)$ is bounded for small values of ρ and the functional equation shows that this conclusion extends to every finite interval. It follows then that $g(\rho)$ is continuous and satisfies a Lipschitz condition of order one. Let us set

$$\alpha = \frac{j}{n}\rho, \beta = \frac{j+1}{n}\rho, \qquad j = 0, 1, 2, \cdots, n-1,$$

and add the inequalities obtaining

$$|g(\rho)| \leqq \sum_{j=0}^{n-1}\left|g\left(\frac{j+1}{n}\rho\right) - g\left(\frac{j}{n}\rho\right)\right| \leqq \frac{n}{\rho}\left|g\left(\frac{\rho}{n}\right)\right|\sum_{j=0}^{n-1}\left[1 + B\left|g\left(\frac{j}{n}\rho\right)\right|\right]\frac{\rho}{n}.$$

Passing to the limit with n gives

$$(25.3.18) \qquad |g(\rho)| \leqq |a|\int_0^\rho [1 + B|g(\sigma)|]\,d\sigma.$$

Repeated substitution shows that $g(\rho)$ satisfies (25.3.17) as asserted.

Suppose now that we have two solutions $g(\rho)$ and $h(\rho)$. The functional equation shows that

$$|g(\rho) - h(\rho)| = |(g(2^{-k}\rho); 2^k) - (h(2^{-k}\rho); 2^k)|.$$

By (25.3.17)

$$|g(2^{-k}\rho)|, |h(2^{-k}\rho)| \leqq B^{-1}[\exp(2^{-k}\rho B|a|) - 1] < 2^{1-k}\rho|a|$$

for all large k. Hence by Lemma 25.3.3

$$|g(\rho) - h(\rho)| < 2^k e^{2B\rho|a|}|g(2^{-k}\rho) - h(2^{-k}\rho)|$$

$$= e^{2B\rho|a|}\rho\left|\frac{g(2^{-k}\rho)}{2^{-k}\rho} - \frac{h(2^{-k}\rho)}{2^{-k}\rho}\right| \to 0$$

as $k \to \infty$. This proves that $h(\rho) \equiv g(\rho) \equiv f(\rho a)$.

It remains to prove (25.3.15) when P_4 holds. Actually all that is needed is that for the particular choice of p, q in question

$$F(2^{-k}p, 2^{-k}q) = F(2^{-k}q, 2^{-k}p)$$

for a sequence of k's tending to infinity. If this condition is satisfied we have

$$F((2^{-k}p; 2^k), (2^{-k}q; 2^k)) = (F(2^{-k}p, 2^{-k}q); 2^k).$$

To prove the desired relation we have then merely to prove that the difference

$$\delta_k(p, q) \equiv (F(2^{-k}p, 2^{-k}q); 2^k) - (2^{-k}(p+q); 2^k)$$

tends to zero as $k \to \infty$. Now for large values of k it follows from P_6 that

$$|F(2^{-k}p, 2^{-k}q) - 2^{-k}(p+q)| < 2^{-k}R\omega(2^{1-k}R)$$

if $\max(|p|, |q|) \leqq R$, so that ultimately

$$2^k \mid F(2^{-k}p, 2^{-k}q) \mid < 3R.$$

Applying Lemma 25.3.3 with

$$x = F(2^{-k}p, 2^{-k}q), \qquad y = 2^{-k}(p+q), \qquad j = k,$$

we see that

$$\mid \delta_k(p, q) \mid \leqq 2^k e^{3BR} \mid F(2^{-k}p, 2^{-k}q) - 2^{-k}(p+q) \mid < e^{3BR} R\omega(2^{1-k}R)$$

and this tends to zero as $k \to \infty$. Hence (25.3.15) holds and this completes the proof of the theorem.

The above theorem may be strengthened as follows.

THEOREM 25.3.2. *Assume the conditions* S, P_1, P_2, P_3, P_5, P_6, *and* P_7. *If* $g(\rho)$ *is a solution of* (25.3.13) *tending to zero with* ρ, *then there exists an* $a \in \Pi$ *such that* (25.3.14) *holds.*

PROOF. Suppose $g(\rho)$ is such a solution and that ρ_0 is so small that $2 \mid g(\rho) \mid \leqq \tau$ for $\rho \leqq \rho_0$. Repeated use of P_6 gives

$$g(\rho) = 2^k g(2^{-k}\rho) + R_k(\rho)$$

where

$$\mid R_k(\rho) - R_{k-1}(\rho) \mid \leqq 2^{k-1} \mid g(2^{-k}\rho) \mid \omega[2 \mid g(2^{-k}\rho) \mid].$$

On the other hand

$$\text{(25.3.19)} \qquad \begin{aligned} \mid g(\rho) \mid &\geqq \{1 - \tfrac{1}{2}\omega[2 \mid g(2^{-1}\rho) \mid]\} 2 \mid g(2^{-1}\rho) \mid \geqq \cdots \\ &\geqq \prod_{j=1}^{\nu} \{1 - \tfrac{1}{2}\omega[2 \mid g(2^{-j}\rho) \mid]\} \mid 2^{\nu} g(2^{-\nu}\rho) \mid \end{aligned}$$

so that

$$\sum_1^{\infty} \mid R_k(\rho) - R_{k-1}(\rho) \mid \leqq \mid g(\rho) \mid \sum_{k=1}^{\infty} \omega[2 \mid g(2^{-k}\rho) \mid] \prod_{j=1}^{k} \{1 - \tfrac{1}{2}\omega[2 \mid g(2^{-j}\rho) \mid]\}^{-1}.$$

The series in the right hand member converges if and only if $\sum \omega[2 \mid g(2^{-k}\rho) \mid]$ converges. Now the terms of this series may be estimated with the aid of (25.3.19). First we note that $\omega[2 \mid g(2^{-k}\rho) \mid] < \frac{1}{2}$. This gives

$$2 \mid g(2^{-k}\rho) \mid \leqq 2(\tfrac{2}{3})^k \mid g(\rho) \mid \leqq \tau(\tfrac{2}{3})^k \equiv \alpha_k.$$

Hence

$$\omega[2 \mid g(2^{-k}\rho) \mid] < \omega(\alpha_k) < 3 \int_{\alpha_k}^{\alpha_{k-1}} \frac{\omega(t)}{t}\, dt$$

so that the series converges if P_7 holds. But this implies that $R_k(\rho)$ tends to a finite limit when $k \to \infty$ and this holds uniformly with respect to ρ in $[0, \rho_0]$. It follows that

$$\lim_{k\to\infty} 2^k g(2^{-k}\rho) \equiv l(\rho) \tag{25.3.20}$$

exists and is a continuous function of ρ in $[0, \rho_0]$. Further setting $l(\rho) = \rho m(\rho)$ one sees that $m(2\rho) = m(\rho)$, and this can serve to define $l(\rho)$ for all positive ρ.

Proceeding in exactly the same manner one shows that

$$\lim_{k\to\infty} 3^k g(3^{-k}\rho) \equiv l_1(\rho)$$

exists and is a continuous function of ρ. Further putting $l_1(\rho) = \rho m_1(\rho)$ one finds that $m_1(3\rho) = m_1(\rho)$. Actually $l(\rho) \equiv l_1(\rho)$. To see this we note first that (25.3.20) implies that $|g(\rho)| < M\rho$ for $0 < \rho < \rho_0$. Hence by P_2 and P_5

$$\begin{aligned} |g(\rho+\sigma) - g(\sigma)| &= |F(g(\rho), g(\sigma)) - F(0, g(\sigma))| \\ &\leqq [1 + B|g(\sigma)|]\,|g(\rho)| < [1 + B|g(\sigma)|]M\rho, \end{aligned}$$

or a Lipschitz condition for $g(\rho)$. Next we consider the continued fraction expansion of (log 3/log 2); let its successive convergents be m_k/n_k so that

$$\frac{\log 3}{\log 2} = \frac{m_k}{n_k} + \frac{\epsilon_k}{n_k^2}, \qquad |\epsilon_k| < 1.$$

We have then

$$3^{n_k} = 2^{m_k + \epsilon_k/n_k}$$

and

$$3^{n_k} g(3^{-n_k}\rho) = 2^{\epsilon_k/n_k} 2^{m_k} g(2^{-\epsilon_k/n_k} 2^{-m_k}\rho).$$

Using the Lipschitz condition we see that

$$3^{n_k} g(3^{-n_k}\rho) = 2^{m_k} g(2^{-m_k}\rho) + O(n_k^{-1}).$$

Letting $k \to \infty$ one obtains $l(\rho) \equiv l_1(\rho)$ as asserted. We have consequently $m(\rho) = m_1(\rho)$ and the double functional equation $m(\rho) = m(2\rho) = m(3\rho)$. From this one concludes that $m(\rho)$ takes on the value $m(\rho_0) \equiv a$ in the set $\{2^{\pm m} 3^{\pm n} \rho_0\}$ which is dense in $[0, \rho_0]$ since log 2 and log 3 are independent over the rational field. Since $m(\rho)$ is continuous this requires that $m(\rho) \equiv a$. This constant, being the limit of elements in the closed set Π must also be in Π. From

$$\lim_{k\to\infty} 2^k \rho^{-1} g(2^{-k}\rho) = a,$$

the convergence being uniform with respect to ρ in the interval $\rho_0/2 \leqq \rho \leqq \rho_0$, one finally concludes that

$$\lim_{\rho\to 0+} \rho^{-1} g(\rho) = a.$$

This completes the proof.

25.4. The canonical orbits. The canonical function $f(p)$ maps the unit vectors in Π in a one-to-one manner on the canonical sub-semi-groups S_a of Π which admit the origin as a limit element:

$$S_a \equiv [f(\rho a), 0 < \rho < \infty]. \tag{25.4.1}$$

When we want to emphasize the geometry of the situation, S_a is denoted by Γ_a instead and is referred to as a *canonical orbit*.

A number of properties of these orbits may be read off from the preceding discussion. Thus Γ_a is tangent to the ray $[\rho a;\ \rho \geqq 0]$ at the origin and has a unique tangent for almost all $\rho > 0$.

When assumption S is replaced by S_0 so that $\mathfrak{P} = E_n$ and $\Pi = \bar{E}_n^+$, the path Γ_a may be determined as the solution of a system of n first order differential equations provided a suitable differentiability condition holds such as P_8.

Here and below coordinates will be denoted by superscripts:

$$p = (p^1, \cdots, p^n), \qquad q = (q^1, \cdots, q^n), \qquad F = (F^1, \cdots, F^n);$$

and differentiation of $F(p; q)$ with respect to the variables p^i, q^j is indicated by subscripts separated by a semi-colon to distinguish between the first and the second set of variables. Thus the symbols $F^i_{j;}$ and $F^i_{;j}$ indicate the first partials of F^i with respect to p^j and q^j respectively and so on. For small values of $|q|$ we have

$$F^i(p^1, \cdots, p^n; q^1, \cdots, q^n) = p^i + \textstyle\sum_{j=1}^n q^j[F^i_{;j}(p^1, \cdots, p^n; 0, \cdots, 0) + \epsilon_{ij}]$$

where $\epsilon_{ij} \to 0$ with $|q|$; here we have used the fact that $F^i(p; 0) = p^i$ by virtue of P_2. Setting $g(\rho) = (g^i(\rho))$ we then get

$$\frac{1}{\sigma}\{F^i[g^1(\rho), \cdots, g^n(\rho); g^1(\sigma), \cdots, g^n(\sigma)] - g^i(\rho)\}$$

$$= \sum_{j=1}^n \frac{g^j(\sigma)}{\sigma}\{F^i_{;j}[g^1(\rho), \cdots, g^n(\rho); 0, \cdots, 0] + \epsilon_{ij}\}.$$

Letting $\sigma \to 0$ and using the initial condition with $a = (a^1, \cdots, a^n)$ we get the system

$$\begin{cases} \dfrac{dg^i(\rho)}{d\rho} = \displaystyle\sum_{j=1}^n a^j F^i_{;j}[g^1(\rho), \cdots, g^n(\rho); 0, \cdots, 0], \\ g^i(0) = 0, \end{cases} \qquad i = 1, 2, \cdots, n. \tag{25.4.2}$$

This system, when it exists, is usually more convenient for the determination and discussion of Γ_a than equations (25.3.13) and (25.3.14).

Let us return to the general case. Every path Γ_a is confined to Π but beyond this very little is known concerning the behavior in the large of Γ_a. Some isolated results are listed as lemmas below. The basic assumptions are S, P_1, P_2, P_3, P_5, and P_6.

LEMMA 25.4.1. *If* $\lim_{\rho\to\infty} f(\rho a)$ *exists for* $a \neq 0$, *then this limit cannot be zero.*

PROOF. The contrary assumption would imply that

$$0 = \lim_{\rho\to\infty} f((\rho + \sigma)a) = \lim_{\rho\to\infty} F[f(\rho a), f(\sigma a)] = f(\sigma a).$$

Thus $f(\sigma a) \equiv 0$ and hence $a = 0$.

LEMMA 25.4.2. *If* $-a \notin \Pi$ *then* $f(\rho a) \neq 0$ *for* $0 < \rho < \infty$.

PROOF. Suppose contrariwise that $f(\rho_0 a) = 0$. Then for $0 < \rho < \rho_0$

$$\begin{aligned} 0 = f(\rho_0 a) &= F[f(\rho a), f((\rho_0 - \rho)a)] \\ &= f(\rho a) + f((\rho_0 - \rho)a) + o(\rho) \end{aligned}$$

by P_6 since $f(\rho a) = \rho a + o(\rho)$ and $f((\rho_0 - \rho)a) = o(1)$ as $\rho \to 0+$. It follows that $f((\rho_0 - \rho)a) = -\rho a + o(\rho)$ and, dividing by ρ and passing to the limit as $\rho \to 0+$, we see that this implies that $-a \in \Pi$ against the assumption.

It should be observed that postulate S assumes that $-a$ and a never belong to Π_0 simultaneously. Hence the only case where $f(\rho a)$ can equal 0 for some $\rho > 0$ is that in which $a \in \Pi \ominus \Pi_0$.

As ρ increases it may possibly happen that Γ_a intersects itself. Suppose that $p = f(\rho a)$, $0 < \rho < \rho_0$, is a simple arc but $f(\rho_0 a) = f((\rho_0 + \omega)a)$ where ω is the least value for which return occurs. We have then for $\rho \geqq \rho_0$

$$\begin{aligned} f((\rho + \omega)a) &= F[f((\rho - \rho_0)a), f((\rho_0 + \omega)a)] \\ &= F[f((\rho - \rho_0)a), f(\rho_0 a)] = f(\rho a) \end{aligned}$$

so that $f(\rho a)$ has the period ω for $\rho \geqq \rho_0$. Thus Γ_a consists of a simple arc Γ_a^1 plus a closed loop Γ_a^2 which is described infinitely often as $\rho \to \infty$. The elements of Γ_a^2 form a group, the unit element of which is $f(m\omega a)$ where $m\omega$ is the least multiple of ω with $m\omega \geqq \rho_0$. We note first that

$$F[f(\rho a), f(m\omega a)] = f(\rho a) = F[f(m\omega a), f(\rho a)], \qquad \rho \geqq \rho_0.$$

Secondly, if $\rho > \rho_0$ we can find a positive integer k such that $k\omega - \rho \geqq \rho_0$. Since $k \geqq m$ we have

$$\begin{aligned} F[f(\rho a), f((k\omega - \rho)a)] &= f(k\omega a) = f(m\omega a) \\ &= F[f((k\omega - \rho)a), f(\rho a)] \end{aligned}$$

so that $f((k\omega - \rho)a)$ is the inverse of $f(\rho a)$. We obviously have

$$F[f(m\omega a), f(m\omega a)] = f(m\omega a)$$

so that $f(m\omega a)$ is an idempotent under the product operation.

While in general looping does not seem to be excluded, there are situations where we can assert that the orbits are simple arcs.

LEMMA 25.4.3. *Suppose that* P_9 *holds and that* $-a \notin \Pi$. *Suppose also that* $\text{Int}(\Gamma_a) \subset \Pi_0$. *Then* Γ_a *is a simple arc.*

PROOF. Suppose the contrary were true. We would then have Γ_a consisting of a simple arc Γ_a^1, corresponding to $0 < \rho < \rho_0$, to which is attached a loop Γ_a^2 described infinitely often, $f(\rho a)$ having the period ω say. The possibility of Γ_a being all loop, that is of $f(\omega a) = 0$, is excluded by Lemma 25.4.2. Hence consider the set $K \equiv [f(\rho a); \frac{1}{2}\rho_0 \leqq \rho \leqq \rho_0 + \omega]$. Since K is a closed continuous arc in Π_0, it is necessarily a compact subset of Π_0. By P_9 there exists a $\delta = \delta(K)$ such that for $c \in K$, $h \in \Pi_0$, and $|h| < \delta$ the equation $F(h, b) = c$ has a unique solution $b = \psi(c, h)$. Let $\epsilon = \epsilon(\delta) < \frac{1}{2}\rho_0$ be so small that $|f(\rho a)| < \delta$ for $0 < \rho < \epsilon$. We have then on the one hand

$$F[f(\rho a), f((\rho_0 - \rho)a)] = f(\rho_0 a),$$

and on the other

$$F[f(\rho a), f((\rho_0 + \omega - \rho)a)] = f((\rho_0 + \omega)a) = f(\rho_0 a).$$

Thus it follows from P_9 that $f((\rho_0 + \omega - \rho)a) = f((\rho_0 - \rho)a)$, $0 < \rho < \epsilon$. But this means that $f(\rho_0 a)$ is not the first multiple point on Γ_a against our assumption. Hence Γ_a is necessarily a simple arc.

Two orbits Γ_a and Γ_b may intersect, but if they do, they have to intersect infinitely often since $f(\rho_0 a) = f(\sigma_0 b)$ implies $f(k\rho_0 a) = f(k\sigma_0 b)$, $k = 2, 3, \cdots$. An illustration is given by the hyperbolical semi-group of sections 8.8 and 8.9; here all orbits intersect.

Let us again return to the general case. The results stated just before Lemma 25.4.3 may be given a sharper form.

LEMMA 25.4.4. *Every ultimately periodic orbit contains one and only one idempotent. Conversely, a canonical orbit containing an idempotent as an interior point is ultimately periodic.*

PROOF. Suppose that $e = f(\eta a)$ and $e_1 = f(\eta_1 a)$, $\eta_1 > \eta$, are idempotents on Γ_a where η is the least value of ρ for which $f(\rho a)$ is idempotent. Then it is easy to show that the sets $[f(\rho a); \eta \leqq \rho \leqq 2\eta]$ and $[f(\rho a); \eta_1 \leqq \rho \leqq 2\eta_1]$ form groups under the product definition having e and e_1 as their respective unit elements. Further, $f(\rho a)$ is periodic with period η for $\rho \geqq \eta$ and periodic with period η_1 for $\rho \geqq \eta_1$. Thus for $k\eta \geqq \eta_1$ both $f(k\eta a)$ and $f(\eta_1 a)$ act as unit elements on each other so that

$$e = f(k\eta a) = F[f(k\eta a), f(\eta_1 a)] = f(\eta_1 a) = e_1.$$

If η and η_1 are commensurable, there exists an ω_0 such that $\eta = n\omega_0$, $\eta_1 = m\omega_0$, $(m, n) = 1$, and a classical argument shows that $f(\rho a)$ is actually periodic for $\rho \geqq \eta$ with period ω_0. On the other hand, if η and η_1 are incommensurable, arbitrary small periods exist; and since $f(\rho a)$ is continuous this implies that

$f(\rho a) \equiv e$ for $\rho \geqq \eta$. In this case Γ_a simply stops at the idempotent. The occurrence of this second case appears doubtful.

There is another interesting possibility.

LEMMA 25.4.5. *If* $\lim_{\rho\to\infty} f(\rho a) \equiv e$ *exists, then* e *is idempotent.*

The proof is obvious. In general no unique limit exists, but there will be convergent subsequences if $[f(\rho a)]$ is compact. The following result is due to J. E. L. Peck (personal communication).

LEMMA 25.4.6. *Suppose that* $\Pi \subset E_n$ *and that* Γ_a *is a bounded canonical orbit in* Π. *Consider all sets* $[f(\rho_n a); 0 < \rho_n < \rho_{n+1}, \rho_n \to \infty]$. *The totality of limit elements of these sets forms an abelian group, the unit element of which is an idempotent associated with* Γ_a.

PROOF. Since the set $\{f(\rho_n a)\}$ is sequentially conditionally compact there exists a subsequence converging to an element g of Π. If $h = \lim_{k\to\infty} f(\sigma_k a)$, then $g \bigcirc h = h \bigcirc g$ is a limit of $\{f((\rho_j + \sigma_k)a)\}$. Further we can find an $x \in \Pi$ such that $g \bigcirc x = h$. Indeed we have merely to consider a subsequence $\{\sigma_{k_j}\}$ such that $\sigma_{k_j} > \rho_j + j$ and take the set $X \equiv \{f((\sigma_{k_j} - \rho_j)a)\}$. Since

$$F[f(\rho_j a), f((\sigma_{k_j} - \rho_j)a)] = f(\sigma_{k_j} a)$$

a subsequence of X will converge to an element x with the desired property. It follows that the limits form an abelian group G_a. If e_a is the unit element of G_a, then e_a is the desired idempotent. It is clearly unique.

In this case there is consequently an idempotent associated with every (bounded) canonical orbit. It may very well happen that all these idempotents are identical. This is evidently the case for the hyperbolical semi-group of section 8.8 where the orbits are loxodromic spirals having $z = 1$ as limit point and this is the only idempotent in the right half-plane. The opposite extreme occurs in the case of the three parameter semi-group Π_3 defined at the end of section 8.4. Here the canonical orbits are found to be straight line segments joining the origin with the points on that part of the hyperbolical paraboloid $x_1 + 1 = x_2 x_3$ which lies in Π_3. Every point on the paraboloid is an idempotent so each orbit defines a distinct idempotent.

It should be observed that the idempotents $\neq 0$ are bounded away from the origin for they are the roots of the equation

$$F(p, p) = p \tag{25.4.3}$$

and by P_6

$$| F(p, p) - p | \geqq | p | [1 - \omega(2 | p |)] > 0$$

as long as $\omega(2 | p |) < 1$.

This brings up the question of what part of Π is covered by canonical orbits. Here we note the following preliminary result.

LEMMA 25.4.7. *Given* $p_0 \in \Pi$ *then a necessary condition for the existence of an* $a \in \Pi$ *such that* $f(a) = p_0$ *is that the equations*

$$F(p_n, p_n) = p_{n-1}, \qquad n = 1, 2, 3, \cdots \tag{25.4.4}$$

have solutions $p_1, p_2, \cdots, p_n, \cdots$ *in* Π *with* $p_n \to 0$. *Conversely if* P_7 *also holds then the condition is sufficient as well as necessary and* $a = \lim_{n\to\infty} 2^n p_n$.

PROOF. The necessity is obvious for if $f(a) = p_0$ then all points $f(2^{-n}a)$ are on Γ_a and we may take $p_n = f(2^{-n}a.)$ Conversely suppose that the condition holds and that P_7 is valid. We can then proceed as in the proof of Theorem 25.3.2. Without restricting the generality we may suppose that $2 \mid p_n \mid < \tau$ for all n. Using P_6 we have then in analogy with (25.3.19) and subsequent formulas

$$2^n \mid p_n \mid \leqq \mid p_0 \mid \prod_{k=1}^{n} [1 - \tfrac{1}{2}\omega(2 \mid p_k \mid)]^{-1} \leqq \mid p_0 \mid \prod_{k=1}^{n} [1 - \tfrac{1}{2}\omega(\alpha_k)]^{-1}$$

and

$$\sum_{n=1}^{\infty} \mid 2^{n-1} p_{n-1} - 2^n p_n \mid \leqq \tfrac{1}{2} \mid p_0 \mid \sum_{n=1}^{\infty} \omega(\alpha_n) \prod_{k=1}^{n} [1 - \tfrac{1}{2}\omega(\alpha_k)]^{-1},$$

where the series on the right converges. It follows that $\lim 2^n p_n \equiv a$ exists and is an element of Π. Set $2^n p_n = a + q_n$. Then $p_0 = (p_n ; 2^n) = (2^{-n}(a + q_n); 2^n)$ and using Lemma 25.3.3 we see that

$$\mid (2^{-n}a; 2^n) - (2^{-n}(a + q_n); 2^n) \mid < \exp [B(\mid a \mid + \mid q_n \mid)] \mid q_n \mid \to 0$$

when $n \to \infty$. But $(2^{-n}a; 2^n) \to f(a)$ so that $f(a) = p_0$ as asserted.

Thus under the assumptions made above a point p of Π lies on a canonical orbit if and only if p has 2^nth roots of all orders tending to zero with $1/n$. The set of such points p is a region R in Π, the *root region*; under suitable assumptions R will contain a full neighborhood of the origin in Π. The following observation is due to R. S. Phillips.

LEMMA 25.4.8. *Let* P_7 *also hold and suppose that* K *is a compact subset of* Π *such that* (1) $K \subset [F(p, p); p \in K]$ *and* (2) $\mid F(p, p) \mid > \mid p \mid$ *for all* $p \in K \ominus \{0\}$. *Then* K *belongs to the root region of* Π.

PROOF. Given $p_0 \in K$, $p_0 \neq 0$, there exists by (1) at least one root of the equation $f(p, p) = p_0$ in K, say p_1. Further by (2) $0 < \mid p_1 \mid < \mid F(p_1, p_1) \mid = \mid p_0 \mid$. Repeating this process, we obtain a sequence $\{p_n\} \subset K$ and satisfying (25.4.4). Since $\mid p_n \mid < \mid p_{n-1} \mid$ we see that $\lim_{n\to\infty} \mid p_n \mid \equiv d$ exists. On the other hand K is compact and hence there exists a subsequence $\{p_{n_k}\}$ tending to a limit, say $q \in K$, and from $p_{n_k-1} = F(p_{n_k}, p_{n_k})$ we see that $\lim_{k\to\infty} p_{n_k-1} \equiv r$ also exists and belongs to K. Finally by continuity we have $r = F(q, q)$ and $\mid q \mid = d = \mid r \mid$. This can happen only if $q = 0 = r$; it follows that $p_n \to 0$. Lemma 25.4.7 now shows that K belongs to the root region of Π.

By restricting ourselves to the case $\Pi = \bar{E}_n^+$ we obtain the following corollary; here the points of $\bar{E}_n^+$ are partially ordered by the convention: $p < q$ if $q - p \in \bar{E}_n^+ \ominus 0$.

COROLLARY. *Let* $\Pi = \bar{E}_n^+$. *Let* $F(p, p)$ *have the following properties*: (1) *the jth coordinate of* $F(p, p)$ *vanishes with the jth coordinate of* p, $j = 1, \cdots, n$, *and* (2) *there exists a parallellopiped* $P\colon 0 \leqq p^j \leqq \pi^j$, *in which* $F(p, p) > p$, $p \neq 0$. *Then* P *belongs to the root region of* Π.

PROOF. If one introduces an equivalent norm in E_n, namely, $| p | = \sum_{j=1}^n | p^j |$, then condition (2) above implies (2) of the lemma. It is obvious that P is compact. Condition (1) of the lemma can be established from

LEMMA 25.4.9. *Let* $f_k(x)$ *be continuous functions of* $x = (x^1, \cdots, x^n)$ *in* P *such that*

$$\begin{aligned} f_k(x^1, \cdots, x^{k-1}, 0, x^{k+1}, \cdots, x^n) &< 0, \\ f_k(x^1, \cdots, x^{k-1}, \pi^k, x^{k+1}, \cdots, x^n) &> 0, \end{aligned} \qquad k = 1, 2, \cdots, n.$$

Then the equations $f_k(x) = 0$ *have at least one common root in* P.

For a proof see F. Severi and G. Scorza Dragoni [1, pp. 134–136]. The lemma is a consequence of Kronecker's existence theorem, see P. Alexandroff and H. Hopf [1, p. 470].

In order to apply this lemma we write $f_k(x) = F^k(x, x) - p_0^k$ where $F^k(x, x)$ is the kth coordinate of $F(x, x)$ and $p_0 = (p_0^1, \cdots, p_0^n)$ is an arbitrary element of $\text{Int}(P)$. On the face $x^k = 0$ the value of $f_k(x)$ is $-p_0^k < 0$ and on the opposite face $x^k = \pi^k$ the value of $f_k(x)$ is greater than or equal to $\pi^k - p_0^k > 0$ since $F(x, x) - x$ is positive in P. Thus the assumptions of Lemma 25.4.9 are satisfied and we see that $F(x, x) = p_0$ has at least one root in P. Consequently the point set $[F(p, p); p \in P]$ contains $\text{Int}(P)$ and since the continuous image of a compact set is again compact, it must contain all of P. Thus condition (1) of Lemma 25.4.8 is satisfied and the corollary is proved.

The set of all idempotents in Π, omitting the origin, will be denoted by Φ in the following. It is clear that $\Phi \cup \{0\}$ is closed relative to Π. Further, if P_6 holds, Φ lies at a positive distance from the origin, in fact it was observed after formula (25.4.3) that

$$d(0, \Phi) \geqq r_0 \quad \text{where} \quad \omega(2r_0) = 1. \tag{25.4.5}$$

This is under the assumption that $\omega(t)$ can take on values exceeding one, if not, $r_0 = +\infty$.

Among the elements of Φ figure the *fixed points* of Π. We say that p *is a fixed point of* Π *if either*

$$F(p, a) = p \quad \textit{or} \quad F(a, p) = p \tag{25.4.6}$$

for every $a \in \Pi$. An example is furnished by the hyperbolical semi-group of the right half-plane where $z = 1$ is a fixed point.

In this connection we mention that the existence of idempotents and fixed points and related questions have been examined in general topological semi-groups. For such questions we refer to the paper of A. D. Wallace [1] which contains an extensive bibliography.

2. Parametric Transformation Semi-Groups

25.5. Further assumptions; suboperative functions. We shall now consider a parameter semi-group of linear bounded transformations in the sense of Definition 8.3.4:

$$\mathfrak{T} = [T(p); p \in \Pi, T(p) \in \mathfrak{E}(\mathfrak{X})]. \tag{25.5.1}$$

We restrict ourselves to the case in which $\Pi \subset E_n$, but it is not necessary that $\Pi = \bar{E}_n^+$, in fact, for our present considerations it suffices that Π satisfies:

S_1. $\text{Int}(\Pi) \equiv \Pi_0$ *is an open simply-connected set in* E_n, *having the origin as a limit point, and* Π_0 *is closed under the product operation.*

The latter shall satisfy some of the assumptions of section 25.2. In particular we shall need P_9, but in the following much stronger form:

P_9^*. *To every compact set* K *in* $\Pi_0 \ominus \Phi$ *there is a* $\delta = \delta(K)$ *such that for* $q \in K$, $h \in S(\delta) \equiv [h; h \in \Pi_0 \cup \{0\}, |h| < \delta]$ *the equations* $F(h, p) = q$ *and* $F(s, h) = q$ *have unique solutions* $p = \psi(q, h)$ *and* $s = \chi(q, h)$ *which are continuous functions of* (q, h). *If* q *is fixed in* K, *if* H, P_q *and* S_q *are corresponding sets under the mappings* $h \to \psi(q, h)$, $h \to \chi(q, h)$, *then the measurability of one of these sets shall imply the measurability of the other two. Further, there exists a positive constant* $C = C(K)$ *such that*

$$C(K)m[H] \leqq m[P_q], m[S_q]. \tag{25.5.2}$$

We shall also need:

P_{10}. *To every* $R > 0$, $\rho > 0$, *there is a positive* $\delta = \delta(R, \rho)$ *such that for fixed* p *with* $p \in \Pi_0 \ominus \Phi$, $|p| < R$, $d(p, \Phi) > \rho$, *the mappings* $h \to F(p, h)$ *and* $h \to F(h, p)$ *are open mappings of the set* $S(\delta) \equiv [h; h \in \Pi_0 \cup \{0\}, |h| < \delta]$ *into* $\Pi_0 \ominus \Phi$.

Various assumptions will be made below concerning $T(p)$ as a function of p. The basic assumption is T_1, the other seven postulates are supplementary and are listed more or less in order of increasing restrictiveness.

T_1. $T(p) \in \mathfrak{E}(\mathfrak{X})$ *for each* $p \in \Pi$, $T(0) = I$, *and*

$$T(p \circ q) = T(p)T(q), \qquad p, q \in \Pi. \tag{25.5.3}$$

T_2. *$T(p)$ is a strongly measurable function of p in Π_0.*
T_3. *$T(p)$ is bounded on each compact subset of $\Pi_0 \ominus \Phi$.*
T_4. *$T(p)$ is continuous in the strong operator topology for $p \in \Pi_0 \ominus \Phi$.*
T_5. *$T(p)$ is continuous at the origin in the strong operator topology.*
T_6. *$T(p)$ is a uniformly measurable function of p in Π_0.*
T_7. *$T(p)$ is continuous in the uniform operator topology for $p \in \Pi_0 \ominus \Phi$.*
T_8. *$T(p)$ is continuous at the origin in the uniform operator topology.*

We shall elucidate the relationship between these various assumptions on $T(p)$ in conjunction with the semi-group property and the assumptions on the product operation. In principle we could also impose conditions of differentiability or analyticity, but such assumptions are not appropriate without similar restrictions on $F(p, q)$.

In analogy with the one-parameter case one would expect T_6 to imply T_7. In the former case the proof was based on Theorem 7.4.1 for subadditive functions. The more general situation requires a corresponding result for *suboperative functions* (cf. section 8.9).

DEFINITION 25.5.1. *A function $\varphi(p)$ on Π to real numbers is said to be suboperative with respect to the product operation if*

$$\varphi(p \bigcirc q) \leqq \varphi(p) + \varphi(q), \qquad p, q \in \Pi. \tag{25.5.4}$$

We shall prove

THEOREM 25.5.1. *A suboperative function $\varphi(p)$, defined on Π_0, which is measurable and $< +\infty$ on $\Pi_0 \ominus \Phi$, is bounded above on compact sets in $\Pi_0 \ominus \Phi$ provided S_1, P_1, P_2, P_3, and P_9^* hold.*

PROOF. The assumptions made will enable us to follow the pattern of the proof of Theorem 7.13.1. Suppose that K is a compact set in $\Pi_0 \ominus \Phi$ and that $\varphi(p)$ is not bounded above on K. Then there exists a sequence $\{p_n\} \subset K$ such that $\varphi(p_n) \geqq 2n$ and $p_n \to p_0 \in K$. Let $\delta = \delta(K)$, $C(K)$ be the positive constants associated with K by assumption P_9^*. Let ρ be so small that the sphere

$$| p - p_0 | < \rho$$

is in $\Pi_0 \ominus \Phi$. We set $S(\delta) = [h; h \in \Pi_0 \cup \{0\}, | h | < \delta]$. For $h \in S(\delta)$, $c \in K$, we can solve the equation $F(h, b) = c$ obtaining $b = \psi(c, h)$. Now this function is a continuous function of (c, h) with $c = \psi(c, 0)$ and K is compact. Consequently given $\epsilon > 0$ there exists a $\delta' \leqq \delta$ such that $| b - c | < \epsilon$ for $h \in S(\delta')$ and all $c \in K$. Without restricting the generality we may assume that

$$\delta + 2\epsilon < | p_0 |, \qquad 2\epsilon < \rho.$$

We set $S_k = [p; | p - p_0 | < k\epsilon]$, $k = 1, 2$. Then $\Pi_0 \ominus \Phi \supset S_2 \supset S_1$ and

$$S_2 \cap S(\delta) = \emptyset.$$

If $n > n(\delta)$ we have $p_n \in S_1$ and $\psi(p_n, h) \in S_2$ for each $h \in S(\delta')$. We have now

$$2n \leqq \varphi(p_n) = \varphi[h \bigcirc \psi(p_n, h)] \leqq \varphi(h) + \varphi[\psi(p_n, h)]$$

so that at least one of the two terms in the last member equals or exceeds n. Set $E_n \equiv [p; \varphi(p) \geqq n, p \in S(\delta') \cup S_2]$. This is a measurable set and (25.5.2) gives us the lower bound $C_1(K)m[S(\delta')]$ for its measure where $C_1(K) = \min [1, C(K)]$. Here $E_n \supset E_{n+1}$ for each n so that if $E = \limsup_{n\to\infty} E_n$, then E is measurable and $m[E] \geqq C_1(K)m[S(\delta')]$. Since $\varphi(p)$ must equal $+\infty$ in E, we are led to a contradiction and conclude that $\varphi(p)$ is bounded above in K.

25.6. Uniform continuity of $T(p)$. We shall now prove an analogue of Theorem 9.3.1.

THEOREM 25.6.1. *If the assumptions* S_1, P_1, P_2, P_3, P_6, *and* P_9^* *hold, then* T_6 *implies* T_7, *that is, an operator function $T(p)$ measurable in the uniform topology is continuous in that topology for $p \in \Pi_0 \ominus \Phi$.*

PROOF. We try to imitate the procedure used in the proof of Theorem 9.3.1. Consider a point p_0 and a sphere K; $|p - p_0| \leqq \rho$ located in $\Pi_0 \ominus \Phi$ and let $S(\delta)$ be defined by P_9^*. Let $h \in S(\delta)$ and $p_1 \in K$. Then

$$T(p_1) - T(p_0) = T(h)\{T[\psi(p_1, h)] - T[\psi(p_0, h)]\}.$$

This identity will be integrated with respect to h over a compact sphere $C \subset \operatorname{Int} S(\delta) \ominus \Phi$, the existence of which is assured by P_6. Since

$$\log \| T(p \bigcirc q) \| \leqq \log \| T(p) \| + \log \| T(q) \|$$

and $T(p)$ is measurable in the uniform operator topology, $\log \| T(p) \|$ is seen to be a measurable suboperative function, defined on Π_0 and $\neq +\infty$. According to the assertion of Theorem 25.5.1, $\log \| T(p) \|$ is bounded above on compact subsets of $\Pi_0 \ominus \Phi$. There is consequently a constant M such that $\| T(h) \| \leqq M$ for all $h \in C$. From

$$[T(p_1) - T(p_0)]m[C] = \int_C T(h)\{T[\psi(p_1, h)] - T[\psi(p_0, h)]\}\, dh$$

one concludes that

$$\| T(p_1) - T(p_0) \| m[C] \leqq \int_C \| T(h) \| \, \| T[\psi(p_1, h)] - T[\psi(p_0, h)] \| \, dh$$

$$\leqq M \int_C \| T[\psi(p_1, h)] - T[\psi(p_0, h)] \| \, dh,$$

provided the last integral has a sense. Since $\psi(p, h)$ is continuous in (p, h) we may suppose that $\psi(p_1, h)$ and $\psi(p_0, h)$ are both located in the sphere K. The integrand is then dominated by $2M(K)$, where $M(K)$ is the upper bound of $\| T(p) \|$ in K. It remains to prove that $T[\psi(p, h)]$ is a measurable function of h for fixed p in K, when h ranges over C. By assumption $T(q)$ is uniformly measurable for $q \in \Pi_0$. The mapping $h \to q = \psi(p, h)$, p fixed in K, carries C into a measurable set $C(p)$, which for small values of $| p - p_0 |$ lies in K, and measurable subsets correspond to measurable subsets under the inverse mapping. It follows that $T(q)$ is a uniformly measurable function of q in $C(p)$ and hence that $T[\psi(p, h)]$ is a uniformly measurable function of h in C. Thus the integral in question has a sense. Now $\psi(p_1, h) \to \psi(p_0, h)$ as $p_1 \to p_0$, uniformly with respect to h in C. In order to show that this implies that the integral tends to zero, we argue as follows. Given an $\eta > 0$ we can find a function $F(q)$ defined on K to $\mathfrak{E}(\mathfrak{X})$, continuous in the uniform operator topology, and such that

$$\int_K \| T(q) - F(q) \| \, dq < \eta.$$

Hence if

$$G(q) \equiv \| T(q) - F(q) \|$$

we have $G(q) > \eta^{1/2}$ in a set $E(\eta)$ the measure of which does not exceed $\eta^{1/2} m[K]$ and $F(q)$ may be so chosen that $G(q) \leqq 2M(K)$, independently of η, for all $q \in K$. Then

$$\int_C \| T[\psi(p_1, h)] - T[\psi(p_0, h)] \| \, dh \leqq \int_C \| T[\psi(p_1, h)] - F[\psi(p_1, h)] \| \, dh$$

$$+ \int_C \| F[\psi(p_1, h)] - F[\psi(p_0, h)] \| \, dh + \int_C \| T[\psi(p_0, h)] - F[\psi(p_0, h)] \| \, dh.$$

Here the integrand of the second integral on the right is a continuous function of h which tends to zero as $p_1 \to p_0$; its limit is consequently zero. The first and the second integrals are of the form

$$\int_C G[\psi(p, h)] \, dh, \qquad \psi(p, h) \in K,$$

and here we can use the properties of $G(q)$. By virtue of (25.5.2) the integrand exceeds $\eta^{1/2}$ in a subset of C the measure of which does not exceed $[C(K)]^{-1} m[K] \eta^{1/2}$ and its value in this set is at most $2M(K)$. It follows that such an integral has a value not exceeding $\{2M(K)[C(K)]^{-1} m[K] + m[C]\} \eta^{1/2}$ and hence

$$\limsup_{p_1 \to p_0} \int_C \| T[\psi(p_1, h)] - T[\psi(p_0, h)] \| \, dh$$

$$\leqq 2\{2M(K)[C(K)]^{-1} m[K] + m[C]\} \eta^{1/2}.$$

Here η is arbitrary so the limit must be zero and, hence,

$$\lim_{p_1 \to p_0} \| T(p_1) - T(p_0) \| = 0$$

as asserted. This completes the proof.

That continuity ordinarily does not spread to the boundary of the parameter semi-group Π_0 we know from the canonical one-parameter case. The following example shows that the fixed points of Π are also possible points of discontinuity. We take $\mathfrak{X} = l_2$, $\Pi = Z_1^+$, $z_1 \bigcirc z_2 = (z_1 + z_2)/(1 + z_1 z_2)$ with $\Phi = \{1\}$ and define

$$(25.6.1) \qquad T(z)\ \{a_n\} = \left\{ \left| \frac{z-1}{z+1} \right|^{1/n} a_n \right\}, \qquad T(1) = \Theta.$$

It is easy to show that $T(z) \in \mathfrak{E}(l_2)$ and $T(z_1 \bigcirc z_2) = T(z_1)T(z_2)$. The multiplier of a_n is <1 in Z_1^+ but tends to 1 when $n \to \infty$, $z \neq 1$, uniformly with respect to z on compact sets in $Z_1^+ \ominus \{1\}$. From this it follows that $\| T(z) \| = 1$, $z \neq 1$, and also that for $z_0 \neq 1$, $T(z) \to T(z_0)$ in the uniform operator topology when $z \to z_0$. Since $\| T(z) - T(1) \| \equiv 1$, the point $z = 1$ is obviously a point of discontinuity in the uniform topology. It should be observed, however, that $T(z)$ is continuous at $z = 1$ in the strong operator topology.

We shall now prove

THEOREM 25.6.2. If S_1, P_1, P_2, P_3, P_9^*, *and* P_{10} *hold, then* T_8 *implies* T_7, *that is, uniform continuity at the origin implies uniform continuity everywhere in* $\Pi_0 \ominus \Phi$.

PROOF. To every $\epsilon > 0$ there exists by assumption a $\delta(\epsilon) > 0$ such that $\| T(h) - I \| < \epsilon$ for $h \in S(\delta(\epsilon)) \equiv [h; h \in \Pi_0 \cup \{0\}, |h| < \delta(\epsilon)]$. We have then, for every $p \in \Pi_0 \ominus \Phi$

$$(25.6.2) \qquad \begin{gathered} \lim \| T(p) - T(p_0) \| = 0, \\ p = h \bigcirc p_0 \quad \text{or} \quad p = p_0 \bigcirc h, \qquad h \in \Pi_0,\ h \to 0. \end{gathered}$$

We say that $T(p)$ is Π-continuous in $\Pi_0 \ominus \Phi$. It is required to show that Π-continuity implies ordinary continuity in $\Pi_0 \ominus \Phi$.

We shall first prove boundedness of $T(p)$ on compact subsets of $\Pi_0 \ominus \Phi$. Let K be such a compact set, let R and ρ be such that $\max [\,|q|; q \in K] < R$ and $0 < \rho < d(\Phi, K)$, and let $\delta(K)$ and $\delta(R, \rho)$ be the quantities defined by assumptions P_9^* and P_{10} respectively. Finally set $E \equiv [p; p \in \Pi_0, |p| < R, d(p, \Phi) > \rho]$. Now by virtue of P_9^* the equation $q = F(s, h)$ has a unique solution $s = \chi(q, h)$ for $q \in K$ and $h \in \Pi_0 \cup \{0\}$, $|h| < \delta(K)$. Since $\chi(q, h)$ is a continuous function of (q, h) with $\chi(q, 0) = q$ and since K is compact, there exists a $\delta(E)$ such that $\chi(q, h) \in E$ for all $q \in K$ and $h \in \Pi_0$ with $|h| < \delta(E)$. Next set $\delta = \min [\delta(\epsilon), \delta(K), \delta(R, \rho), \delta(E)]$ and let $S(\delta) \equiv [h; h \in \Pi_0 \cup \{0\}, |h| < \delta]$. With each point p of E we associate the set $S(p, \delta) \equiv [q; q = F(p, h), h \in \text{Int } S(\delta)]$. Because of P_{10} each set $S(p, \delta)$ is open and contained in $\Pi_0 \ominus \Phi$. It is claimed that the family of sets $[S(p, \delta); p \in E]$ provides a covering of K. For if $q_0 \in K$ and $h_0 \in \text{Int } S(\delta) \subset S(\delta(K)) \cap S(\delta(E))$, then $s_0 = \chi(q_0, h_0) \in E$ and $q_0 = F(s_0, h_0) \in S(s_0, \delta)$. K being compact, there is a finite subcovering of K, say by the sets $E_1, \cdots, E_m$, where $E_j = S(p_j, \delta)$. Hence for every $q \in K$ there is a j such that $\| T(q) - T(p_j) \| \leqq \epsilon \| T(p_j) \|$. Setting $M = (1 + \epsilon) \max. \| T(p_j) \|$, we see that $\| T(q) \| \leqq M$ for all $q \in K$ so that $\| T(q) \|$ is bounded on compact sets.

Next given any point $p_0 \in K$ we wish to show that $T(p)$ is continuous at $p = p_0$. Now $S(p_0, \delta)$ contains a sphere, $|q - q_0| < \sigma$ say, and since $q_0 \in S(p_0, \delta)$, there exists an $h_0 \in \text{Int } S(\delta)$ such that $q_0 = F(p_0, h_0)$. Further, $F(p, h)$ being continuous in (p, h), there exists an $\eta > 0$ such that $|F(p, h) - F(p_0, h_0)| < \sigma$ for $|p - p_0| < \eta$ and $|h - h_0| < \eta$, $h \in S(\delta)$. Now take any $p \in K$ such that $|p - p_0| < \eta$ and consider the set $S(p, \delta)$. If $h \in \text{Int } S(\delta)$ and $|h - h_0| < \eta$, then $|F(p, h) - q_0| < \sigma$ so that the point $F(p, h) \in S(p, \delta) \cap S(p_0, \delta)$. Consequently there is an $h_1 \in \text{Int } S(\delta)$ such that $F(p, h) = F(p_0, h_1) \equiv q_1$. We have then

$$\begin{aligned} \| T(p) - T(p_0) \| &\leqq \| T(p) - T(q_1) \| + \| T(p_0) - T(q_1) \| \\ &= \| T(p) - T(p \circ h) \| + \| T(p_0) - T(p_0 \circ h_1) \| \\ &\leqq \| T(p) \| \, \| T(h) - I \| + \| T(p_0) \| \, \| T(h_1) - I \| \leqq 2M\epsilon \end{aligned}$$

and this proves the assertion.

THEOREM 25.6.3. *Let us assume* S_1, P_1, P_2, P_3, T_1, *and* T_8. *Let* $\mathfrak{T}_1 \subset \mathfrak{T}$ *be the subset of all elements* $T(p)$ *of the form*

$$T(p) = \prod_{k=1}^{m} T(h_k), \qquad h_k \in S(\delta), \tag{25.6.3}$$

where δ *is so small that* $\| T(h) - I \| < 1$ *for* h *in* $S(\delta)$. *Then* $\mathfrak{T}_1$ *is also a semi-group, and* $\mathfrak{T}_1$ *may be embedded in a group* $\mathfrak{G}_1$.

PROOF. Since $\| T(h) - I \| < 1$, Theorem 4.3.1 shows that $T(h)$ has an inverse in $\mathfrak{E}(\mathfrak{X})$ and hence every $T(p)$ which is the product of a finite number of operators $T(h_k)$ will also have an inverse. It is clear that $\mathfrak{T}_1$ is a sub-semi-group of $\mathfrak{T}$. We obtain $\mathfrak{G}_1$ by adjoining all inverses and forming all possible finite products.

If $p \in \Phi$, that is, if p is an idempotent, $p \circ p = p$, and, if $T(p)$ is defined, we have also $[T(p)]^2 = T(p)$ so that $T(p)$ is an idempotent in $\mathfrak{E}(\mathfrak{X})$. If $T(p) \neq I$ we cannot embed $T(p)$ in a group. In general we cannot expect to be able to embed all of $\mathfrak{T}$ in a group when T_8 holds owing to the idempotents. It may happen that the set Φ forms a barrier separating Π_0 into disjoint components. In such a case one would not in general expect to be able to embed that part of $\mathfrak{T}$ which corresponds to a component of Π_0 separated from the origin by Φ.

25.7. Strong continuity of $T(p)$. We now turn to the case in which the basic assumption is T_2, that is $T(p)$ is strongly measurable on $\Pi_0 \ominus \Phi$. We start by proving a result analogous to Lemma 10.2.1. Cf. I. Miyadera [**1**, Theorem 4] where the case $\Pi_0 = E_1^+$ is discussed.

LEMMA 25.7.1. *If* S_1, P_1, P_2, P_3, *and* P_9^* *hold, then* T_2 *implies* T_3.

PROOF. We apply the argument used in proving Lemma 10.2.1. By assumption $\| T(p)x \|$ is a measurable function of p in $\Pi_0 \ominus \Phi$ for each $x \in \mathfrak{X}$. Given a compact set K in $\Pi_0 \ominus \Phi$, the uniform boundedness theorem enables us to conclude that $\| T(p) \|$ is bounded on K if it can be shown that $\sup_{p\in K} \| T(p)x \|$ is finite for each $x \in \mathfrak{X}$. Suppose this were not so. We could then find an $x \in \mathfrak{X}$ and a sequence $\{p_n\}$, $p_n \in K$, $p_n \to p_0 \in K$, such that $\| T(p_n)x \| \geqq n$. Let $S(\delta) = S(\delta(K))$ where $\delta(K)$ is defined by P_9^*. To each p_n and each $h \in S(\delta)$ there is a unique $\chi(p_n, h)$ such that $p_n = F[\chi(p_n, h), h]$. Since $\| T(p)x \|$ is Lebesgue measurable, there exists a measurable subset S of $S(\delta)$ and a finite M such that $m[S] > \frac{1}{2}m[S(\delta)]$ and $\| T(p)x \| \leqq M$ for all $p \in S$. Let $E_n \equiv [q;\, q = \chi(p_n, h),\, h \in S]$; then E_n is measurable and, by (25.5.2), we have $m[E_n] \geqq C(K)m[S] \geqq \frac{1}{2}C(K)m[S(\delta)]$. Since $\chi(p, h)$ is continuous in (p, h) with $\chi(p, 0) = p$, we may suppose, without loss of generality, that δ is so small that the E_n are contained in a bounded subset of $\Pi_0 \ominus \Phi$. Now if $h \in S$ we have

$$n \leqq \| T(p_n)x \| \leqq \| T[\chi(p_n, h)] \| \, \| T(h)x \| \leqq \| T[\chi(p_n, h)] \| M$$

so that $\| T[\chi(p_n, h)] \| \geqq n/M$. Let us now consider the set $F_n \equiv \bigcup_{k>n} E_k$. It is clear that F_n has a finite measure exceeding $\frac{1}{2}C(K)m[S(\delta)]$ and that $\| T(q) \| \geqq n/M$ for all $q \in F_n$. Since $F_n \supset F_{n+1}$ the usual argument shows that $\| T(q) \|$ must be infinite on a subset of $\Pi_0 \ominus \Phi$ of positive measure. This contradicts the fact that $\| T(q) \|$ is finite valued in $\Pi_0 \ominus \Phi$ and thus implies the statement of the lemma.

We can now prove that strong measurability implies strong continuity.

THEOREM 25.7.1. *If* S_1, P_1, P_2, P_3, P_6, *and* P_9^* *hold, then* T_2 *implies* T_4.

PROOF. We can proceed essentially as in the proof of Theorem 25.6.1, from which we take our notation. Since

$$T(p_1)x - T(p_0)x = T(h)\{T[\psi(p_1, h)]x - T[\psi(p_0, h)]x\},$$

we get

$$\| T(p_1)x - T(p_0)x \| \, m[C] \leqq M(C) \int_C \| T[\psi(p_1, h)]x - T[\psi(p_0, h)]x \| \, dh$$

where C is a compact sphere in Int $S(\delta) \ominus \Phi$ and $M(C)$ is the supremum of $\| T(h) \|$ in C which is finite by Lemma 25.7.1. The integrand is bounded by $2M(K) \| x \|$ where $M(K)$ is the supremum of $\| T(q) \|$ on K. One proves next that $T[\psi(p, h)]x$ is a measurable function of h in C for fixed p in K because the mapping $h \to \psi(p, h)$ defines a correspondence between measurable sets. Further one can find a function $F(q, x)$ defined in K to $\mathfrak{X}$, continuous in q, and such that, if $G(q, x) \equiv \| T(q)x - F(q, x) \|$, then $\int_K G(q, x)\, dq < \eta$, a preassigned arbitrarily small positive number, $G(q, x) \leqq 2M(K) \| x \|$ for all $q \in K$, and $G(q, x) > \eta^{1/2}$ in a set of measure $< \eta^{1/2} m[K]$. The remaining details may be left to the reader.

THEOREM 25.7.2. *If* S_1, P_1, P_2, P_3, P_9^*, *and* P_{10} *hold, then* T_5 *implies* T_4, *that is, strong continuity at the origin implies strong continuity everywhere in* $\Pi_0 \ominus \Phi$.

PROOF. Here we follow the pattern of Theorem 25.6.2. For any given $x \in \mathfrak{X}$

and $\epsilon > 0$ there exists a $\delta = \delta(\epsilon, x)$ and for $h \in S(\delta)$ we have $\| T(h) x - x \| < \epsilon$. We have consequently strong one-sided Π-continuity everywhere in $\Pi_0 \ominus \Phi$, that is

$$\lim \| T(p)x - T(p_0)x \| = 0, \qquad p = p_0 \circ h,\ h \in \Pi_0,\ h \to 0.$$

We first show that Π-continuity implies that $\| T(p) \|$ is bounded on compact subsets of $\Pi_0 \ominus \Phi$. It suffices for this purpose to show for each $x \in \mathfrak{X}$ and each compact set $K \subset \Pi_0 \ominus \Phi$ that the supremum of $\| T(q)x \|$ on K is finite. As in the uniform case we can find a finite covering of K, depending now on x, by open sets $S[p_j(x), \delta(x)]$ such that for each $q \in K$ there is a j with $\| T(q)x - T[p_j(x)]x \| < \epsilon \| T[p_j(x)] \|$. Hence $\| T(q)x \| \leqq M(K, x)$ in K. The uniform boundedness theorem then implies that $\| T(q) \| \leqq M(K)$ for all $q \in K$. The rest of the proof also proceeds as in the uniform case and we get

$$\begin{aligned} \| T(p)x - T(p_0)x \| &\leqq \| T(p) \| \, \| T(h)x - x \| + \| T(p_0) \| \, \| T(h_1)x - x \| \\ &\leqq 2M(K)\epsilon. \end{aligned}$$

This proves the theorem.

3. The Lie Theory

25.8. The structural constants. We shall now turn to the infinitesimal properties of $[T(p); p \in \Pi]$. Here we shall require the existence of canonical orbits and strong continuity of $T(p)$ at the origin. For most purposes we have to restrict ourselves to the case in which $\Pi \subset E_n$ and it will simplify matters to assume $\Pi = \bar{E}_n^+$. It is convenient at this point to introduce still another variant of P_9, namely,

P_9°. *There is a $d_0 > 0$ such that to every compact set K in $\Pi_0 \ominus \Phi$ with $d(0, K) < d_0$ there corresponds a $\delta = \delta(K)$, and for $q \in K$, $h \in S(\delta) \equiv [h; h \in \Pi, | h | < \delta]$ the equations $F(h, p) = q$ and $F(s, h) = q$ have unique solutions $p = \psi(q, h)$ and $s = \chi(q, h)$, which are continuous functions of (q, h). If q is fixed in K, if H, P_q, and S_q are corresponding sets under the mappings $h \to \psi(q, h)$, $h \to \chi(q, h)$, then the measurability of one of these sets shall imply the measurability of the other two. Further, there exists a positive constant $C = C(K)$ such that*

$$C(K)m(H) \leqq m[P_q], m[S_q].$$

Since we are interested in differential properties of $T(p)$, it is necessary to impose stronger differentiability properties on $F(p, q)$ than we have used before. Hence in addition to P_1, P_2, P_3, we shall also assume throughout this paragraph the condition

P_{11}. *At every point of $\bar{E}_n^+ \times \bar{E}_n^+$ the n coordinates of $F(p, q)$ have continuous partial derivatives with respect to the coordinates of p and q up to and including the third order.*

By virtue of this assumption we have a Taylor expansion with remainder in the neighborhood of every point of the product space including the origin where one-sided derivatives exist. The expansion at the origin has a very particular form owing to the restrictions imposed on $F(p, q)$ by assumptions P_2 and P_3. To express the results more succinctly we shall use the compact notation for derivatives introduced in section 25.4.

LEMMA 25.8.1. *For $i, j, k, l = 1, 2, \cdots, n$, we have at the origin*

$$(F^i_{j;})_{00} = (F^i_{;j})_{00} = \delta^i_j, \qquad (F^i_{jk;})_{00} = (F^i_{;jk})_{00} = 0, \tag{25.8.1}$$
$$(F^i_{jkl;})_{00} = (F^i_{;jkl})_{00} = 0.$$

PROOF. One differentiates the identities

$$F^i(a, 0) = a^i, \qquad F^i(0, b) = b^i$$

and passes to the limit with a or b.

Writing

$$(F^i_{j;k})_{00} = \alpha^i_{jk} \tag{25.8.2}$$

we obtain

$$F^i(a, b) = a^i + b^i + \sum \alpha^i_{jk} a^j b^k + R_3 \tag{25.8.3}$$

where the remainder is $O(\rho^3)$ as $\rho \to 0$, $\rho = \max(|a|, |b|)$. Postulate P_3 imposes certain relations between the second and third order derivatives.

LEMMA 25.8.2. *For $i, j, k, = 1, \cdots, n$, we have at the origin*

$$(F^i_{j;kl})_{00} + \sum_{\mu=1}^n \alpha^i_{j\mu}\alpha^\mu_{kl} = (F^i_{jk;l})_{00} + \sum_{\mu=1}^n \alpha^i_{\mu l}\alpha^\mu_{jk}. \tag{25.8.4}$$

PROOF. One starts from the identity

$$F[a, F(b, c)] = F[F(a, b), c]$$

and differentiates the ith component with respect to a^j, b^k, c^l. In the result one lets $a, b, c \to 0$ and uses (25.8.1). The result is (25.8.4) and turns out to be independent of the order of differentiation. We shall sketch the steps for the case in which the differentiations are taken in lexicographical order. The results of the three differentiations are respectively

$$F^i_{j;}(\cdot) = F^i_{\mu;}(\cdot)F^\mu_{j;}(a, b),$$
$$F^i_{j;\mu}(\cdot)F^\mu_{k;}(b, c) = F^i_{\mu\nu;}(\cdot)F^\mu_{j;}(a, b)F^\nu_{;k}(a, b) + F^i_{\mu;}(\cdot)F^\mu_{j;k}(a, b),$$
$$F^i_{j;\mu\nu}(\cdot)F^\mu_{k;}(b, c)F^\nu_{;l}(b, c) + F^i_{j;\mu}(\cdot)F^\mu_{k;l}(b, c)$$
$$= F^i_{\mu\nu;l}(\cdot)F^\mu_{j;}(a, b)F^\nu_{;k}(a, b) + F^i_{\mu;l}(\cdot)F^\mu_{j;k}(a, b).$$

Here a dot indicates a function of all three variables and we have used the summation convention. Passing to the limit with a, b, c and using (25.8.1) we obtain formula (25.8.4).

In accordance with the classical theory of Lie groups we define

$$\gamma^i_{jk} = \alpha^i_{jk} - \alpha^i_{kj} \tag{25.8.5}$$

as the *structural constants* of Π. These constants satisfy certain relations which constitute the necessary part of Lie's Third Fundamental Theorem [1, p. 396]:

THEOREM 25.8.1. *For all $i, j, k, l = 1, 2, \cdots, n$, we have*

$$\gamma^i_{kj} = -\gamma^i_{jk}, \tag{25.8.6}$$

$$\sum_{\mu=1}^{n} [\gamma^i_{j\mu}\gamma^\mu_{kl} + \gamma^i_{k\mu}\gamma^\mu_{lj} + \gamma^i_{l\mu}\gamma^\mu_{jk}] = 0. \tag{25.8.7}$$

PROOF. The first relation follows from the definition. In order to obtain the second one we use (25.8.4); here we apply the six operations of the permutation group on three letters to the subscripts j, k, l and multiply the result by $(-1)^t$ where t is the number of transpositions. When the six equations are added, the third order derivatives cancel, since $(F^i_{jk;l})_{00} = (F^i_{kj;l})_{00}$ and $(F^i_{j;kl})_{00} = (F^i_{j;lk})_{00}$, and the remaining products of second order derivatives give (25.8.7) when the terms are collected.

According to Lie these necessary conditions are also sufficient: given a system of n^3 constants γ^i_{jk} satisfying the relations (25.8.6) and (25.8.7) there exists a group (germ) of which they are the structural constants. In our case this would mean that Π can be extended to a group germ Γ containing a full neighborhood of the origin. We shall make no use of this proposition.

25.9. The infinitesimal generators. We now turn back to the canonical function $f(p)$ of Π introduced in section 25.3. Since $f(\rho a) \bigcirc f(\sigma a) = f((\rho + \sigma)a)$, assumption T_1 gives:

THEOREM 25.9.1. *Under the assumptions of Theorem 25.3.1 together with T_1, the set $\mathfrak{T}_a \equiv [T[f(\rho a)]; 0 < \rho < \infty]$ for $a \in \Pi$ is a canonical one-parameter sub-semi-group of $\mathfrak{T}$.*

If $\mathfrak{T}_a$ is continuous in the strong operator topology as a function of ρ, $\rho > 0$, then the discussion in section 10.3 is applicable. The operator $A_o(a)$ defined by

$$\lim_{\eta\to 0+} \eta^{-1}\{T[f(\eta a)]x - x\} \equiv A_o(a)x, \tag{25.9.1}$$

whenever the limit exists, is the infinitesimal operator of the semi-group $\mathfrak{T}_a$; the domain of $A_o(a)$ is denoted by $\mathfrak{D}_o(a)$. The smallest closed linear extension of $A_o(a)$, if it exists, is called the infinitesimal generator of $\mathfrak{T}_a$. It is denoted by $A(a)$ and its domain by $\mathfrak{D}(a)$. As a direct consequence of Theorem 10.5.5 we have

THEOREM 25.9.2. *Under the assumptions of Theorem 25.3.1 together with* T_5, $\mathfrak{T}_a$ *is a semi-group of class* (C_0) *with infinitesimal generator* $\dot{A}(a) = A_o(a)$.

If T_5 holds then every element of the form

$$y \equiv \int_\alpha^\beta T[f(\rho a)]x \, d\rho, \qquad x \in \mathfrak{X}, \tag{25.9.2}$$

belongs to $\mathfrak{D}(a)$, and these elements are dense in $\mathfrak{X}$. However the big problem before us is to find a set of elements, dense in $\mathfrak{X}$, which belongs to every domain $\mathfrak{D}(a)$. This can be done by a suitable modification of constructions used by N. Dunford [4], L. Gårding [1], and I. Gelfand [3] for similar purposes. Cf. Theorem 10.3.4. We can carry through the construction only in E_n, however. We shall consider elements of the form

$$y = K[x] \equiv \int_D K(p)T(p)x \, dp, \qquad x \in \mathfrak{X}. \tag{25.9.3}$$

Here $K(p)$ is a numerical valued kernel with the following properties.

(i) $K(p)$ *is defined in* $\bar{E}_n^+$ *and equals zero for p outside of a domain D, where* $\bar{D} \subset \text{Int } \bar{E}_n^+$ *and D lies within a sufficiently small distance* $d(\Pi)$ *of the origin;*

(ii) $K(p)$ *is of class* $C^{(m)}$, $m \geqq 3$;

(iii) $K(p)$ *and its existing derivatives are integrable over D;*

(iv) If D_η *is a homeomorphic image of D where corresponding points are less than* η *apart, and if* $N(p)$ *stands for* $K(p)$ *or any one of its partial derivatives, then*

$$\int_\Delta |N(p)| \, dp = o(\eta), \qquad \Delta = D \ominus D \cap D_\eta. \tag{25.9.4}$$

We denote the class of all such kernels by $\mathfrak{K}$ and let $\mathfrak{K}[\mathfrak{X}]$ symbolize the set of all elements of the form (25.9.3) with $K(p) \in \mathfrak{K}$. Here $\mathfrak{K}$ depends on the topological semi-group Π and $\mathfrak{K}[\mathfrak{X}]$ depends both on Π and on $[T(p)]$.

For suitable choices of $\{a^j\}$, $\{b^j\}$ the following special kernel will be in $\mathfrak{K}$. We take D to be the cube $a^j < p^j < b^j$, $b^j - a^j = \delta$, $j = 1, \cdots, n$, and define

$$K(p) = K(p; a, b) = (C\delta)^{-n} \exp\left\{- \delta^2 \sum_1^n [(b^j - p^j)(p^j - a^j)]^{-1}\right\}, \tag{25.9.5}$$

where $p \in D$ and

$$C = \int_0^1 \exp\{- [t(1 - t)]^{-1}\} \, dt.$$

Condition (i) requires merely that $a^j > 0$ and that $|b| < d(\Pi)$. Conditions (ii) and (iii) are obviously satisfied and so is (iv) since the integral of $K(p)$ over Δ is dominated by

$$2nC^{-1} \int_0^{\eta/\delta} e^{-1/t} \, dt < (2n\eta/C\delta)e^{-\delta/\eta} \qquad \text{if } 2\eta < \delta,$$

and similar estimates hold for the derivatives. Thus $K(p; a, b) \in \mathfrak{K}$.

THEOREM 25.9.3. *Under the assumptions* S_0, P_1, P_2, P_3, P_5, P_6, P_9°, P_{11}, T_1, *and* T_5 *the set* $[\bigcap_a \mathfrak{D}(a)] \cap [\bigcap_{a,b} \mathfrak{D}[A(a)A(b)]]$ *is dense in* $\mathfrak{X}$ *and contains the set* $\mathfrak{K}[\mathfrak{X}]$.

PROOF. It is convenient to let $S(\delta)$ denote the set $[h; h \in \Pi, |h| < \delta]$, and this we shall do throughout this section.

Our first objective is to show that various mappings defined by $q = F(h, p)$ are well behaved near the origin by virtue of P_2, P_3, P_9°, and P_{11}. It follows from (25.8.1) that $F^i_{;k}(h, p) \to \delta^i_k$ and $F^i_{k;}(h, p) \to \delta^i_k$ as $h, p \to 0$. Consequently there exists a $d_1 > 0$ such that

$$\det [F^i_{;k}(h, p)] \geqq \tfrac{1}{2} \quad \text{and} \quad \det [F^i_{k;}(h, p)] \geqq \tfrac{1}{2} \tag{25.9.6}$$

for $|h|, |p| < d_1$. Since P_6 is also assumed, the relation (25.4.5) shows that $d_2 \equiv d(0, \Phi) > 0$. We now set

$$d \equiv d(\Pi) = \tfrac{1}{2} \min [d_0, d_1, d_2],$$

where d_0 is determined by the condition P_9°; this is the constant $d(\Pi)$ stipulated in the definition of $\mathfrak{K}$.

The integral (25.9.3) will exist for any choice of $K(p)$ in $\mathfrak{K}$ if it can be shown that $T(p)$ is continuous in the strong operator topology for all $p \in \text{Int } S(2d)$. This in turn may be established following the argument used in the proof of Theorem 25.7.2, if it can be shown that P_9^* and P_{10} are valid in $S(2d)$. Now P_9° assures us that P_9^* holds in $S(2d)$. On the other hand it follows from (25.9.6) that the mappings $h \to F(p, h)$ and $h \to F(h, p)$, with p fixed, are three times continuously differentiable, having nonsingular Jacobians for $h, p \in S(2d)$, and hence that these mappings are local homeomorphisms. Thus given any $R < 2d$ and $\rho > 0$, if we choose $\delta(R, \rho)$ so that $|F(p, h)|, |F(h, p)| < 2d$ for $|h| < \delta(R, \rho)$ and $|p| \leqq R$, then these maps are open and take Int $S[\delta(R, \rho)]$ into $\Pi_0 \ominus \Phi$; the conditions of P_{10} are therefore satisfied in $S(2d)$.

The set $\mathfrak{K}[\mathfrak{X}]$ being well defined, we note that it is dense in $\mathfrak{X}$. In fact the subset of $\mathfrak{K}[\mathfrak{X}]$ which is based on the special kernels (25.9.5) is itself dense in $\mathfrak{X}$ since for $p_0 \in \text{Int } S(d)$, $h = (\delta, \cdots, \delta)$, and $x \in \mathfrak{X}$ we have

$$\lim_{p_0 \to 0} \lim_{\delta \to 0+} \int_D K(p; p_0, p_0 + h)T(p)x \, dp = \lim_{p_0 \to 0} T(p_0)x = x. \tag{25.9.7}$$

Let us now take an arbitrary $K(p) \in \mathfrak{K}$ and form the difference quotient

$$\eta^{-1}\{T[f(\eta a)]y - y\} = \eta^{-1} \int_D K(p)\{T[F(f(\eta a), p)]x - T(p)x\} \, dp. \tag{25.9.8}$$

In order to find the limit of this expression as $\eta \to 0+$ we have first to study the mapping H_η defined by $p \to q = F(f(\eta a), p)$ which maps $\bar{D} \subset \text{Int } S(d)$ onto a set $\bar{D}_\eta$. By (25.3.17) $|f(\eta a)| = O(\eta)$ and by P_5

$$|F(f(\eta a), p) - p| < (1 + B|p|)|f(\eta a)| < M\eta,$$

where M depends only upon B and D. It follows that the mapping H_η involves only small distortions so that $\bar{D}_\eta \subset \text{Int } S(d)$ for η sufficiently small. To show that H_η is a homeomorphism we again apply P_9°, this time to the set $\bar{D}_\eta \subset \text{Int } S(d)$. Let η_0 be chosen so that $| f(\eta a) | < \delta(\bar{D}_\eta)$ for $0 < \eta < \eta_0$, where $\delta(\bar{D}_\eta)$ is given by P_9°. Then this condition implies that the map $p \to q = F(f(\eta a), p)$ maps $\bar{D}$ onto $\bar{D}_\eta$ in a one-to-one fashion. Thus H_η is a homeomorphism. It should be noted that $f(\rho a)$ now satisfies the system (25.4.2) and is consequently twice differentiable with respect to ρ. By (25.9.6) the two Jacobians

$$(25.9.9)\qquad \begin{aligned} J(q; p) &= \det [F^i_{;k}(f(\eta a), p)], \\ J(p; q) &= \det [\psi^i_{k;}(q, f(\eta a))] = [J(q; p)]^{-1} \end{aligned}$$

are bounded away from zero and infinity; they are moreover twice continuously differentiable with respect to η.

We can now return to equation (25.9.8), the right member of which may be written as

$$(25.9.10)\qquad \begin{aligned} &\eta^{-1} \int_{D_\eta} K(p)J(p; q)T(q)x\, dq - \eta^{-1} \int_D K(q)T(q)x\, dq \\ &= \int_{D_1} \eta^{-1}[K(p)J(p; q) - K(q)]T(q)x\, dq \\ &\quad + \eta^{-1} \int_{D_2} K(p)J(p; q)T(q)x\, dq - \eta^{-1} \int_{D_3} K(q)T(q)x\, dq \\ &\equiv J_1 + J_2 + J_3, \end{aligned}$$

where $D_1 = D \cap D_\eta$, $D_2 = D_\eta \ominus D \cap D_\eta$, $D_3 = D \ominus D \cap D_\eta$. Now $\| T(q) \|$ has a finite upper bound on $D \cup D_\eta$, M say. This follows from the strong continuity of $T(q)$ on compact subsets of Int $S(2d)$. Consequently $\| J_2 \|$ and $\| J_3 \|$ do not exceed $M \| x \|$ times

$$\eta^{-1} \int_{D_2} | K(p) | J(p; q)\, dq \qquad \text{and} \qquad \eta^{-1} \int_{D_3} | K(q) |\, dq$$

respectively. Here the second expression obviously tends to zero with η by (25.9.4), since H_η is a homeomorphism and the distortion is $O(\eta)$. Reverting to the variable p in the first expression, we obtain $\int | K(p) |\, dp$ extended over the inverse image of D_2 to which (25.9.4) also applies. Hence $\| J_2 \|$ and $\| J_3 \| \to 0$ with η.

In J_1 we substitute $p = \psi(q, f(\eta a))$ and pass to the limit as $\eta \to 0+$. This is permissible by virtue of the differentiability properties of the functions involved. Noting that $D_1 \to D$ we see that $y \in \mathfrak{D}(a)$ and

$$(25.9.11)\qquad \begin{aligned} A(a)y &= \int_D K_1(q; a)T(q)x\, dq, \\ K_1(q; a) &= \frac{\partial}{\partial \eta} \Big\{ K[\psi(q, f(\eta a))]\ J[\psi(q, f(\eta a)); q] \Big\}_{\eta=0}. \end{aligned}$$

It is clear that $K_1(q; a)$ is a linear combination of $K(q)$ and its n first order partials with coefficients which are bounded, twice differentiable functions of q. The coefficients depend also upon a and are uniformly bounded with respect to a, $|a| \leqq 1$. It follows that $A(a)K \in \mathfrak{E}[\mathfrak{X}]$ and

$$\| A(a)K \| \leqq \int_D | K_1(q; a) | \, \| T(q) \| \, dq.$$

We have now to consider the existence of $A(b)A(a)y$. Here $A(a)y$ is given by (25.9.11) and this integral is of type (25.9.3) with $K(p)$ replaced by $K_1(p; a)$. The properties of $K_1(p; a)$ are essentially the same as those of $K(p)$ except for differentiability, but $K_1(p; a) \in C^{(2)}$ at least and this suffices for our needs. The argument given above may consequently be used also to prove the existence of $A(b)A(a)y$. Further we see that $A(b)A(a)K \in \mathfrak{E}(\mathfrak{X})$ and its norm is a bounded function of a and b on the unit sphere in $\bar{E}_n^+$. This completes the proof.

We now define an auxiliary operator $U(a)$ by

$$U(a)x = \lim_{\eta \to 0+} \eta^{-1}[T(\eta a)x - x] \tag{25.9.12}$$

whenever the limit exists.

THEOREM 25.9.4. *Under the assumptions of Theorem 25.9.3, $U(a)y$ exists and equals $A(a)y$ if $y \in \mathfrak{R}[\mathfrak{X}]$.*

PROOF. The existence of $U(a)y$ is proved by the argument used in the proof of Theorem 25.9.3. We have merely to replace $f(\eta a)$ by ηa throughout. Since $f(\eta a)$ is twice differentiable with respect to η, we have $f(\eta a) = \eta a + O(\eta^2)$. Hence

$$\psi(q, f(\eta a)) - \psi(q, \eta a) = O(\eta^2),$$

$$J[\psi(q, f(\eta a); q] - J[\psi(q, \eta a); q] = O(\eta^2),$$

and these relations may be differentiated with respect to η, the derivatives of the right hand members being $O(\eta)$. It follows that the passage to the limit gives the same result in both cases so that $U(a)y = A(a)y$ for $y \in \mathfrak{R}[\mathfrak{X}]$.

THEOREM 25.9.5. *Let $K(p) \in \mathfrak{R}$ be given and set $y = K[x]$. Under the assumptions of Theorem 25.9.3 there exists a $\rho > 0$ and a constant C depending on the kernel K such that for $p_1, p_2 \in S(\rho)$ we have*

$$\| T(p_1)y - T(p_2)y \| \leqq C \, | p_1 - p_2 | \, \| x \| \tag{25.9.13}$$

and the same inequality holds when y is replaced by $A(a)y$, $|a| \leqq 1$.

PROOF. Proceeding as in the proof of Theorem 25.9.3 we start with

$$T(p_1)y - T(p_2)y = \int_D K(p)T[F(p_1, p)]x \, dp - \int_D K(p)T[F(p_2, p)]x \, dp.$$

We then choose $\rho > 0$ so that the mapping $p \to F(h, p)$ is a homeomorphism for each $h \in S(\rho)$, $p \in \bar{D}$ by essentially the same argument as given above for the mapping H_η. It therefore follows that if $p_1, p_2 \in S(\rho)$, then the mapping $F(p_1, p) \to F(p_2, p)$, p ranging over $\bar{D}$, is also a homeomorphism and by P_5

$$| F(p_1, p) - F(p_2, p) | < (1 + B | p |) | p_1 - p_2 | \equiv M | p_1 - p_2 |.$$

We now set $q_1 = F(p_1, p)$, $q_2 = F(p_2, p)$ and introduce q_1, q_2 as new variables of integration. In the first integral $p = \psi(q_1, p_1) \equiv \psi_1$, whereas in the second $p = \psi(q_2, p_2) \equiv \psi_2$. The result then is

$$T(p_1)y - T(p_2)y = \int_{D_1} K(\psi_1)J(\psi_1 ; q)T(q_1)x\, dq_1 - \int_{D_2} K(\psi_2)J(\psi_2 ; q_2)T(q_2)x\, dq_2,$$

where $D_k = p_k \circ D$, $k = 1, 2$. Here D_2 is a homeomorphic image of D_1 involving displacements less then a constant times $| p_1 - p_2 |$. We can then write each integral as the sum of three: one over $D_0 = D_1 \cap D_2$ plus or minus integrals over that part of D_1 which is not in D_2 or vice versa. The two integrals over D_0 we combine into a single one:

$$J \equiv \int_{D_0} [K(\psi_1)J(\psi_1 ; s) - K(\psi_2)J(\psi_2 ; s)]T(s)x\, ds.$$

Here $K(p) \in C^{(3)}$ and will consequently satisfy a Lipschitz condition on compact sets. Thus

$$| K(\psi_1) - K(\psi_2) | \leqq M_1 | \psi_1 - \psi_2 | \leqq M_2 | p_1 - p_2 |.$$

Similarly

$$| J(\psi_1 ; s) - J(\psi_2 ; s) | \leqq M_3 | \psi_1 - \psi_2 | \leqq M_4 | p_1 - p_2 |.$$

Finally $\| T(s) \| \leqq M_5$ for s in $S(\rho) \circ \bar{D}$ with obvious notation. The residual integrals are all of a type to which (25.9.4) applies; *a fortiori* each has a norm not exceeding a constant multiple of $| p_1 - p_2 | \| x \|$. Combining these estimates we see that (25.9.13) holds. To get the corresponding inequality for $A(a)y$ we have merely to replace $K(p)$ by $K_1(p; a)$ in the preceding argument. Since $K_1(p; a) \in C^{(2)}$ we arrive at the same result with, possibly, a larger constant C. The final constant in (25.9.13) will then be the larger of the two constants involved. It should be observed that this constant is a bounded function of a in the unit sphere since $\| K_1(\cdot; a) \|_\infty$ has this property. This completes the proof.

25.10. The fundamental theorems. We recall that to every $a \in \Pi$ there is an infinitesimal generator $A(a)$ with domain $\mathfrak{D}(a)$ dense in $\mathfrak{X}$ and containing $\mathfrak{R}[\mathfrak{X}]$. We shall study the correspondence between a and $A(a)$. In a certain sense it is a homomorphism.

THEOREM 25.10.1. *Under the assumptions of Theorem 25.3.1 together with* T_1 *and* T_5 *we have* $\mathfrak{D}(\alpha a) = \mathfrak{D}(a)$ *for* $\alpha > 0$ *and* $A(\alpha a) = \alpha A(a)$.

This is an immediate consequence of (25.9.1).

For the correspondence under addition we need much more stringent hypotheses and the results obtained from such assumptions are still fairly weak. Throughout the rest of this section we shall assume the postulates S_0, P_1, P_2, P_3, P_5, P_6, P_9°, P_{11}, T_1, and T_5. In the parameter space we shall be restricted to a small neighborhood of the origin, essentially the sphere $S(2d)$ of Theorem 25.9.3, and here P_9° and P_{11} replace P_9^* and P_{10} as we have seen above. It is convenient to keep P_5 and P_6 since our discussion of the canonical function $f(\rho a)$ is based on Theorem 25.3.1 and not on the differential equation (25.4.2).

THEOREM 25.10.2. *For* $y \in \mathfrak{R}[\mathfrak{X}]$

$$A(a_1 + a_2)y = A(a_1)y + A(a_2)y. \tag{25.10.1}$$

PROOF. It is sufficient to prove the corresponding result with A replaced by U since the operators $A(a)$ and $U(a)$ coincide on $\mathfrak{R}[\mathfrak{X}]$ by Theorem 25.9.4. This will be accomplished if it can be shown that

$$\{T[\delta(a_1 + a_2)] - T(\delta a_1) - T(\delta a_2) + I\}y = o(\delta)$$

when $\delta \to 0$. The left member equals

$$\{T[\delta(a_1 + a_2)] - T[F(\delta a_1, \delta a_2)]\}y + [T(\delta a_1) - I][T(\delta a_2) - I]y.$$

By Theorem 25.9.5 the norm of the first term does not exceed

$$C \mid \delta(a_1 + a_2) - F(\delta a_1, \delta a_2) \mid \| x \|$$

which is $O(\delta^2)$ by (25.8.3). Since $y \in \mathfrak{D}(a_2)$, we have $[T(\delta a_2) - I]y = \delta[U(a_2)y + y(\delta)] = \delta[A(a_2)y + y(\delta)]$ where $\| y(\delta) \| \to 0$ with δ. Hence

$$\| [T(\delta a_1) - I][T(\delta a_2) - I]y \| \leqq \delta \| [T(\delta a_1) - I]A(a_2)y \| + \delta \| [T(\delta a_1) - I]y(\delta) \| = o(\delta)$$

by T_5. This proves the assertion.

The terms in (25.10.1) make sense if y is in the intersection of the domains of the three operators involved and one would expect the relation to be true precisely in this set. This will be so if the graph $[[y, A(a)y]; y \in \mathfrak{R}[\mathfrak{X}]]$ is always dense in the graph $[[x, A(a)x]; x \in \mathfrak{D}(a)]$ in $\mathfrak{X} \times \mathfrak{X}$. Perhaps this is the case but the argument tendered by E. Hille in **[15]** to prove the dense graph theorem is not conclusive. In the commutative case one has

$$\{T[f(\delta(a_1 + a_2))] - T[f(\delta a_1)] - T[f(\delta a_2)] + I\}x = o(\delta)$$

for any x in $\mathfrak{D}(a_1) \cap \mathfrak{D}(a_2)$ and (25.10.1) is valid for x in this set.

THEOREM 25.10.3. *Let*

$$a = (a^1, a^2, \cdots, a^n) = a^1 e_1 + a^2 e_2 + \cdots + a^n e_n$$

and set $A(e_k) = A_k$. *Then for* $y \in \mathfrak{R}[\mathfrak{X}]$

$$A(a)y = a^1 A_1 y + a^2 A_2 y + \cdots + a^n A_n y. \tag{25.10.2}$$

This is an immediate consequence of the two preceding theorems. The maximal domain of validity of (25.10.2) is subject to the same remarks as that of (25.10.1).

Normally the n operators A_k may be expected to be linearly independent over the real field when they act in their common domain of definition. However in cases of degeneracy the A_k may be dependent over the reals. For example, let $[T(\xi); \xi > 0]$ be a semi-group of class (C_0) with infinitesimal generator A and define $T[(x^1, x^2)] = T(x^1 + x^2)$ on $\Pi = \bar{E}_2^+$. In this case $A_1 = A_2 = A$. It should also be observed that linear independence over the reals does not imply independence over the complex field.

The infinitesimal generators $[A(a)]$ of $\mathfrak{T}$ form a cone which is closed under addition and multiplication by positive numbers. On the other hand, multiplication by negative numbers is not allowed since A and $-A$ are normally not both infinitesimal generators of bounded semi-groups. In this respect there is a striking difference between the cone of generators of a semi-group and the Lie ring of generators of a group. Another difference will be found below.

We turn now to the analogue of Lie's First Fundamental Theorem (S. Lie and G. Scheffers [1, p. 376]), or, from our point of view, the extension of Theorem 10.3.3 to n-parameter semi-groups.

THEOREM 25.10.4. *For $y \in \mathfrak{R}[\mathfrak{X}]$ and small values of $p \in \bar{E}_n^+$*

$$\frac{\partial}{\partial p^j} T(p)y = \sum_{k=1}^{n} c_j^k(p) T(p) A_k y, \qquad j = 1, 2, \cdots, n, \tag{25.10.3}$$

where the matrix $[c_j^k(p)] \equiv [F^k_{;j}(p, 0)]^{-1}$ tends to the unit matrix as $p \to 0$.

PROOF. The left side of (25.10.3) is the limit as $\sigma \to 0+$ of

$$\sigma^{-1}[T(p + \sigma e_j) - T(p)]y,$$

if this limit exists. Here our first problem is to solve the equation $F(p, h_j) = p + \sigma e_j$ for h_j when σ is small. Using the Taylor expansion we get the conditions

$$\sum_{m=1}^{n} F^i_{;m}(p, 0) h_j^m + O(|h_j|^2) = \delta_j^i \sigma, \qquad i = 1, \cdots, n.$$

Here $F^i_{;m}(p, 0) \to \delta^i_m$ as $p \to 0$ so the system certainly has a solution for small values of $|p|$. We get

$$h_j = \sigma c_j(p) + O(\sigma^2)$$

where $c_j(p) = (c_j^k(p))$ is the vector obtained by solving the linear system

$$\sum_{m=1}^{n} F^i_{;m}(p, 0) c_j^m(p) = \delta_j^i, \qquad i = 1, \cdots, n. \tag{25.10.4}$$

Once this is done we have

$$\sigma^{-1}[T(p + \sigma e_j) - T(p)]y = T(p)\{\sigma^{-1}[T(h_j) - I]y\}.$$

Here

$$\sigma^{-1}[T(h_j) - I]y = \sigma^{-1}\{T[\sigma c_j(p)] - I\}y + \sigma^{-1}\{T(h_j) - T[\sigma c_j(p)]\}y$$

and the first expression on the right tends to $A[c_j(p)]y$ by Theorem 25.9.4 while the second one is $O(\sigma)$ by Theorem 25.9.5 since $|\sigma c_j(p) - h_j| = O(\sigma^2)$. It follows that the derivative in question exists and equals $T(p)\{A[c_j(p)]y\}$. Since

$$c_j(p) = c_j^1(p)e_1 + c_j^2(p)e_2 + \cdots + c_j^n(p)e_n,$$

we have merely to apply Theorem 25.10.3 in order to get the stated result.

Inverting the system of equation (25.10.3), we obtain

COROLLARY. *For* $y \in \mathfrak{R}[\mathfrak{X}]$ *we have*

$$T(p)[A_iy] = \sum_{k=1}^{n} F_{;i}^k(p, 0) \frac{\partial}{\partial p^k} T(p)y, \qquad i = 1, \cdots, n. \tag{25.10.5}$$

We now come to the analogue of Lie's Second Fundamental Theorem (S. Lie and G. Scheffers [1, pp. 390–391]) which involves the structure constants γ_{jk}^i defined by (25.8.5).

THEOREM 25.10.5. *If* $y \in \mathfrak{R}[\mathfrak{X}]$

$$[A_i, A_j]y \equiv (A_iA_j - A_jA_i)y = \sum_{m=1}^{n} \gamma_{ij}^m A_m y. \tag{25.10.6}$$

PROOF. We use (25.10.5) twice, forming

$$\begin{aligned} T(p)[A_iA_jy] &= \sum_{k=1}^{n} F_{;i}^k(p, 0) \frac{\partial}{\partial p^k} [T(p)A_jy] \\ &= \sum_{k=1}^{n} F_{;i}^k(p, 0) \frac{\partial}{\partial p^k} \left\{ \sum_{m=1}^{n} F_{;j}^m(p, 0) \frac{\partial}{\partial p^m} T(p)y \right\}. \end{aligned}$$

Interchanging i and j, subtracting and simplifying, we get

$$T(p)\{[A_i, A_j]y\} = \sum_{m=1}^{n} \Gamma_{ij}^m(p) \frac{\partial}{\partial p^m} T(p)y$$

where

$$\Gamma_{ij}^m(p) = \sum_{k=1}^{n} \{F_{;i}^k(p, 0)F_{k;j}^m(p, 0) - F_{;j}^k(p, 0)F_{k;i}^m(p, 0)\}$$

and this tends to γ_{ij}^m as $p \to 0$ by (25.8.1), (25.8.2), and (25.8.5). Formula (25.10.3) shows that the partial of $T(p)y$ with respect to p^m tends to A_my. It should be observed that the second order partials of $T(p)y$ which arise in the process, but cancel in the subtraction, actually exist since $y \in \mathfrak{R}[\mathfrak{X}]$, and may be found by formal differentiation of (25.10.3).

The second fundamental theorem asserts that each commutator $[A_i, A_j]$ is a linear combination of the n basic infinitesimal generators, but in the semi-group case it does not follow that $[A_i, A_j]$ is also an infinitesimal generator, that is, we cannot always find a vector a in $\bar{E}_n^+$ such that $[A_i, A_j]y = A(a)y$. In particular, we note that if $[A_i, A_j]$ is an infinitesimal generator then $[A_j, A_i]$ cannot be one. Thus in the semi-group case the cone of infinitesimal generators ordinarily does not form a ring under addition and commutator multiplication.

Finally we note the Jacobi identities

$$(25.10.7) \quad [A_i, [A_j, A_k]]y + [A_j, [A_k, A_i]]y + [A_k, [A_i, A_j]]y = 0$$

valid for any $y \in \mathfrak{R}[\mathfrak{X}]$.

CHAPTER XXVI

FUNCTIONS ON VECTORS TO VECTORS

26.1. Orientation. The present chapter is devoted to a systematic development of the theory of functions on vectors to vectors, supplementing the earlier discussion in Chapter III, paragraph 3. Again the emphasis is on analytic functions, with the difference that we now study in much greater detail the power series expansion for such functions. Much of this material depends on results from the theory of functions of several complex variables due in their classical form to F. Hartogs [**1**] and extended by Max Zorn (see section 3.15).

There are two paragraphs: *Differentiable Functions* and *Power Series.* The principal references have already been listed at the end of Chapter III; for a more extensive bibliography the reader is referred to the expository papers of Graves [**3**], Hyers [**1**], Michal [**1**], and Taylor [**7**]. A suitable source for the classical interpolation theory used in section 26.2 is J. F. Steffensen [**1**, §§ 2–4, 19].

This chapter was written in collaboration with Max Zorn to whom most of the new results are due. Zorn has published his investigations in [**1**, **2**, **3**].

1. Differentiable Functions

26.2. Multilinear forms and polynomials. In the following we shall be concerned with functions defined on a complex (B)-space $\mathfrak{X}$ and having its values in another such space $\mathfrak{Y}$. In Chapter II we considered linear functions on $\mathfrak{X}$ to $\mathfrak{Y}$. The extension to multilinear functions is immediate.

Definition 26.2.1. *Suppose $x_1, x_2, \cdots, x_n$ are variables in $\mathfrak{X}$. Then a function $F(x_1, x_2, \cdots, x_n)$ defined for all values of the variables and having values in $\mathfrak{Y}$ is called a symmetric n-linear form, if* (i) *it is linear in each variable separately and* (ii) *it is a symmetric function of the variables. It is said to be continuous if it is continuous in each variable separately.*

Definition 26.2.2. *A function $y = P(x)$ on $\mathfrak{X}$ to $\mathfrak{Y}$ defined for all x is called a polynomial in x of degree m if for all $a, h \in \mathfrak{X}$ and all complex α*

$$P(a + \alpha h) = \sum_{\nu=0}^{m} P_\nu(a, h)\alpha^\nu, \tag{26.2.1}$$

where $P_\nu(a, h)$ are independent of α. The degree is exactly m if $P_m(a, h) \not\equiv \theta$. $P(x)$

is a power or a homogeneous polynomial of degree n if it is a polynomial and $P(\alpha x) \equiv \alpha^n P(x)$ *in* x *and* α.

According to this definition, the zero element is homogeneous of arbitrary degree. This is the only case, however, in which the degree of a power differs from its degree as a polynomial. More precisely expressed: *if* $P(x)$ *is a polynomial of degree exactly* m, *if* $P(x)$ *is homogeneous of degree* n, *and if* $P(x) \not\equiv \theta$, *then* $m = n$. The following simple argument, due to Christine S. Williams, starts from the identity $\alpha^m P(a + \alpha^{-1}h) = \alpha^m P[\alpha^{-1}(h + \alpha a)]$. Using the homogeneity on the right and expanding both sides with the aid of (26.2.1), we see that

$$\sum_{\nu=0}^{m} P_\nu(a, h)\, \alpha^{m-\nu} = \sum_{\nu=0}^{m} P_\nu(h, a)\, \alpha^{m-n+\nu}.$$

Since neither $P_0(a, h)$ nor $P_m(h, a)$ can vanish identically, this identity requires that $m = n$ and

$$P_\nu(a, h) = P_{n-\nu}(h, a), \qquad \nu = 0, 1, 2, \cdots, n. \tag{26.2.2}$$

Returning to general polynomials, we rewrite (26.2.1) as a Newton interpolation polynomial

$$\begin{aligned} P(a + \alpha h) &= \sum_{\nu=0}^{m} \Delta_h^\nu P(a) \binom{\alpha}{\nu}, \\ \Delta_h^\nu P(a) &= \sum_{\mu=0}^{\nu} (-1)^{\nu-\mu} \binom{\nu}{\mu} P(a + \mu h). \end{aligned} \tag{26.2.3}$$

This relation is a simple identity for integral values of $\alpha \leqq m$. Further, $\binom{\alpha}{\nu}$ being a polynomial of degree ν in α, we see that $\sum_{\nu=0}^{m} \Delta_h^\nu P(a)\binom{\alpha}{\nu}$ is a polynomial of degree m in α, equal to $P(a + \alpha h)$ for $\alpha = 0, 1, 2, \cdots, m$. Since $P(a + \alpha h)$ is also a polynomial of degree m in α by (26.2.1), this establishes (26.2.3) for all α and proves that the coefficients $P_\nu(a, h)$ are uniquely determined as linear combinations of the $P(a + \mu h)$, $\mu = 0, 1, 2, \cdots, m$. It follows that $P_\nu(a, h)$ is itself a polynomial in both a and h. By (26.2.1) we see that $P_\nu(a, h)$ is homogeneous of degree ν in h. Further we have

$$P_0(a, h) = P(a), \qquad P_m(a, h) = \frac{1}{m!} \Delta_h^m P(a). \tag{26.2.4}$$

That the last coefficient is actually independent of a will be shown below. If, in addition, $P(x)$ is homogeneous of degree m, then by (26.2.2) we see that $P_\nu(a, h)$ is homogeneous of degree $m - \nu$ in a.

From (26.2.3) we obtain by complete induction that

$$\begin{aligned} &P(a + \alpha_1 h_1 + \cdots + \alpha_n h_n) \\ &\qquad = \sum \cdots \sum \Delta_{h_1}^{\nu_1} \cdots \Delta_{h_n}^{\nu_n} P(a) \binom{\alpha_1}{\nu_1} \cdots \binom{\alpha_n}{\nu_n}, \end{aligned} \tag{26.2.5}$$

where the increments h_μ are arbitrary elements of $\mathfrak{X}$ and the summations with respect to $\nu_1, \cdots, \nu_n$ go from 0 to m. The right side is a polynomial in the n complex variables $\alpha_1, \cdots, \alpha_n$ of total degree nm. Replacing each α_μ by $\lambda\alpha_\mu$ we obtain a polynomial in λ of degree nm which by formula (26.2.1) must reduce to one of degree m. Owing to the arbitrariness of the α's this is possible if and only if each difference of $P(a)$ involving more than m spans h_μ vanishes identically. This leads to the basic

THEOREM 26.2.1. *If $P(x)$ is a polynomial of degree exactly m, then*

$$\Delta^{m+1}_{h_1 h_2 \cdots h_{m+1}} P(x) = \theta \tag{26.2.6}$$

identically in x and in the increments $h_1, \cdots, h_{m+1}$. No difference of order m can vanish identically. Conversely, if $P(x)$ is a polynomial and if a difference of $P(x)$ of order k vanishes identically in x and in the increments, then the degree of $P(x)$ is less than k.

PROOF. (26.2.6) was proved above. If the degree of $P(x)$ is exactly m, then $m!P_m(a, h) = \Delta_h^m P(a)$ does not vanish identically. The converse assertion follows from (26.2.3) which shows that $P(a + \alpha h)$ reduces to a polynomial in α of degree less than k since the identical vanishing of a kth order difference implies the identical vanishing of all differences of order $\geqq k$.

Setting

$$P(x_1, \cdots, x_m) = \frac{1}{m!} \Delta^m_{x_1 \cdots x_m} P(a), \tag{26.2.7}$$

we note that giving a an increment h and taking the difference of span h produces a result which vanishes identically in h. It follows that $P(x_1, \cdots, x_m)$ is actually independent of a. This shows in particular that the coefficient $P_m(a, h)$ is independent of a as asserted above.

It is obvious that any sum of polynomials is a polynomial of a degree not exceeding the highest degree present among the summands. If $P(x)$ is a polynomial in x of exact degree m so is $P(x + c)$ for any fixed c. Further, a polynomial of degree m is the sum of homogeneous polynomials

$$P(x) = \sum_{\nu=0}^{m} P_\nu(x),$$

where $P_\nu(x)$ is homogeneous in x of degree ν. It follows from (26.2.1) that $P_\nu(x) = P_\nu(\theta, x)$ so that $P_\nu(x)$ is uniquely determined.

DEFINITION 26.2.3. *The polar form of a homogeneous polynomial $P(x)$ of degree m is defined by* (26.2.7).

THEOREM 26.2.2. *If $P(x)$ is a homogeneous polynomial of degree n, its polar form is a symmetric n-linear form. In terms of the polar form we have $P(x) = P(x, \cdots, x)$ and*

$$P(\lambda a + \mu b) = \sum_{\nu=0}^{n} \binom{n}{\nu} P(a, \cdots, a, b, \cdots, b)\lambda^{\nu}\mu^{n-\nu}, \tag{26.2.8}$$

where the coefficient of $\lambda^{\nu}\mu^{n-\nu}$ *is a homogeneous polynomial in* a *of degree* ν *and in* b *of degree* $n - \nu$.

PROOF. That the polar form is symmetric follows from (26.2.7). In order to prove that it is linear in x_n we set

$$f(x) = \frac{1}{n!} \Delta^{n-1}_{x_1 \cdots x_{n-1}} P(x).$$

According to Theorem 26.2.1, $f(x)$ is a non-null polynomial in x, all of whose second differences vanish identically. It follows that $f(x)$ is a polynomial of degree one and hence that $g(x) = f(x) - f(\theta)$ is a homogeneous polynomial of degree one. Now

$$\Delta^1_{\alpha x} \Delta^1_{\beta y} f(\theta) = [g(\alpha x + \beta y) - g(\alpha x)] - [g(\beta y) - g(\theta)] = \theta$$

and therefore $g(\alpha x + \beta y) = \alpha g(x) + \beta g(y)$. Consequently $P(x_1, \cdots, x_n) = \Delta^1_{x_n} f(\theta) = g(x_n)$ is linear in x_n and, since it is symmetric, it is linear in each of the x_i's. Further, $P(x)$ being homogeneous, we see from the relations (26.2.1), (26.2.4), and (26.2.7) that

$$P(x, \cdots, x) = \frac{1}{n!} \Delta^n_x P(\theta) = P_n(\theta, x) = P(x).$$

Formula (26.2.8) follows directly from the fact that $P(x_1, \cdots, x_n)$ is a symmetric n-linear form. A converse to this result is given by

THEOREM 26.2.3. *If* $F(x_1, \cdots, x_n)$ *is a symmetric n-linear form, then* $P(x) = F(x, \cdots, x)$ *is a homogeneous polynomial of degree n and* $F(x_1, \cdots, x_n)$ *is again the polar form of* $P(x)$.

PROOF. As in formula (26.2.8) we obtain

$$P(a + \alpha h) = \sum_{\nu=0}^{m} \binom{n}{\nu} F(a, \cdots, a, h, \cdots, h)\, \alpha^{\nu} \tag{26.2.9}$$

where on the right the variable h occurs ν times in the coefficient of α^{ν}. It is obvious from (26.2.9) that $P(x)$ is a homogeneous polynomial of degree n. Suppose now that $P(x_1, \cdots, x_n)$ is the polar form of $P(x)$. Then

$$P\left(\sum_1^n \alpha_i x_i, \cdots, \sum_1^n \alpha_i x_i\right) = P\left(\sum_1^n \alpha_i x_i\right) = F\left(\sum_1^n \alpha_i x_i, \cdots, \sum_1^n \alpha_i x_i\right).$$

If we now expand the two extreme members and compare the coefficients of $\alpha_1\alpha_2 \cdots \alpha_n$ we obtain

$$\sum P(x_{i_1}, \cdots, x_{i_n}) = \sum F(x_{i_1}, \cdots x_{i_n}),$$

each sum being taken over all permutations of $1, \cdots, n$. Because of the symmetry of each of these forms this equality reduces to $P(x_1, \cdots, x_n) = F(x_1, \cdots, x_n)$.

The rest of this section is devoted to questions of continuity.

THEOREM 26.2.4. *A homogeneous polynomial of degree n is continuous if and only if it is bounded in some sphere. It is then bounded in every fixed finite sphere and satisfies a Lipschitz condition of order one uniformly in such a sphere. Moreover, there exists a constant M such that for all x*

$$\| P(x) \| \leqq M \| x \|^n.$$

PROOF. Suppose that $\| P(x) \| \leqq B$ in $\| x - a \| \leqq \rho$. The coefficients $P_k(a, b)$ in the expression for $P(a + \alpha b)$ given by (26.2.1) can obviously be expressed linearly with numerical coefficients in terms of $P(a + \omega^\nu b)$, $\nu = 0, 1, 2, \cdots, n$, where ω is a primitive $(n + 1)$th root of unity. There is consequently a B_1 such that for $\| b \| \leqq \rho$ we have

$$\| P_k(a, b) \| \leqq B_1, \qquad k = 0, 1, \cdots, n.$$

$P_k(a, b)$ being homogeneous of degree k in b, one infers that for all values of b

$$\| P_k(a, b) \| \leqq B_1\rho^{-k} \| b \|^k, \qquad k = 0, 1, \cdots, n, \tag{26.2.10}$$

and this shows that $P(x) = P(a + (x - a))$ is bounded in every finite sphere. Since $P_0(a, b) = P(a)$, the estimate also gives

$$\| P(a + h) - P(a) \| = O[\| h \|].$$

Since (26.2.10) holds uniformly with respect to a in any bounded domain of $\mathfrak{X}$, the Lipschitz condition also holds uniformly in such a domain. For $a = \theta$, $b = x$, $k = n$ we get the desired inequality $\| P(x) \| \leqq M \| x \|^n$.

Conversely, if $P(x)$ is continuous at $x = a$, then $P(x)$ is bounded in some sphere $\| x - a \| \leqq \rho$ and the preceding argument applies.

THEOREM 26.2.5. *A symmetric n-linear form $F(x_1, \cdots, x_n)$ is continuous in each variable separately if and only if it is bounded in the sense that there is a constant M such that $\| F(x_1, \cdots, x_n) \| \leqq M \| x_1 \| \cdots \| x_n \|$ for all values of the variables. It is then a continuous function of $(x_1, \cdots, x_n)$.*

PROOF. The condition is sufficient since

$$\| F(x_1, \cdots, x_n) - F(a_1, \cdots, a_n) \|$$
$$\leqq \sum_{k=1}^{n} \| F(a_1, \cdots, a_{k-1}, x_k - a_k, x_{k+1}, \cdots, x_n) \| \leqq M\rho^{n-1} \sum_{k=1}^{n} \| x_k - a_k \|$$

if $\| a_k \| \leqq \rho$, $\| x_k \| \leqq \rho$ for all k.

To prove the necessity it is enough to consider the case $n = 2$; the general case can then be handled by complete induction. Thus we have a symmetric bilinear form $F(y, x)$ continuous in y and x separately. For each y, $F(y, x)$ is a linear bounded operator in x on $\mathfrak{X}$ to $\mathfrak{Y}$ and hence there exists a constant $M(y)$ such that $\| F(y, x) \| \leqq M(y) \| x \|$. By symmetry, $\| F(y, x) \| \leqq M(x) \| y \|$. In particular for all y, $\| y \| \leqq 1$, we have $\| F(y, x) \| \leqq M(x)$. By the uniform boundedness theorem applied to the operators $[F(y, \cdot), \| y \| \leqq 1]$ there exists an M such that $\| F(y, \cdot) \| \leqq M$ for all y, $\| y \| \leqq 1$. In other words $\| F(y, x) \| \leqq M$ for all x, y of norm one and therefore $\| F(y, x) \| \leqq M \| y \| \| x \|$. Thus $F(y, x)$ is bounded and consequently a continuous function of (y, x) as asserted.

THEOREM 26.2.6. *A homogeneous polynomial is continuous if and only if its polar form is continuous. Conversely, a symmetric n-linear form* $F(x_1, \cdots, x_n)$ *is continuous if and only if the polynomial* $F(x, \cdots, x)$ *is continuous.*

PROOF. It is clear from Theorems 26.2.2 and 26.2.3 that the two parts of this theorem are equivalent. Further it is obvious that $F(x_1, \cdots, x_n)$ continuous in $(x_1, \cdots, x_n)$ implies $F(x, \cdots, x)$ continuous in x. It remains to show that if $P(x)$ is continuous then so is its polar form. However $P(x_1, \cdots, x_n)$ is a fixed linear combination of the $P(\sum_1^n \epsilon_i x_i)$, $\epsilon_i = 0$ or 1. For $\| x_i \| \leqq 1$ we have $\| \sum_1^n \epsilon_i x_i \| \leqq n$. It now follows by Theorem 26.2.4 that $\| P(x_1, \cdots, x_n) \|$ is bounded for $\| x_i \| \leqq 1$ and hence by Theorem 26.2.5 that $P(x_1, \cdots, x_n)$ is continuous.

DEFINITION 26.2.4. *If* $P(x)$ *is a homogeneous continuous polynomial of degree n, the least value of* M *for which* $\| P(x) \| \leqq M \| x \|^n$ *for all* x *is called the bound or the norm of* $P(\cdot)$ *and is denoted by* $\| P \|$. *For a continuous symmetric n-linear form the bound or norm* $\| F \|$ *is the least value of* M *such that* $\| F(x_1, \cdots, x_n) \| \leqq M \| x_1 \| \cdots \| x_n \|$ *for all* $x_1, \cdots, x_n$.

DEFINITION 26.2.5. *A function* $y = f(x)$ *on* $\mathfrak{X}$ *to* $\mathfrak{Y}$ *defined in the set* $\mathfrak{D}$, *is said to be continuous in the sense of Baire if there exists a set of the first category* $\mathfrak{D}_0 \subset \mathfrak{D}$ *such that* $f(x)$ *is continuous in the set* $\mathfrak{D} \ominus \mathfrak{D}_0$.

Our use of the notion of continuity in the sense of Baire will be restricted to the implications of the next three theorems; for the first of these we refer to Kuratowski [**2**, p. 191 et seq.], for the others to S. Mazur and W. Orlicz [**1**, p. 182].

THEOREM 26.2.7. *A convergent sequence of functions, continuous in the sense of Baire, converges to a function having the same property.*

THEOREM 26.2.8. *If a homogeneous polynomial is continuous in the sense of Baire, then it is continuous everywhere. A symmetric n-linear form which is continuous in the sense of Baire with respect to each variable separately is continuous everywhere with respect to the variables jointly.*

Combining these results we obtain:

THEOREM 26.2.9. *A convergent sequence of continuous symmetric n-linear forms converges to a form with the same properties or the limit vanishes identically. Also, a convergent sequence of continuous homogeneous polynomials of fixed degree n converges either to such a polynomial or to the zero element.*

26.3. Differentiability. We return to the question of differentiability already considered in sections 3.16 and 3.17. There it was shown that the nth variation $\delta^n f(x; h)$ of a (G)-differentiable function is homogeneous of degree n in h. We shall now prove that $\delta^n f(x; h)$ is actually a homogeneous polynomial of degree n in h, continuous in h when $f(x)$ is analytic.

Our starting point is the following corollary to Theorem 3.16.1.

THEOREM 26.3.1. *$f(x)$ is (G)-differentiable in the finitely open set $\mathfrak{D}$ if and only if for every $x \in \mathfrak{D}$ and any $h_1, \cdots, h_n \in \mathfrak{X}$ the function $f(x + \sum_1^n \zeta_k h_k)$ is partially differentiable with respect to ζ_k, $k = 1, \cdots, n$, in the open subset of the space Z_n which corresponds to points $x + \sum_1^n \zeta_k h_k$ in $\mathfrak{D}$.*

THEOREM 26.3.2. *Let $f(x)$ be (G)-differentiable in the finitely open set $\mathfrak{D}$. For $x \in \mathfrak{D}$, $\delta_x^h f = \delta f(x; h)$ is linear in h.*

PROOF. We have already noted that $\delta f(x; h)$ is homogeneous of degree one in h; it remains to show the additivity. Consider the function $f(x + \zeta_1 h_1 + \zeta_2 h_2)$ of $\boldsymbol{\zeta} = (\zeta_1, \zeta_2) \in Z_2$. It is partially differentiable and thus by Theorem 3.15.1 (iii)

$$f(x + \zeta_1 h_1 + \zeta_2 h_2) = f(x) + \zeta_1 \left(\frac{\partial f}{\partial \zeta_1}\right)_0 + \zeta_2 \left(\frac{\partial f}{\partial \zeta_2}\right)_0 + o(\| \boldsymbol{\zeta} \|).$$

It is clear that

$$\left\{\frac{\partial}{\partial \zeta_i} f(x + \zeta_1 h_1 + \zeta_2 h_2)\right\}_0 = \delta f(x; h_i), \qquad i = 1, 2;$$

and this gives

$$f(x + \zeta_1 h_1 + \zeta_2 h_2) = f(x) + \zeta_1 \delta f(x; h_1) + \zeta_2 f(x; h_2) + o(\| \boldsymbol{\zeta} \|).$$

Setting $\zeta_1 = \zeta_2 = \zeta$ we obtain

$$f[x + \zeta(h_1 + h_2)] - f(x) = \zeta[\delta f(x; h_1) + \delta f(x; h_2)] + o(| \zeta |)$$

since $o(\| \boldsymbol{\zeta} \|) = o(| \zeta |)$. This shows that

$$\begin{aligned} \delta f(x; h_1 + h_2) &= \lim_{\zeta \to 0} \frac{1}{\zeta} \{ f[x + \zeta(h_1 + h_2)] - f(x) \} \\ &= \delta f(x; h_1) + \delta f(x; h_2) \end{aligned}$$

and the theorem is proved.

In defining (F)-differentiability it is customary to require that $\delta f(x; h)$ exist as a *linear* continuous function of h. However as Theorem 26.3.2 shows, the assumption of linearity is redundant; continuity is implied by condition (i) of Definition 3.16.5. Thus *for fixed x, $\delta f(x; \cdot)$ is a linear bounded operator on $\mathfrak{X}$ to $\mathfrak{Y}$, that is*

$\delta f(x; \cdot) \in \mathfrak{E}(\mathfrak{X}, \mathfrak{Y})$. Consequently $\delta f(x; h)$ is a function on $\mathfrak{D}$ to $\mathfrak{E}(\mathfrak{X}, \mathfrak{Y})$ which, as we shall see later, is actually analytic in x.

THEOREM 26.3.3. *Let $f(x)$ be (G)-differentiable in the finitely open set $\mathfrak{D}$. For every $h \in \mathfrak{X}$, $\delta_x^h f = \delta f(x; h)$ is a (G)-differentiable function of x in $\mathfrak{D}$.*

PROOF. We have

$$\begin{aligned} \delta_x^{h_2}\delta_x^{h_1}f(x) &= \delta_x^{h_2}\left\{\frac{d}{d\zeta}f(x+\zeta h_1)\right\}_{\zeta=0} \\ &= \left\{\frac{\partial}{\partial\zeta_2}\left[\frac{\partial}{\partial\zeta_1}f(x+\zeta_1h_1+\zeta_2h_2)\right]_{\zeta_1=0}\right\}_{\zeta_2=0} \\ &= \left\{\frac{\partial^2}{\partial\zeta_1\partial\zeta_2}f(x+\zeta_1h_1+\zeta_2h_2)\right\}_{0,0}. \end{aligned}$$

Since $f(x + \zeta_1h_1 + \zeta_2h_2)$ is partially differentiable, Theorem 3.15.1 ensures the existence of the higher partials.

We may therefore define the nth variation $\delta^n f(x; h_1, \cdots, h_n)$ of $f(x)$ with increments $h_1, \cdots, h_n$ by

DEFINITION 26.3.1. *We set* $\delta^1 f(x; h_1) = \delta f(x; h_1)$;

$$\delta^{n+1}f(x; h_1, \cdots, h_{n+1}) = \delta_x^{h_{n+1}}[\delta^n f(x; h_1, \cdots, h_n)];\ \delta^n f(x; h) = \delta^n f(x; h, \cdots, h).$$

For the sake of convenience we add the convention that $[\delta^n f(x; h_1, \cdots, h_n)]_{n=0} = f(x)$. We state without proof

THEOREM 26.3.4. *We have*

$$\text{(i)} \qquad \left\{\frac{\partial^n}{\partial\zeta_1\cdots\partial\zeta_n}f\left(x+\sum_{k=1}^n\zeta_kh_k\right)\right\}_{0,\cdots,0} = \delta^n f(x; h_1, \cdots, h_n);$$

$$\text{(ii)} \qquad \left\{\frac{\partial^n}{\partial\zeta_1\cdots\partial\zeta_n}f\left(x+\sum_{k=1}^n\zeta_kh_k\right)\right\}_{h,\cdots,h;0,\cdots,0} = \left\{\frac{d^n}{d\zeta^n}f(x+\zeta h)\right\}_{\zeta=0}.$$

THEOREM 26.3.5. *Let $f(x)$ be (G)-differentiable in the finitely open set $\mathfrak{D}$. Then $\delta^n f(x; h_1, \cdots, h_n)$ is a symmetric n-linear form in $h_1, \cdots, h_n$ which is (G)-differentiable with respect to x for $x \in \mathfrak{D}$. $\delta^n f(x; h)$ is a homogeneous polynomial of degree n in h.*

PROOF. By definition $\delta^n f(x; h_1, \cdots, h_n)$ is a variation with respect to its last argument and therefore by Theorem 26.3.2 it is linear in its last argument. It follows from Theorem 26.3.4 (i) that $\delta^n f(x; h_1, \cdots, h_n)$ is symmetric and hence that it is linear in each h_k. By Theorem 26.3.3 it is (G)-differentiable in x and by Theorem 26.2.3, $\delta^n f(x; h)$ is a homogeneous polynomial of degree n in h.

THEOREM 26.3.6. *If $f(x)$ is analytic in the domain $\mathfrak{D}$, then $\delta^n f(x; h_1, \cdots, h_n)$ is an analytic function of x in $\mathfrak{D}$ for fixed $h_1, \cdots, h_n \in \mathfrak{X}$ and a continuous symmetric n-linear form in $(h_1, \cdots, h_n)$ for fixed $x \in \mathfrak{D}$.*

PROOF. By the previous theorem $\delta^n f(x; h)$, considered as a function of x for fixed h, is (G)-differentiable in $\mathfrak{D}$ and according to Theorem 3.17.1 it is also locally bounded in $\mathfrak{D}$. Thus by Definition 3.17.2, $\delta^n f(x; h)$ is an analytic function of x in $\mathfrak{D}$ for fixed h. This conclusion extends also to $\delta^n f(x; h_1, \cdots, h_n)$ which is the polar form of $\delta^n f(x; h)$ and hence expressible linearly in terms of the functions $\delta^n f(x; \sum_1^n \epsilon_k h_k)$, $\epsilon_k = 0, 1$, all of which are analytic functions of x. On the other hand for fixed $x \in \mathfrak{D}$, $\delta^n f(x; h)$ is a bounded homogeneous function of degree n in h by Theorem 3.17.1. Applying Theorems 26.2.4, 26.2.6, and 26.3.5, we see that its polar form, namely $\delta^n f(x; h_1, \cdots, h_n)$, is a continuous symmetric n-linear form in $(h_1, \cdots, h_n)$.

At this juncture an examination of the differentiability properties of powers is in order.

THEOREM 26.3.7. *A homogeneous polynomial is always (G)-differentiable; it is (F)-differentiable if and only if it is continuous.*

PROOF. The (G)-differentiability of a homogeneous polynomial $P(x)$ is an immediate consequence of Definition 26.2.2. If $P(x)$ is also continuous, then it is analytic and hence (F)-differentiable by Theorem 3.17.1. On the other hand, if $P(x)$ is (F)-differentiable, then it is obviously continuous.

The above theorem suggests a new terminology for homogeneous polynomials, namely *(F)-powers* or *(G)-powers* according as they are (F)-differentiable or merely (G)-differentiable, that is, according as they are continuous or not. By Theorem 26.2.4 we may replace continuity by boundedness in this statement.

The Taylor series of $f(x)$ given in Theorems 3.16.2 and 3.17.1 are expansions in (G)-powers and in (F)-powers respectively and consequently could be described as *(G)-power series* or *(F)-power series* in the respective cases. For $f(x)$ analytic in the open domain $\mathfrak{D}$ it was shown that the series converged uniformly in some sphere about each $x \in \mathfrak{D}$. We shall now show for $f(x)$ merely (G)-differentiable in the finitely open $\mathfrak{D}$ that the series will converge uniformly in some finitely open c-star about each $x \in \mathfrak{D}$. In the following theorem we have taken $x = \theta$.

THEOREM 26.3.8. *Let $f(x)$ be (G)-differentiable in the finitely open set $\mathfrak{D}$. Let $\mathfrak{C}^*(\theta) \subset \mathfrak{D}$ be a finitely open c-star about θ, and let $\mathfrak{S}$ be a compact subset of $\mathfrak{C}^*(\theta)$ which is contained in a finite-dimensional subspace of $\mathfrak{X}$. Then there exist quantities ϵ and M, $0 < \epsilon < 1$, $0 < M$, and a finitely open c-star $\mathfrak{C}_0^*(\theta) = \mathfrak{C}$ such that* (i) $\mathfrak{S} \subset (1 - \epsilon)\mathfrak{C} \equiv \mathfrak{C}(\epsilon) \subset \mathfrak{C} \subset \mathfrak{C}^*(\theta)$, (ii) $\| f(x) \| \leqq M$ *in* $\mathfrak{C}$, *and* (iii)

$$\sum_{m=0}^{\infty} \frac{1}{n!} \sup_{x \in \mathfrak{C}(\epsilon)} \| \delta^n f(\theta, x) \| < \infty.$$

PROOF. The function $\varphi(x) = \max_{|\zeta| \leqq 1} \| f(\zeta x) \|$ is well defined when $x \in \mathfrak{C}^*(\theta)$. If $\mathfrak{X}_n$ is any finite-dimensional linear subspace of $\mathfrak{X}$, then $f(x)$ is continuous in $\mathfrak{X}_n \cap \mathfrak{C}^*(\theta) \equiv \mathfrak{C}_n$ since each x in $\mathfrak{C}_n$ is of the form $\sum_1^n \zeta_k h_k$ and $f(\sum_1^n \zeta_k h_k)$ is a partially differentiable function of $\zeta_1, \cdots, \zeta_n$ for $\sum_1^n \zeta_k h_k$ in $\mathfrak{C}_n$. However this implies that $\varphi(x)$ is also continuous in $\mathfrak{C}_n$. In order to see this, let $\{x_k\} \subset \mathfrak{C}_n$ and $x_k \to x_0 \in \mathfrak{C}_n$ as $k \to \infty$. There exist complex numbers ζ_k and ζ_0 of absolute value one such that $\varphi(x_k) = \| f(\zeta_k x_k) \|$ and $\varphi(x_0) = \| f(\zeta_0 x_0) \|$. Now

$$\limsup \varphi(x_k) = \limsup \| f(\zeta_k x_k) \| \leqq \max_{|\zeta| \leqq 1} \| f(\zeta x_0) \| = \varphi(x_0).$$

On the other hand, $\varphi(x_k) \geqq \| f(\zeta_0 x_k) \|$ whence

$$\liminf \varphi(x_k) \geqq \| f(\zeta_0 x_0) \| = \varphi(x_0)$$

and, finally, $\lim \varphi(x_k) = \varphi(x_0)$.

This being established, we consider the subset $\mathfrak{D}(x_0)$ of $\mathfrak{C}^*(\theta)$ in which $\varphi(x) < \varphi(x_0) + 1$, $x_0 \in \mathfrak{S}$. This set is finitely open since its intersection with any $\mathfrak{X}_n$ is relatively open; it is a c-star about θ from the definition of $\varphi(x)$; and it contains x_0. $\mathfrak{S}$ being compact and contained in a finite-dimensional subspace of $\mathfrak{X}$, there exists a finite subset $x_1, \cdots, x_n \in \mathfrak{S}$ such that the corresponding sets $\mathfrak{D}(x_1), \cdots, \mathfrak{D}(x_n)$ suffice to cover $\mathfrak{S}$. We set $\mathfrak{C} = \bigcup \mathfrak{D}(x_k)$, $M = 1 + \max_k \varphi(x_k)$. Then $\mathfrak{C}$ is a finitely open c-star about θ contained in $\mathfrak{C}^*(\theta)$, containing $\mathfrak{S}$, and such that $\| f(x) \| \leqq M$ in $\mathfrak{C}$. Suppose that $\mathfrak{S}$ is contained in the n-dimensional linear subspace $\mathfrak{X}_n$; since $\mathfrak{C} \cap \mathfrak{X}_n$ is relatively open, we can find an ϵ, $0 < \epsilon < 1$, such that $\mathfrak{S} \subset (1 - \epsilon)\mathfrak{C} \equiv \mathfrak{C}(\epsilon)$. For $x \in \mathfrak{C}(\epsilon)$ we can use formula (3.16.3) with x replaced by θ, h by x, and ρ' by $1/(1 - \epsilon)$. It follows that $\sup_{x \in \mathfrak{C}(\epsilon)} \| \delta^n f(\theta, x) \| \leqq M(1 - \epsilon)^n n!$. This completes the proof.

We shall now summarize the main results of sections 3.16, 3.17, and the present section in the following two theorems. It is suggestive for purposes of comparison to formulate these results in the new terminology.

THEOREM 26.3.9. *If $f(x)$ is defined and (G)-differentiable in the finitely open set $\mathfrak{D} \subset \mathfrak{X}$, then to every $x \in \mathfrak{D}$ there is a (G)-power series $\sum_{n=0}^{\infty} P_n(x, h)$ where $P_n(x, h)$ is a (G)-power in h of degree n and a (G)-differentiable function of x in $\mathfrak{D}$. The functions $P_n(x, h)$ are uniquely determined by $f(x)$; $P_n(x, h) = \delta^n f(x;h)/n!$. We have*

$$f(x + h) = \sum_{n=0}^{\infty} P_n(x, h),$$

valid if h is in the c-star in $\mathfrak{D}$ about x. To each $x \in \mathfrak{D}$ there is a finitely open c-star $\mathfrak{C}$ about x in which $f(x)$ is bounded and the series converges uniformly and normally in the sense that $\sum_{n=0}^{\infty} \sup_{x+h \in \mathfrak{C}} \| P_n(x, h) \|$ converges.

THEOREM 26.3.10. *If $f(x)$ is defined and analytic in the open domain $\mathfrak{D}$, then to every $x \in \mathfrak{D}$ there is an (F)-power series $\sum_{n=0}^{\infty} P_n(x, h)$ where $P_n(x, h)$ is an (F)-power in h of degree n and an analytic function of x in $\mathfrak{D}$. $P_n(x, h) = \delta^n f(x; h)/n!$ and we have*

$$f(x + h) = \sum_{n=0}^{\infty} P_n(x, h),$$

valid if h is in the c-star in $\mathfrak{D}$ about x. To each $x \in \mathfrak{D}$ there is a sphere about x in which $f(x)$ is bounded and the series converges uniformly and normally.

26.4. The Taylor series of an (L)-analytic function. When the underlying space is a commutative (B)-algebra and $f(x)$ is (L)-analytic, then the Taylor series expansion takes on a particularly simple form. This is the content of the following theorem which may be thought of as a converse to Theorem 3.19.1.

THEOREM 26.4.1. *Let $\mathfrak{B}$ be a complex commutative Banach algebra with a unit element. If $f(x)$ is (L)-analytic in the sphere $\| x - a \| < \rho$, then there exist unique elements a_n in $\mathfrak{B}$, $a_n n! = \delta^n f(a; e) = f^{(n)}(a)$, and we have*

$$f(x) = \sum_{n=0}^{\infty} a_n (x - a)^n, \tag{26.4.1}$$

valid for all x in $\| x - a \| < \rho$.

PROOF. An (L)-analytic function is *a fortiori* analytic in the sense of Definition 3.17.2. Consequently by Theorem 26.3.6 its first variation $\delta f(x; h) = hf'(x)$ is an analytic function of x and the same is true of $f'(x) = \delta f(x; e)$. We shall show that $f'(x)$ is also (L)-analytic. To this end we set $f_2(x, k) = \delta f'(x; k)$. Clearly

$$\delta^2 f(x; h, k) = \lim_{\zeta \to 0} \frac{h}{\zeta} [f'(x + \zeta k) - f'(x)] = hf_2(x, k).$$

Since the second variation is symmetric in h and k we have

$$\delta^2 f(x; h, k) = hf_2(x, k) = kf_2(x, h).$$

Putting $h = e$, we see that $f_2(x, k) = kf_2(x, e) = kf''(x)$ where we have set $f_2(x, e) = f''(x)$. Since $f_2(x, k) = \delta f'(x; k)$ and $f'(x)$ is (F)-differentiable, it follows that

$$\| f'(x + k) - f'(x) - kf''(x) \| = o(\| k \|).$$

Hence $f''(x)$ is actually the derivative of $f'(x)$ and the latter is (L)-analytic. From this one concludes that $f(x)$ has derivatives of all orders with $\delta^n f(x; h) = f^{(n)}(x)h^n$ and that $f(x)$ may be expanded in a Taylor series

$$f(x) = \sum_{n=0}^{\infty} a_n (x - a)^n, \qquad n! a_n = f^{(n)}(a),$$

convergent for all x in $\| x - a \| < \rho$.

2. POWER SERIES

26.5. (G)-power series. Let $P_n(x)$, $n = 0, 1, 2, \cdots$, be a given sequence of (G)-powers in x, the degree of $P_n(x)$ being n. We shall study the (G)-power series

$$\sum_{n=0}^{\infty} P_n(x), \tag{26.5.1}$$

its region of convergence and the properties of its sum.

We start from a local point of view which enables us to determine the cross sections of the region of convergence, denoted by $\mathfrak{C}[P_n]$, with planes through its center $x = \theta$.

DEFINITION 26.5.1. *Let $u \in \mathfrak{X}$ and $\| u \| = 1$ and put*

$$\mu(u) = \limsup_{n\to\infty} \| P_n(u) \|^{1/n}, \qquad \rho(u) = 1/\mu(u).$$

We state without proof

THEOREM 26.5.1. (i) *If the terms of* (26.5.1) *are bounded for $x = x_0$, then the series is absolutely convergent for $x = \zeta x_0$, $|\zeta| < 1$.* (ii) *If $x = \zeta u$, $\| u \| = 1$, then the series converges absolutely for $|\zeta| < \rho(u)$ and diverges for $|\zeta| > \rho(u)$.*

COROLLARY. *The closure of the region of convergence of a (G)-power series is a c-star about θ (or reduces to θ).*

The properties of $\rho(u)$ appear to be very complicated even in cases which look fairly simple on the surface. As an example we may take $\mathfrak{X} = l_2$, the space of sequences $x = \{\alpha_n\}$, with $\| x \| = \{\sum |\alpha_n|^2\}^{1/2}$, and $\mathfrak{Y} = Z_1$, the space of complex numbers with the usual norm. Then

$$f(x) = \sum_{n=1}^{\infty} (n\alpha_n)^n, \qquad x = \{\alpha_1, \alpha_2, \cdots, \alpha_n, \cdots\}$$

is an (F)-power series as is easily verified. Here the function $\rho(u)$ has the property that its lower and upper limit functions are identically 0 and $+\infty$ respectively. Indeed, suppose that $u = \{\alpha_n\}$ and $\| u \| = 1$. To any given $\epsilon > 0$ we may find an integer N and two numbers A and B near unity such that

$$\sum_{N+1}^{\infty} |\alpha_n|^2 < \epsilon^2$$

and

$$u_1 = \{A\alpha_1, A\alpha_2, \cdots, A\alpha_N, 0, 0, \cdots\},$$
$$u_2 = B\{\alpha_1, \alpha_2, \cdots, \alpha_N, (N+1)^{-2/3}, (N+2)^{-2/3}, \cdots\}$$

are unit vectors. Here $\rho(u_1) = \infty$ and $\rho(u_2) = 0$, further

$$\| u - u_1 \| \leqq A - 1 + \epsilon, \qquad \| u - u_2 \| \leqq 2BN^{-1/6} + |B - 1| + \epsilon,$$

so that u_1 and u_2 are as close to u as we please. This proves the assertion.

A satisfactory theory is obtainable in the case in which there exists a finitely open domain of convergence. This assumption implies certain restrictions on $\rho(u)$ (essentially lower semi-continuity) which, however, we shall not investigate here. We start with a simple lemma.

LEMMA 26.5.1. *If a (G)-power $P(x)$ satisfies $\| P(x) \| \leqq M$ in a c-star $\mathfrak{C}^*(x_0) = x_0 + H$, then it satisfies the same inequality in the c-star $\mathfrak{C}^*(\theta) = \zeta x_0 + H$ for $|\zeta| \leqq 1$.*

PROOF. Let $h_0 \in H$ and consider the function $P(\zeta x_0 + h_0)$ for $|\zeta| \leqq 1$. It is holomorphic for all such values of ζ since it is a polynomial in ζ. By the principle of the maximum $\| P(\zeta x_0 + h_0) \|$ reaches its largest value on the unit circle, say for $\zeta = \zeta_0$. But the homogeneity of $P(x)$ gives

$$\| P(\zeta_0 x_0 + h_0) \| = \| P(x_0 + \zeta_0^{-1} h_0) \| \leqq \max_{h \in H} \| P(x_0 + h) \| \leqq M$$

since H is a c-star. This proves the lemma.

THEOREM 26.5.2. *If the (G)-power series (26.5.1) converges in the finitely open c-star* $\mathfrak{C}^*(x_0) = x_0 + H$, *it will also converge in* $\mathfrak{C}^*(\theta) = \zeta x_0 + H$, $|\zeta| \leqq 1$.

PROOF. Let $h \in H$ and consider the linear subspace $[\zeta_1 x_0 + \zeta_2 h] = [x_0 + (\zeta_1 - 1)x_0 + \zeta_2 h]$. It intersects $\mathfrak{C}^*(x_0)$ in a relatively open set $\mathfrak{D}_0$ corresponding to an open set Δ in Z_2. Since $(1, \zeta_2) \in \Delta$ for $|\zeta_2| \leqq 1$, we can find an $\epsilon > 0$ such that the closed bicylinder $\Gamma_2 : |\zeta_1 - 1| < \epsilon$, $|\zeta_2| \leqq 1 + \epsilon$ is in Δ. Substituting $x = \zeta_1 x_0 + \zeta_2 h$ into the power series, one obtains a series of homogeneous polynomials in (ζ_1, ζ_2) convergent in Δ. By Theorem 3.15.2 (ii) this implies that the terms of the series are uniformly bounded in Γ_2. Thus there is a finite M such that $\| P_k(x) \| \leqq M$ for x in the c-star $\mathfrak{C}_1^*(x_0)$: $x_0 + \eta_1 x_0 + \eta_2 h$ with $(\eta_1 + 1, \eta_2) \in \Gamma_2$, that is $|\eta_1| \leqq \epsilon$, $|\eta_2| \leqq 1 + \epsilon$. By Lemma 26.5.1 the inequality $\| P_k(x) \| \leqq M$ holds also in the c-star $\mathfrak{C}_1^*(\theta)$: $\zeta x_0 + \eta_1 x_0 + \eta_2 h$, $|\zeta| \leqq 1$, $|\eta_1| \leqq \epsilon$, $|\eta_2| \leqq 1 + \epsilon$. Using the homogeneity of $P_k(x)$ or referring to Theorem 3.15.2 again, we see that the power series is absolutely convergent in the c-star $\rho\mathfrak{C}_1^*(\theta)$ for any $\rho < 1$. This c-star, for ρ sufficiently near to one, contains all points of the form $\zeta_1 x_0 + \zeta_2 h$, $|\zeta_1| \leqq 1$, $|\zeta_2| \leqq 1$. Since $h \in H$ is arbitrary, it follows that the power series converges in the finitely open c-star $\zeta x_0 + H$, $|\zeta| \leqq 1$, as asserted.

COROLLARY. *If a (G)-power series converges in an open set, it converges in a neighborhood of* θ.

THEOREM 26.5.3. *If the (G)-power series (26.5.1) converges in a finitely open set* $\mathfrak{D}$, *it will converge uniformly on discs, that is, for fixed* $x \in \mathfrak{D}$, $h \in \mathfrak{X}$ *the series*

$$\sum_0^\infty P_k(x + \zeta h)$$

will be uniformly convergent for $|\zeta| \leqq \rho' < \rho(x, h)$.

PROOF. Consider the series $\sum_0^\infty P_k(\zeta_1 x + \zeta_2 h)$; it is a series of homogeneous polynomials in (ζ_1, ζ_2) which converges in an open set Δ in Z_2, containing all points $(1, \zeta_2)$ with $|\zeta_2| < \rho(x, h)$. The subset defined by $|\zeta_2| \leqq \rho' < \rho(x, h)$ will be a region of uniform convergence by Theorem 3.15.2 (iii).

THEOREM 26.5.4. *If a sequence of functions* $f_n(x)$ *converges uniformly on each of the discs* $x + \zeta h$, $|\zeta| \leqq \rho' < \rho(x, h)$, *of a finitely open set* $\mathfrak{D}$, *and if the functions* $f_n(x)$ *are (G)-differentiable in* $\mathfrak{D}$, *then the limit will also be (G)-differentiable in* $\mathfrak{D}$.

PROOF. If $f(x) = \lim_{n\to\infty} f_n(x)$, we have to show that $f(x_0 + \zeta h)$ is differentiable with respect to ζ for $|\zeta| < \rho(x_0, h)$. This follows from Theorem 3.11.6. The same theorem also yields:

THEOREM 26.5.5. *Under the conditions of the preceding theorem we have for* $x \in \mathfrak{D}$, $h \in \mathfrak{X}$

$$\delta^k f(x; h) = \lim_{n\to\infty} \delta^k f_n(x; h).$$

Combining Theorems 26.5.3 and 26.5.4 we obtain:

THEOREM 26.5.6. *If a (G)-power series converges in a finitely open set* $\mathfrak{D}$, *its sum will be a (G)-differentiable function in* $\mathfrak{D}$.

In the preceding theorem we may replace strong convergence by weak.

THEOREM 26.5.7. *If a (G)-power series converges weakly to a limit in a finitely open set, then its weak sum is (G)-differentiable.*

PROOF. By assumption the series $\sum_0^\infty y^*[P_n(x)]$ converges in $\mathfrak{D}$ for every $y^* \in \mathfrak{Y}^*$. We note that $y^*[P_n(x)]$ is a (G)-power in x with values in Z_1 and the series $\sum_0^\infty y^*[P_n(x)]$ converges in the sense of the strong metric of Z_1 (that is, in the sense of Cauchy). By Theorem 26.5.6, $\sum_0^\infty y^*[P_n(x)]$ is (G)-differentiable in $\mathfrak{D}$ and if $f(x)$ is the weak sum of the series $\sum_0^\infty P_n(x)$, we have consequently that $y^*[f(x)]$ is (G)-differentiable for every y^*. This is equivalent to $y^*[f(x + \zeta h)]$ being a holomorphic function of ζ in $|\zeta| < \rho(x, h)$ for every y^* and this implies that $f(x + \zeta h)$ is holomorphic in the same domain. It follows that $f(x)$ is (G)-differentiable in $\mathfrak{D}$ (in the strong sense). We may however go still further:

THEOREM 26.5.8. *Let* $\sum_0^\infty P_k(x)$ *be a (G)-power series. If the series* $\sum_0^\infty y^*[P_k(x)]$ *converges for each* $y^* \in \mathfrak{Y}^*$ *and for all* x *belonging to a finitely open set* $\mathfrak{D}$, *then the (G)-power series converges in norm for each* $x \in \mathfrak{D}$.

PROOF. For fixed $x_0 \in \mathfrak{D}$, $h \in \mathfrak{X}$, and $y^* \in \mathfrak{Y}^*$, $y^*[P_k(\zeta_1 x_0 + \zeta_2 h)]$ is a homogeneous polynomial in (ζ_1, ζ_2) with values in Z_1. By hypothesis $\sum_0^\infty y^*[P_k(\zeta_1 x_0 + \zeta_2 h)]$ converges in an open set $\Delta \subset Z_2$ containing the point (1, 0), and Δ does not depend on our choice of y^*. Let Δ_0 be a bounded neighborhood of (1, 0) such that $\bar{\Delta}_0 \subset \Delta$. Theorem 3.15.2 (ii) applied to the numerical case asserts that there exists a constant $M(y^*)$ such that

$$| y^*[P_k(\zeta_1 x_0 + \zeta_2 h)] | \leqq M(y^*)$$

for all $k = 0, 1, 2, \cdots$ and $(\zeta_1, \zeta_2) \in \Delta_0$. By the uniform boundedness theorem there exists an M independent of y^* such that $\| P_k(\zeta_1 x_0 + \zeta_2 h) \| \leqq M$, again for all $k = 0, 1, 2, \cdots$ and $(\zeta_1, \zeta_2) \in \Delta_0$. Hence $\| P_k(\zeta_1 x_0 + \zeta_2 h) \| \leqq M(1 - \epsilon)^k$ in the set $(1 - \epsilon)\Delta_0$, which for small enough ϵ, $0 < \epsilon < 1$, still contains the point (1, 0). It follows that $\sum_0^\infty P_k(x_0)$ converges absolutely and hence in the norm.

We come now to the question of the structure of the finitely open set of convergence of a (G)-power series. Theorem 26.5.2 has a bearing on this problem. We have further:

THEOREM 26.5.9. *The largest finitely open set in which a (G)-power series converges is a c-star about θ.*

PROOF. We assume the existence of a finitely open set of convergence. It is clear that the union of all such sets is a finitely open set of convergence and that it is the largest such set. It remains to prove that it is a c-star about θ. Denoting this largest set by $\mathfrak{G}_0[P_k]$, we assume that $x_0 \in \mathfrak{G}_0[P_k]$, $x_0 \neq \theta$. In the comments to Definition 3.16.2 it was observed that to every point of a finitely open set there is a finitely open c-star about that point which is also in the set. Let $\mathfrak{C}^*(x_0) = x_0 + H$ be a c-star in $\mathfrak{G}_0[P_k]$ about x_0. By Theorem 26.5.2, $\mathfrak{G}_0[P_k]$ then contains the finitely open c-star $\mathfrak{C}^*(\theta) = \zeta x_0 + H$, $|\zeta| \leqq 1$. In particular, we see that $x_0 \in \mathfrak{G}_0[P_k]$ implies $\zeta x_0 \in \mathfrak{G}_0[P_k]$, for $|\zeta| \leqq 1$, that is, $\mathfrak{G}_0[P_k]$ is a c-star.

DEFINITION 26.5.2. *A subset $\mathfrak{D}$ of $\mathfrak{X}$ is called c-convex if, Δ being any bounded open set of complex numbers and Γ its boundary, the assumption that $x + \Gamma h$ is contained in $\mathfrak{D}$ implies that $x + \Delta h$ lies in $\mathfrak{D}$.*

THEOREM 26.5.10. *$\mathfrak{G}_0[P_k]$ is c-convex.*

PROOF. Suppose that Δ is a bounded open set of complex numbers and Γ is its boundary; let x_0 and h_0 be such that the set $\mathfrak{S} \subset \mathfrak{G}_0[P_k]$ where $\mathfrak{S} = x_0 + \zeta h_0$, $\zeta \in \Gamma$. Since $\mathfrak{S}$ lies in a two-dimensional linear subspace and is compact there, Theorem 26.3.8 applies. By Theorem 26.5.9 $\mathfrak{G}_0[P_k]$ is itself a finitely open c-star about θ. Consequently there exist a finitely open c-star $\mathfrak{C}$ about θ and an ϵ, $0 < \epsilon < 1$, such that (i) $\mathfrak{S} \subset (1 - \epsilon)\mathfrak{C} \subset \mathfrak{C} \subset \mathfrak{G}_0[P_k]$ and (ii) $\sum M_k$ converges where $M_k = \sup \| P_k(x) \|$ for $x \in (1 - \epsilon)\mathfrak{C}$. Consider now the set H of all elements $h \in \mathfrak{X}$ such that $x_0 + \zeta h_0 + h \in (1 - \epsilon)\mathfrak{C}$ for every $\zeta \in \Gamma$. It is claimed that H is finitely open. To see this, intersect H by a finite-dimensional linear subspace $\mathfrak{X}_n$ and we may assume without loss of generality that $\mathfrak{X}_n$ contains x_0 and h_0. Let $h \in H \cap \mathfrak{X}_n = H_n$ and let $(1 - \epsilon)\mathfrak{C} \cap \mathfrak{X}_n = \mathfrak{C}_n$. Then $x_0 + \zeta h_0 + h \in \mathfrak{C}_n$ for every $\zeta \in \Gamma$. Suppose we could find a sequence $\{h_\nu\} \subset \mathfrak{X}_n$ and a sequence $\{\zeta_\nu\} \subset \Gamma$ such that $h_\nu \to h$ and $x_0 + \zeta_\nu h_0 + h_\nu \in \mathfrak{X}_n - \mathfrak{C}_n$. Without loss of generality we may assume that $\zeta_\nu \to \zeta_0 \in \Gamma$. We have then $x_0 + \zeta_\nu h_0 + h_\nu \to x_0 + \zeta_0 h_0 + h$, an element of the relatively open set $\mathfrak{C}_n$. This involves a contradiction and we see that for all large values of ν we must have $x_0 + \zeta_\nu h_0 + h \in \mathfrak{C}_n$. It follows that H_n is relatively open and that H is finitely open.

We now consider $P_k(x_0 + \zeta h_0 + h)$ for $h \in H$. But when $\zeta \in \Gamma$ we have $x_0 + \zeta h_0 + h \in (1 - \epsilon)\mathfrak{C}$ so that $\| P_k(x_0 + \zeta h_0 + h) \| \leqq M_k$. By the principle of the maximum, this inequality holds also for $\zeta \in \Delta$. It follows that $\sum \| P_k(x) \|$ converges uniformly for $x \in x_0 + \Delta h_0 + H$ which is a finitely open set. Hence $x_0 + \Delta h_0 + H \in \mathfrak{G}_0[P_k]$. In particular, $x_0 + \zeta h_0 \in G_0[P_k]$ for $\zeta \in \Delta$ and the theorem is proved.

Let $\mathfrak{D}$ be a finitely open set in $\mathfrak{X}$. We can always find a finitely open c-convex c-star about θ containing $\mathfrak{D}$ and if $\mathfrak{D} \neq \mathfrak{X}$, there are infinitely many such stars which form a partially ordered system under the relation of inclusion. The intersection of any finite number of such stars is a star with the same property; for an infinite system the intersection is still a c-convex c-star but may possibly fail to be finitely open. Let $\mathfrak{T}$ denote the intersection and $\mathfrak{J}$ the "finite interior" of $\mathfrak{T}$, that is, the set of all points of $\mathfrak{T}$ belonging to finitely open subsets of $\mathfrak{T}$. It is clear that $\mathfrak{D} \subset \mathfrak{J}$ and $\mathfrak{J}$ is finitely open. It is not so obvious that $\mathfrak{J}$ is also a c-convex c-star. To see this, suppose that $x_0 \in \mathfrak{J} \subset \mathfrak{T}$ so that $\zeta x_0 \in \mathfrak{T}$ for $|\zeta| \leqq 1$ and there is a finitely open set H containing θ such that $x_0 + H \subset \mathfrak{J}$. It follows that $\zeta x_0 + \zeta H \subset \mathfrak{T}$ for $|\zeta| < 1$. But the left side of this inclusion is a finitely open set containing ζx_0; thus $\zeta x_0 \in \mathfrak{J}$ and $\mathfrak{J}$ is a c-star. The c-convexity is proved by the same type of argument. The assumption that $x_0 + \Gamma h_0 \subset \mathfrak{J}$ implies the existence of a finitely open set H such that $x_0 + \Gamma h_0 + H \subset \mathfrak{J}$ or $(x_0 + H) + \Gamma h_0 \subset \mathfrak{J} \subset \mathfrak{T}$. Since $\mathfrak{T}$ is c-convex, this implies that $(x_0 + H) + \Delta h_0 \subset \mathfrak{T}$. But $x_0 + \Delta h_0 + H$ is a finitely open set containing $x_0 + \Delta h_0$ and thus belonging to $\mathfrak{J}$. This proves the c-convexity of $\mathfrak{J}$.

Thus to every system of finitely open c-convex c-stars containing $\mathfrak{D}$ there is a unique smallest set with the same properties which is contained in all of them, namely the finite interior of the intersection of the sets in the system. In particular there is a unique smallest set for the system of *all* finitely open c-convex c-stars containing $\mathfrak{D}$. We denote this minimal star by $\mathfrak{C}[\mathfrak{D}]$. The two preceding theorems then imply

THEOREM 26.5.11. *If $\mathfrak{D}$ is a given finitely open set in $\mathfrak{X}$ and if a (G)-power series $\sum P_k(x)$ converges in $\mathfrak{D}$, then $\mathfrak{C}[\mathfrak{D}] \subset \mathfrak{G}_0[P_k]$.*

Another consequence is the following

THEOREM 26.5.12. *If $f(x)$ is (G)-differentiable in a finitely open c-star $\mathfrak{D}$ about θ, then it may be continued as a (G)-differentiable function into all of $\mathfrak{C}[\mathfrak{D}]$.*

26.6. (F)-power series. Let $\{P_k(x)\}$ be a given sequence of (F)-powers, the degree of $P_k(x)$ being k, and consider the (F)-power series $\sum_0^\infty P_k(x)$. This is of course also a (G)-power series so the results of section 26.5 apply. Owing to the continuity of the terms, sharper results may be proved, however.

THEOREM 26.6.1. *If $\mathfrak{C}[P_k]$ is the region of convergence of an (F)-power series and if $\mathfrak{G}[P_k] \equiv \operatorname{Int} \mathfrak{C}[P_k]$ is non-void, then $\mathfrak{G}[P_k]$ is a c-convex c-star about θ.*

PROOF. As in Theorems 26.5.9 and 26.5.10, let $\mathfrak{G}_0[P_k]$ denote the largest finitely open set of convergence for the (F)-power series. Then clearly $\mathfrak{G}[P_k] \subset \mathfrak{G}_0[P_k] \subset \mathfrak{C}[P_k]$ and $\mathfrak{G}[P_k] = \operatorname{Int} \mathfrak{G}_0[P_k]$. Consequently we need only to show that the interior of a c-convex c-star set about θ is again of this type. Suppose that $\mathfrak{G}[P_k]$ is non-vacuous and contains the point x_0. Then there is an open sphere $[x_0 + H]$ about x_0 contained in $\mathfrak{G}_0[P_k]$. Since $\mathfrak{G}_0[P_k]$ is c-star about θ we see that $\zeta(x_0 + H)$, $|\zeta| \leqq 1$, belongs to $\mathfrak{G}_0[P_k]$ so that ζx_0, $|\zeta| \leqq 1$, is interior to $\mathfrak{G}_0[P_k]$ and therefore lies in $\mathfrak{G}[P_k]$. Thus $\mathfrak{G}[P_k]$ is c-star about θ. On the other hand suppose that there is a bounded open subset Δ of the complex plane with boundary Γ such that for some x_0 and h_0 the point set $[x_0 + \zeta h_0\,;\, \zeta \in \Gamma]$ belongs to $\mathfrak{G}[P_k]$. Then for each $\zeta_0 \in \Gamma$ there is a maximum positive number $\rho(\zeta_0)$ such that $x_0 + \zeta_0 h_0 + h \in \mathfrak{G}_0[P_k]$ if $\|h\| < \rho(\zeta_0)$. The set Γ being compact, $\inf \rho(\zeta_0) \equiv \rho > 0$. Thus $x_0 + \zeta h_0 + h \in \mathfrak{G}_0[P_k]$ when $\zeta \in \Gamma$, $\|h\| < \rho$. Since $\mathfrak{G}_0[P_k]$ is c-convex we see that $x_0 + \zeta h_0 + h \in \mathfrak{G}_0[P_k]$ for all $\zeta \in \Delta$ and $\|h\| < \rho$. Thus the set $[x_0 + \zeta h_0\,;\, \zeta \in \Delta]$ is interior to $\mathfrak{G}_0[P_k]$ and therefore belongs to $\mathfrak{G}[P_k]$. This proves the c-convexity of $\mathfrak{G}[P_k]$.

It is natural to expect that an (F)-power series should converge to an analytic function in $\mathfrak{G}[P_k]$, but this simple proposition is by no means easy to prove. The first stage of the proof is given in

THEOREM 26.6.2. *If $\mathfrak{G}[P_k]$ is non-vacuous, it contains a non-vacuous open c-star about θ in which the series converges uniformly to an analytic function.*

PROOF. If $\mathfrak{G}[P_k]$ is non-void, the sum of the series, $f(x)$ say, is a (G)-differentiable function of x in $\mathfrak{G}[P_k]$ by Theorem 26.5.6 and $f(x)$ may be expanded in a convergent Taylor series about each point of $\mathfrak{G}[P_k]$. In particular, the Maclaurin series is seen to be identical to the given (F)-power series, that is, $\delta^k f(\theta; x) = k!P_k(x)$ for all k and all x. We can then appeal to Theorem 26.3.8 which asserts the existence of a finitely open c-star $\mathfrak{C}$ about θ such that $\sum_k M_k$ converges where M_k is the supremum of $\| P_k(x) \|$ for x in $\mathfrak{C}$. $P_k(x)$ being continuous, we may replace $\mathfrak{C}$ by its closure $\bar{\mathfrak{C}}$ without changing the value of the supremum. It is desired to show that $\bar{\mathfrak{C}}$ contains an open sphere. Let $\mathfrak{C}_n$ denote the set of all points x such that $(1/n)x \in \mathfrak{C}$, $n = 2, 3, \cdots$. Every point $x \in \mathfrak{X}$ belongs to some $\bar{\mathfrak{C}}_n$ since $\mathfrak{C}$ is finitely open and contains θ. It follows that $\bar{\mathfrak{C}}_n \subset \bar{\mathfrak{C}}_{n+1}$ and hence that $\bigcup \bar{\mathfrak{C}}_n = \mathfrak{X}$. This shows that some set $\bar{\mathfrak{C}}_n$ contains an open sphere since $\mathfrak{X}$ is not of the first category in itself. It follows that $\bar{\mathfrak{C}}$ contains an open sphere, $\mathfrak{S}: \| x - a \| < \rho$ say. We have then $\| P_k(x) \| \leqq M_k$ for $x \in \mathfrak{S}$. By Lemma 26.5.1 the same inequality is valid in the open c-star $\mathfrak{C}_0: \{\zeta a + h\}$, $| \zeta | \leqq 1$, $\| h \| \leqq \rho$. In $\mathfrak{C}_0$ the (F)-power series converges uniformly to $f(x)$; the terms being continuous, $f(x)$ is also continuous in $\mathfrak{C}_0$ and consequently analytic there.

We note in passing the following theorem, the proof of which is left to the reader:

THEOREM 26.6.3. *If the terms of an (F)-power series are locally uniformly bounded in an open set $\mathfrak{D}$, then the series converges locally uniformly in $\mathfrak{D}$ and vice versa. The sum of the series is analytic in $\mathfrak{D}$.*

We come now to the main theorem:

THEOREM 26.6.4. *The sum of an (F)-power series is analytic in $\mathfrak{G}[P_k]$.*

PROOF. We already know that the sum $f(x)$ is analytic in some neighborhood of $x = \theta$. We now take $a \in \mathfrak{G}[P_k]$, $a \neq \theta$, and consider the Taylor series

$$f(a + h) = \sum_{n=0}^{\infty} \frac{1}{n!} \delta^n f(a; h)$$

which is a (G)-power series in h by Theorem 26.3.9. As such it converges in the largest finitely open c-convex c-star about a contained in the finite interior of $\mathfrak{G}[P_k]$. In particular, it will converge in a non-void open c-star about a, that is, it converges for h in a non-void open c-star about θ. We shall prove that this series is actually an (F)-power series in h. To this end we note that

$$\delta^n f(a; h) = \sum_{k=n}^{\infty} \delta^n P_k(a; h),$$

the convergence of the series being ensured by Theorem 26.5.5 for all values of n. Here $\delta^n P_k(a; h)$ is a homogeneous polynomial in h of degree n, continuous in h by Theorem 26.3.7. Thus the terms are (F)-powers in h of fixed degree n. By Theorem 26.2.9 the sum of the series $\delta^n f(a; h)$ is then either identically θ or else an (F)-power in h of degree n. Thus the Taylor series for $f(a + h)$ is actually an (F)-power series in h. The series being convergent in an open set containing $h = \theta$, Theorem 26.6.2 asserts that it converges uniformly in some neighborhood of $h = \theta$. Hence $f(a + h)$ is an analytic function of h for small values of $\| h \|$, that is, $f(x)$ is analytic everywhere in $\mathfrak{G}[P_k]$.

THEOREM 26.6.5. *If $a \in \mathfrak{G}[P_k]$, then there exists an open c-star $\mathfrak{C}$ about θ containing a, such that $\sum_k M_k$ converges where M_k is the supremum of $\| P_k(x) \|$ in $\mathfrak{C}$.*

PROOF. This is the analogue for (F)-differentiable functions of Theorem 26.3.8 and the same type of proof applies. Since $f(x)$ is now continuous in $\mathfrak{G}[P_k]$, the function $\varphi(x) = \max \| f(\zeta x) \|$, $| \zeta | \leqq 1$, is also a continuous function in $\mathfrak{G}[P_k]$. The set $\mathfrak{C}_0$ of points x such that $\varphi(x) < \varphi(a) + 1$ is an open c-star about θ containing $x = a$. If ϵ is a small positive number, $\mathfrak{C} = (1 - \epsilon)\mathfrak{C}_0$ still contains $x = a$ and is an open c-star about θ. From Theorem 3.16.3 we get $\| \delta^k f(\theta; x) \| \leqq k![1 + \varphi(a)](1 - \epsilon)^k$ for x in $\mathfrak{C}$, so that $M_k \leqq [1 + \varphi(a)](1 - \epsilon)^k$ and the convergence of the series $\sum M_k$ is obvious.

COROLLARY. *The (F)-power series $\sum_k P_k(x)$ converges locally uniformly in $\mathfrak{G}[P_k]$.*

The preceding theorems give a reasonably good qualitative description of $\mathfrak{G}[P_k]$. We can obtain some quantitative results related to the function $\rho(u)$ of Definition 26.5.1.

DEFINITION 26.6.1. *Let* $\rho_a = \inf [\rho(u); \| u \| = 1]$ *and put*

$$\rho_u = 1/\mu_u, \qquad \mu_u = \limsup_{n\to\infty} \| P_n \|^{1/n}.$$

We call ρ_a the radius of absolute and ρ_u the radius of uniform convergence of $\sum_k P_k(x)$.

We recall that $\| P_n \|$ is the norm of $P_n(x)$ in the sense of Definition 26.2.4 and is finite if and only if $P_n(x)$ is an (F)-power. For a true (G)-power series we have always $\rho_u = 0$. The justification of the terminology is given by:

THEOREM 26.6.6. *For the (F)-power series $\sum_0^\infty P_n(x)$ we have $0 < \rho_u \leqq \rho_a$ if and only if $\mathfrak{G}[P_k]$ is non-void. The series is absolutely convergent for $\| x \| < \rho_a$; this is the largest open sphere with center at θ contained in $\mathfrak{G}[P_k]$. On every spherical surface $\| x \| = \rho$ with $\rho_a < \rho$ there are points where the series diverges. It is uniformly convergent for $\| x \| < (1 - \epsilon)\rho_u$, $\epsilon > 0$, and fails to converge uniformly on any spherical surface $\| x \| = \rho$ with $\rho_u < \rho$.*

PROOF. Suppose that $\rho_a > 0$. Then for $x = \alpha u$, $\| u \| = 1$,

$$\limsup_{n\to\infty} \| P_n(x) \|^{1/n} = \| x \| / \rho(u) \leqq \| x \| / \rho_a,$$

so that the series converges absolutely in the sphere $\| x \| < \rho_a$; $\mathfrak{G}[P_k]$ is non-void and contains this sphere. On the other hand, from the definition of the infimum it follows that for every $\epsilon > 0$ there is a point u on the unit sphere such that $\rho(u) < \rho_a + \epsilon$. Hence if $| \alpha | > \rho_a + \epsilon$, $\limsup_n \| P_n(\alpha u) \|^{1/n} > 1$ and the series diverges for $x = \alpha u$. Thus every spherical surface $\| x \| = \rho$ with $\rho > \rho_a$ contains infinitely many points where the series diverges because the terms do not tend to zero. It follows that $\| x \| < \rho_a$ is the largest open sphere with center at θ contained in $\mathfrak{G}[P_k]$. The converse follows from Theorem 26.5.1.

If $\rho_u > 0$ and $\| x \| < (1 - \epsilon)\rho_u$, then

$$\| P_n(x) \| \leqq \| P_n \| \| x \|^n \leqq [\rho_u(1 - \epsilon/2)]^{-n}[\rho_u(1 - \epsilon)]^n < (1 - \epsilon/2)^n$$

for $n \geqq n(\epsilon)$ and the uniform convergence of the series for $\| x \| < (1 - \epsilon)\rho_u$ is evident. On the other hand, the definition of the supremum implies that for every $\epsilon > 0$ the inequality $\| P_n \| \geqq [\rho_u(1 + \epsilon)]^{-n}$ holds for infinitely many values of n. For each such n we can find a u_n on the unit sphere such that $\| P_n(u_n) \| \geqq [\rho_u(1 + 2\epsilon)]^{-n}$ and if ρ is given, $\rho_u < \rho$, ϵ may be chosen so small that

$$\| P_n(\rho u_n) \| = \rho^n \| P_n(u_n) \| > \rho^n[\rho_u(1 + 2\epsilon)]^{-n} \equiv A^n$$

where $A > 1$. This implies that the series cannot converge uniformly on the spherical surface $\| x \| = \rho$ when $\rho > \rho_u$. Theorem 26.6.2 shows that $\rho_u > 0$ whenever $\rho_a > 0$. This completes the proof.

We may very well have $\rho_u < \rho_a$. This is shown by the following modification of the example used in section 26.5. We take $\mathfrak{X} = l_2$, $\mathfrak{Y} = Z_1$ and

$$f(x) = \sum_1^\infty (\alpha_n)^n, \qquad x = \{\alpha_1, \alpha_2, \cdots, \alpha_n, \cdots\}.$$

This is an (F)-power series and a simple computation shows that $\rho_u = 1$ and $\rho_a = \infty$.

26.7. A theorem on analyticity. We now return to the problem of determining when a (G)-differentiable function is actually (F)-differentiable. The earliest result in this direction is due to A. E. Taylor (1937) who showed that the added assumption of continuity was sufficient for this purpose. We have already shown in section 3.17 that continuity can be replaced by local boundedness; and we now show that continuity can also be replaced by continuity in the sense of Baire, a result due to Max Zorn [1]. For the argument we shall need two simple lemmas which we state without proof.

LEMMA 26.7.1. *If $f(x)$ and $g(x)$ are Baire continuous, so are $\alpha f(x) + \beta g(x)$ and $f(\alpha x + a)$, where α, β, and a are fixed, $\alpha, \beta \in Z_1$, $a \in \mathfrak{X}$.*

LEMMA 26.7.2. *If $f(x)$ is Baire continuous for $\|x - a\| < \rho$ and $F(x) = f(x)$ or θ according as $\|x - a\| < \rho$ or $\geqq \rho$, then $F(x)$ is continuous in the sense of Baire for all x.*

THEOREM 26.7.1. *If $f(x)$ is defined as a (G)-differentiable function, continuous* *in the sense of Baire, in an open set* $\mathfrak{D}$*, then* $f(x)$ *is (F)-differentiable and hence analytic in* $\mathfrak{D}$.

PROOF. The basic idea of the proof is to show that the Baire continuity of $f(x)$ with respect to x extends to the variations $\delta^n f(x; h_1, \cdots, h_n)$ with respect to all $(n + 1)$ variables. Since these variations are symmetric n-linear forms of $h_1, \cdots, h_n$, Baire continuity implies ordinary continuity with respect to the increments by virtue of Theorem 26.2.8; this in turn implies that $\delta^n f(x; h)$ is an (F)-power in h and the Taylor series is an (F)-power series with a positive radius of absolute convergence so that $f(x + h)$ is analytic in h, for small $\| h \|$ and $x \in \mathfrak{D}$. This makes $f(x)$ analytic in $\mathfrak{D}$.

We start with the first variation. Let $a \in \mathfrak{D}$ be fixed and let $\mathfrak{S}$: $\| x - a \| < \rho$ be a sphere in $\mathfrak{D}$. We define

$$F(x) = \begin{cases} f(x), \ x \in \mathfrak{S}, \\ \theta, \ x \in \mathfrak{X} \ominus \mathfrak{S}; \end{cases}$$

$$F_n(x, h) = n\left[F\left(x + \frac{1}{n}h\right) - F(x)\right], \qquad n = 1, 2, 3, \cdots.$$

The functions are continuous in the sense of Baire with respect to both x and h by Lemmas 26.7.1 and 26.7.2. Further

$$\lim_{n\to\infty} F_n(x, h) = \delta f(x; h), \qquad x \in \mathfrak{S}, h \in \mathfrak{X}.$$

Theorem 26.2.7 then implies that $\delta f(x; h)$ is Baire continuous with respect to x (in $\mathfrak{S}$ to start with but the extension to all of $\mathfrak{D}$ is immediate) and $h \in \mathfrak{X}$. As already observed, this makes $\delta f(x; h)$ continuous in h for x fixed in $\mathfrak{D}$. We can now apply the same type of argument to $\delta f(x; h_1)$. It is a (G)-differentiable, Baire continuous function of x in $\mathfrak{D}$ and a continuous function of h_1. It follows that $\delta^2 f(x; h_1, h_2)$ is continuous in the sense of Baire with respect to all three variables and hence continuous in the ordinary sense with respect to h_1 and h_2. In the same manner we prove that all variations $\delta^n f(x; h_1, \cdots, h_n)$ are continuous in the sense of Baire with respect to all variables and hence continuous with respect to $h_1, \cdots, h_n$ so that $\delta^n f(x; h)$ is also a continuous function of h for $x \in \mathfrak{D}$. The Taylor series

$$f(x + h) = \sum_0^\infty \frac{1}{n!}\delta^n f(x; h), \qquad x \in \mathfrak{D},$$

converges in a neighborhood of $h = \theta$ and is an (F)-power series in h. By Theorem 26.6.4 it is then an analytic function of h in this neighborhood of $h = \theta$. Hence $f(x)$ is analytic in $\mathfrak{D}$ and the theorem is proved.

BIBLIOGRAPHY

Alaoglu, L.

1. *Weak topologies of normed linear spaces.* Annals of Math., (2) 41 (1940) 252–267.

Alaoglu, L., and **Birkhoff, G.**

1. *General ergodic theorems.* Proc. Nat. Acad. Sci., 25 (1939) 628–630.
2. *General ergodic theorems.* Annals of Math., (2) 41 (1940) 293–309.

Alexandroff, P., and **Hopf, H.**

1. *Topologie.* I. J. Springer, Berlin, xiii + 636 pp.

Alexiewicz, A., and **Orlicz, W.**

1. *Analytic operations in real Banach spaces.* Studia Math., 14 (1953) 57–78.

Andersen, A. F.

1. *Studier over Cesàro's Summabilitetsmetode.* Thesis. J. Gjellerup, Copenhagen, 1921, 100 pp.

Arens, R. F.

1. *On a theorem of Gelfand and Neumark.* Proc. Nat. Acad. Sci., 32 (1946) 237–239.

Arens, R. F., and **Singer, I. M.**

1. *Function values as boundary integrals.* Proc. Amer. Math. Soc., 5 (1954) 735–745.

Arnold, I.

1. *Ideale in kommutativen Halbgruppen.* Rec. Math., 36 (1929) 401–407.

Austin, D. C.

1. *On the existence of the derivative of the Markoff transition probability functions.* Proc. Nat. Acad. Sci., 41 (1955) 224–226.

Baer, R.

1. *Radical ideals.* Amer. J. Math., 65 (1943) 537–568.

Baker, B. B., and **Copson, E. T.**

1. *The mathematical theory of Huygens' principle.* Oxford Univ. Press, 1939, vi + 155 pp.

Baker, G. A., Jr.

1. *Einstein numbers.* Amer. Math. Monthly, 61 (1954) 39–41.

Banach, S.

1. *Sur les fonctionnelles linéaires* II. Studia Math., 1 (1929) 223–239.
2. *Théorie des opérations linéaires.* Warszawa, 1932, vii + 254 pp.
3. *Die Theorie der Operationen und ihre Bedeutung für die Analysis.* C. R. Congrès Internat. Math., Oslo, 1936. I, 261–268.
4. *Über homogene Polynome in* (L^2). Studia Math., 7 (1938) 36–44.

Bergström, V.

1. *Beiträge zur Theorie der endlichdimensionalen Moduln und der diophantischen Approximationen.* Dissertation, Lund, 1935, 97 pp.

Bernstein, F.

1. *Die Integralgleichung der elliptischen Thetanullfunktion.* Sitzungsber. preuss. Akad. Wiss., 1920, 735–747.

Bernstein, S.

1. *Demonstration du théorème de Weierstrass, fondée sur le calcul des probabilités.* Comm. Soc. Math. Charkow, (2) 13 (1912) 1–2.

Beurling, A.

1. *Sur les intégrales de Fourier absolument convergentes et leur application à une transformation fonctionnelle.* Neuvième Congrès Math. Scand., Helsingfors, 1938, 345–366.

Birkhoff, G. (Cf. **Alaoglu, L.**)

1. *Integration of functions with values in a Banach space.* Trans. Amer. Math. Soc., 38 (1935) 357–378.
2. *Moore-Smith convergence in general topology.* Annals of Math., (2) 38 (1937) 39–56.
3. *Analytical groups.* Trans. Amer. Math. Soc., 43 (1938) 61–101.
4. *An ergodic theorem for general semi-groups.* Proc. Nat. Acad. Sci., 25 (1939) 625–627.
5. *Lattice theory.* Amer. Math. Soc. Coll. Publ., XXV. New York, 1948, xiv + 283 pp.

Bochner, S.

1. *Vorlesungen über Fouriersche Integrale.* Akad. Verlagsges., Leipzig, 1932, viii + 229 pp.
2. *Integration von Funktionen, deren Werte die Elemente eines Vektorraumes sind.* Fund. Math., 20 (1933) 262–276.
3. *Completely monotone functions in partially ordered spaces.* Duke Math. J., 9 (1942) 519–526.
4. *Diffusion equations and stochastic processes.* Proc. Nat. Acad. Sci., 35 (1949) 368–370.

Bochner, S., and **Phillips, R. S.**

1. *Absolutely convergent Fourier expansions for non-commutative normed rings.* Annals of Math., (2) 43 (1942) 409–418.

Bochner, S., and **Taylor, A. E.**

1. *Linear functionals on certain spaces of abstract-valued functions.* Annals of Math., (2) 39 (1938) 913–944.

Bohnenblust, H. F., and **Sobczyk, A.**

1. *Extensions of functionals on complex linear spaces.* Bull. Amer. Math. Soc., 44 (1938) 91–93.

Bourbaki, N.

1. *Sur les espaces de Banach.* C.R. Acad. Sci. Paris, 206 (1938) 1701–1704.

Burgess, D. C. J.

1. *Abstract moment problems with applications to the l^p and L^p spaces.* Proc. London Math. Soc., (3) 4 (1954) 107–128.

Butzer, P. L.

1. *Über den Approximationsgrad des Identitätsoperators durch Halbgruppen von linearen Operatoren und Anwendungen auf die Theorie der singulären Integrale.* Math. Annalen, 133 (1957) 410–425.

Calkin, J. W.

1. *Two-sided ideals and congruences in the ring of bounded operators in Hilbert space.* Annals of Math., (2) 42 (1941) 839–873.

Carlson, F.

1. *Sur une classe de série de Taylor.* Thèse. Upsala, 1914, 76 pp.
2. *Sur les séries de coefficients binominiaux.* Nova Acta R. Soc. Sci. Upsaliensis, (4) 4, No. 3 (1915) 61 pp.

Cartan, E.

1. *La théorie des groupes finis et continus et l'analysis situs.* Mémorial Sci. Math., 42. Gauthier-Villars, Paris, 1930, 60 pp.

Caton, W. B., and **Hille, E.**

1. *Laguerre polynomials and Laplace integrals.* Duke Math. J., 12 (1945) 217–242.

Clarkson, J. A.

1. *Uniformly convex spaces.* Trans. Amer. Math. Soc., 40 (1936) 396–414.

Clifford, A. H. (Cf. **Michal, A. D.**)

1. *Matrix representations of completely simple semigroups.* Amer. J. Math., 64 (1942) 327–342.

Cohen, L. W.

1. *On the mean ergodic theorem.* Annals of Math., (2) 41 (1940) 505–509.

Cooper, J. L. B.

1. *One-parameter semi-groups of isometric operators in Hilbert space.* Annals of Math. (2) 48 (1947) 827–842.

Cooper, R.

1. *The converse of the Cauchy-Hölder inequality and the solutions of the inequality* $g(x + y) \leqq g(x) + g(y)$. Proc. London Math. Soc., (2) 26 (1927) 415–432.

Copson, E. T. (Cf. **Baker, B. B.**)

1. *Some applications of Marcel Riesz's integrals of fractional order.* Proc. R. Soc. Edinburgh, A 61 (1943) 260–272.

Courant, R., and **Hilbert, D.**

1. *Methoden der mathematischen Physik,* I. J. Springer, Berlin, 1931, xiv + 469 pp.

Cramér, H.

1. *Random variables and probability distributions.* Cambridge Univ. Press, 1937, viii + 121 pp.
2. *Mathematical methods of statistics.* Almqvist and Wiksell, Uppsala, 1945, and Princeton Univ. Press, 1946, xvi + 575 pp.

Dahlgren, L.

1. *A theorem on translations by Hille, and its interpretation from the point of view of probability.* Skand. Aktuarietidskrift, 33 (1950) 184–192.

van Dantzig, D.

1. *Zur topologischen Algebra.* I. Math. Annalen, 107 (1932) 587–626.

Day, M. M.

1. *The spaces L^p with $0 < p < 1$.* Bull. Amer. Math. Soc., 46 (1940) 816–823.
2. *Some more uniformly convex spaces.* Bull. Amer. Math. Soc., 47 (1941) 504–507.
3. *Ergodic theorems for abelian semi-groups.* Trans. Amer. Math. Soc., 51 (1942) 399–412.
4. *Means for the bounded functions and ergodicity of the bounded representations of semi-groups.* Trans. Amer. Math. Soc., 69 (1950) 276–291.

Devinatz, A.

1. *A note on semi-groups of unbounded self-adjoint operators.* Proc. Amer. Math. Soc., 5 (1954) 101–102.

Dickson, L. E.

1. *On semi-groups and the general isomorphism between infinite groups.* Trans. Amer. Math. Soc., 6 (1905), 205–208.

Dieudonné, J.

1. *Sur le théorème de Banach.* La Revue Scientifique (Paris), 79 (1941) 642–643.
2. *Sur le théorème de Lebesgue-Nikodym* (V). Canadian J. Math., 3 (1951) 129–139.

Doetsch, G.

1. *Handbuch der Laplace-Transformation,* I. Birkhäuser, Basel, 1950, 581 pp.

Doob, J. L.

1. *Stochastic processes depending on a continuous parameter.* Trans. Amer. Math. Soc., 42 (1937) 107–140.
2. *Topics in the theory of Markoff chains.* Trans. Amer. Math. Soc., 52 (1942) 37–64.
3. *Markoff chains— denumerable case.* Trans. Amer. Math. Soc., 58 (1945) 455–473.

Dubreil, P.

1. *Contributions à la théorie des demi-groupes.* Mém. Acad. Sci., Inst. de France, 63, No. 3 (1941) 52 pp.

Dunford, N.

1. *Integration in general analysis.* Trans. Amer. Math. Soc., 37 (1935) 441–453; 38 (1935) 600–601.
2. *Integration and linear operations.* Trans. Amer. Math. Soc., 40 (1936) 474–494.
3. *Uniformity in linear spaces.* Trans. Amer. Math. Soc., 44 (1938) 305–356.

4. *On one parameter groups of linear transformations.* Annals of Math., (2) 39 (1938) 569–573.
5. *An ergodic theorem for n-parameter groups.* Proc. Nat. Acad. Sci., 25 (1939) 195–196.
6. *A mean ergodic theorem.* Duke Math. J., 5 (1939) 635–646.
7. *Spectral theory.* Bull. Amer. Math. Soc., 49 (1943), 637–651.
8. *Spectral theory* I. *Convergence to projections.* Trans. Amer. Math. Soc., 54 (1943) 185–217.
9. *Direct composition of Banach spaces.* Boletin de la Sociedad Matematica Mexicana, 3 (1946) 1–12.

Dunford, N., and **Hille, E.**

1. *The differentiability and uniqueness of continuous solutions of addition formulas.* Bull. Amer. Math. Soc., 50 (1944) 668; 53 (1947) 799–805.

Dunford, N., and **Schwartz, J.**

1. *Convergence almost everywhere of operator averages.* Proc. Nat. Acad. Sci., 41 (1955) 229–231.

Dunford, N., and **Segal, I. E.**

1. *Semi-groups of operators and the Weierstrass theorem.* Bull. Amer. Math. Soc., 52 (1946) 911–914.

Dye, H. A., and **Phillips, R. S.**

1. *Groups of positive operators.* Canadian J. Math., 8 (1956) 462–486.

Eberlein, W. F.

1. *Weak compactness in Banach spaces.* I. Proc. Nat. Acad. Sci., 33 (1947) 51–53.
2. *Abstract ergodic theorems.* Proc. Nat. Acad. Sci., 34 (1948) 43–47.
3. *Abstract ergodic theorems and weak almost periodic functions.* Trans. Amer. Math. Soc., 67 (1949) 217–240.

Eidelheit, M.

1. *Zur Theorie der konvexen Mengen in linearen normierten Räumen.* Studia Math., 6 (1936) 104–111.

Evans, G. C.

1. *Functionals and their applications. Selected topics, including integral equations.* Amer. Math. Soc. Coll. Publ., V_1. New York, 1918, xii + 136 pp.

Fantappiè, L.

1. *Le calcul des matrices.* C. R. Acad. Sci. Paris, 186 (1928) 619–621.
2. *I funzionali analitici.* Mem. R. Accad. Naz. dei Lincei, (6) 3 (1930) 453–683.
3. *Überblick über die Theorie der analytischen Funktionale und ihre Anwendungen.* Jahresber. D.M.V., 43 (1933) 1–25.
4. *Nuovi fondamenti della teoria dei funzionali analitici.* Atti Accad. Italia, Mem. Cl. Sci. Fis., Mat., Nat., 12 (1942) 617–706.

Favard, J.

1. *Sur les multiplicateurs d'interpolation.* J. Math. Pures et Appliqués, (9) 23 (1944) 219–247.

Feller, W.

1. *Neuer Beweis für die Kolmogoroff—P. Lévysche Charakterisierung der unbeschränkt teilbaren Verteilungsfunktionen.* Bull. Internat. Acad. Yougoslave des Sci., Cl. Sci. Math. et Nat., 32 (1939) no. 13, 8 pp. [From Rad, 263 (1939) 95–112].
2. *On the integro-differential equations of purely discontinuous Markoff processes.* Trans. Amer. Math. Soc., 48 (1940) 488–515, 58 (1945) 474.
3. *The parabolic differential equations and the associated semi-groups of transformations.* Annals of Math., (2) 55 (1952) 468–519.
4. *On a generalization of Marcel Riesz' potentials and the semi-groups generated by them.* Proc. R. Physiogr. Soc. Lund, 21 (1952) 73–81.

5. *Semi-groups of transformations in general weak topologies.* Annals of Math., (2) 57 (1953) 287–308.
6. *On the generation of unbounded semi-groups of bounded linear operators.* Annals of Math., (2) 58 (1953) 166–174.

Fréchet, M.

1. *Une définition fonctionnelle des polynomes.* Nouv. Ann. Math., (4) 9 (1909) 145–162.
2. *Sur les fonctionnelles continues.* Ann. École Norm. Sup., (3) 27 (1910) 193–216.
3. *La notion de différentielle dans l'analyse générale.* Ann. École Norm. Sup., (3) 42 (1925) 293–323.
4. *Les espaces abstraits.* Gauthier-Villars, Paris, 1928, xi + 296 pp.
5. *Les polynomes abstraits.* J. de Math., (9) 8 (1929) 71–92.
6. *Recherches théoriques modernes sur le calcul des probabilités.* 2. *Méthode des fonctions arbitraires. Theorie des événements en chaîne dans le cas d'un nombre fini d'états possible.* [Tome I, fasc. 3 of É. Borel et al., *Traité du calcul des probabilités*, Gauthier-Villars, Paris, 1938, x + 315 pp.]

Fremberg, N. E.

1. *Proof of a theorem of M. Riesz concerning a generalization of the Riemann-Liouville integral.* Proc. R. Physiogr. Soc. Lund, 10 (1945) 265–276.
2. *A study of generalized hyperbolic potentials with some physical applications.* Dissertation. Comm. Sem. Math. Univ. Lund, 7 (1946) 1–100.

Frobenius, G.

1. *Über die cogredienten Transformationen der bilinearen Formen.* Sitzungsber. K. preuss. Akad. Wiss. Berlin, 1896:1, 7–16.

Frobenius, G., and Schur, I.

1. *Über die Äquivalenz der Gruppen linearer Substitutionen.* Sitzungsber. K. preuss. Akad. Wiss. Berlin, 1906, 209–217.

Fuglede, B.

1. *A commutativity theorem for normal operators.* Proc. Nat. Acad. Sci., 36 (1950) 35–40.

Fukamiya, M.

1. *On one-parameter groups of operators.* Proc. Imp. Acad. Sci. Tokyo, 16 (1940) 262–265.

Gantmacher, V.

1. *Über schwache totalstetige Operatoren.* Rec. Math., 7 (1940) 301–308.

Garabedian, H. L., Hille, E., and Wall, H. S.

1. *Formulations of the Hausdorff inclusion problem.* Duke Math. J., 8 (1941) 193–213.

Gårding, L.

1. *Note on continuous representations of Lie groups.* Proc. Nat. Acad. Sci., 33 (1947) 331–332.
2. *The solution of Cauchy's problem for two totally hyperbolic linear differential equations by means of Riesz integrals.* Annals of Math., (2) 48 (1947) 785–826.
3. *Linear hyperbolic partial differential equations with constant coefficients.* Acta Math., 85 (1951) 1–62.

Gâteaux, R.

1. *Fonctions d'une infinité des variables indépendantes.* Bull. Soc. Math. France, 47 (1919) 70–96.
2. *Sur les fonctionnelles continues et les fonctionnelles analytiques.* Bull. Soc. Math. France, 50 (1922) 1–21.

Gelfand, I.

1. *Sur un lemme de la théorie des espaces linéaires.* Comm. Soc. Math. Kharkoff, (4) 13 (1936) 35–40.
2. *Abstrakte Funktionen und lineare Operatoren.* Rec. Math., N.S. 4 (1938) 235–284.
3. *On one-parametrical groups of operators in a normed space.* C. R. (Doklady) Acad Sci. URSS, N.S. 25 (1939) 713–718.

4. *Normierte Ringe.* Rec. Math., N.S. 9 [51] (1941) 3–24.
5. *Zur Theorie der Charaktere der Abelschen topologischen Gruppen.* Rec. Math., N.S. 9 [51] (1941) 49–50.
6. *Über absolut konvergente trigonometrische Reihen und Integrale.* Rec. Math., N.S. 9 [51] (1941) 51–66.

Gelfand, I., and Neumark, M.
1. *On the imbedding of normed rings into the rings of operators in Hilbert space.* Rec. Math., N.S. 12 [54] (1943) 197–213.

Gelfand, I., Raikov, D., and Šilov, G.
1. *Commutative normed rings.* Uspeki Mat. Nauk, N.S. 2 (1946) 48–146.

Gelfand, I., and Šilov, G.
1. *Über verschiedene Methoden der Einführung der Topologie in die Menge der maximalen Ideale eines normierten Ringes.* Rec. Math., N.S. 9 [51] (1941) 25–39.

Giorgi, G.
1. *Nuove osservazioni sulle funzioni delle matrici.* Rendiconti R. Accad. Naz. dei Lincei, (6) 8 (1928) 3–8.

Graves, L. M. (Cf. **Hildebrandt, T. H.**)
1. *Riemann integration and Taylor's theorem in general analysis.* Trans. Amer. Math. Soc., 29 (1927) 163–177.
2. *Implicit functions and differential equations in general analysis.* Trans. Amer. Math. Soc., 29 (1927) 514–552.
3. *Topics in the functional calculus.* Bull. Amer. Math. Soc., 41 (1935) 641–662.

Hadamard, J.
1. *Essais sur l'étude des fonctions données par leur developpement de Taylor.* J. de Math., (4) 8 (1892) 101–186.
2. *Sur un problème mixte aux dérivées partielles.* Bull. Soc. Math. France, 31 (1903) 208–224.
3. *Principe de Huygens et prolongement analytique.* Bull. Soc. Math. France, 52 (1924) 241–278.
4. *Le principe de Huygens.* Bull. Soc. Math. France, 52 (1924) 610–640.

Hahn, H.
1. *Über lineare Gleichungssysteme in linearen Räumen.* J. f. Math., 157 (1927) 214–229.

Halmos, P. R.
1. *Measure theory.* D. Van Nostrand, New York, 1950, xi + 304 pp.

Halmos, P. R., Lumer, G., and Schäffer, J. J.
1. *Square roots of operators.* Proc. Amer. Math. Soc., 4 (1953) 142–149.

Hardy, G. H., Littlewood, J. E., and Pólya, G.
1. *Inequalities.* Cambridge Univ. Press, 1934, xii + 314 pp.

Hartogs, F.
1. *Zur Theorie der analytischen Funktionen mehrerer unabhängiger Veränderlichen, insbesondere über die Darstellung derselben durch Reihen, welche nach Potenzen einer Veränderlichen fortschreiten.* Math. Annalen, 62 (1906) 1–88.

Hausdorff, F.
1. *Summationsmethoden und Momentfolgen.* I, II. Math. Z., 9 (1921) 74–109, 280–299.

Helly, E.
1. *Über lineare Funktionaloperationen.* Sitzungsber. Akad. Wiss. Wien, Math.-nat. Kl. IIa, 121 (1912) 265–297.
2. *Über Systeme linearer Gleichungen mit unendlich vielen Unbekannten.* Monatshefte f. Math. u. Phys. 31 (1921) 60–91.

Henstock, R., and Macbeath, A. M.
1. *On the measure of sum sets.* (I) *The theorems of Brunn, Minkowski, and Lusternik.* Proc. London Math. Soc., (3) 3 (1953) 182–194.

Highberg, I. E.

1. *Polynomials in abstract spaces.* Thesis, Cal. Inst. of Techn. 1936.
2. *A note on abstract polynomials in complex spaces.* J. de Math., (9) 16 (1937) 307–314.

Hilbert, D. (Cf. **Courant, R.**)

1. *Wesen und Ziele einer Analysis der unendlich vielen unabhängigen Variabeln.* Rend. Circ. Mat. Palermo, 27 (1909) 59–74.

Hildebrandt, T. H.

1. *Lesbesgue integration in general analysis.* Bull. Amer. Math. Soc., 33 (1927) 646.
2. *Über lineare vollstetige Transformationen.* Acta Math., 51 (1928) 311–318.
3. *On bounded linear functional operations.* Trans. Amer. Math. Soc., 36 (1934) 868–875.
4. *Integration in abstract spaces.* Bull. Amer. Math. Soc., 59 (1953) 111–139.

Hildebrandt, T. H., and **Graves, L. M.**

1. *Implicit functions and their differentials in general analysis.* Trans. Amer. Math. Soc., 29 (1927) 127–153.

Hille, E. (Cf. **Caton, W. B., Dunford, N.,** and **Garabedian, H. L.**)

1. *Note on Dirichlet's series with complex exponents.* Annals of Math., (2) 25 (1924) 261–278.
2. *A class of reciprocal functions.* Annals of Math., (2) 27 (1926) 427–468.
3. *On Laplace integrals.* 8[de] Skand. Mat. Kongr. Stockholm 1934, 216–227.
4. *Notes on linear transformations,* I. Trans. Amer. Math. Soc., 39 (1936) 131–153.
5. *On semi-groups of transformations in Hilbert space.* Proc. Nat. Acad. Sci., 24 (1938) 159–161.
6. *Analytical semi-groups in the theory of linear transformations.* 9ème Congrès Math. Scand. Helsingfors 1938, 135–145.
7. *Notes on linear transformations.* II. *Analyticity of semi-groups.* Annals of Math., (2) 40 (1939) 1–47.
8. *Remarks concerning group spaces and vector spaces.* Comp. Math., 6 (1939) 375–381.
9. *Representation of one-parameter semi-groups of linear transformations.* Proc. Nat. Acad. Sci., 28 (1942) 175–178.
10. *On the analytical theory of semi-groups.* Proc. Nat. Acad. Sci., 28 (1942) 421–424.
11. *On the theory of characters of groups and semi-groups in normed vector rings.* Proc. Nat. Acad. Sci., 30 (1944) 58–60.
12. *Remarks on ergodic theorems.* Trans. Amer. Math. Soc., 57 (1945) 246–269.
13. *Functional analysis and semi-groups.* Amer. Math. Soc. Coll. Publ., XXXI, New York, 1948, xii + 528 pp.
14. *Les semi-groupes linéaires.* C. R. Acad. Sci. Paris, 227 (1949) 35–37.
15. *Lie theory of semi-groups of linear transformations.* Bull. Amer. Math. Soc., 56 (1950) 89–114.
16. *On the differentiability of semi-group operators.* Acta Sci. Math. Szeged, 12 B (1950) 19–24.
17. *On the generation of semi-groups and the theory of conjugate functions.* Proc. R. Physiogr. Soc. Lund, 21:14 (1951) 130–142.
18. *A note on Cauchy's problem.* Annales Soc. Polonaise de Math., 25 (1952), 56–68.
19. *Une généralisation du problème de Cauchy.* Annales Institut Fourier (Grenoble), IV, Année 1952 (1953) 31–48.
20. *Le problème abstrait de Cauchy.* Rend. Sem. Mat. Univ. e Politec. Torino, 12 (1953) 95–103.
21. *The abstract Cauchy problem and Cauchy's problem for parabolic differential equations.* J. d'Analyse Math., 3 (1954) 81–196.
22. *Sur un théorème de perturbation.* Rend. Sem. Mat. Univ. e Politec. Torino, 13 (1954) 169–184.

23. *Perturbation methods in the study of Kolmogoroff's equations.* Proc. Int. Congr. Math. Amsterdam, 1954, III, 365–376.

Hille, E., and **Tamarkin, J. D.**

1. *On the theory of linear integral equations.* I. Annals of Math., (2) 31 (1930) 479–528.
2. *On the summability of Fourier series.* II. Annals of Math., (2) 34 (1933) 329–348.
3. *On the summability of Fourier series.* III. Math. Annalen, 108 (1933) 525–577.
4. *Questions of relative inclusion in the domain of Hausdorff means.* Proc. Nat. Acad. Sci., 19 (1933) 573–577.
5. *On moment functions.* Proc. Nat. Acad. Sci., 19 (1933) 902–908.
6. *On the theory of Laplace integrals.* I, II. Proc. Nat. Acad. Sci., 19 (1933) 908–912; 20 (1934) 140–144.
7. *On the absolute integrability of Fourier transforms.* Fund. Math., 25 (1935) 329–352.

Hille, E., and **Zorn, M.**

1. *Open additive semi-groups of complex numbers.* Annals of Math., (2) 44 (1943) 554–561.

Hopf, E.

1. *Ergodentheorie.* Ergebnisse d. Math., 5:2, J. Springer, Berlin, 1937, v + 83 pp.

Hopf, H. See **Alexandroff, P.**

Hukahura, M.

1. *Sur le fonction $S(x)$ de M. E. Kamke.* Jap. J. Math., 17 (1941) 289–298.

Hyers, D. H.

1. *Linear topological spaces.* Bull. Amer. Math. Soc., 51 (1945) 1–21.

Jacobson, N.

1. *Structure theory of simple rings without finiteness assumptions.* Trans. Amer. Math. Soc., 57 (1945) 228–245.
2. *The radical and semi-simplicity for arbitrary rings.* Amer. J. Math., 67 (1945) 300–320.
3. *Some remarks on one sided inverses.* Proc. Amer. Math. Soc., 1 (1950) 352–355.

James, R. C.

1. *A non-reflexive Banach space isometric with its second conjugate space.* Proc. Nat. Acad. Sci., 37 (1951) 174–177.

Kakutani, S.

1. *Ein Beweis des Satzes von M. Eidelheit über konvexe Mengen.* Proc. Imp. Acad. Tokyo, 13 (1937) 93–94.
2. *Iteration of linear operations in complex Banach spaces.* Proc. Imp. Acad. Tokyo, 14 (1938) 295–300.
3. *Weak topology and regularity of Banach spaces.* Proc. Imp. Acad. Tokyo, 15 (1939) 169–173.
4. *Weak topology, bicompact sets, and the principle of duality.* Proc. Imp. Acad. Tokyo, 16 (1940) 63–67.
5. *Concrete representations of abstract (L)-spaces and the mean ergodic theorem.* Annals of Math., (2) 42 (1941) 523–537.
6. *Concrete representations of abstract (M)-spaces.* Annals of Math., (2) 42 (1941) 994–1024.
7. *Ergodic theory.* Proc. Internat. Congr. Math. Cambridge 1950, II, 128–142.
8. *A proof of Schauder's theorem.* J. Math. Soc. Japan, 3 (1951) 228–231.

Kakutani, S., and Klee, V. L., Jr.

1. *The finite topology of a linear space*, Arch. Math., 14 (1963) 55–58.

Kamke, E.

1. *Über die eindeutige Bestimmtheit der Integrale von Differentialgleichungen*, II. Sitzungsber. Akad. Wiss. Heidelberg, Math.-nat. Kl., 1931, no. 17, 15 pp.

Kaplansky, I.

1. *Topological rings.* Amer. J. of Math., 69 (1947) 153–183.

Kato, T.

1. *On the semi-groups generated by Kolmogoroff's differential equations.* J. Math. Soc. Japan, 6 (1954) 1–15.

Kelley, J. L.

1. *Note on a theorem of Krein and Milman.* J. Osaka Inst. Sci. Techn., I, 3 (1951) 1–2.

Kendall, D. G.

1. *Bernstein polynomials and semi-groups of operators.* Math. Scand., 2 (1954) 185–186.

Kendall, D. G., and **Reuter, G. E. H.**

1. *Some pathological Markov processes with a denumerable infinity of states and the associated semi-groups of operators on l.* Proc. Int. Congr. Math. Amsterdam 1954, III, 377–415.

Kerner, M.

1. *Die Differentiale in der allgemeinen Analysis.* Annals of Math., (2) 34 (1933) 546–572.

Khintchine, A.

1. *Deduction nouvelle d'une formule de M. Paul Lévy.* Bull. Univ. d'état à Moscou. Série Int., Section A, 1 (1937) 1–5.
2. *Contribution à l'arithmétique des lois de distribution.* Bull. Univ. d'état à Moscou. Série Int., Section A, 1 (1937) 6–17.

Kitagawa, T.

1. *On the theory of linear translatable functional equations and Cauchy's series.* Jap. J. Math., 13 (1937) 233–332.

Klee, V. L., Jr.

1. *Convex sets in linear spaces.* Duke Math. J., 18 (1951) 443–466.

Kober, H.

1. *On a theorem of Schur and on fractional integrals of purely imaginary order.* Trans. Amer. Math. Soc., 50 (1941) 160–174.

Kolmogoroff, A. N.

1. *Über die analytischen Methoden der Wahrscheinlichkeitsrechnung.* Math. Annalen, 104 (1931) 415–458.
2. *Sulla forme generale di un processo stocastico omogeneo.* (*Un problema di Bruno di Finetti.*) Atti Accad. Naz. dei Lincei, Rend., (6) 15 (1932) 805–808, 866–869.
3. *Zur Normierbarkeit eines allgemeinen topologischen Raumes.* Studia Math., 5 (1934) 29–33.
4. *On the differentiability of the transition probabilities in stationary Markov processes with a denumerable number of states.* (Russian.) Moskov. Gos. Univ. Ucenye Zapiski, 148, Mat. 4 (1951) 53–59.

Krein, M.

1. *Sur quelques questions de la géométrie des ensembles convexes situés dans un espace linéaire normé et complet.* C.R. Acad. Sci. URSS, 14 (1937) 5–7.

Krein, M. and **Milman, D.**

1. *On the extreme points of regularly convex sets.* Studia Math., 9 (1940) 133–138.

Kuratowski, C.

1. *Topologie* I. Warszawa-Wrocław, 1948, xii + 452 pp.

La Salle, J. P.

1. *Pseudo-normed linear spaces.* Duke Math. J., 8 (1941) 131–135.

Ledermann, W. See **Reuter, G. E. H.**

Lefschetz, S.

1. *Algebraic topology.* Amer. Math. Soc. Coll. Publ., XXVII. New York, 1942, vi + 389 pp.

Le Roux, J.

1. *Recherches sur les équations aux dérivées partielles.* J. de Math., (5) 9 (1903) 403–455.

Lévy, Paul

1. *Théorie de l'addition des variables aléatoires.* [*Monographies des probabilités*, Fasc. 1.] Gauthier-Villars, Paris, xvii + 328 pp.

Lie, S., and **Scheffers, G.**

1. *Vorlesungen über continuierliche Gruppen mit geometrischen und anderen Anwendungen.* Teubner, Leipzig, 1893, xii + 810 pp.

Lindelöf, E. (Cf. **Phragmén, E.**)

1. *Sur l'application de la théorie des résidus au prolongement analytique des séries de Taylor.* J. de Math., (5) 9 (1903) 213–221.
2. *Quelques remarques sur la croissance de la fonction $\zeta(s)$.* Bull. Sci. Math., (2) 32 (1908) 341–356.
3. *Démonstration nouvelle d'un théorème fondamental sur les suites des fonctions monogènes.* Bull. Soc. Math. France, 41 (1913) 171–178.

Liouville, J.

1. *Mémoire sur le calcul des différentielles à indices quelconques.* J. École Polytechn., 13 (1832) cah. 21, 71–162.

Littlewood, J. E. See **Hardy, G. H.**

Loomis, L. H.

1. *An introduction to abstract harmonic analysis.* D. Van Nostrand, New York, 1953, x + 190 pp.

Lorch, E. R.

1. *On a calculus of operators in reflexive vector spaces.* Trans. Amer. Math. Soc., 45 (1939) 217–234.
2. *The spectrum of linear transformations.* Trans. Amer. Math. Soc., 52 (1942) 238–248.
3. *The theory of analytic functions in normed abelian vector rings.* Trans. Amer. Math. Soc., 54 (1943) 414–425.
4. *The structure of normed abelian rings.* Bull. Amer. Math. Soc., 50 (1944) 447–463.

Lorenzen, P.

1. *Abstrakte Begründung der multiplikativen Idealtheorie.* Math. Z., 45 (1939) 533–553.

Martin, R. S. (Cf. **Michal, A. D.**)

1. *Contributions to the theory of functionals.* Thesis, Cal. Inst. Techn., 1932.

Mazur, S.

1. *Über konvexe Mengen in linearen normierten Räumen.* Studia Math., 4 (1933) 70–84.
2. *Sur les anneaux linéaires.* C.R. Acad. Sci. Paris, 207 (1938) 1025–1027.

Mazur, S., and **Orlicz, W.**

1. *Grundlegende Eigenschaften der polynomischen Operationen.* Studia Math., 5 (1934) 50–68, 179–189.

Michal, A. D.

1. *General differential geometries and related topics.* Bull. Amer. Math. Soc., 45 (1939) 529–563.

Michal, A. D., and **Clifford, A. H.**

1. *Fonctions analytiques implicites dans des espaces vectoriels abstraits.* C.R. Acad. Sci. Paris, 197 (1933) 735–737.

Michal, A. D., and **Martin, R. S.**

1. *Some expansions in vector space.* J. de Math., (9) 13 (1934) 69–91.

Milman, D.

1. *On some criteria for the regularity of spaces of the type (B).* C.R. Acad. Sci. URSS, 20 (1938) 243–246.

Minkowski, H.

1. *Geometrie der Zahlen.* Teubner, Berlin-Leipzig, 1910 (1896) viii + 256 pp.

Miyadera, I.

1. *On one-parameter semi-groups of operators.* J. Math. Tokyo, 1 (1951) 23–26.
2. *Generation of a strongly continuous semi-group of operators.* Tôhoku Math. J., (2) 4 (1952) 109–114.

3. *On the generation of a strongly ergodic semi-group of operators*. Tôhoku Math. J., (2) 6 (1954) 38–52.

4. *On the generation of strongly ergodic semi-groups of operators*. II. Tôhoku Math. J., (2) 6 (1954) 231–242.

Moore, E. H., and **Smith, H. L.**

1. *A general theory of limits*. Amer. J. Math., 44 (1922) 102–121.

Müntz, H. Ch.

1. *Umkehrung bestimmter Integrale und absolute Approximation*. Math. Z., 21 (1924) 96–110.

Murray, F. J.

1. *On complementary manifolds and projections in spaces L_p and l_p*. Trans. Amer. Math. Soc., 41 (1937) 138–152.

Nagumo, M.

1. *Einige analytische Untersuchungen in linearen metrischen Ringen*. Jap. J. Math., 13 (1936) 61–80.

Nathan, D. S.

1. *One-parameter groups of transformations in abstract vector spaces*. Duke Math. J., 1 (1935) 518–526.

von Neumann, J.

1. *Zur Theorie der Darstellungen kontinuierlicher Gruppen*. Sitzungsber. preuss. Akad. Wiss., 1927, 76–90.

2. *Über die analytischen Eigenschaften von Gruppen linearer Transformationen und ihrer Darstellungen*. Math. Z., 30 (1929) 3–42.

3. *Zur Algebra der Funktionaloperationen und Theorie der normalen Operatoren*. Math. Annalen 102 (1929) 370–427.

4. *Über einen Satz von M. H. Stone*. Annals of Math., (2) 33 (1932) 567–573.

Neumark, M., See **Gelfand, I.**

Nevanlinna, F., and **R.**

1. *Über die Eigenschaften analytischer Funktionen in der Umgebung einer singulären Stelle oder Linie*. Acta Soc. Sci. Fennicæ, 50 (1922) no. 5, 46 pp.

Nielsen, J.

1. *Om Strukturen af en Vektormodul med endelig Basis. Med Anvendelse paa Diofantiske Approximationer*. Mat. Tidsskrift, B, 1932, 29–42.

Nörlund, N. E.

1. *Vorlesungen über Differenzenrechnung*. J. Springer, Berlin, 1924, ix + 551 pp.

2. *Leçons sur les séries d'interpolation*. Gauthier-Villars, Paris, 1926, vii + 236 pp.

Orlicz, W. (Cf. **Alexiewicz, A.,** and **Mazur, S.**)

1. *Beiträge zur Theorie der Orthogonalentwicklungen*. I, II. Studia Math., 1 (1929) 1–39, 241–255.

2. *Über unbedingte Konvergenz in Funktionenräumen*, I. Studia Math., 4 (1933) 33–37.

Perlis, S.

1. *A characterization of the radical of an algebra*. Bull. Amer. Math. Soc., 48 (1942) 128–132.

Pettis, B. J.

1. *On integration in vector spaces*. Trans. Amer. Math. Soc., 44 (1938) 277–304.

2. *A note on regular Banach spaces*. Bull. Amer. Math. Soc., 44 (1938) 420–428.

3. *A proof that every uniformly convex space is reflexive*. Duke Math. J., 5 (1939) 249–253.

Phillips, R. S. (Cf.. **Bochner, S.,** and **Dye, H. A.**)

1. *Integration in a convex linear topological space*. Trans. Amer. Math. Soc., 47 (1940) 114–145.

2. *On linear transformations*. Trans. Amer. Math. Soc., 48 (1940) 516–541.

3. *On weakly compact subsets of a Banach space*. Amer. J. Math., 65 (1943) 108–136.

4. *On one-parameter semi-groups of linear transformations.* Proc. Amer. Math. Soc., 2 (1951) 234–237.
5. *A note on ergodic theory.* Proc. Amer. Math. Soc., 2 (1951) 663–670.
6. *Spectral theory for semi-groups of linear operators.* Trans. Amer. Math. Soc., 71 (1951) 393–415.
7. *On the generation of semi-groups of linear operators.* Pacific J. Math., 2 (1952) 343–369.
8. *Perturbation theory for semi-groups of linear operators.* Trans. Amer. Math. Soc., 74 (1953) 199–221.
9. *An inversion formula for Laplace transforms and semi-groups of linear operators.* Annals of Math., 59 (1954) 325–356.
10. *A note on the abstract Cauchy problem.* Proc. Nat. Acad. Sci., 40 (1954) 244–248.
11. *Semi-groups of operators.* Bull. Amer. Math. Soc., 61 (1955) 16–33.
12. *The adjoint semi-group.* Pacific J. Math., 5 (1955) 269–283.
13. *Dissipative hyperbolic systems.* Trans. Amer. Math. Soc., 86 (1957).

Phragmén, E., and **Lindelöf, E.**
1. *Sur une extension d'un principe classique de l'analyse et sur quelques propriétés des fonctions monogènes dans le voisinage d'un point singulier.* Acta Math., 31 (1908) 381–406.

Picard, E.
1. *Sur la théorie des groupes et les surfaces algébriques.* Rend. Circ. Mat. Palermo, 9 (1895) 244–255.

Pierpont, J.
1. *Theory of functions of real variables.* II. Ginn, Boston, 1912, xiii + 645 pp.

Pitt, H. R. (Cf. **Wiener, N.**)
1. *General Tauberian theorems.* Proc. London Math. Soc., (2) 44 (1938) 243–288.

Plessner, A.
1. *Eine Kennzeichnung der totalstetigen Funktionen.* J. f. Math., 160 (1929) 26–32.

Pólya, G. (Cf. **Hardy, G. H.**)
1. *Über die Funktionalgleichung der Exponentialfunktion im Matrizenkalkül.* Sitzungsber. preuss. Akad. Wiss., 1928, 96–99.
2. *Untersuchungen über Lücken und Singularitäten von Potenzreihen.* Math. Z., 29 (1929) 549–640.
3. *Aufgaben* 105, 106. Jahresbericht D.M.V., 40 (1931) *80, 81.*

Pólya, G., and **Szegö, G.**
1. *Aufgaben und Lehrsätze aus der Analysis.* I. J. Springer, Berlin, 1925, xvi + 338 pp.

Pontrjagin, L.
1. *Topological groups.* Princeton Univ. Press, 1939, ix + 299 pp.

Post, E. L.
1. *Generalized differentiation.* Trans. Amer. Math. Soc., 32 (1930) 723–781.

Price, G. B.
1. *The theory of integration.* Trans. Amer. Math. Soc., 47 (1940) 1–50.

Raikov, D. A. (Cf. **Gelfand, I.**)
1. *Harmonic analysis on commutative groups with a Haar measure and the theory of characters.* Trav. Inst. Math. Stekloff, 14 (1945) 86 pp.

Rees, D.
1. *On semi-groups.* Proc. Cambridge Phil. Soc., 36 (1940) 387–400.

Rellich, F.
1. *Störungstheorie der Spektralzerlegung.* V. Math. Annalen, 118 (1942) 462–484.

Reuter, G. E. H., and **Ledermann, W.**
1. *On the differential equations for the transition probabilities of Markov processes with enumerably many states.* Proc. Cambridge Phil. Soc., 49 (1953) 247–262.

Rickart, C. E.

1. *Integration in a convex topological space*. Trans. Amer. Math. Soc., 52 (1942) 498–521.
2. *The singular elements of a Banach algebra*. Duke Math. J., 14 (1947) 1063–1077.

Riemann, B.

1. *Versuch einer allgemeinen Auffassung der Integration und Differentiation*. Math. Werke, Leipzig, 1876, 331–344.

Riesz, F.

1. *Untersuchungen über Systeme integrierbarer Funktionen*. Math. Annalen, 69 (1910) 449–497.
2. *Les systèmes d'équations linéaires à une infinité d'inconnues*. Gauthier-Villars, Paris, 1913, vi + 182 pp.
3. *Über lineare Funktionalgleichungen*. Acta Math., 41 (1918) 71–98.
4. *Über die linearen Transformationen des komplexen Hilbertschen Raumes*. Acta Litt. ac Sci., Szeged, 5 (1930) 23–54.
5. *Some mean ergodic theorems*. J. London Math. Soc., 13 (1938) 274–278.

Riesz, M.

1. *Sur les maxima des formes bilinéaires et sur les fonctionnelles linéaires*. Acta Math., 49 (1926) 465–497.
2. *Sur les fonctions conjuguées*. Math. Z., 27 (1928) 218–244.
3. *L'intégrale de Riemann-Liouville et le problème de Cauchy pour l'équation des ondes*. Conf. Réunion Internat. Math. Paris 1937. Gauthier-Villars, Paris 1939, 153–170.
4. *Intégrales de Riemann-Liouville et potentiels*. Acta Litt. ac Sci., Szeged, 9 (1938) 1–42, 116–118.
5. *L'intégrale de Riemann-Liouville et le problème de Cauchy*. Acta Math., 81 (1949) 1–223.

Romanoff, N. P.

1. *One-parameter operator groups and semi-groups, and new methods in the theory of prime numbers based on examination of them*. (In Russian.) Siberian Physico-Technological Institute of the University of Tomsk. Tomsk, 1942, 42 pp. [Mimeographed.]
2. *On one-parameter groups of linear transformations* I. Annals of Math., (2) 48 (1947) 216–233.

Rosenbaum, R. A.

1. *Sub-additive functions*. Duke Math. J., 17 (1950) 227–247.

Schauder, J.

1. *Über lineare vollstetige Funktionaloperationen*. Studia Math., 2 (1930) 183–196.

Schur, I. (Cf. **Frobenius, G.**)

1. *Über lineare Transformationen in der Theorie der unendlichen Reihen*. J. f. Math., 151 (1921) 79–111.
2. *Über die stetigen Darstellungen der allgemeinen linearen Gruppe*. Sitzungsber. preuss. Akad. Wiss., 1928, 100–124.

Schwartz, J. See **Dunford, N.**

Schwartz, L.

1. *Un lemme sur la dérivation des fonctions vectorielles d'une variable réelle*. Ann. Inst. Fourier (Grenoble), 2 (1951) 17–18.

Scorza-Dragoni, G. See **Severi, F.**

Segal, I. E. (Cf. **Dunford, N.**)

1. *The group algebra of a locally compact group*. Trans. Amer. Math. Soc., 61 (1947) 69–105.

de Séguier, J. A.

1. *Éléments de la théorie des groupes abstraits*. Gauthier-Villars, Paris, 1904, ii + 176 pp.

Severi, F., and Scorza-Dragoni, G.

1. *Lezioni di analisi*, Vol. 3. Cesare Zuffi, Bologna, 1951, vi + 255 pp.

Sierpiński, W.

1. *Sur les fonctions convexes mesurables.* Fund. Math., 1 (1920) 125–129.

Šilov, G. (Cf. **Gelfand, I.**)

1. *On the extension of maximal ideals.* C. R. (Doklady) Acad. Sci. URSS, N. S. 29 (1940) 83–84.
2. *On normed rings possessing one generator.* Rec. Math., 21 (1947) 25–47.
3. *On the decomposition of a commutative normed ring in a direct sum of ideals.* Rec. Math., 32 (1953) 353–364.

Silva, J. S.

1. *As funções analíticas e a análise funcional.* Dissertation. Instituto para á Alta Cûltura, Lisbon, 1948; Portugaliae Math., 9 (1950) 1–130.

Silvermann, L. L., and **Tamarkin, J. D.**

1. *On the generalization of Abel's theorem for certain definitions of summability.* Math. Z., 29 (1928) 161–170.

Smith, H. L. See **Moore, E. H.**

Smith, P. A.

1. *Foundations of the theory of Lie groups with real parameters.* Annals of Math., (2) 44 (1943) 481–513.
2. *Foundations of Lie groups.* Annals of Math., (2) 48 (1947) 29–42.

Šmulian, V.

1. *On multiplicative linear functionals in certain special normed rings.* C. R. (Doklady) Acad. Sci. URSS, N. S. 26 (1940) 13–16.

Sobczyk, A. See **Bohnenblust, H. F.**

Šreĭder, Yu. A.

1. *The structure of maximal ideals in rings of measures with convolutions.* (Russian.) Mat. Sbornik, N. S. 27 (69) (1950) 297–318.

Steffensen, J. F.

1. *Interpolation.* Williams and Wilkins, Baltimore, 1927, ix + 248 pp.

Stone, M. H.

1. *Linear transformations in Hilbert space* III. Proc. Nat. Acad. Sci., 16 (1930) 172–175.
2. *On one parameter unitary groups in Hilbert space.* Annals of Math., (2) 33 (1932) 643–648.
3. *Linear transformations in Hilbert space and their applications to analysis.* Amer. Math. Soc. Coll. Publ., XV. New York, 1932, viii + 622 pp.
4. *Applications of the theory of Boolean rings to general topology.* Trans. Amer. Math. Soc., 41 (1937) 375–481.
5. *A general theory of spectra* I. Proc. Nat. Acad. Sci., 26 (1940) 280–283.
6. *On a theorem of Pólya.* J. Indian Math. Soc., 12 (1948) 1–7.
7. *Boundedness properties in function-lattices.* Canadian J. Math., 1 (1949) 176–186.

Suchumlinov, G.

1. *Analytische Funktionale.* [Russian, German summary.] Bull. Univ. d'état à Moscou. Sér. Internat., Section A, 1 (1937) fasc. 2, 20 pp.

Suschkewitsch, A.

1. *Über die endlichen Gruppen ohne das Gesetz der eindeutigen Umkehrbarkeit.* Math. Annalen, 99 (1928) 30–50.
2. *Über die Matrizendarstellung der verallgemeinerten Gruppen.* Comm. Soc. Math. Kharkoff, (4) 6 (1933) 27–38.

Szász, O.

1. *Generalization of S. Bernstein polynomials to the infinite interval.* J. Research Nat. Bureau of Standards, 45 (1950) 239–245.

Szegö, G. (Cf. **Pólya, G.**)

1. *Lösung der Aufgaben* 105, 106. Jahresber. D. M. V., 43 (1933) *10–11, 15–16.*
2. *Orthogonal polynomials.* Amer. Math. Soc. Coll. Publ., XXIII. New York, 1939, ix + 401 pp.

Sz.-Nagy, B.

1. *Über messbare Darstellungen Liescher Gruppen.* Math. Annalen, 112 (1936) 286–296.
2. *On semi-groups of self-adjoint transformations in Hilbert space.* Proc. Nat. Acad. Sci., 24 (1938) 559–560.
3. *Spektraldarstellungen linearer Transformationen des Hilbertschen Raumes.* Ergebnisse d. Math., Vol. 5, No. 5. J. Springer, Berlin, 1942, iv + 80 pp.

Tamarkin, J. D. See **Hille, E.**, and **Silverman, L. L.**

Taylor, A. E.

1. *Analytic functions in general analysis.* Ann. R. Scuola Norm. Sup. Pisa, (2) 6 (1937) 277–292.
2. *The resolvent of a closed transformation.* Bull. Amer. Math. Soc., 44 (1938) 70–74.
3. *Additions to the theory of polynomials in normed linear spaces.* Tôhoku Math. J., 44 (1938) 302–318.
4. *Biharmonic functions in abstract spaces.* Amer. J. Math., 60 (1938) 416–422.
5. *Linear operations which depend analytically upon a parameter.* Annals of Math., (2) 39 (1938) 574–593.
6. *On the properties of analytic functions in abstract spaces.* Math. Annalen, 115 (1938) 466–484.
7. *Analysis in complex Banach spaces.* Bull. Amer. Math. Soc., 49 (1943) 652–669.
8. *Conjugations in complex linear spaces.* Univ. of California Publ. in Math., N. S. 2 (1944) 85–102.
9. *Spectral theory of closed distributive operators.* Acta Math., 84 (1950) 189–224.

Titchmarsh, E. C.

1. *Introduction to the theory of Fourier integrals.* Oxford Univ. Press, 1937, x + 390 pp.

Tricomi, F.

1. *Sulle trasformazioni funzionali lineari commutabili con la derivazione.* Comm. Math. Helvetici, 8 (1935–1936) 70–87.

Tukey, J. W.

1. *Some notes on the separation of convex sets.* Portugaliae Math., 3 (1942) 95–102.

Tychonoff, A.

1. *Ein Fixpunktsatz.* Math. Annalen, 111 (1935) 767–776.

de la Vallée-Poussin, Ch. J.

1. *Leçons sur l'approximation des fonctions d'une variable réelle.* Gauthier-Villars, Paris, 1919, vi + 151 pp.

Visser, C.

1. *On the iteration of linear operations in a Hilbert space.* Nederl. Akad. Wetensch., 41 (1938) 487–495.

Volterra, V.

1. *Leçons sur les fonctions de lignes.* Gauthier-Villars, Paris, 1913, vi + 230 pp.

Vulich, B.

1. *Sur les opérations linéaires multiplicatives.* C. R. (Doklady) Acad. Sci. URSS, 41 (1943) 142–144.
2. *Sur la représentation analytique d'opérations linéaires multiplicatives.* C. R. (Doklady) Acad. Sci. URSS, 41 (1943) 187–190.

Wall, H. S. See **Garabedian, H. L.**

Wallace, A. D.

1. *The structure of topological semi-groups.* Bull. Amer. Math. Soc., 61 (1955) 95–112.

Walsh, J. L.

1. *Interpolation and approximation by rational functions in the complex domain*. Amer. Math. Soc. Coll. Publ., XX. New York, 1935, x + 382 pp.

Wehausen, J. V.

1. *Transformations in linear topological spaces*. Duke Math. J., 4 (1938) 157–169.

Weil, A.

1. *Sur les espaces à structure uniforme et sur la topologie générale*. Actualités Sci. et Ind., 551. Hermann, Paris, 1937, 39 pp.
2. *L'intégration dans les groupes topologiques et ses applications*. Actualités Sci. et Ind., 869. Hermann, Paris, 1938, 158 pp.

Weyl, H.

1. *Bemerkungen zum Begriff des Differentialquotienten gebrochener Ordnung*. Vierteljahrschrift Naturforsch. Ges. Zürich, 62 (1917) 296–302.

Widder, D. V.

1. *The Laplace transform*. Princeton Univ. Press, 1941, x + 406 pp.

Wiener, N.

1. *Note on a paper of M. Banach*. Fund. Math., 4 (1923) 136–143.
2. *The operational calculus*. Math. Annalen, 95 (1926) 557–584.
3. *Tauberian theorems*. Annals of Math., (2) 33 (1932) 1–100.
4. *The ergodic theorem*. Duke Math. J., 5 (1939) 1–18.

Wiener, N., and Pitt, H. R.

1. *On absolutely convergent Fourier-Stieltjes transforms*. Duke Math. J., 4 (1938) 420–436.

Yosida, K.

1. *On the group embedded in the metrical complete ring*. Jap. J. Math., 13 (1936) 7–26, 459–472.
2. *Mean ergodic theorems in Banach spaces*. Proc. Imp. Acad. Sci. Tokyo, 14 (1938) 292–294.
3. *On the differentiability and the representation of one-parameter semi-group of linear operators*. J. Math. Soc. Japan, 1 (1948) 15–21.
4. *An operator-theoretical treatment of temporally homogeneous Markoff process*. J. Math. Soc. Japan, 1 (1949) 244–253.
5. *On Cauchy's problem in the large for wave equation*. Proc. Japan Acad., 28 (1952) 396–403.
6. *Semi-group theory and the integration problem of diffusion equations*. Proc. Internat. Congr. Math. Amsterdam 1954, II, 16 pp.
7. *An operator-theoretical integration of the wave equation*. J. Math. Soc. Japan, 8 (1956) 79–92.

Zorn, M. A. (Cf. **Hille, E.**)

1. *Characterization of analytic functions in Banach spaces*. Annals of Math., (2) 46 (1945) 585–593.
2. *Gâteaux differentiability and essential boundedness*. Duke Math. J., 12 (1945) 579–583.
3. *Derivatives and Fréchet differentials*. Bull. Amer. Math. Soc., 52 (1946) 133–137.

Zygmund, A.

1. *Trigonometrical series*. Warszawa, 1935, iv + 332 pp.

INDEX

The numbers refer to pages. Names in the References are not included.

Other Titles in This Series

(Continued from the front of this publication)

3.1 **G. A. Bliss,** Fundamental existence theorems, 1913

3.2 **E. Kasner,** Differential-geometric aspects of dynamics, 1913

2 **E. H. Moore,** Introduction to a form of general analysis

M. Mason, Selected topics in the theory of boundary value problems of differential equations

E. J. Wilczyński, Projective differential geometry, 1910

1 **H. S. White,** Linear systems of curves on algebraic surfaces

F. S. Woods, Forms on noneuclidean space

E. B. Van Vleck, Selected topics in the theory of divergent series and of continued fractions, 1905